W9-CCK-964

World – Political

160°W
140°W
120°W
100°W
80°W
80°N
ARCTIC OCEAN
ALASKA
(U.S.)
60°N
CANADA
Aleutian Islands
(U.S.)
NORTH
AMERICA
St. Pierre
and Miquel
(FR.)
40°N
UNITED STATES
PACIFIC OCEAN
Bermuda
(U.K.)
ATLANT
OCEAN
Midway Islands
(U.S.)
THE
CARIBBEAN
MEXICO
20°N
Hawaii (U.S.)
Caribbean Sea
AUSTRALIA
AND OCEANIA
GUYANA
SURINAME
VENEZUELA
FRENCH GUIAN
(FR.)
See inset below
COLOMBIA
KIRIBATI
Equator
0°
Galápagos Islands
(ECUADOR)
ECUADOR
Tokelau
(N.Z.)
Cook Islands
(N.Z.)
PERU
LATIN AMERICA
Wallis and Futuna (FR.)
BRAZIL
SAMOA
French Polynesia
(FR.)
American Samoa
(U.S.)
TONGA
BOLIVIA
20°S
Niue
(N.Z.)
CHILE
PARAGUAY
Easter I.
(CHILE)
Pitcairn Islands
(U.K.)
URUGUAY
PACIFIC OCEAN
ARGENTINA
40°S
Falkland Islands
(U.K.)
South Ge
60°S

The Caribbean

90°W
80°W
70°W
60°W
FLORIDA
(U.S.)
0
250
500 Miles
0
250
500 Kilometers
Gulf of Mexico
THE
BAHAMAS
ATLANTIC OCEAN
Turks & Caicos
(U.K.)
20°N
CUBA
Virgin Is.
(U.K.)
Anguilla (U.K.)
St. Barthélemy (FR.)
ANTIGUA AND BARBUDA
Cayman Islands
(U.K.)
Puerto Rico
(U.S.)
MEXICO
JAMAICA
HAITI
DOMINICAN
REPUBLIC
Montserrat (U.K.)
Guadeloupe (FR.)
BELIZE
Virgin Is. (U.S.)
St. Martin (FR. & NETH.)
DOMINICA
GUATEMALA
ST. KITTS AND NEVIS
Martinique (FR.)
HONDURAS
Caribbean Sea
ST. LUCIA
ST. VINCENT AND
THE GRENADINES
BARBADOS
EL SALVADOR
Aruba (NETH.)
Netherlands
Antilles
(NETH.)
GRENADA
NICARAGUA
TRINIDAD
AND
TOBAGO
10°N
COSTA
RICA
PACIFIC
OCEAN
PANAMA
VENEZUELA
GUYANA
COLOMBIA
90°W
80°W

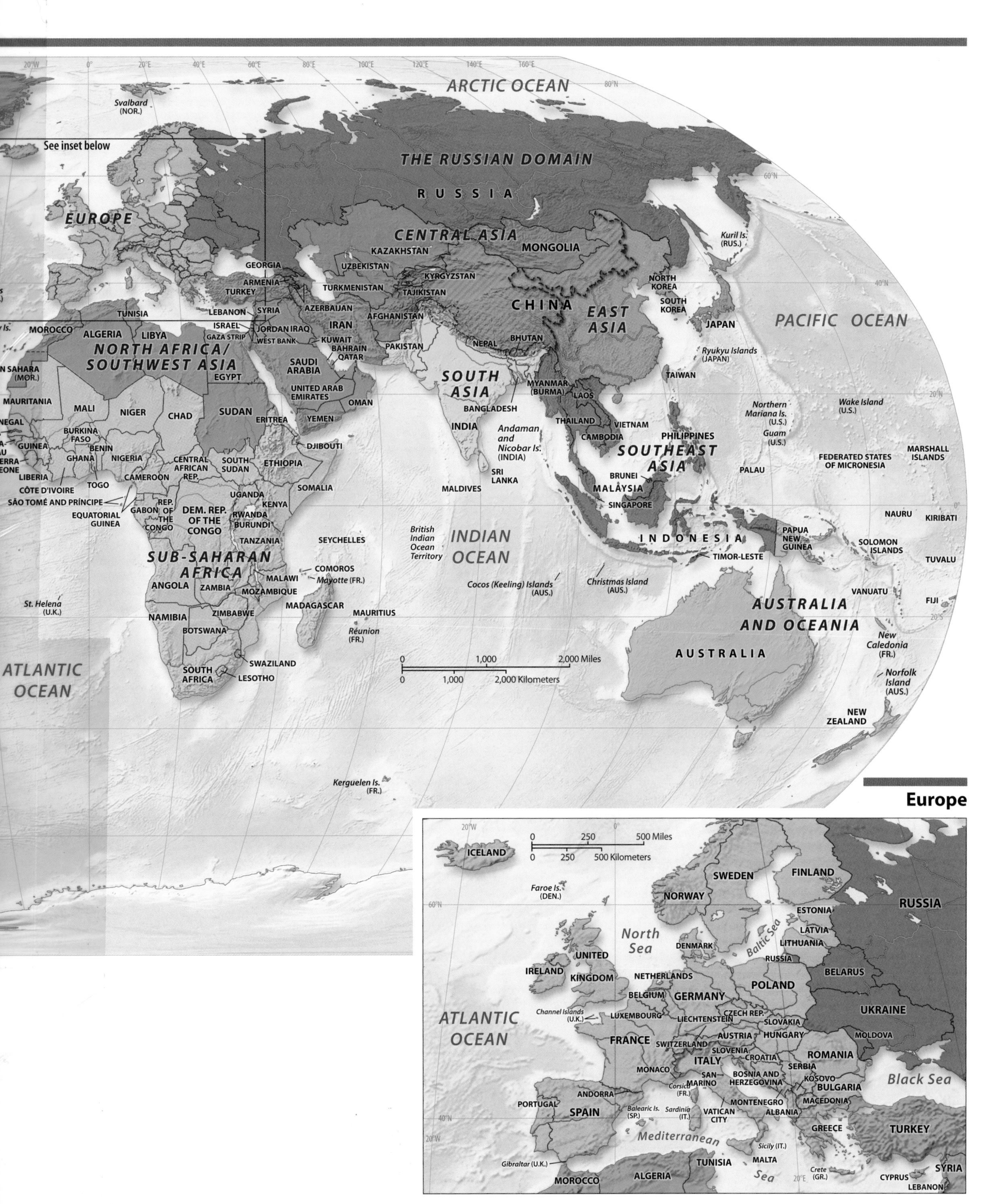

ARCTIC OCEAN
PACIFIC OCEAN
INDIAN OCEAN
ATLANTIC OCEAN
See inset below
EUROPE
THE RUSSIAN DOMAIN
RUSSIA
CENTRAL ASIA
EAST ASIA
SOUTH ASIA
SOUTHEAST ASIA
NORTH AFRICA/ SOUTHWEST ASIA
SUB-SAHARAN AFRICA
AUSTRALIA AND OCEANIA
Svalbard (NOR.)
KAZAKHSTAN
MONGOLIA
UZBEKISTAN
GEORGIA
ARMENIA
TURKEY
TURKMENISTAN
KYRGYZSTAN
TAJIKISTAN
AZERBAIJAN
CHINA
NORTH KOREA
SOUTH KOREA
JAPAN
Kuril Is. (RUS.)
Ryukyu Islands (JAPAN)
TAIWAN
TUNISIA
LEBANON
SYRIA
ISRAEL
GAZA STRIP
WEST BANK
JORDAN
IRAQ
IRAN
KUWAIT
BAHRAIN
QATAR
AFGHANISTAN
PAKISTAN
NEPAL
BHUTAN
MOROCCO
ALGERIA
LIBYA
EGYPT
SAUDI ARABIA
UNITED ARAB EMIRATES
OMAN
YEMEN
MYANMAR (BURMA)
LAOS
BANGLADESH
INDIA
THAILAND
VIETNAM
CAMBODIA
PHILIPPINES
MAURITANIA
MALI
NIGER
CHAD
SUDAN
ERITREA
DJIBOUTI
BURKINA FASO
GUINEA
BENIN
GHANA
NIGERIA
LIBERIA
TOGO
CÔTE D'IVOIRE
SÃO TOMÉ AND PRÍNCIPE
EQUATORIAL GUINEA
CAMEROON
CENTRAL AFRICAN REP.
SOUTH SUDAN
ETHIOPIA
SOMALIA
UGANDA
KENYA
GABON
REP. OF THE CONGO
DEM. REP. OF THE CONGO
RWANDA
BURUNDI
TANZANIA
SEYCHELLES
COMOROS
Mayotte (FR.)
MALAWI
ANGOLA
ZAMBIA
MOZAMBIQUE
MADAGASCAR
NAMIBIA
ZIMBABWE
BOTSWANA
MAURITIUS
Réunion (FR.)
SWAZILAND
SOUTH AFRICA
LESOTHO
St. Helena (U.K.)
Andaman and Nicobar Is. (INDIA)
SRI LANKA
MALDIVES
British Indian Ocean Territory
Cocos (Keeling) Islands (AUS.)
Christmas Island (AUS.)
BRUNEI
MALAYSIA
SINGAPORE
INDONESIA
PAPUA NEW GUINEA
TIMOR-LESTE
PALAU
Northern Mariana Is. (U.S.)
Guam (U.S.)
Wake Island (U.S.)
FEDERATED STATES OF MICRONESIA
MARSHALL ISLANDS
NAURU
KIRIBATI
SOLOMON ISLANDS
TUVALU
VANUATU
FIJI
New Caledonia (FR.)
Norfolk Island (AUS.)
AUSTRALIA
NEW ZEALAND
Kerguelen Is. (FR.)
0 1,000 2,000 Miles
0 1,000 2,000 Kilometers
Europe
0 250 500 Miles
0 250 500 Kilometers
ICELAND
Faroe Is. (DEN.)
SWEDEN
FINLAND
NORWAY
RUSSIA
ESTONIA
LATVIA
LITHUANIA
North Sea
Baltic Sea
DENMARK
UNITED KINGDOM
IRELAND
NETHERLANDS
BELGIUM
GERMANY
POLAND
BELARUS
Channel Islands (U.K.)
LUXEMBOURG
LIECHTENSTEIN
CZECH REP.
SLOVAKIA
UKRAINE
AUSTRIA
HUNGARY
MOLDOVA
FRANCE
SWITZERLAND
SLOVENIA
CROATIA
ROMANIA
MONACO
ITALY
SERBIA
SAN MARINO
BOSNIA AND HERZEGOVINA
KOSOVO
BULGARIA
Black Sea
Corsica (FR.)
ANDORRA
MONTENEGRO
MACEDONIA
PORTUGAL
SPAIN
Balearic Is. (SP.)
Sardinia (IT.)
VATICAN CITY
ALBANIA
GREECE
TURKEY
Mediterranean Sea
Sicily (IT.)
MALTA
Gibraltar (U.K.)
TUNISIA
Crete (GR.)
CYPRUS
SYRIA
LEBANON
MOROCCO
ALGERIA

Help Students Understand the Tensions Between Global Forces & Local Diversity

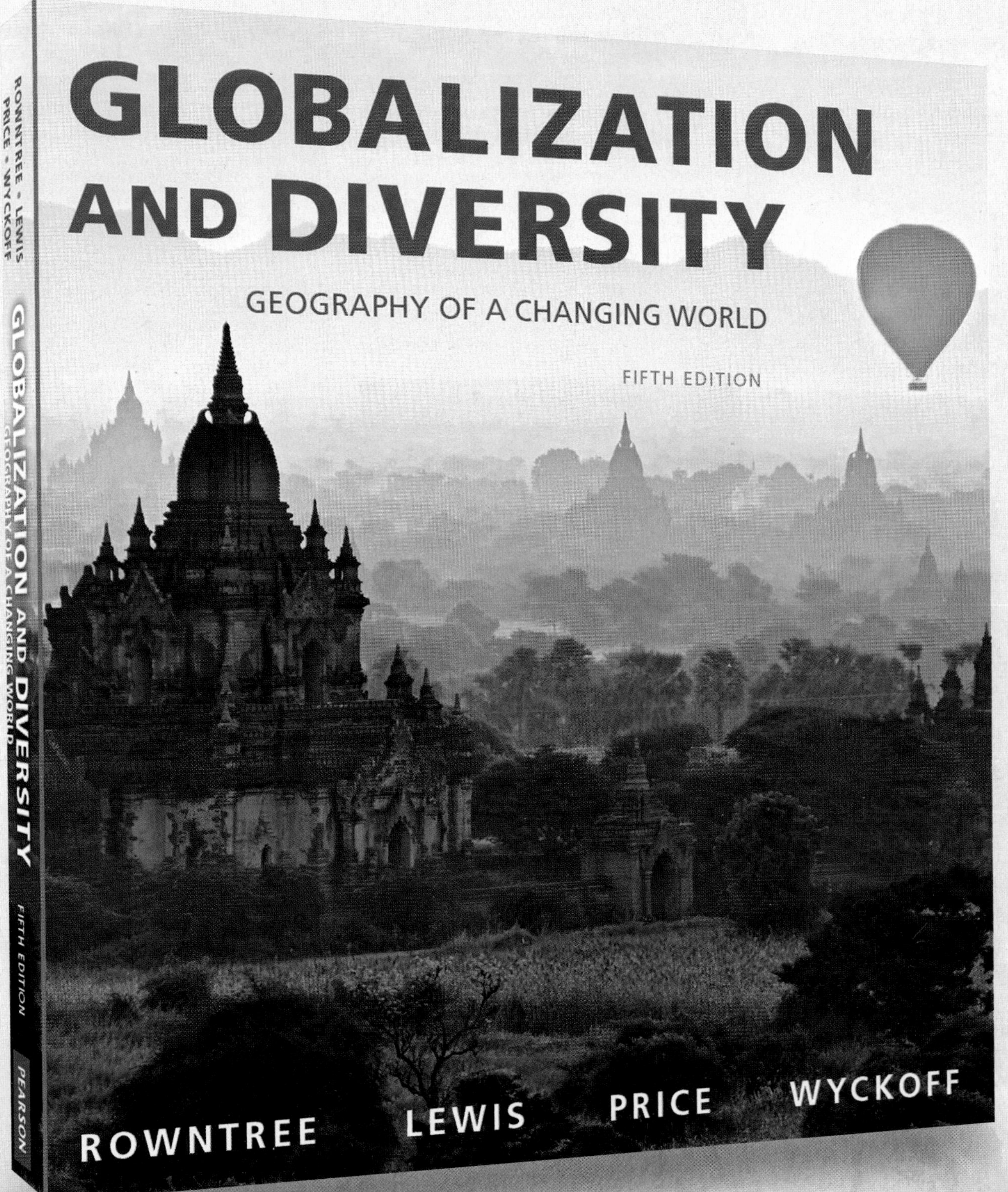

PEARSON

Global Forces & Local Diversity

NEW! *Everyday Globalization* features explore everyday products and commodities that we commonly use and consume, tracing global vs local links and the implications of consumption and behavior. Critical thinking questions ask students to reflect on their experience, behavior, and interaction with these products and commodities.

EVERYDAY GLOBALIZATION

The Rainforest and Your Chocolate Fix

Your chocolate bar comes from the tropical rainforest, and satisfying your sweet tooth could be either destroying or saving the rainforest, depending on how the cocoa was grown. Cocoa, chocolate's main ingredient, comes from cacao trees, which grow exclusively in equatorial rainforests—mainly in Ghana and other African countries, but also in the Amazon Basin of South America. Cacao trees prefer the shade of higher rainforest trees, which is good news. But to meet the ever-increasing demand for chocolate, cacao is also cultivated for short periods of time in the full sunlight of newly cleared rainforest plots. That's the bad news—because this method of cacao farming is a major factor in the destruction of African rainforests.

So what's a rainforest-loving chocolate lover to do? Easy: Take an extra 30 seconds and read the candy bar label to see whether there's any mention of shade-grown and/or sustainably farmed cacao trees. After that, it's up to you.

1. Identify other foods you eat that come from tropical rainforests, and describe how their cultivation affects the forests.
2. What are the different ways you eat chocolate, and where is that cocoa grown?

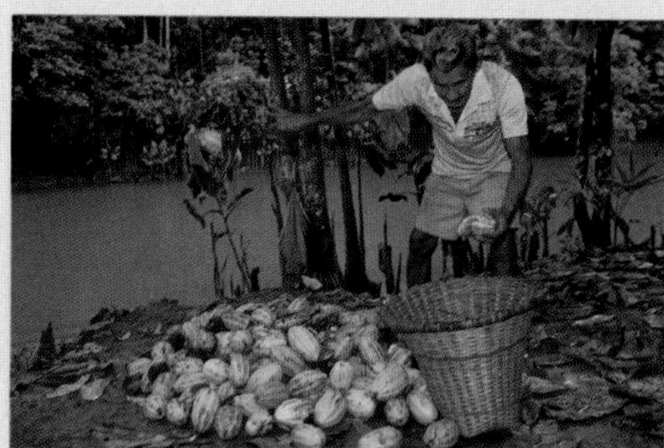

Figure 2.3.1 Sustainable Cocoa This farmer sorts cocoa pods harvested on a sustainable plantation in Brazil's Amazon region.

EVERYDAY GLOBALIZATION

Popping Pills from Israel

Every year U.S. doctors write more than 2.5 *billion* prescriptions for generic pharmaceuticals. Few people realize how many of these drugs are actually manufactured in Southwest Asia—specifically, Israel. When you reach for that generic antibiotic (amoxicillin), painkiller (oxycodone), or anti-inflammatory (naproxen), you may well be taking pills manufactured halfway around the world. Israel is home to seven research universities and a host of companies that focus on the biological sciences and innovations in the pharmaceutical industry.

The largest player in Israel's generic drug industry is Teva Pharmaceutical Industries (Figure 7.4.1). The company estimates that it manufactures 73 *billion* tablets a year and that one in six generic prescriptions in the United States is filled with a Teva (Hebrew for "nature") product. Today Teva is the largest global manufacturer of generic pharmaceuticals, as well as an innovative producer of its own proprietary drugs. The result is that Israel has emerged as one of the planet's key focal points in an industry that seems destined to grow along with the world's insatiable demand for affordable pharmaceuticals.

1. For the American public, describe some of the benefits and drawbacks of depending on a global geography of prescription drugs.
2. Visit a local pharmacy and select two over-the-counter medications. Can you find out who manufactured them and where they came from?

Figure 7.4.1 Teva Headquarters, Petah Tikva, Israel Employing thousands of skilled workers, Teva Pharmaceuticals Industries produces both the world's largest volume of generic drugs and a growing array of its own patented pharmaceuticals.

EXPLORING GLOBAL CONNECTIONS

The Libyan Highway to Europe

Google Earth (MG) Virtual Tour Video
http://goo.gl/Mb0mHp

Revolutions bring many unintended consequences. When Libyan dictator Muammar al-Qaddafi was overthrown in 2011, few experts believed it would dramatically reorient and enhance one of the world's most diverse flows of refugees. The newly formed Libyan Highway has truly international implications that reach from Syria and Nigeria to Italy and Sweden (Figure 7.2.1).

A Highway for Refugees All of the critical variables in the creation of the highway fell into place in 2014. First and foremost, Libya itself ceased to truly exist as multiple political forces vied for power, essentially ending any effective control over the country. Migrants and smugglers were free to make trip arrangements without much fear of government interference.

Second, an unregulated extralegal industry designed around transporting desper-

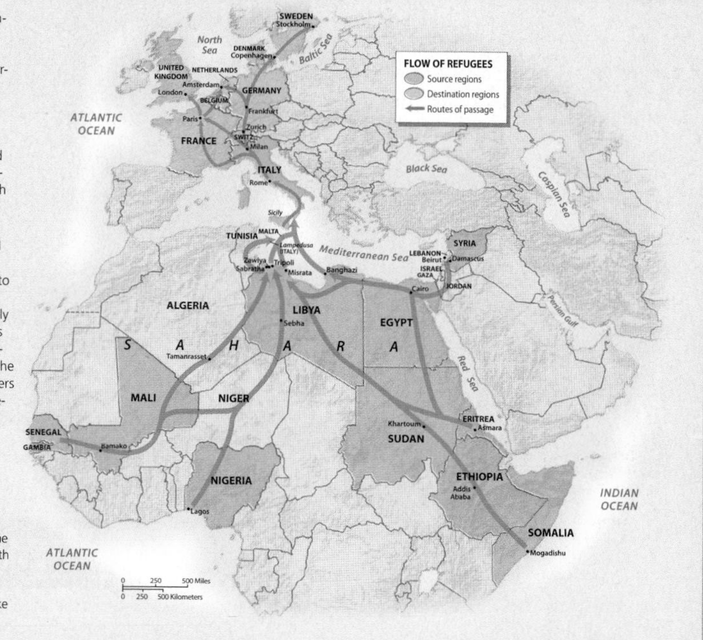

Figure 7.2.1 Libyan Highway to Europe The map shows some of the overland routes across North Africa that converge on Libyan ports, as well as general routes across the Mediterranean that take desperate migrants to Europe.

UPDATED! *Exploring Global Connections* features explore the often-surprising connections between places and people around the world. Mobile-ready Quick Response (QR) codes link to narrated **Google Earth Virtual Tour Videos** that explore landscapes related to each feature.

The Critical Issues & Work of Geography

GEOGRAPHERS AT WORK

Tracking Conflict from Space

Figure 1.4.1 **Susan Wolfinbarger**

As an undergraduate at Eastern Kentucky University, Susan Wolfinbarger took a world regional geography class, and was mesmerized: "There are so many things you learn in geography, and the methods of analysis can be applied to different careers and research." Years later, with a PhD in Geography from the Ohio State University, Wolfinbarger directs the Geospatial Technologies Project at the American Association for the Advancement of Science (AAAS) (Figure 1.4.1). Her group uses high-resolution satellite imagery to track conflicts and document issues of global concern, such as human rights abuses and damage to cultural heritage sites.

Most people have used Google Earth satellite images to look at places. Wolfinbarger's team employs a time series of such images in order to assess events such as destruction of villages. Interpreting images and quantifying findings is a challenge, but, she says, "Geography taught me not just mapping but statistics and surveying . . . it gave me a great toolkit to apply to any topic." Much of her analysis is used by human rights organizations such as the European Court of Human Rights and the Inter-American Court of Human Rights.

Wolfinbarger's team analyzed the increase in roadblocks in the Syrian city of Aleppo (Figure 1.4.2). Roadblocks demonstrate a decline in the circulation of people and goods in this densely settled city, which is a major problem. The Geospatial Technologies Project has also documented heritage sites at risk from damage and looting, especially in the Southwest Asia, and is developing training materials so that others can use this technology.

Geographers are at the cutting edge of applying satellite imagery to a broad spectrum of human rights issues. Wolfinbarger notes, "There are a lot of ways that geographers can contribute to things happening in the world, and a lot of opportunities out there other than academic jobs. Everyone wants a geographer!"

Figure 1.4.2 **Monitoring Aleppo** This image shows the city of Aleppo in May 2013, where over 1000 roadblocks were detected. Roadblocks are an indicator of ongoing conflict and potential humanitarian concerns because they restrict the movement of people and goods throughout the city. In a nine-month period from September 2012 to May 2013, the number of roadblocks doubled.

1. Suggest ways that satellite imagery could be used to document not just conflict but environmental change.
2. Government agencies are constantly developing and using satellite technology. How might a citizen or non-governmental group in your city or state use this kind of analysis?

NEW! *Geographers At Work* features look at how geography is practiced in the real world, profiling active geographers who are using the unique tools and techniques of geography. These features emphasize the diverse issues and places that geographers explore, emphasizing the different career and research opportunities of geography, and the interesting and important real-world problems that contemporary geography addresses.

UPDATED! *Working Toward Sustainability* features explore how the theme of sustainability plays out across world regions, looking at initiatives and positive outcomes of environmental, cultural, and economic sustainability. Mobile-ready Quick Response (QR) codes link to narrated **Google Earth Virtual Tour Videos** that explore landscapes related to each feature.

NEW & UPDATED! Expanded coverage of **Climate Change, Sustainability, Gender Issues, Food, Art, Music, Film, Sport, and Geopolitics** in each regional chapter.

WORKING TOWARD SUSTAINABILITY

Women and Water in the Developing World

Google Earth Tour Sahel region Africa https://goo.gl/oz3E31

Women and children bear the burden of water problems in most developing countries. Not only are children the most vulnerable to waterborne diseases, but also adult females (mothers, aunts, grandmothers, and older siblings) are the major caregivers for these sick children, adding yet another time-consuming task to their already busy days.

Further, women and older girls are the primary conveyers of water from wells or streams to their village homes. Every person requires about 5 gallons (18 liters) of water per day for their hydration, cooking, and sanitation needs; consequently, this amount (multiplied by the number of people in a family) must be carried each day from source to residence. In addition, women and children are responsible for supplying water for kitchen gardens that provide the family's food. At a global level, the water source for about a third of the developing world's rural population is more than half a mile (1 km) away from residences. To meet water needs, women spend about 25 percent of their day carrying water. A recent United Nations study estimated that in Sub-Saharan Africa about 40 billion hours a year are spent collecting and carrying water, the same amount of time spent in 1 year by France's entire workforce.

Besides the time expenditure, water is heavy, and most of it is carried by hand. In Africa, 40-pound (151-liter) jerry cans are common; in northwest India, women and girls balance several 5-gallon (19-liter) containers on their heads to lessen the number of trips made (Figure 2.4.1). (Note that 40 pounds is about the weight of the suitcase you check with the airlines on a typical trip. Try carrying it on your head through the airport parking lot someday.) After years of carrying water, women commonly suffer from chronic neck and back problems, many of which complicate childbirth. Additionally, girls' water-carrying responsibilities often interfere with their schooling, resulting in a high dropout rate and furthering female illiteracy in rural villages.

Figure 2.4.1 **Women in India carrying water on their heads**

Toward a Solution: The Wello WaterWheel After studying the water-carrying issue in semiarid northwestern India, Cynthia Koenig, a recent engineering graduate from the University of Michigan, invented the Wello WaterWheel, a barrel-like 13-gallon (50-liter) rolling water container that greatly reduces women's water-carrying duties (Figure 2.4.2).

Figure 2.4.2 **Woman using Wello WaterWheel**

Previously in that part of India, women and girls were spending 42 hours per week carrying water back and forth; with the Wello WaterWheel, that has been reduced to only 7 hours a week. Using this time-saving device has also reduced the school dropout rate for young girls in the region. Currently, Wello, which is a nonprofit organization, can deliver a WaterWheel from its factory in Mumbai to a rural Indian family for a mere $20. In the last year, thousands of Wello WaterWheels have been purchased by international aid organizations and donated to villages in Rajasthan, moving them closer to a sustainable existence.

1. List the social costs incurred when the responsibility for providing water falls to the women and children of a village.
2. List the probable social benefits to a village where clean water is readily available instead of requiring transport over long distances by women and children.

Structured to Facilitate Learning

PHYSICAL GEOGRAPHY AND ENVIRONMENTAL ISSUES

The region's vulnerability to water shortages is likely to increase in the early 21st century as growing populations, rapid urbanization, and increasing demands for agricultural land consume limited supplies.

POPULATION AND SETTLEMENT

Many settings within the region continue to see rapid population growth. These demographic pressures are particularly visible in fragile, densely settled rural zones as well as in fast-growing large cities.

CULTURAL COHERENCE AND DIVERSITY

Islam continues to be a vital cultural and political force within the region, but increasing fragmentation within that world has led to more culturally defined political instability.

GEOPOLITICAL FRAMEWORK

The Arab Spring uprisings in the early 2010s jolted the geopolitical status quo in Tunisia, Egypt, Libya, Yemen, and Bahrain. Internal instability and the growth of ISIL have produced extensive bloodshed in Syria and Iraq. Prospects for peace between Israel and the Palestinians remain murky, and Iran's growing political role is seen by many as a threat both within and beyond the region.

ECONOMIC AND SOCIAL DEVELOPMENT

Unstable world oil prices and unpredictable geopolitical conditions have discouraged investment and tourism in many countries. The pace of social change, especially for women, has quickened, stimulating diverse regional responses.

The Critical Themes of Geography Following two unique introductory chapters, each regional chapter is organized into five thematic sections: **Physical Geography and Environmental Issues, Population and Settlement, Cultural Coherence and Diversity, Geopolitical Framework,** and **Economic and Social Development.**

UPDATED! Region-specific *Learning Objectives* set up a structured learning path in the book and MasteringGeography, framing the major learning goals of each chapter.

Learning Objectives *After reading this chapter you should be able to:*

7.1 **Explain how latitude and topography produce the region's distinctive patterns of climate.**

7.2 **Describe how the region's fragile, often arid setting shapes contemporary environmental challenges.**

7.3 **Describe four distinctive ways in which people have learned to adapt their agricultural practices to the region's arid environment.**

7.4 **Summarize the major forces shaping recent migration patterns within the region.**

7.5 **List the major characteristics and patterns of diffusion of Islam.**

7.6 **Identify the key modern religions and language families that dominate the region.**

7.7 **Identify the role of cultural variables in understanding key regional conflicts in North Africa, Israel, Syria, Iraq, and the Arabian Peninsula.**

7.8 **Summarize the geography of oil and gas reserves in the region.**

7.9 **Describe traditional roles for Islamic women and provide examples of recent changes.**

UPDATED! *Review Questions and Key Terms* at the end of each section help students check their comprehension of key concepts as they read.

Review

7.1 Describe the climatic changes you might experience as you travel on a line from the eastern Mediterranean coast at Beirut to the highlands of Yemen. What are some of the key climatic variables that explain these variations?

7.2. Discuss five important human modifications of the Southwest Asian and North African environment, and assess whether these changes have benefited the region.

KEY TERMS Arab Spring, sectarian violence, ISIL (Islamic State of Iraq and the Levant) Islamic fundamentalism, Islamism, culture hearth, Organization of Petroleum Exporting Countries (OPEC), Maghreb, Levant, salinization, fossil water, hydropolitics, choke point

Data-Rich, Visual Explorations of Earth's People & Places

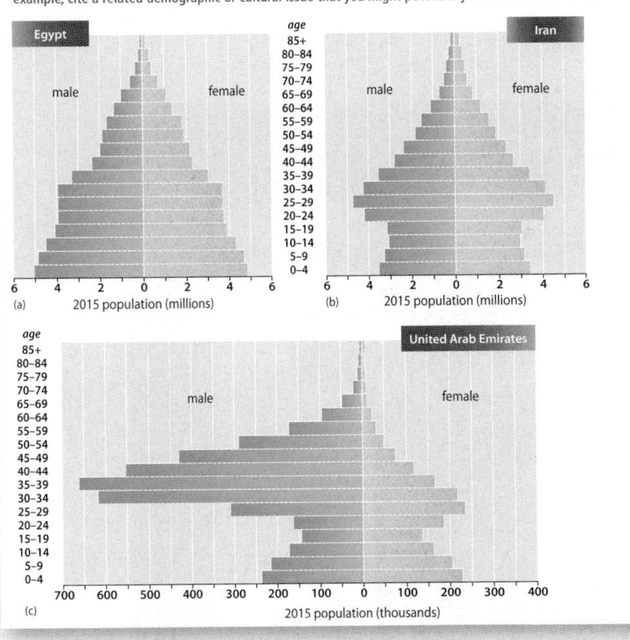

Figure 7.27 Population Pyramids: Egypt, Iran, and United Arab Emirates, 2015 Three distinctive demographic snapshots highlight regional diversity: (a) Egypt's above-average growth rates differ sharply from those of (b) Iran, where a focused campaign on family planning has reduced recent family sizes. (c) Male immigrant laborers play a special role in skewing the pattern within the United Arab Emirates. **Q: For each example, cite a related demographic or cultural issue that you might potentially find in these countries.**

NEW! Visual Analysis Questions within each chapter section give students more opportunity to stop and practice visual analysis, data analysis, and critical thinking as they read.

NEW! End-of-Chapter *Review* features provide highly-visual and interactive reviews of each chapter, organized around learning outcomes and incorporating satellite-based imagery, photos, and GIS-built maps. This active-review section revisits the key issues from the region at multiple spatial scales, links to constantly updated resources at the **Author Blogs**, and presents students with **NEW conceptual, visual, & Data Analysis activities**.

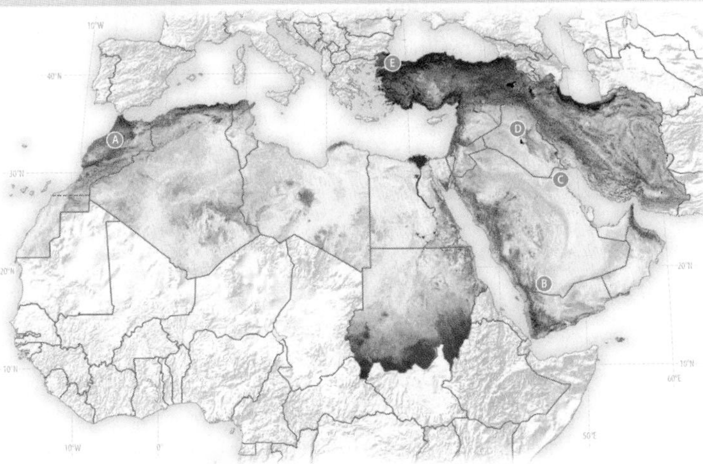

Review

Physical Geography and Environmental Issues

7.1 Explain how latitude and topography produce the region's distinctive patterns of climate.

7.2 Describe how the region's fragile, often arid setting shapes contemporary environmental challenges.

7.3 Describe four distinctive ways in which people have learned to adapt their agricultural practices to the region's arid environment.

Many nations within the region face significant environmental challenges and growing pressures on limited supplies of agricultural land and water. The results, from the eroded soils of the Atlas Mountains to the overworked garden plots along the Nile, illustrate the environmental price paid when population growth outstrips the ability of the land to support it.

1. If populations outstrip water supplies in North Africa's oasis settlements, how might residents adjust?
2. List ways in which modern technology might address water shortages across the region. Are there limits or challenges to this approach?

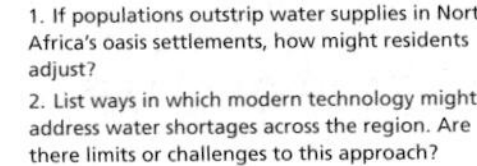

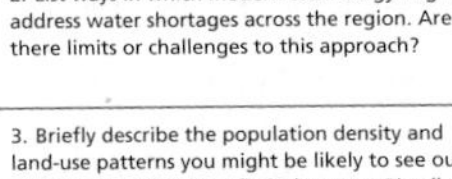

Population and Settlement

7.4 Summarize the major forces shaping recent migration patterns within the region.

The population geography of Southwest Asia and North Africa is strikingly uneven. Areas with higher rainfall or access to exotic water often have very high physiological population densities, whereas nearby arid zones remain almost empty of settlement.

3. Briefly describe the population density and land-use patterns you might be likely to see out the plane window on a flight between Riyadh (Saudi Arabia) and San'a (Yemen).
4. How might very low population densities impose special problems for maintaining effective political control across all portions of nations such as Saudi Arabia, Libya, and Algeria?

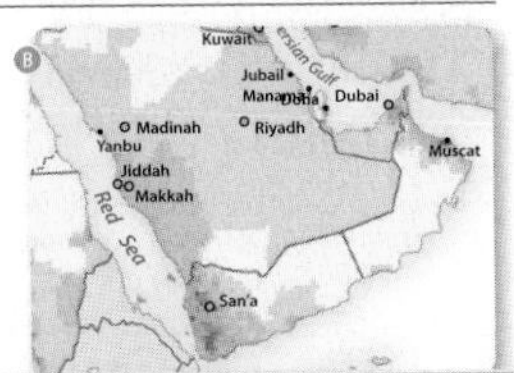

Cultural Coherence and Diversity

7.5 List the major characteristics and patterns of diffusion of Islam.

7.6 Identify the key modern religions and language families that dominate the region.

7.7 Identify the role of cultural variables in understanding key regional conflicts in North Africa, Israel, Syria, Iraq, and the Arabian Peninsula.

Culturally, the region remains the hearth of Christianity, the spatial and spiritual core of Islam, and the political and territorial focus of modern Judaism. In addition, important sectarian divisions within religious traditions (especially the schism between Sunnis and Shiites), as well as long-standing linguistic differences, continue to shape the local cultural geographies and regional identities.

5. Why is Islam both a powerful unifying and a divisive cultural force in the region?
6. Why does Saudi Arabia remain such a pivotal part of the Islamic world?

Geopolitical Framework

7.8 Summarize the geography of oil and gas reserves in the region.

Political conflicts have disrupted economic development. Civil wars, sectarian violence, conflicts between states, and regional tensions work against initiatives for greater cooperation and trade. Perhaps most important, the region must deal with the conflict between modernity and more fundamentalist interpretations of Islam.

7. How likely is it that the cultural and religious divisions in Iraq will be healed in 5–10 years?
8. Work with other students in the class to organize a debate on whether a renewed oil boom in the Iraqi economy might spur *greater* or *reduced* levels of sectarian violence within the country.

Economic and Social Development

7.9 Describe traditional roles for Islamic women and provide examples of recent changes.

Abundant reserves of oil and natural gas, coupled with the global economy's continuing reliance on fossil fuels, ensure that the region will remain prominent in world petroleum markets. Also likely are moves toward economic diversification and integration, which may gradually draw the region closer to Europe and other participants in the global economy.

9. What are likely to be the chief drivers of economic growth in settings such as Istanbul, Turkey, in the next 10–20 years?
10. Write an essay comparing and contrasting the challenges of producing sustained economic growth in Turkey and Saudi Arabia between 2020 and 2030.

KEY TERMS

Arab League *(p. 242)*
Arab Spring *(p. 219)*
brain drain *(p. 248)*
choke point *(p. 224)*
culture hearth *(p. 220)*
domestication *(p. 226)*
exotic river *(p. 228)*
Fertile Crescent *(p. 226)*
fossil water *(p. 224)*
Greater Arab Free Trade Area (GAFTA) *(p. 245)*
Hajj *(p. 233)*
hydropolitics *(p. 224)*
ISIL (Islamic State of Iraq and the Levant; also ISIS or Islamic State) *(p. 220)*
Islamic fundamentalism *(p. 220)*
Islamism *(p. 220)*
Levant *(p. 220)*
Maghreb *(p. 220)*
medina *(p. 229)*
monotheism *(p. 233)*
Organization of the Petroleum Exporting Countries (OPEC) *(p. 220)*
Ottoman Empire *(p. 234)*
Palestinian Authority (PA) *(p. 241)*
pastoral nomadism *(p. 226)*
physiological density *(p. 226)*
Quran *(p. 233)*
salinization *(p. 223)*
sectarian violence *(p. 219)*
Shiite *(p. 233)*
Suez Canal *(p. 239)*
Sunni *(p. 233)*
theocratic state *(p. 233)*
transhumance *(p. 226)*

DATA ANALYSIS

Health care is often considered a basic human right in more developed portions of the world, but large parts of Southwest Asia and North Africa are poorly served by health-care providers. The World Health Organization (WHO) gathers data on the number of physicians per 1000 population, which can be used as a measure of access to health care as well as social development. According to recent data, the United States had about 2.5 physicians per 1000 and Germany about 3.9. Go to the WHO website (www.who.int) and access the data/interactive atlas page on physicians per 1000 population.

1. Make your own data table and map showing the regional pattern of health-care access across Southwest Asia and North Africa.
2. In a few sentences, summarize the general patterns and trends you see. How would you explain some of the major variations you observe across the region?
3. Compare the pattern you see for physicians with the map in the text on childhood mortality (Figure 7.44). What similarities and differences do you see? How might these two indicators be a good measure of future social development? How might they predict political stability?

MasteringGeography™

Looking for additional review and test prep materials? Visit the Study Area in MasteringGeography™ to enhance your geographic literacy, spatial reasoning skills, and understanding of this chapter's content by accessing a variety of resources, including MapMaster interactive maps, geoscience animations, videos, *In the News* RSS feeds, flashcards, web links, self-study quizzes, and an eText version of *Globalization and Diversity*.

Authors' Blogs

Scan to visit the **Author's Blog for field notes, media resources, and chapter updates**

Scan to visit the **GeoCurrents Blog**

Continuous Learning Before, During, and After Class

BEFORE CLASS

Mobile Media & Reading Assignments Ensure Students Come to Class Prepared.

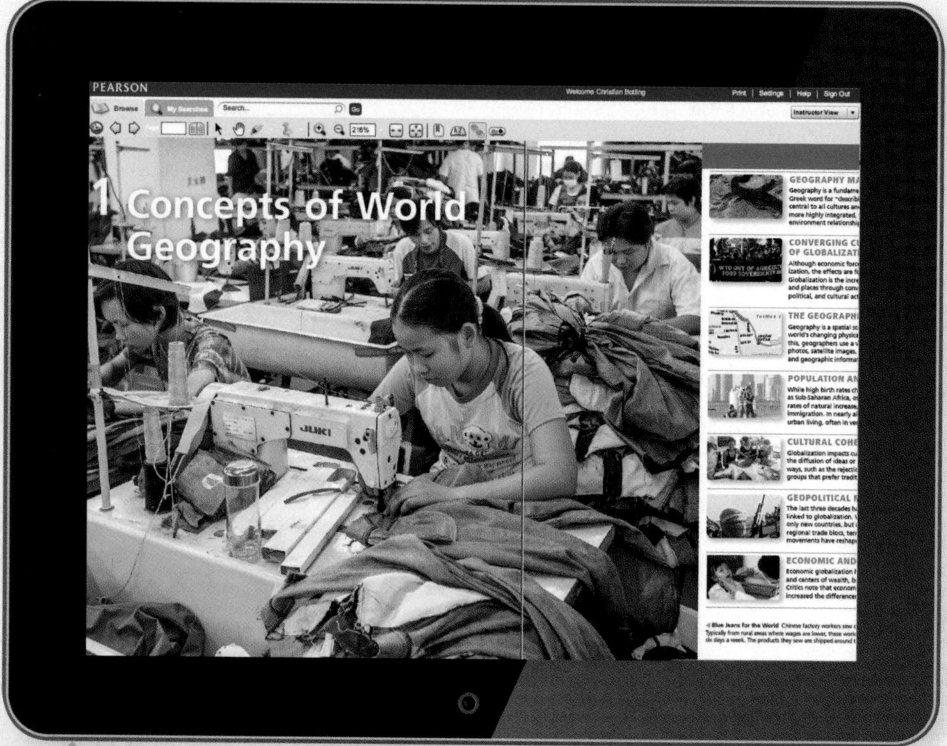

NEW! Dynamic Study Modules personalize each student's learning experience. Created to allow students to acquire knowledge on their own and be better prepared for class discussions and assessments, this mobile app is available for iOS and Android devices.

Pearson eText in MasteringGeography gives students access to the text whenever and wherever they can access the internet. eText features include:

- Now available on smartphones and tablets.
- Seamlessly integrated videos and other rich media.
- Fully accessible (screen-reader ready).
- Configurable reading settings, including resizable type and night reading mode.
- Instructor and student note-taking, highlighting, bookmarking, and search.

Pre-Lecture Reading Quizzes are easy to customize & assign

Reading Questions ensure that students complete the assigned reading before class and stay on track with reading assignments. Reading Questions are 100% mobile ready and can be completed by students on mobile devices.

with MasteringGeography™

DURING CLASS

Learning Catalytics™ & Engaging Media

What has Teachers and Students excited? Learning Cataltyics, a 'bring your own device' student engagement, assessment, and classroom intelligence system, allows students to use their smartphone, tablet, or laptop to respond to questions in class. With Learning Cataltyics, you can:

- Assess students in real-time using open ended question formats to uncover student misconceptions and adjust lecture accordingly.
- Automatically create groups for peer instruction based on student response patterns, to optimize discussion productivity.

> ***"My students are so busy and engaged answering Learning Catalytics questions during lecture that they don't have time for Facebook."***
>
> Declan De Paor, *Old Dominion University*

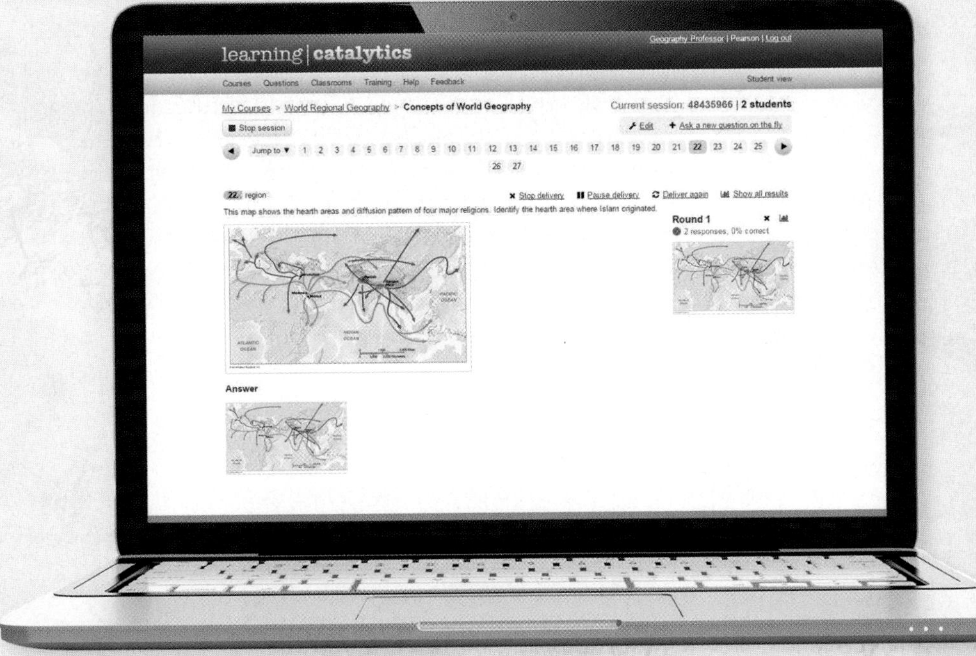

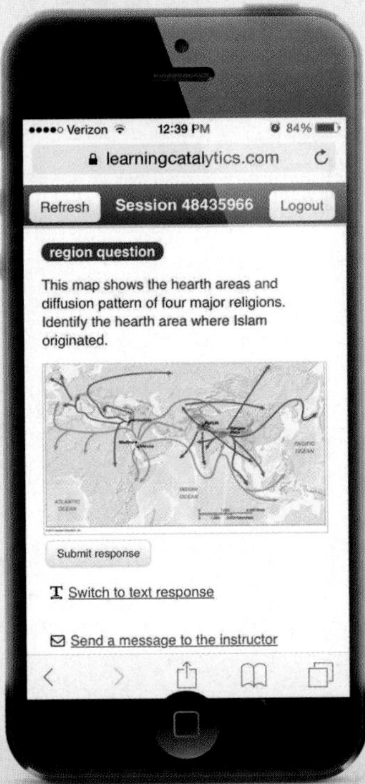

Enrich Lecture with Dynamic Media

Teachers can incorporate dynamic media from MasteringGeography into lecture, such as Videos, MapMaster Interactive Maps, and Geoscience Animations.

Mastering Geography™

MasteringGeography delivers engaging, dynamic learning opportunities—focusing on course objectives and responsive to each student's progress—that are proven to help students absorb world regional geography course material and understand challenging geography processes and concepts.

AFTER CLASS

Easy to Assign, Customizable, Media-Rich, and Automatically Graded Assignments

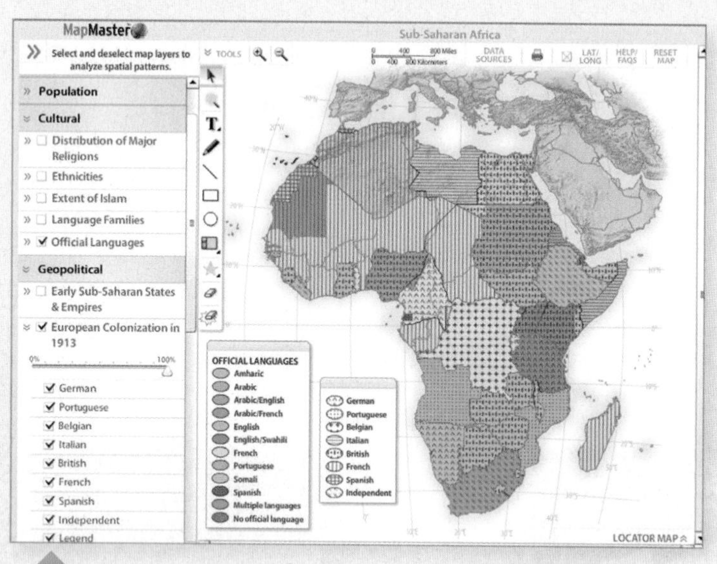

UPDATED! MapMaster Interactive Map Activities are inspired by GIS, allowing students to layer various thematic maps to analyze spatial patterns and data at regional and global scales. This tool includes zoom and annotation functionality, with hundreds of map layers leveraging recent data from sources such as NOAA, NASA, USGS, United Nations, and the CIA.

NEW! Geography Videos from such sources as the BBC and *The Financial Times* are now included in addition to the videos from Television for the Environment's *Life and Earth Report* series in **MasteringGeography**. Approximately 200 video clips for over 30 hours of footage are available to students and teachers and **MasteringGeography**.

NEW! Google Earth Virtual Tour Videos enhance *Exploring Global Connections* and *Working Toward Sustainability* features with brief, mobile-ready, narrated video explorations of landscapes related to each feature.

www.masteringgeography.com

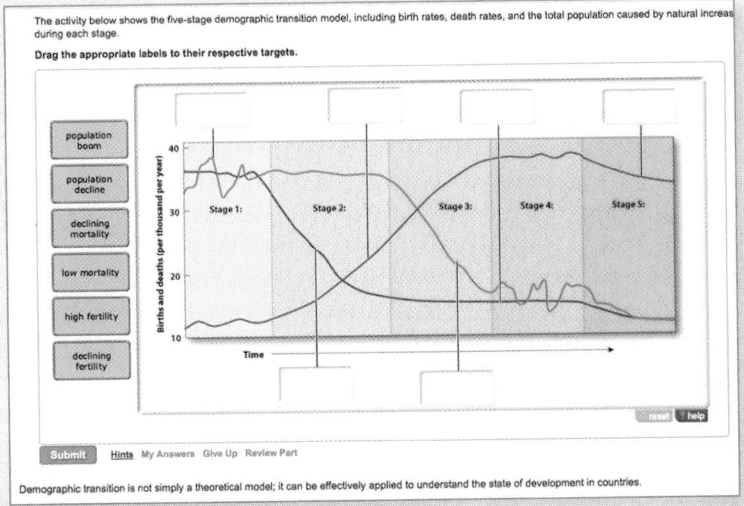

NEW! GeoTutors Highly visual and data-rich coaching items with hints and specific wrong answer feedback help students master the toughest topics in geography.

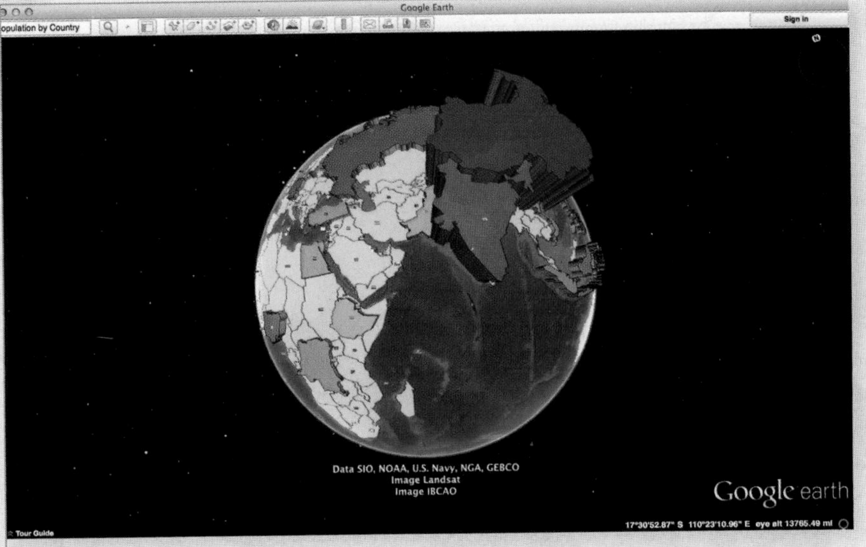

UPDATED! Encounter (Google Earth) activities provide rich, interactive explorations of regional geography concepts, allowing students to visualize spatial data and tour distant places on the virtual globe.

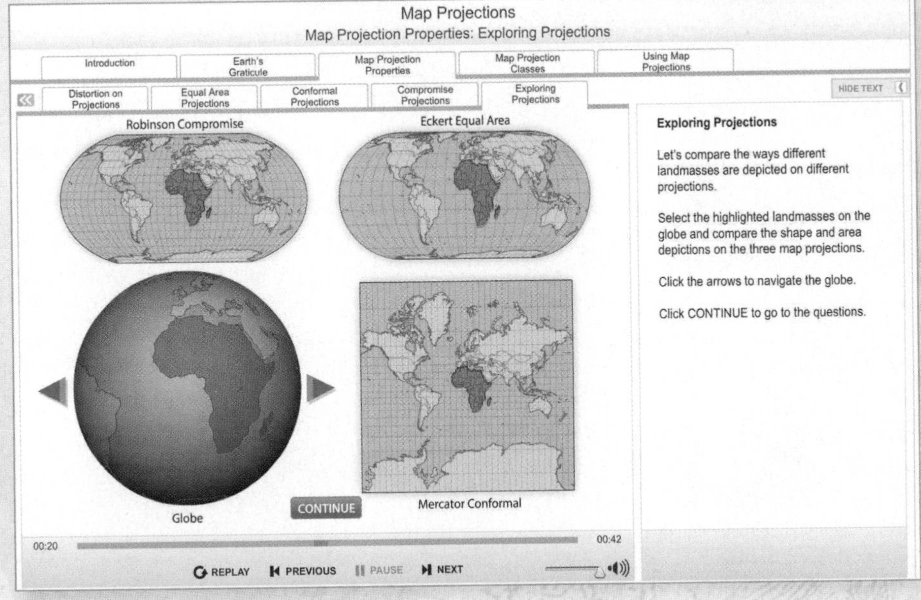

Map Projections interactive tutorial media helps reinforce and remediate students on the basic yet challenging introductory map projection concepts.

GLOBALIZATION

AND DIVERSITY

GEOGRAPHY OF A CHANGING WORLD

FIFTH EDITION

LES ROWNTREE
University of California, Berkeley

MARTIN LEWIS
Stanford University

MARIE PRICE
George Washington University

WILLIAM WYCKOFF
Montana State University

PEARSON

Senior Geography Editor: Christian Botting
Executive Product Marketing Manager: Neena Bali
Senior Field Marketing Manager: Mary Salzman
Program Manager: Anton Yakovlev
Director of Development: Jennifer Hart
Development Editor: Veronica Jurrgena
Marketing Assistant: Ami Sampat
Senior Content Producer: Tim Hainley
Team Lead, Program Management: Kristen Flathman
Team Lead, Project Management: David Zielonka
Project Manager: Nicole Antonio
Full Service/Composition: Cenveo Publisher Services
Full Service Project Manager: Mary Tindle, Cenveo Publisher Services
Cartography and Illustrations: International Mapping
Image Lead Manager, Rights Management: Rachel Youdelman
Photo Researcher: Kristin Piljay
Design Manager: Mark Ong
Interior and Cover Designer: Jeff Puda Design
Procurement Specialist: Maura Zaldivar-Garcia
Cover Photo Credit: Getty/WIN-Initiative

Credits and acknowledgments borrowed from other sources and reproduced, with permission, in this textbook appear on the appropriate page within the text or on the credits page beginning on page C-1.

Copyright © 2017, 2014, 2011, 2008, 2004 Pearson Education, Inc. All rights reserved. Manufactured in the United States of America. This publication is protected by Copyright, and permission should be obtained from the publisher prior to any prohibited reproduction, storage in a retrieval system, or transmission in any form or by any means: electronic, mechanical, photocopying, recording, or likewise. To obtain permission(s) to use material from this work, please submit a written request to Pearson Education, Inc., Permissions Department, 1 Lake Street, Department 1G, Upper Saddle River, NJ 07458.

Many of the designations used by manufacturers and sellers to distinguish their products are claimed as trademarks. Where those designations appear in this book, and the publisher was aware of a trademark claim, the designations have been printed in initial caps or all caps.

Library of Congress Cataloging-in-Publication Data
Names: Rowntree, Lester, 1938- author.
Title: Globalization and diversity : geography of a changing world / Les Rowntree, University of California, Berkeley, Martin Lewis, Stanford University, Marie Price, George Washington University, William Wyckoff, Montana State University.
Description: Fifth edition. | Hoboken, NJ : Pearson, [2017]
Identifiers: LCCN 2015040429 | ISBN 9780134117010 (Student Edition)
Subjects: LCSH: Economic geography. | Globalization.
Classification: LCC HF1025 .G59 2017 | DDC 330.9--dc23 LC record available at http://lccn.loc.gov/2015040429

About Our Sustainability Initiatives

Pearson recognizes the environmental challenges facing this planet, as well as acknowledges our responsibility in making a difference. This book is carefully crafted to minimize environmental impact. The binding, cover, and paper come from facilities that minimize waste, energy consumption, and the use of harmful chemicals. Pearson closes the loop by recycling every out-of-date text returned to our warehouse.

Along with developing and exploring digital solutions to our market's needs, Pearson has a strong commitment to achieving carbon-neutrality. As of 2009, Pearson became the first carbon- and climate-neutral publishing company, having reduced our absolute carbon footprint by 22% since then. Pearson has protected over 1,000 hectares of land in Columbia, Costa Rica, the United States, the UK and Canada. In 2015, Pearson formally adopted The Global Goals for Sustainable Development, sponsoring an event at the United Nations General Assembly and other ongoing initiatives. Pearson sources 100% of the electricity we use from green power and invests in renewable energy resources in multiple cities where we have operations, helping make them more sustainable and limiting our environmental impact for local communities.

The future holds great promise for reducing our impact on Earth's environment, and Pearson is proud to be leading the way. We strive to publish the best books with the most up-to-date and accurate content, and to do so in ways that minimize our impact on Earth. To learn more about our initiatives, please visit **https://www.pearson.com/social-impact/sustainability/environment.html**

PEARSON

www.pearsonhighered.com

2 16

Brief Contents

Contents

Forestview Chinese Restaurant
BANK
Forestview Chinese Restaurant
ORIENT HARVEST
隆豐行
ORIENT HARVEST
FACTORY PRICE

BOR•04415

14 AUSTRALIA AND OCEANIA 452

Preface

Globalization and Diversity: Geography of a Changing World, Fifth Edition, is an issues-oriented textbook for college and university world regional classes that explicitly recognizes the vast geographic changes taking place because of globalization. With this focus we join the many scholars who consider globalization to be the most fundamental reorganization of the world's socioeconomic, cultural, and geopolitical structure since the Industrial Revolution. That provides the point of departure and thematic structure for our book.

As geographers, we think it essential for our readers to understand and critique two interactive themes: the consequences of converging environmental, cultural, political, and economic systems inherent to globalization and the persistence—and even expansion—of geographic diversity and differences in the face of globalization. These two opposing forces, homogenization and diversification, are reflected in our book's title, *Globalization and Diversity*.

NEW TO THE FIFTH EDITION

- **Everyday Globalization** illustrates how globalization permeates every aspect of one's life, even the most mundane and taken-for-granted, such as one's food, clothing, cell phones, and music.
- **Geographers at Work** introduce readers to 14 professional geographers their research, fieldwork, teaching, and lives, including how and when they decided to make Geography the focus of their lives.
- **Google Earth Virtual Tour Videos** link via Quick Response (QR) codes from *Working Toward Sustainability* and *Exploring Global Connections* features, providing mobile-ready, on-the-go virtual tours of the geography and places discussed in the sidebar.
- Chapter opening pages introduce readers to key themes and characteristics of the regions with large panoramic photographs, a selection of visual and brief textual previews of the chapter sections, and a real-world vignette.
- **Visual questions** integrated with select figures in each chapter section give students opportunities to apply critical thinking skills and perform visual analysis.
- **End-of-chapter Review** sections provide a highly-visual summary and active-learning review of each chapter, with integrated maps, photos, and illustrations, critical thinking questions, key terms, a Data Analysis exercise, and QR links to **two Author Blogs**.
- The **Author Blogs** discuss everything from current events to author travels and field research. Both blogs are graphically rich with innovative maps and photos, extending the print book with dynamically updated information and data on current events from around the globe.

New to Chapter 1: Concepts of World Geography

- **Geography Matters**. New discussion of fundamental geographic concepts, including areal differentiation, regions, and the cultural landscape.
- **Geographer's Toolbox**. New discussion of latitude and longitude, map projections, scale, chorographic maps, aerial photos, remote sensing, and GIS.
- A new and expanded discussion of *Global Migration and Settlement*.
- Demographic transition revised. Following the lead of professional demographers, a fifth stage has been added to the traditional demographic transition model to account for the current very low natural population rates in developed countries.
- **The Nation-State Revisited**. A critical view of the traditional nation-state concept sets the scene for regional material on post- and neocolonial tensions, microregionalism, ethnic separatism, migrant enclaves, and multicultural nationalism.

New to Chapter 2: Physical Geography and the Environment

- New graphics and content on tectonic plate boundaries.
- An expanded and graphically rich section on climate controls. This expanded section explains the climate controls of solar energy, latitude, land-water interactions, global pressure and wind systems, and topography.
- Updated and expanded section on climate change and global warming. Drawing upon the latest data from the Intergovernmental Panel on Climate Change's *Fifth Assessment Report* this section presents not just the latest data about climate change and global warming, but also the complex international negotiations on limiting CO_2 emissions from the 2015 Paris meeting.
- A new section on global energy issues. Linked to the previous material on climate change and global warming, this new section discusses the geography of global energy resources, both renewable and nonrenewable, including material on hydraulic fracturing ("fracking").
- Revised and expanded material on bioregions and biodiversity. A more detailed cartographic depiction of biomes and bioregions is complemented by a fuller discussion of the world's ecological diversity, as well as the issues faced in protecting those environments around the globe.

Organization

Globalization and Diversity opens with two substantive introductory chapters that provide the geographic fundamentals of both human and physical geography. Chapter 1: "Concepts of World Geography" begins by providing readers with background on the geographic dimensions of globalization, including a section on the costs and benefits of globalization according to proponents and opponents. Next is an introduction to the discipline of geography and its major concepts, which leads into a section called "The Geographer's Toolbox," where students are informed about such matters as map-reading, cartography, aerial photos, remote sensing, and GIS. This initial chapter concludes with a discussion of the concepts and tabular data that are used throughout the regional chapters.

Chapter Two: "Physical Geography and the Environment" builds an understanding of physical geography and environmental issues with discussions of geology; environmental hazards; weather, climate, and global warming; energy; hydrology and water stress; and global bioregions and biodiversity.

Each regional chapter is structured around five geographic themes:

- **Physical Geography and Environmental Issues**, in which we not only describe the physical geography of each region, but also environmental issues, including climate change and energy.
- **Population and Settlement**, where we examine the region's demography, migration patterns, land use, and settlement, including cities.
- **Cultural Coherence and Diversity** covers the traditional topics of language and religion, but also examines the ethnic and cultural tensions resulting from globalization. Gender issues and popular culture topics such as sports and music are also included in this section.
- **Geopolitical Framework** examines the political geography of the region, taking on such issues as postcolonial tensions, ethnic conflicts, separatism, micro-regionalism, and global terrorism.
- **Economic and Social Development**, where explores each region's economic framework at both local and global scales and examines such social issues as health, education, and gender inequalities.

CHAPTER FEATURES

- **Structured learning path**. Every chapter begins with an explicit set of learning objectives to provide students with the larger context of each chapter. Review questions after each section allow students to test their learning. Each chapter ends with an innovative, graphically rich "Review" section, where students are asked to apply what they have learned from the chapter using an active-learning framework.
- **Comparable regional maps**. Of the many maps in each regional chapter, many are constructed on the same themes and with similar data so that readers can easily draw comparisons between regions. Most regional chapters have maps of physical geography, climate, environmental issues, population density, migration, language, religion, and geopolitical issues.
- **Other chapter maps pertinent to each region**. The regional chapters also contain many additional maps illustrating important geographic topics such as global economic issues, social development, and ethnic tensions.
- **Comparable regional data sets in appendices**. Two thematic tables related to each regional chapter facilitate comparisons between regions and provide insights into the characteristics of each region. The first table provides population data on a number of issues, including fertility rates and proportions of the population under 15 and over 65 years of age, as well as net migration rates for each country within the region. The second table presents economic and social development data for each country, including gross national income per capita, gross domestic product growth, life expectancy, percentage of the population living on less than $2 per day, child mortality rates, and the international gender inequality index.
- **Sidebar essays**. Each chapter has four sidebars that expand on geographic themes:
  - ***Geographers at Work*** profiles active geographers, exploring their lives, education, and field work.
  - ***Working Toward Sustainability*** feature case studies of sustainability projects throughout the world, emphasizing positive environmental and social initiatives and their results. Each includes a QR link to an online Google Earth Virtual Tour Video.
  - ***Exploring Global Connections*** investigate the many ways in which activities in different parts of the world are linked so that students understand that in globalized world regions are neither isolated nor discrete. Each includes a QR link to an online Google Earth Virtual Tour Video.
  - ***Everyday Globalization*** illustrate the many ways that globalization permeates one's everyday life, from food, to clothing, to cell phones, to music.

ACKNOWLEDGEMENTS

We have many people to thank for the conceptualization, writing, rewriting, and production of *Globalization and Diversity*, Fifth Edition. First, we'd like to thank the thousands of students in our world regional geography classes who have inspired us with their energy, engagement, and curiosity; challenged us with their critical insights; and demanded a textbook that better meets their need to understand the contemporary geography of their dynamic and complex world.

Next, we are deeply indebted to many professional geographers and educators for their assistance, advice, inspiration, encouragement, and constructive criticism as we labored through the different stages of this book. Among the many who provided invaluable comments on various drafts and editions of *Globalization and Diversity* or who worked on supporting print or digital material are:

Gilian Acheson, *Southern Illinois University, Edwardsville*
Joy Adams, Humboldt State University
Dan Arreola, Arizona State University
Bernard BakamaNume, Texas A&M University
Brad Baltensperger, Michigan Technological University
Max Beavers, Samford University
Laurence Becker, Oregon State University
Dan Bedford, Weber State University

James Bell, University of Colorado
Katie Berchak, University of Louisiana, Lafayette
William H. Berentsen, University of Connecticut
Kevin Blake, Kansas State University
Mikhail Blinnikov, St. Cloud State University
Michelle Brym, University of Central Oklahoma
Karl Byrand, University of Wisconsin, Sheboygan County
Michelle Calvarese, California State University, Fresno
Craig Campbell, Youngstown State University
G. Scott Campbell, College of DuPage
Elizabeth Chacko, George Washington University
Philip Chaney, Auburn University
Xuwei Chen, Northern Illinois University
David B. Cole, University of Northern Colorado
Amanda Coleman, Northeastern State University
Malcolm Comeaux, Arizona State University
Jonathan C. Comer, Oklahoma State University
Jeremy Crampton, George Mason University
Kevin Curtin, University of Texas at Dallas
James Curtis, California State University, Long Beach
Dydia DeLyser, California State University, Fullerton
Francis H. Dillon, George Mason University
Jason Dittmer, Georgia Southern University
Jerome Dobson, University of Kansas
Caroline Doherty, Northern Arizona University
Vernon Domingo, Bridgewater State College
Roy Doyon, Ball State University
Dawn Drake, Missouri Western State University
Jane Ehemann, Shippensburg University
Steven Ericson, University of Alabama
Chuck Fahrer, Georgia College and State University
Dean Fairbanks, California State University, Chico
Emily Fekete, University of Kansas
Caitie Finlayson, Florida State University
Colton Flynn, University of Arkansas, Fort Smith
Doug Fuller, University of Miami
Douglas Gamble, University of North Carolina, Wilmington
Sherry Goddicksen, California State University, Fullerton
Sarah Goggin, Cypress College
Reuel Hanks, Oklahoma State University
Megan Hoberg, Orange Coast College & Golden West College
Steven Hoelscher, University of Texas, Austin
Erick Howenstine, Northeastern Illinois University
Tyler Huffman, Eastern Kentucky University
Peter J. Hugil, Texas A&M University
Eva Humbeck, Arizona State University
Shireen Hyrapiet, Oregon State University
Drew Kapp, University of Hawaii, Hilo
Ryan S. Kelly, University of Kentucky
Richard H. Kesel, Louisiana State University
Cadey Korson, Kent State University
Rob Kremer, Front Range Community College
Robert C. Larson, Indiana State University
Alan A. Lew, Northern Arizona University
Elizabeth Lobb, Mt. San Antonio College
Catherine Lockwood, Chadron State College
Max Lu, Kansas State University
Luke Marzen, Auburn University
Kent Matthewson, Louisiana State University
Daniel McGowin, Auburn University
James Miller, Clemson University
Bob Mings, Arizona State University
Wendy Mitteager, SUNY, Oneonta
Sherry D. Morea-Oakes, University of Colorado, Denver
Anne E. Mosher, Syracuse University
Julie Mura, Florida State University
Tim Oakes, University of Colorado
Nancy Obermeyer, Indiana State University
Karl Offen, University of Oklahoma
Thomas Orf, Las Positas College
Kefa Otiso, Bowling Green State University
Joseph Palis, University of North Carolina
Jean Palmer-Moloney, Hartwick College
Bimal K. Paul, Kansas State University
Michael P. Peterson, University of Nebraska, Omaha
Richard Pillsbury, Georgia State University
Brandon Plewe, Brigham Young University
Jess Porter, University of Arkansas at Little Rock
Patricia Price, Florida International University
Erik Prout, Texas A&M University
Claudia Radel, Utah State University
David Rain, George Washington University
Rhonda Reagan, Blinn College
Joshua Regan, Western Connecticut State University
Kelly Ann Renwick, Appalachian State University
Craig S. Revels, Portland State University
Pamela Riddick, University of Memphis
Scott M. Robeson, Indiana State University
Paul A. Rollinson, Southwest Missouri State University
Jessica Salo, University of Northern Colorado
Yda Schreuder, University of Delaware
Kathy Schroeder, Appalachian State University
Kay L. Scott, University of Central Florida
Patrick Shabram, South Plains College
Duncan Shaeffer, Arizona State University
Dimitrii Sidorov, California State University, Long Beach
Susan C. Slowey, Blinn College
Andrew Sluyter, Louisiana State University
Christa Smith, Clemson University
Joseph Spinelli, Bowling Green State University
William Strong, University of Northern Alabama
Philip W. Suckling, University of Northern Iowa
Curtis Thomson, University of Idaho
Suzanne Traub-Metlay, Front Range Community College
James Tyner, Kent State University
Nina Veregge, University of Colorado
Jonathan Walker, James Madison University
Fahui Wang, Louisiana State University
Gerald R. Webster, University of Alabama
Keith Yearman, College of DuPage
Emily Young, University of Arizona
Bin Zhon, Southern Illinois University, Edwardsville
Henry J. Zintambia, Illinois State University
Sandra Zupan, University of Kentucky

In addition, we wish to thank the many publishing professionals who have made this book possible. We start with Christian Botting,

Pearson Senior Editor for Geography, Meteorology, and Geospatial Technologies, a consummate professional and good friend whose leadership, high standards, and enduring patience has been laudable, inspiring, and necessary. Next in line is Program Manager Anton Yakovlev, for his daily attention to production matters as well as his graceful and diplomatic interaction with four demanding and often cranky authors. Many thanks, as well, to Nicole Antonio, our Pearson Project Manager. Veronica Jurgena, our outstanding Development Editor, whose insights, guidance, and encouragement (much of it coming from her education as a geographer) has been absolutely crucial to this revision. Also to be thanked are a number of behind-the-curtain professionals: Kristin Piljay, photo researcher; Rachel Youdelman, Manager, Rights Management; Mary Tindle, Cenveo Project Manager, for somehow turning thousands of pages of manuscript into a finished product; and Kevin Lear, *International Mapping* Senior Project Manager, for his outstanding work creating and revising our maps. Thanks are due to Gloriana Sojo for her timely production of all data tables.

Not to be overlooked are the 14 professional geographers who allowed us to pry into their personal and professional lives so we could profile them in the new Geographers at Work sidebars. They are:

Fenda Akiwumi, University of South Florida
Holly Barcus, Macalester College
Sarah Blue, Texas State University
Laura Brewington, East-West Center
Karen Culcasi, West Virginia University
Corrie Drummond Garcia, USAID
M Jackson, University of Oregon
P.P. Karah, University of Kentucky
Weronika Kusek, Northern Michigan University
Lucia Lo, York University
Rachel Silvey, University of Toronto
Dmitry Streletskiy, The George Washington University
Gregory Veeck, Western Michigan University
Susan Wolfinbarger, American Association for the Advancement of Science

Last, the authors want to thank that special group of friends and family who were there when we needed you most—early in the morning and late at night; in foreign countries and at home; when we were on the verge of tears and rants, but needed lightness and laughter; for your love, patience, companionship, inspiration, solace, enthusiasm, and understanding. Words cannot thank you enough: Magdalena Cooper, Meg Conkey, Rob, Joseph, and James Crandall, Marie Dowd, Evan and Eleanor Lewis, Karen Wigen, and Linda, Tom, and Katie Wyckoff.

Les Rowntree

Martin Lewis

Marie Price

William Wyckoff

About The Authors

Les Rowntree is currently a Research Associate at the University of California, Berkeley, where he writes about global and local environmental issues. This career change comes after 35 years teaching both Geography and Environmental Studies at San Jose State University. As an environmental geographer, Dr. Rowntree's interests focus on international environmental issues, biodiversity conservation, and climatic change. He sees world regional geography as way to engage and inform students by providing them with the conceptual tools to critically and constructively assess the contemporary world. His current writing projects include a natural history book and website about California's Coast Ranges, and several essays on different European environmental topics. Along with these writings he maintains an assortment of web-based blogs and websites.

Martin Lewis is a Senior Lecturer in History at Stanford University, where he teaches courses on global geography. He has conducted extensive research on environmental geography in the Philippines and on the intellectual history of world geography. His publications include *Wagering the Land: Ritual, Capital, and Environmental Degradation in the Cordillera of Northern Luzon, 1900–1986* (1992), and, with Karen Wigen, *The Myth of Continents: A Critique of Metageography* (1997). Dr. Lewis has traveled extensively in East, South, and Southeastern Asia. His most recent book, co-written with Asya Pereltsvaig, is *The Indo-European Controversy: Facts and Fallacies in Historical Linguistics* (2015). In April 2009, Dr. Lewis was recognized by *Time* magazine as one of American's most favorite lecturers.

Marie Price is a Professor of Geography and International Affairs at George Washington University. A Latin American specialist, Dr. Price has conducted research in Belize, Mexico, Venezuela, Panama, Cuba, and Bolivia. She has also traveled widely throughout Latin America and Sub-Saharan Africa. Her studies have explored human migration, natural resource use, environmental conservation, and sustainability. She is a nonresident fellow of the Migration Policy Institute, a nonpartisan think tank that focuses on migration issues, and is President of the American Geographical Society of the American Geographical Society. Dr. Price brings to *Globalization and Diversity* a special interest in regions as dynamic spatial constructs that are shaped over time through both global and local forces. Her publications include the co-edited book *Migrants to the Metropolis: The Rise of Immigrant Gateway Cities* (2008) and numerous academic articles and book chapters.

William Wyckoff is a geographer in the Department of Earth Sciences at Montana State University specializing in the cultural and historical geography of North America. He has written and co-edited several books on North American settlement geography, including *The Developer's Frontier: The Making of the Western New York Landscape* (1988), *The Mountainous West: Explorations in Historical Geography* (1995) (with Lary M. Dilsaver), *Creating Colorado: The Making of a Western American Landscape 1860–1940* (1999), and *On the Road Again: Montana's Changing Landscape* (2006). His most recent book, entitled *How to Read the American West: A Field Guide*, appeared in the Weyerhaeuser Environmental Books series and was published in 2014 by the University of Washington Press. A World Regional Geography instructor for 26 years, Dr. Wyckoff emphasizes in the classroom the connections between the everyday lives of his students and the larger global geographies that surround them and increasingly shape their future.

Digital & Print Resources

This edition provides a complete world regional geography program for students and teachers.

FOR STUDENTS & TEACHERS

MasteringGeography™ with Pearson eText.
The Mastering platform is the most widely used and effective online homework, tutorial, and assessment system for the sciences. It delivers self-paced coaching activities that provide individualized coaching, focus on course objectives, and are responsive to each student's progress. The Mastering system helps teachers maximize class time with customizable, easy-to-assign, and automatically graded assessments that motivate students to learn outside of class and arrive prepared for lecture. MasteringGeography offers:

- **Assignable activities** that include GIS-inspired MapMaster™ interactive map activities, *Encounter* Google Earth Explorations, Video activities, Geoscience Animation activities, Map Projections activities, GeoTutor coaching activities on the toughest topics in geography, Dynamic Study Modules that customize the student's learning experience, book questions and exercises, reading quizzes, Test Bank questions, and more.
- **A student Study Area** with GIS-inspired MapMaster™ interactive maps, videos, Geoscience Animations, web links, glossary flashcards, "In the News" RSS feeds, Google Earth Virtual Tour videos, chapter quizzes, PDF downloads of outline maps, an optional Pearson eText and more.

Pearson eText gives students access to the text whenever and wherever they can access the Internet. Features of Pearson eText include:

- Now available on smartphones and tablets. Seamlessly integrated videos and other rich media.
- Fully accessible (screen-reader ready).
- Configurable reading settings, including resizable type and night reading mode.
- Instructor and student note-taking, highlighting, bookmarking, and search. **www.masteringgeography.com**

Television for the Environment Earth Report Geography Videos on DVD (0321662989).
This three-DVD set helps students visualize how human decisions and behavior have affected the environment and how individuals are taking steps toward recovery. With topics ranging from the poor land management promoting the devastation of river systems in Central America to the struggles for electricity in China and Africa, these 13 videos from Television for the Environment's global *Earth Report* series recognize the efforts of individuals around the world to unite and protect the planet.

FOR STUDENTS

Goode's World Atlas, 23rd Edition (0133864642).
Goode's World Atlas has been the world's premiere educational atlas since 1923. It features over 260 pages of maps, from definitive physical and political maps to important thematic maps that illustrate the spatial aspects of many important topics. The 23rd edition includes over 160 pages of digitally-produced reference maps, as well as new thematic maps on global climate change, sea level rise, CO_2 emissions, polar ice fluctuations, deforestation, extreme weather events, infectious diseases, water resources, and energy production, and more.

Pearson's Encounter Series.
provides rich, interactive explorations of geoscience concepts through GoogleEarth™ activities, covering a range of topics in regional, human, and physical geography. For those who do not use MasteringGeography, all chapter explorations are available in print workbooks, as well as in online quizzes at **www.mygeoscienceplace.com**, accommodating different classroom needs. Each exploration consists of a worksheet, online quizzes, and a corresponding Google Earth™ KMZ file.

- *Encounter World Regional Geography* Workbook and Website by Jess C. Porter (0321681754)
- *Encounter Human Geography* Workbook and Website by Jess C. Porter (0321682203)
- *Encounter Physical Geography* Workbook and Website by Jess C. Porter and Stephen O'Connell (0321672526)

Dire Predictions: Understanding Climate Change, 2nd Edition (0133909778 by Michael E. Mann and Lee R. Kump).
Periodic reports from the Intergovernmental Panel on Climate Change (IPCC) evaluate the risk of climate change brought on by humans. In just over 200 pages, this practical text presents and expands upon the IPCC's essential findings in a visually stunning and undeniably powerful way to the lay reader. Scientific findings that provide validity to the implications of climate change are presented in clear-cut graphic elements, striking images, and understandable analogies. The **Second Edition** covers the latest climate change data and scientific consensus from the IPCC Fifth Assessment Report and integrates mobile media links to online media. The text is also available in various eText formats, including an eText upgrade option from MasteringGeography courses.

FOR INSTRUCTORS

Instructor Resource Manual (Download) (0134142667).
The *Instructor Resource Manual* follows the new organization of the main text. It includes a sample syllabus, chapter learning objectives, lecture outlines, a list of key terms, and answers to the textbook's review and end-of-chapter questions. Discussion questions, classroom activities, and advice about how to integrate visual supplements (including MasteringGeography and Learning Catalytics resources) are integrated throughout the chapter lecture outlines.

TestGen/Test Bank (Download) (0134142683).
TestGen is a computerized test generator that lets instructors view and edit Test Bank questions, transfer questions to tests, and print tests in a variety of customized formats. This Test Bank includes approximately 1,500 multiple-choice, true/false, and short-answer/essay questions. Questions are correlated with the book's learning objectives, the revised U.S. National Geography Standards, chapter-specific learning outcomes, and Bloom's Taxonomy to help teachers better map the assessments against both broad and specific teaching and learning objectives. The Test Bank is also available in Microsoft Word® and Blackboard formats.

Instructor Resource DVD (0134142780).
The *Instructor Resource DVD* provides a collection of resources to help instructors make efficient and effective use of their time. All digital resources can be found in one well-organized, easy-to-access place. The IRC DVD includes:

- All textbook images as JPEGs, PDFs, and PowerPoint™ Presentations
- Pre-authored Lecture Outline PowerPoint™ Presentations, which outline the concepts of each chapter with embedded art and can be customized to fit instructors' lecture requirements
- CRS "Clicker" Questions in PowerPoint™ format, which correlate to the book's learning objectives, the U.S. National Geography Standards, chapter-specific learning outcomes, and Bloom's Taxonomy
- The TestGen software and *Test Bank* questions
- Electronic files of the *Instructor Resource Manual* and *Test Bank*

This Instructor Resource content is also available completely online via the Instructor Resources section of MasteringGeography and **www.pearsonhighered.com/irc**.

LEARNING CATALYTICS

Learning Catalytics™ is a "bring your own device" student engagement, assessment, and classroom intelligence system. With Learning Catalytics, you can

- Assess students in real time, using open-ended tasks to probe student understanding.
- Understand immediately where students are and adjust your lecture accordingly.
- Improve your students' critical thinking skills.
- Access rich analytics to understand student performance.
- Add your own questions to make Learning Catalytics fit your course exactly.
- Manage student interactions with intelligent grouping and timing.

Learning Catalytics is a technology that has grown out of 20 years of cutting-edge research, innovation, and implementation of interactive teaching and peer instruction. Available integrated with MasteringGeography.
www.learningcatalytics.com
www.masteringgeography.com

1 Concepts of World Geography

GEOGRAPHY MATTERS

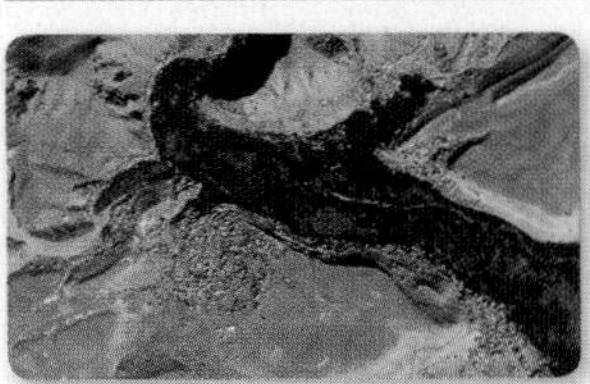

Geography is a fundamental science with its roots in the Greek word for "describing the Earth." This discipline is central to all cultures and helps us better understand a more highly integrated, yet diverse world in which human–environment relationships are constantly changing.

CONVERGING CURRENTS OF GLOBALIZATION

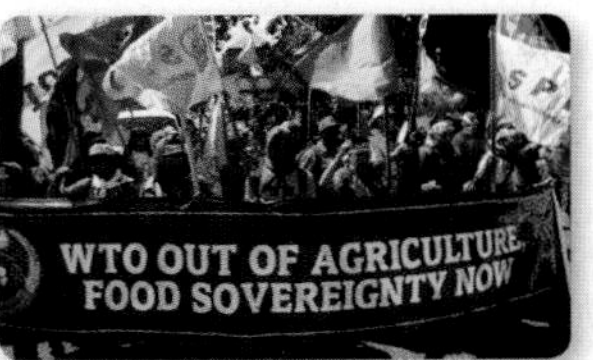

Although economic forces may drive many aspects of globalization, the effects are found in all facets of land and life. Globalization is the increasing interconnectedness of people and places through converging economic, technological, political, and cultural activities.

THE GEOGRAPHER'S TOOLBOX

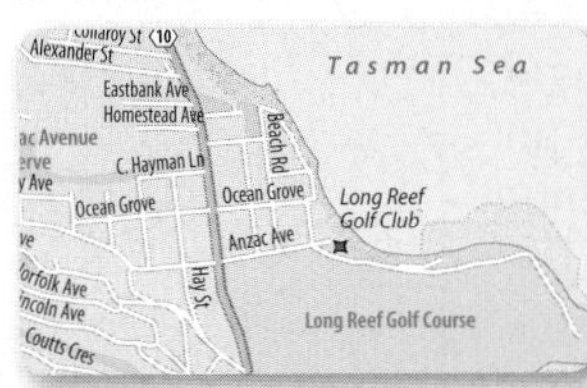

Geography is a spatial science that describes and analyzes the world's changing physical and human environments. To do this, geographers use a variety of tools such as maps, aerial photos, satellite images, global positioning systems (GPS), and geographic information systems (GIS).

POPULATION AND SETTLEMENT

While high birth rates characterize some world regions such as Sub-Saharan Africa, other areas such as Europe have low rates of natural increase, with most growth coming from immigration. In nearly all countries, the trend is for more urban living, often in very large cities.

CULTURAL COHERENCE AND DIVERSITY

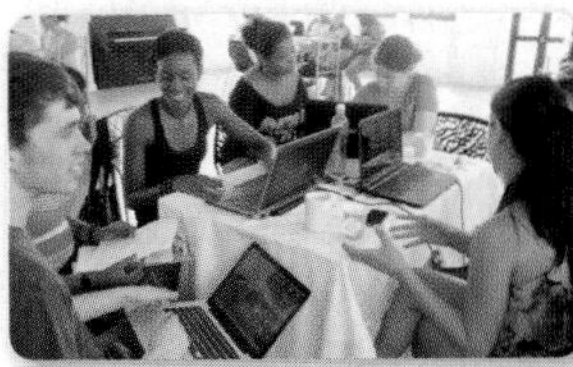

Globalization impacts culture in anticipated ways, such as the diffusion of ideas or practices, and in unanticipated ways, such as the rejection of introduced changes by cultural groups that prefer traditional ways of doing things.

GEOPOLITICAL FRAMEWORK

The last three decades have seen rapid geopolitical changes linked to globalization. With the end of the Cold War, not only new countries, but also new political actors such as regional trade blocs, terrorist networks, and ethnic separatist movements have reshaped the world map.

ECONOMIC AND SOCIAL DEVELOPMENT

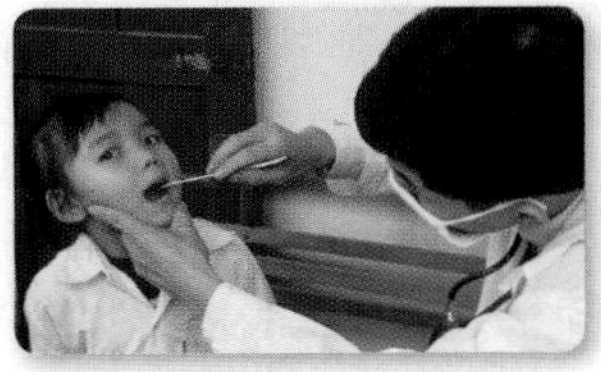

Economic globalization has created new world trade patterns and centers of wealth, but not for all people in all places. Critics note that economic and social disparities have actually increased the differences between rich and poor.

◀ Chinese factory workers sew denim jeans in the city of Shenzhen, China. Typically from rural areas where wages are lower, these workers live in factory-owned dorms and work six days a week. The products they sew are shipped around the world.

Denim jeans are an iconic American cultural symbol found all over the world—yet very few jeans are made in the United States anymore. The story of the humble blue work pants created for gold miners in 1850s California, but later reimagined by high-end design houses in Europe and assembled in sweatshops in China and Mexico, is indicative of the long-term and uneven workings of globalization. We can think of jeans as an assemblage of materials and processes. Cotton grown in the United States, India, Uzbekistan, or Australia makes its way to textile houses in Turkey and Pakistan that convert it into denim. The bolts of denim are shipped to scores of countries for cutting and sewing, usually places where labor costs are low or trade agreements give preference to particular markets. One of the largest producers of blue jeans is China. In places such as Shenzhen in southern China, men and women toil in large factories sewing pant legs or putting in rivets. The long hours of work with this dyed fabric stain the workers' fingers blue. The finished products are then packaged and shipped all over the world or sold to Chinese consumers. After years of wear, a pair of jeans may even end up in the recycled clothing market or sold in bundles to hawkers in African cities such as Lusaka or Accra. Such global patterns of production and trade are increasingly the norm.

Blue jeans are also a cultural product, which can be a comfortable work pant or a high-end status symbol. In some cultures, mostly men wear jeans, which are associated with modern Western values; meanwhile women might wear more traditional attire. Styles and colors of jeans vary from place to place. The meaning people give to the blue jeans, and the decisions about who can wear them and when, underscore the diverse cultural practices at play with this ordinary garment. A single pair of blue jeans has environmental impacts as well. A study by Levi Strauss & Co. determined that 3000 liters of water were consumed during the life of one pair of 501 jeans: Half of the water went to growing the cotton, less than half went to customers washing jeans, and a small fraction was used to manufacture them.

Globalization and Diversity investigates these global patterns and interactions through the lens of geography. The analysis is by world regions, which invites consideration of long-term cultural and environmental practices that characterize and shape these distinct areas. Yet we contend that globalization—the increasing interconnectedness of people and places through converging economic, political, and cultural activities—is one of the most important forces reshaping the world today. Pundits say globalization is like the weather: It's everywhere, all the time. It is a ubiquitous part of our lives and landscapes that is both beneficial and negative, depending on our needs and point of view. While some people in some places embrace the changes brought by globalization, others resist and push back, seeking refuge in traditional habits and places. Thus, globalization's impact is highly uneven across space, which invites the need for a geographic (or spatial) understanding. As you will see in the pages that follow, geographers, who study places and phenomena around the globe and seek to explain the similarities and differences among places, are uniquely suited to analyze the impacts of globalization in different countries and world regions.

Globalization—connecting people and places through converging economic, political, and cultural activities—is one of the most important forces reshaping the world today.

As a counterpoint to globalization, **diversity** refers to the state of having different forms, types, practices or ideas, as well as the inclusion of distinct peoples, in a particular society. We live on a diverse planet with a mix of languages, cultures, environments, political ideologies, and religions that influence how people in particular localities view the world. At the same time, the intensification of communication, trade, travel, and migration that result from global forces have created many more settings in which people from vastly different backgrounds live, work, and interact. For example, in metropolitan Toronto, Canada's largest city, over half of the area's 5.5 million residents were born in another country. Increasingly, modern diversifying societies must find ways to build social cohesion among distinct peoples. Confronting diversity can challenge a society's tolerance, trust, and sense of shared belonging. Yet, diverse societies also stimulate creative exchanges and new understandings that are beneficial, building greater inclusion. The regional chapters that follow provide examples of the challenges and opportunities that diverse societies in an interconnected world experience today. We begin by introducing the discipline of geography and then examine this ongoing diversity in the context of globalization from a geographer's perspective.

Learning Objectives

After reading this chapter you should be able to:

1.1 **Describe the conceptual framework of world regional geography.**

1.2 **Identify the different components of globalization, including controversial aspects, and list several ways in which globalization is changing world geographies.**

1.3 **Summarize the major tools used by geographers to study Earth's surface.**

1.4 **Explain the concepts and metrics used to document changes in global population and settlement patterns.**

1.5 **Describe the themes and concepts used to study the interaction between globalization and the world's cultural geographies.**

1.6 **Explain how different aspects of globalization have interacted with global geopolitics from the colonial period to the present day.**

1.7 **Identify the concepts and data important to documenting changes in the economic and social development of more and less developed countries.**

Geography Matters: Environments, Regions, Landscapes

Geography is a foundational discipline, inspired and informed by the long-standing human curiosity about our surroundings and how we are connected to the world. The term *geography* has its roots in the Greek word for "describing the Earth," and this discipline is central to all cultures and civilizations as humans explore their world, seeking natural resources, commercial trade, military advantage, and scientific knowledge about diverse environments. In some ways, geography can be compared to history: Historians describe and explain what has happened over time, whereas geographers describe and explain the world's spatial dimensions—how it differs from place to place.

Given the broad scope of geography, it is no surprise that geographers have different conceptual approaches to investigating the world. At the most basic level, geography can be broken into two complementary pursuits: *physical* and *human geography*. **Physical geography** examines climate, landforms, soils, vegetation, and hydrology. **Human geography** concentrates on the spatial analysis of economic, social, and cultural systems.

Figure 1.1 Rio Negro Settlement in the Amazon Basin A young child plays with butterflies attracted to fish drying on wooden racks. Settlers in the Amazon Basin have often relied on the vast rivers of this region for food and transport.

A physical geographer, for example, studying the Amazon Basin of Brazil, might be interested primarily in the ecological diversity of the tropical rainforest or the ways in which the destruction of that environment changes the local climate and hydrology. A human geographer, in contrast, would focus on the social and economic factors explaining the migration of settlers into the rainforest or the tensions and conflicts over resources between new migrants and indigenous peoples. Both human and physical geographers share an interest in human–environment dynamics, asking how humans transform the physical environment and how the physical environment influences human behaviors and practices. Thus, they learn that Amazon residents may depend on fish from the river and plants from the forest for food (Figure 1.1) but raise crops for export and grow products such as black pepper or soy, rather than wheat, because wheat does poorly in humid tropical lowlands.

Another basic division in geography is the focus on a specific topic or theme as opposed to analyzing a specific place or a region. The theme approach is termed **thematic** or **systematic geography**, while the regional approach is called **regional geography**. These two perspectives are complementary and by no means mutually exclusive. This textbook, for example, utilizes a regional scheme for its overall organization, dividing Earth into 12 separate world regions. It then presents each chapter thematically, examining the topics of environment, population and settlement, cultural differentiation, geopolitics, and socioeconomic development in a systematic way. In doing so, each chapter combines four kinds of geography: physical, human, thematic, and regional geography.

Areal Differentiation and Integration

As a spatial science, geography is charged with the study of Earth's surface. A central theme of that responsibility is describing and explaining what distinguishes one piece of the world from another. The geographical term for this is **areal differentiation** (*areal* means "pertaining to area"). Why is one part of Earth humid and lush, while another, just a few hundred kilometers away, is arid (Figure 1.2)?

Geographers are also interested in the connections between different places and how they are linked. This concern is one of **areal integration**, or the study of how places interact with one another. An

Figure 1.2 Areal Differentiation This satellite photo of oasis villages on the southern slope of Morocco's Atlas Mountains is a classic illustration of areal differentiation, or how landscapes can differ significantly within short distances. The dark green bands are irrigated date palm and vegetable fields, watered by rivers that rise in the high mountains and then flow southward into the Sahara Desert. Since irrigated fields near the rivers are precious land, the village settlements are nearby in the dry areas.

example is the analysis of how and why the economies of Singapore and the United States are closely intertwined, even though the two countries are situated in entirely different physical, cultural, and political environments. Questions of areal integration are becoming increasingly important because of the new global linkages inherent in globalization.

Global to Local All systematic inquiry has a sense of *scale*, whatever the discipline. In biology, some scientists study the very small units such as cells, genes, or molecules, while others take a larger view, analyzing plants, animals, or whole ecosystems. Geographers also work at different scales. While one may concentrate on analyzing a local landscape—perhaps a single village in southern China—another might focus on the broader regional picture, examining all of southern China. Other geographers do research on a still larger global scale, perhaps studying emerging trade networks between southern India's center of information technology in Bangalore and North America's Silicon Valley, or investigating how the Indian monsoon might be connected to and affected by the Pacific Ocean's El Niño phenomenon. But even though geographers may work at different scales, they never lose sight of the interactivity and connectivity among local, regional, and global scales. They will note the ways that the village in southern China might be linked to world trade patterns or how the late arrival of the monsoon could affect agriculture and food supplies in Bangladesh.

The Cultural Landscape: Space into Place

Humans transform space into distinct places that are unique and heavily loaded with meaning and symbolism. This diverse fabric of *placefulness* is of great interest to geographers because it tells us much about the human condition throughout the world. Places can tell us how humans interact with nature and how they interact among themselves; where there are tensions and where there is peace; where people are rich and where they are poor.

A common tool for the analysis of place is the concept of the **cultural landscape**, which is the tangible, material expression of human settlement, past and present. Thus, the cultural landscape visually reflects the most basic human needs—shelter, food, and work. Additionally, the cultural landscape acts to bring people together (or keep them apart) because it is a marker of cultural values, attitudes, and symbols. As cultures vary greatly around the world, so do cultural landscapes (Figure 1.3).

Increasingly, however, we see the uniqueness of places being eroded by the homogeneous landscapes of globalization—shopping malls, fast-food outlets, business towers, theme parks, and industrial complexes. Understanding the forces behind the spread of these homogenized landscapes is important because they tell us much about the expansion of global economies and cultures. Although a modern shopping mall in Hanoi, Vietnam, may simply seem familiar to someone from North America, this new landscape represents yet another component of globalized world culture that has been implanted into a once remote and distinctive city.

Regions: Formal and Functional

The human intellect seems driven to make sense of the universe by lumping phenomena together into categories that emphasize similarities. Biology has its taxa of living organisms, while history marks off eras and periods of time. Geography, too, organizes information about

Figure 1.3 The Cultural Landscape Despite globalization, the world's landscapes still have great diversity, as shown by this village and its surrounding rice terraces on the island of Luzon, Philippines. Geographers use the cultural landscape concept to better understand how people interact with their environment.

the world into units of spatial similarity called **regions**—each a contiguous bounded territory that shares one or many common characteristics.

Sometimes, the unifying threads of a region are physical, such as climate and vegetation, resulting in a regional designation like the *Sahara Desert* or *Siberia*. Other times, the threads are more complex, combining economic and social traits, as in the use of the term *Rust Belt* for parts of the northeastern United States that have lost industry and population. Think of a region as spatial shorthand that provides an area with some signature characteristic that sets it apart from surrounding areas. In addition to delimiting an area, generalizations about society or culture are often embedded in these regional labels.

Geographers designate two types of regions: formal and functional. **Formal regions** take their name from the fact that they are defined by some aspect of physical form, such as a climate type or mountain range, such as Appalachia. Cultural features, such as the dominance of a particular language or religion, can also be used to define formal regions. Belgium, for example, can be divided into Flemish-speaking Flanders and French-speaking Wallonia. Many of the maps in this book denote formal regions. In contrast, a **functional region** is one where a certain activity (or cluster of activities) takes place. The earlier example of North America's Rust Belt is such a region because it encompasses a triangle from Milwaukee to Cincinnati to Syracuse, where manufacturing dominated through the 1960s and then experienced steady decline as factories shut down and people left (Figure 1.4). Geographers designate functional regions to show changing regional associations such as the spatial extent of a sports team's fan base or the commuter shed of a major metropolitan area such as Los Angeles. Delimiting such regions can be valuable for marketing, planning transportation, or thinking about the ways that people identify with an area.

Regions can be defined at various scales. In this book, we divide the world into 12 *world regions* based on formal characteristics such as physical features, language groups, and religious affiliations, but also relying on functional characteristics such as trade groups and regional associations (Figure 1.5). Some of these regional groupings are in common use, such as Europe or East Asia. Understandings and characteristics of these regions have often evolved over centuries. But the boundaries of these regions do shift. For example, during the Cold War

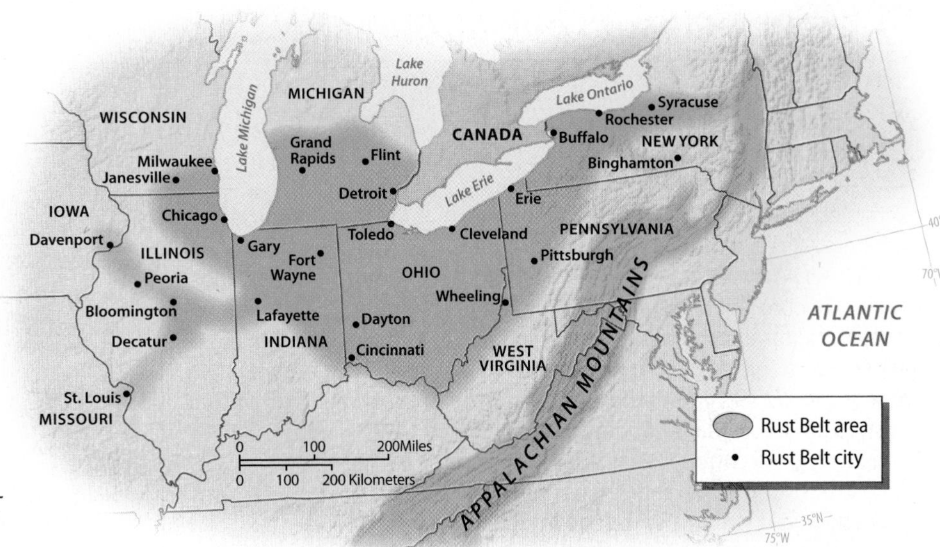

Figure 1.4 U.S. Rust Belt The rust belt is an example of a functional region. It is delimited to show an area that has lost manufacturing jobs and population over the last four decades. By constructing this region, a set of functional relationships is highlighted. **Q: In what formal and functional regions do you live?**

it made sense to divide Europe into east and west, with eastern Europe closely linked to the Soviet Union. With the 1991 collapse of the Soviet Union and the expansion of the European Union in the 2000s, that divide became less meaningful. Working at the world regional scale invariably creates regions that are not homogeneous, with some states fitting better into regional stereotypes than others. Yet understanding world regional formations is an important way to explore the impact of globalization on environments, cultures, politics, and development.

Review

1.1 Explain the difference between areal differentiation and areal integration.

1.2 How is the concept of the cultural landscape related to areal differentiation?

1.3 How do functional regions differ from formal regions?

KEY TERMS areal differentiation, areal integration, cultural landscape, region, geography, human geography, formal region, functional region, physical geography, regional geography, thematic geography (systematic geography)

Converging Currents of Globalization

One of the most important features of the 21st century is **globalization**—the increasing interconnectedness of people and places. Once-distant regions and cultures are now increasingly linked through commerce, communications, and travel. Although earlier forms of globalization existed, especially during Europe's colonial period, the current degree of planetary integration is stronger than ever. In fact, many observers argue that contemporary globalization is the most fundamental reorganization of the world's socioeconomic structure

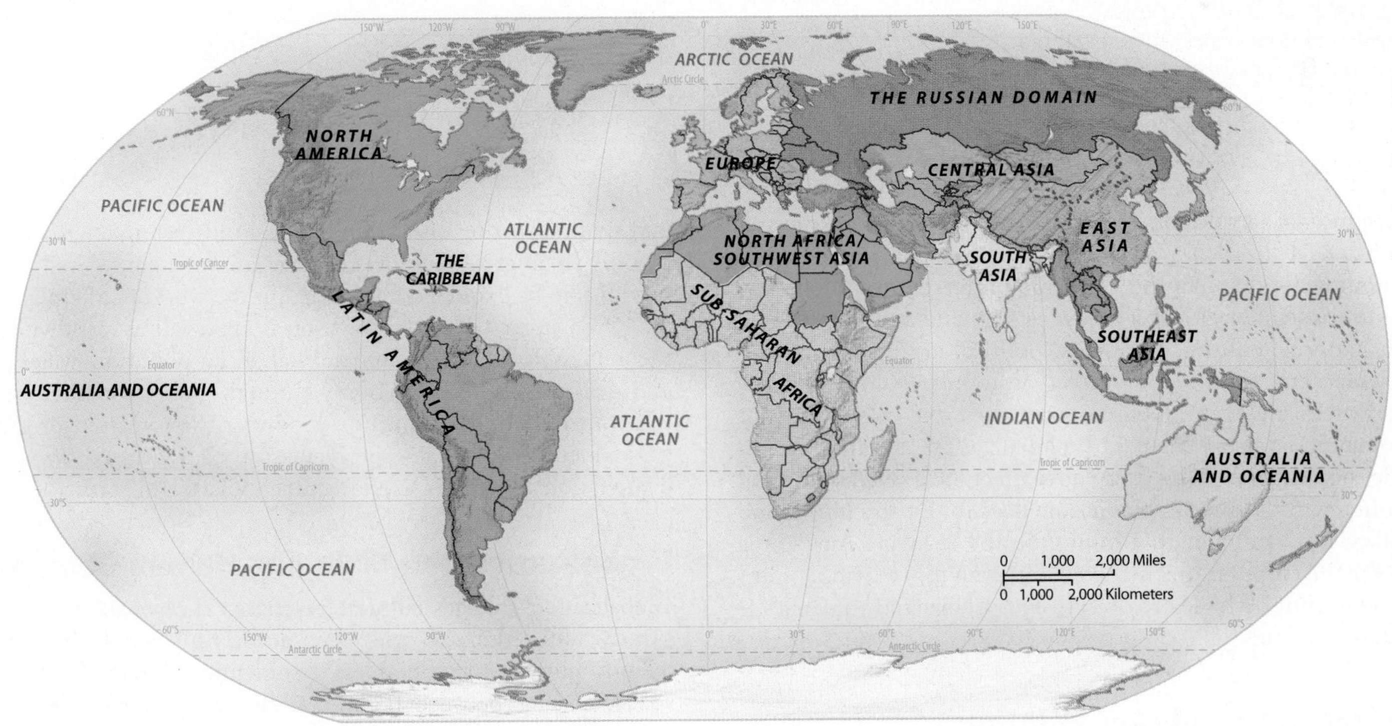

Figure 1.5 World Regions The boundaries shown here are the basis for the 12 regional chapters in this book. Countries or areas within countries that are treated in more than one chapter are designated on the map with a striped pattern. For example, western China is discussed in both Chapter 10, on Central Asia, and Chapter 11, on East Asia. Also, three countries on the South American continent are discussed as part of the Caribbean region because of their close cultural similarities to the island region.

EXPLORING GLOBAL CONNECTIONS

A Closer Look at Globalization

Google Earth (MG) Virtual Tour Video: http://goo.gl/5uPpKb

Globalization comes in many shapes and forms as it connects far-flung people and places. Many of these interactions are common knowledge, such as the global reach of multinational corporations. Others may be rather surprising. Who would expect to find Australian firefighters dowsing California wildfires as they migrate between Southern and Northern Hemisphere fire seasons? Would you predict that Saudi investors are leasing large tracts of land in Ethiopia to grow cotton, sugarcane, and palm oil for export to the Arabian Peninsula, while many Ethiopians struggle with food insecurity?

Indeed, global connections are ubiquitous and often complex—so much so that understanding the many different shapes, forms, and scales of these interactions is a key component of the study of world geography. To complement that study, each chapter of this book contains an *Exploring Global Connections* sidebar that presents a globalization case study.

The Chapter 7 sidebar, for example, explains how and why migrants are leaving Syria for neighboring countries and destinations in Europe (Figure 1.1.1). Record numbers of people from Southwest Asia and Sub-Saharan Africa make this perilous journey, and each year thousands die crossing in overcrowded boats. Other examples include international protections for Antarctica (Chapter 2); crisis mapping in Haiti (Chapter 5); South Korean investment in Africa (Chapter 11); and Southeast Asia's resurging opium trade (Chapter 13). Many of these sidebars are illustrated with Google Earth virtual tour videos.

1. Consider complex global connections based on your own experiences. For example, what food from another part of the world did you buy today, and how did it get to your store?
2. Now choose a city or a rural settlement in a completely different part of the world, and suggest ways in which globalization affects the lives of people in that place.

Figure 1.1.1 Conflict in Syria More than 4 million Syrians have left their country over the past five years due to civil war and insecurity. While most reside in refugee camps in bordering countries, increasingly refugees from Syria are resettling in Europe.

since the Industrial Revolution (see *Exploring Global Connections: A Closer Look at Globalization*).

Economic activities may be the major force behind globalization, but the consequences affect all aspects of land and life: Human settlement patterns, cultural attributes, political arrangements, and social development are all undergoing profound change. Because natural resources are now global commodities, the planet's physical environment is also affected by globalization. Financial decisions made thousands of miles away now affect local ecosystems and habitats, often with far-reaching consequences for Earth's health and sustainability. For example, gold mining in the Peruvian Amazon is profitable for the corporations involved and even for individual miners, but it may ruin biologically diverse ecosystems and threaten indigenous communities.

The Environment and Globalization

The expansion of a globalized economy is creating and intensifying environmental problems throughout the world. Transnational firms conducting business through international subsidiaries disrupt ecosystems around the globe with their incessant search for natural resources and manufacturing sites. Landscapes and resources previously used by only small groups of local peoples are now considered global commodities to be exploited and traded in the world marketplace.

On a larger scale, globalization is aggravating worldwide environmental problems such as climate change, air pollution, water pollution, and deforestation. Yet it is only through global cooperation, such as the United Nations treaties on biodiversity protection or greenhouse gas reductions, that these problems can be addressed. Environmental degradation and efforts to address it are discussed further in Chapter 2.

Globalization and Changing Human Geographies

Globalization changes cultural practices. The spread of a global consumer culture, for example, often accompanies globalization and frequently hurts local economies. It sometimes creates deep and serious social tensions between traditional cultures and new, external global culture. Television shows and movies available via satellite, Facebook, Twitter, and online videos implicitly promote Western values and culture that are then imitated by millions throughout the world (Figure 1.6).

Fast-food franchises are changing—some would say corrupting—traditional diets, with explosive growth in most of the world's

Figure 1.6 Global Communications The effects of globalization are everywhere, even in remote villages in developing countries. Here, in a small village in southwestern India, a rural family earns a few dollars a week by renting out viewing time on its globally linked television set.

cities. Although these foods may seem harmless to North Americans because of their familiarity, they are an expression of deep cultural changes for many societies and are also generally unhealthy and environmentally destructive. Yet some observers contend that even multinational corporations have learned to pay attention to local contexts. **Glocalization** (which combines globalization with locale) is the process of modifying an introduced product or service to accommodate local tastes or cultural practices. For example, a McDonald's in Japan may serve shrimp burgers along with Big Macs.

Although the media give much attention to the rapid spread of Western consumer culture, nonmaterial culture is also dispersed and homogenized through globalization. Language is an obvious example—American tourists in far-flung places are often startled to hear locals speaking an English made up primarily of movie or TV clichés. However, far more than speech is involved, as social values also are dispersed globally. Changing expectations about human rights, the role of women in society, and the intervention of nongovernmental organizations are also expressions of globalization that may have far-reaching effects on cultural change.

In return, cultural products and ideas from around the world greatly impact U.S. culture (Figure 1.7). The large and diverse immigrant population in the United States has contributed to heightened cultural diversity and exchange. The internationalization of American food and music and the multiple languages spoken in American cities are all expressions of globalization.

Globalization also clearly influences population movements. International migration is not new, but increasing numbers of people from all parts of the world are now crossing national boundaries, legally and illegally, temporarily and permanently (Figure 1.8). Migration from Latin America, the Caribbean, and Asia has transformed the ethnic and racial makeup of the United States, and migration from Africa and Asia has transformed western Europe. Countries such as Japan and South Korea, long perceived as ethnically homogeneous, now have substantial immigrant populations. Even several relatively poor countries, such as Ghana and the Ivory Coast, have large numbers of immigrants coming from even poorer countries, such as Burkina Faso and Mali. Although international migration is curtailed by the laws of every country—much more so, in fact, than the movement of goods or capital—it is still rapidly mounting, propelled in part by the uneven economic development associated with globalization (discussed in more detail later in the chapter).

Geopolitics and Globalization

Globalization also has important geopolitical components. To many, an essential dimension of globalization is that it is not restricted by territorial or national boundaries. For example, the creation of the United Nations (UN) following World War II was a step toward creating an international governmental structure in which all nations could find representation. The simultaneous emergence of the Soviet Union as a military and political superpower led to a rigid division into Cold War blocs that slowed further geopolitical integration. However, with the peaceful end of the Cold War in the early 1990s, the former communist countries of eastern Europe and the Soviet Union were opened almost immediately to global trade and cultural exchange, changing those countries immensely (Figure 1.9).

Figure 1.7 Global Culture in the United States The multilingual welcome offered by a public library in Montgomery County, Maryland not only illustrates the many different languages spoken in the suburbs of Washington, DC, but also reminds us that expressions of globalization are found throughout North America.

Figure 1.8 International Migration Globalization in its many different forms is connected to the largest migration in human history as people are drawn to centers of economic activity in hopes of a better life. But along with the pull forces that lure people to new places are the forces of civil strife, environmental deterioration, and economic collapse that push migrants out of their homelands. This truckload of African migrants is crossing the Sahara to the Mediterranean shore, where many will attempt to illegally enter Europe through Spain or Italy. **Q: What international groups are found in your city?**

Figure 1.9 End of the Cold War The peaceful end of the Cold War in 1991 greatly facilitated global economic expansion and jump-started cultural and political globalization. In 1989, Germans celebrate the opening of the Berlin Wall that had divided East and West Berlin since 1961.

A significant international criminal element is another globalization outcome and includes terrorism (discussed later in this chapter), drugs, pornography, slavery, and prostitution, which require international coordination and agreements to address (Figure 1.10). Some of the world's most remote places, such as the mountains of northern Burma and the valleys of southern Afghanistan, are thoroughly integrated into the circuits of global exchange through the production of opium that is central to the world heroin trade. Even areas that do not directly produce drugs are involved in their global sale and transshipment; many Caribbean economies are becoming reoriented to drug transshipments and the laundering of drug money. Prostitution, pornography, and gambling have also emerged as highly profitable global businesses. Over the past decades, for example, parts of eastern Europe have become major sources of both pornography and prostitution, finding a lucrative, but morally questionable, niche in the new global economy.

Further, there is a strong argument that globalization—almost by definition—has weakened the political power of individual states by strengthening regional economic and political organizations, such as the European Union and the World Trade Organization. In some world regions, a weakening of traditional state power has led to stronger local

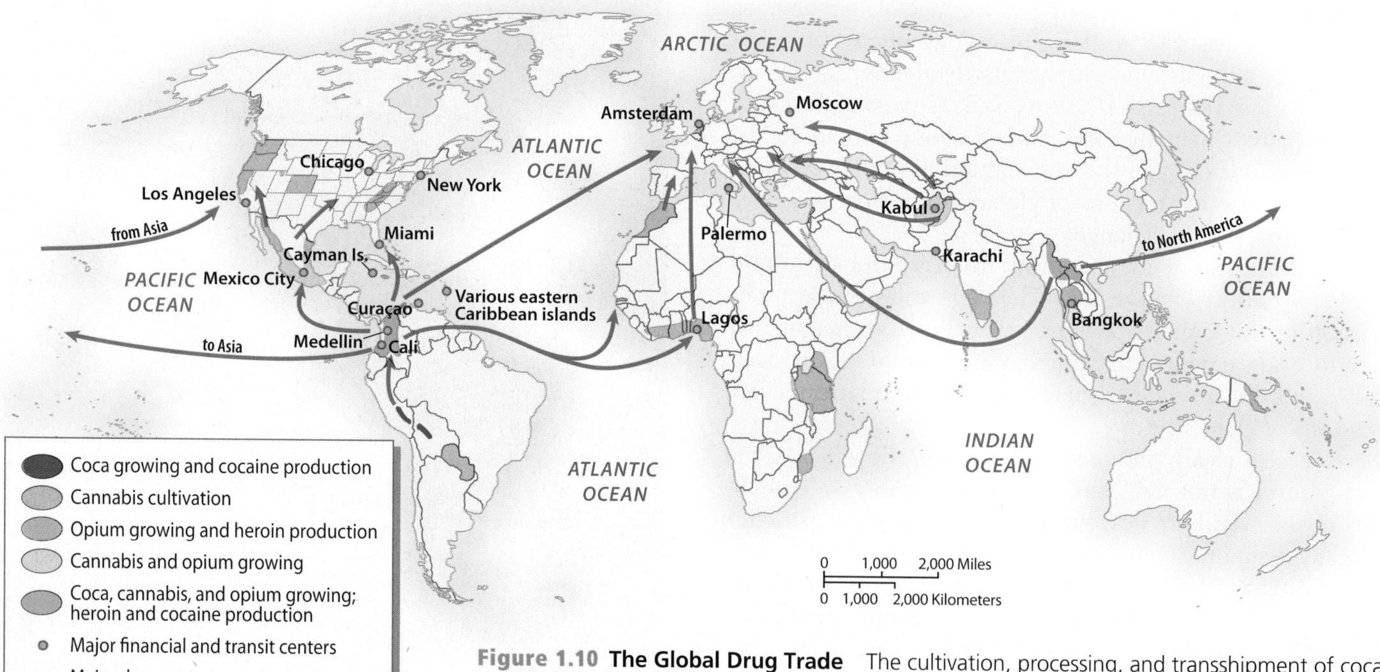

Figure 1.10 The Global Drug Trade The cultivation, processing, and transshipment of coca (cocaine), opium (heroin), and cannabis (marijuana) are global issues. The most important cultivation centers are Colombia, Mexico, Afghanistan, and northern Southeast Asia, and the major drug financing centers are located mostly in the Caribbean, the United States, and Europe. Nigeria and Russia also play significant roles in the global transshipment of illegal drugs.

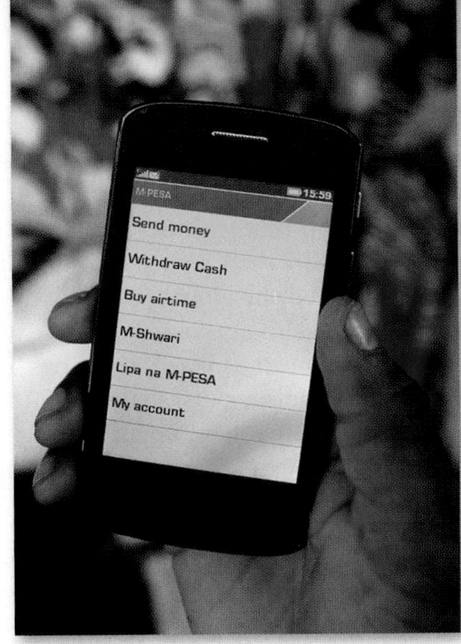

Figure 1.11 Global Use of Cell Phones Mobile technologies have revolutionized the way people communicate, acquire information, and interact in a globalized world. Today cell phones are used for far more than simply talking. In Nairobi, Kenya, the majority of the city's adult population now uses M-Pesa, a cell-phone-based money transfer service, to pay for everything from street food to rides on the city's privately owned minibuses.

and separatist movements, as illustrated by the turmoil on Russia's southern borders or the plethora of separatist organizations in Europe.

Economic Globalization and Uneven Development Outcomes

Most scholars agree that the major component of globalization is the economic reorganization of the world. Although different forms of a world economy have existed for centuries, a well-integrated and truly global economy is primarily the product of the past several decades. Attributes of this system, while familiar, bear repeating:

- Global communication systems that instantaneously link all regions and most people on the planet (Figure 1.11)
- Transportation systems capable of moving goods quickly by air, sea, and land
- Transnational business strategies that have created global corporations more powerful than many sovereign nations
- New and more flexible forms of capital accumulation and international financial institutions that make 24-hour trading possible
- Global agreements that promote free trade among countries
- Market economies and private enterprises that have replaced state-controlled economies and services
- An abundance of global goods and services created to fulfill consumer demand—real or imagined (Figure 1.12)
- Economic disparities between rich and poor regions and countries that drive people to migrate, both legally and illegally, in search of a better life
- An army of international workers, managers, and executives who give this powerful economic force a human dimension

As a result of this global reorganization, economic growth in some areas of the world has been unprecedented during recent decades; China is a good example, with an average annual growth rate of 8.6 percent from 2010 to 2014. But not everyone profits from economic globalization, as continuing wage gaps within China indicate, nor have all world regions shared equally in the benefits. Additionally, the global recession of 2008–2010 demonstrated that economic interconnectivity also increases economic vulnerability, as illustrated by the collapse of financial institutions in Iceland or the decline in remittances to Mexico.

Thinking Critically About Globalization

Globalization, especially its economic aspects, is one of today's most contentious issues. Supporters believe that it results in a greater economic efficiency that will eventually lead to rising prosperity for the entire world. In contrast, critics claim that globalization largely benefits those who are already prosperous, leaving most of the world poorer than before as the rich and powerful exploit the less fortunate.

Economic globalization is generally applauded by corporate leaders and economists and has substantial support within both major political parties in the United States. In fact, moderate and conservative politicians in most countries generally support free trade and other aspects of economic globalization. Opposition to economic globalization is widespread in the labor and environmental movements as well as among many student groups worldwide. Hostility toward globalization is sometimes loudly expressed, as in the Occupy Wall Street demonstrations or marches at World Bank and World Trade Organization meetings (Figure 1.13).

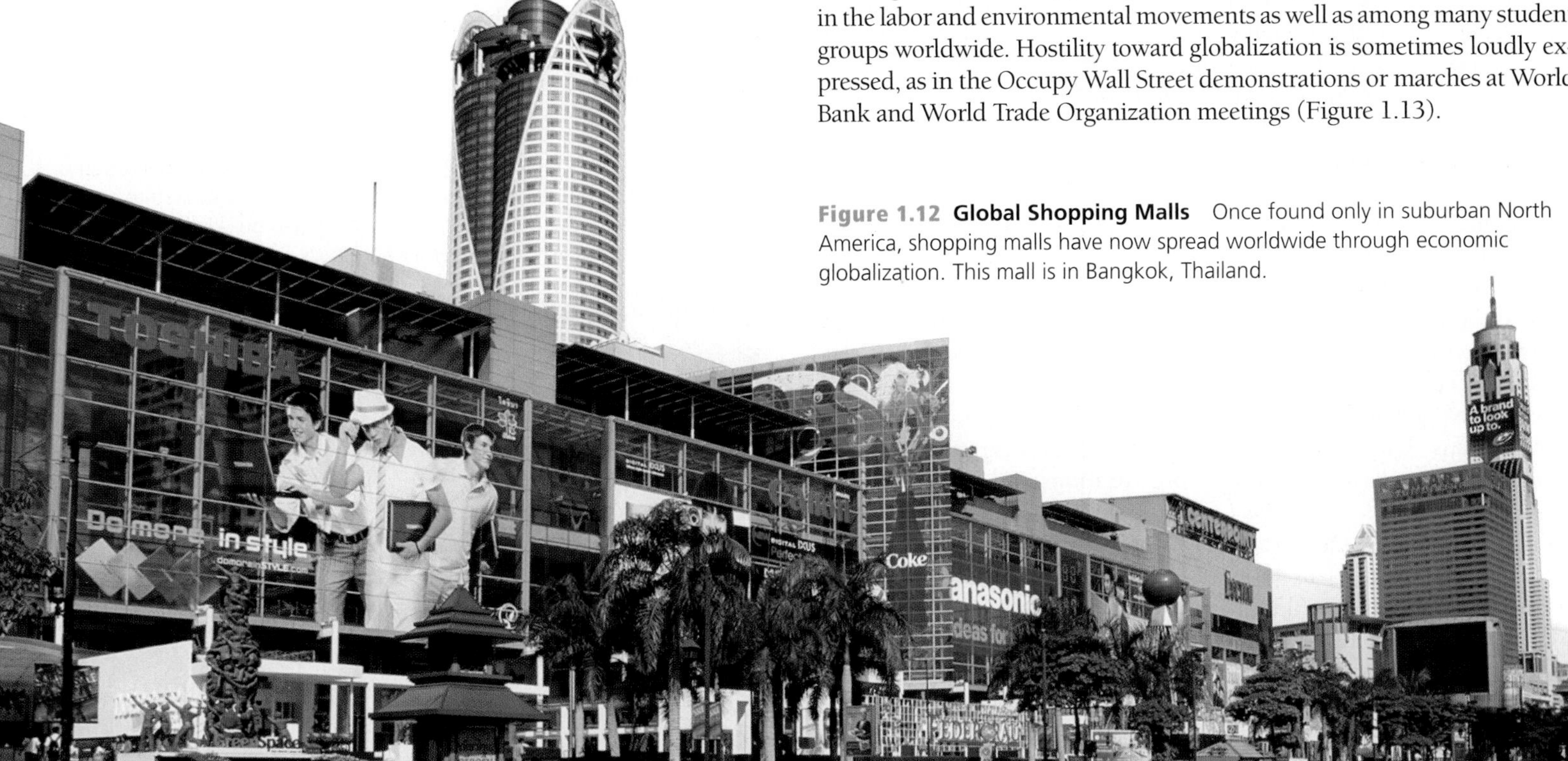

Figure 1.12 Global Shopping Malls Once found only in suburban North America, shopping malls have now spread worldwide through economic globalization. This mall is in Bangkok, Thailand.

Figure 1.13 Protests Against Globalization Meetings of international groups such as the World Trade Organization (WTO) and International Monetary Fund (IMF) commonly draw protesters against economic globalization. This group of protestors at a WTO meeting in Bali, Indonesia are demanding more local autonomy in food systems.

Pro-globalization Arguments Advocates argue that globalization is a logical and inevitable expression of contemporary international capitalism and that it benefits all nations and all peoples. Economic globalization can work wonders, they contend, by enhancing competition, allowing capital to flow to poor areas, and encouraging the spread of beneficial new technologies and ideas. As countries reduce their barriers to trade, inefficient local industries must become more efficient in order to compete with the new flood of imports, thereby enhancing overall national productivity. Those that cannot adjust will most likely go out of business, making the global marketplace more efficient.

Every country and region of the world, moreover, should concentrate on those activities for which it is best suited in the global economy. Enhancing such geographic specialization, pro-globalizers argue, creates a more efficient world economy. Such economic restructuring is made increasingly possible by the free flow of capital to areas having the greatest opportunities. By making access to capital more readily available throughout the world, economists contend, globalization should eventually result in a certain global **economic convergence**, implying that the world's poorer countries will gradually catch up with the more advanced economies.

American journalist Thomas Friedman, an influential advocate of economic globalization, argues that the world has not only shrunk, but also become economically "flat" such that financial capital, goods, and services can move freely from place to place. For example, the need to attract foreign capital forces countries to adopt new economic policies. Friedman describes the great power of the global "electronic herd" of bond traders, currency speculators, and fund managers who either direct money to or withhold it from developing economies, resulting in economic winners and losers (Figure 1.14).

The pro-globalizers also strongly support the large multinational organizations that facilitate the flow of goods and capital across international boundaries. Three such organizations are particularly important: the World Bank, the International Monetary Fund (IMF), and the World Trade Organization (WTO). The World Bank's primary function is to make loans to poor countries so that they can invest in infrastructure and build more modern economic foundations. The IMF is concerned with making short-term loans to countries in financial difficulty—those having trouble, for example, paying interest on the loans previously taken. The WTO, a much smaller organization than the other two, works to reduce trade barriers between countries to enhance economic globalization. It also tries to mediate between countries and trading blocs that are engaged in trade disputes.

To support their claims, pro-globalizers argue that countries that have embraced the global economy have generally enjoyed more economic success than those that have sought economic self-sufficiency. The world's most isolated countries, Burma (Myanmar) and North Korea, are economic disasters with little growth and rampant poverty, whereas those that have opened themselves to global forces during the same period, such as Singapore and Thailand, have seen rapid growth and substantial reductions in poverty.

Critics of Globalization Virtually all of the claims of pro-globalizers are strongly contradicted by globalization's critics. Opponents often argue that globalization is not a "natural" process, but rather the product of an explicit economic policy promoted by free trade advocates, capitalist countries (mainly the United States, but also Japan and the countries of Europe), financial interests, international investors, and multinational firms.

Further, mounting evidence that globalization of the world economy is creating greater inequality between rich and poor refutes the "trickle-down" model of development benefiting all people in all regions. On a global scale, the richest 20 percent of the world's people consume 86 percent of the world's resources, whereas the poorest 80 percent use only 14 percent. The growing inequality of this age of globalization is apparent on both global and national scales. The world's wealthiest countries have grown much richer over the past two decades, while many of the poorest countries have lost ground. And even within developed countries such as the United States, the wealthiest 1 percent of the population has reaped almost all of the gains that globalization has offered, while the remaining 99 percent has seen real income decline as wages have remained static and jobs have been lost to outsourcing (Figure 1.15).

Opponents also contend that globalization promotes free-market, export-oriented economies at the expense of localized, sustainable activities. World forests, for example, are increasingly cut for export timber, rather than serving local needs. As part of their economic structural adjustment package, the World Bank and the IMF often encourage developing countries to expand resource exports to

Figure 1.14 The Electronic Herd One component of globalization is the rapid movement of capital within the global economic system, creating financial hotspots and stampedes as money moves quickly from place to place. This image is of traders in the Hong Kong Stock Exchange.

Figure 1.15 U.S. Employment and Globalization The fact that U.S. manufacturing jobs have moved offshore to lower-wage countries is one criticism of globalization. While true to some extent, this job loss is also the result of other kinds of change in world and domestic economies. These job seekers are in Rochester Hills, Michigan.

earn more hard currency to pay off their foreign debts. This strategy, however, usually leads to overexploitation of local resources. Critics also note that the IMF often requires developing countries to adopt programs of fiscal austerity that entail substantially cutting public spending for education, health, and food subsidies. By adopting such policies, sceptics warn, poor countries will end up even more impoverished than before.

Furthermore, anti-globalizers contend that the "free-market" economic model commonly promoted for developing countries is not the one that Western industrial countries used for their own economic development. In Germany, France, and even to some extent the United States, governments historically have played a strong role in directing investment, managing trade, and subsidizing chosen sectors of the economy.

Those who challenge globalization worry that the entire system—with its instantaneous transfers of vast sums of money over nearly the entire world on a daily basis—is inherently unstable. British author and noted globalization critic John Gray, for example, argues that the same "electronic herd" that Friedman applauds is dangerously susceptible to "stampedes." International managers of capital tend to panic when they think their funds are at risk; when they do so, the entire intricately linked global financial system can quickly become destabilized, leading to a crisis of global proportions. The rapid downturn of the global economy in late 2008 seems to support that assertion.

A Middle Position Many experts argue that both the anti-globalization and the pro-globalization stances are exaggerated. Those in the middle ground argue that economic globalization is indeed unavoidable and that, despite its promises and pitfalls, globalization can be managed at both the national and the international levels to reduce economic inequalities and protect the natural environment. These experts stress the need for strong, yet efficient national governments, supported by international institutions (such as the UN, World Bank, and IMF) and globalized networks of environmental, labor, and human rights groups.

Unquestionably, globalization is one of the most important issues of the day—and certainly one of the most complicated. While this book does not pretend to resolve the controversy, nor does it take a position, it does encourage readers to reflect on these critical points as they apply to each world region.

Diversity in a Globalizing World

As globalization progresses, many observers foresee a world far more uniform and homogeneous than today's. The optimists among them imagine a universal global culture uniting all humankind into a single community untroubled by war, ethnic strife, or resource shortages—a global utopia of sorts.

A more common view is that the world is becoming blandly homogeneous as different places, peoples, and environments lose their distinctive character and become indistinguishable from their neighbors. Yet even as globalization generates a certain degree of homogenization, the world is still a highly diverse place (Figure 1.16). You can still find marked differences in culture (language, religion, architecture, foods, and other attributes of daily life), economy, and politics—as well as in the physical environment. Such diversity is so vast that it cannot readily be extinguished, even by the most powerful forces of globalization. Diversity may be difficult for a society to live with, but it also may be dangerous to live without. Nationality, ethnicity, cultural distinctiveness—all are defining expressions of humanity that are nurtured in distinct places.

In fact, globalization often provokes a strong reaction on the part of local people, making them all the more determined to maintain what is distinctive about their way of life. Thus, globalization is understandable only if we also examine the diversity that continues to characterize the world and, perhaps most important, the tension between these two forces: the homogenizing power of globalization and the reaction against it, often through demands for protecting cultural diversity.

The politics of diversity demand increasing attention as we try to understand global terrorism, ethnic identity, religious practices, and political independence. Groups of people throughout the world seek self-rule of territory they can call their own. Today most wars are fought *within* countries, not *between* them. As a result, our interest in

Figure 1.16 Shopping in Isfahan Young women shop in the grand bazaar in Isfahan, Iran, in preparation for Eid al-Fitr, the celebration at the end of Ramadan. While few places are beyond the reach of globalization, it is also true that distinct cultures, traditions, and landscapes exist in the world's various regions.

geographic diversity takes many forms and goes far beyond simply celebrating traditional cultures and unique places. People have many ways of making a living throughout the world, and it is important to recognize this fact as the globalized economy becomes increasingly focused on mass-produced retail goods. Furthermore, a stark reality of today's economic landscape is uneven outcomes: While some people and places prosper, others suffer from unrelenting poverty (Figure 1.17). To analyze these patterns of unevenness and change, the next section considers the tools used by geographers to better know the world.

Figure 1.17 Landscape of Economic Inequality The geography of diversity takes many expressions, one of which is economic unevenness, as depicted in this photo from the City of Makati in The Philippines where squatter settlements of the poor contrast with high-rise office buildings and apartments of the more affluent.

Review

1.4 Provide examples of how globalization impacts the culture of a place or region.

1.5 Describe and explain five components of economic globalization.

1.6 Summarize three elements of the controversy about globalization.

KEY TERMS globalization, glocalization, economic convergence

The Geographer's Toolbox: Location, Maps, Remote Sensing, and GIS

Geographers use many different tools to represent the world in a convenient form for examination and analysis. Different kinds of images and data are needed to study vegetation change in Brazil or mining activity in Mongolia; population density in Tokyo or language regions in Europe; religions practiced in the Southwest Asia or rainfall distribution in southern India. Knowing how to display and interpret information in map form is part of a geographer's skill set. In addition to traditional maps, today's modern satellite and communications systems provide an array of tools not imagined 50 years ago.

Latitude and Longitude

To navigate their way through daily tasks, people generally use a mental map of *relative locations* to locate specific places in terms of their relationship to other landscape features. The shopping mall is near the highway, for example, or the college campus is near the river. In contrast, map makers use *absolute location*, often called a mathematical location, which draws on a universally accepted coordinate system that gives every place on Earth a specific numerical address based on latitude and longitude. The absolute location for the Geography Department at the University of Oregon, for example, has the mathematical address of 44 degrees, 02 minutes, and 42.95 seconds north and 123 degrees, 04 minutes, and 41.29 seconds west. This is written 44° 02' 42.95" N and 123° 04' 41.29" W.

Lines of **latitude**, often called parallels, run east–west around the globe and are used to locate places north and south of the equator (0 degrees latitude). In contrast, lines of **longitude**, called meridians, run from the North Pole (90 degrees north latitude) to the South Pole (90 degrees south latitude). Longitude values locate places east or west of the **prime meridian**, located at 0 degrees longitude at the Royal Naval Observatory in Greenwich, England (just east of London) (Figure 1.18). The equator divides the globe into northern and southern hemispheres, whereas the prime meridian divides the world into eastern and western hemispheres; these latter two hemispheres meet at 180 degrees longitude in the western Pacific Ocean. The International Date Line, where each new solar day begins, lies along much of 180 degrees longitude, deviating where necessary to ensure that small Pacific island nations remain on the same calendar day.

Each degree of latitude measures 60 nautical miles or 69 land miles (111 km) and is made up of 60 minutes, each of which is 1 nautical mile (1.15 land miles). Each minute has 60 seconds of distance, each approximately 100 feet (30.5 meters).

From the equator, parallels of latitude are used to mathematically define the tropics: the Tropic of Cancer at 23.5 degrees north and the

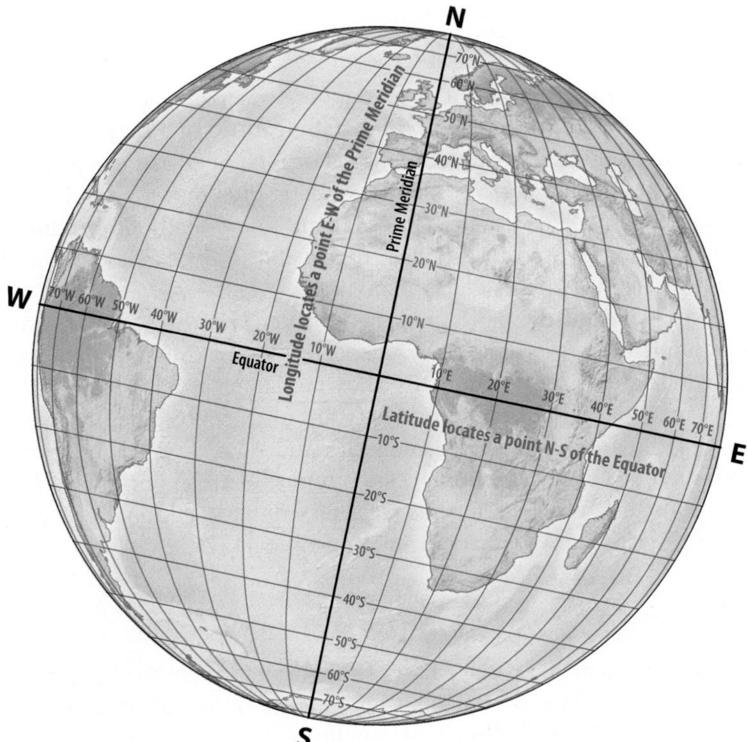

Figure 1.18 Latitude and Longitude Latitude locates a point between the equator and the poles and is designated as so many degrees north or south. Longitude locates a point east or west of the prime meridian, located just east of London, England. **Q: What are the latitude and longitude of your school?**

Tropic of Capricorn at 23.5 degrees south. These latitude lines denote where the Sun is directly overhead at noon on the solar solstices in June and December. The Arctic and Antarctic circles, at 66.5 degrees north and south latitude, respectively, mathematically define the polar regions.

Global Positioning Systems (GPS) Historically, precise measurements of latitude and longitude were determined by a complicated method of celestial navigation, based on one's location relative to the Sun, Moon, planets, and stars. Today absolute location on Earth (or in airplanes above Earth's surface) is determined through satellite-based **global positioning systems (GPS)**. These systems use time signals sent from your location to a satellite and back to your receiver to calculate precise coordinates of latitude and longitude. GPS was first used by the U.S. military in the 1960s and then made available to the public in the later decades of the 20th century. Today GPS guides airplanes across the skies, ships across oceans, private autos on the roads, and hikers through wilderness areas. While most smartphones use locational systems based on triangulation from cell-phone towers, some smartphones are capable of true satellite-based GPS accurate to 3 feet (or 1 meter).

Map Projections

Because the world is a sphere, mapping the globe on a flat piece of paper creates inherent distortions in the latitudinal, or north–south, depiction of Earth's land and water areas. Cartographers (those who make maps) have tried to limit these distortions by using various **map projections**, defined as the different ways to project a spherical image onto a flat surface. Historically, the Mercator projection was the projection of choice for maps used for oceanic exploration. However, a glance at the inflated Greenlandic and Russian landmasses shows its weakness in accurately depicting high-latitude areas (Figure 1.19). Over time, cartographers have created literally hundreds of different map projections in their attempts to find the best ways to map the world while limiting distortions.

For the last several decades, cartographers have generally used the Robinson projection for maps and atlases. In fact, several professional cartographic societies tried unsuccessfully in 1989 to ban projections such as the Mercator because of their spatial distortions. As in many other professional publications, in this book we use only the Robinson projection for our world maps.

(a)

(b)

Animation
Map Projections
http://goo.gl/vRjKDJ

Figure 1.19 Map Projections Cartographers have long struggled with how best to accurately map the world given the distortions inherent in transferring features on a round globe to a flat piece of paper. Early mapmakers commonly used the Mercator projection (a), which distorts features in the high latitudes, but worked fairly well for seagoing explorers. (b) This map is the Robinson projection, developed in the 1960s and now the industry standard because it minimizes cartographic distortion.

Map Scale

All maps must reduce the area being mapped to a smaller piece of paper. This reduction involves the use of **map scale**, or the mathematical ratio between the map and the surface area being mapped. Many maps note their scale as a ratio or fraction between a unit on the map and the same unit in the area being mapped, such as 1:63,360 or 1/63,360. This means that 1 inch on the map represents 63,360 inches on the land surface; thus, the scale is 1 inch equals 1 mile. Although 1:63,360 is a convenient mapping scale to understand, the amount of surface area that can be mapped and fitted on a letter-sized sheet of paper is limited to about 20 square miles. At this scale, mapping 100 square miles would produce a bulky map 8 feet square. Therefore, the ratio must be changed to a larger number, such as 1:316,800. This means that 1 inch on the map now represents 5 miles (8 km) of distance on land.

Based on the **representative fraction**, the ratio between the map and the area being mapped, maps are categorized as having either large or small scales (Figure 1.20). It may be easy to remember that large-scale maps make landscape features like rivers, roads, and cities *larger*, but because the features are larger, the maps must cover *smaller* areas. Conversely, small-scale maps cover *larger* areas but must then make landscape features *smaller*. A bit harder to remember is that the larger the second number of the representative fraction—the 63,360 in the fraction 1:63,360 or the 100,000 in the fraction 1:100,000, for example—the smaller the scale of the map.

Map scale is probably the easiest to interpret when it is a **graphic** or **linear scale**, which visually depicts distance units such as feet, meters, miles, or kilometers on a horizontal bar. Most of the maps in this book are small-scale maps of large areas; thus, the graphic scale is in miles and kilometers. Distances between two points on the map can be calculated by making two tick marks on a piece of paper held next to the points and then measuring the distance between the two marks on the linear scale.

Map Patterns and Map Legends

Maps depict everything from the most basic representation of topographic and landscape features to complicated patterns of population, migration, economic conditions, and more. A map can be a simple *reference map* showing the location of certain features or a **thematic map** displaying data such as rainfall patterns or income distribution. Most of the maps in this text are thematic maps illustrating complicated spatial phenomena. Every map has a **legend** that provides information on the categories used in the map, their values (when relevant), and other symbols that may need explanation.

One type of thematic map used often in this book is the **choropleth map** in which color shades represent different data values, with darker shades generally showing larger average values. Per capita income and population density are often represented by these maps, with data divided into categories and then mapped by spatial units such as countries, provinces, counties, or neighborhoods. The category breaks and spatial units selected can have a dramatic impact on the patterns shown in a choropleth map (Figure 1.21).

Aerial Photos and Remote Sensing

Although maps are a primary tool of geography, much can be learned about Earth's surface by deciphering patterns on aerial photographs taken from airplanes, balloons, or satellites. Originally available only in black and white, today these images are digital and can exploit visible light (like a photograph) or other light wavelengths such as infrared that are not visible to the human eye.

Even more information about Earth comes from electromagnetic images taken from aircraft or satellites, termed **remote sensing** (Figure 1.22). This technology has many applications, including monitoring the loss of rainforests, tracking the biological health of crops and woodlands, and even measuring changes in ocean surface temperatures. Remote sensing is also central to national defense issues, such as monitoring the movements of troops or the building of missile sites in hostile countries. In simple terms, aerial photographs are merely images taken from balloons, airplanes, or satellites, whereas remote sensing gathers electromagnetic data that then must be processed and interpreted by computer software to produce images of Earth's surface.

Figure 1.20 Small- and Large-Scale Maps A portion of Australia's east coast north of Sydney is mapped at two scales: (a) one at a small scale and (b) the other at a large scale. Note the differences in distance depicted on the linear scales of the two maps. There is more close-up detail in the large-scale map, but it covers only a small portion of the area mapped at a small scale.

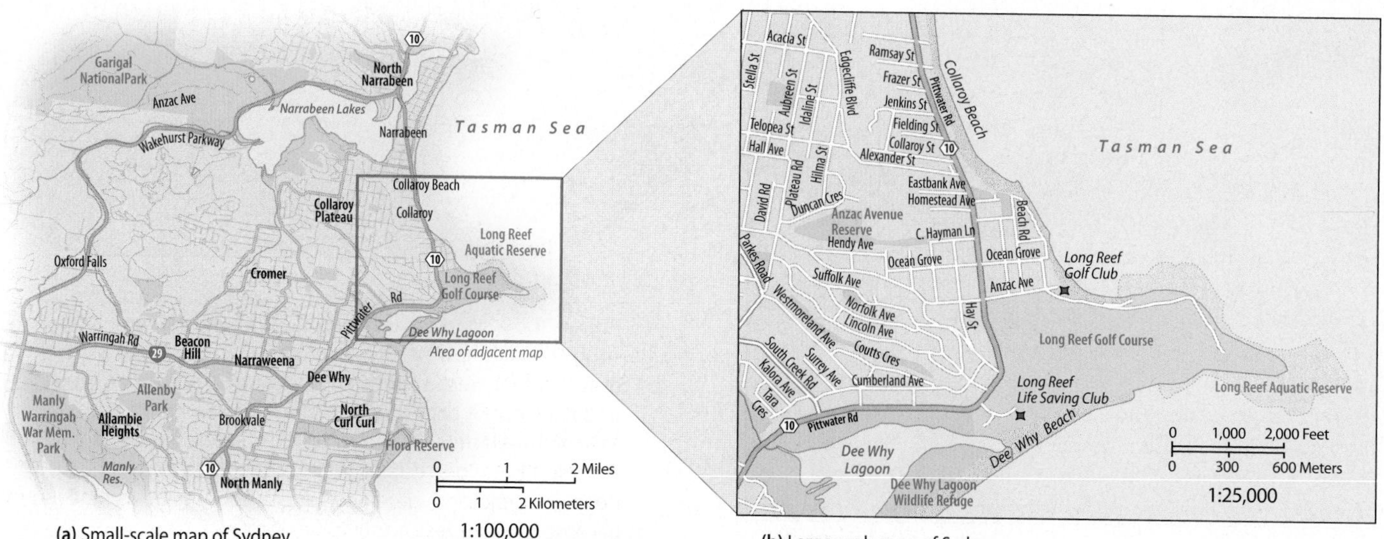

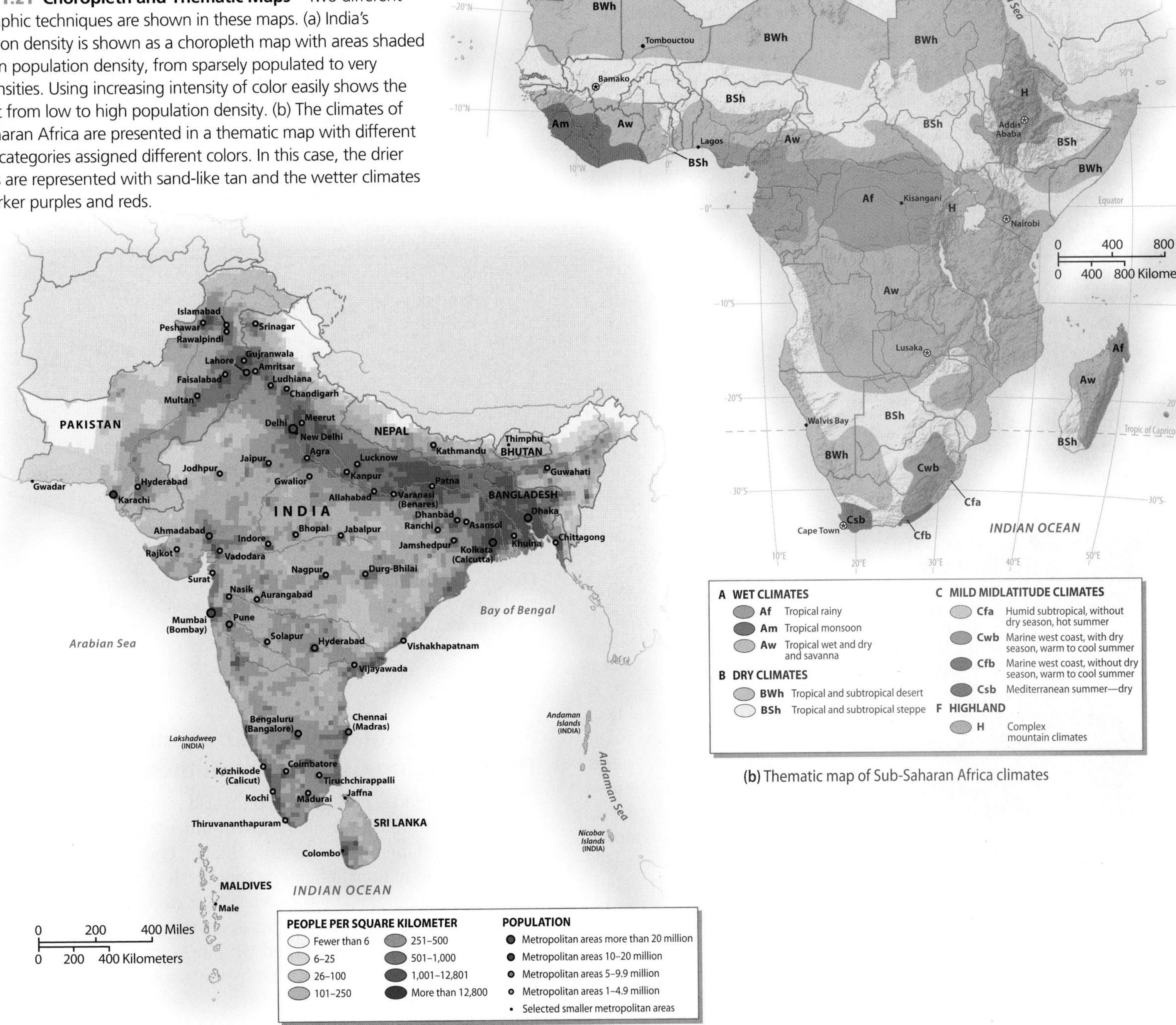

Figure 1.21 Choropleth and Thematic Maps Two different cartographic techniques are shown in these maps. (a) India's population density is shown as a choropleth map with areas shaded based on population density, from sparsely populated to very high densities. Using increasing intensity of color easily shows the gradient from low to high population density. (b) The climates of Sub-Saharan Africa are presented in a thematic map with different climate categories assigned different colors. In this case, the drier climates are represented with sand-like tan and the wetter climates with darker purples and reds.

(a) Choropleth map of South Asia population density

(b) Thematic map of Sub-Saharan Africa climates

The Landsat satellite program launched by the United States in 1972 is a good example of both the technology and the uses of remote sensing. These satellites collect data simultaneously in four broad bands of electromagnetic energy, from visible through near-infrared wavelengths, that is reflected or emitted from Earth. Once these data are processed by computers, they display a range of images, as illustrated in Figure 1.22. The resolution on Earth's surface ranges from areas 260 feet (80 meters) square down to 98 feet (30 meters) square.

Commercial satellites such as GeoEye and DigitalGlobe now provide high-resolution satellite imagery down to 1.5 feet (or 0.5 meters) square. This means that a car, small structure, or group of people would be easily seen, but not an individual person. Of course, cloud cover often compromises the continuous coverage of many parts of the world.

Geographic Information Systems (GIS)

Vast amounts of computerized data from sources such as maps, aerial photos, remote sensing, and census data are brought together in **geographic information systems (GIS)**. The resulting spatial databases are used to analyze a wide range of issues. Conceptually, GIS can be considered a computer system for producing a series of overlay maps showing spatial patterns and relationships (Figure 1.23). A GIS map, for example, might combine a conventional map with data on toxic waste sites, local geology, groundwater flow, and surface hydrology

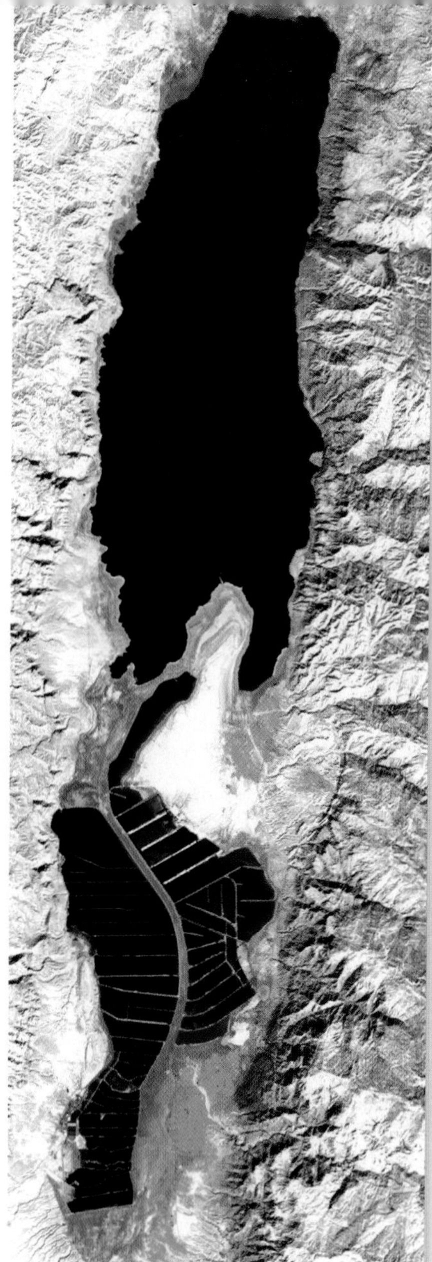

Figure 1.22 Remote Sensing of the Dead Sea This NASA satellite image of the Dead Sea, the lowest spot on Earth at 1300 feet (400 meters) below sea level, uses false color remote sensing to capture different elements of the environment. Black is deeper water, with light blue showing shallow waters. The green areas along the shoreline are irrigated crops, while the white areas are salt evaporation ponds.

to determine the source of pollutants appearing in household water systems.

Although GIS dates back to the 1960s, it is only in the last several decades—with the advent of desktop computer systems and remote sensing data—that GIS has become absolutely central to geographic problem-solving. It plays a central role in city planning, environmental science, public health, and real-estate development, to name a few of the many activities using these systems.

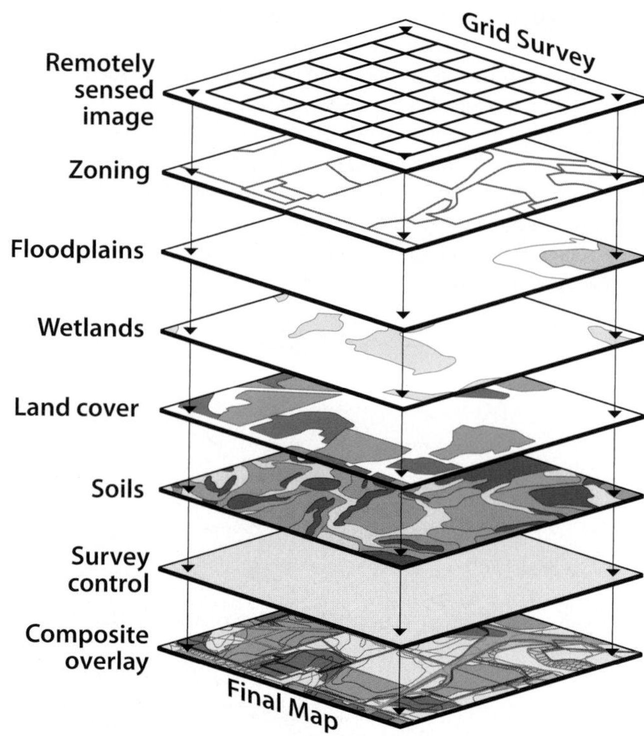

Figure 1.23 GIS Layers Geographic information systems (GIS) maps usually consist of many different layers of information that can be viewed and analyzed separately or as a composite overlay. This is a typical environmental planning map where different physical features (such as wetlands and soils) are combined with zoning regulations.

Themes and Issues in World Regional Geography

Following two introductory chapters, this book adopts a regional perspective, grouping all of Earth's countries into a framework of 12 world regions (see Figure 1.5). We begin with a region familiar to most of our readers—North America—and then move on to Latin America, the Caribbean, Sub-Saharan Africa, Northwest Africa and Southwest Asia, Europe, the Russian Domain, and the different regions of Asia, before concluding with Australia and Oceania. Each regional chapter employs the same five-part thematic structure—physical geography and environmental issues, population and settlement, cultural coherence and diversity, geopolitical framework, and economic and social development. The concepts and data central to each theme are discussed in the following sections.

Physical Geography and Environmental Issues: The Changing Global Environment

Chapter 2 provides background on world physical and environmental geography, outlining the global environmental elements fundamental to human settlement—landforms, climate, energy, hydrology, and vegetation. In the regional chapters, the physical geography sections explain the environmental issues relevant to each world region, covering topics such as climate change, sea-level rise, acid rain, energy and resource issues, deforestation, and wildlife conservation. Each regional chapter addresses specific environmental problems, but also discusses policies and plans to resolve those issues (see *Working Toward Sustainability: A Concept with Many Meanings*).

Review

1.7 Explain the difference between latitude and longitude.

1.8 What does a map's scale tell us? List two ways to portray map scale.

1.9 What is a choropleth map, and what might it depict?

1.10 Describe how remote sensing differs from aerial photos.

KEY TERMS latitude (parallels), longitude (meridians), prime meridian, global positioning systems (GPS), map projection, map scale, representative fraction, graphic (linear) scale, thematic map, legend, choropleth map, remote sensing, geographic information systems (GIS), sustainability

WORKING TOWARD SUSTAINABILITY

Meeting the Needs of Future Generations

Google Earth (MG) Virtual Tour Video http://goo.gl/AEbLtq

The idea of **sustainability** seems to be everywhere, as we hear about sustainable cities, agriculture, forestry, businesses—even sustainable lifestyles. With so many different uses of the word, it is appropriate to revisit its original definition.

Sustainable has two main roots: The first means to endure and to maintain something at a certain level so that it lasts. The second refers to something that can be upheld or defended, such as a *sustainable idea* or *action*. Resource management has long used terms such as *sustained-yield forestry* to refer to timber practices where tree harvesting is attuned to the natural rate of forest growth so that the resource is not exhausted, but can renew itself over time.

Moral and ethical dimensions were added to this traditional usage in 1987 when the UN World Commission on Environment and Development addressed the complicated relationship between economic development and environmental deterioration. The commission stated that "sustainable development is development that meets the needs of the present without compromising the ability of future generations to meet their own needs." This cautionary message expands the notion of sustainability from a narrow focus on managing a specific resource, such as trees or grass, to include the whole range of human "needs," both now and in the future. Fossil fuels, for example, are finite, so we should not greedily consume them now without considering their availability for future generations. Similar cautions apply to sustainable use of all other resources—air, water, soil, genetic biodiversity, wildlife habitats, and so on.

Sustainably utilizing a specific resource, however, can be extremely difficult because it requires knowing the total amount of the resource in question and the current rate of consumption and then estimating the needs of future generations. These challenges have given rise to the new field of *sustainability science*, which emphasizes measuring and quantifying these factors. Because of these measurement difficulties, many researchers suggest that sustainability is better thought of as a process, rather than an achievable state.

Figure 1.2.1 Bogotá's Bikers Abundant bike lanes and an innovative bus system make Bogotá a model for a bike-friendly city with efficient public transportation. In 2003, the mayor introduced the "day without cars" campaign to demonstrate how this congested city could function without automobiles, a practice that still continues.

In the following chapters, we explore the different ways people are thinking about and working toward environmental and resource sustainability worldwide. Examples include green public transportation in Bogotá (Chapter 4; see Figure 1.2.1); desalinization in the desert in Dubai (Chapter 7); and sustainable tea and coffee production in China (Chapter 11). Many of these sidebars link to Google Earth virtual tour videos.

1. Does your college or community have a sustainability plan? If so, what are the key elements?
2. How might the concept of sustainability differ for a city or town in India or China compared to a U.S. city? Browse the Internet to see what you can learn about sustainability programs in other cities.

Population and Settlement: People on the Land

Currently, Earth has more than 7.3 billion people, with demographers (those who study human populations and population change) forecasting an increase to 9.7 billion by 2050. Most of this increase will take place in Sub-Saharan Africa, North Africa and Southwest Asia, and Australia and Oceania (Figure 1.24). In contrast, the regions of Europe, the Russian Domain, and East Asia will likely experience no demographic growth between now and 2050. Population concerns vary, with some countries trying to slow population growth, such as Bangladesh, while others worry about population decline, like Ukraine.

Population is a complex topic, but several points may help to focus the issues:

- The current rate of population growth is now half the peak rate experienced in the early 1960s, when the world population was around 3 billion. At that time, talk of a "population bomb" and "population explosion" was common, as scholars and activists voiced concern about what might happen if such high growth rates continued. Still, even with today's slower growth, demographers predict that over 2 billion more people will be added by 2050, with much of this growth taking place in the world's poorest countries.
- Population planning takes many forms, from the fairly rigid two-child policy of China to slow population growth to the family-friendly policies of no-growth countries that would like to increase their natural birth rates. Over half of the world's married women use modern contraceptive methods, which has contributed to slower growth (Figure 1.25).
- Not all attention should be focused on natural growth because migration is increasingly a significant cause of growth in some countries. International migration is often driven by a desire for a better life in a new country. Although much international migration is to developed countries in Europe, North America, and Oceania, there are comparable flows of migrants moving between developing countries, such

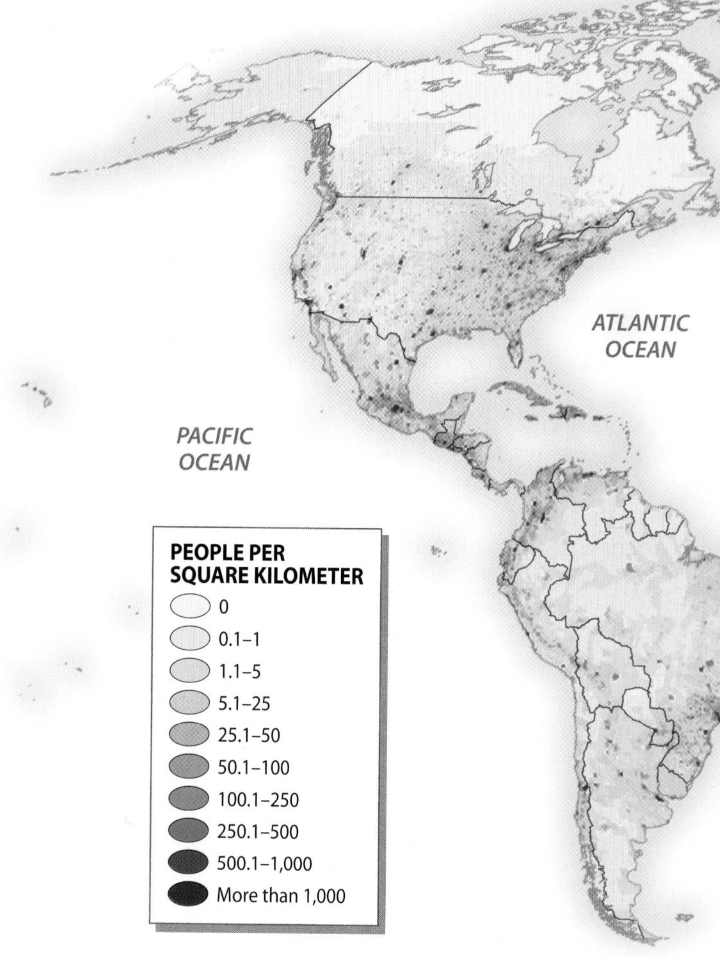

Figure 1.24 World Population This map emphasizes the world's different population densities. East and South Asia stand out as the most populated regions, with high densities in Japan, eastern China, northern India, and Bangladesh. In North Africa and Southwest Asia, population clusters are often linked to the availability of water for irrigated agriculture, as is apparent with the population cluster along the Nile River. Higher population densities in Europe, North America, and other countries are usually associated with large cities, their extensive suburbs, and nearby economic activities.

as flows from South Asia to Southwest Asia or immigration within Latin America and Sub-Saharan Africa. In addition, the UN estimates that 60 million people were displaced as a result of civil strife, political persecution, and environmental disasters in 2015, the largest number every recorded. This includes both internally displaced people and refugees who have left their country of origin.

- The greatest migration in human history is going on now as millions of people move from rural to urban places. In 2009, a landmark was reached when demographers estimated that for the first time more than half the world's population lived in towns and cities.

Population Growth and Change

Geographers make use of a variety of ways to define the population characteristics of a region. The most common measures and models are described here. Because of the central importance of demography in shaping localities, each regional chapter has a table of population indicators for the countries of that region, located in an appendix at the back of the book. Table 1.1 includes key population indicators for the world's 10 largest countries by total population size in 2015. Keep in mind that one-third of the world's 7.3 billion people live in two countries—China and India. The next largest country is the United States (319 million), followed by Indonesia (252 million) and Brazil (203 million). Combined, these 10 countries account for 60 percent of the world's population.

Population size alone tells only part of the story. **Population density**, for the purposes of this text, is the average number of people per square kilometer. Thus, China is the world's largest country demographically, but the population density of India, the second largest country, is more than twice that of China. Bangladesh's population density is far greater still at 1218 people per square kilometer.

Population densities differ considerably across a large country and between rural and urban areas, making the gross national figure a bit misleading. Many of the world's largest cities, for example, have densities of more than 30,000 people per square mile (10,300 per square kilometer), with the central areas of São Paulo and Shanghai

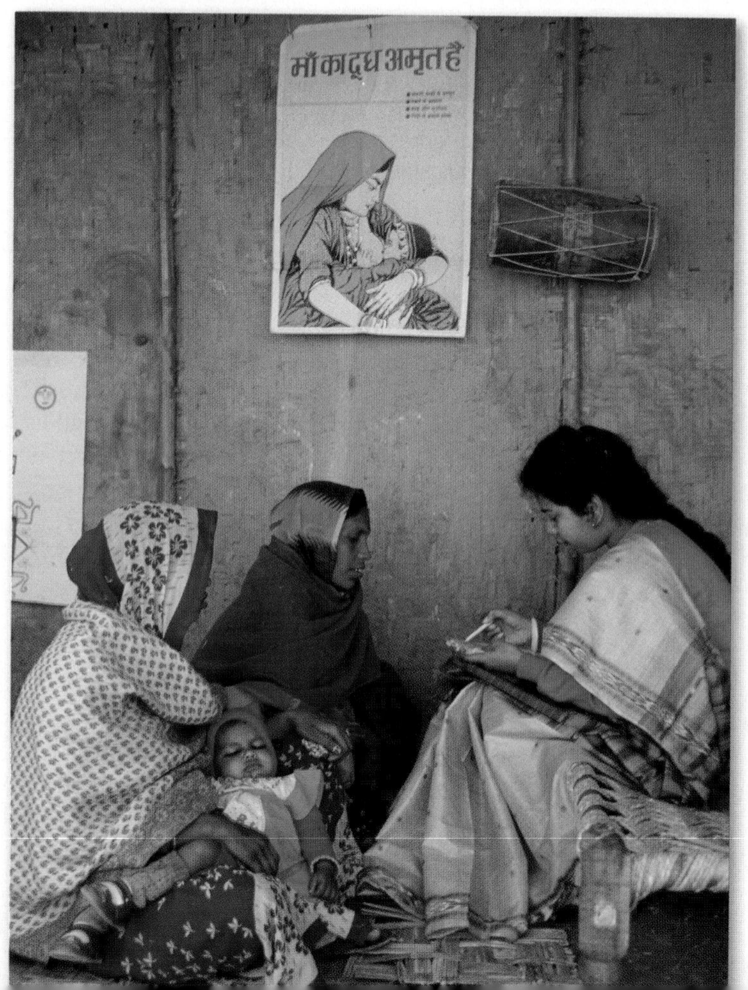

Figure 1.25 Family Planning Many countries with fast-growing populations attempt to slow growth through government clinics, like this one in Agra, India, that offer women advice on family planning and modern contraception methods.

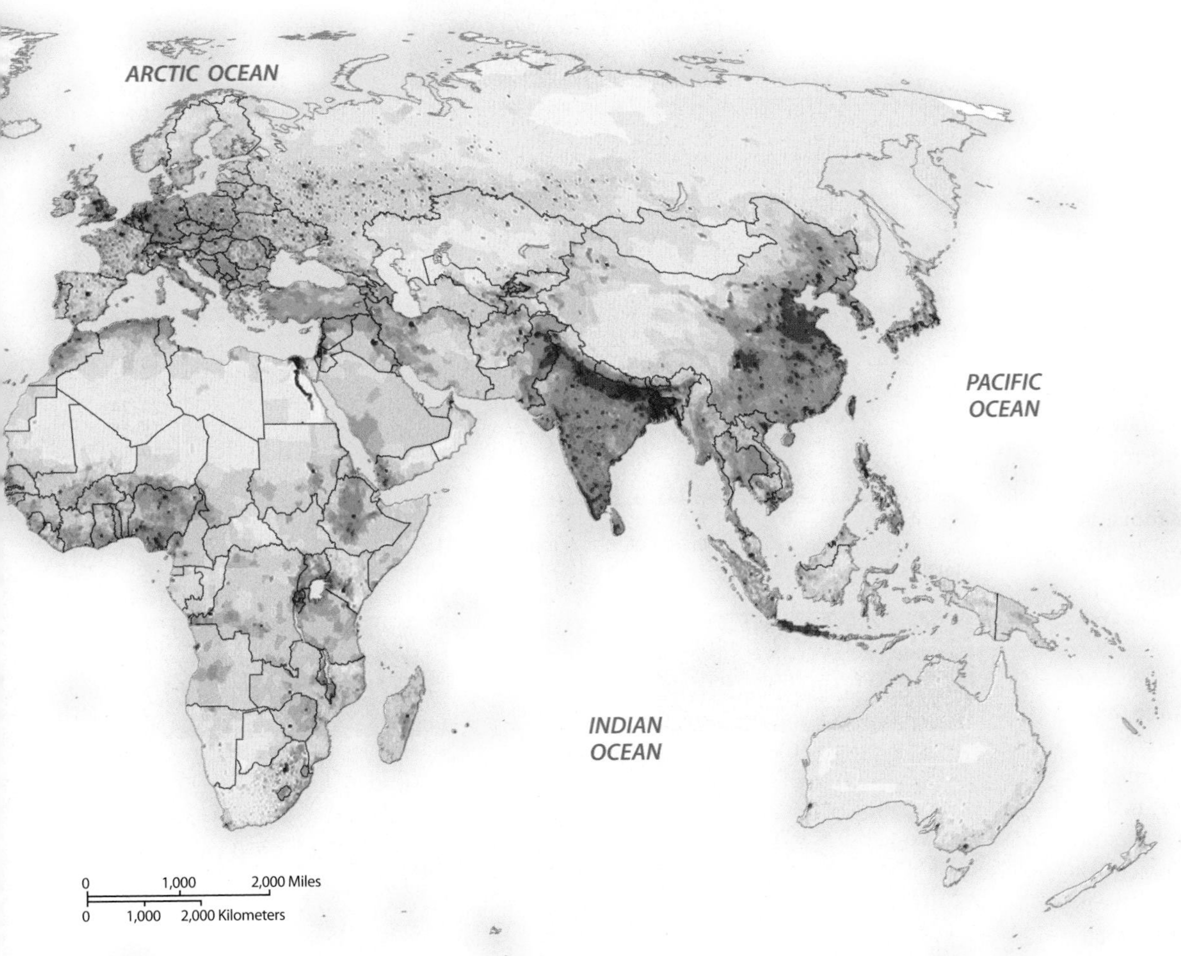

The statistics in Table 1.1 might seem daunting, but this information is crucial to understanding general population trends, overall growth rates, and patterns of settlement among the countries that make up various world regions.

Natural Population Increase A common starting point for measuring demographic change is the **rate of natural increase (RNI)**, which provides the annual growth rate for a country or region as a percentage. This statistic is produced by subtracting the number of deaths from the number of births in a given year. Important to remember is that population gains or losses through migration are not considered in the RNI.

The RNI is a small number with major consequences. It can be positive, as in the case of Nigeria, or negative, as in the case of Japan. China's RNI is 0.5, whereas India's is 1.5. Yet if those rates are maintained, China's population will double in 140 years, whereas India's will double in 47 years. This is why demographers are confident that India will surpass China as the largest country in the next decade or so. The country with the highest RNI on Table 1.1 is Nigeria at 2.5. If Nigeria maintains that rate, it will double its size in 28 years. Countries with a rate close to zero are demographically stable, but countries with persistent negative rates will experience slow declines in population unless immigration occurs.

easily twice as dense because of the prevalence of high-rise apartment buildings. In contrast, most North American cities have densities of fewer than 10,000 people per square mile (3,800 per square kilometer), due largely to a cultural preference for single-family dwellings on individual urban lots.

Table 1.1 POPULATION INDICATORS

Country	Population (millions) 2013	Population Density (per square kilometer)[1]	Rate of Natural Increase (RNI)	Total Fertility Rate	Percent Urban	Percent < 15	Percent > 65	Net Migration (Rate per 1000)
China	1,371.9	145	0.5	1.7	55	17	10	0
India	1,314.1	421	1.4	2.3	32	29	5	–1
United States	321.2	35	0.5	1.9	81	19	15	3
Indonesia	255.7	138	1.5	2.6	54	29	5	–1
Brazil	204.5	24	0.9	1.8	86	24	7	0
Pakistan	199.0	236	2.3	3.8	38	36	4	–2
Nigeria	181.8	191	2.5	5.5	50	43	3	0
Bangladesh	160.4	1,203	1.4	2.3	23	33	5	–3
Russia	144.3	9	0.0	1.8	74	16	13	2
Japan	126.9	349	–0.2	1.4	93	13	26	1

Source: Population Reference Bureau, *World Population Data Sheet, 2015.*

[1]World Bank, *World Development Indicators,* 2015.

Total Fertility Rate Population change is impacted by the **total fertility rate (TFR)**, which is the average number of live births a woman has in her lifetime. The TFR is a good indicator of a country's potential for growth. Clearly, women do not have 1.6 or 5.6 children; rather, women in some countries on average have 1 to 2 children versus 5 to 6 children, which means the potential for population growth is very different. A TFR of 2.1 is considered the **replacement rate** and suggests that it takes two children per woman, with a fraction more to compensate for infant and child mortality, to maintain a stable population. Where infant mortality is high, a country's actual replacement rate could be higher—say, 3.0. In 1970, the global TFR was 4.7, but by 2013 that rate was nearly cut in half to 2.5. Around the world, total fertility rates have been coming down for the last four decades as women move to cities, become better educated, work outside the home, control their fertility with modern contraception, and receive better medical care for themselves and their infants.

Half the countries listed in Table 1.1 have a below-replacement TFR, meaning that over time their natural growth will slow as fewer children are born, and in some cases population will decline if immigration does not occur. Even India's current TFR is 2.4, a dramatic change from 5.5 in 1970. India will still grow for many decades to come, but the potential for growth has been reduced as Indian women have smaller families. The countries with the highest total fertility rates are in Sub-Saharan Africa, where the average is about 5. Nigeria's TFR is slightly higher at 5.5.

Population Age Structure Another important indicator of a population's relative youthfulness, and its potential for growth, is the percentage of the population under age 15. Currently, 26 percent of the world's population is younger than age 15. However, in fast-growing Sub-Saharan Africa, that figure is 43 percent. This is another indicator of the population growth that will continue in this region for at least another generation. In contrast, only 16 percent of the populations of East Asia and Europe are under 15, suggesting slower growth and shrinking family sizes.

The other end of the age spectrum is also important, and it is measured by the percentage of a population over age 65. Just 8 percent of the world's population is over age 65, yet the percentage is twice that in many developed countries. Japan distinguishes itself in this regard with 26 percent of its population over 65; in contrast, only 13 percent of its population is under 15, indicating that the average age of the population is increasing as well. An aging population is significant when calculating a country's need to provide social services for its senior citizens and pensioners. It also has implications for the size of the overall workforce that supports retired and elderly individuals.

Population Pyramids The best graphic indicator of a population's age and gender structure is the **population pyramid**, which depicts the percentage of a population (or, in some cases, the raw number) that is male or female in different age classes, from young to old (Figure 1.26). If a country has higher numbers of young people than old, the graph has a broad base and a narrow tip, thus taking on a pyramidal shape that commonly forecasts rapid population growth. In contrast, slow-growth or no-growth populations are top-heavy, with a larger number of seniors than younger age classes.

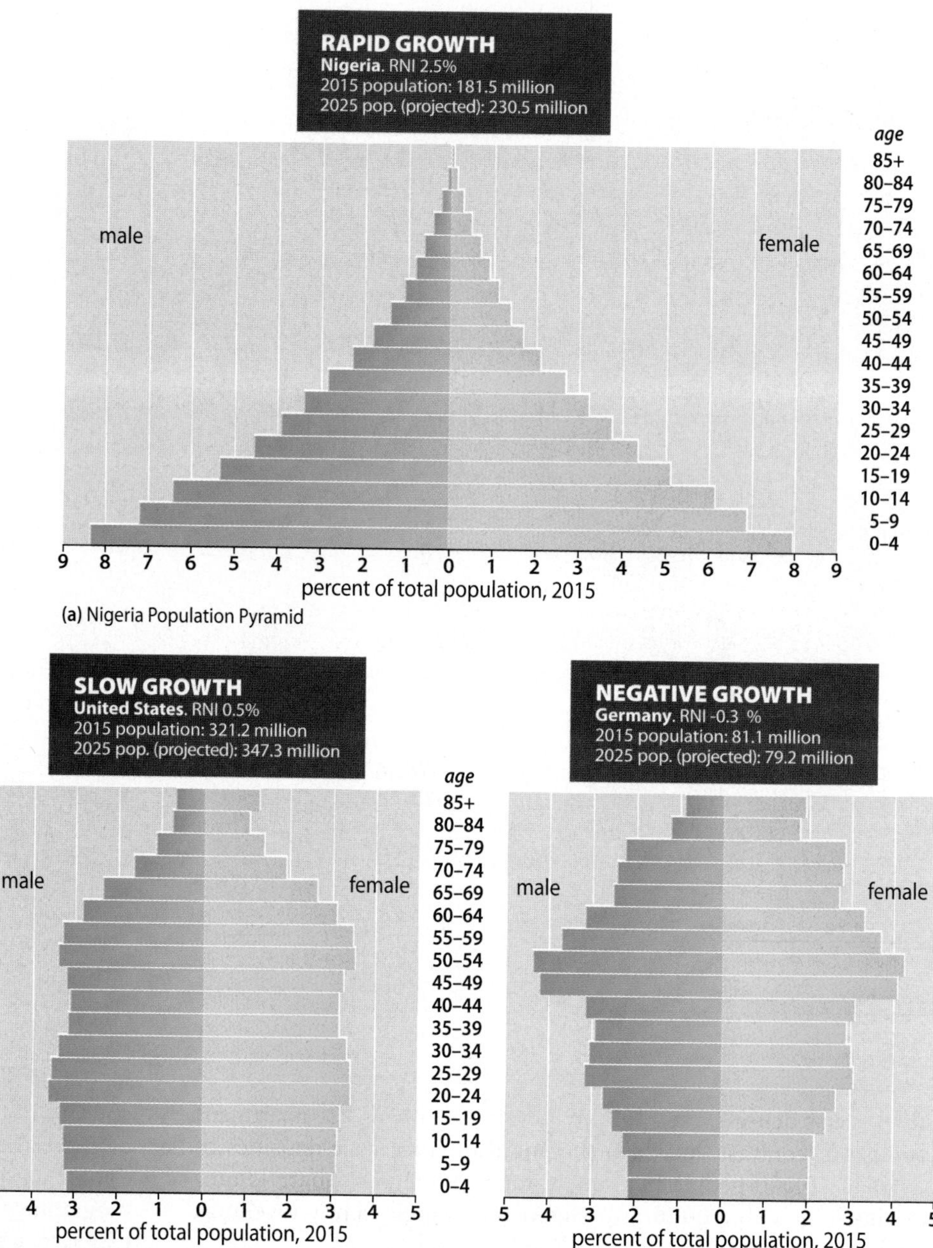

Figure 1.26 Population Pyramids The term *population pyramid* comes from the shape of the graph representing a rapidly growing country such as (a) Nigeria, when data for age and sex are plotted as percentages of the total population. The broad base illustrates the high percentage of young people in the country's population, which indicates that rapid growth will probably continue for at least another generation. This pyramidal shape contrasts with the narrow bases of slow- and negative-growth countries, such as (b) the United States and (c) Germany, which have fewer children and people in the childbearing years and a larger proportion of the population over age 65. **Q: Think of two example countries that fit into each of these three categories: rapid growth, slow growth, and negative growth.**

Not only are population pyramids useful for comparing different population structures around the world at a given point in time, but also they capture the structural changes of a population as it transitions from fast to slow growth. In addition, population pyramids display gender differences within a population, showing whether or not there is a disparity in the numbers of males and females. In the mid-20th century, for example, population pyramids for those countries that fought in World War II (such as the United States, Germany, the former Soviet Union, and Japan) showed a distinct deficit of males, indicating those lost to warfare. Similar patterns are found today in countries experiencing widespread conflict and civil unrest.

Cultural preferences for one sex or another, such as the preference for male infants in China and India, show up in population pyramids when there are more male children than female. Because of their utility in displaying population structures graphically, comparative population pyramids are found throughout the regional chapters of this book.

Life Expectancy A demographic indicator containing information about health and well-being in a society is *life expectancy*, which is the average number of years a typical male or female in a specific country can be expected to live. Life expectancy generally has been increasing around the world, indicating that conditions supporting life and longevity are improving. To illustrate, in 1970 the average life expectancy figure for the world was 58 years, whereas today it is 71. Some countries, such as Bangladesh, Iran, and Nepal, have seen average life expectancies increase by 20 years or more since 1970.

Because a large number of social factors—such as health services, nutrition, and sanitation—influence life expectancy, many researchers use life expectancy as a surrogate measure for development. When this figure is improving, it indicates that other aspects of development are occurring. Thus in the appendix for this book, life expectancy figures are shown in a separate table of development indicators for each world region.

The Demographic Transition The historical record suggests that population growth rates have slowed over time. More specifically, in Europe and North America, population growth slowed as countries became increasingly industrialized and urbanized. From these historical data, demographers generated the **demographic transition model**, a conceptualization that tracks the changes in birth rates and death rates over time. Birth rates are the number of annual births in a country per 1000 people, and death rates are the annual number of deaths per 1000 people. When annual birth rates exceed death rates, natural increase occurs. Originally, this model had four stages; however, a fifth stage is now commonly added to illustrate that many countries have slowed still further to a no-growth point (Figure 1.27).

In the original demographic transition model, Stage 1 is characterized by both high birth rates and high death rates, resulting in a very low rate of natural increase. Historically, this stage is associated with Europe's preindustrial period, a time that predated common public health measures such as water and sewage treatment, an understanding of disease transmission, and the most fundamental aspects of modern medicine. Not surprisingly, death rates were high and life expectancy was short. Tragically, these conditions are still found today in some parts of the world.

In Stage 2, death rates fall dramatically, while birth rates remain high, thus producing a rapid rise in the RNI. In both historical and contemporary times, this decrease in death rates is commonly associated with the development of public health measures and modern medicine. Additionally, one of the assumptions of the demographic transition model is that these health services become increasingly available only after some degree of economic development and urbanization takes place.

However, even as death rates fall and populations increase, it takes time for people to respond with lower birth rates, which happens in Stage 3. This, then, is the transitional stage in which people become aware of the advantages of smaller families in an urban and industrial setting, contrasted with the earlier need for large families in rural, agricultural settings or where children worked at industrial jobs (both legally and illegally).

Then in Stage 4, a low RNI results from a combination of low birth rates and low death rates. Until recently, this stage was assumed to be the static end point of change for developing and urbanizing populations. This, however, does not seem to be the case; in many highly urbanized developed countries, such as those in Europe, the death rate now exceeds the birth rate. As a result, the RNI falls below a replacement level, expressed as a negative number. This negative growth state can be considered a fifth stage of the traditional

Figure 1.27 Demographic Transition As a country industrializes, its population moves through the five stages in this diagram, known as the *demographic transition model*. In Stage 1, population growth is low because high birth rates are offset by high death rates. Rapid growth takes place in Stage 2, as death rates decline. Stage 3 is characterized by a decline in birth rates. The transition was initially thought to end with low growth once again in Stage 4, resulting from a relative balance between low birth rates and low death rates. But with many developed countries now showing no natural growth, demographers have recently added a fifth stage to the traditional model to show no growth or even negative natural growth.

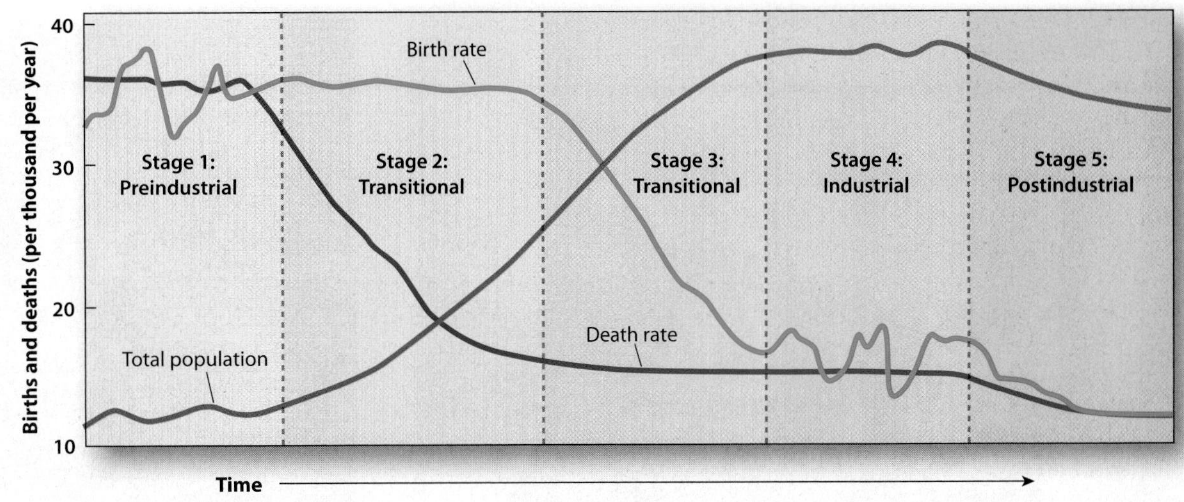

demographic transition model. Of the world's largest countries, Japan falls into this category.

Remember, though, that the RNI is just that—the rate of *natural* increase—and does not capture a country's growth from immigration. For example, even if RNI is negative, a country may demographically grow or stabilize due to immigration from other countries.

Global Migration and Settlement

Never before have so many people been on the move, either from rural areas to cities or across international borders. Today more than 230 million people live outside the country of their birth and thus are officially designated as immigrants by the UN and other international agencies. Much of this international migration is directly linked to the new globalized economy because the majority of these migrants live either in the developed world or in developing countries with vibrant industrial, mining, or petroleum extraction economies. In the oil-rich countries of the United Arab Emirates, Qatar, and Saudi Arabia, the labor force is composed primarily of foreign migrants, especially from South Asia (Figure 1.28). The top six destination countries, which account for 40 percent of the world's immigrants, are major industrial or mining economies: United States, Russia, Germany, Saudi Arabia, United Kingdom, and United Arab Emirates. Within these countries, and for most other destinations, migrants are drawn to the opportunities found in major metropolitan areas.

Not all migrants move for economic reasons. War, persecution, famine, and environmental destruction cause people to flee to safe havens elsewhere. Accurate data on refugees are often difficult to obtain for several reasons (such as individuals illegally crossing international boundaries or countries deliberately obscuring the number for political reasons), but UN officials estimate that there are currently 60 million refugees or internally displaced persons. More than half of these people are in Africa and Southwest Asia. The conflict in Syria has displaced over half of that country's population. Most of the 11.6 million displaced people are scattered within Syria, but 4 million live outside the territory, mostly in Turkey, Lebanon, Jordan, and Iraq (Figure 1.29).

Net Migration Rates The amount of immigration (people entering a country) and emigration (those leaving a country) is measured by the **net migration rate**. A positive figure means that a country's population is growing because of migration, whereas a negative number means more people are leaving. As with other demographic indicators, the net migration rate is provided for the number of migrants per 1000 of a base population. Returning to Table 1.1, only the United States, Russia and Japan show positive net migration rates. Three states (Indonesia, Pakistan, and Bangladesh) have negative rates, and five (China, India, Brazil, Nigeria) have rates at zero, meaning the numbers of people entering and leaving in a particular year cancel each other out. This does not mean that these countries do not produce immigrants—both India and China have large populations overseas—but for that particular year, incoming and outgoing flows equaled each other.

Countries with some of the highest net migration rates depend heavily on migrants for their labor force, such as the United Arab Emirates, Kuwait, and Oman. Countries with the highest negative migration rates include those in conflict (Syria at –11) and Pacific island nations such as Samoa (–24) and Micronesia (–19) with relatively small populations and weak economies.

Settlement in an Urbanizing World The focal points of today's globalizing world are cities—the fast-paced centers of deep and widespread economic, political, and cultural change. Because of this vitality and the options cities offer to impoverished and uprooted rural peoples, they are magnets for migration. The scale and rate of growth of some world cities are absolutely staggering. Between natural growth and in-migration, Mumbai (Bombay) is expected to add over 7 million people by 2020, which, assuming growth is constant throughout the period (perhaps a questionable assumption), would mean that this urban area would add over 10,000 new people each week. The same projections show Lagos, Nigeria, which currently has the highest annual growth of any megacity, adding almost 15,000 residents per week.

Based on data on the **urbanized population**, which is the percentage of a country's population living in cities, 53 percent of the

Figure 1.28 Global Workforce South Asia laborers working on a construction site in Doha, Qatar. Many of the Persian Gulf countries rely upon vast numbers of contract laborers, mostly from South Asia, to provide the labor force necessary to build these modern cities and serve the populations living there.

Figure 1.29 Refugee Camps A Kurdish woman from Syria walks with her baby at a refugee camp near Suruc, Turkey. Suruc is located close to the Syrian border and this refugee camp is one of the largest ones in Turkey. The crisis in Syria has displaced millions. In 2015 it was estimated that 4 million Syrians had left their homeland.

world's population is urbanized. Further, demographers predict that the world will be 60 percent urbanized by 2025. Urbanization rates vary by region. Sub-Saharan Africa may be rapidly urbanizing, but in countries such as Ethiopia, Kenya, and Malawi, more than three-quarters of the population still lives in rural areas. The population of India, with its megacities of over 10 million people, is still two-thirds rural.

Generally speaking, most countries with high rates of urbanization are also more developed and industrialized because manufacturing tends to cluster around urban centers. We know that urbanization is a major demographic reality, but it is important to remember that 3 billion people live in rural settings that are also being transformed by globalization.

Review

1.11 What is the rate of natural increase (RNI), and how can it be a negative number?

1.12 Explain a high versus a low total fertility rate, and give examples.

1.13 Describe and explain the demographic transition model.

1.14 How is a population pyramid constructed, and what kind of information does it convey?

KEY TERMS population density, rate of natural increase (RNI), total fertility rate (TFR), replacement rate, population pyramid, demographic transition model, net migration rate, urbanized population

Cultural Coherence and Diversity: The Geography of Change and Tradition

Social scientists often say that culture binds together the world's diverse social fabric. If this is true, one glance at the daily news suggests this complex global tapestry could be unraveling because of widespread cultural tensions and conflict. As noted earlier, the recent rise of global communication systems (TV, films, smartphones, and the Internet) have diffused stereotypical Western culture at a rapid pace. Although some cultures accept these new influences willingly, others resist and push back against these new forms of cultural imperialism through local protests, censorship, and even terrorism. Still others use this technology to advance their own cultural or political agendas.

The geography of cultural cohesion and diversity studies tradition and change, new cultural forms produced by interactions between cultures, gender issues, and global languages and religions.

Culture in a Globalizing World

The dynamic changes connected with globalization have blurred traditional definitions of culture. A very basic definition provides a starting point. **Culture** is learned, not innate, behavior shared by a group of people, empowering them with what is commonly called a "way of life."

Culture has both abstract and material dimensions: speech, religion, ideology, livelihood, and value systems, but also technology, housing, foods, dress, and music. Even something like sports can have deep cultural meaning. Think of how billions of people watch the World Cup and support their "national" teams with near-religious devotion. These varied expressions of culture are relevant to the study of world regional geography because they tell us much about the way people interact with their environment, with one another, and with the larger world (Figure 1.30). Not to be overlooked is that culture is dynamic and ever changing, not static. Thus, culture is a process, not a condition—an abstract, yet useful concept that is constantly adapting to new circumstances. As a result, there are often tensions between the conservative, traditional elements of a culture and the newer forces promoting change.

When Cultures Collide Cultural change often takes place within the context of international tensions. Sometimes, one cultural system will replace another; at other times, resistance by one group to another's culture will stave off change. More commonly, however, a newer, hybrid form of culture results from an amalgamation of two cultural traditions. Historically, colonialism was the most important perpetuator of these cultural collisions; today globalization in its varied forms is a major vehicle for cultural tensions and change (see *Everyday Globalization: Common Cultural Exchanges*).

The active promotion of one cultural system at the expense of another is called **cultural imperialism**. The most severe examples occurred during the colonial period, when European cultures spread worldwide and overwhelmed, eroded, and even replaced indigenous cultures. During this period, Spanish and Portuguese cultures spread widely in Latin America, French culture diffused into parts of Africa and Southeast Asia, and British culture was imprinted on North America as well as much of South Asia and Sub-Saharan Africa. New languages were mandated, new educational systems were implanted, and new administrative institutions replaced the old. Foreign dress styles, diets, gestures, and organizations were added to existing cultural systems. Many vestiges of colonial culture are still evident today. In India, the makeover was so complete that pundits say, with only slight exaggeration, that "the last true Englishman will be an Indian."

Today's cultural imperialism is seldom linked to an explicit colonizing force, but more often comes as a byproduct of economic globalization. Though many expressions of cultural imperialism carry a Western (even U.S.) tone—such as McDonald's, MTV, KFC, Marlboro cigarettes, and the widespread use of English as the dominant language of the Internet—these facets result more from a search for new consumer markets than from deliberate efforts to spread modern U.S. culture throughout the world.

Figure 1.30 U.S. Popular Culture The clothing and appearance of this young couple give clues to their membership in Brooklyn, New York, hipster culture.

EVERYDAY GLOBALIZATION

Common Cultural Exchanges

Globalization is so ubiquitous that it's often taken for granted. What you're wearing was probably made overseas, since 98 percent of all U.S. apparel is imported. Your shirt could be made in China, Bangladesh, Thailand, Haiti, Mexico, or India, all of which are major manufacturing centers for the world's clothing. Even some "Made in the U.S.A." clothing might be pushing the truth a bit by being produced in the U.S. commonwealth countries of Puerto Rico and the Northern Mariana Islands in the far western Pacific. However, if you paid $300 or so for your jeans, they could be made in the United States—most probably in Los Angeles, where 30 different apparel firms turn out designer jeans.

The point is that globalization is not just about multinational corporations doing business all over the world; it is everywhere in your daily life, from what you eat to what you wear to the smartphone in your hand to the coffee you drink. Chances are that whatever it is involves a complex world geography.

We illustrate this idea in each chapter: how your chocolate bar may or may not come from the tropical rainforest (Chapter 2); why the National Basketball Association is so successful globally (Chapter 3); where carnival is celebrated outside of Latin America and the Caribbean (Chapter 5); and which countries build the huge ships that bring commodities and finished goods to ports around the world (Chapter 11). One way that U.S. college students experience the world is through study-abroad programs, which can be important opportunities to learn about cultures other than your own (Figure 1.3.1).

1. How has globalization changed higher education in the United States?
2. Identify a commonplace item or activity in your life that has an interesting backstory involving globalization.

Figure 1.3.1 Cultural Exchange Through Study Abroad College students from the United States and Panama work together on a research project investigating urban sustainability in the historic colonial core of Panama City.

The reaction against cultural imperialism is **cultural nationalism.** This is the process of protecting and defending a cultural system against diluting or offensive cultural expressions, while at the same time actively promoting national and local cultural values. Often cultural nationalism takes the form of explicit legislation or official censorship that simply outlaws unwanted cultural traits. For example, France has long fought the Anglicization of its language by banning "Franglais" in official governmental French, thereby exorcising commonly used words such as *weekend, downtown, chat,* and *happy hour.* France has also sought to protect its national music and film industries by legislating that radio DJs play a certain percentage of French songs and artists each broadcast day (40 percent currently). Many Muslim countries limit Western cultural influences by restricting or censoring international TV, an element they consider the source of many undesirable cultural influences. Most Asian countries are also increasingly protective of their cultural values, and many are demanding changes to tone down the sexual content of MTV and other international TV networks.

Cultural Hybrids The most common product of cultural collision is the blending of forces to form a new, synergistic form of culture, a process called **cultural syncretism** or **cultural hybridization.** To characterize India's culture as British, for example, is to grossly oversimplify

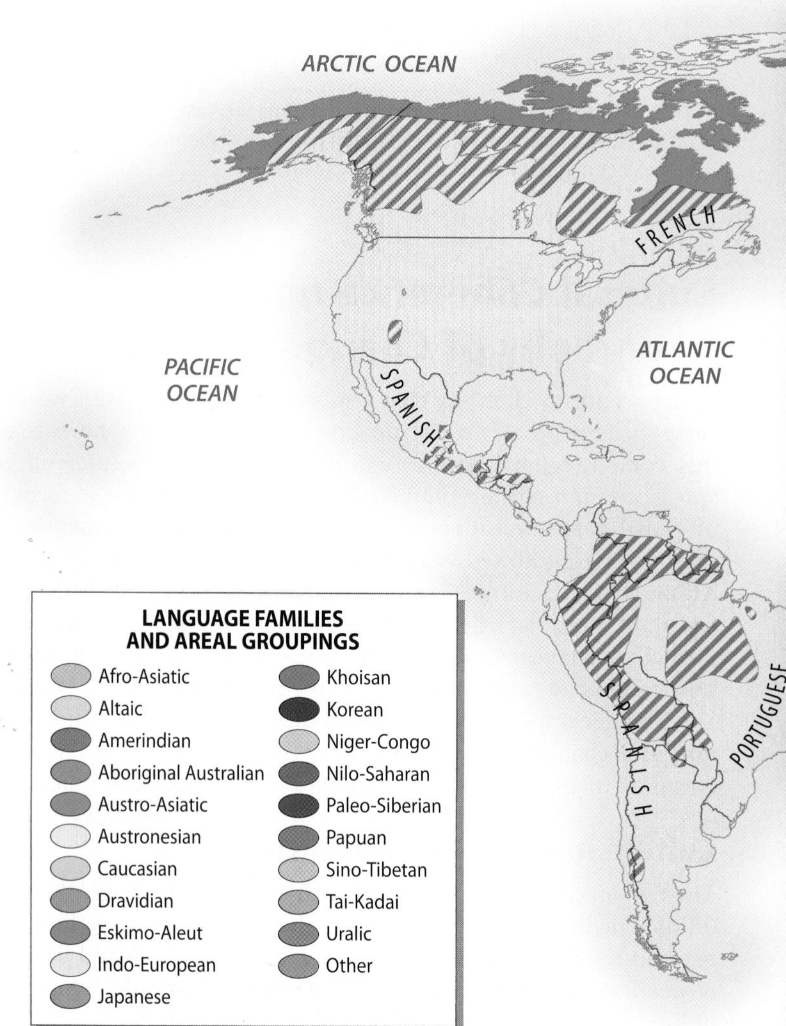

Figure 1.31 Bolivian Hip-Hop Bolivian rappers perform in the highland town of Tiwanaku. Rapping in Spanish and Aymara, these young artists sing about political change and vent their anger over historical oppression and exploitation, while adopting many elements of U.S. hip-hop culture.

and exaggerate England's colonial influence. Instead, Indians have adapted many British traits to their own circumstances, infusing them with their own meanings. India's use of English, for example, has produced a unique form of "Indlish" that often befuddles visitors to South Asia. Nor should we forget that India has added many words to our English vocabulary—*khaki, pajamas, veranda*, and *bungalow*, among others. Clearly, both the Anglo and the Indian cultures have been changed by the British colonial presence in South Asia.

Other examples of cultural hybrids abound: Australian-rules football, Bolivian hip-hop music, fusion cuisine, Tex-Mex fast food, and on and on (Figure 1.31).

Language and Culture in Global Context

Language and culture are so intertwined that often language is the major characteristic that defines one cultural group and differentiates it from another (Figure 1.32). Furthermore, because language is the primary means for communication, it folds together many other aspects of cultural identity, such as politics, religion, commerce, and customs. Language is fundamental to cultural cohesiveness and distinctiveness, for language not only brings people together, but also sets them apart from nonspeakers of that language. Therefore, language is an important component of national or ethnic identity as well as a means for creating and maintaining boundaries for group and regional identity.

Because most languages have common historic (and even prehistoric) roots, linguists have grouped the thousands of languages found throughout the world into a handful of *language families*, based on common ancestral speech. For example, about half of the world's people speak languages of the Indo-European family, a large group that includes not only European languages such as English and Spanish, but also Hindi and Bengali, the dominant languages of South Asia.

Within language families, smaller units also give clues to the common history and geography of peoples and cultures. Language branches and groups (also called *subfamilies*) are closely related subsets within a language family, usually sharing similar sounds, words, and grammar. Well known are the similarities between German and English and between French and Spanish.

Additionally, individual languages often have distinctive *dialects* associated with specific regions and places. Think of the distinctions among British, Canadian, and Jamaican English or the

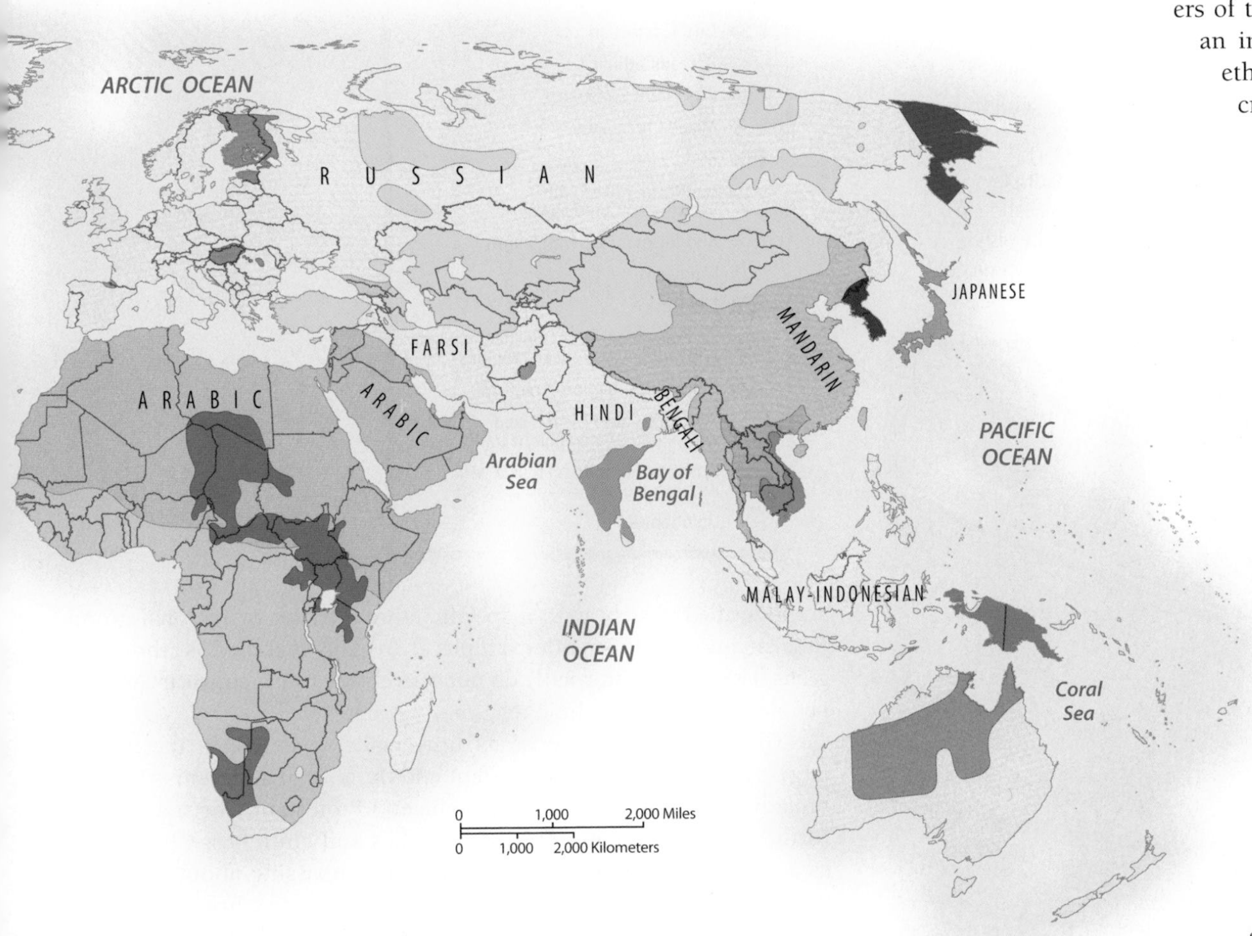

Figure 1.32 World Language Families Most languages of the world belong to a handful of major language families. About 50 percent of the world's population speaks a language belonging to the Indo-European language family, which includes languages common to Europe and Russia, but also major languages in South Asia, such as Hindi. They are in the same family because of their linguistic similarities. The next largest family is the Sino-Tibetan family, which includes languages spoken in China, the world's most populous country. **Q: What languages, other than English, are spoken in your community?**

city-specific dialects that set apart New Yorkers from residents of Dallas, Berliners from inhabitants of Munich, or Parisians from villagers of rural France.

When people from different cultural groups cannot communicate directly in their native languages, they often employ a third language to serve as a common tongue, a **lingua franca**. Swahili has long served that purpose for speakers of the many tribal languages of eastern Africa, and French was historically the lingua franca of international politics and diplomacy. Today English is increasingly the common language of international communications, science, and air transportation (Figure 1.33).

The Geography of World Religions

Another important defining trait of cultural groups is religion (Figure 1.34). Indeed, in this era of a comprehensive global culture, religion has become increasingly important in defining cultural identity. Recent ethnic violence and unrest in far-flung places such as the Balkans, Iraq, Syria, and Burma illustrate the point.

Universalizing religions, such as Christianity, Islam, and Buddhism, attempt to appeal to all peoples, regardless of location or culture. These religions usually have a proselytizing or missionary program that actively seeks new converts throughout the world. In contrast are **ethnic religions**, which are identified closely with a specific ethnic, tribal, or national group. Judaism and Hinduism, for example, are usually regarded as ethnic religions because they normally do not actively seek new converts; instead, people are born into ethnic religions.

Figure 1.33 Chinese and English This road sign in Shanghai displays two of the world's most popular languages, Chinese and English. Mandarin Chinese is spoken by about 12 percent of the world's population. English is the global language of commerce, transportation, and science. An estimated 2 billion people know some English.

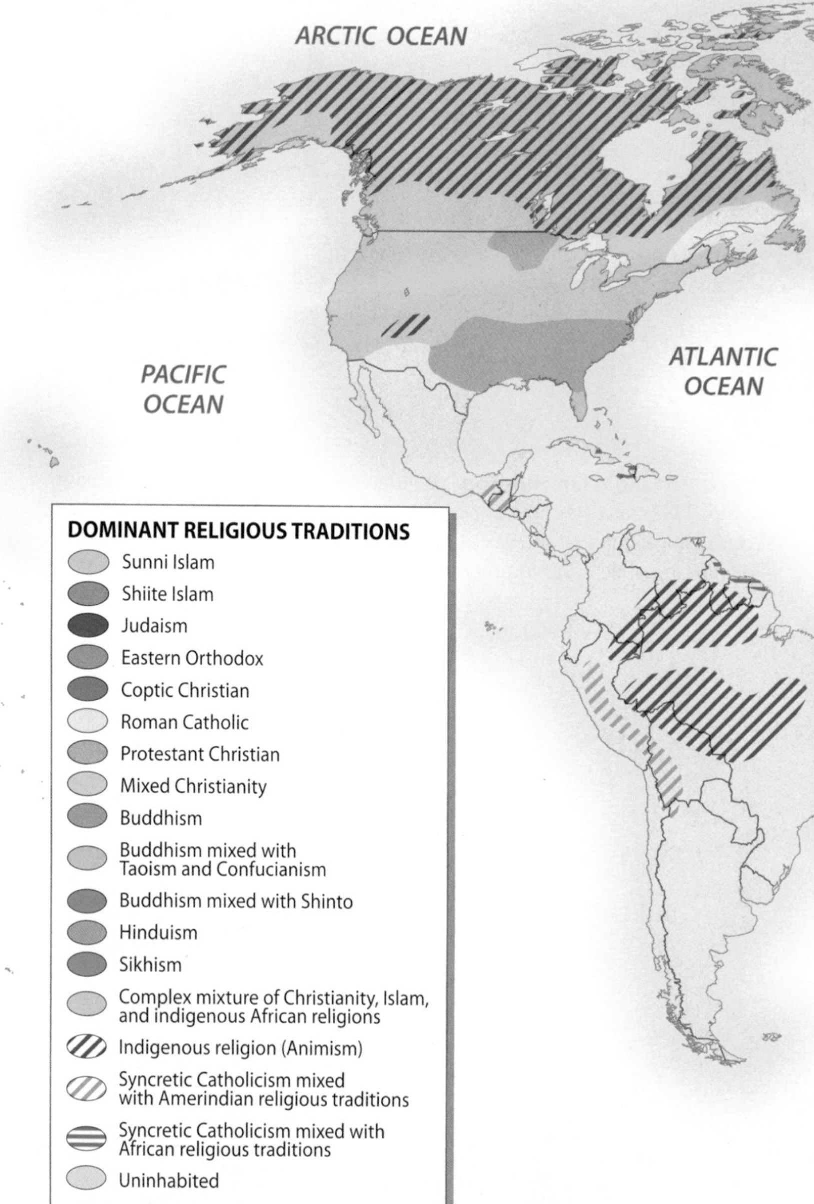

Christianity, because of its universalizing ethos, is the world's largest religion in both areal extent and number of adherents. Though broadly divided into Roman Catholic and Protestant Christianity and further fragmented into myriad branches and churches, Christianity as a whole has 2.1 billion adherents, encompassing about one-third of the world's population. The largest numbers of Christians can be found in Europe, Africa, Latin America, and North America. Islam, which has spread from its origins on the Arabian Peninsula east to Indonesia and the Philippines, has about 1.3 billion members.

Although not as severely fragmented as Christianity, Islam should not be considered a homogeneous religion because it is also split into separate groups. *Shi'a Islam* constitutes about 11 percent of the total Islamic population and represents a majority in Iran and southern Iraq, while the more dominant *Sunni Islam* is found from the Arab-speaking

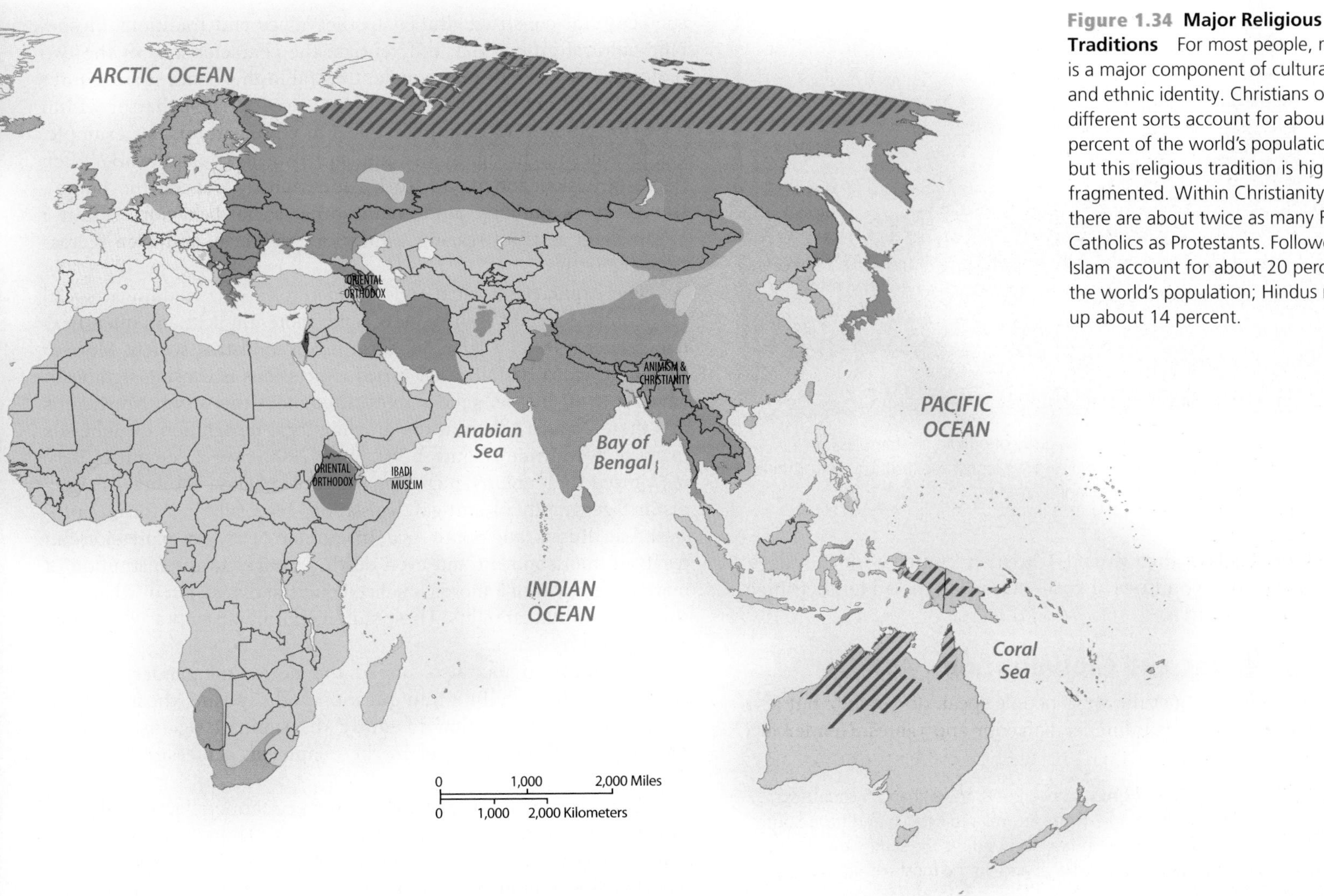

Figure 1.34 Major Religious Traditions For most people, religion is a major component of cultural and ethnic identity. Christians of different sorts account for about 34 percent of the world's population, but this religious tradition is highly fragmented. Within Christianity, there are about twice as many Roman Catholics as Protestants. Followers of Islam account for about 20 percent of the world's population; Hindus make up about 14 percent.

lands of North Africa to Indonesia. Probably in response to Western influences connected to globalization, both the Shi'a and the Sunni branches of Islam are currently experiencing fundamentalist revivals in which proponents seek to maintain purity of faith, separate from these Western influences.

Judaism, the parent religion of Christianity, is also closely related to Islam. Although tensions are often high between Jews and Muslims, these two religions, along with Christianity, actually share historical and theological roots in the Hebrew prophets and leaders. Judaism now numbers about 14 million adherents, having lost perhaps one-third of its total population to the systematic extermination of Jews during World War II.

Hinduism, which is closely linked to India, has about 900 million adherents. Outsiders often regard Hinduism as polytheistic because Hindus worship many deities. Most Hindus argue, however, that all of their faith's gods are merely representations of different aspects of a single divine, cosmic unity. Historically, Hinduism is linked to the caste system, with its segregation of peoples based on ancestry and occupation. However, because India's democratic government is committed to reducing the social distinctions among castes, the connections between religion and caste are now much less explicit than in the past.

Buddhism, which originated as a reform movement within Hinduism 2500 years ago, is widespread in Asia, extending from Sri Lanka to Thailand and from Mongolia to Vietnam (Figure 1.35). There are two major branches of Buddhism: *Theravada*, found throughout Southeast Asia and Sri Lanka, and *Mahayana*, found in Tibet and East Asia. In its spread, Buddhism came to coexist with other faiths in certain areas, making it difficult to accurately estimate the number of its adherents. Estimates of the total Buddhist population range from 350 million to 900 million people.

In some parts of the world, religious practice has declined significantly, giving way to **secularism**, in which people consider themselves either nonreligious or outright atheistic. Though secularism is difficult to measure, social scientists estimate that about 1.1 billion people fit into this category worldwide. Perhaps the best example of secularism comes from the former communist lands of Russia and eastern Europe where, historically, there was overt hostility between government and church from the time of the 1917 Russian Revolution. Since the demise of Soviet control in the 1990s, however, many of these countries have experienced religious revivals.

Secularism has also grown more pronounced in western Europe. Although France was historically, and to some extent still is culturally, a Roman Catholic country, more people in France attend Muslim

Figure 1.35 Buddhist Landscape An array of buildings—temples, monasteries, and shrines—produces a distinctive landscape throughout Southeast Asia, like Chiang Mai in northern Thailand.

mosques on Fridays than attend Christian churches on Sundays. Japan and the other countries of East Asia are also noted for their high degree of secularization.

Culture, Gender, and Globalization

Culture includes not just the ways people speak or worship, but also embedded practices that influence behavior and values. **Gender** is a sociocultural construct, linked to the values and traditions of specific cultural groups that differentiate the characteristics of the two biological sexes, male and female. Central to this concept are **gender roles**, the cultural guidelines that define appropriate behavior within a specific context. In traditional tribal or ethnic groups, for example, gender roles might rigidly distinguish between women's work (often domestic tasks) and men's work (done mostly outside the home). Gender roles similarly guide many other social behaviors within a group, such as child rearing, education, marriage, and even recreational activities.

The explicit and often rigid gender roles of a traditional social unit contrast greatly with the less rigid, more implicit, and often flexible gender roles of a large, modern, urban industrial society. More to the point, globalization in its varied expressions is causing significant changes to traditional gender roles throughout the world. Nowhere is this more apparent than in the growing legal recognition of same-sex marriage worldwide (Figure 1.36). Since 2000, over 25 countries have recognized such unions, including the United States. Yet there is also a distinct geography of anti-gay legislation, especially in Africa, Southwest Asia, Russia, and South Asia. In extreme cases, gay expression can result in imprisonment and even death. Changes to the institution of marriage are part of a more globalized cultural discussion of what constitutes basic human rights. These shifting norms are embraced by some and rejected by others.

Globalization has also spread the notion of gender equality around the globe, calling into question and exposing those cultural groups and societies that blatantly discriminate against women. This topic is discussed later in the chapter as a measure of social development.

There are gender dimensions to the economic effects of globalization in many developed countries. In the United States, for example, male workers have suffered more from unemployment than have females as industrial and technology jobs have been outsourced

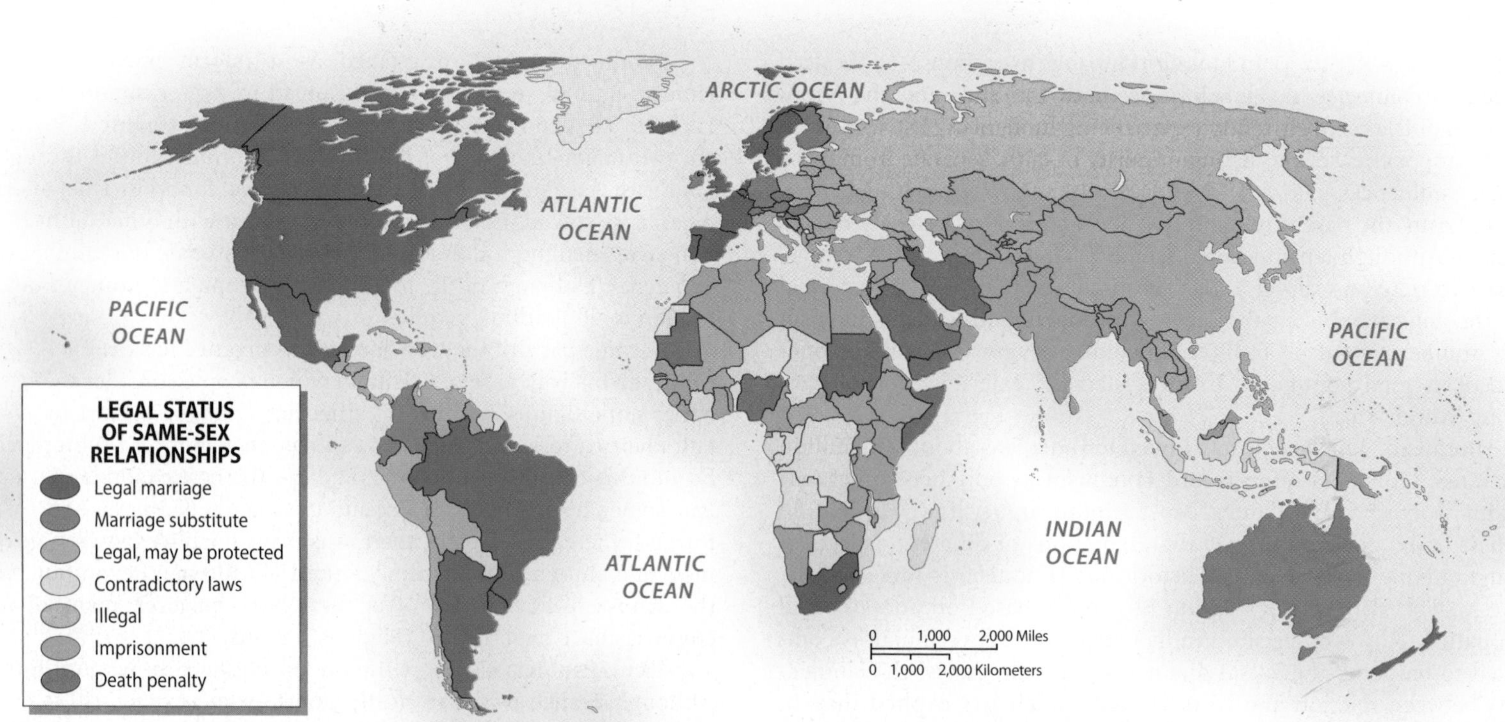

Figure 1.36 Mapping Gay Rights Since 2000, more than 25 countries have recognized same-sex marriage. From Australia to Mexico and from South Africa to Ireland, a major cultural shift has occurred. At the same time, there are countries where gay expression is illegal and, in the most extreme cases, punishable by death.

to China and India. Consequently, in many households women are emerging as primary income earners, while men have taken on new roles in domestic activities.

Review

1.15 Define cultural imperialism and cultural hybridization, and give an example of each.

1.16 What is a lingua franca? Provide two examples.

1.17 Describe the geographies of the two branches of Islam.

1.18 Discuss the patterns of acceptance and exclusion shown in Figure 1.36 with regard to gay rights.

KEY TERMS culture, cultural imperialism, cultural nationalism, cultural syncretism, cultural hybridization, lingua franca, universalizing religion, ethnic religion, secularism, gender, gender roles

Geopolitical Framework: Unity and Fragmentation

The term **geopolitics** is used to describe the close link between geography and politics. More specifically, geopolitics focuses on the interactivity between political power and territory at all scales, from the local to the global. Unquestionably, one of the global characteristics of the last several decades has been the speed, scope, and character of political change in various regions of the world; thus, discussions of geopolitics are central to world regional geography.

With the demise of the Soviet Union in 1991 came opportunities for self-determination and independence in eastern Europe and Central Asia, resulting in fundamental changes to economic, political, and even cultural alignments. Religious freedom helped drive national identities in some new Central Asian republics, whereas eastern Europe was primarily concerned with new economic and political links to western Europe. Russia itself still wavers perilously between different geopolitical pathways. Russia's justification for taking over parts of Ukraine in 2014 is based on the fact that ethnic Russians live there. Meanwhile, these acts have been condemned internationally as an affront to state sovereignty (Figure 1.37). All of these topics are discussed further in Chapters 8, 9, and 10.

The Nation-State Revisited

A map of the world shows an array of nearly 200 countries ranging in size from microstates like Vatican City and Andorra to huge territorial and multiethnic states such as Russia, the United States, Canada, and China. All of these countries are regulated by governmental systems, ranging from democratic to autocratic. Commonly, these different forms of government share a concern with **sovereignty**, defined geopolitically as a government's ability to control activities within its territorial borders.

The notion of sovereignty is closely linked to the concept of the **nation-state**. Here *nation* describes a large group of people with shared sociocultural traits, such as language, religion, and shared identity. The word *state* refers to a political entity (a government) that has delimited territorial boundaries, control over its internal space, and recognition by other political entities. France and England are often cited as the archetypal examples of a nation-state. Contemporary countries such as Albania, Egypt, Bangladesh, Japan, and the two Koreas are more modern examples of countries where there is close overlap between nation and state. The related term *nationalism* is the sociopolitical expression of identity and allegiance to the shared values and goals of the nation-state.

Globalization, however, has weakened the vitality of the nation-state concept because today most of the world's countries have a questionable fit with the traditional definition of nation-state. International migration has led to many countries with large populations of ethnic minorities who may not share the national culture of the majority. In England, for example, large numbers of South Asians form their own communities, speak their own languages, practice their own religions, and dress to their own standards. Similarly, France is home to a mosaic of peoples from its former colonial lands in Africa and Asia. Canada and the United States also have large immigrant populations who are legal citizens of the political state but have changed the very nature of the national culture by their presence. The fact that the United States, Canada, Germany, and other countries officially embrace cultural diversity and declare themselves multicultural states underscores these changes.

Decentralization and Devolution Also residing within many nation-states are groups of people who seek autonomy from the central government and argue for the right to govern themselves. This autonomy can range from the simple decentralization of power from a central government to smaller governmental units, as is the case with U.S. states or French departments. At the far end of the spectrum is outright political separation and full governmental autonomy, termed *devolution*. As an illustration, the citizens of Scotland held a referendum in 2014 to separate from England; although the referendum failed, 45 percent of the electorate chose independence. Other separatist movements are found among French-speaking people of Québec Province in Canada, the Catalonians and Basques of Spain, and the more radical groups of native Hawaiians who seek autonomy from the United States (Figure 1.38).

Not to be overlooked are political organizations that have eclipsed the power of traditional political states. This is certainly the case

Figure 1.37 Russian Troops in Crimea In March 2014, Russian troops quickly took over military bases in Crimea, claiming the Ukrainian territory as part of Russia. While relatively bloodless, fighting continues in eastern Ukraine along the border with Russia.

Figure 1.38 Ethnic Separatism A major aspect of contemporary geopolitics is the way ethnic groups are demanding recognition, autonomy, and often independence from larger political units. These Basque women in southwestern France are protesting the outlawing of a Basque youth group that French and Spanish authorities suspected of aiding Basque terrorists.

for the 28 member states of the European Union, a topic discussed in Chapter 8. Finally, some cultural groups lack political voice and representation due to the way political borders have been drawn. In Southwest Asia, the Kurdish people have long been considered a nation without a state because they are divided by political borders among Turkey, Syria, Iraq, and Iran (Figure 1.39).

Colonialism, Decolonialization, and Neocolonialism

One of the overarching themes in world geopolitics is the waxing and waning of European colonial power in the Americas, the Caribbean, Asia, and Africa. **Colonialism** consists of the formal establishment of rule over a foreign population. A colony has no independent standing in the world community, but instead is seen only as an appendage of the colonial power. The historic Spanish presence and rule over parts of the United States, Latin America, and the Caribbean is an example. Generally speaking, the main period of colonialization by European countries was from 1500 through the mid-1900s, with the major players being England, Belgium, the Netherlands, Spain, and France (Figure 1.40).

Decolonialization refers to the process of a colony gaining (or, more correctly, regaining) control over its own territory and establishing a separate, independent government. As was the case with the Revolutionary War in the United States, this process often involves violent struggle. Similar wars of independence became increasingly common in the mid-20th century, particularly in South Asia, Southeast Asia, and Africa. Consequently, most European colonial powers recognized the inevitable and began working toward peaceful disengagement from their colonies. British rule ended in South Asia in 1947, and in the late 1950s and early 1960s, Britain and France granted independence to their African colonies. This period of European colonialism symbolically closed in 1997 when England turned over Hong Kong to China.

However, decades and even centuries of colonial rule are not easily erased. The influences of colonialism are still found in the culture, government, educational systems, and economic life of the former colonies, evident in the many contemporary manifestations of British culture in India and the continuing Spanish influences in Latin America.

In the 1960s, the term **neocolonialism** came into popular usage to characterize the many ways that newly independent states, particularly those in Africa, felt the continuing control of Western powers, especially in economic and political matters. To receive financial aid from the World Bank, for example, former colonies were often required to revise their internal economic structures to become better integrated with the emerging global system. This economic restructuring may have seemed necessary from a global perspective, but the dislocations caused at national and local scales led critics of globalization to characterize these external influences as no better than the formal control by colonial powers.

Global Conflict and Insurgency

As mentioned earlier, challenges to a centralized political state or authority have long been part of global geopolitics as rebellious and separatist groups seek independence, autonomy, and territorial control.

Figure 1.39 A Nation Without a State Not all nations or large cultural groups control their own political territories. The Kurdish people of Southwest Asia occupy a large cultural territory that lies in four different political states—Turkey, Iraq, Syria, and Iran. As a result of this political fragmentation, the Kurds are considered a minority in each of these four countries. **Q: Suggest issues that might result from the Kurds lacking a political state.**

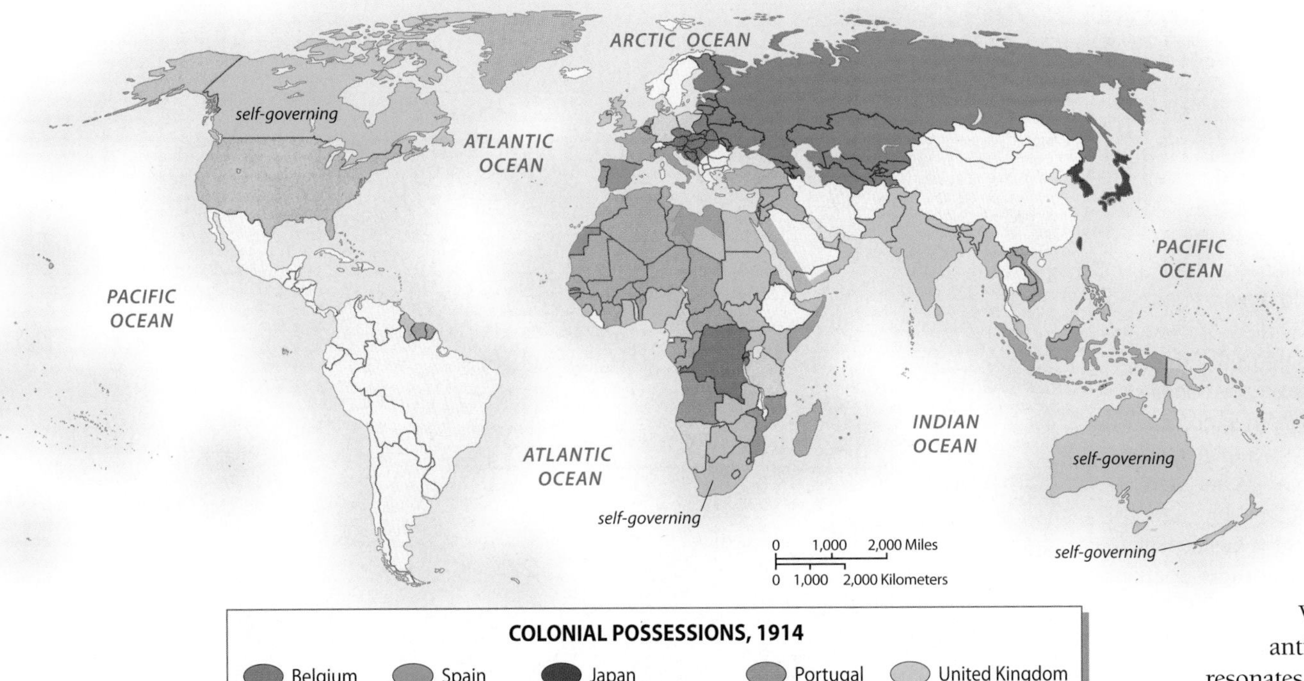

Figure 1.40 The Colonial World, 1914 This map shows the extent of colonial power and territory just prior to World War I. At that time, most of Africa was under colonial control, as were Southwest Asia, South Asia, and Southeast Asia. Australia and Canada were very closely aligned with England. Also note that in Asia, Japan had colonial control of the Korean Peninsula.

These actions are termed **insurgency**. Armed conflict has also been part of this process; the American and Mexican revolutions were both successful wars for independence fought against European colonial powers. **Terrorism**, which can be defined as violence directed at nonmilitary targets, has also been common, albeit to a far lesser degree than today.

Until the September 2001 terrorist attacks by Al Qaeda, terrorism was usually directed at specific local targets and committed by insurgents with focused goals. The Irish Republican Army (IRA) bombings in Great Britain and Basque terrorism in Spain are illustrations. The attacks on the World Trade Center and the Pentagon (as well as the thwarted attack on the U.S. Capitol), however, went well beyond conventional geopolitics as a small group of religious extremists attacked the symbols of Western culture, finance, and power. Experts believe that Al Qaeda's goal was less about disrupting world commerce and politics and more about displaying the strength of their own convictions and power. Regardless of motives, those acts of terrorism underscore the need to expand our conceptualization of the linkages between globalization and geopolitics.

Many experts argue that global terrorism is both a product of and a reaction to globalization. Unlike earlier geopolitical conflicts, the geography of global terrorism is not defined by a war between well-established political states. Instead, the Al Qaeda terrorists appear to belong to a web of small, well-organized cells located in many different countries. Boko Haram, a Muslim extremist group in Nigeria, has terrorized villages and kidnapped schoolchildren and is linked to Al Qaeda. Similarly Al-Shabaab, based in Somalia and responsible for attacks in Kenya, is affiliated with Al Qaeda.

The terrorist group Islamic State of Iraq and the Levant (ISIL; also known as ISIS or Islamic State in Iraq and Syria, or as Daesh) controls significant territory in Iraq and Syria and, unlike other terrorists groups, has the stated goal of forming a modern-day caliphate or a fundamentalist Islamic state in Southwest Asia (Figure 1.41). ISIL's use of terror, kidnapping, public executions, extortion, and social media has gained it international recognition and condemnation—as well as converts to its extremist cause. ISIL has taken advantage of the political vacuum left by years of conflict in Syria and Iraq that weakened those states andfostered the anti-Western, anti-secular, and anti-globalization sentiment that resonates in some parts of the world. Increasingly, geospatial tools, such as satellite imagery, are employed to track such groups and the chaos they cause to people, infrastructure, and cultural heritage sites (See *Geographers at Work: Tracking Conflict from Space*).

The U.S. State Department lists 59 groups as foreign terrorist organizations. While most of these groups are clustered in North Africa and Southwest and Central Asia, this list also includes insurgent groups in all other regions of the world.

The military responses to global terrorism and insurgency involve several components, ranging from the neutralization of terrorist activities, known as counterterrorism, to **counterinsurgency**. The latter is a more complicated, multifaceted strategy that combines military warfare with social and political service activities, designed to win over the local population and deprive insurgents of a political base. Counterinsurgency activities include, first, clearing and then holding territory held by insurgents, followed by building schools, medical clinics, and a viable economy. These nonmilitary activities are often referred to as nation building, since the goal is to replace the insurgency with

Figure 1.41 ISIL Heavily armed Islamic State fighters travel in the back of pickup trucks across the Iraqi desert in this propaganda video released by ISIL.

GEOGRAPHERS AT WORK

Tracking Conflict from Space

Figure 1.4.1 Susan Wolfinbarger

As an undergraduate at Eastern Kentucky University, Susan Wolfinbarger took a world regional geography class, and was mesmerized: "There are so many things you learn in geography, and the methods of analysis can be applied to different careers and research." Years later, with a PhD in Geography from the Ohio State University, Wolfinbarger directs the Geospatial Technologies Project at the American Association for the Advancement of Science (AAAS) (Figure 1.4.1). Her group uses high-resolution satellite imagery to track conflicts and document issues of global concern, such as human rights abuses and damage to cultural heritage sites.

Most people have used Google Earth satellite images to look at places. Wolfinbarger's team employs a time series of such images in order to assess events such as destruction of villages. Interpreting images and quantifying findings is a challenge, but, she says, "Geography taught me not just mapping but statistics and surveying . . . it gave me a great toolkit to apply to any topic." Much of her analysis is used by human rights organizations such as the European Court of Human Rights and the Inter-American Court of Human Rights.

Wolfinbarger's team analyzed the increase in roadblocks in the Syrian city of Aleppo (Figure 1.4.2). Roadblocks demonstrate a decline in the circulation of people and goods in this densely settled city, which is a major problem. The Geospatial Technologies Project has also documented heritage sites at risk from damage and looting, especially in the Southwest Asia, and is developing training materials so that others can use this technology.

Figure 1.4.2 Monitoring Aleppo This image shows the city of Aleppo in May 2013, where over 1000 roadblocks were detected. Roadblocks are an indicator of ongoing conflict and potential humanitarian concerns because they restrict the movement of people and goods throughout the city. In a nine-month period from September 2012 to May 2013, the number of roadblocks doubled.

Geographers are at the cutting edge of applying satellite imagery to a broad spectrum of human rights issues. Wolfinbarger notes, "There are a lot of ways that geographers can contribute to things happening in the world, and a lot of opportunities out there other than academic jobs. Everyone wants a geographer!"

1. Suggest ways that satellite imagery could be used to document not just conflict but environmental change.
2. Government agencies are constantly developing and using satellite technology. How might a citizen or non-governmental group in your city or state use this kind of analysis?

a viable social, economic, and political fabric more complementary to the larger geopolitical state. This is the strategy recently employed by the United States in Iraq and Afghanistan.

Review

1.19 Why is it common to use two different concepts—nation and state—to describe political entities?

1.20 Distinguish colonialism from neocolonialism.

1.21 Describe the differences between counterterrorism and counterinsurgency.

KEY TERMS geopolitics, sovereignty, nation-state, colonialism, decolonialization, neocolonialism, insurgency, terrorism, counterinsurgency

Economic and Social Development: The Geography of Wealth and Poverty

The pace of global economic change and development has accelerated in the past several decades, rising rapidly at the start of the 21st century and then slowing precipitously in 2008 as the world fell into an economic recession. Most countries are now gradually recovering from the depths of the global recession, although major economies such as Brazil and Russia have experienced weaker growth rates in the last couple of years. If nothing else, this recent global recession and its unsteady recovery have highlighted the overarching question of whether the benefits of economic globalization outweigh the negative aspects.

Responses vary considerably, depending on one's point of view, occupation, career aspirations, and socioeconomic status. Anyone attempting to understand the contemporary world needs a basic understanding of global economic and social development. To that end, each regional chapter contains a substantive section on that topic, drawing on the concepts discussed below.

Economic development is considered desirable because it generally brings increased prosperity to people, regions, and nations. Following conventional thinking, this economic development usually translates into social improvements such as better health care, improved educational systems, higher wages, and longer life expectancies. One of the most troubling expressions of global economic growth, however, has been the geographic unevenness of prosperity and social improvement. That is, while some regions and places in the world prosper, others languish and even fall further behind the more developed countries. As a result, the gap between rich and poor regions has actually increased over the past several decades in many regions. This economic and social unevenness has, unfortunately, become one of the signatures of globalization. In addition, the numbers of people living in *extreme poverty*, defined as those living on less than $1.25 a day, have declined since the 1990s, but if one considers the population living on less than $2 a day, the poverty measure used by the World Bank and the UN, about 2.5 billion people in the world still struggle for existence at this level. Many of these people live in Sub-Saharan Africa, South Asia, and Southeast Asia (Figure 1.42).

These inequities are problematic because they are intertwined with political, environmental, and social issues. For example, political instability and civil strife within a state are often driven by the economic disparity between a poor peripheral area and an affluent industrial core—between the haves and have-nots. Such instability, in turn, can strongly influence international economic interactions.

Figure 1.42 Living on Less than $2 a Day The World Bank uses two measures of global poverty. Subsisting on less than $1.25 a day, the definition of extreme poverty, is a reality for roughly 1 billion people; fortunately, the numbers of people at this level have decreased. Poverty is defined as life on less than $2 a day. Unfortunately, there are still 2.5 billion people in this category. Surviving by picking through the rubbish dumps of Manila, Philippines, these children and adults are the faces of global poverty.

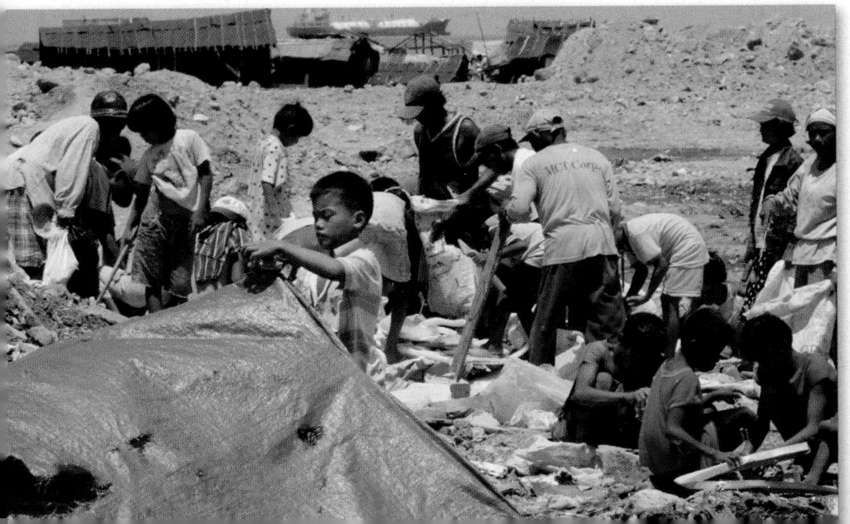

More and Less Developed Countries

Until the later 20th century, economic development was centered in North America, Japan, Europe, and Australia, with most of the rest of the world gripped in poverty. This uneven distribution of economic power led scholars to devise a **core–periphery model** of the world. According to this scheme, these countries and regions constituted the global economic *core*, centered for the most part in the Northern Hemisphere, whereas most of the areas in the Southern Hemisphere made up a less developed *periphery*. Although oversimplified, this core–periphery dichotomy does contain some truth. All the G8 countries—the exclusive club of the world's major industrial nations, made up of the United States, Canada, France, England, Germany, Italy, Japan, and Russia—are located in the Northern Hemisphere. (China—unquestionably an industrial power located in the Northern Hemisphere—is currently excluded from the G8.) Many critics contend that the developed countries achieved their wealth primarily by exploiting the poorer countries of the southern periphery, historically through colonial relationships and today through various forms of neocolonialism and economic imperialism.

Following this core–periphery model, much has been made of "north–south tensions," a phrase that distinguishes the rich and powerful countries of the Northern Hemisphere from the poor and less powerful countries of the Southern Hemisphere. However, this model demands revision because over recent decades the global economy has grown much more complicated. A few former colonies of the periphery or "south"—most notably, Singapore—have become very wealthy, while a few northern countries—notably, Russia—have experienced very uneven economic growth since 1989, with some parts of the country actually seeing economic declines. Additionally, the developed Southern Hemisphere countries of Australia and New Zealand never fit into the north–south division. For these reasons, many global experts conclude that the designation *north–south* is outdated and should be avoided.

Third world is another term often erroneously used as a synonym for the developing world. Historically, the term was part of the Cold War vocabulary used to describe countries that were not part of either the capitalist Westernized first world or the communist second world dominated by the Soviet Union and China. Thus, in its original sense *third world* signified a political and economic orientation (capitalist vs. communist), not a level of economic development. With the Soviet Union's demise and China's considerably changed economic orientation, *third world* has lost its original political meaning. In this book, we prefer relational terms that capture a complex spectrum of economic and social development—*more developed country (MDC)* and *less developed country (LDC)*. This global pattern of MDCs and LDCs can be inferred from a map of gross national income, divided into four categories from low-income countries to high-income ones (Figure 1.43).

Indicators of Economic Development

The terms *development* and *growth* are often used interchangeably when referring to international economic activities. There is, however, value in keeping them separate. *Development* has both qualitative and quantitative dimensions. When we talk about economic development, the term usually implies structural changes, such as a shift from agricultural to manufacturing activity that also involves changes in the

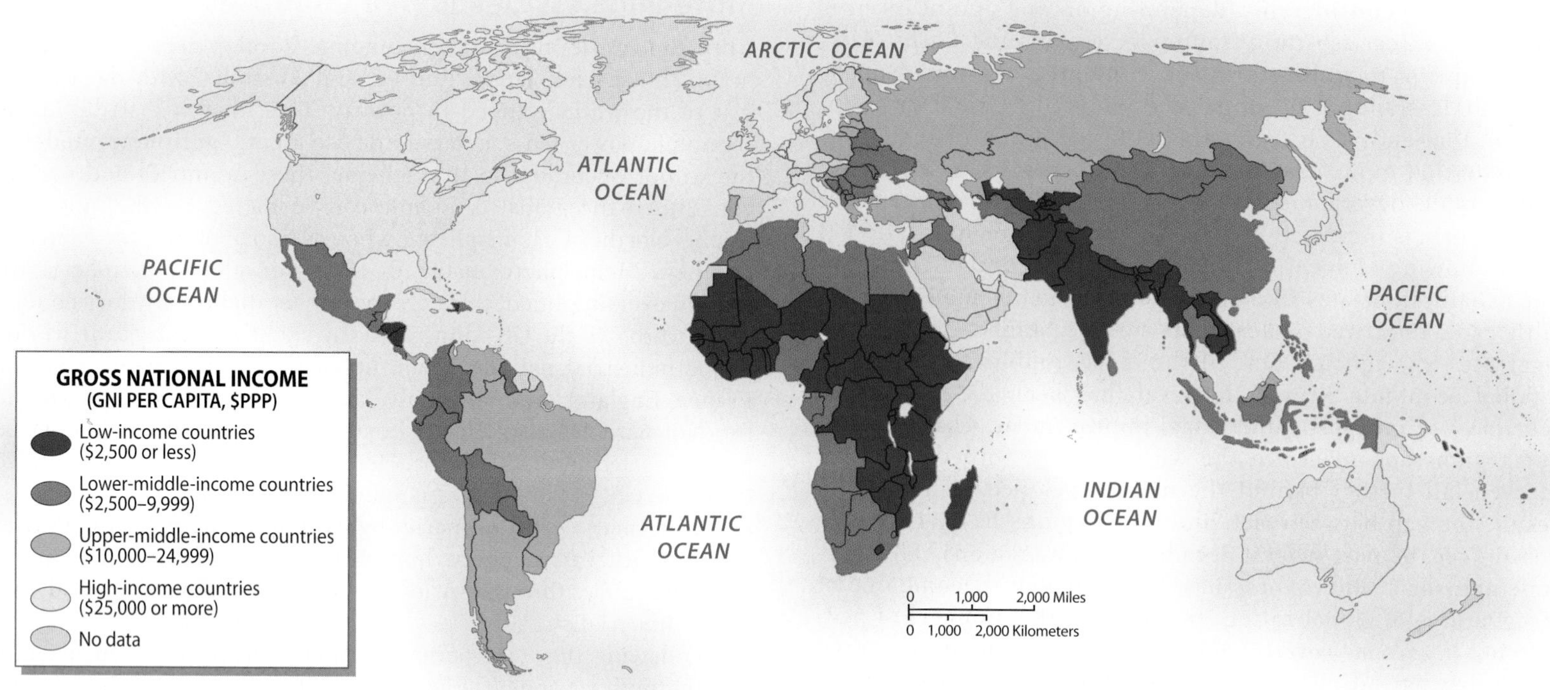

Figure 1.43 More and Less Developed Countries Based on GNI per capita, purchasing power parity (PPP) adjusted, you can see the global pattern of more and less developed countries (MDCs and LDCs). Sub-Saharan Africa and South Asia stand out as the regions with the greatest number of low-income countries.

allocation of labor, capital, and technology. Along with these changes are assumed improvements in standard of living, education, and political organization. The structural changes experienced by Southeast Asian countries such as Thailand and Malaysia in the past several decades capture this process.

Growth, in contrast, is simply the increase in the size of a system. The agricultural or industrial output of a country may grow, as it has for India in the past decade, and this growth may—or may not—have positive implications for development. Many growing economies, in fact, have actually experienced increased poverty with economic expansion. When something grows, it gets bigger; when it develops, it improves. Critics of the world economy often say that we need less growth and more development.

Each of the regional chapters refers to a table of development indicators found in the appendix. Table 1.2 highlights the development indicators used throughout the book for the 10 largest countries.

Table 1.2 DEVELOPMENT INDICATORS

Country	GNI per capita, PPP 2013	GDP Average Annual %Growth 2009–13	Human Development Index (2013)[1]	Percent Population Living Below $2 a Day	Life Expectancy (2015)[2]	Under Age 5 Mortality Rate (1990)	Under Age 5 Mortality Rate (2013)	Youth Literacy (%pop ages 15–24)	Gender Inequality Index (2013)[3,1]
China	11,850	8.7	.719	18.6	75	49	13	100	0.202
India	5,350	6.9	.586	59.2	68	114	53	81	0.563
United States	53,750	2.1	.914	–	79	11	8	–	0.262
Indonesia	9,270	6.2	.684	43.3	71	82	29	99	0.500
Brazil	14,750	3.1	.744	6.8	75	58	14	99	0.441
Pakistan	4,840	3.1	.537	50.7	66	122	86	71	0.563
Nigeria	5,360	5.4	.504	82.2	52	214	117	66	–
Bangladesh	3,190	6.2	.558	76.5	71	139	41	80	0.529
Russia	24,280	3.5	.778	< 2	71	27	10	100	0.314
Japan	37,550	1.6	.890	–	83	6	3	–	0.138

Source: World Bank, *World Development Indicators, 2015*

[1]United Nations, *Human Development Report, 2014.*

[2]Population Reference Bureau, *World Population Data Sheet, 2015.*

[3]Gender Equality Index—A composite measure reflecting inequality in achievements between women and men in three dimensions: reproductive health, empowerment and the labor market that ranges between 0 and 1. The higher the number, the greater the inequality.

Gross Domestic Product and Income A common measure of the size of a country's economy is the **gross domestic product (GDP)**, the value of all final goods and services produced within its borders. Table 1.2 shows GDP average annual growth for 2009–2013. In this five-year period, most countries saw growth in GDP, especially in the less developed world. But this period also captures growth that occurred after the recession. Compare the average annual growth of the United States with that of Indonesia for 2009–2013. The U.S. growth rate of 2.1 percent is far lower than Indonesia's 6.2 percent rate. But the U.S. economy is far larger and more diversified, and in the high-income United States, people have far more resources than they do in Indonesia, a lower-middle-income country. In general, the less developed countries shown in this table have higher growth rates than the more developed countries, with China having the highest annual growth rate at 8.7 percent.

When GDP is combined with net income from outside a country's borders through trade and other forms of investment, this constitutes a country's **gross national income (GNI)**. Although these terms are widely used, both GDP and GNI are incomplete and sometimes misleading economic indicators because they completely ignore nonmarket economic activity, such as bartering and household work, and do not take into account ecological degradation or depletion of natural resources. For example, if a country were to clear-cut its forests—an activity that would probably limit future economic growth—this resource usage would actually increase the GNI for that particular year, but then GNI would likely decline. Diverting educational funds to purchase military weapons might also increase a country's GNI in the short run, but its economy would likely suffer in the future because of its less-well-educated population. In other words, GDP and GNI are a snapshot of a country's economy at a specific moment in time, not a reliable indicator of continued vitality or social well-being.

Comparing Incomes and Purchasing Power

Because GNI data vary widely among countries, **gross national income (GNI) per capita** figures are used, allowing large and small economies to be compared, regardless of population size. An important qualification to these GNI per capita data is the concept of adjustment through **purchasing power parity (PPP)**, which takes into account the value of goods that can be purchased with the equivalent of one international dollar in a particular country. An international dollar has the same purchasing power parity over all GNI and is set at a designated U.S. dollar value. Thus, if a country's food costs are lower than the U.S. cost, the per capita purchasing power for that country increases. PPP was created to adjust comparisons between countries because (for example) an income of $10,000 in Mexico can purchase more basic goods than the same amount of money in Norway.

In Table 1.2, the United States has the strongest GNI (PPP), followed by Japan at $37,550. Without taking PPP into consideration, Japan's GNI per capita would be far greater, over $46,000. The lower PPP figure for Japan reflects the relatively higher cost of living in that country. The country with the lowest GNI (PPP) is Bangladesh at $3190. In this case, purchasing power per capita GNI increases because the cost of basic goods is far less in this developing country. Crude GNI per capita would be closer to $1000 per person.

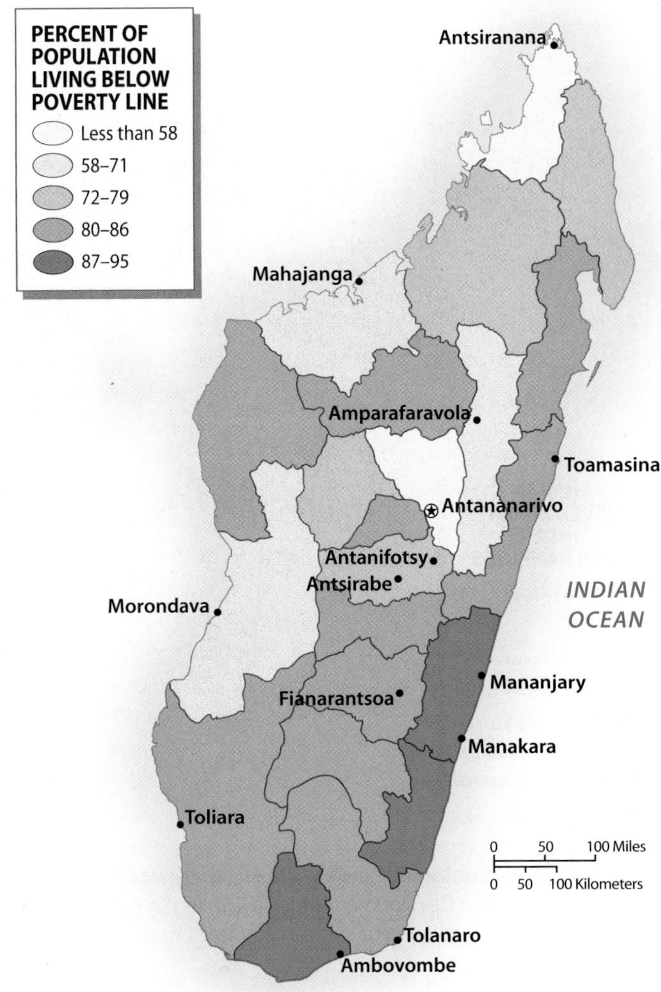

Figure 1.44 Poverty Mapping This map of Madagascar shows that the highest rates of poverty are in the central highlands, where the country's population is concentrated, and parts of the east coast. Poverty rates are lower around the capital and the northern lowlands. Mapping poverty at this scale is used to decide where scarce resources could be better spent to improve overall development. **Q: Why are urban poverty rates lower in cities compared to rural areas?**

Measuring Poverty

As noted earlier, the international definition of *poverty* is living on less than $2 per day, and extreme poverty is living on less than $1.25 per day. While the cost of living varies greatly around the world, the UN usually uses unadjusted per capita income figures when measuring poverty. Table 1.2 shows that 82 percent of Nigeria's population lives in poverty, followed by 76 percent of Bangladesh's population and 59 percent of India's. In contrast, this measure of poverty does not exist for Japan and the United States. While poverty data are usually presented at the country level, the World Bank and other agencies compile data at the sub-state level in order to better understand the poverty landscape within a country. The patterns of poverty in Madagascar provide an instructive example, with areas in the east having higher poverty rates than areas around the capital and in the far north (Figure 1.44). Poverty mapping at this scale helps governments and development agencies decide which areas of a country require more aid and investment in order to reduce poverty.

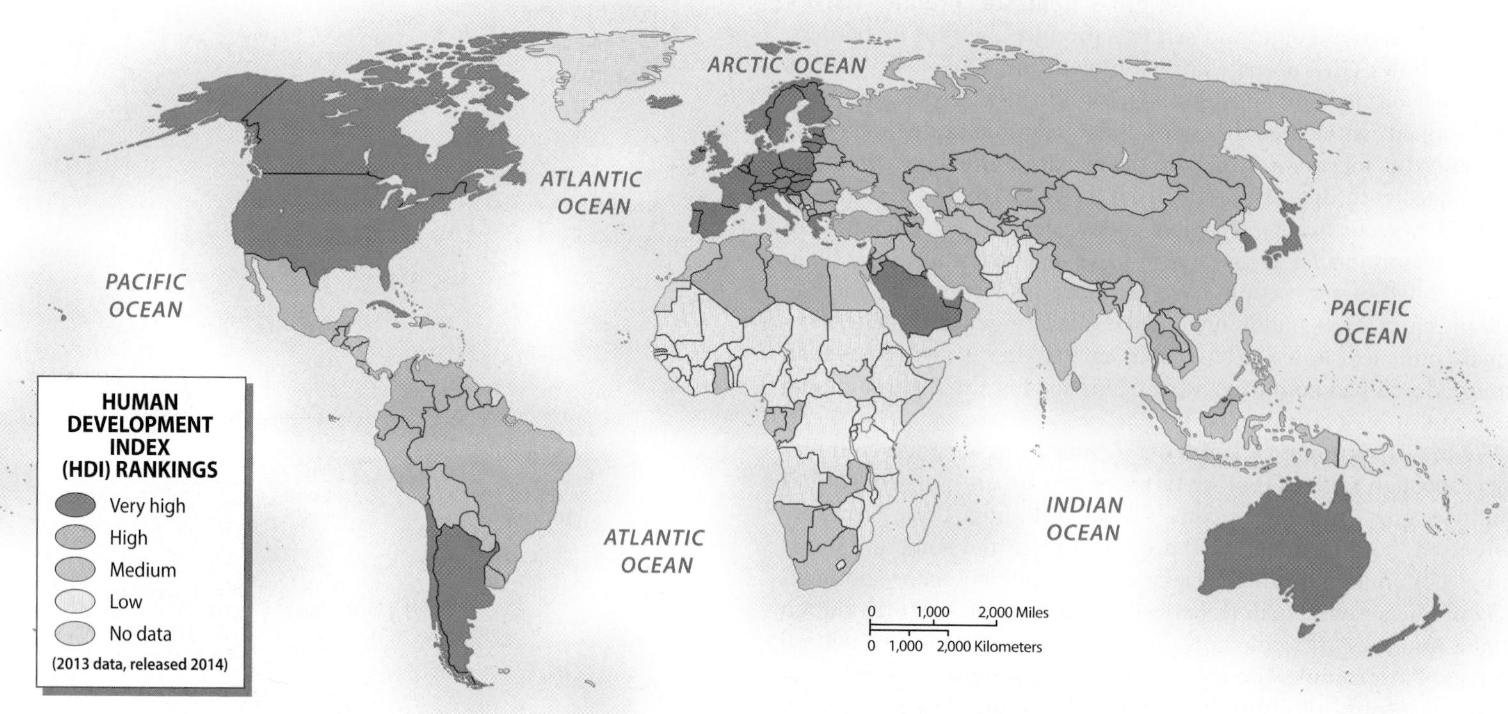

Figure 1.45 Human Development Index This map depicts the most recent rankings assigned to four categories that make up parts of the Human Development Index (HDI). In the numerical tabulation, Norway, Australia, and Switzerland have the highest rankings, while several African countries have the lowest scores.

Indicators of Social Development

Although economic growth is a major component of development, equally important are quality-of-life measures. As noted earlier, the standard assumption is that economic development will spill over into the social infrastructure, leading to improvements in life expectancy, child mortality, **gender inequality**, and education. Even some of the world's poorest countries have experienced significant improvements in all these measures. Much of the foreign development aid since 2000 goes to tracking and improving these development indicators.

The Human Development Index For the past three decades, the UN has tracked social and economic development in the world's countries through the **Human Development Index (HDI)**, which combines data on life expectancy, literacy, educational attainment, gender inequality, and income (Figure 1.45). A 2014 analysis ranks the 187 countries that provided data to the UN from high to low, with Norway at the top and Australia in second place, followed by Switzerland, the Netherlands, and the United States. Sub-Saharan countries score the lowest, including Niger, Democratic Republic of the Congo, Central African Republic, Chad, and Sierra Leone. Countries with high or very high human development have HDI scores of .700 or better; Norway's score is .944 (Figure 1.46). The low human development states score .540 or lower.

Although the HDI is criticized for using national data that miss the diversity of development within a country, overall, it conveys a reasonably accurate sense of a country's human and social development.

Figure 1.46 Norwegian Development Currently the top-ranked country in the Human Development Index, Norway boasts high income (GNI per capita PPP of over $65,000), universal access to education and health care, low income inequality, and long life expectancy. Yet it is a relatively small country of 5 million with substantial oil and gas reserves. Norwegians also have a high cost of living and pay relatively high taxes. This image is a shopping district in Oslo.

Thus, we include HDI data in our development indicator tables for each regional chapter.

Child Mortality Another widely used indicator of social development is data on *under age five mortality*, which is the number of children in that age bracket who die per 1000 children. Aside from the tragedy of infant death, child mortality also reflects the wider conditions of a society, such as the availability of food, health services, and public sanitation. If those factors are lacking, children under age five suffer most; therefore, their death rate is taken as an indicator of whether a country has the necessary social infrastructure to sustain life (Figure 1.47). In the social development table for each region, child mortality data are given for two points in time, 1990 and 2013, to indicate whether the social structure has improved in the intervening years.

Every country has seen improvements over 23 years; especially significant in Table 1.2 are the improvements in China and Bangladesh. States with the lowest rates of child mortality are the most developed countries.

Youth Literacy Reading and writing are crucial in today's world, yet current data show that many adults in the developing world lack these skills. Of those who cannot read or write, two-thirds of them are women. The World Bank focuses its resources on measuring and improving youth literacy (people ages 15–24). The hope is that literacy will increase dramatically for young adults and that disparities between males and females will disappear. Table 1.2 shows that Nigeria and Pakistan have the lowest rates of youth literacy, while in China, Indonesia, and Brazil virtually all youth are literate.

Figure 1.47 Child Mortality The mortality rate of children under the age of five is an important indicator of social conditions such as health services, food availability, and public sanitation. This child is being examined in Nguyen Province, Vietnam.

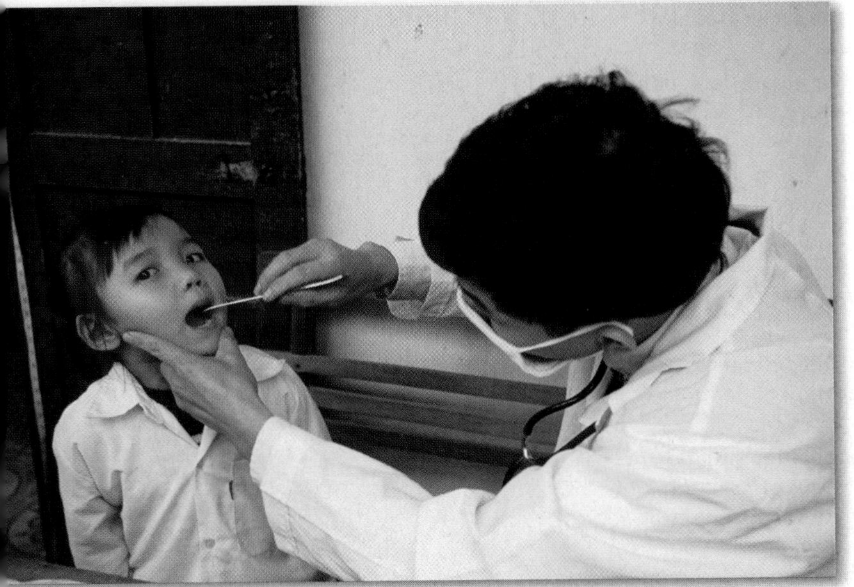

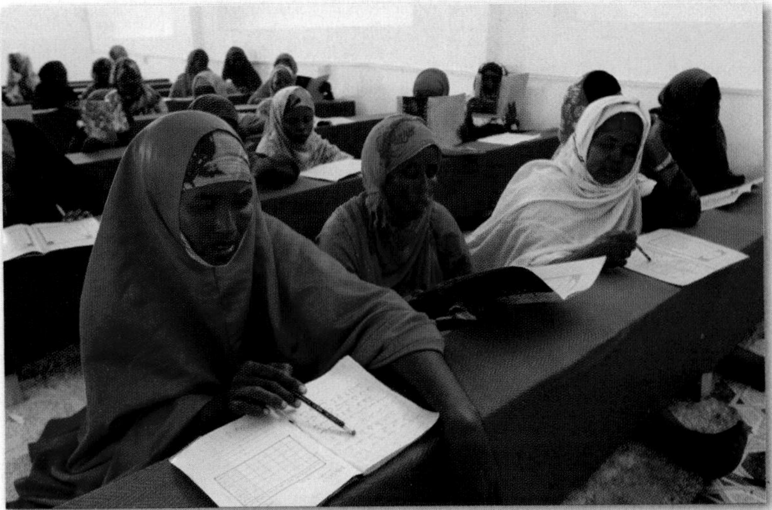

Figure 1.48 Women and Literacy Gender inequalities in education lead to higher rates of illiteracy for women. However, where there is gender equality in education, female literacy has several positive outcomes for society. Educated women have a higher participation rate in family planning, which usually results in lower birth rates and better child survival rates. These women and girls are in a literacy class in a refugee camp outside of Mogadishu, Somalia.

Gender Inequality Discrimination against women takes many forms, from not allowing them to vote to discouraging school attendance (Figure 1.48). Given the importance of this topic, the UN calculates gender inequality among countries in order to measure the relative positions of women and men in terms of employment, empowerment, and reproductive health (measured by maternal mortality and adolescent fertility). The UN index ranges from 0 to 1, which expresses the highest level of gender inequality. In 2014, Slovenia had the lowest score for inequality with 0.021, while Yemen had the highest at 0.733, indicating very high gender inequality.

Some countries may register reasonably high on the HDI, which is positive, yet also receive a relatively high gender inequality score (which is not so good). Qatar, for example, is a rich country that uses assets from its oil resources to provide many social benefits to its citizens, which explains its high HDI ranking, while at the same time its conservative Muslim culture produces a gender inequality score of .524 and a rank of 113 among countries in that index. Table 1.2 shows Pakistan and India with scores of .563, sharing the rank of 127, which means relatively high gender inequality. Readers are advised to look carefully at the development indicators data and note these kinds of contradictions and inconsistencies.

Review

1.22 Explain the difference between GDP and GNI.

1.23 What is PPP, and why is it useful?

1.24 How does the UN measure gender inequality? Explain why this is a useful metric for social development.

KEY TERMS core–periphery model, gross domestic product (GDP), gross national income (GNI), gross national income (GNI) per capita, purchasing power parity (PPP), gender inequality, Human Development Index (HDI)

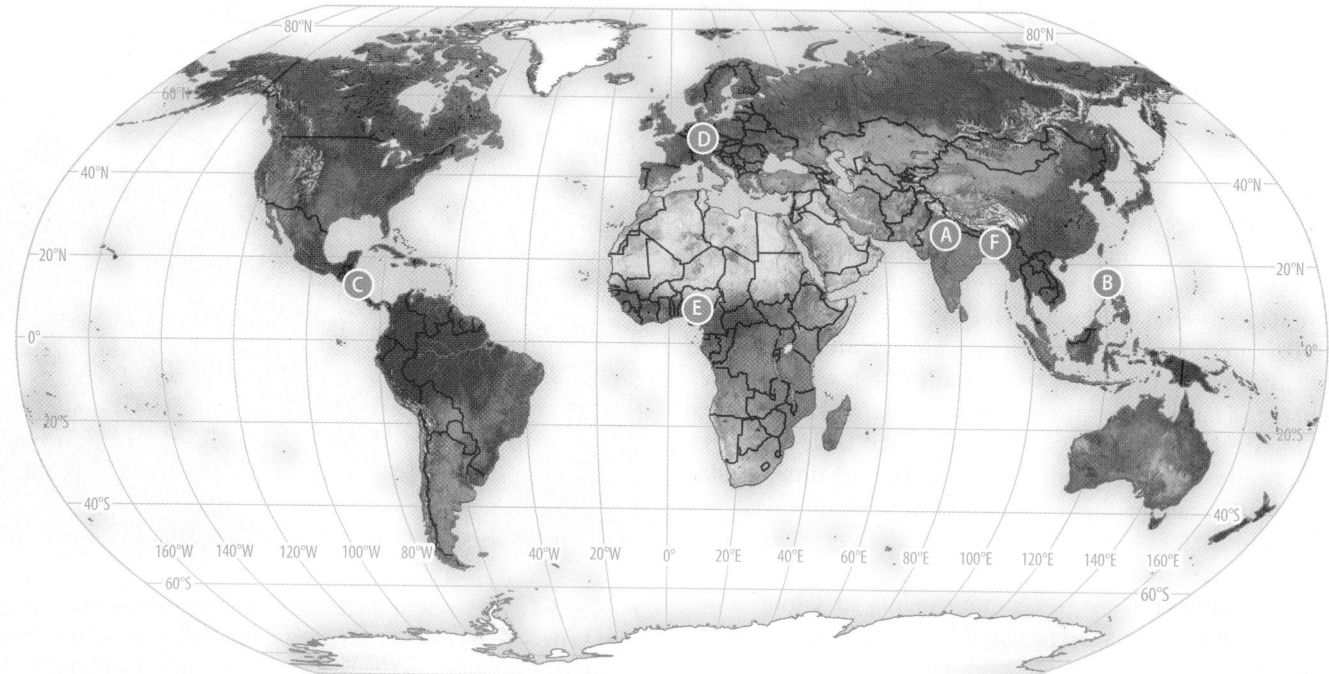

Review

Converging Currents of Globalization

1.1 Describe the conceptual framework of world regional geography.

1.2 Identify the different components of globalization, including controversial aspects, and list several ways in which globalization is changing world geographies.

Globalization appears everywhere, all the time, affecting all aspects of world geography with its economic, cultural, and political connectivity (or lack of it). Despite fears that globalization will produce a homogeneous world, a great deal of diversity is still apparent—often as economic unevenness, other times as pushback against globalization.

1. Why are so many U.S. call centers located in India?

2. How has globalization affected—for better or worse—the local economy in your area?

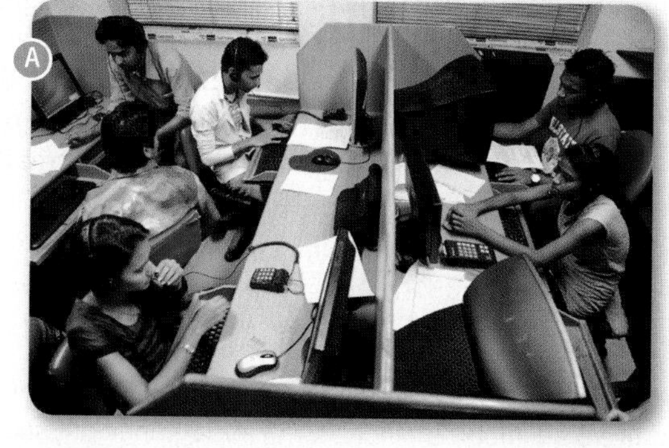

Geographer's Toolbox

1.3 Summarize the major tools used by geographers to study Earth's surface.

Geography describes and explains Earth's varied landscapes and environments. This can be done conceptually in different ways, by using physical or human geography and thematic or regional geography—or by combining these approaches. Geographers use a variety of tools, from paper maps to computer models, drawing on information gathered on the ground and by satellites high above Earth's surface, at all different scales, large and small.

3. Using the landscape as your guide, describe the economy and social structure of this village on the island of Luzon, Philippines.

4. How are geographic tools used by city and county planners in your community?

Population and Settlement

1.4 Explain the concepts and metrics used to document changes in global population and settlement patterns.

Human populations around the world are growing either quickly or slowly as a function of natural increase as well as widely different patterns of in- and out-migration. Urbanization is also a major factor in settlement patterns as people continue to move from rural to urban locales.

5. What are the reasons for the differences in population density in this portion of Central America?

6. Do a brief study of the population situation in your area of the country in terms of its natural growth, out-migration, in-migration, and resulting growth or decline. Then explain why your area's population is either growing or declining.

Cultural Coherence and Diversity

1.5 Describe the themes and concepts used to study the interaction between globalization and the world's cultural geographies.

Culture is learned behavior and includes a wide range of both tangible and intangible behaviors and objects, from language to house architecture, from gender roles to sports. Globalization is altering the world's cultural geography, producing new cultural hybrids in many places. In other places, people resist change with different kinds of cultural nationalism that protect (or even resurrect) traditional ways of life.

7. This photo was taken in downtown Prague, in the Czech Republic. What are the signs of global versus traditional culture?

8. Give five examples of cultural hybridization in your local area.

D

Geopolitical Framework

1.6 Explain how different aspects of globalization have interacted with global geopolitics from the colonial period to the present day.

Varying political systems, ranging from dictatorships to democracies, provide the world with a dynamic geopolitical framework that is stable in some places and filled with tension and violence in others. The traditional concept of the nation-state is currently challenged by separatism, insurgency, and even terrorism.

9. What are the reasons behind terrorist attacks in Nigeria? What groups are responsible, and what are their goals?

10. Working in a small group, choose an African country that was a European colony in the early 20th century, and trace its geopolitical geography and history over the last century.

E

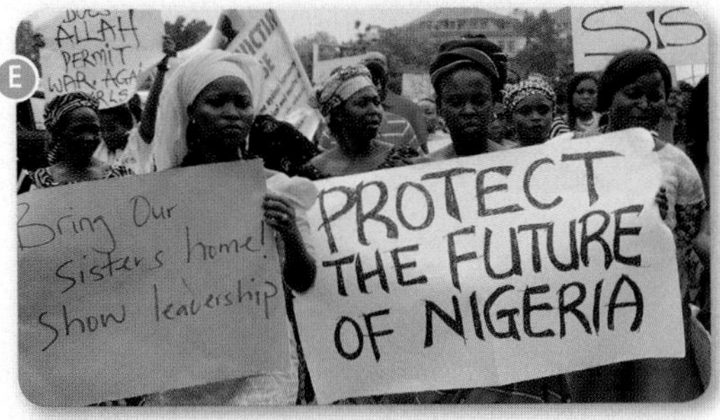

Economic and Social Development

1.7 Identify the concepts and data important to documenting changes in the economic and social development of more and less developed countries.

Although economic globalization proponents argue that all people in all places gain from expanded world commerce, that does not seem to be the case. Instead, there are places that profit, while others lose out—resulting in global patterns of economic disparity. Social development in terms of health care and education also remains highly uneven, but many countries have experienced gains in these measures.

11. Many clothing factories have sprung up in Bangladesh in the last decade. Why? Make a list of the positive and negative aspects of this development from a local perspective.

12. In the latest Human Development Index study, the United States ranks number 5 in the world. Why doesn't the United States rank higher?

F

KEY TERMS

areal differentiation *(p. 5)*
areal integration *(p. 5)*
choropleth map *(p. 16)*
colonialism *(p. 32)*
core–periphery model *(p. 35)*
counterinsurgency *(p. 33)*
cultural imperialism *(p. 25)*
cultural landscape *(p. 6)*
cultural nationalism *(p. 26)*
cultural syncretism or hybridization *(p. 26)*
culture *(p. 25)*
decolonialization *(p. 32)*
demographic transition model *(p. 23)*
economic convergence *(p. 12)*
ethnic religion *(p. 28)*
formal region *(p. 6)*
functional region *(p. 6)*
gender *(p. 30)*
gender inequality *(p. 38)*
gender roles *(p. 30)*
geographic information systems (GIS) *(p. 17)*
geography *(p. 5)*
geopolitics *(p. 31)*
globalization *(p. 7)*
global positioning systems (GPS) *(p. 15)*
glocalization *(p. 9)*
graphic (linear) scale *(p. 18)*
gross domestic product (GDP) *(p. 37)*
gross national income (GNI) *p. 37)*
gross national income (GNI) per capita *(p. 37)*
Human Development Index (HDI) *(p. 38)*
human geography *(p. 5)*
insurgency *(p. 33)*
latitude (parallels) *(p. 14)*
legend *(p. 16)*
lingua franca *(p. 28)*
longitude (meridians) *(p. 14)*
map projection *(p. 15)*
map scale *(p. 16)*
nation-state *(p. 31)*
neocolonialism *(p. 32)*
net migration rate *(p. 24)*
physical geography *(p. 5)*
population density *(p. 20)*
population pyramid *(p. 22)*
prime meridian *(p. 14)*
purchasing power parity (PPP) *(p. 37)*
rate of natural increase (RNI) *(p. 21)*
region *(p. 6)*
regional geography *(p. 5)*
remote sensing *(p. 16)*
replacement rate *(p. 22)*
representative fraction *(p. 16)*
secularism *(p. 29)*
sovereignty *(p. 31)*
sustainability *(p. 19)*
terrorism *(p. 33)*
thematic geography (systematic geography) *(p. 5)*
thematic map *(p. 16)*
total fertility rate (TFR) *(p. 22)*
universalizing religion *(p. 28)*
urbanized population *(p. 24)*

DATA ANALYSIS

http://goo.gl/O28Ueu

The tables in this chapter show data for the world's 10 largest countries. But what are the world's next 10 largest countries, and where are they located? What are their per capita income levels? Are these economies growing or contracting? You can answer these questions and others by going to the website of the World Bank (http://wdi.worldbank.org) and accessing Table 1.1 for the 2015 development indicators.

1. Review the first column, and make a table of the next 10 largest countries. In which world regions are these countries located?
2. After selecting the countries, compare their gross national incomes with their purchasing power parity at a per capita basis. Based on your findings, would you consider these countries more developed or less developed, and why?
3. Compare the population densities of these countries. Some social scientists have argued that population density is a problem that can contribute to higher levels of poverty. Is there any correlation between population density and overall levels of development?

MasteringGeography™

Looking for additional review and test prep materials? Visit the Study Area in MasteringGeography™ to enhance your geographic literacy, spatial reasoning skills, and understanding of this chapter's content by accessing a variety of resources, including MapMaster interactive maps, geoscience animations, videos, *In the News* RSS feeds, flashcards, web links, self-study quizzes, and an eText version of *Globalization and Diversity*.

Authors' Blogs

Scan to visit the **Author's Blog for field notes, media resources, and chapter updates**

https://gad4blog.wordpress.com/category/globalization-and-world-geography/

Scan to visit the **GeoCurrents Blog**

http://geocurrents.info/

2 Physical Geography and the Environment

GEOLOGY: A RESTLESS EARTH

Earth's surface is comprised of numerous tectonic plates that slowly move about, driven by convection cells deep within the mantle. This movement not only provides shape to the world but also causes hazardous earthquakes and volcanoes.

GLOBAL CLIMATES: ADAPTING TO CHANGE

Our world has a wide variety of climate regions, ranging from polar to tropical. These climates, however, are changing because of human-caused global climate change, with problematic consequences.

BIOREGIONS AND BIODIVERSITY: THE GLOBALIZATION OF NATURE

Cloaks of natural vegetation vary greatly from place to place on Earth, creating novel ecosystems that have been highly altered by human activities.

WATER: A SCARCE WORLD RESOURCE

Clean freshwater is a necessity of life, but global supplies are scarce, with increasing water stress and shortages in many areas of the world.

GLOBAL ENERGY: THE ESSENTIAL RESOURCE

Fossil fuels—coal, oil, and natural gas—currently dominate global energy usage, emitting climate-changing atmospheric gases. Renewable energy sources—water, wind, solar, and biomass—currently make up only 9 percent of the world's energy usage.

◀ Located on the Colorado Plateau along the Arizona-Utah border, Monument Valley is a Navajo Tribal Park, the Navajo Nation's designation for a national park. Many of the buttes reach 1000 feet (300 meters) above the valley floor, and are composed of three geologic layers: shale, sandstone, and conglomerate rock

The immense physical diversity of Earth, with its varied climates, deep oceans, towering mountain ranges, dry deserts, and wet tropics, makes it unique in our solar system. Other planets are too warm (Venus) or too cold (Mars), but Earth is the Goldilocks planet with just the right temperature range for life. In turn, life forms of all different sorts—plant, animal, and human—have interacted with the physical environment to produce the diverse landscapes and habitats that make Earth our home. Monument Valley, Arizona, illustrates the crucial environmental relationships that support life on Earth: geology, water, and vegetation. Though seemingly uninhabited, the valley is part of the lands of the Navajo Nation, and the dusty dirt roads winding through the sparse vegetation reminds us that Earth is a human (and humanized) planet. Thus, a necessary starting point for the study of world regional geography is knowing more about Earth's physical environment—its geology, climate, diverse life forms, hydrology, and energy resources.

Life forms of all different sorts interact with the physical environment to produce the diverse landscapes and habitats that make Earth our home.

Learning Objectives

After reading this chapter you should be able to:

2.1 **Describe those aspects of tectonic plate theory responsible for shaping Earth's surface.**

2.2 **Identify on a map those parts of the world where earthquakes and volcanoes are hazardous to human settlement.**

2.3 **List and explain the factors that control the world's weather and climate, and use these to describe the world's major climate regions.**

2.4 **Define the greenhouse effect and explain how it is related to anthropogenic climate change.**

2.5 **Summarize the major issues underlying international efforts to address climate change.**

2.6 **Locate on a map and describe the characteristics of the world's major bioregions.**

2.7 **Name some threats to Earth's biodiversity.**

2.8 **Identify the causes of global water stress.**

2.9 **Describe the world geography of fossil fuel production and consumption.**

2.10 **List the advantages and disadvantages of the different kinds of renewable energy.**

Geology: A Restless Earth

The world's continents, separated by vast oceans, are made up of an array of high mountains, deep valleys, rolling hills, and flat plains created over time by geologic processes originating deep within our planet and then sculpted on the surface by everyday processes such as wind, rain, and running water. Not only does this physical landscape give Earth its unique character, but this geologic fundament also affects a wide range of human activities, creating resources in many places, but posing daunting challenges in others with destructive earthquakes and volcanic eruptions (Figure 2.1).

Plate Tectonics

The starting point for understanding geologic processes is the theory of **plate tectonics**, which states that Earth's outer layer, the **lithosphere**, consists of large geologic platforms, or plates, that move very slowly across its surface. Driving the movement of these plates is a heat exchange deep within Earth; Figure 2.2 illustrates this complicated process.

On top of these plates sit continents and ocean basins; however, note in Figure 2.3 that the world's continents and oceans are not identical to the underlying plates, but rather have different margins and boundaries. This is important because most earthquakes and volcanoes and their associated hazards are found along these plate boundaries. This map also shows that there are different types of tectonic plate boundaries linked to the underlying convection cells. **Convergent plate boundaries** are those where plates move toward one another, whereas **divergent plate boundaries** are those where plates move apart. **Transform plate boundaries** are characterized by two plates grinding laterally past one another.

Along convergent boundaries, one plate often sinks below another, creating a **subduction zone**. Deep trenches characterize these zones where the ocean floor has been pulled downward by sinking plates. Subduction zones exist off the west coast of South America, off

Figure 2.1 Mt. Bromo in the Morning The volcanic landscape of Bromo Tengger Semeru National Park in East Java, Indonesia, illustrates the dramatic diversity of Earth's physical geography. This area became a national park in 1982.

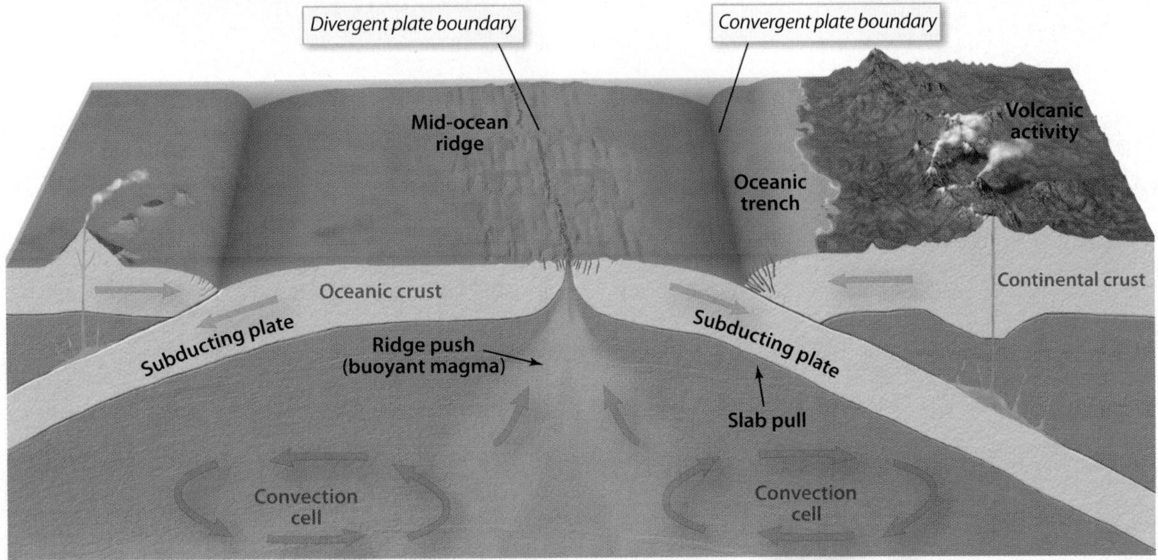

Figure 2.2 Tectonic Plate Theory The driving force behind tectonic plate theory are the convection cells resulting from heat differences within Earth's mantle. These cell circulate slowly, and in different directions, producing surface movement in the crustal tectonic plates. New plate material reaches the surface in the mid-oceanic ridges, then moves away slowly from these divergent boundaries. As the plate material cools it tends to sink, creating subduction zones in convergent plate boundaries.

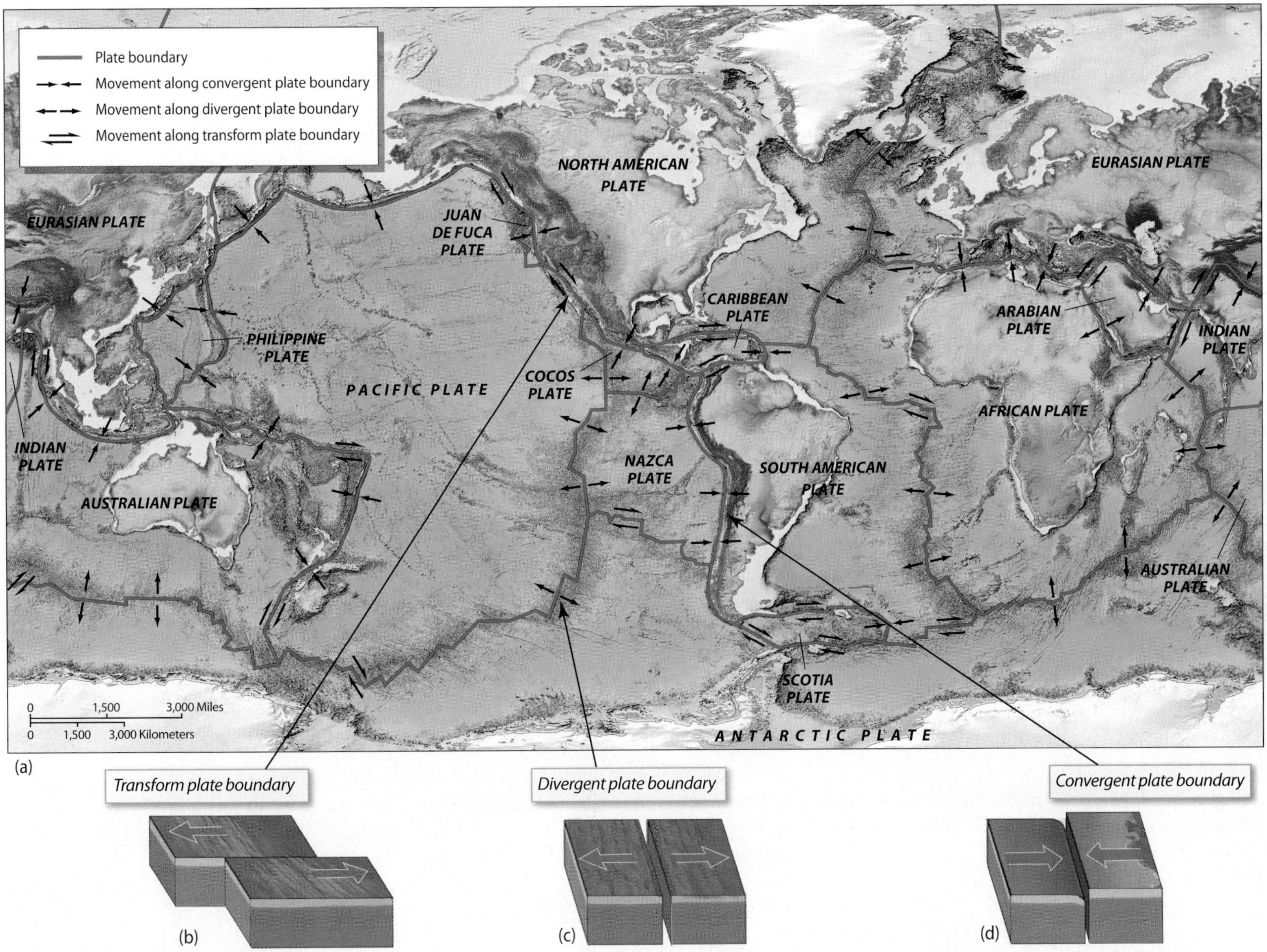

Figure 2.3 Tectonic Plate Boundaries This world map shows the global distribution of the major tectonic plates, along with the general direction of plate movement. As well, the different categories of plate boundaries are shown. Note that continental boundaries do not always coincide with plate boundaries. Put differently, continents are not the same as tectonic plates; instead, to simplify, continents ride on top of the plates.

Figure 2.4 Chile's Subduction Zone Earthquakes The South American country of Chile is particularly vulnerable to tectonic subduction zone earthquakes. Here, a security worker examines a car damaged in the 8.2 magnitude Iquique earthquake of April 2014; four years earlier Chile was struck by a massive 8.8 magnitude earthquake that killed 500 people. More recently, in mid-September 2015, a 8.3 magnitude earthquake devastated Illapel, Chile.

Figure 2.5 Iceland's Divergent Plate Boundary Volcanic activity is common in Iceland because of its location on the divergent tectonic boundary that bisects the Atlantic Ocean. This photo is from a recent eruption on the Holuhraun Fissure near the Bardarbunga Volcano.

the northwest coast of North America, offshore of eastern Japan, and near the Philippines, where the Mariana Trench is the deepest point of the world's oceans at 35,000 feet (10,700 meters) below the surface. These subduction zones are also the locations of Earth's most powerful earthquakes, often with accompanying tsunamis, as evidenced by the magnitude 8.2 earthquake that displaced thousands in Chile in 2014 (Figure 2.4) and the magnitude 9.0 earthquake that devastated coastal Japan in 2011. These zones are also sites for many volcanoes, such as the 130 active volcanoes in Indonesia, located along the Sunda Trench. On convergent boundaries where plates collide rather than subduct, towering mountains are formed. The best-known of these are is the Himalayas that stretch across Asia.

Where plates diverge, magma from Earth's interior often flows to the surface, creating mountain ranges and active volcanoes. In the North Atlantic, Iceland lies on the divergent plate boundary that bisects the Atlantic Ocean (Figure 2.5). But at other divergent boundaries, deep depressions—called **rift valleys**—are formed. An example is the area occupied by the Red Sea between northern Africa and Saudi Arabia.

Figure 2.6 Transform Tectonic Boundary This 3-D map shows the many different earthquake faults in the San Francisco Bay Area associated with the San Andreas fault system, which is a large transform tectonic boundary separating the oceanic Pacific Plate from the continental North American Plate.

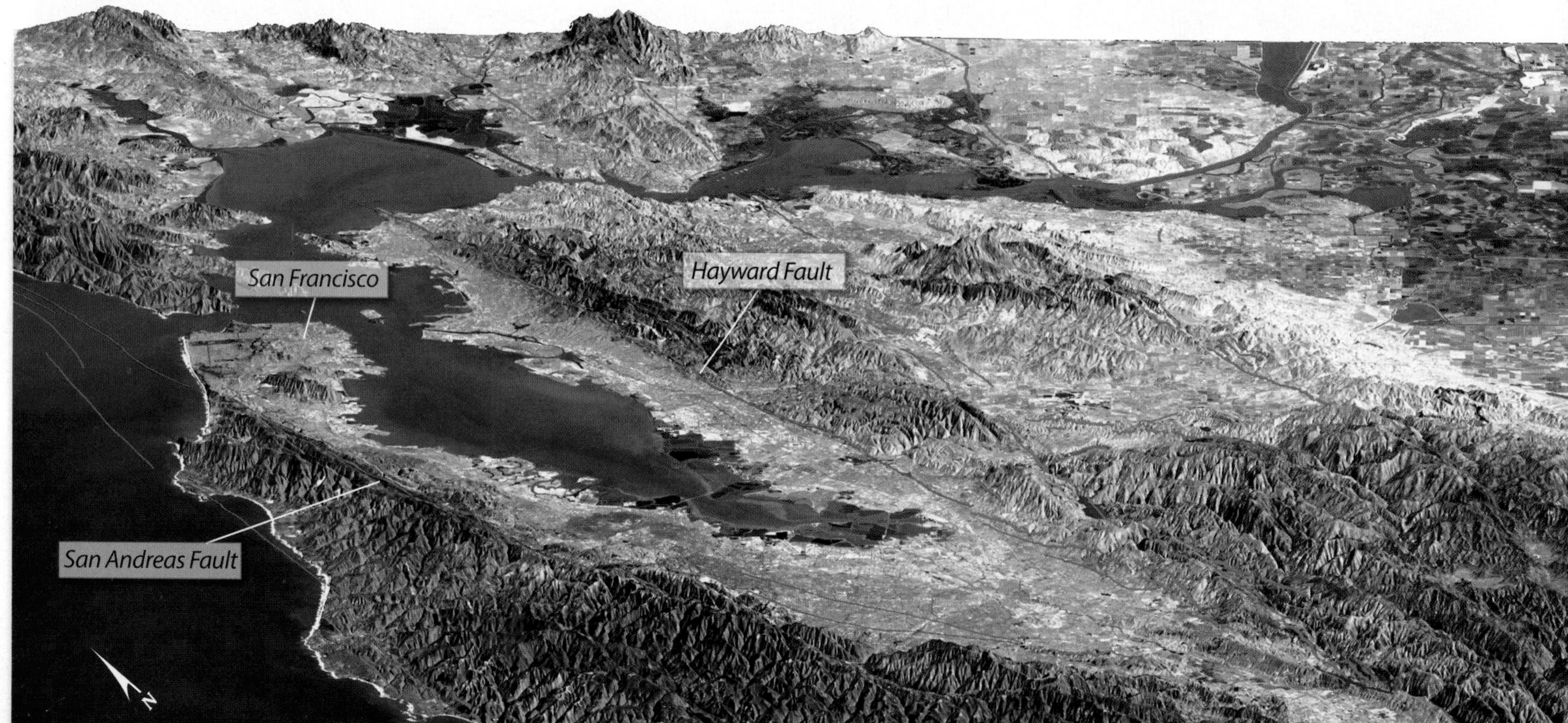

In western North America, coastal California lies atop two different plates—the Pacific and the North American plates. The infamous San Andreas Fault traversing coastal California forms this plate boundary. The San Andreas is a **transform fault** (Figure 2.6), with the eastern edge of the Pacific Plate moving laterally northward at a rate of several inches each year, pushing sideways past the North American Plate. The nearness of San Francisco and Los Angeles to the San Andreas Fault makes these two urban areas—like many others around the world—vulnerable to destructive earthquakes.

Geologic evidence suggests that some 250 million years ago all the world's land masses were tightly consolidated into a supercontinent centered on present-day Africa. Over time, this supercontinent, called **Pangaea**, was broken up as convection cells moved the tectonic plates apart. A hint of this former continent can be seen in the jigsaw-puzzle fit of South America with Africa and of North America with Europe.

Geologic Hazards

Although extreme weather events like floods and tropical storms typically take a higher toll of human life each year, earthquakes and volcanoes can significantly affect human settlement and activities (Figure 2.7). Nearly 20,000 people died in March 2011 from the combination of an earthquake and a tsunami in coastal Japan, and a year earlier (January 2010) over 230,000 people were killed in a magnitude 7.0 earthquake in Haiti. The vastly different effects of these two quakes underscore the fact that vulnerability to geologic hazards differs considerably around the world, depending on local building standards, population density, housing traditions, and the effectiveness of search, rescue, and relief organizations.

In addition to earthquakes, volcanic eruptions occur along both divergent plate boundaries and subduction zones (see Figure 2.7) and can also cause major destruction. But because volcanoes usually provide an array of warnings before they erupt, the loss of life from volcanoes is generally a fraction of that from earthquakes. In the 20th century, an estimated 75,000 people were killed by volcanic eruptions, whereas approximately 1.5 million died in earthquakes.

Unlike earthquakes, volcanoes provide some benefits to people. In Iceland, New Zealand, and Italy, geothermal activity produces energy to heat houses and power factories. In other parts of the world, such as the islands of Indonesia, volcanic ash has enriched soil fertility for agriculture. Additionally, local economies benefit from tourists attracted by scenic volcanic landscapes in such places as Hawaii, Japan, and the Pacific Northwest (Figure 2.8).

(a)

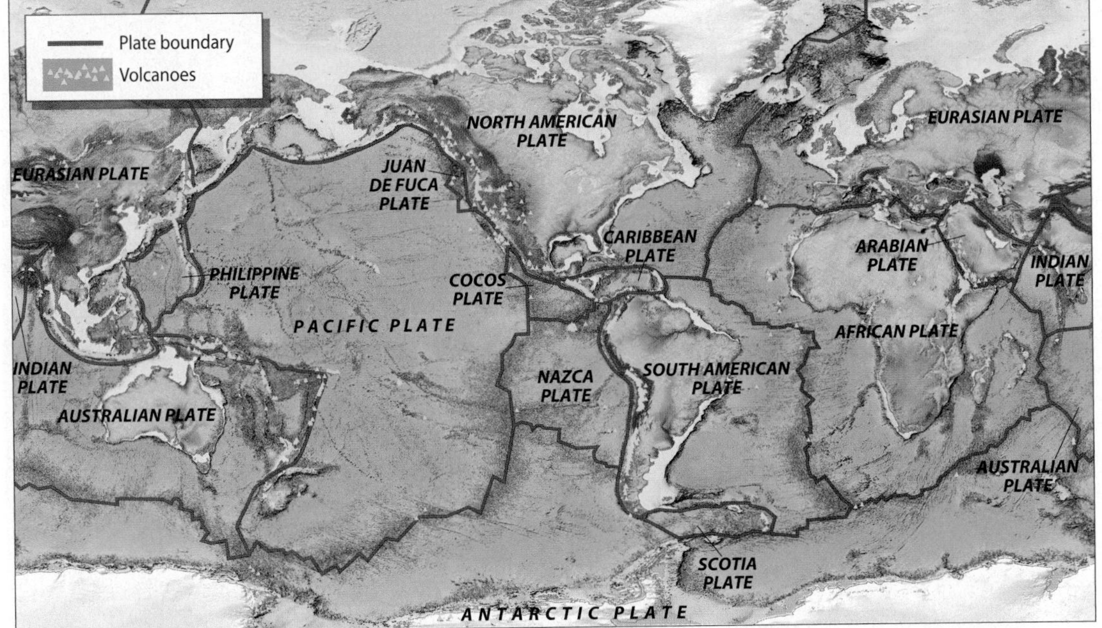

(b)

Figure 2.7 The Geography of Earthquakes and Volcanoes (a) Most, but not all, earthquakes take place near tectonic plate boundaries. Further, most of the strongest and most devastating earthquakes are located near converging, subduction zone boundaries. (b) While there's a strong correlation between the distribution of volcanoes, tectonic plate boundaries, and earthquakes, there are many places in the world where volcanoes are found far removed from plate boundaries. The island volcanoes of Hawaii are an example.

Figure 2.8 **Seattle and Mt Rainier** Picturesque as it may be, Mt Rainier is a classic subduction zone volcano conveying silent warning that Seattle should expect a strong earthquake sometime in the future. The last major quake along the Cascadia subduction zone was in 1700, and estimated to be between 8.7 and 9.2 on the Richter Magnitude. A similar earthquake today would cause considerable damage in both Seattle and Portland, Oregon.

Review

2.1 Sketch and describe the different kinds of tectonic plate boundaries. What causes their differences?

2.2 Where are most of the world's earthquakes and volcanoes? Why are they located where they are?

KEY TERMS plate tectonics, lithosphere, convergent plate boundary, divergent plate boundary, transform plate boundary, subduction zone, rift valley, transform fault, Pangaea

Global Climates: Adapting to Change

Many human activities are closely tied to weather and climate. Farming depends on certain conditions of sunlight, temperature, and precipitation to produce the world's food, while urban transportation systems are often disrupted by extreme weather events like snowstorms, typhoons, and even heat waves. Furthermore, a severe weather event in one location can affect far-flung places. Reduced harvests due to drought in Russia's grain belt, for example, ripple through global trade and food supply systems, with serious consequences worldwide.

Aggravating these interconnections is global climate change. Just what the future holds is not entirely clear, but even if the long-term forecast has some uncertainty, there is little question that all forms of life—including humans—must adapt to vastly different climatic conditions by the middle of the 21st century (see *Geographers at Work: Jerilynn M. Jackson, Climate Change Scientist and Arctic Adventurer*).

Climate Controls

The world's climates differ significantly from place to place and seasonally with highly varying patterns of temperature and precipitation (rain and snow) that can be explained by physical processes referred to as climate controls.

Solar Energy The Sun's heating of Earth and its atmosphere is the most important factor affecting world climates. Not only does solar energy cause temperature differences between warmer and colder regions, but it also drives other important climate controls such as global pressure systems, winds, and ocean currents.

Incoming short-wave solar energy, called **insolation**, passes through the atmosphere and is absorbed by Earth's land and water surfaces. As these surfaces warm, they **reradiate** heat back into the lower atmosphere as infrared, long-wave energy. This reradiating energy, in turn, is absorbed by water vapor and other atmospheric gases such as carbon dioxide (CO_2), creating the envelope of warmth that makes life possible on our planet. Because there is some similarity between this heating process and the way a garden greenhouse traps warmth from the Sun, this natural process of atmospheric heating is called the **greenhouse effect** (Figure 2.9). Without this process, Earth's climate would average about 60°F (33°C) colder, resulting in conditions much like Mars.

Latitude Because Earth is a tilted sphere with the North Pole facing away from the Sun for half of the year and toward the Sun for

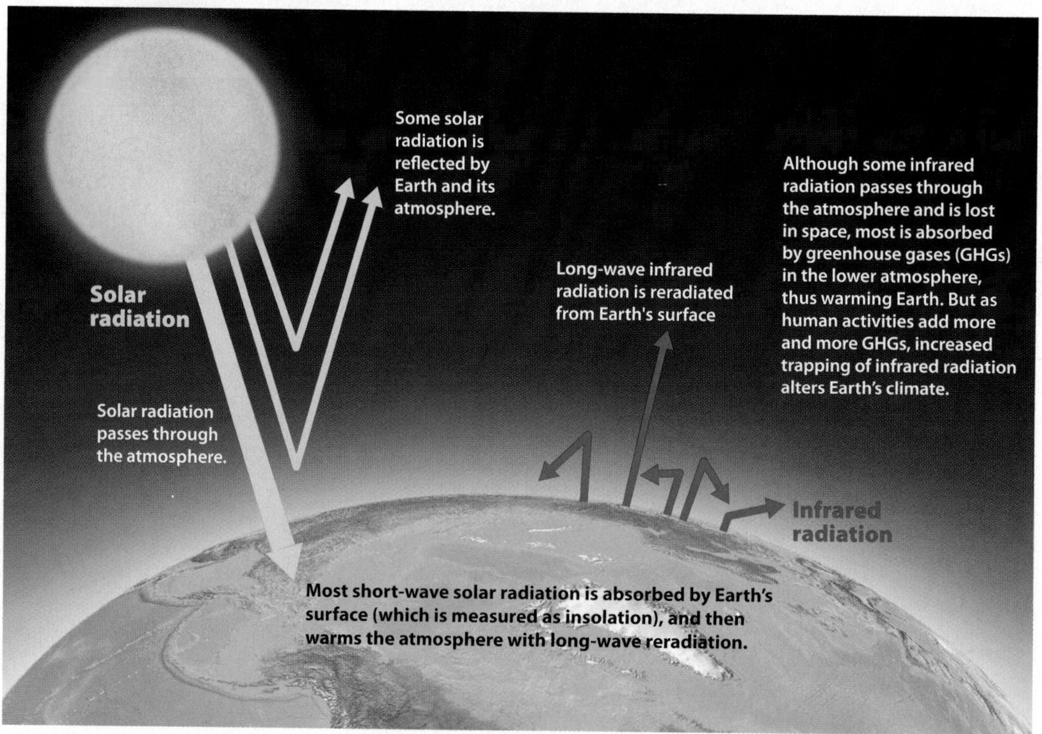

Figure 2.9 **Solar Energy and the Greenhouse Effect** The greenhouse effect is the trapping of solar radiation in the lower atmosphere, resulting in a warm envelope surrounding Earth. As shown here, most incoming short-wave solar radiation is absorbed by land and water surfaces, then rereadiated into the atmosphere as long-wave infrared radiation. It is this long-wave radiation absorbed by greenhouse gases—both natural and human-generated—that warms the lower atmosphere and affects Earth's weather and climate.

GEOGRAPHERS AT WORK

Glaciers and Climate Change

Climate change scientist and Arctic adventurer **M Jackson** is a role model for mixing serious science with global adventuring. M leads National Geographic Society expeditions (Figure 2.1.1), lecturing on arctic science during tours, and recently wrote a book, *While Glaciers Slept: Being Human in a Time of Climate Change*.

Glaciers and Geography After completing her master's degree in environmental science at the University of Montana, Jackson traveled to Turkey as a Fulbright Scholar to study glaciers on the Turkish/Iranian border. What she observed led to more questions—and to the study of geography. "Glaciologists told me that if I really wanted to understand glaciers within a greater system, I should do so from a geographic perspective. Geography provides an avenue to view something from many different angles: the social sciences, the humanities, the natural sciences. And it's the geographer's job to combine these different elements into a whole picture of what is happening . . . I love that."

In addition to her NGS expeditions, M is working on her PhD in geography, doing fieldwork on the impacts of glacier loss in Iceland. Both activities allow her to "sell" geography to undergraduates: "If a student is leaning more toward natural science or lab work, or if a person is interested in social theory, or wants to learn GIS or computer skills, geography is a major where the student can be selective, but add these bits into a greater toolkit. The size and strength of that toolkit will help immensely on the job market."

"If you look at something in just one way, you can't get a full view," continues M. "With geography, you can go in different directions, or just synthesize everything together."

1. Explain the impacts of human-induced climate change on glaciers, and list the possible consequences for people and their communities.
2. What are the advantages of understanding human geography when researching natural phenomena, such as glaciers, forests, or oceans?

Figure 2.1.1 M Jackson takes a break from her glaciology work in Iceland.

the other half, maximum solar radiation at a given location occurs seasonally (with summer marking the season of peak insolation for each hemisphere), and insolation strikes the surface at a true right angle only in the tropics. Therefore, solar energy is more intense and effective at heating land and water at low latitudes than at higher latitudes. Not only does this difference in solar intensity result in warmer tropical climates, contrasted with the cooler middle and high latitudes, but also heat accumulates on a larger scale in these equatorial regions (Figure 2.10). This heat is then distributed away from the tropics through global pressure and wind systems, ocean currents, subtropical typhoons or hurricanes, and even midlatitude storms (Figure 2.11).

Interactions Between Land and Water Land and water areas differ in their ability to absorb and reradiate insolation; thus, the global arrangement of oceans and land areas is a major influence on world climates. More solar energy is required to heat water than to heat land, so land areas heat and cool faster than do bodies of water (Figure 2.12). This explains why the temperature extremes of hot summers and cold winters are found in the continental interiors, such as the Great Plains of North America, while coastal areas experience more moderate winters and cooler summers. These coastal-inland temperature differences also occur at smaller scales, often within just hundreds of miles of each other. In coastal San Francisco, for example,

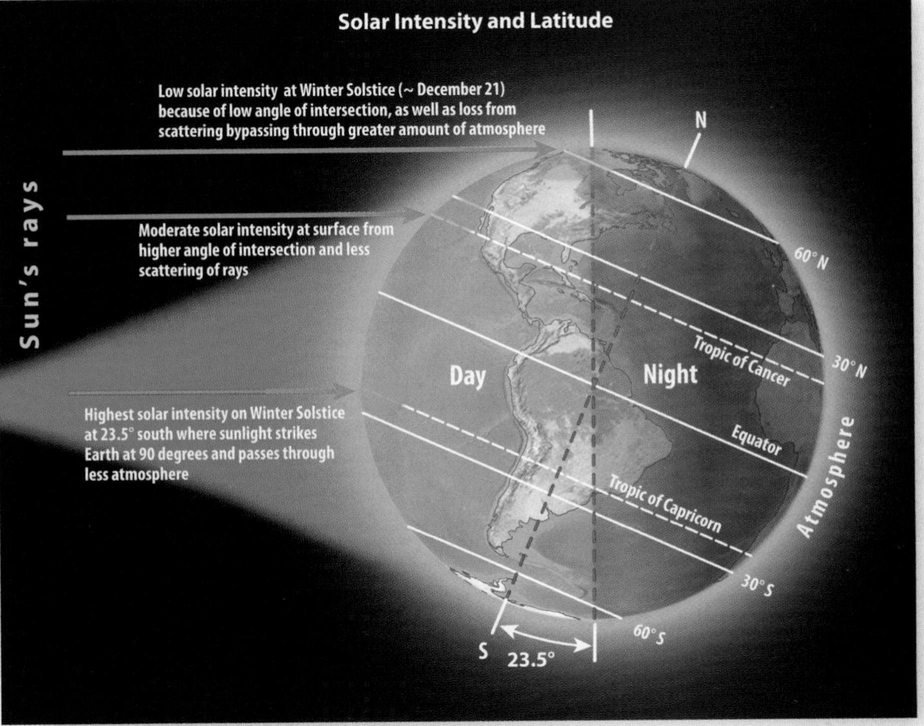

Figure 2.10 Solar Intensity and Latitude Because of Earth's curvature, solar radiation is more intense and more effective at warming the surface in the tropics than at higher latitudes. The resulting heat buildup in the equatorial zone energizes global wind and pressure systems, ocean currents, and tropical storms.

Figure 2.11 Typhoon Pam Atmospheric heat imbalances between the equatorial zone and the midlatitudes produce massive tropical cyclones called typhoons in the Pacific and hurricanes in the Atlantic. Both are capable of widespread damage through high winds, coastal flooding from waves, and heavy rainfall. This is a satellite image of Typhoon Pam, a major storm that devastated many Pacific islands in March 2015.

the average July maximum temperature is 60°F (15.6°C), whereas 80 miles (129 km) away in California's inland capital, Sacramento, the average maximum July temperature is 92.4°F (33.5°C).

The term **continental climate** describes inland climates with hot summers and cold winters, while locations where oceanic influences dominate are referred to as **maritime climates**. The island countries of Southeast Asia and the British Isles in Europe are good examples of areas with maritime climates, while interior North America, Europe, and Asia have continental climates.

Global Pressure Systems The uneven heating of Earth due to latitudinal differences and the arrangement of oceans and continents produces a regular pattern of high and low pressure cells (Figure 2.13). These cells drive the movement of the world's wind and storm systems because air (in the form of wind) moves from high to low pressure. The interaction between high- and low-pressure systems over the North Pacific, for example, produces storms that are carried by winds onto the North American continent. Similar processes in the North Atlantic produce winter and summer weather for Europe. Farther south, over the subtropical zones, large cells of high pressure cause very different conditions. The subsidence (sinking) of warm air moving in from the equatorial regions causes the great desert areas at these latitudes. These high-pressure areas expand during the warm summer months, producing the warm, rainless summers of Mediterranean climate areas in Europe and California. In the low latitudes, summer heating of the oceans also spawns the strong tropical storms known as typhoons in Asia and as hurricanes in North America and the Caribbean.

Global Wind Patterns There are several different wind patterns that strongly influence Earth's weather and climate. At the global level are the **polar and subtropical jet streams**, powerful atmospheric rivers of eastward-moving air that affect storms and pressure systems in both the Northern and the Southern hemispheres (Figure 2.13). These jets are products of Earth's rotation and global temperature differences. The two polar jets (north and south) are the strongest and the most variable, flowing 23,000–39,000 feet (7–12 km) above the surface at speeds reaching 200 miles per hour (322 km/h). The subtropical jets are usually higher and somewhat weaker. While the northern jet stream has a major effect on the weather of North America and Europe, steering storms across the continents, the southern polar jet circles the globe in the sparsely populated areas near Antarctica.

Nearer Earth's surface are continent-scale winds that, as mentioned earlier, flow from high to low pressure areas. Good examples are the **monsoon winds** of Asia and North America; summer monsoons bring welcome rainfall to the dry areas of interior India and the Southwest United States (Figure 2.14).

Topography Weather and climate are affected by topography—an area's surface characteristics—in two ways: Cooler temperatures are found at higher elevations, and precipitation patterns are strongly influenced by topography.

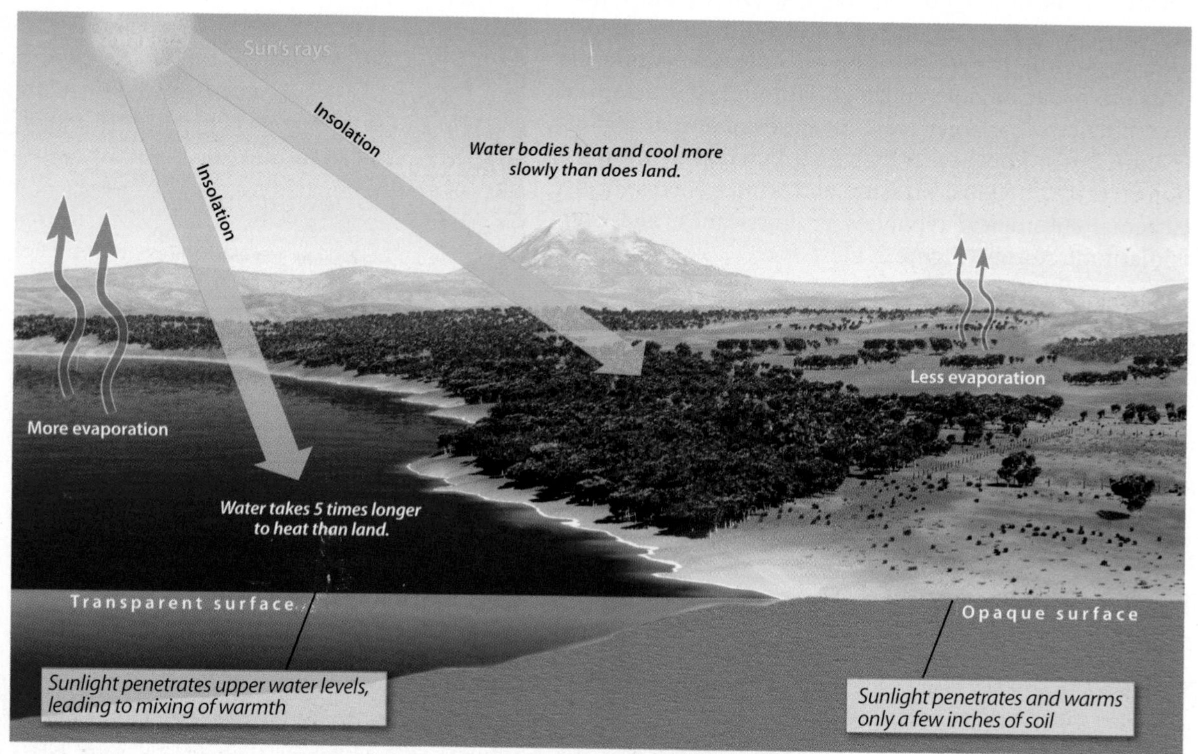

Figure 2.12 Differential Heating of Land and Water Land heats and cools faster than does water through incoming and outgoing solar radiation. This is why inland temperatures are usually both warmer in the summer and colder in the winter than coastal locations near the ocean.

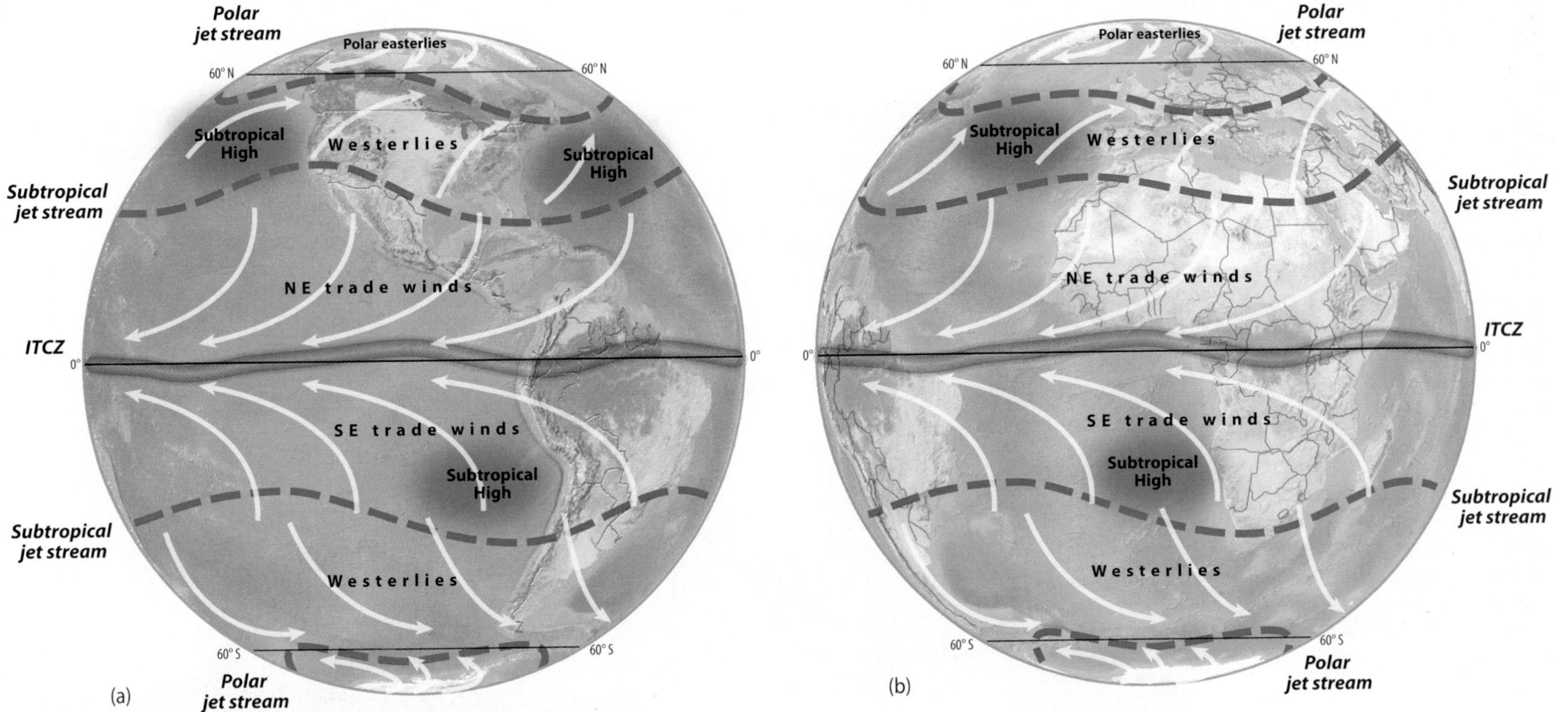

Figure 2.13 Global Pressure Systems and Winds Two jet streams, the Polar and Subtropical Jets are found in each hemisphere, northern and southern. In circling Earth, these jets often change position as they steer storms and air masses. The subtropical high pressure cells are large areas of subsiding air that energize the mid-latitude westerly winds and the tropical trade winds. The Inter-Tropical Convergence Zone (ITCZ) is a belt of low pressure that circles Earth and results from strong solar radiation in the equatorial zone.

Because the lower atmosphere is heated by solar energy reradiated from Earth's surface, air temperatures are warmer closer to the surface and become cooler with altitude. As a general rule, the atmosphere cools by 3.5°F for every 1000 feet gained in elevation (.05°C per 100 meters). This is called the **environmental lapse rate**. To illustrate, on a typical summer day in Phoenix, Arizona, at an elevation 1100 feet (335 meters), the temperature often reaches 100°F (37.7°C). Just 140 miles away, in the mountains of northern Arizona at 7100 feet (2160 meters) in the small town of Flagstaff, the temperature is a pleasant 79°F (26°C). This difference of 21°F (11.7°C) results from 6000 feet (1825 meters) of elevation; this can be easily calculated by multiplying elevation in thousands (6) by the environmental lapse rate (3.5°F).

Figure 2.14 Summer Monsoon in Southwest North America As the U.S. Southwest warms during the northern hemisphere summer, this heating creates thermal lows over the interior that draw in moist air from the Gulfs of Mexico and California, resulting in cloudiness, thunderstorms, and much-needed rainfall from July to September.

If an area has rugged topography, this can wring moisture out of clouds when moist air masses cool as they are forced up and over mountain ranges in what is called the **orographic effect** (Figure 2.15). Cooler air cannot hold as much moisture as warm air, resulting in precipitation. Note that the rising air mass cools (and warms upon descending) faster than a hypothetical stable or nonmoving air mass, for which the change in temperature with elevation is measured by the environmental lapse rate. More specifically, an air mass moving up a mountain slope will cool at 5.5°F per 1000 feet (1°C per 100 meters) of elevation; this is referred to as the **adiabatic lapse rate**.

This process explains the common pattern of wet mountains and nearby dry lowlands. These dry areas are said to be in the **rain shadow** of the adjacent mountains. Rainfall lessens as downslope winds warm (the opposite of upslope winds, which cool), thus increasing an air mass's ability to retain moisture and deprive nearby lowlands of precipitation. Rain shadow areas are common in the mountainous areas of western North America, Andean South America, and many parts of South and Central Asia.

Climate Regions

Even though the world's weather and climate vary greatly from place to place, areas with similarities in temperature, precipitation, and seasonality can be mapped into global climate regions (Figure 2.16). Before going further, it is important to note the difference between these two terms. Weather is the short-term, day-to-day expression of atmospheric processes; that is, weather can be rainy, cloudy, sunny, hot, windy, calm, or stormy, all within a short time period. As a result, weather is measured at regular intervals each day, usually hourly. These data are then compiled over a 30-year period to generate statistical averages that describe the

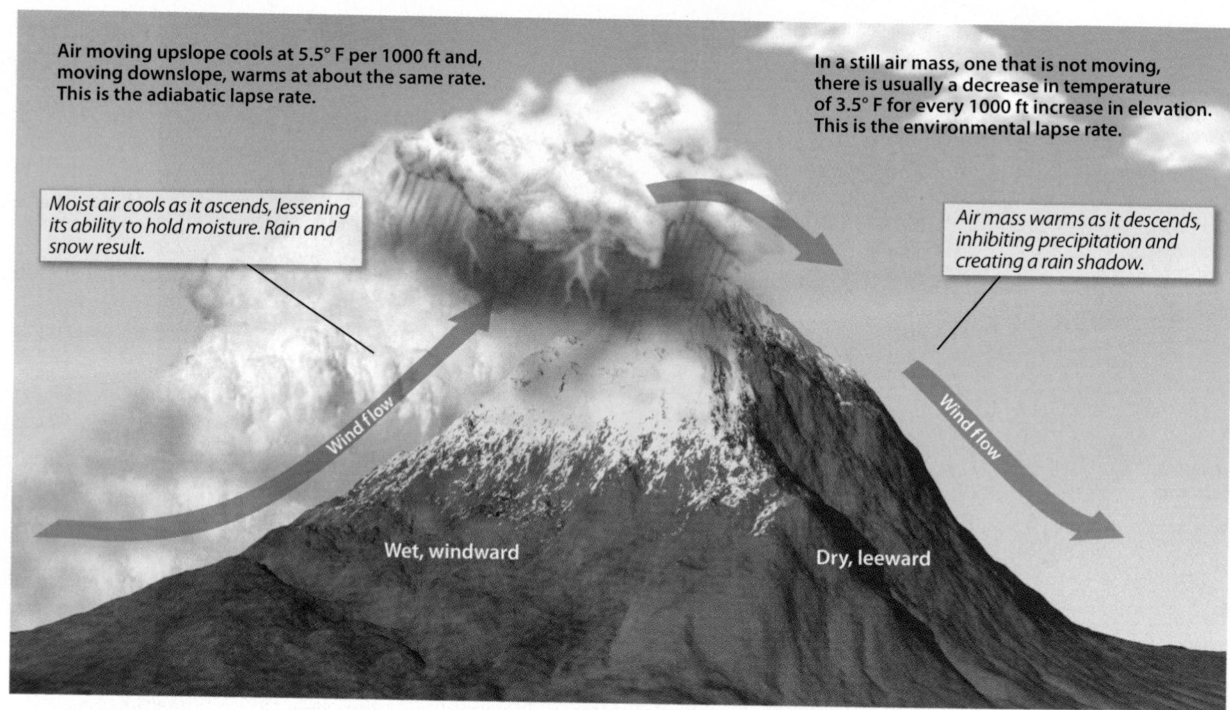

Figure 2.15 The Orographic Effect Upland and mountainous areas are usually wetter than the adjacent lowland areas because of the orographic effect. This results from the cooling of rising air over higher topography, and as the air mass cools it loses its ability to hold moisture, resulting in rain and snowfall. In contrast, the leeward or downwind side of the mountains is drier because downslope air masses warm, thus increasing their ability to retain moisture. These dry areas on the downwind side of mountains are called rain shadows **Q: After reviewing the concept of the orographic effect, look at a map of the world and locate at least five different areas where the orographic effect would be found.**

typical meteorological conditions of a specific place, which is the climate. Simply stated, *weather* is the short-term expression of atmospheric processes, and **climate** is the long-term average from daily weather measurements. As pundits like to say, climate is what you expect and weather is what you get.

We use a standard scheme of climate types throughout this text, and each regional chapter contains a map showing the different climates of that region. In addition, these maps contain **climographs**, which are graphic representations of monthly average temperatures and precipitation. Two lines for temperature data are presented on each climograph: The upper line plots average high temperatures for each month, while the lower line shows average low temperatures. Besides these temperature lines, climographs contain bar graphs depicting average monthly precipitation. The total amount of rainfall and snowfall is important, as is the seasonality of precipitation. Figure 2.16 contains climographs for Tokyo, Japan, and Cape Town, South Africa. In looking at these two climographs, remember that the seasons are reversed in the Southern Hemisphere.

Global Climate Change

Human activities connected to economic development have caused significant **climate change** worldwide over the last century, resulting in warmer temperatures, melting ice caps, rising sea levels, and more extreme weather events (Figure 2.17). More important, unless international action is taken soon to limit atmospheric pollution, these climate changes will produce a challenging world environment by mid-century. Rainfall patterns may change, so that agricultural production in traditional breadbasket areas such as the U.S. Midwest and Canadian prairies may be threatened; low-lying coastal

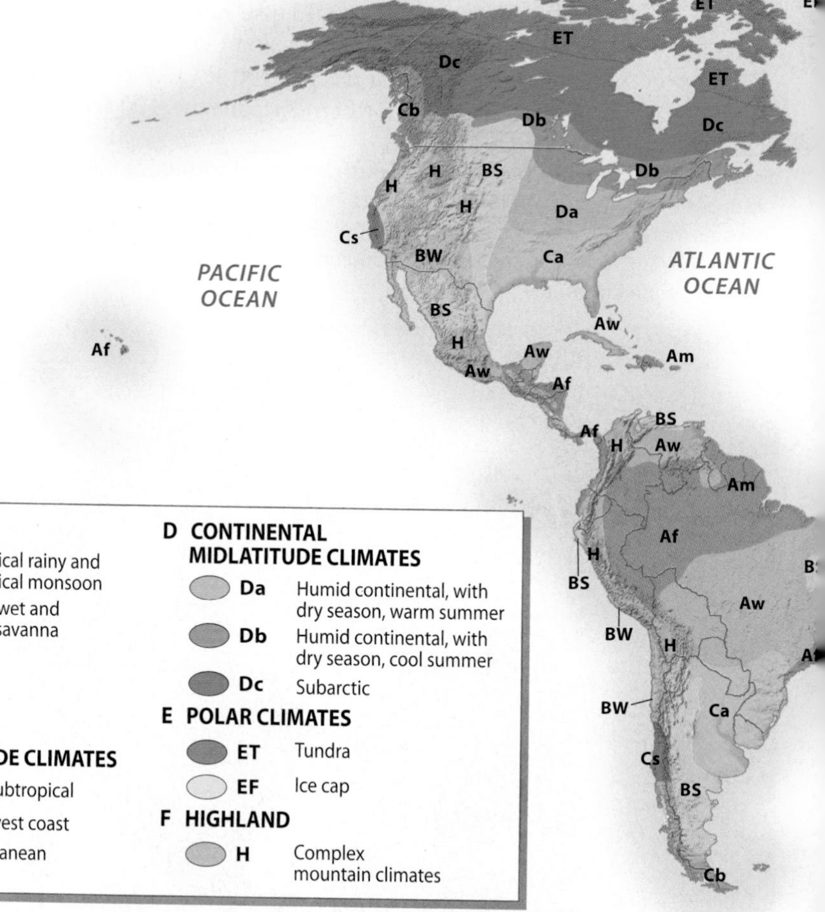

Figure 2.16 Global Climate Regions A standard scheme, called the Köppen system, named after the Austrian geographer who devised the plan in the early 20th century, is used to describe the world's diverse climates. Combinations of upper- and lower-case letters describe the general climate type, along with precipitation and temperature characteristics. Specifically, the *A* climates are tropical, the *B* climates are dry, the *C* climates are generally moderate and are found in the middle latitudes, and the *D* climates are associated with continental and high-latitude locations

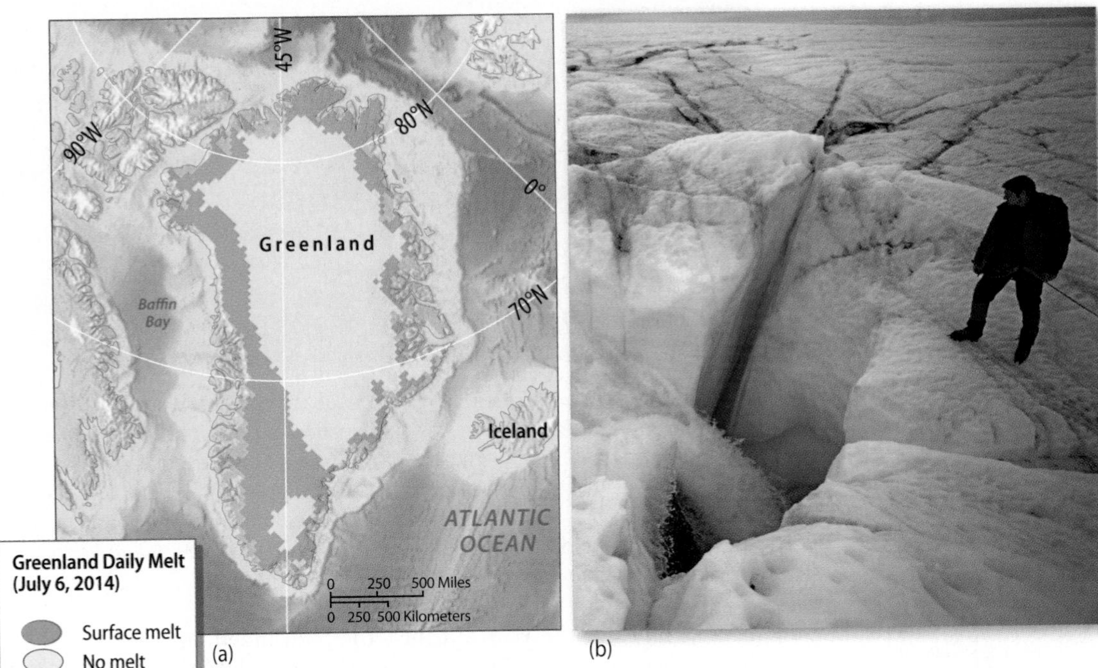

Figure 2.17 Greenland's Melting Ice Cap (a) One problematic component of global warming is the melting of mountain glaciers and arctic ice caps, since the meltwater is causing ocean levels to rise significantly. (b) In this photo a glaciologist studies the surface melting of the Greenland ice cap, observing a river of surface meltwater pouring into a feature known as a moulin.

settlements in places like Florida and Bangladesh will be flooded as sea levels rise; increased heat waves will cause higher human death tolls in the world's cities; and clean water will become increasingly scarce in many areas of the world.

Causes of Climate Change As mentioned earlier, the natural greenhouse effect provides Earth with a warm atmospheric envelope; this warmth comes from incoming and outgoing solar radiation that is trapped by an array of such natural constituents as water vapor, carbon dioxide (CO_2), methane (CH_4), and ozone (O_3). Although the composition of these natural **greenhouse gases (GHGs)** has varied somewhat over long periods of geologic time, it has been relatively stable since the last ice age ended 20,000 years ago (more detail on GHGs is found online in *MasteringGeography*).

However, with the widespread consumption of coal and petroleum associated with global industrialization, a huge increase in atmospheric carbon dioxide and methane has occurred. As a result, the natural greenhouse effect has been greatly magnified by **anthropogenic** or human-generated GHGs, trapping increased amounts of Earth's long-wave reradiation and thus warming the atmosphere and changing our planet's climates. Figure 2.18 shows that in 1860 atmospheric CO_2 was measured at 280 parts per million (ppm); today it is 400 ppm. More troubling is that these CO_2 emissions are forecast to reach 450 ppm by 2020, a level at which climate scientists predict irrevocable climate change.

Although the complexity of the global climate system leaves some uncertainty about exactly how the world's climates may change, climate scientists using high-powered computer models are reaching consensus on what can be expected. These computer models predict that average global temperatures will increase 3.6°F (2°C) by 2020, a temperature change of the same

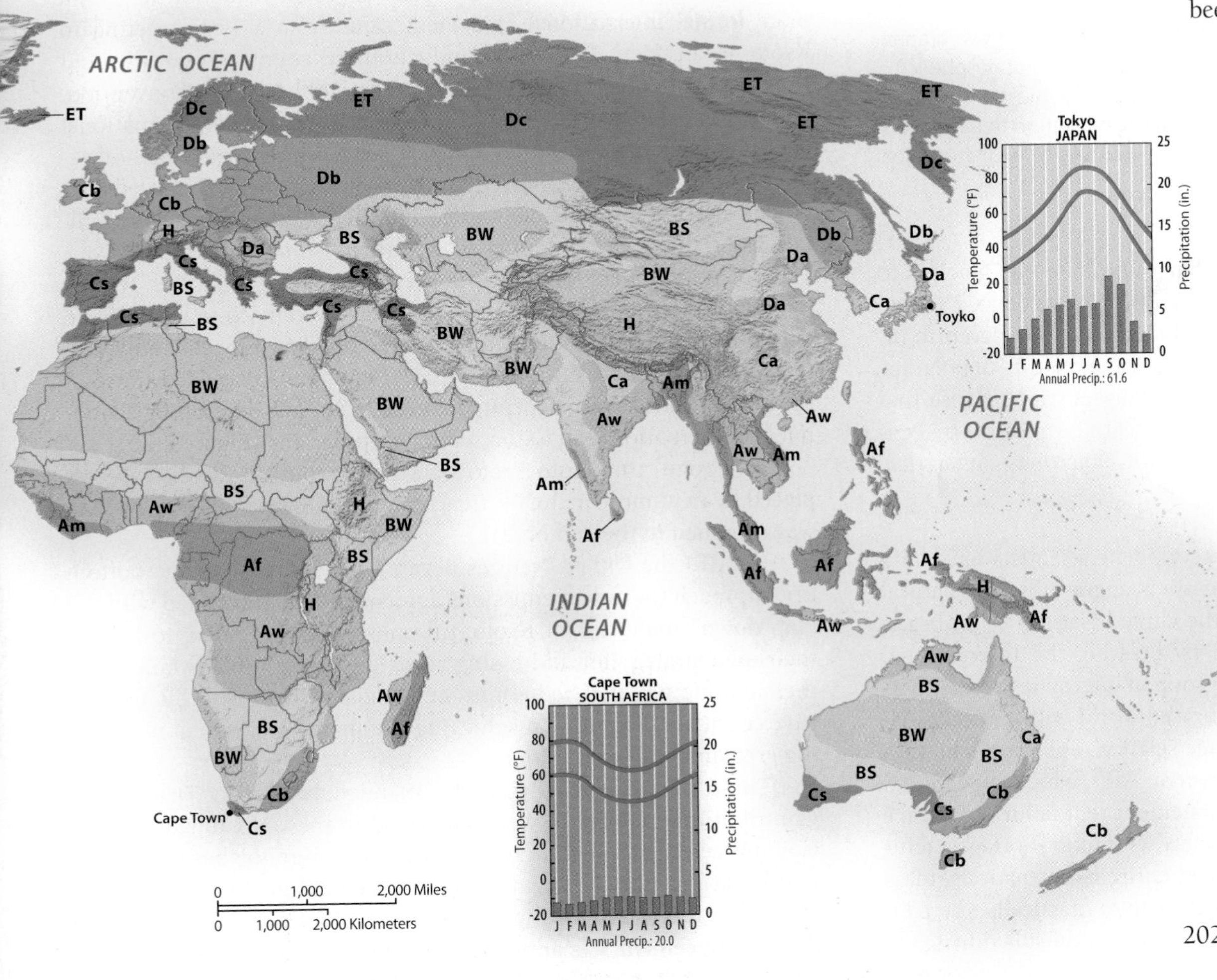

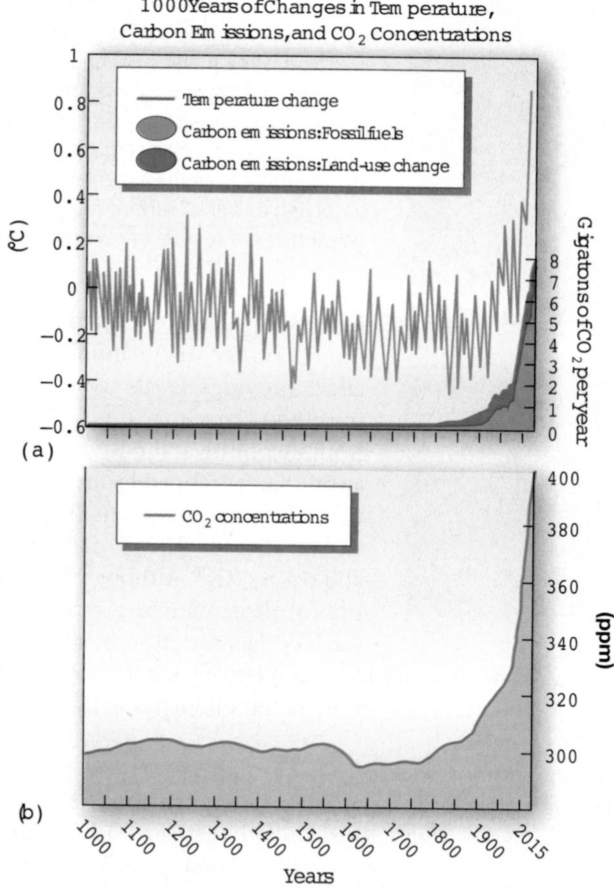

Figure 2.18 Global Increase of CO_2 and Temperature These two graphs show the strong relationship between (b) the recent increase in atmospheric CO_2 and (a) the increase the average annual temperature for the world. The graphs go back 1000 years and show both CO_2 and temperature to have been relatively stable until the industrial period, when the burning of fossil fuels (coal and oil) began on a large scale.

magnitude as the amount of cooling that caused ice-age glaciers to cover much of Europe and North America 30,000 years ago. Further, without international policies to limit emissions, this temperature increase is projected to double by 2100. The resulting melting of polar ice caps, ice sheets, and mountain glaciers will cause a sea-level rise that currently is estimated to be on the order of 4 feet (1.4 meters) by century's end (Figure 2.19; see *Exploring Global Connections: Antarctica, the Science Continent*).

International Efforts to Limit Emissions Climate scientists have long expressed concern about fossil fuel emissions aggravating the natural greenhouse effect; as a result, in 1988 the United Nations (UN) began coordinating study of global warming by creating the International Panel on Climate Change (IPCC). This group of international scientists was (and is still) charged with providing the world with periodic Assessment Reports (ARs) of climate change science. The panel's first report, AR1, came out in 1990; the most recent, AR5, which appeared in late 2014, included a strongly worded statement that failure to reduce atmosphere emissions could threaten society with food shortages, refugee crises, the flooding of major cities and entire island nations, mass extinctions of plants and animals, and a climate so drastically altered it might become dangerous for people to work or play outside during the

Figure 2.19 Sea Level Rise One consequence of global warming will be a rise in the world's sea level due to a combination of polar ice cap melting and the thermal expansion of warmer ocean water. Forecasts are that at the current rate of climate warming, sea levels will rise about 4 feet (1.4 meters) by the year 2100. This rise will cause considerable flooding of low-lying coastal areas throughout the world. This photo shows that rising sea levels due to flooding are already affecting the large delta region of Bangladesh.

hottest times of the year. Further, the "continued emission of greenhouse gases will cause further warming and long-lasting changes in all components of the climate system, increasing the likelihood of severe, pervasive and irreversible impacts for people and ecosystems."

International efforts to reduce atmospheric emissions began in 1992, shortly after the IPCC's first report, when 167 countries meeting in Rio de Janeiro, Brazil, signed the Rio Convention, in which they agreed to voluntarily limit their GHG emissions. However, because none of the Rio signatories reached its emission reduction targets, a more formal international agreement came from a 1997 meeting in Kyoto, Japan. Here 30 Western industrialized countries agreed to cut back their emissions to 1990 levels by 2012. Unlike the Rio Convention, which was voluntary, this **Kyoto Protocol** had the force of international law, with penalties for those countries not reaching their emission reduction targets. At that point in time, the 30 signatories produced over 60 percent of the world's emissions, and there were no emissions limitations on the large developing economies of China and India.

But not all went well with the Kyoto Protocol. First, the world's largest polluter at that time, the United States, refused to ratify Kyoto because of concerns about injuring the struggling U.S. economy. Additionally, by 2008 China's yearly GHG emissions exceeded those of the United States, underscoring the need to include developing economies in international emission reduction programs (Figure 2.20).

By design, the Kyoto Protocol was to expire in 2012, to be replaced by new, more inclusive treaty. But progress was slow and Kyoto was extended to the end of 2015.

In 2014 the United Nations began a new, very different "bottom up" approach toward an emissions agreement that contrasted with the "top down" mandates of Kyoto. All countries were asked to submit their own strategy for addressing climate change in plans tailored to their unique national economies and social structures. This process gave countries flexibility in achieving the shared global goal of reducing greenhouse gas emissions.

These national plans, called "Intended Nationally Determined Contributions" (INDCs), were submitted to the UN Climate Change Committee during the first half of 2015, and were the basis for negotiations at the Climate Change Conference held in Paris, France that December. At that meeting a new international greenhouse gas reduction agreement was approved; this plan now serves as the world's

EXPLORING GLOBAL CONNECTIONS

Antarctica, the Science Continent

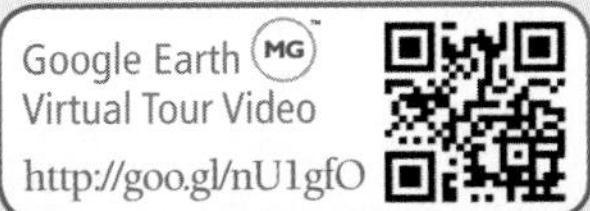

In this era of climate change, all eyes are on the polar regions, for it is there, both north and south, that our future lies. Melting ice caps will dictate how much—and how fast—sea levels rise, flooding lowlands and islands worldwide. Freshwater from melting polar ice caps and glaciers could also dramatically change ocean circulation patterns, which, in turn, might drastically alter Earth's midlatitude weather and climate.

While globalization is often criticized because it increases economic competition between countries, Antarctica embodies international cooperation at its finest, as 72 nations have joined together in the Antarctica Treaty System (ATS) to protect the huge continent for scientific study. Even more amazing was that this international agreement was reached in 1959, during the darkest days of the Cold War between the world's superpowers: the Soviet Union (now Russia) and the United States. At that time, 12 countries (including both superpowers) signed the ATS, which protects Antarctica from territorial claims, mining, and military bases and also establishes the continent as a center for international scientific research. Today the ATS has 50 full-member countries and 22 associate-member countries.

Physical Geography Antarctica, which includes the South Pole, is the world's fifth largest continent, twice as large as Australia. Antarctica is huge, but many other superlatives also apply: It's the coldest, driest, windiest, and highest continent. The average yearly temperature at the South Pole is –70°F (–57°C), and on average only 6.5 inches (16.5 cm) of precipitation fall each year. Winds of 50–60 miles per hour (80–95 km/h) blow constantly, resulting in terrifically low wind-chill factors. Only on the Antarctic Peninsula do summer temperatures get above freezing.

The geography of Antarctica is fairly simple: Two major ice sheets dominate, one in the east and the other in the west, separated by a large mountain range. The ice sheets that cover most of Antarctica have an average depth of 7900 feet (2400 meters) but reach 16,500 feet (5000 meters) in the thickest part. Mt. Vinson is the highest point above sea level at 16,144 feet (4892 meters). In addition, several giant coastal ice shelves extend into nearby ocean waters (Figure 2.2.1).

Laboratory on Ice Antarctica has no indigenous people—no natives have ever lived there. In fact, the only familiar life forms on the entire continent are penguins. This makes Antarctica cleaner and purer than any other place on Earth—perfect for investigating Earth's atmosphere without the effects of air pollution.

Shortly after the ATS was signed, the British Antarctica Survey began to monitor atmospheric ozone, resulting in cooperative science on the infamous ozone hole. This is a depletion of the ozone layer in the upper atmosphere, which normally protects life on Earth from too much harmful ultraviolet radiation from the Sun. Scientists found a hole in the ozone layer over Antarctica that was growing at an alarming rate due to ozone reactions with certain chemicals that drifted into the upper atmosphere. This research led to another landmark in international cooperation, the Montreal Protocol of 1989, which banned ozone-depleting sprays and refrigerants. It is still upheld today as a model of international scientific and legal cooperation.

Today much of the international science being done in Antarctica is focused on climate change. Ice cores from the continent's gigantic ice sheets, for example, provide invaluable information about past climates going back some 800,000 years, almost seven times longer than the ice cores extracted from Greenland.

Not only does Antarctica provide clues to Earth's past climates, but the continent is also a critical player in contemporary climate change scenarios because of the potential for the melting and collapse of the West Antarctica Ice Sheet (Figure 2.2.2). Should that happen, Earth's sea levels could rise by at least 10 feet (33 meters), devastating coastal settlements worldwide. Such a disaster, however, would not happen overnight, for scientists think that such melting may take at least several decades to occur.

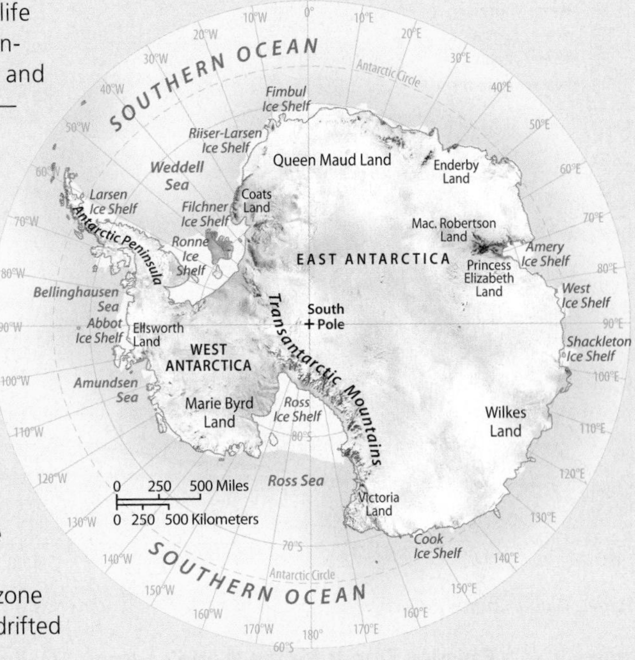

Figure 2.2.1 Antarctica The East and West Antarctica Ice Sheets that make up much of the continent, as well as the Transantarctic Mountains separating the two ice sheets. McMurdo Station, the American Antarctic research center, is southeast of the Ross Ice Shelf, on the small point north of Victoria Land.

1. What is the highest elevation in Antarctica? Is it part of an ice sheet?
2. How do scientists measure melting of the Antarctic ice sheets?

Figure 2.2.2 West Antarctica Ice Sheet The edge of the ice shelf is in the Ross Sea, with the Transantarctic Mountains in the background.

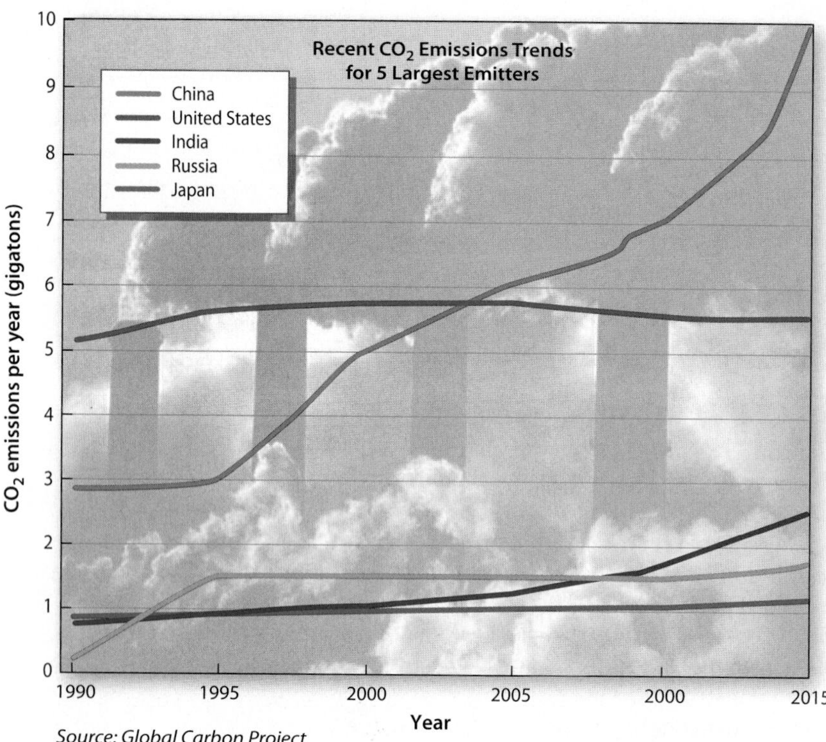

Figure 2.20 Emission Trends for the World's Largest CO_2 Emitters China's yearly CO_2 emissions continue to grow as hundreds of new coal-fired power plants come online. In contrast, emissions in the United States have stabilized recently because more power plants have switched from coal to natural gas as the price of that cleaner source of energy has become increasingly competitive. The flat line of Russia's emissions after the collapse of the Soviet Union is somewhat of a mystery and may be a reporting problem. **Q: What are the similarities and differences between the CO_2 emission reduction plans of China, the United States, and India?**

Figure 2.21 India's CO_2 Emissions India recently became the third-largest CO_2 emitter after China and the United States, with an emissions trend forecast to grow significantly over the next decade. Politically, India appears reluctant to take major steps in reducing its emissions for fear it will inhibit the country's economic development. This photo shows India's largest coal-burning power plant in Munda, a power plant producing energy and emissions from coal imported from Indonesia.

blueprint for addressing the challenges of climate change. While the new Paris Agreement does not solve the climate change problem, it does create a new and promising pathway for the world to follow.

Key components of the Paris Agreement are:

- It is an inclusive, international agreement, approved by 195 countries, covering the economic spectrum from developed to developing economies.
- The signatories are committed to reducing their emissions as presented in their 2015 INDCs. Additionally, each country must assess and revise their INDC every five years with the goal of reducing further their GHG emissions.
- Countries commit to the goal of zero net emissions as soon as possible. This strategy combines emission reductions with carbon offsets, such as planting more trees to store carbon. An important part of this commitment is flexibility in achieving zero net emissions so that developing countries, such as India and Brazil, can move at their own pace toward this goal (Figure 2.21). One concern is the issue of transparency—how to monitor efforts and hold countries accountable for their commitments.
- Developed countries will contribute to a fund of $100 billion by 2020 to assist poorer countries mitigate and adapt to climate change. Low-lying island nations endangered by sea level rise are possible candidates for this aid.

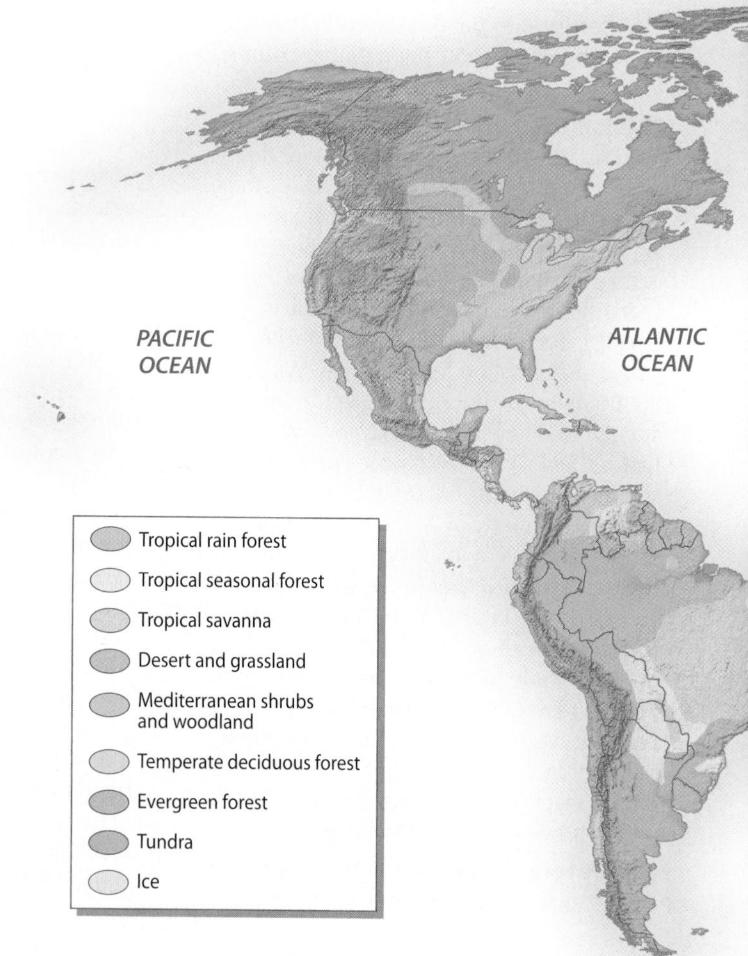

Review

2.3 Define insolation and reradiation, and describe how these interact to warm Earth.

2.4 List the similarities and differences between maritime and continental climates. What causes the differences?

2.5 Explain how topography affects weather and climate.

2.6 What is the natural greenhouse effect? How has it been changed by human activities?

KEY TERMS insolation, reradiate, greenhouse effect, continental climate, maritime climate, polar jet stream, subtropical jet stream, monsoon wind, environmental lapse rate, orographic effect, adiabatic lapse rate, rain shadow, climate, climograph, climate change, greenhouse gases (GHGs), anthropogenic, Kyoto Protocol

Bioregions and Biodiversity: The Globalization of Nature

One aspect of Earth's uniqueness is the rich diversity of plants and animals covering its continents and oceans. This **biodiversity** can be thought of as the green glue that binds together geology, climate, hydrology, and life. Like climate regions, the world's biological resources can be broadly categorized into **bioregions**—areas defined by natural characteristics such as similar plant and animal life (see the bioregion photo essay, Figures 2.22 to 2.26).

Humans obviously are very much a part of this interaction. Not only are we evolutionary products of the African tropical savanna—a specific bioregion—but also our long human prehistory includes the domestication of plants and animals that led to modern agriculture and later our global food systems.

This human activity, however, combined with later urbanization and industrialization, has taken an immense toll on nature to the point where it is questionable whether natural vegetation bioregions truly exist anymore. Instead, cultivated fields have replaced grasslands and woodlands; forests have been logged for wood products (Figure 2.27); and wildlife has been hunted for food and fur or subjected to habitat destruction. As a result, Earth's natural world has become a distinctly human, globalized one, dominated by **novel ecosystems** that are completely new to Earth. Three pressing issues outlined below make biodiversity and bioregions an important part of world regional geography.

Figure 2.22 Bioregions of the World Although global vegetation has been greatly modified by clearing land for agriculture and settlements, as well as by cutting forests for lumber and paper pulp, there are still recognizable patterns to the world's bioregions, from tropical forests to arctic tundra. An important point is that each bioregion has its own unique array of ecosystems; also important is that these natural resources are used, abused, and conserved differently by humans.

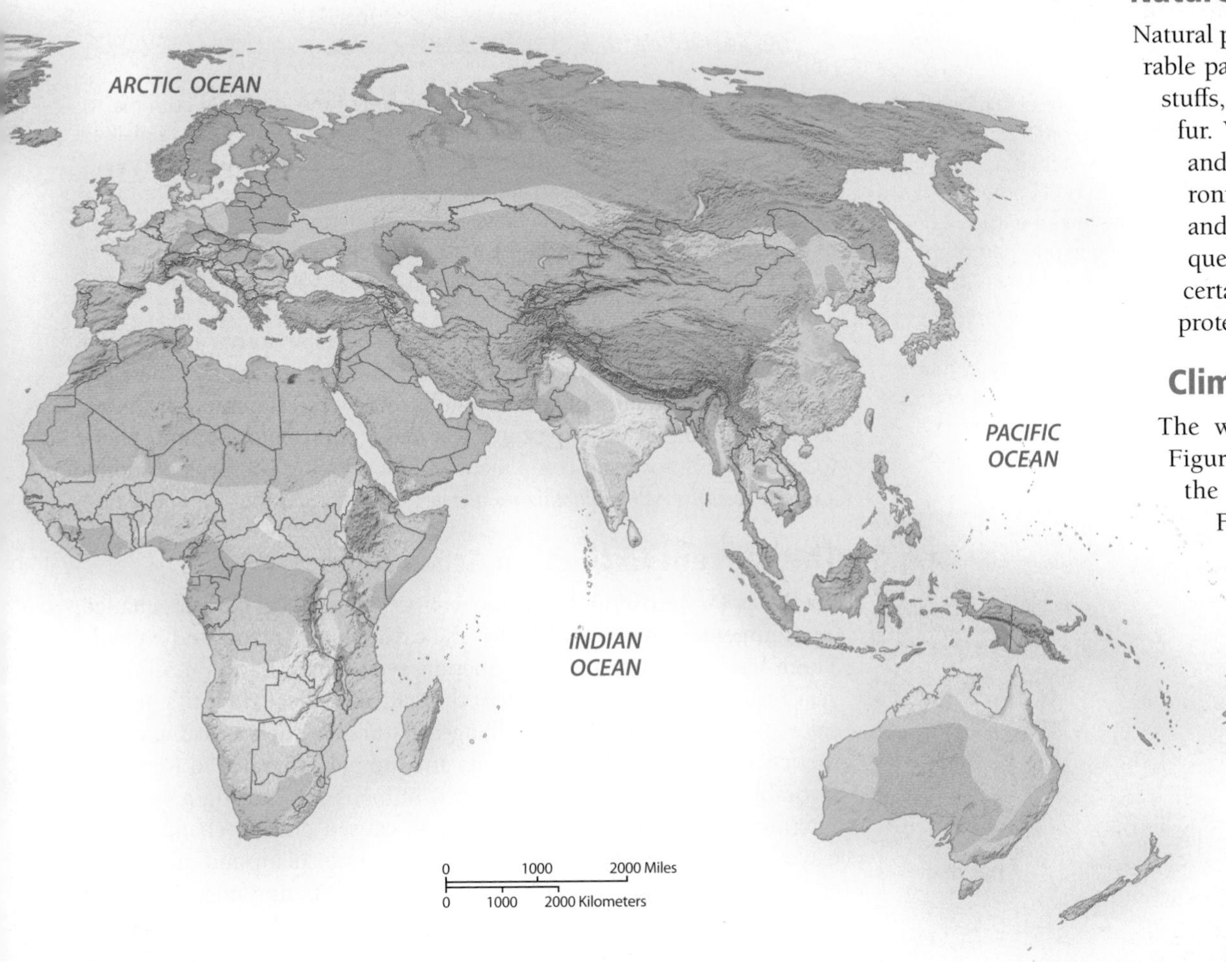

Nature and the World Economy

Natural plant and animal products are an inseparable part of the world economy, be they foodstuffs, wood products, or animal meat and fur. While most of this world trade is legal and appropriately regulated as to its environmental consequences, much is illegal and has detrimental environmental consequences. Examples are the illegal logging of certain rainforest trees and the poaching of protected animal species.

Climate Change and Nature

The world map of natural bioregions (see Figure 2.22) shows a close relationship with the map of global climate regions (see Figure 2.16) because temperature and precipitation—the two major components of climate—are also the two most important influences on flora and fauna. Today, however, global climate change is causing possibly irreversible changes in the world's bioregions because plants and animals cannot adapt to the rapid changes in climate. There is another important linkage as well: Vegetation takes up and stores carbon during growth and then releases CO_2 as it ages and dies. Long-lived forest trees, for example,

(a)

(b)

Figure 2.23 Tropical Rainforest and Savanna (a) The tropical rainforest consists of a rich ecosystem of plants and animals adapted to differing levels of sunlight through the multiple layers of vegetation. This ecosystem is adapted to heavy rainfall throughout the year. This tropical rainforest is in Borneo, Indonesia. (b) The tropical savanna ecosystem is less dense than the true tropical rainforest because it's adapted to less rainfall falling in only part of the year. As a result the vegetation is often sparse, with widely spaced trees in a grassland. This photo is from the famed Serengeti National Park in Tanzania.

Figure 2.24 Tropical Forest Destruction Humans have cleared away much of the tropical rainforest to create grasslands for cattle, plantation crops, and, on a smaller scale, for family subsistence farms, such as this one in southern Laos. Burning the rainforest not only clears an area for crops and pasture but also transfers nutrients from the vegetation to the soil.

(a)

(b)

Figure 2.25 Desert and Steppe (a) Poleward of the tropics, in both the Northern and Southern Hemispheres, are true desert areas of sparse rainfall. On the global climate region map these are the areas of the BW climate found in Asia (such as the Gobi Desert shown here), Africa, North America, South America, and Australia. (b) Lush grasslands characterize the steppe bioregion, as depicted here in Mongolia. These areas are found poleward of the true deserts where rainfall is commonly between 10 and 15 inches (254–381 mm).

are good storehouses of carbon, making responsible forest management an important part of atmospheric emission reduction plans. Therefore, the widespread practice of cutting and burning tropical rainforests to clear land for farming or cattle pastures is a major contributor of atmospheric CO_2 and a controversial component of a developing country's economic strategy (see *Everyday Globalization: The Rainforest and Your Chocolate Fix*).

The Current Extinction Crisis

Finally, not to be overlooked is the reality that as global climates change, environmental stresses will hasten the extinction of plants and animals. There have been five major extinction events over Earth's 4.5-billion-year history, all natural events that dramatically affected Earth's biological evolution. When climates changed naturally (such as at the end of the ice age 20,000 year ago), plants and animals migrated to find more favorable conditions. Today, however, humans are causing a sixth extinction event with more rapid, human-caused climate change; highways, cities, and farmland now form barriers to plant and animal migration, further aggravating habitat loss and extinction rates. Biologists estimate that as a result of habitat destruction and other changes we are losing several dozen species every day, resulting in an extinction rate that could see as much as 50 percent of Earth's species gone by 2050—reducing Earth's genetic resources, with potentially disastrous results.

(a)

(b)

Figure 2.26 Evergreen and Deciduous Forests (a) Evergreen trees usually have needles rather than leaves and keep their foliage throughout the year, which is why they're referred to as "evergreen". Most evergreens are cone-bearing, consist of softer interior wood, and thus are called "softwoods". This photo is of an evergreen forest in British Columbia, Canada. (b) Contrasting to evergreens are the deciduous trees that drop their leaves during the winter season, often after providing a colorful display (as in this photo from Maine) that results from the tree's slowing physiology as it prepares to hibernate.

EVERYDAY GLOBALIZATION

The Rainforest and Your Chocolate Fix

Your chocolate bar comes from the tropical rainforest, and satisfying your sweet tooth could be either destroying or saving the rainforest, depending on how the cocoa was grown. Cocoa, chocolate's main ingredient, comes from cacao trees, which grow exclusively in equatorial rainforests—mainly in Ghana and other African countries, but also in the Amazon Basin of South America. Cacao trees prefer the shade of higher rainforest trees, which is good news. But to meet the ever-increasing demand for chocolate, cacao is also cultivated for short periods of time in the full sunlight of newly cleared rainforest plots. That's the bad news—because this method of cacao farming is a major factor in the destruction of African rainforests.

So what's a rainforest-loving chocolate lover to do? Easy: Take an extra 30 seconds and read the candy bar label to see whether there's any mention of shade-grown and/or sustainably farmed cacao trees. After that, it's up to you.

1. Identify other foods you eat that come from tropical rainforests, and describe how their cultivation affects the forests.
2. What are the different ways you eat chocolate, and where is that cocoa grown?

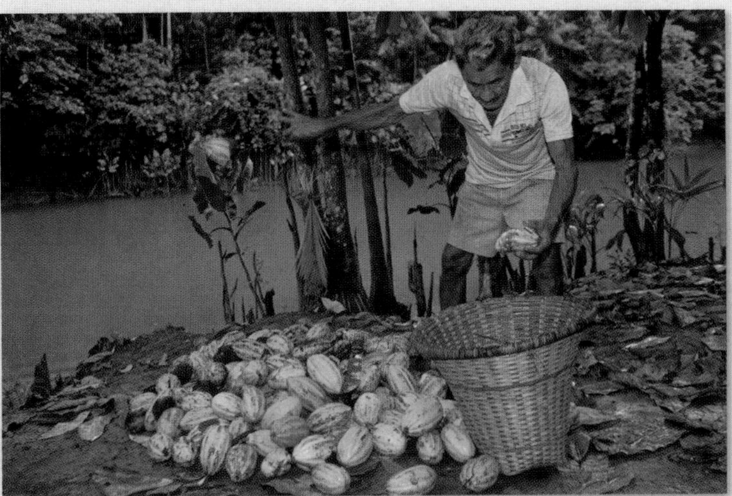

Figure 2.3.1 Sustainable Cocoa This farmer sorts cocoa pods harvested on a sustainable plantation in Brazil's Amazon region.

Review

2.7 Use a map to locate and describe the bioregions found in the world's tropical climates.

2.8 Describe the different kinds of forests found in the midlatitudes.

2.9 What is the difference between an evergreen and a deciduous forest?

KEY TERMS biodiversity, bioregion

Figure 2.27 Clear Cut Forests The evergreen, softwood forests of the Northern Hemisphere are a major source of the world's lumber and paper pulp needs, and are often clear cut, as shown in this aerial photo from British Columbia, Canada, for the efficiency of harvesting as well as to facilitate replanting of nursery species. These forests, however, also store vast amounts of CO_2, much of which is released back into the atmosphere during cutting and milling. Further, because the visual impact of clear-cutting are often unacceptable, the protection (or harvesting) of these northern forests are the foci of numerous environmental controversies.

Water: A Scarce World Resource

Water is central to all life, yet it is unevenly distributed around the world—plentiful in some areas, while distressingly scarce in others. As a result, around 1 billion people lack access to safe and reliable water sources. Water issues are not due simply to the distribution of diverse global climates that produce wet or dry conditions; they are also caused by a range of complex socioeconomic factors at all scales, local to global.

At first glance, Earth is indeed the water planet, with more than 70 percent of its surface area covered by oceans. But 97 percent of the total global water budget is saltwater, with only 3 percent freshwater. Of that small amount of freshwater, almost 70 percent is locked up in polar ice caps and mountain glaciers. Additionally, groundwater accounts for almost 30 percent of the world's freshwater. This leaves less than 1 percent of the world's water in more accessible surface rivers and lakes.

Another way to conceptualize this limited amount of freshwater is to think of the total global water supply as 100 liters, or 26 gallons. Of that amount, only 3 liters (0.8 gallon) is freshwater; and of that small supply, a mere 0.003 liter, or only about half a teaspoon, is readily available to humans.

Water planners use the concept of **water stress** and scarcity to map where water problems exist and also to predict where future problems will occur (Figure 2.28). These water stress data are generated by calculating the amount of freshwater available in relation to current and future population needs. Northern Africa stands out as the region of highest water stress; hydrologists predict that three-quarters of Africa's population will experience water shortages by 2025. Other problem areas are China, India, much of Southwest Asia, and even several countries in humid Europe. Although climate change may actually increase rainfall in some parts of the world, scientists forecast that this will in general aggravate global water problems because of higher evaporation rates and increased water usage due to warmer temperatures.

Water Sanitation

Where clean water is not available, people use polluted water for their daily needs, resulting in a high rate of sickness and even death. More specifically, the United Nations reports that over half of the world's hospital beds are occupied by people suffering from illnesses linked to contaminated water. Further, more people die each year from polluted water than are killed in all forms of violence, including wars. This toll from polluted water is particularly high for infants and children, who have not yet developed any sort of resistance to or tolerance for contaminated water. The United Nations Children's Fund (UNICEF) reports that nearly 4000 children die each day from unsafe water and lack of basic sanitation facilities.

Water Access

By definition, when a resource is scarce, access is problematic, and these hardships take many forms in the case of water. Women and children, for example, often bear the daily burden of providing water for family use, and this can mean walking long distances to pumps and wells and then waiting in long lines to draw water (see *Working Toward Sustainability: Women and Water in the Developing World*). Given the amount of human labor involved in providing water for crops, it is not surprising that some studies have shown that in certain areas people expend as many calories of energy irrigating their crops as they gain from the food itself.

Ironically, some recent international efforts to increase people's access to clean water have gone astray and have actually aggravated access problems instead. Historically, domestic water supplies have been public resources, organized and regulated—either informally by common consent or more formally as public utilities—resulting in free or low-cost water. In recent decades, however, the World Bank and the International Monetary Fund have promoted the privatization of water systems as a condition for providing loans and economic aid to developing countries. The agencies' goals have been laudable, trying to ensure that water is clean and healthful. However, the means have been controversial because typically the international engineering

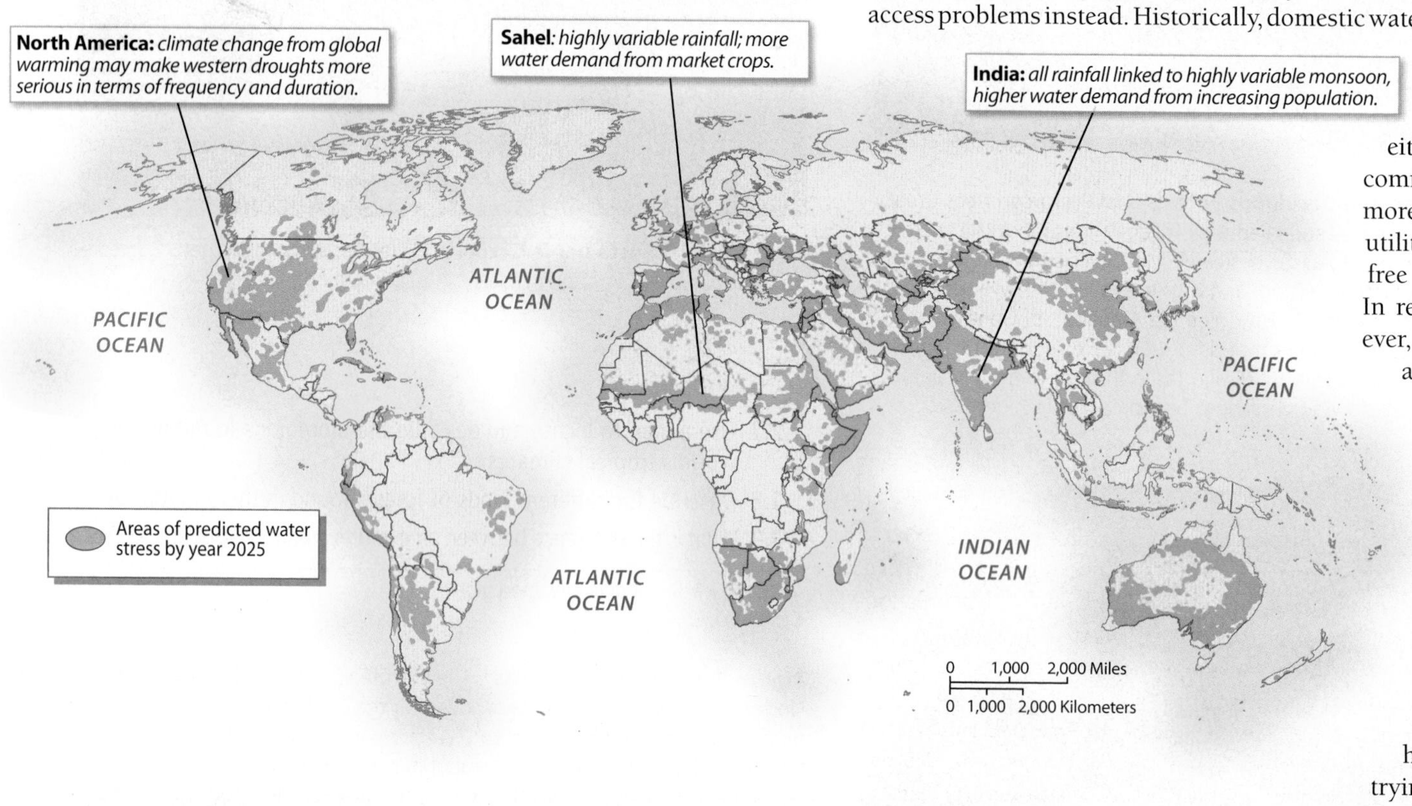

Figure 2.28 Global Water Stress This world map shows where water planners forecast there will be serious water problems in the year 2025. Although drought and highly variable rainfall regimes are major causes of water stress, there can be other socioeconomic factors that limit supplies and access to water. That said, many of the water stress areas on this map, such as western North America, the Mediterranean, the Sahel of Africa, and South Asia are linked to reduced or unreliable rainfall resulting from global warming and climate change.

WORKING TOWARD SUSTAINABILITY

Women and Water in the Developing World

Google Earth Tour (MG)
Sahel region Africa
https://goo.gl/oz3E31

Women and children bear the burden of water problems in most developing countries. Not only are children the most vulnerable to waterborne diseases, but also adult females (mothers, aunts, grandmothers, and older siblings) are the major caregivers for these sick children, adding yet another time-consuming task to their already busy days.

Further, women and older girls are the primary conveyers of water from wells or streams to their village homes. Every person requires about 5 gallons (18 liters) of water per day for their hydration, cooking, and sanitation needs; consequently, this amount (multiplied by the number of people in a family) must be carried each day from source to residence. In addition, women and children are responsible for supplying water for kitchen gardens that provide the family's food. At a global level, the water source for about a third of the developing world's rural population is more than half a mile (1 km) away from residences. To meet water needs, women spend about 25 percent of their day carrying water. A recent United Nations study estimated that in Sub-Saharan Africa about 40 billion hours a year are spent collecting and carrying water, the same amount of time spent in 1 year by France's entire workforce.

Besides the time expenditure, water is heavy, and most of it is carried by hand. In Africa, 40-pound (151-liter) jerry cans are common; in northwest India, women and girls balance several 5-gallon (19-liter) containers on their heads to lessen the number of trips made (Figure 2.4.1). (Note that 40 pounds is about the weight of the suitcase you check with the airlines on a typical trip. Try carrying it on your head through the airport parking lot someday.) After years of carrying water, women commonly suffer from chronic neck and back problems, many of which complicate childbirth. Additionally, girls' water-carrying responsibilities often interfere with their schooling, resulting in a high dropout rate and furthering female illiteracy in rural villages.

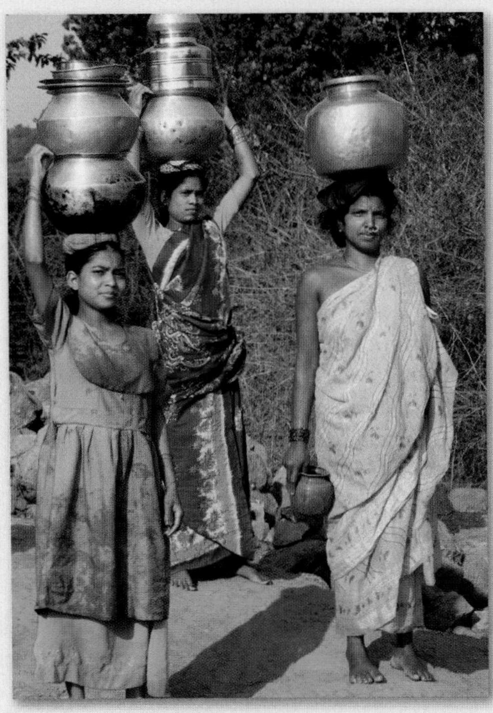

Figure 2.4.1 Women in India carrying water on their heads

Toward a Solution: The Wello WaterWheel

After studying the water-carrying issue in semiarid northwestern India, Cynthia Koenig, a recent engineering graduate from the University of Michigan, invented the Wello WaterWheel, a barrel-like 13-gallon (50-liter) rolling water container that greatly reduces women's water-carrying duties (Figure 2.4.2).

Figure 2.4.2 Woman using Wello WaterWheel

Previously in that part of India, women and girls were spending 42 hours per week carrying water back and forth; with the Wello WaterWheel, that has been reduced to only 7 hours a week. Using this time-saving device has also reduced the school dropout rate for young girls in the region. Currently, Wello, which is a nonprofit organization, can deliver a WaterWheel from its factory in Mumbai to a rural Indian family for a mere $20. In the last year, thousands of Wello WaterWheels have been purchased by international aid organizations and donated to villages in Rajasthan, moving them closer to a sustainable existence.

1. List the social costs incurred when the responsibility for providing water falls to the women and children of a village.
2. List the probable social benefits to a village where clean water is readily available instead of requiring transport over long distances by women and children.

firms that have upgraded rudimentary water systems by installing modern water treatment and delivery technology have increased the costs of water delivery to recoup their investment. Although the people may now have access to cleaner and more reliable water, in many cases the price is higher than they can afford, forcing them to either do without or go to other, unreliable and polluted sources.

In Cochabamba, Bolivia, for example, the privatization of the water system 15 years ago resulted in a 35 percent average increase in water costs. The people responded with demonstrations that became tragically violent. Eventually, the water system was returned to public control. Today, however, 40 percent the city's population is still without a reliable water source.

Review

2.10 How much water is there on Earth, and how much is available for human usage? Use in your answer the concept that Earth's water budget is just 100 liters.

2.11 Describe the three major issues that cause water stress.

2.12 Where in the world are the areas of the most severe water stress?

KEY TERMS water stress

Global Energy: The Essential Resource

The world runs on energy. While sunlight provides the natural world with its driving force, providing energy for the modern human world is much more complicated because of the uneven distribution of energy resources, the complex technologies required, and, not least, the economic, environmental, and geopolitical dynamics involved in finding, exploiting, and transporting these essential resources (Figure 2.29).

Nonrenewable and Renewable Energy

Energy resources are commonly categorized as either nonrenewable or renewable. **Nonrenewable energy** is consumed at a higher rate than it is replenished, which is the case with the world's oil, coal, uranium, and natural gas resources—fossil fuels that took millennia to create. In contrast, **renewable energy** depends on natural processes that are constantly renewed—namely, water, wind, and solar energy. Currently, 91 percent of the world is powered by nonrenewable energy. While renewable energy powers the remaining 9 percent, this energy sector is increasing rapidly because of the world's concerns about reducing atmospheric emissions. Oil, coal, and natural gas are considered "dirty" fuels because of their carbon content, whereas renewable energy is generally "clean," with no GHG emissions resulting from its usage. Among the dirty fuels, coal adds the most harmful emissions, with natural gas emitting 60 percent less CO_2 than coal.

At a global level, oil and coal are the major fuels, currently making up 36 and 33 percent of all fossil fuel usage, respectively; natural gas trails at 26 percent, with nuclear far behind at 5 percent. Recent trends show an increase in natural gas at coal's expense because of environmental concerns; however, with the recent price decrease for North American coal, some countries have actually increased their coal consumption.

Fossil Fuel Reserves, Production, and Consumption

As discussed at the start of this chapter, plate tectonics and other complex geologic forces have shaped ancient landscapes over millions of years. As a result, fossil fuels are not evenly distributed around the world, but instead are clustered into specific geologic formations in specific locations, resulting in a complicated international pattern of supply and demand. Table A2.1 in the Appendix shows the varied world geography of energy reserves, production, and consumption—where the resources lie, who is mining and producing them, and which countries are primary consumers of coal, oil, and gas. More details on this global energy geography appear in each of the regional chapters of this text.

Because of the high degree of technological difficulty involved in mining fossil fuels, the energy industry uses the concept of **proven reserves** of oil, coal, and gas to refer to deposits that are possible to mine and distribute under current economic and technological conditions. An important component of this definition is "current," since economic, regulatory, and technological conditions change rapidly; as they do, this expands or contracts the amount of a resource deemed feasible to mine. If, for example, the price of oil on the global market is high, then oil reserves with relatively high drilling costs (such as in the Arctic) can be produced at a profit. Conversely, if the market price of oil is low, then only easily accessible reserves qualify as "proven."

Not only do the market prices for fossil fuels change over time, but also drilling and mining technologies become more efficient, changing the amount of proven reserves. A good example is the recent development of **hydraulic fracturing** or **fracking**, as it is called, a relatively new oil and gas mining technique that forces oil and gas out of shale rock (Figure 2.30). Currently, fracking in North America has increased the world's oil and gas supplies to the point where traditional global oil suppliers (Russia, Saudi Arabia, and Venezuela) are threatened. In early 2015, the response of these traditional oil countries was to let the price of oil fall in the hope of bankrupting and closing down North American oil shale operations. By late 2015, however, that had not happened, with oil prices remaining low.

Figure 2.29 North American Oil Transport Hydraulic fracturing ("fracking") has produced a North American oil glut that benefits consumers with low fuel prices but has created transportation difficulties in moving oil safely from inland production areas to refineries located primarily in coastal areas, a locational artifact from when the United States imported most of its oil. Although pipelines carry much North American oil, railroads are moving increasing amounts, resulting in railroad traffic jams, oil spills, and noise and traffic problems for railroad towns. The oil train shown in this photo is creeping through the suburbs of Kansas City, Missouri.

Renewable Energy

As mentioned earlier, renewable energy sources can be naturally replenished at a faster rate than their energy is consumed. Wind and solar power are prime examples. The list of renewable energy sources includes not just wind and solar power, but also hydroelectric, geothermal (from Earth's interior heating), tidal currents, and biofuels, which use the carbon in plants as their power source.

Nevertheless, despite the widespread availability of sunlight, wind, and biofuels, renewable energy provides only a fraction of the world's power, at 9 percent. Some countries, however, make considerable use of renewables in their power grid. Iceland, for example, is blessed with bountiful supplies of both water and thermal resources and generates fully 95 percent of its power usage from renewable sources.

Of the large industrial economies, Germany leads the way with over 20 percent of its power coming from renewables, primarily through extensive wind and solar power stations (Figure 2.31). China, the world's largest consumer of energy, reportedly drives a quarter of its massive economy with renewable energy. Unlike Germany, most of China's renewable energy comes from hydropower, although the country is rapidly expanding its wind and solar facilities.

Despite the attractiveness of wind and solar power, significant issues must be resolved before their usage can expand. Both wind and solar, for example, are intermittent sources of power, since the Sun does not always shine or the wind blow. To compensate for these lulls, commercial wind and solar facilities are usually mandated to have gas-fired generators as backups to produce power at a moment's notice.

The intermittent nature of large-scale wind and solar generation also requires that national power grids perform a tricky balancing act between energy supply and demand. Power surges generated by sunny and windy periods must somehow be assimilated into a power grid built long ago under the assumption of steady and continual inputs of power from gas- or coal-fired power plants.

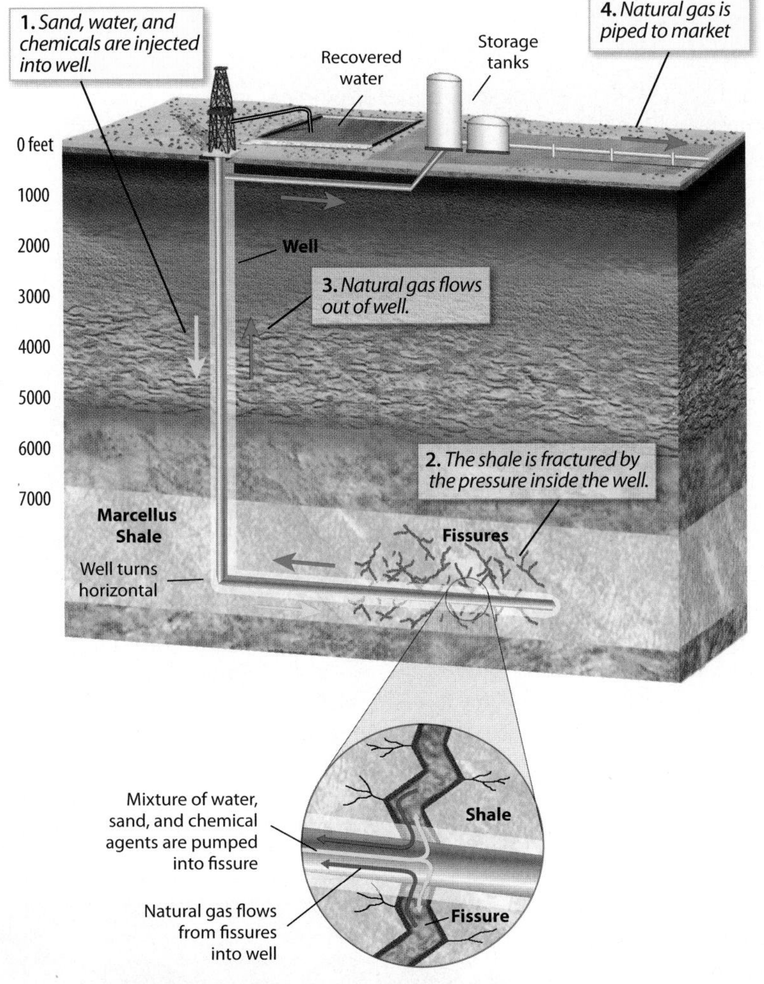

Figure 2.30 Hydraulic Fracturing This graphic illustrates the different components of hydraulic fracturing (fracking) for oil and gas. What it does not show are the environmental issues associated with fracking, which include the high amount of freshwater used, the possibility of polluting local ground water, problems in disposing of waste water, local noise and nuisance issues, and the possibility fracking could cause more earthquakes in certain geologic structures.

Finally, renewable power can be more costly to develop and implement than fossil fuels because renewables lack the same degree of economic subsidies and tax incentives as are enjoyed by oil, coal, gas, and nuclear power. At a global level, it is estimated that traditional fossil fuels receive six times the financial support through governmental incentives that renewables do. When new technologies serve the public interest, they usually receive considerable economic support from national governments, and while this has happened in several notable cases (China and Germany), these subsidies need to become more widespread before renewable energy can compete on a level playing field with fossil fuels.

Figure 2.31 Germany's Renewable Energy Program Germany is on course to become the world's first large industrial economy to be powered primarily by renewable energy. The first step, taken almost a decade ago, was to subsidize its solar industry by offering attractive incentives to households that converted to solar power. The second stage has been to expand the country's wind power sector, a technology that currently supplies most of the country's renewable energy. This photo shows how German homeowners have used the steep roofs of the country's traditional architecture to maximize solar panel placement.

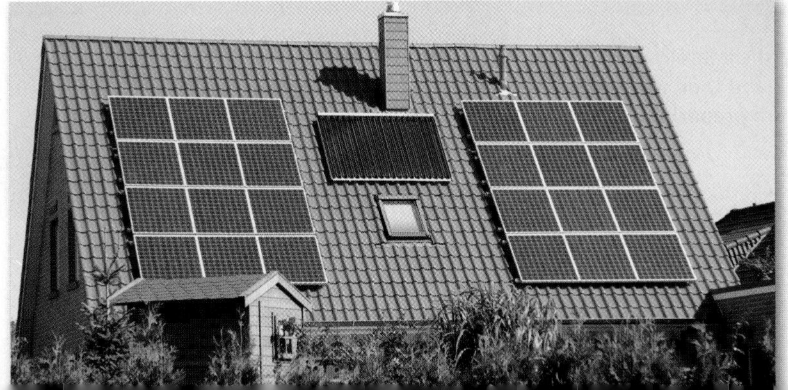

Figure 2.32 Small Solar in Africa Many villages in Africa are moving to small-scale solar systems to generate energy that was formerly produced by gas or diesel generators. Not only is solar less expensive than fossil fuels, but it's more adaptable to specific needs, as is shown here where a women has a small business charging cell phones from a single solar panel.

Energy Futures

Global energy demand is forecast to increase 40 percent by 2030 as the large developing economies of China, India, and Brazil industrialize further. Not to be overlooked is the challenge of providing power to the 25 percent of the world's population that currently lacks access to an energy grid (Figure 2.32). In contrast, energy demand in the developed world (the United States, Europe, Japan) is expected to increase at a much slower rate than in the developing world, and perhaps even to level off, because of technological advancements in energy efficiency.

Forecasts assume that fossil fuel reserves are adequate to meet future energy demands, but the expansion of renewable technologies may actually decrease the demand for coal, oil, and natural gas. Also important is whether unconventional drilling techniques, primarily hydraulic fracturing, will continue to expand. In the United States, which has extensive deposits of the shale oil and gas ripe for fracking, this means first finding solutions to the many environmental issues associated with this new technology. If this is done (and opinions vary on whether these environmental issues can be resolved satisfactorily), the United States might find itself in the enviable position of satisfying most of its energy needs with domestic oil and gas instead of relying on its traditional import partners—Mexico, Canada, Saudi Arabia, and Venezuela.

China and Russia also have extensive oil and gas shales that could be exploited by fracking, and if they, too, are successful, this could bring further changes to the current global energy geography.

Review

2.13 What is meant by fossil fuel proven reserves?

2.14 List the three countries using the most oil, coal, and natural gas.

2.15 What are some of the problems associated with renewable energy?

2.16 What is fracking? How might fracking change the world's energy geography?

KEY TERMS nonrenewable energy, renewable energy, proven reserves, hydraulic fracturing (fracking)

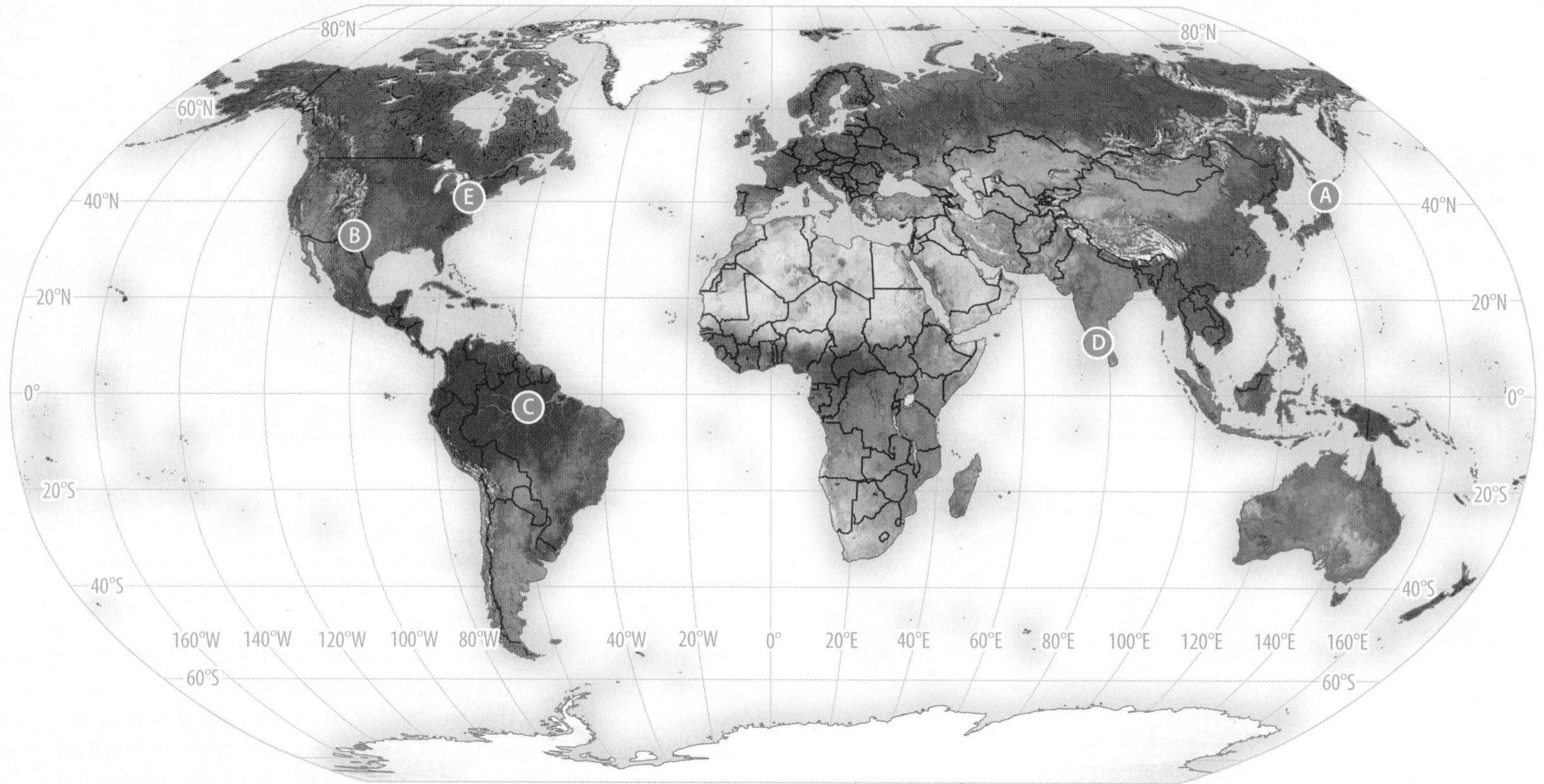

Review

Geology: A Restless Earth

2.1 Describe those aspects of tectonic plate theory responsible for shaping Earth's surface.

2.2 Identify on a map those parts of the world where earthquakes and volcanoes are hazardous to human settlement.

The arrangement of tectonic plates on Earth is responsible for diverse global landscapes. The motion of these plates also causes earthquake and volcanic hazards that threaten the safety of millions of people, particularly in the large cities of North America, South America, and Asia.

1. Compile a list of cities with populations larger than half a million that are located near convergent plate boundaries, making them vulnerable to damage from a major earthquake. Then note those cities near subduction zones where volcanic explosions and tsunamis could be a hazard.
2. Choose one of those cities, and then go on the Internet to gather information about how that city is reducing its earthquake or volcanic hazard vulnerability in terms of planning and disaster preparation.

Global Climates: Adapting to Change

2.3 List and explain the factors that control the world's weather and climate, and use these to describe the world's major climate regions.

2.4 Define the greenhouse effect and explain how it is related to anthropogenic climate change.

2.5 Summarize the major issues underlying international efforts to address climate change.

Current climate change results from additional greenhouse gases (GHGs) emitted into the atmosphere and is a byproduct of human activities, both past and present. Historically, the more developed countries of Europe and North America were the major GHG producers; today, however, the developing economies of China and India have become the major polluters.

3. How does the 2015 Paris Agreement on reducing global greenhouse emissions differ from the earlier Kyoto Protocol? Give some specific examples.
4. Compile a list of impacts of climate change in different parts of the world, and then investigate how people in those areas are preparing for and adapting to those threats.

Bioregions and Biodiversity: The Globalization of Nature

2.6 Locate on a map and describe the characteristics of the world's major bioregions.

2.7 Name some threats to Earth's biodiversity.

Plants and animals throughout the world face an extinction crisis because of habitat destruction from a variety of human activities. Tropical forests are a focus of these problems because they contain the more plant and animal species of any bioregion and are capable of both storing and emitting large amounts of carbon dioxide

5. Describe the different activities that lead to biodiversity loss. Provide examples from specific countries.

6. Discuss how bioregions and climate regions are interconnected. Provide specific examples.

C

Water: A Scarce World Resource

2.8 Identify the causes of global water stress.

Water, a necessity for all life, is becoming an increasingly scarce resource in the world and will cause serious water stress problems in Sub-Saharan Africa, Southwest Asia, and western North America.

7. After reviewing the world map of water stress, discuss the different kinds of water problems found in different climate regions.

8. Consult the Internet to become acquainted with the controversy over large dams in South Asia by listing their benefits and liabilities.

D

Global Energy: The Essential Resource

2.9 Describe the world geography of fossil fuel production and consumption.

2.10 List the advantages and disadvantages of the different kinds of renewable energy.

Fossil fuels (oil, coal, and natural gas) dominate the world's energy picture, with renewable energy (wind, solar, water, and biomass) currently providing only a fraction of the world's energy needs. While energy demand has leveled off (and even decreased) in more developed countries, it continues to rise in China, India, and other developing economies.

9. In what parts of the world are the possibilities for wind and solar energy highest? Besides the environmental conditions for renewable energy, also consider the different kinds of energy usage in those areas.

10. Currently, the United States relies on vast amounts of imported oil to meets its energy needs. Although some say that North American fracking could end this dependence on foreign oil, others disagree. Acquaint yourself with this debate, and then make a case for one side or the other.

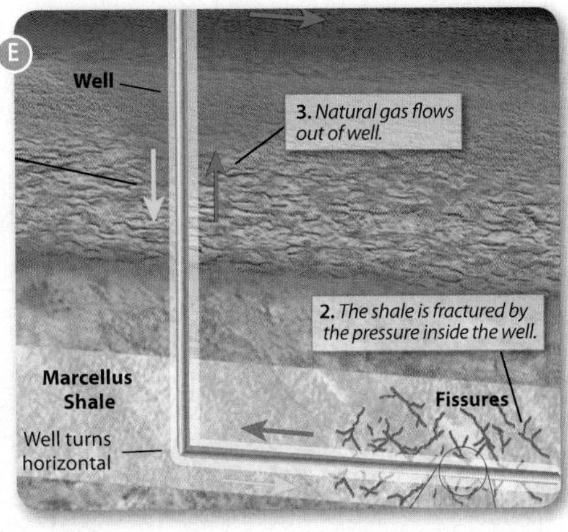

KEY TERMS

adiabatic lapse rate *(p. 51)*
anthropogenic *(p. 53)*
biodiversity *(p. 57)*
bioregion *(p. 57)*
climate *(p. 52)*
climate change *(p. 52)*
climograph *(p. 52)*
continental climate *(p. 50)*
convergent plate boundary *(p. 44)*
divergent plate boundary *(p. 44)*
environmental lapse rate *(p. 51)*
greenhouse effect *(p. 48)*
greenhouse gases (GHGs) *(p. 53)*
hydraulic fracturing (fracking) *(p. 62)*
insolation *(p. 48)*
Kyoto Protocol *(p. 54)*
lithosphere *(p. 44)*
maritime climate *(p. 50)*
monsoon wind *(p. 50)*
nonrenewable energy *(p. 62)*
novel ecosystems *(p. 57)*
orographic effect *(p. 51)*
Pangaea *(p. 47)*
plate tectonics *(p. 44)*
polar jet stream *(p. 50)*
proven reserves *(p. 62)*
rain shadow *(p. 51)*
renewable energy *(p. 62)*
reradiate *(p. 48)*
rift valley *(p. 46)*
subduction zone *(p. 44)*
subtropical jet stream *(p. 50)*
transform plate boundary *(p. 44)*
transform fault *(p. 47)*
water stress *(p. 60)*

DATA ANALYSIS

http://goo.gl/Bl3Ox

Emissions per capita are a useful measure of each country's contribution to climate change. Compare a rich country where people drive gas-guzzling SUVs and live in huge houses requiring lots of energy for heating and cooling with a not-so-rich country where people use public transportation and live in modest energy-efficient apartments. If the two countries have the same population, then clearly the first country would have higher per capita emissions, right? Reality, however, is not so simple. Go to the World Bank website (http://data.worldbank.org) and access data on CO_2 emissions (metric tons per capita). List countries emitting less than .5 ton, and those emitting over 15 tons per person. Rank the countries in both categories from lowest to highest.

1. Do rich countries emit more CO_2 than less-wealthy countries? (Use GNI per capita from tables in the appendix of this text as a measure of wealth for each country.)
2. Does a country's energy mix—renewables vs. fossil fuels—influence emissions per capita? Do countries with abundant fossil fuel resources produce more emissions than those importing fuel? (You can find a country's energy mix and resources in The World Factbook, https://www.cia.gov.)
3. Does climate make a difference? Do countries in colder climates use more energy and emit more CO_2 than those in warmer climates?
4. Write a paragraph on why per capita emissions differ so much. Then browse the Internet for articles on CO_2 emissions per capita, and refine your explanations.

MasteringGeography™

Looking for additional review and test prep materials? Visit the Study Area in MasteringGeography™ to enhance your geographic literacy, spatial reasoning skills, and understanding of this chapter's content by accessing a variety of resources, including MapMaster interactive maps, geoscience animations, videos, *In the News RSS* feeds, flashcards, web links, self-study quizzes, and an eText version of *Globalization and Diversity.*

Authors' Blogs

Scan to visit the **Author's Blog for field notes, media resources, and chapter updates**

https://gad4blog.wordpress.com/category/global-environment/

Scan to visit the **GeoCurrents Blog**

http://www.geocurrents.info/category/physical-geography

3 North America

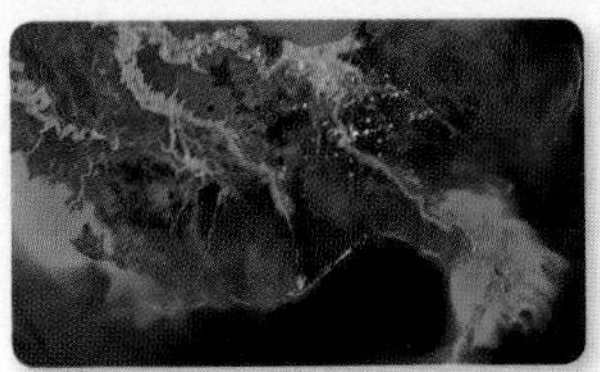

PHYSICAL GEOGRAPHY AND ENVIRONMENTAL ISSUES

Stretching from Texas to the Yukon, the North American region is home to an enormously varied natural setting and to an environment that has been extensively modified by human settlement and economic development.

POPULATION AND SETTLEMENT

Settlement patterns in North American cities reflect the diverse needs of an affluent, highly mobile population. Sprawling suburbs are designed around automobile travel and mass consumption, while many traditional city centers struggle to redefine their role within the decentralized metropolis.

CULTURAL COHERENCE AND DIVERSITY

Cultural pluralism remains strong in North America. Currently, more than 47 million immigrants live in the region, more than double the total in 1990. The tremendous growth in Hispanic and Asian immigrants since 1970 has fundamentally reshaped the region's cultural geography.

GEOPOLITICAL FRAMEWORK

Cultural pluralism continues to shape political geographies in the region. Immigration policy remains hotly contested in the United States, and Canadians confront persistent regional and native peoples' rights issues.

ECONOMIC AND SOCIAL DEVELOPMENT

North America's economy recovered in many settings after the harsh economic downturn between 2007 and 2010. Still, persisting poverty and many social issues related to gender equity, aging, and health care challenge the region today.

◀ Toronto's large and vibrant Chinatown caters to a sizable immigrant population as well as to a variety of visitors from around the globe.

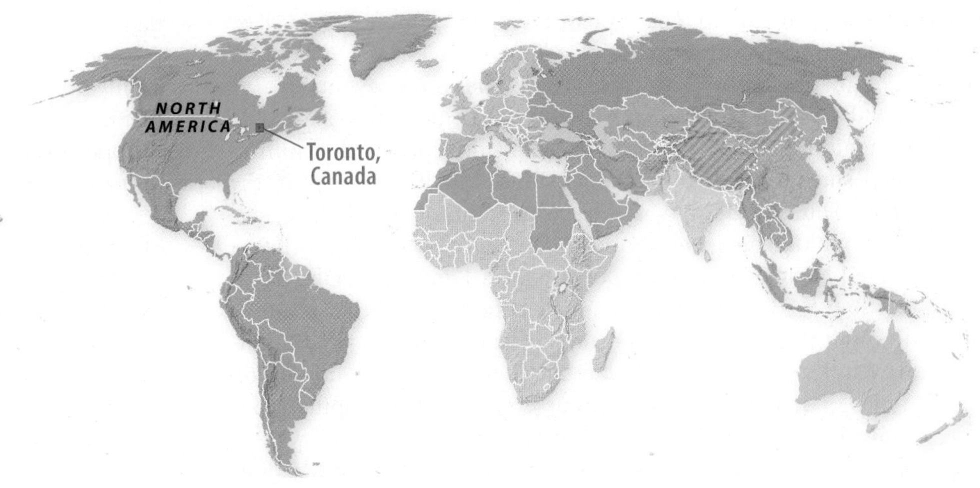

Walk the streets of Toronto and you will hear a veritable symphony of languages. Some experts call Canada's largest metropolis (5.8 million) the "most multicultural city in the world." More than 45 percent of Toronto's residents are foreign-born. Most newer immigrants are from diverse Asian countries, but South Asians, Chinese, and Filipinos often mingle with Latinos, Portuguese, and Italians. The result is a quintessentially globalized North American city that reflects the region's enduring pull on the world's immigrants.

North America, a wealthy region with two highly urbanized, mobile populations, helps drive the processes of globalization.

Toronto's economic success has been a powerful draw. Many immigrants are young, savvy, skilled, and upwardly mobile. Some choose to live near downtown, preferring the high-rise condo lifestyle adjacent to Lake Ontario. Older, traditional inner-city communities such as Chinatown (just west of downtown) also remain vibrant, while other immigrants have found newer homes on the northeast and northwest fringes of the city.

Globalization has fundamentally refashioned Toronto as well as many other portions of North America. Large foreign-born populations are found in many North American settings. Tourism brings in millions of additional foreign visitors and billions of dollars, which are spent everywhere from Las Vegas to Disney World. In more subtle ways, North Americans see globalization in their everyday lives. They eat ethnic foods, enjoy the sounds of salsa and Senegalese music, and surf the Internet from one continent to the next.

Learning Objectives *After reading this chapter you should be able to:*

3.1 Describe North America's major landform and climate regions.

3.2 Identify key environmental issues facing North Americans and connect these to the region's resource base and economic development.

3.3 Analyze map data to identify and trace major migration flows in North American history.

3.4 Explain the processes that shape contemporary urban and rural settlement patterns.

3.5 List the five phases of immigration shaping North America, and describe the recent importance of Hispanic and Asian immigration.

3.6 Provide examples of major cultural homelands (rural) and ethnic neighborhoods (urban) within North America.

3.7 Contrast the development of the distinctive federal political systems in the United States and Canada, and identify each nation's current political challenges.

3.8 Discuss the role of key location factors in explaining why economic activities are located where they are in North America.

3.9 List and explain contemporary social issues that challenge North Americans in the 21st century.

Globalization is a two-way street, and North American capital, culture, and power are ubiquitous. By any measure of multinational corporate investment and global trade, the region's influence far outweighs its population of 355 million residents. North American automobiles, consumer goods, information technology, and investment capital circle the globe along with its music, movies, sports, and fashion.

North America is a culturally diverse and resource-rich region that has seen tremendous human modification of its landscape and extraordinary economic development over the past two centuries (Figure 3.1). As a result, North America is one of the world's wealthiest regions, with two highly urbanized, mobile populations helping drive the processes of globalization and consuming resources at the highest rates on Earth. Indeed, the region exemplifies a **postindustrial economy** shaped by modern technology, innovative information services, and a popular culture dominating both North America and the world beyond.

Politically, North America is home to the United States, arguably the last remaining global superpower. Such status brings the country onto center stage in times of global tensions, whether they are in the Middle East, South Asia, or West Africa. In addition, North America's largest metropolitan area, New York City (22 million people), is home to the United Nations and other global political and financial institutions. North of the United States, Canada is the region's other political unit. Although slightly larger in area than the United States (3.83 million square miles [9.97 million square kilometers] versus 3.68 million square miles [9.36 million square kilometers]), Canada's population is only about 11 percent of that of the United States.

The United States and Canada are commonly referred to as "North America," but that regional terminology can be confusing. As a physical feature, the North American continent includes Mexico, Central America, and often the Caribbean. Culturally, however, the U.S.–Mexico border seems a better dividing line, although the growing Hispanic presence in the southwestern United States, as well as ever-closer economic links across the border, makes even that regional division problematic. In addition, while Hawaii is a part of the United States (and included in Chapter 3), it is also considered a part of Oceania (and discussed in Chapter 14). Finally, Greenland (population 56,000), which often appears on the North American map, is actually an autonomous territory within the Kingdom of Denmark and is mainly known for its valuable, but diminishing, ice cap.

Contemporary North America displays both the bounty and the price of economic development. On one hand, the region shares the benefits of modern agriculture, globally competitive industries, excellent transportation and communications infrastructure, and two of the most highly urbanized societies in the world (Figure 3.2). The cost of development, however, is high: Native populations were all but eliminated by European settlers, and forests have been logged, valuable soils eroded, numerous species extinguished, and natural resources often wasted.

Nevertheless, economic growth has vastly improved the standard of living for many North Americans, who enjoy high rates of consumption and varied amenities that are the envy of the less developed world. Amid this material abundance, however, are continuing differences in income and quality of life. Poor rural and inner-city populations struggle to match the affluence of wealthier neighbors.

North America's cultural fabric also defines the region, including common processes of colonization, a heritage of Anglo dominance, and a shared belief in representative democracy and individual freedom. But the region's history has also juxtaposed Native Americans and a global assortment of immigrants in fresh ways, resulting in two societies unlike any others.

Figure 3.1 North America The North American region plays a key role in globalization. It also contains two of the world's most highly urbanized and culturally diverse populations. With 355 million people and extensive economic development, North America is also one of the largest consumers of natural resources on the planet.

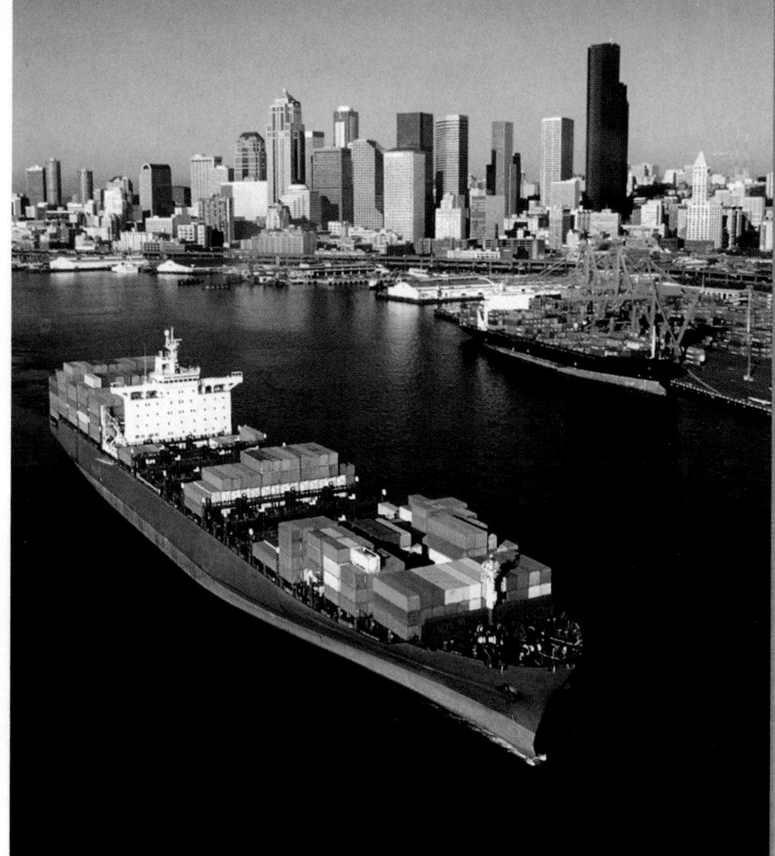

Figure 3.2 Container Shipping, Port of Seattle Major ports such as Seattle are key links in facilitating global trade and economic connections between North America and the rest of the world. Standard-sized container-shipping modules can easily be stored, stacked, and moved in such settings.

Physical Geography and Environmental Issues: A Vulnerable Land of Plenty

North America's physical and human geographies are enormously diverse. In the past decade, the region has also witnessed a dizzying array of natural disasters and environmental hazards that suggest the close connections between the region's complex physical setting and its human population. In 2012, for example, much of the East Coast was slammed by Hurricane Sandy (Figure 3.3). The storm rearranged the coastline of New Jersey, flooded lower Manhattan, and inflicted $50–$60 billion in damage. In 2014 and 2015 much of California experienced its most extreme drought in memory. Farmers were hit hard, many municipalities had to restrict water consumption, and the state's river systems experienced severe water shortages. These events remind us how the costs and impacts of both "natural" and "human" environmental disasters are inevitably intertwined with a region's broader cultural, social, and economic characteristics.

A Diverse Physical Setting

The North American landscape is dominated by interior lowlands bordered by more mountainous topography in the western part of the region (see Figure 3.1). In the eastern United States, extensive coastal plains stretch from southern New York to Texas and include a sizable portion of the lower Mississippi Valley. The Atlantic coastline is complex, made up of drowned river valleys, bays, swamps, and low barrier islands. The nearby Piedmont region, a transition zone between nearly flat lowlands and steep mountain slopes, consists of rolling hills and low mountains that are older and less easily eroded than the lowlands. West and north of the Piedmont are the Appalachian Highlands, an internally complex zone of higher and rougher country reaching altitudes of 3000–6000 feet (900–1830 meters). To the southwest, Missouri's Ozark Mountains and the Ouachita Plateau of northern Arkansas resemble portions of the southern Appalachians.

Much of the North American interior is a vast lowland, extending east–west from the Ohio River valley to the Great Plains and north–south from west central Canada to the lower Mississippi near the Gulf of Mexico (Figure 3.4). Glacial forces, particularly north of the Ohio and Missouri rivers, once actively carved and reshaped the landscapes of this lowland zone, including the environmentally complex Great Lakes Basin.

In the West, mountain-building (including large earthquakes and volcanic eruptions), alpine glaciation, and erosion produce a regional topography quite unlike that of eastern North America. The Rocky Mountains reach more than 10,000 feet (3050 meters) in height and stretch from Alaska's Brooks Range to northern New Mexico's Sangre de Cristo Mountains (Figure 3.5). West of the Rockies, the Colorado Plateau is characterized by colorful sedimentary rock eroded into spectacular buttes and mesas. Nevada's sparsely settled basin and range country features north–south mountain ranges alternating with structural basins with no outlet to the sea. North America's western border

Figure 3.3 Coastal Damage from Hurricane Sandy When Hurricane Sandy made landfall in New Jersey in October 2012, the storm inflicted heavy damage on this portion of the Atlantic City boardwalk.

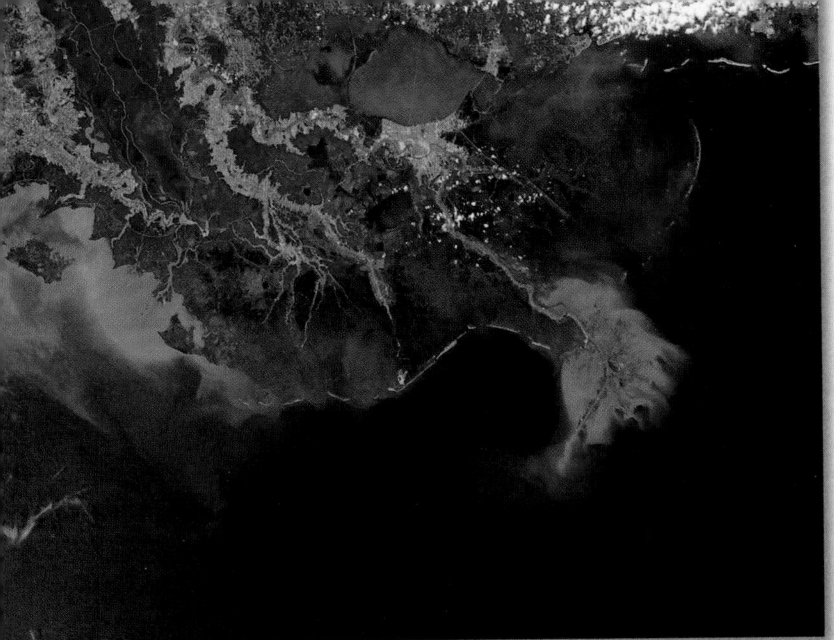

Figure 3.4 Satellite View of the Lower Mississippi Valley This view of the Mississippi Delta shows how sediments from the interior lowlands have accumulated in the area. Note the sediment plume (right) extending into the Gulf of Mexico.

is marked by the mountainous and rain-drenched coasts of southeast Alaska and British Columbia; the Coast Ranges of Washington, Oregon, and California; the lowlands of Puget Sound (Washington), Willamette Valley (Oregon), and Central Valley (California); and the complex uplifts of the Cascade Range and Sierra Nevada.

Patterns of Climate and Vegetation

North America's climates and vegetation are diverse, mainly due to the region's size, latitudinal range, and varied terrain (Figure 3.6). As the climographs for Dallas, Texas, and Columbus, Ohio, suggest, much of North America south of the Great Lakes and east of the Rockies is characterized by a long growing season, 30 to 60 inches (75 to 150 cm) of precipitation annually, and a deciduous broadleaf forest (later cut down and replaced by crops). From the Great Lakes north, the coniferous evergreen forest, or **boreal forest**, dominates the continental interior. Near Hudson Bay and across harsher northern tracts, trees give way to **tundra**, a mixture of low shrubs, grasses, and flowering herbs that briefly flourish in the short growing season of the high latitudes. Drier continental climates from west Texas to Alberta feature large seasonal ranges in temperature and unpredictable precipitation, averaging from 10 to 30 inches (25 to 75 cm) annually. The soils of much of this region are fertile and originally supported **prairie** vegetation, dominated by tall grasslands in the East and by short grasses and scrub vegetation in the West.

Western North American climates and vegetation are greatly complicated by the region's mountain ranges. The Rocky Mountains and the intermontane interior exhibit typical midlatitude seasonal variations, but topography greatly modifies climate and vegetation patterns. Many arid interior settings lie in the dry rain shadow of the Cascade Range and Sierra Nevada. Farther west, marine west coast climates dominate north of San Francisco, while a dry-summer Mediterranean climate occurs across central and southern California.

Figure 3.5 Rocky Mountains This spectacular lake in Alberta's Banff National Park reveals the characteristic signatures of alpine glaciation that are found in many portions of the Rocky Mountain region, both in the United States and in Canada. Many glaciers are retreating in response to global climate change.

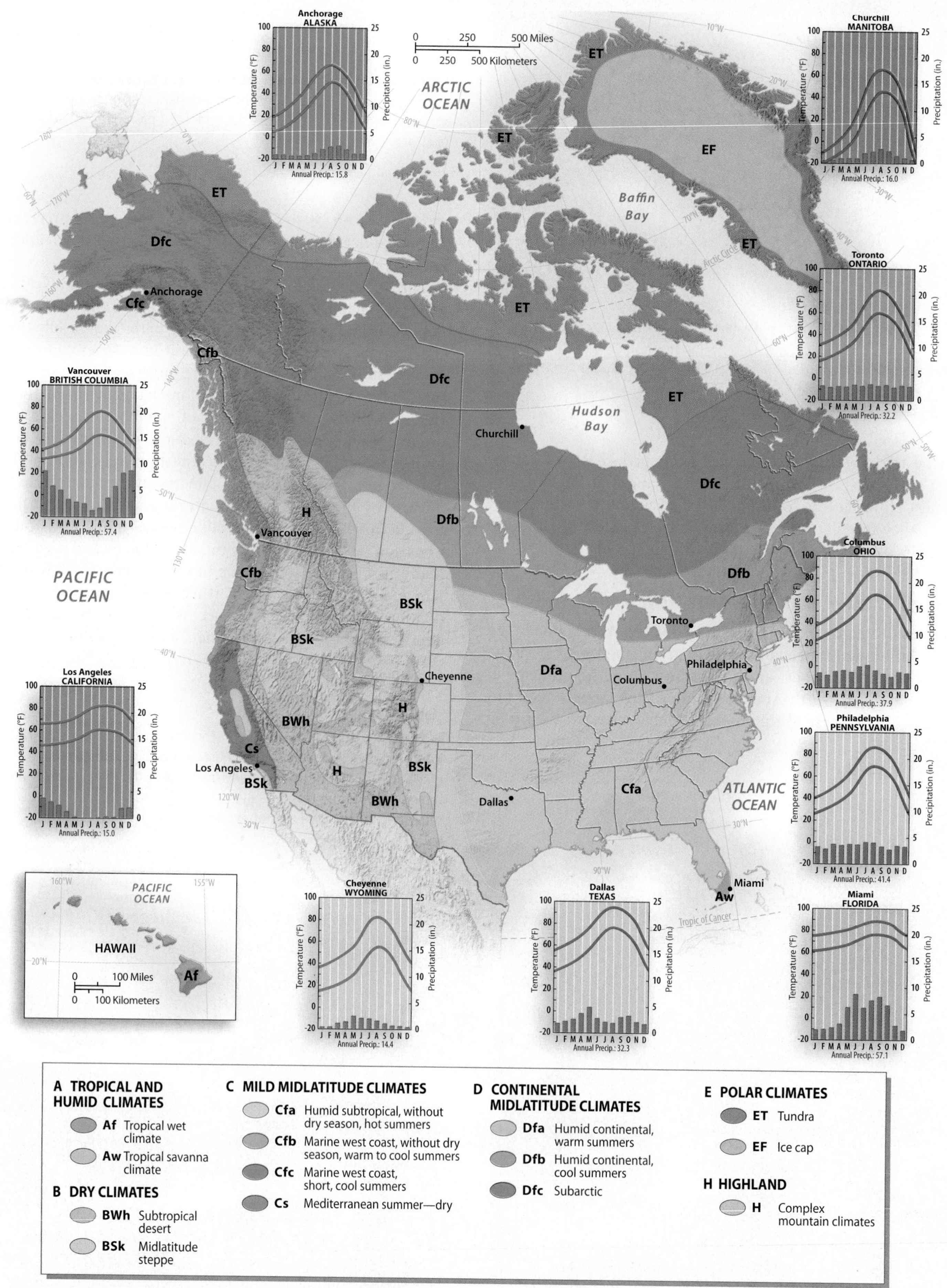

Figure 3.6 Climate Map of North America North American climates include everything from tropical savanna (Aw) to tundra (ET) environments. Most of the region's best farmland and densest settlements lie in the mild (C) or continental (D) midlatitude climate zones.

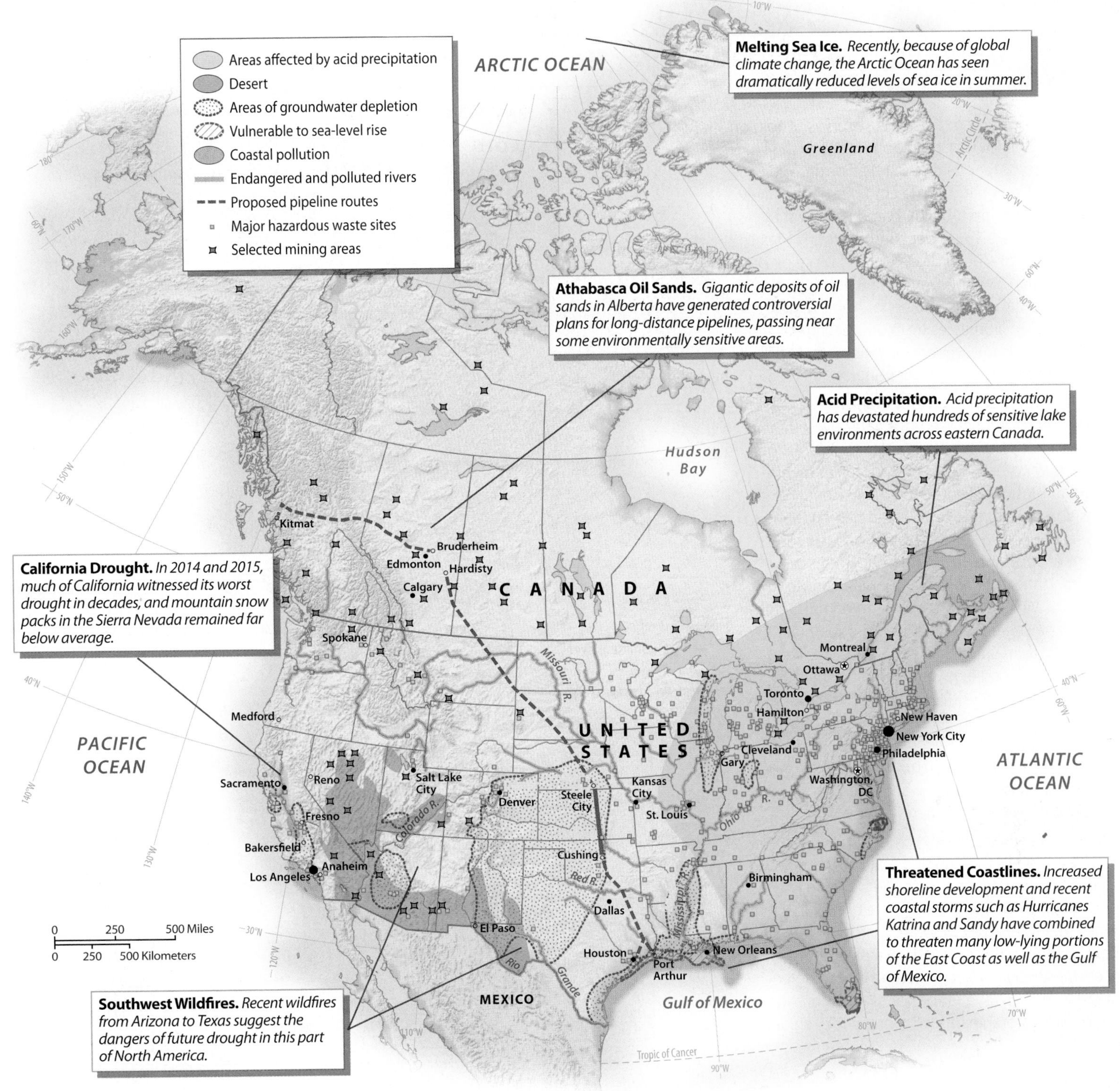

Figure 3.7 Environmental Issues in North America Acid rain damage is widespread in regions downwind from industrial source areas. Elsewhere, extensive water pollution, cities with high levels of air pollution, and zones of accelerating groundwater depletion pose health dangers and economic costs to residents of the region. Since 1970, however, both Americans and Canadians have become increasingly responsive to the dangers posed by these environmental challenges.

The Costs of Human Modification

North Americans have modified their physical setting in many ways. Processes of globalization and accelerated urban and economic growth have transformed the region's landforms, soils, vegetation, and climate. Indeed, problems such as acid rain, nuclear waste storage, groundwater depletion, and toxic chemical spills are manifestations of a way of life unimaginable only a century ago (Figure 3.7).

Transforming Soils and Vegetation While North America has been occupied by indigenous peoples for thousands of years, the arrival of Europeans dramatically affected the region's flora and fauna as countless new species were introduced, including wheat, cattle, and horses. As the number of settlers increased, forest cover was removed from millions of acres. Grasslands were plowed under and replaced with grain and forage crops not native to the region. Widespread soil erosion was increased

WORKING TOWARD SUSTAINABILITY

Greening the Colorado River Delta

A quiet environmental success story is unfolding along the lower Colorado River near the U.S.–Mexico border (Figure 3.1.1). For decades, the Colorado Delta had been a continental sacrifice zone, the victim of dam building and poor water management in the Southwest and northern Mexico. The completion of Hoover Dam (1937), Morelos Dam (1950), and Glen Canyon Dam (1963) helped dry up seasonal water flows into the delta, transforming more than 1.5 million acres (600,000 hectares) of precious wetlands into sunbaked mud flats. For decades, the delta was starved for water.

Creating a Green Coalition But thanks to a broad coalition of environmental groups and government agencies on both sides of the border, thousands of new cottonwood trees are taking root, along with willows and mesquite (Figure 3.1.2). Seasonal birds are revisiting the region's resurrected habitat, and there is real hope that the delta may one day return to its former splendor. Environmental groups based in the United States, such as the Sonoran Institute, the Defenders of Wildlife, and the Environmental Defense Fund, have partnered with Mexican groups, such as Pronatura, to make the case for the delta. Their actions have included legal battles (suing the U.S. government over lost habitat for endangered bird species), small-scale restoration initiatives demonstrating the potential for regreening the region (providing more than 800 acres of new trees and native plants), and intense lobbying efforts with water agencies throughout the Southwest and northern Mexico.

The New Agreement Those combined initiatives have paid off, even amid the ongoing reality of drought in this part of North America. Both the U.S. and Mexican governments, as well as a triumvirate of regional water agencies (the Metropolitan Water District of Southern California, the Southern Nevada Water Authority, and the Central Arizona Project), worked out an agreement to (1) increase the efficiency of existing irrigation and water storage infrastructure in the region,

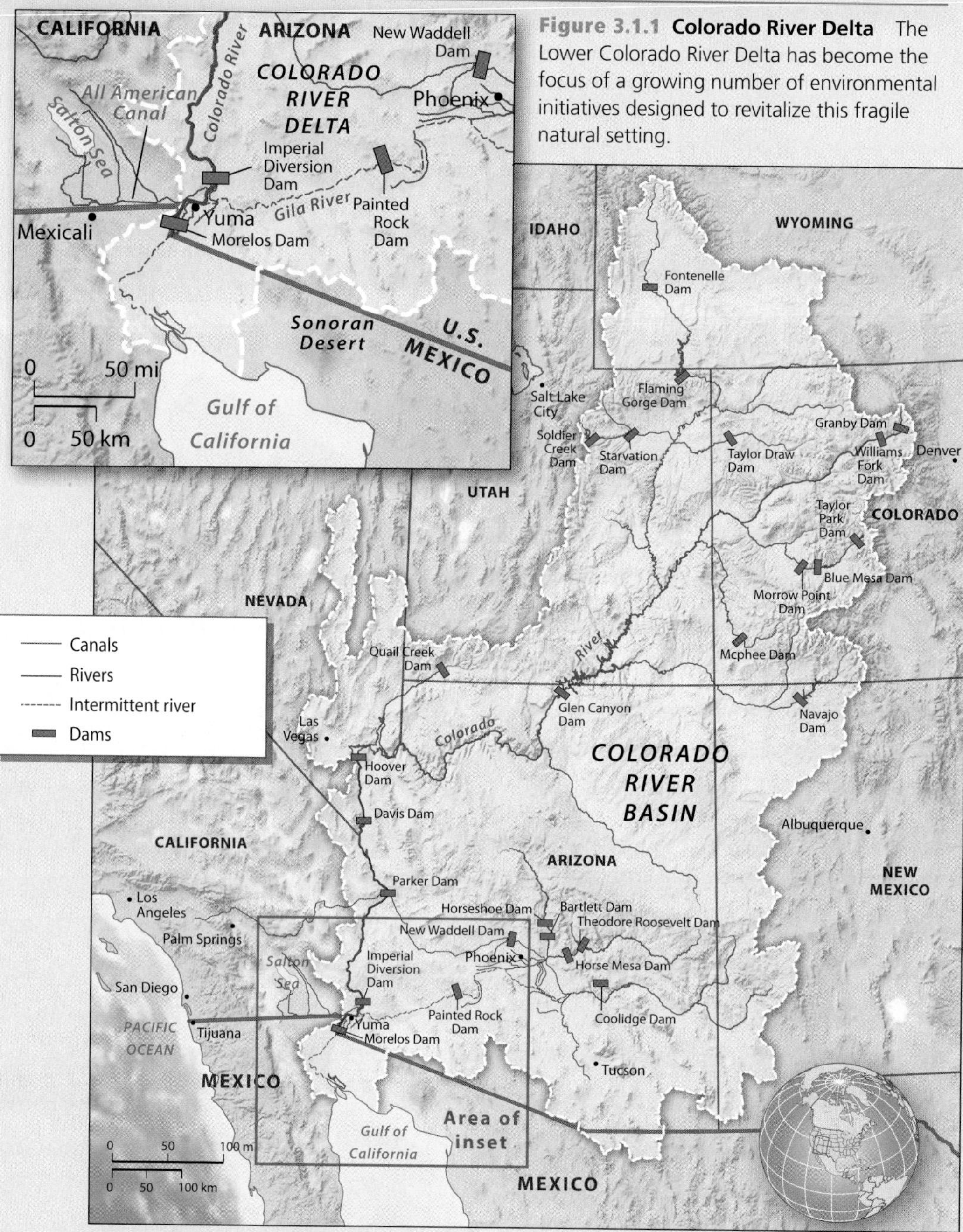

Figure 3.1.1 Colorado River Delta The Lower Colorado River Delta has become the focus of a growing number of environmental initiatives designed to revitalize this fragile natural setting.

by unsustainable cropping and ranching practices, and many areas of the Great Plains and South suffered lasting damage. Globalization has also brought new plants and animals to North America. For example, kudzu, a plant native to Japan, was introduced across the South to control erosion, but today it is a major invasive pest, literally the star of an ecological horror movie that envelops trees, barns, and utility poles in its grasp (Figure 3.8).

Managing Water North Americans consume huge amounts of water. While conservation efforts and technology have slightly reduced per capita rates of water use over the past 25 years, city dwellers still use an average of more than 175 gallons daily. Metropolitan areas such as New York City struggle with outdated municipal water supply systems. Beneath the Great Plains, the waters of the Ogallala Aquifer are being depleted; center-pivot irrigation systems have steadily lowered water tables by as much as 100 feet (30 meters) in the past 50 years. The fluctuating flows of the Colorado River are chronically frustrating to residents in the rapidly growing southwestern United States, although creative approaches to water management are being utilized to improve the river's flows, especially in its lower reaches (see *Working Toward Sustainability: Greening the Colorado River Delta*).

Altering the Atmosphere North Americans modify the very air they breathe, changing local and regional climates as well as the composition of the atmosphere. For example, built-up metropolitan areas create an **urban heat island** effect, in which development associated with cities often produces nighttime temperatures some 9 to 22°F (5 to 12°C)

Figure 3.1.2 Restored Riparian Habitat, Colorado River Delta The two views offer images taken (a) before and (b) after a recent effort to engineer controlled flows of freshwater into the Colorado Delta to help restore riparian vegetation.

(2) release a huge pulse of water (more than 100,000 acre-feet) through the delta to replicate a natural flood, and (3) provide future base flows (about 50,000 acre-feet of water annually) until 2017 to maintain the ecological lifeline to delta plants and wildlife.

A Fragile Rebirth During the spring of 2014, the pulse of water released through the lower Colorado system triggered a massive ecological renaissance: Satellite imagery verified a 40 percent increase in riparian vegetation, water tables have risen, and bird populations are increasing. But experts warn that the recovery is fragile and may be ephemeral. If severe drought conditions return to the Colorado Basin, annual base flows are not guaranteed past 2017. Furthermore, the tentative spirit of international cooperation that made recent progress possible may evaporate if more water shortages loom. But for now, the delta's pale green beauty has partly returned, eloquent testimony to what is possible when water's long-term ecological value is more fully appreciated in the complex calculus of resource management.

1. Describe another setting in North America where an international border potentially complicates an important environmental issue.
2. How might future plans for restoring the Colorado Delta go astray? Offer one scenario for an unexpected challenge, and describe how planning agencies might respond to it.

Google Earth (MG) Virtual Tour Video http://goo.gl/cQEuLP

warmer than those of nearby rural areas. At the local level, industries, utilities, and automobiles contribute carbon monoxide, sulfur, nitrogen oxides, hydrocarbons, and particulates to the urban atmosphere. While the region's worst offenders are U.S. cities such as Los Angeles and Houston, Canadian cities such as Toronto, Hamilton, and Edmonton experience significant air-quality problems. Overall air quality in urban North America has improved in the past few decades, but a 2014 American Lung Association report estimated that about 47 percent of U.S. residents still live in places with unsafe levels of air pollution, including high levels of particulates as well as elevated levels of ozone.

North America remains plagued by **acid rain**—industrially produced sulfur dioxide and nitrogen oxides in the atmosphere that combine with precipitation to damage forests, poison lakes, and kill fish. Many atmospheric pollution producers—industrial plants, power-generating facilities, and motor vehicles—are located in the Midwest and southern Ontario, and prevailing winds transport pollutants and deposit damaging acid rain and snow across the Ohio Valley, the Northeast, and eastern Canada (see Figure 3.7).

Growing Environmental Awareness

Many U.S. and Canadian environmental initiatives have addressed local and regional problems. For example, the improved water quality of the Great Lakes over the past 30 years is an achievement to which both nations contributed and that benefits both. Tougher air-quality standards have also reduced certain types of emissions in many North American cities.

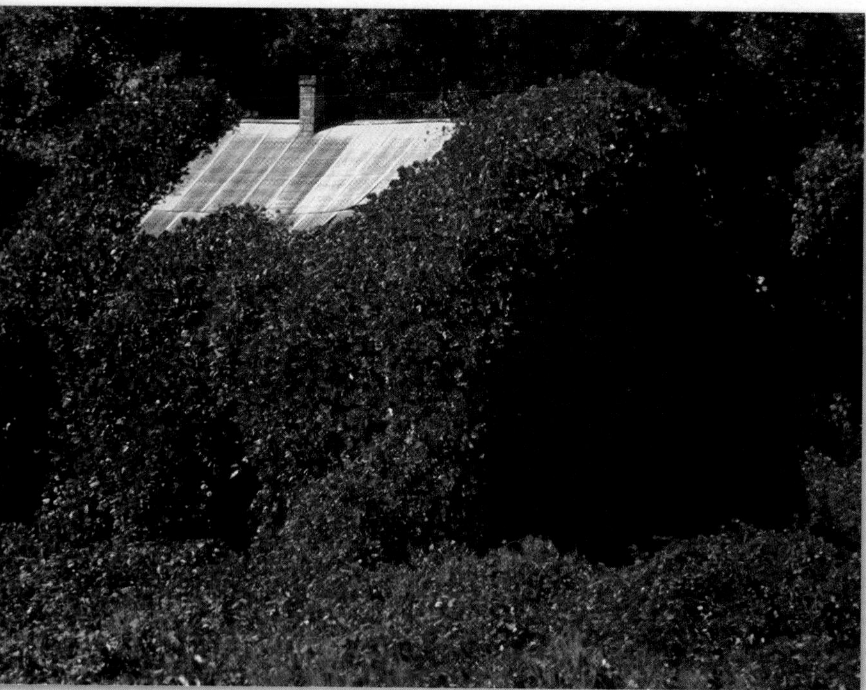

Figure 3.8 Kudzu A native plant of East Asia, kudzu has spread widely in the Southeast, often as an unwanted, invasive plant.

Perhaps most important, North America is increasingly supporting green industries and technologies. The growing popularity of **sustainable agriculture** exemplifies the trend, where organic farming principles, limited use of chemicals, and an integrated plan of crop and livestock management combine to offer both producers and consumers environmentally friendly alternatives. Elsewhere, American homebuilders are integrating alternative energy technology into their housing projects. For example, Lennar Corporation, the nation's second-largest homebuilder, now installs solar panels on many of its new houses in the Southwest; residents enjoy the energy savings while the company receives state and federal tax breaks for adopting the technology.

The Shifting Energy Equation

The region's energy consumption remains extremely high (the United States is still the source of almost 20 percent of Earth's greenhouse gas emissions). More incentives for energy efficiency may lead to lower per capita consumption in the future. The growing technological and economic appeal of **renewable energy sources**, such as hydroelectric, solar, wind, and geothermal, are likely to fundamentally rework North America's economic geography in coming years as policymakers, industrial innovators, and consumers are drawn to their enduring availability and potentially lower environmental costs (Figure 3.9).

At the same time, recent evidence suggests that North American oil, natural gas, coal, and tar sands may offer relatively abundant fossil fuel–based energy resources. New discoveries and drilling technologies have fundamentally shifted the continent's energy equation. The Bakken formation in North Dakota and Montana, thanks to new oil extraction methods, may one day produce more than 15–20 billion barrels of oil, making it one of the planet's great energy reserves (on a par with Alaska's North Slope field). The International Energy Agency predicts the United States will become a major exporter of natural gas by 2020. Canada alone has the world's third-largest proven oil reserves: About 170 billion barrels of oil can be recovered just from its rich oil sands. Overall, even with its high rates of consumption, North America may be an oil export region by 2035. Yet fluctuating global energy prices, such as the dramatic drop in crude oil in 2015, may cloud such long-term projections, especially if a slowdown in the energy economy discourages investments in both alternative sources (such as wind and solar) and shale-based fossil fuels.

Many environmental issues also complicate the clean development of North America's untapped fossil fuels. Moving fossil fuels involves huge investments and risks. Plans for expanding coal exports from the Rocky Mountains (especially from Wyoming and Montana) to the Pacific Coast (for export to Asia) have stirred protests. Controversial energy

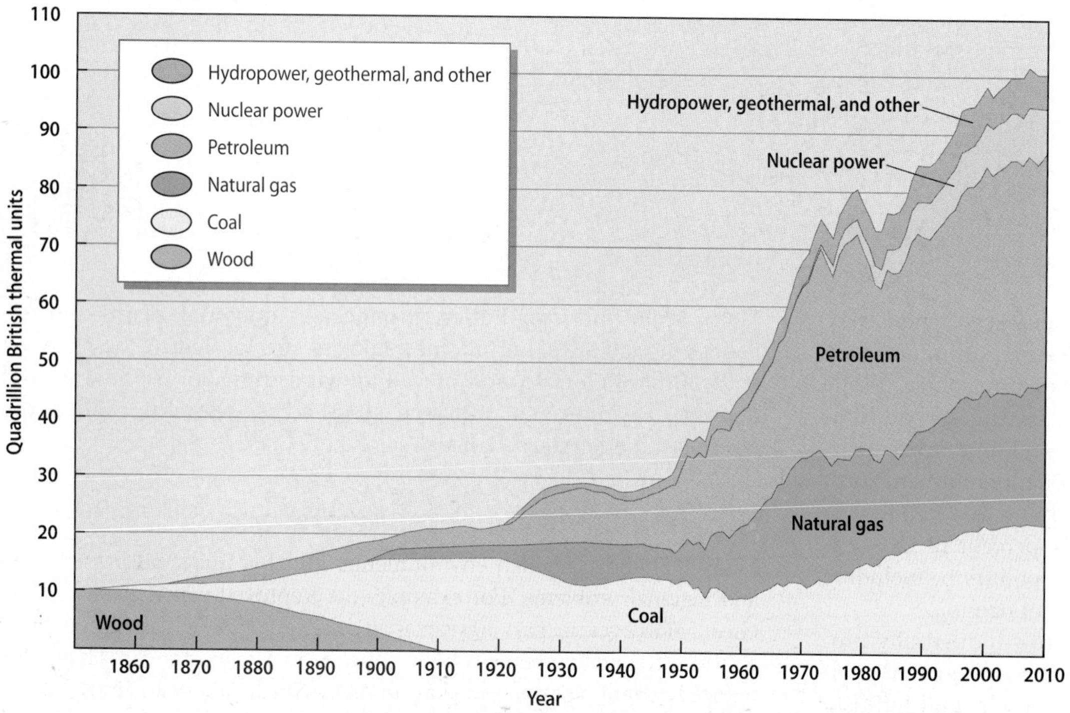

Figure 3.9 U.S. Energy Consumption The growing popularity of fossil fuels is evident in U.S. energy consumption during the late 19th century as coal, oil, and then natural gas supplanted wood consumption. Nuclear power and other renewable energy sources are poised to play a larger role during the 21st century.
Q: Approximately what percentage of U.S. energy consumption is currently accounted for by nuclear power and renewable energy sources?

pipelines such as the Keystone XL project (Alberta to the Gulf of Mexico) and the Northern Gateway Pipeline (Alberta to the Pacific Coast), designed to tap into Canada's rich Athabasca oil sands, also illustrate the tensions between increasing energy production (and job creation) and the potentially dangerous environmental consequences of moving fossil fuels long distances (see Figure 3.7). In addition, the growing use of **fracking** (hydraulic fracturing, a drilling technology that injects a mix of water, sand, and chemicals underground in order to release natural gas) in settings from North Dakota to Pennsylvania may lead to polluted groundwater and potentially hazardous environmental conditions for nearby residents.

Climate Change and North America

In 2014, more than 400,000 people from all around the United States participated in the People's Climate March through New York City (Figure 3.10). The marchers, including New York City Mayor Bill de Blasio, pressed for faster national and global responses to climate change, which has already profoundly reshaped many North American settings. After the unparalleled warmth from 2012 through 2015 (some of the hottest years on record in the United States), many experts predict that more extreme temperatures and precipitation events are likely in the near future. The accompanying droughts and wildfires cost billions of dollars.

Projected impacts of climate change for North America suggest varying regional patterns. The U.S. Southwest, parts of Texas, and southern California may be drier, while northern states may see more precipitation. Many coastal localities along the Atlantic Ocean and the Gulf of Mexico will be especially vulnerable to rising sea levels and more intense coastal storms.

In the high latitudes, changes in arctic temperatures, sea ice, and sea levels have increased coastal erosion, affected whale and polar bear populations, and stressed traditional ways of life for Inuit residents. At the same time, a more ice-free Arctic Ocean enhances the potential for more commercial shipping and resource development, and longer Canadian growing seasons may allow for more high-latitude crop production (see *Exploring Global Connections: Climate Change Brings Luxury Cruises to the Fabled Northwest Passage*).

Western North American mountains are especially sensitive to climate change. Expanding bark beetle populations survive in milder winters and are infesting pine forests. Many of the region's spectacular alpine glaciers are rapidly disappearing. Earlier spring melting of mountain snow packs also impacts downstream fisheries, farms, and metropolitan areas that depend on these seasonal water resources.

Many government agencies have begun to respond to the reality of climate change. Most dramatically, municipal planners are increasingly considering the probable impacts of changes on urban water supplies, flooding hazards, health, and disaster management. Gradually, states and provinces are also developing longer-term planning documents to help future North American residents adjust to the reality of climate change.

Review

3.1 Describe North America's major landform regions and climates, and suggest ways in which the region's physical setting has shaped patterns of human settlement.

3.2 Identify the key ways in which humans have transformed the North American environment since 1600.

3.3 Identify four key environmental problems that North Americans face in the early 21st century.

KEY TERMS postindustrial economy, boreal forest, tundra, prairie, urban heat island, acid rain, sustainable agriculture, renewable energy source, fracking

Population and Settlement: Reshaping a Continental Landscape

The North American landscape is the product of human settlement extending back in time for at least 12,000–25,000 years. The pace of change for much of that period was modest and localized, but the last 400 years have witnessed an extraordinary transformation as Europeans, Africans, Asians, and Central and South Americans arrived in the region, disrupted Native American peoples, and created dramatically new patterns of human settlement. Today the region is home to 355 million people—some of the world's most affluent and highly mobile populations (Table A3.1).

Figure 3.10 Climate Change March, New York City, September 2014 More than 400,000 people participated in the 2014 Climate Change March in downtown New York City.

EXPLORING GLOBAL CONNECTIONS

Climate Change Brings Luxury Cruises to the Fabled Northwest Passage

Google Earth (MG) Virtual Tour Video
http://goo.gl/1ZQEWb

It has been a long wait. But thanks to global climate change, large luxury cruise ships will be plying the Arctic Ocean beginning in 2016, exploring a newly thawed and quite lucrative global connection between Anchorage and New York City. Following the so-called Northwest Passage around the northern perimeter of North America, the high-latitude tourists—for a mere $20,000 per person—can enjoy the pleasures of the polar region, including arctic wildlife, native villages, and iceberg-studded ocean vistas (Figure 3.2.1).

A Fabled Passage The fabled Northwest Passage has long been shrouded in myth and tragedy. Notions of an ice-free passage between the Atlantic and Pacific oceans and ideas about an open polar sea sparked the imagination of European explorers for generations. In the late 16th century, English explorers Martin Frobisher (1576) and John Davis (1585) probed the polar seas from the east between Greenland and Baffin Island. Two centuries later, Denmark's Vitus Bering (1741) and Great Britain's James Cook (1778) braved the Bering Sea and briefly entered the Arctic Ocean from the west. Later 19th-century expeditions ended in death and starvation, the most famous being the ice-choked voyage of John Franklin (sailing with the HMS *Erebus* and the HMS *Terror*) in the 1840s (Figure 3.2.2). Norwegian Roald Amundsen's incredibly dangerous three-year trek (1903–1906) finally accomplished the deed.

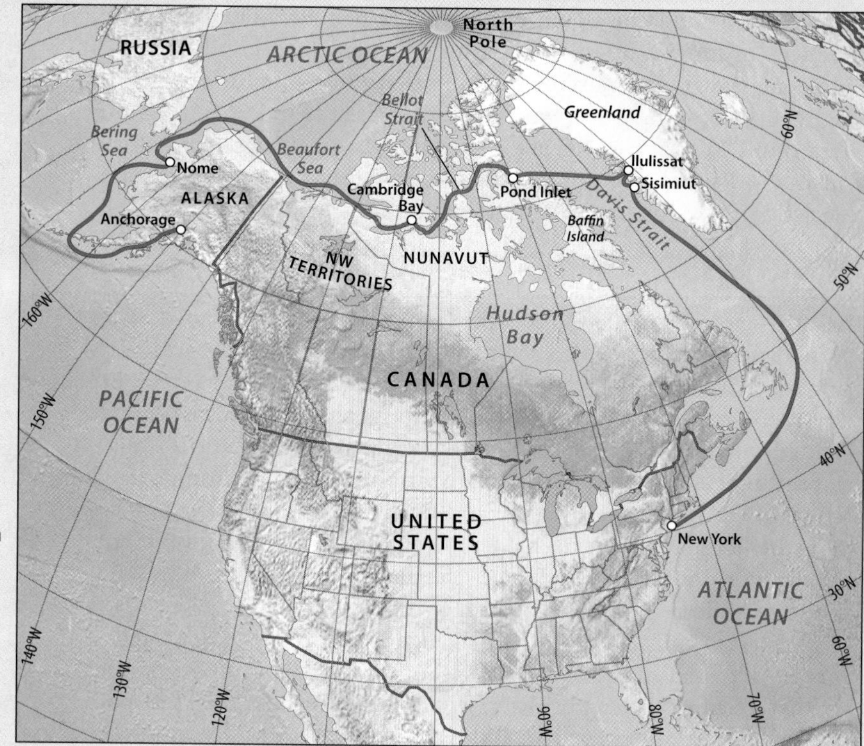

Figure 3.2.1 The Fabled Northwest Passage Modern cruise ships may increasingly ply arctic waters as they follow the fabled Northwest Passage through ice-free polar seas.

Polar Touring with Panache But today's luxury cruise-ship industry and a more ice-free Arctic have changed all that. By 2016, the high-end Crystal Cruises line plans the first

Modern Spatial and Demographic Patterns

Large metropolitan areas (including both central cities and suburbs) dominate North America's population geography, producing uneven regional patterns of settlement (Figure 3.11). Canada's "Main Street" corridor contains most of that nation's urban population, led by Toronto (5.8 million) and Montreal (3.9 million). **Megalopolis**, the largest settlement cluster in the United States, includes Baltimore/Washington, DC (8.6 million), Philadelphia (6.5 million), New York City (22 million), and Boston (7.6 million). Beyond these two core areas, other sprawling urban centers cluster around the southern Great Lakes (Chicago, 9.7 million), in various parts of the South (Dallas, 7.4 million), and along the Pacific Coast (Los Angeles, 17.9 million; Vancouver, 2.4 million) (Figure 3.12).

North America's population has increased greatly since European colonization. Before 1900, high birth rates produced large families, and large numbers of new immigrants arrived in the region. In Canada, a population of fewer than 300,000 Native Americans and Europeans in the 1760s grew to an impressive 3.2 million a century later. For the United States, a late colonial (1770) total of around 2.5 million increased over tenfold to more than 30 million by 1860. Both countries saw even higher rates of immigration in the late 19th and 20th centuries, although birth rates gradually fell after 1900. After World War II, birth rates rose once again in both countries, resulting in the "baby boom" generation born between 1946 and 1965. Today, however, rates of natural increase in North America are below 1 percent annually, and the overall population is growing older, particularly in states such as Iowa (Figure 3.13). Still, the region attracts immigrants. These growing numbers, along with higher birth rates among immigrant populations (exemplified by Texas), recently led experts to increase long-term population projections (Figure 3.13). The U.S. Census Bureau's predictions of a 2050 population of 464 million (423 million in the United States and 41 million in Canada) may prove conservative.

Occupying the Land

When Europeans began occupying North America more than 400 years ago, they were not settling an empty land; the region had been populated for at least 12,000–25,000 years by peoples as culturally diverse as the Europeans who conquered them. Native

Figure 3.2.2 The Failed Franklin Expedition, 1840s This 19th-century artist sketch of the *Terror*, a ship in the failed Franklin Expedition of the 1840s, captures some of the icy challenges encountered when navigating a colder arctic environment.

true luxury-ship commercial crossing of the Northwest Passage. Frobisher, Franklin, and Amundsen would be impressed: In a mere 32 days, the sumptuous 900-passenger *Crystal Serenity*—complete with spa services, onboard casino, and putting green—can make the trip between Anchorage, Alaska, and New York Harbor. The itinerary calls for numerous stops at Inuit villages, onboard lectures by polar experts, an optional round of golf at the world's northernmost nine-hole course, and opportunities for sea kayaking and polar bear sightings.

Rough Seas Ahead? Just as in earlier generations, this new era of northern exploration has its skeptics. While economic development officials tout the potential of new business opportunities for indigenous peoples, some critics worry about the challenges when hundreds of wealthy tourists stream off the decks of a cruise ship into a small arctic village. Cruise lines are already planning on limiting the size of onshore parties and have agreed to remove every bit of garbage they generate, but time may reveal longer-term impacts.

Other observers worry that the ice-free Arctic may not always live up to its temperate expectations—that is, long-term climate trends do not always operate predictably, and a colder- and icier-than-normal summer is always possible. No one wants to see an ice-bound luxury cruise ship, especially the Canadian Coast Guard, which is already engaged in training exercises to prepare for the worst-case scenario. Even the *Crystal Serenity* has an alternative itinerary, should the ice thicken: One contingency plan takes Arctic wannabees through the Panama Canal. But the longer-term trend is clear. Both polar bears and Inuit villagers should be advised: Plan on more 21st-century visitors to come.

1. List the potential positive and negative consequences of mass polar tourism for Inuit villages.
2. Make a brief list of five potential onboard lecture topics that might be of interest to passengers on the *Crystal Serenity*.

Americans migrated to North America from northeast Asia in multiple waves and dispersed across the region, adapting in diverse ways to its many natural environments. Cultural geographers estimate Native American populations in 1500 CE at 3.2 million for the continental United States and 1.2 million for Canada, Alaska, Hawaii, and Greenland.

North America's native peoples met many different fates, reducing their numbers by more than 90 percent following European settlement. Some groups were exterminated by disease and war; others were expelled from their homelands and relocated on reservations, both in Canada and in the United States. For example, as a consequence of the Indian Removal Act of 1830, thousands of Native Americans were driven out of the Southeast in the notorious "Trail of Tears" trek to Indian Territory (later Oklahoma). Some Native Americans also mixed with European populations, losing parts of their cultural identity in the process. Today the majority of the region's native peoples actually live in cities, often far removed from the land of their ancestors.

Who replaced Native North Americans? The first stage of European settlement created a series of colonies between 1600 and 1750, mostly in the coastal regions of eastern North America (Figure 3.14). These regionally distinct societies were anchored in the north by the French settlement of the St. Lawrence Valley and extended south along the Atlantic Coast, including several separate English colonies. Scattered developments along the Gulf Coast and in the Southwest also appeared before 1750.

The second stage in the Europeanization of North America (1750–1850) featured settlement of better agricultural lands in the eastern half of the continent. Pioneers surged westward across the Appalachians following the end of the American Revolution (1783) and a series of Indian conflicts, finding the Interior Lowlands almost ideal for agricultural settlement. Southern Ontario, or Upper Canada, was also opened to development after 1791.

The third stage in North America's settlement accelerated after 1850 through 1910. During this period, most of the region's remaining agricultural lands were settled by a mix of native-born and immigrant farmers. In the American West, settlers were attracted by opportunities in California, Oregon, Utah, and the Great Plains. In Canada, thousands occupied southern portions of Manitoba, Saskatchewan, and Alberta.

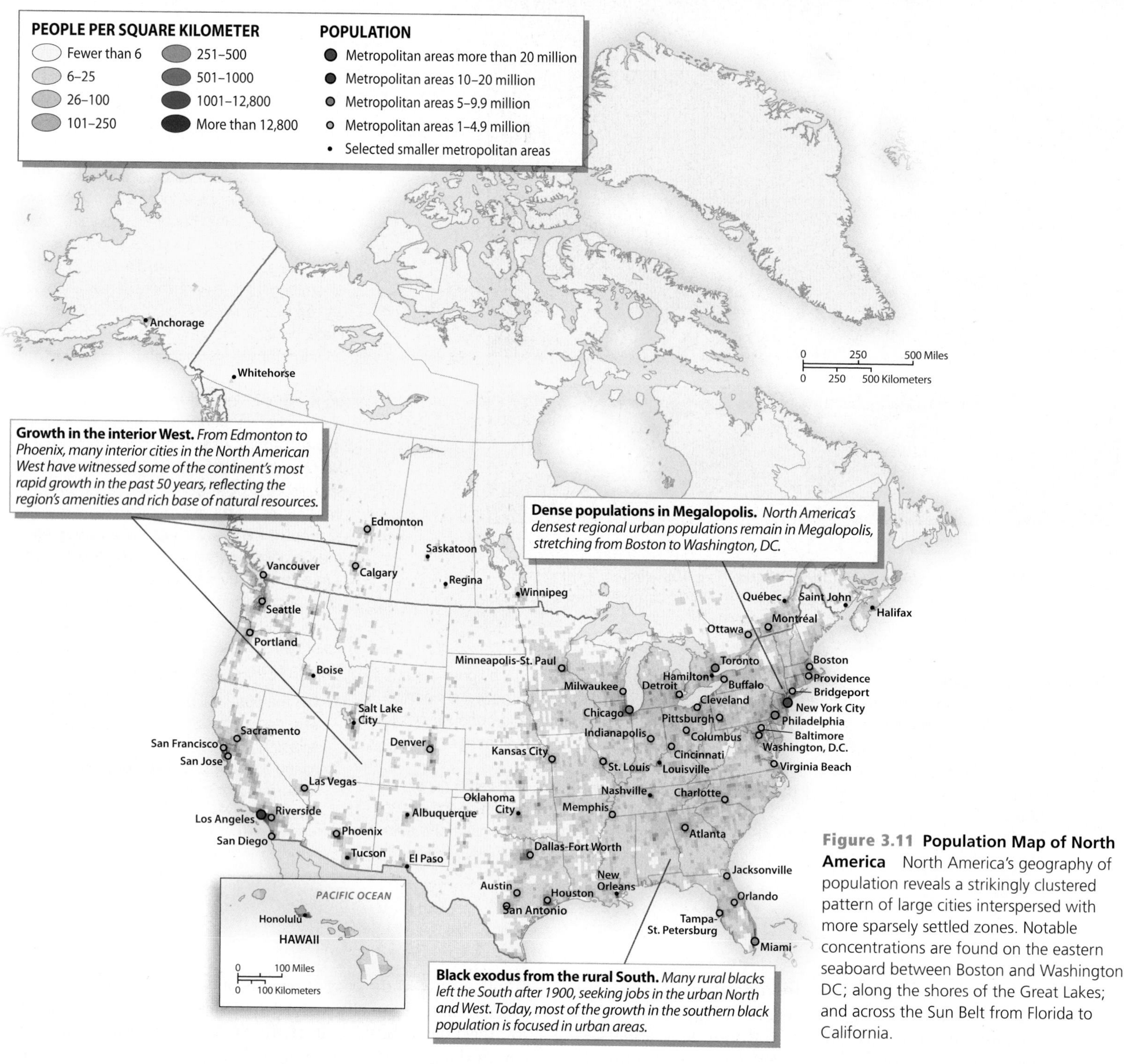

Figure 3.11 Population Map of North America North America's geography of population reveals a strikingly clustered pattern of large cities interspersed with more sparsely settled zones. Notable concentrations are found on the eastern seaboard between Boston and Washington, DC; along the shores of the Great Lakes; and across the Sun Belt from Florida to California.

Discovery of gold and silver led to development in areas such as Colorado, Montana, and British Columbia's Fraser Valley. Incredibly, in a mere 160 years, much of the North American landscape was occupied as expanding populations sought new land to settle and as the global economy demanded resources to fuel its growth.

North Americans on the Move

From the legendary days of Davy Crockett and Calamity Jane to the 20th-century sojourns of John Steinbeck and Jack Kerouac, North Americans have been on the move. Indeed, almost one in every five Americans moves annually, suggesting that the region's people are quite willing to change residence in order to improve their income or their quality of life. Several trends dominate the picture.

Westward-Moving Populations The most persistent regional migration trend has been the tendency for people to move west. By 1990, more than half of the U.S. population lived west of the Mississippi River, a dramatic shift from colonial times. Since 1990, some of the fastest-growing areas have been in the American West (including the states of Arizona and Nevada) and in the western Canadian provinces of Alberta and British Columbia. This trend was fueled by new job creation in high-technology, energy, and service industries as well as by the region's scenic, recreational, and retirement attractions.

Black Exodus from the South African Americans have generated distinctive patterns of interregional migration. Most blacks remained economically tied to the rural South after the Civil War, but by the early 20th century many African Americans migrated because of

Figure 3.12 Chicago Skyline Chicago's spectacular downtown skyline along the Lake Michigan shoreline is a classic example of a large North American central business district (CBD).

declining demands for labor in the agricultural South and growing industrial job opportunities in cities in the North and West. Boston, New York, Philadelphia, Detroit, Chicago, Los Angeles, and Oakland became major destinations for southern black migrants. Since 1970, however, more blacks have moved from North to South. Sun Belt jobs and federal civil rights guarantees now attract many northern urban blacks to growing southern cities. The net result is still a major change from 1900, when more than 90 percent of African Americans lived in the South; today only about 55 percent of the nation's 46 million blacks reside within the region.

Rural-to-Urban Migration Another continuing trend in North American migration (and a growing global phenomenon) has taken people from the country to the city. Two centuries ago only 5 percent of North Americans lived in urban areas (cities of more than 2500 people), whereas today more than 80 percent of the North American population is urban. Shifting economic opportunities account for much of the transformation: As farm mechanization reduced the demand for labor, many young people left for new employment opportunities in the city.

Growth of the Sun Belt South Particularly since the 1970s, southern states from the Carolinas to Texas have grown much more rapidly than states in the Northeast and Midwest. During the 1990s, Georgia, Florida, Texas, and North Carolina all grew by more than 20 percent. The South's expanding economy, modest living costs, adoption of air conditioning, attractive recreational opportunities, and appeal to snow-weary retirees have all contributed to its growth. Dallas–Fort Worth, Houston, and Atlanta have enjoyed some of the nation's fastest metropolitan growth rates since 2000 (Figure 3.15).

Nonmetropolitan Growth During the 1970s, some areas in North America beyond its large cities witnessed significant population gains, including rural settings that had previously lost population. Selectively, this pattern of **nonmetropolitan growth**—in which people leave large cities for smaller towns and rural areas—continues today. The growing retiree population in both Canada and the United States is part of this trend, but a substantial number are younger *lifestyle migrants*. Given our electronically connected world, they find or create employment in affordable smaller cities and rural settings that are rich in amenities and often removed from perceived urban problems.

Settlement Geographies: The Decentralized Metropolis

North American cities are characterized by **urban decentralization**, in which metropolitan areas sprawl in all directions and suburbs take on many of the characteristics of traditional downtowns. Although both Canadian and U.S. cities have experienced decentralization, the impact has been particularly profound in the United States, where inner-city problems, poor public transportation, widespread automobile ownership, and fewer regional-scale planning initiatives encourage middle-class urban residents to move beyond the central city.

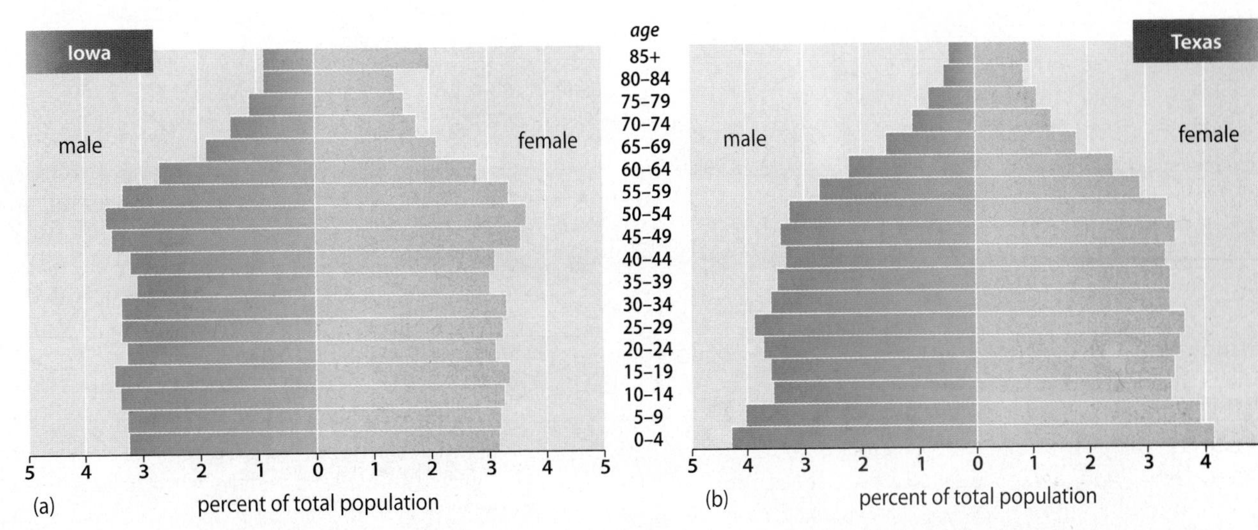

Figure 3.13 Population Pyramids: (a) Iowa and (b) Texas Iowa's aging population stands in contrast to the larger proportion of young people in Texas, reflecting that state's higher birth rate and sizable influx of young immigrants. **Q: How is the baby boom generation reflected in the Iowa pyramid?**

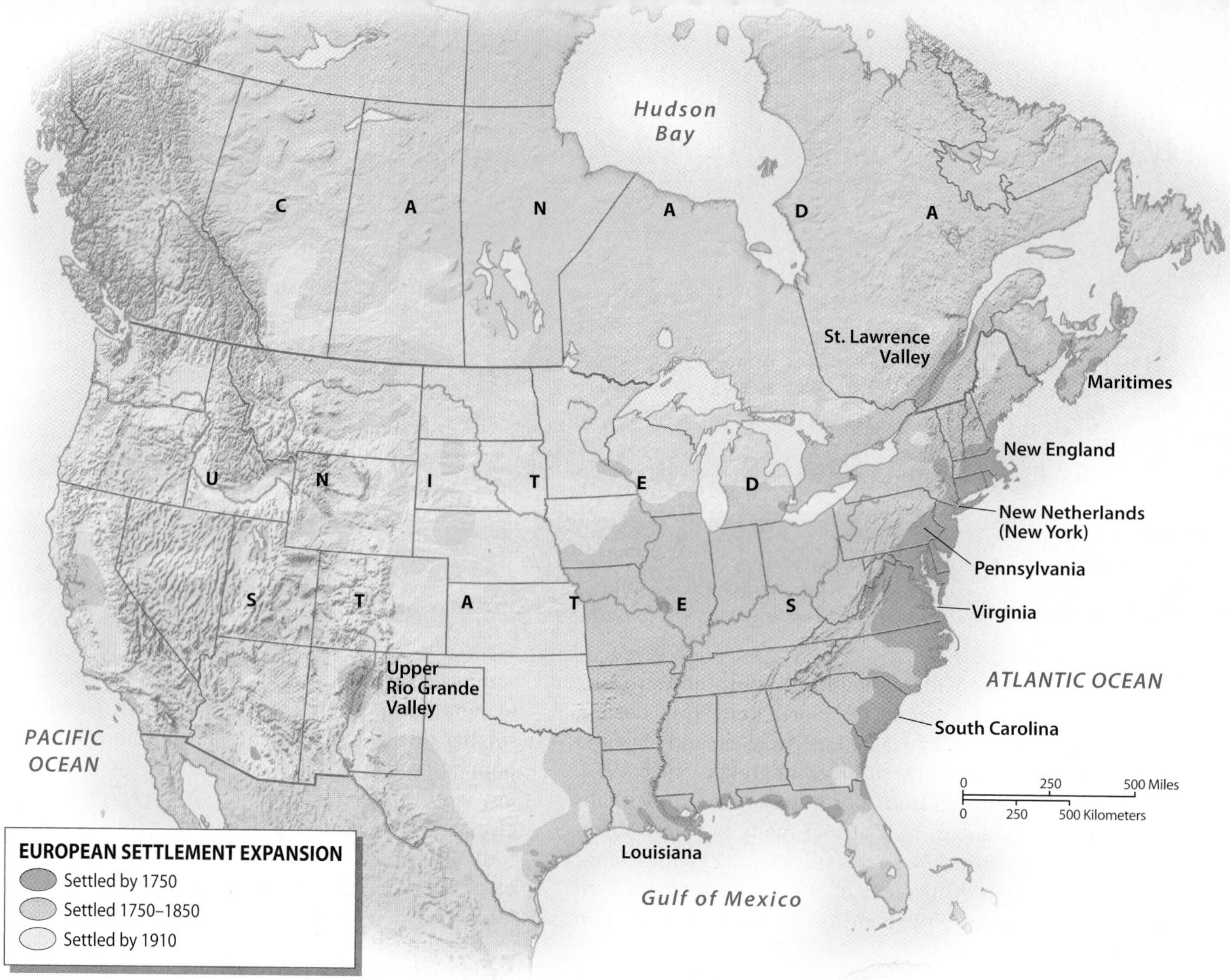

Figure 3.14 European Settlement Expansion Sizable portions of North America's East Coast and the St. Lawrence Valley were occupied by Europeans before 1750. The most remarkable surge of settlement occurred during the next century as Europeans opened vast areas of land and dramatically disrupted Native American populations.

Figure 3.15 Downtown Houston Houston's energy-based economy made it one of North America's fastest-growing cities between 1990 and 2014.

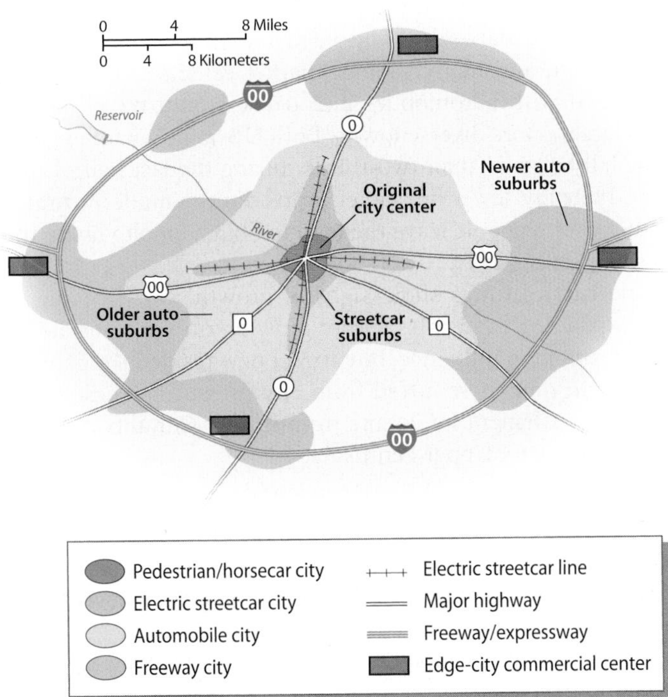

Figure 3.16 Growth of the American City Many U.S. cities became increasingly decentralized as they moved through the pedestrian/horsecar, electric streetcar, automobile, and freeway eras. Each era left a distinctive mark on metropolitan America, including the recent growth of edge cities on the urban periphery.

Historical Evolution of U.S. Cities Changing transportation technologies decisively shaped the evolution of the city in the United States (Figure 3.16). The pedestrian/horsecar city (pre-1888) was compact, essentially limiting urban growth to a 3- or 4-mile-diameter ring around downtown. The invention of the electric trolley in 1888 expanded the urbanized landscape farther into new "streetcar suburbs" that extended outward along streetcar lines, often for 5 to 10 miles from the city center. The biggest technological revolution came after 1920, with the mass production of cars. The automobile city (1920–1945) continued the expansion of middle-class suburbs. Post–World War II growth in the outer city (1945 to the present) promoted more decentralized settlement along commuter routes as built-up areas appeared 40 to 60 miles from downtown.

Urban decentralization also reconfigured land-use patterns, producing metropolitan areas today that are strikingly different from early 20th-century cities. In the city of 1920, urban land uses were generally organized in rings around a highly focused central business district (CBD) containing much of the city's retail and office functions. Residential districts beyond the CBD were added as the city expanded, with higher-income groups seeking more desirable locations on the outside edge of the urbanized area.

Today's suburbs, however, feature a mix of peripheral retailing (commercial strips, shopping malls, and big-box stores), industrial parks, office complexes, and entertainment facilities. These nodes of activity, called **edge cities**, have fewer functional connections with the central city than they have with other suburban centers. Tysons Corner, Virginia, located west of Washington, DC, is an excellent example of the edge-city landscape on the expanding periphery of a North American metropolis (Figure 3.17).

The Consequences of Sprawl As suburbanization increased in the 1960s and 1970s, many inner cities, especially in the Northeast and Midwest, suffered absolute losses in population and a shrinking tax base. Poverty rates average almost three times those of nearby suburbs. Unemployment rates remain above the national average. Central cities in the United States also remain places of racial tension, the product of decades of discrimination, segregation, and poverty.

Even with these challenges, inner-city landscapes enjoy selective improvement via **gentrification**, which involves higher-income residents replacing (some would say displacing) the lower-income residents of central-city neighborhoods, improving deteriorated inner-city landscapes, and constructing shopping complexes, sports and entertainment attractions, and convention centers in selected downtown locations. Seattle's Pioneer Square, Toronto's Yorkville, and Baltimore's Harborplace exemplify how such public and private investments shape the city (Figure 3.18). Many city planners and developers involved in such efforts also advocate **new urbanism**, an urban design movement stressing higher-density, mixed-use, pedestrian-scaled neighborhoods where residents can walk to work, school, and entertainment.

Settlement Geographies: Rural North America

The region's rural cultural landscapes trace their origins to early European settlement. Over time, these immigrants from Europe showed a clear preference for a dispersed rural settlement pattern as

Figure 3.17 Tysons Corner, Virginia North America's edge-city landscape is nicely illustrated by Tysons Corner, Virginia. Far from a traditional metropolitan downtown, this sprawling complex of suburban offices and commercial activities reveals how and where many North Americans will live their lives in the 21st century.

Figure 3.18 **Baltimore's Harborplace Development** Baltimore's Harborplace exemplifies inner-city redevelopment and includes a mix of residential and public space, offices, restaurants, and entertainment venues.

they created new farms. In portions of the United States settled after 1785, the federal government surveyed and sold much of the land. Surveys were organized around the simple, rectangular pattern of the federal government's township-and-range survey system (Canada's system is similar), which offered a convenient method of dividing and selling the public domain in 6-mile-square townships (Figure 3.19).

Commercial farming and technological changes further transformed the settlement landscape. Railroads opened corridors of development, provided access to markets for commercial crops, and helped to establish towns. By 1900, several transcontinental lines spanned North America, radically transforming the farm economy and the pace of rural life. After 1920, however, even greater change accompanied the arrival of the automobile, farm mechanization, and better rural road networks. The need for farm labor declined with mechanization, and many smaller market centers withered as farmers equipped with automobiles and trucks could travel farther and faster to larger, more diverse towns. Both U.S. and Canadian farm populations fell by more than two-thirds during the last half of the 20th century. Typically, fewer but larger farms dot the modern rural scene, and many young people leave the land to obtain employment, often in more urban settings.

Some rural settings show signs of growth, experiencing the effects of expanding edge cities. Other growing rural settings lie beyond direct metropolitan influence, but attract new residents seeking amenity-rich environments removed from city pressures. These trends are shaping the settlement landscape from British Columbia's Vancouver Island to Michigan's Upper Peninsula.

Figure 3.19 **Minnesota Settlement Patterns** The regular rectangular look of this Minnesota landscape is a common cultural landscape feature across North America. In the United States, the township-and-range survey system stamped such predictable patterns across vast portions of the North American interior.

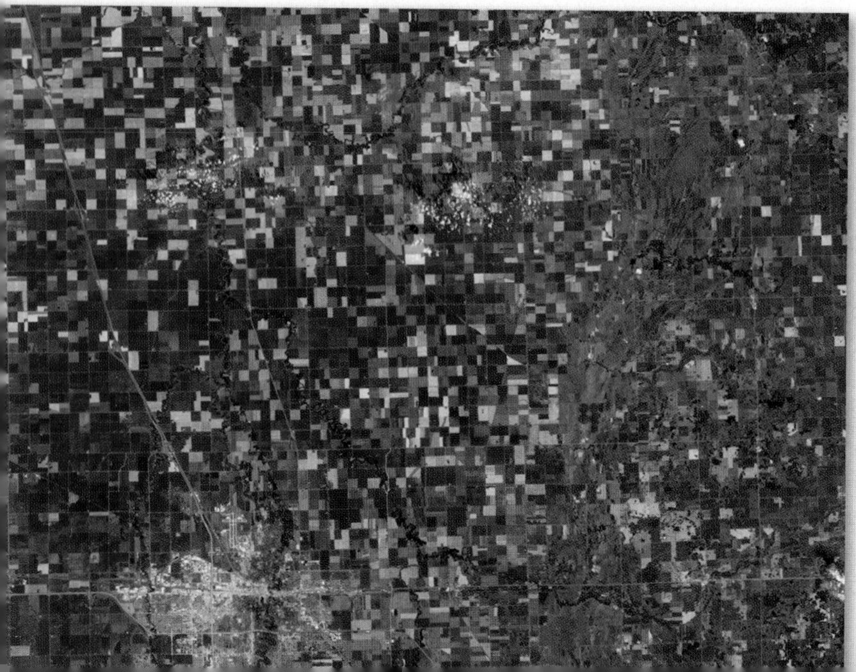

Review

3.4 Describe the dominant North American migration flows since 1900.

3.5 Sketch and discuss the principal patterns of land use within the modern U.S. metropolis, including (a) the central city and (b) the suburbs/edge city. How have forces of globalization shaped North American cities?

KEY TERMS Megalopolis, nonmetropolitan growth, urban decentralization, edge city, gentrification, new urbanism

Cultural Coherence and Diversity: Shifting Patterns of Pluralism

North America's cultural geography exerts global influence. At the same time, it is internally diverse. History and technology have produced a contemporary North American cultural force that is second to none in the world. Yet the region is also a collection of different peoples who retain part of their traditional cultural identities, celebrate their varied roots, and acknowledge the region's multicultural character.

The Roots of a Cultural Identity

Powerful historical forces formed a common dominant culture within North America. Although both the United States (1776) and Canada (1867) became independent from Great Britain, they remained closely tied to their Anglo roots. Key Anglo legal and social institutions solidified core values that North Americans shared with the British and, eventually, with one another. Traditional Anglo beliefs emphasized representative government, separation of church and state, liberal individualism, privacy, pragmatism, and social mobility. From those shared foundations, particularly within the United States, consumer culture blossomed after 1920, producing a shared set of experiences oriented around convenience, consumption, and the mass media.

But North America's cultural unity coexists with pluralism—the persistence and assertion of distinctive cultural identities. Closely related is the concept of **ethnicity**, in which people with a common background and history identify with one another, often as a minority group within a larger society. For Canada, the French colonization of Québec and the enduring power of its native peoples complicate its modern cultural geography. The greater diversity of ethnic groups within the United States produced different cultural geographies on both local and regional scales.

Peopling North America

North America is a region of immigrants. Decisively displacing Native Americans in most portions of the region, immigrant populations created a new cultural geography of ethnic groups, languages, and religions. Though small in number, early migrants had considerable cultural influence. Over time, immigrant groups and their changing destinations produced a varied cultural geography across North America. Also varying among groups were the pace and degree of **cultural assimilation**, the process in which immigrants are absorbed by the larger host society.

Migration to the United States Variations in the number and source regions of migrants produced five distinctive chapters in U.S. history (Figure 3.20). In Phase 1 (prior to 1820), English and African influences dominated. Slaves, mostly from West Africa, added cultural influences in the South. Northwest Europe served as the main source region of immigrants between 1820 and 1870 (Phase 2). During this phase, however, Irish and Germans dominated the flow and provided more cultural variety.

As Figure 3.20 shows, immigration reached a much higher peak around 1900, when almost 1 million foreigners entered the United States *annually*. During Phase 3 (1870–1920), the majority of immigrants were southern and eastern Europeans. Political strife and poor economies in Europe existed during this period, and news of available land and expanding industrialization in the United States offered an escape. By 1910, almost 14 percent of the nation was foreign-born. Very few of these immigrants, however, settled in the job-poor U.S. South, creating a cultural divergence that still exists.

Between 1920 and 1970 (Phase 4), more immigrants came to the United States from neighboring Canada and Latin America, but overall totals fell sharply, a function of more-restrictive federal immigration policies (the Quota Act of 1921 and the National Origins Act of 1924), the Great Depression, and the disruption caused by World War II.

Immigration has sharply increased since 1970 (Phase 5), and now annual arrivals surpass those of a century ago (see Figure 3.20). Most legal migrants since 1970 have originated in Latin America or Asia. In 2000, about 60 percent of immigrants were Hispanics, and only 20 percent were Asians, but by 2010 the balance shifted, with 36 percent from diverse Asian settings and only about 30 percent from Latin America. The post-1970 surge was due to economic and political instability abroad, a growing post-war American economy, and loosening immigration laws. Undocumented immigration, particularly from Mexico, rose after 1970, but since 2008 the pace has slowed appreciably, mostly because of fewer job opportunities in the United States and a more rigorous U.S. border patrol. Today the United States is home to about 11–12 million undocumented immigrants.

The U.S. Hispanic population continues to grow, fundamentally reshaping North America's cultural and economic geography (Figure 3.21). An estimated 12 million Mexican-born residents (about 10 percent of Mexico's population) now live in the United States.

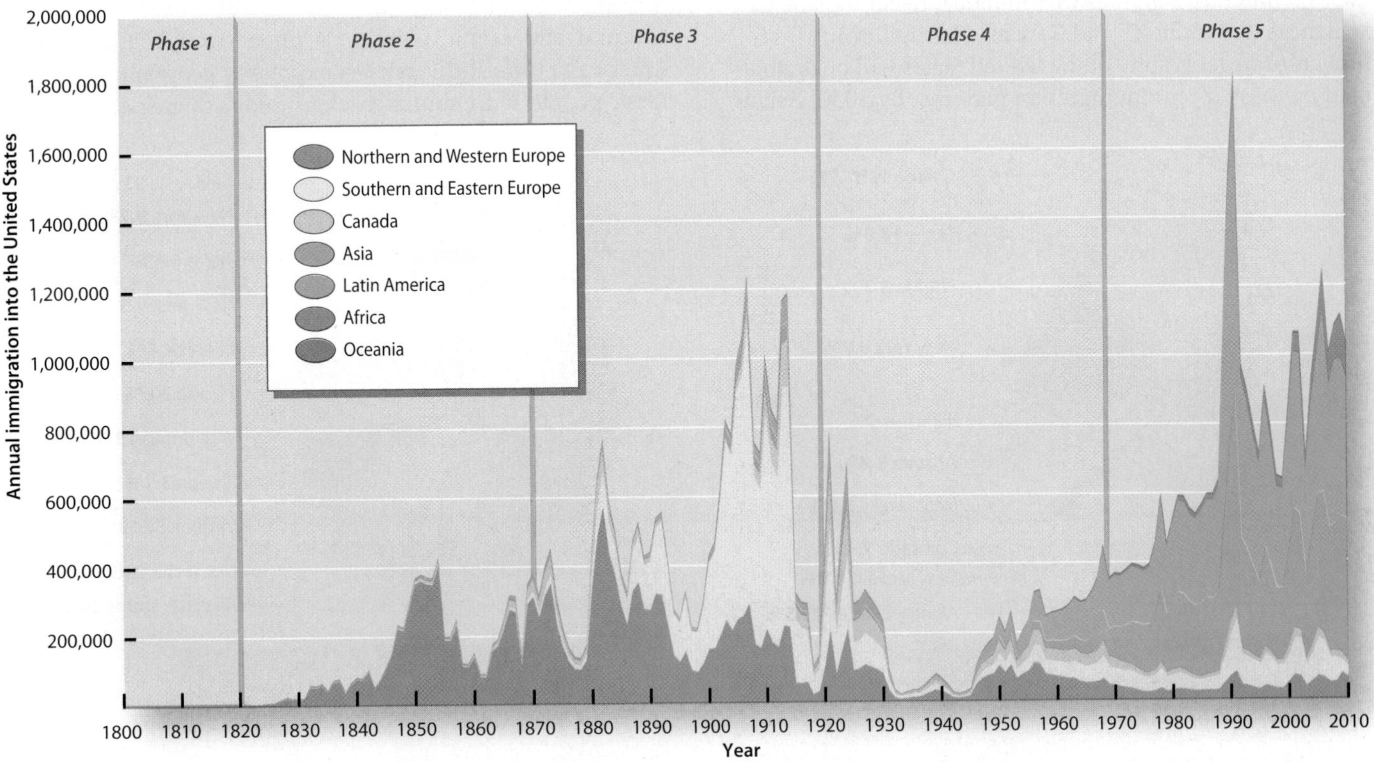

Figure 3.20 U.S. Immigration, by Year and Group Annual immigration rates peaked around 1900, declined in the early 20th century, and then surged again, particularly beginning in 1970. The source areas of these migrants have also shifted. Note the decreased role Europeans currently play and the growing importance of Asians and Latin Americans.
Q: What were the dominant source regions of the U.S. population in 1850, 1910, and 1980? Why did they change?

Figure 3.21 Latino Farm Workers near Salinas, California Latinos make up the vast majority of California's 650,000 farm workers. These migrant workers are picking strawberries in the Salinas Valley. Although the days are long, farm worker unions now mandate accessible toilets in the workplace (distant right).

In the next 25 years, however, most of the projected increase in the U.S. Hispanic population will be fueled by births within the country, rather than by new immigrants. Almost half of U.S. Hispanics live in California (27 percent of California's population is foreign-born) or Texas, but they are increasingly moving to other states such as South Carolina, Wisconsin, and Kansas (Figure 3.22). Across the Great Plains, Hispanic immigrants are bringing new churches, taquerías, and school-aged children to many once-dying communities.

In percentage terms, migrants from Asia constitute the fastest-growing immigrant groups, and various Asian ethnicities, both native and foreign-born, account for 6 percent of the U.S. population. Chinese is the third most commonly spoken language in the United States (behind English and Spanish). Indian immigrants make up one of the wealthiest and best-educated groups. California remains a key entry point for migrants and is home to one-third of the nation's Asian population, while another 15% now live in the mid-Atlantic states of New York and New Jersey (Figure 3.22). The largest Asian groups in the United States include Chinese (4.0 million), Filipinos (3.4 million), Asian Indians (3.2 million), Vietnamese (1.7 million), and Koreans (1.7 million).

The future cultural geography of the United States will be dramatically redefined by these recent immigration patterns. By 2050, Asians may total almost 10 percent of the U.S. population, and almost one in three Americans will be Hispanic. Indeed, it is likely that the U.S. non-Hispanic white population will achieve minority status by that date (Figure 3.23).

The Canadian Pattern French arrivals in the St. Lawrence Valley dominated the early European immigration to Canada. After 1765, many migrants came from Britain, Ireland, and the United States. Canada experienced the same surge and reorientation in migration flows seen in the United States around 1900. Between 1900 and 1920, more than 3 million foreigners ventured to Canada, an immigration rate far higher than that of the United States, given Canada's smaller population. Eastern Europeans, Italians, Ukrainians, and Russians dominated. Today about 60 percent of Canada's immigrants are Asians, and its 21 percent foreign-born population is among the highest in the developed world. In Vancouver, for example, about 40 percent of the population is foreign-born, creating a city of diverse ethnic neighborhoods, especially reflecting the city's strong ties to Asia.

Culture and Place in North America

Cultural and ethnic identity is often strongly tied to place. North America's cultural diversity is expressed geographically in two ways. First, people with similar backgrounds congregate near one another

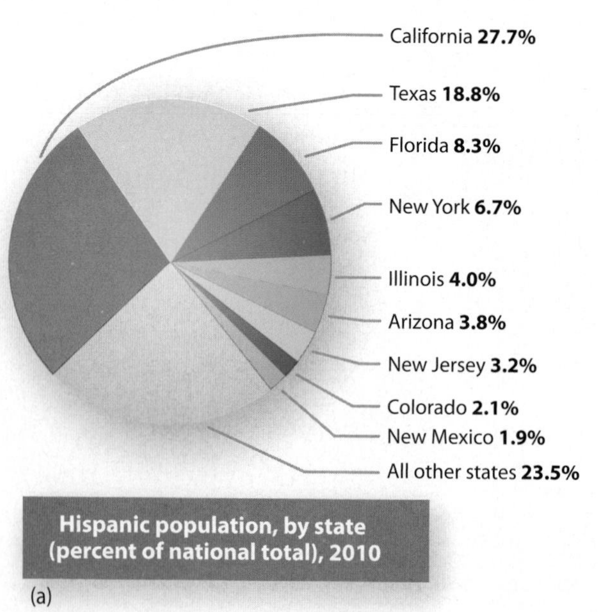

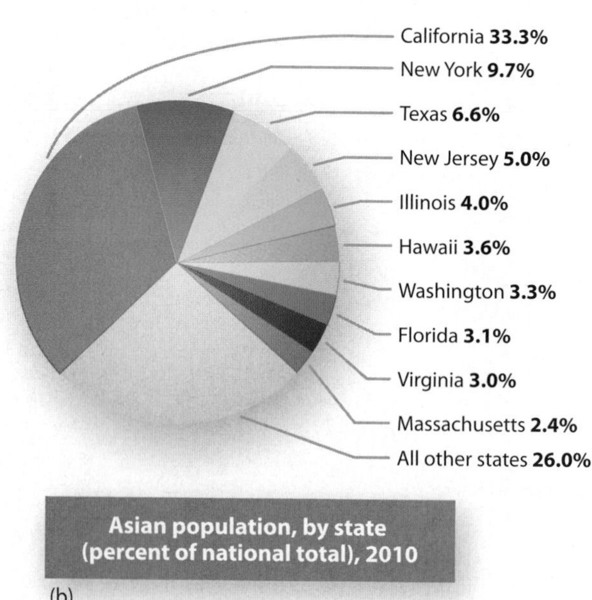

Figure 3.22 Distribution of U.S. Hispanic and Asian Populations, by State, 2010 California, Texas, and Florida (a) claim more than half of the nation's Hispanic population, but a growing number of Hispanics are locating elsewhere. California alone (b) is still home to one-third of the country's Asian population. **Q: Outside of the West and South, why do you think New York, New Jersey, and Illinois are also important destinations for these immigrants?**

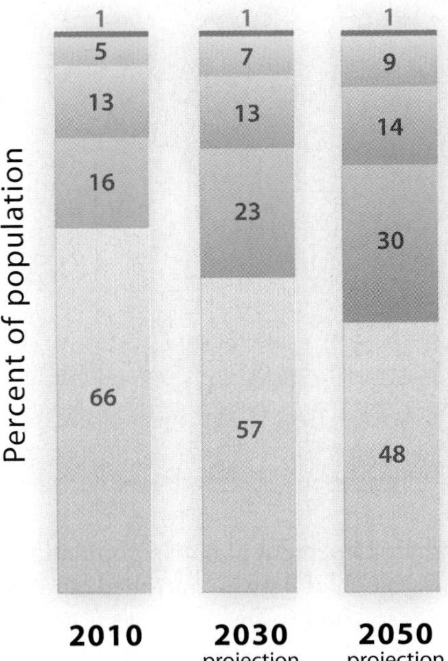

Figure 3.23 Projected U.S. Ethnic Composition, 2010 to 2050 By the middle of the 21st century, almost one in three Americans will be Hispanic, and non-Hispanic whites will assume minority status amid an increasingly diverse U.S. population.

and derive meaning from the territories they occupy together. Second, these distinctive cultures leave their mark—artifacts, habits, language, and values—on the everyday landscape. Boston's Italian North End simply looks and feels different from nearby Chinatown, and rural French Québec is a world away from a Hopi village in Arizona.

Cultural Homelands French-Canadian Québec is an excellent example of a **cultural homeland**: It is a culturally distinctive settlement in a well-defined geographic area, and its ethnicity has survived over time, stamping the landscape with an enduring personality (Figure 3.24). About 80 percent of Québec's population speaks French, and language remains the "cultural glue" that holds the homeland together. Policies adopted after 1976 strengthened the French language within the province by requiring French instruction in schools and national bilingual programming by the Canadian Broadcasting Corporation (CBC). Many Québécois feel that the greatest cultural threat comes not from Anglo-Canadians, but rather from recent immigrants to the province. Southern Europeans and Asians in Montreal, for example, show little desire to learn French, preferring instead to put their children in English-speaking private schools.

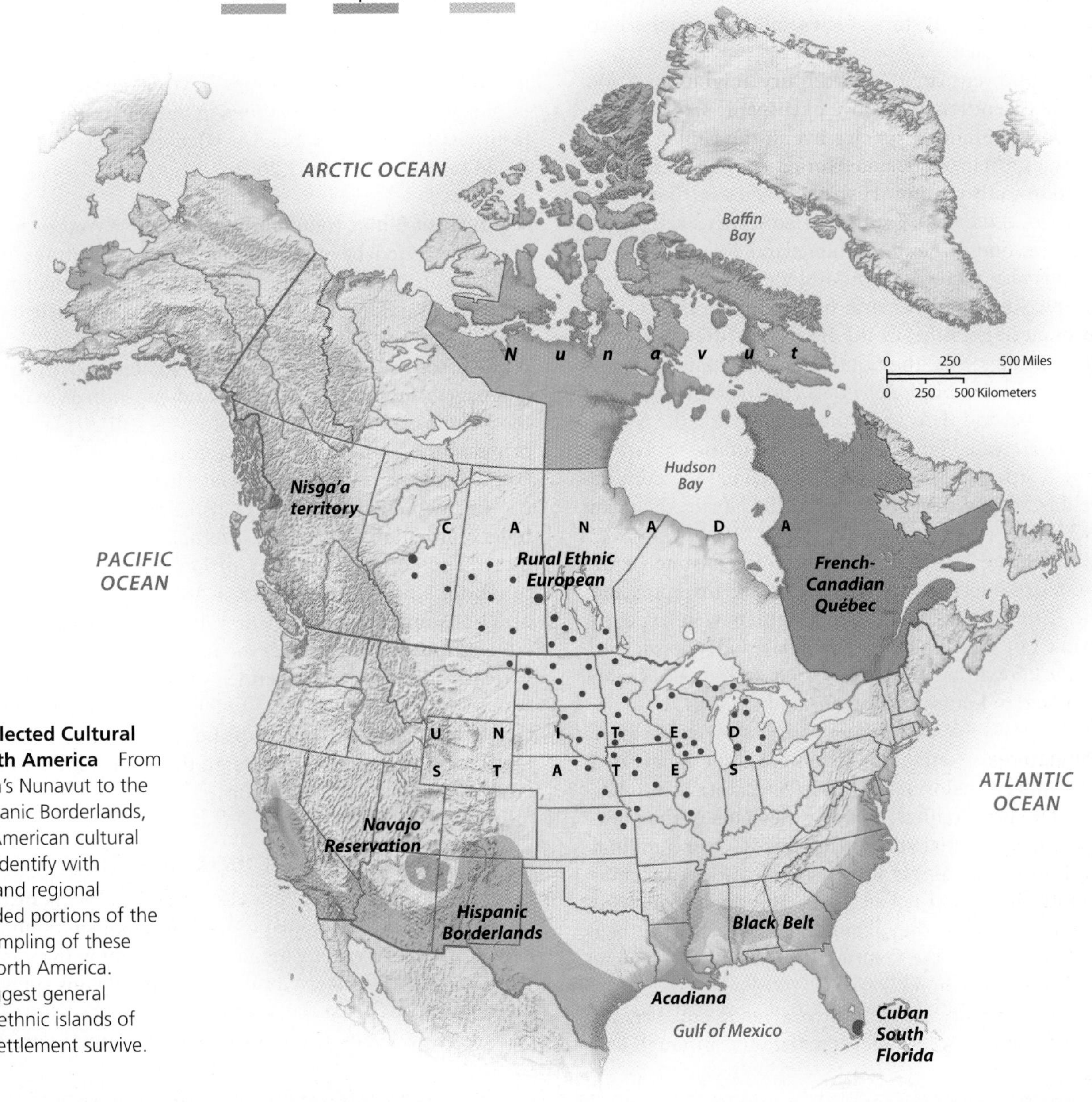

Figure 3.24 Selected Cultural Regions of North America From northern Canada's Nunavut to the Southwest's Hispanic Borderlands, different North American cultural groups strongly identify with traditional local and regional homelands. Shaded portions of the map display a sampling of these regions across North America. Dotted areas suggest general locations where ethnic islands of rural European settlement survive.

Figure 3.25 African-American Rap Artists, Denver, Colorado This colorful mural on Denver's west side celebrates the musical contributions of (left to right) Tupac Shakur, Biggie Smalls, Easy-E, and Krayzie Bone.

Another well-defined cultural homeland is the Hispanic Borderlands (see Figure 3.24). It is similar in size to French-Canadian Québec, but significantly larger in total population and more diffuse in its cultural and political expression. Historical roots of the homeland are deep, extending back to the 17th century, when Spaniards opened the region to the European world. Spanish place names, earth-toned Catholic churches, and traditional Hispanic settlements dot the rolling highlands of northern New Mexico and southern Colorado. From California to Texas, other historical sites and place names also reflect this rich Hispanic legacy.

Unlike Québec, however, large 20th-century migrations from Latin America brought an entirely new wave of Hispanic settlement to the Southwest. About 55 million Hispanics live in the United States, with about half in California, Texas, and Florida combined. Indeed, in 2015, Hispanics outnumbered non-Hispanic whites in California. New York City, Chicago, and Cuban South Florida serve as key points of Hispanic influence beyond this cultural homeland.

African Americans also retain a cultural homeland in the South, but it has become less important because of outmigration (see Figure 3.24). Dozens of rural counties in the Black Belt still have large African-American majorities, and the South remains home to many black folk traditions, including black spirituals and the blues, musical forms now popular far beyond their rural origins. Outside the South, African Americans have created large, vibrant communities in largely urban settings in the Northeast, Midwest, and West, and their cultural influence has indelibly shaped these portions of the North American landscape (Figure 3.25).

Another rural homeland, Acadiana, is a zone of persisting Cajun culture in southwestern Louisiana (see Figure 3.24). This homeland was founded in the 18th century, when French settlers were expelled from eastern Canada (an area known as Acadia) and relocated to Louisiana. Nationally known today through food and music, Cajun culture is strongly linked to Louisiana's bayous and swamps.

Native American Signatures Native Americans are also strongly tied to their homelands. Indeed, many native peoples maintain intimate relationships with their surroundings, weaving elements of the natural environment into their material and spiritual lives. Over 5 million Native Americans, Inuits, and Aleuts live in North America, claiming allegiance to more than 1100 tribal bands. Although many Native Americans now live in cities, they retain close contact with their homelands. Place names, landscape features, and family ties cement this connection between people and place.

Particularly in the American West and the Canadian and Alaskan North, native peoples also control sizable reservations, although less than 25 percent of native populations reside on reservations. The largest block of native-controlled land in the lower 48 states is the Navajo Reservation in the Southwest. About 300,000 people claim allegiance to the Navajo Nation. To the north, Canada's self-governing Nunavut Territory (population about 35,000) is another reminder of the enduring presence of native cultural influence and emergent political power within the region (see Figure 3.24).

Although these homelands preserve traditional ties to the land, they are also settings for pervasive poverty, health problems, and increasing cultural tensions (Figure 3.26a). Within the United States, many Native American groups have taken advantage of the special legal status of their reservations to build gambling casinos and tourist facilities that bring in much-needed capital but also challenge traditional lifestyles (Figure 3.26b).

A Mosaic of Ethnic Neighborhoods North America's cultural mosaic is characterized by smaller-scale ethnic signatures that shape both rural and urban landscapes (see Figure 3.24). During settlement of the agricultural interior, immigrants often established close-knit communities. Among others, German, Scandinavian, Slavic, Dutch, and Finnish settlements took shape, held together by common origins, languages, and religions. Rural landscapes in Wisconsin, Minnesota, the Dakotas, and the Canadian prairies still display these cultural imprints in the form of folk architecture, distinctive settlement patterns, ethnic place names, and rural churches.

Ethnic neighborhoods are also a part of the urban landscape and reflect both global-scale and internal North American migration patterns. The ethnic geography of Los Angeles is an example of both economic and cultural forces at work (Figure 3.27). Because most of its economic expansion took place during the 20th century, the city's ethnic patterns reflect the movements of more recent migrants. African-American communities on the city's south side (Compton and Inglewood) are a legacy of black migration out of the South. Hispanic (East Los Angeles) and Asian (Alhambra and Monterey Park) neighborhoods are a reminder that about 40 percent of the city's population is foreign-born.

Patterns of North American Religion

Distinctive religious traditions also shape North America's cultural geography. Reflecting its colonial roots, Protestantism dominates within the United States, accounting for about 60 percent of the population (Figure 3.28). In some settings, hybrid American religions sprang from broadly Protestant roots. By far the most successful is the Church of Jesus Christ of Latter-Day Saints (Mormons), which claims more than 6 million

Figure 3.26 Native American Landscapes (a) A Navajo girl and her grandfather stand in front of the family home in the Navajo Nation. (b) Near Albuquerque, New Mexico, the Isleta Pueblo Indian Reservation operates the Isleta Resort and Casino, providing many new jobs for local native residents.

North American members, concentrated in Utah and Idaho. Although many traditional Catholic neighborhoods have lost population in the urban Northeast, Catholic numbers are growing in the West and South, reflecting both domestic migration patterns and higher rates of Hispanic immigration and births. Almost 40 percent of the Canadian population is Protestant, with the United Church of Canada claiming large numbers of followers (see Figure 3.28). French-Canadian Québec is a bastion of Catholic tradition and makes Canada's population (43 percent) distinctly more Catholic than that of the United States (24 percent).

Millions of other North Americans practice religions outside of the Protestant and Catholic traditions or are unaffiliated with traditional religions. Orthodox Christians congregate in the urban Northeast, where many Greek, Russian, and Serbian Orthodox communities were established between 1890 and 1920. Ukrainian Orthodox churches also dot the Canadian prairies of Alberta, Saskatchewan, and Manitoba. More than 5 million Jews live in North America, concentrated in East and West Coast cities. Many Muslims (6 million), Buddhists (1 million), and Hindus (1 million) also live in the United States.

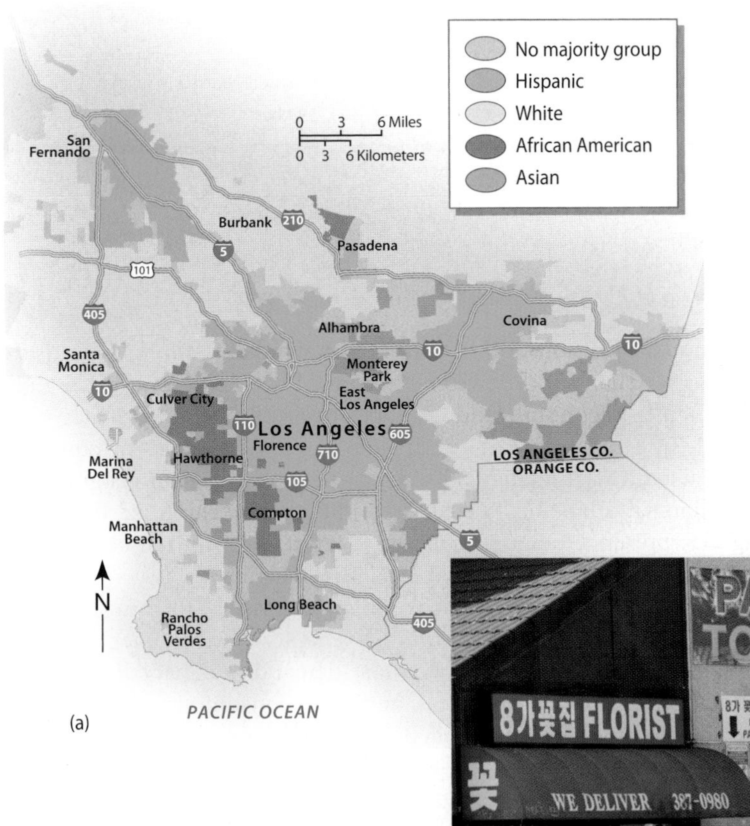

(b)

Figure 3.27 Ethnicity in Los Angeles (a) In many portions of Los Angeles, different ethnic signatures overlap. (b) A growing Latino population is taking over this portion of Koreatown.

The Globalization of American Culture

Simply put, North America's cultural geography is becoming more global at the same time that global cultures are becoming more North American (influenced particularly by the United States). But cultural globalization processes are complex; rather than simple flows of foreign influences into North America or of U.S. cultural dominance invading every traditional corner of the globe, the story of 21st-century cultural globalization increasingly mixes influences that flow in many directions at once, resulting in new hybrid cultural creations.

North Americans: Living Globally More than ever, North Americans in their everyday lives are exposed to people from beyond the region. With more than 47 million foreign-born migrants living across North America, diverse global influences mingle in new ways. Millions of international visitors come to the region annually, both for business and for pleasure. In U.S. colleges and universities, more than 760,000 international students (more than half from Asia) add global flavor to the classroom.

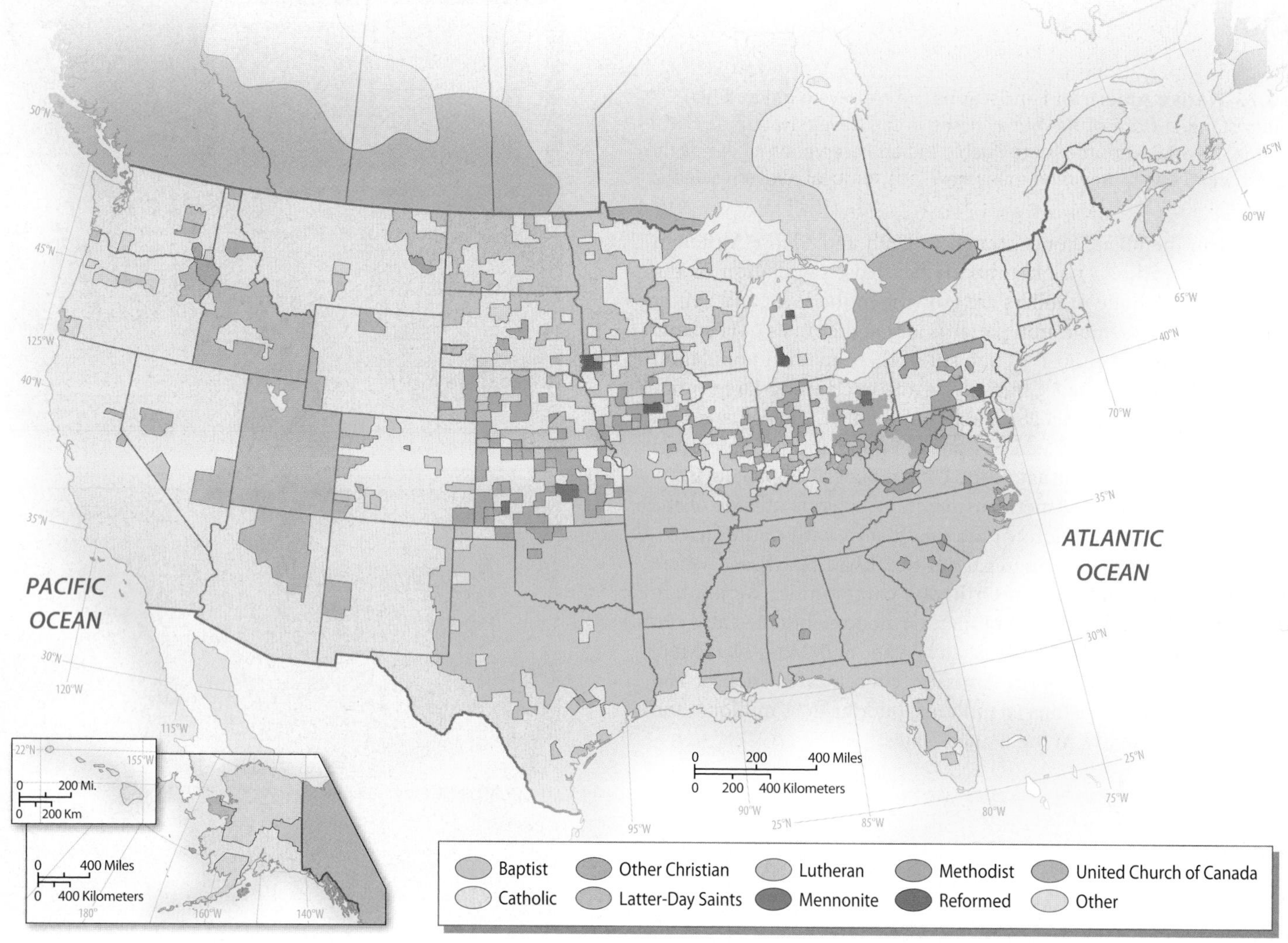

Figure 3.28 Christian Denominations of North America Although many parts of North America feature great religious diversity, Roman Catholicism or various Protestant denominations dominate select regions. Portions of rural Utah and Idaho dominated by the Mormon faith display some of the West's highest concentrations of any single religion.

Globalization presents challenges for North Americans. In the United States, one key issue revolves around the English language, which some have described as the "social glue" holding the nation together. Increasing use of **Spanglish**, a hybrid combination of English and Spanish spoken by Hispanic Americans, illustrates the complexities of North American globalization. Spanglish, an example of "code switching," where a speaker alternates between two or more languages, includes interesting hybrids such as *chatear*, which means "to have an online conversation."

North Americans are going global in other ways. By 2015, most Americans and Canadians had Internet access, launching far-reaching journeys in cyberspace. Social media such as Facebook and Twitter have for many North Americans redefined the kinds of communities and networks that shape their daily lives. The popularity of ethnic cuisine has peppered the region with a bewildering variety of Cuban, Ethiopian, Basque, and South Asian eateries (Figure 3.29). *Gucci, Brioni,* and *Prada* are household words for millions who keep their eye on European fashion, while German techno bands, Gaelic instrumentals, and Latin rhythms have become the soundtrack of daily life. Professional athletes from around the world migrate to North America to compete in basketball, baseball, and soccer (see *Everyday Globalization: The NBA Goes Global*). Indeed, from acupuncture and massage therapy to soccer and New Age religions, North Americans tirelessly borrow, adapt, and absorb the larger world around them.

The Global Diffusion of U.S. Culture In parallel fashion, U.S. culture has forever changed the lives of billions of people beyond the region. Although the economic and military power of the United States were notable by 1900, it was not until after World War II that the country's popular culture fundamentally reshaped global human geographies. The Marshall Plan and Peace Corps initiatives exemplified the growing U.S. presence on the world stage, even as European colonialism waned. Perhaps most critical was the marriage between growing global demands for consumer goods and the rise of the multinational corporation (discussed later in this chapter), which was superbly structured to meet and cultivate those demands. Global corporate advertising, distribution networks, and mass consumption bring Cokes and Big Macs to Moscow and Beijing, golf courses to Thai jungles, and Mickey and Minnie Mouse to Tokyo and Paris. Millions of people, particularly the young, are attracted by the North American emphasis on individualism, consumption, youth, and mobility.

But challenges to U.S. cultural control illustrate the varied consequences of globalization. Hollywood's dominance within the global film industry has declined dramatically as filmmakers have built their own movie businesses in India, Latin America, West Africa, China, and elsewhere. As worldwide use of the Internet has grown, the online dominance of English-speaking users has dramatically declined. Active resistance to U.S. cultural influence is also notable; examples

EVERYDAY GLOBALIZATION

The NBA Goes Global

Switch on the latest college or professional basketball game, or examine some of the NBA's team rosters; without a doubt, this quintessentially North American sport has gone global. Much of this worldly spin on the game came from former NBA Commissioner David Stern, who promoted televised games overseas (the NBA now broadcasts in 43 languages) and helped recruit talented international players. Stern recognized that an increasingly interconnected world was ready for basketball and that international talent could energize the American game.

Between 2007 (60 players from 28 countries) and 2015 (101 players from 37 countries), the NBA's growing global harvest was impressive. Recent top players have included brothers Marc and Pau Gasol (Spain), Dirk Nowitski (Germany), Hedo Türkoglu (Turkey), and Manu Ginóbili (Argentina). The global flow of players destined for the North American court is dominated by Europe, but the reach for talent extends deep into Latin America and Sub-Saharan Africa (Figure 3.3.1). The trend seems likely to continue as the game's global popularity—and lucrative NBA salaries—grow.

1. Name another American sport that has "gone global," and identify two players from foreign countries who exemplify the pattern.
2. In addition to basketball, what is a sport you enjoy playing or watching? Where did it originate and how did it spread to your community?

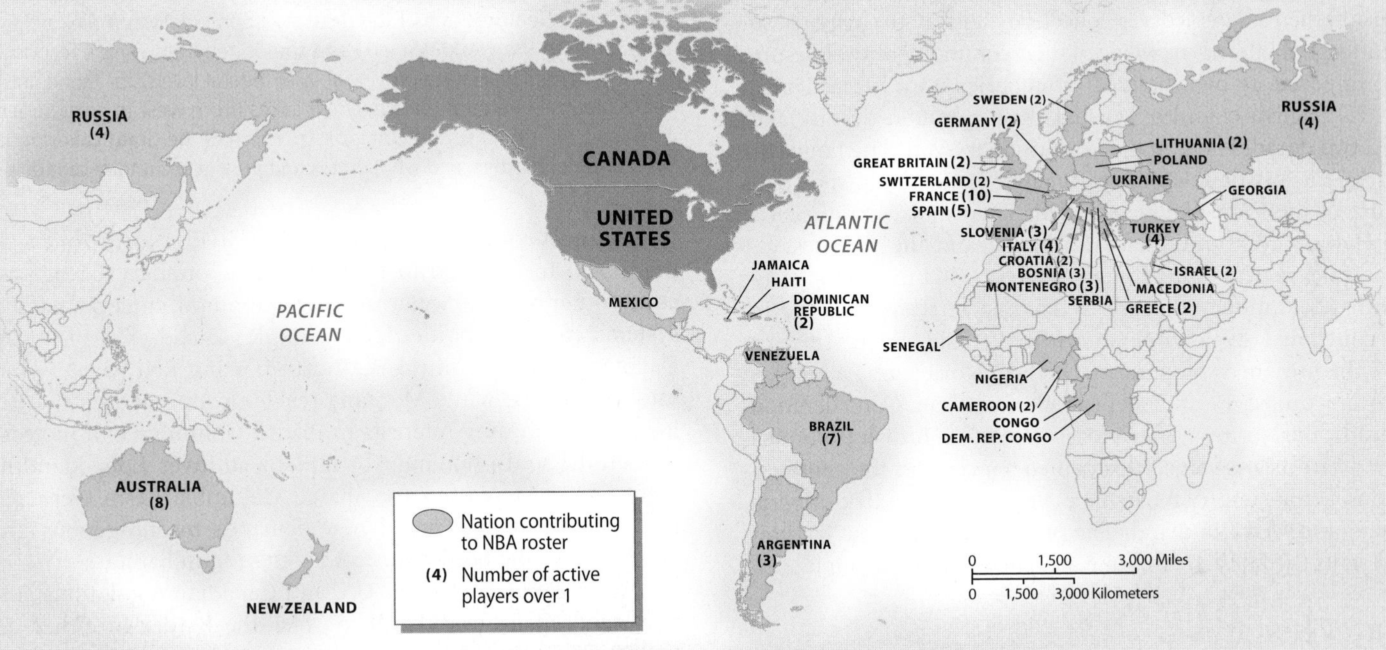

Figure 3.3.1 International Players in North American Professional Basketball, 2015 This map shows the national origins of active NBA players born outside the United States. Tall, sharpshooting immigrants from Europe and Latin America dominate the pattern, but émigrés from Senegal to Turkey have been fitted with NBA jerseys.

Figure 3.29 Indian Restaurant, New York City New York City's large South Asian population supports a wide diversity of small businesses.

include Canadian government efforts to limit U.S. popular culture in the country's radio, television, and film industries and French criticism of U.S. dominance in such media as the Internet. Iran continues to ban satellite dishes and many U.S. films, although illegal copies of top box-office hits often pierce national borders.

Review

3.6 Identify distinctive eras of immigration in U.S. history. How do they compare with those of Canada?

3.7 Identify four enduring North American cultural regions, and describe their key characteristics.

KEY TERMS ethnicity, cultural assimilation, cultural homeland, Spanglish

Geopolitical Framework: Patterns of Dominance and Division

North America is home to two of the world's largest states. Their creation, however, was neither simple nor preordained, but rather the result of historical processes that might have created quite a different North American map. Once established, these two states have coexisted in a close relationship of mutual economic and political interdependence.

Creating Political Space

The United States and Canada have very different political roots. The United States broke cleanly and violently from Great Britain. Canada, in contrast, was a country of convenience, born from a peaceful separation from Britain and then assembled as a collection of distinctive regional societies that only gradually acknowledged their common political destiny.

Europe imposed its own political boundaries on a future United States. The 13 English colonies, sensing their common destiny after 1750, united two decades later in the Revolutionary War. The Louisiana Purchase (1803) nearly doubled the national domain, and by the 1850s the remainder of the West had been added. The acquisition of Alaska (1867) and Hawaii (1898) rounded out what became the 50 states.

Canada was created under quite different circumstances. After the American Revolution, England's remaining territories in the region were controlled by administrators in British North America, and in 1867 the provinces of Ontario, Québec, Nova Scotia, and New Brunswick were united in an independent Canadian Confederation. Within a decade, the Northwest Territories, Manitoba, British Columbia, and Prince Edward Island joined this confederation, and the continental dimensions of the country took shape. Later infilling added Alberta, Saskatchewan, and Newfoundland. The creation of Nunavut Territory (1999) represents the latest change in Canada's political geography.

Continental Neighbors

Geopolitical relationships between Canada and the United States have always been close: Their common 5525-mile (8900-km) boundary requires both nations to pay attention to one another. During the 20th century, the two countries lived largely in harmony. In 1909, the Boundary Waters Treaty created the International Joint Commission, an early step in the common regulation of cross-boundary issues involving water resources, transportation, and environmental quality. The St. Lawrence Seaway (1959) opened the Great Lakes region to better global trade connections. With the signing of the Great Lakes Water Quality Agreement (1972) and the U.S.–Canada Air Quality Agreement (1991), the two nations have joined more formally in cleaning up Great Lakes pollution and in reducing acid rain in eastern North America. In 2012, these environmental agreements were updated, opening the way for new cooperative cleanup efforts between the two countries (Figure 3.30).

Close political ties also have strengthened trade. The United States receives about three-quarters of Canada's exports and supplies almost two-thirds of its imports. Conversely, Canada accounts for roughly 20 percent of U.S. exports and 15 percent of its imports. A bilateral Free Trade Agreement, signed in 1989, paved the way five years later for the larger **North American Free Trade Agreement (NAFTA)**, which extended the alliance to Mexico. Paralleling the success of the European Union (EU), NAFTA has forged the world's largest trading bloc, including more than 450 million consumers and a huge free trade zone stretching from beyond the Arctic Circle to Latin America.

Figure 3.30 Satellite Image of the Great Lakes North America's Great Lakes region features one of the most environmentally complex political boundaries in the world. Canada and the United States share responsibility (at a variety of local, state/provincial, and federal levels) for managing the ecological health of the five Great Lakes (from west to east: Superior, Michigan, Huron, Erie, and Ontario). **Q: Which of the Great Lakes define international borders, and what states and provinces do they separate?**

Political conflicts occasionally still divide North Americans (Figure 3.31). In addition to intricate trans-boundary Great Lakes concerns, other regional water issues are common, since so many drainage systems cross the border. For example, Canada has protested North Dakota's plans to control the north-flowing Red River (which leads into Manitoba), while Montana residents are nervous that Canadian logging and mining interests in British Columbia will increase pollution on the south-flowing North Flathead River. Long-standing agreements on dams within the shared Columbia Basin were also being renegotiated in 2014 amid new demands by indigenous groups on both sides of the border for expanded salmon habitat.

More generally, tighter U.S. and Canadian regulations since 2009 have made it more difficult to cross the border in either direction. Reflecting security concerns in the United States, the world's longest "open border" now sees more surveillance drones and border agents than ever before. Persons crossing the border now must present a passport or other approved form of identification, just as they would on the border with Mexico.

Agricultural and natural resource issues occasionally cause controversy between the two neighbors. The appearance of mad cow disease in Canadian livestock has curtailed exports to the United States and elsewhere. Canadian wheat and potato growers have been periodically accused of dumping their products into U.S. markets, thus depressing prices and profits for U.S. farmers. Similar issues have arisen in the logging industry, although a 2006 bilateral agreement has lessened tensions. Finally, the controversy over the completion of the Keystone XL Pipeline has frustrated various resource and environmental constituencies in both nations.

The Legacy of Federalism

The United States and Canada are **federal states** in that both nations allocate considerable political power to subnational units of government. Other nations, such as France, have traditionally been **unitary states**, in which power is centralized at the national level. Federalism leaves many political decisions to local and regional governments and often allows distinctive cultural and political groups to be recognized within a country. The U.S. Constitution (1787) limited

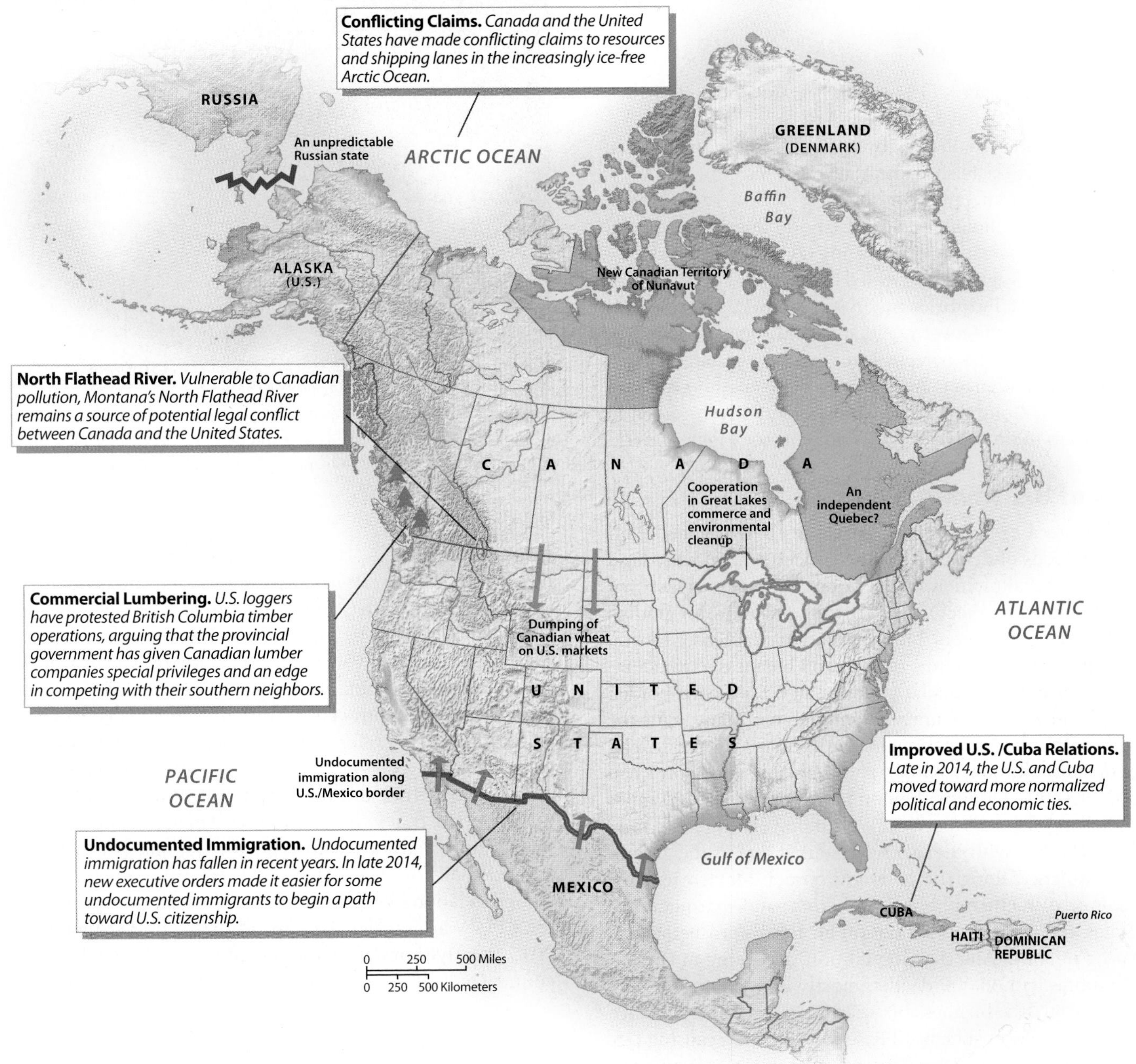

Figure 3.31 Geopolitical Issues in North America Although Canada and the United States share a long and peaceful border, many political issues still divide the two countries. In addition, internal political conflicts cause tensions, particularly in multicultural Canada.

centralized authority, giving all unspecified powers to the states or the people. In contrast, the Canadian Constitution (1867) created a federal state under a parliamentary system, giving most powers to central authorities. Ironically, the evolution of the United States produced an increasingly powerful central government, while Canada's geopolitical balance of power shifted toward more provincial autonomy and a relatively weak national government.

Québec's Challenge The political status of Québec remains a major issue in Canada (see Figure 3.31). Economic disparities between the Anglo and French populations have reinforced cultural differences between the two groups, with the French Canadians often suffering when compared with their wealthier neighbors in Ontario. Beginning in the 1960s, a separatist political party in Québec (the Parti Québecois) increasingly voiced French-Canadian concerns. When the party won provincial elections in 1976, it declared French the official language of Québec. Formal provincial votes over the question of Québec's independence were held in 1980 and 1995. Both measures failed. Since then, support for separation has ebbed in favor of a more modest strategy of increased "autonomy" within Canada.

Native Peoples and National Politics Another challenge to federal political power comes from North American Indian and Inuit populations in both countries. Within the United States, Native Americans asserted their political power in the 1960s, marking a decisive turn away from assimilation policies. Passage of the Indian Self-Determination and Education Assistance Act of 1975 has increased Native Americans' control of their economic and political destiny. The Indian Gaming Regulatory Act (1988) offered potential economic independence for many tribes. In 2014, Indian gaming operations (primarily gambling casinos) nationally netted tribes about $27 billion. In the western American interior, where Native Americans control roughly 20 percent of the land, tribes are also solidifying their hold on resources, reacquiring former reservation acreage, and participating in political interest groups, such as the Native American Fish and Wildlife Society and the Council of Energy Resource Tribes. In Alaska, native peoples acquired title to 44 million acres (18 million hectares) of land in 1971 under the Alaska Native Claims Settlement Act.

In Canada, ambitious challenges by native peoples have yielded dramatic results. Canada established the Native Claims Office in 1975. Agreements with native peoples in Québec, Yukon, and British Columbia turned over millions of acres of land to aboriginal control and increased native participation in managing remaining public lands. By far the most ambitious agreement created Nunavut out of the eastern portion of the Northwest Territories in 1999 (Figure 3.32), representing a new level of native self-government in North America. Nunavut is home to 35,000 people (85 percent Inuit) and is the largest territorial/provincial unit in Canada. Agreements between the Canadian Parliament and British Columbia tribes (the Nisga'a) have resulted in similar moves toward more native self-government (see Figure 3.24).

Figure 3.32 Life in Nunavut This woman navigates the unpaved streets of Pond Inlet on an ATV. The small hamlet is a rugged outpost on the northern end of Baffin Island, part of Canada's Nunavut Territory.

The Politics of U.S. Immigration

Immigration policies are hotly contested in the United States. Four key issues are the focus of debate. The first concerns how many legal immigrants should be allowed into the country. Some suggest that sharply reduced numbers protect American jobs and allow for gradual assimilation of existing foreigners, but others argue that looser restrictions could actually boost economic growth and business expansion.

A second major issue, particularly along the U.S.–Mexico border, is tightening daily flows of undocumented immigrants. Many argue the country's southern border is a national security issue. Recent federal legislation mandates additional border patrol agents, and more than 20,000 officers presently monitor the boundary. More than 700 miles (1125 km) of fencing also have been built or improved (Figure 3.33).

Third, U.S. relations with Mexico have soured due to the growth of drug-related violence along the common border. Mexico remains the leading source of methamphetamine, heroin, and marijuana for the United States and is a key transit nation for northward-bound cocaine originating in South America. In addition, according to Human Rights Watch, more than 60,000 deaths, mostly in northern Mexico, were tied to the illegal drug business between 2006 and 2013. Violence has spilled north into places such as El Paso and Phoenix, causing U.S. officials to worry that the Mexican government has lost effective political control of its northern border.

Finally, there is no political consensus on a policy to deal with existing undocumented workers. In late 2014, the Obama administration issued a series of executive orders (bypassing Congress) that effectively delayed deportation for many undocumented immigrants and also offered a path to citizenship for immigrants who had been in the United States for at least five years and were parents of citizens or legal permanent residents. While many Republicans criticized the plan because it granted a path to amnesty for some undocumented residents, others felt it did not go far enough in addressing the needs of millions more who remain in the country and who seek American citizenship.

A Global Reach

The geopolitical reach of the United States, in particular, extends far beyond the region's borders.

World War II and its aftermath forever redefined the U.S. role in world affairs. The United States emerged from the conflict as the world's dominant political power. It also developed multinational political and military agreements, such as the North Atlantic Treaty Organization (NATO) and the Organization of American States (OAS). Conflicts in Korea (1950–1953) and Vietnam (1961–1975) pitted U.S. political interests against communist attempts to extend control beyond the Soviet Union and China. Even as the Cold War faded during the late 1980s, the global political reach of the United States expanded. Direct involvement in conflicts within Central America, the Middle East, Serbia, and Kosovo exemplified the country's global agenda. Recent controversial wars in Iraq (2003–2011) and Afghanistan (2001–2015) offer further evidence of America's global political presence. The growing U.S. involvement in efforts to defeat ISIL (Islamic State of Iraq and the Levant, also known as ISIS) in 2014 and 2015 demonstrates the continuing American presence in that troubled region. Tensions with Iran, Russia, and North Korea are also a reminder of how the global scene remains unpredictable, while improved relations with Cuba since 2014 mark a profound shift from the Cold War era. Defense expenditures of more than $580 billion in 2015 (nearly as much as the rest of the world combined) suggest the country will continue to play a highly visible role in global affairs.

Figure 3.33 International Border North America's southwestern landscape is boldly bifurcated by an increasingly hardened international border that separates the United States and Mexico.

Review

3.8 How do the political origins of the United States and Canada differ?

3.9 What are four of the key issues surrounding U.S. immigration policy?

KEY TERMS North American Free Trade Agreement (NAFTA), federal state, unitary state

Economic and Social Development: Geographies of Abundance and Affluence

North America possesses the world's most powerful economy and its wealthiest population. Its 355 million people consume huge quantities of global resources but also produce some of the world's most sought-after manufactured goods and services. The region's human capital—the skills and diversity of its population—has enabled North Americans to achieve high levels of economic development (Table A3.2).

An Abundant Resource Base

North America is blessed with numerous natural resources that provide diverse raw materials for development. Indeed, the direct extraction of natural resources still makes up 3 percent of the U.S. economy and more than 6 percent of the Canadian economy. Some of these resources are exported to global markets, while other raw materials are imported to the region, as discussed in the previous section.

Agriculture remains a dominant land use across much of the region (Figure 3.34), but in a highly commercialized, mechanized, and specialized form that emphasizes efficient transportation, global markets, and large capital investments in farm machinery. Agriculture employs only a small percentage of the labor force in both the United States (1 percent) and Canada (2 percent), and the number of farms has sharply dropped, while average farm size has steadily risen.

The geography of North American farming represents the combined impacts of (1) diverse environments; (2) varied continental and global markets; (3) historical patterns of settlement and agricultural evolution; and (4) the role of **agribusiness**, or corporate farming. Agribusiness involves large-scale business enterprises that control closely integrated segments of food production, from farm to grocery store. In the Northeast, dairy operations and truck farms take advantage of proximity to major cities in Megalopolis and southern Canada. Corn, soybeans, and livestock production dominate the Midwest and western Ontario. To the south, the old Cotton Belt has been largely replaced by subtropical specialty crops; poultry, catfish, and livestock production; and commercial logging. Extensive grain-growing operations stretch from Kansas to Saskatchewan and Alberta, while irrigation allows agricultural production in the far West, depending on surface and groundwater resources. Indeed, California, nourished by large agribusiness operations in the irrigated Central Valley, accounts for more than 10 percent of the U.S. farm economy.

North Americans produce and consume huge quantities of other natural resources. The region consumes 40 percent more oil than all of the European Union, and production of fossil fuels is on the rise. Key areas of oil and gas production are the Gulf Coast, the Central Interior (North Dakota became a major producing state after 2000), Alaska's North Slope, and central Canada (especially Alberta's oil sands). The most abundant fossil fuel in the United States is coal (27 percent of the world's total), but its relative importance in the overall energy economy declined in the 20th century as industrial technologies changed and environmental concerns grew.

North America also remains a major producer of metals, although global competition, rising extraction costs, and environmental concerns pose challenges for this sector of the economy.

Creating a Continental Economy

The timing of European settlement in North America was critical in its rapid economic transformation. The region's abundant resources came under the control of Europeans possessing new technologies that reshaped the landscape and reorganized its economy. By the 19th century, North Americans actively contributed to those technological changes. New natural resources were developed in the interior, and new immigrant populations arrived in large numbers. In the 20th century, although natural resources remained important, industrial innovations and more service-sector jobs added to the economic base and extended the country's global reach.

Connectivity and Economic Growth North America's economic success was a function of its **connectivity**, or how well its different locations became linked with one another through vastly improved transportation and communications networks. Those links greatly facilitated the interaction between locations and dramatically reduced the cost of moving people, products, and information, thereby laying the foundation for urbanization, industrialization, and the commercialization of agriculture.

Technological breakthroughs revolutionized North America's economic geography between 1830 and 1920. By 1860, more than 30,000 miles (48,000 km) of railroad track had been laid in the United States, and the network grew to more than 250,000 miles (400,000 km) by 1910. Midwestern and Plains farmers found ready markets for their products in cities hundreds of miles away. Industrialists collected raw materials from faraway places, processed them, and shipped manufactured goods to their final destinations. The telegraph brought similar changes to information: Long-distance messages flowed across eastern North America by the late 1840s, and 20 years later, undersea cables linked the region to Europe, another milestone in the process of globalization.

Transportation and communications systems were modernized further after 1920. Automobiles, mechanized farm equipment, paved highways, commercial air links, national radio broadcasts, and dependable transcontinental telephone service reduced the cost of distance across North America. Perhaps most important, the region has taken the lead in the global information age, integrating computer, satellite, telecommunications, and Internet technologies in a web of connections that facilitates the flow of knowledge both within the region and beyond.

The Sectoral Transformation Changes in employment structure signaled North America's economic modernization just as surely as its increasingly interconnected society. **Sectoral transformation** refers to the evolution of a nation's labor force from one dependent on the *primary* sector (natural resource extraction) to one with more employment in the *secondary* (manufacturing or industrial), *tertiary* (services), and *quaternary* (information processing) sectors. For example, agricultural mechanization reduced demand for primary-sector workers but opened new opportunities in the growing industrial sector. In the 20th century,

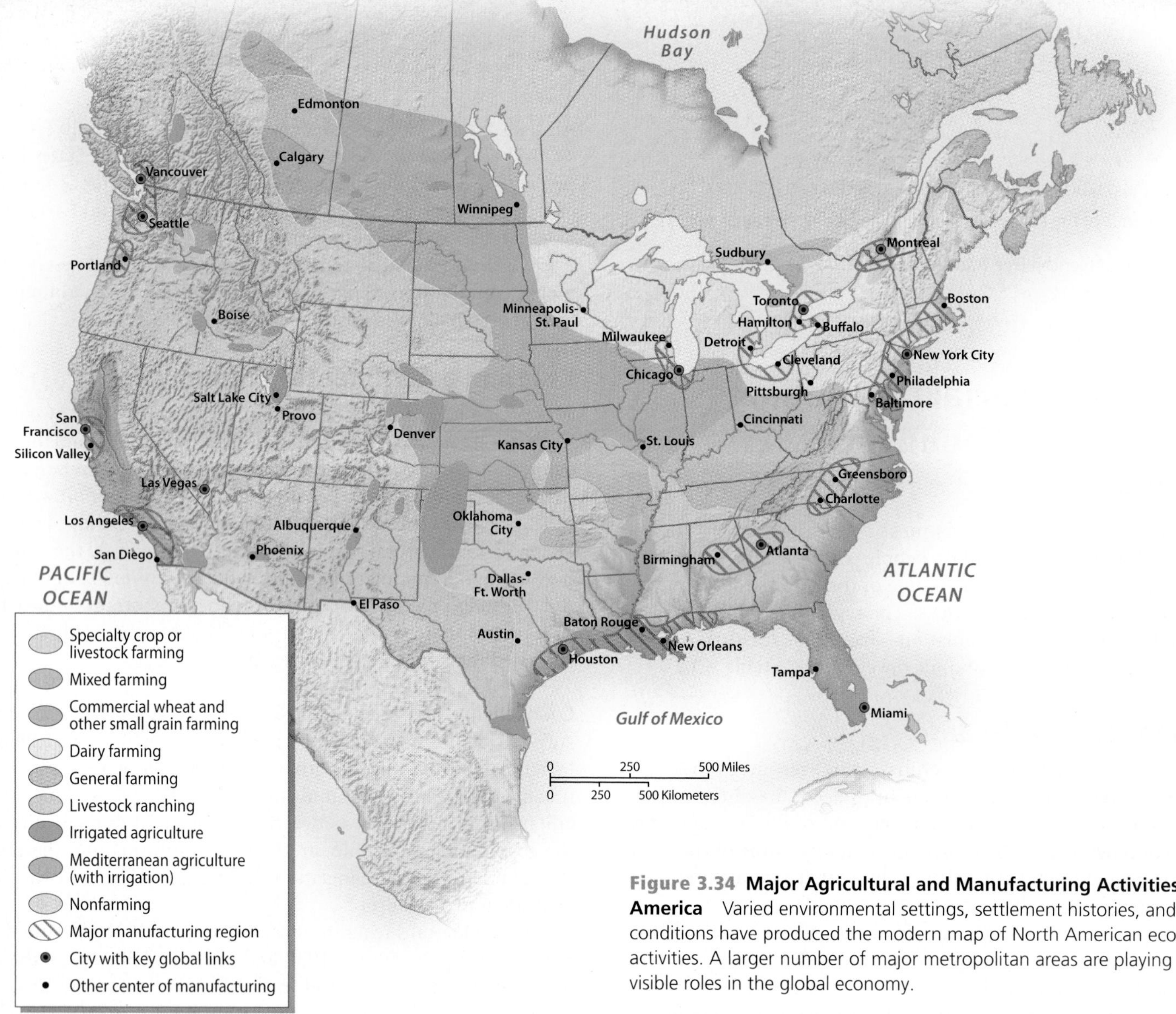

Figure 3.34 Major Agricultural and Manufacturing Activities of North America Varied environmental settings, settlement histories, and economic conditions have produced the modern map of North American economic activities. A larger number of major metropolitan areas are playing increasingly visible roles in the global economy.

new services (trade, retailing) and information-based activities (education, data processing, research) created other employment opportunities. Today the tertiary and quaternary sectors employ more than 70 percent of the U.S. and Canadian labor forces.

Regional Economic Patterns The locations of North America's industries show important regional patterns, influenced by various **location factors** that explain *why* an economic activity is located where it is and *how* patterns of economic activity are shaped. Patterns of industrial location illustrate the concept (see Figure 3.34). The historical manufacturing core includes Megalopolis (Boston, New York, Philadelphia, Baltimore, and Washington, DC), southern Ontario (Toronto and Hamilton), and the industrial Midwest. The region's proximity to *natural resources* (farmland, coal, and iron ore); its increasing *connectivity* (canals and railroad networks, highways, air traffic hubs, and telecommunications centers); its ready supply of *productive labor*; and a growing national, then global, *market demand* for its industrial goods encouraged continued *capital investment* within this industrial core. Traditionally, the core dominated in steel, automobiles, machine tools, and agricultural equipment and played a key role in financial and insurance services.

But in the last half of the 20th century, industrial- and service-sector growth shifted to the South and West. Cities of the South's Piedmont manufacturing belt (Greensboro to Birmingham) grew after 1960, partly because lower labor costs and Sun Belt amenities attracted new investment. North Carolina's "research triangle" area, encompassing Raleigh, Durham, and Chapel Hill, is the nation's third-largest biotech cluster, behind California and Massachusetts. The Gulf Coast industrial region is strongly tied to nearby fossil fuels that provide raw materials for the energy-refining and petrochemical industries (Figure 3.35).

The varied West Coast industrial region stretches from Vancouver, British Columbia, to San Diego, California (and beyond into northern Mexico), demonstrating the increasing importance of Pacific Basin trade. Large western aerospace firms also reflect the role of *government spending* as a location factor. Silicon Valley is now a leading region of manufacturing exports (Figure 3.36), and its proximity to Stanford, Berkeley, and other universities demonstrates the importance of *access to innovation and research* for many fast-changing high-technology industries. Silicon Valley's location also shows the advantages of *agglomeration economies*, in which companies with similar, often integrated manufacturing operations locate near one another. Smaller places such as Provo, Utah, and Austin, Texas, specialize in high-technology industries and demonstrate the growing role of *lifestyle amenities* in shaping industrial location decisions, both for entrepreneurs and for skilled workers attracted to such amenities.

North America and the Global Economy

Together with Europe, Japan, and China, North America is a key player in the global economy and is home to a growing number of

Figure 3.35 Gulf Coast Petroleum Refining Petroleum-related manufacturing has transformed many Gulf Coast settings. Much of Houston's 20th-century growth was fueled by the dramatic expansion of oil-related industries. The port of Houston remains a major center of North America's refining and petrochemical operations.

Figure 3.36 Silicon Valley The high-technology industrial landscape of California's Silicon Valley contrasts sharply with the look of traditional manufacturing centers. Here similar industries form complex links, benefiting from their proximity to one another and to nearby universities such as Stanford and Berkeley.

truly "global cities" that serve as key connecting points and decision-making centers in the world economy (see Figure 3.34). When the global economy is thriving, North America benefits, but in periods of international instability, globalization means that the region is more vulnerable to economic downturns.

Creating the Modern Global Economy The United States, with Canada's firm support, played a formative role in creating much of the new global economy and in shaping its key institutions. In 1944, allied nations met at Bretton Woods, New Hampshire, to discuss economic affairs. Under U.S. leadership, the group set up the International Monetary Fund (IMF) and the World Bank and gave these global organizations the responsibility for defending the world's monetary system. The United States was also the driving force for the creation (in 1948) of the General Agreement on Tariffs and Trade (GATT), renamed the **World Trade Organization (WTO)** in 1995. Its 161 member states are dedicated to reducing global trade barriers. The United States and Canada also participate in the **Group of Eight (G8)**, a collection of economically powerful countries (Japan, Germany, Great Britain, France, Italy, and sometimes Russia are the other members) that regularly meets to discuss key global economic and political issues.

Attracting Skilled Immigrants North America's role in the global economy attracts thousands of skilled workers from other countries, adding to the region's supply of human capital. Statistics gathered by the U.S. Department of Homeland Security point to the unique contributions of highly skilled immigrants. H-1B visas are granted to special "temporary skilled workers" to encourage computer programmers, doctors, and other professionals to work in the United States. Many other immigrants become urban entrepreneurs, starting new businesses that cater to their own communities as well as to larger metropolitan populations (Figure 3.37). Whether it is the Chinese in Vancouver and Toronto or the Cubans in Miami, immigrants in many of North America's largest, most global cities have made huge capital and human investments in their adopted communities. Almost 30 percent of the Korean-born and 20 percent of the Iranian-born populations in the United States are self-employed, a strong indicator of business ownership. The pattern offers a powerful reminder that the economic evolution of both the United States and Canada remains intimately connected to skilled immigrant populations (see *Geographers at Work: Lucia Lo and Toronto's Chinese Entrepreneurs*).

Doing Business Globally Patterns of capital investment and corporate power place North America at the center of global trade and money flows. The region attracts inflows of foreign capital, both as investments in North American companies and as foreign direct investment (FDI) by international companies. Multinational corporations based in North America also directly invest in foreign countries.

But the geography of 21st-century multinational corporations illustrates recent changes in broader patterns of globalization. One sign of the times can be measured by the home countries of the world's largest multinational companies. The 2014 *Forbes* list of "Global 2000" companies (largest as measured by revenues, assets, profits, and market value) included 674 from Asia, 629 from North America, and 506 from Europe. Many of these same multinationals are making huge investments of their own in the less developed world, from Africa to Southeast Asia, bypassing North American control altogether. The late-20th-century, top-down model of multinational corporate control and investment, traditionally based in North America, Europe, and Japan, is being replaced by a more globally distributed model of corporate control. This new model has many origins, many destinations, and new patterns of labor, capital, production, and consumption.

North Americans directly experience the consequences of these shifts in global capitalism. People in the United States are increasingly reacting to corporate **outsourcing**, a business practice that transfers portions of a company's production and service activities to lower-cost settings, often located overseas. Millions of jobs in manufacturing, textiles, semiconductors, and electronics have effectively migrated to places such as China, India, and Mexico, as those localities offer low-cost, less regulated settings for production, for both local and foreign firms. The results are complex: North American consumers benefit from cheap imports, but they may find their own jobs threatened in the corporate restructurings that make such bargains possible.

Enduring Social Issues

Profound socioeconomic problems shape North America's human geography. Despite the region's continental wealth, great differences persist between rich and poor. Race, particularly within the United States, continues

GEOGRAPHERS AT WORK

Toronto's Chinese Entrepreneurs

International migrants tend to relocate for political or economic reasons, but Lucia Lo, a York University geographer studying Canada's Chinese immigrant businesses, focuses on the "place and space dimension," noting that "every social phenomenon has a spatial manifestation, a geographical component." What she finds satisfying about geography is "integrating the various disciplines, rather than studying just economics, politics, or history." Lo uses a variety of data sources to explore how Chinese immigrants in Toronto secure business capital, where they locate, and how their businesses impact the city's economy. Her innovative approach combines statistics collected on Canadian businesses with on-the-ground case studies of particular immigrant entrepreneurs, both in Toronto's central city and in the suburbs. Says Lo, "It's always interesting to talk to people about life stories, but we also need some numbers and techniques—mapping, GIS, and so on—to make the picture complete. With quantitative and qualitative data, one can inform the other, raising new questions."

Immigrants Reshape the City What has Lo found? Toronto's Chinese immigrant entrepreneurs have successfully tapped into both local and global sources of capital in financing a variety of small and (increasingly) larger businesses (Figure 3.4.1). They tend to be younger and more educated than non-Chinese entrepreneurs, and they have extensively decentralized both their businesses and their residences into the suburbs. Overall, these immigrants have had a strongly positive impact on Toronto's economy. Lo has also examined some of the challenges immigrants face in utilizing available social services, especially in the outer suburbs, and how they successfully navigate both mainstream and ethnic networks. Lo's work provides policymakers and immigrant entrepreneurs with empirical evidence to facilitate future immigrant business formation.

1. Give examples of both quantitative data and qualitative information, and explain how combining these could improve a research project.

Figure 3.4.1 Suburban Toronto Store This Chinese-owned shop in the Toronto suburb of Richmond Hill offers customers an assortment of dried seafood, including shark fins, abalone, scallops, and sea cucumbers.

2. Identify and describe an immigrant-owned business in your own community or in a nearby town. What particular advantages or disadvantages might they have?

to be an issue of overwhelming importance. Both nations also face concerns related to gender inequity and challenges for aging populations.

Wealth and Poverty The regional landscape displays contrasting scenes of wealth and poverty. Elite northeastern suburbs, gated California neighborhoods, upscale shopping malls, and posh alpine ski resorts are expressions of private, exclusive landscapes characterizing wealthy North America. In contrast, substandard housing, abandoned property, aging infrastructure, and unemployed workers illustrate the gap between rich and poor. Rural poverty remains a major social issue in the Canadian Maritimes, Appalachia, the Deep South, the Southwest, and agricultural California. While many of the poorest Americans live in central cities, poverty is also moving to the suburbs. A 2013 Brookings Institute report noted that more poor people live in suburbs than in central cities and that poverty rates are growing much more rapidly in suburbs than in either inner cities or rural areas. Overall, about 13–16 percent of the U.S. and Canadian populations live in poverty. In the United States, about 27 percent of the country's African-American and Hispanic populations live below the poverty line.

Access to Education Education is also a major public policy issue in Canada and the United States. Although politics hampers consensus, most public officials agree that more investment in education can only improve North America's chances for competing successfully in the global marketplace. Despite steady improvements in graduation rates in both countries, dramatically lower numbers in many poor urban and rural districts suggest ongoing educational challenges. Race plays a key role: American whites are twice as likely to hold a college degree as African Americans.

Gender, Culture, and Politics Since World War II, both the United States and Canada have seen great improvements in the role that women play in society. However, the **gender gap** is yet to be closed when it comes to differences in salary, working conditions, and political power. Women make up more than half of the North American workforce and are often more educated than men, but they still earn only about 78 cents for every dollar

Figure 3.37 Immigrant Entrepreneurs These Vietnamese merchants own the Tien Hung Complete Oriental Foods and Gifts store in Orlando, Florida.

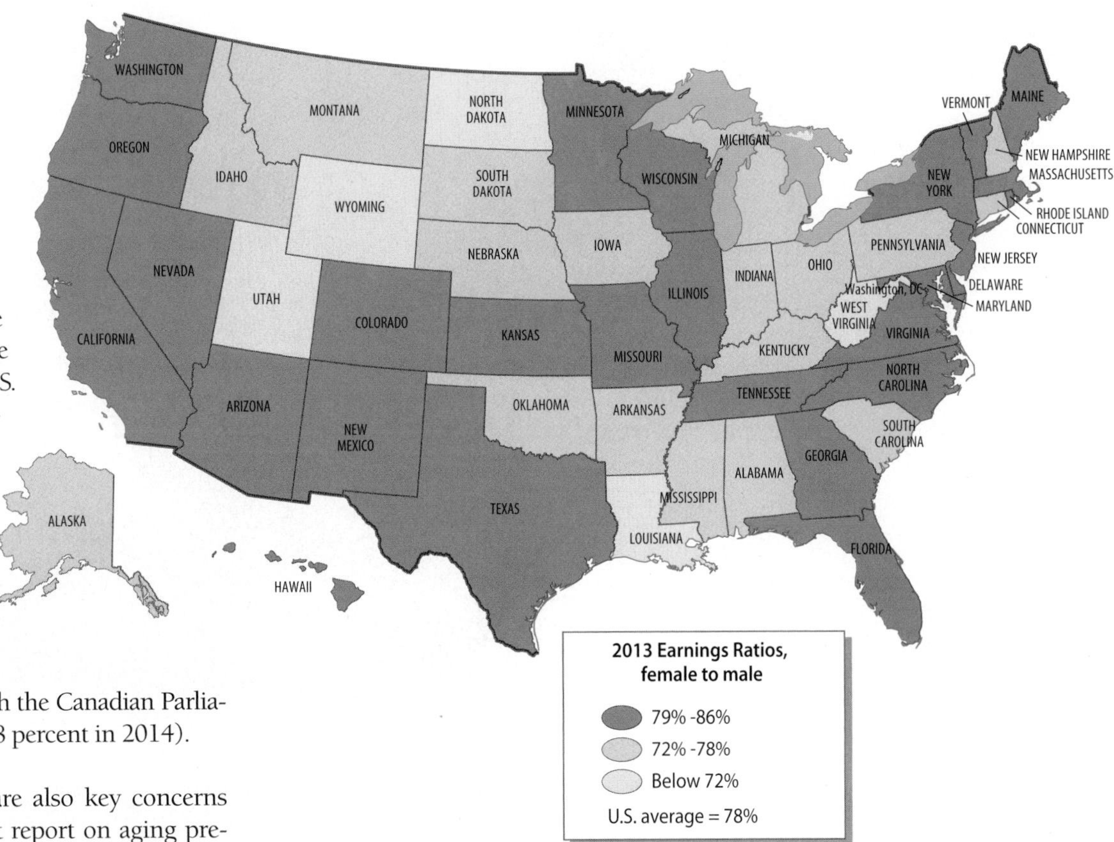

Source, AAUW

Figure 3.38 Earnings Ratios, by Gender, United States This map shows the relative median annual earnings for women versus men, by state. Note the relatively higher earnings for women in the Northeast and in California versus selected southern and midwestern settings. **Q: What variables might help explain how your own state fits into the larger national pattern?**

that men earn (Figure 3.38). Women also head the vast majority of poorer single-parent families in the United States, and more than 40 percent of all U.S. births are to unwed mothers. Canadian women, particularly single mothers working full-time, are also greatly disadvantaged, averaging only about 70 percent of the salaries of Canadian men.

Although women have played critical roles in deciding recent national elections, political power remains largely in male hands. Canadian women have voted since 1918 and U.S. women since 1920, but females in the early 21st century remain clear minorities in both the Canadian Parliament (25 percent in 2014) and the U.S. Congress (18 percent in 2014).

Health Care and Aging Aging and health care are also key concerns within a region of graying baby boomers. A recent report on aging predicted that 20 percent of the U.S. population will be older than 65 by 2030. Poverty rates are also higher for seniors. With fewer young people to support their parents and grandparents, officials debate the merits of reforming social security programs. Whatever the outcome of such debates, the geographical consequences of aging are already abundantly clear: Whole sections of the United States—from Florida to southern Arizona—have become increasingly oriented around retirement (Figure 3.39).

Health care remains a key issue in both countries. Both systems are costly by global standards: Canadians spend about 12 percent of their gross domestic product (GDP) on health care, and costs are even higher in the United States (over 15 percent of GDP). Canada offers an enviable system of government-subsidized universal health care to its residents (who pay higher taxes to fund it). The United States is gradually moving toward more universal coverage (although largely within a system of private insurers) through the Patient Protection and Affordable Care Act signed into law in 2010. Despite ongoing issues related to costs and access, the long life spans and low childhood mortality rates (see Table A3.2) of the region suggest that both countries reap many rewards from these modern health-care systems.

Yet rising rates of chronic diseases associated with aging (heart disease, cancer, and stroke are the three leading causes of death) will continue to pressure both health-care systems. In addition, hectic lives are often oriented around dining out at fast-food eateries, contributing to rapidly growing rates of obesity (the average American consumed 603 more calories daily in 2000 than in 1980). Almost two-thirds of adult Americans are overweight, often leading to heart disease and diabetes. Other strains on the region's health care include alcohol-related health problems and binge drinking, often linked with assault and domestic and sexual abuse. The cost of treating HIV/AIDS victims is another critical health-care issue, particularly among poorer black and Hispanic populations.

Figure 3.39 Tomorrow's Baby Boom Landscape? Hundreds of golf resorts and retirement communities have been built across North America's Sun Belt since 1980 and now cater increasingly to the baby boom generation. This is an aerial view of Sun City, Arizona.

Review

3.10 Define sectoral transformation. How does it help explain economic change in North America?

3.11 Cite five types of location factors, and illustrate each with examples from your local economy.

3.12 What common social issues are faced by both Canadians and Americans? How do they differ?

KEY TERMS agribusiness, connectivity, sectoral transformation, location factor, World Trade Organization (WTO), Group of Eight (G8), outsourcing, gender gap

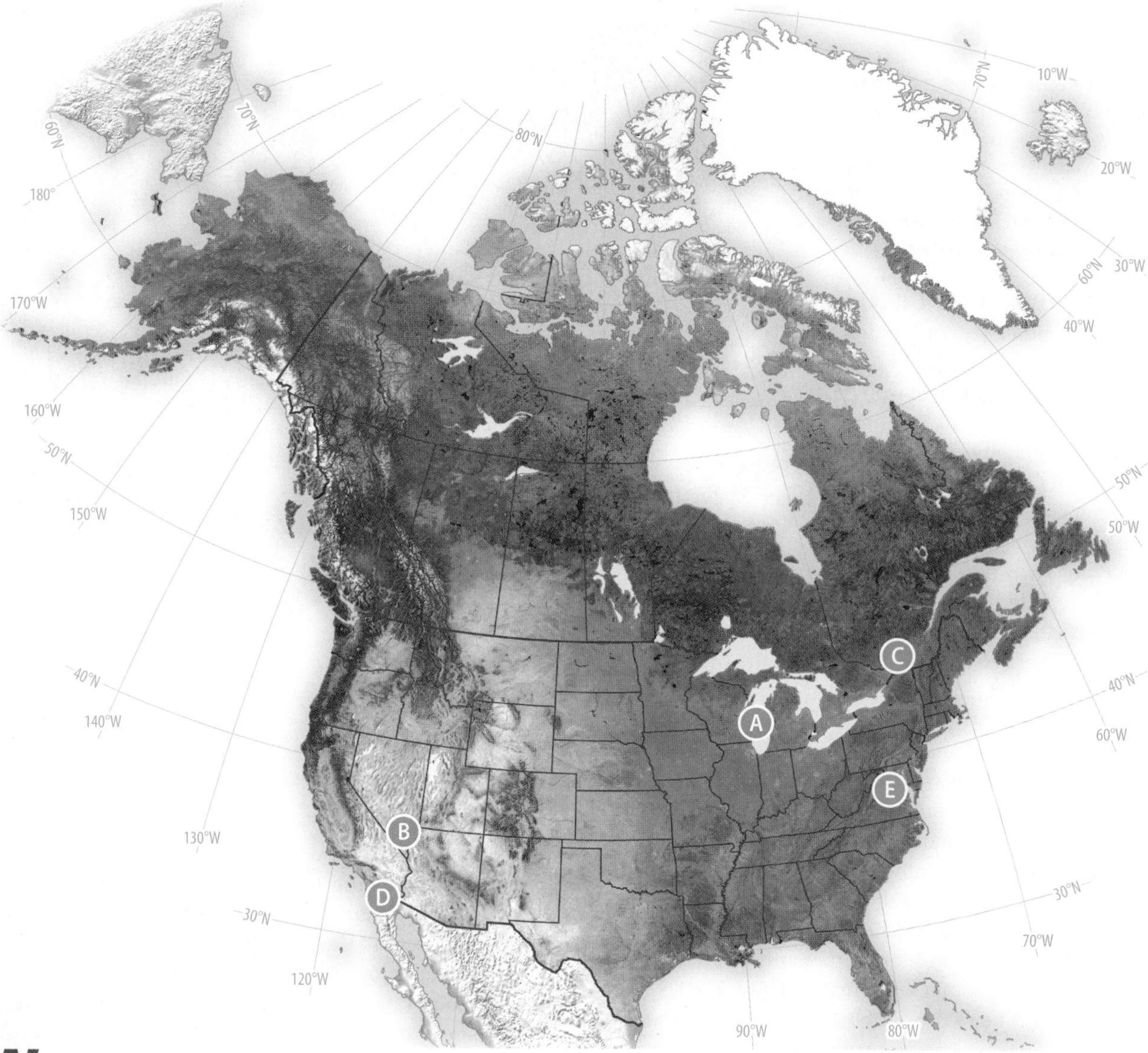

Review

Physical Geography and Environmental Issues

3.1 Describe North America's major landform and climate regions.

3.2 Identify key environmental issues facing North Americans and connect these to the region's resource base and economic development.

North Americans have reaped the natural abundance of their region, and in the process they have transformed the environment, created a highly affluent society, and extended their global economic, cultural, and political reach. North America's affluence has come with a considerable price tag, and today the region faces significant environmental challenges, including soil erosion, acid rain, and air and water pollution.

1. The yellow squares on the map indicate major hazardous waste sites. Why are so many sites concentrated along major rivers and near the Great Lakes?

2. Can you identify key hazardous waste sites in your area? What are the sources of the waste? Have these sites been cleaned up?

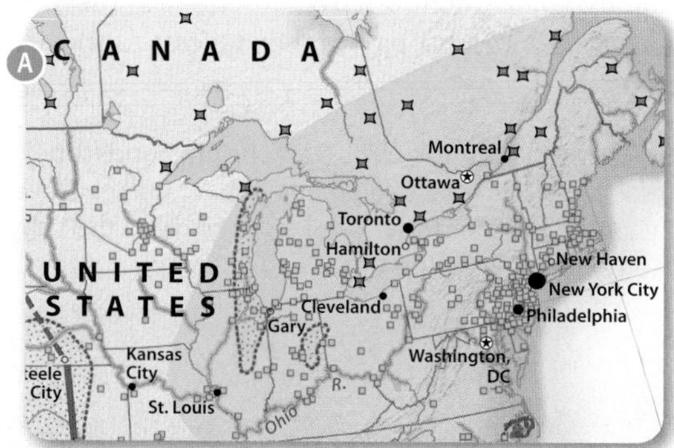

Population and Settlement

3.3 Analyze map data to identify and trace major migration flows in North American history.

3.4 Explain the processes that shape contemporary urban and rural settlement patterns.

In a remarkably short time, a unique mix of varied cultural groups from around the world has contributed to the settlement of a huge and resource-rich continent that is now one of the world's most urbanized regions.

3. What are some of the reasons for the rapid growth of Las Vegas since 1980?

4. In this harsh desert (especially vulnerable to future drought), describe a path to sustainability for maintaining southern Nevada's population in the next 50–100 years.

Cultural Coherence and Diversity

3.5 List the five phases of immigration shaping North America and describe the recent importance of Hispanic and Asian immigration.

3.6 Provide examples of major cultural homelands (rural) and ethnic neighborhoods (urban) within North America.

North Americans produced two societies that are closely intertwined but face distinctive national political and cultural issues. The Canadian identity remains problematic as it works through the persistent challenges of its multicultural character and the costs and benefits of its proximity to its dominating continental neighbor.

5. Why does one still find large numbers of French speakers in the Canadian province of Québec?

6. Is the French language likely to retain its cultural vitality in Québec over the next century? What key challenges does it face?

Geopolitical Framework

3.7 Contrast the development of the United States' and Canada's distinctive federal political systems, and identify each nation's current political challenges.

Canada and the United States enjoy a close political relationship, but several issues—often related to environmental quality, resource claims, and trade—still produce tensions between the two countries.

7. What are the key characteristics of a political "borderlands" zone such as this one along the Mexico–California border?

8. Immigration remains a key issue for North America. Organize a class debate on the pros and cons of sharply curtailing immigration in the future. What about opening up the region to even larger flows of immigrants?

Economic and Social Development

3.8 Discuss the role of key location factors in explaining why economic activities are located where they are in North America.

3.9 List and explain contemporary social issues that challenge North Americans in the 21st century.

North America displays great regional affluence, but enduring issues of poverty, health care, and gender equity continue to challenge both countries in the 21st century.

9. What national and global economic trends are illustrated by edge-city settings such as Tysons Corner, Virginia?

10. Identify an edge city or peripheral suburban shopping area near you. What economic activities are emphasized in these settings?

KEY TERMS

acid rain *(p. 75)*
agribusiness *(p. 95)*
boreal forest *(p. 71)*
connectivity *(p. 95)*
cultural assimilation *(p. 85)*
cultural homeland *(p. 87)*
edge city *(p. 83)*
ethnicity *(p. 84)*
federal state *(p. 92)*
fracking *(p. 77)*
gender gap *(p. 98)*
gentrification *(p. 83)*
Group of Eight (G8) *(p. 97)*
location factor *(p. 96)*
Megalopolis *(p. 78)*
new urbanism *(p. 83)*
nonmetropolitan growth *(p. 81)*
North American Free Trade Agreement (NAFTA) *(p. 92)*
outsourcing *(p. 97)*
postindustrial economy *(p. 68)*
prairie *(p. 71)*
renewable energy source *(p. 76)*
sectoral transformation *(p. 95)*
Spanglish *(p. 90)*
sustainable agriculture *(p. 76)*
tundra *(p. 71)*
unitary state *(p. 92)*
urban decentralization *(p. 81)*
urban heat island *(p. 74)*
World Trade Organization (WTO) *(p. 97)*

DATA ANALYSIS

http://goo.gl/CH6Ysy

Every decade, the Census Bureau gathers and summarizes an enormous amount of data for the United States. These data are used by planners and government agencies to forecast future needs for public infrastructure and social services. Age and sex distributions for cities and states can provide real insights into the social and economic characteristics of these settings. Population pyramids are convenient ways to visualize these characteristics (see Figure 3.13). Go to the Census Bureau's website (www.census.gov) and access the summaries and predictions of state populations.

1. Examine the 2010 and 2030 (projected) pyramids for Florida and Utah. Describe major similarities and differences for both years. Write a paragraph that summarizes reasons for these differences.
2. Select two additional states that display quite different population structures. Write a paragraph that summarizes and explains these differences.
3. From the point of view of a planner or budget expert, explain how the different population structures in the states you selected might impact future expenditures and trends in economic development in 2030 and beyond.

MasteringGeography™

Looking for additional review and test prep materials? Visit the Study Area in MasteringGeography™ to enhance your geographic literacy, spatial reasoning skills, and understanding of this chapter's content by accessing a variety of resources, including MapMaster interactive maps, geoscience animations, videos, *In the News* RSS feeds, flashcards, web links, self-study quizzes, and an eText version of *Globalization and Diversity*

Authors' Blogs

Scan to visit the
Author's Blog for field notes, media resources, and chapter updates

https://gad4blog.wordpress.com/category/north-america/

Scan to visit the
GeoCurrents Blog

http://www.geocurrents.info/category/place/north-america

4 Latin America

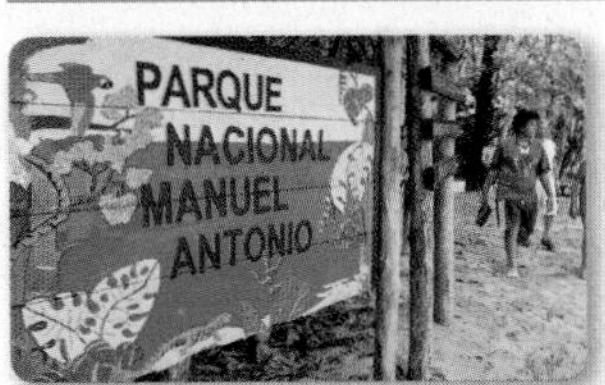

PHYSICAL GEOGRAPHY AND ENVIRONMENTAL ISSUES

Tropical forests in Latin America, especially in the Amazon Basin, are one of the planet's greatest reserves of biological diversity. How it will be managed is critical with increasing pressure to extract mineral wealth, build roads and dams, and convert forests into farms or pasture.

POPULATION AND SETTLEMENT

Latin America is the most urbanized region of the developing world, with 78 percent of the population in cities. Four megacities (>10 million people) are found here. Yet the region also has high emigration, especially to North America.

CULTURAL COHERENCE AND DIVERSITY

Amerindian activism is on the rise. Indigenous peoples from Central America to the Andes and the Amazon demand territorial and cultural recognition while Latin America is on a global stage as Brazil hosts the 2016 Olympics.

GEOPOLITICAL FRAMEWORK

As Latin American governments mark 200 years of independence from Spain, most are fully democratic. Recent elections have seen liberal democrats and populists gain power, promising to reduce income inequality through government programs. Women are also political actors in Latin America, holding nearly one-quarter of all seats in national parliaments.

ECONOMIC AND SOCIAL DEVELOPMENT

Economic growth, increased trade, and fewer people in extreme poverty are all positive trends for the region, but serious income inequality persists. Government programs such as Brazil's Bolsa Familia seek to address both social and economic well-being for poor families. Meanwhile, heightened violence and insecurity, especially in Central America, have more people living in fear and trying to leave.

◀ A monumental infrastructure project, the expansion of the Panama Canal with a new set of locks will increase the number and size of ships that use the canal when it opens in 2016. On the left side of the image, the new far larger locks are being constructed. To the right are the 100-year-old Gatun Locks with ships moving through them. In the background stretches Gatun Lake, a reservoir that was created to move ships across the country and to manage the water supply needed to run canal.

A powerful symbol of Latin America's role in global trade is the Panama Canal. Built by the United States, it opened in 1914. Since 2000, the canal has been controlled and managed by Panamanians. In 2016, a new set of locks will open, allowing for the passage of more and much larger ships. The size of the old locks limited access to vessels no more than 100 feet wide and 1000 feet long, the so-called Panamax vessels. The Panama Canal Authority, which manages the canal, anticipates total cargo levels will increase by 3 percent a year, doubling the 2005 tonnage by 2025. Ports around North America and Europe have also retrofitted to adjust for the arrival of larger post-Panamax ships. The timing of this expansion is especially important, considering potential competition from Arctic sea routes opening up as a consequence of climate change.

Yet just 400 miles north in Nicaragua a new project is under way, financed by Chinese billionaire Wang Jing. In 2014, the Nicaraguan Canal was approved, and the Hong Kong Nicaraguan Canal Development Investment (HKND) Group was formed. This canal is optimistically estimated to cost $50 billion and to be constructed in five years. If completed, it will be 170 miles (270 km) long and contain two locks. Workers have begun to build access roads to support the project, and local protests have resulted. The environmental implications for Lake Nicaragua, coastal zones, and tropical forests will be substantial. The economic justification for a second canal is questionable, and many doubt the project will be completed. But the Nicaraguan government envisions greater prosperity through global trade and has entered a long-term and complex partnership with China to build this canal.

Beginning with Mexico and extending to the tip of South America, Latin America has a regional unity that stems largely from its shared colonial history, rather than from different levels of development seen today. More than 500 years ago, Spain and Portugal began their conquest of the Americas and left lasting marks in Latin America: Officially, two-thirds of the population speaks Spanish; the rest speak Portuguese. Iberian architecture and town design add homogeneity to the colonial landscape. The largest concentration of Roman Catholics worldwide lives in the region, and in 2013 an Argentine priest became Pope Francis, the first Pope from the Americas.

> Through colonialism, immigration, and trade, the forces of globalization are embedded in Latin America's landscape.

European culture blended with those of various Amerindian peoples, who maintain a strong presence in Bolivia, Peru, Ecuador, Guatemala, and southern Mexico. After the initial colonial conquest, other cultural groups from Africa and Asia were added to this mix of native and Iberian peoples, making it one of the world's most racially diverse regions.

The concept of Latin America as a distinct region has been popularly accepted for nearly a century. The region's boundaries are straightforward, beginning at the Rio Grande (called the Rio Bravo in Mexico) and ending at Tierra del Fuego (Figure 4.1). French geographers coined the term *Latin America* in the 19th century to distinguish the Spanish- and Portuguese-speaking republics of the Americas plus Haiti from the English-speaking territories. There is nothing particularly "Latin" about the area, other than the predominance of Romance languages. The term stuck because it encompassed different colonial histories, while also offering a clear cultural distinction from Anglo-America, the region referred to as North America in this book.

This chapter defines Latin America as the Spanish- and Portuguese-speaking countries of Central and South America, including Mexico. This division emphasizes the important Amerindian and Iberian influences affecting mainland Latin America and separates it from the unique colonial and demographic history of the Caribbean and the Guianas, discussed in Chapter 5.

Learning Objectives

After reading this chapter you should be able to:

4.1 Explain the relationships among elevation, climate, and agricultural production, especially in tropical highland areas.

4.2 Identify the major environmental issues of Latin America and how countries are addressing them.

4.3 Summarize the demographic issues impacting this region, such as rural-to-urban migration, urbanization, smaller families, and emigration.

4.4 Describe the cultural mixing of European and Amerindian groups in this region and indicate where Amerindian cultures thrive today.

4.5 Explain the global reach of Latino culture through immigration, sport, music, and television.

4.6 Describe the Iberian colonization of the region and how it affected the formation of today's modern states.

4.7 Identify the major trade blocs in Latin America and how they are influencing development.

4.8 Summarize the significance of primary exports from Latin America, especially agricultural commodities, minerals, wood products, and fossil fuels.

4.9 Describe the neoliberal economic reforms that have been applied to Latin America and how they have influenced the region's development.

Roughly equal in area to North America, Latin America has a much larger and faster-growing population of 585 million people. Its most populated state, Brazil, has over 200 million people, making it the world's fifth-largest country. The next largest state, Mexico, has a population of nearly 130 million. Collectively, many Latin American states fall into the middle-income category and support a significant middle class. Nearly 80 percent of the population lives in cities. Yet poverty and income inequality remain major concerns, as one in ten people in the region is estimated to live on less than $2 per day.

Through colonialism, immigration, and trade, the forces of globalization have been embedded in the Latin American landscape (Figure 4.2).

Figure 4.1 Latin America Roughly equal in size to North America, Latin America supports a larger population and far greater ecological diversity. The 17 countries in this region share a history of Iberian colonization. Three-quarters of the region's 585 million people live in cities, making it the most urbanized region of the developing world. It is noted for production of primary exports and manufactured goods, although rates of economic development vary greatly among states. Latin America's subregions include Central America (Guatemala, El Salvador, Honduras, Nicaragua, Costa Rica, and Panama), the Andean States (Colombia, Ecuador, Peru, and Bolivia), and the Southern Cone (Chile, Argentina, Uruguay, and southern Brazil).

The early Spanish Empire concentrated on extracting precious metals, sending galleons laden with silver and gold back across the Atlantic. The Portuguese became important producers of dyewoods, sugar products, gold, and later coffee. In the late 19th and early 20th centuries, exports to North America and Europe fueled the region's economy. Most countries specialized in one or two products: bananas and coffee, meat and wool, wheat and corn, petroleum and copper. Such a primary export tradition led to an unhealthy economic dependence; economists argued in the 1960s that Latin American economies were too specialized and faced unequal terms of trade that inhibited overall development.

Since then, the countries of the region have industrialized, urbanized, and diversified their production, yet they continue to be major producers of primary goods for world markets. Neoliberal policies that encourage foreign investment, export production, and privatization have been adopted by many states, with mixed results; some states achieved impressive economic growth, but disparities between rich and poor increased. Intraregional trade within Latin America, stimulated by Mercosur (the Southern Cone Common Market) and impacted by the North American Free Trade Agreement (NAFTA; see Chapter 3) and the Central American Free Trade Agreement (CAFTA), is an indicator of heightened economic integration in the Western Hemisphere. In 2014, a dozen Brazilian cities hosted the World Cup, and Rio de Janeiro hosts the Olympics in 2016.

Despite the region's growing industrial capacity, extractive industries will continue to prevail, in part because of the area's impressive resource base. Latin America is home to Earth's largest rainforest, the greatest river by volume, and massive reserves of natural gas, oil, and copper. With its extensive territory, tropical location, and relatively low population density (Latin America has half the population of India in nearly seven times the area), the region is also recognized as one of the world's great reserves of biological diversity. How this diversity will be managed in the face of global demand for natural resources is an increasingly important question for the region.

Figure 4.2 **Spanish Colonial Influence** Throughout Latin America, the Iberian influence is still seen in towns and cities established by Spain and Portugal. Typically on a grid pattern, the center is a market plaza with nearby church and municipal buildings. This is the village of Mucuchies, Venezuela, which was established by the Spanish in the 17th century.

Physical Geography and Environmental Issues: Neotropical Diversity and Urban Degradation

Much of Latin America is characterized by its tropicality. Travel posters of the region showcase lush forests and brightly colored parrots. The diversity and uniqueness of the **neotropics** (tropical ecosystems of the Western Hemisphere) have long attracted naturalists eager to understand their unique flora and fauna. It is no accident that Charles Darwin's theory of evolution was inspired by his two-year journey in tropical America. Even today scientists throughout the region work to understand complex ecosystems, discover and protect new species, and interpret the impact of human settlement, especially in neotropical forests.

Not all of the region is tropical. Important population centers lie below the Tropic of Capricorn—most notably Buenos Aires, Argentina, and Santiago, Chile. Much of northern Mexico, including the city of Monterrey, is north of the Tropic of Cancer. Highlands and deserts exist throughout the region. Yet Latin America's tropical climate and vegetation define the region's image. Given its large size and relatively low population density, Latin America has not experienced the same levels of environmental degradation witnessed in East Asia and Europe. Huge areas remain relatively untouched, supporting an incredible diversity of plant and animal life. Throughout the region, national parks offer some protection to unique plant and animal communities. A growing environmental movement in countries such as Costa Rica and Brazil has yielded both popular and political support for "green" initiatives. In short, Latin Americans have entered the 21st century with a real opportunity to avoid many of the environmental mistakes seen in other world regions. At the same time, global market forces are driving governments to exploit minerals, fossil fuels, forests, shorelines, transportation routes, and soils. The region's biggest resource management challenge is to balance the economic benefits of extraction with the principles of sustainable development. Another major challenge is to improve the environmental quality of Latin American cities.

Western Mountains and Eastern Lowlands

Latin America is a region of diverse landforms, including high mountains, extensive upland plateaus, and vast river basins. The movement of tectonic plates explains much of the region's basic topography, including the formation of its geologically young western mountain ranges, such as the Andes and the Volcanic Axis of Central America (see Figure 4.1). For example, Villarrica volcano in Chile is one of the most active in the Andes, with eruptions in 2015. This area is also prone to earthquakes that threaten people and damage property. In February 2010, for example, a massive magnitude 8.8 earthquake struck off the coast near the Chilean city of Concepción, killing some 400 people and unleashing tsunami warnings across the Pacific. In contrast, the Atlantic side of South America is characterized by humid lowlands interspersed with large upland plateaus called *shields*. Across these lowlands meander some of the great rivers of the world, including the Amazon, Plata, and Orinoco.

Historically, the most important areas of settlement in tropical Latin America were not along the major rivers, but across its shields, plateaus, and fertile mountain valleys. In these places, the combination of arable land, mild climate, and sufficient rainfall produced the region's most productive agricultural areas and its densest settlement. The Mexican Plateau, for example, is a massive upland area ringed by the Sierra Madre mountains. The Valley of Mexico is located at the plateau's southern end. Similarly, the elevated and well-watered basins of Brazil's southern mountains provide an ideal setting for agriculture. These especially fertile areas can support high population densities, so it is not surprising that the region's two largest cities, Mexico City and São Paulo, emerged in these settings. The Latin American highlands also lend a special character to the region. Lush tropical valleys nestled below snow-covered mountains hint at diverse ecosystems found near one another. The most dramatic of these highlands, the Andes, runs like a spine down the length of the South American continent.

Figure 4.3 Bolivian Altiplano The Altiplano is an elevated plateau straddling the Bolivian and Peruvian Andes. Displayed is picturesque Laguna Canapa in Bolivia with the Andean peaks towering in the background. This high and windswept land is the home of many Amerindian peoples.

The Andes From northwestern Venezuela to Tierra del Fuego, the Andes are relatively young mountains that extend nearly 5000 miles (8000 km). They are an ecologically and geologically complex mountain chain, with some 30 peaks higher than 20,000 feet (6000 meters). Created by the collision of oceanic and continental plates, the Andes are a series of folded and faulted sedimentary rocks with intrusions of crystalline and volcanic rock. Many rich veins of precious metals and minerals are found in these mountains. In fact, the initial economic wealth of many Andean countries came from mining silver, gold, tin, copper, and iron.

The lengthy Andean chain is typically divided into northern, central, and southern components. In Colombia, the northern Andes split into three distinct mountain ranges before merging near the border with Ecuador. High-altitude plateaus and snow-covered peaks distinguish the central Andes of Ecuador, Peru, and Bolivia. The Andes reach their greatest width here. Of special interest is the treeless high plain of Peru and Bolivia, the **Altiplano**. The floor of this elevated plateau ranges from 11,800 feet (3600 meters) to 13,000 feet (4000 meters) in altitude, limiting its usefulness for grazing. Two high-altitude lakes—Titicaca on the Peruvian and Bolivian border and the smaller Poopó in Bolivia—are located in the Altiplano, as are many mining sites (Figure 4.3). The highest peaks are found in the southern Andes, shared by Chile and Argentina, including the Western Hemisphere's highest peak, Aconcagua, at almost 23,000 feet (7000 meters).

The Uplands of Mexico and Central America The Mexican Plateau and the Volcanic Axis of Central America are the most important Latin American uplands in terms of settlement, as many major cities are found here. The Mexican Plateau is a large, tilted block with its highest elevations, about 8000 feet (2500 meters), in the south around Mexico City and its lowest, just 4000 feet (1200 meters), at Ciudad Juárez. The southern end of the plateau, the Mesa Central, contains several flat-bottomed basins interspersed with volcanic peaks that have long been significant areas for agricultural production (Figure 4.4). It also contains Mexico's megalopolis—a concentration of the largest population centers, such as Mexico City and Puebla.

Along Central America's Pacific coast lies the Volcanic Axis, a chain of volcanoes that stretches from Guatemala to Costa Rica. It is a handsome landscape of rolling green hills, elevated basins with sparkling lakes, and volcanic peaks. More than 40 volcanoes, many still active, have produced a rich volcanic soil that yields a wide variety of domestic and export crops. Most of Central America's population is also concentrated in this zone, in the capital cities or surrounding rural villages. The bulk of the agricultural land is tied up in large holdings that produce beef, cotton, and coffee for export. However, in terms of numbers, most of the farms are small subsistence properties that produce corn, beans, squash, and assorted fruits.

The Shields South America has three major **shields**—large upland areas of exposed crystalline rock that are similar to upland plateaus found in Africa and Australia. The Brazilian and Patagonian shields (the Guiana Shield will be discussed in Chapter 5) vary in elevation from 600 to 5000 feet (200 to 1500 meters). The Brazilian Shield is larger and more important in terms of natural resources and settlement. Far from a uniform land surface, this shield covers much of Brazil from the Amazon Basin in the north to the Plata Basin in the south. In the southeast corner of the plateau is São Paulo, the largest urban conglomeration in South America. The other major population centers are on the coastal edge of the plateau, where large protected bays made the sites of Rio de Janeiro and Salvador attractive to Portuguese colonists. Finally, the Paraná basalt plateau on the southern end of the Brazilian Shield is famous for its fertile red soils (*terra roxa*), which yield coffee, oranges, and soybeans. So fertile is this area that the economic rise of São Paulo is attributed to the expansion of commercial agriculture, especially coffee, into this area.

The Patagonian Shield lies in the southern tip of South America. Beginning south of Bahia Blanca and extending to Tierra del Fuego, the region to this day is sparsely settled and hauntingly beautiful. It is treeless, covered by scrubby steppe vegetation, and home to wildlife such as the guanaco (Figure 4.5). Sheep were introduced to Patagonia in the late 19th century, spurring a wool boom. More recently, offshore oil production has renewed the economic importance of Patagonia.

River Basins Three great river basins drain the Atlantic lowlands of South America: the Amazon, Plata, and Orinoco. The Amazon drains an area of roughly 2.3 million square miles (5.9 million square kilometers), making it the largest river system in the world by volume and area and the second largest by length. The Amazon Basin

Figure 4.4 Mexico's Mesa Central Mexico's elevated central plateau has long been the demographic and agricultural core of the country. This image shows a variety of agave grown in Jalisco, used for tequila production. Tequila, a traditional drink in Mexico, has a growing export market.

Figure 4.5 Patagonian Wildlife Guanacos, native to South America, thrive on the steppe vegetation found throughout Patagonia. The population fell dramatically due to hunting and competition with introduced livestock, but guanacos thrive today in protected areas such as Torres del Paine in Chile.

is home to the world's largest rainforest; annual rainfall is more than 60 inches (150 centimeters) everywhere in the basin and close to 100 inches (250 centimeters) in the basin's largest city, Belem. The mighty Amazon drains eight countries, but two-thirds of the watershed is within Brazil. Active settlement of the Brazilian portion of the Amazon since the 1960s has boosted the population. Today some 34 million people live in the Amazon Basin, which is equal to 8 percent of the total population in South America. The basin's development—most notably through towns, roads, dams, farms, and mines—is forever changing what was viewed as a vast tropical wilderness just a half century ago. Brazil's government has plans to build 30 new dams in its portion of the Amazon. Perhaps the most contested dam is Belo Monte on the Xingu River, a tributary of the Amazon (Figure 4.6).

The region's second-largest watershed, the Plata Basin, begins in the tropics and discharges into the Atlantic in the midlatitudes near Buenos Aires. Several major rivers make up this system: the Paraná, the Paraguay, and the Uruguay. Unlike the Amazon Basin, much of the Plata Basin is now economically productive through large-scale mechanized agriculture, especially soybean production. The basin contains several major dams, including the region's largest hydroelectric plant, the Itaipú on the Paraná, which generates electricity for all of Paraguay and much of southern Brazil. As agricultural output in the watershed grows, sections of the Paraná River have been canalized and dredged to enhance the river's capacity for barge and boat traffic.

The third-largest basin by area is the Orinoco in northern South America. Although its watershed is only one-seventh the size of the Amazon watershed, the Orinoco has a discharge roughly equal to that of the Mississippi River. The Orinoco River meanders through much of southern Venezuela and part of eastern Colombia, giving character to the sparsely settled tropical grasslands called the *Llanos*. Since the colonial era, these grasslands have supported large cattle ranches. Although cattle are still important, the Llanos are also a dynamic area of petroleum production for both Colombia and Venezuela.

Climate and Climate Change in Latin America

In tropical Latin America, average monthly temperatures in settings such as Managua (Nicaragua), Quito (Ecuador), and Manaus (Brazil) show little variation (see the climographs in Figure 4.7). Precipitation patterns, however, are variable and create distinct wet and dry seasons. In Managua, for example, January is typically a dry month, and October is a wet one. The tropical lowlands of Latin America, especially east of the Andes, are usually classified as tropical humid climates that support forest or savanna, depending on the amount of rainfall. The region's desert climates are found along the Pacific coasts of Peru and Chile and in Patagonia, northern Mexico, and the state of Bahia in Brazil. Thus, a city such as Lima, Peru, which is clearly in the tropics, averages only 1.5 inches (4 centimeters) of rainfall a year due to the extreme aridity of the Peruvian coast. Some sections of the Atacama Desert of Chile get no measurable rainfall (Figure 4.8). Yet the discovery of resources such as nitrates in the 19th century and copper in the 20th century made this hyper-arid region a source of conflict among Chile, Bolivia, and Peru.

Midlatitude climates, with hot summers and cold winters, prevail in Argentina, Uruguay, and parts of Paraguay and Chile (see the climographs for Buenos Aires and Punta Arenas in Figure 4.7). Recall that the midlatitude temperature shifts in the Southern Hemisphere are the opposite of those in the Northern Hemisphere (cold Julys and warm Januarys). In the mountain ranges, complex climate patterns result from changes in elevation. To appreciate how humans adapt to tropical mountain ecosystems, the concept of **altitudinal zonation**, which is the relationship between cooler temperatures at higher elevations and changes in vegetation, is important.

Altitudinal Zonation First described in the scientific literature by Alexander von Humboldt in the early 1800s, altitudinal zonation has practical applications that are intimately understood by the region's

Figure 4.6 Amazonian Dam An early phase in the construction of the Belo Monte Hydroelectric Project on the Xingu River in Brazil is shown. It is located in Para state near the town of Altimira. When completed, Belo Monte will be the world's third-largest dam, generating more than 11,000 megawatts of electricity. It will include two dams, two canals, two reservoirs, and a system of dikes.

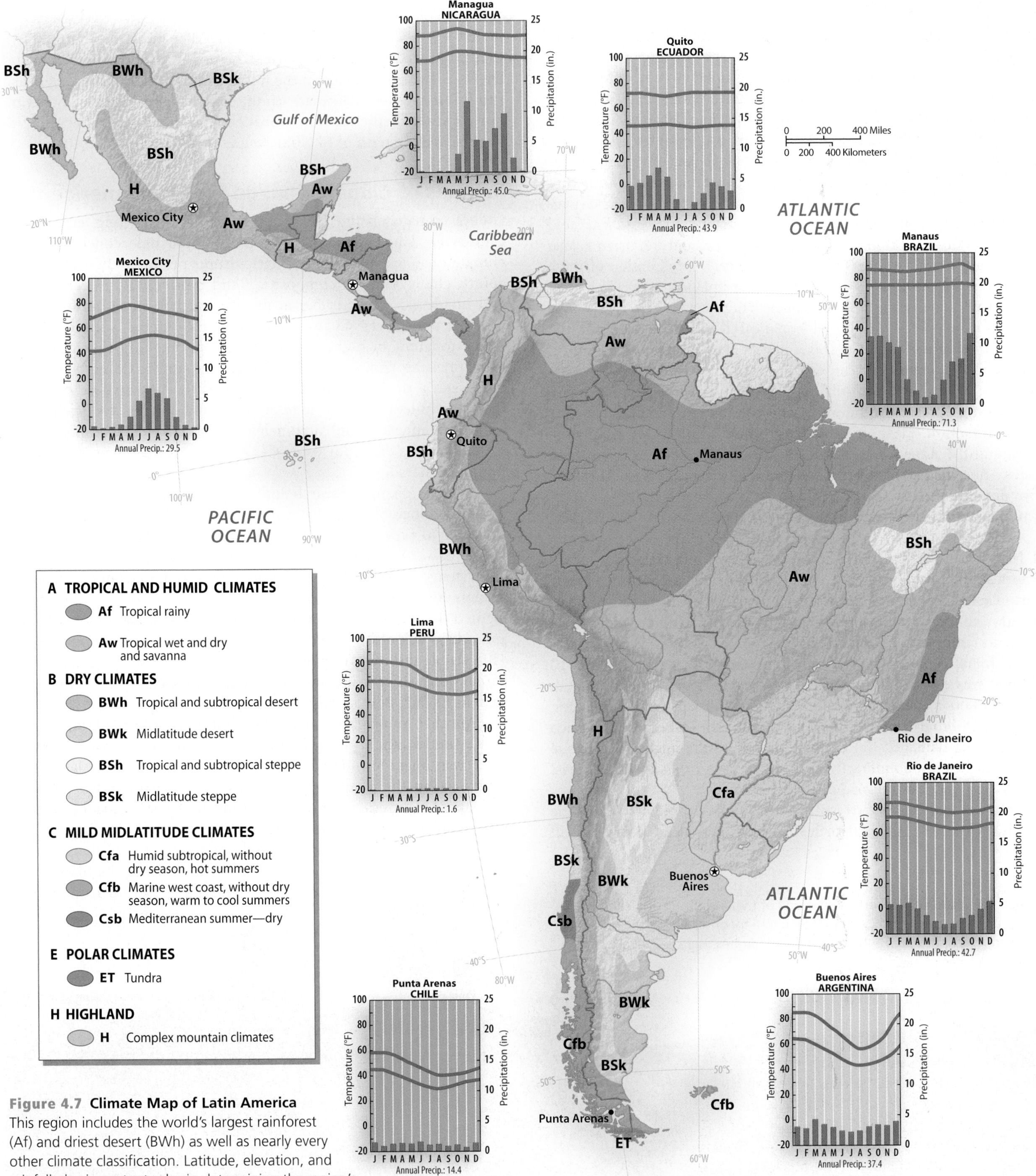

Figure 4.7 Climate Map of Latin America This region includes the world's largest rainforest (Af) and driest desert (BWh) as well as nearly every other climate classification. Latitude, elevation, and rainfall play important roles in determining the region's climates. Note the contrast in rainfall patterns between humid Quito and arid Lima.

native inhabitants. Humboldt systematically recorded declines in temperature as he ascended to higher elevations, a phenomenon known as the **environmental lapse rate**. According to Humboldt, temperature declines approximately 3.5°F for every 1000 feet in higher elevation, or 6.5°C for every 1000 meters. Humboldt also noted changes in vegetation by elevation, demonstrating that plant communities common to the midlatitudes could thrive in the tropics at higher elevations. These different altitudinal zones are commonly termed the *tierra caliente* (hot land), from sea level to 3000 feet (900 meters); the *tierra templada* (temperate land), at 3000–6000 feet (900–1800 meters); the *tierra fría* (cold land), at 6000–12,000 feet (1800–3600 meters); and the *tierra helada* (frozen land), above 12,000 feet (3600 meters). Exploitation of these zones allows agriculturists, especially in the uplands, access to a great diversity of domesticated and wild plants (Figure 4.9).

The concept of altitudinal zonation is most relevant for the Andes, the highlands of Central America, and the Mexican Plateau. For example, traditional Andean farmers might use the high pastures of the Altiplano for grazing llamas and alpacas, the *tierra fría* for growing potatoes and quinoa, and the lower temperate zone for corn production. All the great precontact civilizations, especially the Incas and the Aztecs, systematically extracted resources from these zones, thus ensuring a diverse and abundant resource base. Yet these complex ecosystems are extremely fragile and have become important areas of research on the effects of climate change in the tropics.

El Niño One of the most studied weather phenomena in Latin America, **El Niño** (referring to the Christ child), occurs when a warm Pacific current arrives along the normally cold coastal waters of Ecuador and Peru in December, around Christmastime. This change in ocean temperature, which happens every few years, produces torrential rains, signaling the arrival of an El Niño event. The 2009–2010 El Niño was especially bad for Latin America; scores of people were killed by floods or storms attributed to El Niño–related disturbances. Devastating floods occurred in Peru and Brazil. In Peru, heavy rains and flooding damaged the railroad leading to the ancient Incan site of Machu Picchu, temporarily limiting access to this popular tourist destination until the railroad could be rebuilt.

The less-talked-about result of El Niño is drought. While the Pacific coasts of South and North America experienced record rainfall in the 1997–1998 El Niño, Colombia, Venezuela, northern Brazil, Central America, and Mexico battled drought. In addition to crop and livestock losses estimated in the billions of dollars, hundreds of brush and forest fires left their mark on the region's landscape. Scientists are not yet sure whether or how global climate change will affect the frequency and strength of El Niño cycles.

Impacts of Climate Change for Latin America

Global climate change has both immediate and long-term implications for Latin America. Of greatest immediate concern is how climate change will influence agricultural productivity, water availability, changes in the composition and productivity of ecosystems, and incidence of vector-borne diseases such as malaria and, especially, dengue fever. Changes attributable to global warming are already apparent in higher-elevation regions, making these concerns more pressing. For example, coffee growers in the Colombian Andes have seen a decline in productivity over the past five years, which they attribute to higher temperatures and longer dry spells. The long-term effects of global climate change on lowland tropical forest systems is less clear; for example, some areas may experience more rainfall, others less.

Climate change research indicates that highland areas are particularly vulnerable to global warming. Tropical mountain systems are projected to experience temperature increases of 2 to 6°F (1 to 3°C) as well as lower rainfall. This will raise the altitudinal limits of various ecosystems, impacting the range of crops and arable land available to farmers and pastoralists. Research over the past 50 years has documented the dramatic retreat of Andean glaciers; some no longer exist, and others will cease to exist in the next 10 to 15 years (Figure 4.10). This visible indicator of global warming also has pressing human repercussions, since many mountain communities, as well as large cities such as La Paz, Bolivia, rely on water from glacial runoff.

Another immediate concern brought on by warmer temperatures is the sudden rise in dengue fever, a mosquito-borne virus. Dengue fever was once considered relatively uncommon in highland Latin America, but the number of cases has risen sharply in the past decade. Tens of thousands now suffer from its fever, headache, nausea, joint pain, and, in rare cases, external and internal bleeding that can be fatal. In Latin America, the sudden rise in cases of dengue fever suggests that warmer highland temperatures have placed millions more at risk.

Figure 4.8 Atacama Desert This is one of the driest places on earth, with almost no vegetation; many visitors liken it to a moonscape. Yet the soils of the Atacama contain a wealth of copper and nitrates. Here is the Valley of the Moon in northern Chile.

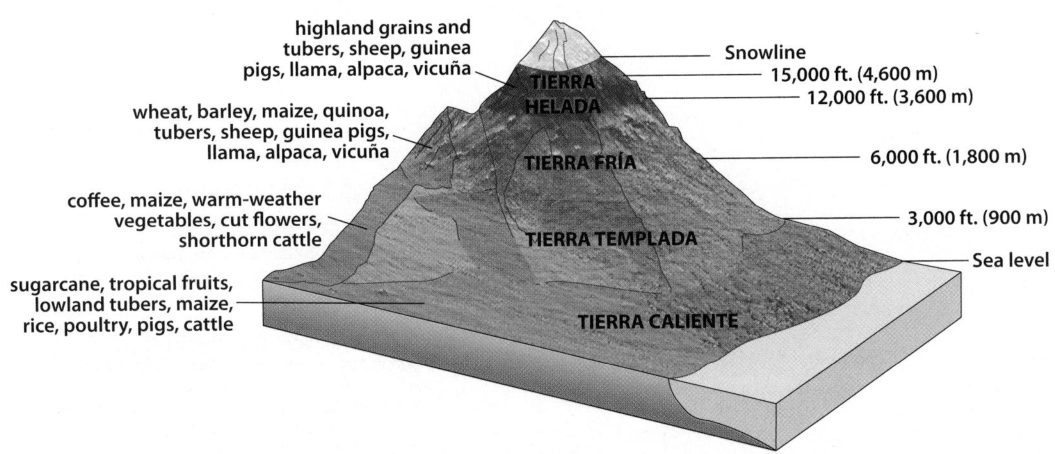

Figure 4.9 Altitudinal Zonation Tropical highland areas support a complex array of ecosystems. In the *tierra fría* zone (6000 to 12,000 feet, or 1800 to 3700 meters), for example, midlatitude crops such as wheat and barley can be grown. The diagram depicts the range of crops and animals found at different elevations in the Andes. **Q: Quinoa, an Andean grain grown in the *tierra fría*, has become globally popular in the last two decades. Where else might quinoa be grown?**

Environmental Issues: The Destruction and Conservation of Forests

Perhaps the environmental issue most commonly associated with Latin America is deforestation (Figure 4.11). The Amazon Basin and portions of the eastern lowlands of Central America and Mexico still maintain unique and impressive stands of tropical forest. Other woodland areas, such as the Atlantic coastal forests of Brazil and the Pacific forests of Central America, have nearly disappeared as a result of agriculture, settlement, and ranching. The coniferous forests of northern Mexico are also falling, in part because of a bonanza for commercial logging stimulated by NAFTA. In Chile, the ecologically unique evergreen rainforest (the Valdivian forest) in the midlatitudes is being cleared for wood chip exports to Asia.

The loss of tropical rainforests is most critical in terms of biological diversity. Tropical rainforests cover only 6 percent of Earth's landmass, but at least 50 percent of the world's species are found in this biome. Moreover, the Amazon contains the largest undisturbed stretches of rainforest in the world. Unlike Southeast Asian forests, where hardwood extraction drives deforestation, Latin American forests are usually seen as an agricultural frontier. State governments divide areas in an attempt to give land to the landless and reward political elites. Thus, forests are cut and burned, with settlers and politicians carving them up to create permanent settlements, slash-and-burn plots, or large cattle ranches. In addition, some tropical forest cutting has been motivated by the search for gold (Brazil, Venezuela, and Costa Rica) and the production of coca leaf for cocaine (Peru, Bolivia, and Colombia).

Brazil has been criticized more than other countries for its Amazon forest policies. During the past 40 years, one-fifth of the Brazilian Amazon has been deforested. In states such as Rondônia in western Brazil, close to 60 percent of the land has been deforested (Figure 4.12). What most alarms environmentalists and forest dwellers (Indians and rubber tappers) is the dramatic increase in the rate of rainforest clearing since 2000, estimated at nearly 8000 square miles (20,000 square kilometers) per year. The increased rates of deforestation in the Brazilian Amazon are due to the expansion of industrial mining and logging, the growth of corporate farms, the development of new road networks, the incidence of human-ignited wildfires, and continued population growth. Under the Advance Brazil program started in 2000, some US$40 billion has gone to new highways, railroads, gas lines, hydroelectric projects, power lines, and river canalization projects that will reach into remote areas of the basin. In an effort to slow deforestation rates, the Brazilian government has created new conservation areas, many of them alongside the "arc of deforestation"—a swath of agricultural development along the southern edge of the Amazon Basin (see Figure 4.11). Yet Brazil's Forest Code, as revised in 2012, has reduced the amount of "forest reserve" that private landholders must maintain. Many conservationists fear this will lead to more forest clearing and fragmentation.

(a) In 1966, Chacaltaya Glacier was small but still existed.

(b) In 2009, Chacaltaya Glacier has only a small remaining area on the far right of the image.

Figure 4.10 Andean Glacial Retreat The rapid disappearance of Chacaltaya Glacier in the Bolivian Andes is a visible reminder of warming conditions in high-altitude tropical zones. Seasonal melt from this glacier, which has long been an important water source for La Paz, Bolivia, has declined dramatically.

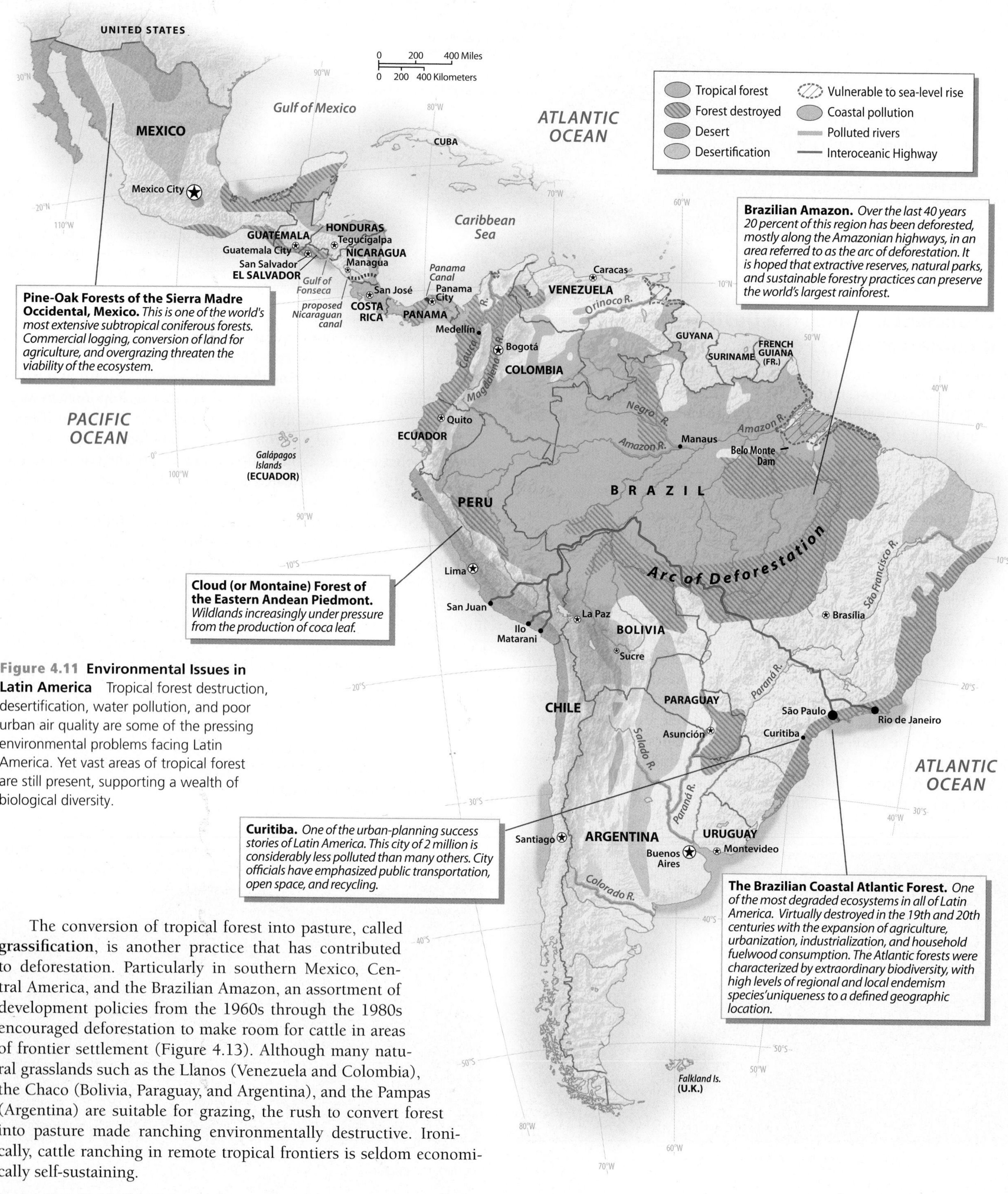

Figure 4.11 Environmental Issues in Latin America Tropical forest destruction, desertification, water pollution, and poor urban air quality are some of the pressing environmental problems facing Latin America. Yet vast areas of tropical forest are still present, supporting a wealth of biological diversity.

The conversion of tropical forest into pasture, called **grassification**, is another practice that has contributed to deforestation. Particularly in southern Mexico, Central America, and the Brazilian Amazon, an assortment of development policies from the 1960s through the 1980s encouraged deforestation to make room for cattle in areas of frontier settlement (Figure 4.13). Although many natural grasslands such as the Llanos (Venezuela and Colombia), the Chaco (Bolivia, Paraguay, and Argentina), and the Pampas (Argentina) are suitable for grazing, the rush to convert forest into pasture made ranching environmentally destructive. Ironically, cattle ranching in remote tropical frontiers is seldom economically self-sustaining.

Protecting Lands for Future Generations Latin America has more nationally protected lands than any other developing region. The areas

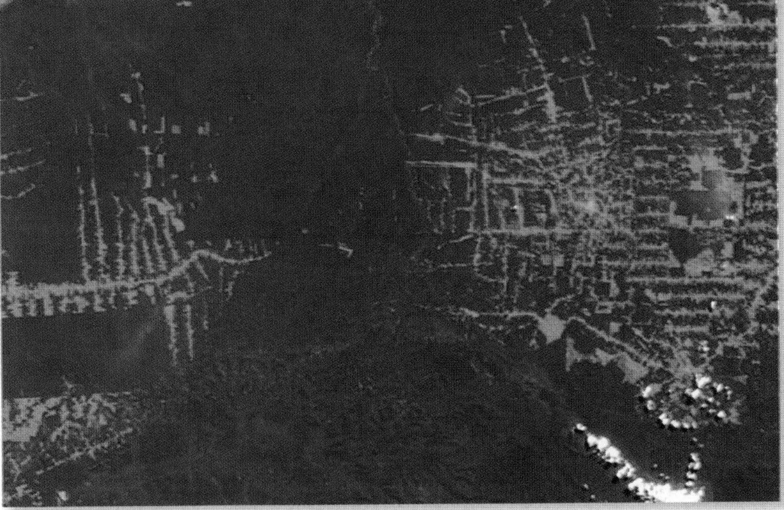

(a) July 30, 2000

(b) August 2, 2010

Figure 4.12 Tropical Forest Settlement in the Amazon These satellite images of Rondônia, Brazil, illustrate the dramatic change in forest cover in just 10 years, between (a) 2000 and (b) 2010, near the settlement of Buritis and road BR-364. Intact forest is dark green, whereas cleared areas are light green (crops) or tan (bare ground). Typically, the first clearings appear off roads, forming a fishbone pattern. Over time, as more forest is cleared and settlements grow, the fishbone pattern collapses into a mosaic of pasture, farmland, and forest fragments.

designated as national parks, nature reserves, wildlife sanctuaries, and scientific reserves with limited public access went from 10 percent of the territory in 1990 to 20 percent in 2010, according to World Bank estimates. Brazil's protected land went from just 9 percent of the national territory to 26 percent in 20 years. Although conservationists complain that many of these areas are "paper parks" with limited real protection, many countries in the region have used the conservation of forests and other lands as a means to attract tourists.

Costa Rica is a Latin American pioneer in creating national parks and promoting ecotourism. In the 1970s, Costa Rican conservationist Mario Boza successfully lobbied for the creation of national parks in response to the rampant destruction of forests to expand coffee and banana plantations and cattle pasture. By 1990, almost 20 percent of the territory had been protected—by then, nearly all the unprotected lands had been cleared for agriculture or settlement. Drawn by Costa Rica's impressive natural beauty, Pacific and Caribbean beaches, volcanoes, and biodiversity, about 2.4 million international tourists visit the country each year (Figure 4.14). The parks are accessible to Costa Ricans at a reduced fee, whereas tourists pay higher park entrance fees to support conservation and park maintenance.

Figure 4.13 Converting Forest into Pasture Cattle graze in northern Guatemala's Petén region. Clearing of this tropical forest lowland began in the 1960s and continues today. Ranching is a status-conferring occupation in Latin America with serious ecological costs. The beef produced from this region is for domestic and export markets.

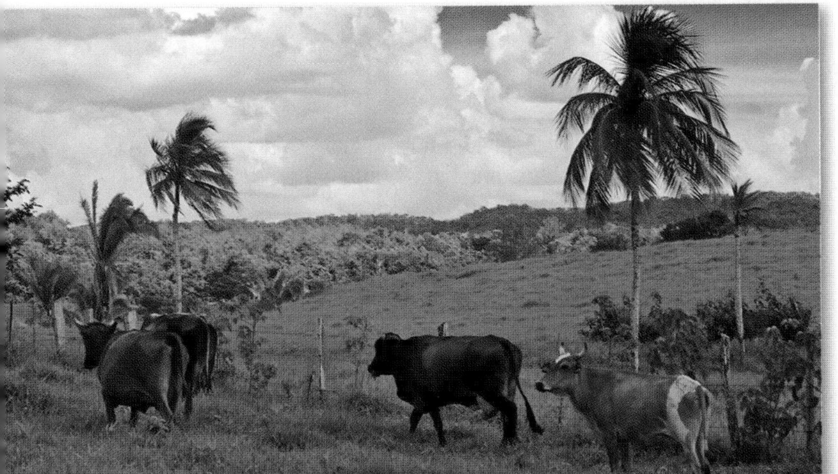

Urban Environmental Challenges

For most Latin Americans, air pollution, water availability and quality, and garbage removal are the pressing environmental problems of everyday life. Consequently, many environmental activists in the region focus their efforts on making urban environments cleaner by introducing "green" legislation and calling people to action. In this most urbanized region of the developing world, city dwellers do have better access to water, sewers, and electricity than their counterparts in Asia and Africa. Moreover, the density of urban settlement seems to encourage the widespread use of mass transportation; both public and private bus and van routes make getting around cities fairly easy. However, the usual environmental problems that come from dense urban settings ultimately require expensive remedies, such as new

Figure 4.14 Costa Rican National Park Hugging the Pacific coast, the tropical forest and beaches of Manuel Antonio National Park make it a popular destination for both Costa Ricans and international tourists. The pressures to develop tropical coasts are strong, which makes creating protected areas an urgent need.

Figure 4.15 Air Pollution in Santiago Smog blankets Santiago, with the Andes in the background. Thermal inversion layers form during the winter months (May through August), trapping pollutants near ground level and causing a spike in pollution-related health problems. Reducing vehicular traffic and greatly expanding public transportation have improved air quality.

power plants and modernized sewer and water lines. The money for such projects is never enough, due to currency devaluation, inflation, and foreign debt. Because many urban dwellers tend to reside in unplanned squatter settlements, servicing these communities with utilities after they are built is difficult and costly.

Air Pollution Most major cities, but especially the capitals Santiago and Mexico City, suffer from air pollution (see *Working Toward Sustainability: Greening Transport and Expanding Access in Bogotá*). Air pollution is not just an aesthetic issue—the health costs of breathing such contaminated air are significant, as elevated death rates due to heart disease, asthma, influenza, and pneumonia suggest. The burden of air pollution is not evenly distributed among city residents, as the elderly, the very young, and the poor are more likely to suffer the negative health effects of contaminated air. Fortunately, both cities have taken steps to address this vexing problem.

Santiago, a prosperous city of nearly 7 million in Chile's Central Valley, has an elevation of 1700 feet (520 meters). Although not as high as Mexico City, this basin setting regularly produces thermal inversions, when warm air traps a layer of cold air near the surface. This trapped surface layer becomes filled with engine exhaust, industrial pollution, garbage, and even fecal matter (Figure 4.15). The inversion layers happen year-round but can be especially bad in the winter months from May through August, often forcing schools to suspend all sports when smog emergencies are called. Santiago officials began addressing this problem in the late 1980s by restricting vehicular traffic. On a given weekday, 20 percent of all buses, taxis, and cars are prohibited from driving, based on license plate numbers. During smog emergencies, up to 40 percent of vehicles can be restricted. In addition, numerous private buses were replaced by a large fleet of clean-running public buses. These buses, combined with the city's subway system, can move over 2 million riders a day. By 2010, air quality had noticeably improved, and public support for restrictive measures and public transport had solidified.

Mexico City's smog has been so bad that most visitors today have no idea that mountains surround them. Air quality has been a major issue for Mexico City since the 1960s, driven in part by the city's unusually high rate of growth. (Between 1950 and 1980, the city's annual rate of growth was 4.8 percent.) It is difficult to imagine a better setting for creating air pollution. The city sits in a bowl 7400 feet (2200 meters) above sea level, and thermal inversions regularly form. Steps were finally taken in the late 1980s to reduce emissions: Unleaded gas is now widely available for the 4 million cars in the metropolitan area, and cars manufactured for the Mexican market must have catalytic converters. Also, some of the worst polluting factories in the Valley of Mexico have closed. In the last few years, the mayor of Mexico City has expanded a low-emissions bus system, eliminating thousands of tons of carbon monoxide. In 2007, the decision was made to close the elegant Paseo de Reforma to traffic on Sunday mornings and open it to bike riders. This change was so popular that now bike lanes have been introduced to some downtown areas in an effort to encourage bike ridership. For longer commutes, a suburban train system is being built that will complement the existing subway system. The payoff is real: Mexico City no longer ranks among the most polluted cities in the world, and it seems to have cut most of its pollutants by at least half.

Water Providing access to clean and reliable freshwater is also a challenge for Latin America's large cities. When Vicente Fox was president of Mexico, he declared water scarcity and water quality a national security issue—not just for the capital, but also for the entire country. Ironically, it was the abundance of water that initially made Mexico City attractive for settlement. Large shallow lakes once filled the valley, but over the centuries most were drained to expand agricultural land. As surface water became scarce, wells were dug to tap the basin's massive freshwater aquifer. Today approximately 70 percent of the metropolitan area's water is drawn from the valley's aquifer. There is troubling evidence that the aquifer is being overdrawn and at risk of contamination, especially in areas where unlined drainage canals can leak pollutants into the surrounding soil, which then leach into the aquifer. To reduce reliance on the aquifer, the city now pumps water nearly a mile uphill from more than 100 miles (160 km) away.

Andean cities such as Bogotá, Quito, and La Paz are increasingly experiencing water scarcity and rationing. Some of this is due to increased demands on aging water systems brought about by population growth. However, changes in precipitation patterns due to El Niño years or global climate change make these large urban centers especially vulnerable. La Paz, for example, gets much of its water from glacial runoff. A major Bolivian glacier, Chacaltaya, has lost 80 percent of its area in the past 20 years. Thus, as average temperatures increase in the highlands and glaciers recede, there is widespread concern about future drinking-water supplies in this metropolitan area of nearly 2 million people.

 **REVIEW**

4.1 Describe the major ecosystems in Latin America and how humans have adapted to and modified these different ecosystems.

4.2 Summarize some of the major environmental issues impacting this region and how different countries have tried to address them.

KEY TERMS neotropics, Altiplano, shield, altitudinal zonation, environmental lapse rate, El Niño, grassification

WORKING TOWARD SUSTAINABILITY

Greening Transport and Expanding Access in Bogotá

Google Earth (MG) Virtual Tour Video http://goo.gl/NqzYx7

Most major Latin American cities are over four centuries old and were designed for pedestrians and carriage traffic, not automobiles. As these cities exploded in size in the 20th century, many observers lamented that automobile dependence had destroyed urban life in Latin America. An infusion of cars took over public space, contaminated air, and saddled people with slow commutes. Yet innovative leaders in the region's cities dreamed of something different.

Mass Transit Buses For the 9 million people in Bogotá, Colombia, a new urban phase began in 2000 with the opening of a rapid-transit bus system called TransMilenio. TransMilenio replicates some aspects of the highly regarded transit system in Curitiba, Brazil, long considered Latin America's "Green City." Using large-capacity articulated buses with dedicated bus lanes and rapid-loading platforms, the bright red vehicles now dominate the city's main arteries (Figure 4.1.1). Today about 1400 buses are in use, plus several hundred smaller feeder buses, moving 1.5 million people per day. Older, more polluting buses were pulled from the streets. Extensive coverage, the use of smart cards that integrate transfers from one bus to another, and a fixed fare of less than a dollar make this an accessible system. Rapid-transit bus systems with dedicated lanes are also much less expensive than subways or light rail, an important consideration for developing countries.

TransMilenio was just one component of a broader vision for Bogotá that focused on increased social integration, improved mobility, enhanced public space, and a human scale of design. Guided by Mayor Enrique Peñalosa in the late 1990s, the plan emphasized creating new pedestrian zones, revitalized parks and sidewalks, bike paths, and a more integrated and efficient public transport system. Better public transport made driving in private cars less attractive. Moreover, based on license plate numbers, the city prohibited private cars from driving during rush hour two days a week.

Figure 4.1.2 Bogotá's Bikers A cyclist on a bike path pedals past a nun. Abundant bike lanes make Bogotá Latin America's most bike-friendly city. In 2000, the mayor introduced the "day without cars" campaign to demonstrate how this congested city could function without automobiles.

Figure 4.1.1 TransMilenio in Bogotá Gliding along the busy arteries of Bogotá are large red articulated buses that make up the TransMilenio system. Operating along designated lanes and with large fast-loading platforms, the system moves 1.5 million people a day.

Mile-High Biking Bogotá's planners intentionally included bicycles in the transportation system. In the past decade, bike lanes have been built throughout the city, and bike stations are found at suburban TransMilenio bus stops (Figure 4.1.2). Bogotá is at an elevation of 8600 feet (2600 meters), but it is relatively flat with springlike temperatures year-round, so the bicycle is a practical and clean form of urban transport. Today Bogotá is considered Latin America's most bike-friendly city—and one of the most bike-friendly big cities in the world. Biking is a popular sport in Colombia, and the practice of closing some streets for weekend biking has existed for decades. To reinforce the value of biking and walking, Bogotá held its first Car Free Day in 2000; on that day, no private cars and no trucks could be used in the city. This weekday event was so popular that it is now held annually.

So what is Bogotá like today? Congestion, especially in the downtown, has been greatly reduced. Also, the city's air quality has visibly improved. Bicycles are definitely more prevalent, and the city is filled with lively and popular public spaces.

1. Consider the density of Bogotá. Why would a bus system work well here?
2. What are the advantages of designated bus lanes?

Figure 4.16 Population Map of Latin America This map highlights the concentration of population in urban and coastal settlements. Population density in central and southern Mexico, as well as in Central America, is quite high. In South America, the majority of people live on or near the coasts, leaving the interior of the continent lightly populated.

Population and Settlement: The Dominance of Cities

Latin America did not have great river basin civilizations like those in Asia. In fact, the great rivers of the region are surprisingly underused as areas of settlement or corridors for transportation. The major population clusters of Central America and Mexico are in the interior plateaus and valleys, whereas the interior lowlands of South America are relatively empty. Historically, the highlands supported most of this region's population during the pre-Hispanic and colonial eras. In the 20th century, population growth and migration to the Atlantic lowlands of Argentina and Brazil, along with continued growth of Andean coastal cities such as Guayaquil, Barranquilla, and Maracaibo, have reduced the demographic importance of the highlands. Major highland cities such as Mexico City, Guatemala City, Bogotá, and La Paz still dominate their national economies, but the majority of large cities are on or near the coasts (Figure 4.16).

Like the rest of the developing world, Latin America has experienced dramatic population growth. In 1950, its population totaled 150 million people, which equaled the population of the United States at that time. By 1995, the population had tripled to 450 million; in comparison, the U.S. population only reached 300 million in 2006. Latin America outpaced the United States because infant mortality fell and life expectancy soared, while birth rates remained higher than that of the United States. In 1950, Brazilian life expectancy was only 43 years; by the 1980s, it was 63, and now it is 75. Four countries account for 70 percent of the region's population: Brazil with 205 million, Mexico with 127 million, Colombia with 48 million, and Argentina with 42 million (Table A4.1). In addition, more than three-quarters of those in Latin America are urban residents.

Patterns of Rural Settlement

Although the majority of people live in cities, some 125 million people do not. Throughout the region, a distinct rural lifestyle exists, especially among peasant subsistence farmers. In Brazil alone, more than 35 million people live in rural areas. Interestingly, the absolute number of people living in rural areas today is roughly equal to the number in the 1960s. Yet rural life has definitely changed. Along with subsistence agriculture, highly mechanized, capital-intensive farming occurs in most rural areas. The links between rural and urban areas are much improved, making rural areas less isolated. Also, as international migration increases, many rural communities are directly connected to cities in

North America and Europe, with immigrants sending back remittances and supporting hometown associations. This is especially evident in rural Mexico and Central America. The rural landscape is divided by extremes of poverty and wealth, and the root of social and economic tension in the countryside is the uneven distribution of arable land.

Rural Landholdings Historically, the control of land in Latin America was the basis for political and economic power. Colonial authorities granted large tracts of land to the colonists, who were also promised the services of Indian laborers. These large estates typically took up the best lands along the valley bottoms and coastal plains. The owners were often absentee landlords, spending most of their time in the city and relying on a mixture of hired and slave labor to run their rural operations. Passed down from one generation to the next, many estates can trace their ownership back a century or more. The establishment of large blocks of estate land meant that peasants were denied territory of their own, so they were forced to work for the estates. This long-observed practice of maintaining large estates is called **latifundia**.

Although the pattern of estate ownership is well documented, peasants have always farmed small plots for their own use. This practice of **minifundia** can lead to permanent or shifting cultivation. Small farmers typically plant a mixture of crops for subsistence as well as for trade. Peasant farmers in Colombia or Costa Rica, for example, grow corn, fruits, and various vegetables alongside coffee bushes that produce beans for export. Strains on the minifundia system occur when rural populations grow and land becomes scarce, forcing farmers to divide their properties into smaller and less productive parcels or seek out new parcels on steep slopes.

Much of the turmoil in 20th-century Latin America surrounded the issue of land ownership, with peasants demanding its redistribution through the process of **agrarian reform**. Governments have addressed these concerns in different ways. The Mexican Revolution in 1910 led to a system of communally held lands called *ejidos*. In the 1950s, Bolivia crafted agrarian reform policies that led to the government appropriating estate lands and redistributing them to small farmers. As part of the Sandinista revolution in Nicaragua in 1979, lands were taken from the political elite and converted into collective farms. In 2000, President Hugo Chavez ushered in a new era of agrarian reform in Venezuela, and Bolivian President Evo Morales introduced an agrarian reform program in 2006 to give land title to indigenous communities in the eastern lowlands. Each of these programs has met with resistance and proved to be politically and economically difficult. Eventually, the path chosen by most governments has been to make frontier lands available to land-hungry peasants.

Agricultural Frontiers The creation of agricultural frontiers served several purposes: providing peasants with land, tapping unused resources, and filling in blank spots on the map with settlers. Several frontier colonization efforts are noteworthy. Peru developed its *Carretera Marginal* (Perimeter Highway) in an effort to lure colonists into cloud forests and rainforests of eastern Peru some four decades ago. Most recently, the Interoceanic Highway, completed in 2013, links ports in Peru with those in Brazil and opens up new areas of settlement in the Amazonian territories of these countries (Figure 4.17). Bolivia, Colombia, and Venezuela have all had agricultural frontier schemes in their interior lowland tropical plains that attracted more large-scale investors than peasant farmers. Guatemala built roads to open the Petén region.

The opening of the Brazilian Amazon for settlement was the region's most ambitious frontier colonization scheme. In the 1960s, Brazil began its frontier expansion by constructing several new Amazonian highways, a new capital (Brasília), and state-sponsored mining operations. The Brazilian military directed the opening of the Amazon to provide an outlet for landless peasants and to extract the region's many resources. However, the generals' plans did not deliver as intended. Government-promised land titles, agricultural subsidies, and credits were slow to reach small farmers. Instead, too much money went to subsidizing large cattle ranches through tax breaks and improvement deals in which "improved" meant cleared forestland. Today five times more people live in the Amazon than in the 1960s; thus, continued human modification of this region is inevitable (see purple arrows in Figure 4.18). Yet most of the people in the Brazilian Amazon live in large cities such as Manaus and Belém.

The Latin American City

A quick glance at the population map of Latin America shows a concentration of people in cities (see Figure 4.16). One of the most significant demographic shifts has been the movement out of rural areas to cities, which began in earnest in the 1950s. Just one-quarter of the region's population was urban in 1950; the rest lived in small villages and the countryside. Today the pattern is reversed, with three-quarters of the population living in cities. In the most urbanized countries, such as Argentina, Chile, Uruguay, and Venezuela, more than 85 percent of the population lives in cities (Table A4.1). This preference for urban life is attributed to cultural as well as economic factors; under Iberian rule, people residing in cities had higher social status and greater economic opportunity. Initially, only Europeans were allowed to live in the colonial cities, but this exclusivity was not strictly enforced. Over the centuries, colonial cities became the hubs for transportation and communication, making them the primary centers for economic and social activities.

Latin America is noted for high levels of **urban primacy**, a condition in which a country has a *primate city* three to four times larger

Figure 4.17 Amazon Mining Town Mazuco, a small Peruvian gold-mining town, is part of a resource boom that is bringing in settlers and bringing down forest. This community is also served by the new Interoceanic Highway.

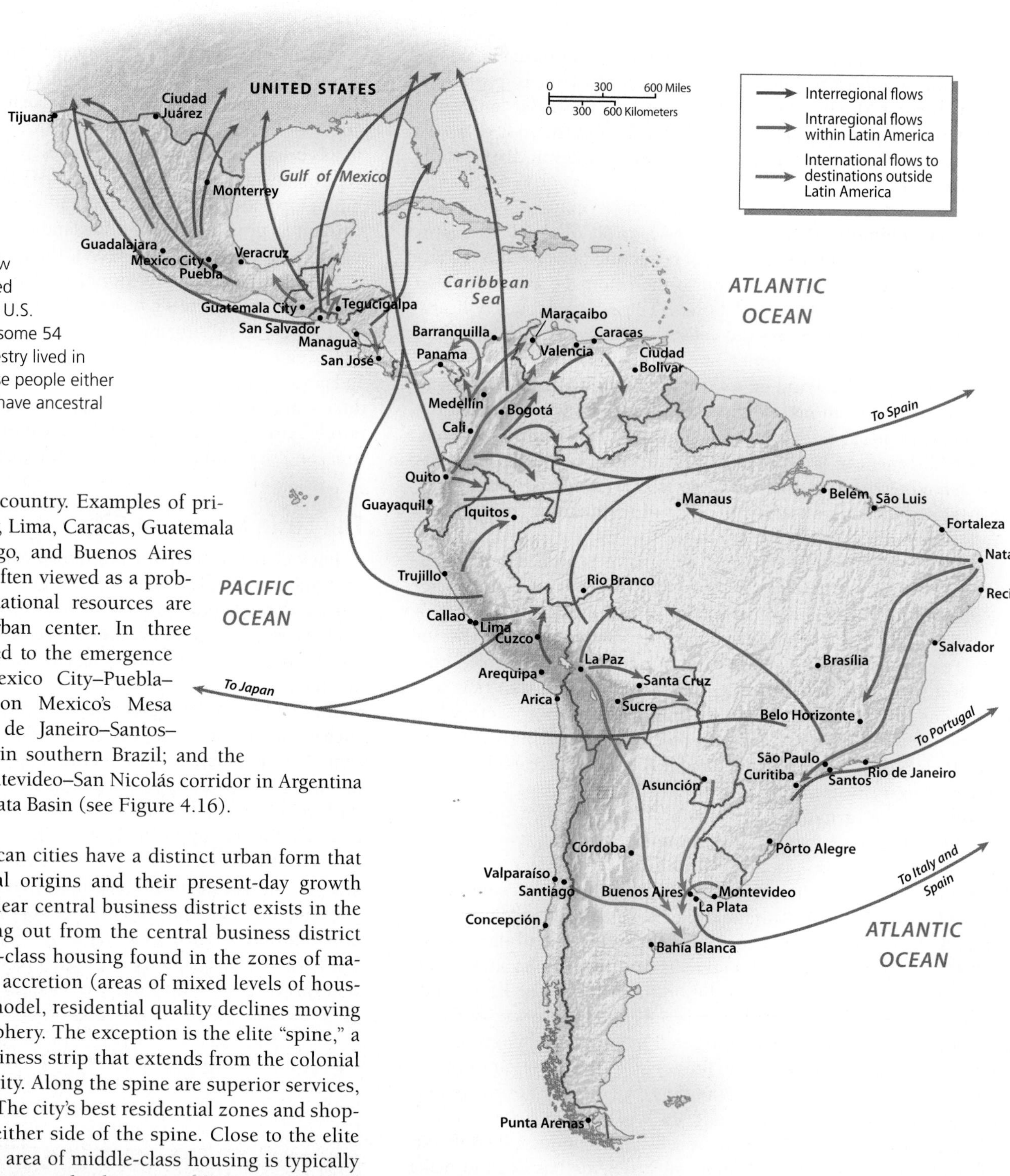

Figure 4.18 Major Latin American Migration Flows Internal, intraregional, and international migrations have opened frontier zones (purple arrows) and created transnational communities (red and blue arrows). Over the past three decades, the flow of Latin Americans to the United States has grown. In 2013, the U.S. Census Bureau estimated that some 54 million people of Hispanic ancestry lived in the United States. Most of these people either were born in Latin America or have ancestral ties to Latin America.

than any other city in the country. Examples of primate cities are Mexico City, Lima, Caracas, Guatemala City, Panama City, Santiago, and Buenos Aires (Figure 4.19). Primacy is often viewed as a problem because too many national resources are concentrated into one urban center. In three cases, urban growth has led to the emergence of a megalopolis: the Mexico City–Puebla–Toluca–Cuernavaca area on Mexico's Mesa Central; the Niterói–Rio de Janeiro–Santos–São Paulo–Campinas axis in southern Brazil; and the Rosario–Buenos Aires–Montevideo–San Nicolás corridor in Argentina and Uruguay's lower Rio Plata Basin (see Figure 4.16).

Urban Form Latin American cities have a distinct urban form that reflects both their colonial origins and their present-day growth (Figure 4.20). Usually, a clear central business district exists in the old colonial core. Radiating out from the central business district is older middle- and lower-class housing found in the zones of maturity and *in situ* (natural) accretion (areas of mixed levels of housing and services). In this model, residential quality declines moving from the center to the periphery. The exception is the elite "spine," a newer commercial and business strip that extends from the colonial core to newer parts of the city. Along the spine are superior services, roads, and transportation. The city's best residential zones and shopping malls are usually on either side of the spine. Close to the elite residential sector, a limited area of middle-class housing is typically found. Most major urban centers also have a *periférico* (a ring road or beltway highway) encircling the city. Industry is located in isolated areas of the inner city and in larger industrial parks outside the ring road.

Straddling the *periférico* is a zone of peripheral **squatter settlements**, where many of the urban poor live in self-built housing on land that does not belong to them. Services and infrastructure are extremely limited: Roads are unpaved, water is often trucked in, and sewer systems are nonexistent. The dense ring of squatter settlements that encircles Latin American cities reflects the speed and intensity with which these zones were created. In some cities, more than one-third of the population lives in these self-built homes of marginal or poor quality. These kinds of dwellings are found throughout the developing world, yet the practice of building homes on the "urban frontier" has a longer history in Latin America than in most Asian and African cities. The combination of a rapid inflow of migrants, the inability of governments to meet pressing housing needs, and the eventual official recognition of many of these neighborhoods with land titles and utilities meant that this housing strategy was seldom discouraged. Each successful squatter settlement on the urban edge encouraged more.

Population Growth and Movement

Latin America's high growth rates throughout the 20th century are attributed to natural increases as well as to immigration in the early part of the century. The 1960s and 1970s were decades of tremendous growth, resulting from high fertility rates and increasing life expectancy. In the 1960s, for example, a Latin America woman typically had six or seven children. By the 1980s, family sizes were half as big. Today the total fertility rate (TFR) for the region is 2.2, which is only slightly higher than replacement value (see Table A4.1). Several factors explain this: more urban families, which tend to be smaller than rural ones; increased participation of women in the workforce; higher education levels of women; state support of family planning; and better access to birth control.

Even with family sizes shrinking—and in Brazil, Costa Rica, Uruguay, and Chile falling below replacement value—built-in potential for continued growth exists because of the relative demographic youth of these countries. The average percentage of the population below age 15 is 27 percent. In North America, that same group is 19 percent of the population, and in Europe it is just 16 percent. This means that a proportionally larger segment of the population has yet to enter the childbearing years.

The population pyramids of two countries, Uruguay and Guatemala, contrast the profile of a country that has a stable population size with that of a demographically growing state (Figure 4.21). Uruguay is a small but prosperous country with a high human development index ranking and relatively little poverty. Uruguayan women average two children, which is slightly below replacement level. Life expectancy is also high, but most population projections have the country growing very slowly between now and 2050. In contrast, Guatemala has a wider-based population pyramid and is considerably poorer. Total fertility rates have declined in Guatemala but are still considered high at 3.8. Due to its youthful population and increased life expectancy, Guatemala's population is expected to double between 2010 and 2050.

In addition to natural increase, waves of immigrants into Latin America and migrant streams within Latin America have influenced population size and patterns of settlement. Beginning in the late 19th century, new immigrants from Europe and Asia added to the region's size and ethnic complexity. Important population shifts within countries have also occurred in recent decades, as illustrated by the growth of Mexican border towns and the demographic expansion of the Bolivian plains. In an increasingly globalized economy, even more Latin Americans live and work outside the region, especially in the United States and Europe.

European Migration After Latin American countries gained independence from Spain and Portugal in the 19th century, their new leaders sought to develop economically through immigration. Firmly believing that "to govern is to populate," many countries set up immigration offices in Europe to attract hard-working peasants to till the soils and "whiten" the ***mestizo*** (people of mixed European and Indian ancestry) population. The Southern Cone countries of Argentina, Chile, Uruguay, Paraguay, and southern Brazil were the most successful in attracting European immigrants from the 1870s until the depression of the 1930s. During this period, some 8 million Europeans arrived (more than came during the entire colonial period), with Italians, Portuguese, Spaniards, and Germans being the most numerous.

Asian Migration Less well known than the European immigrants to Latin America are the Asian immigrants, who also arrived during the late 19th and 20th centuries. Although considerably fewer, over time they established an important presence in the large cities of Brazil, Peru, Argentina, and Paraguay. Beginning in the mid-19th century, Chinese and Japanese laborers were contracted to work on the coffee estates

Figure 4.19 Primacy in Buenos Aires This capital city with a metropolitan population of over 13 million is the economic and cultural hub of Argentina. The bustling ceremonial boulevard, 9 de Julio Avenue, cuts through the downtown and commemorates Argentine Independence Day (July 9). The obelisk built in the early 20th century is an iconic structure for the city. Buenos Aires is both a primate city and a megacity (over 10 million people).

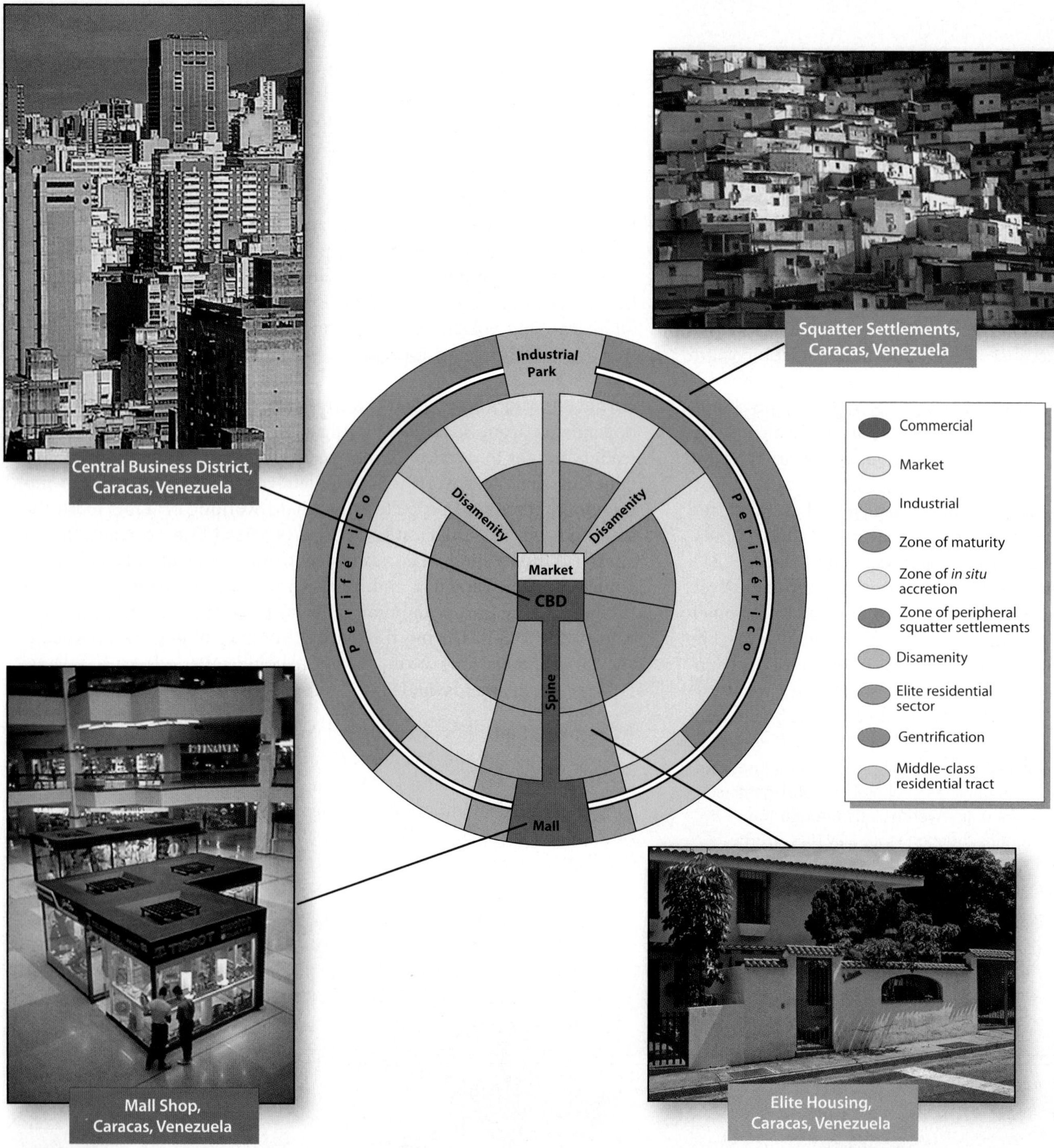

Figure 4.20 Latin American City Model This urban model highlights the growth of Latin American cities and the class divisions within them. Although the central business district, elite spine, and residential sectors may have excellent access to services and utilities, life in the zone of peripheral squatter settlements (variously called *ranchos*, *favelas*, *barrios jovenes*, or *pueblos nuevos*) is much more difficult. In many Latin American cities, one-third of the population resides in squatter settlements. **Q: How does the Latin American city model compare with that of North America in terms of where the rich and the poor live? What are the factors that drive urban growth?**

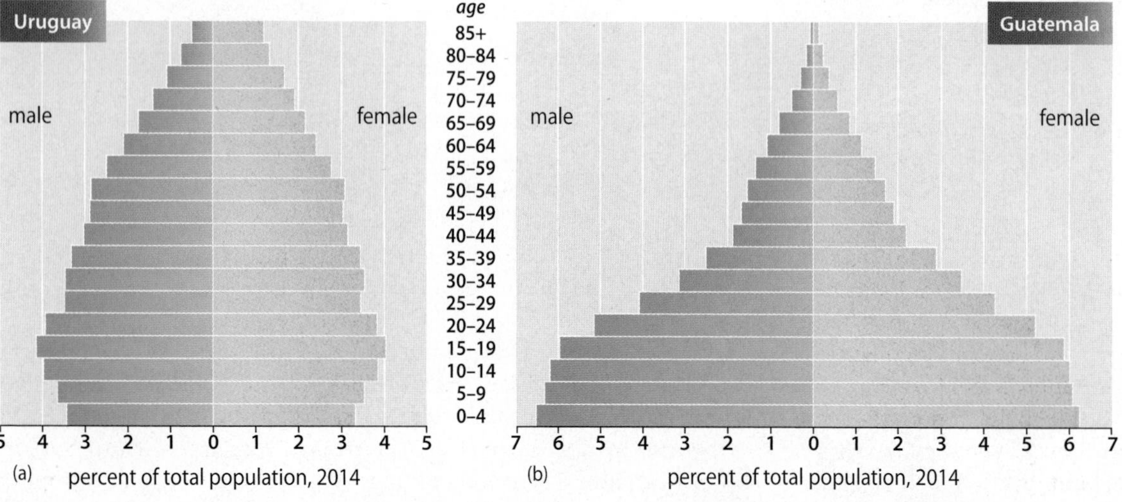

Figure 4.21 Population Structure of Uruguay and Guatemala These two pyramids contrast the population structure of (a) the more developed and demographically stable Uruguay with (b) that of the youthful and rapidly growing Guatemala. The average Uruguayan woman has 2 children, whereas Guatemalan women have a total fertility rate (TFR) of 3.8. Due in part to this difference, Uruguay is projected to have about the same population size in 2050, but Guatemala is projected to be twice as large, with 30 million people.

Figure 4.22 Japanese-Brazilians Brazilian youth of Japanese ancestry perform in Curitiba, Brazil, to mark the 100th anniversary of the beginning of Japanese immigration to Brazil. In 1908, the first Japanese immigrants arrived as agricultural workers, choosing Brazil as a destination after countries such as the United States and Canada banned Japanese immigration. Today there are over 1.3 million ethnic Japanese in Brazil, especially in the states of São Paulo and Paraná, and they have distinguished themselves as large-scale farmers and urban professionals.

in southern Brazil and the sugar estates and coastal mines of Peru. Over time, these Asian immigrants became prominent members of society; for example, Alberto Fujimori, a son of Japanese immigrants, was president of Peru from 1990 to 2000.

Between 1908 and 1978, one-quarter of a million Japanese immigrated to Brazil; today the country is home to 1.3 million people of Japanese descent (Figure 4.22). As a group, the Japanese have been closely associated with the expansion of soybean and orange production. Increasingly, second- and third-generation Japanese have taken professional and commercial jobs in Brazilian cities; many have married outside their ethnic group and are losing their fluency in Japanese. South America's economic turmoil in the 1990s encouraged many ethnic Japanese to emigrate to Japan in search of better wages. Nearly one-quarter of a million ethnic Japanese, mostly from Brazil and Peru, left South America in the 1990s and now work in Japan.

Latino Migration and Hemispheric Change Migration within Latin America and between Latin America and North America has had a significant impact on both sending and receiving communities. Within Latin America, international migration is shaped by shifting economic and political realities. For example, Venezuela's oil wealth attracted Colombian immigrants, who tended to work as domestics or agricultural laborers. Argentina has long been a destination for Bolivian and Paraguayan laborers. Nicaraguans seek employment in Costa Rica. And farmers in the United States have depended on Mexican laborers for more than a century.

Political turmoil has also sparked international migration. The bloody civil wars in El Salvador and Guatemala in the 1980s, for example, sent waves of refugees into Mexico and the United States. Violence in northern Central America is again driving people toward Mexico and the United States. In the summer of 2014, nearly 60,000 unaccompanied minors from Guatemala, El Salvador, and Honduras overwhelmed border facilities in the United Sates as these youth sought to reunite with family in the United States and to find refuge from gang-driven violence in Central America. Although some of these youth were able to remain, many were returned to Central America. Moreover, Central American countries actively dissuaded people from sending their youth to the north because of the many dangers involved in the illegal crossing.

Presently, Mexico is the country of origin for the most legal immigrants to the United States. There are 54 million Hispanics in the United States; two-thirds of them claim Mexican ancestry, including approximately 12 million who were born in Mexico. Mexican labor migration to the United States dates back to the late 1800s, when relatively unskilled labor was recruited to work in agriculture, mining, and railroads. Mexican immigrants are concentrated in California and Texas, but increasingly they are found throughout the country. Although Mexicans continue to have the greatest presence among Latinos in the United States, the number of immigrants from El Salvador, Guatemala, Nicaragua, Colombia, Ecuador, and Brazil has steadily grown. Most U.S. Hispanics have ancestral ties with peoples from Latin America and the Caribbean (see Chapter 5 on Caribbean migration).

Today Latin America is seen as a region of emigration, rather than one of immigration. Both skilled and unskilled workers from the region are an important source of labor in North America, Europe, and Japan. Many of these immigrants send monthly **remittances** (monies sent back home) to sustain family members. Peaking at nearly US$70 billion in 2008, remittances fell to $61 billion by 2013, indicating the lingering impact of the economic recession. In particular, remittance income to Mexico (Latin America's largest recipient) declined as many Mexicans returned to their country or were removed from the United States, which has stepped up deportation efforts over the last decade. The economic significance of remittances will be discussed again at the end of the chapter.

 **REVIEW**

4.3 How have policies such as agrarian reform and frontier colonization impacted the patterns of settlement and primary resource extraction in the region?

4.4 What are the historical and economic explanations for urban dominance and urban primacy in Latin America?

4.5 Demographically, Latin America has grown much faster than North America. What factors contribute to its faster growth, and is this growth likely to continue?

KEY TERMS latifundia, minifundia, agrarian reform, urban primacy, squatter settlement, *mestizo*, remittance

Cultural Coherence and Diversity: Repopulating a Continent

The Iberian colonial experience (1492 to the 1800s) imposed a political and cultural coherence on Latin America that makes it recognizable today as a world region. Yet this was not a simple transplanting of Iberia across the Atlantic. Instead, a process unfolded in which European and Indian traditions blended as indigenous groups were added into either the Spanish or the Portuguese empires. Indian cultures have shown remarkable resilience in some areas, as evidenced by the survival of Amerindian languages. However, the prevailing pattern was one of forced assimilation in which European religion, languages, and political organization were imposed on surviving Amerindian societies. Later, other cultures—especially more than 10 million African slaves—added to the cultural mix of Latin America, the Caribbean, and North America. The legacy of the African slave trade is examined in greater detail in Chapters 5 and 6. For Latin America, perhaps the single most important factor in the dominance of European culture was the demographic collapse of native populations.

The Decline of Native Populations

It is difficult to grasp the enormity of cultural change and human loss due to this encounter between the Americas and Europe. Throughout the region, archaeological sites are reminders of the complexity of Amerindian civilizations prior to European contact. Dozens of stone temples found throughout Mexico and Central America, where the Mayan and Aztec civilizations flourished, attest to the ability of these societies to thrive in the area's tropical forests and upland plateaus. The Mayan city of Tikal flourished in the lowland forests of Guatemala, supporting tens of thousands, before its mysterious collapse centuries before the arrival of Europeans (Figure 4.23). In the Andes, the complexity of Amerindian civilizations can be seen in political centers such as Cuzco and Machu Picchu. The Spanish, too, were impressed by the sophistication and wealth they saw around them, especially in Tenochtitlán, where Mexico City sits today. Tenochtitlán, the political and ceremonial center of the Aztecs, supported a complex metropolitan area with some 300,000 residents. The largest city in Spain at the time was considerably smaller.

The Demographic Toll The most telling indicators of the impact of European expansion in Latin America are demographic. It is widely believed that the precontact Americas had 54 million inhabitants; by comparison, western Europe in 1500 had approximately 42 million people. Of the 54 million, about 47 million were in what is now Latin America, and the rest were in North America and the Caribbean. By 1650, after a century and a half of colonization, the indigenous population was one-tenth its precontact size. The human tragedy of this population loss is hard to comprehend. The relentless elimination of 90 percent of the native population was largely caused by epidemics of influenza and smallpox, but warfare, forced labor, and starvation due to the collapse of food-production systems also contributed to the rapid population decline.

Indian Survival Presently, Mexico, Guatemala, Ecuador, Peru, and Bolivia have the largest indigenous populations. Not surprisingly, these areas had the densest native populations at the time of European contact. Indigenous survival also occurs in isolated settings where the workings of national and global economies are slow to break through, such as in eastern Panama, the Miskito Coast of Honduras, and the roadless sections of western Amazonia.

In many cases, Indian survival comes down to one key resource—land. Indigenous peoples who are able to maintain a territorial home, formally through land title or informally through long-term occupancy, are more likely to preserve a distinct ethnic identity. Because of this close association between identity and territory, native peoples are increasingly insisting on a recognized space within their countries. Some of Panama's indigenous groups have organized territories called *comarcas*, where they assert local authority and have limited autonomy. The *comarca* of Guna Yala, on Panama's Caribbean coast, is the recognized territory of some 40,000 Guna (Figure 4.24). These efforts to define indigenous territory are seldom welcomed by the state, but they are occurring throughout the region.

Patterns of Ethnicity and Culture

The Indian demographic collapse enabled Spain and Portugal to reshape Latin America into a European likeness. However, instead of a neo-Europe rising in the tropics, a complex ethnic blend evolved. Within the first years of contact, unions between European sailors and Indian women began the process of racial mixing that over time became a defining feature of the region. The courts of Spain and Portugal officially discouraged racial mixing, but such positions could not be realistically enforced in the colonial territories.

Generations of intermarriage led to four broad racial categories: *blanco* (European ancestry), *mestizo* (mixed ancestry), *indio* (Indian ancestry), and *negro* (African ancestry). The *blancos* (or Europeans) continue to be well represented among the elites, yet the vast majority of Latin Americans are of mixed racial ancestry. *Dia de la Raza*, the region's observance of Columbus Day, recognizes the emergence of a new *mestizo* race as the legacy of European conquest. Throughout Latin America, more than other regions of the world, miscegenation (or racial mixing) is the norm, making the process of mapping racial or ethnic groups especially difficult.

Figure 4.23 Tikal, Guatemala This ancient Mayan city, located in the lowland forests of the Petén, was part of a complex network of cities located in the Yucatan and northern Guatemala. At its height, Tikal supported over 100,000 people before its collapse in the late 10th century. Today it is a major tourist destination.

Figure 4.24 Panama's Guna A Guna woman with her four children in front of her home on one of the San Blas islands in Panama. The Guna (formerly known as the Kuna) have maintained their territory of Guna Yala in eastern Panama since the 1920s.

Languages Roughly two-thirds of Latin Americans are Spanish speakers, and one-third speak Portuguese. These colonial languages were so widespread by the 19th century that they were the unquestioned languages of government and education for the newly independent Latin American republics. In fact, until recently, many countries actively discouraged, and even repressed, Indian tongues. It took a constitutional amendment in Bolivia in the 1990s to legalize native-language instruction in primary schools and to recognize the country's multiethnic heritage (more than half the population is Indian, and Quechua, Aymara, and Guaraní are widely spoken) (Figure 4.25).

Because Spanish and Portuguese dominate, there is a tendency to overlook the influence of indigenous languages in the region. Mapping the use of native languages, however, reveals important areas of Indian resistance and survival. In the Central Andes of Peru, Bolivia, and southern Ecuador, more than 10 million people still speak Quechua and Aymara, along with Spanish. In Paraguay and lowland Bolivia, there are 4 million Guaraní speakers, and in southern Mexico and Guatemala at least 6 to 8 million speak Mayan languages. Small groups of native-language speakers are found scattered throughout the sparsely settled interior of South America and the more isolated forests of Central America. However, many of these languages have fewer than 10,000 speakers.

Blended Religions Like language, the Roman Catholic faith appears to have been imposed on the region without challenge. Most countries report 90 percent or more of their population as Catholic. Every major city has dozens of churches, and even the smallest village maintains a graceful church on its central square. In some countries, such as El Salvador, Brazil, and Uruguay, a sizable portion of the population practices Pentecostalism, an evangelical form of Christianity, but the Catholic core of this region remains intact (see *Exploring Global Connections: The Catholic Church and the Argentine Pope*). Exactly what native peoples absorbed of the Christian faith is complex. Throughout Latin America, **syncretic religions**—blends of different belief systems—enabled animist practices to be included in Christian worship. These blends took hold and endured, in part because Christian saints were easy replacements for pre-Christian gods and because the Catholic Church tolerated local variations in worship as long as the process of conversion was under way. The Mayan practice of paying tribute to spirits of the underworld seems to be replicated today in Mexico and Guatemala through the practice of building small cave shrines to favorite Catholic saints and leaving offerings of fresh flowers and fruits. One of the most celebrated religious symbols in Mexico is the Virgin of Guadalupe—a dark-skinned virgin seen by an Indian shepherd boy—who became the patron saint of Mexico.

The Global Reach of Latino Culture

Latin American culture, vivid and diverse as it is, is widely recognized throughout the world. Whether it is the sultry pulse of the tango or the fanaticism with which Latinos embrace soccer as an art form, aspects of Latin American culture have been absorbed into a globalizing world culture. In the arts, Latin American writers such as Jorge Luis Borges, Gabriel García Marquez, and Isabel Allende have obtained worldwide recognition. In terms of popular culture, musical artists such as Colombia's Shakira and Brazil's hip-hop samba singer Max de Castro have international audiences. Through music, literature, art, and even *telenovelas* (soap operas), Latino culture is being transmitted to an eager worldwide audience.

Telenovelas Popular nightly soap operas are a mainstay of Latin American television. These tightly plotted series are filled with intrigue and double-dealing. Unlike their U.S. counterparts, the dramas end, usually after 100 episodes. Once standard fare for the working class, many telenovelas take hold and absorb an entire nation. During particularly popular episodes, the streets are noticeably calm as millions catch up on the lives of their favorite heroines. Brazil, Venezuela, and Mexico each produce scores of telenovelas, but the Mexican ones are international megahits.

Televisa, a Mexican production agency, has aggressively marketed its inventory of soap operas to an eager global public. Mexican telenovelas are avidly watched in countries as diverse as Croatia, Russia, China, South Korea, Iran, the United States, and France as well as throughout Latin America. Predictably scripted as Mexican Cinderella stories, these sagas of poor underclass women (often domestics) falling in love with members of the elite, battling jealous rivals, and ultimately emerging triumphant seem to resonate with fans around the world. In addition to their broad appeal, telenovelas are big business, perhaps Mexico's largest cultural export. Hollywood and Mumbai grind out movies, but much of Mexico's entertainment industry is geared toward producing this popular art form.

Soccer Perhaps the quintessential global sport, soccer has a fanatical following throughout much of the world. Yet it is Latin America, and especially South America, where *fútbol* is considered a cultural necessity. It is still largely a male game. Although girls are beginning to play, young boys and men are constantly seen playing on fields, beaches, and blacktops, especially in late afternoons and on weekends. The great soccer stadiums of Buenos Aires (Bombonera) and Rio de Janeiro (Maracaña)

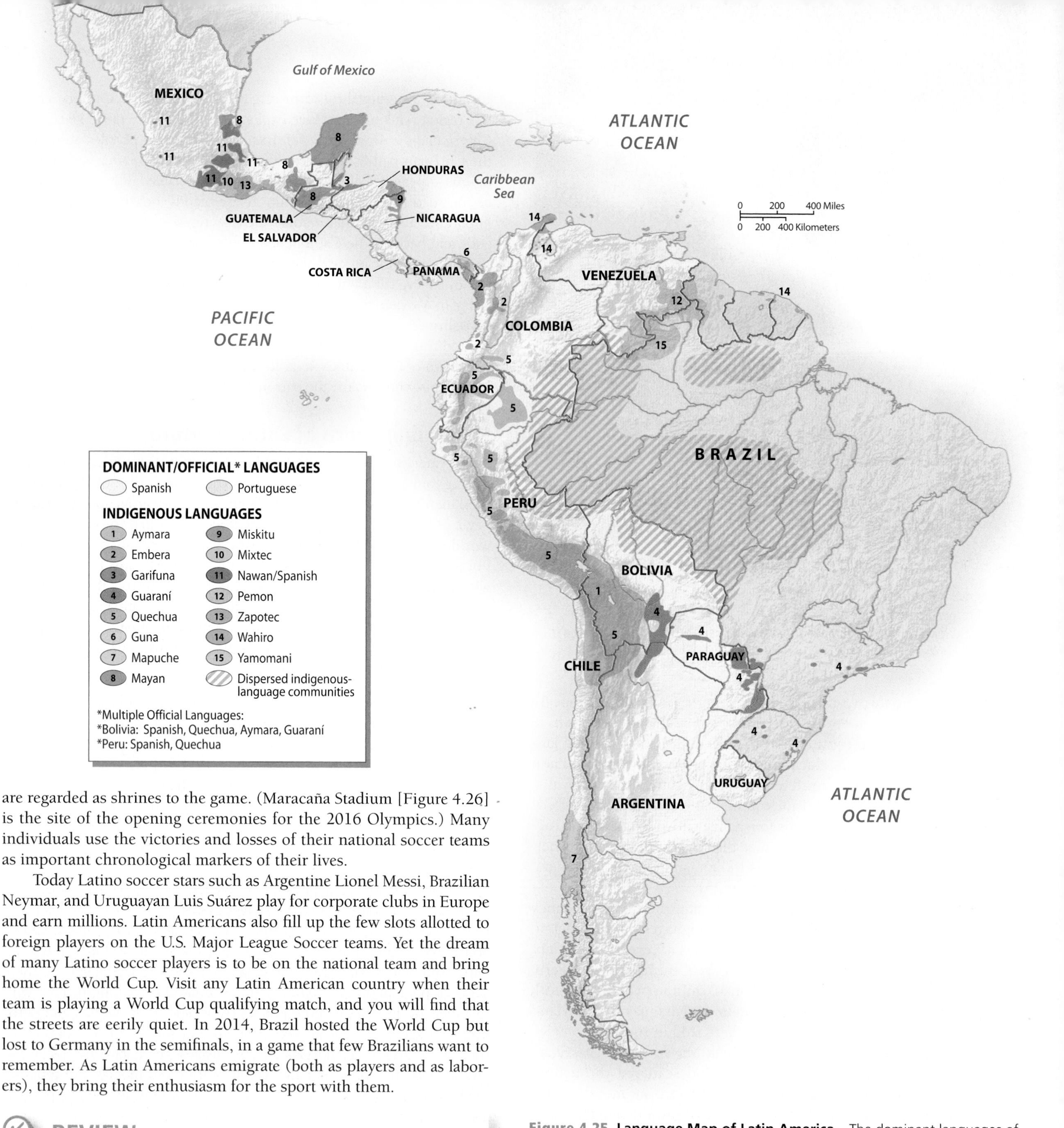

DOMINANT/OFFICIAL* LANGUAGES

Spanish | Portuguese

INDIGENOUS LANGUAGES

1 Aymara
2 Embera
3 Garifuna
4 Guaraní
5 Quechua
6 Guna
7 Mapuche
8 Mayan
9 Miskitu
10 Mixtec
11 Nawan/Spanish
12 Pemon
13 Zapotec
14 Wahiro
15 Yamomani
Dispersed indigenous-language communities

*Multiple Official Languages:
*Bolivia: Spanish, Quechua, Aymara, Guaraní
*Peru: Spanish, Quechua

are regarded as shrines to the game. (Maracaña Stadium [Figure 4.26] is the site of the opening ceremonies for the 2016 Olympics.) Many individuals use the victories and losses of their national soccer teams as important chronological markers of their lives.

Today Latino soccer stars such as Argentine Lionel Messi, Brazilian Neymar, and Uruguayan Luis Suárez play for corporate clubs in Europe and earn millions. Latin Americans also fill up the few slots allotted to foreign players on the U.S. Major League Soccer teams. Yet the dream of many Latino soccer players is to be on the national team and bring home the World Cup. Visit any Latin American country when their team is playing a World Cup qualifying match, and you will find that the streets are eerily quiet. In 2014, Brazil hosted the World Cup but lost to Germany in the semifinals, in a game that few Brazilians want to remember. As Latin Americans emigrate (both as players and as laborers), they bring their enthusiasm for the sport with them.

REVIEW

4.6 What factors contributed to racial mixing in Latin America, and where are the areas of strongest Amerindian survival?

4.7 What are the cultural legacies of Iberia in Latin America, and how are they expressed?

KEY TERMS syncretic religion

Figure 4.25 Language Map of Latin America The dominant languages of Latin America are Spanish and Portuguese. Nevertheless, there are significant areas in which native languages still exist and, in some cases, are recognized as official languages. Smaller language groups exist in Central America, the Amazon Basin, and southern Chile. **Q: What does this language map tell us about the patterns of Amerindian survival and endurance in Latin America?**

Geopolitical Framework: Redrawing the Map

Latin America's colonial history, more than its current condition, unifies this region geopolitically. For the first 300 years after the arrival of Columbus, Latin America was a territorial prize sought by various European countries, but the contest was effectively won by Spain and Portugal. By the 19th century, the independent states of Latin America had formed, but they continued to experience foreign influence, and sometimes overt political pressure, especially from the United States. At various times, a more neutral hemispheric vision of American relations and cooperation has held sway, represented by the formation of the **Organization of American States (OAS)**. The present OAS was officially formed in 1948, but its origins date to 1889. Yet there is no doubt that U.S. policies toward trade, economic assistance, political development, and sometimes military intervention are often seen as undermining the independence of these states.

Today the geopolitical influence of the United States in the region is declining, especially in South America. In many South American countries, trade with the European Union, China, and Japan is as important as, if not more important than, trade with the United States. For example, Brazil's largest trading partner is now China. Brazil's own influence in the region and the world is rising. It is the world's seventh-largest economy and one of the world's rapidly advancing BRIC countries (Brazil, Russia, India, and China). Trade blocs such as Mercosur and UNASUR are reshaping patterns of trade and political engagement.

Iberian Conquest and Territorial Division

When Christopher Columbus claimed the Americas for Spain, the Spanish became the first active colonial agents in the Western Hemisphere. In contrast, the Portuguese presence in the Americas was the result of the **Treaty of Tordesillas** in 1493–1494. By that time, Portuguese navigators had charted much of the coast of Africa in an attempt to find a water route to the Spice Islands (Moluccas) in Southeast Asia. With the help of Columbus, Spain sought a western route to the Far East. When Columbus discovered the Americas, Spain and Portugal asked the Pope to settle how these new territories should be divided. Without consulting other European powers, the Pope divided the Atlantic world in half: The eastern half, containing the African continent, was awarded to Portugal; the western half, with most of the Americas, was given to Spain. The line of division established by the treaty actually cut through the eastern part of South America, placing it under Portuguese rule. The treaty was never recognized by the French, English, or Dutch, who also claimed territory in the Americas, but it did provide the legal justification for the creation of Portuguese Brazil. This state would later become the largest and most populous in Latin America (Figure 4.27).

Six years after the treaty was signed, Portuguese navigator Alvares Cabral accidentally reached the coast of Brazil on a voyage to southern Africa. The Portuguese soon realized that this territory was on their side of the Tordesillas line. Initially, they were unimpressed by what Brazil had to offer; there were no spices or major native settlements. Over time, they came to appreciate the utility of the coast as a provisioning site as well as a source for brazilwood, used to produce a valuable dye. Portuguese interest in the territory intensified in the late 16th century, with the development of sugar estates and the expansion of the slave trade, and in the 17th century, with the discovery of gold in the Brazilian interior.

Spain, in contrast, aggressively pursued the conquest and settlement of its new American territories from the very start. After discovering little gold in the Caribbean, by the mid-16th century Spain directed its energy toward developing the silver resources of Central Mexico and the Central Andes (most notably Potosí in Bolivia). Gradually, the economy diversified to include some agricultural exports, such as cacao (for chocolate) and sugar, as well as a variety of livestock. In terms of foodstuffs, the colonies were virtually self-sufficient. Manufacturing, however, was forbidden in the Spanish-American colonies in order to keep them dependent on Spain.

Figure 4.26 Maracaña Stadium Rio de Janeiro's soccer stadium, a sports icon for fans around the world, is the site of the 2016 Olympic Opening Ceremonies. The final match of the 2014 World Cup, in which Germany beat Argentina to win the cup, was held there as well.

EXPLORING GLOBAL CONNECTIONS

The Catholic Church and the Argentine Pope

Google Earth (MG) Virtual Tour Video http://goo.gl/KpSJhI

In 2013, a new pope was selected, and for the first time the spiritual leader of nearly 1.2 billion Roman Catholics was a man born in the Americas. Pope Francis, formerly Bishop Jorge Mario Bergoglio, is the son of Italian immigrants and was born in Buenos Aires, Argentina. As a religious leader and a member of the Jesuit Order, he earned a reputation for his humility and devotion to the poor. Now, as leader of the Catholic Church, he oversees a vast global network of churches, schools, missions, and clergy that look to his guidance from Rome.

Shifting Catholic Demographics Pope Francis's selection brings to the fore the gradual but dramatic demographic shift in the world's Catholic population. In 1900, the clear majority of the world's Catholics resided in Europe, but today less than one-quarter live there. For some time, the demographic core of the Catholic Church has been in Latin America. Worldwide, Brazil has the largest Catholic population (150 million), followed by Mexico (106 million), the Philippines (75 million), and then the United States (75 million)—helped in part by the large influx of immigrants from Latin America (Figure 4.2.1). Half the world's Catholics are in the Americas. Sub-Saharan Africa—particularly the Democratic Republic of Congo and Nigeria—is another region where the numbers of Catholics are growing quickly, due in part to missionary work and colonial legacies.

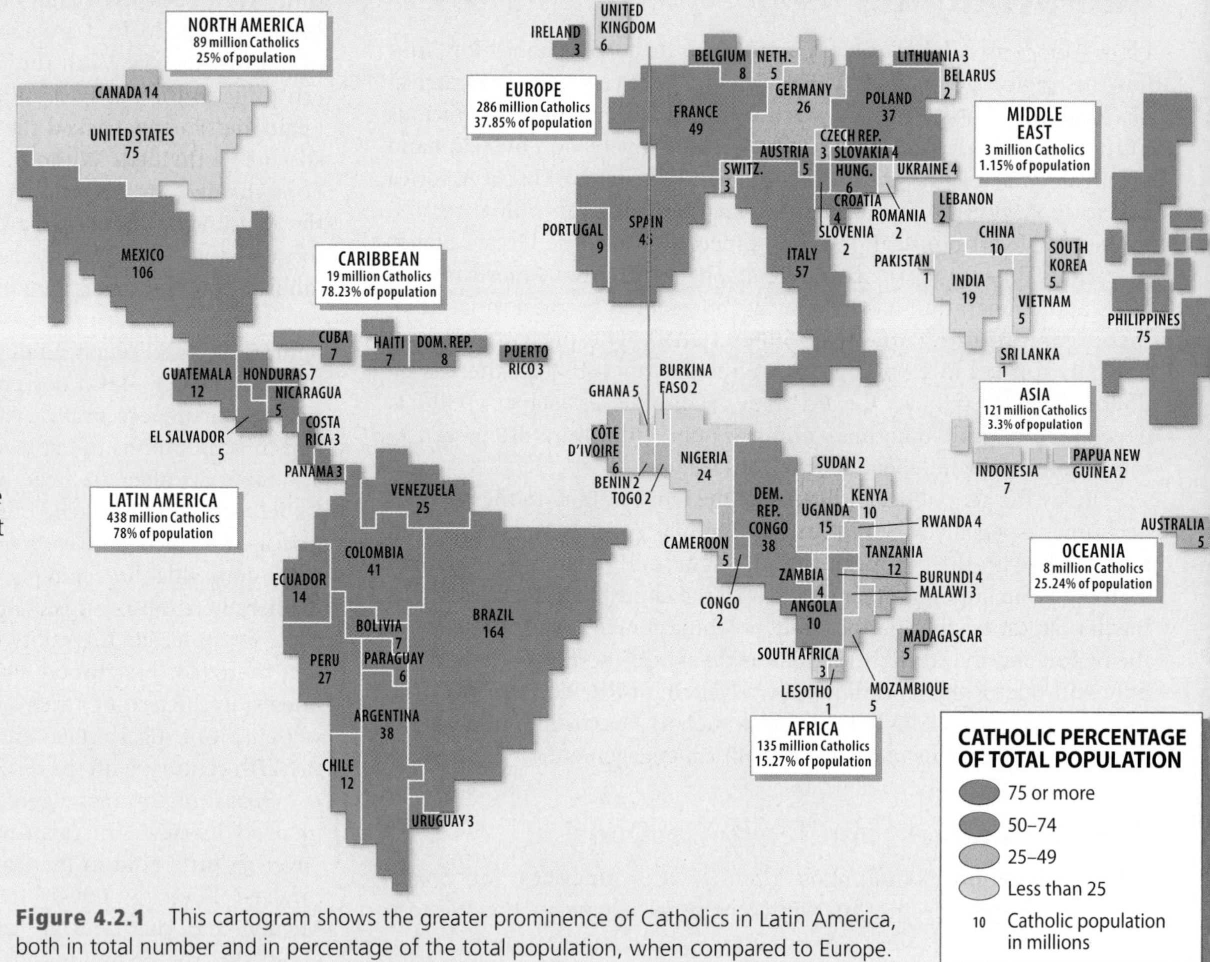

Figure 4.2.1 This cartogram shows the greater prominence of Catholics in Latin America, both in total number and in percentage of the total population, when compared to Europe. The number of Catholics in Sub-Saharan Africa is also growing. A cartogram is a thematic map in which area is distorted to reflect the relative value of what is mapped. In this case, the larger the country, the greater the number of Catholics.

Speaking for the Poor and the Environment Although the Pope is foremost a spiritual leader, he is clearly one with political clout. In his teachings and actions, he stresses reducing poverty and ensuring a clean and safe planet for future generations. In 2015, the Pope visited Ecuador, Bolivia, and Paraguay, some of Latin America's poorest countries that also face serious problems of environmental degradation (Figure 4.2.2). Many Latin Americans are thrilled at the selection of Pope Francis and the recognition it implies. Ironically, the future of the Catholic Church may lie more in the Southern Hemisphere than in the Northern.

1. What processes explain the growth of the Catholic faith in Latin America and Sub-Saharan Africa?
2. Given population trends, where is membership in the Catholic Church likely to grow?

Figure 4.2.2 Pope Francis in Ecuador Pope Francis waves to enthusiastic followers from his popemobile in El Quinche, Ecuador, during a visit in 2015. A native of Argentina, he has been an outspoken advocate for social justice and better protection of the global environment.

Figure 4.27 Shifting Political Boundaries (a) The evolution of Latin American political boundaries began with the 1494 Treaty of Tordesillas, which gave much of the Americas to Spain and a slice of South America (Brazil) to Portugal. The larger Spanish territory was gradually divided into viceroyalties and *audiencias*, which formed the basis for many modern national boundaries. (b) The 1830 borders of these newly independent states were far from fixed. Bolivia lost its access to the coast, Peru gained much of Ecuador's Amazon, and Mexico was stripped of its northern territory by the United States.

Revolution and Independence Not until the rise of revolutionary movements between 1810 and 1826 was Spanish authority on the mainland challenged. Ultimately, European elites born in the Americas gained control, displacing leaders loyal to the crown. In Brazil, the evolution from Portuguese colony to independent republic was a slower and less violent process that spanned eight decades (1808–1889). In the 19th century, Brazil was declared a separate kingdom from Portugal, with its own king, and later became a republic.

The territorial division of Spanish and Portuguese America into administrative units provided the legal basis for the modern states of Latin America (see Figure 4.27). The Spanish colonies were first divided into two viceroyalties (the administrative units of New Spain and Peru), and later subdivisions became the basis for the modern states. (In the 18th century, the Viceroyalty of Peru, which included all of Spanish South America, was divided to form three viceroyalties: La Plata, Peru, and New Granada.) Unlike Brazil, which evolved from a colony into a single republic, the former Spanish colonies experienced fragmentation in the 19th century.

Today the former Spanish mainland colonies include 16 states (plus 3 Caribbean islands), with a total population of over 400 million. If the Spanish colonial territory had remained a single political unit, it would now have the third-largest population in the world, following China and India.

Persistent Border Conflicts As the colonial administrative units turned into states, it became clear that the territories were not clearly demarcated, especially the borders that stretched into the sparsely populated South American interior. This would later become a source of conflict as new states struggled to define their territorial limits. Numerous border wars erupted in the 19th and 20th centuries, and the map of Latin America had to be redrawn many times. Some notable conflicts were the War of the Pacific (1879–1882), in which Chile expanded to the north and Bolivia lost its access to the Pacific; warfare between Mexico and the United States in the 1840s, resulting in the present border under the Treaty of Hidalgo (1848); and the bloody War of the Triple Alliance (1864–1870), in which Argentina, Brazil, and Uruguay together defeated Paraguay's claim on the upper Paraná River Basin. In the 1980s, Argentina lost a war with Great Britain over control of the Falkland, or Malvinas, Islands in the South Atlantic. And as recently as 1998, Peru and Ecuador fought over a disputed boundary in the Amazon Basin (Figure 4.28).

The Trend Toward Democracy Most of Latin America's 17 countries have celebrated or will soon celebrate their bicentennials. Compared with most of the rest of the developing world, Latin Americans have been independent for a long time. Yet political stability is not a characteristic of the region. Among these countries, some 250 constitutions have been written since independence, and military takeovers have been alarmingly frequent. Since the 1980s, however, the trend has been toward democratically elected governments, the opening of markets, and broader public participation in the political process. Where dictators once outnumbered elected leaders, by the

Figure 4.28 Geopolitics and Trade Blocs in Latin America Of the five economic trade blocs shown, Mercosur and NAFTA are the most dynamic. As UNASUR develops, it could fold Mercosur and the Andean Community into one common market. In 2004, members of the Central American Common Market entered into an agreement called CAFTA (Central American Free Trade Agreement), which soon came to include the Dominican Republic and henceforth has been known as CAFTA-DR. **Q: How could the growth and strength of trade blocs impact how Latin America functions as a region?**

1990s each country in the region had a democratically elected president. (Cuba, the one exception, will be discussed in Chapter 5.)

Democracy may not be enough for the millions frustrated by the slow pace of political and economic reform. In survey after survey, Latin Americans reveal their dissatisfaction with politicians and governments. Many of the democratic leaders have also been free-market reformers who are quick to eliminate state-backed social safety nets, such as food subsidies, government jobs, and pensions. Many of the poor and members of the middle class have doubts about whether this brand of neoliberal democracy can make their lives better. Even in more prosperous Chile, widespread student protests from 2011 to 2015 demanding lower-cost university tuition and fairer access to and funding for secondary education have left politicians scrambling. One Chilean student leader is a charismatic geography student, Camila Vallejo, who has gained international recognition for her ability to mobilize university students and to shut down a city (Figure 4.29). Under such conditions, it is not surprising that left-leaning politicians have won presidential elections in Brazil, Bolivia, Nicaragua, Ecuador, Peru, and Venezuela. These leaders have not rolled back neoliberal trade reforms, but many have tried to improve social services and reduce income inequalities.

Figure 4.29 Student Protests in Chile Camila Vallejo (center) and other Chilean student leaders lead a march of some 50,000 students in Santiago over the growing costs of higher education in Chile. The Confederation of Chilean Students has organized massive marches for educational reform since 2011.

Regional Organizations

At the same time that democratically elected leaders struggle to address the pressing needs of their countries, political developments at the supranational and subnational levels pose new challenges to their authority. The most discussed **supranational organizations** (governing bodies that include several states) are the trade blocs, the newest one being **UNASUR—The Union of South American Nations. Subnational organizations** (groups that represent areas or people within a state) form along ethnic or ideological lines or can support organized crime. Subnational organizations can have positive or destabilizing impacts. Examples include native groups that seek territorial recognition (such as the Guna in Panama); insurgent groups (such as the FARC [Revolutionary Armed Forces of Colombia] or the Zapatistas in Mexico) that challenge the authority of the state; and more recently drug cartels such as Los Zetas and Sinaloa, which have terrorized Mexico with extreme violence since 2006.

Trade Blocs Beginning in the 1960s, countries formed regional trade alliances in an effort to promote internal markets and reduce trade barriers. The Latin American Integration Association (formerly LAFTA), the Central American Common Market (CACM), and the Andean Group have existed for decades, but their ability to influence economic trade and growth has been limited at best. In the 1990s, **Mercosur (the Southern Cone Common Market)** and NAFTA emerged as supranational structures that could influence development (Figure 4.28). For Latin America, lessons learned from Mercosur in particular led Brazil to propose UNASUR in 2008, uniting virtually all of South America.

NAFTA took effect in 1994 as a free trade area that would gradually eliminate tariffs and ease the movement of goods among the member countries (Mexico, the United States, and Canada). NAFTA has increased intraregional trade, but it has provoked considerable controversy about costs to the environment and to employment (see Chapter 3). NAFTA did prove, however, that a free trade area combining industrialized and developing states was possible. In 2004, the United States, five Central American countries—Guatemala, El Salvador, Nicaragua, Honduras, and Costa Rica—and the Dominican Republic signed **CAFTA (Central American Free Trade Agreement)**. Like NAFTA, CAFTA aims to increase trade and reduce tariffs among member countries. The treaty was fully ratified in 2009, but whether such a treaty will lead to more economic development in Central America is a much-debated question.

Mercosur was formed in 1991 with Brazil and Argentina—the two largest economies in South America—and the smaller states of Uruguay and Paraguay as members. Since its formation, trade among these countries has grown tremendously, so much so that eventually Chile, Bolivia, Peru, Ecuador, and Colombia joined the group as associate members, and Venezuela received ratification as a full member in 2012. Mercosur's success is significant in two ways: It reflects the growth of these economies and also their willingness to put aside old rivalries (especially long-standing antagonisms between Argentina and Brazil) for the economic benefits of cooperation.

In 2008, Brazil initiated the formation of the 12-member Union of South American Nations (UNASUR). Some see this as an assertion of Brazil's greater political and economic clout in the region. UNASUR includes every South American country except French Guiana, which is a territory of France. It has formally organized with a permanent secretariat and has responded to political crises in Bolivia (2008), Ecuador (2010), Paraguay (2012), and Colombia (2014). Significantly, unlike NAFTA or CAFTA, UNASUR was a Brazilian-led effort, not one led by the United States. Brazil's lead in this initiative underscores its larger geopolitical ambitions to influence South American development and to secure a permanent seat on the United Nations Security Council. Brazil's ambitions have not always been well received by its neighbors, but the strengthening of UNASUR suggests a shift in the region's geopolitical alignment and the potential to form a common market more like the European Union.

Insurgencies and Drug Cartels Guerilla groups such as the FARC in Colombia have controlled large territories within their countries through the support of those loyal to the cause and through the use of such tactics as theft, kidnapping, and violence. The FARC, along with the ELN (National Liberation Army), gained wealth and weapons through the drug trade. The level of violence in Colombia has escalated further with the rise of paramilitary groups—armed private groups that terrorize those sympathetic to insurgency. The paramilitary groups have been blamed for hundreds of politically motivated murders each year. As many as 2.5 million Colombians have been internally displaced by violence since the late 1980s, most fleeing rural areas for towns and cities. Fortunately, after more than a decade of negotiations, in 2014 unilateral cessation of FARC hostilities was agreed to, and the situation has improved considerably. While the level of violence has fallen, Colombia remains the world's largest cocaine producer, followed by Peru and Bolivia (see *Geographers at Work: Development Work in Postconflict Colombia*).

Drug cartels and gangs in states as diverse as Mexico, Guatemala, El Salvador, Honduras, and Brazil have been blamed for increasing violence and lawlessness. The spike in violence and corruption in Mexico has been especially destabilizing. Profiting from the illegal production and/or shipment of cocaine, marijuana, methamphetamine, and heroin, the cartels generate billions of dollars. Beginning in 2006, the Mexican government brought in the army to quell the violence, kidnapping, and intimidation perpetrated by cartel groups, especially in the border region, but now extending throughout Mexico and into Central America (Figure 4.30). Human rights groups estimate that some 120,000 cartel-related murders occurred from 2006 to 2013 along with many thousands of instances of "disappeared" persons. Stemming the violence (rather than the flow of drugs) was one of the biggest issues

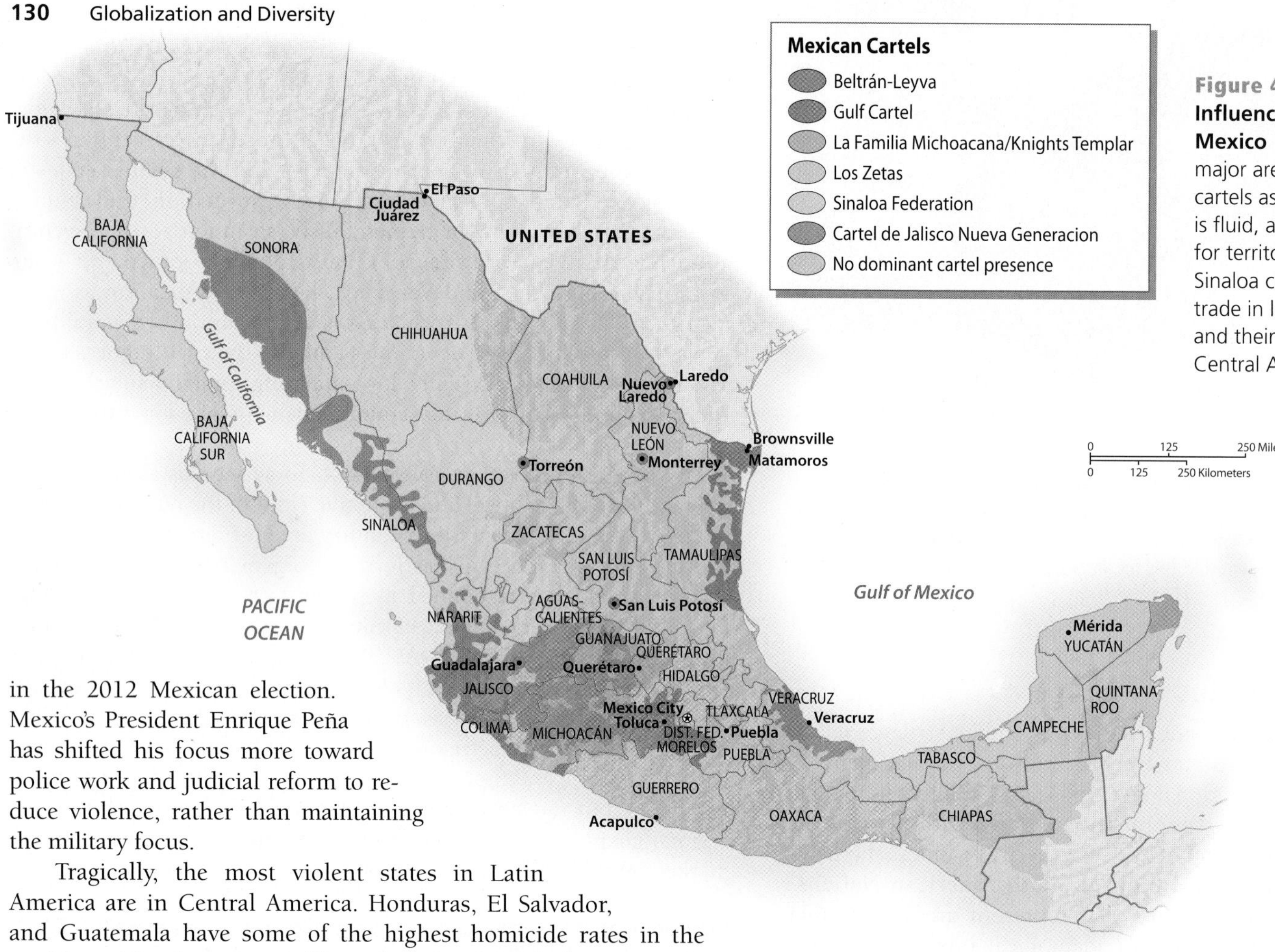

Mexican Cartels

- Beltrán-Leyva
- Gulf Cartel
- La Familia Michoacana/Knights Templar
- Los Zetas
- Sinaloa Federation
- Cartel de Jalisco Nueva Generacion
- No dominant cartel presence

Figure 4.30 Mapping the Influence of Drug Cartels in Mexico This map suggests the major areas of influence of Mexican cartels as of 2012. Cartel influence is fluid, and groups often compete for territories. Los Zetas and the Sinaloa cartels control the drug trade in large swaths of Mexico, and their influence crosses into Central America as well.

in the 2012 Mexican election. Mexico's President Enrique Peña has shifted his focus more toward police work and judicial reform to reduce violence, rather than maintaining the military focus.

Tragically, the most violent states in Latin America are in Central America. Honduras, El Salvador, and Guatemala have some of the highest homicide rates in the world outside of war zones. These three impoverished countries have long been in the transshipment zone of cocaine produced in Colombia, Peru, and Bolivia. More recently, however, Mexico's Sinaloa and Los Zetas cartels have been active in the isthmus, paying locals in drugs, creating drug-processing centers, and both directly and indirectly driving up the murder rate. The high levels of violence in Central America contributed to the surge in unaccompanied minors crossing through Mexico and into the United States in 2014.

REVIEW

4.8 How did Latin America's colonization by Iberia lead to the formation of the region's modern states?

4.9 Explain how trade blocs are reshaping the region's geopolitics.

KEY TERMS Organization of American States (OAS), Treaty of Tordesillas, supranational organization, UNASUR (Union of South American Nations), subnational organization, Mercosur (Southern Cone Common Market), Central American Free Trade Agreement (CAFTA)

Economic and Social Development: Focusing on Neoliberalism

Most Latin American economies fit into the broad middle-income category set by the World Bank. Clearly part of the developing world, Latin American people are much better off than those in Sub-Saharan Africa, South Asia, and most of China. Still, the economic contrasts are sharp, both between states and within them. Generally, the Southern Cone states (including southern Brazil and excluding Paraguay) and Mexico are the richest. The poorest countries in terms of per capita purchasing power parity (PPP) are Nicaragua, Bolivia, and Honduras. Although per capita incomes in Latin America are well below levels of developed countries, the region has witnessed steady improvements in various social indicators, such as life expectancy, child mortality, and literacy. Also, some small states, such as Costa Rica, do very well in the human development index (see Table A4.2).

The economic engines of Latin America are its two largest countries, Brazil and Mexico. According to the International Monetary Fund, in 2015 Brazil was the world's 7th-largest economy and Mexico was the 15th-largest, based on gross domestic product (GDP). The region has also seen reductions in extreme poverty, as the percentage of people living on less than \$2 per day dropped from 22 percent in 1999 to 12 percent in 2008. Today about one in ten Latin Americans lives on less than \$2 a day.

Yet the path toward economic development in Latin America has been a volatile one. In the 1960s, Brazil, Mexico, and Argentina all seemed poised to enter the ranks of the developed world. Multilateral agencies such as the World Bank and the Inter-American Development Bank loaned money for big development projects: continental highways, dams, mechanized agriculture, and power plants. All sectors of the economy were radically transformed. Agricultural production increased with the application of "green revolution" technology and mechanization (see Chapter 12). State-run industries reduced the need for imported goods,

GEOGRAPHERS AT WORK

Development Work in Post-Conflict Colombia

Corrie Drummond Garcia, who works for the U.S. Agency for International Development (USAID), says that her geography studies "provided a background in several disciplines that can be applicable in many types of jobs." She now works in Colombia in the field of post-conflict development, coordinating with USAID partners (NGOs, municipal governments, and contractors) and monitoring community infrastructure or agricultural projects (Figure 4.3.1). What fascinates Drummond Garcia about Colombia is its regional diversity: "It is a country with a strong sense of national pride, but also strong regional identities."

Building a Career Drummond Garcia discovered geography as an undergraduate at Bucknell University where she had a professor who could "take what is happening in the world on any particular day and bring it to your attention in ways you never thought about . . . it made me totally rethink what geography was and how it affected my analysis of the world." She later brought these analytic skills to a stint with the Peace Corps in El Salvador, where she found that she enjoyed working in Latin America with marginalized groups. After earning her MA in Geography at George Washington University and working for the Pan American Development Foundation, she joined USAID, working in Washington DC, Haiti, and now Colombia.

Post-Conflict Development After decades of violence, Colombia is in a post-conflict phase where many rural areas need development or redevelopment. An overarching concern is Colombia's displaced population, estimated to be at least 2.5 million. Legislation enacted in 2011 assists victims of violence and helps those uprooted by violence return to their land, though many have been displaced for more than a decade.

During the conflict years, political dysfunction was high, and some groups thrived in the intentional chaos. The challenge now is to build trust and find local solutions. One creative development strategy is to design educational or cultural parks on reclaimed sites once associated with violent episodes. Actively involving local groups in planning and reclaiming sites is the key to community healing and development. Drummond Garcia says, "With regard to conflict, each community has experienced it differently . . . we go into a community without any preconceived notions, and build solutions together."

1. What is unique about development efforts in a post-conflict situation?
2. What skills do geographers bring to the field of development?

Figure 4.3.1 Development worker, Corrie Drummond Garcia, at work Here she is wearing protective netting as she visits a beekeeping project in rural Colombia.

and the service sector ballooned as a result of new government and private-sector jobs. In the end, most Latin American countries made the transition from predominantly rural and agrarian economies, dependent on one or two commodities, to more economically diversified and urbanized countries with mixed levels of industrialization.

The modernization dreams of Latin American countries were trampled in the 1980s, when debt, currency devaluation, hyperinflation, and falling commodity prices undermined the region's aspirations. By the 1990s, most Latin American governments had radically changed their economic development strategies. State-run national industries and tariffs were jettisoned for policy reforms that emphasized privatization, direct foreign investment, and free trade, collectively labeled **neoliberalism**. Through tough fiscal policy, increased trade, privatization, and reduced government spending, most countries saw their economies grow and poverty decline. In aggregate, the economies of the region averaged an annual growth rate of 3.8 percent from 2009 to 2013. However, sporadic economic downturns have made these neoliberal policies highly unpopular with the masses, at times causing major political and economic turmoil. Much of the economic growth in the last few years is attributed to a boom in primary exports, which are still critical to the region's economic well-being.

Primary Export Dependency

Historically, Latin America's abundant natural resources were its wealth. In the colonial period, silver, gold, and sugar generated great wealth for the colonists. With independence in the 19th century, a series of export booms introduced commodities such as bananas, coffee, cacao, grains, tin, rubber, copper, wool, and petroleum to an expanding world market. One legacy of this export-led development was a tendency to specialize in one or two major commodities, a pattern that continued into the 1950s. During that decade, 90 percent of Costa Rica's export earnings came from bananas and coffee, 70 percent of Nicaragua's came from coffee and cotton, 85 percent of Chile's came from copper, and half of Uruguay's came from wood. Even Brazil generated 60 percent of its export earnings from coffee in 1955. However, by 2000, coffee accounted for less than 5 percent of the country's exports, even though Brazil remained the world leader in coffee production (see *Everyday Globalization: Good Morning Coffee*).

Agricultural Production Since the 1960s, the trend in Latin America has been to diversify and mechanize agriculture. Nowhere is this more evident than in the Plata Basin, which includes southern Brazil, Uruguay, northern Argentina, Paraguay, and eastern Bolivia. Soybeans, used for oil and animal feed, transformed these lowlands in the 1980s and early 1990s.

EVERYDAY GLOBALIZATION

Good Morning Coffee

Many Americans begin their day with coffee, at an annual rate of 10 pounds per person. We look to the tropics, particularly Latin America, for our morning fix. Brazil has been the world leader in coffee production for over a century. Until 1990, Colombia was the second-largest producer but has been surpassed by Vietnam. Today Latin America accounts for 58 percent of the world's coffee production.

Coffee is an ideal commodity for global trade because it does not spoil. Unlike other major agricultural exports, coffee is labor-intensive, and the International Coffee Organization estimates that coffee growing employs 26 million people worldwide. The berries are handpicked; preparing the beans for shipment is also labor-intensive (Figure 4.4.1). Small family farms do not produce great wealth, but in the highland tropics coffee is often the most profitable legal crop. Increasingly, **fair trade** coffee is promoted to help small growers receive a higher price for their crop by cutting out some of the middlemen. When consumers understand the connection between their daily cup of coffee and the people who produce it, rural livelihoods in Latin America can be more economically and environmentally sustainable.

1. What are the environmental and economic advantages of fair trade coffee?
2. If you drink coffee, find out what country it comes from and how it is produced.

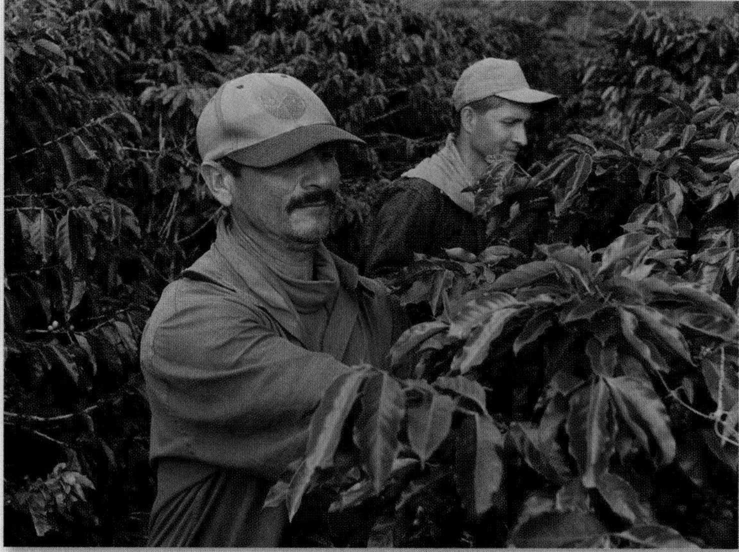

Figure 4.4.1 Harvesting Colombian Coffee Pickers in the department of Risaralda harvest coffee by hand. Colombia is noted for its quality coffee, but its harvests have declined in the last decade due to climate change and rural violence.

Brazil is now the second-largest producer of soy in the world (following the United States) and the world's largest soy exporter. Argentina is the third-largest, and production is still increasing. Between the late 1990s and 2010, soy production tripled in Argentina. The speed with which the Plata and Amazon basins are being converted into soy fields alarms many; this conversion is eliminating forest and savanna, negatively impacting biodiversity and increasing greenhouse gas emissions. But with soy prices high, the rush to plant continues (Figure 4.31).

Similar large-scale agricultural frontiers exist along the Pacific slope of Central America (cotton and some tropical fruits) and in the Central Valley of Chile and the foothills of Argentina (wine and fruit production). In northern Mexico, water supplied from dams along the Sierra Madre Occidental has turned the valleys in Sinaloa into intensive agricultural centers of fruits and vegetables for U.S. consumers. The relatively mild winters in northern Mexico allow growers to produce strawberries and tomatoes during the winter months.

In each of these cases, the agricultural sector is capital-intensive. By using machinery, hybrid crops, chemical fertilizers, and pesticides, many corporate farms are extremely productive and profitable. What these operations fail to do is employ many rural people, which is especially problematic in countries where one-third or more of the population depends on agriculture for its livelihood. Interestingly, a few traditional Amerindian foods, such as quinoa, are gaining consumers thanks to a growing appetite for organic and healthful foods. Recently, Peru and Bolivia have experienced a boom in quinoa production and exports, with much of the crop being grown in small and medium-sized highland farms. Bolivian President Evo Morales even declared 2012 the "year of quinoa" in an effort to promote traditional "Indian" foods.

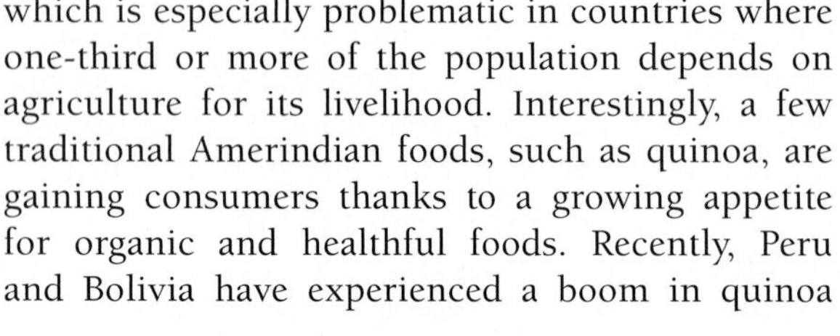

Mining and Forestry The mining of silver, zinc, copper, iron ore, bauxite, and gold is an economic mainstay for many countries in the region. Moreover, many commodity prices reached record levels in the last decade, boosting foreign exchange earnings. Chile is the world leader in copper production, far surpassing the next two largest producers, Peru and the United States. Mexico is the global leader in silver production as of 2013, Peru is the second-leading producer, and Bolivia is sixth. Peru is Latin America's top gold producer.

Lithium has gained world interest. This soft, silver-white metal is used to make lightweight batteries, like those in cell phones and laptops. It is also a key metal for electric car batteries. Today the largest producer of lithium is Chile, but the world's largest reserves are in Bolivia under the Solar de Uyuni in the Altiplano. These reserves are so immense that Bolivia has been dubbed the Saudi Arabia of lithium.

Figure 4.31 Soy Production in Brazil Fartura Farm in the state of Mato Grosso, Brazil, embodies the large-scale industrial agriculture that has transformed much of South America into one of the world's largest producers and exporters of soy products.

But it remains to be seen how and under what terms this critical resource will be extracted from this remote region.

Logging is another important, and controversial, resource-based activity. Several countries rely on plantation forests of introduced species of pine, teak, and eucalyptus to supply domestic fuel wood, pulp, and board lumber. These plantation forests grow single species and fall far short of the complex ecosystems occurring in natural forests. Still, growing trees for paper or fuel reduces the pressure on other forested areas. Leaders in plantation forestry are Brazil, Venezuela, Chile, and Argentina. Considered one of Latin America's economic stars, Chile exports timber and wood chips to add to its mineral export earnings. Thousands of hectares of exotics (especially pine) have been planted, systematically harvested, and cut into boards or chipped for wood pulp. East Asian capital is heavily invested in this sector of the Chilean economy. Expansion of the wood chip business, however, led to a dramatic increase in the logging of native forests.

The Energy Sector The oil-rich countries of Mexico, Venezuela, Ecuador, and now Brazil are able to meet most of their own fuel needs and also earn vital revenues from oil exports. In 2014, Mexico was the 9th-largest producer of oil in the world, and Brazil and Venezuela were top-20 producers. Of the three, Venezuela is the most dependent on energy revenues, earning up to 90 percent of its foreign exchange from petroleum and natural gas products. The largest new oil discovery in recent years has been off the coast of Brazil. In the past decade, oil production in Brazil has doubled, moving the country into the ranks of oil exporters and drawing increased foreign investment. Although Latin American oil producers do not receive as much attention as those in the Middle East, they are significant. Venezuela was one of the five founding members of the Organization of the Petroleum Exporting Countries (OPEC), and Ecuador joined the group in 1973.

Natural gas production is also on the rise in this region. Venezuela and Bolivia have the largest proven reserves of natural gas in Latin America, but Mexico and Argentina are by far the largest producers. In recent years, Argentina's natural gas production has been boosted by new finds in Patagonia. In 2012, Argentina made headlines when President Christina Fernandez de Kirchner seized majority control of YPF, the major producer of oil and natural gas in Argentina, from a Spanish-owned company, claiming frustration over unnecessary declines in output. Argentina is especially dependent on natural gas for its urban markets and exports; the production declines produced shortages and price increases in the domestic liquefied natural gas market that required action.

In the area of biofuels, Brazil offers a story of sweet success. In the 1970s, when oil prices skyrocketed, then oil-poor Brazil decided to convert its abundant sugarcane harvest into ethanol. Over the years, even when oil prices plummeted, Brazil continued to invest in ethanol, building mills and a distribution system that delivered ethanol to gas stations. One of Brazil's major technological successes was inventing flex-fuel cars that run on any combination of ethanol and gasoline. At the time, Brazil was motivated by its limited oil reserves, but today, with interest in biofuels growing as a way to reduce CO_2 emissions, Brazil's support of its ethanol program looks visionary.

Thanks to these innovations, Latin America's energy mix has changed considerably since the 1970s (Figure 4.32). Forty years ago, 60 percent of the region's energy was supplied by oil and 20 percent by wood fuel. By 2010, the energy supply was more diverse and cleaner, most notably through the growth in natural gas, hydroelectricity, and biofuels (or bagasse). That said, Latin America's energy consumption has also increased fivefold from 1970 to 2010 due to population increase, urban growth, improved transportation, and greater economic activity.

Latin America in the Global Economy

During the 1990s, Latin American governments and the World Bank became champions of neoliberalism as a sure path to economic development. Neoliberal policies accentuate the forces of globalization by turning away from policies that emphasize state intervention and self-sufficiency. Most Latin American political leaders have embraced neoliberalism and the benefits that come with it, such as increased trade, greater direct foreign investment, and more favorable terms for debt repayment. However, there are signs of discontent with neoliberalism throughout the region. Recent protests in Brazil and Chile reflect popular anger against trade policies that seem to benefit only the elite.

Maquiladoras and Foreign Investment The growth in foreign investment and the presence of foreign-owned factories are examples of neoliberalism. The Mexican assembly plants, called **maquiladoras**, that line the border with the United States are characteristic of manufacturing systems in an increasingly globalized economy. Construction of these plants began in the 1960s as part of a border industrialization program, and by 2000 there were over 3000 maquiladoras along the border. With NAFTA, foreign-owned manufacturing plants are no longer restricted to the border zone and are increasingly being built near the population centers of Monterrey, Puebla, and Veracruz. Aguascalientes, in central Mexico, has emerged as the country's auto city, although most of the cars are produced there by foreign companies and are destined for export. The city of Chihuahua, a four-hour drive from the U.S.–Mexico border, is Mexico's center for aerospace (Figure 4.33). In the last decade, over three dozen aerospace plants have opened there, providing parts for the booming U.S. airplane manufacturing business. Labor costs average $6 per hour in Chihuahua, far cheaper than U.S. labor. And although Mexican wages are higher than those in China, northern Mexico is still an attractive location because of its proximity to the U.S. border and its membership in NAFTA.

Considerable controversy on both sides of the border surrounds this form of industrialization. Organized labor in the United States complains that well-paying manufacturing jobs are being lost to low-cost competitors, whereas environmentalists point out serious

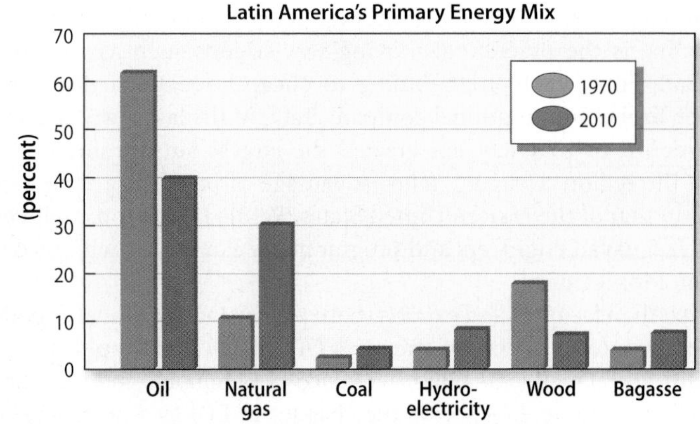

Figure 4.32 Latin America's Energy Mix, 1970 and 2010 As the region's energy sources have increased and diversified, there is much less reliance on wood and more reliance on natural gas. **Q: What could explain the decline in wood fuel for this region?**

Figure 4.33 Chihuahua's Aerospace Industry Mexican workers at the Hawker Beechcraft plant in Chihuahua assemble jet airplane parts for export to the United States. Chihuahua has the largest concentration of aerospace engineers and technicians in Mexico.

industrial pollution resulting from lax government regulation of the maquiladoras. Mexicans worry that these plants are poorly integrated into the rest of the economy and that many factories choose to hire young, unmarried women because they are viewed as docile laborers.

Other Latin American states are attracting foreign companies through tax incentives and low labor costs. Assembly plants in Honduras, Guatemala, and El Salvador are drawing foreign investors, especially in the apparel industry. A recent report from El Salvador claims that not one of its apparel factories has a union. Making goods for major American labels, many Salvadoran garment workers complain that they do not make a living wage, work 80-hour weeks, and will lose their jobs if they become pregnant. The situation in Costa Rica, which has been a major computer chip manufacturer for Intel since 1998, is quite different. With a well-educated population, low crime rate, and stable political scene, Costa Rica is now attracting other high-tech firms. Hopeful officials claim that Costa Rica is transitioning from a banana republic (bananas and coffee were the country's long-standing exports) to a high-tech manufacturing center. As a result, the Costa Rican economy averaged 4.3 percent annual growth from 2010 to 2014.

Uruguay is another small country with a well-educated population that has recently emerged as the leader in Latin American **outsourcing** operations. Outsourcing, most commonly associated with India, is the practice of moving service jobs such as tech support, data entry, and programming to cheaper locations. Partnered with the Indian multinational company Tata, in the last few years TCS Iberoamerica in Uruguay has created the largest outsourcing operation in the region. Uruguay takes advantage of being in a time zone similar to that of the eastern United States. While India's top engineers sleep, Uruguayan engineers and programmers can serve their customers from Montevideo.

Growth in factories and exports (both primary and secondary goods) is fueled by foreign direct investment (FDI). In 2013, the top three FDI recipients were Brazil ($81 billion), Mexico ($42 billion), and Colombia ($16 billion). Figure 4.34 shows the changes in FDI as a percentage of GDP from 1990 to 2011. For nearly every country in the region, the value of FDI in terms of the percentage of GDP went up, except for Ecuador. Much of this investment came from Europe and Asia. In 2012, China became Brazil's largest trading partner. China and Brazil—the so-called BRIC nations, along with Russia and India—have recognized their global strategic partnership. Not only does Brazil export grains, minerals, and energy resources, but also it has signed an agreement that will allow Brazilian airplane maker Embraer to manufacture and sell its regional jets in China.

Remittances Another important indicator that reflects the integration of Latin American workers into global labor markets is remittances. Scholars debate whether this flow of capital can actually lead to sustained development or whether it is simply a survival strategy of last resort. World Bank research shows that remittances sharply dropped during the global economic recession that began in 2008 but by 2013 had reached $61 billion (below the 2008 peak of $69 billion). Remittances are growing on average by 5–10 percent per year, so many economists project that these funds will continue to be a major source of capital for the region.

The economic impact of remittances on a per capita basis is real (see Figure 4.34). Mexico is the regional leader, receiving over $23 billion in remittance income in 2012 (equivalent to over $200 per capita). But for smaller countries such as El Salvador and Honduras, remittances contribute far more to the domestic economy. El Salvador, a country of about 6.4 million people, received $4 billion in remittances in 2012, or nearly $630 per capita. For many Latinos, remittances are the surest way to alleviate poverty, even though they depend on an international migration system that is constantly changing and includes both legal and illegal channels of movement.

Dollarization During the 1990s, as Latin American governments faced various financial crises, many began to consider the economic benefits of **dollarization**, a process by which a country adopts—in whole or in part—the U.S. dollar as its official currency. In a totally dollarized economy, the U.S. dollar becomes the only medium of exchange, and the country's national currency ceases to exist. This was the radical step taken by Ecuador in 2000 to address the dual problems of currency devaluation and hyperinflation rates of more than 1000 percent annually. El Salvador adopted dollarization in 2001 as a means to reduce the cost of borrowing money. Dollarization is not a new idea; back in 1904, Panama dollarized its economy the year after it gained independence from Colombia. Until 2000, however, Panama was the only fully dollarized state in Latin America.

A more common strategy in Latin America is limited dollarization, in which U.S. dollars circulate and are used alongside the country's national currency. Limited dollarization exists in many countries around the world, but most notably in Latin America as a type of insurance because the region's economies are prone to currency devaluation and hyperinflation. Many banks in Latin America, for example, allow customers to maintain accounts in dollars to avoid the problem of capital flight should a local currency be devalued. Other countries keep their national currency but peg its value one-for-one to the dollar; this was the innovative strategy adopted by Argentina in 1991, although it led to a serious financial crisis in 2001 and was eventually stopped. Dollarization, partial or full, tends to reduce inflation, eliminate fears of currency devaluation, and reduce the cost of trade by eliminating currency conversion costs.

Dollarization has its drawbacks. The obvious one is that a country no longer has control of its monetary policy, making it reliant on the decisions of the U.S. Federal Reserve. Foreign governments do not have to ask permission to dollarize their economies. At the same

Figure 4.34 Global Linkages: Foreign Investment and Remittances Foreign investors and immigrants are responsible for significant increases in the amount of capital flowing into Latin America. As the map indicates, most countries saw increases in foreign direct investment between 1990 and 2011. Immigrants working abroad sent $61 billion to the region, providing much-needed capital to many poor households. In the cases of Guatemala, El Salvador, and Honduras, remittances totaled more than $300 per capita.

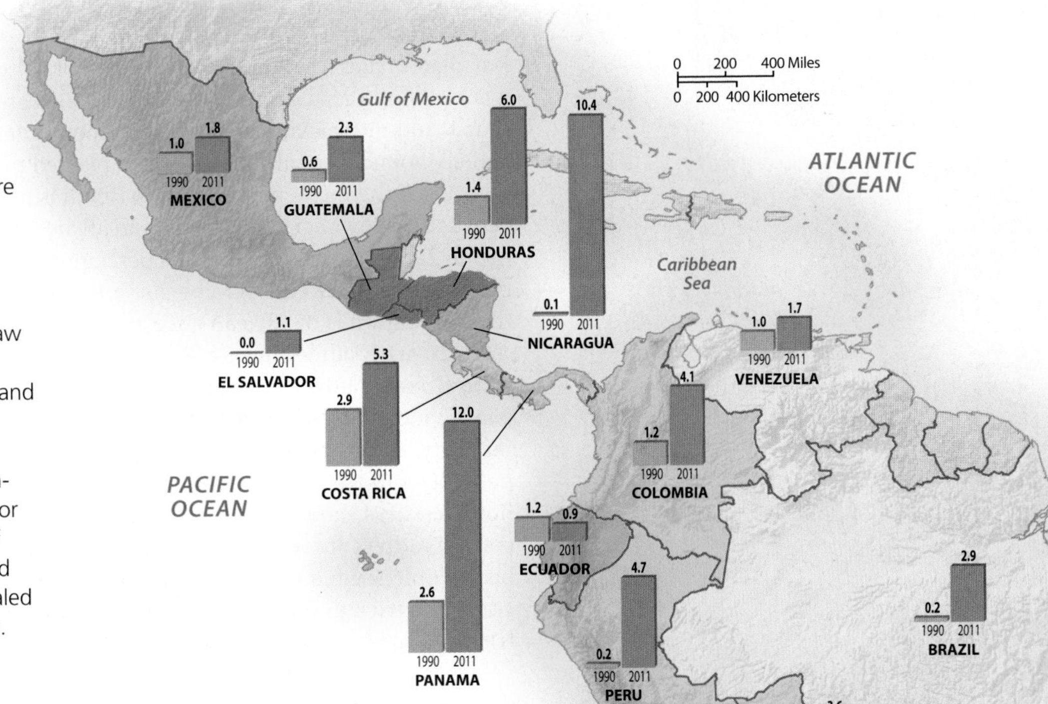

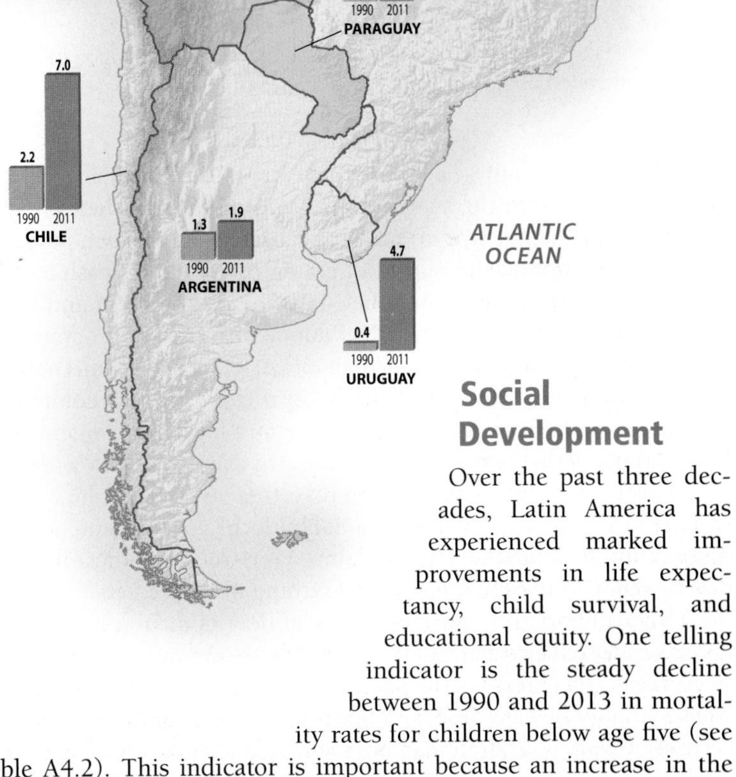

time, the United States insists that all its monetary policies be based exclusively on domestic considerations, regardless of the impact such decisions may have on foreign countries. The political impact of eliminating a national currency is serious. The case of Ecuador is instructive. In 1999, when President Jamil Mahuad announced his plan to dollarize the economy to head off hyperinflation, he was quickly forced out of office by a coalition of military and Indian activists. When Vice President Gustavo Naboa became president and the economic situation worsened, the country's political leadership went ahead with dollarization. In short, dollarization may help in a time of economic duress, but it is not a popular policy.

The Informal Sector Even in prosperous capital cities, a short drive to the urban periphery shows large neighborhoods of self-built housing filled with street traders and family-run workshops. Such activities make up the **informal sector**, which is the provision of goods and services without government regulation, registration, or taxation. Most people in the informal economy are self-employed and receive no wages or benefits except the profits they clear. The most common informal activities are housing construction (in many cities, one-third of all residents live in self-built housing), manufacturing in small workshops, street vending, transportation services (messenger services, bicycle delivery, and collective taxis), garbage picking, street performing, and even paid waiting in line (Figure 4.35).

No one is sure how big this economy is, in part because it is difficult to separate formal activities from informal ones. Visit Lima, Belém, Guatemala City, or Guayaquil, and it is easy to get the impression that the informal economy *is* the economy. From self-help housing that dominates the landscape to hundreds of street vendors who crowd the sidewalks, it is impossible to avoid. The informal sector has some advantages—hours are flexible, children can work with their parents, and there are no bosses. As important as this sector may be, however, widespread dependence on it signals Latin America's poverty, not its wealth. It reflects the inability of the formal economies of the region, especially in industry, to provide enough jobs for the many people seeking employment.

Social Development

Over the past three decades, Latin America has experienced marked improvements in life expectancy, child survival, and educational equity. One telling indicator is the steady decline between 1990 and 2013 in mortality rates for children below age five (see Table A4.2). This indicator is important because an increase in the number of children younger than five years surviving suggests that basic nutritional and health-care needs are being met. We can also conclude that resources are being used to sustain women and their children. Despite economic downturns, the region's social networks have been able to lessen the negative effects on children.

A combination of government policies and grassroots and nongovernmental organizations (NGOs) plays a fundamental role in contributing to social well-being. In the past few years, conditional cash transfer programs, such as **Bolsa Familia**, have reduced extreme poverty.

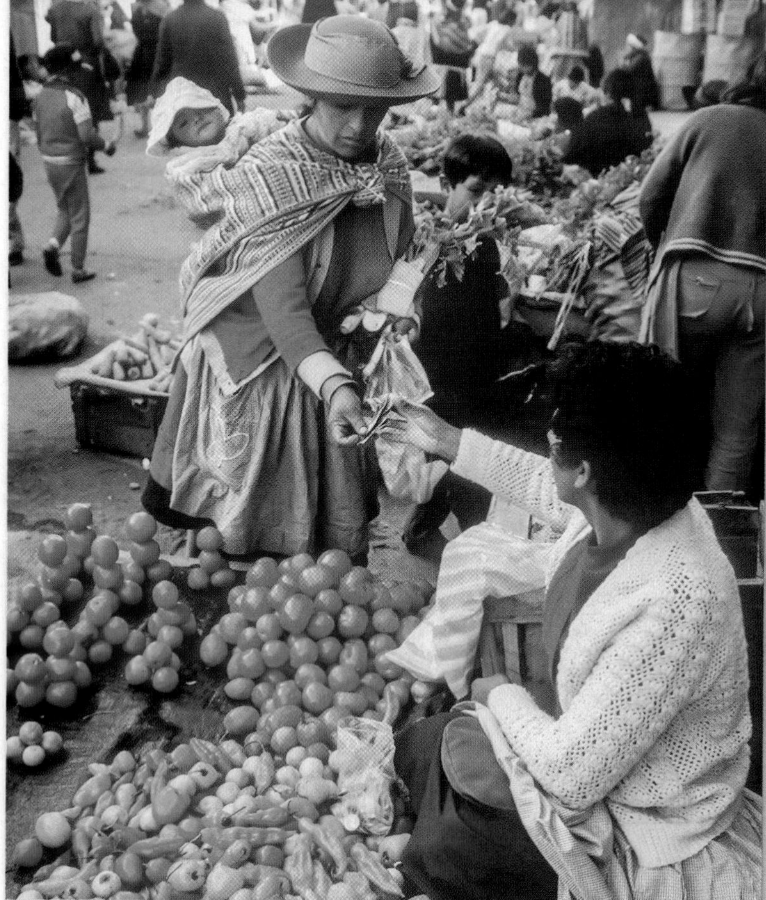

Figure 4.35 Peruvian Street Vendors Street vendors sell produce in Huancayo, Peru. Street vending plays a critical role in the distribution of goods and the generation of income. It is representative of the informal sector in Latin America.

Poor Brazilian families who qualify for Bolsa Familia receive a monthly check from the state but are required to keep their children in school and take them to clinics for health checkups. Such programs have both the immediate impact of giving poor families cash and the long-term impact of improving the educational attainment and health care of their children. Mexico has adopted a similar program. For states with far fewer resources than Brazil or Mexico, international humanitarian organizations, church organizations, and community activists provide many services that state and local governments cannot. Catholic Relief Services and Caritas, for example, work with rural poor throughout the region to improve their water supplies, health care, and education. Other groups lobby local governments to build schools and recognize squatters' claims. Grassroots organizations also develop cooperatives that market everything from sweaters to cheeses.

Other important indicators for social development are life expectancy, gender educational equity, and access to improved water sources. In aggregate, 94 percent of the people in the region have access to an adequate amount of water from an improved source, slightly more girls receive secondary education than boys (Figure 4.36), and life expectancy (for both men and women) is 75 years. Masked by these combined data are extreme variations between rural and urban areas, between regions, and along racial and ethnic lines.

Race and Inequality There is much to admire about race relations in Latin America. The complex racial and ethnic mix that was created in Latin America fostered tolerance for diversity. That said, Indians and blacks are more likely to be counted among the poor of the region. More than ever before, racial discrimination is a major political issue in Brazil. Reports of organized killings of street children, most of them Afro-Brazilian, make headlines. For decades, Brazil championed its vision of a color-blind racial democracy. Residential segregation by race is rare in Brazil, and interracial marriage is common, but certain patterns of social and economic inequality seem best explained by racial discrimination.

Assessing racial inequalities in Brazil is problematic. The Brazilian census asks few racial questions, and all are based on self-classification. In the 2000 census, less than 11 percent of the population called itself black. Some Brazilian sociologists, however, claim that more than half the population is of African ancestry, making Brazil the second-largest "African state" after Nigeria. Racial classification is always highly subjective and relative, but some patterns support the existence of racism. Evidence from northeastern Brazil, where Afro-Brazilians are the majority, shows death rates approaching those of some of the world's poorest countries. Throughout Brazil, blacks suffer higher rates of homelessness, landlessness, illiteracy, and unemployment than others. To address this problem, various affirmative action measures have been implemented (along with the Bolsa Familia program). From federal ministries to public universities, various quota systems are being tried to improve the condition of Afro-Brazilians.

In areas of Latin America where Indian cultures are strong, indicators of low socioeconomic position are also present. In most countries, areas where native languages are widely spoken regularly correspond with areas of persistent poverty. In Mexico, the Indian south lags behind the booming north and Mexico City. Prejudice is embedded in the language. To call someone an *indio* (Indian) is an insult in Mexico. In Bolivia, women who dress in the Indian style of full, pleated skirts and bowler hats are called *cholas*, a descriptive term referring to the rural *mestizo* population that suggests a backwardness and even cowardice. No one of high social standing, regardless of skin color, would ever be called a *chola* or *cholo*.

Figure 4.36 School Children in Panama Uniformed public school children walk to school in Panama City. Latin American states have seen steady improvements in youth literacy, with 97 percent of youth (between the ages of 15 and 24) being literate. Per capita expenditures on education and access to postsecondary education still lag behind levels in Europe and North America.

PERCENT OF SEATS HELD BY WOMEN IN NATIONAL PARLIAMENTS, 2014

5–14.9 | 15–24.9 | 25–34.9 | 35–44.9 | 45–54.9

Countries that have or have had women presidents

In Latin America 28.5% of seats are held by women in national parliaments. By comparison, in the United States 19.3% of congressional seats are held by women. In Canada the figure is 25.1%.

Figure 4.37 Women's Participation in National Politics Women are active players in Latin American politics. On average, 25 percent of seats in national parliaments are held by women. In Brazil only 9 percent of the seats are held by women, but in Mexico and Argentina 37 percent were held by women in 2014. Moreover, six countries have had or have female presidents. (World Bank Development Indicators, 2015.)

It is difficult to separate status divisions based on class from those based on race. From the days of conquest, being European meant an immediate elevation in status over the Indian, African, and *mestizo* populations. Race does not necessarily determine one's economic standing, but it certainly influences it. Amerindian people, however, are politically asserting themselves. For example, Evo Morales was inaugurated as president of Bolivia in 2006, making him the first Indian leader in that country.

The Status of Women Many contradictions exist with regard to the status of women in Latin America. Many Latina women work outside the home. In most countries, the formal figures are between 30 and 40 percent of the workforce, not far off from many European countries, but lower than in the United States. Legally speaking, women can vote, own property, and sign for loans, although they are less likely to do so than men, reflecting the society's patriarchal (male-dominated) tendencies. Even though Latin America is predominantly Catholic, divorce is legal and family planning is promoted. In most countries, however, abortion remains illegal.

Overall, access to education in Latin America is good, compared to other developing regions, and thus illiteracy rates tend to be low. Rates of adult illiteracy are slightly higher for women than for men, but usually by only a few percentage points. Throughout higher education in Latin America, male and female students are equally represented today. Consequently, women are regularly employed in the fields of education, medicine, and law.

The biggest changes for women are the trends toward smaller families, urban living, and educational parity with men. These factors have greatly improved the participation of women in the labor force. In the countryside, however, serious inequalities remain. Rural women are less likely to be educated and tend to have larger families. In addition, they are often left to care for their families alone, as husbands leave in search of seasonal employment. In most cases, the conditions facing rural women have been slow to improve.

Women are increasingly playing an active role in politics. In 1990, Nicaragua elected Latin America's first woman president, Violeta Chamorro, the owner of an opposition newspaper. Nine years later, Panamanians voted Mireya Moscoso into power. In 2005, South America had its first woman president of a nation: Dr. Michelle Bachelet, a pediatrician and single mother, took the oath of office in Chile. She was reelected president for a second term in 2013. Brazilian President Dilma Rousseff took office in 2011, so the seventh-largest economy in the world is run by a woman. As shown in Figure 4.37, many Latin American states have larger percentages of women in seats of national parliaments than does the United States.

Across the region, women and indigenous groups are active organizers and participants in cooperatives, small businesses, and unions and are elected to national office. In a relatively short period, they have won a formal place in the economy and a political voice. Moreover, evidence suggests that this trend will continue.

REVIEW

4.10 How has the export of primary products (food, fiber, and energy) shaped the economies of Latin America?

4.11 What explains some of the positive indicators of social development in Latin America?

KEY TERMS neoliberalism, fair trade, maquiladora, outsourcing, dollarization, informal sector, Bolsa Familia

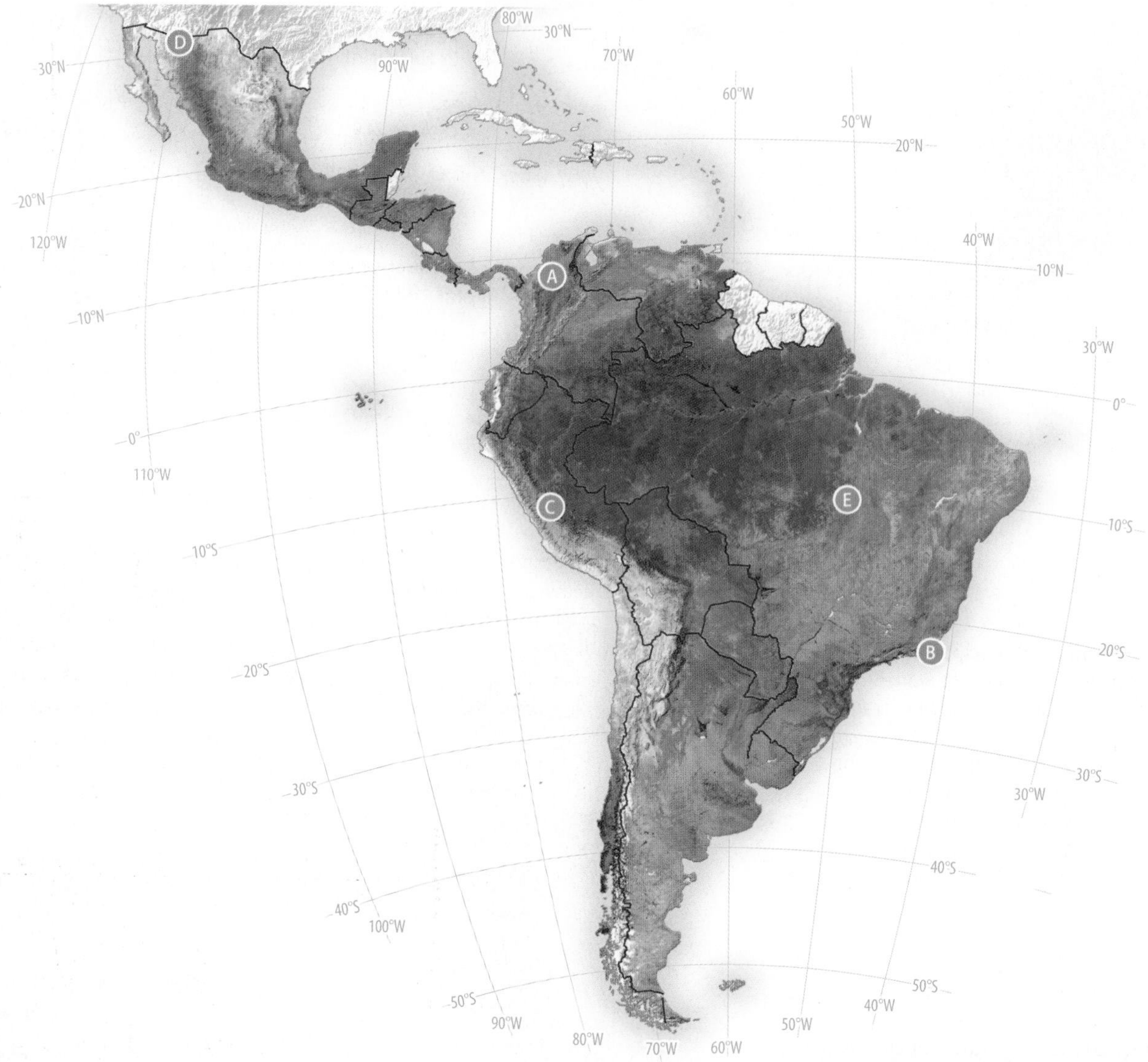

Review

Physical Geography and Environmental Issues

4.1 Explain the relationships among elevation, climate, and agricultural production, especially in tropical highland areas.

4.2 Identify the major environmental issues of Latin America and how countries are addressing them.

Compared to Europe and Asia, this region is still rich in natural resources and relatively lightly populated. Yet growing populations and increasing trade in primary resources put pressure on the environment. The relentless cutting of tropical forests and construction of new dams are of particular concern. In terms of urban environments, Latin American cities have made strides in improving air quality.

1. How have changes in urban transportation planning made Latin American cities greener?

2. Provide some examples of resource conservation in Latin America. Are these practices working?

Population and Settlement

4.3 Summarize the demographic issues impacting this region, such as rural-to-urban migration, urbanization, smaller families, and emigration.

Unlike in other developing areas, three-quarters of Latin Americans live in cities. This shift started early and reflects a cultural bias toward urban living with roots in the colonial past. Cities are large and combine aspects of the formal industrial economy along with the informal one. Grinding poverty in rural areas drives people to live in cities or to emigrate in search of employment abroad.

3. Explain where this area might be with regard to the Latin American urban model (Figure 4.20).

4. Why has Latin America become a region of emigration? Is this pattern likely to continue?

Cultural Coherence and Diversity

4.4 Describe the cultural mixing of European and Amerindian groups in this region and indicate where Amerindian cultures thrive today.

4.5 Explain the global reach of Latino culture through immigration, sport, music, and television.

Latin America and the Caribbean were the first world regions to be fully colonized by Europe. In the process, perhaps 90 percent of the native population died from disease, cruelty, and forced resettlement. The slow demographic recovery of native peoples and the steady arrival of Europeans and Africans resulted in an unprecedented level of racial and cultural mixing. Today Amerindian activism is on the rise as indigenous groups seek territorial and political recognition.

5. What factors explain the language patterns in this area of Latin America?

6. How do religious practices in this region reflect both globalization and diversity?

C

Geopolitical Framework

4.6 Describe the Iberian colonization of the region and how it affected the formation of today's modern states.

4.7 Identify the major trade blocs in Latin America and how they are influencing development.

Most Latin American states have been independent for 200 years, yet during that time they experienced dependent political and trade relations with Europe and North America that limited the region's overall development. Today Latin America, especially Brazil, is exerting more geopolitical influence. Within the region, new political actors are emerging—from indigenous groups to women—who challenge old ways of doing things.

7. What international agreements have shaped land-use patterns in this area?

8. How might the evolution of UNASUR impact development within South America?

D

Economic and Social Development

4.8 Summarize the significance of primary exports from Latin America, especially agricultural commodities, minerals, wood products, and fossil fuels.

4.9 Describe the neoliberal economic reforms that have been applied to Latin America and how they have influenced the region's development.

Latin American governments were early adopters of neoliberal economic policies. Some states prospered, whereas others faltered, sparking popular protests against the negative effects of neoliberalism and globalization. Since 2000, most states have experienced economic growth, driven in part by higher commodity prices. In addition, extreme poverty has declined, and social indicators of development are improving.

9. What commodity is this, and how is its rise in production changing agricultural practices and patterns of trade?

10. Do you think neoliberal policies are increasing or decreasing income inequality in the region? Why?

E

KEY TERMS

agrarian reform *(p. 117)*
Altiplano *(p. 107)*
altitudinal zonation *(p. 108)*
Bolsa Familia *(p. 135)*
Central American Free Trade Agreement (CAFTA) *(p. 129)*
dollarization *(p. 134)*
El Niño *(p. 110)*
environmental lapse rate *(p. 110)*
fair trade *(p. 132)*
grassification *(p. 112)*
informal sector *(p. 135)*
latifundia *(p. 117)*
maquiladora *(p. 133)*
Mercosur (Southern Cone Common Market) *(p. 129)*
mestizo *(p. 119)*
minifundia *(p. 117)*
neoliberalism *(p. 131)*
neotropics *(p. 106)*
Organization of American States (OAS) *(p. 125)*
outsourcing *(p. 134)*
remittance *(p. 121)*
shield *(p. 107)*
squatter settlement *(p. 118)*
subnational organization *(p. 129)*
supranational organization *(p. 129)*
syncretic religion *(p. 123)*
Treaty of Tordesillas *(p. 125)*
UNASUR (Union of South American Nations) *(p. 129)*
urban primacy *(p. 117)*

DATA ANALYSIS

http://goo.gl/rJDjBc

Coffee production is extremely important for many Latin American countries, and the region remains the world's top producer of coffee. Coffee is a valued commodity grown on large estates and small farms in the tropics. For many rural families, the coffee bean provides access to international markets and income. The International Coffee Organization (ICO) maintains statistics on coffee production worldwide. Consult the ICO's website (http://www.ico.org) and access the coffee production figures from 2011 to 2014.

1. Which countries were the top producers in 2011? Which countries were the top producers in 2014? What political or economic factors might explain the changes?
2. Which world regions have seen the biggest increase in coffee production? Which world regions have seen the biggest decline?
3. Write a paragraph explaining how significant increases or decreases in coffee production could impact the physical environment and economic development potential of these regions.

MasteringGeography™

Looking for additional review and test prep materials? Visit the Study Area in MasteringGeography™ to enhance your geographic literacy, spatial reasoning skills, and understanding of this chapter's content by accessing a variety of resources, including MapMaster interactive maps, geoscience animations, videos, *In the News* RSS feeds, flashcards, web links, self-study quizzes, and an eText version of *Globalization and Diversity.*

Authors' Blogs

Scan to visit the **Author's Blog for field notes, media resources, and chapter updates.**

http://gad4blog.wordpress.com/category/latin-america/

Scan to visit the **GeoCurrents Blog**

http://geocurrents.info/category/place/latin-america

5 The Caribbean

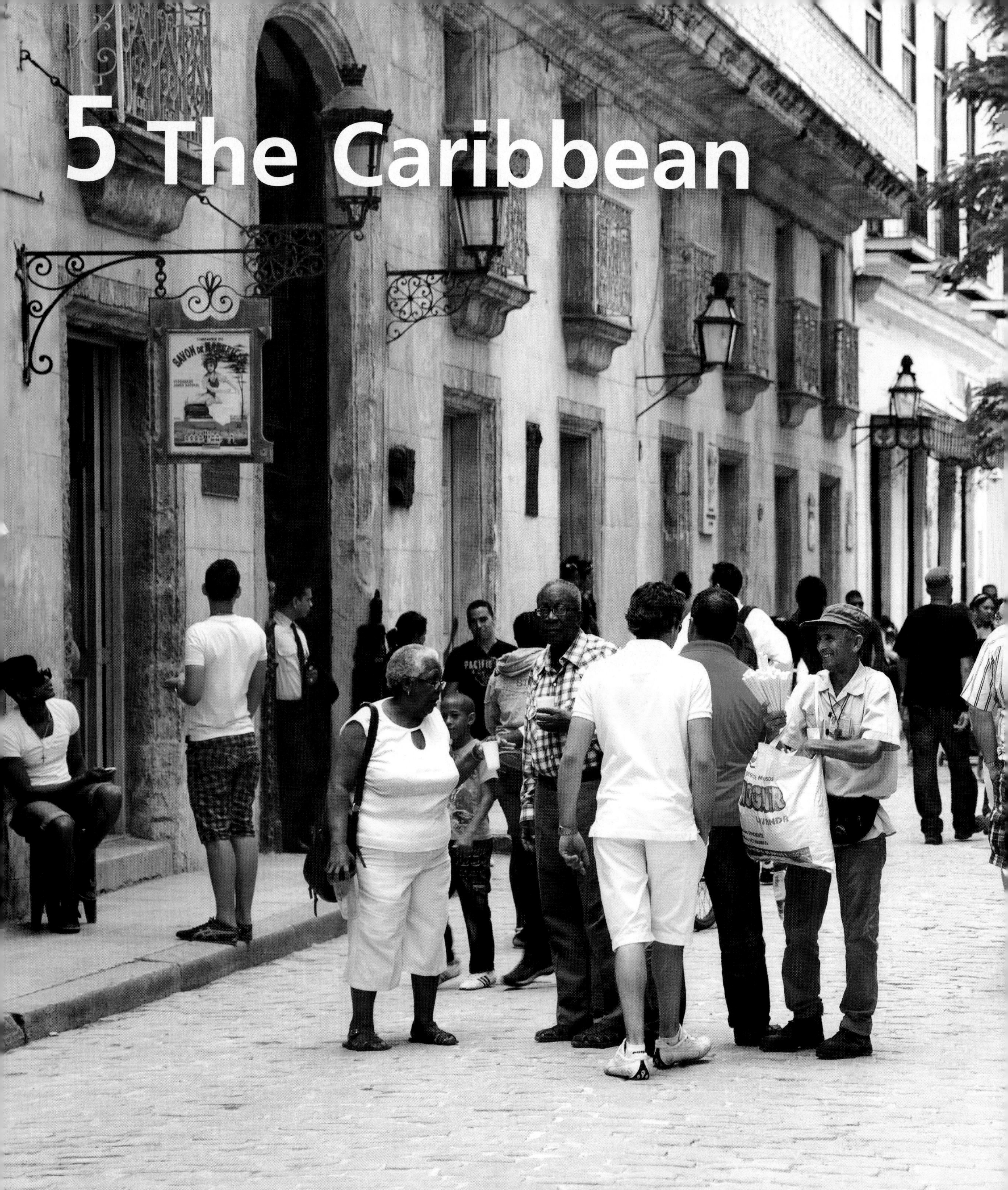

PHYSICAL GEOGRAPHY AND ENVIRONMENTAL ISSUES

Climate change threatens the Caribbean, with the potential for stronger and more frequent hurricanes, loss of land due to rising sea levels, and destruction of coral reefs. A devastating 2010 earthquake flattened Port-au-Prince, the capital of Haiti, initiating the region's worst natural disaster in decades. Rebuilding the capital city has been slow.

POPULATION AND SETTLEMENT

Having experienced its demographic transition, the region now has slow population growth. In addition, large numbers of Caribbean peoples have emigrated from the region in search of economic opportunity, sending back billions of dollars.

CULTURAL COHERENCE AND DIVERSITY

Creolization—the blending of African, European, and Amerindian elements—led to many unique Caribbean expressions of culture, such as rara, reggae, and steel drum bands. Caribbean-styled celebrations of carnival have diffused with the movement of people from the region to Europe and North America.

GEOPOLITICAL FRAMEWORK

The first area of the Americas to be extensively explored and colonized by Europeans, the region has seen many rival European claims and, since the early 20th century, has experienced strong U.S. influence. Many Caribbean territories have been independent states for 50 years or less.

ECONOMIC AND SOCIAL DEVELOPMENT

Environmental, locational, and economic factors make tourism a vital component of this region's economy, particularly in Puerto Rico, Cuba, the Dominican Republic, Jamaica, and The Bahamas. Offshore manufacturing and banking are also significant in the region's modern economic development.

◀ Tourism has been growing in Cuba for the last 25 years, but it is about to grow more. With the normalization of relations between Cuba and the United States, more U.S. citizens will have an opportunity to visit the Caribbean's largest and most distinct island. Here Cubans and tourists mingle in the streets of Old Havana.

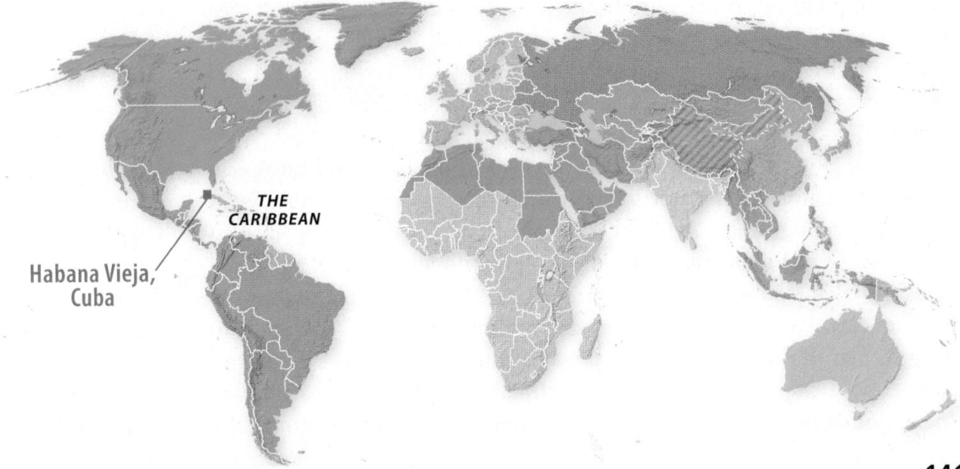

Cuba is the Caribbean's largest country. Yet as a socialist country in which the state controls most aspects of the economy, it is an anomaly. Short on cash and basic materials since the early 1990s, Cuba has steadily grown its tourism economy so that nearly 3 million people visit the island annually. Drawn to the island's historic architecture, white sandy beaches, iconic 1950s America–era cars, and rich cultural and political history, tourists from Europe, Canada, and other parts of Latin America regularly visit. With the normalization of relations between the United States and Cuba in December 2014, for the first time in five decades large numbers of Americans may also visit the island, provided such travel constitutes an educational activity or supports the Cuban people—requirements that can be easily met with a slight tweaking of tourism packages. Many speculate what this means for Cuba's future. Part of the island's appeal is its authenticity—missing are the chain hotels, restaurants, and clothing stores found in other Caribbean tourist destinations. It was only in 2010 that President Raúl Castro, the younger brother of Fidel Castro, permitted private entrepreneurs (*cuentapropistas*) to own and operate small restaurants, barbershops, family-run hotels, hardware stores, and even religious article shops (for both Christian and Santería worshipers). Although international tourists visit, foreign travel is restricted for most Cubans, and relatively low incomes limit access to modern technology. Cubans, and their aging revolutionary leaders, continue to reinvent their country in response to global economic and political forces. Although quite distinct, in the future the island may more closely resemble its Caribbean neighbors.

The Caribbean is distinguished from the largely Iberian-influenced Latin America by its more diverse colonial history and strong African imprint.

Learning Objectives *After reading this chapter you should be able to:*

5.1 Differentiate between island and rimland environments and the environmental issues that affect these areas.

5.2 Summarize the demographic shifts in the Caribbean as population growth slows, urbanization intensifies, emigration abroad continues, and a return migration begins.

5.3 Identify the demographic and cultural implications of the transfer of African peoples to the Caribbean and the creation of a neo-African society in the Americas.

5.4 Explain why European colonists so aggressively sought control of the Caribbean and why independence in the region came about more gradually than in neighboring Latin America.

5.5 Describe how the Caribbean is linked to the global economy through offshore banking, emigration, and tourism.

5.6 Suggest reasons why the Caribbean does better in social indicators of development than in economic indicators.

The Caribbean was the first region of the Americas to be extensively explored and colonized by Europeans. Yet its modern regional identity is unclear, often merged with Latin America, but also viewed as apart from it. Today the region is home to 45 million people, scattered across 26 countries and dependent territories. They range from the small British dependency of Turks and Caicos, with 36,000 people, to the island of Hispaniola, split between Haiti and the Dominican Republic, with over 21 million. In addition to the Caribbean islands, Belize of Central America and the three Guianas—Guyana, Suriname, and French Guiana—of South America are considered part of the Caribbean. For historical and cultural reasons, the peoples of these mainland states identify with the island nations and are thus included in this chapter (Figure 5.1).

Historically, the Caribbean was a battleground where rival European powers competed for territorial control of these tropical lands. As in many developing areas, external control of the Caribbean by foreign governments and companies produced highly dependent and inequitable economies, characterized by reliance on slave labor, plantation agriculture, and the prosperity that sugar promised in the 17th and 18th centuries. In the early 1900s, the United States took over as the dominant geopolitical force in the region, maintaining what some have called a neocolonial presence. Over time, other products became economically important, as did the international tourist industry. Increasingly, governments in the region have sought to diversify their economies to reduce their dependence on agriculture and tourism.

The basis for treating the Caribbean as a distinct world region lies within its particular cultural and economic history. Culturally, the region can be distinguished from the largely Iberian-influenced mainland of Latin America by its more diverse European colonial history and strong African imprint. The dominance of export-oriented plantation agriculture also explains many of the social, economic, and environmental patterns in the region.

Still a developing area, most Caribbean states have achieved life expectancies in the 70s, low child mortality rates, and high literacy rates. Millions of tourists view the Caribbean as an international playground for sun, sand, and fun. However, there is another Caribbean that is far poorer and economically more dependent than the one portrayed on travel posters. Haiti, by most measures the Western Hemisphere's poorest country, has 10.5 million people. The Dominican Republic has 10.7 million, and Cuba has 11.2 million. Each of these major Caribbean nations suffers from serious economic problems and widespread poverty.

The majority of Caribbean people are poor, living in the shadow of North America's vast wealth. The concept of **isolated proximity** has

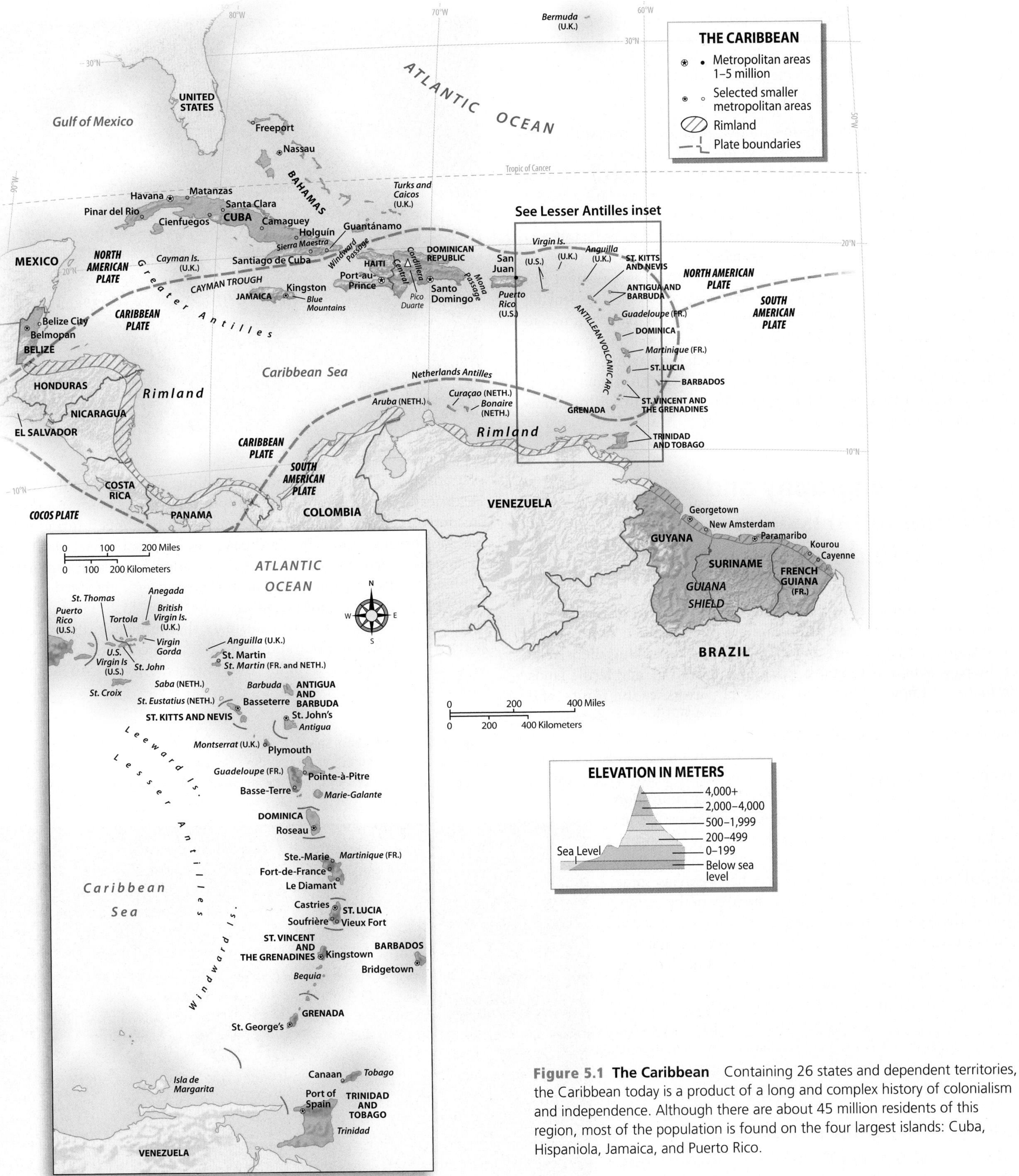

Figure 5.1 The Caribbean Containing 26 states and dependent territories, the Caribbean today is a product of a long and complex history of colonialism and independence. Although there are about 45 million residents of this region, most of the population is found on the four largest islands: Cuba, Hispaniola, Jamaica, and Puerto Rico.

Figure 5.2 Carnival Drummer A steel pan drummer performs while his drum cart is pushed through the streets during carnival in Port of Spain, Trinidad. Steel drums, created in Trinidad in the 1940s from discarded oil drums from a U.S. military base, have become the iconic sound for the region.

been used to explain the region's unique position in the world. The *isolation* of the Caribbean sustains the area's cultural diversity (Figure 5.2) and also explains its limited economic opportunities. Caribbean writers note that this isolation fosters a strong sense of place and a tendency to focus inward. Yet the relative *proximity* of the Caribbean to North America (and, to a lesser extent, Europe) ensures its international connections and economic dependence.

Through the years, the Caribbean has evolved as a distinct but economically marginal world region, characterized by workers leaving the region in search of employment and foreign companies drawn to the Caribbean for its cheap labor. The economic well-being of most Caribbean countries is uncertain. Despite such uncertainty, the people of the Caribbean maintain an enduring cultural richness and attachment to place that may explain a growing countercurrent of immigrants back to the region.

Physical Geography and Environmental Issues: Paradise Undone

Tucked between the Tropic of Cancer and the equator, with year-round temperatures averaging in the high 70s, the hundreds of islands and picturesque waters of the Caribbean have often inspired comparisons to paradise. Columbus began the tradition by describing the islands of the New World as the most marvelous, beautiful, and fertile lands he had ever known, filled with flocks of parrots, exotic plants, and friendly natives. Writers today are still lured by the sea, sands, and swaying palms of the Caribbean.

Ecologically speaking, it is difficult to picture a landscape more completely altered than that of the Caribbean. For nearly five centuries, the destruction of forests and the unrelenting cultivation of soils resulted in the extinction of many endemic (native) Caribbean plants and animals, including various shrubs and trees, songbirds, large mammals, and monkeys. This severe depletion of biological resources helps explain some of the present economic and social instability of the region. Most of the region's environmental problems are associated with agricultural practices, soil erosion, excessive reliance on wood and charcoal for fuel, and the threat of global climate change. The devastating Port-au-Prince earthquake of 2010 underscored how quickly a place such as Haiti can become undone due to a major natural disaster. However, because many countries rely on tourism as a vital source of income, the region has also experienced a growth in protected areas, both on the land and in maritime areas.

Island and Rimland Landscapes

It is the Caribbean Sea itself—the body of water enclosed between the *Antillean* islands (the arc of islands that begins with Cuba and ends with Trinidad) and the mainland of Central and South America—that links the states of the region. Historically, the sea connected people through its trade routes and sustained them with its marine resources of fish, green turtle, manatee, lobster, and crab. The Caribbean Sea is noted for its clarity and biological diversity, but it has never supported large-scale commercial fishing because the quantities of any one species are not large. Sea surface temperatures range from 73° to 84°F (23° to 29°C), over which forms a warm tropical marine air mass that influences daily weather patterns. This warm water and tropical setting continue to be key resources for the region, as millions of tourists visit the Caribbean each year (Figure 5.3).

Figure 5.3 Caribbean Sea Noted for its calm turquoise waters, steady breezes, and treacherous shallows, the Caribbean Sea has both sheltered and challenged sailors for centuries. This aerial photograph shows English Caye off the coast of Belize.

Figure 5.4 Post-earthquake Reconstruction in Haiti The magnitude 7.0 earthquake that struck Port-au-Prince on January 12, 2010, was one of the worst natural disasters in the region's history, killing over 200,000 and leveling over 150,000 structures. The Iron Market in downtown Port-au-Prince, where thousands of vendors work, was one of the first major public structures to be rebuilt after the earthquake.

The arc of islands that stretches across the Caribbean Sea is its most distinguishing physical feature. The Antillean islands are divided into the Greater and Lesser Antilles. The majority of the region's population lives on these islands. The **rimland** (the Caribbean coastal zone of the mainland) includes Belize and the Guianas as well as the Caribbean coast of Central and South America (see Figure 5.1). In contrast to the islands, the rimland has low population densities.

Most of the islands, with the exception of Cuba, are on the Caribbean tectonic plate, wedged between the South American and North American plates. Generally, this is not one of the most tectonically active zones, although earthquakes and volcanic eruptions do happen. In January 2010, a magnitude 7.0 earthquake leveled Port-au-Prince in one of the most tragic natural disasters to strike the Caribbean.

The Enriquillo Fault, near the densely settled and extremely poor capital city of Haiti, had been inactive for more than a century. When it violently shifted in 2010, the epicenter of the resulting earthquake was just a few miles from the city. The disaster affected nearly 3 million people, as homes were rendered unsafe and water and electricity supplies were disrupted. Shockingly, over 200,000 people died and 1 million people were left homeless. Many of the government agencies that would have assisted in the relief response were also destroyed. The tragedy of the Haitian earthquake was compounded by the state's poverty and corruption, as most buildings were not built to standards that could withstand an earthquake of this magnitude. The international community, as well as the large diaspora of Haitians living abroad, immediately offered financial aid and assistance (see *Exploring Global Connections: Crisis Mapping in Haiti After the Earthquake*). More than $12 billion in humanitarian and development aid and debt relief were pledged in the wake of the crisis. Slowly rebuilding has occurred, including a new two-lane highway between Port-au-Prince and Gonaives, a new airport, and hundreds of new schools (Figure 5.4). The last of the tent cities that housed over 1 million people immediately after the earthquake were finally taken down. But housing remains a problem, and Haiti's large impoverished population still struggles.

Greater Antilles The four large islands of Cuba, Jamaica, Hispaniola (shared by Haiti and the Dominican Republic), and Puerto Rico make up the **Greater Antilles**. These islands contain the bulk of the region's population, arable lands, and large mountain ranges. Many people are knowledgeable about the Caribbean coasts but are surprised to learn that Pico Duarte in the Cordillera Central of the Dominican Republic is more than 10,000 feet (3000 meters) tall, Jamaica's Blue Mountains top 7000 feet (2100 meters), and Cuba's Sierra Maestra is more than 6000 feet (1800 meters) tall. Historically, the mountains of the Greater Antilles were of little economic interest because plantation owners preferred the coastal plains and valleys. However, the mountains were an important refuge for runaway slaves and subsistence farmers and thus are important in the region's cultural history.

The best farmlands are found in the Greater Antilles, especially in the central and western valleys of Cuba, where limestone contributes to the formation of a fertile red clay soil (locally called *mantanzas*), and in Jamaica, where a gray or black soil type called *rendzinas* is found. Surprisingly, given the region's agricultural history, many other soils on these islands are nutrient-poor, heavily leached, and acidic.

Lesser Antilles The **Lesser Antilles** form a double arc of small islands stretching from the Virgin Islands to Trinidad. Smaller in size and population than the Greater Antilles, they were important early footholds for rival European colonial powers. The islands from St. Kitts to Grenada form the inner arc of the Lesser Antilles. These mountainous islands, with peaks ranging from 4000 to 5000 feet (1200 to 1500 meters), have volcanic origins. Erosion of the island peaks and accumulation of ash from volcanic eruptions formed small pockets of arable soils, although the terrain's steepness limits agricultural development. Some of these volcanically active islands are tapping into clean geothermal energy. For example, western Guadeloupe's La Bouillante power station generates 15 megawatts of geothermal energy.

Just east of this volcanic arc are the low-lying islands of Barbados, Antigua, Barbuda, and the eastern half of Guadeloupe. Covered in limestone that overlays volcanic rock, these lands were much more inviting for agriculture. In particular, such soils were ideal for growing sugarcane, making these islands important early settings for the plantation economy that diffused throughout the region.

Rimland States The rimland consists of the coastal zone of the mainland, beginning with Belize and extending along the coast of Central America to northern South America. Unlike the rest of the Caribbean, the rimland states of Belize and the Guianas still contain significant amounts of forest cover. As on the islands, agriculture in these states is closely tied to local geology and soils. Much of low-lying Belize is limestone. Sugarcane is the dominant crop in the drier north, whereas citrus is produced in the wetter central portion of the state. The Guianas are characterized by the rolling hills of the Guiana Shield, whose crystalline rock is responsible for the area's overall poor soil quality. Thus, most agriculture in the Guianas occurs on the narrow coastal plain, with sugarcane and rice as the primary crops. French Guiana, an overseas territory of France, relies mostly on French subsidies but exports shrimp and timber. It is also home to the European Space Center at Kourou (Figure 5.5).

EXPLORING GLOBAL CONNECTIONS

Crisis Mapping in Haiti After the Earthquake

Google Earth (MG) Virtual Tour Video http://goo.gl/6naaRN

In response to the 2010 Haitian earthquake, social media, humanitarian organizations, and crisis mappers joined forces in a new and unique way that has changed how governments and civil societies respond to complex humanitarian crises. One of the leaders in the crisis-mapping movement is Patrick Meier, who was a key player in assembling the crisis-mapping team for Haiti. In 2013, Meier was a National Geographic Emerging Explorer blogging about the Haitian experience.

Crisis Mapping Crisis mapping is the leveraging of mobile devices (texts and tweets), open-source applications, participatory maps, satellite imagery, and crowdsourced event data for rapid responses to complex humanitarian crises. Humanitarian workers need precise, real-time information that localities in crisis are often unable to provide. Working through an African-created platform called Ushahidi, crisis mappers assembled at Tufts University, just outside of Boston, gathering tweets and text messages from Haiti (with translations provided by Haitians living in the United States). New global connections were forged, resulting in maps used by first responders that saved lives.

Two free open-source mapping platforms were critical in moving crisis mapping forward: Ushahidi and Open Street Map. Ushahidi was developed by African bloggers who sought to report on post-election violence in Kenya in 2008 that was not covered by the media. Ushahidi (Swahili for "witness") relies on a Google Web-based map interface that plots acts of violence sent by crowd-sourced text messages. In the case of Haiti, Open Street Map was incorporated into the platform in order to construct an extremely detailed and interactive map that people could use in the field and drill down to individual reports (Figure 5.1.1). Key to the project's success was a team of crisis mappers (initially students at Tufts University) and translators who scanned for tweets. Later, through collaboration with Haiti's largest mobile phone provider, a texting number

Figure 5.5 Kourou, French Guiana The European Space Agency launches rockets from its center in Kourou, French Guiana. This French territory, near the equator and on the coast, makes an ideal launching site. In this photo, an unmanned Ariane rocket is prepared for takeoff.

Caribbean Climate and Climate Change

Much of the Antillean islands and rimland receives more than 80 inches (200 centimeters) of rainfall annually, which is enough to support tropical forests. Average temperatures are typically highs of 80 degrees and lows of 70. Distinct dry areas exist as well, such as the rain-shadow basin in western Hispaniola (Figure 5.6).

As in many other tropical lowlands, seasonality in the Caribbean is defined by changes in rainfall more than temperature. Although rain falls throughout the year, for much of the region the rainy season is from July to October. This is also when unstable atmospheric conditions can cause hurricanes to form. During the slightly cooler months of December through March, rainfall declines (see the climographs in Figure 5.6 for Havana, Port-au-Prince, and Bridgetown). This time of year corresponds with the peak tourist season.

The Guianas have a different rainfall cycle. On average, these territories receive more rain than the Antillean islands. In Cayenne, French Guiana, an average of 126 inches (320 centimeters) falls each year (see the climograph in Figure 5.6). Unlike the Antilles, the Guianas experience a brief dry period from September to October, while January tends to be a wet period. The Guianas also differ from the rest of the region because they are not affected by hurricanes.

Hurricanes Each year several **hurricanes** pound the Caribbean, as well as Central and North America, with heavy rains and fierce winds. Beginning in July, westward-moving low-pressure disturbances form off the coast of West Africa, picking up moisture and speed as they move across the Atlantic. Usually no more than 100 miles across, these disturbances achieve hurricane status when wind speeds reach 74 miles per hour. Hurricanes may take several paths through the region, but typically enter through the Lesser Antilles. They then curve north or northwest and collide with the Greater Antilles, Central America, Mexico, or southern North America before moving to the northeast and dissipating in the Atlantic Ocean. The hurricane zone

Live Mapping Haiti
Bas Peu De Choses
Saint Gerard
Pacot
Deprez
Bolosse

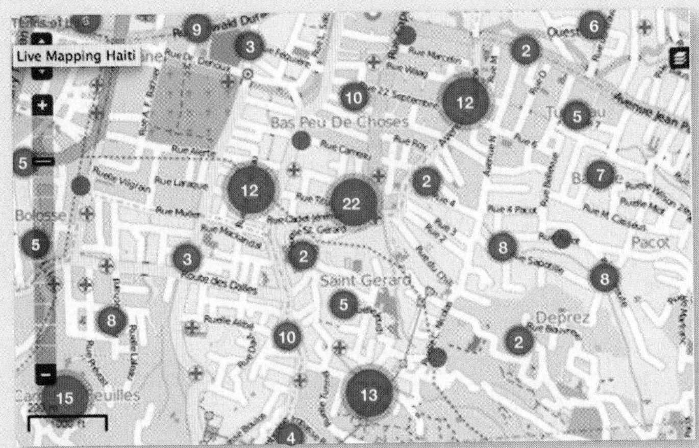

Figure 5.1.1 Crisis Mapping for Port-au-Prince A portion of the map created using Open Street Map in the days after the Haitian earthquake. The circles represent the number of individual reports for a particular area.

was set up so that anyone in Haiti could text urgent needs. As thousands of texts poured in, Haitians in the United Sates were mobilized to translate the texts from Haitian Creole to English so that the mappers could add the geo-referenced information to the map. As the real-time map grew, so did the number of contributors and users. The U.S. Coast Guard and Marines and various humanitarian groups on the ground in Port-au-Prince relied almost exclusively on its output.

Future Crisis Mapping Since the Haiti experience with crisis mapping, similar efforts have been used in response to earthquakes in Nepal and tsunamis in the Philippines. Various organizations of crisis-mapping volunteers have formed to respond to future events. As Patrick Meier likes to say, "To map the world is to know it. But to map the world live is to change it before it's too late."

Source: Adapted from www.newswatch.nationalgeographic.com, How Crisis Mapping Saved Lives in Haiti, July 2, 2012.

1. What are the advantages of an open-source mapping platform when responding to humanitarian crises such as the one experienced in Haiti?
2. Find out whether there is a crisis-mapping organization at your college or university.

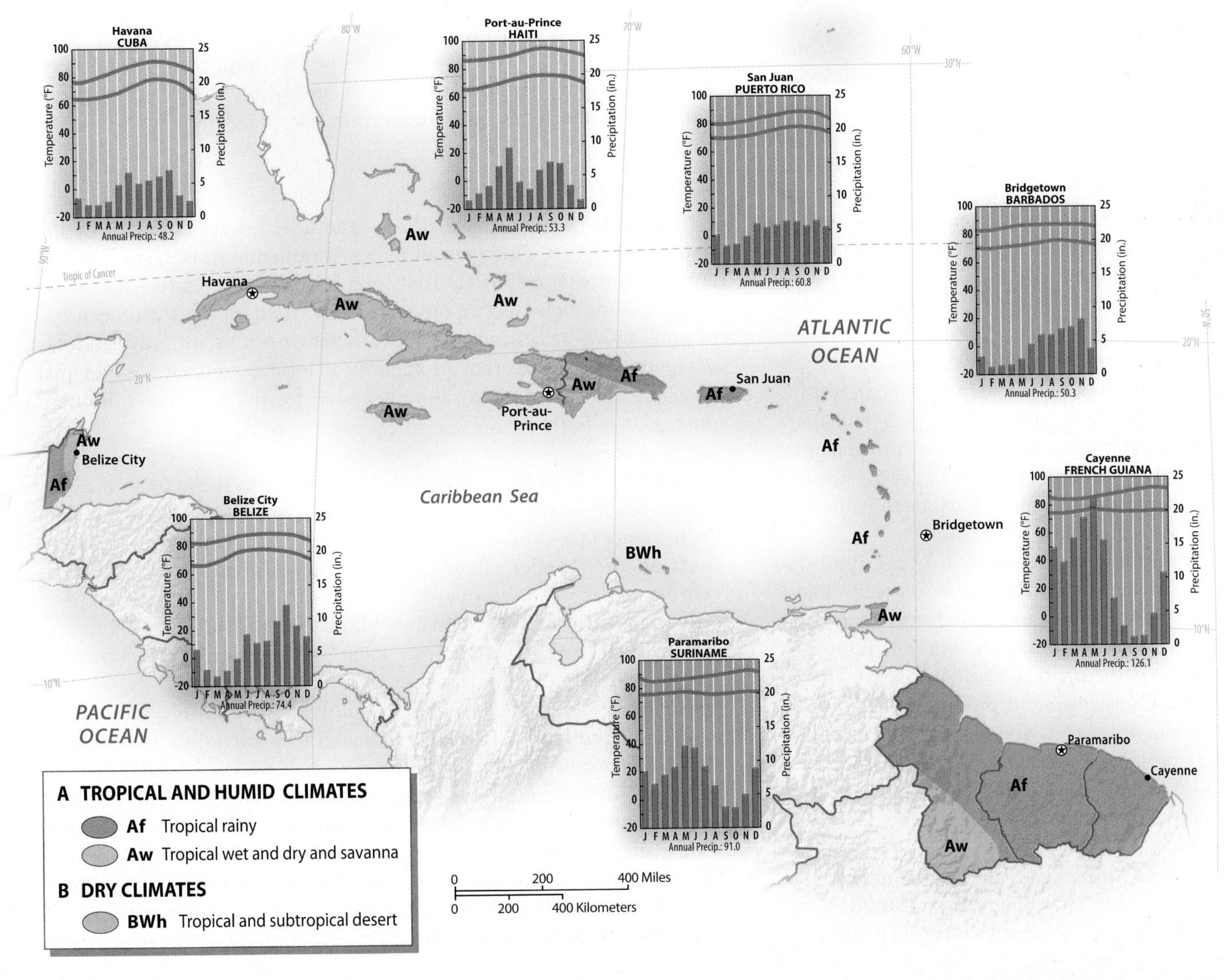

Havana
CUBA
Annual Precip.: 48.2
Port-au-Prince
HAITI
Annual Precip.: 53.3
San Juan
PUERTO RICO
Annual Precip.: 60.8
Bridgetown
BARBADOS
Annual Precip.: 50.3
Cayenne
FRENCH GUIANA
Annual Precip.: 126.1
Belize City
BELIZE
Annual Precip.: 74.4
Paramaribo
SURINAME
Annual Precip.: 91.0
Temperature (°F)
Precipitation (in.)
J F M A M J J A S O N D
Tropic of Cancer
ATLANTIC OCEAN
Caribbean Sea
PACIFIC OCEAN
Havana
Port-au-Prince
San Juan
Belize City
Bridgetown
Paramaribo
Cayenne
Aw
Af
BWh
A TROPICAL AND HUMID CLIMATES
Af Tropical rainy
Aw Tropical wet and dry and savanna
B DRY CLIMATES
BWh Tropical and subtropical desert
0 200 400 Miles
0 200 400 Kilometers

Figure 5.6 Climate Map of the Caribbean Most of the region is classified as having either a tropical wet (Af) or a tropical savanna (Aw) climate. Temperature varies little across the region, as shown by the relatively straight temperature lines. Important differences in total rainfall and the timing of the dry season distinguish different localities. In the Guianas, for example, the dry season is September through October, whereas the drier months for the islands are December through March.

lies just north of the equator on both the Pacific and Atlantic sides of the Americas. Typically, a half-dozen to a dozen hurricanes form each season and move through the region, causing limited damage.

There are, of course, exceptions, and most longtime residents of the Caribbean have felt the full force of at least one major hurricane in their lifetimes. The destruction caused by these storms is not just from the high winds, but also from the heavy downpours, which can cause severe flooding and deadly coastal storm surges. Modern tracking equipment has improved hurricane forecasting and reduced the number of fatalities, primarily through early evacuation of areas in a hurricane's path. This has saved lives but cannot reduce the damage to crops, forests, or infrastructure. In 2012, Hurricane Sandy struck Jamaica and eastern Cuba, removing roofs, downing power lines, and flooding rivers before moving up the East Coast of the United States (Figure 5.7). Similarly, in 2010, Antigua in the Lesser Antilles was hit hard by Hurricane Earl, which destroyed homes and caused serious flooding.

Figure 5.7 Hurricane Sandy Residents walk through the wreckage of Santiago de Cuba after Hurricane Sandy moved through this east Cuban town in October 2012. In addition to fallen trees and missing roofs, 11 Cubans were killed by this fierce storm, which later pounded the New York metropolitan area.

Climate Change Of all the issues facing the Caribbean, one of the most difficult to address is climate change. The Caribbean is not a major contributor of greenhouse gases, but this maritime region is extremely vulnerable to the negative effects of climate change, including sea-level rise, increased intensity of storms, variable rainfall leading to both floods and droughts, and loss of biodiversity (both in forests and in coral reefs). The scientific consensus is that climate change could cause a sea-level rise of 3 to 10 feet (1 to 3 meters) in this century. In terms of land loss due to inundation, the low-lying Bahamas would be the most impacted country in the region, losing nearly 30 percent of its land with a 10-foot (3-meter) sea-level rise. In terms of people affected by inundation, Suriname, French Guiana, Guyana, Belize, and The Bahamas would be the most severely impacted: Just a 3-foot (1-meter) sea-level rise would be devastating because most of the population lives near the coast. With a sea-level rise of 10 feet (3 meters), 30 percent of Suriname's population and 25 percent of Guyana's would be displaced (Figure 5.8).

In addition to land loss and population displacement due to sea-level rise, other concerns include changes in rainfall patterns, leading to declines in agricultural yields and freshwater supplies, and more intense storms—especially hurricanes—that destroy infrastructure and cause other problems. All these changes would negatively affect tourism and thus the gross domestic income of Caribbean countries. Some of the worst-case scenarios are catastrophic.

In terms of biodiversity, continued warming of ocean temperatures will further negatively impact the Caribbean's coral reefs. These reefs, particularly those of the rimland, are already threatened by water pollution and subsistence fishing practices. There is mounting evidence of coral bleaching and die-off due to higher sea temperatures. Coral reefs are biologically diverse and productive ecosystems that function as nurseries for many marine species. Healthy reefs also serve as barriers to protect populated coastal zones as well as mangroves and wetlands. As the reefs become more ecologically vulnerable, so, too, do the human populations that depend on the many benefits that the reefs provide.

Throughout the Caribbean, protecting the environment and preparing for the effects of climate change are increasingly being recognized not as a luxury, but as a question of economic livelihood. In fact, the Caribbean Community and Common Market (CARICOM) has monitored the threat of climate change for over a decade. To address the issue of greenhouse gases regionally, Guyana entered into an innovative agreement with Norway in 2009. Norway provided Guyana with an initial payment of $30 million into its Reducing Emissions from Deforestation and Forest Degradation (REDD) development fund. Norway has continued supporting the project, but disputes over fund management and program implementation have left the project's future uncertain after 2015.

Environmental Issues

Climate change is both a medium- and a long-term concern for the Caribbean region. However, other environmental issues, such as soil erosion and deforestation, have preoccupied the region due to its long-standing dependence on agriculture. Also, as the Caribbean has become more urbanized and more reliant on tourism, governments have come to realize that protection of local ecosystems is not just good for the environment, but also good for the overall economy of the region (Figure 5.8).

Agriculture's Legacy of Deforestation Prior to the arrival of Europeans, much of the Caribbean was covered in tropical forests. The great clearing of these forests began on European-owned plantations on the smaller islands of the eastern Caribbean in the 17th century and spread westward. The island forests were removed not only to make room for sugarcane, but also to provide the fuel necessary to turn the cane juice into sugar and the lumber for housing, fences, and ships. Primarily, however, tropical forests were removed because they were seen as unproductive; the European colonists valued cleared land. The newly exposed tropical soils easily eroded and ceased to be productive after several harvests, a situation that led to two distinct land-use strategies. On the larger islands of Cuba and Hispaniola, as well as on the mainland, new lands were constantly cleared and older ones abandoned or fallowed in an effort to keep up sugar production. On the smaller islands where land was limited, such as Barbados and Antigua, labor-intensive efforts to conserve soil and maintain fertility were employed. In either case,

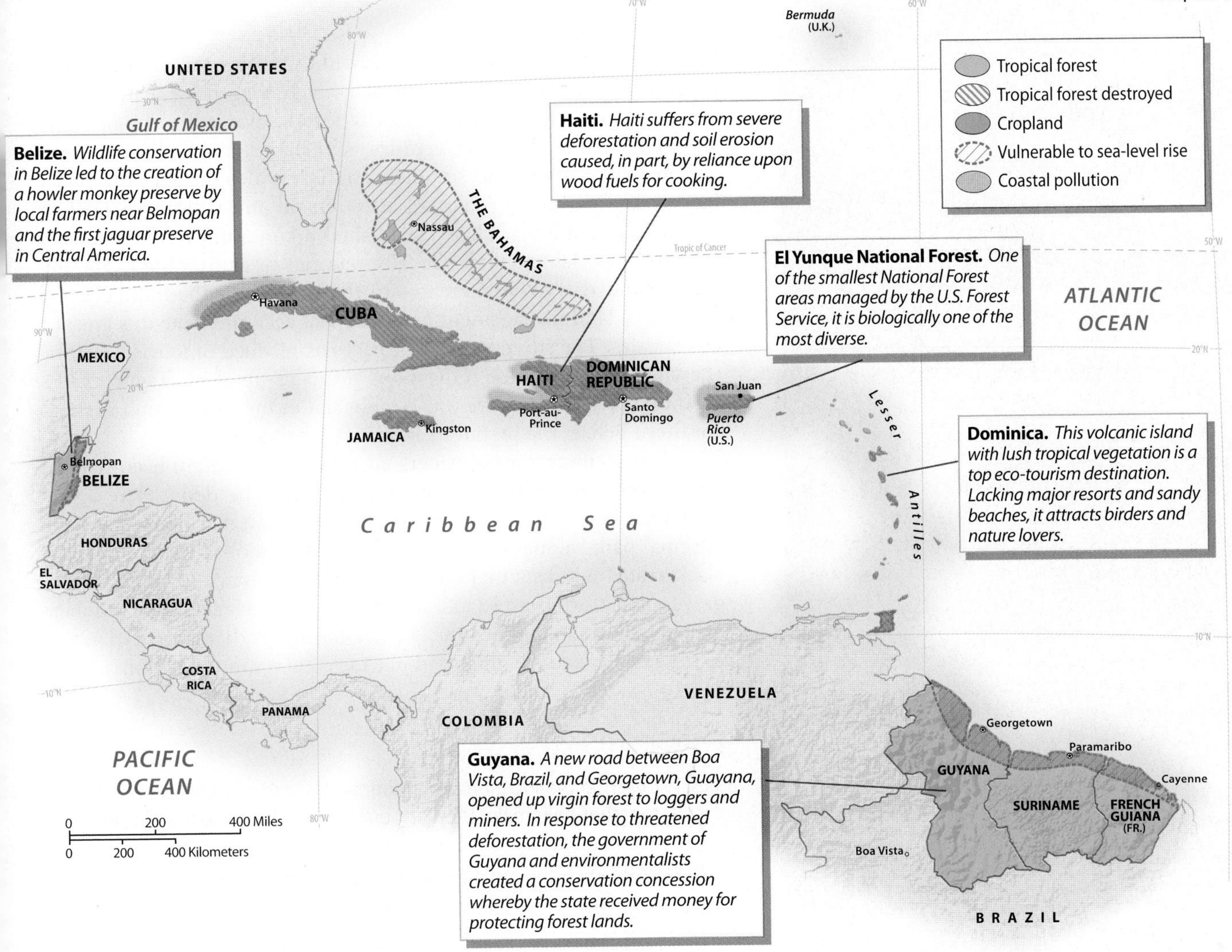
Tropical forest
Tropical forest destroyed
Cropland
Vulnerable to sea-level rise
Coastal pollution
Belize. *Wildlife conservation in Belize led to the creation of a howler monkey preserve by local farmers near Belmopan and the first jaguar preserve in Central America.*
Haiti. *Haiti suffers from severe deforestation and soil erosion caused, in part, by reliance upon wood fuels for cooking.*
El Yunque National Forest. *One of the smallest National Forest areas managed by the U.S. Forest Service, it is biologically one of the most diverse.*
Dominica. *This volcanic island with lush tropical vegetation is a top eco-tourism destination. Lacking major resorts and sandy beaches, it attracts birders and nature lovers.*
Guyana. *A new road between Boa Vista, Brazil, and Georgetown, Guayana, opened up virgin forest to loggers and miners. In response to threatened deforestation, the government of Guyana and environmentalists created a conservation concession whereby the state received money for protecting forest lands.*
Bermuda (U.K.)
UNITED STATES
Gulf of Mexico
THE BAHAMAS
Nassau
Tropic of Cancer
ATLANTIC OCEAN
Havana
CUBA
MEXICO
HAITI
DOMINICAN REPUBLIC
Port-au-Prince
Santo Domingo
San Juan
Puerto Rico (U.S.)
JAMAICA
Kingston
Lesser Antilles
Belmopan
BELIZE
Caribbean Sea
HONDURAS
EL SALVADOR
NICARAGUA
COSTA RICA
PANAMA
COLOMBIA
VENEZUELA
Georgetown
Paramaribo
Cayenne
GUYANA
SURINAME
FRENCH GUIANA (FR.)
Boa Vista
BRAZIL
PACIFIC OCEAN
0 200 400 Miles
0 200 400 Kilometers
80°W
70°W
60°W
50°W
90°W
30°N
20°N
10°N

Figure 5.8 Environmental Issues in the Caribbean Most of the island forests were removed long ago for agriculture or fuelwood, and soil erosion is a chronic problem. Coastal pollution is serious around the largest cities and industrial zones. The forest cover of the rimland states, however, is largely intact. As tourism becomes increasingly important for the Caribbean, efforts to protect the beaches and reefs, along with the fauna and flora, are growing. Areas vulnerable to sea-level rise are noted on this map.

the island forests were replaced by a landscape devoted to crops for world markets.

Haiti's problems with deforestation were accentuated throughout the 20th century as a destructive cycle of environmental and economic impoverishment was established. Half of Haiti's people are peasants who work small hillside plots and seasonally labor on larger estates. As the population grew, people sought more land. They cleared the remaining hillsides, subdivided their plots into smaller units, and abandoned the practice of fallowing land in an effort to eke out an annual subsistence. When the heavy tropical rains came, the exposed and easily eroded mountain soils were washed away. As sediments collected in downstream irrigation ditches and behind dams, agriculture suffered, electricity production declined, and water supplies were degraded. Deforestation was further aggravated by reliance of many poor Haitians on charcoal (made from trees) for household needs. It is estimated that less than 3 percent of Haiti remains forested (Figure 5.9).

While Haiti has lost most of its forest cover, on Jamaica and Cuba nearly one-third of the land is still forested. In the case of the Dominican Republic, about 40 percent of the country is forested, and more than half of Puerto Rico is forested. On these more forested islands, the decline in agriculture has caused some fields to be abandoned and forests to recover. In the case of Puerto Rico, a territory of the United States, its national forests—such as El Yunque on the

Google earth

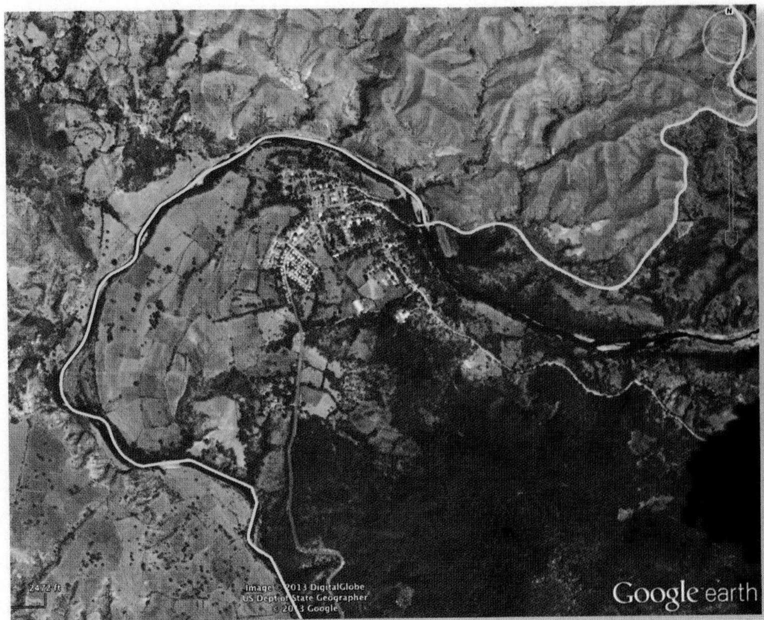

Figure 5.9 Differences in Deforestation Hillsides that were once forested become denuded as trees are cleared for agriculture and fuel production, especially in Haiti. **Q: Which side of the yellow border is Haiti? How could you compare land use in Haiti with that in the Dominican Republic?**

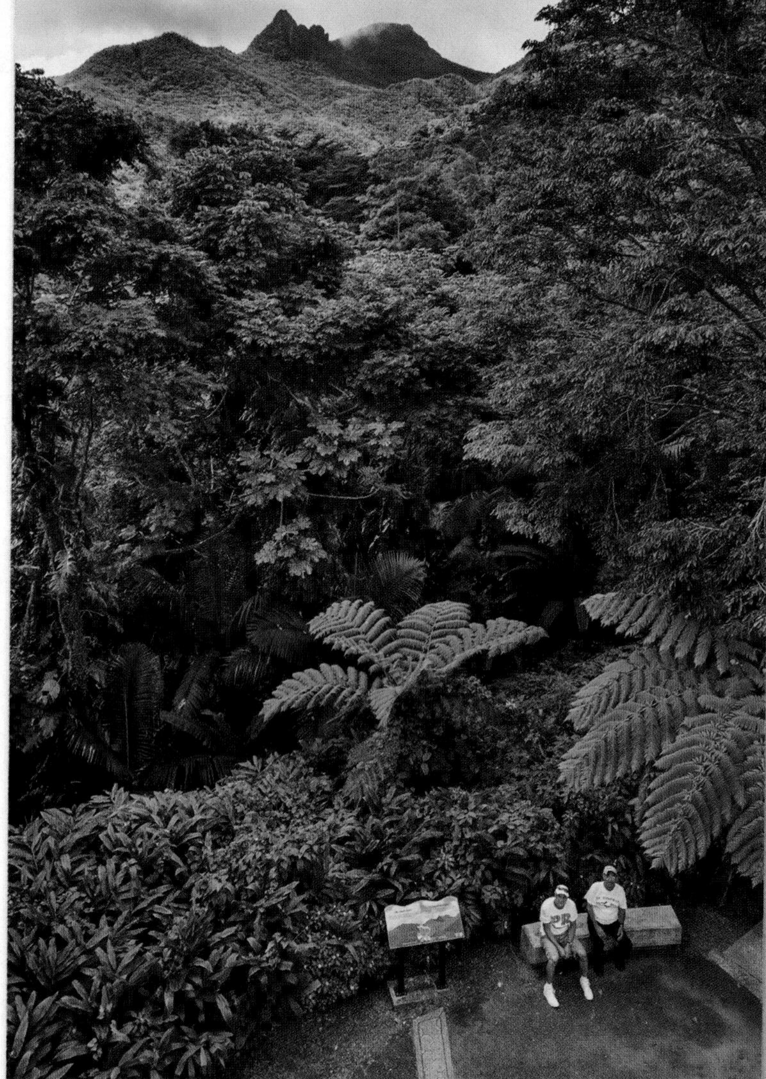

Figure 5.10 El Yunque National Forest Puerto Rico's largest remaining rainforest is in El Yunque National Forest. Visitors enjoy the park's hiking trails and rich biological diversity.

eastern side of the island—contribute greatly to the biological diversity protected by the U.S. National Forest System (Figure 5.10).

Energy Needs and Innovations With the exception of Trinidad and Tobago, which exports oil and liquefied natural gas, the Caribbean states are net importers of oil and highly dependent on foreign sources for their energy needs. The region has some oil refineries that process crude oil shipments into petroleum for domestic consumption and even some for export, but dependence on foreign energy and volatile oil pricing make the small economies of this region vulnerable.

Not surprisingly, Caribbean nations have a growing interest in renewable energy. In many ways, wind energy has long been important for the Caribbean economy. After all, the entire colonial enterprise depended on the trade winds to move commodities and people across the Atlantic. Commercial wind energy is relatively new is gaining popularity. Puerto Rico just opened a major new wind farm on its southern coast near the city of Ponce. The Puerto Rican government intends to generate 12 percent of its energy from renewable sources by 2015. Similarly, the Los Cocos wind farm in the Dominican Republic is a major investment in wind power to serve the country's growing electricity needs. The region's potential for solar power is also excellent.

Conservation Efforts In general, biological diversity and stability are less threatened in the rimland states than in the rest of the Caribbean. Thus, current conservation efforts could produce important results. Even though much of Belize was selectively logged for mahogany in the 19th and 20th centuries, healthy forest cover still supports a diversity of mammals, birds, reptiles, and plants. Public awareness of the negative consequences of deforestation is also greater now. Belize has established many protected areas. In the mid-1980s, villagers in Bermudian Landing, Belize, established a community-run sanctuary for black howler monkeys (locally referred to as baboons). The villagers banded together to maintain the habitat for the monkeys and commit to land management practices that accommodate this gregarious species. The success of the project has resulted in tourists visiting the villages to see these indigenous primates up close (Figure 5.11). In 1986, a jaguar preserve was established in the Cockscomb Basin in southern Belize, the first of its kind in Centrl America.

Slowly, the territorial waters surrounding the Caribbean nations have gained protection, although more could be done. Here again, Belize has been a leader in creating over a dozen marine reserves and national parks along its barrier reef and outer atolls. The country has also created a substantial coastal wildlife sanctuary to protect mangroves. The Caribbean island of Bonaire, which attracts large numbers of scuba divers, maintains the Bonaire Marine Park, recognized as one of the most effectively managed marine reserves in the region.

Review

5.1 Describe the locational, environmental, and climate factors that together help make the Caribbean a major international tourist destination.

5.2 What environmental issues currently affect the Caribbean? Describe the risks and possible solutions.

KEY TERMS isolated proximity, rimland, Greater Antilles, Lesser Antilles, hurricane

Figure 5.11 Protecting Habitat and Wildlife Tourists visit the Community "Baboon" Sanctuary in Bermudian Landing, Belize. The sanctuary is a community-run project to preserve the habitat and increase the number of black howler monkeys (locally called baboons). The sanctuary, established in 1985, attracts domestic and foreign visitors.

Population and Settlement: Densely Settled Islands and Rimland Frontiers

Caribbean population density is generally quite high and, as in neighboring Latin America, increasingly urban. Eighty-five percent of the region's population is concentrated on the four islands of the Greater Antilles (Figure 5.12). Add to this Trinidad's 1.3 million and Guyana's 780,000, and most of the population of the Caribbean is accounted for by six countries and one U.S. territory (Puerto Rico).

In terms of total population, few people inhabit the Lesser Antilles; nevertheless, some of these small island states are densely settled. The island of Barbados is an extreme example. With only 166 square miles (430 square kilometers) of territory, it has over 1700 people per square mile (660 per square kilometer). Bermuda, which is one-third the size of the District of Columbia, has more than 3000 people per square mile (1200 per square kilometer). Population densities on St. Vincent, Martinique, and Grenada, while not as high, are still more than 700 people per square mile (270 per square kilometer). Because arable land is scarce on some islands, access to land is a basic resource problem for many island states. The growth in the region's population, coupled with the scarcity of land, has forced many people into cities or abroad.

In contrast to the islands, the larger rimland states are lightly populated: Guyana and Suriname average 10 people per square mile (3 per square kilometer) and Belize 41 people per square mile (16 per square kilometer). These areas are sparsely settled in part because the relatively poor quality and accessibility of arable land made them less attractive to colonial enterprises.

Demographic Trends

Prior to European contact with the New World, diseases such as smallpox, influenza, and malaria did not exist in the Americas. As discussed

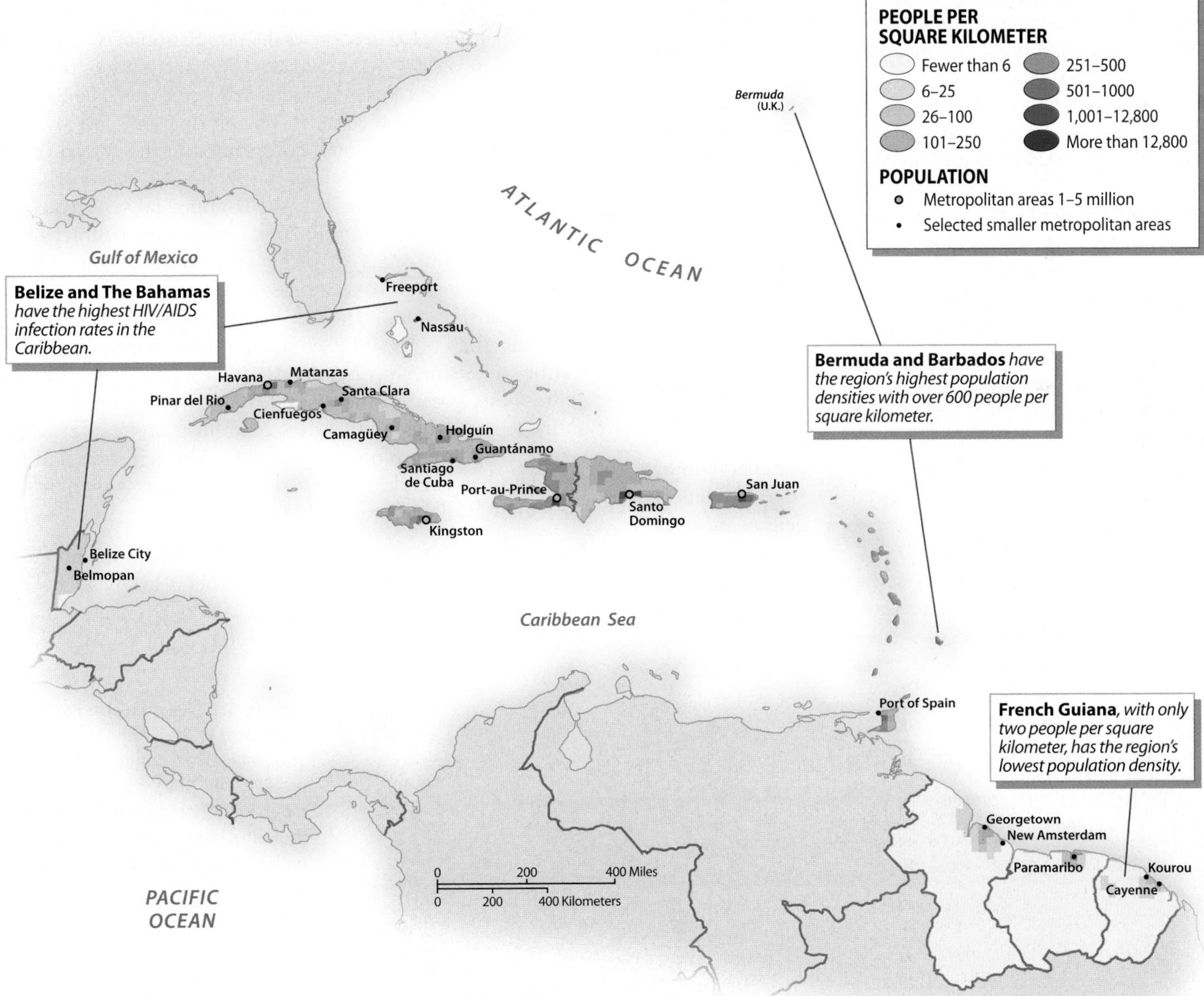

Figure 5.12 Population of the Caribbean The major population centers are on the islands of the Greater Antilles. The pattern here, as in the rest of Latin America, is a tendency toward greater urbanism. The largest metropolitan areas in the region are San Juan, Santo Domingo, and Havana; each has over 2 million residents. In comparison, the rimland states are very lightly settled.

in Chapter 4, these diseases contributed to the demographic collapse of Amerindian populations. Epidemics spread quickly in the Caribbean, and within 50 years of Columbus's arrival, the indigenous population was virtually gone. Only the name *Caribbean* suggests that a Carib people once inhabited the region. Initially, European planters experimented with white indentured labor to work on sugar plantations. However, newcomers from Europe were especially vulnerable to malaria in the lowland Caribbean; typically, half died during the first year of settlement. Those who survived were considered "seasoned." In contrast, Africans had prior exposure to malaria and thus some immunity. They, too, died from malaria, but at much lower rates. This is not to argue that malaria caused slavery in the region, but it did strengthen the economic rationale for it.

During the years of slave-based sugar production, mortality rates were extremely high due to disease, inhumane treatment, and malnutrition. Consequently, the only way to maintain population levels was to continue to import African slaves. With the end of slavery in the mid- to late 19th century and the gradual improvement of health and sanitary conditions on the islands, populations began to increase naturally. In the 1950s and 1960s, many states achieved peak growth rates of 3.0 percent or higher, causing population totals and densities to soar. Over the past 30 years, however, growth rates have come down or stabilized. As noted earlier, the current population of the Caribbean is 45 million. The population is now growing at an annual rate of 1.1 percent, and the population projected for 2025 is 49 million (see Table A5.1).

Fertility Decline and Longer Lives The most significant demographic trends in the Caribbean are the decline in fertility and the increase in life expectancy. Cuba, Puerto Rico, and Trinidad and Tobago have the region's lowest rates of natural increase (0.3). In socialist Cuba, thanks to the education of women, combined with the availability of birth control and abortion, the average woman has 1.7 children (compared to 2.0 in the United States). Yet in capitalist Puerto Rico and Trinidad and Tobago, low rates of natural increase have also been achieved, along with a total fertility rate of 1.6. In general, educational improvements, urbanization, and a preference for smaller families have contributed to slower growth rates. Even states with relatively high total fertility rates, such as Haiti, have seen declines in family size. Haiti's total fertility rate fell from 5.8 in 1970 to 3.4 in 2014.

Figure 5.13 provides a stark contrast in the population profiles of Cuba and Haiti. Although both are poor Caribbean countries, Haiti has the classic broad-based pyramid of a developing country, where more than one-third of the population is under the age of 15. Also, there are very few old people, due to Haiti's relatively low life expectancy (63 years). In contrast, Cuba's population pyramid is more diamond-shaped, bulging in the 35- to 49-year-old age cohort. Here the impact of the Cuban revolution and socialism is evident. Family size fell sharply after education improved and modern contraception became readily available. With better health care, Cuba's population also lives longer, having nearly the same life expectancy as people in the United States (78 years). Cuba has 13 percent of its population over 65 and just 17 percent under 15; thus, it has an extremely low rate of natural increase, similar to many developed countries.

The Impact of HIV/AIDS The rate of HIV/AIDS infection in the Caribbean has come down in the past decade, but it is still twice that of North America, making the disease an important regional issue. Although the infection rates are nowhere near those in Sub-Saharan Africa (see Chapter 6), 1.0 percent of the Caribbean population between the ages of 15 and 49 was living with HIV/AIDS in 2012—or roughly 250,000 people. In Haiti, one of the locations where AIDS was detected earliest, 1.8 percent of the population between the ages of 15 and 49 is infected with the virus. The infection rate is 1.8 percent in Jamaica, 2.3 percent in Belize, and 2.8 percent in The Bahamas.

Poverty, gender inequality, misinformation, and stigma attached to people infected with HIV contributed to the spread of the disease in the 1990s and 2000s. Added to this are the considerable movement among the island nations and the commercial sex trade, encouraged by a strong tourism economy. Now nearly every country has launched educational campaigns to bring awareness up and infection rates down. Reflecting global patterns, heterosexual sex is the main route of HIV transmission in the Caribbean, often associated with paid sex. More than half of the infected population is female. In 2001, the Pan-Caribbean Partnership Against HIV/AIDS (PANCAP) formed to help address the spread of the disease. PANCAP negotiated for lower-cost antiretroviral drugs, so that now two-thirds of the infected population receive these life-prolonging therapies. Through state and regional efforts, mother-to-child transmission prevention is the norm, condoms are widely available, and testing is easily done.

Emigration Driven by the region's limited economic opportunities, a pattern of emigration to other Caribbean islands, North America, and Europe began in the 1950s. Over more than 50 years, a **Caribbean diaspora**—the economic flight of Caribbean peoples

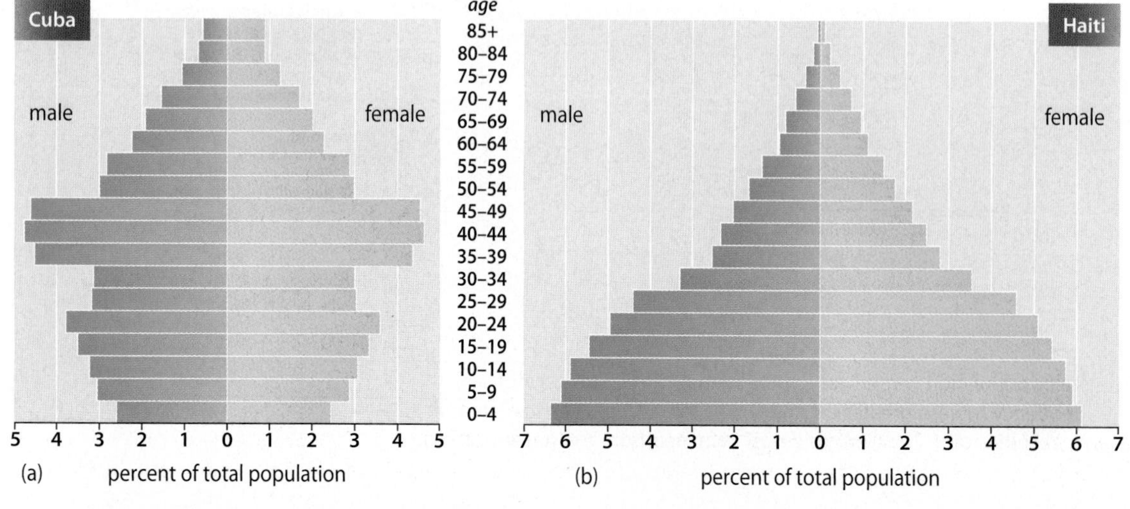

Figure 5.13 Population Pyramids of Cuba and Haiti Although neighbors, Cuba and Haiti have extremely different population profiles. (a) Cuba's population is stable and older, with a notable decline in family size. (b) Haiti's population is much younger and growing, which is reflected in its broad-based pyramid.

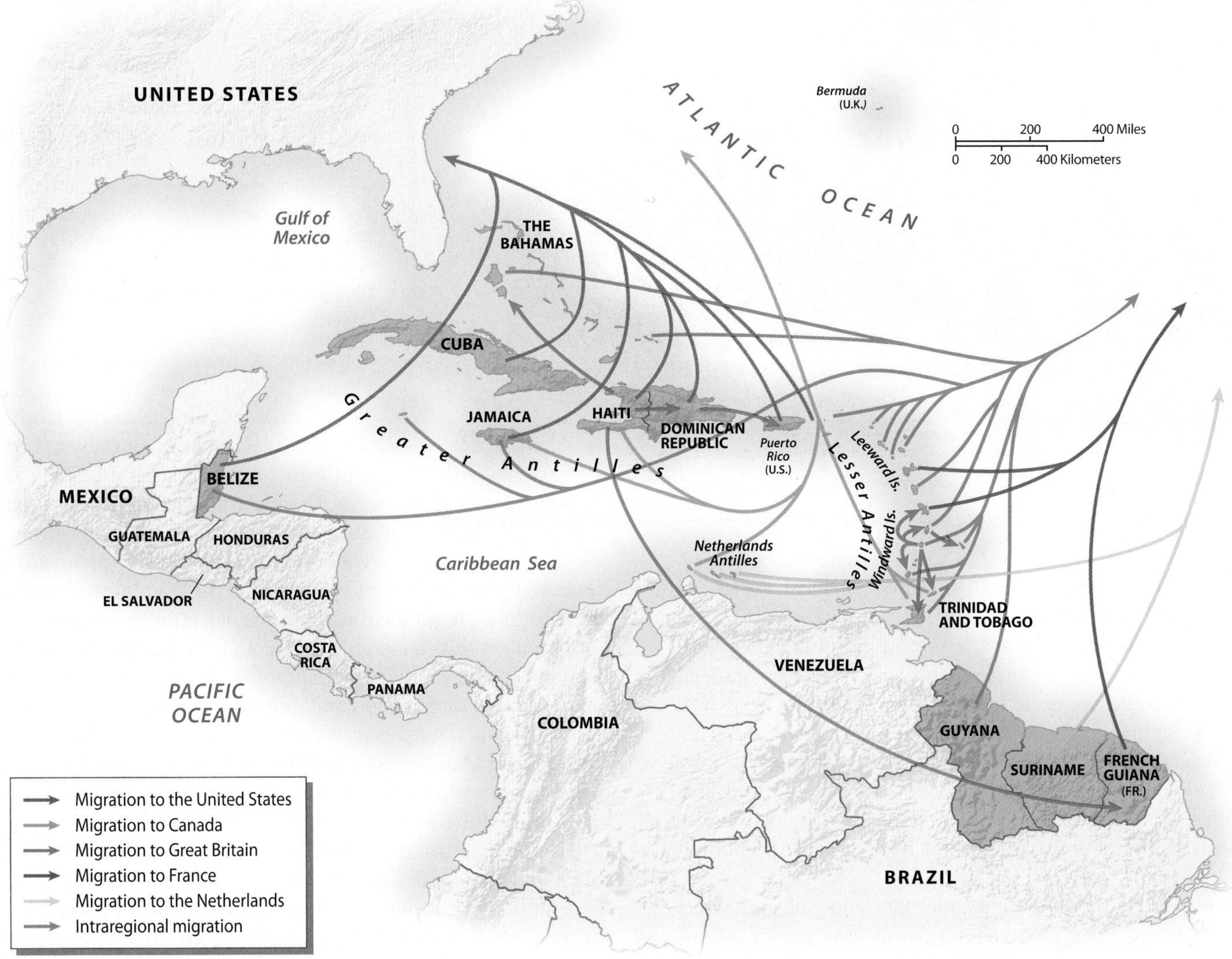

Figure 5.14 Caribbean Diaspora Emigration has long been a way of life for Caribbean peoples. With relatively high education levels, but limited professional opportunities, migrants from the region head to North America, the United Kingdom, France, and the Netherlands. Intraregional migrations also occur between Haiti and the Dominican Republic and between the Dominican Republic and Puerto Rico. **Q: Compare this map with Figure 5.22 on Caribbean languages. What is the relationship between migration flows and the languages spoken in the origin and destination countries?**

across the globe—has become a way of life for much of the region (Figure 5.14). Barbadians generally choose to move to England. In contrast, one out of every three Surinamese has moved to the Netherlands, with most residing in Amsterdam. As for Puerto Ricans, only slightly more live on the island than reside on the U.S. mainland. In the 1980s, roughly 10 percent of Jamaica's population legally emigrated to North America (some 200,000 to the United States and 35,000 to Canada). Cubans have made the city of Miami their destination of choice since the 1960s; today they are a large percentage of that city's population.

The Caribbean region has one of the world's highest negative rates of net migration at –4.0. This means that for every 1000 people, 4 leave the Caribbean each year for other world regions. Individual countries have much higher rates: Guyana has a rate of –15 per 1000, Puerto Rico has a rate of –13, and Martinique has a rate of –10 (see Table A5.1). The economic implications of this labor-related migration are significant and will be discussed later.

The Rural–Urban Continuum

Initially, plantation agriculture and subsistence farming shaped Caribbean settlement patterns. Low-lying arable lands were dedicated to export agriculture and controlled by the colonial elite. Only small amounts of land were set aside for subsistence production. Over time, villages of freed or runaway slaves were established, especially in remote island interiors. However, the vast majority of people lived on estates as owners, managers, or slaves. Cities were formed to serve the administrative and social needs of the colonizers, but most were small, containing just a fraction of a colony's population. The colonists who linked the Caribbean to the world economy saw no need to develop major urban centers.

Plantation America Anthropologist Charles Wagley coined the term **plantation America** to designate a cultural region that extends from midway up the coast of Brazil through the Guianas and the Caribbean into the southeastern United States. Ruled by a European elite dependent on an African labor force, this society was primarily coastal and produced agricultural exports. It relied on **mono-crop production** (production of a single commodity, such as sugar) under a plantation system that concentrated land in the hands of elite families. Such a system created rigid class lines and formed a multiracial society in which people with lighter skin were privileged. The term *plantation America* is meant to describe not a race-based division of the Americas, but rather

Figure 5.15 Tobacco Plantation This woodcut from the 1840s depicts slaves harvesting tobacco in Cuba while a white supervisor looks on smoking a cigar. Commodities such as tobacco and sugar were profitable for estate owners, but the work was arduous and done by slaves. Several million Africans were enslaved and forcibly relocated to the region to produce these commodities.

a production system that brought about specific ecological, social, and economic relations (Figure 5.15).

Even today the structure of Caribbean communities reflects this plantation legacy. Many of the region's subsistence farmers are descendants of former slaves and continue to farm their small plots and work part-time as wage laborers on estates, especially in Haiti. The social and economic patterns generated by slavery still mark the landscape. Rural communities tend to be loosely organized, labor is temporary, and small farms are scattered on available pockets of land. Because men tend to leave home for seasonal labor, female-headed households are common.

Caribbean Cities Since the 1960s, the mechanization of agriculture, offshore industrialization, and population growth have caused a surge in rural-to-urban migration. Cities have grown accordingly, and today 66 percent of the region is classified as urban. Of the large countries, Puerto Rico is the most urban (99 percent) and Haiti the least (53 percent). Caribbean cities are not large by world standards, as only five have 1 million or more residents: Santo Domingo, Havana, Port-au-Prince, San Juan, and Kingston.

Like their counterparts in Latin America, the Spanish Caribbean cities were laid out on a grid with a central plaza. Vulnerable to raids by rival European powers and pirates, these cities were usually walled and extensively fortified. The oldest continuously occupied European city in the Americas is Santo Domingo in the Dominican Republic, settled in 1496. Today it is a metropolitan area of 2.9 million people. *Merenque*—fast-paced, highly danceable music that originated in the Dominican Republic—is the soundtrack that pulses through the metropolis day and night. As rural migrants poured into the city over the last four decades in search of employment and opportunity, the city steadily grew. In 2009, a high-speed Metro opened in Santo Domingo. It now has two lines and over 30 stations to help reduce the city's crushing traffic (Figure 5.16).

The region's second-largest city is metropolitan San Juan, estimated at 2.6 million. It, too, has a renovated colonial core (Figure 5.17) that is dwarfed by the modern sprawling city, which supports the island's largest port. San Juan is the financial, political, manufacturing, and tourism hub of Puerto Rico. With its highways, high rises, shopping malls, and ever-present shoreline, it is an interesting blend of Latin American, North American, and Caribbean urbanism.

Strategically situated on Cuba's north coast at a narrow opening to a natural deep-water harbor, Havana emerged as the most important colonial city in the region, serving as a port for all incoming and outgoing Spanish ships. Consequently, Old Havana possesses a handsome collection of colonial architecture, especially from the 18th and 19th centuries, and is a UNESCO World Heritage Site. The modern city is more sprawling, with a mix of Spanish colonial and Soviet-inspired concrete apartment blocks. It is also a city that had to reinvent

Figure 5.16 Santo Domingo Metro Passengers load onto Metro cars in downtown Santo Domingo, Dominican Republic. The Metro, which opened in 2009, received technical support from the Metro in Madrid, Spain.

itself when subsidies from the former Soviet Union stopped flowing (see *Working Toward Sustainability: Urban Agriculture in Havana*).

Other colonial powers left their mark on the region's cities. For example, Paramaribo, the capital of Suriname, has been described as a tropical extension of Holland, without tulips. In the British and French colonies, a preference for wooden whitewashed cottages with shutters was evident. Yet these port cities tended to be unplanned afterthoughts, built to serve the rural estates and small colonial bureaucracies. Most have grown dramatically over the past 40 years. These cities are no longer small ports for agricultural exports; increasingly, their focus is on welcoming cruise ships and sun-seeking tourists.

Caribbean cities and towns do have their charms and reflect a blend of cultural influences. Throughout the region, houses are often simple structures made of wood, brick, or stucco, raised off the ground a few feet to avoid flooding, and painted in pastels. Most people still get around by foot, bicycle, motorbike, or public transportation; neighborhoods are filled with small shops and services within easy walking distance (Figure 5.18). Streets are narrow, and the pace of life is markedly slower than in North America and Europe. Even when space is tight in town, most settlements are near the sea and its cooling breezes. An afternoon or evening stroll along the waterfront is a common activity.

Review

5.3 What are the major demographic trends for this region, and what factors explain these patterns?

5.4 How did the long-term reliance on a plantation economy influence settlement patterns in the Caribbean?

KEY TERMS Caribbean diaspora, plantation America, mono-crop production

Figure 5.17 Old San Juan Tourists roam the cobbled streets of Old San Juan, a UNESCO World Heritage Site. Many of the 18th- and 19th-century structures of this port city have been handsomely restored.

Figure 5.18 Caribbean Motorbikes A woman rides her motorbike past the whitewashed Dutch colonial-styled buildings in Paramaribo, Suriname. Across the Caribbean, motorbikes and bicycles are popular urban transportation options.

Cultural Coherence and Diversity: A Neo-Africa in the Americas

Linguistic, religious, and ethnic differences abound in the Caribbean. The presence of several former European colonies, millions of descendants of ethnically distinct Africans and indentured workers from India and China, and isolated Amerindian communities on the mainland challenge any notion of cultural coherence.

Common historical and cultural processes hold this region together. In particular, this section focuses on three cultural influences shared throughout the Caribbean: the European colonial presence, African influences, and the mix of European and African cultures termed *creolization*.

The Cultural Impact of Colonialism

The arrival of Columbus in 1492 triggered a devastating chain of events that depopulated much of the Caribbean islands within 50 years. A combination of Spanish brutality, enslavement, warfare, and disease changed the densely settled islands, which supported up to 3 million Caribs and Arawaks, into an uninhabited territory ready for the colonizer's hand. By the mid-16th century, as rival European states competed for Caribbean territory, the lands they fought for were virtually empty. In many ways, this simplified their task, as they did not have to recognize native land claims or work amid Amerindian societies. Instead, the colonizers reorganized the Caribbean territories to serve a plantation-based production system. The critical missing element was labor. Once slave labor from Africa, and later contract labor from Asia, were secured, the small Caribbean colonies became surprisingly profitable.

Creating a Neo-Africa The introduction of African slaves to the Americas began in the 16th century and continued into the 19th century. This forced migration of Africans to the Americas was only part of a much more complex **African diaspora**—the forced removal of Africans from their native areas. The slave trade also crossed the Sahara to include North Africa and linked East Africa with a slave trade in Southwest Asia (see Chapter 6). The best-documented slave route was the transatlantic one—at least 10 million Africans landed in

WORKING TOWARD SUSTAINABILITY

Urban Agriculture in Havana

Google Earth (MG) Virtual Tour Video http://goo.gl/NiDBtY

Many cities around the world have seen a renewed interest in urban gardening as a way to build community unity, reduce food insecurity, improve nutrition, generate income, and enhance urban environments by converting brown spaces into green ones. Cuba is a global leader in urban agriculture, and these farming efforts are especially evident in metropolitan Havana. Scattered throughout this city of 2.3 million are thousands of small and large plots where urban residents are producing vegetables on raised beds; harvesting fruit trees; and raising rabbits, chickens, and goats for meat, eggs, and milk (Figure 5.2.1). Although the Cuban context is unique—a socialist planned economy with fixed prices and limited exposure to market forces—some of the successes of Havana farmers are transferable to other cities in the world.

Gardening by Necessity In 1989, Cuba's government officially recognized the potential for urban gardens as a means to address the pressing food shortages provoked by drastic cuts in food and energy supports from the Soviet Union. The Ministry of Agriculture responded by creating the first coordinated urban agriculture program, providing access to land, especially small urban lots; extension services for training and research; supply stores; and sales outlets. In addition to government actions, nongovernmental organizations from Germany, Canada, and the United States were consulted for best urban agricultural practices and innovative organic techniques. From the start, intensive organic farming techniques and the use of biological agents for pest control were emphasized—in part due to the expense of imported fertilizers and pesticides.

The producers on these smaller farms have use rights to the land for an indefinite period, can choose the crops they grow, and can sell their products at market prices. Residents interested in growing food on empty city lots or other open spaces (say, public parks) have free use of the land as long as they keep it in production. An Urban Agriculture Department was formed to change city laws so that gardeners would have legal priority for all unused space and also to set up consulting centers in each of the city's administrative districts. These consulting centers are a key innovation of the program, providing tools, seeds, compost, and advice for a population that before these reforms was unaccustomed to farming.

Havana as a Garden City These efforts have transformed Havana, improving access to a quantity and variety of fresh foods, creating new forms of self-employment, and forging new green spaces. Empty plots are now filled with raised beds growing eggplants, tomatoes, or strawberries that are tended by local residents. Residents have better access to fresh eggs, milk, and meats than before the initiative began. Even portions of Havana's *Parque Metropolitano* have been converted into food-producing plots. Today food shortages are less severe in Havana because of urban and rural agricultural practices, as well as other market-based reforms that have stimulated economic growth and small-scale entrepreneurship. Interestingly, a new form of tourism has emerged as international visitors book special tours to view Havana's agricultural innovations. In this respect, Havana's urban gardeners have much to share with a rapidly urbanizing world.

1. What factors led the residents of Havana to become leaders in urban agriculture?
2. Beyond increasing the food supply, what are the other advantages of growing food in cities?

the Americas, and it is estimated that another 2 million died en route. More than half of these slaves were sent to the Caribbean(Figure 5.19).

This influx of slaves, combined with the elimination of nearly all the native inhabitants, remade the Caribbean as the area with the greatest concentration of relocated African people in the Americas. The African source areas extended from Senegal to Angola, and slave purchasers intentionally mixed tribal groups in order to weaken ethnic identities. Consequently, intact transfer of religion and languages into the Caribbean rarely occurred; instead, languages, customs, and beliefs were blended.

Maroon Societies Communities of runaway slaves—termed **maroons** in English, *palenques* in Spanish, and *quilombos* in Portuguese—offer interesting examples of African cultural diffusion across the Atlantic. Hidden settlements of escaped slaves existed wherever slavery was practiced. While many of these settlements were short-lived, others have endured and allowed for the survival of African traditions, especially farming practices, house designs, community organization, and language.

The maroons of Suriname still show clear links to West Africa. Whereas other maroon societies gradually blended into local populations, to this day the Suriname maroons maintain a distinct identity. Six tribes formed, ranging in size from a few hundred to 20,000 (Figure 5.20). Living relatively undisturbed for 200 years, these rainforest inhabitants fashioned a rich ritual life for themselves, involving prophets, spirit possession, and witch doctors. Recent pressures to modernize and extract resources have placed Suriname's maroons in direct conflict with the state and private business.

African Religions Linked to maroon societies, but more widely diffused, was the transfer of African religious and magical systems to the Caribbean. These patterns, another reflection of neo-Africa in the Americas, are most closely associated with northeastern Brazil and the Caribbean. Millions of Brazilians practice the African-based religions of Umbanda, Macuba, and Candomblé, along with Catholicism. Likewise, Afro-religious traditions in the Caribbean have evolved into unique forms that have clear ties to West Africa. The most widely practiced are Voodoo (also Vodoun) in Haiti, Santería in Cuba, and Obeah in Jamaica. Each of these religions has its own priesthood and unique pattern of worship. Their impact is considerable; the father-and-son dictators of Haiti, the Duvaliers, were known to hire Voodoo priests to scare off government opposition.

Indentured Labor from Asia By the mid-19th century, most colonial governments in the Caribbean had begun to free their slaves. Fearful of labor shortages, they sought **indentured labor** (workers contracted to labor on estates for a set period of time, often several years) from South, Southeast, and East Asia.

The legacy of these indentured arrangements is clearest in Suriname, Guyana, and Trinidad and Tobago. In Suriname, a former Dutch colony, more than one-third of the population is of South Asian descent, and 16

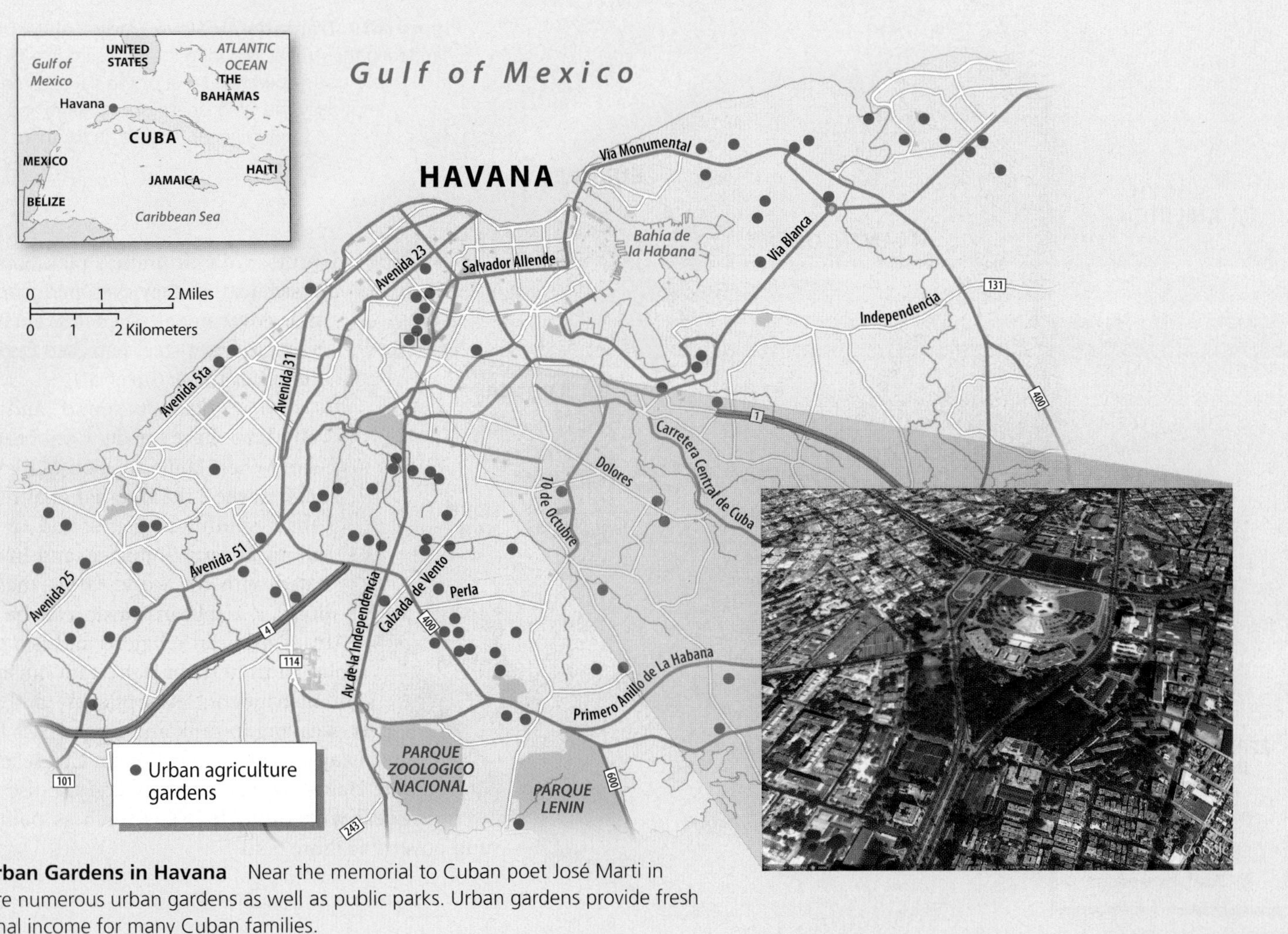

Figure 5.2.1 Urban Gardens in Havana Near the memorial to Cuban poet José Marti in central Havana are numerous urban gardens as well as public parks. Urban gardens provide fresh food and additional income for many Cuban families.

percent is Javanese (from Indonesia). Guyana and Trinidad were British colonies, and most of their contract labor came from India. Today half of Guyana's population and 40 percent of Trinidad and Tobago's claim South Asian ancestry. Hindu temples are found in the cities and villages, and many families speak Hindi at home. The current prime minister of Trinidad and Tobago, Kamla Persad-Bissessar, is of Indian ancestry (Figure 5.21).

Most of the former English colonies have Chinese populations of not more than 2 percent. Once these East Asian immigrants fulfilled their agricultural contracts, they often became merchants and small-business owners, positions they still hold in Caribbean society. Cuba and Suriname have the largest ethnic Chinese populations in the region. Moreover, Suriname has experienced a substantial surge in Chinese immigration, with some reports suggesting that recent Chinese arrivals may account for 10 percent of the country's population.

Creolization and Caribbean Identity

Creolization consists of the blending of African, European, and some Amerindian cultural elements into the unique cultural systems found in the Caribbean. The Creole identities that have formed over time are complex; they illustrate the dynamic cultural and national identities of the region. Today Caribbean writers (V. S. Naipaul, Derek Walcott, and Jamaica Kinkaid), musicians (Bob Marley, Ricky Martin, and Juan Luís Guerra), and athletes (Dominican baseball player David Ortiz and Jamaican sprinter Usain Bolt) are internationally recognized. Collectively, these artists and athletes represent their individual islands and Caribbean culture as a whole.

Language The dominant languages in the region are European: Spanish (25 million speakers), French (11 million), English (7 million), and Dutch (0.5 million) (Figure 5.22). Yet these figures tell only part of the story. In Cuba, the Dominican Republic, and Puerto Rico, Spanish is the official language and is universally spoken. As for the other countries, local variants of the official language, especially in spoken form, can be difficult for a nonnative speaker to understand. In some cases, completely new languages have emerged. In the islands of Aruba, Bonaire, and Curaçao, Papiamento (a trading language that blends Dutch, Spanish, Portuguese, English, and African languages) is the lingua franca, with use of Dutch declining. Similarly, French Creole, or *patois*, in Haiti has been given official status as a distinct language. In practice, French is used in higher education, government, and the courts, but *patois* (with clear African influences) is the language of the street, the home, and oral tradition.

With the independence of Caribbean states in the 1960s, Creole languages became politically and culturally charged with national meaning. While most formal education is taught using standard language forms, the richness of vernacular expression and its capacity to

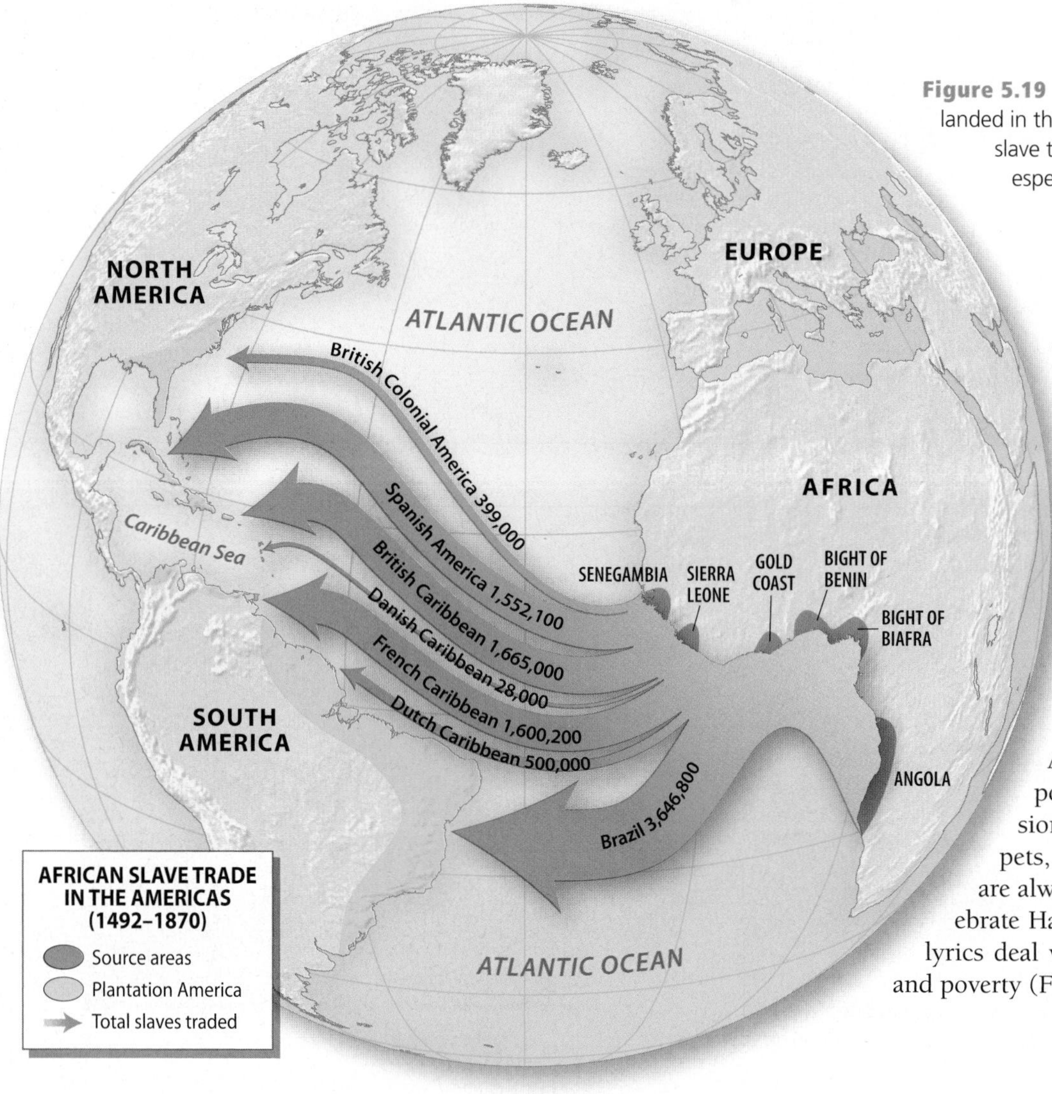

Figure 5.19 Transatlantic Slave Trade At least 10 million Africans landed in the Americas during the four centuries in which the Atlantic slave trade operated. Most of the slaves came from West Africa, especially the Gold Coast (now Ghana) and the Bight of Biafra (now Nigeria). Angola, in southern Africa, was also an important source area.

instill a sense of identity are appreciated. Locals rely on their ability to switch from standard to vernacular forms of speech. Thus, a Jamaican can converse with a tourist in standard English and then switch to a Creole variant when a friend walks by, effectively excluding the outsider from the conversation. This ability to switch is evident in many cultures, but it is widely used in the Caribbean.

Music The rhythmic beats of the Caribbean might be the region's best-known product. This small area is the home of reggae, calypso, merengue, rumba, zouk, and scores of other musical forms. The roots of modern Caribbean music reflect a combination of African rhythms with European forms of melody and verse. These diverse influences, coupled with a long period of relative isolation, sparked distinct local sounds. As movement among the Caribbean population increased, especially during the 20th century, musical traditions were blended, but characteristic sounds remained.

The famed steel pan drums of Trinidad were created from oil drums discarded from a U.S. military base there in the 1940s. The bottoms of the pans are pounded with a sledgehammer to create a concave surface that produces different tones. During carnival (a pre-Lenten celebration), racks of steel pans are pushed through the streets by dancers while drummers play. So skilled are these musicians that they even perform classical music, and government agencies encourage troubled teens to learn steel pan (see *Everyday Globalization: Caribbean Carnival*).

The distinct sound and the ingenious rhythms have made Caribbean music popular. When Jamaican Bob Marley and the Wailers crashed the international music scene with their soulful reggae sound, it was the lyrics about poverty, injustice, and freedom that resonated with the world. More than good dancing music, Caribbean music can be closely tied to Afro-Caribbean religions and is a popular form of political protest. In Haiti, rara music mixes percussion instruments, saxophones, and bamboo trumpets, weaving in funk and reggae bass lines. The songs are always performed in French Creole and typically celebrate Haiti's African ancestry and the use of Voodoo. The lyrics deal with difficult issues, such as political oppression and poverty (Figure 5.23).

Figure 5.20 Maroons in Suriname A maroon with white clay rubbed on her face dances at the Black People's Day celebration in Paramaribo. Surinamese maroons (whose heritage is from runaway slave communities) recognize the first Sunday in January as Black People's Day to express their shared transatlantic cultural heritage.

Figure 5.21 South Asian Influences Trinidad and Tobago's Prime Minister, Kamla Persad-Bissessar, attends a celebration to mark the 165th anniversary of Indian Arrival Day in Trinidad. The Prime Minister is of Indian descent.

Sports: From Baseball to Béisbol

Latin Americans are known for their love of soccer, but baseball is the dominant sport for much of the Caribbean. A byproduct of early U.S. influence in the region, baseball is the sport of choice in Cuba, Puerto Rico, and the Dominican Republic. Even socialist Cuba embraces baseball with a fervor that would humble many U.S. fans. Recently, several Cuban baseball stars have defected to the United States; players such as Jose Abreu, Alex Guerrero, and Aroldis Chapman landed multimillion-dollar contracts and appreciative fans. Over one-quarter of Major League Baseball (MLB) players are born outside the United States. The Dominican Republic sends the most players to the major leagues, accounting for 10 percent of all players in 2014.

The Dominican Republic became a talent pipeline for MLB due to a complex mix of boyhood dreams, economic inequality, and greed. This small country has produced many baseball legends, and over the decades franchises have invested millions in training camps there. In the past two decades, however, Dominican pride in its baseball prowess has been tinged by the realities of a merciless feeder system that depends on impoverished kids, performance-enhancing drugs, fake documents, and scouts who skim a percentage of the signing bonuses. Yet the reality is that more and more young boys, who can sign contracts at age 16, see their future in baseball rather than in schooling. Even a modest signing bonus of $10,000 to $20,000 can build a nice home for a boy's family (Figure 5.24).

San Pedro de Macoris, not far from Santo Domingo, epitomizes this field of dreams. It is a place of cane fields, kids on bicycles with bats and gloves, sugarcane factories, dusty baseball diamonds, and large homes of former players, such as George Bell, Pedro Guerrero, and Sammy Sosa. These houses are silent testaments to what is possible through baseball. In an effort to clean up baseball's image, MLB has officials in the Dominican Republic investigating drug use and fraudulent papers. However, as long as there are families pushing their teenage boys and a talent pool that delivers, this transnational system is self-perpetuating.

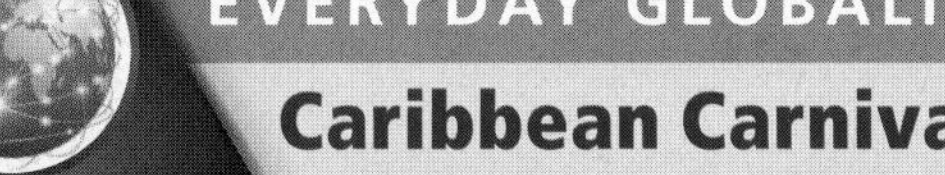

EVERYDAY GLOBALIZATION

Caribbean Carnival

Modern carnival originated in Christian Lenten beliefs and African musical traditions. The term is Latin in origin, a reference to giving up eating meat (*carnivale* means to put meat away) in observance of Lent (the 40-day period prior to Easter). In the Caribbean, former slaves imbued carnival with special meaning because it became their opportunity to break the monotony of their daily lives.

Today carnival is celebrated on nearly every Caribbean island as a national street party that can go on for weeks and attract thousands of tourists. Many official carnivals are no longer tied to Lent, but are scheduled at other times of the year. Carnival celebrations have also followed the Caribbean diaspora to new settings in North America and Europe. One of North America's biggest carnivals is in Toronto, where a large Caribbean immigrant population maintains the tradition every July (Figure 5.3.1). In London, Birmingham, and Leicester, Caribbean carnivals are celebrated annually. As a cultural measure of globalization, the diffusion of carnival shows that a great party transfers easily!

1. Why do you think carnival became an important expression of Caribbean identity?
2. Outside the Caribbean, where else have you witnessed or experienced this kind of celebration?

Figure 5.3.1 Carnival in Toronto Revelers march in the Toronto Caribbean Carnival Parade held each summer in Canada's largest and most diverse city. Toronto has long been an important destination for Caribbean migrants.

Review

5.5 What kinds of African influences exist in the Caribbean, and how do they express themselves?

5.6 What is creolization, and how does it explain different cultural influences and patterns found in the Caribbean?

KEY TERMS African diaspora, maroon, indentured labor, creolization

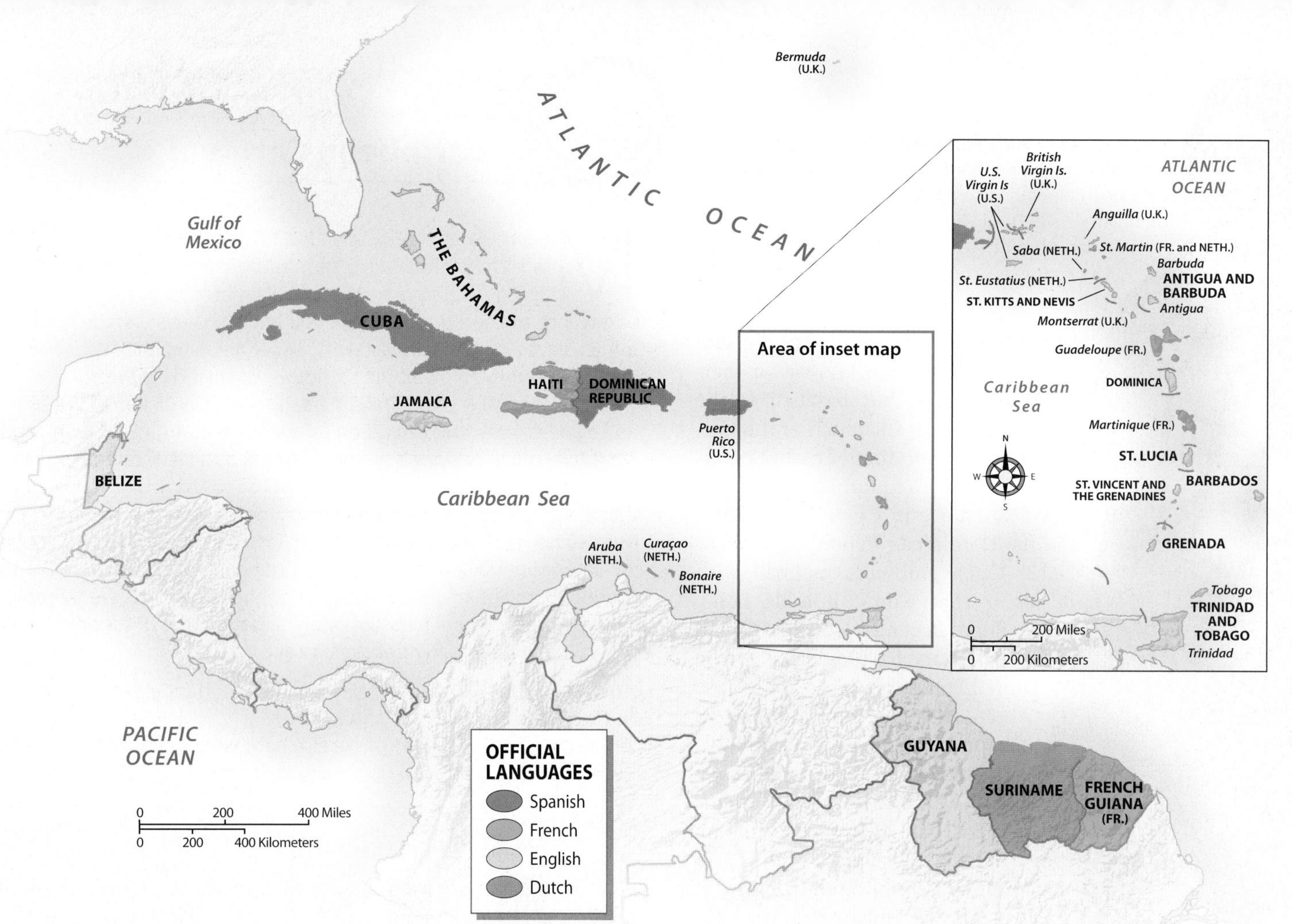

Bermuda (U.K.)
ATLANTIC OCEAN
Gulf of Mexico
THE BAHAMAS
CUBA
HAITI
DOMINICAN REPUBLIC
JAMAICA
Puerto Rico (U.S.)
BELIZE
Caribbean Sea
Aruba (NETH.)
Curaçao (NETH.)
Bonaire (NETH.)
Area of inset map
PACIFIC OCEAN
0 200 400 Miles
0 200 400 Kilometers
OFFICIAL LANGUAGES
Spanish
French
English
Dutch
GUYANA
SURINAME
FRENCH GUIANA (FR.)

British Virgin Is. (U.K.)
U.S. Virgin Is (U.S.)
ATLANTIC OCEAN
Anguilla (U.K.)
Saba (NETH.)
St. Martin (FR. and NETH.)
Barbuda
St. Eustatius (NETH.)
ANTIGUA AND BARBUDA
ST. KITTS AND NEVIS
Antigua
Montserrat (U.K.)
Guadeloupe (FR.)
Caribbean Sea
DOMINICA
Martinique (FR.)
ST. LUCIA
ST. VINCENT AND THE GRENADINES
BARBADOS
GRENADA
Tobago
TRINIDAD AND TOBAGO
Trinidad
0 200 Miles
0 200 Kilometers

Figure 5.22 Language Map of the Caribbean Because this region has no significant Amerindian population (except on the mainland), the dominant languages are European: Spanish, French, English, and Dutch. However, many of these languages have been creolized, making it difficult for outsiders to understand them. **Q: Where is English spoken in the Caribbean, and what does this tell us about the early colonization of this region?**

Figure 5.23 Haiti's Rara Music Performed in procession, rara music is sung in *patois*. Considered the music of the poor, it is used to express risky social commentary. This rara band performs at a folk festival in Washington, DC.

Figure 5.24 Caribbean Baseball A young Cuban batter takes aim during a pickup game in rural Cuba. Several Caribbean states have adopted baseball as their national sport—most notably, the Dominican Republic and Cuba.

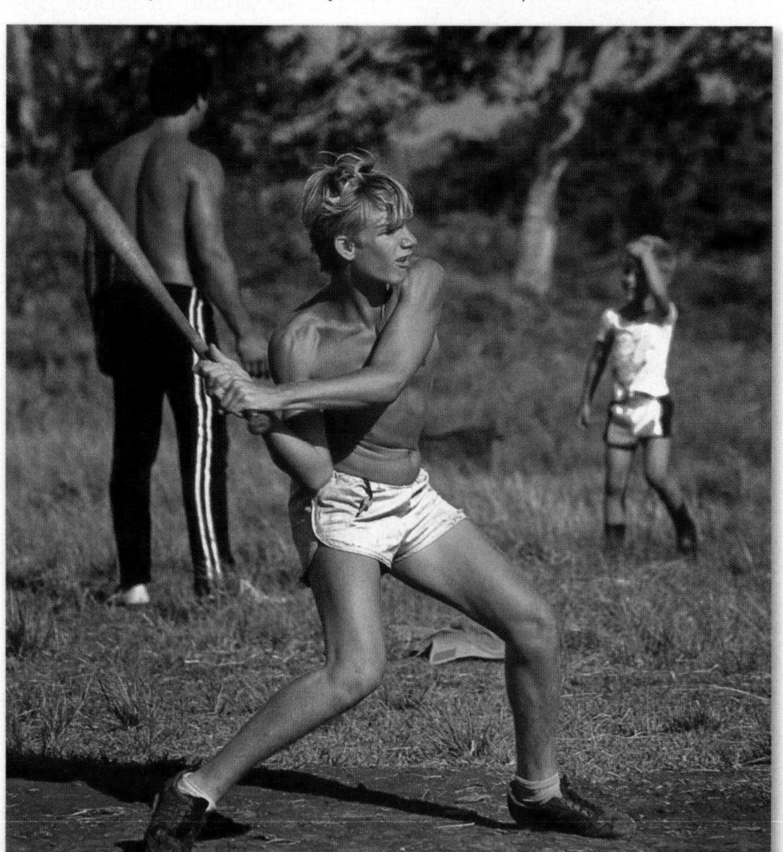

Geopolitical Framework: Colonialism, Neocolonialism, and Independence

Caribbean colonial history is a patchwork of competing European powers fighting over profitable tropical territories. By the 17th century, the Caribbean had become an important proving ground for European ambitions. Spain's grip on the region was slipping, and rival European nations felt confident that they could gain territory by gradually pushing Spain out. Many territories, especially islands in the Lesser Antilles, changed European rulers several times.

Europeans viewed the Caribbean as a strategically located, profitable region in which to produce sugar, rum, and spices. Geopolitically, rival European powers also felt that their presence in the Caribbean limited Spanish authority there. However, Europe's geopolitical dominance in the Caribbean began to diminish by the mid-19th century, just as the U.S. presence increased. Inspired by the **Monroe Doctrine**, which claimed that the United States would not tolerate European military involvement in the Western Hemisphere, the U.S. government made it clear that it considered the Caribbean to be within its sphere of influence. This view was highlighted during the Spanish–American War in 1898. Even though several English, Dutch, and French colonies remained after this date, the United States indirectly (and sometimes directly) asserted its control over the region, bringing in a period of **neocolonialism**. Neocolonialism is the indirect control of one country or region by another through economic and cultural domination, rather than through direct military or political control as occurs under colonialism.

In an increasingly global age, however, even neocolonial interests can be short-lived or sporadic. The Caribbean has not attracted the level of private foreign investment seen in other regions, and as its strategic importance in a post–Cold War era fades, new geopolitical forces are shaping the region. Taiwan began wooing small Caribbean islands in the 1990s with strategic investments, in the hope of winning United Nations votes for its cause. Not surprisingly, China has invested still more money in the region, in part to convince nations that supported Taiwan, such as Dominica and Grenada, to switch their support to China.

Life in "America's Backyard"

To this day, the United States exerts considerable influence in the Caribbean, which was commonly referred to as "America's backyard" in the early 20th century. The stated foreign policy objectives were to free the region from European authority and encourage democratic governance. Yet time and again, American political and economic ambitions undermined those goals. President Theodore Roosevelt made his priorities clear with imperialistic policies that extended the influence of the United States beyond its borders. Policies and projects such as the construction of the Panama Canal and the maintenance of open sea-lanes benefited the United States, but did not necessarily support social, economic, or political gains for the Caribbean people. The United States later offered benign-sounding development packages, such as the Good Neighbor Policy (1930s), the Alliance for Progress (1960s), and the Caribbean Basin Initiative (1980s). The Caribbean view of these initiatives has been wary at best. Rather than feeling liberated, many residents believe that one kind of political dependence was traded for another—colonialism for neocolonialism.

In the early 1900s, the role of the United States in the Caribbean was overtly military and political. The Spanish–American War (1898) secured Cuba's freedom from Spain and also resulted in Spain giving up the Philippines, Puerto Rico, and Guam to the United States; the latter two are still U.S. territories. The U.S. government also purchased the Danish Virgin Islands in 1917, renaming them the U.S. Virgin Islands and developing the harbor of St. Thomas. French, English, and Dutch colonies were tolerated as long as these allies recognized U.S. supremacy in the region. Outwardly against colonialism, the United States had become much like an imperial force.

One privilege of an empire is the ability to impose one's will, by force if necessary. When a Caribbean state refused to abide by U.S. trade rules, U.S. Navy vessels would block its ports. Marines landed, and U.S.-backed governments were installed throughout the Caribbean Basin. These were not short-term engagements: U.S. troops occupied the Dominican Republic from 1916 to 1924, Haiti from 1913 to 1934, and Cuba from 1906 to 1909 and 1917 to 1922 (Figure 5.25). Even today the United States maintains several important military bases in the region, including Guantánamo in eastern Cuba.

Critics of U.S. policy in the Caribbean complain that business interests overwhelm democratic principles when foreign policy is determined. The U.S. banana companies that settled the coastal plain of the Caribbean rimland operated as if they were independent states. Sugar and rum manufacturers from the United States bought the best lands in Cuba, Haiti, and Puerto Rico. Meanwhile, truly democratic institutions remained weak, and there was little improvement in social development. True, exports increased, railroads were built, and port facilities were improved, but levels of income, education, and health remained dreadfully low throughout the first half of the 20th century.

The Commonwealth of Puerto Rico Puerto Rico is both within the Caribbean and apart from it because of its status as a commonwealth of the United States. Throughout the 20th century, various Puerto Rican independence movements sought to separate the island from the United States. Even today Puerto Ricans are divided about their island's political future. At the same time, Puerto Rico depends on U.S. investment and welfare programs; U.S. food stamps are a major source of income for many Puerto Rican families. Commonwealth status also means that Puerto Ricans can freely move between the island and the U.S. mainland, a right they actively assert. In other ways, Puerto Ricans symbolically display their independence. For example, they support their own "national" sports teams and send a Miss Puerto Rico to international beauty pageants. In 2012, a controversial island referendum resulted in the majority of residents voting for a change in political status, with their preference being for statehood. There are no plans in the U.S. Congress to change the political status of the island (Figure 5.26).

Puerto Rico led the Caribbean in the transition from an agrarian economy to an industrial one, beginning in the 1950s. For some U.S. officials, Puerto Rico became the model for the rest of the region. Puerto Rican President Muñoz Marín advocated an industrialization program called "Operation Bootstrap." Drawn by tax incentives and cheap labor, hundreds of U.S. textile and clothing firms relocated to Puerto Rico. Over the next two decades, 140,000 industrial jobs were added, resulting in a marked increase in per capita gross national product. In the 1970s, when Puerto Rico faced stiff competition from Asian clothing manufacturers, the government encouraged petrochemical and pharmaceutical plants to relocate to the island. By the 1990s, Puerto Rico was one of the most industrialized places in the region, with a significantly higher per capita income than its neighbors. By the 2000s many of the U.S. tax benefits associated with establishing factories in Puerto Rico had ceased to exist and manufacturing declined. Consequently, there are growing signs of underdevelopment, including widespread poverty, extensive out-migration, low rates of education, and serious debt problems.

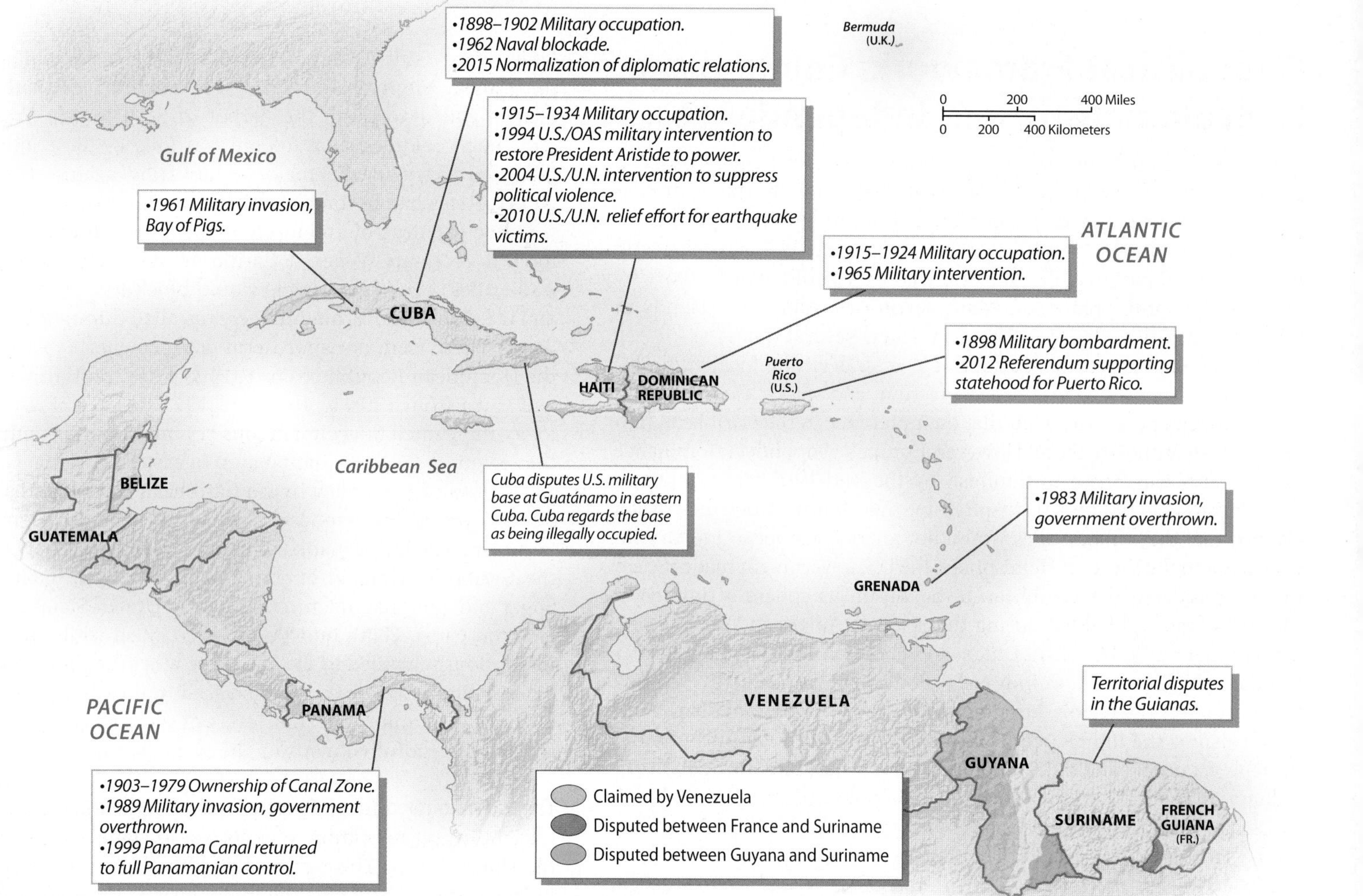

•1898–1902 Military occupation.
•1962 Naval blockade.
•2015 Normalization of diplomatic relations.
Bermuda (U.K.)
0 200 400 Miles
0 200 400 Kilometers
•1915–1934 Military occupation.
•1994 U.S./OAS military intervention to restore President Aristide to power.
•2004 U.S./U.N. intervention to suppress political violence.
•2010 U.S./U.N. relief effort for earthquake victims.
Gulf of Mexico
•1961 Military invasion, Bay of Pigs.
ATLANTIC OCEAN
•1915–1924 Military occupation.
•1965 Military intervention.
CUBA
•1898 Military bombardment.
•2012 Referendum supporting statehood for Puerto Rico.
HAITI
DOMINICAN REPUBLIC
Puerto Rico (U.S.)
BELIZE
Caribbean Sea
Cuba disputes U.S. military base at Guatánamo in eastern Cuba. Cuba regards the base as being illegally occupied.
•1983 Military invasion, government overthrown.
GUATEMALA
GRENADA
Territorial disputes in the Guianas.
VENEZUELA
PACIFIC OCEAN
PANAMA
GUYANA
•1903–1979 Ownership of Canal Zone.
•1989 Military invasion, government overthrown.
•1999 Panama Canal returned to full Panamanian control.
Claimed by Venezuela
Disputed between France and Suriname
Disputed between Guyana and Suriname
SURINAME
FRENCH GUIANA (FR.)

Figure 5.25 Geopolitical Issues in the Caribbean: U.S. Military Involvement and Regional Disputes The Caribbean was regarded as the geopolitical backyard of the United States, and U.S. military occupation was a common occurrence in the first half of the 20th century. Border and ethnic conflicts also exist—most notably, in the Guianas.

Cuba and Geopolitics The most profound challenge to U.S. authority in the region came from Cuba and its superpower ally, the former Soviet Union. In the 1950s, a revolutionary effort began in Cuba, led by Fidel Castro against the pro-American Batista government. Cuba's economic productivity had soared, but its people were still poor, uneducated, and increasingly angry. The contrast between the lives of average sugarcane workers and the foreign elite was sharp. Castro, who took power in 1959, had tapped a deep vein of Cuban resentment against six decades of American neocolonialism.

After Castro's government nationalized U.S. industries and took ownership of all foreign-owned properties, the United States responded by refusing to buy Cuban sugar and ultimately ending diplomatic relations with the state. Various U.S. embargoes (laws forbidding trade with a particular country) against Cuba have existed for five decades. When Cuba established strong diplomatic relations with the Soviet Union in 1960, at the height of the Cold War, the island state became a geopolitical enemy of the United States. With the Soviet Union financially and militarily backing Castro, a direct U.S. invasion of Cuba was too risky. The fall of 1962 produced one of the most dangerous episodes of the Cold War, when Soviet missiles were discovered on Cuban soil. Ultimately, the Soviet Union removed its weapons; in return, the United States promised not to invade Cuba.

Even with the end of the Cold War and the loss of financial support from the Soviet Union, Cuba managed to reinvent itself by

Figure 5.26 Puerto Rican Referendum for Statehood Supporters of the pro-statehood New Progressive party hold placards supporting the vote for Puerto Rico to become the 51st state. The referendum passed in November 2012, but the island is still deeply divided over its relationship with the United States. Meanwhile, the U.S. Congress has yet to take any action regarding Puerto Rican statehood.

I LOVE THE UNITED STATES OF AMERICA
NO MORE COLONY
PR. STATE 51 NOW
"PUERTO RICO, NO PODEMOS ESPERAR MÁS"
VOTA NO

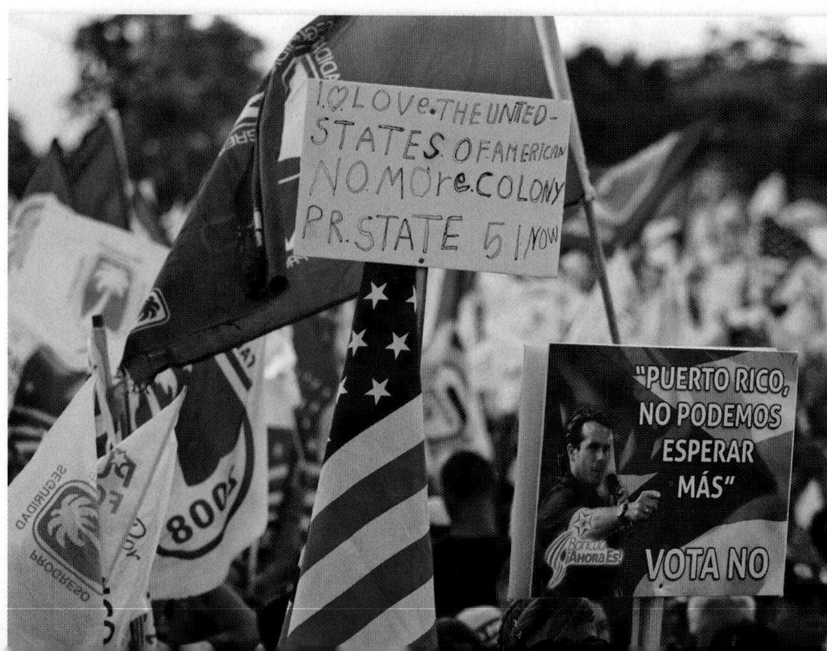

growing its tourism sector and courting foreign investment, especially from Spain, and oil from Venezuela.

In many ways, Cuba is entering a new political era. In 2008, Fidel Castro, then 82 and in poor health, left office, and his younger brother, Raúl, assumed the duties of president. Raúl Castro has been more willing to encourage private enterprise in Cuba and has expanded the availability of licenses that allow self-employed individuals and small businesses to operate legally in the country for the first time in decades. In 2013, he announced that he would not seek a third term in office and praised the appointment of the first vice president, Miguel Díaz-Canel. Many believe that Díaz-Canel, a much younger man in his 50s, will lead Cuba in the future. The most dramatic change was the December 2014 announcement that President Obama and President Castro had agreed to restore full diplomatic relations between the two countries. The United States agreed to ease restrictions on remittances, travel, and banking, while Cuba agreed to allow more Internet access. As of 2015, the trade embargo remains in place, as ending it would require an act of Congress.

Independence and Integration

Given the repressive colonial history of the Caribbean, it is no wonder that the struggle for political independence began more than 200 years ago. Haiti was the second colony in the Americas to gain independence, in 1804, after the United States in 1776. However, political independence in the region has not guaranteed economic independence. Many Caribbean states struggle to meet the basic needs of their people. Today some Caribbean territories maintain their colonial status as an economic asset. For example, the French territories of Martinique, Guadeloupe, and French Guiana are overseas departments of France; residents have full French citizenship and social welfare benefits.

Independence Movements Haiti's revolutionary war began in 1791 and ended in 1804. During this conflict, the island's population was cut in half by casualties and emigration; ultimately, the former slaves became the rulers. Independence, however, did not allow this crown jewel of the French Caribbean to prosper. Slowed by economic and political problems, Haiti was ignored by the European powers and never fully accepted by the Latin American countries on the mainland once they gained their own political independence in the 1820s.

Several revolutionary periods followed in the 19th century. In the Greater Antilles, the Dominican Republic finally gained independence in 1844, after taking control of the territory from Spain and Haiti. Cuba and Puerto Rico were freed from Spanish colonialism in 1898, but their independence was weakened by greater U.S. involvement. The British colonies also faced revolts, especially in the 1930s; it was not until the 1960s that independent states emerged from the English Caribbean. First, the larger colonies of Jamaica, Trinidad and Tobago, Guyana, and Barbados gained independence. Other British colonies followed throughout the 1970s and early 1980s. Suriname, the only Dutch colony on the rimland, became a self-governing territory in 1954 but remained part of the Netherlands until 1975, when it declared itself an independent republic.

Limited Regional Integration Perhaps the most difficult task facing the Caribbean is increasing economic integration. Scattered islands, a divided rimland, different languages, and limited economic resources hinder the formation of a meaningful regional trade bloc. Economic cooperation is more common between groups of islands with a shared colonial background than between, for example, former French and English colonies.

During the 1960s, the Caribbean began to experiment with regional trade associations as a means to improve its economic competitiveness. The goal of regional cooperation was to raise employment, increase intraregional trade, and ultimately reduce economic dependence. The countries of the English Caribbean took the lead in this development strategy. In 1963, Guyana proposed an economic integration plan with Barbados and Antigua. In 1972, the integration process intensified with the formation of the **Caribbean Community and Common Market (CARICOM)**. Representing the former English colonies, CARICOM proposed an ambitious regional industrialization plan and the creation of the Caribbean Development Bank to assist the poorer states. As important as CARICOM is as an economic symbol of regional identity, it has produced limited improvements in intraregional trade. It has 13 full member states—all of the English Caribbean and French-speaking Haiti. Other dependencies, such as Anguilla, Turks and Caicos, Bermuda, and the British Virgin Islands, are associate members. Still, the predominance of English-speaking territories in CARICOM illustrates the deep linguistic divides in the Caribbean.

The dream of regional integration as a way to produce a more stable, self-sufficient Caribbean has never been realized. One scholar of the region argues that a limiting factor is a "small-islandist ideology." Islanders tend to keep their backs to the sea, oblivious to the needs of neighbors. At times, such isolationism results in suspicion, distrust, and even hostility toward nearby states. Yet economic necessity dictates engagement with partners outside the region. This peculiar status of isolated proximity expresses itself in the Caribbean's uneven social and economic development trends.

Review

5.7 Which countries have had colonial or neocolonial influences in the Caribbean, and why have they engaged with the region?

5.8 Describe the obstacles to Caribbean political or economic integration.

KEY TERMS Monroe Doctrine, neocolonialism, Caribbean Community and Common Market (CARICOM)

Economic and Social Development: From Cane Fields to Cruise Ships

Collectively, Caribbean peoples, although poor by U.S. standards, are economically better off than most residents of Sub-Saharan Africa, South Asia, and even China. Despite periods of economic stagnation, social gains in education, health, and life expectancy are significant (Table A5.2). Historically, the Caribbean's links to the world economy were through tropical agricultural exports, but several specialized industries—such as tourism, offshore banking, and assembly plants—have challenged agriculture's dominance. These industries grew because of the region's proximity to North America and Europe, abundant cheap labor, and policies that created a nearly tax-free environment for foreign-owned companies. Unfortunately, growth in these industries does not employ all the region's displaced rural laborers, so the lure of jobs outside the region is still strong.

From Fields to Factories and Resorts

Agriculture once dominated the economic life of the Caribbean. However, decades of unstable crop prices and a decline in special trade agreements with former colonial states have produced more hardship than prosperity. Ecologically, the soils are overworked, and there are no frontier areas into which to expand production except for areas of the rimland. Moreover, agricultural prices have not kept pace with rising production costs, so wages and profits remain low. With the exception of a few mineral-rich territories, such as Trinidad, Guyana, Suriname, and Jamaica, most countries have tried to diversify their economies, relying less on their soils and more on manufacturing and services.

Comparing export figures over time illustrates the shift away from mono-crop dependence. In 1955, Haiti earned more than 70 percent of its foreign exchange through coffee exports; by 1990, coffee accounted for only 11 percent of its export earnings. Similarly, in 1955, the Dominican Republic earned close to 60 percent of its foreign exchange through sugar, but 35 years later sugar accounted for less than 20 percent of the country's foreign exchange, and pig iron exports nearly equaled exports of sugar.

Sugar The economic history of the Caribbean cannot be separated from the production of sugarcane. Even relatively small territories such as Antigua and Barbados yielded fabulous profits because there was no limit to the demand for sugar in the 18th century. Once considered a luxury crop, it became a popular necessity for European and North American laborers by the 1750s. It sweetened tea and coffee and made jams a popular spread for stale bread. In short, it made the meager and bland diets of ordinary people tolerable and also boosted caloric intake. Distilled into rum, sugar produced a popular intoxicant. Though it is difficult to imagine today, consumption of a pint of rum a day was not uncommon in the 1800s. The Caribbean and Latin America still produce the majority of the world's rum (Figure 5.27).

Sugarcane is still grown throughout the region for domestic consumption and export. Its economic importance has declined, however, mostly due to increased competition from corn and sugar beets grown in the midlatitudes. The Caribbean and Brazil are the world's major sugar exporters. Until 1990, Cuba alone accounted for more than 60 percent of the value of world sugar exports and earned 80 percent of its foreign exchange through sugar production. However, Cuba's dominance in sugar exports had more to do with its subsidized and guaranteed markets in Eastern Europe and the Soviet Union than with exceptional productivity. Since the breakup of the Soviet Union in 1991, the value and volume of the Cuban sugar harvest have plummeted.

Figure 5.27 Bacardi Rum and San Juan Rum is the quintessential Caribbean beverage. Made from sugar cane, it is been an important regional export for five centuries. The Bacardi Factory shown here in San Juan, Puerto Rico, is a popular tourist destination.

Assembly-Plant Industrialization Another important Caribbean development strategy has been to invite foreign investors to set up assembly plants and thus create jobs. This was first tried successfully in Puerto Rico in the 1950s and was copied throughout the region. Today the main driver of the Puerto Rican economy is manufacturing, especially pharmaceuticals, textiles, petrochemicals, and electronics. However, since the signing of NAFTA and CAFTA (see Chapter 4), Puerto Rico has faced increased competition from other states with even lower wages. The 1996 decision of the U.S. Congress to phase out many of the tax exemptions also undercut Puerto Rico's ability to maintain its specialized industrial base. Consequently, since 2006 the Puerto Rican economy has been in recession. Manufacturing is still important, but over one-third of Puerto Ricans live below the poverty line.

Through the creation of **free trade zones (FTZs)**—duty-free and tax-exempt industrial parks for foreign corporations—the Caribbean has become an increasingly attractive location for assembling goods for North American consumers. The Dominican Republic took advantage of tax incentives and guaranteed access to the U.S. market offered through the Caribbean Basin Initiative, and the country now has 50 FTZs. The majority of them are clustered around the outskirts of Santo Domingo and Santiago, the two largest cities (Figure 5.28). The most frequent investors in these zones are U.S. and Canadian firms, followed by Dominican, South Korean, and Taiwanese firms. Traditional manufacturing on the island was tied to sugar refining, whereas production in the FTZs focuses on garments and textiles. These manufacturing centers now account for three-quarters of the country's exports.

Growth in manufacturing depends on national and international policies supporting export-led development through foreign investment. Certainly, new jobs are being created, and national economies are diversifying in the process, but critics believe that foreign investors gain more than the host countries. Most goods are assembled from imported materials, so there is little development of national suppliers. Wages may be higher than local averages, but they are still low compared to those in the developed world—sometimes just a few dollars a day.

Offshore Banking and Online Gambling The rise of offshore banking in the Caribbean began with The Bahamas in the 1920s. **Offshore banking** centers appeal to foreign banks and corporations by offering specialized services that are confidential and tax exempt. These offshore banks make money through registration fees, not taxes. The Bahamas was so successful in developing this sector that by 1976 the country was the world's third-largest banking center. Its dominance began to decline due to corruption concerns linked to money laundering and competition from the Caribbean, Hong Kong, and Singapore. In the 1990s, the Cayman Islands emerged as the region's leader in financial services—and one of the leading financial centers globally. With a population of 45,000, this crown colony of Britain has some 50,000 registered companies and a per capita purchasing power parity of nearly $50,000. Yet with growing competition from Asia, the Pacific islands, and even Europe, the Caymans has slipped from one of the top 5 banking centers to one of the top 50.

An estimated $20–$30 trillion is hidden in offshore tax havens all over the world; the Caribbean is just one location where corporations and rich individuals park their money. Each of the Caribbean offshore banking centers tries to develop specialized financial services to attract clients, such as functional operations, insurance, and trusts. Bermuda,

Figure 5.28 Free Trade Zones in the Dominican Republic One sign of globalization is the increase in duty-free and tax-exempt industrial parks in the Caribbean. The Dominican Republic, also a member of the Central American Free Trade Association, has 50 FTZs with foreign investors from the United States, Canada, South Korea, and Taiwan.
Q: Where are FTZs clustered in the Dominican Republic, and what might explain this pattern?

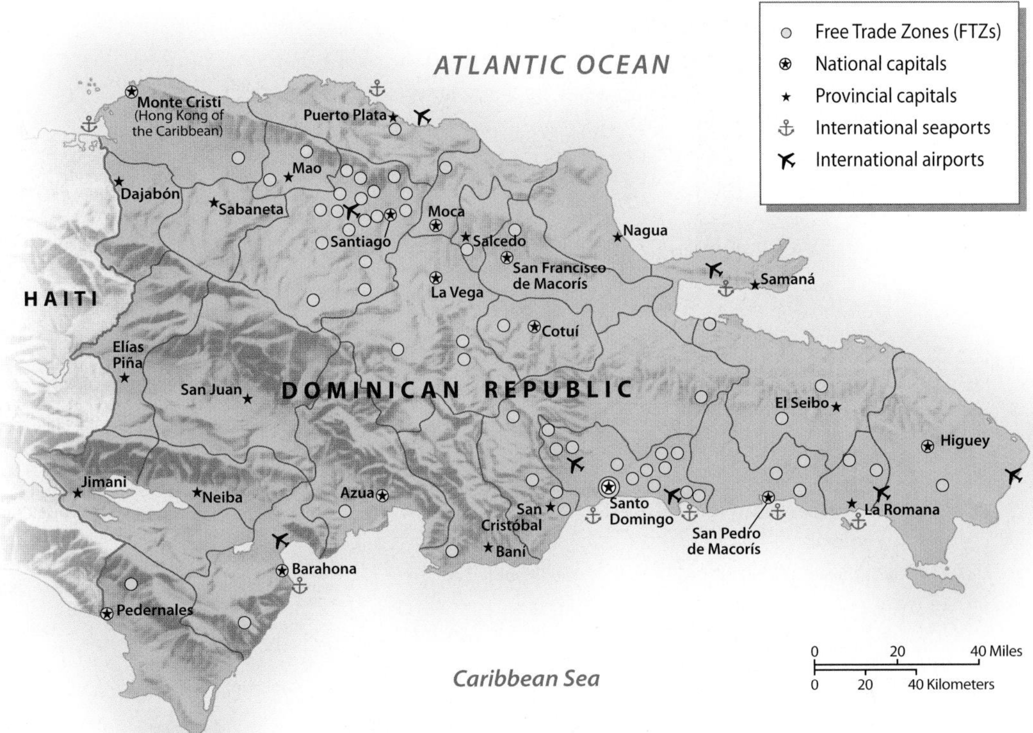

for example, is a global leader in the reinsurance business, which makes money from underwriting part of the risk of other insurance companies (Figure 5.29). The Caribbean is an attractive location for such services because of proximity to the United States (home of many of the registered firms), client demand for these services in different countries, and advances in telecommunications that make this industry possible. The resource-poor islands of the region see financial services as a way to bring foreign capital to state treasuries. Envious of the economic success of The Bahamas, Bermuda, and the Cayman Islands, countries such as Antigua, Aruba, Barbados, and Belize have also established international banking services. Barbados, for example, is the preferred tax haven for Canadians who desire to bank offshore.

Online gambling is the newest industry for the microstates of the Caribbean. Antigua and St. Kitts were the region's leaders, beginning legal online gambling services in 1999. Other states soon followed; as of 2003, Dominica, Grenada, Belize, and the Cayman Islands had gambling domain sites. In 2007, the WTO deemed U.S. restrictions imposed on overseas Internet gambling sites to be illegal. The tiny nation of Antigua is currently seeking $3 billion in compensation from the United States for lost revenue due to illegal restrictions placed on Antigua's business.

Meanwhile, in the face of a lucrative business opportunity, efforts to legalize Internet gambling in the United States moved into high gear. By 2013, the governors of Delaware, New Jersey, and Nevada had signed laws to allow online poker in their states. Other cash-strapped states have followed suit, seeing the taxation of online gambling as an attractive revenue source. For the Caribbean, however, this suggests that its era as an online gambling destination may soon be eclipsed.

Tourism Environmental, locational, and economic factors converge to support tourism in the Caribbean. The earliest visitors to this tropical sea admired its clear and sparkling turquoise waters. By the 19th century, wealthy North Americans were fleeing winter to enjoy the healing warmth of the Caribbean during the dry season. Developers later realized that the simultaneous occurrence of the Caribbean's dry season and the Northern Hemisphere's winter was ideal for beach resorts. By the 20th century, tourism was well established, with both destination resorts and cruise lines. By the 1950s, the leader in tourism was Cuba, with The Bahamas a distant second. Castro's rise to power, however, eliminated this sector of the Cuba's economy for nearly three decades and opened the door for other islands to develop tourism economies.

Six countries or territories hosted two-thirds of the 21 million international tourists who came to the Caribbean in 2011: the Dominican Republic, Puerto Rico, Cuba, The Bahamas, Jamaica, and Aruba (Figure 5.30).

Figure 5.29 Financial Services in Bermuda Front Street in Hamilton, Bermuda, is a reflection of the territory's ties to the United Kingdom and its prosperity. Tourism and financial services in the reinsurance business explain Bermuda's wealth.

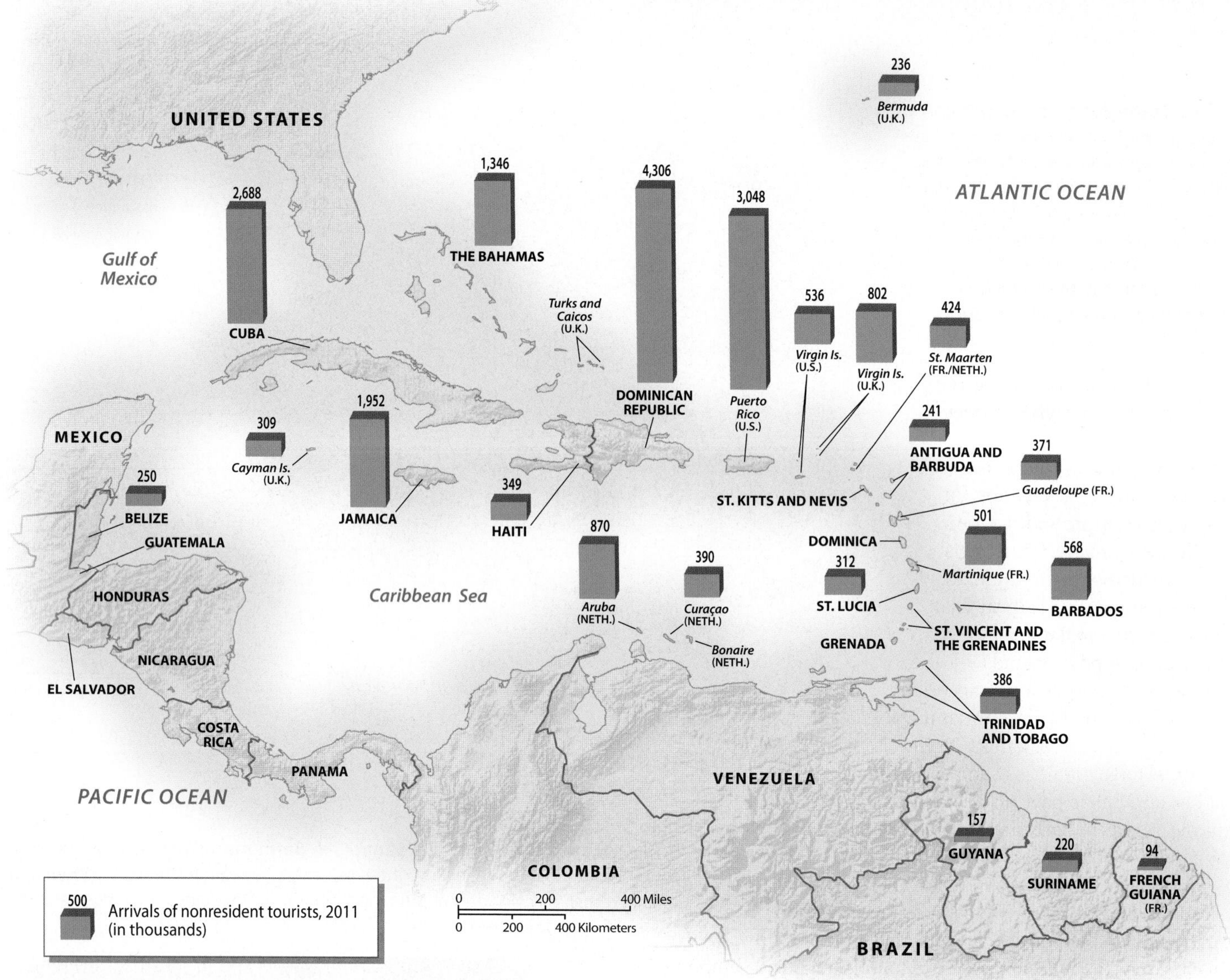

Figure 5.30 Global Linkages: International Tourism The Caribbean is directly linked to the global economy through tourism. Each year more than 21 million tourists come to the islands, mostly from North America, Latin America, and Europe. The most popular destinations are the Dominican Republic, Puerto Rico, Cuba, Jamaica, The Bahamas, and Aruba. **Q: What might explain the lower levels of tourism in the Guianas compared to the rest of the region?**

Puerto Rico's tourist sector began to grow with commonwealth status in 1952. San Juan is now the largest home port for cruise lines and the second-largest cruise-ship port in the world in terms of total visitors. The Dominican Republic is the region's largest tourist destination, receiving 4 million visitors in 2011, many of them Dominican nationals who live overseas. Since 1980, tourist receipts have increased 20-fold, making tourism a leading foreign exchange earner at more than $4 billion.

The Bahamas attributes most of its economic development and high per capita income to tourism. With more than 1.3 million hotel guests in 2011 and 1.7 million cruise-ship passengers, The Bahamas is another major tourist hub. Some 30 percent of Bahamians are employed in tourism, and tourism represents nearly half the country's GDP (Figure 5.30).

After years of neglect, Cuba has revived tourism in an attempt to earn badly needed foreign currency. Tourism represented less than 1 percent of the national economy in the early 1980s. By 2011, 2.7 million tourists (mostly Canadians and Europeans) poured onto the island, bringing in $2.5 billion in tourism receipts. Conspicuous in their absence were U.S. visitors, forbidden to travel to Cuba as "tourists" because of the U.S.-imposed sanctions. But with the normalization of U.S.-Cuban relations in 2014, there has been a growth in "educational" tours to Cuba, which means that more U.S. citizens can visit this country (see *Geographers at Work: Educational Tourism in Cuba*).

As important as tourism is for the larger states, it is often the principal source of income for smaller ones. The Virgin Islands, Barbados, Turks and Caicos, and, Belize all significantly depend on international tourists. To show how quickly this sector can grow, consider this example: When Belize began promoting tourism in the early 1980s, it had just 30,000 arrivals per year. An English-speaking country close to North America, Belize specialized in ecotourism that showcased its interior tropical forests and coastal barrier reef. By the 1990s, the number of tourists topped 300,000, and tourism employed one-fifth of the workforce. Belize City became a port of call for day visitors from cruise ships in 2000, making it the fastest-growing tourist port in the Caribbean. Yet the influx of day visitors to this impoverished coastal town of 60,000 has done little to improve the city's infrastructure or high unemployment.

For more than four decades, tourism has been the foundation of the Caribbean economy, but the industry has grown more slowly in recent years, compared to other tourist destinations in the Middle East, southern Europe, and even Central America. Americans seem to favor domestic destinations, such as Hawaii, Florida, and Las Vegas, or more "exotic"

GEOGRAPHERS AT WORK

Educational Tourism in Cuba

Unlike many people who discovered geography later on, **Sarah Blue**, an associate professor at Texas State University, says that she started college as a geography major: "I wanted to do something about pollution. Then I realized that geography was so much more than that." She first visited Cuba in 1996 to improve her Spanish-language skills and learn to dance salsa, then returned to study topics such as the role of remittances in the Cuban economy and Cuba's influence on medical practices throughout the developing world.

When the Obama administration allowed people-to-people tourism in 2013, Blue started Candela Cuba Tours, a company that leads educational trips to Cuba, as a way to give back to Cuban colleagues. Her tours incorporate food production, culture, music, education, health care, and the economy (Figure 5.4.1). Blue's clients stay in family-run establishments, eating breakfast with their Cuban hosts. (These *casas particulares*, licensed by the Cuban government, allow Cubans to rent rooms in their homes to benefit directly from international tourism and educational tours.)

Changing Perceptions The recent change in U.S.–Cuba diplomatic relations means that U.S. visitors to Cuba will likely increase, especially through educational tourism. Blue notes that these developments "will boost Cuba's economy, but it's not going to change things radically. Really, what's opening up are opportunities for Americans." Her geography background helps her analyze possible changes to Cuban society. Says Blue, "A colleague once said to me, 'History is about the past, but geography is about the future.' I love that!"

Blue's American clients are surprised by how open, friendly, and happy Cubans are; they also appreciate the different ways that Cubans approach problems and view the role of government. These insights offer a poignant juxtaposition, leading visitors to meaningful reflection and deeper understanding about society and place. "Traveling is fundamental to geography and understanding different places," notes Blue. "The more you see other places in the world, the more you can reflect on your own home."

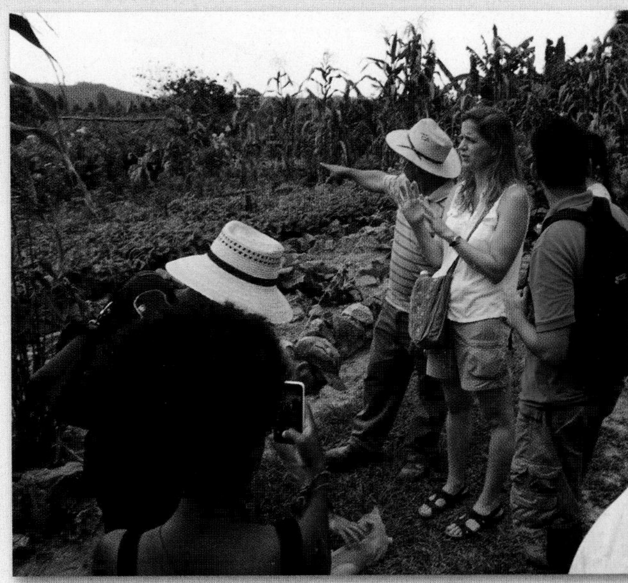

Figure 5.4.1 Cuba Tourism Sarah Blue discusses food production on the island with a group of American tourists.

1. How is people-to-people tourism different from regular tourism? Have you or someone you know engaged in this kind of tourist experience?
2. Consider how globalization is changing Cuba today. Do you think Cuba can keep its distinctive qualities in the future?

settings, such as Costa Rica. European tourists also seem to be either sticking closer to home or venturing to new locations, such as Dubai on the Persian Gulf and Goa in India. Increasingly, tourists are opting to experience the Caribbean from the decks of cruise ships rather than land-based resorts. This trend undermines the local benefits of tourism, directing capital to large cruise lines rather than island economies (Figure 5.31).

Tourism-led growth has other drawbacks. It heavily depends on the overall health of the world economy and current political affairs. Thus, if North America experiences a recession or international tourism declines due to terrorism concerns, the flow of tourist dollars to the Caribbean dries up. Where tourism is on the rise, local resentment may build as residents confront the differences between their own lives and those of the tourists. There is also a serious problem of **capital leakage**, which is the huge gap between gross income and the total tourist dollars that remain in the Caribbean. Because many guests stay in hotel chains or cruise ships with corporate headquarters outside the region, leakage of profits is inevitable. On the plus side, tourism tends to promote stronger environmental laws and regulations. Countries quickly learn that their physical environment is the foundation for success. And though tourism does have its costs (higher energy and water consumption and demand for more imports), it is environmentally less destructive than traditional export agriculture—and currently more profitable.

Social Development

The Caribbean's record for economic growth is inconsistent, but measures of social development are generally strong. For example, most Caribbean peoples have an average life expectancy of more than 73 years (see Table A5.2). Literacy levels are high, with near parity in terms of school enrollment by gender. Indeed, high levels of educational attainment and out-migration have markedly lowered the natural increase rate over the past 30 years, which now hovers around 1 percent.

Figure 5.31 Caribbean Cruising A massive cruise ship docks in St. John's harbor in Antigua, an island of the Lesser Antilles. The cruise business brings many visitors to these small islands, but they spend relatively little time there.

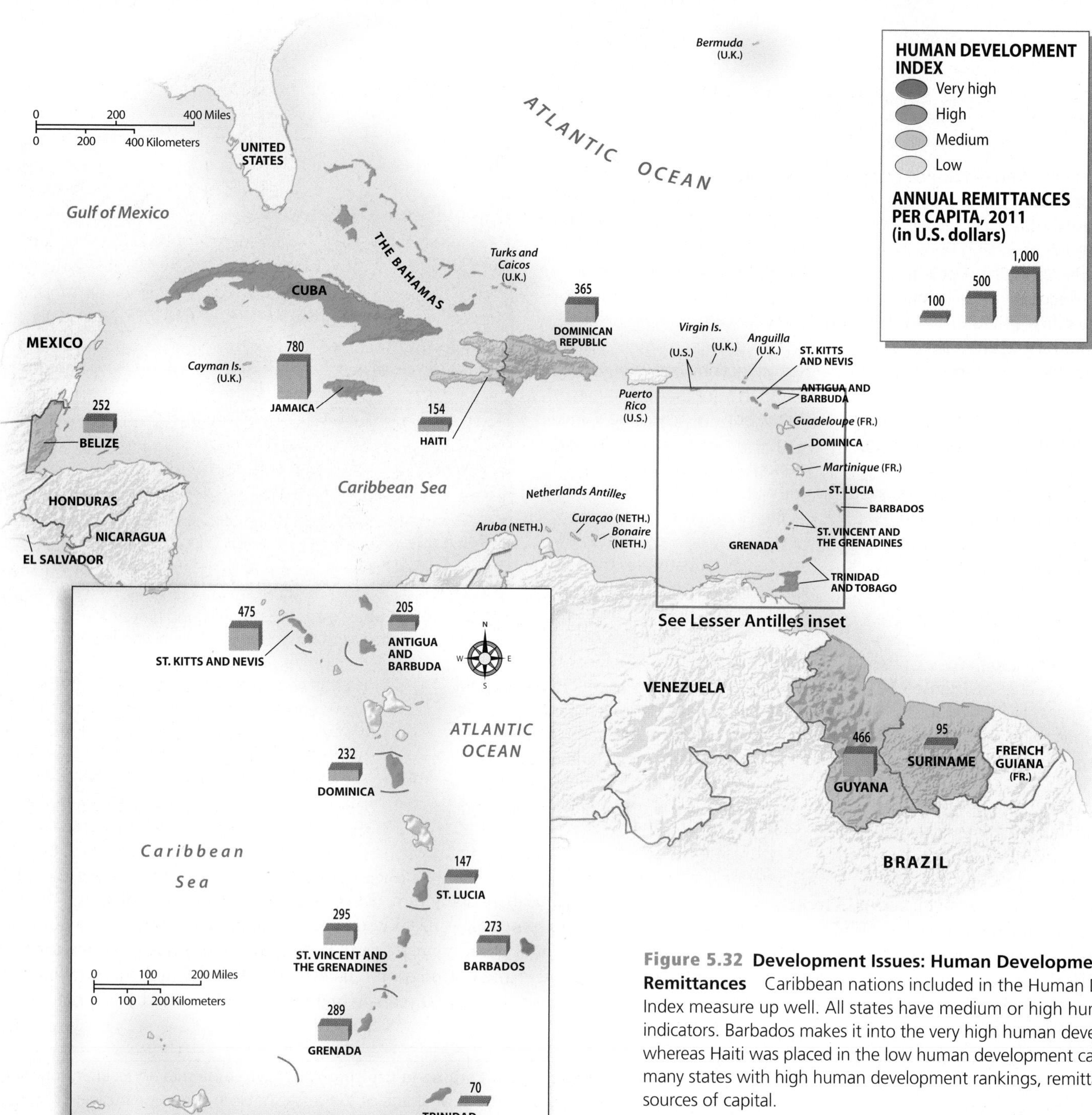

Figure 5.32 Development Issues: Human Development and Remittances Caribbean nations included in the Human Development Index measure up well. All states have medium or high human development indicators. Barbados makes it into the very high human development category, whereas Haiti was placed in the low human development category in 2013. For many states with high human development rankings, remittances are significant sources of capital.

These demographic and social indicators explain why Caribbean nations fare well on the Human Development Index (Figure 5.32). Most fall within the high and medium human development categories. Barbados, ranked 38th in the world in 2013, is in the very high human development category, while Haiti has the lowest ranking, 161st, placing it in the category of low human development. Figure 5.32 also shows that many of these well-ranked states, especially Jamaica, Guyana, and St. Kitts and Nevis, have significant annual per capita inflows of **remittances**, monies sent back home by migrants working overseas. It has been argued that remittances have become extremely important in boosting the overall level of social and economic development in the region. Yet despite real social gains, many inhabitants are chronically underemployed, poorly housed, and perhaps overly dependent on foreign remittances. For rich and poor alike, the temptation to leave the region in search of better opportunities remains.

Gender, Politics, and Culture

The matriarchal (female-dominated) basis of Caribbean households is often singled out as a distinguishing characteristic of the region. The rural custom of men leaving home for seasonal employment tends to nurture strong, self-sufficient female networks. Women typically run the local street markets. With men absent for long stretches, women tend to make household and community decisions. Although this gives women local power, it does not always imply higher status. In rural areas, female status is often undermined by the relative exclusion of women from the cash economy—men earn wages, while women provide subsistence.

As Caribbean society urbanizes, more women are being employed in assembly plants (the garment industry, in particular, prefers to hire women), in data-entry firms, and in tourism. With new employment opportunities, female participation in the labor force has surged; in countries such as Barbados, Haiti, Jamaica, Puerto Rico, and Trinidad and Tobago, more than 40 percent of the workforce is female. Increasingly, women are the principal earners of cash and are more likely to complete secondary education than men. There are also signs of greater political involvement by women. In recent years, Jamaica, Dominica, Trinidad and Tobago, and Guyana have all had female prime ministers.

Today slightly more women than men migrate to the United States, and many seek employment in the globalized care economy as health-care workers, nannies, and eldercare aides. Many feminist scholars argue that the care industry has increasingly become the domain of immigrant women of color throughout North America and Europe. This segmentation of the labor market is driven by a complex mix of income inequalities, racial and gender preferences, and the demands and relatively low status of care work in general.

An estimated 300,000 to 500,000 Caribbean-born health-care professionals, mostly women, work in the United States. More than half are home health-care aides, and another third are registered nurses or technicians (Figure 5.33). Many of these women are reliable senders of remittances. Yet it is also true that these female migrants may leave their own families and children behind due to work demands and visa stipulations. Many Caribbean governments worry about this trend, but it is not one that can be easily addressed.

Education Many Caribbean states have excelled in educating their citizens. Literacy is the norm, and most people receive at least a high school diploma. In many respects, Cuba's educational accomplishments are the most impressive, given the size of the country and its high illiteracy rates in the 1960s. Today nearly all adults are literate. Hispaniola is the obvious contrast to Cuba's success. Although the Dominican Republic has made strides in improving adult literacy (88 percent of adults are literate), half of all Haitian adults are illiterate. Political stability and economic growth have helped the Dominican Republic better its social conditions over the past decade. In fact, many Haitians have crossed the border into the Dominican Republic because conditions, although far from ideal, are much better there than in their homeland.

Figure 5.33 Caribbean Health-Care Worker A Jamaican nurse with her colleagues in a Massachusetts hospital.

Education is expensive for these nations, but it is considered essential for development. Ironically, many states express frustration about training professionals for the benefit of developed countries in a phenomenon called **brain drain**. Brain drain occurs throughout the developing world, especially between former colonies and the mother countries. In the early 1980s, the prime minister of Jamaica complained that 60 percent of his country's newly university-trained workers left for the United States, Canada, and Britain, representing a subsidy to these economies far greater than the foreign aid Jamaica received from them. A World Bank study of skilled migrants revealed that 40 percent of Caribbean migrants living abroad were college educated. For countries such as Guyana, Grenada, Jamaica, St. Vincent and the Grenadines, and Haiti, over 80 percent of the college-educated population will emigrate. No other world region has proportionately this many educated people leaving. Given the small population of many Caribbean territories, each professional person lost to emigration can negatively impact local health care, education, and enterprise. However, despite the high outflow of professionals, many countries are more recently experiencing a return migration of Caribbean peoples from North America and Europe as a **brain gain**. This term refers to the potential of returnees to contribute to the social and economic development of their home country with the experiences they have gained abroad.

Labor-Related Migration Given the region's high educational rates and limited employment opportunities, Caribbean countries have seen their people emigrate for decades. After World War II, better transportation and political developments in the Caribbean produced a surge of migrants to North America. This trend began with Puerto Ricans going to New York in the early 1950s and intensified in the 1960s, with the arrival of nearly half a million Cubans. Since then, large numbers of Dominicans, Haitians, Jamaicans, Trinidadians, and Guyanese have also migrated to North America, typically settling in Miami, New York, Los Angeles, and Toronto. A substantial number of Caribbean migrants go to the United Kingdom, France, and the Netherlands.

Crucial in this exchange of labor is the flow of cash remittances. Immigrants are expected to send something back, especially when immediate family members are left behind. Collectively, these remittances add up; an estimated $3 billion is sent annually to the Dominican Republic by immigrants in the United States, making remittances the country's second leading source of income. Jamaicans and Haitians remit nearly $2 billion annually to their countries. Governments and individuals alike depend on these transnational family networks. Families carefully select the member most likely to succeed abroad, in the hope that money will flow back and help build a base for future immigrants. Labor-related migration has become a standard practice for tens of thousands of households in the region as well as a clear expression of the globalization of labor flows.

Review

5.9 As the Caribbean has shifted out of dependence on agriculture, what other economic sectors have emerged in the region?

5.10 What explains the relatively high levels of social development in the Caribbean, given the region's relative poverty?

KEY TERMS free trade zone (FTZ), offshore banking, capital leakage, remittances, brain drain, brain gain

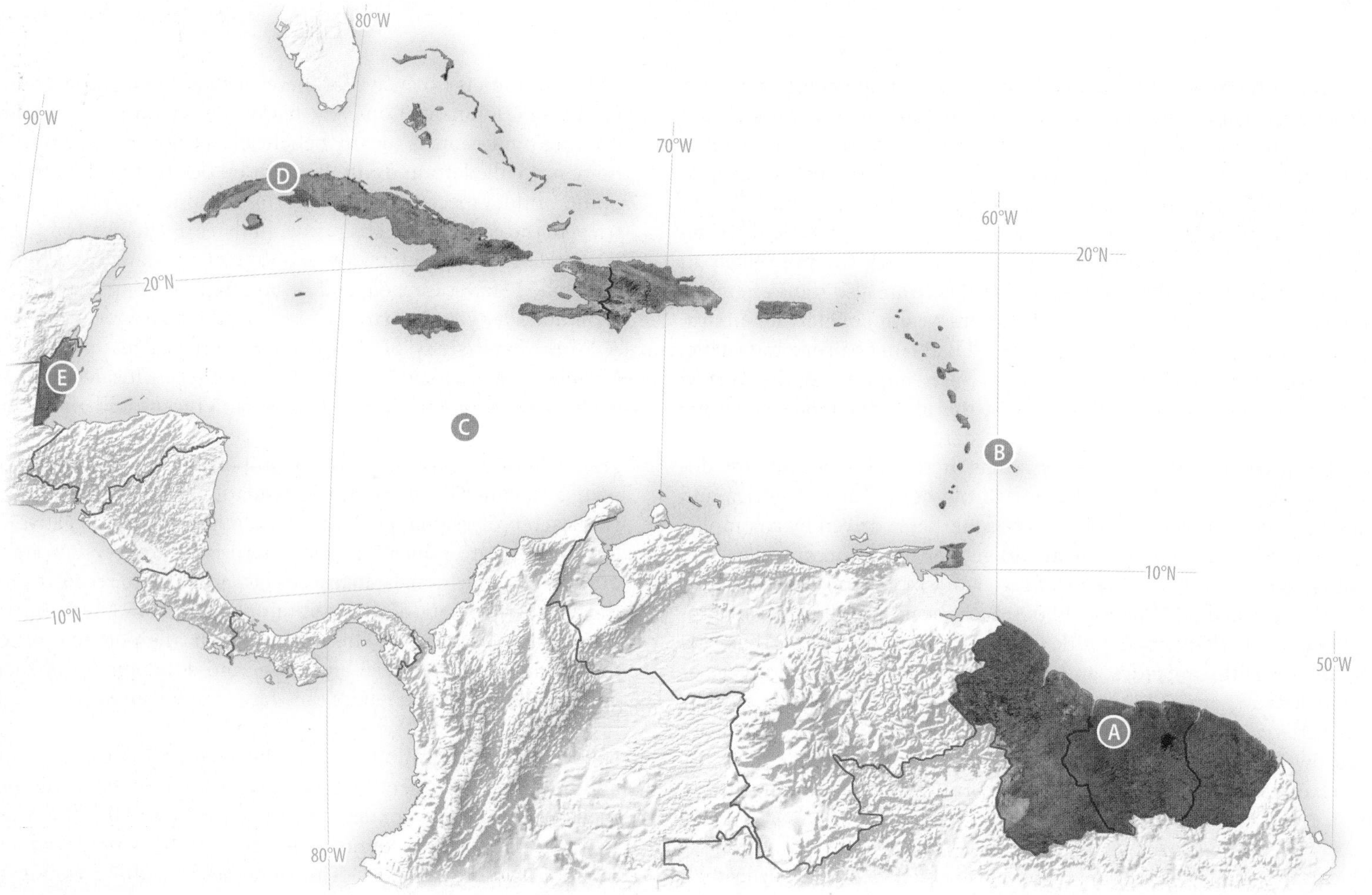

Review

Physical Geography and Environmental Issues

5.1 Differentiate between island and rimland environments and the environmental issues that affect these areas.

This tropical region has been exploited to produce export commodities such as sugar, coffee, and bananas. The region's warm waters and mild climate attract millions of tourists. Yet serious problems with deforestation, soil erosion, and water contamination have degraded urban and rural environments. Climate change poses a serious threat to the region, with the likelihood of more intense hurricanes and sea-level rise.

1. What factors contributed to the maintenance of forest in this Caribbean state?

2. How will climate change impact the Caribbean over the next century?

Population and Settlement

5.2 Summarize the demographic shifts in the Caribbean as population growth slows, urbanization intensifies, emigration abroad continues, and a return migration begins.

Population growth in the Caribbean has slowed over the past two decades; the average woman now has two to three children. Life expectancy is quite high. Most Caribbean people live in cities, but Caribbean cities are not large by world standards. The Caribbean region is noted for high rates of emigration, especially among the highly skilled, who settle in North America and Europe.

3. What factors might explain the slow population growth in the Caribbean country of Barbados?

4. Figure 5.14 shows migration flows within and beyond the Caribbean. What are the reasons behind particular destination preferences?

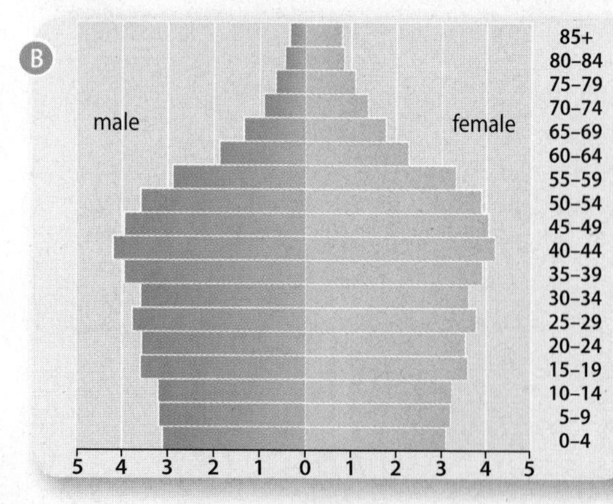

Cultural Coherence and Diversity

5.3 Identify the demographic and cultural implications of the forced transfer of African peoples to the Caribbean and the creation of a neo-African society in the Americas.

The Caribbean was forged through European colonialism and the enslaved labor of millions of Africans. Creolization has resulted in many unique cultural expressions in music, language, and religion. Some view the Caribbean as a neo-Africa, in which African peoples, cultures, and even some agricultural practices dominate, especially in isolated maroon communities.

5. What religious practices transferred from Africa to the Caribbean, and how did they diffuse?

6. What explains the language diversity in the Caribbean?

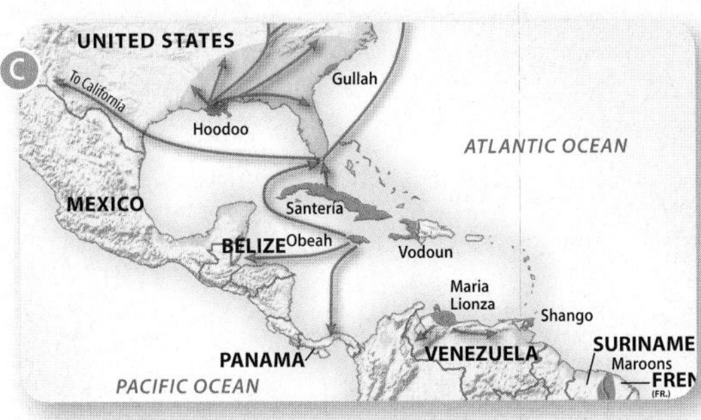

Geopolitical Framework

5.4 Explain why European colonists so aggressively sought control of the Caribbean and why independence in the region came about more gradually than in neighboring Latin America.

Today the region contains 26 independent countries and territories. Even when the Cold War ended, Cuba remained a geopolitical hot spot. Recent developments in U.S.–Cuban relations may bring political and economic change to the Caribbean's largest country. While U.S. influence is strong in this region, other countries such as Venezuela, China, and the United Kingdom also exert influence in the Caribbean.

7. This protected harbor was vital for Spain's control of the Caribbean. Where is this location, and why was this port city important for Spain?

8. Why have U.S. actions in the region been considered neocolonial? What other countries are influencing the region today?

Economic and Social Development

5.5 Describe how the Caribbean is linked to the global economy through offshore banking, emigration, and tourism.

5.6 Suggest reasons why the Caribbean does better in social indicators of development than in economic indicators.

The Caribbean has gradually shifted from being an exporter of primary agricultural resources (especially sugar) to a service and manufacturing economy. Employment opportunities in assembly plants, tourism, and offshore banking have replaced jobs in agriculture. The region's strides in social development, especially in education, health, and the status of women, distinguish it from other developing areas.

9. What environmental, economic, and locational factors contribute to the strength of the tourism economy in the Caribbean?

10. Is reliance on remittances a sign of the Caribbean's isolation from or integration into the global economy?

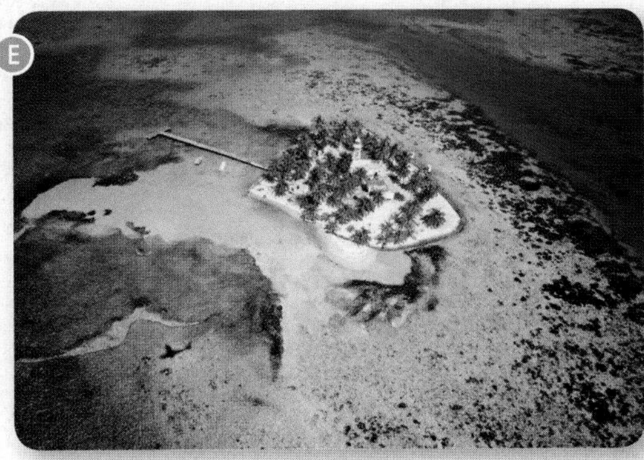

KEY TERMS

African diaspora *(p. 155)*
brain drain *(p. 169)*
brain gain *(p. 169)*
capital leakage *(p. 167)*
Caribbean Community and Common Market (CARICOM) *(p. 163)*
Caribbean diaspora *(p. 152)*
creolization *(p. 157)*
free trade zone (FTZ) *(p. 164)*
Greater Antilles *(p. 145)*
hurricane *(p. 146)*
indentured labor *(p. 156)*
isolated proximity *(p. 142)*
Lesser Antilles *(p. 145)*
maroon *(p. 156)*
mono-crop production *(p. 153)*
Monroe Doctrine *(p. 161)*
neocolonialism *(p. 161)*
offshore banking *(p. 164)*
plantation America *(p. 153)*
remittances *(p. 168)*
rimland *(p. 145)*

DATA ANALYSIS

http://goo.gl/qEpia

Remittances are an important source of revenue for many Caribbean countries. It is important to know not only the total amount a country receives from remittances, but also where the money comes from. Visit the website of the World Bank (http://econ.worldbank.org) and search for data on migration and remittances. Select the most recent bilateral remittance matrix to download a spreadsheet listing countries receiving remittances across the top row and source countries for remittances down the first column.

1. Begin by picking five or six Caribbean countries. Be sure to mix both large and small countries.
2. Create a table with the major source countries and the amounts of remittances sent to each of your selected countries. Now map these patterns.
3. Which countries are the most important sources of remittances, and why? Which countries receive relatively few remittances, and why?
4. What does your map tell you about the major destinations for Caribbean migrants? Write a summary that compares two or more of your selected countries and outlines cultural, geopolitical, and economic factors explaining the patterns you identified.

MasteringGeography™

Looking for additional review and test prep materials? Visit the Study Area in MasteringGeography™ to enhance your geographic literacy, spatial reasoning skills, and understanding of this chapter's content by accessing a variety of resources, including MapMaster interactive maps, geoscience animations, videos, *In the News* RSS feeds, flashcards, web links, self-study quizzes, and an eText version of *Globalization and Diversity*.

Authors' Blogs

Scan to visit the **Author's Blog for field notes, media resources, and chapter updates**

http://gad4blog.wordpress.com/category/latin-america/

Scan to visit the **GeoCurrents Blog**

http://geocurrents.info/category/place/latin-america

6 Sub-Saharan Africa

PHYSICAL GEOGRAPHY AND ENVIRONMENTAL ISSUES

Wood is a main energy source for this region. The Green Belt Movement has led to the planting of millions of trees by rural women throughout the region. In areas such as the Sahel, policy changes that provided ownership or incentives for the protection of trees have increased tree cover.

POPULATION AND SETTLEMENT

Sub-Saharan Africa is demographically young and growing. With 950 million people, its rate of natural increase is 2.6, making it the fastest-growing world region in terms of population. It is also the region hit hardest by HIV/AIDS, which has lowered overall life expectancies in some countries.

CULTURAL COHERENCE AND DIVERSITY

Religious life is important in this region, with large and growing numbers of Muslims and Christians. With a few notable exceptions, religious diversity and tolerance have been distinctive features of this region. However, religious conflict, especially in the Sahel region, has been on the rise.

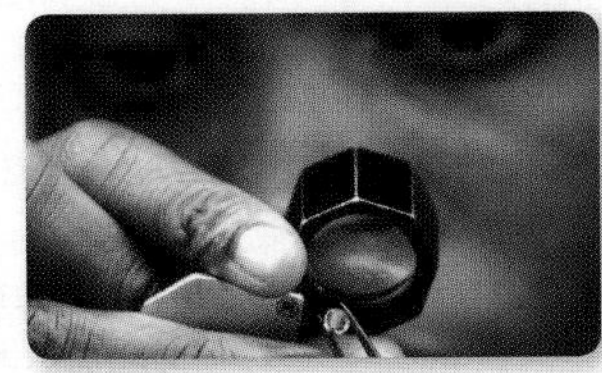

GEOPOLITICAL FRAMEWORK

Most countries gained independence in the 1960s. Many ethnic conflicts emerged as governments struggled for national unity within boundaries drawn by European colonialists. Al Qaeda and ISIL also have footholds in the region, sparking turmoil. There are significant numbers of internally displaced people and refugees.

ECONOMIC AND SOCIAL DEVELOPMENT

The UN's Millennium Development Goals to reduce extreme poverty by 2015 were not met by most states of the region, but progress was made in terms of education, life expectancy, and economic growth. International assistance has increased, and growing global demand for natural resources has led to Asian, European, and North American investment in the extraction of the region's metals and fossil fuels.

◀ The view from Johannesburg's 50-story Carlton Centre (Sub-Saharan Africa's tallest building) shows a sprawling global city that is one of the economic engines for the continent. In the distance are the gold tailings, many of them being reprocessed for trace amounts of gold, as silent reminders of the resource that sparked the growth of this city in the 1880s.

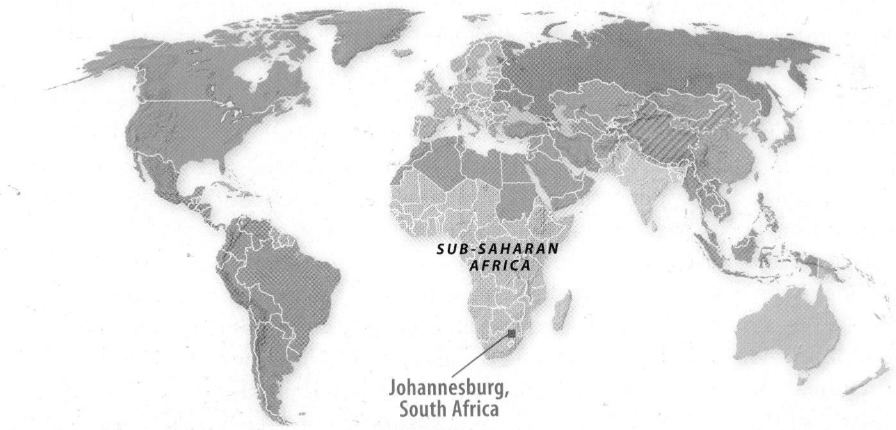

Greater Johannesburg sprawls across a vast plateau studded with the tailings from South African gold mines. With the discovery of gold in 1886, the city grew as thousands searching for riches poured in from around the world. Johannesburg continues to be South Africa's economic engine, and the greater metropolitan area is home to one out of seven South Africans. Sub-Saharan Africa's third-largest metropolitan area, Jozi, as it is often called, boasts some of the region's best infrastructure. Yet as in many African cities struggling to meet the power needs of its rapidly growing population, rolling blackouts have been a problem for the city and for South African industry.

Downtown Johannesburg today reflects the complex history of this global city. Streets are named for English, Afrikaans, and Black African heroes such as Albertina Sisulu, a leader in the struggle against apartheid. Two downtown bridges, one named for Queen Elizabeth and the other for Nelson Mandela, cross the railroad tracks. Nearby is a monument to barrister Mahatma Ghandi—the leader in India's nonviolent independence movement—who practiced law and advocated for civil rights in Johannesburg in the early 1900s. During the apartheid years, Johannesburg was the downtown for white South Africans. Black South African workers were trained in from distant townships such as Soweto, located beyond the gold tailings that can be seen in the distance.

Over the past two decades, the downtown has been transformed. Mostly abandoned by white elites and businesses that have moved to Sandton, the newer commercial center to the north, it is now occupied by black South Africans and black-owned businesses. Government programs to promote black entrepreneurship have produced a black middle class and even many black millionaires who live in Jozi, but unemployment and inequality remain challenges for South Africa.

Since 2000, most of the national economies of Sub-Saharan Africa have been growing faster than their populations, a positive indicator of development.

Learning Objectives

After reading this chapter you should be able to:

6.1 Describe the major ecosystems in the region and how humans have adapted to living in them.

6.2 Outline the environmental issues that challenge Africa south of the Sahara.

6.3 Explain the region's rapid demographic growth and describe the differential impact of diseases such as HIV/AIDS and Ebola on the region.

6.4 Describe the relationship between ethnicity and conflict in this region and the strategies for maintaining peace.

6.5 Summarize various cultural influences of African peoples within the region and globally.

6.6 Trace the colonial history of the region and link colonial policies to postindependence conflicts in the region.

6.7 Assess the roots of African poverty and explain why many of the fastest-growing economies in the world today are in Sub-Saharan Africa.

6.8 Explain how reductions in conflicts can improve educational and social development outcomes in the region.

In general, Africa south of the Sahara is poorer and more rural, and its population is much younger, when compared to Latin America and the Caribbean. Over 950 million people reside in this region, which includes 48 states and 1 territory (Reunion, off the coast of Madagascar). Demographically, this is the world's fastest-growing region (with a 2.6 percent rate of natural increase); in most countries, nearly half the population (43 percent) is under 15 years of age. Income levels are extremely low: 41 percent of the population lives in extreme poverty, surviving on less than $1.25 per day. Life expectancy is only 57 years. Such statistics and the all-too-frequent negative headlines about violence, disease, and poverty might lead to despair. Yet this is also a region of resilience, and many Africans are optimistic about the future. Local and international nongovernmental organizations (NGOs) and various government agencies are improving the quality of life in many parts of the region. In the process, many countries have reduced infant mortality, expanded basic education, and increased food production in the past two decades. One of the most transformational changes has been the rapid diffusion of cell phones, along with innovative applications that improve communication and information sharing. Since 2000, most of the national economies of the region have been growing faster than their populations, which is a positive indicator of development.

Sub-Saharan Africa—that portion of the African continent lying south of the Sahara Desert—is a commonly accepted world region, sharing similar livelihood systems and colonial experiences (Figure 6.1). No common religion, language, philosophy, or political system ever united the area, but loose cultural bonds developed from a variety of lifestyles and idea systems that evolved here. The impact of outsiders also helped to determine the region's identity. Slave traders from Europe, North Africa, and Southwest Asia treated Africans as property; up until the mid-1800s, millions of Africans were taken from the region and sold into slavery. In the late 1800s, the entire African continent was divided by European colonial powers, imposing political boundaries that remain to this day. In the postcolonial period, which began in the 1960s, most countries of the region faced many of the same economic and political challenges.

Figure 6.1 Sub-Saharan Africa This vast world region, which includes 48 states and 1 territory, is often categorized as being divided into the western, central, eastern, and southern subregions (see Table A6.1). The rainforests, tropical savannas, and deserts of Sub-Saharan Africa are home to 950 million people. Much of the region consists of broad plateaus ranging in elevation from 1600 to 6500 feet (500 to 2000 meters). Although the population is growing rapidly, the overall population density is low. Considered one of the least developed world regions, Africa south of the Sahara remains an area rich in natural resources.

In the context of this particular regional division, the major question is how to treat North Africa. Some scholars argue for considering the African continent as one world region. Regional organizations such as the African Union are modern expressions of this continental unity. However, North Africa is more closely linked, both culturally and physically, to Southwest Asia. Arabic is the dominant language and Islam is the leading religion of North Africa. Consequently, North Africans feel more closely connected to the Arab hearth in Southwest Asia than to the Sub-Saharan world.

This chapter focuses on the states south of the Sahara. We preserve political boundaries when delimiting the region so that the Mediterranean states of North Africa, along with Western Sahara and Sudan, are discussed with Southwest Asia (Chapter 7). Before South Sudan's independence from Sudan in 2011, Sudan was Africa's largest state by area. In the more populous and powerful north, Muslim leaders have crafted an Islamic state that is culturally and politically oriented toward North Africa and Southwest Asia, while the new country of South Sudan has more in common with the Christian and animist groups in Sub-Saharan Africa. South Sudan, along with the Sahelian states of Mauritania, Mali, Niger, and Chad, forms the northern boundary of the countries discussed in this chapter.

The region is culturally complex, with dozens of languages spoken in some states. Consequently, most Africans understand and speak several languages. Ethnic identities do not follow the political divisions of Africa, sometimes resulting in deadly ethnic conflict, such as in Rwanda in the mid-1990s and in Nigeria today. Nevertheless, throughout the region peaceful coexistence between distinct ethnic groups is the norm. The cultural significance of European colonizers cannot be ignored: European languages, religions, educational systems, and political ideas were adopted and modified. Yet the daily rhythms of life are far removed from the industrial world. Most Africans still engage in subsistence and cash-crop agriculture (Figure 6.2). The influence of African peoples outside the region is significant; in historical times, the slave trade transferred African peoples, religious systems, and musical traditions throughout the Western Hemisphere. Even today, African-based religious systems are widely practiced in the Caribbean and Latin America.

African economies are growing: From 2009 to 2013, the region's annual growth averaged 4.2 percent, primarily due to high commodity prices, robust domestic demand, rising exports, and steady remittance flows. Yet according to the World Bank, the region's economic output in 2013 was just over 2 percent of global output, even though the region contains 13 percent of the world's population. In 2014, Nigeria became the region's largest economy, followed by South Africa.

Many experts feel that Sub-Saharan Africa has benefited little from its integration (both forced and voluntary) into the global economy. Slavery, colonialism, and export-oriented mining and agriculture served the needs of consumers outside the region, while failing to improve domestic food supplies, infrastructure, and standards of living. Ironically, many of the scholars and politicians concerned by the lack of benefit from economic integration also worry that negative global attitudes about the region have produced a pattern of neglect. Private capital investment in Sub-Saharan Africa lags far behind that in other developing regions, but it is growing, especially in oil-rich states such as Angola, Chad, and Equatorial Guinea. And China has become an increasingly significant investor in the region.

The past decade has witnessed a surge in philanthropic outreach to the region. The Bill and Melinda Gates Foundation, which launched in 2000, has directed billions of dollars to support health care, disease prevention, education, and poverty reduction in Sub-Saharan Africa. Many observers believe that through a combination of internal reforms, better governance, foreign assistance, and foreign investment in infrastructure and technology, social and economic gains are possible for Sub-Saharan Africa. Such investments in social development are paying off, as a dramatic decline in child mortality was achieved between 1990 and 2013.

Physical Geography and Environmental Issues: The Plateau Continent

Sub-Saharan Africa is the largest landmass straddling the equator. Its physical environment is vast in scale and remarkably beautiful, dominated by extensive areas of geologic uplift that formed huge elevated plateaus. The highest elevations are found on the continent's eastern edge, where the **Great Rift Valley** forms a complex upland area of lakes, volcanoes, and deep valleys. In contrast, lowlands prevail in West Africa (see Figure 6.1). Despite this region's immense biodiversity, vast water resources, and wealth of precious minerals, it also has relatively poor soils, widespread disease, and vulnerability to drought.

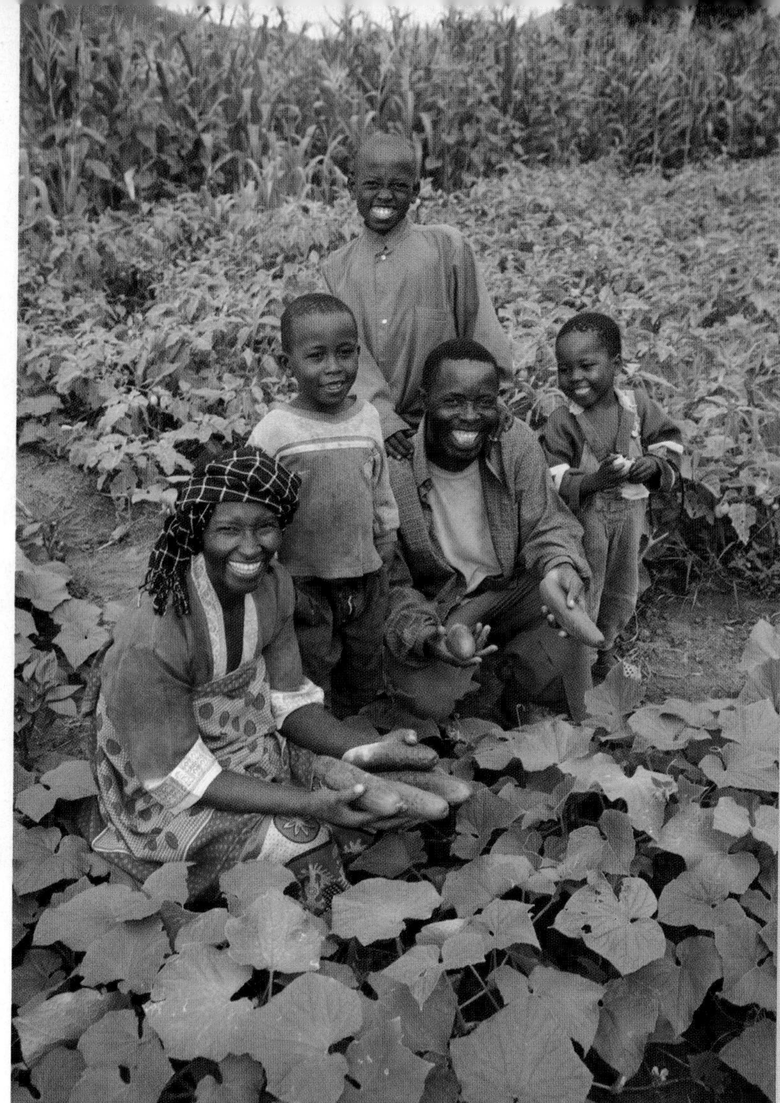

Figure 6.2 Subsistence Farmers in Tanzania A family from Iringa, Tanzania, pose in their garden. Many African households grow food and tend animals for their own consumption, selling any surplus at local markets.

Plateaus and Basins

A series of plateaus and elevated basins dominates the African interior and forms much of the region's unique physical geography (see Figure 6.1). Generally, elevations increase toward the south and east; most of southern and eastern Africa lies well above 2000 feet (600 meters), and sizable areas sit above 5000 feet (1500 meters). These areas are typically referred to as High Africa; Low Africa includes western Africa and much of central Africa. Steep escarpments form where plateaus abruptly end, as illustrated by the majestic Victoria Falls on the Zambezi River (Figure 6.3). Much of southern Africa is rimmed by a landform called the **Great Escarpment** (a high cliff separating two comparatively level areas), which begins in southwestern Angola and ends in northeastern South Africa. Such landforms proved to be a barrier to European colonial settlement in the interior of the continent and, in part, explain the prolonged colonization process.

Figure 6.3 Victoria Falls The Zambezi River descends over Victoria Falls. A fault zone in the African plateau explains the existence of the 360-foot (110-meter) drop. The Zambezi has never been important for transportation, but it is a vital source of hydroelectricity for Zimbabwe, Zambia, and Mozambique.

Though Sub-Saharan Africa is an elevated landmass, it has few significant mountain ranges. The one extensive area of mountainous topography is in Ethiopia, which lies in the northern portion of the Rift Valley zone. Receiving heavy rains in the wet season, the Ethiopian Plateau forms the headwaters of several important rivers—most notably the Blue Nile, which joins the White Nile at Khartoum, Sudan. Volcanic mountains, some quite tall, are located in the southern half of the Rift Valley, such as Tanzania's Mount Kilimanjaro—the region's highest peak at 19,000 feet (5900 meters) (Figure 6.4). However, the high plateaus are dominated by deep valleys rather than dramatic mountain ranges.

Watersheds Africa south of the Sahara does not have the broad, alluvial lowlands that influence patterns of settlement throughout other regions. The four major river systems are the Congo, Niger, Nile, and Zambezi (the Nile is discussed in Chapter 7). Smaller rivers—such as the Orange in South Africa; the Senegal, which divides Mauritania and Senegal; and the Limpopo in Mozambique—are locally important but drain much smaller areas. Ironically, most people believe that Sub-Saharan Africa suffers from water scarcity and tend to discount the size and importance of the watersheds (or catchment areas) drained by these river systems.

The Congo River (also called the Zaire River) is the region's largest watershed in terms of both drainage area and volume of river flow. It is second only to South America's Amazon River in terms of annual flow. The Congo flows across a relatively flat basin more than 1000 feet (300 meters) above sea level, meandering through Africa's largest tropical forest, the Ituri (Figure 6.5). Entry from the Atlantic into the Congo Basin is prevented by a series of rapids and falls, making the Congo River only partially navigable. Despite these limitations, the Congo River has been the major corridor for travel within the Republic of the Congo and the Democratic Republic of the Congo (formerly Zaire); the capitals of these countries, Brazzaville and Kinshasa, rest on opposite sides of the river to form a metropolitan area of nearly 10 million people.

The Niger River is the critical water source for all of West Africa, but especially for the arid countries of Mali and Niger. Beginning in the humid Guinea highlands, the Niger flows to the northeast and then spreads out to form a huge inland delta in Mali before making a great bend southward at the margins of the Sahara near Gao. On the banks of the Niger River are the capitals of Mali (Bamako) and Niger (Niamey) as well as the historical city of Tombouctou (Timbuktu). After flowing through the Sahel, the Niger returns to the humid lowlands of Nigeria, where the Kainji Reservoir temporarily blocks its flow to produce electricity for Africa's most populous state. At the river's end lies the Niger Delta, which is also the center of Nigeria's oil industry. This fertile delta region, home to ethnic groups such as the Igbo and Ogoni, is extremely poor. For decades, conflict in the region has centered on who benefits from the region's oil and on the serious environmental degradation that has resulted from oil and gas extraction.

The considerably smaller Zambezi River begins in Angola and flows east, spilling over an escarpment at Victoria Falls before finally reaching Mozambique and the Indian Ocean. More than other rivers in the region, the Zambezi is a major supplier of commercial energy.

Sub-Saharan Africa's two largest hydroelectric installations are located on this river.

Soils With a few major exceptions, Sub-Saharan Africa's soils are relatively infertile. Generally speaking, fertile soils are young soils, deposited in recent geologic time by rivers, volcanoes, glaciers, or windstorms. In older soils—especially those located in moist tropical environments—natural processes tend to wash out most plant nutrients over time. Over most of the region, the agents of soil renewal are largely absent.

Portions of Sub-Saharan Africa are, however, noted for their natural soil fertility, and not surprisingly, these areas support denser settlement. Some of the most fertile soils are in the Great Rift Valley, made productive by the area's volcanic activity. The population densities of rural Rwanda and Burundi, for example, are partially explained by the highly productive soils. The same can be said for highland Ethiopia, which supports the region's second-largest population. The Lake Victoria lowlands and central highlands of Kenya are also noted for their sizable populations and productive agricultural bases.

The drier grassland and semidesert areas feature a soil type called *alfisols*. High in aluminum and iron content, these red soils have greater fertility than comparable soils found in wetter zones. This helps to explain why farmers tend to plant in drier areas, such as the Sahel, despite the risk of drought. Many agronomists suggest that with irrigation, the southern African countries of Zambia and Zimbabwe could greatly increase commercial grain production on these soils.

Climate and Vegetation

Most of Sub-Saharan Africa lies in the tropical latitudes. Only the far south of the continent extends into the subtropical and temperate belts. Much of the region averages high temperatures from 70 to 80°F (22 to 28°C) year-round (Figure 6.6). Rainfall, more than temperature, determines the different vegetation belts that characterize the region. Addis Ababa, Ethiopia, and Walvis Bay, Namibia, have similar average temperatures (see the climographs in Figure 6.6), but Addis Ababa is in the moist highlands and receives nearly 50 inches (127 cm) of rainfall annually, whereas Walvis Bay is in the Namibian Desert and receives less than 1 inch (2.5 cm).

Figure 6.5 Congo River Africa's largest river by volume, the mighty Congo River flows through the Ituri rainforest in the Democratic Republic of the Congo.

The three main biomes of the region are tropical forests, savannas, and deserts.

Tropical Forests The center of Sub-Saharan Africa falls in the tropical wet climate zone. The world's second largest expanse of equatorial rainforest, the Ituri, lies in the Congo Basin, extending from the Atlantic Coast of Gabon two-thirds of the way across the continent, including the northern portions of the Republic of the Congo and the Democratic Republic of the Congo. Conditions here are constantly warm to hot, and precipitation falls year-round (see the climograph for Kisangani in Figure 6.6).

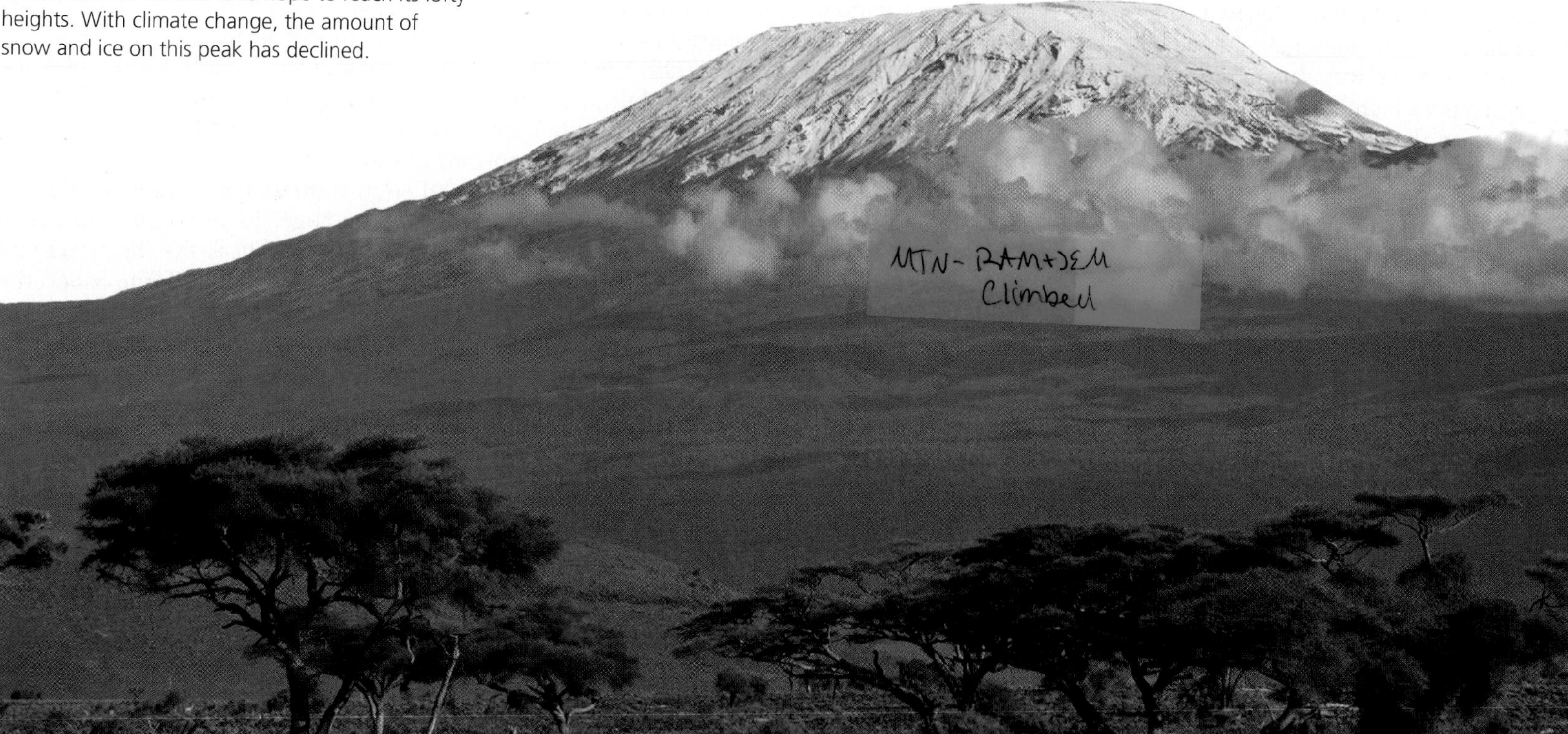

Figure 6.4 Mount Kilimanjaro Rising from the tropical plains and capped in snow, Africa's highest peak is a popular destination for tourists who hope to reach its lofty heights. With climate change, the amount of snow and ice on this peak has declined.

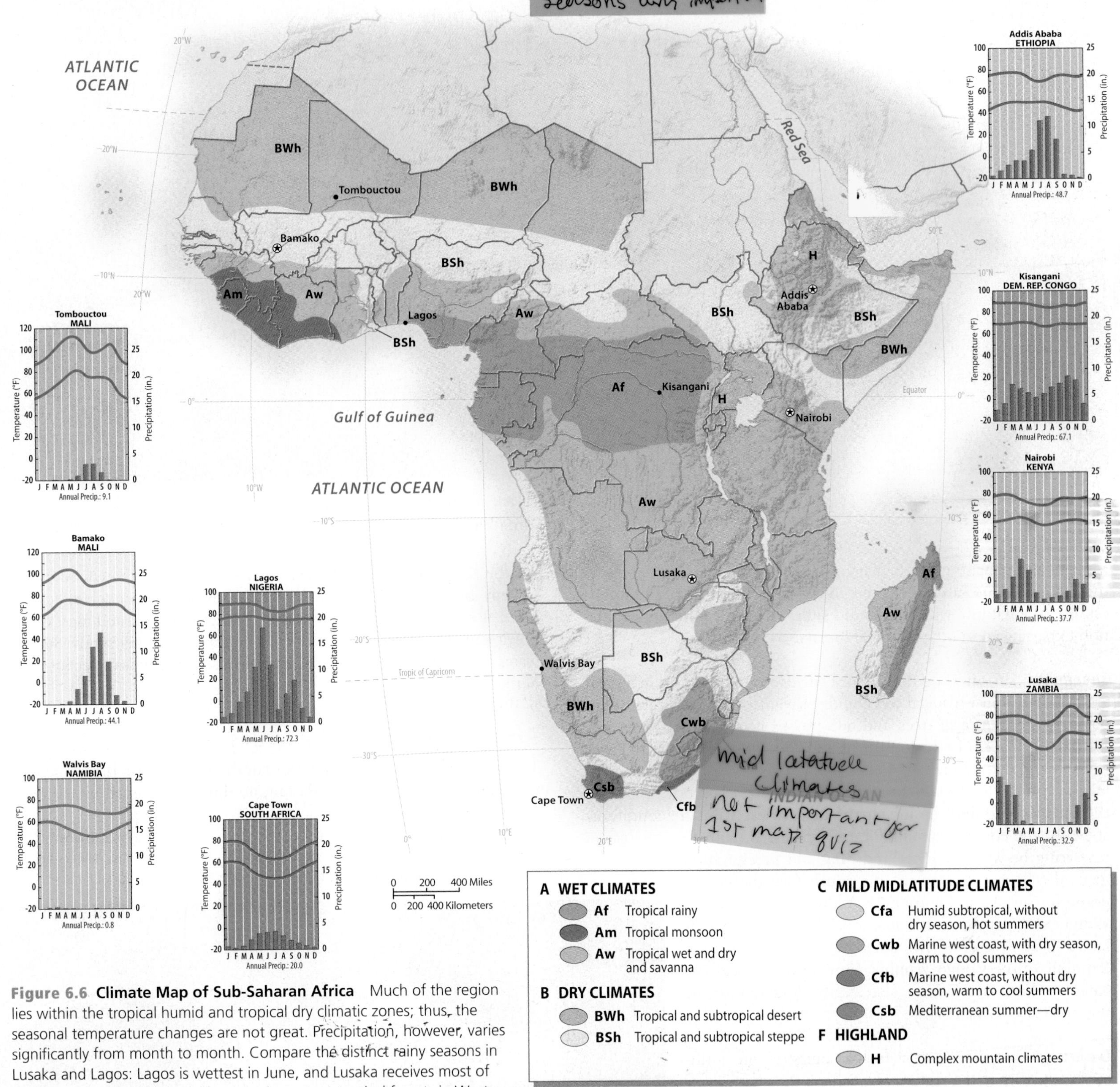

A WET CLIMATES
- **Af** Tropical rainy
- **Am** Tropical monsoon
- **Aw** Tropical wet and dry and savanna

B DRY CLIMATES
- **BWh** Tropical and subtropical desert
- **BSh** Tropical and subtropical steppe

C MILD MIDLATITUDE CLIMATES
- **Cfa** Humid subtropical, without dry season, hot summers
- **Cwb** Marine west coast, with dry season, warm to cool summers
- **Cfb** Marine west coast, without dry season, warm to cool summers
- **Csb** Mediterranean summer—dry

F HIGHLAND
- **H** Complex mountain climates

Figure 6.6 Climate Map of Sub-Saharan Africa Much of the region lies within the tropical humid and tropical dry climatic zones; thus, the seasonal temperature changes are not great. Precipitation, however, varies significantly from month to month. Compare the distinct rainy seasons in Lusaka and Lagos: Lagos is wettest in June, and Lusaka receives most of its rain in January. Although there are important tropical forests in West and Central Africa (coinciding with the tropical wet and monsoon climate zones), vegetation in much of the region is tropical savanna.

Commercial logging and agricultural clearing have degraded the western and southern fringes of the Ituri, but much of this vast forest is still intact. The Ituri has, so far, fared much better than Southeast Asia and Latin America in terms of tropical deforestation. Major national parks such as Okapi and Virunga have been created in the Democratic Republic of the Congo. Virunga National Park—one of the oldest in Africa—sits on the eastern fringe of the Ituri forest and is home to the endangered mountain gorillas. Poor infrastructure and political chaos in the Democratic Republic over the past two decades have made large-scale logging impossible, but have also made conservation difficult. Due to regional conflict, parks such as Virunga have been repeatedly taken over by rebel groups, and park rangers have

Figure 6.7 African Savannas Buffalo gather at a watering hole in Zambezi National Park in Zimbabwe. The savannas of southern Africa are a noted habitat for the region's large mammals, such as buffalo, elephant, zebra, and lion.

been killed; poaching has become a means of survival for people in the region. In the future, it seems likely that Central Africa's rainforest, and the wildlife within it, could suffer the same kind of degradation experienced in other tropical forests.

Savannas Surrounding the Central African rainforest belt in a great arc lie Africa's vast tropical wet and dry savannas. Savannas are dominated by a mixture of trees and tall grasses in the wetter zones immediately adjacent to the forest belt and by shorter grasses with fewer trees in the drier zones. North of the equatorial belt, rain generally falls only from May to October. The farther north you travel, the less the total rainfall and the longer the dry season. Climatic conditions south of the equator are similar, only reversed, with the wet season occurring between October and May and precipitation generally decreasing toward the south (see the climograph for Lusaka in Figure 6.6). A larger area of wet savanna exists south of the equator, with extensive woodlands located in southern portions of the Democratic Republic of the Congo, Zambia, Zimbabwe, and eastern Angola. These savannas are also a critical habitat for the region's large fauna (Figure 6.7).

Deserts Major deserts exist in the southern and northern boundaries of the region. The Sahara, the world's largest desert and one of its driest, crosses the landmass from the Atlantic coast of Mauritania all the way to the Red Sea coast of Sudan. A narrow belt of desert extends to the south and east of the Sahara, wrapping around the **Horn of Africa** (the northeastern corner that includes Somalia, Ethiopia, Djibouti, and Eritrea) and pushing as far as eastern and northern Kenya. An even drier zone is found in southwestern Africa. In the striking red dunes of the Namib Desert of coastal Namibia, rainfall is a rare event, although temperatures are usually mild (Figure 6.8). Inland from the Namib lies the Kalahari Desert. Because it receives slightly more than 10 inches (25 cm) of rain a year, most of the Kalahari is not dry enough to be classified as a true desert. Its rainy season, however, is brief, and most of the precipitation is immediately absorbed by the underlying sands. Surface water is thus scarce, giving the Kalahari a desert-like appearance for most of the year.

Africa's Environmental Issues

The prevailing perception of Africa south of the Sahara is one of environmental scarcity and degradation, no doubt reinforced by televised

Figure 6.8 Namib Desert Arid Namibia has one of the region's lowest population densities, yet its coastal dunes attract foreign tourists. Hikers are seen scaling the orange-colored dunes in Namib-Naukluft National Park.

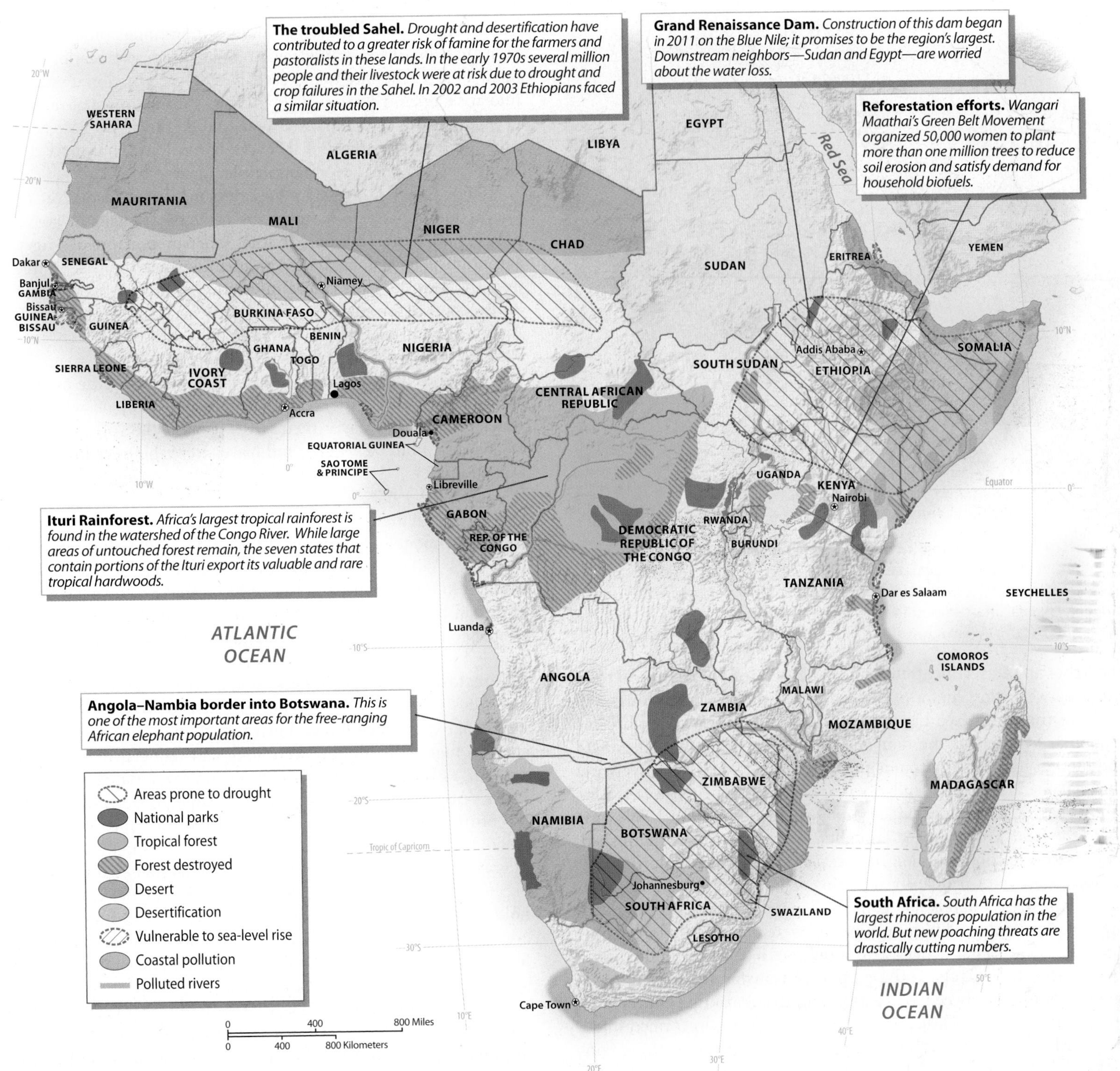

Figure 6.9 Environmental Issues in Sub-Saharan Africa Given the immense size of this world region, it is difficult to generalize about environmental problems. Dependence on trees for fuel places strain on forests and wooded savannas. In semiarid regions such as the Sahel, population pressures and land-use practices have led to desertification. Yet Sub-Saharan Africa also supports the most impressive array of wildlife, especially large mammals, on Earth.

images of drought-ravaged regions and starving children. Single explanations such as rapid population growth and colonial exploitation cannot fully capture the complexity of Africa's environmental issues or the ways that people have adapted to living in marginal ecosystems with debilitating tropical diseases. Because most of Sub-Saharan Africa's population is rural, earning its livelihood directly from the land, sudden environmental changes can have devastating effects on household income and food consumption.

As Figure 6.9 illustrates, deforestation and **desertification**—the expansion of desert-like conditions as a result of human-induced degradation—are commonplace. Sub-Saharan Africa is also vulnerable to drought, most notably in the Horn of Africa, parts of southern Africa, and the Sahel. Many scientists fear that drought will become more frequent and prolonged under global climate change. Yet the region is also home to some of the most impressive wildlife conservation in the world, which is a source of pride and revenue for many African nations.

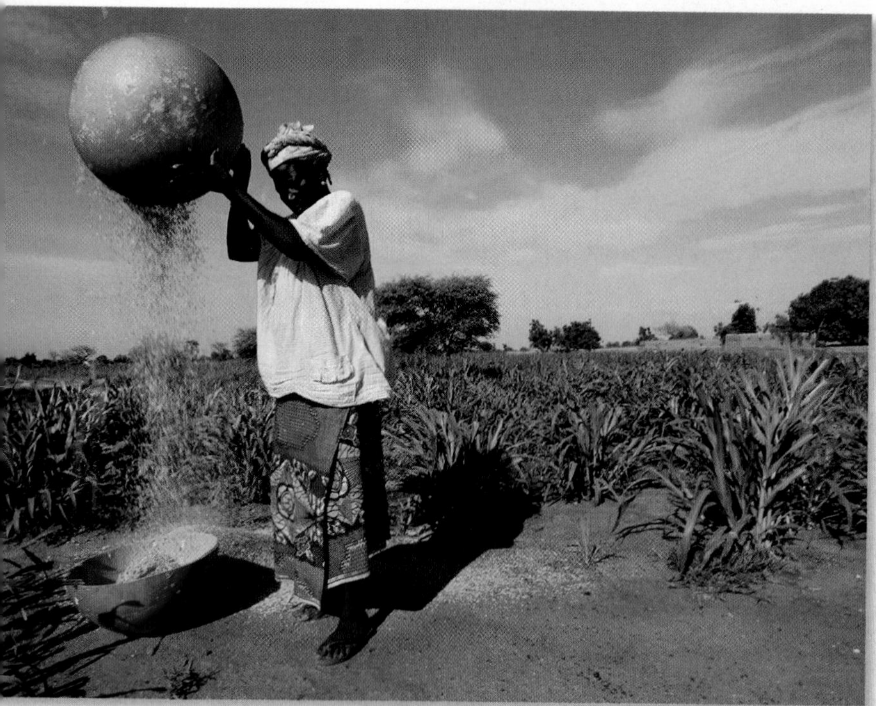

Figure 6.10 The Sahel in Bloom A woman prepares millet grains grown near the city of Maradi, Niger. The soils of the Sahel are fertile, and peasant farmers can produce a surplus with adequate rain. However, in times of drought, crop failures can lead to famine in this region.

The Sahel and Desertification The **Sahel** is a zone of ecological transition between the Sahara to the north and the wetter savannas and forest in the south (see Figure 6.9). In the 1970s, the Sahel became a popular symbol for the dangers of unchecked population growth and human-induced environmental degradation when a relatively wet period came to an abrupt end. Six years of drought (1968–1974) were followed by a second prolonged drought during the mid-1980s, ravaging the land. During these droughts, rivers in the area diminished, and desert-like conditions began to expand southward. Unfortunately, tens of millions of people lived in this area, and farmers and pastoralists whose livelihoods had come to depend on the more abundant precipitation of the relatively wet period were temporarily forced out.

Life in the Sahel depends on a delicate balance of limited rain, drought-resistant plants, and a pattern of **transhumance** (the movement of animals between wet-season and dry-season pasture). What appears to be desert wasteland in April or May is transformed into a lush garden of millet, sorghum, and peanuts after the drenching rains of June. Relatively free of the tropical diseases found in the wetter zones to the south, Sahelian soils are quite fertile, which helps to explain why people continue to live there despite the unreliable rainfall patterns (Figure 6.10).

The main practices cited in Sahel desertification are the expansion of agriculture and overgrazing, leading to the loss of natural vegetation and declines in soil fertility. For example, French colonial authorities forced villagers to grow peanuts as an export crop, a policy continued by the newly independent states of the region. However, peanuts tend to deplete several key soil nutrients, which means that peanut fields are often abandoned after a few years as cultivators move on to fresh sites. Harvesting this crop also turns up the soil at the onset of the dry season, leading to accelerated wind erosion of valuable topsoil.

Overgrazing by livestock, another traditional product of the region, has also been implicated in Sahelian desertification. Development agencies hoping to increase livestock production dug deep wells into areas previously unused by herders through most of the year. The new supplies of water allowed year-round grazing in places that, over time, could not withstand it. Large barren circles around each new well began to appear even on satellite images.

Some Sahelian areas are experiencing some vegetation recovery thanks to simple actions taken by farmers, changes in government policy, and better rainfall. In the Sahelian portion of Niger, local agronomists have documented an unanticipated increase in tree cover over the past 35 years. More interesting still, increases in tree cover have occurred in some of the most densely populated rural areas. After a drought in 1984, farmers began to actively protect rather than clearing them from their fields, including the nitrogen-fixing *goa* tree, which had disappeared from many villages. During the rainy season, the goa tree loses its leaves, so it does not compete with crops for water or sun. The leaves themselves fertilize the soil. Sahelian farmers also use branches, pods, and leaves from the trees for fuel and for animal fodder.

Until the 1990s, all trees were considered property of the state of Niger, thus giving farmers little incentive to protect them. Since then, the government has recognized the value of allowing individuals to own trees. Villages that protect their trees are much greener and more resilient in times of drought than villages that do not. The Sahel is still poor and prone to drought, but as the case of Niger shows, relatively simple conservation practices can have a positive impact.

Deforestation Although Sub-Saharan Africa still contains extensive forests, much of the region is either grasslands or agricultural lands that were once forest. Lush forests that once existed in places such as highland Ethiopia were long ago reduced to a few remnant patches, as local populations rely on such woodlands for their daily needs. Tropical savannas, which cover large portions of the region to the north and south of the tropical rainforest zone, are dotted with woodlands. For many people of the region, savannah deforestation is of greater local concern than the commercial logging of the rainforest because **biofuels**—wood or wood-derived charcoal used for household energy needs, especially cooking—are the leading source of energy for many rural settlements. Loss of woody vegetation has resulted in extensive hardship, especially for women and children who must spend many hours a day looking for wood.

In some countries, village women have organized into community-based NGOs to plant trees and create greenbelts to meet ongoing fuel needs. Kenya boasts one of the most successful efforts, led by Wangari Maathai. Maathai's Green Belt Movement has nurseries in 4000 local communities. Since the group's beginning in 1977, millions of trees have been planted. In those areas, village women now spend less time collecting fuel, and local environments have improved. Kenya's success has drawn interest from other African countries, spurring a Pan-African Green Belt Movement largely organized through NGOs interested in biofuel generation, protection of the environment, and empowerment of women. In 2004, Maathai was awarded a Nobel Peace Prize for her contribution to sustainable development, democracy, and peace. She died in 2011, but the Green Belt Movement remains a powerful force in the region.

Destruction of tropical rainforests through logging is most evident in the fringes of Central Africa's Ituri (see Figure 6.9). Given its

WORKING TOWARD SUSTAINABILITY

Can Bamboo Reduce Deforestation in Africa?

Google Earth (MG) Virtual Tour Video http://goo.gl/1PaV3K

For the near future, wood fuels will be a major contributor to Africa's energy needs. But what can be done to reduce the pressure on the region's forests beyond planting more trees, as the Green Belt Movement has done? Growing bamboo may be a solution. Bamboo grows on every continent and in a tremendous variety of climates, with several species native to Africa. The International Network for Bamboo and Rattan (INBAR) has several pilot programs in Ethiopia and Ghana to promote bamboo as an alternative to wood and charcoal.

Bamboo's Advantages Bamboo is a fast-growing grass, and once it matures (after five to six years), it can be harvested annually for the decades-long life of the plant. Bamboo burns more cleanly than wood and can be converted into charcoal, a preferred cooking fuel. Beyond fuel needs, bamboo is a perennial plant that helps prevent soil erosion when planted on hillsides or along riverbanks. More important, it does not consume much water, a critical resource in drought-prone countries such as Ethiopia. Throughout Asia, bamboo is commercially harvested as a lightweight, but remarkably strong building and flooring material used mostly for domestic markets. Bamboo floors and furniture are also marketed as sustainable or "green" alternatives to hardwoods in many developed countries. However, bamboo has never been commercially important for Sub-Saharan Africa.

Changing Attitudes and Landscapes To change this, INBAR provides bamboo seedlings, trains people to manage bamboo forests, and teaches villagers to build kilns to convert bamboo into charcoal. Some of the technology, such as energy-efficient cooking stoves fueled by bamboo charcoal, comes from China, and cultural resistance persists, especially in Ghana, where wood fuel is more available. Also, bamboo won't grow in the region's driest climates, where biofuels are most needed.

But bamboo is being adopted in countries such as Ethiopia and Mozambique as an inexpensive and sustainable solution to meet pressing demands for fuel and construction materials (Figure 6.1.1). Sustainable jobs are being created as well: Inexpensive bamboo boats are being built in Ethiopia to ferry tourists, while in Ghana a government program promoting the use of local materials such as bamboo to construct schools, market stalls, public toilets, and affordable housing provides employment and reduces the importation of building materials.

1. Which countries in the region could benefit from increased bamboo production?
2. Besides fuel, what are the other uses of bamboo?

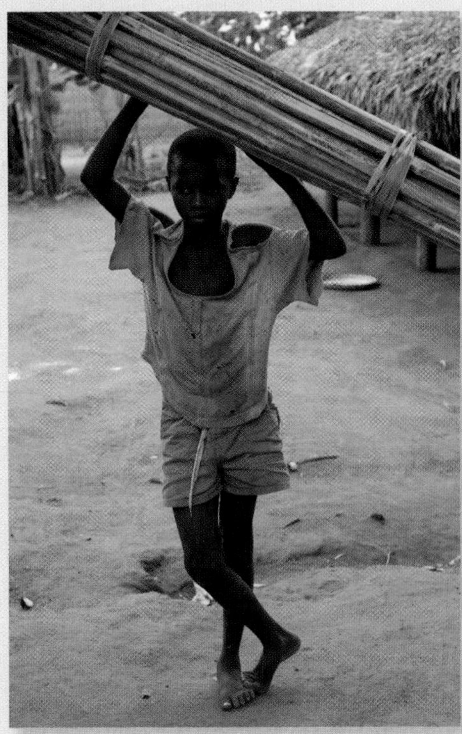

Figure 6.1.1 Building with Bamboo A young boy in Mozambique carries bamboo poles used as construction materials. Many believe that fast-growing bamboo is a better alternative to reliance on wood fuels and lumber.

vast size and the relatively small number of people living there, however, the Ituri is less threatened than other forest areas. Two smaller rainforests—one along the Atlantic coast from Sierra Leone to western Ghana and the other along the eastern coast of the island of Madagascar—have nearly disappeared due to commercial logging and agricultural clearing. Madagascar's western dry forests have also suffered serious degradation in the past three decades. Deforestation in Madagascar is especially worrisome because the island forms a unique environment with a large number of native species—most notably, the charismatic lemurs. In order to address the biofuel needs in such areas, NGOs are experimenting with the introduction of bamboo (see *Working Toward Sustainability: Can Bamboo Reduce Deforestation in Africa?*).

Energy Issues The people of Sub-Saharan Africa suffer from serious energy shortages. At the same time, foreign investors are actively developing the region's supplies of oil and natural gas, mostly for export. Many Sub-Saharan states have oil and natural gas; major producers such as Nigeria and Angola are even members of the Organization of Petroleum Exporting Countries (OPEC; Figure 6.11). More recently, countries such as the Ivory Coast, Tanzania, and Mozambique have developed their natural gas reserves for domestic consumption and export. Yet for most Africans, wood and charcoal (labeled as combustible renewables) account for the majority of total energy production. Figure 6.11 shows the 21 states for which the World Bank estimates what percentage of national energy production these biofuels contribute. Even though Angola is a major oil producer, biofuels supply more than half of the country's energy. In Nigeria, biofuels account for 84 percent of the national energy supply. That figure is over 90 percent in large countries such as Ethiopia and the Democratic Republic of the Congo. This is why energy production places tremendous strain on forests and vegetation and why so many countries are developing alternatives such as hydroelectricity and solar power. Another environmental issue for the majority who rely on biofuels is the smoke that fills homes and causes respiratory problems.

Developing oil and natural gas reserves is not a sure path to economic development, and some observers even call it a curse for Sub-Saharan Africa. In Nigeria, politicians and oil executives have prospered from oil revenue. But in the Niger Delta, where oil was first extracted

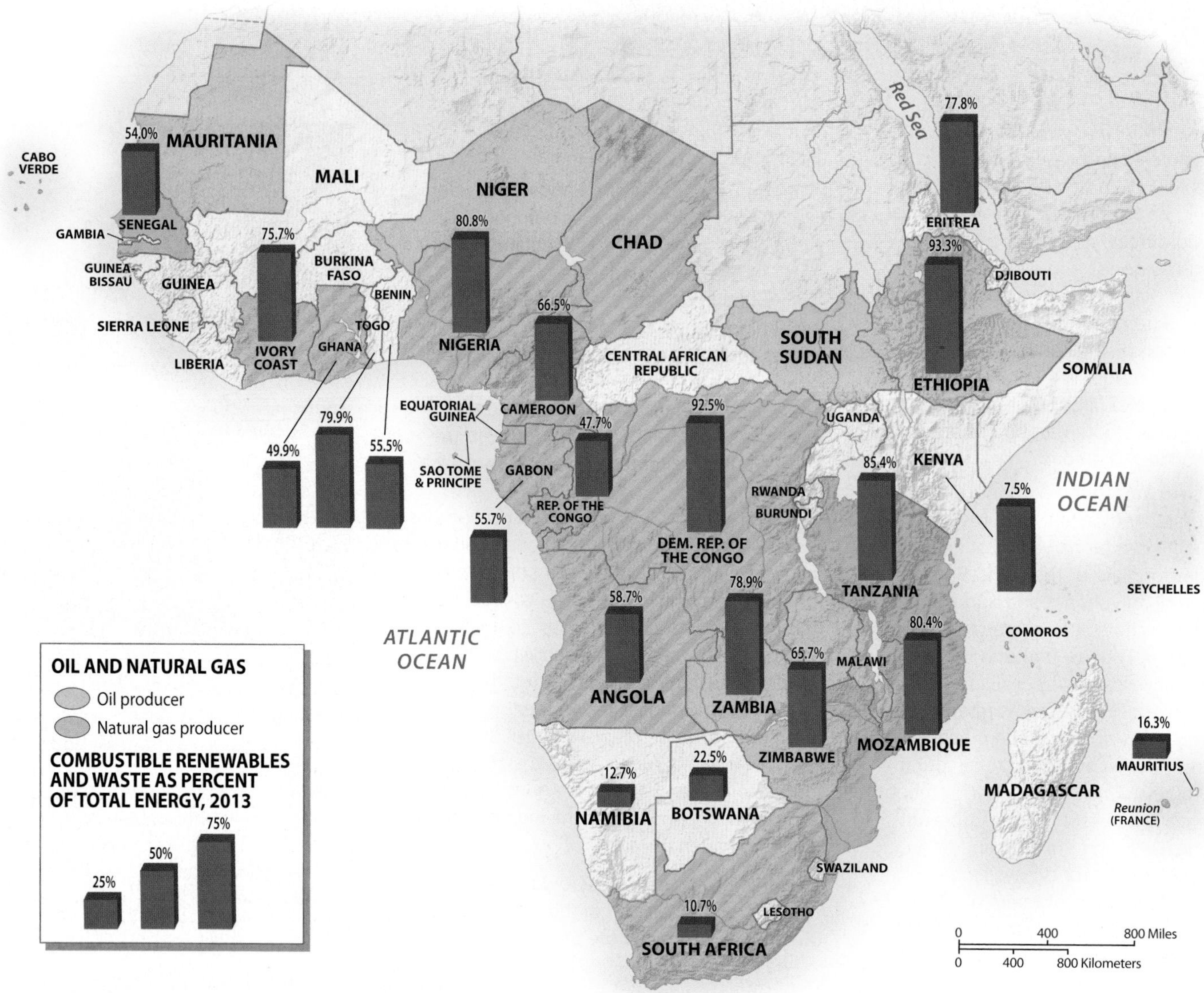

Figure 6.11 Energy Production in Sub-Saharan Africa The region has many states that produce oil and natural gas. Two of the largest producers, Nigeria and Angola, are OPEC members. Yet many states still obtain most of their total energy from burning wood and agricultural waste. **Q: Which states in this region are less dependent on biofuels, and why?**

over 50 years ago, many places lack roads, electricity, and schools. Moreover, careless and unregulated oil extraction has grossly degraded the delta ecosystem (Figure 6.12). As geographer Michael Watts has observed about the delta, oil has been "a dark tale of neglect and unremitting misery." Not all oil production leads to such misery, but Nigeria is a cautionary tale about the limits of oil's ability to foster development.

Wildlife Conservation Sub-Saharan Africa is famous for its wildlife. No other world region has such abundance and diversity of large mammals. The survival of wildlife here reflects, to some extent, the historically low human population density and the fact that sleeping sickness and other diseases have kept people and their livestock out of many areas. In addition, many African peoples have developed various ways of successfully coexisting with wildlife, and about 12 percent of the region is included in nationally protected lands.

However, as is true elsewhere in the world, wildlife is declining in much of Sub-Saharan Africa. The most noted wildlife reserves are in East Africa (Kenya and Tanzania) and southern Africa (South Africa, Zimbabwe, Namibia, and Botswana). These reserves are vital for wildlife protection and are major tourist attractions. Wildlife reserves in southern Africa now seem to be the most secure, and in Zimbabwe elephant populations are considered too large for the land to sustain. Yet throughout the region, population pressure, political instability, and poverty make the maintenance of large wildlife reserves difficult, even though many countries benefit from wildlife tourism.

In 1989, a worldwide ban on ivory trade was imposed as part of the Convention on International Trade in Endangered Species (CITES). Although several African states, such as Kenya, lobbied hard for the ban, others, including Zimbabwe, Namibia, and Botswana, complained that their herds were growing and the sale of ivory helped to fund conservation efforts. Conservationists feared that lifting the ban would bring on a new wave of poaching and illegal trade. However, in the late 1990s, the ban was lifted so that some southern African states could sell down their inventories of elephant ivory confiscated from poachers, and limited sales continued. The last legal auction of elephant ivory was in 2008; officials are reluctant to hold more auctions due to a corresponding increase in poaching.

Today the rhinoceros is especially threatened. The illegal market for rhino horn is lucrative, with most of the demand coming from Vietnam and China. Wildly valued for its questionable medicinal

Figure 6.12 Oil Pollution in the Niger Delta Thousands of Nigerians hack into oil pipelines to steal crude oil, refine it, and then sell it locally or abroad. This informal practice leaves the delta horribly polluted and cuts deeply into Nigeria's national oil production.

properties, ground rhino horn can fetch $65,000 per kilo in the black market. According to geographer Elizabeth Lunstrum, 80 percent of Sub-Saharan Africa's rhinos are in South Africa, and half of those are in one place—Kruger National Park (Figure 6.13), which borders Mozambique. In 2008, rangers noted a spike in shot and dehorned rhinos, and by 2013, 1000 rhinos were reported killed in South Africa, 600 of them in Kruger National Park. Many poachers stage their attacks from Mozambique, which makes this an international as well as a domestic issue. With poaching rates still rising, South African officials are deploying drones and military personnel in an attempt to stop the slaughter. Even with these resources, protecting these endangered animals is proving difficult.

Figure 6.13 South Africa's Rhinos A white rhinoceros moves across the savannas of Kruger National Park, home to more than half the world's rhinos. A sudden spike in poaching since 2008 has led to dramatic losses. In 2015, it was estimated that four rhinos a day were being killed. The fate of the rhino, a protected and endangered species, in the wild is being challenged by this latest wave of poaching.

Climate Change and Vulnerability in Sub-Saharan Africa

Global climate change poses extreme risks for Sub-Saharan Africa due to the region's poverty, recurrent droughts, and overdependence on rain-fed agriculture. Sub-Saharan Africa is the lowest emitter of greenhouse gases in the world, but it is likely to experience greater-than-average human vulnerability to climate change because of the region's limited ability to either respond or adapt to environmental changes. Most vulnerable are arid and semiarid regions such as the Sahel and the Horn of Africa, some grasslands, and the coastal lowlands of West Africa and Angola.

Climate change models suggest that parts of highland East Africa and equatorial Central Africa may receive more rainfall in the future. Thus, some lands that are currently marginal for farming might become more productive. These effects are likely to be offset, however, by declining agricultural productivity in the Sahel and in the grasslands of southern Africa, especially in Zambia and Zimbabwe. Drier grassland areas could deplete wildlife populations, which are a major factor behind the growing tourist economy. As in Latin America, higher temperatures in the tropics may result in the expansion of vector-borne diseases such as malaria and dengue fever into the highlands, where they have until now been relatively rare. Given the region's relatively high elevations, the negative consequences of a rising sea level would be felt mostly on the West African coast (Senegal, Gambia, Sierra Leone, Nigeria, Cameroon, and Gabon).

Even without the threat of climate change, famine stalks many areas of Africa. The Famine Early Warning Systems (FEWS) network monitors food insecurity throughout the developing world, but especially in Sub-Saharan Africa. By tracking rainfall, vegetation cover, food production, food prices, and regional conflict, the network maps food security along a continuum from food secure to famine. Figure 6.14 shows the status of

Figure 6.14 Food Insecurity in West Africa Anticipating areas of food insecurity, based on the timing and amount of rainfall and changes in vegetation/crop cover, is the mission of the FEWS network. Since the late 1980s, FEWS has mapped areas of potential famine, especially in the Sahelian region of Sub-Saharan Africa. (Famine Early Warning System (FEWS) Network, April to June 2013 Report, U.S. Agency for International Development)

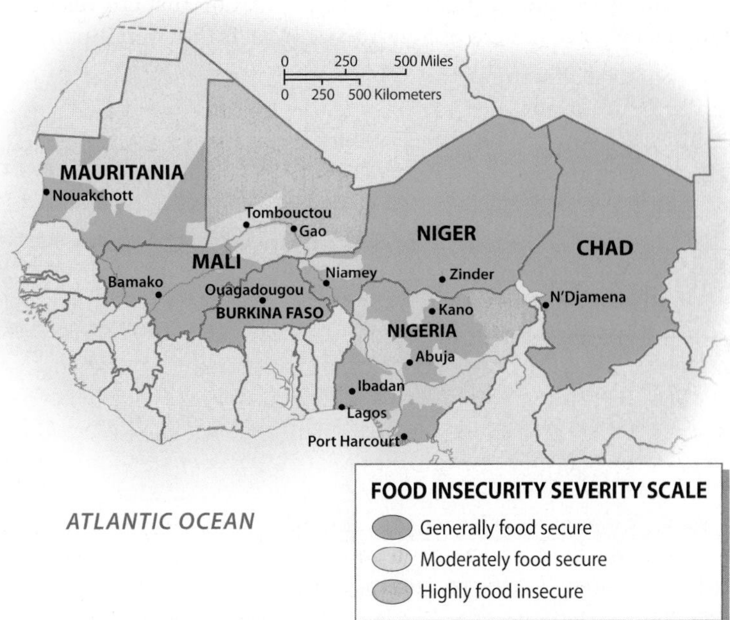

food security in West Africa during 2013; food insecurity was greatest in Mali and northeast Nigeria, in part due to conflict in these areas. When early warning systems are taken seriously, prompt relief efforts can reduce famine and mortality rates.

Review

6.1 What economic and environmental factors contribute to reliance on biofuels in this region?

6.2 Summarize the factors that make Sub-Saharan Africans especially vulnerable to climate change.

KEY TERMS Great Rift Valley, Great Escarpment, Horn of Africa, desertification, Sahel, transhumance, biofuel

Population and Settlement: Young and Restless

Sub-Saharan Africa's population is growing quickly. By 2050, the population is expected to reach 2 billion, double the current population. It is also a young population, with 43 percent of the people under age 15, compared to just 16 percent for more developed countries. Only 3 percent of the region's population is older than age 65. Families tend to be large, with the average woman having five children (Table A6.1). However, child and maternal mortality rates are also high, although child mortality rates have declined substantially in the past two decades. The most troubling indicator for the region is its low life expectancy, which dropped to 50 years in 2008 (in part due to the AIDS epidemic) and is currently estimated at 57 years. Life expectancy in other developing nations is much better: India's is 68 years and China's is 75. The growth of cities is also a major trend in this world region. In 1980, an estimated 23 percent of the population lived in cities; now the figure is 38 percent.

Behind these demographic facts lie complex differences in settlement patterns, livelihoods, belief systems, and access to health care. Although the region is experiencing rapid population growth, Sub-Saharan Africa is not densely populated. The entire region holds 950 million people—roughly half the population that crowds into the much smaller land area of South Asia. In fact, the overall population density of the region (38 people per square kilometer or 99 people per square mile) is similar to that of the United States (33 people per square kilometre or 86 people per square mile). Just six states—Nigeria, Ethiopia, the Democratic Republic of the Congo, South Africa, Tanzania, and Kenya—account for half of the region's population. Some states (such as Rwanda and Mauritius) have very high population densities, whereas others (Namibia and Botswana) are sparsely settled. However, population density is not correlated with overall development. Mauritius is a densely settled island nation that is well governed and relatively prosperous, and the same could be said for the arid and sparsely settled state of Botswana. Population density, however, does provide an indicator of the relative population pressures of states in the region, so it is included in Table A6.1.

Demographic Trends and Disease Challenges

The demographic profile of the region is changing. One positive change is the decline in child mortality due to greater access to primary health care and new disease prevention efforts. Gone are the days when 1 child in 5 did not live past his or her fifth birthday. Today the child mortality figure is closer to 1 in 10—still high by world standards, but a considerable improvement. Also, life expectancy figures bottomed out in the 2000s due to the devastating impact of HIV/AIDS and are now on the rise. Finally, like people in other world regions, Africans are moving to cities, which usually leads to fertility declines. Family size in South Africa, one of the more urbanized large countries, is half the regional average.

Figure 6.15 compares the population pyramids of Ethiopia and South Africa. Ethiopia has the classic broad-based pyramid of a demographically growing and youthful country. Most Ethiopian women have four children, and the rate of natural increase is 2.3 percent. There are nearly even numbers of men and women, and only 4 percent of the people are over the age of 65 (life expectancy is now 64). In contrast, the South African population pyramid tapers down, reflecting the country's smaller family size (the average woman has two or three children). One unusual aspect in this graphic is the smaller number of women in their 30s and early 40s compared to men the same age. This is due to the disproportionate impact of AIDS on women in Africa (which will be discussed later). South Africa has more people over the age of 65 (6 percent), but its life expectancy is 61 years, less than that of Ethiopia. This, too, is attributable to the AIDS epidemic, which hit

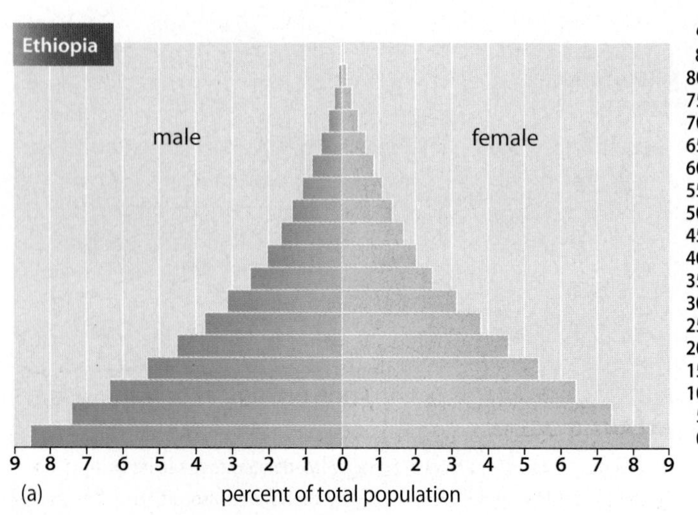

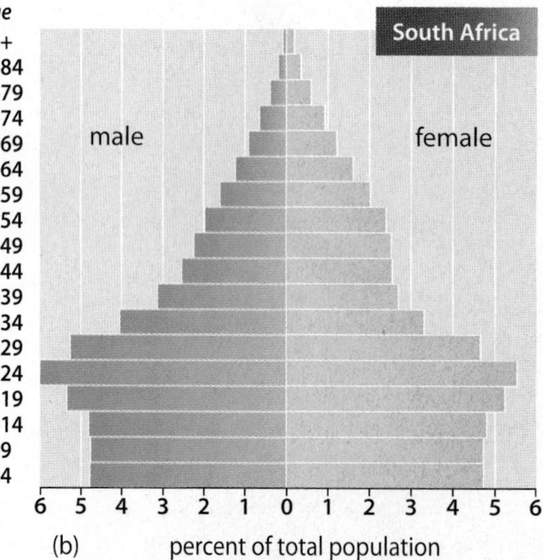

Figure 6.15 Ethiopian and South African Population Pyramids The contrasting demographic profiles of (a) the more rural and rapidly growing Ethiopia and (b) the more urbanized and slower-growing South Africa are shown here. A lost generation of South African women in their 30s is also evident, due to the disproportionate impact of HIV/AIDS on the women of South Africa.

Figure 6.16 Large Families The average total fertility rate for Africa south of the Sahara is five children per woman.

southern Africa with deadly force in the 1990s. Thankfully, infection rates are now on the decline in many states.

Family Size A continued preference for large families is the basis for the region's demographic growth. In the 1960s, many areas in the developing world had total fertility rates (TFRs) of 5.0 or higher. Today Sub-Saharan Africa, at 5.1, is the only region with such a high TFR. A combination of cultural practices, rural lifestyles, child mortality, and economic realities encourages large families (Figure 6.16). Yet average family sizes are coming down; as recently as 1996, the regional TFR was 6.0.

Throughout the region, large families guarantee a family's lineage and status. Even now, most women marry young, typically as teenagers, maximizing their childbearing years. Demographers often point to the limited formal education available to women as another factor contributing to high fertility. Religious affiliation has little bearing on the region's fertility rates; Muslim, Christian, and animist communities all have similarly high birth rates.

The everyday realities of rural life make large families an asset. Children are an important source of labor; from tending crops and livestock to gathering wood, they add more to the household economy than they take. Also, for the poorest places in the developing world such as Sub-Saharan Africa, children are seen as social security: When parents' health falters, they expect their grown children to care for them.

National policies began shifting in the 1980s. For the first time, government officials argued that smaller families and slower population growth were needed for social and economic development. Other factors are bringing down the rate of natural increase. As African states slowly become more urban, there is a corresponding decline in family size—a pattern seen throughout the world. Tragically, declines in natural increase were also occurring as a result of AIDS.

The Disease Factor: Malaria, HIV/AIDS, and Ebola

Historically, the hazards of malaria and other tropical diseases such as sleeping sickness limited European settlement in the tropical portions of Sub-Saharan Africa. It was only in the 1850s, when European doctors discovered that a daily dose of quinine could protect against malaria, that the balance of power in Africa radically shifted. Explorers immediately began to penetrate the interior of the continent, while merchants and expeditionary forces moved inland from the coast. The first imperial claims soon followed, culminating in colonial division of Africa in the 1870s (discussed later in this chapter).

Malaria Malaria has been a scourge in this region for centuries. Transmitted from infected individuals to others via the anopheles mosquito, malaria causes high fever, severe headache, and, in the worst cases, death. The World Health Organization estimates that 200 million people contract malaria each year, resulting in 600,000 deaths. The majority of infections and deaths occur in Sub-Saharan Africa. Since 2000, African governments, NGOs, and foreign aid sources have increased spending to reduce the threat of infection. Presently, a malaria vaccine does not exist, but research in this area is promising. Medication helps in many cases but is not reliable over the long term, and insecticide use has led to mosquito resistance to the chemicals. The most effective tool to reduce infection has been the distribution of insecticide-treated mosquito nets to millions of African homes. That, along with rapid diagnostic tests and access to medication once infected, has cut infections and related deaths by one-third since 2000.

Malaria and poverty are closely related in Sub-Saharan Africa, with many of the poorest tropical countries experiencing higher infection rates. West and Central Africa are the areas hardest hit. The Democratic Republic of the Congo and Nigeria, along with India, account for 40 percent of worldwide malaria cases, and 40 percent of malaria deaths occur in these two African nations.

HIV/AIDS Now in its fourth decade, HIV/AIDS has been one of the deadliest epidemics in modern human history, yet it is beginning to be tamed. This is especially welcome news for Sub-Saharan Africa, home to 70 percent of the 35 million people living with HIV/AIDS. Human immunodeficiency virus (HIV) is the virus that can lead to acquired immunodeficiency syndrome (AIDS). The human body cannot get rid of HIV, but antiretroviral drugs can suppress it and keep it from becoming AIDS.

The HIV/AIDS virus is thought to have originated in the forests of the Congo, possibly crossing over from chimpanzees to humans sometime in the late 1950s. Yet it was not until the 1980s that the impact of the disease was first felt. In Sub-Saharan Africa, as in much of the developing world, the virus is transmitted mostly by unprotected heterosexual activity or from mother to child during the birth process or through breastfeeding. A long-standing pattern of seasonal male labor migration helped to spread the disease. So, too, did lack of education, inadequate testing early on in the epidemic, and disempowerment of women. Consequently, women bear a disproportionate burden of the HIV/AIDS epidemic. They account for approximately 60 percent of HIV infections, and they are usually the caregivers for those who are infected. Until the late 1990s, many African governments were unwilling to acknowledge publicly the severity of the situation or to discuss frankly the measures necessary for prevention.

Southern Africa is ground zero for the AIDS epidemic; the countries with the highest HIV prevalence (South Africa, Swaziland, Lesotho, Botswana, Namibia, Mozambique, Zambia, Zimbabwe, and Malawi) are all located there. In South Africa, the most populous state in southern Africa, 7 million people (nearly one person in five age 15–49) are infected with HIV/AIDS. The rates of infection in neighboring

Figure 6.17 HIV/AIDS Activism in South Africa School children march through Cape Town to celebrate the anniversary of the establishment of a program to distribute antiretroviral drugs to AIDS patients. The program, run by Doctors Without Borders, began distributing these life-prolonging drugs in 2001.

Botswana, Lesotho, and Swaziland are even higher. Infection rates in other African states are lower, but still high by world standards. In Kenya, an estimated 6 percent of the 15–49 age group has HIV or AIDS. By comparison, only 0.6 percent of the same age group in North America is infected.

The social and economic implications of this epidemic have been profound. Life expectancy rates tumbled—in a few places even dropping to the early 40s. AIDS typically hits the portion of the population that is most economically productive. The time lost to care for sick family members and the outlay of workers' compensation benefits have reduced economic productivity and overwhelmed public services in hard-hit areas. The disease makes no class distinctions: Countries are losing both peasant farmers and educated professionals (doctors, engineers, and teachers). Many countries struggle to care for millions of children orphaned by AIDS.

After three devastating decades, there are finally hopeful signs. Prevention measures are widely taught, and treatment with a mix of available drugs means that HIV infection is now manageable and not a death sentence. Due to international financial support and national outreach efforts, Sub-Saharan Africa now has many more health facilities that offer HIV testing and counseling. Some 15 million people receive life-prolonging drugs. Prevention services in prenatal clinics provide the majority of pregnant HIV-positive women with antiretroviral drugs to prevent transmission of the virus to their babies. Political activism and changes in sexual practices, driven by educational campaigns, more condom use, and higher rates of male circumcision, have prevented hundreds of thousands of new HIV cases (Figure 6.17).

Ebola The 2014–2015 Ebola outbreak in West Africa attracted international attention due to the highly contagious and deadly nature of this disease—once infected, an individual can die within a week or two without intensive medical care. The 2014 outbreak was the largest Ebola epidemic in history, affecting multiple countries in West Africa and resulting in some 30,000 confirmed cases. A rare disease, Ebola was first identified in 1976 along the Ebola River in the Democratic Republic of the Congo. There had been isolated outbreaks before, but the 2014 outbreak was by far the largest and deadliest (Figure 6.18), although as of late 2015 there were no new cases in Liberia and Sierra Leone.

International organizations such as Medecins Sans Frontieres (Doctors Without Borders) and various government agencies contributed money, medical personnel, and equipment to fight this epidemic, fearing that it could spread and potentially infect millions. There is no medication to treat Ebola, and it is critical that people not come into contact with the blood or bodily fluids of those infected. This often meant quarantining infected people, developing stringent procedures for health-care personnel, conducting education efforts, and abandoning traditional burial practices to reduce the spread of infection. International cooperation and the efforts on the ground in Sierra Leone, Liberia, and Guinea worked in the 2014–2015 outbreak, and worst-case scenarios were not realized. How such diseases spread and are treated has much to do with settlement patterns, quality of infrastructure, and availability of health care in this poor but developing region.

Patterns of Settlement and Land Use

Because of the dominance of rural settlements in Sub-Saharan Africa, people are widely scattered throughout the region (Figure 6.19). Population concentrations are highest in West Africa, highland East Africa, and the eastern half of South Africa. The first two areas have some of the region's best soils, and native systems of permanent agriculture developed there. In South Africa, the more densely settled east is a result of an urbanized economy based on mining as well as the forced concentration of black South Africans into eastern homelands.

Figure 6.18 Ebola Outbreak in West Africa Liberia, Sierra Leone and Guinea were the states most impacted by the recent Ebola outbreak. In Liberia, cases were found throughout the country but especially in the the capital city of Monrovia. Sierra Leone's cases were also more concentrated near the cities of Freetown and Kenama. Guinea, which had fewer deaths, had a concentration of rural cases near the Liberia border.

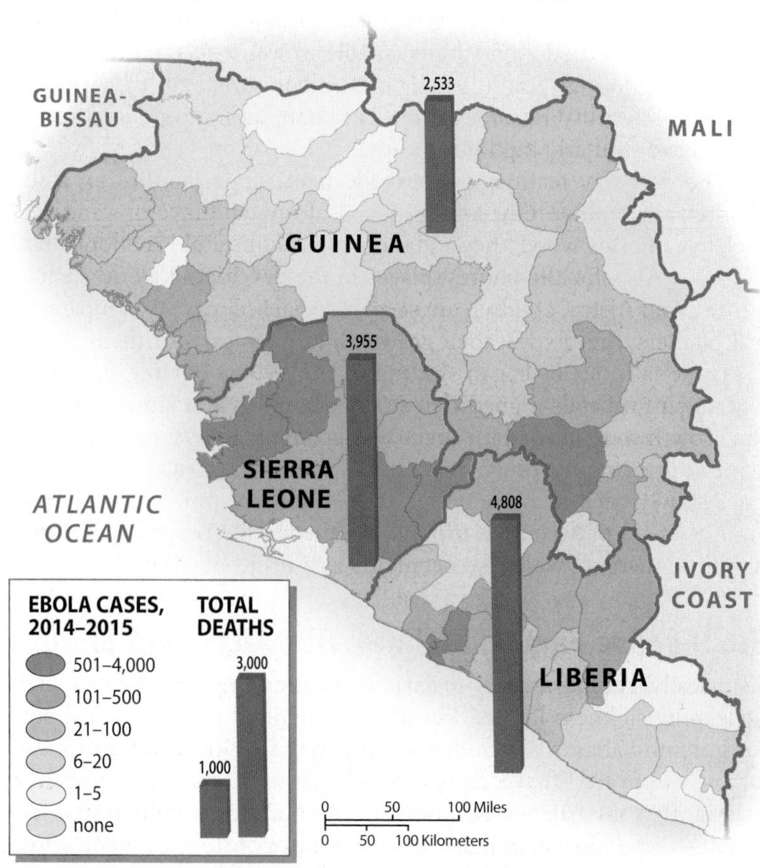

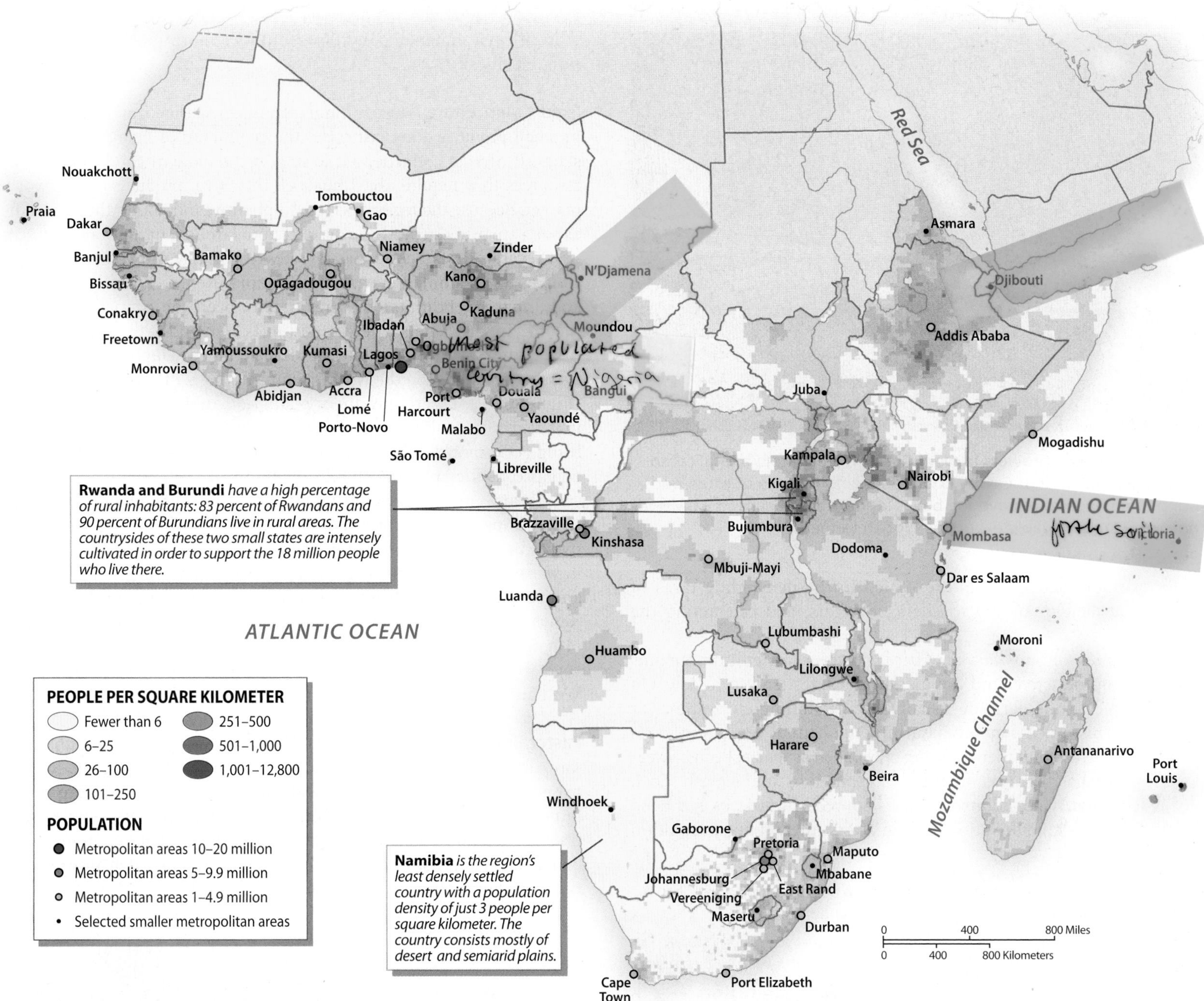

Figure 6.19 Population Map of Sub-Saharan Africa The majority of people in this region live in rural areas. Some of these rural zones, however, are densely settled, such as West Africa and the East African highlands. Major urban centers, especially in South Africa and Nigeria, support millions. There is only one megacity in the region with more than 10 million residents (Lagos), but more than two dozen cities have more than 1 million residents. **Q: What factors contribute to the extremely low population density in the southwest corner of the continent?**

As more Africans move to cities, settlement patterns are becoming more concentrated. Towns that were once small administrative centers for colonial elites have grown into major cities. The region even has its own megacity, Lagos, estimated at 13 to 17 million residents. Throughout the continent, African cities are growing faster than rural areas. Before examining the Sub-Saharan urban scene, however, a more detailed discussion of rural subsistence is needed.

Agricultural Subsistence The staple crops over most of Sub-Saharan Africa are millet, sorghum, and corn (maize) as well as a variety of tubers and root crops such as yams. Irrigated rice is widely grown in West Africa and Madagascar. Geographer Judith Carney, in her book *Black Rice*, documents how African slaves introduced rice cultivation to the Americas. Corn, in contrast, was introduced to Africa from the Americas through the slave trade and quickly grew to become a basic food. Wheat and barley are grown in the higher elevations of Ethiopia and South Africa. Intermixed with subsistence foods are a variety of export crops—coffee, tea, rubber, bananas, cacao, cotton, and peanuts—grown in distinct ecological zones, often in some of the best soils.

In areas that support high annual crop yields, population densities are greater. In parts of humid West Africa, for example, the yam became the dominant subsistence crop. The Igbos' mastery of yam farming enabled them to produce more food and live in denser permanent settlements in the southeastern corner of present-day Nigeria.

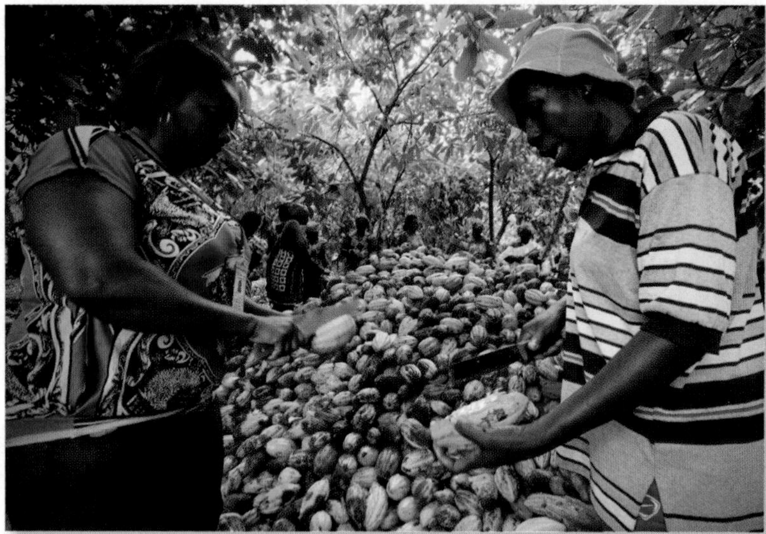

Figure 6.20 Cacao Growers in Ivory Coast Women from a local cacao farmers association in the Ivory Coast work among a pile of cacao pods, slicing them open to remove the beans inside. Cacao beans (the basis for chocolate) are a valuable export for Ivorian farmers.

Much of traditional Igbo culture is tied to the demanding tasks of clearing the fields, tending the delicate plants, and celebrating the harvest.

Over much of Sub-Saharan Africa, however, agriculture remains relatively unproductive, and population densities tend to be low. On poorer tropical soils, farming usually entails shifting cultivation (or **swidden**). This process involves burning the natural vegetation to release fertilizing ash and then planting crops such as maize, beans, sweet potatoes, bananas, papayas, manioc, yams, melons, and squash. Each plot is temporarily abandoned once its source of nutrients has been exhausted. Swidden cultivation is often a very finely tuned adaptation to local environmental conditions, but it cannot support high population densities.

Export Agriculture Agricultural exports, whether from large estates or small producers, are critical to the economies of many Sub-Saharan states. If African countries are to import the modern goods and energy resources they require, they must sell their own products on the world market. Because the region has few competitive industries, the bulk of its exports are primary products derived from farming, mining, and forestry.

Several African countries rely heavily on one or two export crops. Coffee, for example, is vital for Ethiopia, Kenya, Rwanda, Burundi, and Tanzania. Peanuts have historically been the primary source of income in the Sahel, whereas cotton is tremendously important for the Central African Republic and South Sudan. Ghana and the Ivory Coast have long been the world's main suppliers of cacao for chocolate (Figure 6.20), Liberia produces plantation rubber, and many farmers in Nigeria specialize in palm oil. The export of such products can bring good money when commodity prices are high, but when prices collapse, as they periodically do, economic devastation may follow.

Nontraditional agricultural exports that depend on significant capital inputs and refrigerated air transport have emerged in the last two to three decades. One industry is floriculture for the plant and cut flower industry. Here the highland tropical climate of Kenya, Ethiopia, and South Africa is advantageous. After Colombia in South America, Kenya is the largest exporter of cut flowers in the world, and most of its exports go to Europe. Similarly, the European market for fresh vegetables and fruits in the winter is being met by some producers in West and East Africa.

Pastoralism Animal husbandry is extremely important in Sub-Saharan Africa, particularly in semiarid zones. Camels and goats are the principal animals in the Sahel and the Horn of Africa (Figure 6.21), but cattle are most common farther south. Many African peoples have traditionally specialized in cattle raising and are often tied in mutually beneficial relationships with neighboring farmers. Such **pastoralists** typically graze their stock on the stubble of harvested fields during

Figure 6.21 Pastoralists in the Horn of Africa Two Ethiopian girls lead a herd of camels, which are well suited for the arid conditions of eastern Ethiopia and Somalia. Pastoralists rely on their camels for milk, transport, and trade.

the dry season and then move them to drier uncultivated areas during the wet season, when the pastures turn green. Farmers thus have their fields fertilized by the manure of the pastoralists' stock, while the pastoralists find good dry-season grazing. At the same time, the nomads can trade their animal products for grain and other goods of the sedentary world. Several pastoral peoples of East Africa, particularly the Masai of the Tanzanian–Kenyan borderlands, are noted for their extreme reliance on cattle and general (but never complete) independence from agriculture.

Large expanses of Sub-Saharan Africa have been off-limits to cattle because of infestations of **tsetse flies**, which spread sleeping sickness to cattle, humans, and some wildlife. Environments containing brush or woodland necessary for tsetse fly survival can support large numbers of wild animals immune to sleeping sickness, but vulnerable cattle cannot thrive. Tsetse fly eradication programs are currently reducing the threat, and cattle raising is spreading into areas where it was previously restricted. This benefits African herders but may threaten the continued survival of many wild animals. When people and their stock move into new areas in large numbers, wildlife almost inevitably declines.

Urban Life

Sub-Saharan Africa and South Asia are the least urbanized world regions, although in both regions cities are growing at twice the national growth rates. More than one-third of Sub-Saharan people live in urbanized areas. One of the consequences of this surge in city living is urban sprawl. Rural-to-urban migration, industrialization, and refugee flows are forcing the cities of the region to absorb more people and use more resources. As in Latin America, the tendency is toward urban primacy—the condition in which one major city dominates and is at least three times larger than the next largest city. Nairobi, Kenya's capital, was a city of 250,000 in 1960 but now has 3.5 million residents and is considered the hub of transportation, finance, and communication for all of East Africa. In the last decade, Nairobi has become a high-tech superstar, with half the city's population using the Internet and many start-up companies being created. Despite such a robust embrace of technology, Nairobi still has many unemployed and impoverished residents. Close to the downtown and bordering a golf course is the slum of Kibera, with some 200,000 residents (Figure 6.22). Here garbage lines the streets, crime is rampant, and housing is crude and crowded. Municipal officials throughout the region struggle to build enough roads and provide electricity, water, trash collection, and employment for so many people in such rapidly growing places.

Figure 6.22 Kibera Slum One of Nairobi's largest slums close to the downtown, Kibera is home to at least 200,000 people. Kibera, one of East Africa's oldest and most studied slums, is experiencing the dual process of upgrading and resident relocation.

European colonialism greatly influenced urban form and development in the region. Although a very small percentage of the population lived in cities, Africans did have an urban tradition prior to the colonial era. Ancient cities, such as Axum in Ethiopia, thrived 2000 years ago, while in the Sahel major trans-Saharan trade centers, such as Tombouctou (Timbuktu) and Gao, have existed for more than a millennium. In East Africa, an urban trading culture rooted in Islam and the Swahili language emerged. West Africa, however, had the most-developed precolonial urban network, reflecting both native and Islamic traditions. It also supports some of the region's largest cities today.

West African Urban Traditions The West African coastline is dotted with cities, from Dakar, Senegal, in the far north to Lagos, Nigeria, in the east. Half of Nigerians live in cities, and in 2015 the country had eight metropolitan areas with populations of more than 1 million. Historically, the Yoruba cities in southwestern Nigeria have been the best documented. Developed in the 12th century, cities such as Ibadan were walled and gated, with a palace encircled by large rectangular courtyards at its core. An important center of trade for an extensive surrounding area, Ibadan was also a religious and political center. Lagos was another Yoruba settlement. Founded on a coastal island on the Bight of Benin, most of the modern city has spread onto the nearby mainland. Its coastal setting and natural harbor made this relatively small native city attractive to colonial powers. When the British took control in the mid-19th century, the city grew in size and importance. When Nigeria gained its independence from the United Kingdom in 1960, Lagos was a city of 1 million; today it is Sub-Saharan Africa's largest city.

Most West African cities are hybrids, combining Islamic, European, and national elements such as mosques, Victorian architecture, and streets named after independence leaders. Accra, the capital of Ghana, is home to nearly 2.6 million people. Originally settled by the Ga people in the 16th century, it became a British colonial administrative center by

Figure 6.23 Elite Accra Neighborhood The Legon area east of downtown Accra boasts the city's nicest homes, private schools, and the University of Ghana.

the late 1800s. The modern city is being transformed through neoliberal policies introduced in the 1980s to attract international corporations. Increased foreign investment in financial and producer services led to the creation of a "Global Central Business District (CBD)" on the east side of the city, away from the "National CBD." Here foreign companies clustered in areas with secure land title, new modern roads, parking, and airport access. Upper-income gated communities have also formed near the Global CBD, with names such as Trasacco Valley, Airport Hills, and Buena Vista (Figure 6.23). Accra, like other cities in the region, is rapidly changing due to an influx of foreign capital. The result is highly segregated urban spaces reflecting a global phase in urban development, in which world market forces, rather than colonial or national ones, drive the change.

Urban Industrial South Africa The major cities of southern Africa, unlike those of West Africa, are colonial in origin. Most of these cities, such as Lusaka, Zambia, and Harare, Zimbabwe, grew as administrative or mining centers. South Africa is one of the most urbanized states in the entire region, and it is certainly the most industrialized. The foundations of South Africa's urban economy rest largely on its incredibly rich mineral resources (diamonds, gold, chromium, platinum, tin, uranium, coal, iron ore, and manganese). Eight of its metropolitan areas have more than 1 million people; the largest of them are Johannesburg, Durban, and Cape Town.

The form of South African cities continues to be imprinted by the legacy of **apartheid** (an official policy of racial segregation that shaped social relations in South Africa for nearly 50 years). Even though apartheid was abolished in 1994, it is still evident in the landscape. Under apartheid rules, South African cities were divided into residential areas according to racial categories: white, **coloured** (a South African term describing people of mixed African and European ancestry), Indian (South Asian), and African (black). Whites occupied the largest and most desirable portions of the city. Blacks were crowded into the least desired areas, called **townships**, such as Soweto outside of Johannesburg and Gugulethu outside of Cape Town. Today blacks, coloureds, and Indians can legally live anywhere they want. Yet the economic differences between racial groups, as well as deep-rooted animosity, hinder residential integration.

Even in relatively urbanized South Africa, the challenge of accommodating new urban residents is constant. The largely black community of Diepsloot, north of Johannesburg, illustrates how even planned communities can become densely settled in short order. Diepsloot is a postapartheid project where the government divided land into lots and eventually built several thousand houses. Figure 6.24 illustrates the increased settlement density of a Diepsloot neighborhood over a decade. The relatively large lots were gradually filled in with small shacks built with scrap metal, wood, and plastic. The original homes have access to running water and sewage, but the informal dwellings surrounding them do not. Later, small shops were added and trees were planted, but much of the population lacks formal employment. This area, which was planned for a few thousand people, is now a community of over 150,000.

Review

6.3 Explain the factors that contribute to the region's high population growth rates and rapid urbanization.

6.4 How have infectious diseases impacted population trends, and what are governments and aid organizations doing to fight diseases such as HIV/AIDS and malaria?

6.5 What are the major rural livelihoods in this world region?

KEY TERMS swidden, pastoralist, tsetse fly, apartheid, coloured, township

(a)

(b)

Figure 6.24 Satellite Images of Diepsloot, South Africa In just over a decade, black settlers poured into the planned community of Diepsloot, filling in single-family-home lots with multiple structures. They also added businesses, planted trees, and vastly increased the area's population density.

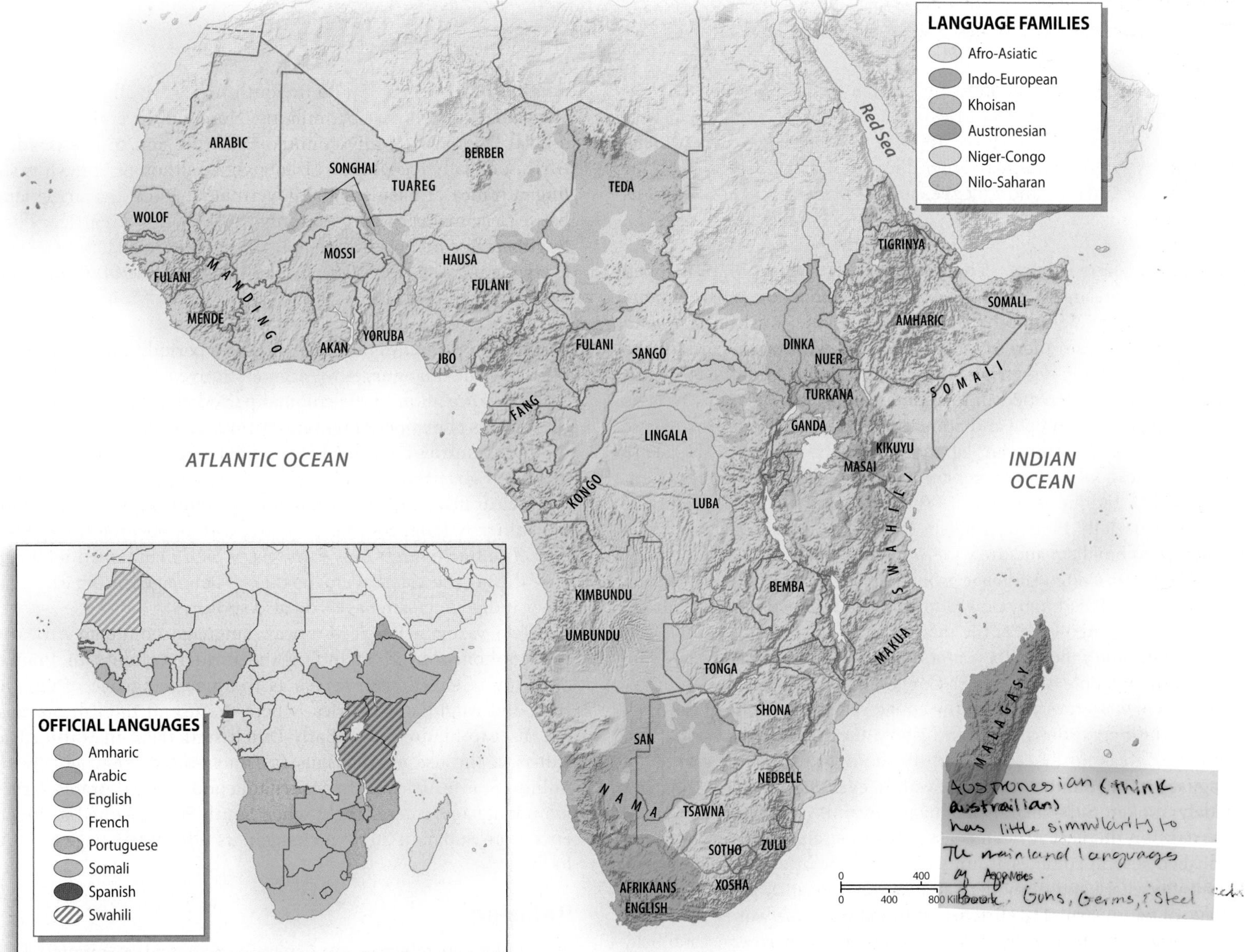

Figure 6.25 Language Map of Sub-Saharan Africa: Language Families and Official Languages Mapping the region's languages is a complex task. Some languages, such as Swahili, have millions of speakers, while other languages are spoken by a few hundred people living in isolated areas. Six language families are represented in the region. Among these families are scores of individual languages (see the labels on the map). Because most modern states have many native languages, the colonial language often became the "official" language. English and French are the most common official languages in the region (see inset). **Q: Consider the distribution of language families in this region. What does this pattern tell us about this region's interaction with peoples from other regions?**

Cultural Coherence and Diversity: Unity Through Adversity

No world region is culturally homogeneous, but most have been partially unified in the past through widespread systems of belief and communication. The lack of traditional cultural and political coherence across Sub-Saharan Africa is not surprising, considering the region's huge size. Sub-Saharan Africa is more than four times larger than Europe or South Asia. Had foreign imperialism not impinged on the region, it is quite possible that western Africa and southern Africa would have developed into their own distinct world regions.

An African identity south of the Sahara was created through a common history of slavery and colonialism as well as through struggles for independence and development. More telling is the fact that the region's people often define themselves as African, especially to the outside world. No one will deny that Sub-Saharan Africa is poor. Yet the cultural expressions of its people—its music, dance, and art—are joyous. Africans share a resilience and optimism that visitors to the region often comment on. The cultural diversity of the region is obvious, yet there is unity among the people, drawn from surviving many adversities.

Language Patterns

In most Sub-Saharan countries, as in other former colonies, the multiple languages spoken reflect tribal, colonial, and national affiliations. Indigenous languages, many from the Bantu subfamily, are often localized to relatively small rural areas. More widely spoken African trade languages, such as Swahili and Hausa, serve as a lingua franca over broader areas. Overlaying native languages are Indo-European (French and English) and Afro-Asiatic (Arabic and Somali) languages. Figure 6.25 illustrates the complex pattern of language families and major languages found in Africa today. Comparing the larger map with the inset showing current "official" languages indicates that most African countries are multilingual, which can be a source of tension within states. In Nigeria, for example, the official language is English, yet there are millions of Hausa, Yoruba, Igbo (or Ibo), Ful (or Fulani), and Efik speakers as well as speakers of dozens of other languages.

African Language Groups Three of the six language groups mapped in Figure 6.25 (Niger-Congo, Nilo-Saharan, and Khoisan) are unique to the region, while the other three (Afro-Asiatic, Austronesian, and Indo-European) are more closely associated with other parts of the world. Afro-Asiatic languages, especially Arabic, dominate North Africa and are understood in Islamic areas of Sub-Saharan Africa as well. Amharic in Ethiopia and Somali in Somalia are also Afro-Asiatic languages. The Austronesian language family is limited to the island of Madagascar, which many believe was first settled by seafarers from present-day Indonesia some 1500 years ago. Indo-European languages, especially French, English, Portuguese, and Afrikaans, are a legacy of colonialism and are widely spoken today.

The Niger-Congo language group is by far the most important one in the region. This linguistic group, which originated in West Africa, includes Mandingo (one of the many Mande languages), Yoruba, Ful (or Fulani), and Igbo (or Ibo), among others. Around 3000 years ago, the Bantu, a people of the Niger-Congo language group, began to expand out of western Africa into the equatorial zone. In one of the most far-ranging migrations in human history, they introduced agriculture into large areas in the central and southern parts of the region. One language in the Bantu subfamily, Swahili, eventually became the most widely spoken Sub-Saharan language. Swahili originated as a trade language on the East African coast, where merchants from Arabia established several colonies around 1100 CE. A hybrid society grew in a narrow coastal band of modern Kenya and Tanzania, speaking a language of Bantu structure enriched with many Arabic words. Although Swahili became the primary language only in this narrow coastal belt, it spread far into the interior as the language of trade. After independence was achieved, Kenya, Tanzania, and Uganda adopted Swahili as an official language, along with English. Swahili, with some 100 million speakers, is a lingua franca for East Africa and parts of Central Africa. It has generated a fairly extensive literature and is often studied in other regions of the world.

Language and Identity Historically, ethnic identity and linguistic affiliation have been highly unstable over much of Sub-Saharan Africa. The tendency was for new language groups to form when people threatened by war fled to less-settled areas, where they often mixed with peoples from other places. In such situations, new languages arise quickly, and divisions between groups are blurred. Nevertheless, distinct tribes formed, consisting of a group of families or clans with common kinship, language, and definable territory. European colonial administrators were eager to establish a fixed social order to better control native peoples. During this process, a flawed cultural map of Sub-Saharan Africa evolved. Some tribes were artificially divided, meaningless names were applied, and cultural areas were often misinterpreted.

Social boundaries between different ethnic and linguistic groups have become more stable in recent years, and some individual languages have become particularly important for communication on a national scale. Wolof in Senegal; Mandingo in Mali; Mossi in Burkina Faso; Yoruba, Hausa, and Igbo in Nigeria; Kikuyu in central Kenya; and Zulu, Xhosa, and Sotho in South Africa are all nationally significant languages spoken by millions. None, however, has the status of official language for any country. With the end of apartheid, South Africa officially recognized 11 languages, although English is still the lingua franca of business and government. Indeed, a single language has a clear majority status in only a handful of countries. The more linguistically homogeneous states include Somalia (where virtually everyone speaks Somali) and the very small states of Rwanda, Burundi, Swaziland, and Lesotho.

European Languages In the colonial period, European countries used their own languages for administrative purposes in their African empires. Education in the colonial period also stressed literacy in the language of the imperial power. Postindependence, most Sub-Saharan African countries continued to use the languages of their former colonizers for government and higher education. Few of these new states had a clear majority language that they could employ, and picking any one minority tongue would have met with opposition from other peoples. The one exception is Ethiopia, which maintained its independence during the colonial era. Its official language is Amharic, although other indigenous languages are also spoken.

Two vast blocks of European language dominate Africa today: Francophone Africa, including the former colonies of France and Belgium, where French serves as the main language of administration, and Anglophone Africa, where the use of English prevails (see the inset to Figure 6.25). Early Dutch settlement in South Africa resulted in the use of Afrikaans (a Dutch-based language) by several million South Africans. In Mauritania and Eritrea, Arabic serves as a main language. Interestingly, when South Sudan gained its independence from Sudan in 2011, it changed its official language from Arabic to English.

Religion

Native African religions are generally classified as animist. This is a somewhat misleading catchall term used to classify all local faiths that do not fit into one of the handful of "world religions." Most animist religions are centered on the worship of nature and ancestral spirits, but within the animist tradition there is great internal diversity. Classifying a religion as animist says more about what it is not than about what it actually is.

Both Christianity and Islam entered Sub-Saharan Africa early in their histories but advanced slowly for many centuries. Since the beginning of the 20th century, both religions have spread rapidly—more rapidly, in fact, than in any other part of the world. However, tens of millions of Africans still follow animist beliefs, and many others combine animist practices and ideas with their observance of Christianity or Islam.

The Introduction and Spread of Christianity Christianity came first to northeastern Africa. Kingdoms in both Ethiopia and central Sudan were converted by 300 CE—the earliest conversions outside the Roman Empire. The peoples of northern and central Ethiopia adopted the Coptic form of Christianity and thus historically have looked to Egypt's Christian minority for their religious leadership (Figure 6.26). At present, roughly half of the population in both Ethiopia and Eritrea is Coptic Christian; most of the rest is Muslim, but some animist communities still exist, especially in Ethiopia's western lowlands.

Figure 6.26 Eritrean Christians at Prayer Coptic Christians gather for an Easter celebration in Asmara, Eritrea. Half of the populations in Eritrea and Ethiopia belong to the Coptic Church, which has ties to Egypt's Christian minority.

European settlers and missionaries introduced Christianity to other parts of Sub-Saharan Africa beginning in the 1600s. The Dutch, who began to colonize South Africa at that time, brought their Calvinist Protestant faith. Later European immigrants to South Africa brought Anglicanism and other Protestant beliefs as well as Catholicism. Most black South Africans eventually converted to one or another form of Christianity as well. In fact, churches in South Africa were instrumental in the long fight against white racial supremacy. Religious leaders such as Bishop Desmond Tutu were outspoken critics of the injustices of apartheid and worked to bring down the system.

Elsewhere in Africa, Christianity came with European missionaries, most of whom arrived after the mid-1800s. As was true in the rest of the world, missionaries had little success where Islam had preceded them, but they eventually made numerous conversions in animist areas. As a general rule, Protestant Christianity prevails in former British colonies, while Catholicism is more important in former French, Belgian, and Portuguese territories. In the postcolonial era, African Christianity has spread out, at times taking on a life of its own, independent from foreign missionary efforts. Still active in the region are various Pentecostal, Evangelical, and Mormon missionary groups, mostly from the United States. It is difficult to map the distribution of Christianity in Africa, however, because it has spread irregularly across the non-Islamic portion of the region.

The Introduction and Spread of Islam Islam began to advance into Sub-Saharan Africa 1000 years ago (Figure 6.27). Berber traders from North Africa and the Sahara introduced the religion to the Sahel, and by 1050 the Kingdom of Tokolor in modern Senegal emerged as the first Sub-Saharan Muslim state. Somewhat later, the ruling class of the powerful Mande-speaking trading empires of Ghana and Mali converted as well. In the 14th century, the emperor of Mali astounded the Muslim world when he and his royal court made the pilgrimage to Mecca, bringing with them so much gold that they set off a brief period of high inflation throughout Southwest Asia.

Mande-speaking traders, whose networks spanned the Sahel to the Gulf of Guinea, gradually introduced the religion to other areas of West Africa. Many peoples remained committed to animism, however, and Islam made slow and unsteady progress. Today orthodox Islam prevails through most of the Sahel. Farther south, Muslims are mixed with Christians and animists, but their numbers continue to grow, and their practices tend to be orthodox as well (Figure 6.28).

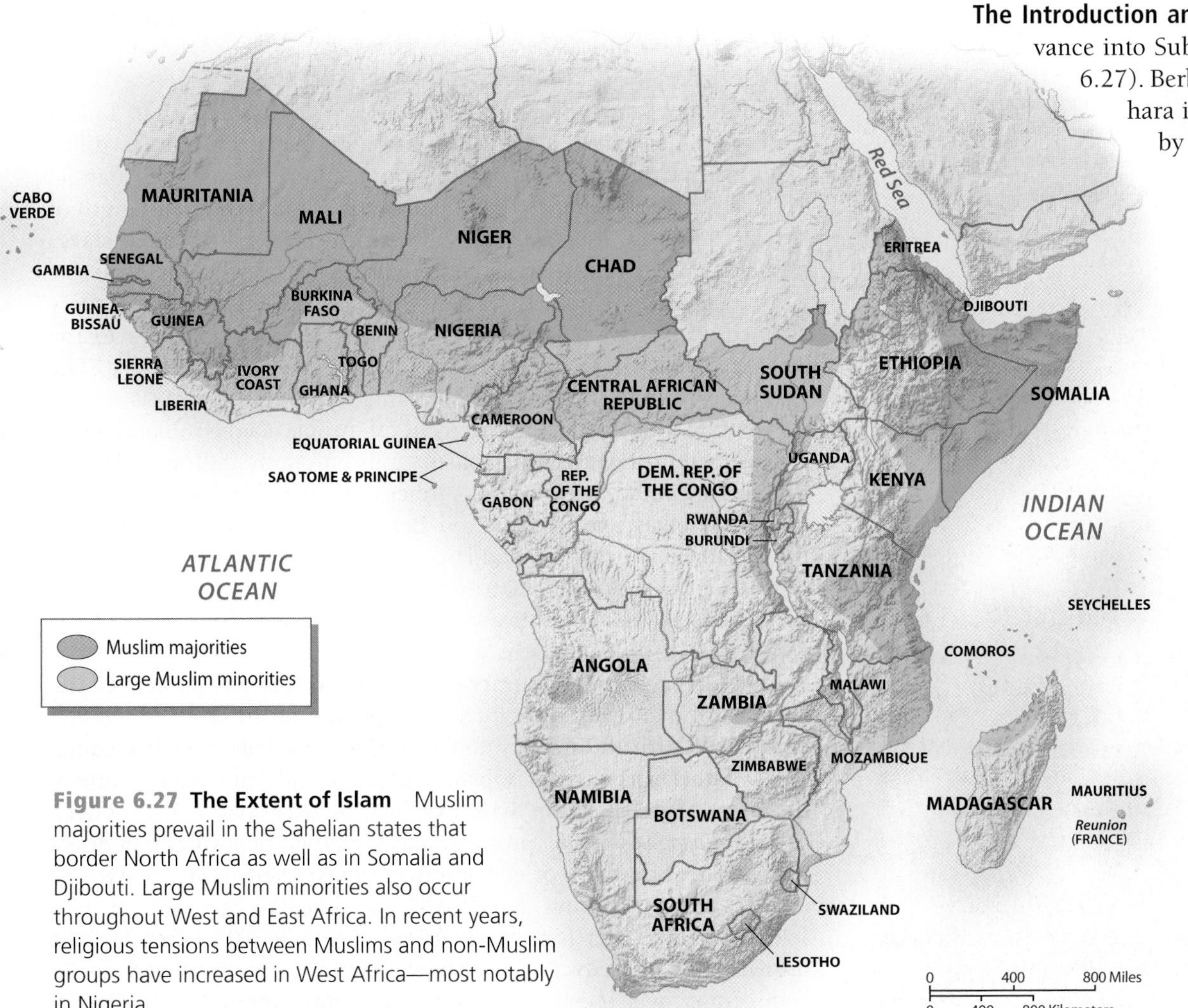

Figure 6.27 The Extent of Islam Muslim majorities prevail in the Sahelian states that border North Africa as well as in Somalia and Djibouti. Large Muslim minorities also occur throughout West and East Africa. In recent years, religious tensions between Muslims and non-Muslim groups have increased in West Africa—most notably in Nigeria.

Interaction Between Religious Traditions The southward spread of Islam from the Sahel, coupled with the northward spread of Christianity from the port cities, has generated a complex religious frontier across much of West Africa. In Nigeria, the Hausa are firmly Muslim, while the southeastern Igbo are largely Christian. The Yoruba of the southwest are divided between Christian and Muslim. In the more remote parts of Nigeria, animist traditions remain strong. Despite this religious diversity, religious conflict in Nigeria has been relatively rare until recently. In 2000, seven northern Nigerian states imposed Muslim sharia laws, which has triggered intermittent violence ever since, especially in the northern city of Kaduna. In the last few years, an armed jihadist group called Boko Haram formed in northeastern Nigeria, escalating levels of violence in the north and south through kidnapping and killing. Initially claiming affiliation with al Qaeda, in 2015 Boko Haram claimed a formal allegiance with the Islamic State of Iraq and the Levant (ISIL, also called ISIS; see Chapter 7). The Nigerian military launched an offensive in 2013 to dislodge this group from Borno Province in northeastern Nigeria, but Boko Haram has been a difficult organization to control.

Religious conflict historically has been far more acute in northeastern Africa, where Muslims and Christians have struggled against each other for centuries. Such a clash eventually led to the creation of the region's newest state when South Sudan separated from the country of Sudan in 2011. Islam was introduced to Sudan in the 1300s by an invasion of Arabic-speaking pastoralists who destroyed the indigenous Coptic Christian kingdoms of the area. Within a few hundred years, northern and central Sudan had become completely Islamic. The southern equatorial province of Sudan, where tropical diseases and extensive wetlands prevented Arab advances, remained animist or converted to Christianity under British colonial rule.

In the 1970s, the Arabic-speaking Muslims of northern and central Sudan began to build an Islamic state. Experiencing both religious discrimination and economic exploitation, the Christian and animist peoples of the south launched a massive rebellion. Fighting became intense in the 1980s, with the government generally controlling the main towns and roads and the rebels maintaining power in the countryside. A peace was brokered in 2003, and as part of the peace agreement southern Sudan was promised an opportunity to vote on secession from the north in 2011. The vote took place, and the new nation was formed with Juba as its capital. Yet this landlocked territory is still not at peace. A power struggle between rival factions of Dinka and Nuer began in 2013 and has led to some 2 million displaced people, 200,000 refugees, and thousands dead. In 2015, after two years of civil war, President Salva Kiir signed a peace accord. It is likely that conflict, which is now more ethnically driven than religious, will continue in South Sudan.

Sub-Saharan Africa is a land of religious vitality. Both Christianity and Islam are spreading rapidly, and devotional activities are part of the daily flow of life in cities and rural areas. Animism continues to have widespread appeal as well, so that new and syncretic (blended) forms of religious expression are also emerging. With such a diversity of faiths, it is fortunate that religion is not typically the cause of overt conflict in the region.

Figure 6.28 West African Muslims Residents of Djenne, Niger, gather in an open market in front of the city's great mosque. Much of the Sahelian region converted to Islam more than six centuries ago.

Globalization and African Culture The slave trade that linked Africa to the Americas and Europe set in motion paths of cultural diffusion that transferred African peoples and cultures across the Atlantic. Tragically, slavery damaged the demographic and political strength of African societies, especially in West Africa, from which most slaves were taken. An estimated 12 million Africans were shipped to the Americas as slaves from the 1500s until 1870 (Figure 6.29). Slavery affected all of Sub-Saharan Africa, sending Africans not just to the Americas, but also to Europe, North Africa, and Southwest Asia. The vast majority, however, worked on plantations in the Americas.

Out of this tragic displacement of people came a blending of African cultures with Amerindian and European cultures. African rhythms are at the core of various American musical styles, from rumba to jazz, the blues, and rock and roll. Brazil, Latin America's largest country, is claimed to be the second-largest "African state" (after Nigeria) because of its huge Afro-Brazilian population. Thus, the forced migration of Africans as slaves had a huge cultural influence on many areas of the world.

So, too, have contemporary movements of Africans influenced the cultures of many world regions. Perhaps one of the most celebrated persons of African ancestry today is U.S. President Barack Obama, whose father was Kenyan. Obama's heritage and upbringing embody the forces of globalization. In Kenya, he is hailed as part of the modern African *diaspora*—young professionals (and their offspring) who leave the continent for work or education and make their mark somewhere else. In popular culture, South African comedian Trevor Noah was selected to take over Jon Stewart's *Daily Show* in 2015 when Stewart retired. This comedy "news" show's brand of political and social commentary will inevitably become more international with a South African at the helm.

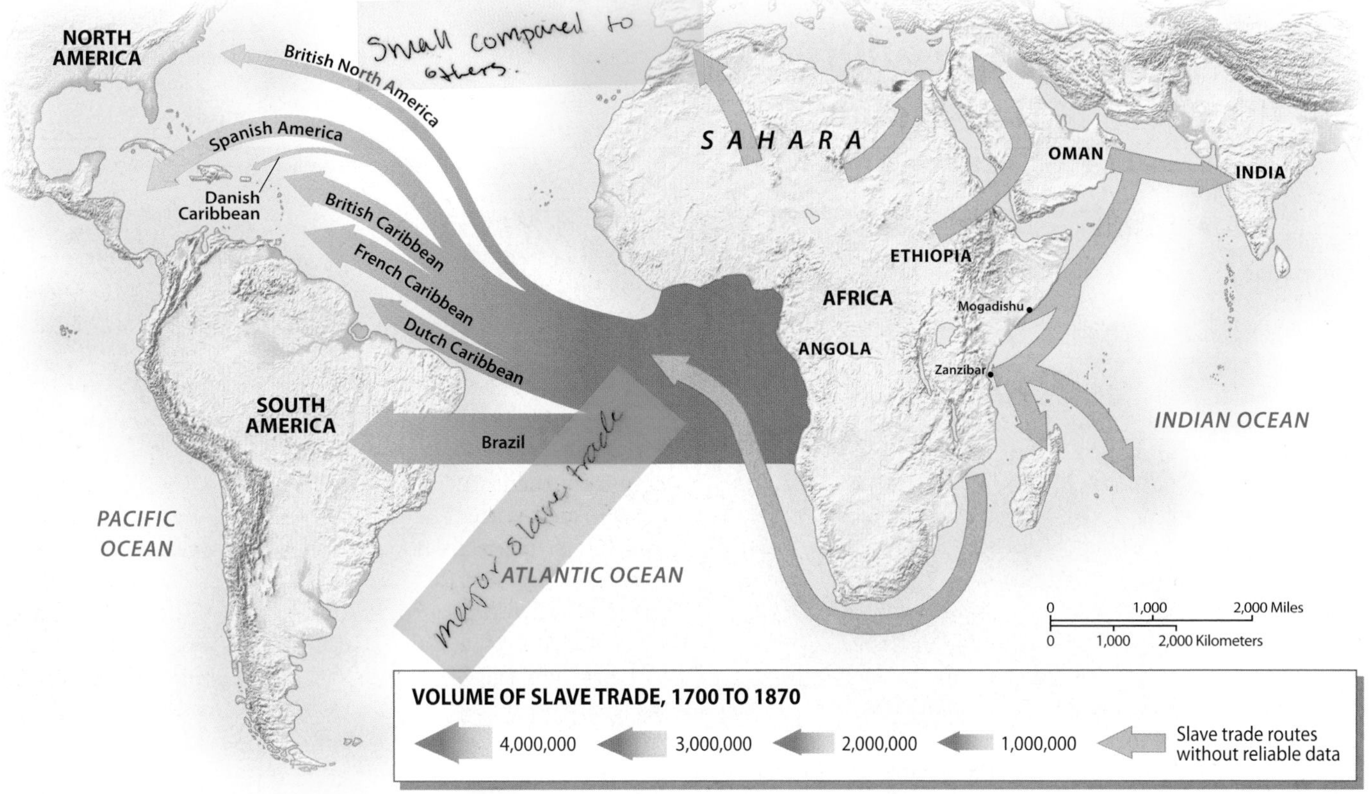

Figure 6.29 African Slave Trade Sub-Saharan societies were devastated by the slave trade. From ship logs, it is estimated that 12 million Africans were shipped to the Americas to work as slaves on sugar, cotton, and rice plantations; the majority went to Brazil and the Caribbean. Other slave routes existed, although the data on them are less reliable. Africans from south of the Sahara were used as slaves in North Africa. Others were traded across the Indian Ocean into Southwest Asia and South Asia.

Popular culture in Africa, like everywhere else in the world, is a dynamic mixture of global and local influences. Kwaito, a popular music form in South Africa, sounds a lot like rap from the United States. A closer listen, however, reveals an incorporation of local rhythms, lyrics in Zulu and Xhosa, and themes about life in postapartheid townships. Zola, a kwaito superstar, also costarred in the Oscar-winning film *Tsotsi*, about gang life and hardship in Johannesburg. Nigeria, however, is the center of Africa's film industry, second only to India in the number of films produced. "Nollywood" films are popular all over Africa, but some directors have global ambitions (*Exploring Global Connections: The Reach of Nollywood*).

Music in West Africa Nigeria is also the musical center of West Africa, with a well-developed and cosmopolitan recording industry. Modern Nigerian styles such as juju, highlife, and Afro-beat are influenced by jazz, rock, reggae, and gospel, and they are driven by an easily recognized African sound.

Farther up the Niger River lies the country of Mali. Bamako, the capital, is a music center that has produced scores of recording artists. Many Malian musicians descend from a caste of musical storytellers performing on either the traditional *kora* (a cross between a harp and a lute) or the guitar. The musical style is strikingly similar to that of blues from the Mississippi Delta, so much so that Ali Farka Touré, who was one of Africa's most renowned musicians, is still referred to as the Bluesman of Africa. Each January, not far from Timbuktu, music fans from West Africa and Europe have gathered for the Festival in the Desert. In this remote Saharan locale, a celebration of Malian music and Touareg nomadic culture has drawn together tourists from Europe, African musicians, and nomads (Figure 6.30). Even with the recent conflict in Mali (which will be discussed later), the festival was held in 2012. Fighting later intensified, and the military took control of the government in Bamako in March 2012, so the future of this cultural event—as well as that of the country overall—is uncertain.

Contemporary African music can be both commercially and politically important. Nigerian singer Fela Kuti (1938–1997) was an influential musician and a voice of political conscience for Nigerians struggling for true democracy. From an elite family and educated in England, Fela borrowed from jazz, traditional, and popular music to produce the Afro-beat sound of the 1970s. The music was irresistible,

Figure 6.30 Festival in the Desert The Touareg band Igbayen plays at the Mali Festival in the Desert. This annual winter festival in the oasis town of Essakane, Mali, draws thousands of Malian musicians, Touareg nomads, and Western tourists.

EXPLORING GLOBAL CONNECTIONS

The Reach of Nollywood

Google Earth (MG) Virtual Tour Video
http://goo.gl/9SM9ER

Africa's undisputed film capital is Lagos, Nigeria. Called Nollywood, the Nigerian movie industry currently grinds out 2500 films a year (50 films a week) and employs more than 1 million people. Nollywood makes more films than Hollywood. Relying on relatively inexpensive digital video technology, most of these movies are shot in a few days and with budgets of $10,000 to $20,000. The typical themes of religion, ethnicity, corruption, witchcraft, the spirit world, violence, and injustice resonate with African audiences. The films are almost always shot on location—in city streets, office buildings, and homes or in the countryside. Nollywood films can be bloody and exploitive; they can also be overtly evangelical, promoting Christianity over indigenous faiths. Many are acted in English, but there are also movies for Yoruba, Igbo, and Hausa speakers (Figure 6.2.1).

Figure 6.2.1 Location Filming in Nigeria A film crew prepares to shoot a scene in Lagos. Nicknamed Nollywood, Nigeria produces more films each year than Hollywood.

The DVD Market Rather than being released in theaters, most Nollywood movies go directly to DVD and are rarely viewed beyond Africa. The shelf life of these movies is rather short; production companies need to make their money back quickly, before pirated copies undermine profits. However, a $20,000 film can earn $500,000 in DVD sales in a couple of weeks. Consequently, film distribution remains tightly controlled by Igbo businessmen, sometimes called the Alaba cartel for their distribution center on the outskirts of Lagos. But this is beginning to change as an African middle-class population wants to go to theaters and directors want to create higher-quality films for Nigeria and beyond.

The Queen of Nollywood Omotola Jalade-Ekeinde, popularly referred to as *OmoSexy*, is the reigning queen of Nollywood. This talented actress and singer has made over 300 feature films and is also a philanthropist known for her Youth Empowerment Program in Nigeria and her work throughout Africa for the UN (Figure 6.2.2). In the plot of *Ijé,* a film partially shot in Los Angeles, Omotola is charged with the murder of three men, including her husband. The story is told through her sister, who comes from Nigeria to visit her in a U.S. jail. Such transnational intrigue is part of Nollywood's trajectory as it reaches out to the world.

Figure 6.2.2 Queen of Nollywood Nigerian actress Omotola Jalade-Ekeinde was recognized by *Time* magazine as one of the 100 most influential people in 2013. One of Nigeria's top movie stars, she is also active in charity work around the world.

1. What technological changes allowed Nigeria to increase its movie production?
2. Have you ever seen a Nigerian movie? If so, how does it compare with films made in North America?

but his searing lyrics also attracted attention. Acutely critical of the military government, he sang of police harassment, the inequities of the international economic order, and even Lagos's infamous traffic. Singing in English and Yoruba, Fela transmitted his message to a larger audience, but he also became a target of state harassment. He died in 1997 from complications related to AIDS, yet his music and politics later became the subject of the award-winning Broadway musical *Fela!*

Pride in East African Runners Ethiopia and Kenya have produced many of the world's greatest distance runners. Abebe Bikila won Ethiopia's and Africa's first Olympic gold medal, running barefoot at the Rome games in 1960. Since then, nearly every Olympic Games has yielded medals for Ethiopia and Kenya. At the 2012 London Olympics, Kenyan runners won 11 medals and Ethiopians 7. These states, along with South Africa, were the top medal winners for Sub-Saharan Africa in London.

Running is a national pastime in Kenya and Ethiopia, where elevation—Addis Ababa sits at 7300 feet (2200 meters) and Nairobi at 5300 feet (1600 meters)—increases oxygen-carrying capacity. Past medalists Haile Gebrselassie and Derartu Tulu are national celebrities in Ethiopia, where they are idolized by the country's youth. Tulu, the first black African woman to win a gold medal in distance running, is a forceful voice for women's rights in a country where women are discouraged from putting on running shorts. In the 2012 Olympics, Ethiopian women Tiki Gelana and Tirunesh Dibaba won gold in the women's marathon and 10,000-meter race, respectively (Figure 6.31).

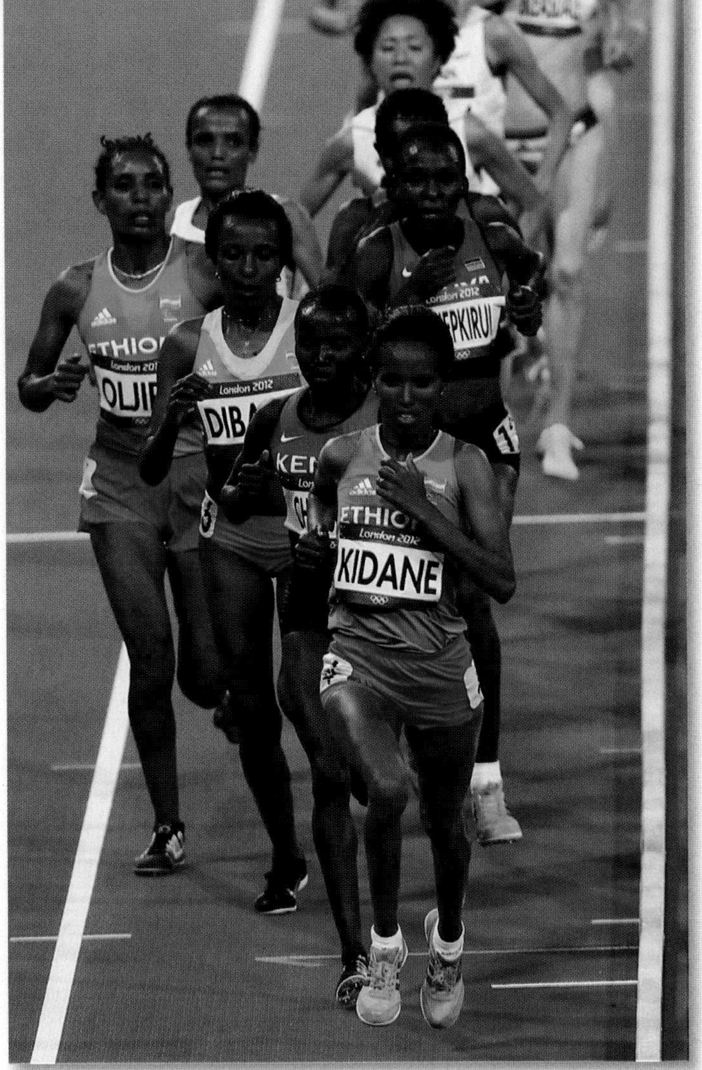

Figure 6.31 East African Distance Runners Ethiopian and Kenyan runners lead the pack in the 10,000-meter race at the 2012 London Olympics. Ethiopian Tirunesh Dibaba (third from the front) took gold, and Kenyans Sally Jepkosgei Kipyego and Vivian Jepkemoi Cheruiyot took silver and bronze.

Review

6.6 What are the dominant religions of Sub-Saharan Africa, and how have they diffused throughout the region?

6.7 Describe the ways in which African peoples have influenced world regions beyond Africa.

Geopolitical Framework: Legacies of Colonialism and Conflict

The duration of human settlement in Sub-Saharan Africa is unmatched by any other region. Evidence shows that humankind originated there, and many diverse ethnic groups have formed in the region over the past few thousand years. Although ethnic conflicts have occurred, cooperation and coexistence among different peoples have also continued over centuries.

Some 2000 years ago the Kingdom at Axum arose in northern Ethiopia and Eritrea, strongly influenced by political models derived from Egypt and Arabia. The first wholly indigenous African states were founded in the Sahel around 700 CE. Over the next several centuries, a variety of other states emerged in West Africa. By the 1600s, the states located near the Gulf of Guinea took advantage of the opportunities presented by the slave trade—namely, selling slaves to Europeans (Figure 6.32).

Thus, prior to European colonization, Sub-Saharan Africa presented a complex mosaic of kingdoms, states, and tribal societies. The arrival of Europeans forever changed patterns of social organization and ethnic relations. As Europeans rushed to carve up the continent to serve their imperial ambitions, they set up various administrations that heightened ethnic tensions and promoted hostility. Many of the region's modern conflicts can trace their roots back to the colonial era, especially the drawing of political boundaries.

European Colonization

Unlike the relatively rapid colonization of the Americas, Europeans needed centuries to gain effective control of Sub-Saharan Africa. Portuguese traders arrived along the coast of West Africa in the 1400s, and by the 1500s they were established in East Africa as well. Initially, the Portuguese made large profits, converted a few local rulers to Christianity, established several defensive trading posts, and gained control over the Swahili trading cities of the east. They stretched themselves too thin, however, and failed in many of their colonizing activities. Only along the coasts of modern Angola and Mozambique, where a sizable population of mixed African and Portuguese peoples emerged, did Portugal maintain power. Along the Swahili, or eastern, coast, the Portuguese were eventually expelled by Arabs from Oman, who then established their own trade-based empire in the area.

One reason for the Portuguese failure was Sub-Saharan Africa's disease environment. With no resistance to malaria and other tropical diseases, roughly half of all Europeans who remained on the African mainland died within a year. Protected both by their armies and by the diseases of their native lands, African states were able to maintain an upper hand over European traders and adventurers well into the 1800s. Unlike in the Americas, where European conquest was facilitated by the introduction of Old World diseases that devastated native populations (see Chapters 4 and 5), in Sub-Saharan Africa endemic disease limited European settlement until the mid-19th century.

Also in the early 1800s, two small territories were established in West Africa as a haven for freed and runaway slaves (Figure 6.33). The American Colonization Society set up a territory in 1822 to settle former African American slaves; by 1847, it was the independent free state of Liberia. Sierra Leone served a similar function for ex-slaves from the British Caribbean, but it remained a British protectorate until the 1960s. Despite the intentions behind the creation of these territories, they were colonies. Liberia, in particular, was imposed on existing indigenous groups who viewed their new "African" leaders with contempt.

The Scramble for Africa In the 1880s, European colonization of the region quickly accelerated, leading to the so-called scramble for Africa. By this time, the colonists had practices to reduce malaria transmission, and after the invention of the machine gun, no African state could long resist European force.

As the colonization of Africa intensified, tensions among the colonizing forces of Britain, France, Belgium, Germany, Italy, Portugal, and Spain mounted. Rather than risk war, 13 countries convened in Berlin at the invitation of German Chancellor Bismarck in 1884, at a gathering known as the **Berlin Conference**. During the conference,

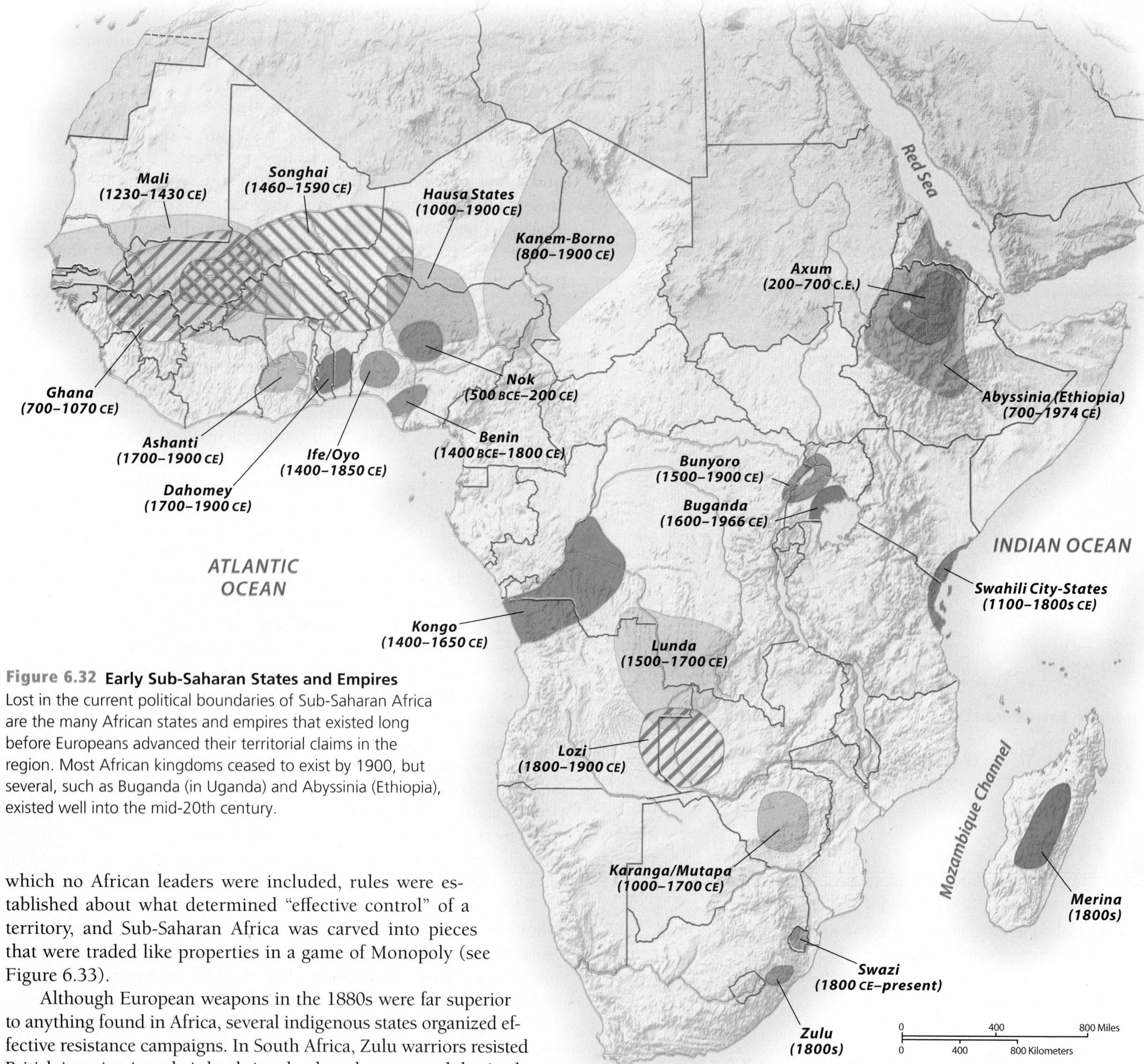

Figure 6.32 Early Sub-Saharan States and Empires Lost in the current political boundaries of Sub-Saharan Africa are the many African states and empires that existed long before Europeans advanced their territorial claims in the region. Most African kingdoms ceased to exist by 1900, but several, such as Buganda (in Uganda) and Abyssinia (Ethiopia), existed well into the mid-20th century.

which no African leaders were included, rules were established about what determined "effective control" of a territory, and Sub-Saharan Africa was carved into pieces that were traded like properties in a game of Monopoly (see Figure 6.33).

Although European weapons in the 1880s were far superior to anything found in Africa, several indigenous states organized effective resistance campaigns. In South Africa, Zulu warriors resisted British invasion into their lands in what have been termed the Anglo-Zulu Wars (1879–1896). Eventually, European forces prevailed everywhere except Ethiopia. The Italians had conquered the Red Sea coast and the far northern highlands (modern Eritrea) by 1890 and quickly set their sights on the large Ethiopian kingdom called Abyssinia, which had been vigorously expanding for several decades. In 1896, however, Abyssinia defeated the invading Italian army, earning the respect of the European powers. In the 1930s, fascist Italy launched a major invasion of the country, by this time renamed Ethiopia, to redeem its earlier defeat and quickly prevailed with the help of poison gas and aerial bombardment. However, Ethiopia had regained its freedomby 1942.

Although Germany was a principal instigator of the scramble for Africa, it lost its own colonies after suffering defeat in World War I. Britain and France then partitioned most of Germany's African empire between themselves. Figure 6.33 shows the region's colonial status in 1913, prior to Germany's territorial loss.

While the Europeans were cementing their rule over Africa, South Africa was inching toward political independence, at least for its white population. One of the oldest colonies in Sub-Saharan Africa, in 1910 South Africa became the first to obtain its political independence from Europe. However, its formalized system of discrimination and racism hardly made it a symbol of liberty. Ironically, as the Afrikaners tightened their political and social control over the nonwhite population

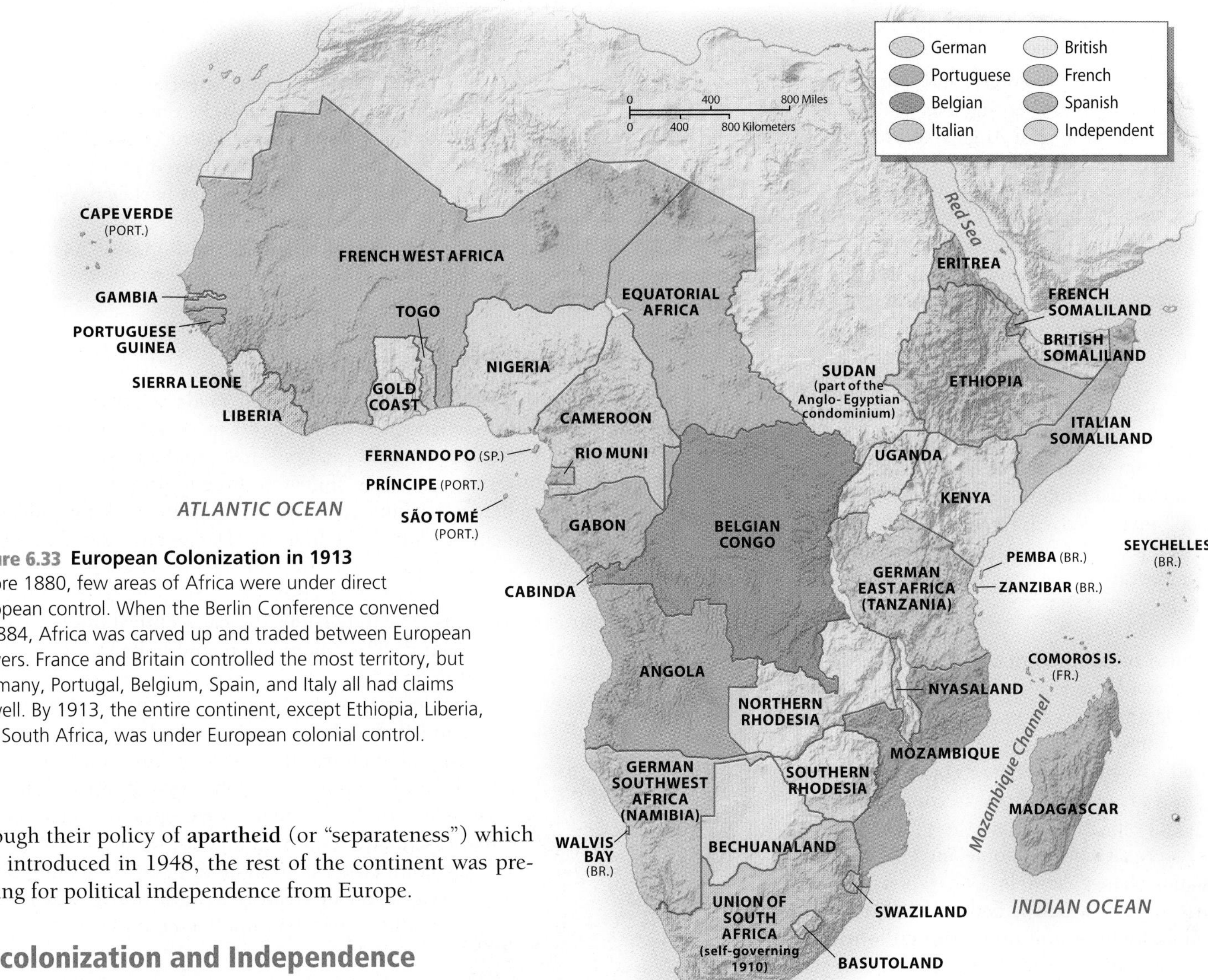

Figure 6.33 European Colonization in 1913
Before 1880, few areas of Africa were under direct European control. When the Berlin Conference convened in 1884, Africa was carved up and traded between European powers. France and Britain controlled the most territory, but Germany, Portugal, Belgium, Spain, and Italy all had claims as well. By 1913, the entire continent, except Ethiopia, Liberia, and South Africa, was under European colonial control.

through their policy of **apartheid** (or "separateness") which was introduced in 1948, the rest of the continent was preparing for political independence from Europe.

Decolonization and Independence

Decolonization of Sub-Saharan Africa happened rather quickly and peacefully, beginning in 1957. Independence movements, however, had sprung up throughout the continent, some dating back to the early 1900s. Workers' unions and independent newspapers became voices for African discontent and the hope for freedom.

By the late 1950s, political demands from within Sub-Saharan Africa and changing attitudes within Europe made it clear to leaders in Britain, France, and Belgium that they could no longer maintain their African empires. (Italy had already lost its colonies during World War II, and Britain gained Somalia and Eritrea.) Once started, the decolonization process moved rapidly. By the mid-1960s, virtually the entire region had achieved independence. In most cases, the transition was relatively peaceful and smooth, with the exception of southern Africa.

Dynamic African leaders put their mark on the region during the early decades after independence. Kenya's Jomo Kenyatta, the Ivory Coast's Felix Houphouët-Boigny, Tanzania's Julius Nyerere, Ghana's Kwame Nkrumah, and others became powerful father figures who molded their new nations (Figure 6.34). President Nkrumah's vision for Africa was the most expansive. After helping to secure independence for Ghana in 1957, his ultimate aspiration was African political unity. Although never realized, his dream set the stage for the founding of the Organization of African Unity (OAU) in 1963, which was renamed the **African Union** **(AU)** in 2002. The AU is a continent-wide organization whose main role has been to mediate disputes between neighbors. Certainly in the 1970s and 1980s, the AU was a constant voice of opposition to South Africa's minority rule, and it continues to intervene in some of the region's ethnic conflicts and humanitarian emergencies.

Southern Africa's Independence Battles Independence did not come easily to southern Africa. In Southern Rhodesia (modern-day Zimbabwe), the problem was the presence of some 250,000 white residents, most of whom owned large farms. Unwilling to see power pass to the country's black majority, then some 6 million strong, these settlers declared themselves the rulers of an independent, white-supremacist state in 1965. The black population continued to resist, however, and in 1978 the Rhodesian government was forced to give up power. Renamed Zimbabwe, the country was henceforth ruled by the black majority, although the remaining whites still form an economically privileged community. Since the mid-1990s, disputes over government land reform (splitting up the large commercial farms mostly owned by whites and giving the land to black farmers) and President Robert Mugabe's strongman politics have resulted in serious racial and political tensions as well as the collapse of the country's economy.

In the former Portuguese colonies, independence came violently. Unlike the other imperial powers, Portugal refused to hand over its colonies in the 1960s, so the people of Angola and Mozambique turned to armed resistance. The most powerful rebel movements adopted a socialist orientation and received support from the Soviet Union and Cuba. When a new government came to power in 1974, however, Portugal withdrew suddenly from its African colonies. At this point, Marxist regimes quickly took over in both Angola and Mozambique. The United States, and especially South Africa, responded to this perceived threat by supplying arms to rebel groups opposing the new governments. Fighting dragged on for three decades in Angola and Mozambique; their respective countrysides became so heavily laden with land mines that it was hard to farm. With the end of the Cold War, however, outsiders lost their interest in continuing these conflicts, and sustained efforts began to negotiate peace settlements. Mozambique has been at peace since the mid-1990s. After several failed attempts, Angola's army signed a peace treaty with rebels in 2002 that ended a 27-year conflict in which more than 300,000 Angolans died and 3 million more were displaced. With peace, Angola's impressive oil reserves have come on line, led by Chinese investment.

Apartheid's Demise in South Africa While fighting continued in the former Portuguese zone, South Africa underwent a remarkable transformation. From 1948 through the 1980s, the ruling Afrikaners' National Party was firmly committed to white supremacy. Under apartheid, only whites enjoyed real political freedom, whereas blacks were denied even citizenship in their own country—technically, they were citizens of "homelands."

The first major change came in 1990 when South Africa withdrew from Namibia, which it had controlled as a protectorate (a dependent political unit under the authority of another state) since the end of World War I. South Africa now stood alone as Africa's single white-dominated state. A few years later, the leaders of the Afrikaner-dominated political party decided they could no longer resist internal and international pressure for change. In 1994, free elections were held, and Nelson Mandela, a black leader who had been imprisoned for 27 years by the old regime, emerged as the new president from the African National Congress (ANC) party. Black and white leaders pledged to put the past behind them and work together to build a new, multiracial South Africa. Since then, orderly elections have been held; South Africans elected Thabo Mbeki for two terms (1999–2009), and Jacob Zuma is currently serving his second term after winning reelection in 2014.

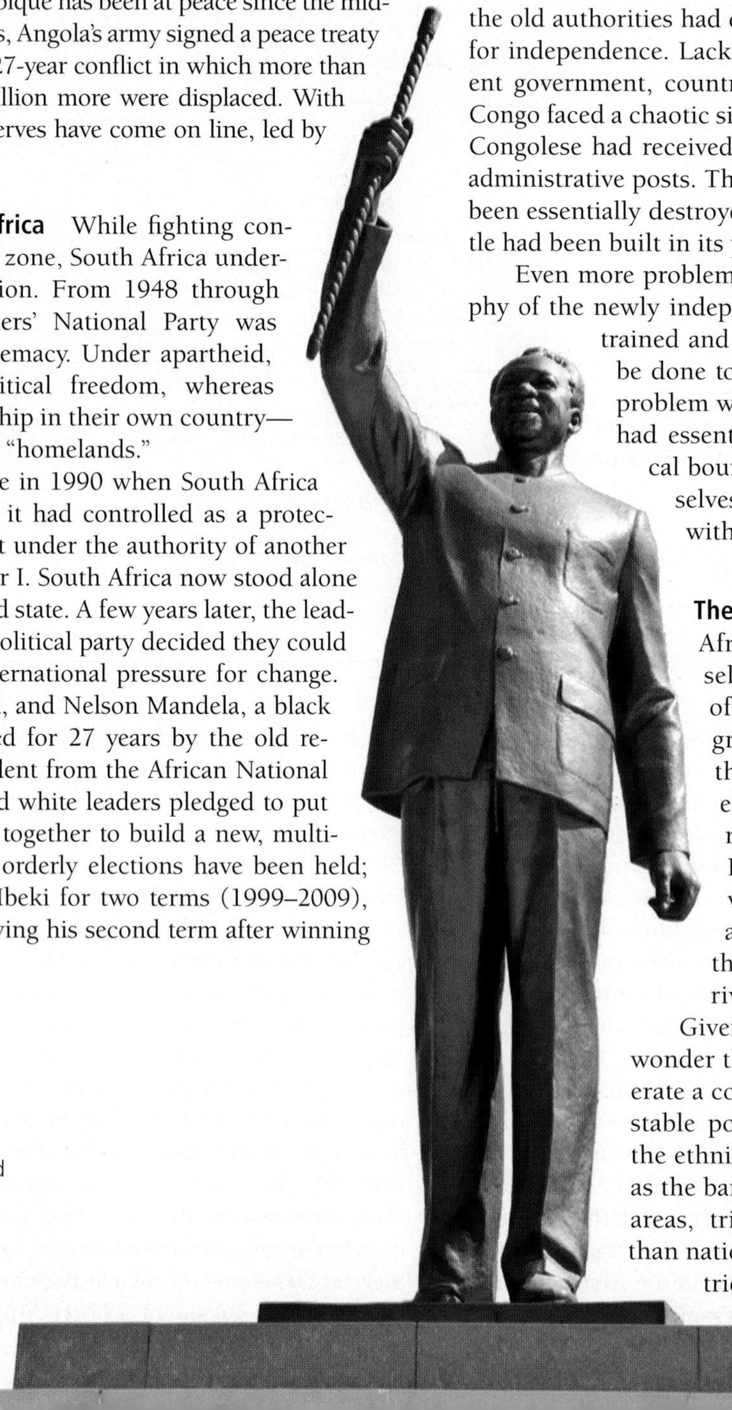

Figure 6.34 Julius Nyerere An independence leader from 1961 until he retired from the presidency in 1985, Julius Nyerere is Tanzania's founding father. This statue is located in Dodoma, the official capital since 1996. Dar es Salaam, the former capital, is still the largest city and has many government offices.

Unfortunately, the legacy of apartheid is not so easily erased. Residential segregation is illegal, but neighborhoods are still sharply divided along racial lines. Under the multiracial political system, a black middle class emerged, but most blacks remain extremely poor (and most whites remain prosperous). Violent crime has increased, and rural migrants and immigrants have poured into South African cities, producing a xenophobic anti-immigrant backlash, especially in greater Johannesburg. Because the political change was not matched by significant economic transformation, the hopes of many people have been frustrated.

Enduring Political Conflict

Although most Sub-Saharan countries made a relatively peaceful transition to independence, virtually all of them immediately faced a difficult set of institutional and political problems. In several cases, the old authorities had done almost nothing to prepare their colonies for independence. Lacking an institutional framework for independent government, countries such as the Democratic Republic of the Congo faced a chaotic situation from the beginning. Only a handful of Congolese had received higher education, let alone been trained for administrative posts. The indigenous African political framework had been essentially destroyed by colonization, and in most cases very little had been built in its place.

Even more problematic in the long run was the political geography of the newly independent states. Civil servants could always be trained and administrative systems built, but little could be done to rework the region's basic political map. The problem was the fact that the European colonial powers had essentially ignored indigenous cultural and political boundaries, both in dividing Africa among themselves and in creating administrative subdivisions within their own imperial territories.

The Tyranny of the Map All over Sub-Saharan Africa, different ethnic groups found themselves forced into the same state with peoples of different linguistic and religious backgrounds, many of whom had recently been their enemies. At the same time, several larger ethnic groups of the region found their territories split between two or more countries. The Hausa people of West Africa, for example, were divided between Niger (formerly French) and Nigeria (formerly British), each of which they had to share with several former ethnic rivals.

Given the imposed political boundaries, it is no wonder that many African countries struggled to generate a common sense of national identity or establish stable political institutions. **Tribalism**, or loyalty to the ethnic group rather than to the state, has emerged as the bane of African political life. Especially in rural areas, tribal identities are usually more important than national ones. Because nearly all of Africa's countries inherited an inappropriate set of colonial borders, some observers thought

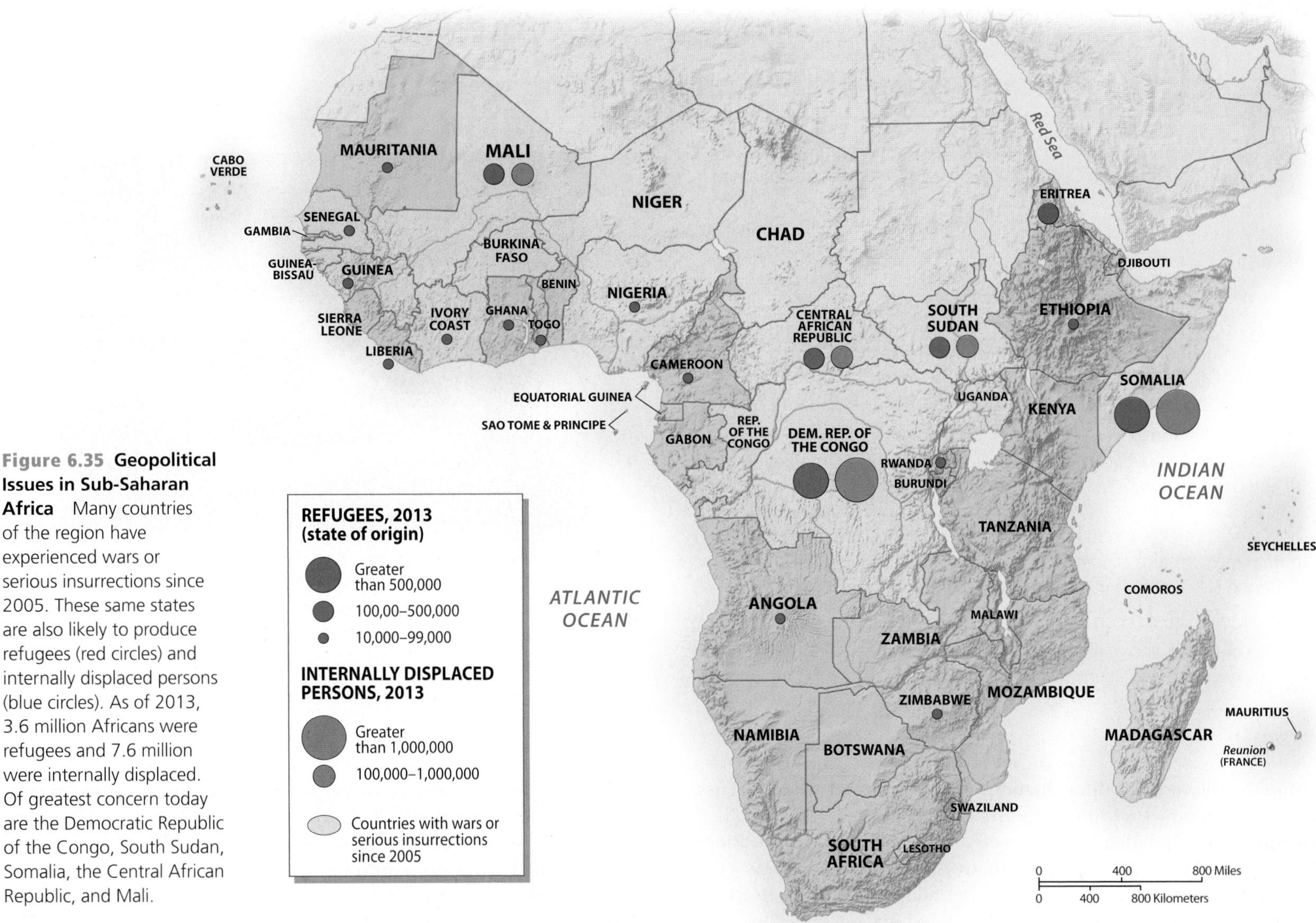

Figure 6.35 Geopolitical Issues in Sub-Saharan Africa Many countries of the region have experienced wars or serious insurrections since 2005. These same states are also likely to produce refugees (red circles) and internally displaced persons (blue circles). As of 2013, 3.6 million Africans were refugees and 7.6 million were internally displaced. Of greatest concern today are the Democratic Republic of the Congo, South Sudan, Somalia, the Central African Republic, and Mali.

they would have been better off drawing a new political map based on indigenous identities. However, such a strategy was impossible, as all the leaders of the newly independent states realized. Any new territorial divisions would have created winners and losers and thus would have resulted in even more conflict. Moreover, because ethnicity in Sub-Saharan Africa was traditionally fluid and because many ethnic groups were intermixed, it would have been difficult to generate a clear-cut system of division. Finally, most African ethnic groups were considered too small to form viable countries. With such complications in mind, the new African leaders, meeting in 1963 to form the Organization of African Unity, agreed that colonial boundaries should remain. The violation of this principle, they argued, would lead to pointless wars between states and endless civil struggles within them.

Despite the determination of Africa's leaders to build their new nations within existing political boundaries, challenges to the states began soon after independence. Figure 6.35 maps the ethnic and political conflicts that have disabled parts of Africa since 2005. The human cost of this turmoil is several million refugees and internally displaced persons. **Refugees** are people who flee their state because of a well-founded fear of persecution based on race, ethnicity, religion, or political orientation. According to the United Nations, nearly 3.6 million Sub-Saharan Africans were considered refugees at the end of 2013, with Somalia accounting for one-third of that total. Added to this figure are another 7.6 million **internally displaced persons (IDPs)**. IDPs have fled from conflict but still reside in their country of origin. The Democratic Republic of the Congo has the largest number of IDPs (3 million), followed by Somalia (1.1 million) and the Central African Republic (0.9 million). The number of IDPs in Mali and South Sudan is also growing. These people are not technically considered refugees, but they may receive some assistance and/or protection from the UN High Commissioner for Refugees.

Ethnic Conflicts In the 1990s, nearly two-thirds of the states in the region were experiencing serious ethnic conflict and, in the case of Rwanda, even genocide. As Figure 6.35 suggests, much of the conflict since 2005 occurred in the Sahelian states and Central Africa. Fortunately, in the past decade peace has returned to Sierra Leone, Liberia, the Ivory Coast, and Angola, states that produced large numbers of refugees in the 1990s and early 2000s.

Many observers attribute the cycles of violence in Sierra Leone and Liberia to the availability of diamonds as a means of financing conflict. Although the relationship between resources and conflict is

EVERYDAY GLOBALIZATION

The African Origins of the Diamond Engagement Ring

The tradition of a diamond engagement ring stems from the remarkable advertising efforts of DeBeers, a South African firm that dominated southern Africa's enormous diamond market throughout the 20th century. DeBeers Consolidated Mines was established in 1888 and steadily expanded to include diamond mines in Botswana, Namibia, and Canada, eventually controlling 90 percent of the global diamond trade (Figure 6.3.1).

DeBeers' genius was threefold: It expanded the supply of quality diamonds, controlled the global market, and convinced a growing middle class that diamonds were proof of love. After sales to Europe slowed in the 1920s, DeBeers began marketing in the United States, convincing American suitors to spend a month's wages on a diamond ring.

"Conflict Diamonds" DeBeers' lock on diamonds unraveled in the 1990s when Russia became a major diamond producer, marketing outside of the DeBeers commodity chain. The idea of "conflict diamonds" also emerged at that time, tainting the image of African diamonds. A certification process makes blood diamonds less of a concern today, but DeBeers' dominance has also slipped; the company now accounts for 40 percent of diamond sales, although southern Africa is still a major producer.

1. How did globalization weaken DeBeers' control of the diamond market?
2. Are diamond rings still popular among the newly engaged? Describe any diamond ring advertisements you have seen recently in the media.

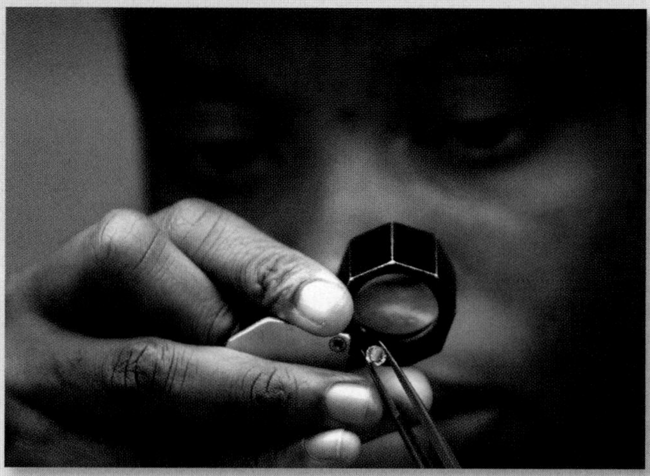

Figure 6.3.1 The Diamond Trade A young diamond cutter checks a 2-carat diamond she is polishing in Gaborone, Botswana. In 2008, the government of Botswana launched its own diamond-trading company in partnership with DeBeers in order to retain more jobs and income in this diamond-producing country.

complex, the term **conflict diamonds** was employed when discussing the diamond trade in West Africa in the 1990s (see *Everyday Globalization: The African Origins of the Diamond Engagement Ring*). One result of the public concern about conflict diamonds was a certification scheme, the Kimberly Process, adopted in 2002. Its aim is to keep conflict diamonds out of the global market and thus avoid tainting the image of the diamond business. In the Ivory Coast, where conflict began as violence that spilled over from Liberia in 2002, a peace deal was brokered in 2007 between the New Forces rebel group in the north and the government-controlled south.

The deadliest ethnic and political conflict in the region has been in the Democratic Republic of the Congo. It is estimated that between 1998 and 2010, 5.4 million people died there, although many of the deaths were from war-induced starvation and disease, rather than bullets or machetes. Current UN figures show half a million refugees are living outside the country and 3 million IDPs within it. In 1996 and 1997, a loose alliance of armed groups from Rwanda (led by Tutsis) and Uganda joined forces with other militias in the Congo and marched their way across the country, installing Laurent Kabila as president. Under Kabila's rocky and ruthless leadership, which ended in his assassination in 2001, rebel groups again invaded from Uganda and Rwanda and soon controlled the northern and eastern portions of the country, while the Kinshasa-based government loosely controlled the western and southern portions.

With Kabila's death, his son Joseph took power and signed a peace accord with the rebels in 2002. In 2003, rebel leaders were made part of a transitional government, and an unsteady peace was in place, with help from the UN, the AU, and Western donors. Remarkably, when elections were held in 2006, Joseph Kabila was elected president, and he was reelected in 2011. Yet Sub-Saharan Africa's largest state in terms of territory has only limited experience with democracy, its civil service barely functions, corruption is rampant, and there are few roads and little working infrastructure for the nearly 75 million people who live there. Moreover, armed groups scattered throughout the country continue to commit serious crimes, including mass rapes, torture, and murders. Due to years of conflict, the formal economy is small, and the informal economy dominates. The level of violence is certainly lower now, but this is still one of the region's troubled countries.

A low-intensity ethnic conflict in northern Mali heated up in 2012 as a Touareg-based National Movement for the Liberation of Azawad (MNLA) proclaimed independence from the Bamako-based government in the south, largely controlled by Mande-speaking groups. The MNLA had been formed in 2011, partly by armed Touareg fighters returning from Libya after the fall of Gaddafi's regime as part of the Arab Spring (see Chapter 7). As fighting intensified, a military coup in the capital of Bamako removed President Toure from office in March 2012. Leaders of the coup were dissatisfied with the state's inability to fight Touareg rebels emboldened by the arrival of weapons from Libya. The AU has denounced the actions and suspended Mali from the union, but the UN estimates that some half a million people have been uprooted by the ongoing conflict and several million people in the north face food insecurity.

The idea of Touareg independence, or autonomy, has existed for decades. Although this latest conflict seems to be driven by political events in North Africa and ethnic rivalries, others point to ecological pressures brought on by drought and/or climate change that undermine traditional patterns of resource sharing between farmers and pastoralists. There is no definitive connection between resource scarcity and ethnic clashes, but several of the region's current conflicts—from Nigeria to Mali, South Sudan to Somalia—are found in semiarid zones that are becoming more vulnerable to drought and famine.

Secessionist Movements Problematic African political boundaries have occasionally led to attempts by territories to secede and form new states. The Shaba (or Katanga) Province tried to leave what was then Zaire soon after independence. The rebellion was crushed a couple of years after it started, with the help of France and Belgium. Similarly, the Igbo in oil-rich southeastern Nigeria declared an independent state of Biafra in 1967. After a short but brutal war, during which Biafra was essentially starved into submission, Nigeria was reunited.

In 1991, the government of Somalia disintegrated, and the territory has been in civil war ever since. The lack of political control facilitated the rise of piracy; Somali pirates raid vessels and extort ransom from ships from the Gulf of Aden to the Indian Ocean. The territory has been ruled by warlords and their militias, who have informally divided the country into clan-based units. **Clans** are social units constituting branches of a tribe or an ethnic group larger than a family. Early in the conflict, the northern portion of Somalia declared its independence as a new country—Somaliland. Somaliland has a constitution, a functioning parliament, government ministries, a police force, a judiciary, and a president. The territory produces its own currency and passports. Yet no country has recognized this territory. In 1998, neighboring Puntland also declared autonomy but is not seeking independence. Meanwhile, Islamic insurgents with their well-armed militias control the south, around Mogadishu. In the past three years, UN peacekeeping forces and the Kenyan and Ethiopian militaries have been fighting the insurgents, trying to reclaim the cities. The need for stability has been exacerbated by four years of drought, creating a humanitarian emergency. The Somalian government was able to re-form in 2012, but conflict still persists. In June 2013, the Shabab, a violent Islamist group formed just to control Mogadishu, bombed a UN compound in that city (Figure 6.36).

Only two territories in the region have successfully seceded. In 1993, Eritrea gained independence from Ethiopia after two decades of civil conflict. This territorial secession is striking because Ethiopia gave up its access to the Red Sea, making it landlocked. Yet the creation of Eritrea still did not bring about peace. After years of fighting, the transition to Eritrean independence began remarkably well. Unfortunately, border disputes between the two countries erupted in 1998, resulting in the deaths of some 100,000 troops. In 2000, a peace accord was reached, and the fighting stopped. The second example is South Sudan, which gained its independence from Sudan in 2011 after some three decades of violent conflict between the largely Arab and Muslim north and the Christian and animist south. But peace has not come to this new territorial state either, which has experienced a surge in internally displaced people due to ethnic tensions that erupted in 2013. The difficulties experienced by the newly created states of South Sudan and Eritrea suggest that major changes to Africa's political map should not be expected.

Review

6.8 What are the processes behind Sub-Saharan Africa's political map, and why have there been relatively few boundary changes since the 1960s?

6.9 What are the present major conflicts in this region, and where are they occurring?

KEY TERMS Berlin Conference, apartheid, African Union (AU), tribalism, refugee, internally displaced person (IDP), conflict diamonds, clan

Figure 6.36 Somalian Conflict Somali National Government soldiers patrol the streets of Mogadishu after Shabab insurgents attacked the UN compound in that city in June 2013. The country has suffered through more than 25 years of conflict.

Economic and Social Development: The Struggle to Develop

By almost any measure, Sub-Saharan Africa is the poorest world region. According to World Bank estimates, 41 percent of the population lives in extreme poverty, surviving on under $1.25 per day, although in 1993 the figure was 61 percent. Due to poverty and low life expectancy, nearly all the states in the region are ranked at the bottom of the Human Development Index. Some demographically small or resource-rich states, such as Botswana, Equatorial Guinea, Mauritius, the Seychelles, and South Africa, have much higher per capita gross national incomes adjusted for purchasing power parity (GNI-PPP), but the average for the region was about $3300 in 2013 (Table A6.2). By way of comparison, the figure for South Asia, the next poorest region, was $5000.

Since 2000, strong commodity prices, new infrastructure, and improved technology (mobile phone subscriptions cover most of the population) have brightened the region's economic prospects. Over the past 10 years, real income per person grew (20 to 30 percent), whereas in the previous 20 years it actually decreased. The most optimistic views of the region see strengthened democracies, greater civic engagement, less violence, and growing investment (from within and outside the region). American economist Jeffrey Sachs argues that in order for the region to get out of the poverty trap, it will need substantial sums of new foreign aid and investment.

Roots of African Poverty

In the past, observers often attributed Africa's poverty to its colonial history, poorly conceived development policies, and/or corrupt governance.

Those who favored environmental explanations pointed to the region's infertile soils, erratic rainfall patterns, lack of navigable rivers, and virulent tropical diseases as reasons for underdevelopment. The best explanations for the region's poverty now point more to historical and institutional factors than to environmental circumstances.

Numerous scholars have singled out the slave trade for its debilitating effect on Sub-Saharan African economic life. Large areas of the region were depopulated, and many people were forced to flee into poor, inaccessible refuges. Colonization was another blow to Africa's economy. European powers invested little in infrastructure, education, and public health and were instead interested mainly in developing mineral and agricultural resources for their own benefit. Several plantation and mining zones did achieve some prosperity under colonial regimes, but strong national economies failed to develop. In almost all cases, the basic transport and communications systems were designed to link administration centers and zones of extraction directly to the colonial powers, rather than to their own surrounding areas. As a result, after achieving independence, Sub-Saharan African countries faced economic and infrastructural challenges that were as daunting as their political problems.

Failed Development Policies The first decade or so of independence was a time of relative prosperity and optimism for many African countries. Most of them relied heavily on the export of mineral and agricultural products, and through the 1970s commodity prices generally remained high. The region attracted some foreign capital, and in many cases the European economic presence actually increased after decolonization.

In the 1980s, as most commodity prices began to decline, foreign debt began to weigh down many Sub-Saharan countries. By the 1990s, most states were registering low or negative growth rates. Not only was the AIDS crisis raging, but also an economic and debt crisis in the 1980s and 1990s prompted the introduction of **structural adjustment programs** by the International Monetary Fund (IMF) and the World Bank. These programs typically reduce government spending, cut food subsidies, and encourage private-sector initiatives. Yet these same policies caused immediate hardships for the poor, especially women and children, and led to social protest, most notably in cities. Although the region's debt was low compared to that of other developing regions (such as Latin America), as a percentage of economic output, Sub-Saharan Africa's debt was the highest in the world.

Many economists argue that the region's governments enacted counterproductive economic policies and thus brought some of their misery on themselves. Eager to build their own economies and reduce their dependency on the former colonial powers, most African countries followed a course of economic nationalism. More specifically, they set about building steel mills and other forms of heavy industry that were simply not competitive. Local currencies were often maintained at artificially elevated levels, which benefited the elite who consumed imported products, but undercut exports.

Corruption Although prevalent throughout most of the world, corruption seems to have been particularly widespread in several African countries. Part of this is driven by a lack of transparent and representative governance and by a civil service class that lacks both resources and professionalism. According to a poll by an international business magazine, Nigeria ranks as the most corrupt country. (Skeptical observers, however, point out that several Asian nations with highly successful economies, such as China, are also noted for having high levels of corruption, so corruption alone may not be the problem.)

With millions of dollars in loans and aid pouring into the region, officials at various levels have been tempted to take something for themselves. Some African states, such as the Democratic Republic of the Congo, were dubbed *kleptocracies*. A **kleptocracy** is a state in which corruption is so institutionalized that most politicians and government bureaucrats siphon off a huge percentage of the country's wealth. President Mobutu, who ruled the Democratic Republic of the Congo (then Zaire) from 1965 to 1997, was a legendary kleptocrat. While his country was saddled with an enormous foreign debt, he reportedly skimmed several billion dollars from government funds and deposited them in Belgian banks.

Signs of Economic Growth

Most of the economies in the region are growing. From 2000 to 2009, the average annual growth rate for Sub-Saharan Africa was 5.7 percent. This rate has slowed to 4.2 percent since 2009, but it is still far higher than the rates in the 1990s.

Some policy makers argue that domestic and international aid targeted toward reducing extreme poverty has contributed to the region's economic and social development. As of 2015, the percentage of people in extreme poverty has declined, more children are in school, and tremendous strides have been made in combating both HIV/AIDS and malaria. This is still the world's poorest region, with serious problems, but its connections with the world are deepening and in many cases proving beneficial.

One bright spot in the region's economy is the growth in cellular and digital technology. Admittedly, fixed telephone lines are scarce; the regional average is 1 line per 100 people. Cell-phone usage, however, has soared. In 2007, the World Bank estimated 23 cell-phone subscriptions per 100 people in Sub-Saharan Africa; by 2013, the figure had nearly tripled, to 66 per 100 people. Multinational providers now compete for mobile-phone customers. Development specialists and entrepreneurs are exploring many new uses for cell phones and smartphones, with applications to secure not only micro-finance, but also educational tools and updates about health issues or weather patterns (Figure 6.37).

Figure 6.37 Mobile Phones for Africa A woman uses her cell phone in a market in Yaoundé, Cameroon. Cell-phone subscriptions in Sub-Saharan Africa tripled between 2007 and 2013, greatly improving communication.

Figure 6.38 East African Railroad Construction Kenyan workers lay new standard gauge railroad track as part of the Chinese-funded regional railway project that will modernize the rail link between Kenya's port of Mombasa and its capital, Nairobi. Plans to improve railroad links throughout East Africa are under way.

In Kenya, 39 percent of the population uses the Internet; the average for the region as a whole is 17 percent. The website *Africa Good News* (based in South Africa) focuses on the new ways in which Sub-Saharan peoples are developing their communities and engaging with the world.

A Surge in New Infrastructure Limited paved roads and railroads place limits on national economies, but large-scale infrastructural programs are in the works. South Africa is the only African state with a fully developed modern road network. Recently, Kenya inaugurated its first superhighway, an eight-lane, 42-kilometer road from Nairobi to Thika. The highway not only has transformed many of the towns along its route, but also is an expression of growing Chinese investment in this region, because the Chinese firm Wu Yi Co. performed much of the engineering and construction work.

Two major projects to improve regional train networks are under way. In East Africa, billions are being invested to renovate existing railroads, standardize gauges, and construct new lines to link East African cities to the port of Mombasa, Kenya. A new line from Kigali, Rwanda, is now being constructed and will connect to Uganda and then to Kenya. There are also plans to extend the East African rail network into South Sudan, Ethiopia, and the eastern Democratic Republic of the Congo. Much of the engineering and some of the financing for this 3000-mile (4800-km) network comes from China (Figure 6.38). In 2015, seven West African states (the Ivory Coast, Ghana, Togo, Benin, Nigeria, Burkina Faso, and Niger) announced plans to renovate, build, and integrate 1860 miles (3000 km) of railroads to facilitate the export of minerals and other primary products. For landlocked Burkina Faso and Niger, this project could greatly reduce transportation costs.

Finally, major water and energy projects are under way in the region. Ethiopia is constructing the Renaissance Dam on the Blue Nile. When completed in 2017, it will be Africa's largest dam. A project of this scale is not without controversy; the downstream nation of Egypt is deeply concerned that the dam will reduce total water flow into the Nile River. Others complain that the scale of the dam is more than Ethiopia needs—annual energy production is estimated to be 15,000 gigawatt-hours—and that the dam's environmental impacts are not fully understood. The region's potential for solar energy is also great. Sub-Saharan Africa's largest photovoltaic solar power project opened near Kimberly, South Africa, in 2014. The Jasper plant can produce 180,000 megawatt-hours of energy a year, enough to power 80,000 homes (Figure 6.39). This renewable energy project was developed by Solar Reserve, a California-based company that builds solar power stations for utilities around the world. Many other states in the region are interested in following South Africa's lead in solar power and are exploring other ways to encourage development that takes into account indigenous practices (*Geographers at Work: Vision for Sustainable Development in West Africa*).

Links to the World Economy

Sub-Saharan Africa's trade connections with the world are limited, accounting for just over 2 percent of global trade. The overall level of international trade is low, both within the region and outside it. Traditionally, most exports went to the European Union (EU), especially the former colonial powers of England and France. The United States is the second most common destination. That pattern is changing rapidly; China is now the single largest trading partner for the region, although collective trade between the EU nations and Sub-Saharan Africa is greater. Throughout the decade of the 2000s, China's trade with Sub-Saharan Africa grew 30 percent per year on average. During the same decade, India's and Brazil's levels of trade with Sub-Saharan Africa grew more than 20 percent annually.

The rise of China as the largest trading partner of and investor in Sub-Saharan Africa has generated much geopolitical discussion about China's influence in this resource-rich, but developing, world region. In 2013, China's President Xi Jinping visited Tanzania, South Africa, and the Democratic Republic of the Congo, promoting the mutual benefits of Sino-African relations. Estimates of the number of Chinese currently working in Sub-Saharan Africa vary. At the high end, Chinese sources claim over 1 million Chinese migrants have settled in the region; the low end is a quarter that number. Most agree that the top destination is South Africa, where Chinese migrants number somewhere between 200,000 and 400,000.

Aid Versus Investment In many ways, Sub-Saharan Africa is more tightly linked to the global economy through the flow of financial aid and loans than through the flow of goods. As Figure 6.40 shows, for several states (Burundi, Liberia, and Malawi) this aid accounted for more than 20 percent of GNI in 2013. Most of the aid comes from a handful of developed regions (Europe, North America, and Japan). In that same year, foreign development assistance to the region was

Figure 6.39 Solar Power The newly built Jasper Solar Power plant in Postmasburg, South Africa, can power up to 80,000 homes. Built by the U.S. based company Solar Reserve, this plant hopes to be a model for clean, sustainable energy in the region.

GEOGRAPHERS AT WORK

Vision for Sustainable Development in West Africa

Fenda Akiwumi was always interested in the environment and the development of her country, Sierra Leone. Trained as a hydrogeologist in the United Kingdom, Akiwumi worked for over a decade in Sierra Leone's Ministry of Agriculture and Forestry and as a consultant developing and managing water resources. Yet her "eureka moment" came much later when she found a way to bridge the gap between the physical sciences and the social sciences—by studying geography. Says Akiwumi: "One thing you have to understand why geography is so important is that no matter what we do, no matter what field we're in, ultimately it all boils down to people... you have a great opportunity to help people in whatever it is you're doing."

One of the things Akiwumi loved most was fieldwork: constructing wells, regulating wetlands for rice production, and training mostly male technicians in basic hydrogeology. Since mining is a vital part of Sierra Leone's economy, she also worked as a consultant at mine sites. During those years, she often wondered what villagers thought about these projects: "Our countries are embracing these investments without regard to people and places."

The blood-diamond war forced Akiwumi to migrate to the United States in 1991. She taught for a while but wanted a more holistic approach toward development that included the physical environment and cultural geography. Eventually she earned a PhD in geography at Texas State University.

Now an associate professor at the University of South Florida, Akiwumi continues to work and conduct research in West Africa (Figure 6.4.1). She says that her geographical training encourages her to examine human–environment dynamics, particularly the interconnections between customary livelihoods and sacred places (rivers, lakes, and springs with great cultural meaning) that local communities use in ceremonies and rituals. Sierra Leone's government recognizes these spaces, but when conflict over mining rights ensues, these indigenous claims are often ignored. "The mining industry must respect cultural heritage and promote community participation for sustainable, conflict-free mining to exist," argues Akiwumi.

Figure 6.4.1 **Discussing Sustainability** Professor Fenda Akiwumi (in yellow and green dress, fourth from right) speaks with village women in rural Sierra Leone about natural resource practices. Akiwumi now teaches geography at the University of South Florida and conducts research in West Africa.

One of the biggest challenges facing West Africa is developing practices that equitably consider indigenous systems because "colonial policies are still embedded in post-colonial policies and laws." Yet Akiwumi is optimistic about the resilience of the rural people she works with: "Despite external influences and pressures, they find ways to sustain important aspects of their culture."

1. How does development conflict with cultural practices in West Africa?
2. Provide an example of a development project that affects local practices in your area.

valued at $47 billion. Other countries, including Botswana, South Africa, Angola, and Nigeria, receive relatively little aid because their economies are larger and they have mineral or oil wealth.

Although aid is extremely important for many African states, foreign direct investment in the region substantially increased from only $4.5 billion in 1995 to $38 billion in 2013. Yet the overall level of foreign investment remains low when compared to other developing regions. In 2014, the largest recipients of foreign investment were Angola, Nigeria, Mozambique, Ghana, and South Africa. China has been the leading investor in the region at a time when the United States and the EU are more focused on offering aid or fighting terrorism. China wants to secure the oil and ore it needs for its massive industrial economy. In exchange, it offers Sub-Saharan nations money for roads, railways, housing, and schools, with relatively few strings attached. Some African leaders see China as a new kind of global partner, one that wants straight commercial relations without an ideological or political agenda. Angola, a country in which China has invested heavily, is now one of China's top suppliers of oil. Yet not all investments have paid off. A massive real estate development outside of Luanda, Angola, called Nova Cidade de Kilamba, was largely unoccupied in 2012 because the cost of purchasing an apartment was well beyond the means of the average Angolan (Figure 6.41).

Debt Relief Another development strategy making a difference in the region is debt relief. In 1996, the World Bank and the IMF proposed

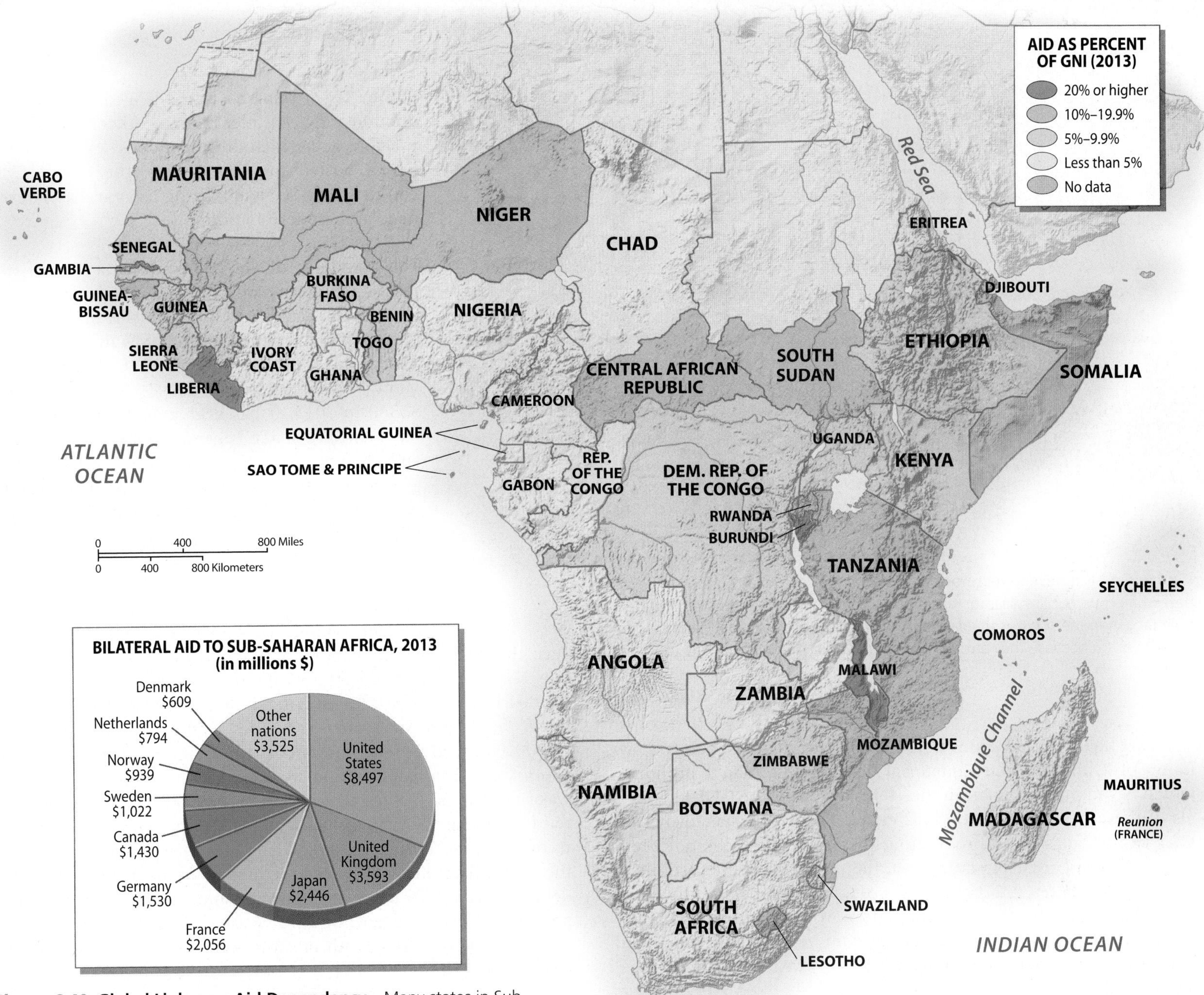

Figure 6.40 Global Linkages: Aid Dependency Many states in Sub-Saharan Africa are dependent on foreign aid as their primary link to the global economy. This figure maps aid as a percentage of gross national income (GNI), which ranges from less than 1 percent to 31.5 percent in Malawi. **Q: Compare the most aid-dependent states in the region with the least aid-dependent ones. What are the differences between these two groups of states?**

reducing debt for heavily indebted poor countries, many of which are in Sub-Saharan Africa. Most Sub-Saharan states are indebted to official creditors such as the World Bank, rather than to commercial banks (as is the case in Latin America and Southeast Asia). Under this World Bank/IMF program, Sub-Saharan countries that are found to have "unsustainable" debt burdens are permitted debt relief. Mauritania, for example, was spending six times more money on debt repayment than it was on health care.

States qualify for different levels of debt relief, depending on their poverty-reduction strategies. Uganda was the first state to qualify for the program in 2000, using money it saved on debt repayment to expand primary schooling. Ghana, which qualified for debt relief in 2004, received a $3.5 billion relief package. Other countries that have benefited from debt reduction are Tanzania, Mozambique, Ethiopia, Mauritania, Mali, Niger, Nigeria, Senegal, Burkina Faso, and Benin. Other African countries may also be able to redirect debt payments toward building infrastructure and improving basic health and educational services.

Economic Differentiation Within Africa

As in most other regions, considerable differences in levels of economic and social development persist in Sub-Saharan Africa. In many respects, the small island nations of Mauritius and the Seychelles have little in common with the mainland. With high levels of per capita GNI, life expectancies averaging in the low 70s, and economies built on tourism, they could more easily fit into the Caribbean were it not for their Indian Ocean location.

Figure 6.41 Chinese Investment in Angola Children play on the basketball courts at the Kilambi Kiaxi housing development, a massive Chinese-built project in the suburbs of Luanda, Angola. Most of these apartments remain vacant because they are far too expensive for Angolan workers to purchase.

Two small African states noted for oil wealth are Gabon (population of 2 million) and Equatorial Guinea (less than 1 million), which began producing oil in the 1990s. In 2013, the GNI-PPP levels of these two states were $17,230 and $23,270, respectively. Yet after two decades of oil production, these revenues have not been invested in the country's citizens; rather, as is often the case, they seem to have fallen into the pockets of a few members of the elite. In contrast, two-thirds of the population of mainland Sub-Saharan Africa subsists on less than $2 per day. Only a few states, mostly in southern Africa, have per capita GNI-PPP levels over $5000 (see Table A6.2).

Given the scale of the African continent, it is not surprising that groups of states have formed trade blocs to facilitate intraregional exchange and development (Figure 6.42). The two most active regional organizations are the **Southern African Development Community (SADC)** and the **Economic Community of West African States (ECOWAS)**. Both were founded in the 1970s but became more important in the 1990s, and each is anchored by one of the region's two largest economies: South Africa and Nigeria. Other regional trade blocs include the Economic Community of Central African States (ECCAS) and the smaller, but more effective East African Community (EAC).

South Africa and SADC South Africa is the most developed large country of Sub-Saharan Africa, with a per capita GNI-PPP of $12,500. Botswana and Namibia, with strong mining economies, also do well in terms of per capita income. Through SADC, there have been efforts to integrate and improve the infrastructure of its member countries. Yet only South Africa has a well-developed and well-balanced industrial economy. It also boasts a healthy agricultural sector, and, more important, it is one of the world's mining superpowers. South Africa remains unchallenged in gold production and is a leader in many other minerals and precious gems, including diamonds. In 2010, South Africa hosted the World Cup, the first African country to do so, symbolizing its arrival as a developed and modern nation.

The Leaders of ECOWAS Nigeria is Africa's most populous country and its largest economy; it is also the core member of ECOWAS. Nigeria has the largest oil reserves in Sub-Saharan Africa, and it is a member of OPEC. Yet despite its natural resources, its per capita GNI-PPP was only $5360 in 2013. It has been argued that oil money has helped to make Nigeria notoriously inefficient and corrupt. A small minority of Nigerians have grown fantastically wealthy, more by manipulating the system than by engaging in productive activities. Eighty-two percent of Nigerians, however, remain trapped in poverty, earning less than $2 per day.

The second and third most populous states in ECOWAS, the Ivory Coast and Ghana, are also important commercial centers that rely on a mix of agricultural and mineral exports. In the mid-1990s, the Ivorian economy began to take off. Boosters within the country called it an emerging "African elephant" (comparing it to the successful "economic tigers" of eastern Asia). However, a destructive civil war that began in 2002 resulted in rebel forces controlling the northern half of the country and over half a million displaced Ivorians. A peace agreement was signed in 2007, but the economic growth of the 1990s has yet to return. Ghana, a former British colony, also began to see economic recovery in the 1990s. In 2001, Ghana negotiated with the IMF and the World Bank for debt relief to reduce its nearly $6 billion foreign debt. Between 2009 and 2013, Ghana maintained an average annual growth rate of 10 percent. In 2011, Ghana also became an emerging oil producer for Africa, with offshore wells being pumped near the city of Takoradi.

East Africa Long the commercial and communications center of East Africa, Kenya experienced economic decline and political tension throughout the 1990s. Yet from 2009 to 2013, its economy averaged 6 percent annual growth, and its per capita GNI-PPP was at $2780. Kenya boasts good infrastructure by African standards, and over 1 million foreign tourists come each year to marvel at its wildlife and natural beauty. Traditional agricultural exports of coffee and tea, as well as nontraditional exports such as cut flowers, dominate the economy.

Kenya is also East Africa's technological leader. In 2009, the government launched plans for Konza City, a walkable technology-oriented development 60 kilometers (37 miles) southeast of Nairobi. The intent is to capture the growing business in outsourced information technology services; Kenya is an English-speaking country where 82 percent of youth (ages 15–24) are literate. If Kenya can avoid political unrest due to ethnic rivalries, it could lead East Africa toward better economic integration.

The political and economic indicators for Kenya's neighbors, Uganda and Tanzania, are also improving, with average annual growth rates at 5.9 and 6.6 percent, respectively (see Table A6.2).

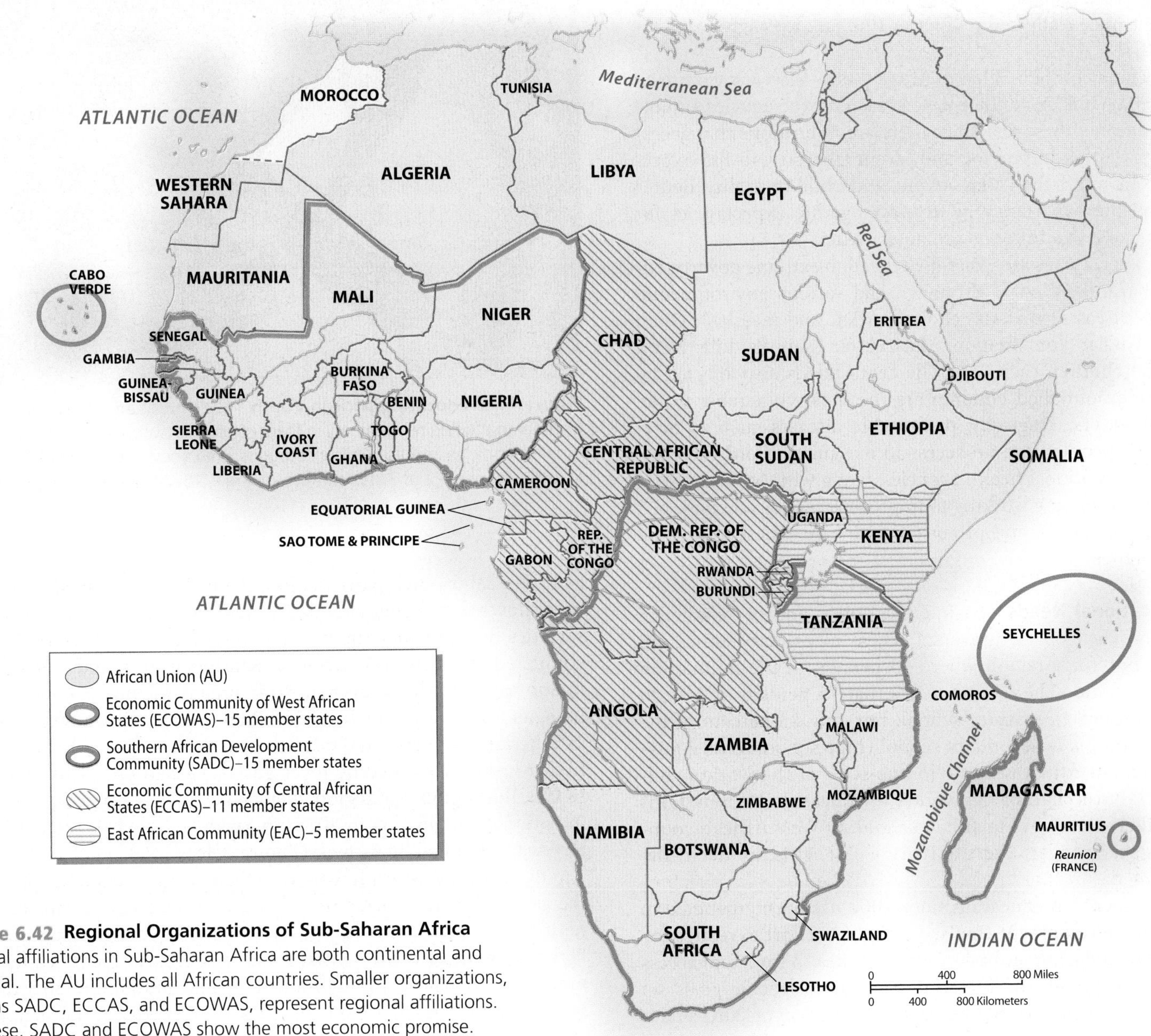

Figure 6.42 Regional Organizations of Sub-Saharan Africa Political affiliations in Sub-Saharan Africa are both continental and regional. The AU includes all African countries. Smaller organizations, such as SADC, ECCAS, and ECOWAS, represent regional affiliations. Of these, SADC and ECOWAS show the most economic promise.

Both countries rely heavily on agricultural exports and mining (especially gold), and both benefited from debt reduction agreements in the 2000s that redirected debt repayment funds to education and health care.

Measuring Social Development

By global standards, measures of social development in Sub-Saharan Africa are extremely low. Yet some positive trends, especially with regard to child survival and youth literacy, are cause for hope (see Table A6.2). Many governments in the region have reached out to the modern African diaspora (economic migrants and refugees who now live in Europe and North America). In other parts of the world, African immigrant organizations have worked to improve schools and health care, and former emigrants have returned to invest in businesses and real estate. The economic impact of remittances in this region is small, but growing.

Child Mortality and Life Expectancy Reductions in child mortality are a surrogate measure for improved social development, because if most children make it to their fifth birthday, it usually indicates adequate primary care and nutrition. A child mortality rate of 200 per thousand means that one child out of five dies before his or her fifth birthday. As Table A6.2 shows, most of the states in the region saw modest to significant improvements in child survival between 1990 and 2013. Eritrea, Liberia, Madagascar, Malawi, and Rwanda actually experienced dramatic gains in child survival rates. However, countries with prolonged conflict (such as Somalia and the Democratic Republic of the Congo) have seen little improvement. In 1990, the regional child mortality rate was 175 per thousand; in 2013, it was down to 92. Thus, high child mortality is still a major concern for the region, but steady reductions in this rate are significant for African families.

Life expectancy for Sub-Saharan Africa is only 57 years. Countries hit hard by HIV/AIDS or conflict have seen life expectancies tumble into the 40s. Despite these statistics, there are indications that access to basic health care is improving and, eventually, so will life expectancies. Keep in mind that high infant- and child-mortality figures depress overall life expectancy figures; average life expectancies for people who make it to adulthood are much better.

Low life expectancies are generally related to extreme poverty, environmental hazards (such as drought), and various environmental and infectious diseases (malaria, cholera, AIDS, and measles). Often these factors work in combination. Malaria, for example, kills half a million African children each year. The death rate is also affected by poverty, as undernourished children are the most vulnerable to the effects of high fevers. Tragically, preventable diseases such as measles occur when people have no access to or cannot afford vaccines. National and international health agencies, along with NGOs such as the Gates Foundation, are working to improve access to vaccines, bed nets (to prevent malaria), and primary health care. These efforts are making a difference.

Figure 6.43 Educating African Youth Students attend classes at a Lutheran theological seminary in Francistown, Botswana. Religious organizations, as well as public institutions, are critical in providing education. In 2013, 96 percent of Botswana's youth (ages 15–24) were literate.

Meeting Educational Needs Basic education is another challenge for the region. The goal of universal access to primary education is a daunting one for a region where 43 percent of the population is under 15 years old. The UN estimates that 75 percent of African children are enrolled in primary school, but only 23 percent of the relevant population is in secondary school (high school or its equivalent). Sub-Saharan Africa is home to one-sixth of the world's children under 15, but to half of the world's uneducated children. Girls are still less likely than boys to attend school. In West African countries such as Chad, Niger, and the Ivory Coast, girls are decidedly underrepresented.

A renewed focus on education since 2000 has been attributed to the **Millennium Development Goals**, a global UN effort to reduce extreme poverty by focusing on basic education, health care, and access to clean water. Although the region did not meet specific goals set for 2015, more government and nonprofit organization resources have been directed to education. More schools are being built across the region, and more children are attending them (Figure 6.43).

Women and Development

Development gains cannot be achieved in Africa unless the economic contributions of African women are recognized. Officially, women are the invisible contributors to local and national economies. In agriculture, women account for 75 percent of the labor that produces more than half the food consumed in the region. Tending subsistence plots, taking in extra laundry, and selling surplus produce in local markets all contribute to household income. Yet because many of these activities are considered part of the informal sector, they are not counted. For many of Africa's poorest people, however, the informal sector is the economy, and within this sector women dominate.

Status of Women The social position of women is difficult to measure for Sub-Saharan Africa. Female traders in West Africa, for example, have considerable political and economic power. By such measures as female labor force participation, many Sub-Saharan African countries show relative gender equality. Also, women in most Sub-Saharan societies do not suffer the kinds of traditional social restrictions encountered in much of South Asia, Southwest Asia, and North Africa; in Sub-Saharan Africa, women work outside the home, conduct business, and own property. In 2006, Ellen Johnson-Sirleaf was sworn in as Liberia's president, making her Africa's first elected female leader. In 2012, she was joined by Joyce Banda, elected president of Malawi. In fact, throughout the region, women occupied 22 percent of all seats in national parliaments in 2012. In Rwanda, over half the parliamentary seats are filled by women (Figure 6.44).

By other measures, however, such as the prevalence of polygamy, the practice of the "bride-price," and the tendency for males to inherit property over females, African women do suffer discrimination. Perhaps the most controversial issue regarding women's status is the practice of female circumcision, or genital mutilation. In Ethiopia, Somalia, and Eritrea, as well as parts of West Africa, the majority of girls are subjected to this practice, which is extremely painful and can have serious health consequences. Yet because the practice is considered traditional, most African states are unwilling to ban it.

Regardless of their social position, most African women still live in remote villages where educational and wage-earning opportunities remain limited and caring for large families is time-consuming and demanding labor. As education levels increase and urban society expands—and as reduced infant mortality provides greater security—we can expect fertility in the region to gradually decrease. Governments can speed up the process by providing birth control information and cheap contraceptives—and by investing more money in women's health and education. As the economic importance of women receives greater attention from national and international organizations, more programs are being directed exclusively toward them.

Building from Within

Surveys reveal that the majority of people in Sub-Saharan Africa are optimistic about their future. Considering many of the real development

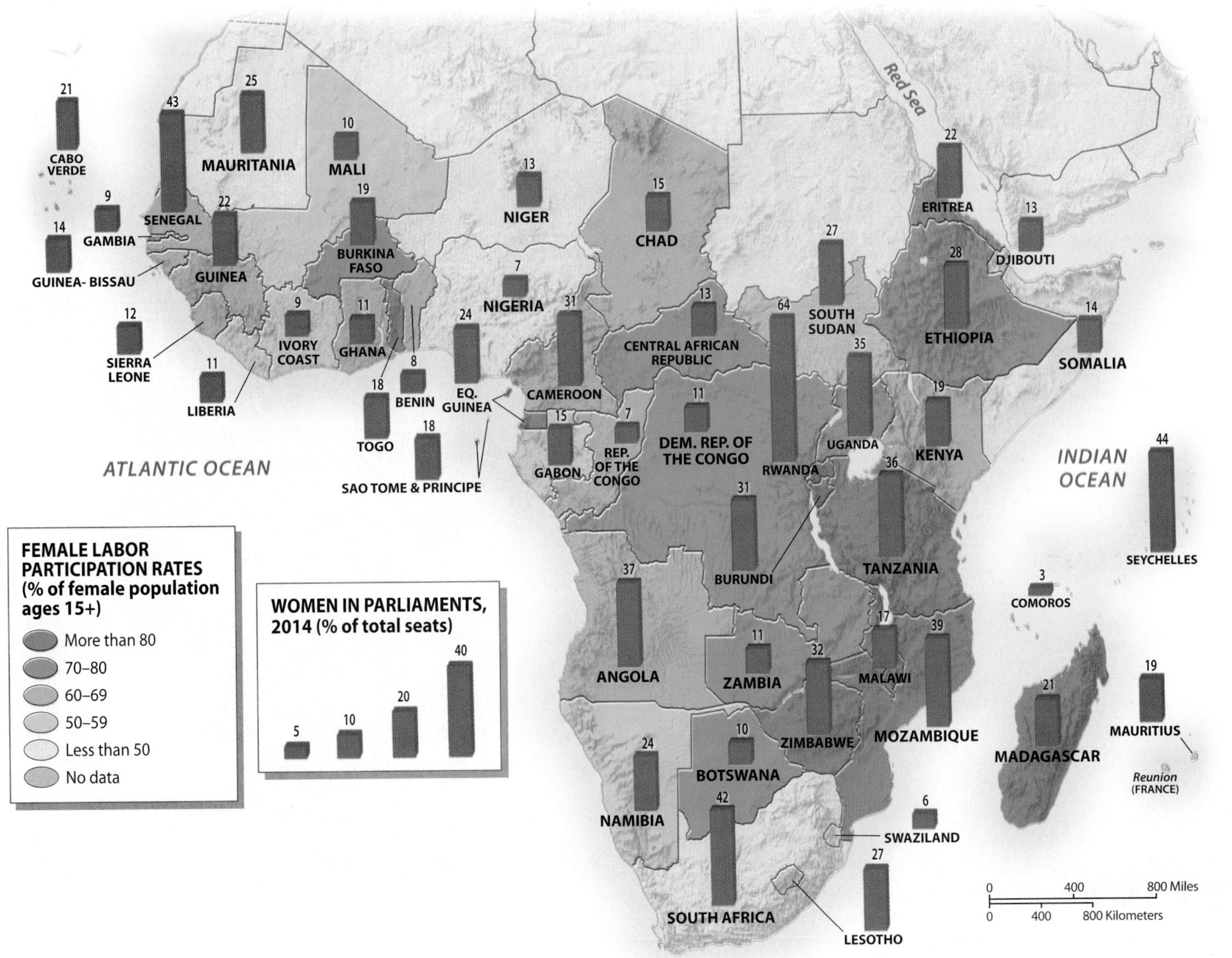

Figure 6.44 Development Issues: African Women in the Workforce and Politics Female participation in the workforce is comparable to that of developed countries, as 70 percent of women over the age of 15 are in the labor force. Another significant change for the region is the increase in women holding seats in national parliaments. The regional average in 2014 was 22 percent, but women held 42 percent of parliamentary seats in South Africa and 64 percent in Rwanda. In contrast, only 7 percent of the parliamentary seats were held by women in Africa's largest country, Nigeria.

hurdles the region faces, this surprises outside observers. Yet considering the levels of conflict, food insecurity, and neglect that Sub-Saharan Africa experienced during the 1990s, perhaps the developments of the past decade are a cause for hope. Most African states have been independent for only half a century. During that time, these countries have shifted from one-party authoritarian states to multiparty democracies. Targeted aid projects to address particular critical indicators, such as mother and child mortality or malaria prevention, have proved their effectiveness. Civil society is also vigorous, from raising the status of women to supporting small businesses with micro-credit loans. Even members of the African diaspora are beginning to return and invest in their countries. The rapid adoption of and adaptation to mobile phone technology in the region demonstrates Sub-Saharan Africans' desire to be better connected with each other and with the world.

Review

6.10 What are the historical, structural, and institutional reasons offered to explain poverty in the region?

6.11 What technological and infrastructural investments are affecting the region's development?

KEY TERMS structural adjustment program, kleptocracy, Southern African Development Community (SADC), Economic Community of West African States (ECOWAS), Millennium Development Goals

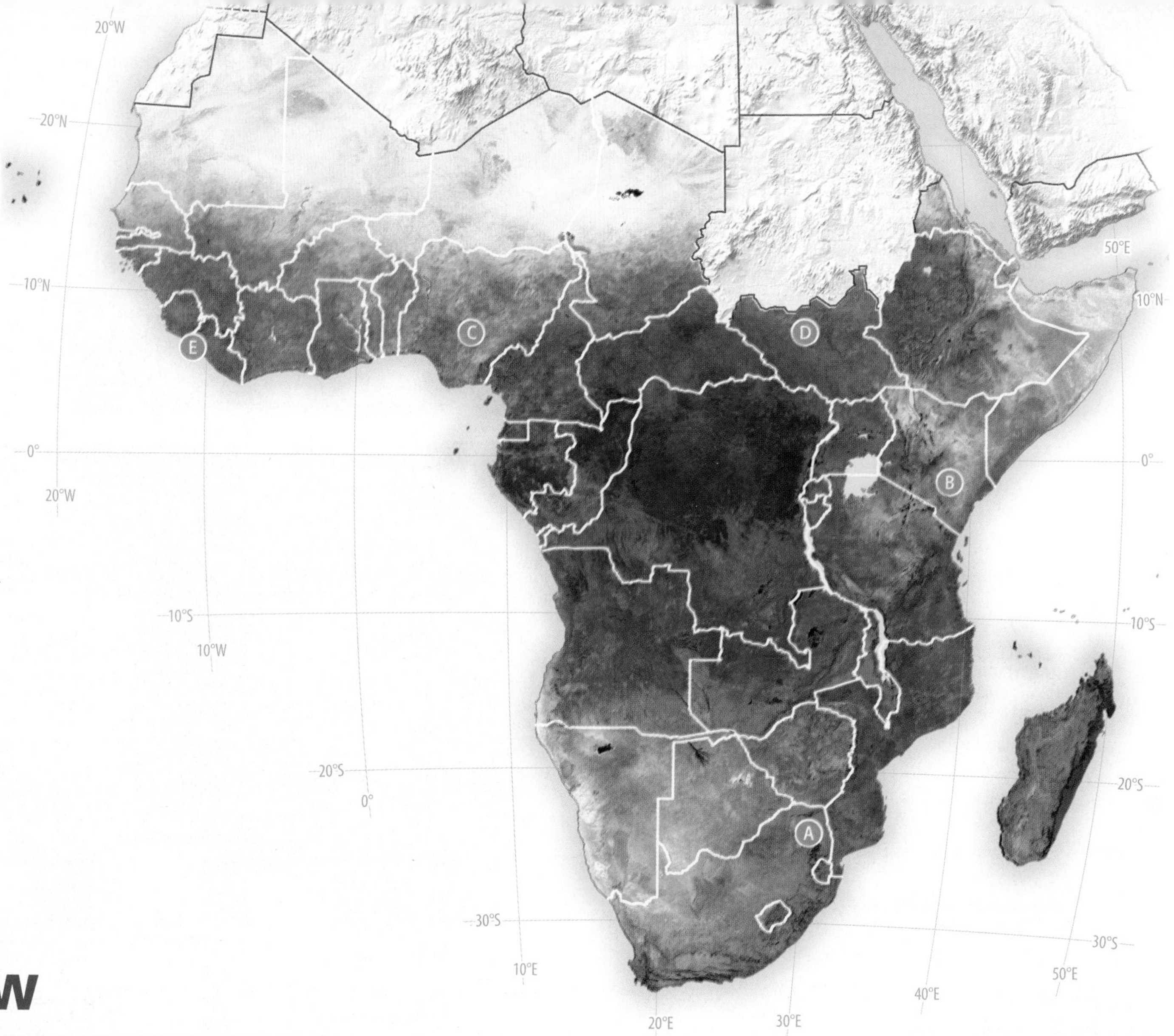

Review

Physical Geography and Environmental Issues

6.1 Describe the major ecosystems in the region and how humans have adapted to living in them.

6.2 Outline the environmental issues that challenge Africa south of the Sahara.

The largest landmass straddling the equator, Africa is called the plateau continent because it is dominated by extensive uplifted plains. Key environmental issues facing this tropical region are desertification, deforestation, and drought. At the same time, the region supports a tremendous diversity of wildlife.

1. Sub-Saharan Africa is noted for its wildlife, especially large mammals. In which Sub-Saharan African country are encounters with large mammals likely to occur, and why?
2. Look at Figure 6.9. In which areas is tropical deforestation occurring, and why?

Population and Settlement

6.3 Explain the region's rapid demographic growth and describe the differential impact of diseases such as HIV/AIDS and Ebola on the region.

With 950 million people, Sub-Saharan Africa is the fastest-growing region in terms of population, with the average woman having five children. Yet it is also the poorest world region, with two-thirds of its people living on less than $2 a day. In addition, it has the lowest average life expectancy, at 57 years.

3. What factors might explain the density of settlement in this region of Africa?
4. Consider how increasing urbanization may affect the overall structure of the population in Sub-Saharan Africa.

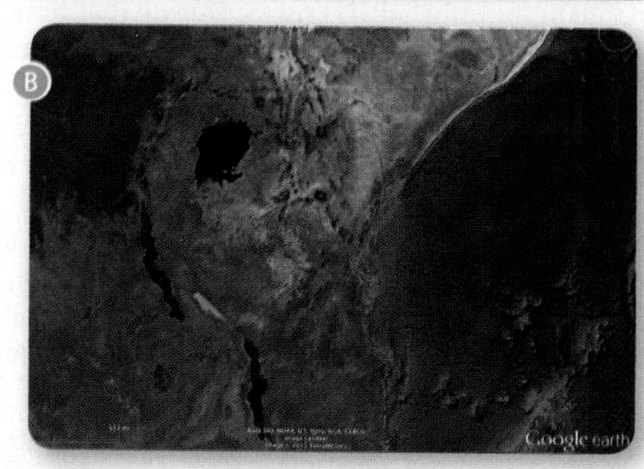

Cultural Coherence and Diversity

6.4 Describe the relationship between ethnicity and conflict in this region and the strategies for maintaining peace.

6.5 Summarize various cultural influences of African peoples within the region and globally.

Sub-Saharan Africa is culturally diverse, where multiethnic and multireligious societies are the norm. With a few exceptions, religious diversity and tolerance have been distinctive features of the region. Most states have been independent for 50 years, and in that time pluralistic but distinct national identities have been forged. Many African cultural expressions, such as music, dance, and religion, have been influential beyond the region.

5. Nigeria is linguistically diverse. What do the distinct language families in this state tell us about the area's settlement?

6. Compare and contrast the role of tribalism in Sub-Saharan Africa with that of nationalism in Europe.

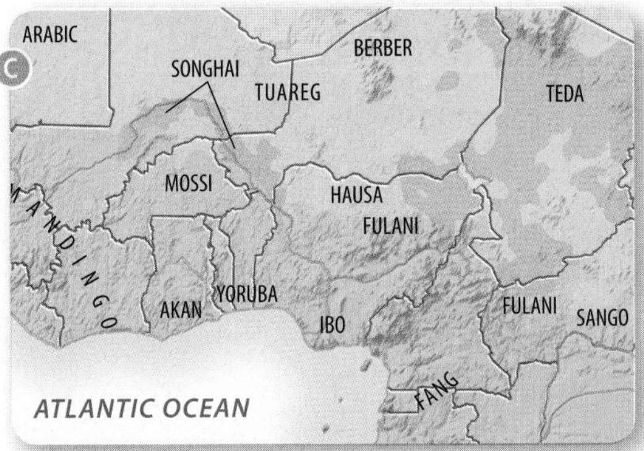

Geopolitical Framework

6.6 Trace the colonial history of the region and link colonial policies to postindependence conflicts.

In the 1990s, many bloody ethnic and political conflicts occurred in the region. Peace now exists in many conflict-ridden areas, such as Angola, Sierra Leone, and Liberia. However, ongoing ethnic and territorial disputes in Somalia, the Democratic Republic of the Congo, South Sudan, and Mali have produced millions of internally displaced persons and refugees.

7. Consider South Sudan. Which regional trade bloc do you think this new nation is likely to join?

8. Historically, how was Sub-Saharan Africa integrated into the global economy? Was its role similar to that of other developing regions?

Economic and Social Development

6.7 Assess the roots of African poverty and explain why many of the fastest-growing economies in the world today are in Sub-Saharan Africa.

6.8 Explain how reductions in conflicts can improve educational and social development outcomes in the region.

Widespread poverty is the region's most pressing concern. Since 2000, Sub-Saharan economies have grown, led in part by higher commodity prices, greater investment, and the end of some of the longest-running conflicts in the region. Social indicators of development are also improving, due to greater attention from the international community and better access to health care and drugs to fight HIV/AIDS.

9. This photograph was taken in Sierra Leone. What economic activity is taking place, and what are its economic and environmental consequences?

10. Compare and contrast the development model put forward by the United States and Europe with that of China. Will Chinese influence in the region alter the course of development for Sub-Saharan Africa?

KEY TERMS

African Union (AU) *(p. 201)*
apartheid *(p. 201)*
Berlin Conference *(p. 199)*
biofuel *(p. 182)*
clan *(p. 205)*
coloured *(p. 192)*
conflict diamonds *(p. 204)*
desertification *(p. 181)*
Economic Community of West African States (ECOWAS) *(p. 210)*
Great Escarpment *(p. 176)*
Great Rift Valley *(p. 176)*
Horn of Africa *(p. 180)*
internally displaced person (IDP) *(p. 203)*
kleptocracy *(p. 206)*
Millennium Development Goals *(p. 212)*
pastoralist *(p. 190)*
refugee *(p. 203)*
Sahel *(p. 182)*
Southern African Development Community (SADC) *(p. 210)*
structural adjustment program *(p. 206)*
swidden *(p. 190)*
township *(p. 192)*
transhumance *(p. 182)*
tribalism *(p. 202)*
tsetse fly *(p. 191)*

DATA ANALYSIS

http://goo.gl/yY2id3

Sub-Saharan Africa has experienced more AIDS-related deaths and holds more people living with HIV/AIDS than any other world region. Country-specific data shed light on the impact of this disease and its geography. Go to the United Nations AIDS info website (http://aidsinfo.unaids.org) and access an interactive map of country-level data for 1990–2014. Scroll over a country to see the data, and adjust the time bar along the bottom to retrieve data from a particular year. The column on the left allows you to select different variables. Pick six to eight countries in Sub-Saharan Africa, including at least one country from its western, eastern, central, and southern subregions.

1. Compare and contrast the number of male adults and the number of female adults living with HIV/AIDS. Where are female numbers higher, and why?
2. New infections are declining, but not in all places. Look at new HIV infections, and compare figures in 1990 and 2000 with figures in 2014. Describe the trends.
3. Now explore AIDS-related deaths (perhaps comparing males and females) and also estimates of AIDS orphans.
4. Use your data to write a couple of paragraphs about how gender, treatment, and loss of life have impacted each subregion and the region as a whole.

MasteringGeography™

Looking for additional review and test prep materials? Visit the Study Area in MasteringGeography™ to enhance your geographic literacy, spatial reasoning skills, and understanding of this chapter's content by accessing a variety of resources, including MapMaster interactive maps, geoscience animations, videos, *In the News* RSS feeds, flashcards, web links, self-study quizzes, and an eText version of *Globalization and Diversity*.

Authors' Blogs

Scan to visit the **Author's Blog for field notes, media resources, and chapter updates**

http://gad4blog.wordpress.com/category/sub-saharan-africa/

Scan to visit the **GeoCurrents Blog**

http://geocurrents.info/category/place/ subsaharan-africa

7 Southwest Asia and North Africa

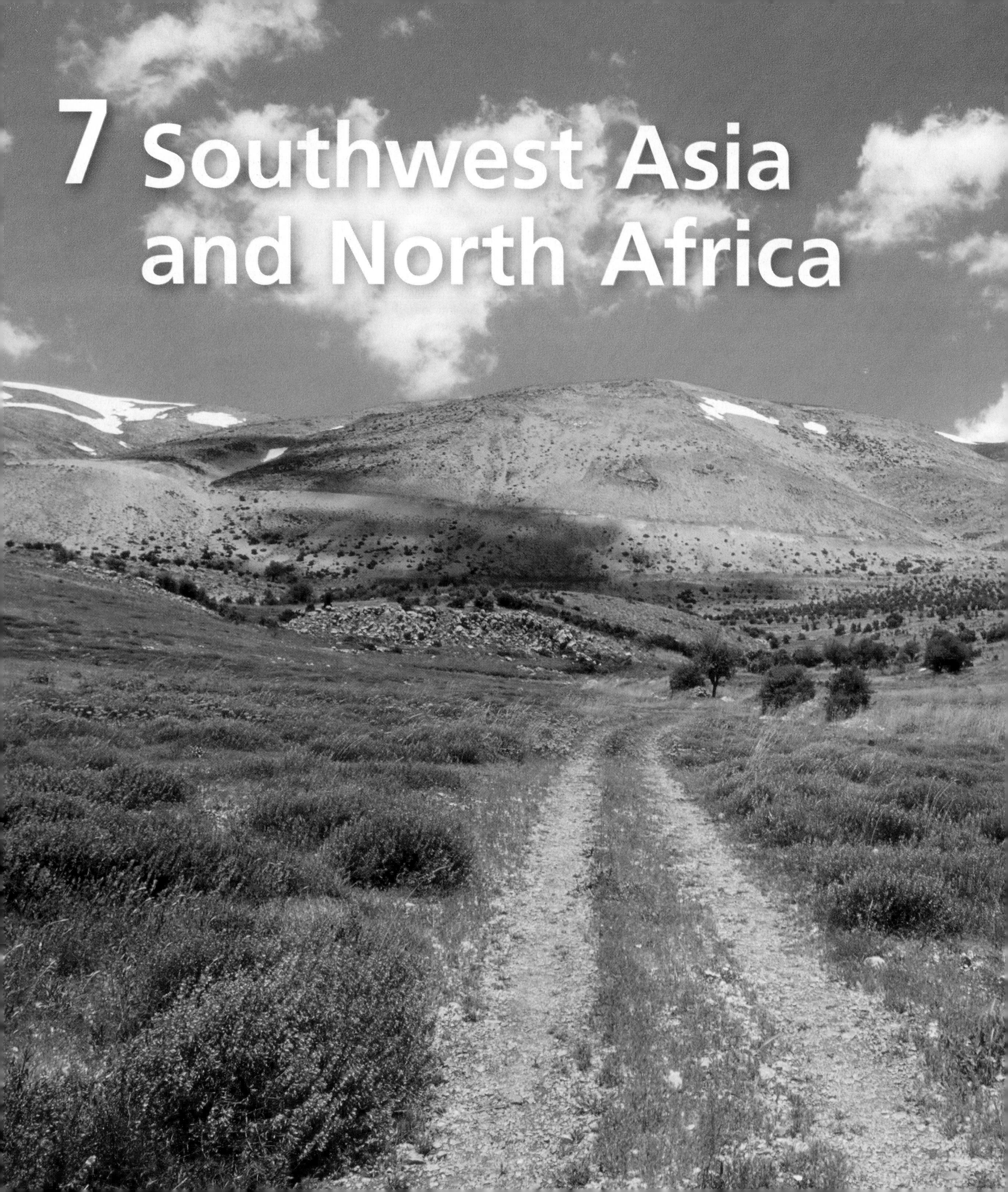

PHYSICAL GEOGRAPHY AND ENVIRONMENTAL ISSUES

The region's vulnerability to water shortages is likely to increase in the early 21st century as growing populations, rapid urbanization, and increasing demands for agricultural land consume limited supplies.

POPULATION AND SETTLEMENT

Many settings within the region continue to see rapid population growth. These demographic pressures are particularly visible in fragile, densely settled rural zones as well as in fast-growing large cities.

CULTURAL COHERENCE AND DIVERSITY

Islam continues to be a vital cultural and political force within the region, but increasing fragmentation within that world has led to more culturally defined political instability.

GEOPOLITICAL FRAMEWORK

The Arab Spring uprisings in the early 2010s jolted the geopolitical status quo in Tunisia, Egypt, Libya, Yemen, and Bahrain. Internal instability and the growth of ISIL have produced extensive bloodshed in Syria and Iraq. Prospects for peace between Israel and the Palestinians remain murky, and Iran's growing political role is seen by many as a threat both within and beyond the region.

ECONOMIC AND SOCIAL DEVELOPMENT

Unstable world oil prices and unpredictable geopolitical conditions have discouraged investment and tourism in many countries. The pace of social change, especially for women, has quickened, stimulating diverse regional responses.

◀ This narrow road takes travelers from Lebanon's Bekaa Valley into the beautiful Mount Lebanon Range northeast of Beirut.

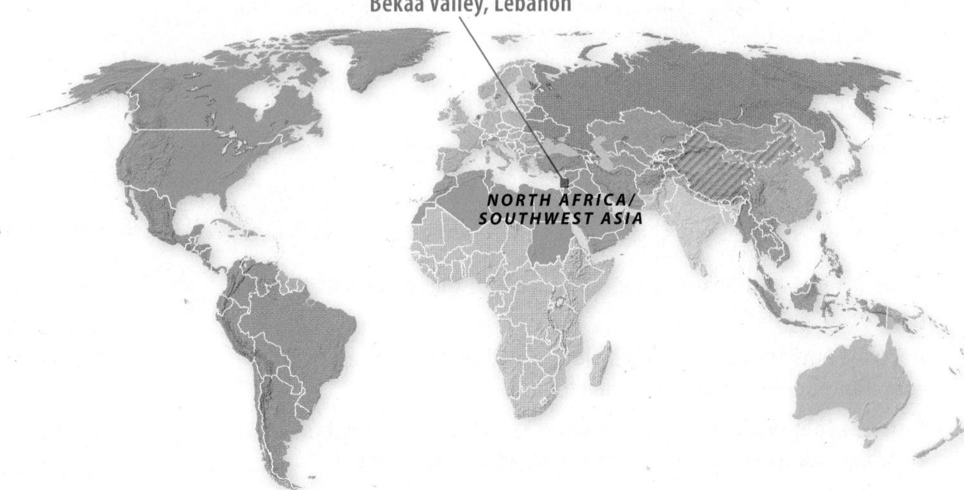

Spring in the Mount Lebanon Range northeast of Beirut can be a beautiful time of year as highland snowbanks retreat, flowers blossom, and upland pastures reveal their first hint of green. Increasingly, these mountains are protected from development and portions of the region are a UNESCO Biosphere Reserve. They remain home to rare trees (the cedars of Lebanon) as well as an amazing collection of mammals (boars, gray wolves, mountain gazelles) and birds (griffon vultures, Eurasian jays). Now a focus of sustainable tourism, the region is a tiny pinpoint of optimism in a region reeling from economic turmoil and political upheaval.

> Diverse languages, religions, and ethnic identities have molded land and life within North Africa and Southwest Asia for centuries.

While settings such as the Mount Lebanon Range are reminders of the region's diversity, broadly shared patterns of climate, culture, and oil resources help define the unifying features of the Southwest Asia and North Africa region. Located at the historic meeting ground of Europe, Asia, and Africa, the region includes thousands of square miles of parched deserts, rugged plateaus, and oasis-like river valleys. It extends 4000 miles (6400 km) between Morocco's Atlantic coastline and Iran's boundary with Pakistan. More than two dozen nations are included within its borders, with the largest populations located in Egypt, Turkey, and Iran (Figure 7.1).

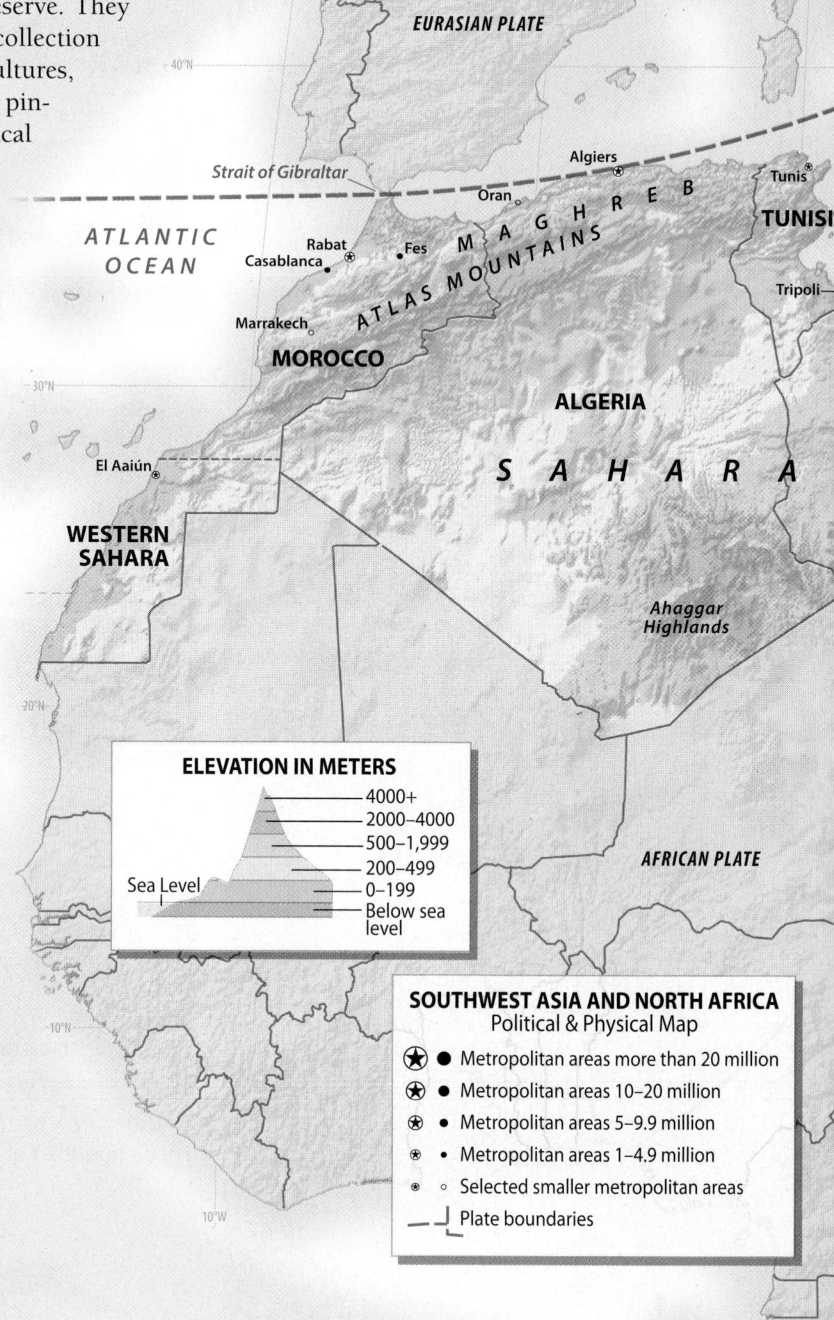

Figure 7.1 Southwest Asia and North Africa This vast region extends from the shores of the Atlantic Ocean to the Caspian Sea. Within its boundaries, major cultural differences and globally important petroleum reserves have contributed to recent political tensions.

Learning Objectives

After reading this chapter you should be able to:

7.1 Explain how latitude and topography produce the region's distinctive patterns of climate.

7.2 Describe how the region's fragile, often arid setting shapes contemporary environmental challenges.

7.3 Describe four distinctive ways in which people have learned to adapt their agricultural practices to the region's arid environment.

7.4 Summarize the major forces shaping recent migration patterns within the region.

7.5 List the major characteristics and patterns of diffusion of Islam.

7.6 Identify the key modern religions and language families that dominate the region.

7.7 Identify the role of cultural variables in understanding key regional conflicts in North Africa, Israel, Syria, Iraq, and the Arabian Peninsula.

7.8 Summarize the geography of oil and gas reserves in the region.

7.9 Describe traditional roles for Islamic women and provide examples of recent changes.

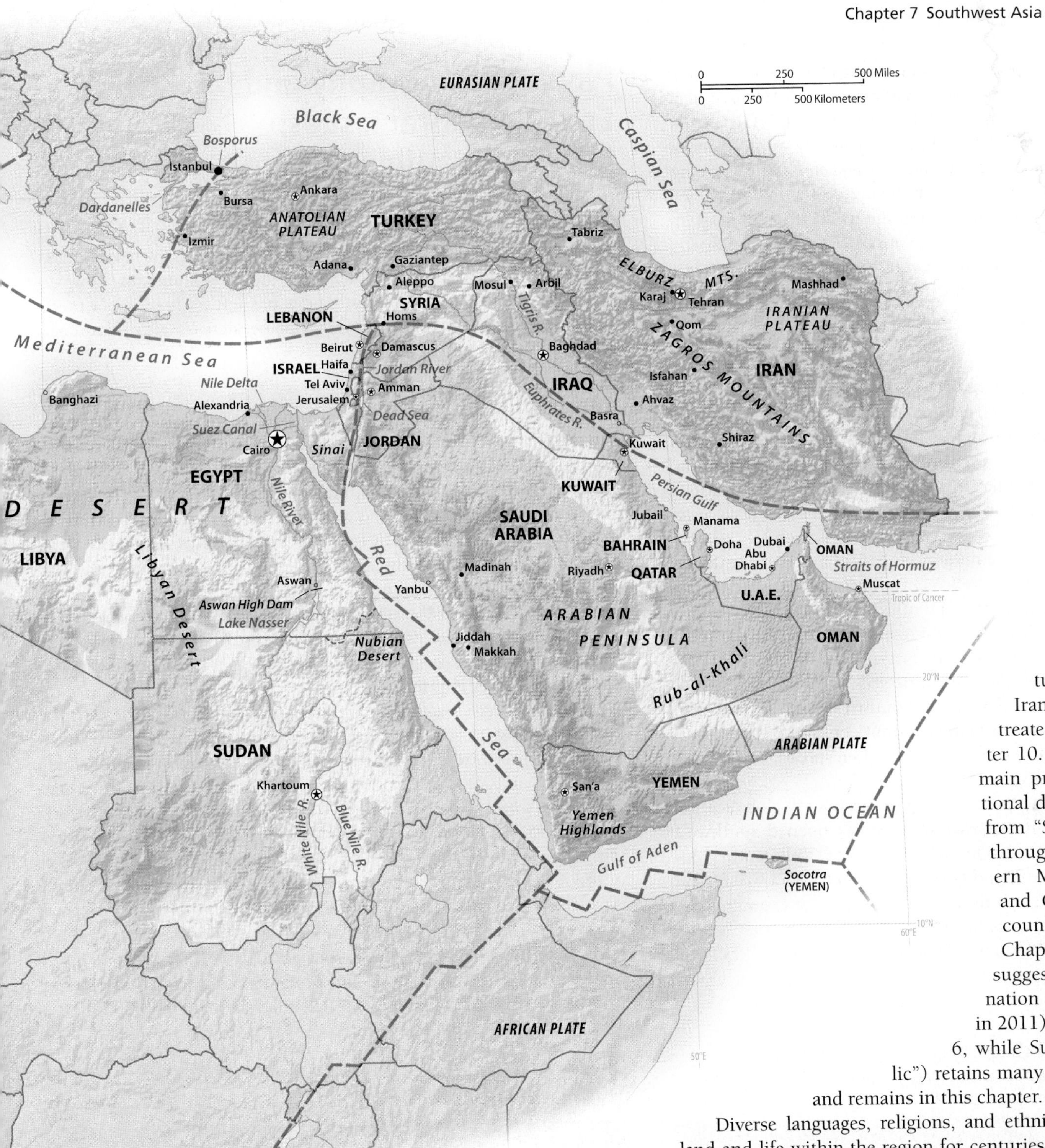

Regional borders also remain unclear. A small piece of northwest Turkey actually sits west of the Bosporus Strait, generally considered the dividing line between Europe and Asia (see Figure 7.1). To the northeast, the Islamic peoples of Central Asia share many cultural ties with Turkey and Iran, but those groups are treated separately in Chapter 10. African borders also remain problematic. The conventional division of "North Africa" from "Sub-Saharan Africa" cuts through the middle of modern Mauritania, Mali, Niger, and Chad. These transitional countries are discussed in Chapter 6. Sudan's split also suggests the complexity. The nation of South Sudan (created in 2011) is also treated in Chapter 6, while Sudan (an "Islamic republic") retains many ties to the Muslim world and remains in this chapter.

Diverse languages, religions, and ethnic identities have molded land and life within the region for centuries, strongly wedding people and place in ways that have had profound social and political implications. One traditional zone of conflict surrounds Israel, where Jewish, Christian, and Islamic peoples have yet to resolve long-standing differences. Since the early 2010s, **Arab Spring** movements—a series of public protests, strikes, and rebellions, often facilitated by social media, that have called for fundamental government and economic reforms—have toppled some governments in the region and pressured others to accelerate political and economic changes, but overall the uprisings failed to produce many democratic reforms or more stable political regimes.

At the same time, cycles of **sectarian violence**—conflicts dividing people along ethnic, religious, and sectarian lines—have repeatedly plagued the region. For example, enduring differences have led to clashes between Jews and Muslims and between varying factions of Islam

"Southwest Asia and North Africa" is both an awkward term and a complex region. The same area is often called the Middle East, but some experts exclude the western parts of North Africa, as well as Turkey and Iran, from such a region. Moreover, the term "Middle East" suggests a European point of view—Lebanon is in the "middle of the east" only from the perspective of the western Europeans who colonized the region and still shape the names we give the world today. Instead, "Southwest Asia and North Africa" offers a straightforward way to describe the general limits of the region.

Syria and Iraq have been especially violent settings for recent instability brought about by the Sunni extremist organization **ISIL (Islamic State of Iraq and the Levant**; also known as ISIS or Islamic State). ISIL has expanded its influence and further destabilized an already war-torn region as it attempts to create a new religious state (a *caliphate*) in the area (Figure 7.2). Some experts estimate that more than 20,000 nonlocal fighters have been recruited into ISIL from around the world (though most are from nearby parts of Southwest Asia and North Africa).

Islamic fundamentalism in the region more broadly advocates a return to traditional practices within Islam. Fundamentalists in any religion advocate a conservative adherence to enduring beliefs within their creed, and they strongly resist change. A related political movement within Islam, known as **Islamism**, challenges the encroachment of global popular culture and blames colonial, imperial, and Western elements for the region's political, economic, and social problems. Islamists resent the role they claim the West has played in creating poverty in their world, and many Islamists advocate merging civil and religious authority and rejecting many characteristics of modern, Western-style consumer culture.

No world region better exemplifies globalization than Southwest Asia and North Africa. The region is a key global **culture hearth**, producing new cultural ideas that subsequently diffuse widely. As an early center for agriculture, civilizations, and major world religions, the region has been a key human crossroads for thousands of years. Important long-distance trade routes have connected North Africa with the Mediterranean and Sub-Saharan Africa. Southwest Asia has also had historical ties to Europe, the Indian subcontinent, and Central Asia. As a result, new ideas within the region have often spread well beyond its bounds.

Particularly within the past century, processes of globalization and the region's strategic importance increasingly opened Southwest Asia and North Africa to outside influences. The 20th-century development of the petroleum industry, largely initiated by U.S. and European investment, had enormous consequences for economic development in oil-rich countries of the region. Many key members of the **Organization of the Petroleum Exporting Countries (OPEC)** are found within the region, and these countries greatly influence global prices and production levels for petroleum.

Figure 7.2 Zaatari Refugee Camp Syria's civil war and the conflict with ISIL have created growing refugee populations. More than 79,000 Syrian refugees lived in Jordan's Zaatari refugee camp as of December 2015.

Physical Geography and Environmental Issues: Life in a Fragile World

In the popular imagination, much of Southwest Asia and North Africa is a land of shifting sand dunes, searing heat, and scattered oases. Although examples of those stereotypes certainly exist, the actual physical setting is much more complex. One theme is dominant, however: A lengthy legacy of human settlement has left its mark on a fragile environment, and the entire region is facing increasingly difficult ecological problems.

Regional Landforms

A quick tour of Southwest Asia and North Africa reveals diverse environmental settings (see Figure 7.1). In North Africa, the **Maghreb** (meaning "western island") includes the nations of Morocco, Algeria, and Tunisia and is dominated near the Mediterranean coastline by the Atlas Mountains. The rugged flanks of the Atlas rise like a series of islands above the narrow coastal plains to the north and the vast stretches of the lower Saharan deserts to the south (Figure 7.3). South and east of the Atlas Mountains, interior North Africa varies between rocky plateaus and extensive desert lowlands. In northeast Africa, the Nile River shapes regional drainage patterns as it flows north through Sudan and Egypt (Figure 7.4).

Southwest Asia is more mountainous than North Africa. In the **Levant**, or eastern Mediterranean region, mountains rise within 20 miles (30 km) of the sea, and the highlands of Lebanon reach heights of more than 10,000 feet (3000 meters). Farther south, the Arabian Peninsula forms a massive tilted plateau, with western highlands higher than 5000 feet (1500 meters) gradually sloping eastward to extensive lowlands in the Persian Gulf area (Figure 7.5). North and east of the Arabian Peninsula lie the two great upland areas of Southwest Asia: the Iranian and Anatolian plateaus (*Anatolia* refers to the large peninsula of Turkey, sometimes called Asia Minor; see Figure 7.1). Both plateaus, averaging 3000–5000 feet (1000–1500 meters) in elevation, are geologically active and prone to earthquakes. One quake near the Iranian city of Bam (2003) claimed more than 30,000 lives.

Figure 7.3 Atlas Mountains The rugged Atlas Mountains dominate a broad area of interior Morocco.

Figure 7.4 Nile Valley This satellite image dramatically reveals the impact of water on the North African desert. Cairo lies at the southern end of the Nile Delta, where it begins to widen toward the Mediterranean Sea. The coastal city of Alexandria sits on the northwest edge of the low-lying delta. Lake Nasser is visible toward the bottom (center) of the image.

Smaller lowlands characterize other portions of Southwest Asia. Narrow coastal strips are common in the Levant, along the southern (Mediterranean Sea) and the northern (Black Sea) Turkish coastlines, and north of Iran's Elburz Mountains near the Caspian Sea. Iraq contains the most extensive alluvial lowlands in Southwest Asia, dominated by the Tigris and Euphrates rivers, flowing southeast to empty into the Persian Gulf. The much smaller Jordan River Valley is a notable lowland that straddles the strategic borderlands of Israel, Jordan, and Syria and drains southward to the Dead Sea (Figure 7.6).

Patterns of Climate

Although the region of Southwest Asia and North Africa is often termed the "dry world," a closer look reveals more complex patterns (Figure 7.7). Both latitude and altitude come into play. Aridity dominates large portions of the region (see the climographs for Cairo, Riyadh, Baghdad, and Tehran in Figure 7.7). A nearly continuous belt of desert land stretches eastward across interior North Africa, through the Arabian Peninsula, and into central and eastern Iran (Figure 7.8). Throughout this zone, plant and animal life adapts to extreme conditions. Deep or extensive root systems allow desert plants to benefit from the limited moisture they receive, and animals efficiently store water, hunt at night, or migrate seasonally to avoid the worst of the dry cycle.

Figure 7.5 Yemen Highlands Steep highlands and narrow valleys dominate much of Yemen's rugged interior.

Elsewhere, altitude and latitude produce a surprising variety of climates. The Atlas Mountains and nearby lowlands of northern Morocco, Algeria, and Tunisia experience a Mediterranean climate, in which dry summers alternate with cooler, wet winters (see the climographs for Rabat and Algiers). In these areas, the landscape resembles those in nearby southern Spain or Italy (Figure 7.9). A second zone of Mediterranean climate extends along the Levant coastline, into the nearby mountains, and northward across sizable portions of northern Syria, Turkey, and northwestern Iran (see the climographs for Jerusalem and Istanbul).

Legacies of a Vulnerable Landscape

The environmental history of Southwest Asia and North Africa reflects both the shortsighted and the resourceful practices of its

Figure 7.6 Jordan Valley This view of the fertile Jordan Valley shows a mix of irrigated vineyards and date palm plantations.

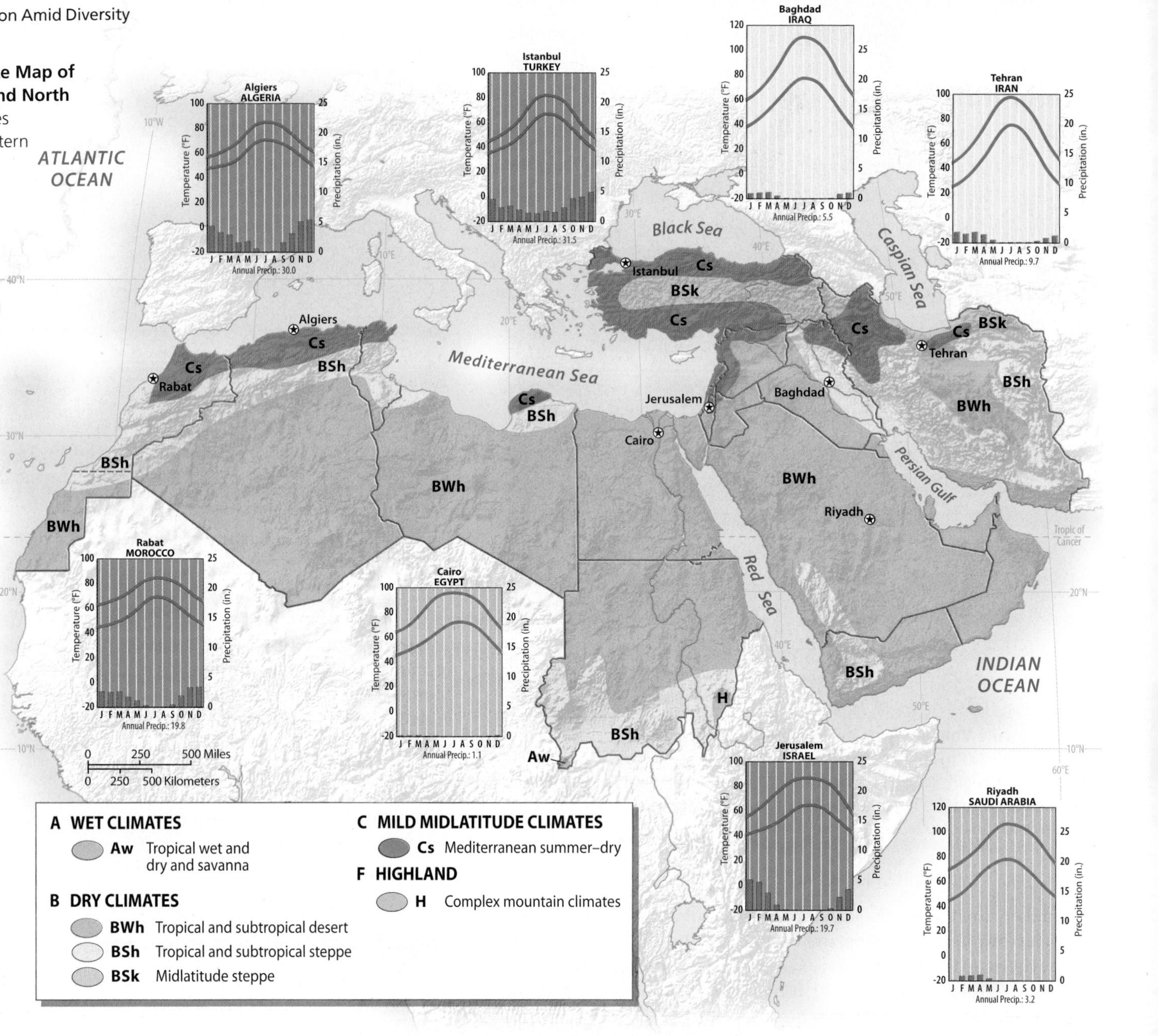

Figure 7.7 Climate Map of Southwest Asia and North Africa Dry climates dominate from western Morocco to eastern Iran. Within these zones, subtropical high-pressure systems offer only limited opportunities for precipitation. Elsewhere, mild midlatitude climates with wet winters are found near the Mediterranean Basin and Black Sea. To the south, tropical savanna climates provide summer moisture to southern Sudan.

A WET CLIMATES
- **Aw** Tropical wet and dry and savanna

B DRY CLIMATES
- **BWh** Tropical and subtropical desert
- **BSh** Tropical and subtropical steppe
- **BSk** Midlatitude steppe

C MILD MIDLATITUDE CLIMATES
- **Cs** Mediterranean summer–dry

F HIGHLAND
- **H** Complex mountain climates

human occupants. Littered with environmental problems, the region reveals the hazards of lengthy human settlement in a marginal land (Figure 7.10). The island of Socotra illustrates the region's fragile and vulnerable environment and suggests how processes of globalization threaten the area's ecological health. Socotra's stony slopes rise out of the shimmering waters of the Indian Ocean southeast of Yemen. Separated for millions of years from the Arabian Peninsula, Socotra's environment evolved in isolation. The island is home to hundreds of plants found nowhere else on Earth. Exotic dragon's blood trees dot the island's dry and rocky hillsides, and dozens of other species have only recently been catalogued by botanists (Figure 7.11).

In 2008, the island was recognized as a United Nations World Natural Heritage Site, and the European Union

Figure 7.8 Arid Iran Only sparse vegetation dots this scene from central Iran, a landscape characterized by isolated mountain ranges and dry interior plateaus.

Figure 7.9 Mediterranean Landscape, Northern Algeria The Mediterranean moisture in northern Algeria produces an agricultural landscape similar to that of southern Spain or Italy. Winter rains create a scene that contrasts sharply with deserts found elsewhere in the region.

advocates the preservation of Socotra's unique biogeography. In the past several years, with UN funding, the Socotra Governance and Biodiversity Project has attempted to regulate the island's tourism and economic development. At the same time, the island's unique plant species remain vulnerable to illegal harvesting. In addition, the Yemeni government has invited international petroleum companies to explore the island's offshore oil and gas potential and has considered developing luxury tourist hotels that would forever change the island's character. Even well-meaning ecotourists on the island have led to increased firewood gathering, coral reef damage, and road construction. Socotra, often cited as the "Galápagos of the Indian Ocean," is an interesting case study for the battle of environmental preservation versus economic development. A recent mixed-use zoning plan and a newly paved airstrip suggest that development is inevitable.

Deforestation and Overgrazing Deforestation remains an ancient regional problem. Although much of the region is too dry for trees, the more humid and elevated lands bordering the Mediterranean once supported heavy forests. Human activities have combined with natural conditions to reduce most of the region's forests to grass and scrub. Mediterranean forests often grow slowly, are highly vulnerable to fire, and usually fare poorly if subjected to heavy grazing. Browsing by sheep and goats in particular has often been blamed for much of the region's forest loss. Deforestation has resulted in a long, slow deterioration of the region's water supplies and in accelerated soil erosion.

Salinization **Salinization**, the buildup of toxic salts in the soil, is another ancient environmental issue in an arid region where irrigation has been practiced for centuries (see Figure 7.10). Hundreds of thousands of acres of the region's once-fertile farmland have been destroyed or damaged by salinization. The problem has been particularly severe in Iraq, where centuries of canal irrigation along the Tigris and Euphrates rivers have seriously degraded land quality. Similar conditions affect central Iran, Egypt, and other irrigated portions of North Africa.

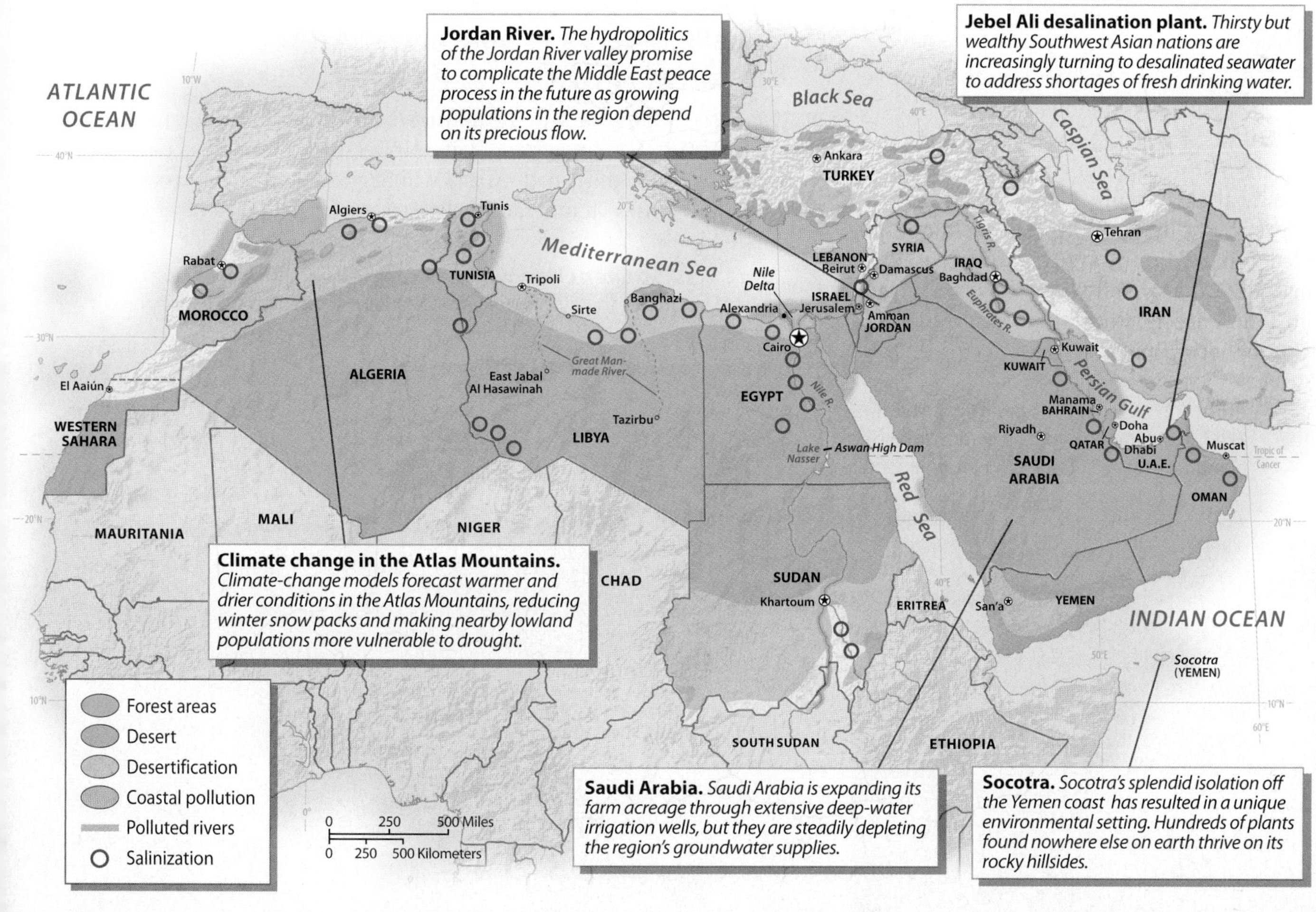

Jordan River. *The hydropolitics of the Jordan River valley promise to complicate the Middle East peace process in the future as growing populations in the region depend on its precious flow.*

Jebel Ali desalination plant. *Thirsty but wealthy Southwest Asian nations are increasingly turning to desalinated seawater to address shortages of fresh drinking water.*

Climate change in the Atlas Mountains. *Climate-change models forecast warmer and drier conditions in the Atlas Mountains, reducing winter snow packs and making nearby lowland populations more vulnerable to drought.*

Saudi Arabia. *Saudi Arabia is expanding its farm acreage through extensive deep-water irrigation wells, but they are steadily depleting the region's groundwater supplies.*

Socotra. *Socotra's splendid isolation off the Yemen coast has resulted in a unique environmental setting. Hundreds of plants found nowhere else on earth thrive on its rocky hillsides.*

Forest areas
Desert
Desertification
Coastal pollution
Polluted rivers
Salinization

0 250 500 Miles
0 250 500 Kilometers

Figure 7.10 Environmental Issues in Southwest Asia and North Africa Growing populations, pressures for economic development, and widespread aridity combine to create environmental hazards across the region. A long history of human activities has contributed to deforestation, irrigation-induced salinization, and expanding desertification. Saudi Arabia's deep-water wells and Egypt's Aswan High Dam are recent technological attempts to expand settlement, but carry a high long-term environmental price tag. **Q: Compare this map with Figure 7.7. What climate types are most strongly associated with desertification?**

Managing Water Water has been managed and manipulated in the region for thousands of years. Traditional systems emphasized directing and conserving surface and groundwater resources at a local scale, but in the past half-century the scope of environmental change has been greatly magnified. One remarkable example is Egypt's Aswan High Dam, completed in 1970 on the Nile River south of Cairo (see Figure 7.10). Increased storage capacity in the upstream reservoir made more water available for agriculture and generates clean electricity. But irrigation has increased salinization because water is not rapidly flushed from fields. The dam has led to more use of costly fertilizers, the infilling of Lake Nasser behind the dam with accumulating sediments, and the collapse of the Mediterranean fishing industry near the Nile Delta, an area previously nourished by the river's silt.

Elsewhere, other water-harvesting strategies have proven useful. **Fossil water**, or water supplies stored underground during earlier and wetter climatic periods, is being mined. For example, Saudi Arabia has invested huge sums to develop deep-water wells, allowing it to greatly expand its food output. Unfortunately, underground supplies are being depleted more rapidly than they are recharged, limiting their long-term sustainability. Elsewhere in the region, seawater desalination is becoming another popular alternative (see *Working Toward Sustainability: Desalination in the Desert at Dubai's Jebel Ali Plant*).

Most dramatically, **hydropolitics**, or the interplay of water resource issues and politics, has raised tensions between countries that share drainage basins. For example, with the help of Chinese capital and engineering expertise, Ethiopia recently built Tekeze Dam on a tributary of the Nile. Already referred to by promoters as "China's Three Gorges Dam in Africa," the controversial project threatens to disrupt downstream fisheries and irrigation in North Africa. Similarly, Sudan's Merowe Dam project (on another stretch of the Nile) has raised major concerns in nearby Egypt.

In Southwest Asia, Turkey's growing development of the upper Tigris and Euphrates rivers (the Southeast Anatolia Project, or GAP), complete with 22 dams and 19 power plants, has raised issues with Iraq and Syria, who argue that capturing "their" water might be considered a provocative political act (Figure 7.12). Turkey has periodically withheld water from Syria by controlling flows along the Euphrates, provoking protests. In addition, a 2013 study based on new NASA satellite data measured the overall water flows within the Tigris–Euphrates Basin and found accelerating losses since 2003. As surface water flows from neighboring nations have declined, Iraqi farmers have dug over 1000 new wells, severely impacting groundwater supplies and lowering regional water tables.

Figure 7.11 Socotra's Dragon's Blood Tree Unique to Socotra, the rare dragon's blood tree reflects the island's environmental isolation. Evolving as a part of the island's ecosystem, the tree survives in the region's dry tropical climate.

Figure 7.12 Southeast Anatolia Project Turkey's Ataturk Dam, one of the world's largest, was completed in 1990 and is a centerpiece of the region's Southeast Anatolia Project.

Water is not only a resource, but also a vital transportation link in the area. The region's physical geography created enduring **choke points**, where narrow waterways are vulnerable to military blockade or disruption. For example, the Strait of Gibraltar (entrance to the Mediterranean), Turkey's Bosporus and Dardanelles, and the Suez Canal have all been key historical choke points within the region (see Figure 7.1). Iran's periodic threat to close the Straits of Hormuz (at the eastern end of the Persian Gulf) to world oil shipments suggests the strategic role water continues to play in the region (Figure 7.13).

Climate Change in Southwest Asia and North Africa

The 2014 report of the 10th Intergovernmental Panel on Climate Change (IPCC) suggested that 21st-century climate changes in Southwest Asia and North Africa will aggravate already existing environmental issues. Temperature changes are predicted to have a greater

Figure 7.13 Straits of Hormuz This satellite view shows the entrance to the Persian Gulf at the Straits of Hormuz, one of the region's most important choke points.

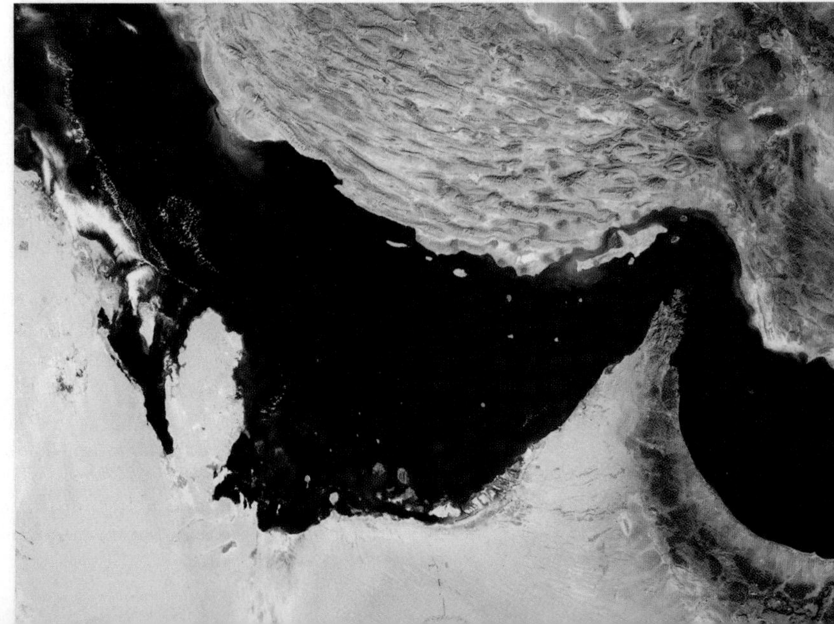

WORKING TOWARD SUSTAINABILITY

Desalination in the Desert at Dubai's Jebel Ali Plant

Google Earth (MG) Virtual Tour Video
http://goo.gl/neuvnb

More than 1.2 billion people globally live in zones of chronic water scarcity, and that number is expected to increase to 1.8 billion by 2025. The dry zones of North Africa and Southwest Asia, especially given many climate-change scenarios, will be getting drier at the same time their populations will be larger and thirstier than ever.

Desalination of seawater is one partial solution to this water crisis in the dry world. Wealthier countries such as Saudi Arabia, the United Arab Emirates, and Israel are making huge investments in new and emerging desalination technologies, which can more affordably and efficiently transform seawater into potable drinking water. The region already accounts for half of the world's desalination plants, and that total will grow in the next 20 years.

The Jebel Ali Plant In Dubai (a part of the United Arab Emirates), one of the world's largest combined desalination and power plants opened in 2013 along the shores of the Persian Gulf at Jebel Ali (Figure 7.1.1). Huge pipes extending into the Gulf can draw in up to a billion gallons of water a day. The seawater is heated by efficient natural gas and diesel fuel boilers to produce steam and provide more than 2000 megawatts of electricity daily. Recent new additions to the plant (to be completed in 2019) by Siemens, a German company, will increase that capacity to more than 2700 megawatts. Plant operators note that the thermal efficiency of their technology is among the best in the world.

Figure 7.1.1 **Jebel Ali Desalination Plant** The United Arab Emirates and many other countries in the region have made large investments in desalination technology. The Jebel Ali plant also supplies electricity to the area.

A Sustainable Approach Eight diesel-powered desalination units, among the world's largest, take the seawater and transform it into drinking water for the area's population. Currently, the plant can produce more than 168 million gallons of drinking water per day. Lime and other chemicals are added to the water to make it more palatable for human consumption. Plant operators can also adjust their power and water output seasonally (higher demand in summer, lower in winter). They claim that the plant will help reduce the region's greenhouse gas emissions and that it symbolizes Dubai's commitment to reduce its carbon footprint even as it takes its increasingly affluent population into the mid-21st century.

1. While the desalination plant in Dubai may work well, what special challenges might there be in constructing similar plants elsewhere in the region?
2. Where does your daily drinking water come from?

impact on the region than changes in precipitation. The already arid and semiarid region will probably remain relatively dry, but warmer average temperatures are likely to have several major consequences:

- Higher overall evaporation rates and lower overall soil moisture across the region will stress crops, grasslands, and other vegetation. Semiarid lands in North Africa's Maghreb region are particularly vulnerable, especially dryland cropping systems that cannot depend on irrigation.
- Warmer temperatures will reduce runoff into rivers, reducing hydroelectric potential and water available for the region's increasingly urban population. Less snow in the Atlas Mountains will stress nearby farmers who depend on meltwater for irrigation.
- More extreme, record-setting summertime temperatures will lead to more heat-related deaths, particularly in cities.

Sea-level changes pose special threats to the Nile Delta. This portion of northern Egypt is a vast, low-lying landscape of settlements, farms, and marshland. Studies that model sea-level changes suggest much of the delta could be lost to inundation, erosion, or salinization. Farmland losses of more than 250,000 acres (100,000 hectares) are quite possible with even modest sea-level changes. The IPCC estimates that a sea-level rise of 3.3 feet (1 meter) could affect 15 percent of Egypt's habitable land and displace 8 million Egyptians in coastal and delta settings. Total losses of $30 billion have been projected for Alexandria because sea-level changes will devastate the city's huge resort industry as well as nearby residential and commercial areas (Figure 7.14).

Experts also estimate broader political and economic costs associated with potential climate changes. Given the region's political instability, even small changes in water supplies, particularly where they might involve several nations, could add to potential conflicts. In addition, wealthier nations such as Israel and Saudi Arabia may have more available resources to plan, adjust, and adapt to climate shifts and extreme events than poorer, less developed countries such as Yemen, Syria, and Sudan.

Figure 7.14 Alexandria, Egypt This beachside view along northern Egypt's low-lying coastline at Alexandria could change significantly if global sea levels rise.

Review

7.1 Describe the climatic changes you might experience as you travel on a line from the eastern Mediterranean coast at Beirut to the highlands of Yemen. What are some of the key climatic variables that explain these variations?

7.2. Discuss five important human modifications of the Southwest Asian and North African environment, and assess whether these changes have benefited the region.

KEY TERMS Arab Spring, sectarian violence, ISIL (Islamic State of Iraq and the Levant) Islamic fundamentalism, Islamism, culture hearth, Organization of Petroleum Exporting Countries (OPEC), Maghreb, Levant, salinization, fossil water, hydropolitics, choke point

Population and Settlement: Changing Rural and Urban Worlds

The human geography of Southwest Asia and North Africa demonstrates the intimate tie between water and life in this part of the world. The pattern is complex: Large areas of the population map remain almost devoid of permanent settlement, whereas lands with available moisture suffer increasingly from problems of crowding and overpopulation (Figure 7.15).

The Geography of Population

Today about 500 million people live in Southwest Asia and North Africa (see Table A7.1). The distribution of that population is strikingly varied (see Figure 7.15). In North Africa, the moist slopes of the Atlas Mountains and nearby better-watered coastal districts support dense populations, a stark contrast to thinly occupied lands southeast of the mountains. Egypt's zones of almost empty desert differ dramatically from crowded, irrigated locations, such as those along the Nile River. In Southwest Asia, many residents live in well-watered coastal and highland settings and in desert localities where water is available from nearby rivers or subsurface aquifers. High population densities are found in better-watered portions of the eastern Mediterranean (Israel, Lebanon, and Syria), Turkey, and Iran. While overall population densities in such countries appear modest, the **physiological density**, which is the number of people per unit area of arable land, is quite high by global standards. Although less than two-thirds of the region's overall population is urban, many nations are dominated by huge cities (for example, Cairo in Egypt, Istanbul in Turkey, and Tehran in Iran) that suffer the same problems of urban crowding found elsewhere in the developing world (Figure 7.16).

Water and Life: Rural Settlement Patterns

Water and life are closely linked across rural settlement landscapes of Southwest Asia and North Africa (Figure 7.17). Indeed, Southwest Asia is one of the world's earliest hearths of **domestication**, where plants and animals were purposefully selected and bred for their desirable characteristics. Beginning around 10,000 years ago, increased experimentation with wild varieties of wheat and barley led to agricultural settlements that later included domesticated animals, such as cattle, sheep, and goats. Much of the early agricultural activity focused on the **Fertile Crescent**, an ecologically diverse zone stretching from the Levant inland through the fertile hill country of northern Syria into Iraq. Between 5000 and 6000 years ago, improved irrigation techniques and increasingly powerful political states encouraged the spread of agriculture into nearby lowlands, such as the Tigris and Euphrates valleys (Mesopotamia) and North Africa's Nile Valley.

Pastoral Nomadism In the drier portions of the region, **pastoral nomadism**, where people move livestock seasonally, is a traditional form of subsistence agriculture. The settlement landscape of pastoral nomads reflects their need for mobility and flexibility as they move camels, sheep, and goats from place to place. Near highland zones such as the Atlas Mountains and the Anatolian Plateau, nomads practice **transhumance**—seasonally moving livestock to cooler, greener high-country pastures in the summer and returning them to valley and lowland settings for fall and winter grazing. Elsewhere, seasonal movements often involve huge areas of desert that support small groups of a few dozen families. Fewer than 10 million pastoral nomads remain in the region today.

Oasis Life Permanent oases exist where high groundwater levels or modern deep-water wells provide reliable water (Figures 7.17 and 7.18). Tightly clustered, often walled villages sit next to small, intensely utilized fields where underground water is applied to tree and cereal crops. In newer oasis settlements, concrete blocks and prefabricated housing add a modern look. Traditional oasis settlements contain families that work their own irrigated plots or, more commonly, work for absentee landowners. While oases are usually small, eastern Saudi Arabia's Hofuf Oasis covers more than 30,000 acres (12,000 hectares). Although some crops are raised for local consumption, expanding world demand for products such as figs and dates increasingly draws even these remote locations into the global economy as products end up on the tables of hungry Europeans or North Americans.

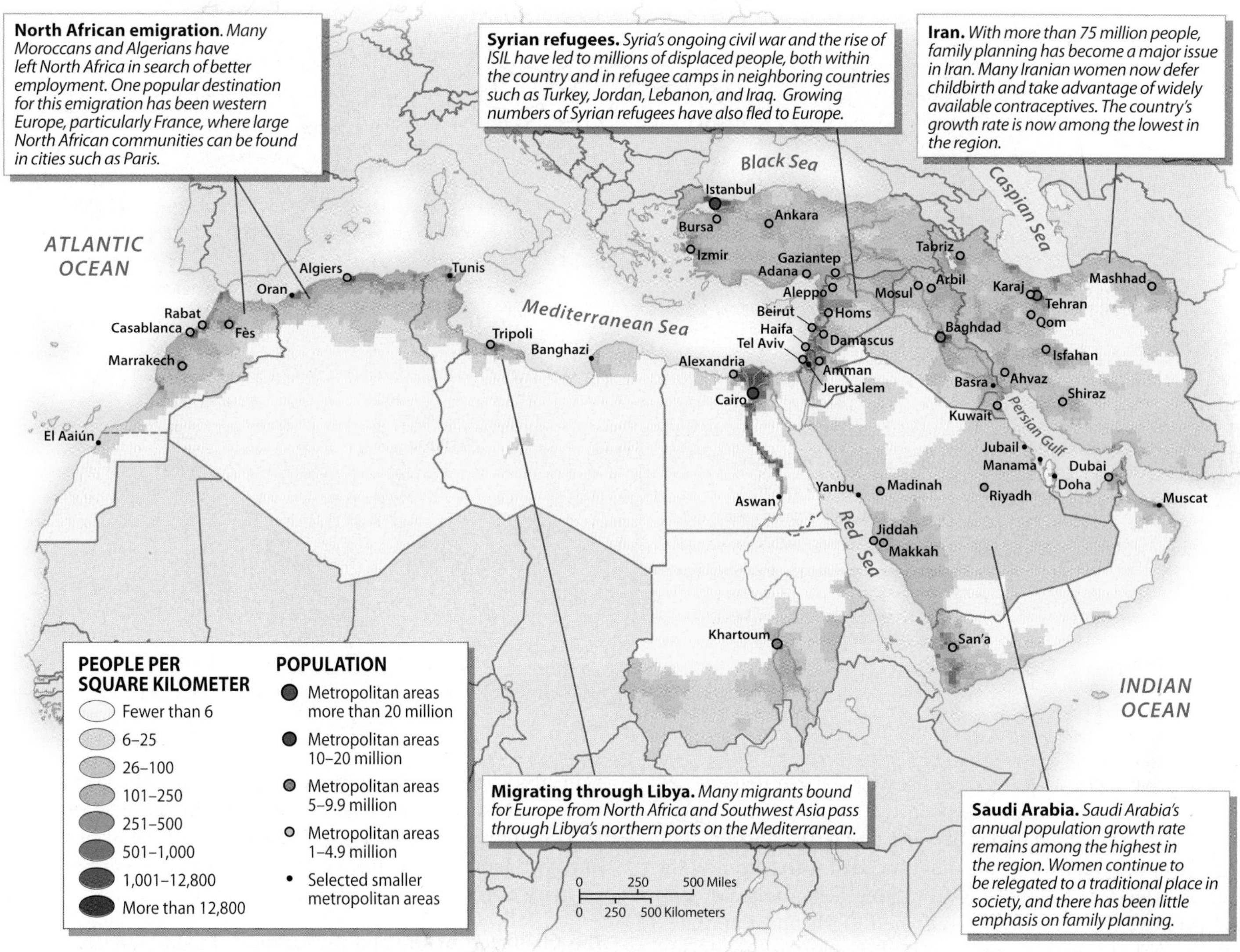

North African emigration. *Many Moroccans and Algerians have left North Africa in search of better employment. One popular destination for this emigration has been western Europe, particularly France, where large North African communities can be found in cities such as Paris.*

Syrian refugees. *Syria's ongoing civil war and the rise of ISIL have led to millions of displaced people, both within the country and in refugee camps in neighboring countries such as Turkey, Jordan, Lebanon, and Iraq. Growing numbers of Syrian refugees have also fled to Europe.*

Iran. *With more than 75 million people, family planning has become a major issue in Iran. Many Iranian women now defer childbirth and take advantage of widely available contraceptives. The country's growth rate is now among the lowest in the region.*

ATLANTIC OCEAN
Black Sea
Caspian Sea
Mediterranean Sea
Red Sea
Persian Gulf
INDIAN OCEAN
Istanbul
Bursa
Ankara
Izmir
Gaziantep
Adana
Aleppo
Tabriz
Mosul
Arbil
Karaj
Tehran
Mashhad
Qom
Isfahan
Beirut
Homs
Haifa
Tel Aviv
Damascus
Amman
Jerusalem
Baghdad
Basra
Ahvaz
Shiraz
Kuwait
Jubail
Manama
Dubai
Doha
Riyadh
Muscat
Yanbu
Madinah
Jiddah
Makkah
San'a
Khartoum
Aswan
Cairo
Alexandria
Tripoli
Banghazi
Tunis
Algiers
Oran
Rabat
Casablanca
Fès
Marrakech
El Aaiún

PEOPLE PER SQUARE KILOMETER
- Fewer than 6
- 6–25
- 26–100
- 101–250
- 251–500
- 501–1,000
- 1,001–12,800
- More than 12,800

POPULATION
- Metropolitan areas more than 20 million
- Metropolitan areas 10–20 million
- Metropolitan areas 5–9.9 million
- Metropolitan areas 1–4.9 million
- Selected smaller metropolitan areas

Migrating through Libya. *Many migrants bound for Europe from North Africa and Southwest Asia pass through Libya's northern ports on the Mediterranean.*

0 250 500 Miles
0 250 500 Kilometers

Saudi Arabia. *Saudi Arabia's annual population growth rate remains among the highest in the region. Women continue to be relegated to a traditional place in society, and there has been little emphasis on family planning.*

Figure 7.15 Population Map of Southwest Asia and North Africa The striking contrasts are clearly evident between large, sparsely occupied desert zones and much more densely settled regions where water is available. The Nile Valley and the Maghreb region contain most of North Africa's people, while Southwest Asian populations cluster in the highlands and along the better-watered shores of the Mediterranean.

Figure 7.16 Istanbul Turkey's largest city is now home to more than 13 million people.

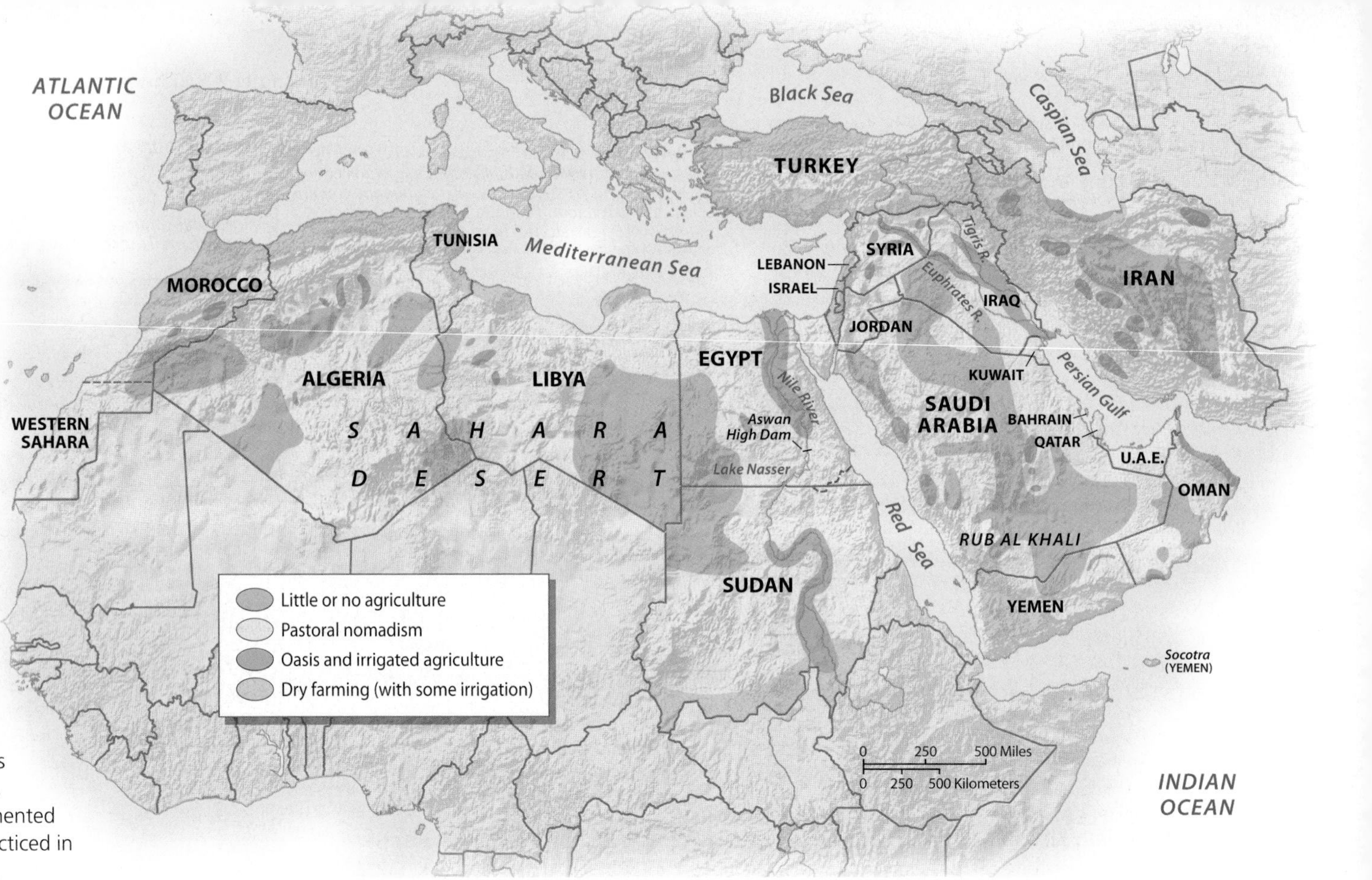

Figure 7.17 Agricultural Regions of Southwest Asia and North Africa Important agricultural zones include oases and irrigated farms where water is available. Elsewhere, dry farming supplemented with irrigation is practiced in midlatitude settings.

Exotic Rivers For centuries, the densest rural settlement of Southwest Asia and North Africa has been tied to its great irrigated river valleys and their seasonal floods of water and fertile nutrients. In such settings, **exotic rivers** transport much-needed water from more humid areas to drier regions suffering from long-term moisture deficits (Figure 7.19). The Nile and the combined Tigris and Euphrates rivers are the largest regional examples of such activity, and both systems have large, densely settled deltas. Similar settlements are found along the Jordan River in Israel and Jordan, in the foothills of the Atlas Mountains, and on the peripheries of the Anatolian and Iranian plateaus. These settings, although capable of supporting sizable populations, are also vulnerable to overuse and salinization. Rural life is also changing in such settings. New dam- and canal-building schemes in Egypt, Israel, Syria, Turkey, and elsewhere are increasing the storage capacity of river systems, allowing for more year-round agriculture.

The Challenge of Dryland Agriculture Mediterranean climates in the region permit dryland agriculture that depends largely on seasonal

Figure 7.18 Oasis Settlement Date palms and irrigated fields shape the landscape around Tinehir, a fertile oasis settlement located in central Morocco.

Figure 7.19 Nile Valley Agriculture Irrigated rice is a major staple in Egypt's fertile Nile Valley.

moisture. These zones include better-watered valleys and coastal lowlands of the northern Maghreb, lands along the shore of the eastern Mediterranean, and favored uplands across the Anatolian and Iranian plateaus. A mix of tree crops, grains, and livestock is raised in these settings. More mechanization, crop specialization, and fertilizer use are also transforming such agricultural settings, following a pattern set earlier in nearby areas of southern Europe. One commercial adaptation of growing regional and global importance is Morocco's flourishing hashish crop. More than 200,000 acres (80,000 hectares) of cannabis are cultivated in the hill country near Ketama in northern Morocco, generating more than $2 billion annually in illegal exports (mostly to Europe) (Figure 7.20).

Many-Layered Landscapes: The Urban Imprint

Cities have played a key role in the human geography of Southwest Asia and North Africa. Indeed, some of the world's oldest urban places are located in the region. Today continuing political, religious, and economic ties link the cities with the surrounding countryside.

A Long Urban Legacy Cities have traditionally been centers of political and religious authority as well as focal points of trade. Urbanization in Mesopotamia (modern Iraq) began by 3500 BCE, and cities such as Eridu and Ur reached populations of 25,000 to 35,000 residents. Similar centers appeared in Egypt by 3000 BCE, with Memphis and Thebes assuming major importance in the middle Nile Valley. By 2000 BCE, however, a different kind of city emerged in the eastern Mediterranean and along important overland trade routes. Beirut, Tyre, and Sidon, all in modern Lebanon, as well as Damascus in nearby Syria, exemplified the growing role of trade in shaping the urban landscape. Expanding port facilities, warehouse districts, and commercial neighborhoods suggested how commerce shaped these urban settlements, and many early Middle Eastern trading towns survive to the present.

Figure 7.20 **Cannabis Fields, Morocco** Bundles of processed hash dry in the sun near Ketama, Morocco, and cannabis fields clothe the nearby hillside. Much of the region's hashish crop is bound for Europe.

Figure 7.21 **Qom Mosque** Qom's Hazrati Masumeh Shrine is annually visited by thousands of faithful Shiites in this sacred Iranian city south of Tehran.

Islam also left an enduring mark on cities because urban centers traditionally served as places of Islamic religious power and education. Both Baghdad and Cairo were seats of religious authority. Urban settlements from North Africa to Turkey felt its influence. Indeed, the Moors carried Islam to Spain, where it shaped the architecture and culture of centers such as Córdoba and Málaga.

The traditional Islamic city features a walled core, or **medina**, dominated by the central mosque and its associated religious, educational, and administrative functions (Figure 7.21). A nearby bazaar, or *suq*, serves as a marketplace where products from city and countryside are traded (Figure 7.22). Housing districts feature a maze of narrow, twisting streets that maximize shade and emphasize the privacy of residents, particularly women. Houses have small windows, frequently are situated on dead-end streets, and typically open inward to private courtyards.

More recently, European colonialism has shaped selected cities. Particularly in North Africa, colonial builders added many architectural features from Great Britain and France. Victorian building blocks, French mansard roofs, suburban housing districts, and wide, European-style commercial boulevards remain landscape features in cities such as Algiers and Cairo.

Signatures of Globalization Since 1950, cities in Southwest Asia and North Africa have become gateways to the global economy. Expanded airports, commercial and financial districts, industrial parks, and luxury tourist facilities all reveal the influence of the global economy. Many cities, such as Algiers and Istanbul, have more than doubled in population in recent years. Crowded Cairo now has more than 15 million residents.

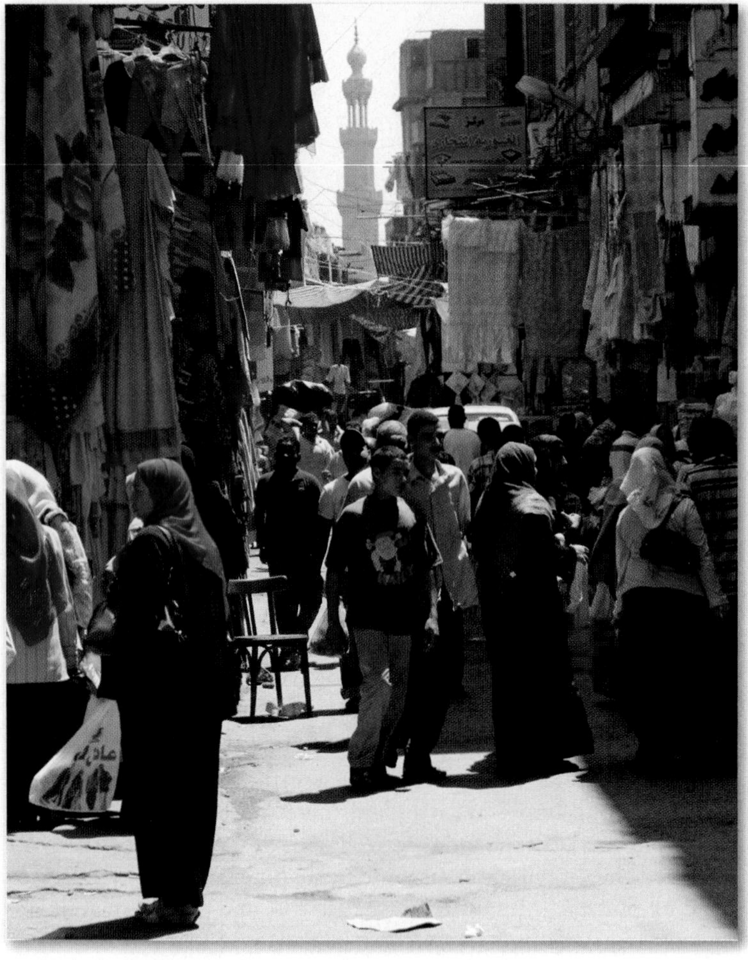

Figure 7.22 Old Cairo The Egyptian capital's narrow, twisting streets are often crowded with shoppers and pedestrians.

Escalating demand for homes has produced ugly, cramped high-rise apartment houses, while elsewhere extensive squatter settlements provide little in the way of quality housing or public services. In New Cairo City east of downtown, large, new high-density suburbs attempt to keep up with the growing demand for urban housing (Figure 7.23).

Figure 7.23 New Cairo City Recent construction east of downtown is creating a very different landscape in suburban Cairo.

Undoubtedly, the most dramatic new urban landscapes are visible in the oil-rich states of the Persian Gulf. Before the 20th century, urban traditions were relatively weak in the area, and even as late as 1950 only 18 percent of Saudi Arabia's population lived in cities. All that has changed, however, and today Saudi Arabia is more urban than many industrialized nations, including the United States. Particularly since 1970, cities such as Dubai (United Arab Emirates), Doha (Qatar), Manama (Bahrain), and Kuwait City (Kuwait) have mushroomed in size, often featuring central-city skylines that feature futuristic architecture (Figure 7.24).

A Region on the Move

Although nomads have crisscrossed the region for ages, entirely new patterns of migration reflect the global economy and recent political events. The rural-to-urban shift seen widely in the less developed world is reworking population patterns across Southwest Asia and North Africa. The Saudi Arabian example is echoed in many other countries. Cities from Casablanca to Tehran are experiencing phenomenal growth, spurred by in-migration from rural areas.

Foreign workers have also migrated to areas within the region that have large labor demands. In particular, Persian Gulf nations support immigrant workforces that often comprise large proportions of the overall population. Over 40 percent of the Gulf nations' total population is made up of foreign-born workers. The influx has major economic, social, and demographic implications in nations such as the United Arab Emirates, where more than 90 percent of the country's private workers are immigrants. Source regions for these immigrants vary, but most come from South Asia and other Muslim countries within and beyond the region (Figure 7.25). In Dubai, Pakistani cab drivers, Filipino nannies, and Indian shop clerks typify a foreign workforce that currently makes up a large majority of the city's 1.4 million inhabitants. Many of these foreign workers in the Gulf send their wages home: India, Egypt,

Figure 7.24 Manama, Bahrain In this aerial view of the Persian Gulf city, Manama's changing skyline is marked by high-rise construction projects.

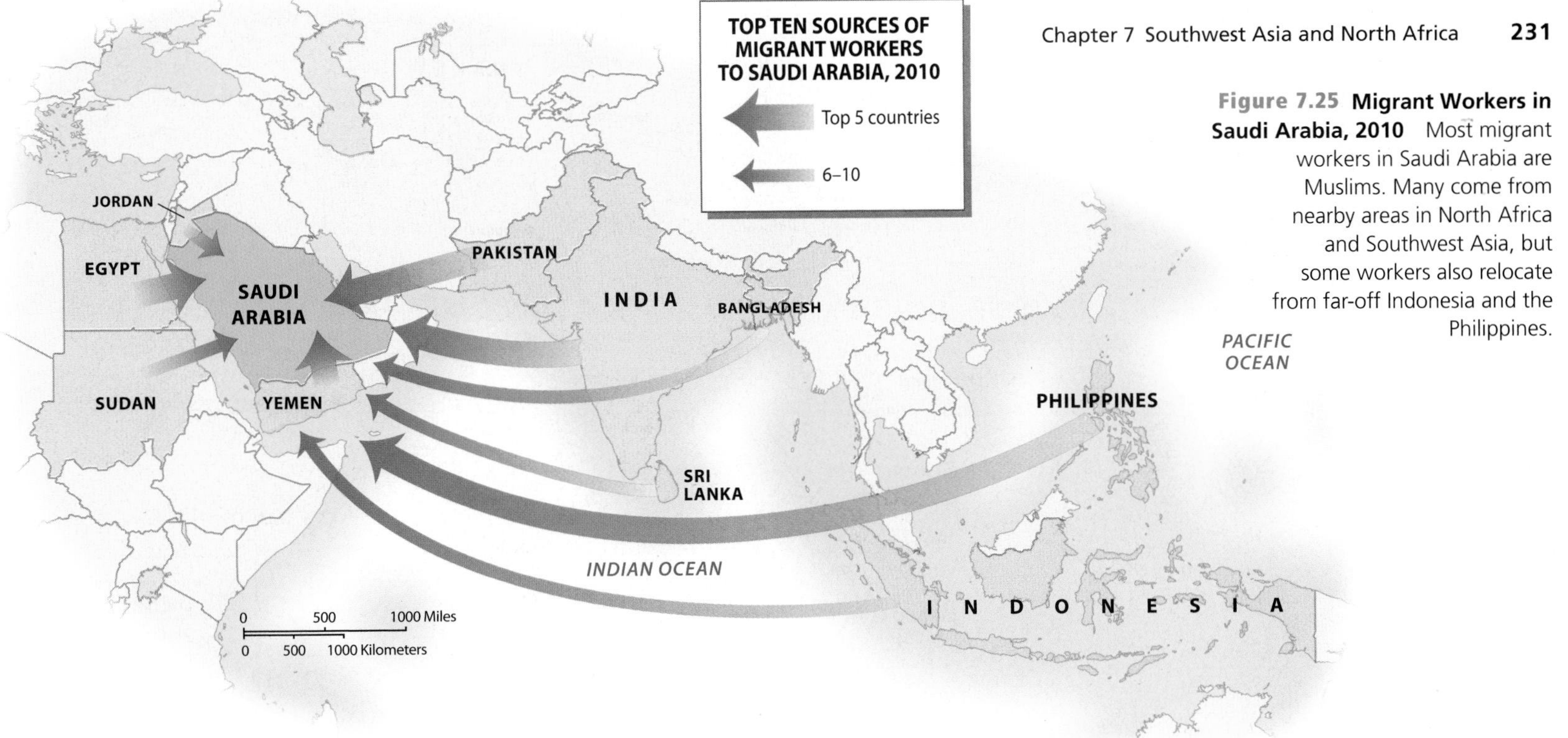

Figure 7.25 Migrant Workers in Saudi Arabia, 2010 Most migrant workers in Saudi Arabia are Muslims. Many come from nearby areas in North Africa and Southwest Asia, but some workers also relocate from far-off Indonesia and the Philippines.

the Philippines, Pakistan, and Bangladesh are often key destinations for these remittances.

Other residents migrate to jobs elsewhere in the world. Because of its strong economy and close location, Europe is a powerful draw. More than 2 million Turkish guest workers live in Germany. Algeria and Morocco also have seen large out-migrations to western Europe, particularly France.

Political instability has also sparked migration. Wealthier residents, for example, have fled nations such as Lebanon, Syria, Iraq, and Iran since the 1980s and today live in cities such as Toronto, Los Angeles, and Paris. More recent political instability has provoked other refugee movements. Since 2003, huge numbers of people in western Sudan's unsettled Darfur region have moved to dozens of refugee camps in nearby Chad. Elsewhere, thousands of displaced Afghans remain in eastern Iran.

Syria's civil war and sectarian conflicts have produced a massive refugee crisis that has displaced over half the country's population. By late 2015, almost 8 million internally displaced persons (IDPs) had left their homes and were living elsewhere in the country. About 4.5 million Syrians were refugees and were living in other countries. The nearby nations of Turkey, Jordan, Lebanon, and Iraq housed the vast majority of these refugees, but growing numbers of Syrians were fleeing the region, bound for opportunities in Europe and beyond (Figure 7.26). By early 2016, more than 150,000 Syrians, for example, had applied for asylum in Germany and many of the more than three million migrants to Europe estimated to arrive there between 2015 and 2017 will be from the same war-torn region. Large numbers of refugees travel via the eastern Mediterranean and through Greece to Europe. Other refugees from the region cross into Europe from Libya (see *Exploring Global Connections: The Libyan Highway to Europe*).

Shifting Demographic Patterns

High population growth remains a critical issue throughout Southwest Asia and North Africa, but the demographic picture is shifting. Uniformly high growth rates in the 1960s have been replaced by more varied regional patterns. For example, women in Tunisia and Turkey now average fewer than three births, representing a large decline in total fertility rates (see Table A7.1). Various factors explain these changes. More urban, consumer-oriented populations opt for fewer children. Many Arab women now delay marriage into their middle 20s and early 30s. Family-planning initiatives are expanding in many countries; programs in Tunisia, Egypt, and Iran have greatly increased access to contraceptive pills, IUDs, and condoms.

Intriguingly, fundamentalist Iran has witnessed a rapid decline in fertility in the past two decades (Figure 7.27). Fertility has fallen since the mid-1970s from an average of 6.6 births to 1.9 births per woman. While Iran's family-planning program was initially dismantled after the fundamentalist revolution in 1979 (it was seen as a Western idea), recent leaders have recognized the wisdom in containing the country's large population.

Still, areas such as the West Bank, Gaza, and Yemen are growing much faster than the world average. Poverty and traditional ways of rural life contribute to large rates of population increase, and even in more urban Saudi Arabia, growth rates remain near 2 percent. The increases result from high birth rates combined with low death rates. In Egypt, even though birth rates may decline, the labor market will need to absorb more than 500,000 new workers annually over the next 10 to 15 years just to keep up with the country's large youthful population (see Figure 7.27).

Review

7.3 Discuss how pastoral nomadism, oasis agriculture, and dryland wheat farming represent distinctive adaptations to the regional environments of Southwest Asia and North Africa. How do these rural lifestyles create distinctive patterns of settlement?

7.4 Describe the distinctive contributions of (a) Islam, (b) European colonialism, and (c) recent globalization to the region's urban landscape.

7.5 Summarize the key patterns and drivers of migration into and out of the region.

KEY TERMS physiological density, domestication, Fertile Crescent, pastoral nomadism, transhumance, exotic river, medina

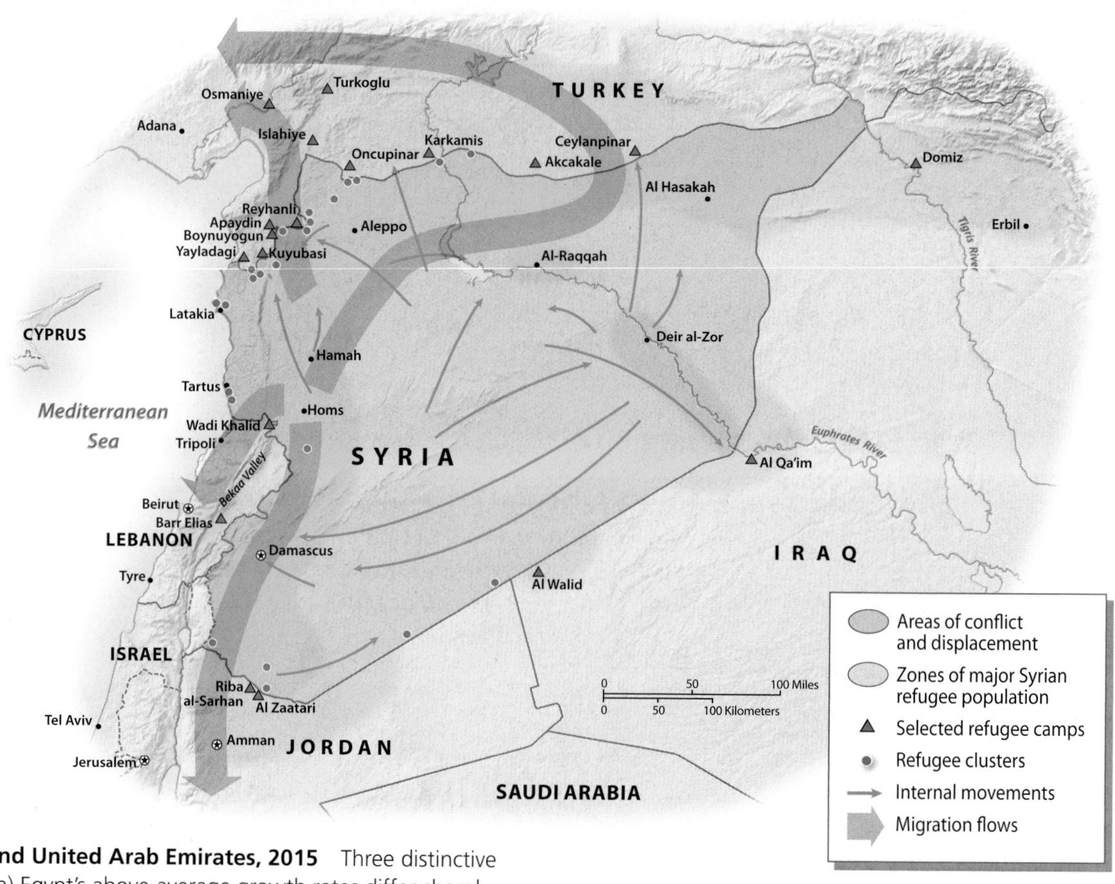

Figure 7.26 Syrian Refugee Zones and Selected Camps, 2015 Neighboring areas of Turkey, Jordan, Iraq, and Lebanon have been inundated with refugees since 2012 and growing numbers are fleeing the region to Europe and beyond.

Figure 7.27 Population Pyramids: Egypt, Iran, and United Arab Emirates, 2015 Three distinctive demographic snapshots highlight regional diversity: (a) Egypt's above-average growth rates differ sharply from those of (b) Iran, where a focused campaign on family planning has reduced recent family sizes. (c) Male immigrant laborers play a special role in skewing the pattern within the United Arab Emirates. **Q: For each example, cite a related demographic or cultural issue that you might potentially find in these countries.**

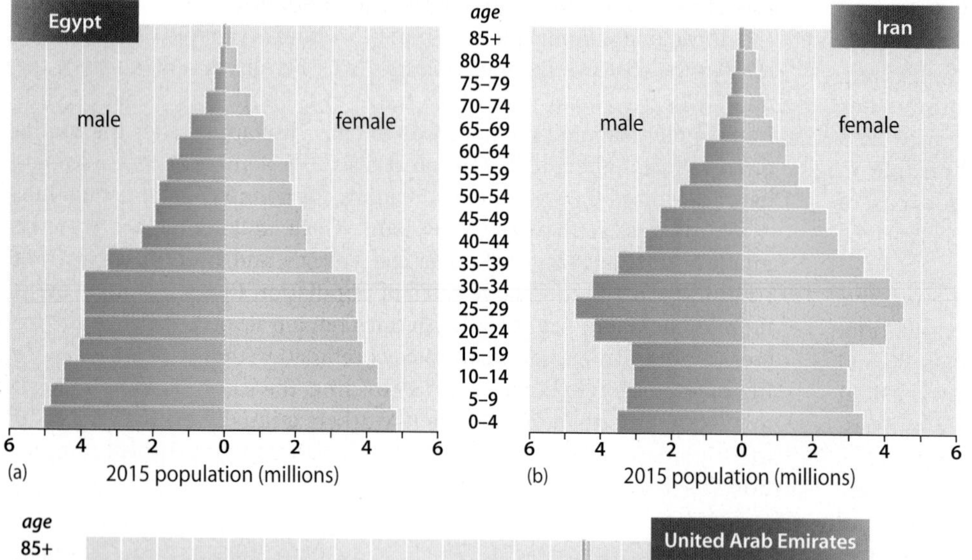

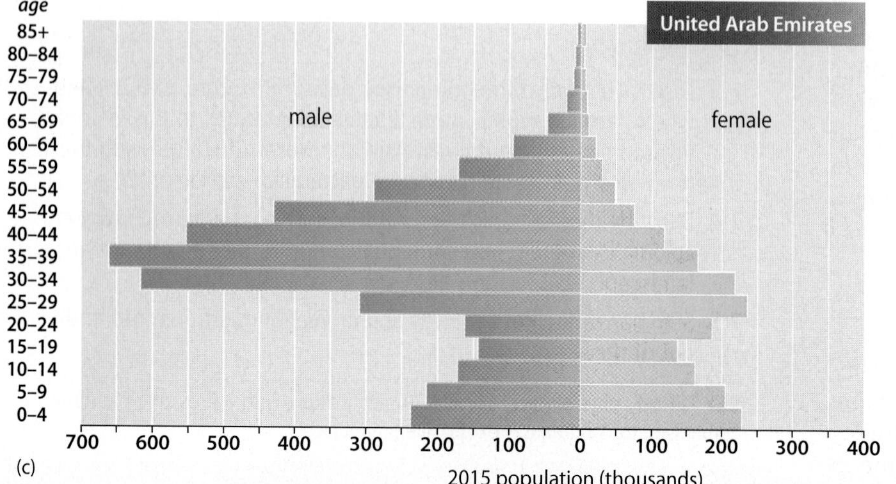

Cultural Coherence and Diversity: Signatures of Complexity

Although Southwest Asia and North Africa remain the heart of the Islamic and Arab worlds, cultural diversity also characterizes the region. Muslims practice their religion in varied ways, often disagreeing strongly on religious views. Elsewhere, other religions complicate cultural geography. Linguistically, Arabic languages are key, but non-Arab peoples, including Persians, Kurds, and Turks, also dominate portions of the region. These cultural geographies can help us understand the region's political tensions and appreciate why many of its residents resist processes of globalization.

Patterns of Religion

Religion is an important part of the lives of most people in Southwest Asia and North Africa. Whether it is the quiet ritual of morning prayers or discussions about current political and social issues, religion remains part of the daily routine of regional residents from Casablanca to Tehran.

Hearth of the Judeo-Christian Tradition Both Jews and Christians trace their religious roots to an eastern Mediterranean hearth. The roots of Judaism lie deep in the past: Some 4000 years ago, Abraham, an early leader in the Jewish tradition, led his people from Mesopotamia to Canaan (modern-day Israel). Jewish history, recounted in the Old Testament of the Bible, focused on a belief in one God (or **monotheism**), a strong code of ethical conduct, and a powerful ethnic identity that continues to the present. During the Roman Empire, many Jews left the eastern Mediterranean to escape Roman persecution. This forced migration, or *diaspora*, took Jews to the far corners of Europe and North Africa. Only in the past century have many of the world's far-flung Jewish people returned to Judaism's place of origin, a process that gathered speed after the creation of Israel in 1948.

Christianity, an outgrowth of Judaism, was based on the teachings of Jesus and his disciples, who lived and traveled in the eastern Mediterranean about 2000 years ago. Although many Christian traditions became associated with European history, forms of early Christianity remain near the religion's hearth. For example, the Coptic Church evolved in nearby Egypt. In Lebanon, Maronite Christians also retain a separate cultural identity.

The Emergence of Islam Islam originated in Southwest Asia in 622 CE, forming another cultural hearth of global significance. Muslims can be found from North America to the southern Philippines, but the Islamic world remains centered on Southwest Asia. Most Southwest Asian and North African peoples still follow its religious teachings. Muhammad, the founder of Islam, was born in Makkah (Mecca) in 570 CE and taught in nearby Medinah (Medina) (Figure 7.28). His beliefs parallel Judeo-Christian traditions. Muslims believe both Moses and Jesus were prophets and that the Hebrew Bible (or Old Testament) and the Christian New Testament, while incomplete, are basically accurate. However, Muslims hold that the **Quran** (or Koran), a book of teachings received by Muhammad from Allah (God), represents God's highest religious and moral revelations to humanity.

Islam offers a blueprint for leading an ethical and religious life. Islam literally means "submission to the will of God," and its practice rests on five essential activities: (1) repeating the basic creed ("There is no god but God, and Muhammad is his prophet"); (2) praying facing Makkah five times daily; (3) giving charitable contributions; (4) fasting between sunup and sundown during the month of Ramadan; and (5) making at least one religious pilgrimage, or **Hajj**, to Muhammad's birthplace of Makkah (Figure 7.29). Islamic fundamentalists also argue for a **theocratic state**, such as modern-day Iran, in which religious leaders (ayatollahs) shape government policy.

A major religious division split Islam almost immediately after the death of Muhammad in 632 CE and endures today. One group, now called **Shiites**, favored passing on religious authority within Muhammad's family, specifically to Ali, his son-in-law. Most Muslims, later known as **Sunnis**, advocated passing down power through established clergy. This group was largely victorious. Ali was killed, and his Shiite supporters went underground. Ever since, Sunni Islam has formed the mainstream branch of the religion, to which Shiite Islam has presented a recurring and sometimes powerful challenge.

Islam quickly spread from the western Arabian Peninsula, following caravan routes and Arab military campaigns as it converted thousands to its beliefs (see Figure 7.28). By the time of Muhammad's death in 632 CE, peoples of the Arabian Peninsula were united under its banner. Shortly thereafter, the Persian Empire fell to Muslim forces, and the Eastern Roman (or Byzantine) Empire lost most of its territory to Islamic influences. By 750 CE, Arab armies swept across North Africa, conquered most of Spain and Portugal, and established footholds in Central and South Asia. By the 13th century, most people in the region were Muslims, and older religions such as Christianity and Judaism became minority faiths.

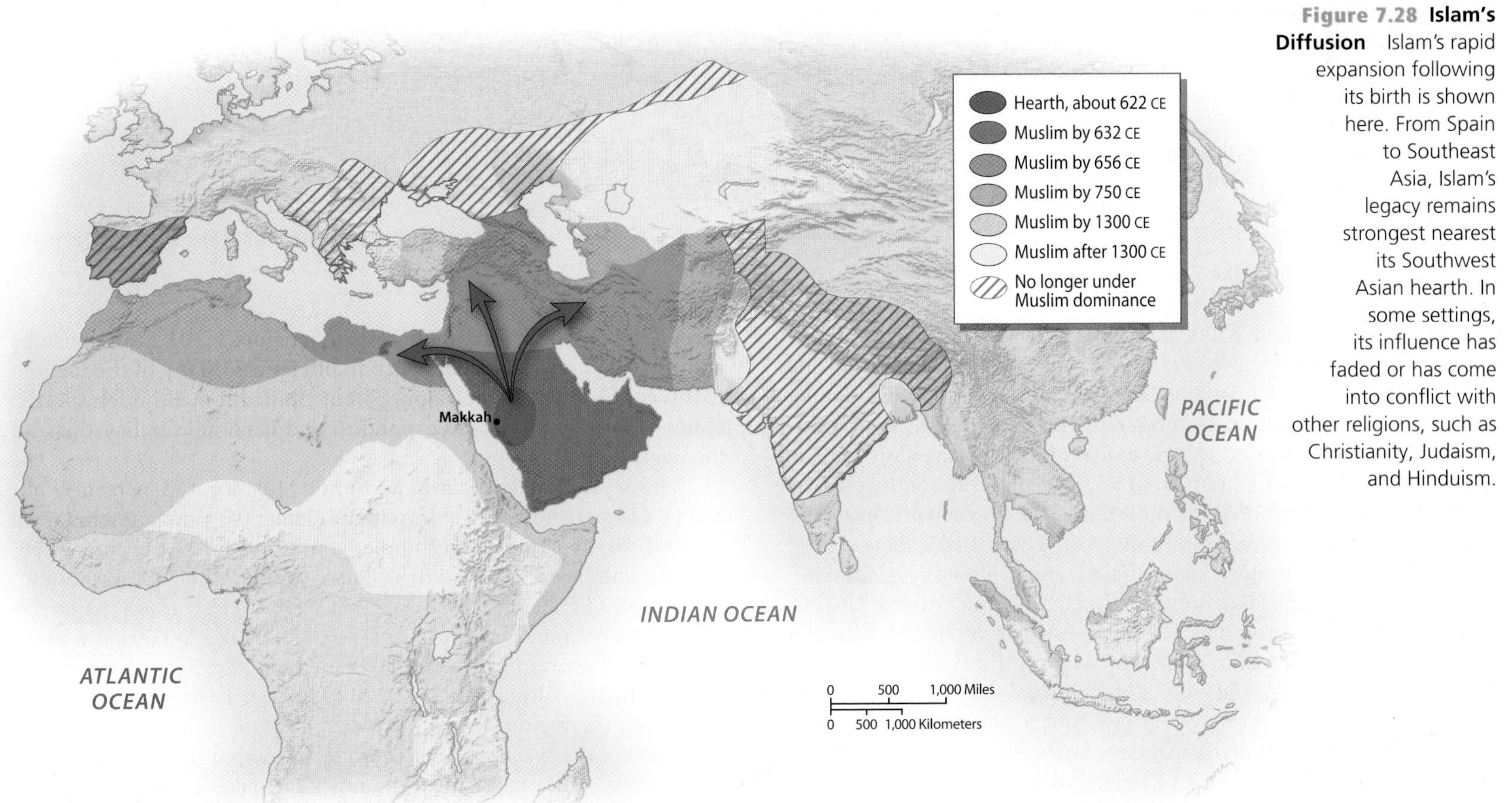

Figure 7.28 Islam's Diffusion Islam's rapid expansion following its birth is shown here. From Spain to Southeast Asia, Islam's legacy remains strongest nearest its Southwest Asian hearth. In some settings, its influence has faded or has come into conflict with other religions, such as Christianity, Judaism, and Hinduism.

EXPLORING GLOBAL CONNECTIONS

The Libyan Highway to Europe

Google Earth (MG) Virtual Tour Video
http://goo.gl/Mb0mHp

Revolutions bring many unintended consequences. When Libyan dictator Muammar al-Qaddafi was overthrown in 2011, few experts believed it would dramatically reorient and enhance one of the world's most diverse flows of refugees. The newly formed Libyan Highway has truly international implications that reach from Syria and Nigeria to Italy and Sweden (Figure 7.2.1).

A Highway for Refugees All of the critical variables in the creation of the highway fell into place in 2014. First and foremost, Libya itself ceased to truly exist as multiple political forces vied for power, essentially ending any effective control over the country. Migrants and smugglers were free to make trip arrangements without much fear of government interference.

Second, an unregulated extralegal industry designed around transporting desper-

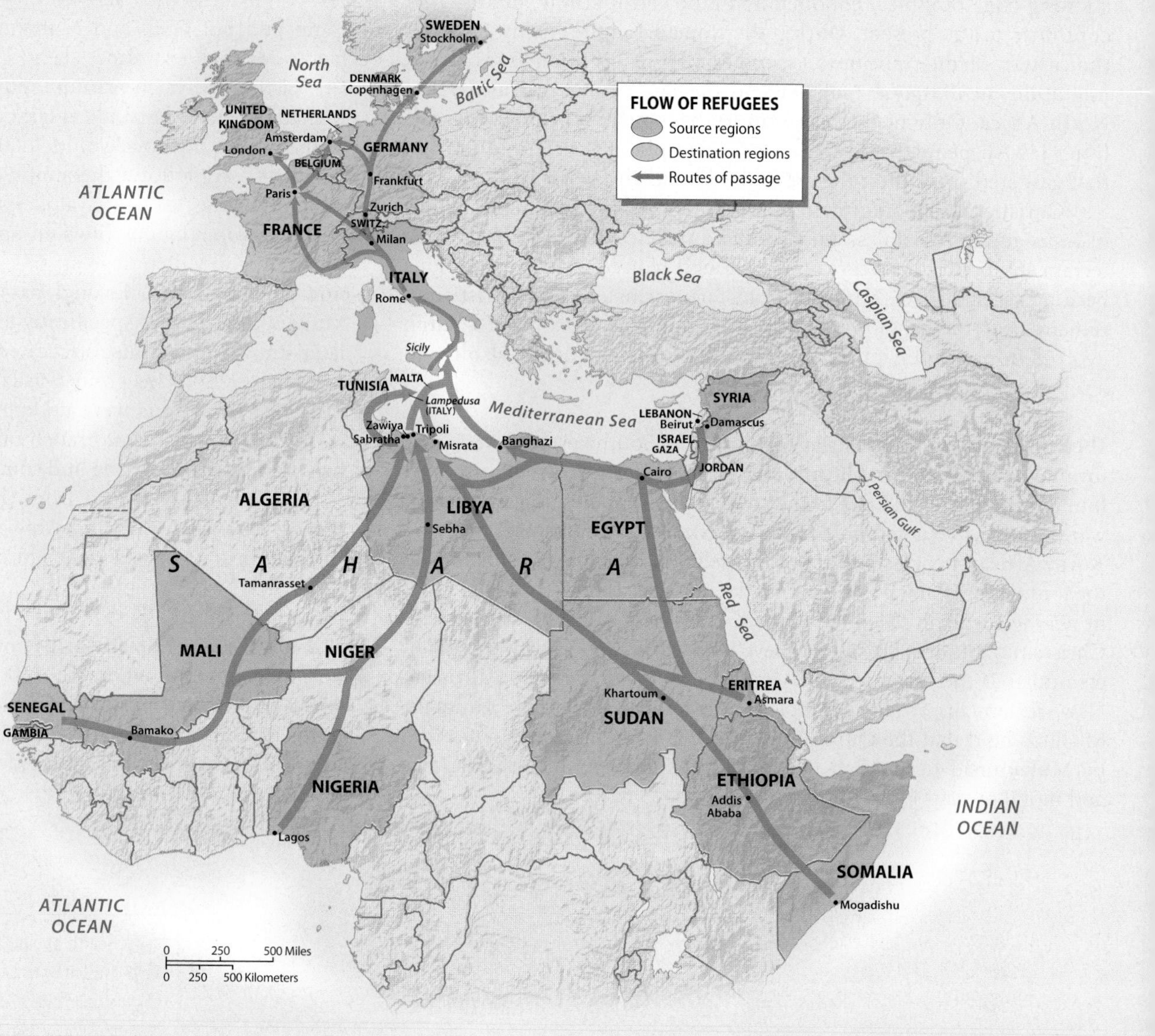

Figure 7.2.1 Libyan Highway to Europe The map shows some of the overland routes across North Africa that converge on Libyan ports, as well as general routes across the Mediterranean that take desperate migrants to Europe.

Between 1200 and 1500, Islamic influences expanded in some areas and contracted elsewhere. The Iberian Peninsula (Spain and Portugal) returned to Christianity in 1492, although Moorish (Islamic) cultural and architectural features remain today. At the same time, Muslims expanded southward and eastward into Africa, while Muslim Turks largely replaced Christian Greek influences in Southwest Asia after 1100. One group of Turks moved into the Anatolian Plateau and conquered the Byzantine Empire in 1453. These Turks soon created the huge **Ottoman Empire** (named after one of its leaders, Osman), which included southeastern Europe (including modern-day Albania, Bosnia, and Kosovo) and most of Southwest Asia and North Africa. It provided a focus of Muslim political power until the empire's disintegration in the late 19th and early 20th centuries.

Modern Religious Diversity Today Muslims form the majority population in all the countries of Southwest Asia and North Africa except Israel, where Judaism dominates (Figure 7.30). Still, divisions within Islam create regional cultural differences. Many of the region's recent conflicts are defined along Sunni–Shiite lines, although specific issues focus more on power, politics, and economic policy than on theological differences.

The region is dominated by Sunni Muslims (73 percent), but Shiites (23 percent) remain important elements in the contemporary cultural mix. In Iraq, for example, southern Shiites (around Najaf, Karbala, and Basra) asserted their cultural and political power following the fall of Saddam Hussein in 2003. Shiites also claim majorities in Iran and Bahrain and form substantial minorities in Lebanon, Saudi Arabia, Yemen, and Egypt. Since 1980, many radicalized Shiite groups have pushed a cultural and political agenda of Islamic fundamentalism across the region. Recently, however, Sunni fundamentalism has also been on the rise, most evident in the growth of ISIL in Iraq, Syria, and beyond. More mainstream Sunnis still make up the majority of

ate refugees expanded across the region. This new geography included overland travel routes across Libya (with checkpoints, smuggler transfer agents, and staging areas) as well as a casual, often chaotic collection of boats (everything from unsafe inflatable rafts to rickety fishing vessels) available to make the short but hazardous journey to Europe (most commonly to southern Italy and Malta) (Figure 7.2.2). The Italian island of Lampedusa (south of Sicily) has been an initial destination for many migrants.

Third, Europe shone as an ever-brighter destination (Germany, Italy, Sweden, and Switzerland have accepted the most asylum seekers) as prospects for decent lives dimmed in West Africa (Mali, Gambia, and Nigeria), the African Horn (Eritrea and Somalia), Southwest Asia (Syria and Gaza), and Libya itself. No surprise that desperate Syrian refugees have dominated recent flows as that country disintegrates. Experts estimate that more than 219,000 people migrated from North Africa to Europe in 2014, most of them along the Libyan Highway. Even reports of disastrous journeys (in 2015, more than 900 people died as an overloaded boat sank) have failed to quell the human migration north. Recently, between 500,000 and 1 million people were camped out in northern Libya (mostly between Benghazi and Tripoli), awaiting passage to a better life. While conditions in these settings were uniformly abysmal, they were typically better than where the refugees began their dangerous journeys.

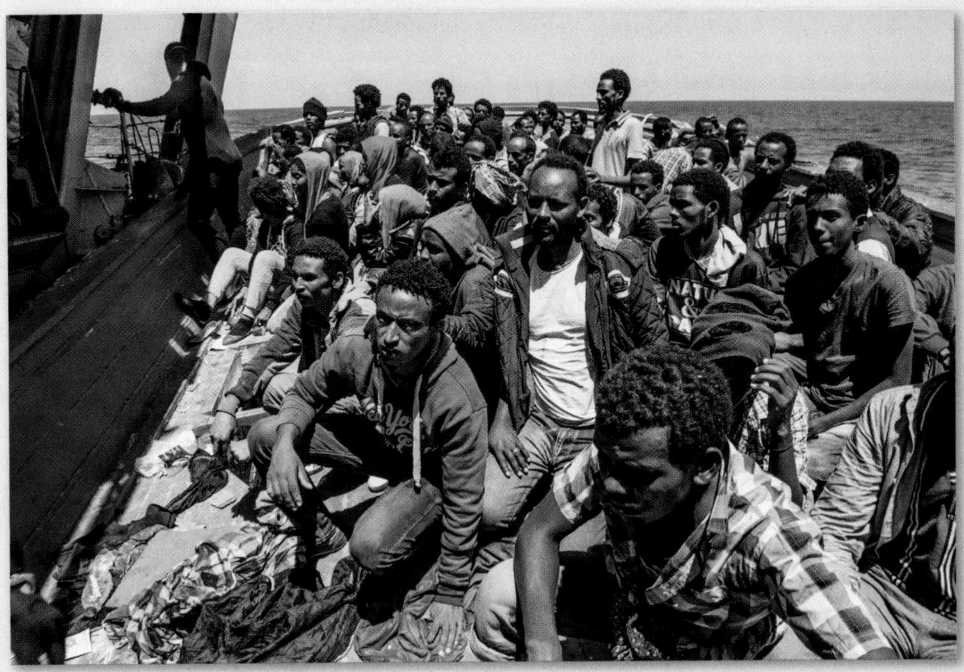

Figure 7.2.2 Refugees on the Mediterranean Sea, 2015 These refugees, rescued in the Straits of Sicily in 2015, were only 30 miles off the coast of Libya. Most of the migrants on this boat were from Eritrea or Syria.

Addressing the Challenge Some officials within the European Union (EU) have asked how to throw an effective roadblock across the Libyan Highway—or at least how to safely control the traffic flow. In the short term, EU nations have agreed to improve patrols and rescue efforts in the Mediterranean to reduce the risk of disasters like the one that took almost 1000 lives in 2015. There is also movement toward a broader resettlement plan for refugees once they fall within EU jurisdiction, an initiative pushed by Italians overwhelmed by migrants. Still, many European interests—especially anti-immigrant political parties—simply want to shut the highway down. In the longer term, however, the Libyan Highway can be controlled only when stable civil authority returns to Libya and when conditions improve in the far-flung localities generating the refugees in the first place. Neither appears likely soon, so for the foreseeable future the hazardous route north will continue to swell with desperate people on the move.

1. From the diverse list of migrant source areas mentioned above, choose two countries and write a paragraph on each that explains why residents of these areas are willing to make the journey.
2. Should Europe welcome or curtail these diverse migrants? Defend your answer.

the region's population and reject its more radical cultural and political precepts, arguing for a more modern Islam that accommodates some Western values and traditions.

While the Sunni–Shiite split is the Muslim world's great divide, other variations of Islam are also practiced in the region. One division separates the mystically inclined form of Islam known as *Sufism* from mainstream traditions. Sufism is prominent in the peripheries of the region, including the Atlas Mountains and across northwestern Iran and portions of Turkey. Elsewhere, the Salafists and Wahhabis, numerous in both Egypt and Saudi Arabia, are austere, conservative Sunnis who adhere to what they see as an earlier, purer form of Islamic doctrine. The Druze of Lebanon practice yet another variant of Islam.

Southwest Asia is also home to many non-Islamic communities. Israel has a Muslim minority (16 percent) dominated by that nation's Jewish population (77 percent). Even Israel's Jewish community is divided between Jewish fundamentalists and more reform-minded Jews. In neighboring Lebanon, a slight Christian (Maronite and Orthodox) majority was in evidence as recently as 1950, but Christian outmigration and higher Islamic birth rates created a nation that today is about 60 percent Muslim.

Jerusalem (Israel's capital) holds special religious significance for several groups and also stands at the core of the region's political problems (Figure 7.31). Indeed, the sacred space of this ancient city remains deeply scarred and divided. Considering just the 220 acres of land within the Old City, Jews pray at the old Western Wall (the site of a Roman-era Jewish temple); Christians honor the Church of the Holy Sepulchre (the burial site of Jesus); and Muslims hold sacred rites in the city's eastern quarter (including the place from which the prophet Muhammad reputedly ascended to heaven). Nearby suburban communities are also contested real estate as Arab and Israeli neighborhoods (including newly built Jewish settlements) uneasily sit next to one another.

Figure 7.29 Makkah Thousands of faithful Muslims gather at the Grand Mosque in central Makkah (Mecca), part of the pilgrimage to this sacred place that draws several million visitors annually. A bevy of hotels and portions of the city's commercial district can be seen in the distance.

Geographies of Language

Although the region is often termed the "Arab World," linguistic complexity creates important cultural divisions across Southwest Asia and North Africa (Figure 7.32).

Semites and Berbers Afro-Asiatic languages dominate the region. Within that family is Arabic, spoken by Semitic peoples from the Persian Gulf to the Atlantic and southward into Sudan. Arabic has religious significance for Muslims because it was the sacred language in which God delivered his message to Muhammad. Although most of the world's Muslims do not speak Arabic, the faithful often memorize prayers in the language, and many Arabic words have entered the other important languages of the Islamic world.

Hebrew, another Semitic language, was reintroduced into the region with the creation of Israel. Hebrew originated in the Levant and was spoken by the ancient Israelites 3000 years ago. Today its modern version survives as the sacred tongue of the Jewish people and is the official language of Israel, although the country's non-Jewish population largely speaks Arabic.

Older Afro-Asiatic languages survive in more remote areas of North Africa. Collectively known as Berber, these languages are related, but not mutually intelligible. Berber speakers are spread across the North African interior, often in isolated rural settings (Figure 7.33). Most Berber languages have never been written, and none has generated a significant literature. Indeed, a Berber-language version of the Quran was not completed until 1999.

Persians and Kurds Although Arabic spread readily through portions of Southwest Asia, much of the Iranian Plateau and nearby mountains is dominated by older Indo-European languages. Here the

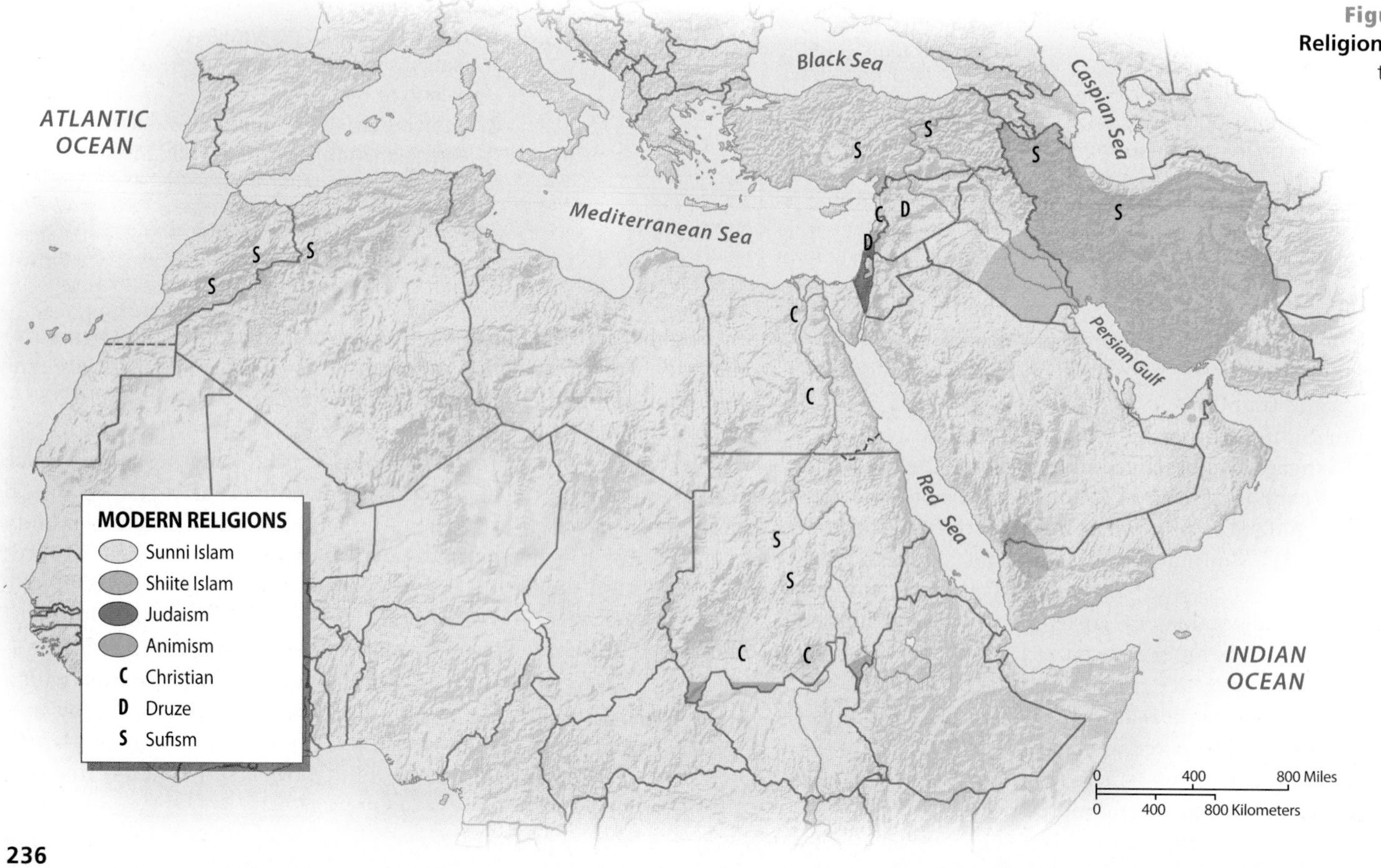

MODERN RELIGIONS
Sunni Islam
Shiite Islam
Judaism
Animism
C Christian
D Druze
S Sufism

ATLANTIC OCEAN
Black Sea
Caspian Sea
Mediterranean Sea
Persian Gulf
Red Sea
INDIAN OCEAN
0 400 800 Miles
0 400 800 Kilometers

Figure 7.30 Modern Religions Islam continues to be the dominant religion across the region. Most Muslims are tied to the Sunni branch, whereas Shiites are found in places such as Iran and southern Iraq. In some locales, however, Christianity and Judaism remain important.

Figure 7.31 Old Jerusalem The historic center of Jerusalem reflects its varied religious legacy. Sacred sites for Jews, Christians, and Muslims are all located within the Old City. The Western Wall, a remnant of the ancient Jewish temple, stands at the base of the Dome of the Rock and Islam's al-Aqsa Mosque.

principal tongue remains Persian, although, since the 10th century, the language has been enriched with Arabic words and written in the Arabic script.

Kurdish speakers of northern Iraq, northwest Iran, and eastern Turkey add further complexity to the regional pattern of languages. Kurdish, also an Indo-European language, is spoken by 10 to 15 million people. The Kurds have a strong sense of shared cultural identity. Indeed, "Kurdistan" is sometimes called the world's largest nation without its own political state. In the 2003 war in Iraq, the Kurds emerged as a cohesive group in the northern part of the country and have now gained more political autonomy. Kurds in eastern Turkey (about 20 percent of that nation's population), while enjoying more individual rights in that increasingly democratic country, have been stifled in their hopes for regional political autonomy.

The Turkic Imprint Turkic languages provide variety across much of modern Turkey and in portions of far northern Iran. Turkic languages are a part of the larger Altaic language family that originated in Central Asia. Turkey remains the largest nation in Southwest Asia dominated by this language family. Tens of millions of people in other countries of Southwest Asia and Central Asia speak related Altaic languages, such as Azeri, Uzbek, and Uyghur.

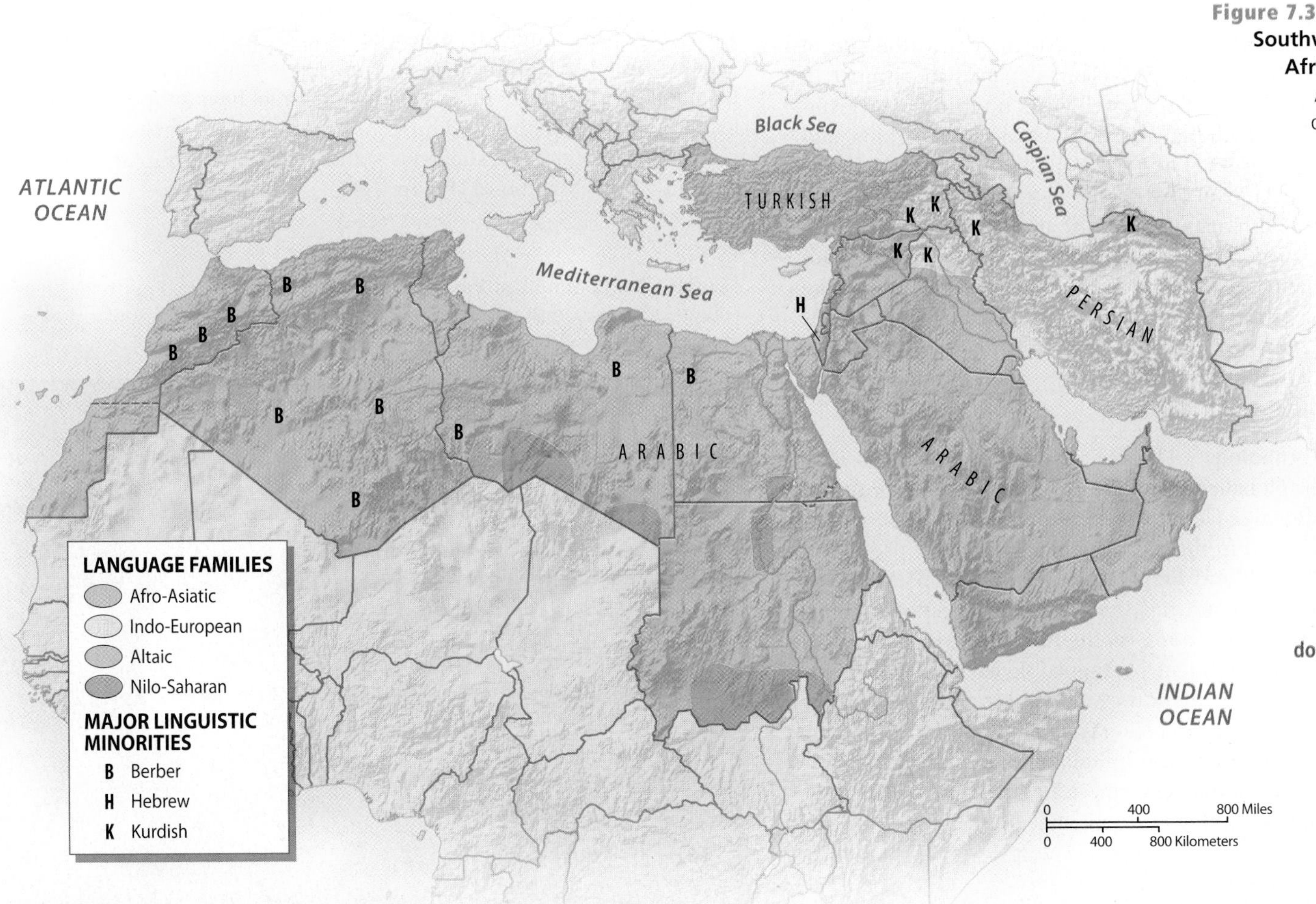

Figure 7.32 Language Map of Southwest Asia and North Africa Arabic, a Semitic Afro-Asiatic language, dominates the region's cultural geography. Turkish, Persian, and Kurdish, however, remain important exceptions, and such differences within the region have had long-lasting political consequences. Israel's more recent reintroduction of Hebrew further complicates the region's linguistic geography. **Q: Cite examples where Islam (see Figure 7.30) dominates in non-Arabic-speaking regions.**

Figure 7.33 Berber Community A shared cultural history and a distinctive language help define the region's Berber community. These Berbers pose outside their tent in the North African interior.

Regional Cultures in Global Context

Many cultural connections tie the region with the world beyond. These global connections are nuanced and complex and are expressed in fascinating, often unanticipated ways.

Islamic Internationalism Islam is geographically and theologically divided, but all Muslims recognize the fundamental unity of their religion. This religious unity extends far beyond Southwest Asia and North Africa. Islamic communities are well established in such distant places as central China, European Russia, central Africa, Indonesia, and the southern Philippines. Today Muslim congregations also are expanding rapidly in the major urban areas of western Europe and North America. Even with its global reach, however, Islam remains centered on Southwest Asia and North Africa, the site of its origins and its holiest places. As Islam expands in number of followers and geographical scope, the religion's tradition of pilgrimage ensures that Makkah will become a city of increasing global significance in the 21st century. The recent global growth of Islamist fundamentalism and Islamism also focuses attention on the region. In addition, the oil wealth accumulated by many Islamic nations is used to sustain and promote the religion. Countries such as Saudi Arabia invest in Islamic banks and economic ventures and make donations to Islamic cultural causes, colleges, and hospitals worldwide. In Southeast Asia, for example, money from Saudi Arabia is funding a growing number of new mosques being built in several Muslim portions of the region.

Globalization and Technology The region also struggles with how its growing role in the global economy is changing traditional cultural values. European colonialism left its own cultural legacy—not only in the architecture still found in the old colonial centers, but also in the widespread use of English and French among the region's Western-educated elite. In oil-rich countries, huge capital investments have had important cultural implications, as the number of foreign workers has grown and as more affluent young people have embraced elements of Western-style music, literature, and clothing. The expansion of Islamic fundamentalism and Islamism is in many ways a reaction to the threat posed by external cultural influences, particularly the supposed evils of European colonialism, American and Israeli power, and local governments and cultural institutions that are seen as selling out to the West.

Technology also shapes cultural and political change. Particularly among the young, millions within and beyond the region found themselves linked by the Internet, cell phones, and various forms of social media during the Arab Spring uprisings of 2011 and 2012 (Figure 7.34). Cell phones, blogs, email, and tweets facilitated the flow of information that helped protesters plan events and coordinate strategies with their allies. Local videos from smartphones and pinhole cameras documented government abuse and often provided (in settings such as Syria) the only proof of widespread state-supported violence. Also, the global diffusion of this information promoted the internationalization of political discourse and made it easier to spread the word about local conflicts and to identify common threads among different protest movements. In a related fashion, the fundamentalist ISIL movement claims considerable success in recruiting new global converts to its cause (including those from Europe and the United States) through the use of social media such as Facebook and Twitter. A related point is that about 60 percent of the region's population is under 30—precisely the group most inclined to use these technologies and often the one most frustrated by unresponsive governments that refuse to change their ways.

The Role of Sports Sports play a hugely important cultural role in everyday life within the region. Soccer rules the day, both as a spectator sport and as an activity that many young people enjoy. Most countries have national football (soccer) associations that also participate in regional and global (FIFA) league competitions. The Union of Arab Football Associations (UAFA), headquartered in Riyadh, Saudi Arabia, offers many opportunities for regional competition, often carried on the Al Jazeera Sports network. A smaller Gulf Cup of Nations competition also sparks spirited rivalries. Large soccer stadiums are a common part of modern urban landscapes, including Kuwait City's Sabah Al-Salem Stadium, which comfortably seats more than 28,000 fans. Recently, Algeria, Tunisia, Iran, and Israel have supported strong teams with high global rankings. As in other parts of the world, regional players can earn superstar status, such as Iraq's goal-shooting veteran Younis Mahmoud (Figure 7.35).

Figure 7.34 Communicating from the Front Lines, Cairo, Egypt A young Egyptian woman talks on a mobile phone in Cairo's Tahrir Square during demonstrations in February 2011.

Figure 7.35 Iraq Soccer Star Younis Mahmoud Emotions run high as Younis Mahmoud scores a goal against Kuwait in the 2013 Gulf Cup tournament.

Review

7.6 Describe the key characteristics of Islam, and explain why distinctive Sunni and Shiite branches exist today.

7.7 Compare the modern maps of religion and language for the region, and identify three major non-Arabic-speaking areas where Islam dominates. Explain why that is the case.

KEY TERMS monotheism, Quran, Hajj, theocratic state, Shiite, Sunni, Ottoman Empire

Geopolitical Framework: Never-Ending Tensions

Geopolitical tensions remain very high in Southwest Asia and North Africa (Figure 7.36). In the Arab Spring rebellions, governments fell in Tunisia (where the regional movement began in late 2010), Egypt, Libya, and Yemen (Figure 7.37); widespread protests shook once-stable states such as Bahrain; and a more protracted civil war erupted in Syria, producing a huge and ongoing refugee crisis. Other countries witnessed shorter, more intermittent demonstrations against state authority. To varying degrees, these uprisings focused broadly on (1) charges of widespread government corruption; (2) limited opportunities for democracy and free elections; (3) rapidly rising food prices; and (4) the enduring reality of widespread poverty and high unemployment, especially for people under 30.

More recently, sectarian conflicts between Sunnis and Shiites have dominated the geopolitical map. Iran (mainly Shiite) and Saudi Arabia (mainly Sunni) have each played major regional roles in bankrolling their respective supporters. For example, Iran supports the Shiite-dominated Iraq government in its struggles against ISIL Sunni extremists as well as Syria's Assad regime (largely made up of Alawites, an offshoot of Shiite Islam). Saudi Arabia, from its perspective, has made it clear that it does not want to see nearby Yemen led by Shiite extremists and began intervening directly in that conflict in 2015.

In addition to these recent conflicts, ongoing issues include the future of Israeli-Palestinian relations. Some of these tensions relate to age-old patterns of cultural geography or to European colonialism because modern boundaries were formed by colonial powers. In addition, geographies of wealth and poverty enter the geopolitical mix: Some residents profit from petroleum resources and industrial expansion, while others struggle to feed their families. The result is a region where the political climate is charged with tension and the sounds of bomb blasts and gunfire remain all-too-common characteristics of everyday life.

The Colonial Legacy

European colonialism arrived relatively late in Southwest Asia and North Africa, but the era left an important imprint on the region's modern political geography. Between 1550 and 1850, the region was dominated by the Ottoman Empire, which expanded from its Turkish hearth to engulf much of North Africa as well as nearby areas of the Levant, the western Arabian Peninsula, and modern-day Iraq. After 1850, Ottoman influences waned, and European colonial dominance grew after the dissolution of the Ottoman Empire in World War I (1918).

Both France and Great Britain were major colonial players within the region. French interests in North Africa included Tunisia and Morocco. French Algeria attracted large numbers of European immigrants. After World War I, France added more colonial territories in the Levant (Syria and Lebanon). The British loosely incorporated places such as Kuwait, Bahrain, Qatar, the United Arab Emirates, and Aden (in southern Yemen) into their empire to help control sea trade between Asia and Europe. Nearby Egypt also caught Britain's attention. Once the European-engineered **Suez Canal** linked the Mediterranean and Red seas in 1869, European banks and trading companies gained more influence over the Egyptian economy. In Southwest Asia, British and Arab forces joined to force out the Turks during World War I. The Saud family convinced the British that a country should be established on the Arabian Peninsula, and Saudi Arabia became fully independent in 1932. Britain divided its other territories into three entities: Palestine (now Israel) along the Mediterranean coast; Transjordan to the east of the Jordan River (now Jordan); and a third zone that became Iraq.

Persia and Turkey were never directly occupied by European powers. In Persia, the British and Russians agreed to establish two spheres of economic influence in the region (the British in the south, the Russians in the north), while respecting Persian independence. In 1935, Persia's modernizing ruler, Reza Shah, changed the country's name to Iran. In Turkey, European powers attempted to divide up the

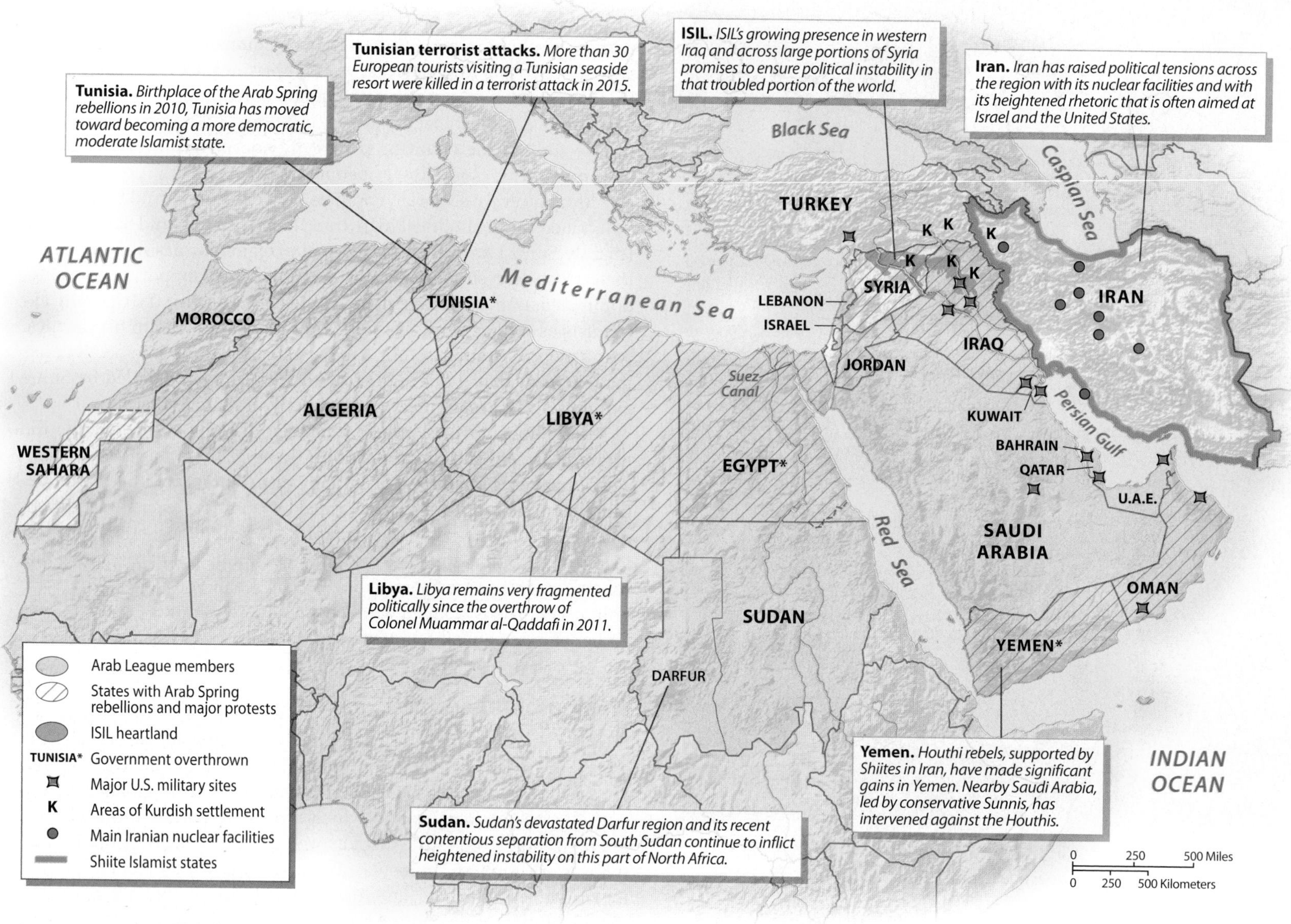

Tunisian terrorist attacks. *More than 30 European tourists visiting a Tunisian seaside resort were killed in a terrorist attack in 2015.*

ISIL. *ISIL's growing presence in western Iraq and across large portions of Syria promises to ensure political instability in that troubled portion of the world.*

Tunisia. *Birthplace of the Arab Spring rebellions in 2010, Tunisia has moved toward becoming a more democratic, moderate Islamist state.*

Iran. *Iran has raised political tensions across the region with its nuclear facilities and with its heightened rhetoric that is often aimed at Israel and the United States.*

Black Sea
Caspian Sea
TURKEY
ATLANTIC OCEAN
Mediterranean Sea
TUNISIA*
MOROCCO
SYRIA
LEBANON
ISRAEL
IRAN
IRAQ
JORDAN
Suez Canal
ALGERIA
LIBYA*
KUWAIT
Persian Gulf
WESTERN SAHARA
BAHRAIN
EGYPT*
QATAR
U.A.E.
Red Sea
SAUDI ARABIA
OMAN
SUDAN
YEMEN*
DARFUR
INDIAN OCEAN

Libya. *Libya remains very fragmented politically since the overthrow of Colonel Muammar al-Qaddafi in 2011.*

Yemen. *Houthi rebels, supported by Shiites in Iran, have made significant gains in Yemen. Nearby Saudi Arabia, led by conservative Sunnis, has intervened against the Houthis.*

Sudan. *Sudan's devastated Darfur region and its recent contentious separation from South Sudan continue to inflict heightened instability on this part of North Africa.*

Arab League members
States with Arab Spring rebellions and major protests
ISIL heartland
TUNISIA* Government overthrown
Major U.S. military sites
K Areas of Kurdish settlement
Main Iranian nuclear facilities
Shiite Islamist states

0 250 500 Miles
0 250 500 Kilometers

Figure 7.36 Geopolitical Issues in Southwest Asia and North Africa Political tensions continue across much of the region. The Arab Spring rebellions shaped subsequent political changes in several settings, and the rise of ISIL has disrupted life in Iraq, Syria, and elsewhere. The Israeli–Palestinian conflict also remains pivotal.

Figure 7.37 Tahrir Square A focal point for political protests, Cairo's Tahrir Square fills with demonstrators at a mass rally in November 2011.

Ottoman Empire following World War I. The successful Turkish resistance to European control was based on new leadership provided by Kemal Ataturk. Ataturk decided to imitate the European countries and establish a modern, culturally unified, secular state.

European colonial powers began withdrawing from Southwest Asian and North African colonies before World War II. By the 1950s, most countries in the region were independent. In North Africa, Britain withdrew troops from Sudan and Egypt in 1956. Libya (1951), Tunisia (1956), and Morocco (1956) achieved independence peacefully during the same era, but the French colony of Algeria became a major problem. Several million French citizens resided there, and France had no intention of simply withdrawing. A bloody war for independence began in 1954, and France finally agreed to an independent Algeria in 1962.

Southwest Asia also lost its colonial status between 1930 and 1960. Iraq became independent from Britain in 1932, but its later instability resulted in part from its imposed borders, which never recognized its cultural diversity. Similarly, the French division of the Levant into Syria and Lebanon (1946) greatly angered local Arab populations and set the stage for future political instability. As a favor to the Lebanese Maronite Christian majority, France carved out a separate Lebanese state from largely Arab Syria, even guaranteeing the Maronites constitutional control of the government. The action created a culturally divided Lebanon as well as a Syrian state that has repeatedly asserted its influence over its Lebanese neighbors.

Modern Geopolitical Issues

The geopolitical instability in Southwest Asia and North Africa continues today. A quick regional transect from the shores of the Atlantic to the borders of Central Asia suggests how these forces are playing out in different settings early in the 21st century.

Across North Africa Varied North African settings have recently witnessed dramatic political changes (see Figure 7.36). In Tunisia, birthplace of the Arab Spring, a moderate Islamist government was elected to replace deposed dictator Zine el-Abidine Ben Ali, but the country has not been immune to terrorist attacks by jihadist extremists, including an attack on British tourists at a seaside resort in 2015.

In nearby Libya, while many cheered the end of Colonel Muammar al-Qaddafi's rule in 2011, two rival militia alliances have split the nation (it has two parliaments). One group is focused in the west (around Tripoli) and another in the northeast at Beida. Libya's political power vacuum has made it a North African stronghold for ISIL sympathisers.

Next door, following the 2011 overthrow of Hosni Mubarak (see Figure 7.37), Egypt remained politically unstable. Parliamentary and presidential elections in 2012 ushered in a brief period of rule by the Muslim Brotherhood, led by President Mohamed Morsi. The government adopted a new constitution, but many ordinary Egyptians felt Morsi moved forward too quickly and ignored many democratically guaranteed civil rights. Also unsettling (particularly to the nation's Coptic Christian population) was the growing visibility of fundamentalist Islamist extremists. In 2013, the Egyptian military staged a coup and ousted Morsi. Since then, some political stability has returned to the country within the shadows of an authoritarian regime strongly influenced by the military.

Sudan also faces daunting political issues. A Sunni Islamist state since a military coup in 1989, Sudan imposed Islamic law across the country, antagonizing both moderate Sunni Muslims and the large non-Muslim (mostly Christian and animist) population in the south. Civil war between the north and south produced more than 2 million casualties (mostly in the south) between 1988 and 2004. A tentative peace agreement was signed in 2005, which opened the way for a successful vote on independence in South Sudan. Even though the two nations officially split in 2011, tensions remain, especially focused on an oil-rich and contested southern border zone between the two countries.

In addition, Sudan's western Darfur region remains in shambles (see Figure 7.36). Ethnicity, race, and control of territory seem to be at the center of the struggle in the largely Muslim region, as a well-armed Arab-led militia group (with many ties to the central government in Khartoum) has attacked hundreds of black-populated villages, killing more than 300,000 people (through violence, starvation, and disease) and driving 2.5 million more from their homes.

The Arab–Israeli Conflict The 1948 creation of the Jewish state of Israel produced another enduring zone of cultural and political tensions within the eastern Mediterranean (Figure 7.38). Jewish migration to Palestine increased after the defeat of the Ottoman Empire in World War I. In 1917, Britain issued the Balfour Declaration, a pledge to encourage the creation of a Jewish homeland in the region. After World War II, the UN divided the region into two states, one to be predominantly Jewish, the other primarily Muslim. Indigenous Arab Palestinians rejected the partition, and war erupted. Jewish forces proved victorious, and by 1949 Israel had actually grown in size. Hundreds of thousands of Palestinian refugees fled from Israel to neighboring countries, where many of them remained in makeshift camps. Under these conditions, Palestinians nurtured the idea of creating their own state on land that had become part of Israel.

Israel's relations with neighboring countries remained poor. Supporters of Arab unity and Muslim solidarity sympathized with the Palestinians, and their antipathy toward Israel grew. Israel fought additional wars in 1956, 1967, and 1973. In territorial terms, the Six-Day War of 1967 was the most important conflict (see Figure 7.38). In this struggle against Egypt, Syria, and Jordan, Israel occupied substantial new territories in the Sinai Peninsula, the Gaza Strip, the West Bank, and the Golan Heights. Israel also annexed the eastern (Muslim) part of the formerly divided city of Jerusalem, arousing particular bitterness among the Palestinians (see Figure 7.31). A peace treaty with Egypt resulted in the return of the Sinai Peninsula in 1982, but tensions focused on other occupied territories under Israeli control. To strengthen its geopolitical claims, Israel built additional Jewish settlements in the West Bank and in the Golan Heights, further angering Palestinian residents.

Palestinians and Israelis began to negotiate a settlement in the 1990s. Preliminary agreements called for a quasi-independent Palestinian state in the Gaza Strip and across much of the West Bank. A tentative agreement late in 1998 strengthened the potential control of the ruling **Palestinian Authority (PA)** in the Gaza Strip and portions of the West Bank (Figure 7.39). But a new cycle of heightened violence erupted late in 2000 as Palestinian attacks against Jews increased and the Israelis continued to build new settlements in occupied lands (especially in the West Bank). Indeed, the Israeli government has pledged to continue its support for new construction, including developments in and near Jerusalem and Bethlehem. Palestinian authorities continue to strongly protest these new West Bank Jewish settlements near Israel's capital.

Adding to the friction is the ongoing construction of an Israeli security barrier, a partially completed series of concrete walls, electronic fences, trenches, and watchtowers designed to effectively separate the Israelis from Palestinians across much of the West Bank region (see Figure 7.39). Israeli supporters of the barrier (to be more than 400 miles long when completed) see it as the only way to protect their citizens from terrorist attacks. Palestinians see it as a land grab, an "apartheid wall" designed to isolate many of their settlements along the Israeli border.

Political fragmentation of the Palestinians adds further uncertainty. In 2006, control of the Palestinian government was split between the Fatah and Hamas political parties. Israelis have long regarded Hamas as an extremist political party, whereas Fatah has shown more willingness to work peacefully with Israel. Hamas gained effective control of the PA within Gaza, and Fatah maintains its greatest influence across the West Bank. Rockets launched into Israel from Hamas-controlled Gaza have repeatedly provoked Israeli counterattacks, decimating the Gaza economy.

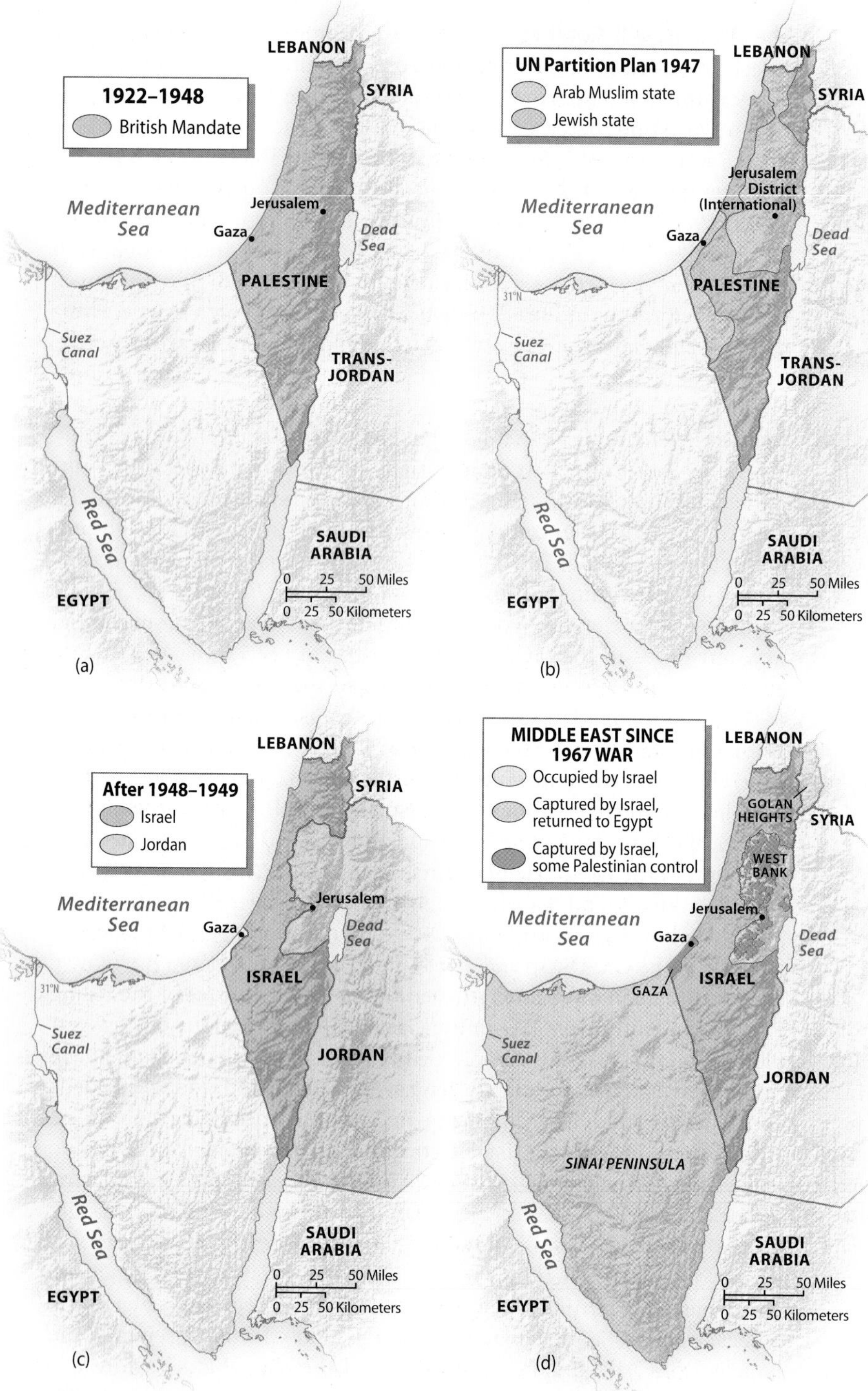

Figure 7.38 Evolution of Israel Modern Israel's complex evolution began with (a) an earlier British colonial presence and (b) a UN partition plan in the late 1940s. (c) Thereafter, multiple wars with nearby Arab states produced Israeli territorial victories in Gaza, the West Bank, and the Golan Heights. (d) Each of these regions continues to be important in Israel's recent relations with nearby states and with resident Palestinian populations.

One thing is certain: Geographical issues will remain at the center of the conflict. Israelis continue their search for secure borders to guarantee their political integrity. Most Palestinians still call for a "two-state solution," in which their autonomy is guaranteed, but a growing minority of Palestinians, frustrated with the stalemate, suggest considering a "one-state solution," in which Israel would be compelled to recognize the Palestinians as equals.

Instability in Syria and Iraq Elsewhere in the region, political instability in Syria erupted into civil war in 2011. Rebel (mostly Sunni Muslim) protests against the autocratic regime of President Bashar Hafez al-Assad (a member of the minority Alawite sect) reached a fever pitch, and government soldiers killed thousands of civilians and used chemical weapons in a series of violent confrontations. The larger regional Arab community reacted against Assad, suspending Syria from the **Arab League** (a regional political and economic organization focused on Arab unity and development; see Figure 7.36) and urging an international solution to the crisis. Since 2014, the presence of ISIL in eastern Syria (see Figure 7.36)—as well as a growing military response to ISIL from moderate Arab states, Kurdish fighters, and U.S.-led bombing raids—has added to Syria's political disintegration. Russian involvement in Syria has also grown, mostly to bolster the Assad regime and to increase Russian influence in the region. By 2015, more than 220,000 deaths were directly related to the violence, and millions of people had fled their homes (see *Geographers at Work: How Do We Define "Middle East"?*).

Neighboring Iraq, another multinational state born during the colonial era, has yet to escape the consequences of its geopolitical origins. When the country was carved out of the British Empire in 1932, it contained the cultural seeds of its later troubles. Iraq remains culturally complex today (Figure 7.40). Most of the country's Shiites live in the lower Tigris and Euphrates river valleys near and south of Baghdad. Indeed, the region near Basra contains some of the world's holiest Shiite shrines. In northern Iraq, the Kurds have their own ethnic identity and political aspirations. Many Kurds want complete independence from Baghdad and have managed to establish a federal region that already enjoys some autonomy from the central Iraqi government. A third major subregion traditionally is dominated by the Sunnis and encompasses part of the Baghdad area as well as territory to the north and west that includes strongholds such as Fallujah and Tikrit. The country's oil fields are mainly located in Shiite- and Kurdish-controlled portions of Iraq, an uneasy fact of resource geography that has long troubled the nation's sizable Sunni population.

When Iraqi leaders assumed control of their new state in 2004, growing sectarian violence between different Iraqi factions threw portions of the nation into civil war. Rival Sunni and Shiite groups forced many Iraqis from their communities. Before largely leaving Iraq in 2011, American troops successfully worked with Iraqi officials to reduce the level of violence. But the growth of ISIL in 2014 (as in nearby Syria) disrupted Iraq's regional balance of power. As ISIL enlarged its sphere of influence in the north and west, increased military responses from the United States and others (especially Shiite-led militias supported by Iran) suggest that Iraq will remain a political battleground haunted by sectarian violence and terrorism for the foreseeable future.

Politics in the Arabian Peninsula Change has also rocked the Arabian Peninsula. In Saudi Arabia, the Al Saud royal family retains its conservative control of the country, although the regime is gradually passing into the hands of younger family members who might be more inclined to democratize the nation's political structure. Saudi Arabia officially supports U.S. efforts to provide stable flows of petroleum and fight terrorism, but beneath the surface, certain elements

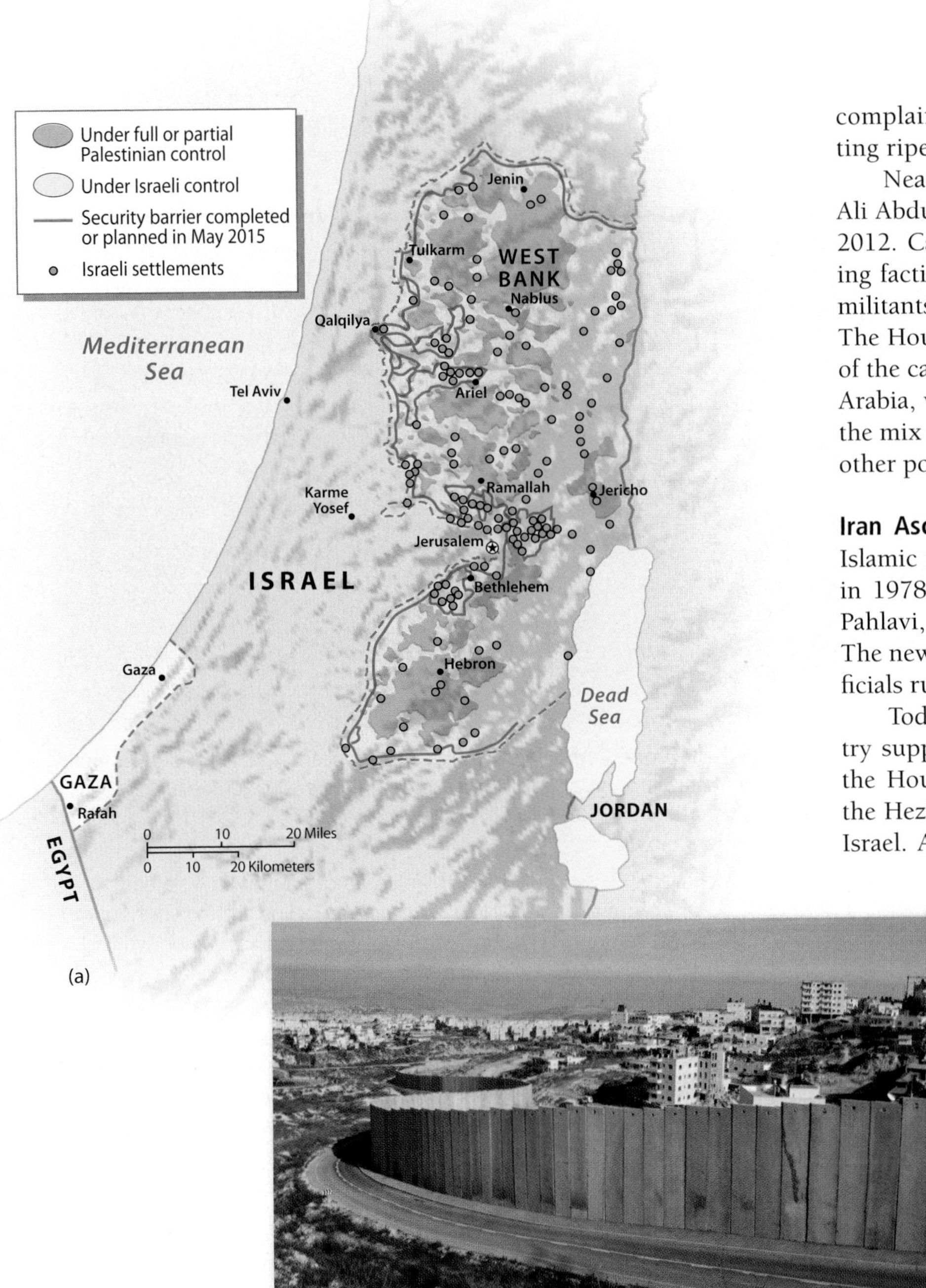

Figure 7.39 West Bank (a) Portions of the West Bank were returned to Palestinian control in the 1990s, but Israel has partially reasserted its authority in some areas and has expanded the construction of its security barrier since 2000. New Israeli settlements are scattered throughout the West Bank in areas still under Israel's nominal control. (b) The photo shows a segment of the Israeli security barrier. **Q: Look carefully at the scale of the map. Measure the approximate distance between Jerusalem and Hebron, and find two local towns in your area that are a similar distance apart.**

of the regime may have financed radically anti-American groups such as Al Qaeda. The Saudi people themselves, largely Sunni Arabs, are torn among an allegiance to their royal family (and the economic stability it brings); the lure of a more democratic, open Saudi society; and an enduring distrust of foreigners, particularly Westerners. The Sunni majority also includes Wahhabi sect members, whose radical Islamist philosophy has fostered anti-American sentiment. In addition, the large number of foreign laborers and the persistent U.S. military and economic presence within the country (a chief complaint of former Al Qaeda leader Osama bin Laden) create a setting ripe for political instability.

Nearby Yemen has been torn apart by political conflict. President Ali Abdullah Saleh was forced from office, and elections were held in 2012. Calls for democratic reforms have been complicated by ongoing factionalism within the country, including the presence of Shiite militants (the Houthis) who maintain close connections with Iran. The Houthis' political gains in Yemen (including their taking control of the capital, San'a) provoked a military response in 2015 from Saudi Arabia, which fears growing Iranian interests in the region. Added to the mix is Al Qaeda's influence (including terrorist training camps) in other portions of the country.

Iran Ascendant? Iran increasingly garners international attention. Islamic fundamentalism dramatically appeared on the political scene in 1978 as Shiite Muslim clerics overthrew Shah Mohammad Reza Pahlavi, an authoritarian, pro-Western ruler friendly to U.S. interests. The new leaders proclaimed an Islamic republic in which religious officials ruled both clerical and political affairs.

Today Iran's influence has grown across the region. The country supports Shiite-allied interests throughout the region (including the Houthis in Yemen, sympathetic regimes in Iraq and Syria, and the Hezbollah movement in Lebanon) and has repeatedly threatened Israel. Adding uncertainty has been Iran's ongoing nuclear development program, an initiative its government claims is related to the peaceful construction of power plants (see Figure 7.36). Both Israel and Arab states such as Saudi Arabia, the United Arab Emirates, and Egypt fear Iran's ascendance. Others in the West (including the United States) have moved toward a negotiated settlement with Iran that allows for limited development of its nuclear capabilities and an end to economic sanctions against the country.

Within Iran, varied political and cultural impulses are evident. Many younger, wealthier, more cosmopolitan Iranians are hopeful that the country will become less isolated on the world stage. Popular interest in fundamentalism has waned, and many Iranians have actually moved toward a more secular lifestyle. At the same time, most Iranians support the nuclear program, arguing that they have the same right as Pakistan or India to develop this resource. Hard-line religious extremists also maintain control of key government positions, and these Shiite clerics continue to harshly criticize Western sanctions and suspected interference in their country.

Tensions in Turkey Turkey has also emerged as a key geopolitical question mark, as it is strategically positioned between diverse, often contradictory geopolitical forces. Many pro-Westerners within Turkey, for example, are committed to joining the EU. To do so, the country has embarked on an active agenda of reforms designed to demonstrate its commitment to democracy. On the other hand, Islamist elements (mainly Sunni) within the country are wary of moving too close to Europe.

Regional issues within the country also remain important. In the east, the Kurds (a key cultural minority in Turkey) have continued to press for more recognition and regional autonomy from the Turkish government (see Figure 7.40). In addition, Syria's political

GEOGRAPHERS AT WORK

How Do We Define "Middle East"?

Figure 7.3.1 Karen Culcasi

Karen Culcasi is no stranger to the Middle East. A geographer at the University of West Virginia, Culcasi regularly teaches courses on the Arab World and takes her students to the region, including explorations of Jordan and the United Arab Emirates (Figure 7.3.2). This allows students the opportunity to experience both the region's diverse natural setting as well as its complex cultural mosaic. The geographic perspective is important, stresses Culcasi: "The spatial element that you don't get [in other disciplines] is powerful." This is particularly true when looking at the politics of this region. "Geopolitics doesn't just happen in a place; the place affects geopolitics," she explains.

Culcasi explores how the "Middle East" has evolved as a regional idea and how it is represented on maps. She has examined old maps and atlases to discover where the term "Middle East" originated (largely from imperial Great Britain), and traveled to the region to ask local residents and experts if they use the term ("Arab Homeland" is more commonly used). She concludes that the "Middle East" is largely an imposed term popularized by European and American politicians to compartmentalize and simplify a complicated mix of peoples and places.

Refugee Experiences More recently, Culcasi has examined the challenges faced by refugees, especially women. She interviews Palestinians, both in the United States and in the region, about their homeland. How do they map its location and describe its character? She is also spending time with Syrian refugee women in Jordan, trying to understand how their lives have been suddenly transformed by this incredible disruption. Whether exploring archives or working with people in a refugee camp, Culcasi credits her geography studies for her research approach. "My professors helped me cultivate a critical perspective and to question assumptions. And on the undergrad level, the diversity and breadth of geography is a benefit."

Figure 7.3.2 Refugee Camp Professor Culcasi's recent research has taken her to Syrian refugee camps in Jordan where she has interviewed women about their recent experiences.

1. On a blank map of Southwest Asia and North Africa, draw a line around your definition of the "Middle East" and then write a paragraph defending your answer. Compare your map with those of classmates.
2. Select a regional term used locally ("New England," "Southern California," "the Panhandle," etc.). Have five friends/classmates identify the area on a blank map of the region and defend their answers. Then summarize and explain their responses.

fragmentation has produced a huge refugee problem in the southern part of the country, and Turkey has closed its border with its troubled neighbor to control the flow of desperate Syrian migrants (see Figure 7.26).

Review

7.8 Describe the role played by the French and British in shaping the modern political map of Southwest Asia and North Africa. Provide specific examples of their lasting legacy.

7.9 Discuss how the Sunni–Shiite split has recently played out in sectarian violence across the region.

7.10 Explain how ethnic differences have shaped Iraq's political conflicts in the past 50 years.

KEY TERMS Suez Canal, Palestinian Authority (PA), Arab League

Economic and Social Development: Lands of Wealth and Poverty

Southwest Asia and North Africa constitute a region of incredible wealth and discouraging poverty (Table A7.2). While some countries enjoy prosperity, due mainly to rich reserves of petroleum and natural gas, other nations are among the world's least developed. Continuing political instability contributes to the region's struggling economy. Civil wars and internal conflicts within Syria, Iraq, and Libya have devastated their economies. Palestinians living in the Gaza and West Bank regions also suffer as political minorities within Israel. Elsewhere, recent economic sanctions have hurt the Iranian economy. Petroleum will no doubt figure significantly in the region's future economy, but some countries in the area have also focused on increasing agricultural output, investing in new industries, and promoting tourism to broaden their economic base.

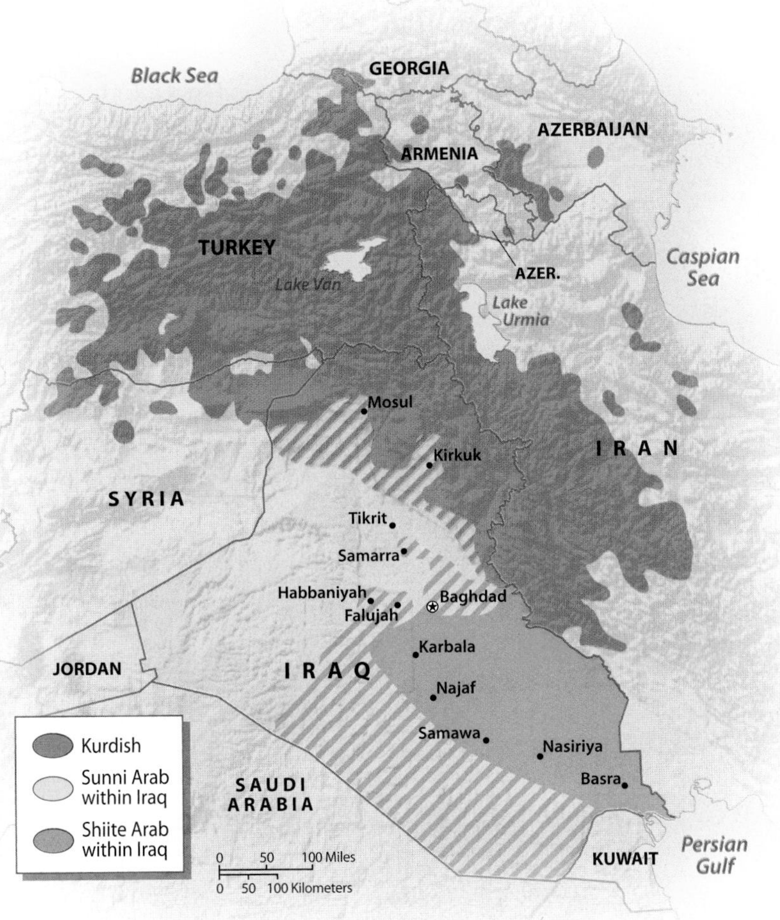

Figure 7.40 Multicultural Iraq Complex colonial origins produced a state with varying ethnic characteristics. Shiites dominate south of Baghdad, Sunnis hold sway in the western triangle zone, and Kurds are most numerous in the north, near oil-rich Kirkuk and Mosul.

The Geography of Fossil Fuels

The striking global geographies of oil and natural gas reveal the region's continuing importance in the world economy as well as the extremely uneven distribution of these resources within the region (Figure 7.41). North African settings (especially Algeria and Libya), as well as the Persian Gulf region, have large sedimentary basins containing huge reserves of oil and gas, whereas other localities (for example, Israel, Jordan, and Lebanon) lie outside zones of major fossil fuel resources. Saudi Arabia, Iran, Iraq, Kuwait, and the United Arab Emirates hold large petroleum reserves, while Iran and Qatar possess the largest regional reserves of natural gas. The distribution of fossil fuel reserves suggests that regional supplies will not be exhausted anytime soon. Overall, with only 7 percent of the world's population, the region holds over half of the world's proven oil reserves. Saudi Arabia's pivotal position, both regionally and globally, is clear: Its 30 million residents live atop almost 20 percent of the planet's known oil supplies.

Global Economic Relationships

Southwest Asia and North Africa share close economic ties with the rest of the world. While oil and gas remain critical commodities that dominate international economic linkages, the growth of manufacturing and tourism is also redefining the region's role in the world.

OPEC's Changing Fortunes Although OPEC does not control global oil and gas prices, it still influences the cost and availability of these pivotal products. While the United States has made some progress toward greater energy independence from the Mideast (via its larger domestic oil and natural gas output), western Europe, Japan, China, and many less industrialized countries still depend on the region's fossil fuels.

Falling energy prices late in 2014 and 2015, however, may be signaling changes for OPEC's fortunes. Falling prices indicate increased global energy production by non-OPEC producers, such as Canada and the United States, suggesting more competition that might decrease OPEC's importance on the global stage. OPEC's move to keep production high indicates the organization (especially Saudi Arabia) is interested in driving high-cost producers (such as North American fracking operations) out of business. In the meantime, however, lower prices have already exerted pressure on the budgets of many OPEC nations within the region. Government programs are being cut, and many residents resent paying more for many services. In 2015, the International Monetary Fund estimated that major Gulf producers ran a combined budget deficit of more than 6 percent of gross domestic product (compared with surpluses of more than 10 percent in recent years).

Other Global and Regional Linkages Beyond key OPEC producers, other trade flows also contribute to global economic integration. Turkey, for example, ships textiles, food products, and manufactured goods to its principal trading partners: Germany, the United States, Italy, France, and Russia. Tunisia sends more than 60 percent of its exports (mostly clothing, food products, and petroleum) to nearby France and Italy. Israeli exports emphasize the country's highly skilled workforce: Products such as cut diamonds, electronics, and machinery parts go to the United States, western Europe, and Japan.

Future interconnections between the global economy and Southwest Asia and North Africa may depend increasingly on cooperative economic initiatives far beyond OPEC. Relations with the European Union (EU) are critical. Since 1996, Turkey has enjoyed close ties with the EU, but recent attempts at full membership in the organization have failed. Other so-called Euro-Med agreements have been signed between the EU and countries across North Africa and Southwest Asia that border the Mediterranean Sea.

Most Arab countries, however, are wary of too much European dominance. In 2005, 17 Arab League members established the **Greater Arab Free Trade Area (GAFTA)**, an organization designed to eliminate all intraregional trade barriers and spur economic cooperation. In addition, Saudi Arabia plays a pivotal role in regional economic development through organizations such as the Islamic Development Bank and the Arab Fund for Economic and Social Development. Many of these financial organizations offer services compliant with Islamic law (*sharia law*). In fact, these Islamic banking assets are projected to grow about 20 percent annually between 2015 and 2018.

Regional Economic Patterns

Remarkable economic differences characterize Southwest Asia and North Africa (see Table A7.2). Some oil-rich countries have prospered greatly since the early 1970s, but in many cases fluctuating oil prices, political disruptions, and rapidly growing populations threaten future economic growth.

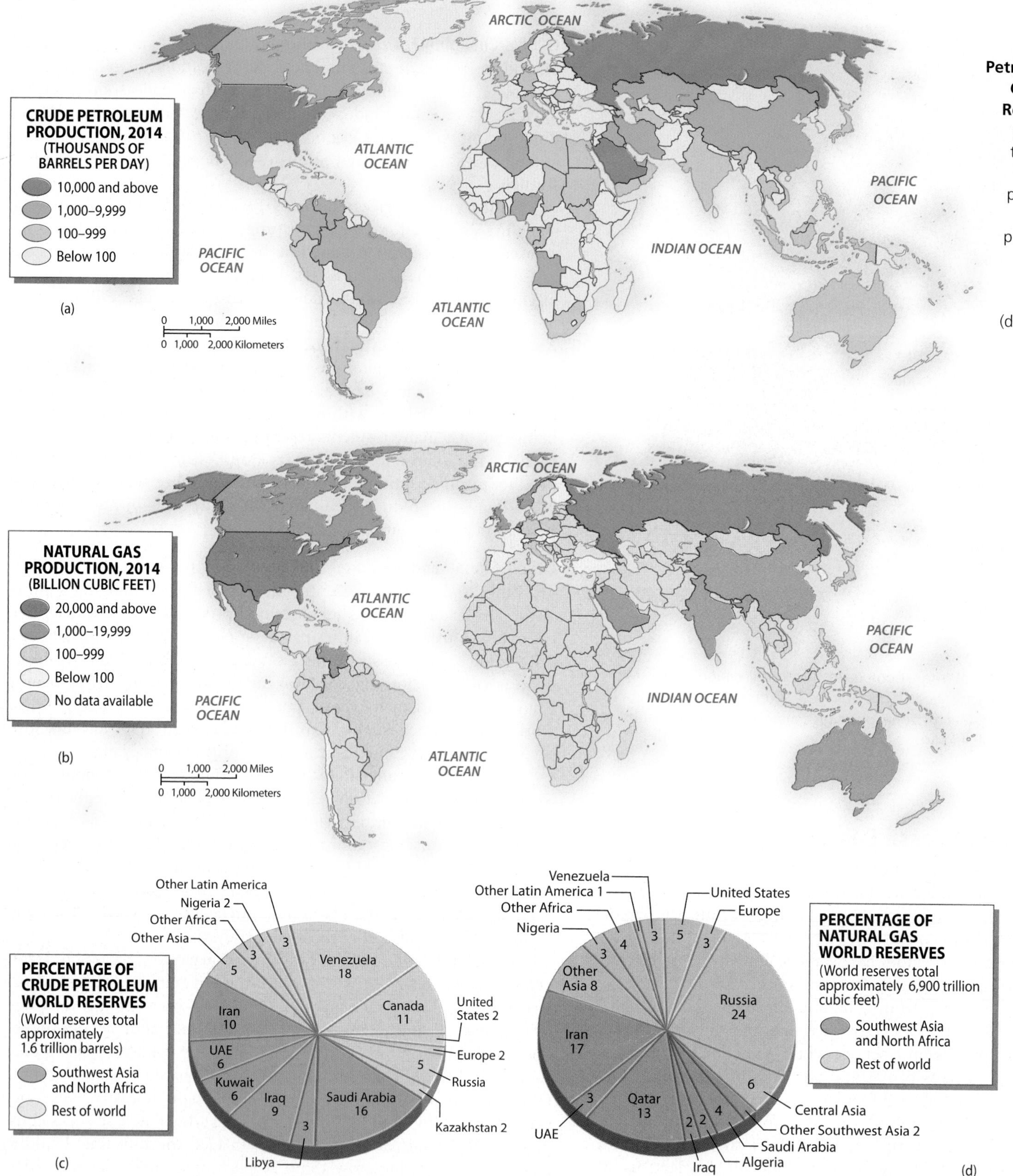

Figure 7.41 Crude Petroleum and Natural Gas Production and Reserves The region plays a central role in the global geography of both (a) crude petroleum production and (b) natural gas production. Abundant regional reserves of both (c) crude petroleum and (d) natural gas suggest that the pattern will continue.

Higher-Income Oil Exporters The richest countries of Southwest Asia and North Africa owe their wealth to massive oil reserves. Nations such as Saudi Arabia, Kuwait, Qatar, Bahrain, and the United Arab Emirates benefit from fossil fuel production as well as from their relatively small populations. Large investments in transportation networks, urban centers, and other petroleum-related industries have reshaped the cultural landscape. The petroleum-processing and -shipping centers of Jubail (on the Persian Gulf) and Yanbu (on the Red Sea) are examples of this commitment to expand the Saudi economic base beyond simple extraction of crude oil (Figure 7.42). Billions of dollars have poured into new schools, medical facilities, low-cost housing, and modernized agriculture, significantly raising the standard of living in the past 40 years.

Still, problems remain. Dependence on oil and gas revenues produces economic pain when prices fall. Such fluctuations in world oil markets will inevitably continue in the future, disrupting construction

Figure 7.42 Yanbu, Saudi Arabia This Google Earth image shows regional roads converging on Yanbu, a petro-city of almost 200,000 people in western Saudi Arabia. The Red Sea is at the lower left.

projects, producing large layoffs of immigrant populations, and slowing investment in the region's economic and social infrastructure. In addition, countries such as Bahrain and Oman must contend with rapidly depleting reserves over the next 20 to 30 years.

Lower-Income Oil Exporters Some countries possess fossil fuel reserves, but different political and economic variables have hampered sustained economic growth. For example, Algerian oil and natural gas overwhelmingly dominate the country's exports, but the past 20 years have brought political instability and shortages of consumer goods. Nearby Libya's political disintegration has had profound economic consequences, sharply reducing oil and gas output and causing severe economic disruptions.

Iraq faces huge challenges. War has crippled much of its already deteriorated infrastructure, and political instability has made rebuilding its economy more difficult. Iraq suffers from high unemployment, more than 20 percent of the population remains malnourished, and only 25 percent of the country is served by dependable electricity. Oil output has increased since 2009, however, suggesting the potential for economic recovery if sectarian conflicts can be contained.

The situation in Iran is also challenging. The country's oil and gas reserves are huge, but Iran is relatively poor, burdened with a stagnating standard of living. Since 1980, fundamentalist leaders have limited international trade in consumer goods and services, fearing the import of unwanted cultural influences. International sanctions on purchases of Iranian oil (related to its nuclear development program) have depressed the economy. Some economic bright spots have appeared, however: The country benefits from energy developments in Central Asia, and literacy rates (particularly for women) have risen, reflecting a new emphasis on rural education.

Prospering Without Oil Some countries, while lacking petroleum resources, have still found paths to economic prosperity. Israel, for example, supports one of the highest standards of living in the region, even with its political challenges (see Table A7.2). Israelis and many foreigners have invested large amounts of capital to create a highly productive industrial base, which produces many products for the global marketplace (see *Everyday Globalization: Popping Pills from Israel*). The country is also a global center for high-tech computer and telecommunications products, known for its fast-paced and highly entrepreneurial business culture that resembles California's Silicon Valley. Israel also has daunting economic problems. Its struggles with the Palestinians and with neighboring states have sapped potential vitality. Defense spending absorbs a large share of total gross national income (GNI), necessitating high tax rates. Poverty and unemployment

EVERYDAY GLOBALIZATION
Popping Pills from Israel

Every year U.S. doctors write more than 2.5 *billion* prescriptions for generic pharmaceuticals. Few people realize how many of these drugs are actually manufactured in Southwest Asia—specifically, Israel. When you reach for that generic antibiotic (amoxicillin), painkiller (oxycodone), or anti-inflammatory (naproxen), you may well be taking pills manufactured halfway around the world. Israel is home to seven research universities and a host of companies that focus on the biological sciences and innovations in the pharmaceutical industry.

The largest player in Israel's generic drug industry is Teva Pharmaceutical Industries (Figure 7.4.1). The company estimates that it manufactures 73 *billion* tablets a year and that one in six generic prescriptions in the United States is filled with a Teva (Hebrew for "nature") product. Today Teva is the largest global manufacturer of generic pharmaceuticals, as well as an innovative producer of its own proprietary drugs. The result is that Israel has emerged as one of the planet's key focal points in an industry that seems destined to grow along with the world's insatiable demand for affordable pharmaceuticals.

1. For the American public, describe some of the benefits and drawbacks of depending on a global geography of prescription drugs.
2. Visit a local pharmacy and select two over-the-counter medications. Can you find out who manufactured them and where they came from?

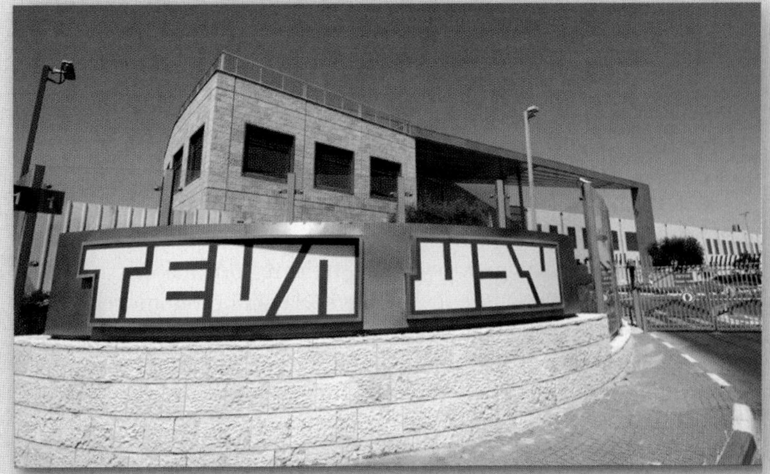

Figure 7.4.1 Teva Headquarters, Petah Tikva, Israel Employing thousands of skilled workers, Teva Pharmaceuticals Industries produces both the world's largest volume of generic drugs and a growing array of its own patented pharmaceuticals.

among Palestinians also remain unacceptably high, both in Gaza (recently devastated by more violence) and in the West Bank.

Turkey has a diversified economy, even though incomes are modest by regional standards. Lacking petroleum, Turkey produces varied agricultural and industrial goods for export. About 24 percent of the population remains in agriculture, and principal commercial products include cotton, tobacco, wheat, and fruit. The industrial economy has grown since 1980, including exports of textiles, processed food, and chemicals. Turkey has also gone high tech: About 44 percent of Turks use the Internet, and the country has been a fertile ground for dozens of global Internet start-up companies that connect well (many are online or virtual gaming enterprises) with younger Turks as well as with the global economy. Turkey also remains a major tourist destination in the region, attracting more than 6 million visitors annually in recent years.

Regional Patterns of Poverty Poorer countries of the region share the problems of the less developed world. For example, Sudan, Egypt, Syria, and Yemen each face unique economic challenges. Sudan's continuing political problems stand in the way of progress. Political disruptions have resulted in major food shortages. The country's transportation and communications systems have seen little new investment, and secondary school enrollments remain very low. On the other hand, Sudan's fertile soils could support more farming, and its share of regional oil prospects suggests petroleum's expanding role in the economy.

Egypt's economic prospects are unclear. Since Mubarak's overthrow in 2011, unemployment has risen, tourism has declined, and foreign investors have remained wary of the nation's uncertain political environment. While leaders have recently been promoting the country's relatively stable political setting, many Egyptians still live in poverty, and the gap between rich and poor continues to widen. Illiteracy is widespread, and the country suffers from the **brain drain** phenomenon as some of its brightest young people leave for better jobs in western Europe or the United States.

Syria once enjoyed both a growing economy and a stable political regime. Now it has neither, as its ongoing civil war and sectarian conflicts have decimated the economy. Millions of people who considered themselves part of the Syrian middle class have been thrown into utter poverty, often as desperate refugees. It will be many years before more normal economic conditions can be restored (Figure 7.43).

Yemen remains the poorest country on the Arabian Peninsula. Positioned far from the region's principal oil fields, Yemen's low per capita GNI puts it on par with nations in impoverished Sub-Saharan Africa. The largely rural country relies mostly on marginally productive subsistence agriculture, and much of its mountain and desert interior lacks effective links to the outside world. Coffee, cotton, and fruits are commercially grown, and modest oil exports bring in needed foreign currency. Overall, however, widespread unemployment and high childhood mortality suggest the failure of the region's inadequate healthcare system combined with challenging social and economic conditions and complex environmental determinants (Figure 7.44). Recent political upheavals have only further dimmed the nation's economic prospects.

Gender, Culture, and Politics: A Woman's Changing World

The role of women in largely Islamic Southwest Asia and North Africa remains a major social issue. Female participation rates in the workforce are among the world's lowest, and large gaps typically exist between levels of education for males and females. In conservative parts of the region, few women work outside the home. Even in parts of Turkey, where Western influences are widespread, it is rare to see rural women selling goods in the marketplace or driving cars in the street. A recent poll on women's rights in the region ranked Egypt, Iraq, and Saudi Arabia among the lowest and Oman, Kuwait, Jordan, and Qatar among the highest.

More orthodox Islamic states impose legal restrictions on the activities of women. In Saudi Arabia, for example, women are not allowed to drive, although growing protests among many younger, educated Saudi females may overturn that ban. In neighboring Qatar, women can vote (and hold public office), but they still need a husband's consent to obtain a driver's license. In Iran, full veiling remains mandatory in more conservative areas, but many wealthier Iranian women have adopted Western dress, reflecting a more secular outlook,

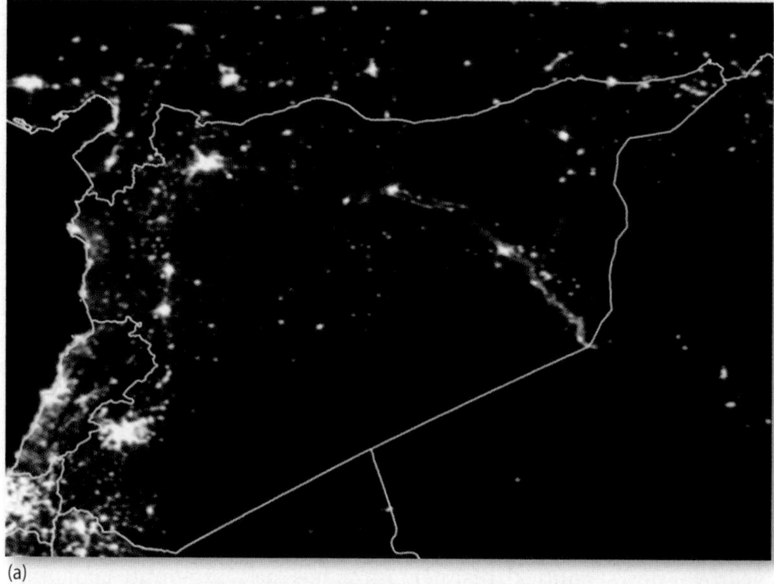

(a)

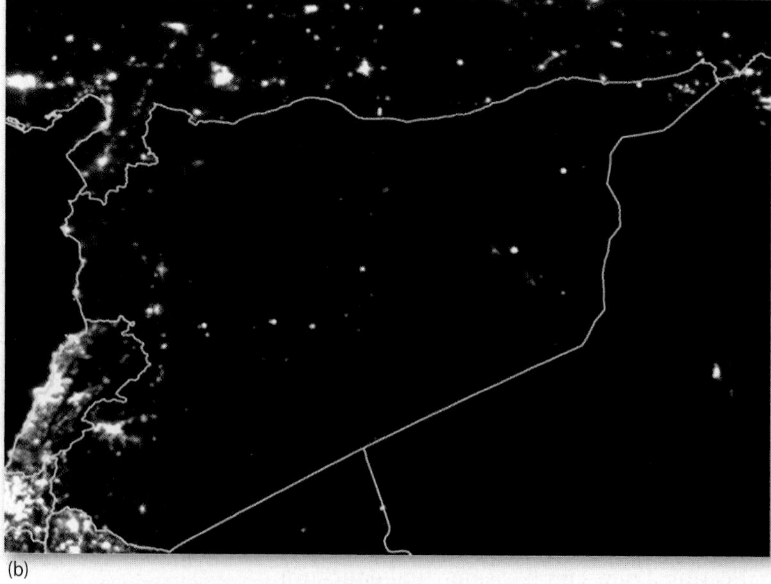

(b)

Figure 7.43 Syria Sinks into Darkness These comparative nighttime satellite views reveal the devastating political and economic impact of the Syrian civil war and the ongoing presence of ISIL. Photo (a) was taken before the conflicts began, photo (b) after hostilities began to rock the country.

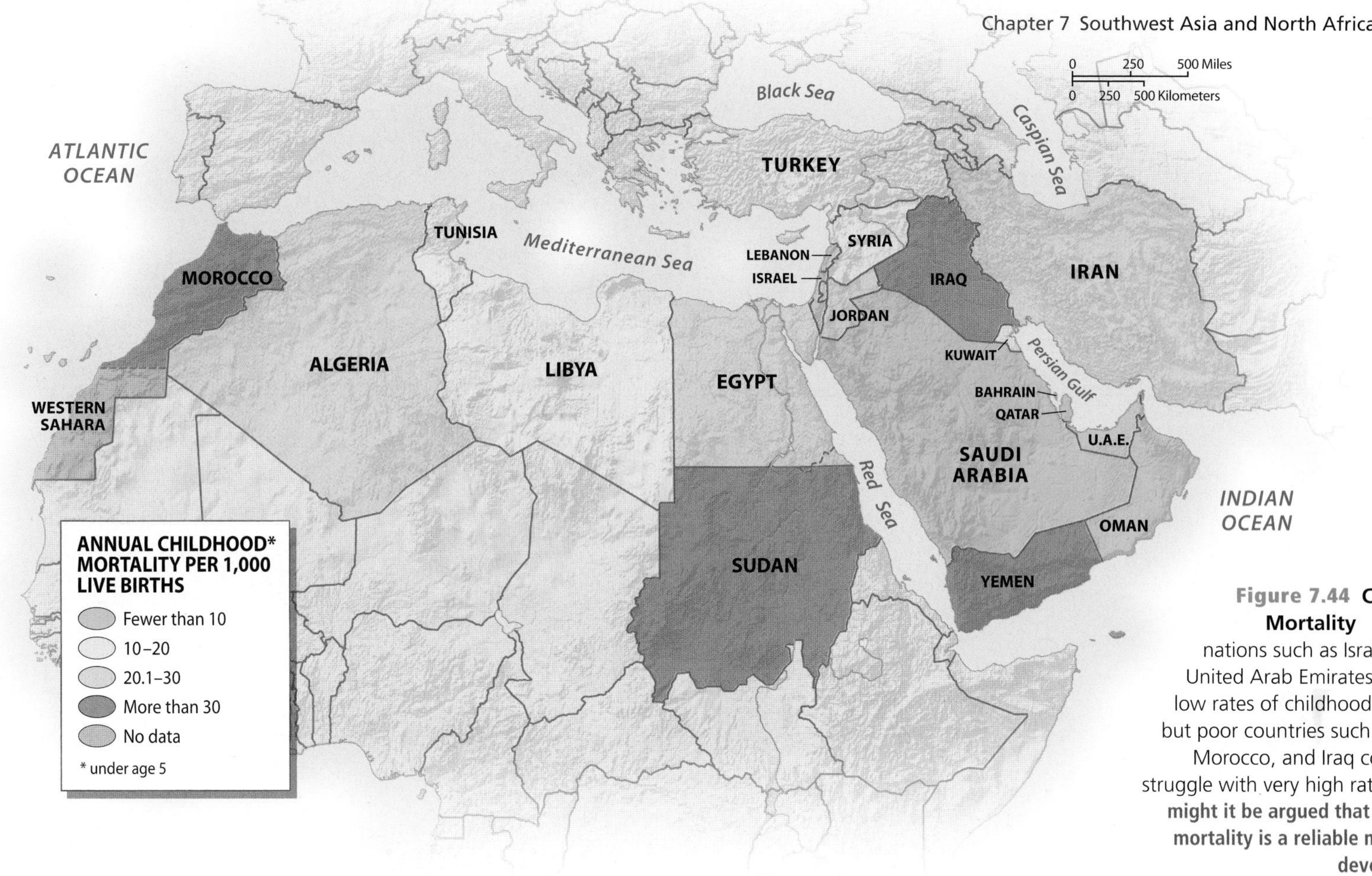

ANNUAL CHILDHOOD* MORTALITY PER 1,000 LIVE BIRTHS

- Fewer than 10
- 10–20
- 20.1–30
- More than 30
- No data

* under age 5

Figure 7.44 Childhood Mortality Wealthier nations such as Israel and the United Arab Emirates have very low rates of childhood mortality, but poor countries such as Sudan, Morocco, and Iraq continue to struggle with very high rates. **Q: Why might it be argued that childhood mortality is a reliable measure of development?**

especially among the young (Figure 7.45). Generally, Muslim women still lead more private lives than men: Much of their domestic space is shielded from the world by walls and shuttered windows, and their public appearances are filtered through the use of the *niqab* (face veil) or *chador* (full-body veil).

In some settings, women's lives are changing, even within norms of more conservative Islamist societies. From Tunisia to Yemen, women widely participated in the Arab Spring rebellions, asserting their new political visibility in very public ways. Kurdish women have also volunteered to fight ISIL, both in Iraq and in Syria. The local consequences of these political and social changes are complex: In liberated Libya, young Islamic women rejoice in their freedom to wear the niqab in public, a practice banned under Qaddafi's rule. Other women have been encouraged to play more active roles in political affairs, including running for office.

Algerian women demonstrate the pattern. Most studies suggest those in the younger generation are more religious than their parents and more likely to cover their heads and bodies with traditional clothing. At the same time, they are more likely to be educated and employed. Today 70 percent of Algeria's lawyers and 60 percent of its judges are women. A majority of university students are women, and women dominate the health-care field. These new social and economic roles help explain why birth rates are declining. Women also have a more visible social position in Israel, except in fundamentalist Jewish communities, where conservative social customs limit women to traditional domestic roles.

Figure 7.45 Iranian Women These fashionably dressed young women in Tehran suggest how Iran's more urban and affluent residents have embraced many elements of Western culture.

Review

7.11 Describe the basic geography of oil reserves across the region, and compare the pattern with the geography of natural gas reserves.

7.12 Identify different strategies for economic development recently employed by nations such as Saudi Arabia, Turkey, Israel, and Egypt. How successful have they been, and how are they related to globalization?

KEY TERMS Greater Arab Free Trade Area (GAFTA), brain drain

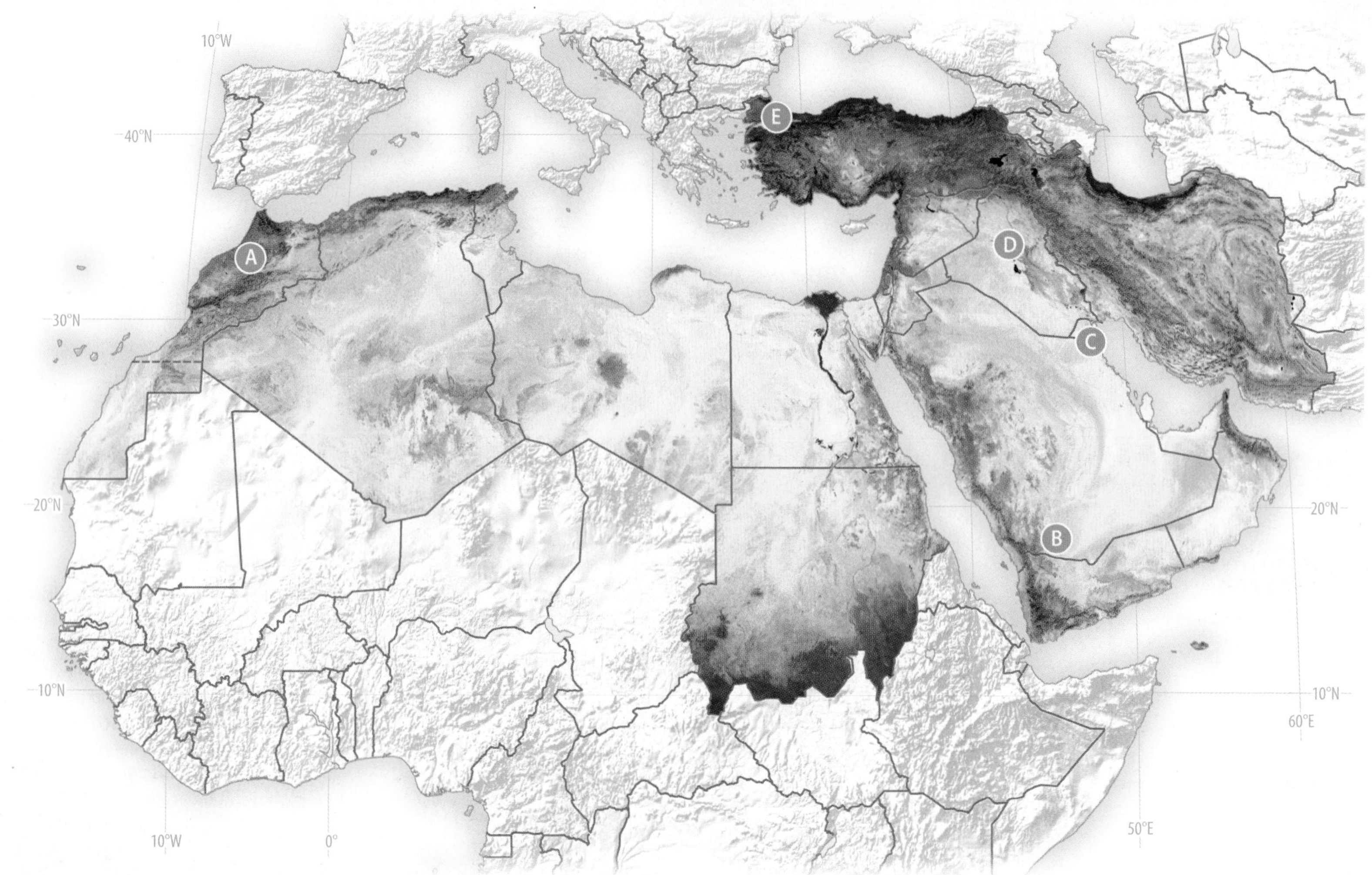

Review

Physical Geography and Environmental Issues

7.1 Explain how latitude and topography produce the region's distinctive patterns of climate.

7.2 Describe how the region's fragile, often arid setting shapes contemporary environmental challenges.

7.3 Describe four distinctive ways in which people have learned to adapt their agricultural practices to the region's arid environment.

Many nations within the region face significant environmental challenges and growing pressures on limited supplies of agricultural land and water. The results, from the eroded soils of the Atlas Mountains to the overworked garden plots along the Nile, illustrate the environmental price paid when population growth outstrips the ability of the land to support it.

1. If populations outstrip water supplies in North Africa's oasis settlements, how might residents adjust?

2. List ways in which modern technology might address water shortages across the region. Are there limits or challenges to this approach?

Population and Settlement

7.4 Summarize the major forces shaping recent migration patterns within the region.

The population geography of Southwest Asia and North Africa is strikingly uneven. Areas with higher rainfall or access to exotic water often have very high physiological population densities, whereas nearby arid zones remain almost empty of settlement.

3. Briefly describe the population density and land-use patterns you might be likely to see out the plane window on a flight between Riyadh (Saudi Arabia) and San'a (Yemen).

4. How might very low population densities impose special problems for maintaining effective political control across all portions of nations such as Saudi Arabia, Libya, and Algeria?

Cultural Coherence and Diversity

7.5 List the major characteristics and patterns of diffusion of Islam.

7.6 Identify the key modern religions and language families that dominate the region.

7.7 Identify the role of cultural variables in understanding key regional conflicts in North Africa, Israel, Syria, Iraq, and the Arabian Peninsula.

Culturally, the region remains the hearth of Christianity, the spatial and spiritual core of Islam, and the political and territorial focus of modern Judaism. In addition, important sectarian divisions within religious traditions (especially the schism between Sunnis and Shiites), as well as long-standing linguistic differences, continue to shape the local cultural geographies and regional identities.

5. Why is Islam both a powerful unifying and a divisive cultural force in the region?

6. Why does Saudi Arabia remain such a pivotal part of the Islamic world?

C

Geopolitical Framework

7.8 Summarize the geography of oil and gas reserves in the region.

Political conflicts have disrupted economic development. Civil wars, sectarian violence, conflicts between states, and regional tensions work against initiatives for greater cooperation and trade. Perhaps most important, the region must deal with the conflict between modernity and more fundamentalist interpretations of Islam.

7. How likely is it that the cultural and religious divisions in Iraq will be healed in 5–10 years?

8. Work with other students in the class to organize a debate on whether a renewed oil boom in the Iraqi economy might spur *greater* or *reduced* levels of sectarian violence within the country.

D

Economic and Social Development

7.9 Describe traditional roles for Islamic women and provide examples of recent changes.

Abundant reserves of oil and natural gas, coupled with the global economy's continuing reliance on fossil fuels, ensure that the region will remain prominent in world petroleum markets. Also likely are moves toward economic diversification and integration, which may gradually draw the region closer to Europe and other participants in the global economy.

9. What are likely to be the chief drivers of economic growth in settings such as Istanbul, Turkey, in the next 10–20 years?

10. Write an essay comparing and contrasting the challenges of producing sustained economic growth in Turkey and Saudi Arabia between 2020 and 2030.

E

KEY TERMS

Arab League *(p. 242)*
Arab Spring *(p. 219)*
brain drain *(p. 248)*
choke point *(p. 224)*
culture hearth *(p. 220)*
domestication *(p. 226)*
exotic river *(p. 228)*
Fertile Crescent *(p. 226)*
fossil water *(p. 224)*
Greater Arab Free Trade Area (GAFTA) *(p. 245)*
Hajj *(p. 233)*
hydropolitics *(p. 224)*
ISIL (Islamic State of Iraq and the Levant; also ISIS or Islamic State) *(p. 220)*
Islamic fundamentalism *(p. 220)*
Islamism *(p. 220)*
Levant *(p. 220)*
Maghreb *(p. 220)*
medina *(p. 229)*
monotheism *(p. 233)*
Organization of the Petroleum Exporting Countries (OPEC) *(p. 220)*
Ottoman Empire *(p. 234)*
Palestinian Authority (PA) *(p. 241)*
pastoral nomadism *(p. 226)*
physiological density *(p. 226)*
Quran *(p. 233)*
salinization *(p. 223)*
sectarian violence *(p. 219)*
Shiite *(p. 233)*
Suez Canal *(p. 239)*
Sunni *(p. 233)*
theocratic state *(p. 233)*
transhumance *(p. 226)*

DATA ANALYSIS

http://goo.gl/oSK5Fa

Health care is often considered a basic human right in more developed portions of the world, but large parts of Southwest Asia and North Africa are poorly served by health-care providers. The World Health Organization (WHO) gathers data on the number of physicians per 1000 population, which can be used as a measure of access to health care as well as social development. According to recent data, the United States had about 2.5 physicians per 1000 and Germany about 3.9. Go to the WHO website (www.who.int) and access the data/interactive atlas page on physicians per 1000 population.

1. Make your own data table and map showing the regional pattern of health-care access across Southwest Asia and North Africa.
2. In a few sentences, summarize the general patterns and trends you see. How would you explain some of the major variations you observe across the region?
3. Compare the pattern you see for physicians with the map in the text on childhood mortality (Figure 7.44). What similarities and differences do you see? How might these two indicators be a good measure of future social development? How might they predict political stability?

MasteringGeography™

Looking for additional review and test prep materials? Visit the Study Area in MasteringGeography™ to enhance your geographic literacy, spatial reasoning skills, and understanding of this chapter's content by accessing a variety of resources, including MapMaster interactive maps, geoscience animations, videos, *In the News* RSS feeds, flashcards, web links, self-study quizzes, and an eText version of *Globalization and Diversity.*

Authors' Blogs

Scan to visit the **Author's Blog for field notes, media resources, and chapter updates**

http://gad4blog.wordpress.com/category/southwest-asia-and-north-africa/

Scan to visit the **GeoCurrents Blog**

http://geocurrents.info/category/place/southwest-asia-and-north-africa

8 Europe

PHYSICAL GEOGRAPHY AND ENVIRONMENTAL ISSUES

Diverse European environments range from subtropical Mediterranean lands to the arctic tundra and from the moderate Atlantic coast to inland continental climates. Europe is also one of the "greenest" world regions, with strong measures regarding pollution, recycling, and renewable energy.

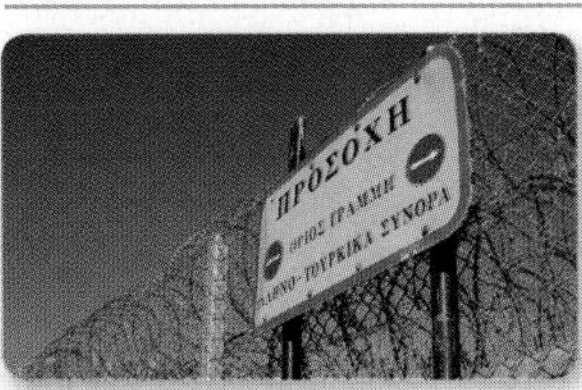

POPULATION AND SETTLEMENT

Europe has very low rates of natural growth and very high rates of internal mobility and international in-migration. Most current international migration consists of refugees from strife-torn Africa and Southwest Asia.

CULTURAL COHERENCE AND DIVERSITY

Europe has a long history of cultural tensions linked to internal differences in language and religion; however, today's tensions are primarily connected with immigration from other world regions.

GEOPOLITICAL FRAMEWORK

Two world wars and a lengthy Cold War divided 20th-century Europe into warring camps, producing an ever-changing map of new states. Although Europe is an integrated and peaceful region, geopolitical tensions linked to micro-nationalism and devolution now dominate.

ECONOMIC AND SOCIAL DEVELOPMENT

For half a century, the European Union (EU) has worked successfully to integrate the region's diverse economies and political systems, making Europe a global superpower. Today, however, internal economic and social issues challenge this unity.

◀ A symbol of Scottish independence, Edinburgh Castle dominates the skyline of Scotland's capital city. The castle was the royal residence from the 12th to early 17th century, after which it became a military garrison. As one of the Scotland's most visited tourist site, the Castle became a UNESCO World Heritage site in 1995.

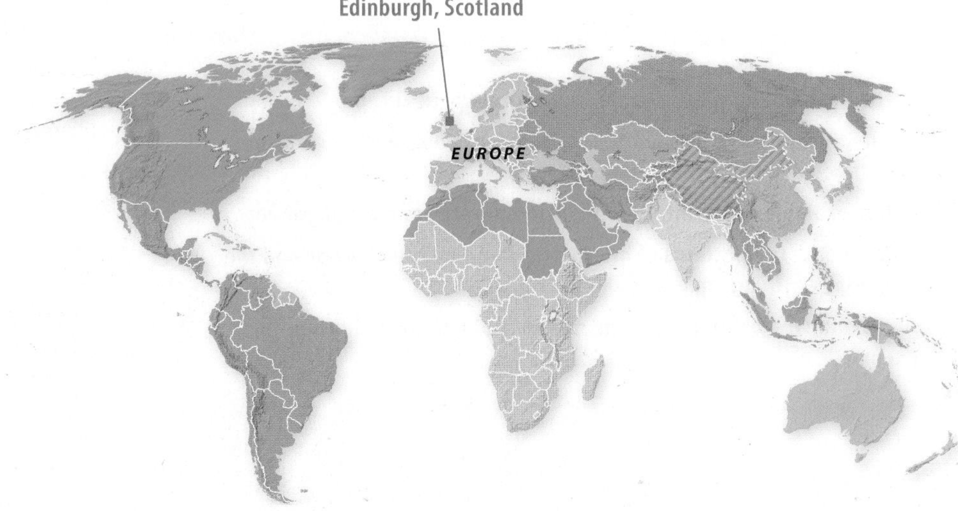

Picturesque Edinburgh, capital of the historical Kingdom of Scotland and today home to the Scottish Parliament, in many ways embodies Europe's complex geographies, past and present. The suffix "burgh" means a fortified and independent place, a fortress town surrounded by protective walls. This concept is not unique to Scotland but has long been a shared concern throughout Europe, expressed in hundreds of place names: Hamburg, Germany; Strasbourg, France; and Burgos, Spain—all clues to the historical search for security on a politically chaotic continent.

In the 1950s, after centuries of nationalistic wars, competition, and conflict, Europe committed itself to an agenda of economic, political, and social integration through the **European Union (EU)**, a supranational organization today made up of 28 countries. Although the EU has been highly successful in uniting and integrating many aspects of European economic and political life, its future is now being called into question with tense geographies of separatism and political devolution.

> Population movements, both within and from outside Europe, have led to new tensions in this world region.

Which brings us back to Edinburgh. Although Scotland narrowly voted against independence in a recent referendum that would have ended its 307-year relationship with England, a strong Scottish separatist movement remains active. As well, and somewhat ironically, the United Kingdom itself is scheduling a referendum on EU membership, testing its voters on the matter of a unified Europe. Other countries with strong anti-EU political parties (Denmark, France, Greece, and Hungary) may follow suit. Similarly, people in smaller regions within existing nation-states (Bretons in France, Walloons in Belgium, and Basques and Catalonians in Spain) may follow Scotland's lead with independence votes of their own (Figure 8.1). This separatist activity raises questions about the survival of an integrated Europe.

Figure 8.1 Basque Separatism The Basque people of northeastern Spain and southwestern France are a distinct cultural group that has long sought autonomy and even independence from Spain and France. At times militant Basque separatists have used violence and terrorism to further their cause. This photo is of a recent Basque demonstration in France showing support for two Basque activists accused by the government of terrorism.

Europe is one of the most diverse regions in the world, encompassing a wide assortment of people and places in an area considerably smaller than North America. More than half a billion people reside in this region, living in 42 different countries that range in size from France, Spain, and Germany to microstates such as Andorra and Monaco (Figure 8.2). Commonly Europe is divided into the four general subregions of western, eastern, southern (or Mediterranean), and northern (or Scandinavian) Europe, terms we use throughout this chapter. Europe is a relatively wealthy and peaceful world region by global standards, yet income and employment disparities, along with population movements both within and from outside Europe, have led to new tensions.

Learning Objectives *After reading this chapter you should be able to:*

- **8.1 Describe the topography, climate, and hydrology of Europe.**
- **8.2 Identify the major environmental issues in Europe as well as measures taken to resolve those problems.**
- **8.3 Provide examples of countries with different rates of natural growth.**
- **8.4 Describe both the patterns of internal migration within Europe and the geography of foreign migration to the region.**
- **8.5 Describe the major languages and religions of Europe and locate these on a map.**
- **8.6 Summarize how the map of European states has changed in the last 100 years.**
- **8.7 Explain why and how Europe was divided during the Cold War and what the geographic implications are today.**
- **8.8 Describe Europe's economic and political integration as driven by the EU.**
- **8.9 Identify the major characteristics of Europe's current economic and social crisis.**

Figure 8.2 Europe Stretching east from Iceland in the Atlantic to Russia, Europe includes 42 countries, ranging in size from large states such as France and Germany to the microstates of Liechtenstein, Andorra, San Marino, and Monaco. Currently, the population of the region is about 531 million. Europe is commonly divided into the four subregions of western, eastern, southern (or Mediterranean), and northern (or Scandinavian) Europe. Tables A8.1 and A8.2 use these subregions for their organizational format.

Physical Geography and Environmental Issues: Human Transformation of a Diverse Landscape

Despite Europe's small size, its environmental diversity is extraordinary. A startling array of landscapes is found within its borders, from the arctic tundra of northern Scandinavia to the semiarid hillsides of the Mediterranean islands, with explosive volcanoes in southern Italy and glaciated seacoasts in Norway and Iceland.

Three factors explain this impressive environmental diversity:

- This western extension of the Eurasian land mass exhibits complex geology.
- Europe's extensive latitudinal range, from the Arctic to the Mediterranean subtropics, affects climate, vegetation, and hydrology (Figure 8.3). However, Europe's high latitudes are modified by the moderating influences of the Atlantic Ocean and its Gulf Stream as well as of the surrounding Baltic, Mediterranean, and Black seas.
- The long history of human settlement, spanning thousands of years, has transformed and modified Europe's landscapes in fundamental ways.

Landform Regions

Europe can be organized into four general landform regions: the European Lowland, forming an arc from southern France to the northeast plains of Poland, but also including southeastern England; the Alpine mountain system, extending from the Pyrenees in the west to the Balkan Mountains of southeastern Europe; the Central Uplands, positioned between the Alps and the European Lowland; and the Western Highlands, which include mountains in Spain and portions of the British Isles and the highlands of Scandinavia (see Figure 8.2). Iceland, unquestionably a part of Europe, yet lying 900 miles (1500 km) west of Norway, has its own unique landforms, straddling two different tectonic plates.

The European Lowland This lowland, also known as the North European Plain, is the unquestioned economic focus of western Europe, with its high population density, intensive agriculture, large cities, and major industrial regions. Though not completely flat, most of this lowland lies below 500 feet (150 meters) in elevation. Many of Europe's major rivers (the Rhine, Loire, Thames, and Elbe) meander across this lowland and form broad estuaries before emptying into the Atlantic. Several of Europe's busiest ports, including London, Le Havre, Rotterdam, and Hamburg, are located in these lowland settings.

The Rhine River delta conveniently divides the unglaciated southern European Lowland from the glaciated plains to the north, which were covered by a Pleistocene ice sheet until about 15,000 years ago. Because of this continental glacier, the northern lowland, including the Netherlands, Germany, Denmark, and Poland, is far less fertile than the unglaciated portions of Belgium and France (Figure 8.4). Rocky clay materials in Scandinavia were eroded and transported south by glaciers. As the climate warmed and the glaciers retreated, piles of glacial debris were left on the plains of Germany and Poland.

The Alpine Mountain System The Alpine mountain system forms the topographic spine of Europe and consists of a series of mountains running west to east from the Atlantic to the Black Sea and the southeastern Mediterranean. These mountain ranges carry distinct regional names, such as the Pyrenees, Alps, Apennines, Carpathians, Dinaric Alps, and Balkans, but they share geologic traits.

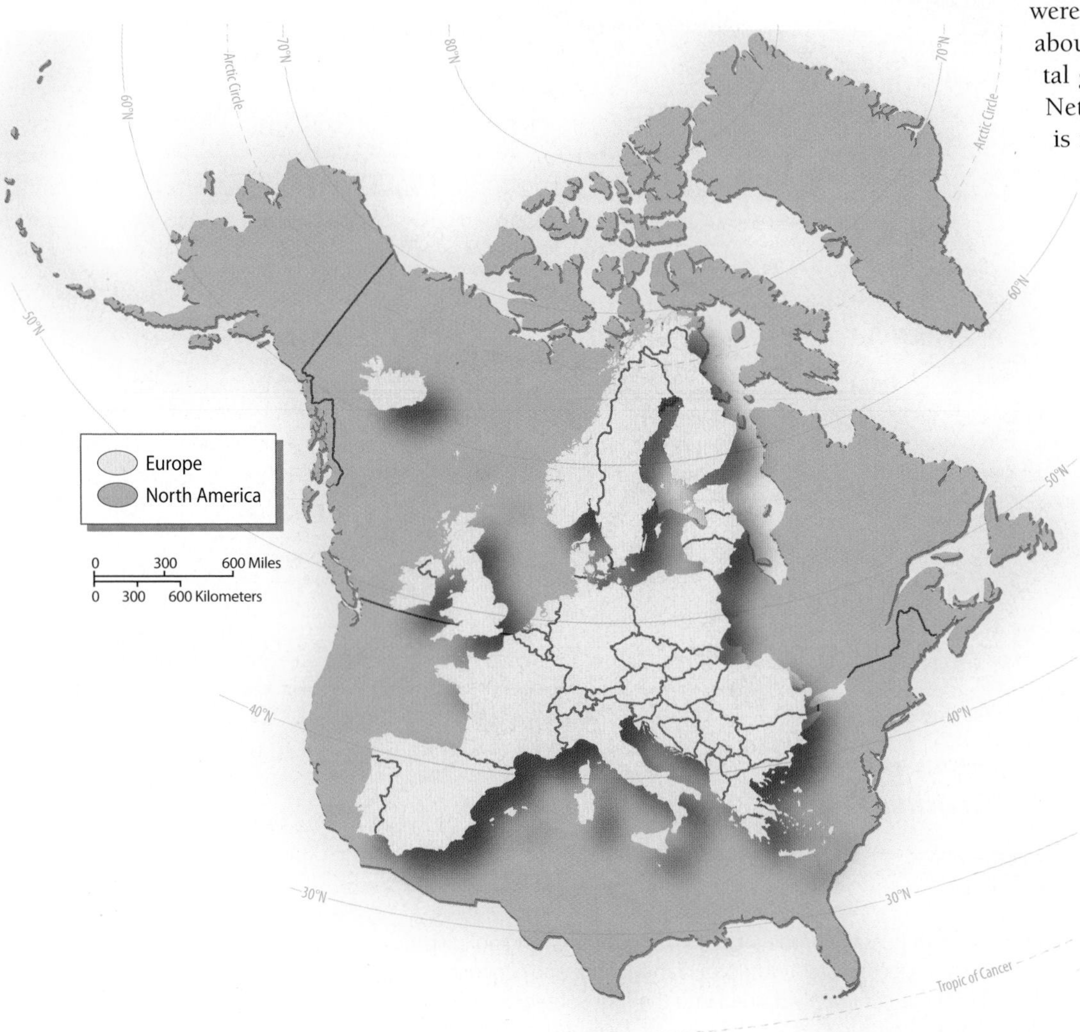

Figure 8.3 Europe's Size and Northerly Location Europe is about two-thirds the size of North America, as shown in this cartographic comparison. Another important characteristic is the northerly location of the region, which affects its climate, vegetation, and agriculture. As depicted, much of Europe lies at the same latitude as Canada; note that even the Mediterranean lands are farther north than the United States–Mexico border. **Q: What parts of Europe are at the same latitude as your location in North America?**

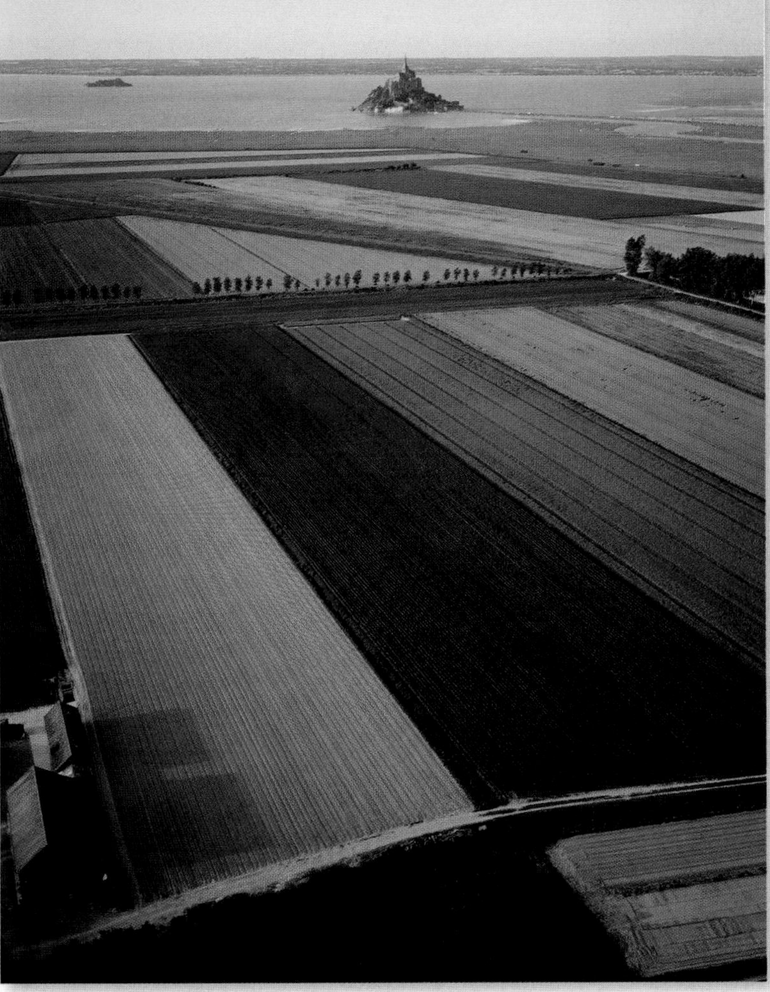

Figure 8.4 The European Lowland Also known as the North European Plain, this large lowland extends from southwestern France to the plains of northern Germany and eastward into Poland. Numerous rivers drain interior Europe by crossing this region, giving rise to large port cities along the coastline. This photo is of the Normandy region in western France. The small island is Mont Saint-Michel, a historic fortified monastery.

The Pyrenees form the political border between Spain and France and include the microstate of Andorra. This rugged range extends almost 300 miles (480 km) from the Atlantic to the Mediterranean. Within the mountain range, glaciated peaks reaching to 11,000 feet (3350 meters) alternate with broad glacier-carved valleys.

The centerpiece of this geologic region is the prototypical mountain range, the Alps, running more than 500 miles (800 km) from France to eastern Austria. These impressive mountains are highest in the west, rising to more than 15,000 feet (4600 meters) at Mt. Blanc on the French-Italian border. In Austria, the Alps are much more subdued, with few peaks exceeding 10,000 feet (3000 meters). Though easily crossed today by car or train through long tunnels and valley-spanning bridges, these mountains have historically formed an important cultural divide between the Mediterranean lands to the south and central and western Europe to the north.

The Apennine Mountains are located south of the Alps, mainly in Italy; the two ranges, however, are physically connected by the hilly coastline of the French and Italian Riviera. Forming Italy's spine, the Apennines are lower and lack the spectacular glaciated peaks and valleys of the true Alps. But farther to the south, the Apennines take on their own distinctive character with the explosive volcanoes of Mt. Vesuvius (just over 4000 feet, or 1200 meters) outside Naples and the much higher Mt. Etna (almost 11,000 feet, or 3350 meters) on the island of Sicily.

To the east, the Carpathian Mountains define the eastern limits of the Alpine system in Europe. They are a plow-shaped upland area extending from eastern Austria to the Iron Gate gorge, which is a narrow passage for Danube River traffic where the borders of Romania and Serbia meet.

Central Uplands In western Europe, a much older highland region occupies an arc between the Alps and the European Lowland in France and Germany. These mountains are much lower in elevation than the Alpine system, with their highest peaks at 6000 feet (1800 meters). Formed about 100 million years ago, much of this upland region is characterized by rolling landscapes about 3000 feet (1000 meters) above sea level.

These uplands are important to western Europe because they contain the raw materials for Europe's industrial areas. In Germany and France, for example, they have provided the iron and coal central to each country's steel industry. To the east, mineral resources from the Bohemian Highlands have also fueled major industrial areas in Germany, Poland, and the Czech Republic.

Western Highlands The Western Highlands define the western edge of the European subcontinent, extending from Portugal in the south, through portions of the British Isles, to the highland backbone of Norway, Sweden, and Finland in the far north. These are Europe's oldest mountains, formed about 300 million years ago.

As with other upland areas that traverse many separate countries, specific names for these mountains differ from country to country. A portion of the Western Highlands forms the highland spine of England, Wales, and Scotland, where picturesque glaciated landscapes are found at modest elevations of 4000 feet (1200 meters) or less. These U-shaped glaciated valleys also appear in Norway's uplands, where they produce a spectacular coastline of **fjords**, or flooded valley inlets, similar to the coastlines of Alaska and New Zealand (Figure 8.5).

Geologically, the far western edge of Europe is found in Iceland, which, as mentioned, is divided by the Eurasian and North American tectonic plates. Like other plate boundaries, Iceland has many active volcanoes that occasionally spew ash into the atmosphere, sometimes causing serious problems for the heavy airline traffic between Europe and North America.

Seas, Rivers, and Ports

In many ways, Europe is a maritime region with strong ties to its surrounding seas. Even landlocked countries such as Austria, Hungary, Serbia, and the Czech Republic have access to the ocean and seas through extensive networks of navigable rivers and canals.

Europe's Ring of Seas Four major seas and the Atlantic Ocean encircle Europe. In the north, the Baltic Sea separates Scandinavia from north-central Europe. Denmark and Sweden have long controlled the narrow Skagerrak and Kattegat straits that connect the Baltic to the North Sea, which is both a major fishing ground and a principal source of Europe's oil and gas, mined from deep-sea drilling platforms.

The English Channel (in French, *La Manche*) separates the British Isles from continental Europe. At its narrowest point, the Dover Straits,

Figure 8.5 Fjord in Norway During the Pleistocene epoch, continental ice sheets and glaciers carved deep U-shaped valleys along what is now Norway's coastline. As the ice sheets melted and sea level rose, these valleys were flooded by Atlantic waters, creating spectacular fjords. Many fjord settlements are accessible only by boat, linked to the outside world by Norway's extensive ferry system.

the channel is only 20 miles (32 km) wide. Although England has regarded the channel as a protective moat, it has primarily been a symbolic barrier, for it deterred neither the French Normans from the continent nor the Viking raiders from the north. Only Nazi Germany found it a formidable barrier during World War II. Since 1993, and after decades of resistance by the English, the British Isles have been connected to France through the 31-mile (50-km) Eurotunnel, with its high-speed rail system carrying passengers, autos, and freight.

Gibraltar guards the narrow straits between Africa and Europe at the western entrance to the Mediterranean Sea, and Britain's stewardship of this passage remains an enduring symbol of a once great sea-based empire. Finally, on Europe's southeastern flanks are the straits of Bosporus and the Dardanelles, the narrows connecting the eastern Mediterranean with the Black Sea. Disputed for centuries, these pivotal waters are now controlled by Turkey. Though these straits are often thought of as the physical boundary between Europe and Asia, they are easily bridged in several places to facilitate truck and train transportation within Turkey and between Europe and Southwest Asia.

Rivers and Ports Europe is a region of navigable rivers, connected by a system of canals and locks that allow inland barge travel from the Baltic and North Seas to the Mediterranean and between western Europe and the Black Sea. Many rivers on the European Lowland—namely the Loire, Seine, Rhine, Elbe, and Vistula—flow into Atlantic and Baltic waters. However, the Danube, Europe's longest river, flows east and south, rising in the Black Forest of Germany only a few miles from the Rhine River and running southeastward to the Black Sea, offering a connecting artery between central and eastern Europe (Figure 8.6). Similarly, the Rhône headwaters rise close to those of the Rhine in Switzerland, yet the Rhône flows southward into the Mediterranean. Both the Danube and the Rhône are connected by locks and canals to the rivers of the European Lowland, making it possible for barge traffic to travel between all of Europe's fringing seas and oceans.

As mentioned, major ports are found at the mouths of most western European rivers, where they serve as transshipment points for inland waterways and rail and truck networks. From south to north, these ports include Bordeaux at the mouth of the Garonne, Le Havre on the Seine, London on the Thames, Rotterdam (the world's largest port in terms of tonnage) at the mouth of the Rhine, Hamburg on the Elbe, and, to the east in Poland, Szczecin on the Oder and Gdansk on the Vistula.

Europe's Climate

Three major climate types characterize Europe (Figure 8.7). Along the Atlantic coast, a moderate and moist **marine west coast climate** modified by oceanic influences dominates. Farther inland, **continental climates** prevail, with hotter summers and colder winters. Finally, a dry-summer **Mediterranean climate** is found in southern Europe, from Spain to Greece.

One of the most important climate controls is that of the Atlantic Ocean. Even though much of Europe is at relatively high latitudes (London, England, for example, is slightly farther north than

Figure 8.6 Danube Barge Traffic In 1992, the Danube River, Europe's longest, was connected by canal to the Rhine River, allowing commercial barge traffic to move throughout Europe and between the North Sea and the Black Sea. Because inland water traffic (IWT) is reportedly 80 percent cheaper than trucking and because industrialization is increasing in eastern Europe, barge traffic has grown considerably in the last decade. Here a tug pushes a barge on the Danube in Serbia.

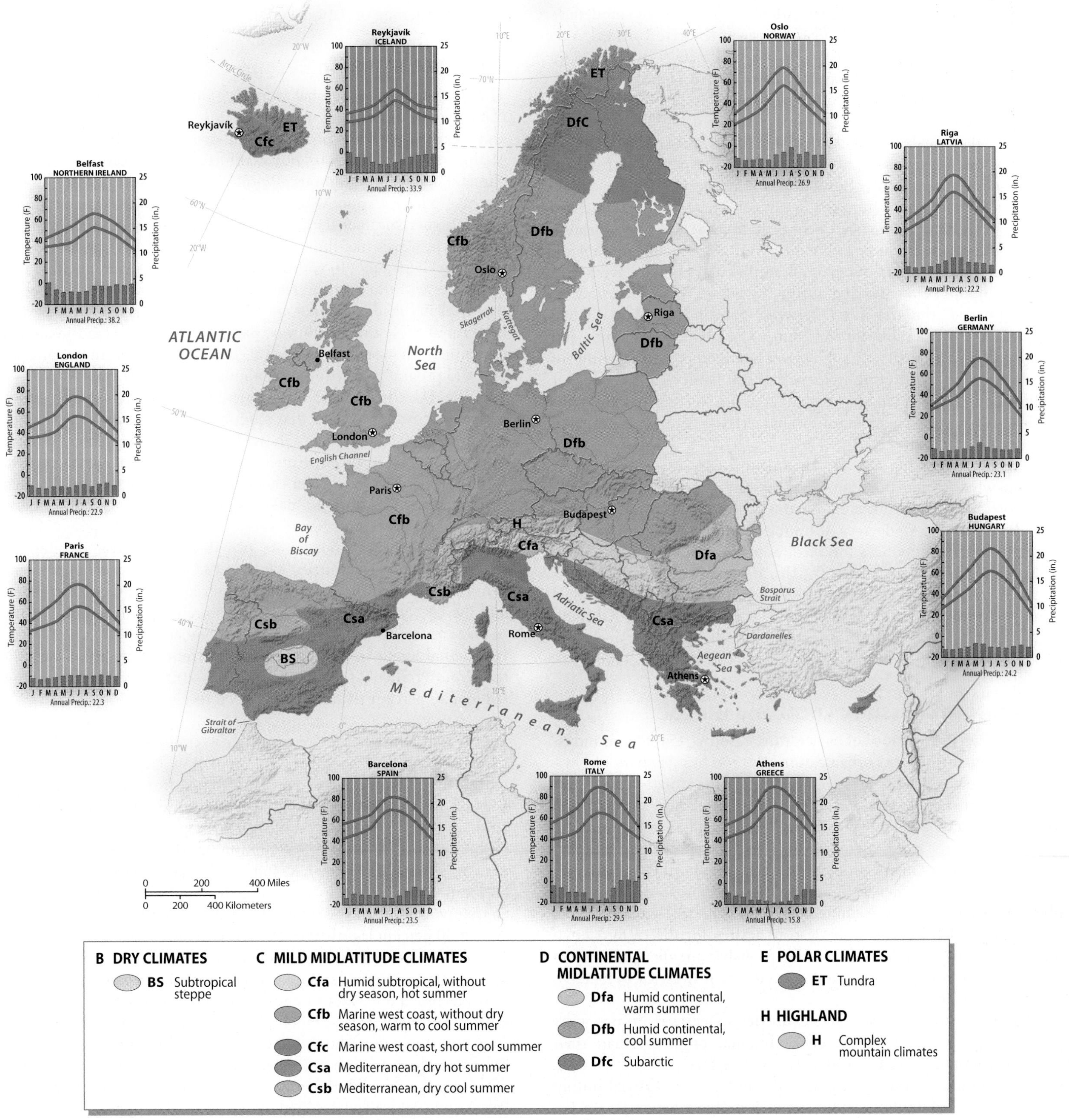

Figure 8.7 Climates of Europe Three major climate zones dominate Europe. Close to the Atlantic Ocean, the marine west coast climate has cool seasons and steady rainfall throughout the year. Farther inland, continental climates have at least one month averaging below freezing, as well as hot summers, with a precipitation maximum occurring during the warm season. Southern Europe has a dry-summer Mediterranean climate. **Q: Where in Europe are there climates similar to the climate where you live in North America?**

Vancouver, British Columbia), the mild North Atlantic Current, which is a continuation of the warmer Atlantic Gulf Stream, moderates coastal temperatures from Iceland and Norway south to Portugal. This maritime influence gives western Europe a climate 5–10°F (3–6°C) warmer than regions at comparable latitudes that lack the moderating influence of a warm ocean current. As a result, in the marine west coast climate region no winter months average below freezing, even though cold rain, sleet, and an occasional blizzard can be common winter events. Summers are often cloudy and overcast, with frequent drizzle and rain as moisture flows in from the ocean.

Inland, far removed from the ocean (or where a mountain chain limits the maritime influence, as in Scandinavia), landmass heating and cooling becomes a strong climatic control, producing hotter summers and colder winters. Indeed, all continental climates average at least one month below freezing during the winter.

In Europe, the transition between maritime and continental climates takes place close to the Rhine River border of France and Germany. Farther north, although Sweden and other nearby countries are close to the moderating influence of the Baltic Sea, their higher latitude coupled with the blocking effect of the Norwegian mountains produces cold winter temperatures characteristic of true continental climates.

The Mediterranean climate has a distinct dry season during the summer, which results from the warm-season expansion of the Atlantic (or Azores) high-pressure area. As this warm air descends between latitudes of 30 and 40 degrees, it inhibits summer rainfall. This same phenomenon also produces the Mediterranean climates of California, western Australia, parts of South Africa, and Chile. These rainless summers may attract tourists from northern Europe, but the seasonal drought is problematic for agriculture. It is no coincidence that traditional Mediterranean cultures, such as the Arab, Moorish, Greek, and Roman civilizations, were major innovators of irrigation technology.

Environmental Issues: Local and Global

Because of its long history of agriculture, resource extraction, industrial manufacturing, and urbanization, Europe has its share of environmental issues (Figure 8.8). Compounding the situation is the fact that pollution rarely stays within political boundaries. Air pollution from England, for example, creates serious acid rain problems in Sweden. Similarly, water pollution from Swiss factories on the upper Rhine River creates major problems downstream for the Netherlands, where Rhine River water is commonly used for urban drinking supplies. As a result of these numerous trans-boundary environmental problems, the EU has taken the lead in addressing the region's environmental issues, and Europe is today probably one the "greenest" of the major world regions, surpassing even North America in its environmental sensibilities.

Until recently, however, the countries of eastern Europe were plagued by far more serious environmental problems than their western neighbors because of that region's history of Soviet control, where economic planning emphasized short-term industrial output at the expense of environmental protection (see Chapter 9). In Poland, for example, industrial effluents reportedly had wiped out all aquatic life in 90 percent of the country's rivers, with damage from air pollution affecting over half of the country's forests. Similar legacies from the Soviet period were reported in the Czech Republic, Romania, and Bulgaria. Today, however, most of these environmental issues in eastern Europe have been resolved through funding from the EU and the strengthening of national environmental laws.

Climate Change in Europe

The fingerprints of global warming are everywhere in Europe, from dwindling sea ice, melting glaciers, and sparse snow cover in arctic Scandinavia to more frequent droughts in the water-starved Mediterranean subregion. Furthermore, the projections for future climate change are ominous: World-class ski resorts in the Alps are forecast to have warmer winters with less snow pack, while in the lowlands, higher summer temperatures will probably produce more frequent heat waves like those in 2003 and 2015, affecting farmers and urban dwellers alike. In addition, rising sea levels from melting polar ice sheets will threaten the Netherlands, where much of the population lives in diked lands that are actually below sea level (Figure 8.9). Because of these threats, Europe has taken a strong stand in addressing climate change and, as a result, has implemented numerous policies and programs to reduce greenhouse gas (GHG) emissions.

The EU entered the 1997 Kyoto Protocol climate negotiations with an innovative scheme that reinforced its philosophy that regional action was superior to that of individual countries. Specifically, the EU set a target of an 8 percent reduction below 1990 GHG emission levels for the EU as a whole. Under this umbrella scheme, several EU member states were required to make significant emissions cuts, whereas others like Greece and Spain were actually allowed increased emissions. The point was to promote growth and industrial development in Europe's poorer countries while at the same time requiring emission reductions in the traditional industrial core of Germany, France, and the UK. Noteworthy is that this umbrella approach has remained in place as the EU has grown from 15 members in 1997, when it made its original Kyoto commitment, to its current 28 member states. For the Kyoto Protocol's second phase (2013–2020), the EU has committed to a 20 percent reduction over 1990 levels.

Energy and Emissions Greenhouse gas emissions are closely linked to a country's energy mix and population size. Not surprisingly, within the EU, the highest emissions come from the member countries having the largest populations and burning the most fossil fuels. Germany, the largest European country by population with almost 83 million people, emits over 900 million metric tons of pollutants each year. It is followed by the UK, Italy, and France, all with populations of about 60 million; all three countries emit about half as much as Germany. Noteworthy is that Germany's current CO_2 emissions are almost 30 percent less than in 1990, the Kyoto Protocol's baseline year.

As for Europe's fuel mix, generally speaking the region runs on fossil fuels—coal, gas, and oil. Although Europe's early industrialization was based on coal, those resources are now running thin in western Europe, and, in fact, much of the EU's emission reductions came from shutting down British and German coal mines during the 1990s. To replace coal, Europe relies heavily on imported gas and oil, much of it from Russia, with the only local supplies coming from the North Sea gas and oil wells developed by the UK and Norway.

Complementing EU emission reduction goals is a policy to increase the region's renewable energy resources to the point where the EU as a whole will be generating 20 percent of its power from hydropower, wind, solar, and biofuels by 2020. This goal seems well within reach given existing hydropower facilities in the Alpine and Scandinavian

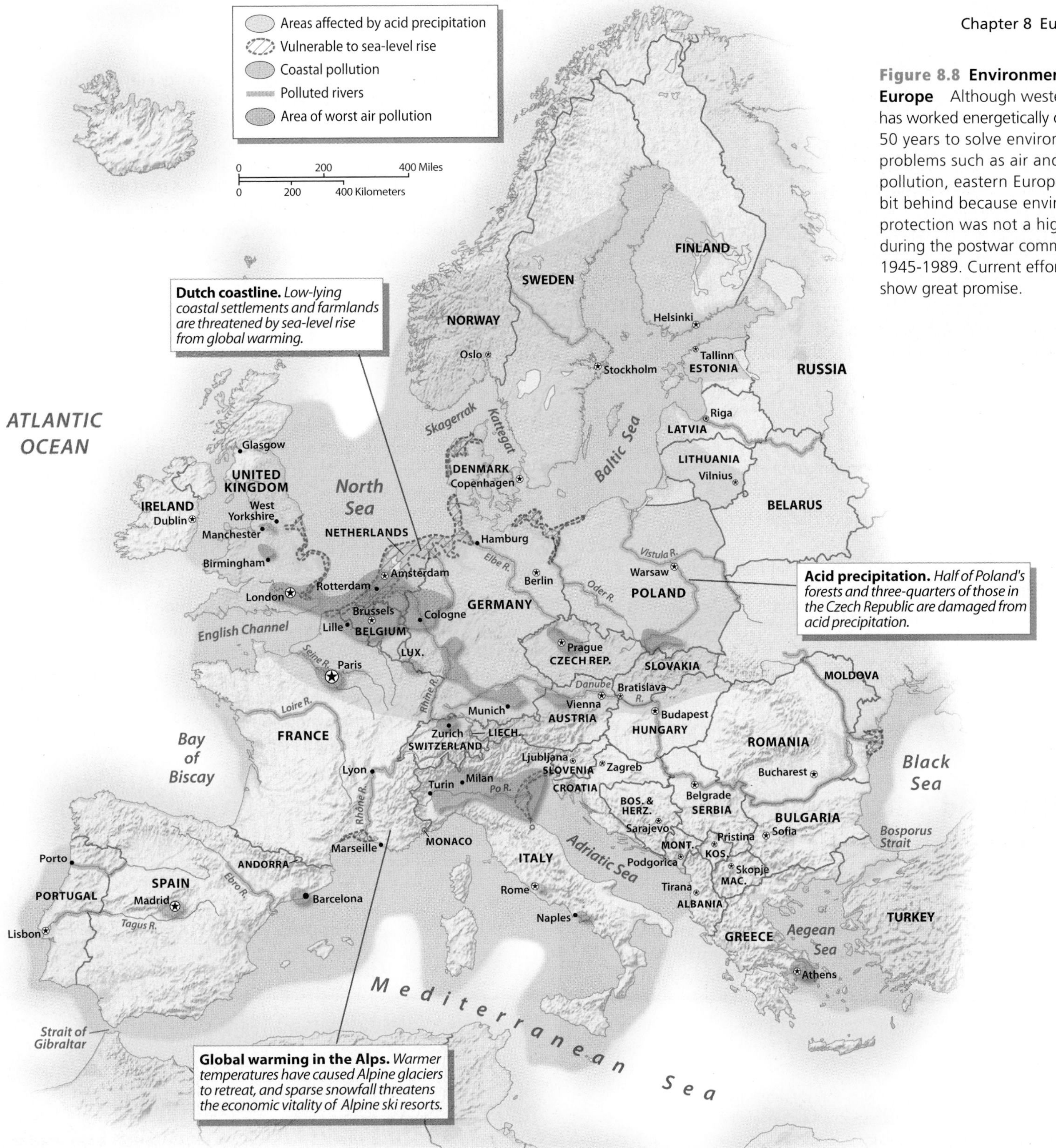

Figure 8.8 Environmental Issues in Europe Although western Europe has worked energetically over the past 50 years to solve environmental problems such as air and water pollution, eastern Europe lags a bit behind because environmental protection was not a high priority during the postwar communist period, 1945-1989. Current efforts, however, show great promise.

countries, coupled with the expansion of wind and solar power in the last few years throughout the EU. Germany, Italy, and Spain already generate well over 20 percent of their energy from renewable resources. Throughout Europe, wind power is the fastest-growing renewable energy segment, supplying 9 percent of the EU's power in 2015, with forecasts for twice that amount by 2020 (Figure 8.10).

The EU's Emission-Trading Scheme As part of its Kyoto Protocol emissions reduction strategy, the EU inaugurated the world's first carbon-trading scheme in 2005. Under this plan, specific yearly emission caps were set for the EU's largest GHG emitters. If these emitters exceeded those caps, they had to either purchase carbon emission equivalences from a source below their own cap or, alternatively, buy credits from the EU carbon market. The goal of this cap-and-trade system was to make business more expensive for companies that pollute, while rewarding those staying under their carbon quota. Although there were considerable problems with this trading scheme in its first decade, these issues had been largely resolved by 2015, making the EU's carbon-trading scheme the world's largest and most successful cap-and-trade plan.

Figure 8.9 Protecting Low-Country Europe from Sea-Level Rise Conceived more than half a century ago, the Dutch Deltaworks were originally built to keep ocean storm surges and Rhine River flooding from the southwestern Netherlands. But now, given the forecasts for sea-level rise from global warming, the Deltaworks must be reengineered and made higher to protect the 50 percent of the Netherlands that lies below the current sea level.

Review

8.1 Name and locate on a map the major lowland and mountainous areas of Europe.

8.2 What and where are the three major climate regions of Europe?

8.3 Describe how inland barge traffic would get from the mouth of the Rhine to the delta of the Danube.

8.4 Explain why and how Europe has been so successful in reducing its CO_2 emissions over the last 20 years.

KEY TERMS European Union (EU), fjord, marine west coast climate, continental climate, Mediterranean climate

Population and Settlement: Slow Growth and Problematic Migration

The major themes of Europe's population and settlement geography are its very low rates of natural growth; its aging population; widespread internal migration that is aggravating population loss in several EU countries; and large streams of legal and extralegal international migration coming from Africa and Southwest Asia. The highly urbanized, industrial, and relatively wealthy core of western Europe, which includes southern England, northern France, Belgium, the Netherlands, and western Germany, is the focus of most migration, both internal and international (Figure 8.11).

Low (or No) Natural Growth

Probably the most striking characteristic of Europe's demography is the lack of natural growth, as death rates exceed birth rates (Table A8.1). Several large countries, notably Germany and Italy, actually have negative natural growth, and their populations could decrease in size over the next decades, with Germany forecast to lose 6 million people by mid-century and Italy 2 million. Numerous smaller European countries (such as Latvia, Lithuania, Bulgaria, Romania, Serbia, and Portugal) are also projected to have smaller populations in the coming decades.

Figure 8.10 Wind Power in Europe Not only is Europe trying to reduce its carbon dioxide emissions, but also it is a world leader in generating renewable energy from wind, sun, and biofuels. This large wind farm is in Denmark.

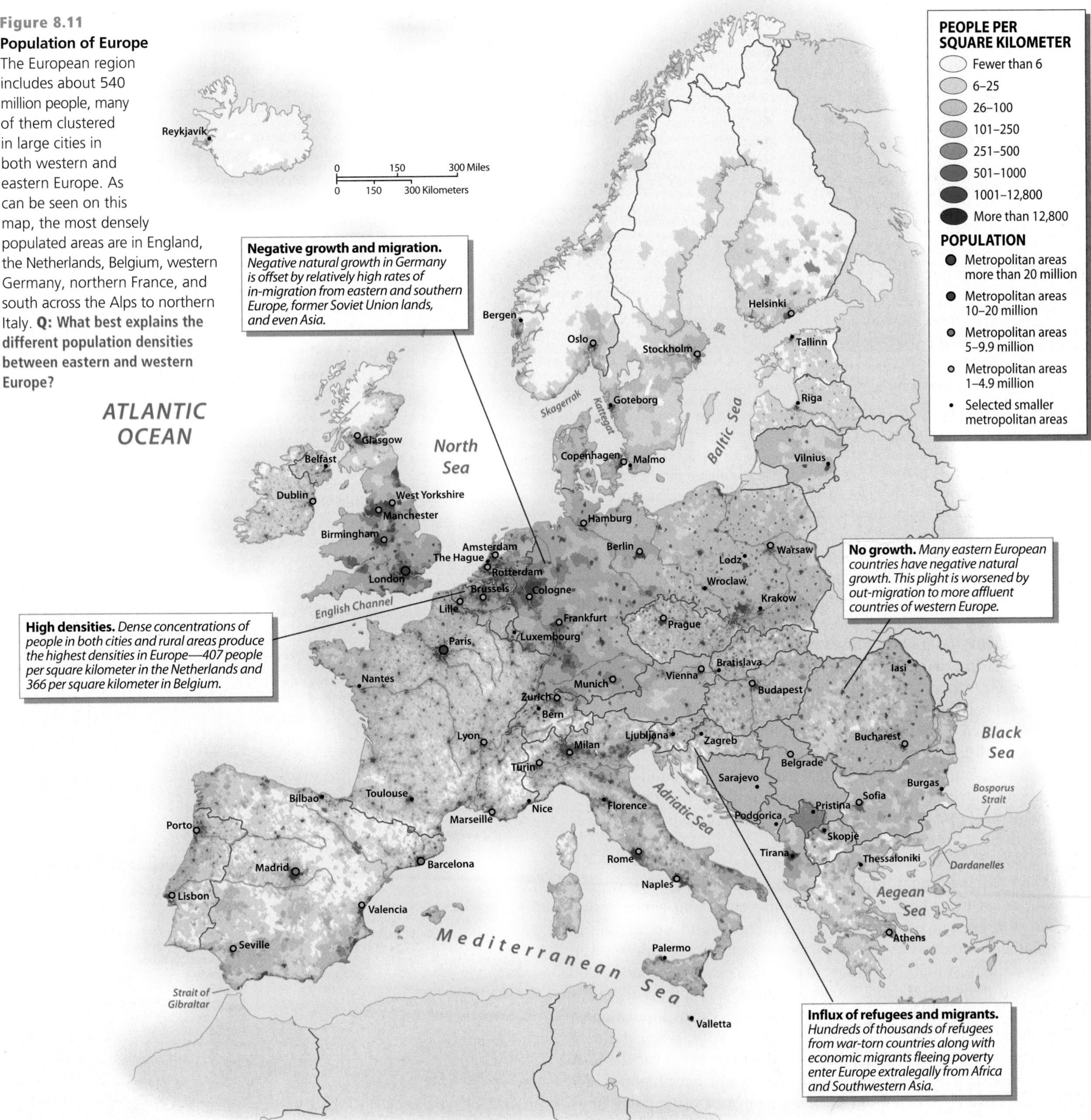

Figure 8.11
Population of Europe
The European region includes about 540 million people, many of them clustered in large cities in both western and eastern Europe. As can be seen on this map, the most densely populated areas are in England, the Netherlands, Belgium, western Germany, northern France, and south across the Alps to northern Italy. **Q: What best explains the different population densities between eastern and western Europe?**

Europe's population, like that of Japan and even the United States, is characterized by the fifth, or postindustrial, stage of the demographic transition (discussed in Chapter 1), in which fertility falls below the replacement level. The consequences of shrinking national populations—labor shortages, smaller internal markets, and reduced tax revenues to support social services (such as retirement pensions) essential for their aging populations—could be significant. The population pyramids for Germany at two dates, 1950 and 2020, illustrate these demographic changes in terms of a shrinking base of young people as well as the overall aging of the population (Figure 8.12).

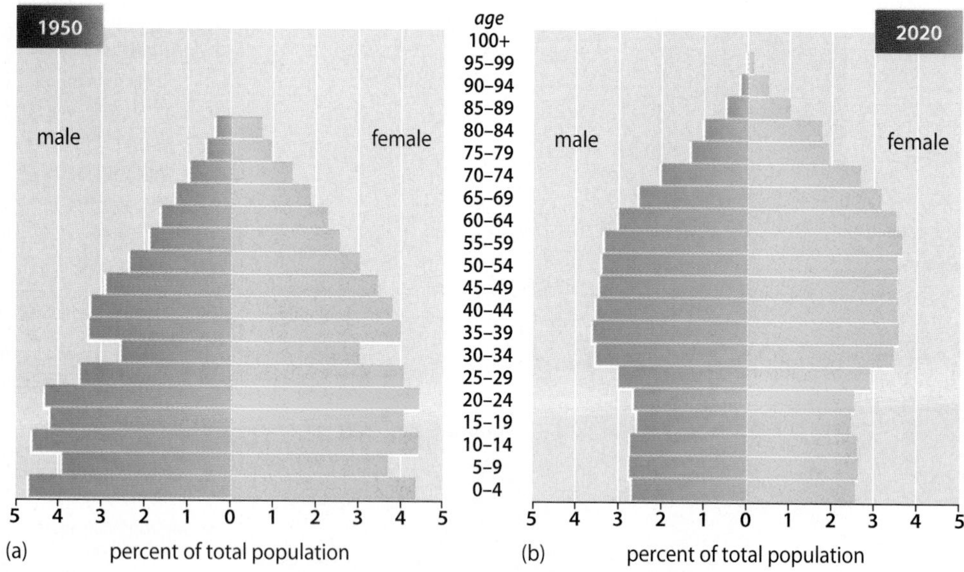

Figure 8.12 Population Pyramids for Germany These two population pyramids illustrate the general aging of Europe, showing (a) Germany's relatively youthful population in 1950, contrasting with (b) Germany's projected population makeup in 2020. A similar pattern would emerge for the other large European countries of France, Spain, Italy, and the United Kingdom.

Pro-growth Policies To address concerns about population loss, many European countries try to promote growth through various programs and policies. These range from bans on abortion and the sale of contraceptives (Hungary) to what are commonly called **family-friendly policies** (Germany, France, and Scandinavia). In these countries, pro-growth policies include full-pay maternity and paternity leaves for both parents, guarantees of continued employment once these leaves conclude, extensive child-care facilities for working parents, outright cash subsidies for having children, and free or low-cost public education and job training for their offspring.

However, important to note is that even with these family-friendly policies, no European country has a total fertility rate above the replacement level of 2.1. Therefore, any population growth taking place is solely through in-migration.

Migration Within Europe Since its origin in 1957, the EU has worked toward the goal of free movement of both people and goods within the larger European community. Consequently, residents of the 28 EU member countries can generally move about as they please. And they are doing just that. In the past decade, for example, some 16,000 Lithuanians moved to Ireland to take advantage of that country's then booming economy, leaving their home country (total population of 3.3 million) to languish economically. More recently, when the Irish boom went bust, almost half of these EU migrants either moved back to Lithuania or went elsewhere in the EU. Recent net migration figures show Ireland is continuing to lose people to out-migration, as are Lithuania and the other two Baltic countries, Latvia and Estonia. Other areas of significant out-migration are the Balkans and the Mediterranean countries; Spain and Greece are two of the largest countries losing population to emigration. As for receiving European migrants, Germany is a favored destination, as are the UK and several smaller countries, including Norway, Luxembourg, and Austria (see *Geographers at Work: Migrants in the Digital Age*).

The Schengen Agreement Underlying this new intra-Europe mobility is a formal legislative act that eroded Europe's historical national borders: the **Schengen Agreement**, named after the city in Luxembourg where it was signed in 1985.

Before Schengen, crossing a European border always involved showing passports and auto insurance papers, car inspection records, and so on at every European border. Today, however, there are either no border stations or only the most cursory formalities as one travels between Schengen countries (Figure 8.13). To older Europeans who knew Europe before Schengen, it's a remarkable experience to freely cross a national border.

Yet today the Schengen Agreement has become increasingly controversial due to a combination of fears over increased terrorism and illegal international migration. Once inside a peripheral Schengen country, such as Italy or Greece, an illegal migrant can theoretically

Figure 8.13 Schengen Border People stroll unhindered by document checks across the Poland–Germany border in the city of Görlitz. Earlier, before Poland became a member of the Schengen Agreement in late 2007, this border was a heavily policed "hard" border between Europe and the non-Schengen countries to the east, where passport and visa checks were rigorously enforced. Currently, these free-access Schengen borders have become focal points for migrants seeking extralegal entry to the core countries of Europe, primarily France and Germany.

GEOGRAPHERS AT WORK

Migrants in the Digital Age

Figure 8.1.1
Dr. Weronika Kusek

Weronika Kusek, an assistant professor of geography at Northern Michigan University, grew up in Poland but did her college and graduate work in the United States. During those years, however, she thought of herself not as a migrant, but as an international student who would surely return to Poland someday. But after 10 years living and working in the United States, she's starting to recalibrate her identity, in part because she studies migrants and how migrant networks have changed dramatically in the digital age.

Everyday Geography Kusek's interest in geography started early, based on a Polish school curriculum that introduces geography in the 4th grade, and reinforced by parents who valued travel and global experiences. She says, "Geography is very applicable to everyday life . . . you interact with people from other cultures, other parts of the world, so a basic understanding of other regions and cultures is extremely important for any career."

Kusek's family hosted an American student while she was in high school; this, in turn, led to her extending a summer visit to Ohio into a trial semester at a college in Toledo. Although she initially planned to return to Poland for college, Kusek received her BA and MA from the University of Toledo, then went on to earn her PhD in Geography from Kent State University.

Ties to Home Based on her fieldwork in England, Kusek says that some migrant groups maintain such strong ties to their homeland through daily phone calls, Skype calls, and other media, that they challenge the traditional process of assimilation into the host culture (Figure 8.1.2). Many Polish migrants in London, for example, spend every holiday in Poland reinforcing ties with family and friends. These new cultural behaviors and geographies contrast significantly with how pre-digital migrant communities interacted with their host cultures.

And what about her own ties to Poland? Just like the people she studies, Kusek Skypes often and makes at least one trip back to Poland each year.

1. Do you know people who are pre-Internet migrants? If so, ask them how they stayed in touch with their home and family.
2. Are you a "digital migrant" of some sort? That is, do you use the Internet to keep in touch with your home?

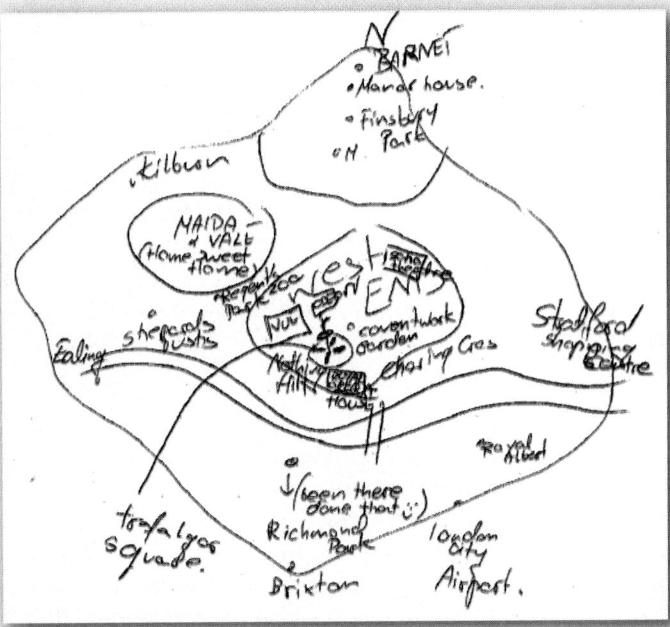

Figure 8.1.2 Mental Map of London One way of learning about a migrant's adaptation to a new place is to ask them to draw a mental map of the area, noting the important parts where they work and live, shop, and recreate. Here's an example from one of Dr. Kusek's London subjects.

move freely across borders, just like EU citizens. True, even without formal border-crossing points, most countries still do maintain some sort of border-policing organization that attempts to limit illegal migration, but those efforts are often ineffective. This fact motivates some Schengen countries to seriously consider returning to the pre-Schengen era of formal border controls. Supporting this cautious mentality is the fact that the UK opted out of Schengen in 1985, drawing on its historical insular location as the rationale. Current conversations about new border controls seem to conflate legitimate concerns about illegal activity (terrorism and organized crime) with equal amounts of neonationalism, cultural xenophobia, and political opportunism, topics that will be discussed later in the chapter.

Legal Migration to Europe Western Europe has long accepted international migrants into its population, particularly in the former colonial powers of Spain, the Netherlands, France, and England, countries that willingly provided their overseas citizens with visas and residential permits. Thus, historically, South Asians came to England, Indonesians to the Netherlands, Africans to France, and South Americans to Spain.

Europe's doors to foreigners were opened further to solve postwar labor shortages as cities and factories rebuilt from the destruction of World War II. The former West Germany, for example, drew heavily on workers from Europe's rural and poorer periphery—Italy, the former Yugoslavia, Greece, and even Turkey—to fill industrial, construction, and service jobs. Later, with the 1991 collapse of the Soviet Union, emigrants from former satellite countries poured into western Europe, seeking relief from the economic chaos in Russia and eastern European countries. This post–Cold War immigration also included refugees from war-torn areas of the former Yugoslavia, particularly Bosnia and Kosovo.

As a result of these different migration streams, foreigners now make up about 5 percent of the EU population; Germany, the largest country, has the highest percentage of foreigners (10 percent), with France and the UK having about 5 percent each. (For comparison, 11.7 percent of the U.S. population are foreigners.)

Extralegal Migration, Leaky Borders, and "Fortress Europe"

While at one level the distinction between legal and illegal migration may seem clear-cut— a migrant either does or does not have the proper entry papers—the situation today is much more complex because of the region's **asylum laws** that protect refugees from global political and ethnic persecution. Current asylum laws stem from Europe's post–World War II humanitarian efforts to care for the

Figure 8.14 Migration into Europe In 2015, close to one million migrants entered Europe by extralegal means. Most were refugees seeking political amnesty from war-torn Syria, Afghanistan, Iraq, and Eritrea, while a smaller number were economic migrants fleeing poverty in their home countries.

region's refugees displaced by the war as well as those displaced by the resulting Cold War that divided the continent politically from 1945 to 1991.

Today, however, Europe is inundated with asylum seekers from afar, primarily from Africa and Southwest Asia. Although it is theoretically possible to apply for asylum in Europe from one's home country, the very nature of persecution often prevents this, leading most asylum seekers to try to enter Europe extralegally. Once (or if) they reach European soil, refugees can then ask authorities for political asylum. In 2015 the number of refugees and migrants reaching Europe was close to one million; a huge increase over the number of extralegal migrants in previous years. This increase is thought to be a result of worsening conditions in war-torn countries, combined with the entry of organized crime groups into human trafficking.

Getting to European soil, however, is not simple, cheap, or safe. Refugees reportedly pay thousands of dollars to smugglers for a risky land or sea journey to Europe. Many die in the process. Even if they reach Europe, they will spend months (often years) in overcrowded relocation camps waiting to make their legal case to European authorities. Again, months (perhaps years) may pass before a decision is reached, since separating those with legitimate claims of persecution from the so-called economic migrants is a difficult and time-consuming task. If the refugees are granted asylum, they will be sent to an accepting European country, where they will face the additional challenges of adapting to a new culture and environment; if they are denied asylum, they will be summarily sent back to their home country.

Sweden and Norway have been the two European countries that, on a per capita basis (of the native population), have accepted the largest number of migrants granted asylum. Germany has accepted the largest total number (more than 40,500 by 2014), with France accepting around 15,000 and the UK about 10,000. Currently, the EU is developing a quota system that will require all 28 member states to accept a specific number of refugees each year. To break the logjam of migrants crowded into relocation camps in Italy, Greece, and Spain, the EU is also working on plans to send migrants to non-Mediterranean countries where they can file their asylum requests. This does not mean that these refugees must stay in those countries; if their asylum claims are validated, they will be able to move freely among the Schengen countries.

To help the EU perimeter countries of Greece, Italy, Malta, and Spain police their borders, the EU has provided funds to strengthen their borders with guards and, in some places, with physical border barriers to inhibit illegal entry. For those with longer memories, these fortifications are disturbingly reminiscent of the Cold War's Iron Curtain that divided Europe into west and east (Figure 8.15).

Today some observers describe Europe as being divided into a geographical system where its perimeter consists of hard borders—a "Fortress Europe," as critics (and anti-immigrant groups) call the plan—while its internal borders are deliberately soft and porous due to the Schengen Agreement. However, until the illegal international migration issue is resolved, those soft internal Schengen borders will become increasingly controversial, challenging earlier political and economic goals of a "Europe without borders."

Landscapes of Urban Europe

Europe is highly urbanized, with almost three-quarters of its population living in cities. In fact, Europe has several microstates that are essentially city-states: Monaco (aka Monte Carlo), Malta, and Vatican City. But urbanization data can produce different landscapes. Belgium, for example, is 99 percent urban, but this comes more from a landscape of connected mid-sized towns than from huge megacities. And no traveler in Iceland would think of that country as 95 percent urban, because outside of its only city, Reykjavik, the landscape is nothing but rural and wild. At the other end of the scale are Bosnia–Herzegovina and Kosovo, Balkan countries that are the only two less than 50 percent urban. Indeed, the rural landscape reinforces that notion.

Europe's largest countries—Germany, France, the UK, and Italy—are more typical of much of the region: three-quarters urban, with numerous large cities scattered among expansive rural landscapes.

The Past in the Present North American visitors often find European cities far more interesting than our own because of the mosaic of both historical and modern landscapes, featuring medieval churches and squares interspersed with high-rise buildings and modern department stores. The imprints of three historical periods can be seen in most European cities: The medieval (900–1500), Renaissance–Baroque (1500–1800), and industrial (1800–present) periods have each left their characteristic traces on the European urban scene. Learning to recognize these stages of historical growth provides visitors to Europe's cities with fascinating insights into both past and present (Figure 8.16).

The **medieval landscape** is one of narrow, winding streets, crowded with three- or four-story masonry buildings with little setback from the street. This is a dense landscape with few open spaces, except around churches and town halls, where public squares or parks are clues to historical medieval open-air marketplaces.

As picturesque as we find medieval-era districts today, they nevertheless present challenges to contemporary inhabitants because of their narrow, congested streets and old housing. Modern plumbing and heating are often lacking, and rooms and hallways are small and cramped compared to present-day standards.

Many cities have enacted legislation to restore and protect their historical medieval landscapes. This movement began in the late 1960s in the Marais area of central Paris and has become increasingly popular throughout Europe as cultures work to preserve the unique sense of place provided by their urban medieval sections. Because restoration costs are high, often these projects lead to a demographic change where low- and fixed-income people are displaced by those able to pay higher rents. Further, historical areas often attract tourists, and with increased foot traffic the array of street-level shops also often changes from neighborhood-serving stores to those catering to tourists. Urban planners use the term *gentrification* to describe these changes to historical districts (see *Working Toward Sustainability: Protecting Europe's Cultural Landscapes*).

In contrast to the cramped and dense medieval landscape, those areas of the city built during the **Renaissance–Baroque period** are much more open and spacious, with expansive ceremonial buildings and squares, monuments, ornamental gardens, and wide boulevards lined with palatial residences. During this period (1500–1800), a new artistic sense of urban planning arose in Europe, resulting in the restructuring of many European cities, particularly the large capitals such as Paris and Vienna where grand boulevards replaced older, more densely settled quarters. These changes were primarily for the benefit of the new urban elite—the royalty and rich merchants.

During the Renaissance–Baroque period, city fortifications limited the outward spread of these growing cities, thus aggravating crowding within. With the advent of assault artillery, European cities were forced to build an extensive system of defensive walls. Once encircled by these walls, the cities could not expand outward. Instead, as the demand for space increased within the cities, a common solution was to add several new stories to the medieval houses.

Industrialization dramatically altered the landscape of European cities. Beginning in the early 19th century, factories clustered together in cities, drawn by their large markets and labor force and supplied by raw materials shipped via barge and railroad. Industrial districts of factories and worker tenements grew up around these transportation lines. In continental Europe, where many cities retained their defensive walls until the late 19th century, the new

WORKING TOWARD SUSTAINABILITY

Protecting Europe's Cultural Landscapes

Europe has a long tradition of protecting its historical monuments, but more recently this preservation consciousness has been expanded to cultural landscapes, both rural and urban. While monument protection laws protect specific points—a church here, a palace there—cultural landscapes, with their diverse arrays of parts, are more difficult to preserve. Several examples illustrate how these landscape laws work—or don't.

World Heritage Sites The "Old Town" of Salzburg, Austria, is a historical and architectural gem by any measure, with its remarkable assemblage of medieval and Baroque buildings. Although many individual buildings were protected as historical monuments under a 19th-century law, in 1967 the city coined Europe's first cultural landscape regulation, one that protected the totality of the inner-city scene with its unique sense of place. This innovative landscape law involved not just preserving existing buildings, but also providing specific guidelines for the appropriate shape, texture, and color of any new buildings in the inner city. Opponents argued that to essentially lock up the urban fabric of a growing city by keeping it from modernizing was unrealistic, but the preservationists won out and were rewarded in 2006 when the inner city was declared a United Nations World Heritage Site. Following the Salzburg model, numerous other European cities—notably Passau, Regensburg, and Dresden, Germany; Bath, England; and Siena, Italy—passed their own cultural landscape protection laws.

Austria has also been a leader in preserving rural cultural landscapes. One of the best examples is legislation protecting the landscape of the Wachau region along the Danube River, a 25-mile-long (40-km-long) valley consisting of several towns, a handful of small wine villages, and thousands of acres of working vineyards (Figure 8.2.1). While the Wachau is unquestionably picturesque, regulating every aspect of its visual cultural landscape is a major administrative challenge. But like Salzburg, the Wachau somehow found a way, and the whole area became a World Heritage Site in 2000.

A still greater challenge—unfortunately, one with a less successful outcome—was an attempt to protect the Elbe River valley in northeastern Germany. Here preservationists linked together the riverine valley landscape with the historical inner-city landscape of Dresden, which had been completely rebuilt with architectural integrity after total destruction in World War II. Initially successful enough to be declared a World Heritage Site in 2004, the area then suffered embarrassment several years later as the first European city to ever lose this coveted status when the Dresden city government built an architecturally modern bridge across the Elbe River to solve a serious interurban traffic problem (Figure 8.2.2).

Figure 8.2.1 The Wachau Region of Austria This picturesque landscape became a World Heritage Site in 2000 after reaching agreement with local residents and landowners on how the cultural landscape should be protected and managed along the 25-mile (40-km) stretch of the Danube River.

Figure 8.2.2 The Waldschlosschen Bridge in Dresden The beautiful Elbe River valley and historic Dresden, Germany, protected its traditional cultural landscape as a World Heritage site until this modern bridge was built to alleviate regional traffic congestion. As a result the area's World Heritage protection was withdrawn in 2009.

This unfortunate delisting of Dresden and the Elbe Valley reminds us that, while there are many benefits to protecting a traditional cultural landscape, there can also be a downside that inhibits modern solutions to urban problems. Bath, England, for example, is asking its citizens to vote on whether completion of a planned modern housing project is worth the risk of losing its World Heritage Site status.

1. Are there protected cultural landscapes near you? What is their significance?
2. Search the Internet to see if you can identify any downsides or disadvantages to becoming a World Heritage Site.

Figure 8.15 Greek Border Fence Because of Greece's location close to the troubled countries of Southwest Asia, that country has been a major entry point to Europe for extralegal migrants from Syria and Iraq, with thousands crossing illegally from Turkey into Greece. In desperation, Greece (with tacit support from the EU) fortified part of its Turkish border with this barrier, creating a precedent for other European countries—most notably Hungary—to do the same to their borders.

industrial districts were often located outside the former city walls, removed from the historical central city. When cities removed these defensive walls, these spaces were commonly converted into ring roads that circled the inner city. Good examples are Vienna, Austria, and Toulouse, France.

Not to be overlooked are the post–World War II changes to European cities as they rebuilt from the war's destruction and adapted to the political and economic demands of the postwar era. As in North American cities, suburban sprawl has become an issue in many European countries as people seek lower-density housing in nearby rural environments. But unlike most North American cities, European urban areas generally have well-developed public transportation systems that offer attractive alternatives to commuting by car.

Figure 8.16 Historical Landscapes This aerial view of Grosseto, Italy, shows how the historical medieval city was encircled by the Renaissance–Baroque fortifications, built to protect the settlement. Today parks and public buildings are located in place of the former walls and moat.

Review

8.5 Which European countries have the highest and lowest rates of natural population increase?

8.6 Which European countries have the highest rates of out-migration? Of in-migration?

8.7 What is the Schengen Agreement, and how is it related to population movement?

8.8 Discuss the reasons behind the current flood of illegal migration to Europe.

8.9 Name three stages of historical urban development still commonly found in European urban landscapes.

KEY TERMS family-friendly policies, Schengen Agreement, asylum laws, medieval landscape, Renaissance–Baroque period

Cultural Coherence and Diversity: A Mosaic of Differences

The rich cultural geography of Europe demands our attention for several reasons. First, the highly varied mosaic of languages, customs, religions, and ways of life that characterize Europe not only strongly shaped regional identities, but also often stoked the fires of conflict. Embers from those historical conflicts still smolder today in several areas.

Second, European cultures played leading roles in globalization as European colonialism brought about changes in languages, religions, political systems, economies, and social values in every corner of the globe. Examples include cricket games in Pakistan, high tea in India, Dutch architecture in South Africa, and the millions of French-speaking inhabitants of equatorial Africa.

Today, however, waves of global culture are spreading back into Europe, and while some Europeans embrace (or passively condone) these changes, others actively resist. In many ways, France, for example, struggles against both global popular culture and the multicultural influences of its large Muslim migrant population (Figure 8.17).

Figure 8.17 Muslims in Europe. Europe has long had a small Muslim population, historically in the Balkans and Spain but more recently in western Europe because of Europe's colonial ties to Asia and Africa. Moreover, the post-war guest worker program led to the creation of Turkish communities in many German cities.These two Muslim schoolgirls in Germany are of Turkish descent, and perhaps are even German citizens. Currently, nationalistic, anti-migrant groups have created concerns about the "Islamization" of Europe from the large numbers of extralegal migrants from Southwest Asia and Africa.

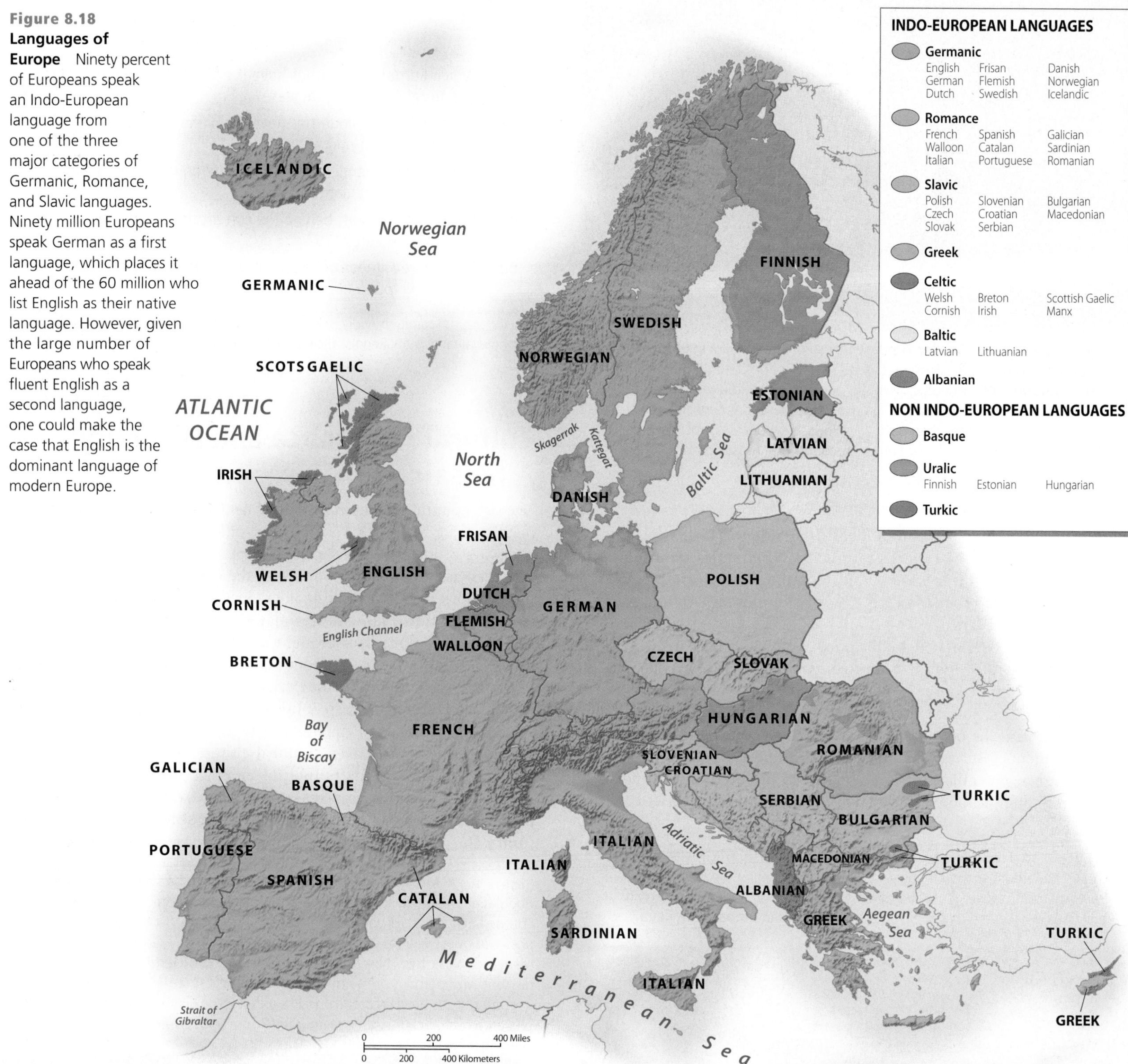

Figure 8.18 Languages of Europe Ninety percent of Europeans speak an Indo-European language from one of the three major categories of Germanic, Romance, and Slavic languages. Ninety million Europeans speak German as a first language, which places it ahead of the 60 million who list English as their native language. However, given the large number of Europeans who speak fluent English as a second language, one could make the case that English is the dominant language of modern Europe.

Geographies of Language

Language has always been an important component of nationalism and group identity in Europe. Today, while some small ethnic groups such as the Irish and the Bretons work hard to preserve their local language, millions of other Europeans are busy learning multiple languages—primarily English—so they can better communicate across cultural and national boundaries.

At the broadest scale, most Europeans speak a language of the Indo-European linguistic family; only in Finland, Estonia, and Hungary are the native languages not Indo-European (Figure 8.18). As their first language, 90 percent of Europe's population speaks a Germanic, Romance, or Slavic language, all of which are linguistic groups within the Indo-European family. Germanic and Romance speakers each number almost 200 million in the European region.

EVERYDAY GLOBALIZATION

English, Europe's New Second Language

English is Europe's new lingua franca, with two-thirds of the region's population having at least a decent working knowledge of the language. This fairly recent development is due to the EU recommending compulsory foreign language study in primary and secondary schools. As a result, English has replaced French, German, and Russian in the race to become Europe's common language (Figure 8.3.1).

A recent EU report found that 94 percent of secondary school pupils and 83 percent of primary school pupils are learning English as their first foreign language. Only in Britain and Ireland is French the top foreign language in schools.

Ironically, English is becoming increasingly popular throughout Europe at a time when the United Kingdom is considering withdrawal from the EU. This prospect would leave the EU with a common language spoken as a native language only by the 4.6 million people of the Irish Republic. And English is only one of the country's two official languages, Irish being the other.

Just so you know, Danes are the most proficient in English (94 percent) and Italians the least, with just 10 percent considering themselves proficient.

1. In what other world regions is English the second language? In regions where English is not the second language, what is?
2. What foreign language have you studied? Have you used it in a foreign country?

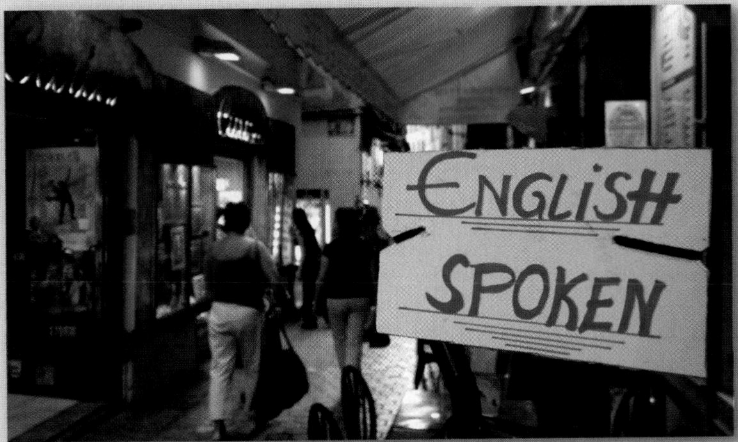

Figure 8.3.1 English Spoken Not only does this sign outside a French store in Paris assure native English speakers that they can do business comfortably in their language but also invites non-French speakers to communicate in Europe's new *lingua franca*.

Although Slavic languages are spoken by 400 million when Russia and its immediate neighbors are included, there are only 80 million Slavic speakers within Europe proper.

Germanic Languages Germanic languages dominate Europe north of the Alps. Today German, claimed by about 90 million people as their mother tongue, is spoken in Germany, Austria, Liechtenstein, Luxembourg, eastern Switzerland, and several small areas in Alpine Italy.

English is the second-largest Germanic language, with about 60 million speakers learning it as their first language. In addition, a large number of Europeans learn English as a second language, with many as fluent in English as are native speakers (see *Everyday Globalization: English, Europe's New Second Language*). Linguistically, English is closest to the Low German spoken along the coastline of the North Sea, which reinforces the theory that an early form of English evolved in the British Isles through contact with the coastal peoples of northern Europe. One distinctive trait of English that sets it apart from German, however, is that almost one-third of the English vocabulary is made up of Romance words brought to England during the Norman French conquest of the 11th century.

Elsewhere in the Germanic linguistic region, Dutch (in the Netherlands) and Flemish (in northern Belgium) together account for another 20 million speakers, with roughly the same number of Scandinavians speaking the closely related languages of Danish, Norwegian, and Swedish. Icelandic is a more distinctive language because of that country's geographic isolation from its Scandinavian roots.

Romance Languages Romance languages, including French, Spanish, and Italian, evolved from the vulgar (or everyday) Latin spoken within the Roman Empire. Today Italian is the most widely used of these Romance languages, with about 60 million Europeans speaking it as their first language. In addition to being spoken in Italy, Italian is an official language of Switzerland and is also spoken on the French island of Corsica.

French is spoken in France, western Switzerland, and southern Belgium (where it is known as *Walloon*). Today there are about 55 million native French speakers in Europe. As with other languages, French has very strong regional dialects.

Spanish also has very strong regional variations. About 25 million people speak Castilian Spanish, the country's official language, which dominates the interior and northern areas of that large country. However, the Catalan form, which some argue is a completely separate language, is found along the eastern coastal fringe, centered on Barcelona, Spain's second-largest city. This distinct language reinforces a strong sense of cultural separateness that has led to the state of Catalonia being given autonomous status within Spain.

Portuguese is spoken by 12 million in Portugal and in the northwestern corner of Spain, although considerably more people speak the language in Brazil, a former Portuguese colony in Latin America. Finally, Romanian represents an eastern outlier of the Romance language family, spoken by 24 million people in Romania. Though unquestionably a Romance language, Romanian also contains many Slavic words.

The Slavic Language Family Slavic speakers are traditionally separated into northern and southern groups, divided by the non-Slavic speakers of Hungary and Romania.

To the north, Polish has 35 million speakers, with Czech and Slovakian speakers totaling about 15 million. As noted earlier, these numbers pale in comparison to the number of northern Slav speakers in nearby Ukraine, Belarus, and Russia, which easily total more than 150 million. Southern Slav languages include three groups: 14 million speakers of Serbian and Croatian (these are now considered separate languages because of the strong political and cultural differences between Serbs and Croats), 11 million Bulgarian or Macedonian speakers, and 2 million Slovenian speakers.

The use of two distinct alphabets further complicates the geography of Slavic languages. In countries with a strong Roman Catholic heritage, such as Poland and the Czech Republic, the Latin alphabet is used. In contrast, countries with close ties to the Orthodox Church—Bulgaria, Montenegro, Macedonia, parts of Bosnia–Herzegovina, and Serbia—use the Greek-derived **Cyrillic alphabet** (Figure 8.19).

Figure 8.19 Cyrillic Alphabet A directional sign in downtown Sofia, Bulgaria, uses both the Cyrillic and the Roman alphabets to guide locals and visitors alike.

Geographies of Religion, Past and Present

Religion is an important component of the geography of cultural coherence and diversity in Europe because many of today's ethnic tensions result from historical religious events. To illustrate, significant cultural borders in the Balkans and eastern Europe are based on the 11th-century split of Christianity into eastern and western churches as well as on the division between Christianity and Islam. Much of the ethnic-cleansing terrorism in the former Yugoslavia during the 1990s was based on these religious differences.

In western Europe, blood is still occasionally shed in Northern Ireland over the tensions resulting from the 17th-century split of Christianity into Catholicism and Protestantism. Understanding these contemporary tensions involves taking a brief look at the historical geography of Europe's religions (Figure 8.20).

The Schism Between Western and Eastern Christianity In southeastern Europe, early Greek missionaries spread Christianity throughout the Balkans and into the lower reaches of the Danube. Because these Greek missionaries refused to accept the control of Roman Catholic bishops in western Europe, there was a formal split with western Christianity in 1054 CE.

This eastern church subsequently splintered into Orthodox sects closely linked to specific nations and states. Today we find Greek Orthodox, Bulgarian Orthodox, and Russian Orthodox churches, all with slightly different rites and rituals.

The Protestant Revolt Besides the division between western and eastern churches, the other great split within Christianity occurred between Catholicism and Protestantism. This division arose in Europe during the 16th century and has divided the region ever since. However, with the exception of "the Troubles" in Northern Ireland, tensions today between these two major groups are far less problematic than in the past, when these religious differences led to several long wars.

Historical Conflicts with Islam Both the eastern and western Christian churches struggled with challenges from the Islamic empires to Europe's south and east. Even though historical Islam was reasonably tolerant of Christianity in its conquered lands, Christian Europe was far less accepting of Muslim imperialism. The first crusade to reclaim Jerusalem from the Turks took place in 1095. After the Ottoman Turks conquered Constantinople in 1453 and gained control over the Bosporus Strait and the Black Sea, they moved rapidly to spread a Muslim empire throughout the Balkans, arriving at the gates of Vienna in the middle of the 16th century. There Christian Europe stood firm militarily and stopped Islam from expanding into western Europe.

Ottoman control of southeastern Europe, however, lasted until the empire's demise in the early 20th century. This historical presence of Islam explains the current coexistence of religions in the Balkans, with intermixed areas of Muslims, Orthodox Christians, and Roman Catholics.

Islam was also the dominant religion and culture of Portugal and most of Spain from the 8th to the 17th century, when the Catholic kingdoms in Spain's northeast expanded their control of the Iberian Peninsula.

A Geography of Judaism Europe has long been a difficult home for the Jews forced to leave Palestine during the Roman Empire. At that time, small Jewish settlements were located in cities throughout the Mediterranean. Later, by 900 CE, about 20 percent of the Jewish population was clustered in the Muslim lands of the Iberian Peninsula, where Islam showed greater tolerance for Judaism than did Christianity. After the Christian reconquest of Iberia, however, Jews once more faced severe persecution and fled from Spain to more tolerant countries in western and central Europe.

One focus for this exodus was the area in eastern Europe that became known as the Jewish Pale. In the late Middle Ages, at the invitation of the Kingdom of Poland, Jews settled in cities and small villages in what are now eastern Poland, Belarus, western Ukraine, and northern Romania. Jews gathered in this region for several centuries in the hope of establishing a true European homeland.

Until emigration to North America began in the 1890s, 90 percent of the world's Jewish population lived in Europe, and most were clustered in the Pale region. Tragically, Nazi Germany devastated this ethnic cluster by focusing its extermination activities on the Pale.

Figure 8.20 Religions of Europe This map shows the divide in western Europe between the Protestant north and the Roman Catholic south. Historically, this distinction was much more important than it is today. Note also the location of the former Jewish Pale, the area that was devastated by the Nazis during World War II. Today ethnic tensions with religious overtones are found primarily in the Balkans, where adherents to Roman Catholicism, Eastern Orthodoxy, and Islam are found in close proximity. **Q: After comparing this map to the one of Europe's languages (Figure 8.18). list those areas where language families and religion appear to be related.**

In 1939, on the eve of World War II, 9.5 million Jews, or about 60 percent of the world's Jewish population, lived in Europe. During the war, German Nazis murdered some 6 million Jews in the horror of the Holocaust. Today fewer than 2 million Jews, about 10 percent of the world population, live in Europe.

Patterns of Contemporary Religion Estimates of religious adherence in contemporary Europe suggest there are 250 million Roman Catholics, fewer than 100 million Protestants, and 13 million Muslims. Generally, Catholics live in the southern half of the region, except for the significant numbers in Ireland and Poland. Protestantism is most widespread in northern Germany (with Catholicism stronger in southern Germany), the Scandinavian countries, and England, and it is intermixed with Catholicism in the Netherlands, Belgium, and Switzerland. Muslims historically are found in Albania, Kosovo, Bosnia–Herzegovina, and Bulgaria. Adding to Europe's Muslim population are postwar migrants from Turkey and northern Africa, with 4.8 million Muslims today in Germany and 4.7 million in France. Between migration and a generally higher natural birth rate, combined with a stagnating number of practicing Christians, Islam is Europe's faster-growing religion.

Because of Europe's long history of religious wars and tensions, the EU's agenda of European unity is explicitly secular, a position that causes its own set of contemporary cultural tensions. For example, the euro, the EU's common currency, has purged all national symbols of Christian crosses and saints, much to the chagrin of countries like Hungary and Poland whose sense of nationalism is inseparable from Catholicism. Perhaps understandably, the EU's secularism is particularly difficult to accept in those former Soviet satellites where their churches were closed and boarded up during the communist period and have only recently reopened as places of gathering and worship.

Despite these tensions, Europe's religious landscape, with its diverse array of cathedrals, churches, monasteries, nunneries, and other religious sites and structures, is a major attraction for travelers and tourists, both international and intra-European.

Because of historical Protestantism's reaction against the ornate cathedrals and statues of the Catholic Church, the landscape of Protestantism in northern Europe is rather sedate and subdued, in contrast to the numerous religious sites of Catholic Europe. The large cathedrals and religious monuments in Britain are associated primarily with the Church of England, which originated with strong historical ties to Catholicism; St. Paul's Cathedral and Westminster Abbey in London are examples. Newer additions to Europe's religious landscape are the Muslim mosques now found in many western European cities (Figure 8.21), with many more being planned to accommodate growing numbers of adherents.

Figure 8.21 Muslim Mosque in Germany The Merkez Mosque in Duisburg, Germany, is said to be the largest mosque in a non-Muslim country. Duisburg, a city of half a million people in the Ruhr industrial area of northwestern Germany, has a population of 85,000 people of Turkish origin.

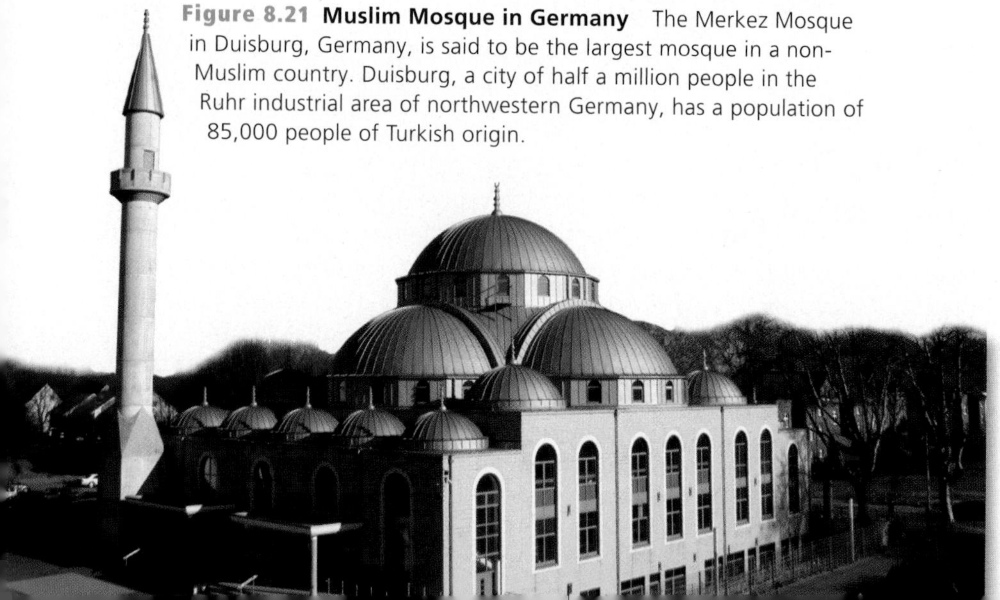

Migrants and Culture

New migration streams from Africa, Asia, and South America are profoundly influencing the dynamic cultural geography of Europe. Unfortunately, in some areas of Europe the products of this recent cultural exchange are highly troubling.

Immigrant clustering, leading to the formation of ethnic neighborhoods and even ghettos, is now common in the cities and towns of western Europe. The high-density apartment buildings of suburban Paris, for example, are home to large numbers of French-speaking Africans and Arab Muslims caught in a web of high unemployment, poverty, and racial discrimination. As a result, cultural struggles, both on the streets and in the courtrooms, are now common in many European countries. For example, in 2004 French leaders drew on the country's constitutional separation of state and religion to ban a key symbol of conservative Muslim life—the head scarf (*hijab*)—for female students in public schools because, officials argued, it interfered with the educational process. And in 2010, full-face veils were banned in public places. Another rationale for the legislation was that traditional Muslim dress inhibited the assimilation process and immigrants should blend into contemporary French society. Other European countries have made their own unique attempts at restricting Muslim dress with a portfolio of rationalizations ranging from concerns about terrorism to the unsafe driving conditions resulting from headscarves and body coverings.

What is clearly at issue here is the social unease about Muslim migrants in what until recently was a relatively homogeneous European culture. A recent Pew Foundation survey asked Europeans about their attitudes toward Muslims. Surprisingly, three-quarters of the French population actually holds favorable views of Muslims. In Germany, home to Europe's largest Muslim population, favorable views are held by 60 percent of the population. But in Spain, Greece, and Poland, less than half the population has positive attitudes toward Muslims. Interestingly, attitudes toward Muslims reflected general political orientation—people who associated with the political right were significantly more unfriendly to Muslims than those aligned with the political center or left.

This political distinction is not surprising, since far-right, neonationalistic political parties throughout Europe share anti-migrant, anti-immigration, anti-asylum positions.

Sports in Europe

Soccer (which Europeans call football) is unquestionably Europe's national sport, played everywhere from sandlots to stadiums, by both women and men, at all levels from family picnics to multiple-level professional leagues (Figure 8.22).

Figure 8.22 Women's Sports in Europe Women's soccer and basketball teams, at both the amateur and the professional levels, are common throughout Europe. Here soccer players from the German (in white) and French (in blue) national teams do battle during the recent Women's World Cup competition.

At the highest pro level, soccer teams draw crowds into stadiums holding 100,000 people. Smaller soccer stadiums seating 30,000–40,000 are common in every European town.

Like many sports throughout the world, soccer is irrevocably linked to globalized culture, with fanatical fans rooting for place-based teams constituted largely of international players lacking any local allegiance. But this contradiction doesn't keep soccer fans from taking their local fandom across Europe's borders to rival towns and cities, where team loyalties sometimes turn violent. Soccer hooliganism, unfortunately, has become a common outlet for Europe's anti-migrant racism and xenophobia.

Aside from homegrown sports like soccer and rugby, Europe has shown some interest in the North American sports of basketball, baseball, and American football. Basketball is unquestionably Europe's favorite American sport, with hoops and courts increasingly common in the region's gyms and playgrounds. Pro leagues at all levels abound for both men and women, with most European cities supporting at least one pro team. It is now common for U.S. Women's National Basketball Association (WNBA) players to spend their off-season playing for a European pro team to augment their modest WNBA salaries. Also common is the increasing number of European basketball players (both male and female) who play for North American college and pro teams (see *Everyday Globalization* in Chapter 3).

Baseball is fairly popular in Europe, having grown from seeds planted by postwar U.S. servicemen into several professional leagues. Today a handful of baseball academies have developed in Germany and France to train athletes who aspire to play Major League Baseball (MLB) in North America. Their goal is to break into MLB, just the way Latin Americans and Japanese did decades ago.

American football remains a novelty in most of Europe, even though the National Football League (NFL) usually schedules a preseason exhibition game in England or Germany. Europeans applaud politely after a touchdown, but most agree that NFL football, with its frequent timeouts, fails to capture the attention of cultures that thrive on soccer's nonstop action.

Review

8.10 Describe the general location within Europe of the three major language groups: Germanic, Romance, and Slavic.

8.11 Summarize the historical distribution within Europe of Catholicism, Protestantism, Judaism, and Islam.

8.12 Discuss the overlap between languages and religion in Europe.

8.13 Which countries have the highest numbers of Muslims? What cultural conflicts have resulted?

KEY TERMS Cyrillic alphabet

Geopolitical Framework: A Dynamic Map

One of Europe's unique characteristics is its dense fabric of 42 independent states within a relatively small area. The ideal of democratic **nation-states** (see Chapter 1 for a discussion of the nation-state concept) arose in Europe and, over time, replaced the fiefdoms and empires ruled by autocratic royalty. France, Italy, Germany, and the UK are major examples.

In many ways, however, Europe's unique geopolitical landscape has been as much problem as promise. Twice in the past century, Europe shed blood to redraw its political borders, and within the past several decades nine new states have appeared—more than half through violent wars. Today, generally speaking, most of Europe's geopolitical hotspots are more about achieving regional autonomy than creating new nation-states (Figure 8.23).

Redrawing the Map of Europe Through War

Two world wars radically reshaped the geopolitical map of 20th-century Europe (Figure 8.24). Although World War I was referred to as the "war to end all wars," it fell far short of solving Europe's geopolitical problems. Instead, according to many experts, the peace treaty actually made another world war unavoidable.

When Germany and Austria–Hungary surrendered in 1918, the Treaty of Versailles set about redrawing the map of Europe with two goals in mind: first, to punish the losers through loss of territory and severe financial reparations and, second, to recognize the nationalistic aspirations of unrepresented peoples by creating several new nation-states. As a result, the new states of Czechoslovakia and Yugoslavia were created. In addition, Poland was reestablished, as were the Baltic states of Finland, Estonia, Latvia, and Lithuania.

Though the goals of the treaty were admirable, few European states were satisfied with the resulting map. New states were resentful when some of their citizens were left outside the redrawn borders. This created an epidemic of **irredentism**, state policies directed toward reclaiming lost territory and peoples.

These imperfect geopolitical solutions were greatly aggravated by the global economic depression of the 1930s, which brought high unemployment, food shortages, and even more political unrest to Europe.

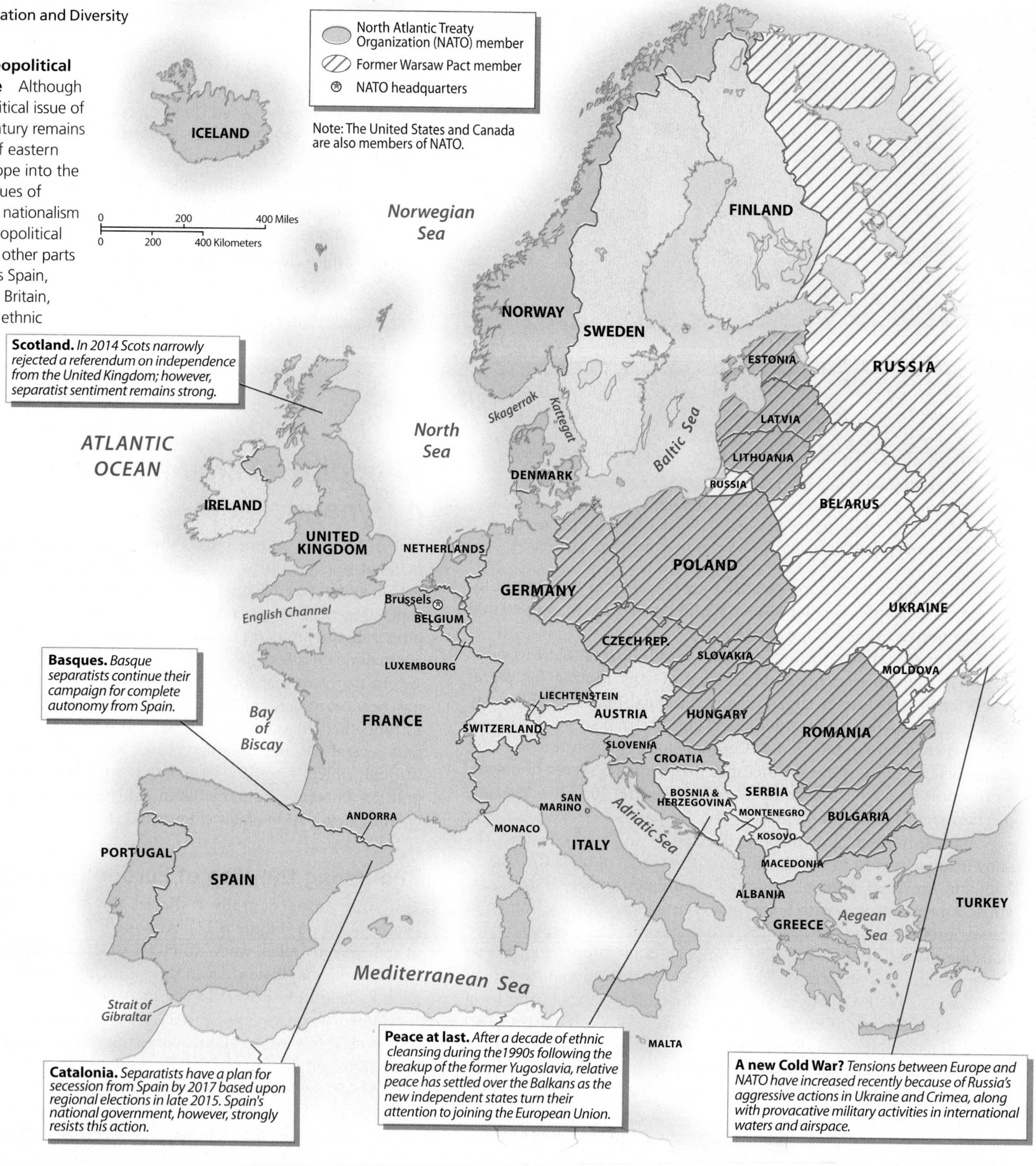
North Atlantic Treaty Organization (NATO) member
Former Warsaw Pact member
NATO headquarters
Note: The United States and Canada are also members of NATO.
0 200 400 Miles
0 200 400 Kilometers
ICELAND
Norwegian Sea
FINLAND
NORWAY
SWEDEN
ESTONIA
RUSSIA
LATVIA
LITHUANIA
RUSSIA
BELARUS
Skagerrak
Kattegat
Baltic Sea
North Sea
DENMARK
ATLANTIC OCEAN
IRELAND
UNITED KINGDOM
NETHERLANDS
GERMANY
POLAND
UKRAINE
Brussels
BELGIUM
English Channel
LUXEMBOURG
CZECH REP.
SLOVAKIA
MOLDOVA
LIECHTENSTEIN
AUSTRIA
HUNGARY
ROMANIA
FRANCE
SWITZERLAND
SLOVENIA
CROATIA
Bay of Biscay
BOSNIA & HERZEGOVINA
SERBIA
SAN MARINO
Adriatic Sea
MONTENEGRO
BULGARIA
ANDORRA
MONACO
KOSOVO
ITALY
MACEDONIA
PORTUGAL
SPAIN
ALBANIA
TURKEY
GREECE
Aegean Sea
Mediterranean Sea
Strait of Gibraltar
MALTA
Scotland. *In 2014 Scots narrowly rejected a referendum on independence from the United Kingdom; however, separatist sentiment remains strong.*
Basques. *Basque separatists continue their campaign for complete autonomy from Spain.*
Catalonia. *Separatists have a plan for secession from Spain by 2017 based upon regional elections in late 2015. Spain's national government, however, strongly resists this action.*
Peace at last. *After a decade of ethnic cleansing during the 1990s following the breakup of the former Yugoslavia, relative peace has settled over the Balkans as the new independent states turn their attention to joining the European Union.*
A new Cold War? *Tensions between Europe and NATO have increased recently because of Russia's aggressive actions in Ukraine and Crimea, along with provacative military activities in international waters and airspace.*

Figure 8.23 Geopolitical Issues in Europe Although the major geopolitical issue of the early 21st century remains the integration of eastern and western Europe into the EU, numerous issues of micro- and ethnic nationalism also engender geopolitical fragmentation. In other parts of Europe, such as Spain, France, and Great Britain, questions of local ethnic autonomy within the nation-state structure challenge central governments.

Three competing ideologies promoted their own solutions to Europe's pressing problems: Western democracy (and capitalism); communism from the Soviet revolution to the east; and a fascist totalitarianism promoted by Mussolini in Italy and Hitler in Germany (see Chapter 1 for descriptions of these political ideologies). With industrial unemployment at record rates in western Europe, public opinion fluctuated wildly between the extremist solutions of far-right fascism and far-left communism and socialism. In 1936, Italy and Germany joined forces through the Rome–Berlin "axis" agreement. As in World War I, this alignment was countered with mutual protection treaties among France, Britain, and the Soviet Union. When an imperialist Japan signed a pact with Germany, the scene was set for a second global war.

Nazi Germany tested western European resolve in 1938 by annexing Austria, the country of Hitler's birth, and then Czechoslovakia, under the pretense of providing protection for ethnic Germans located there. After Germany signed a nonaggression pact with the Soviet Union, Hitler's armies invaded Poland on September 1, 1939. Two days later, France and Britain declared war on Germany. Within a month, the Soviet Union moved into eastern Poland, the Baltic states, and Finland to reclaim territories lost through the peace treaties of World War I.

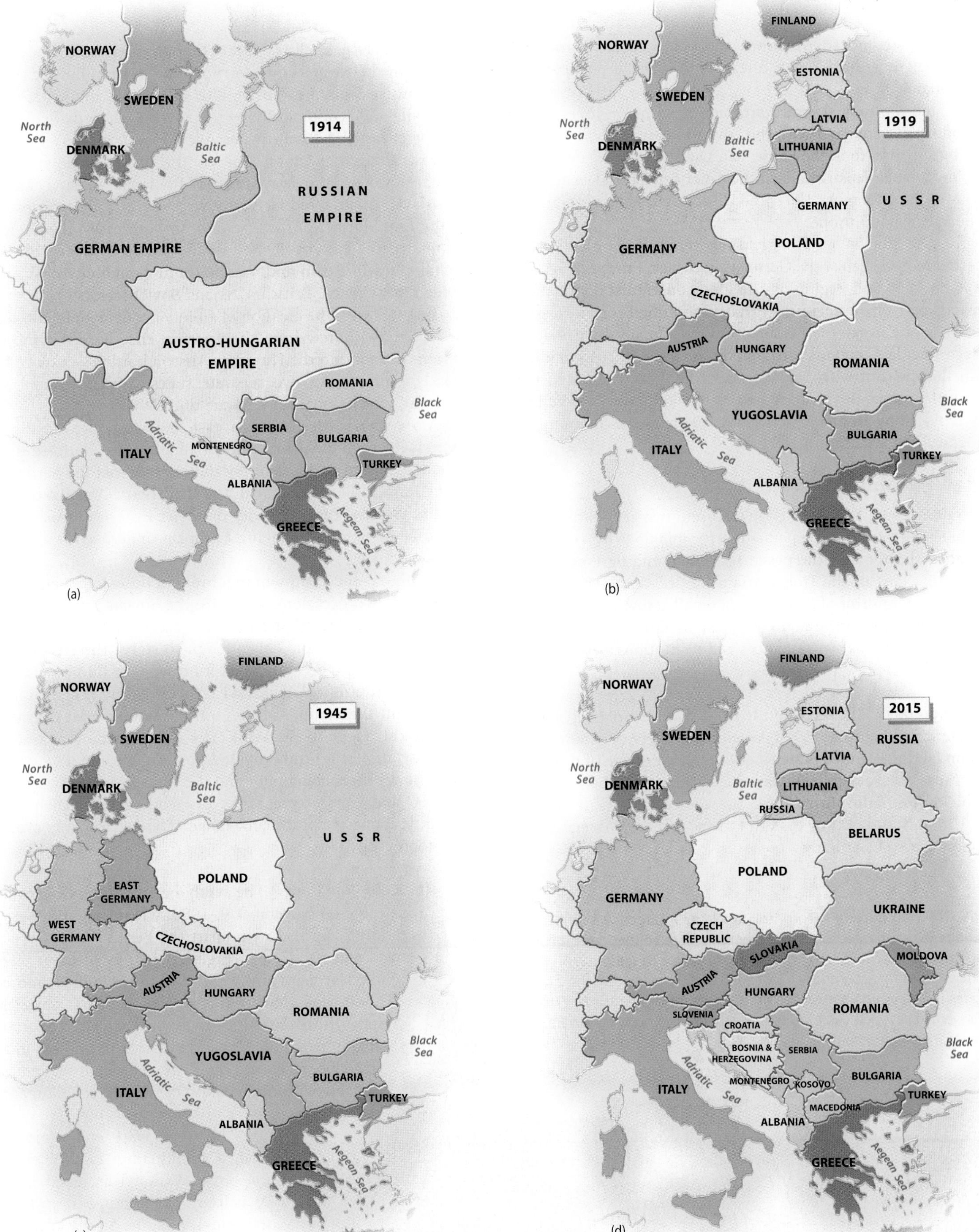

Figure 8.24 A Century of Geopolitical Change (a) At the outset of the 20th century, central Europe was dominated by the German, Austro-Hungarian (or Hapsburg), and Russian empires. (b) Following World War I, these empires were largely replaced by a mosaic of nation-states. (c) More border changes followed World War II, largely as a result of the Soviet Union's turning the area into a buffer zone between itself and western Europe. (d) With the demise of Soviet hegemony in 1990, further political change took place. **Q: Where are the strongest relationships between political change and cultural factors such as language and religion?**

Nazi Germany then moved westward and occupied Denmark, the Netherlands, Belgium, and France, after which it began preparations to invade England.

In 1941, the war took several startling new turns. In June, Hitler broke the nonaggression pact with the Soviet Union and, catching its Red Army by surprise, took the Baltic states and then drove deep into Soviet territory. When Japan attacked the American naval fleet at Pearl Harbor, Hawaii, in December 1941, the United States entered the war in both the Pacific and Europe.

By early 1944, the Soviet army had recovered most of its territorial losses and moved against the Germans in eastern Europe, reaching Berlin in April 1945 and beginning the long communist domination of the eastern part of the region. At that time, Allied forces crossed the Rhine River and began their occupation of Germany. Immediately after Hitler's suicide, Germany surrendered on May 8, 1945, ending the war in Europe. But with Soviet forces firmly entrenched in the eastern part of Europe, the military battles of World War II were immediately replaced by an ideological **Cold War** between communism and democracy that lasted until 1991.

A Divided Europe, East and West

From 1945 until 1991, Europe was divided into two geopolitical and economic blocs, east and west, separated by the infamous **Iron Curtain** that descended shortly after the peace agreement ending World War II (Figure 8.25). East of the Iron Curtain border, the Soviet Union imposed the heavy imprint of communism on all activities—political, economic, military, and cultural. To the west, as Europe rebuilt from the destruction of the war, new alliances and institutions were created to counter the Soviet presence in Europe.

Cold War Geography The seeds of the Cold War were planted at the Yalta Conference of February 1945, when the leaders of Britain, the Soviet Union, and the United States met to plan the shape of postwar Europe. Because the Soviet army was already in eastern Europe and moving quickly on Berlin, Britain and the United States agreed that the Soviet Union would occupy eastern Europe and the Western allies would occupy parts of Germany.

Figure 8.25 The Iron Curtain From 1945 until 1989, Europe was divided politically and physically by the Iron Curtain, which separated the Soviet Union satellite countries of eastern Europe from western Europe. This photo is of the border dividing the former countries of East and West Germany near the city of Vacha. Besides the Iron Curtain itself, the border zone on the eastern side was commonly several miles wide and included military fortifications with severe restrictions on civilian movement.

The larger geopolitical issue, though, was the Soviet desire for a **buffer zone** between its own territory and western Europe. This buffer zone consisted of an extensive bloc of satellite countries, dominated politically and economically by the Soviet Union, that could cushion the Soviet heartland against possible attack from western Europe. In the east, the Soviet Union took control of the Baltic states, Poland, Czechoslovakia, Hungary, Bulgaria, Romania, Albania, and, briefly, Yugoslavia. Austria and Germany were divided into occupied sectors by the four (former) Allied powers. In both cases, the Soviet Union dominated the eastern portion of the country, which contained the capital cities of Berlin and Vienna. Both capital cities, in turn, were divided into French, British, U.S., and Soviet sectors.

In 1955, with the creation of an independent and neutral Austria, the Soviets withdrew from their sector, effectively moving the Iron Curtain eastward to the Hungary–Austria border. Germany, however, quickly evolved into two separate states, West Germany and East Germany, which remained separate until 1990.

Along the border between east and west, two hostile military forces faced each other for almost half a century. Both sides prepared for and expected an invasion by the other across the barbed wire dividing Europe. Both the **North Atlantic Treaty Organization (NATO)** and the **Warsaw Pact** countries of Soviet eastern Europe were armed with nuclear weapons, making Europe a tinderbox for a nightmarish third world war.

Berlin was the flashpoint that brought these forces close to a fighting war on two occasions. In 1948, the Soviets imposed a blockade on the city, denying Western powers access to Berlin across its East German military sector. This attempt to starve the city into submission by blocking food shipments from western Europe was thwarted by a nonstop airlift of food and coal by NATO. Then in 1961, the Soviets built the Berlin Wall to curb the flow of East Germans seeking political refuge in the West. The wall became the concrete-and-mortar symbol of a firmly divided postwar Europe. For several days while the wall was being built and the West agonized over destroying it, NATO and Warsaw Pact tanks and soldiers faced each other with loaded weapons at point-blank range. Though war was avoided, the wall stood until 1989.

The Cold War Thaw The symbolic end of the Cold War in Europe came on November 9, 1989, when East and West Berliners joined forces to rip apart the Berlin Wall with jackhammers and hand tools (Figure 8.26). By October 1990, East and West Germany were officially reunified into a single nation-state. During this period, all other Soviet satellite states, from the Baltic Sea to the Black Sea, also underwent major geopolitical changes that have resulted in a mixed bag of benefits and problems. The Cold War ended completely with the breakup of the Soviet Union at the end of 1991.

The Cold War's end came as much from a combination of problems within the Soviet Union (discussed in Chapter 9) as from rebellion in eastern Europe. By the mid-1980s, the Soviet leadership had advocated for an internal economic restructuring and also recognized the need for a more open dialogue with the West. More recently, unfortunately, the era of

(a) Berlin Wall, 1961

(b) Berlin Wall, 1989

Figure 8.26 The Berlin Wall In August 1961, East Germany built a concrete and barbed wire structure along the border of East Berlin to stem the flow of refugees fleeing communist rule. The wall was the most visible symbol of the Cold War division between east and west until November 1989, when the failing Soviet Union renounced its control over eastern Europe. (a) The extent of the wall zone at the Brandenburg Gate (East Berlin is to the left); (b) Berliners celebrate the end of the wall with East German police, who previously guarded the border zone with shoot-to-kill orders.

cooperation with the West seems to have passed (see *Exploring Global Connections: The New Cold War*).

As a result of the earlier Cold War thaw, the map of Europe changed once again, with the unification of Germany, the peaceful "Velvet Divorce" of Czechs and Slovaks, and the reemergence of the Baltic states. More troublesome was the Balkan area, where the former Yugoslavia violently fractured into a handful of independent states.

The Balkans: Waking from a Geopolitical Nightmare

The Balkans have long been a troubled area with their complex mixture of languages, religions, and ethnic allegiances. Throughout history, these allegiances have led to an often-changing geography of small countries (Figure 8.27). Indeed, the term **balkanization** is used to describe the geopolitical processes of small-scale independence movements based on ethnic fault lines.

Following the fall of the Austro-Hungarian and Ottoman empires in the early 20th century, much of the region was unified under the political umbrella of the former Yugoslavian state. In the 1990s, however, Yugoslavia broke apart as ethnic factionalism and nationalism produced a decade of violence, turmoil, and wars of independence, creating a geopolitical nightmare for Europe, the EU, NATO, and the world. Today, though, despite lingering tensions in several areas, there are signs that the Balkan countries are moving toward a new era of peace and stability. Two Balkan countries, Slovenia and Croatia, are EU members, and several others are candidates for EU membership.

Balkan Wars of Independence In 1990, elections were held in Yugoslavia's different republics over the issue of secession from the mother state. Secessionist parties gained control in Slovenia and Croatia, but Serbian voters opted for continued Yugoslav unification, in what observers considered to be a government-controlled election. Nonetheless, Slovenia and Croatia declared independence in 1991 and Macedonia and Bosnia–Herzegovina in April 1992. When the Yugoslavian army attacked Slovenia, the Balkan situation got Europe's full attention, and a negotiated settlement resulted in Slovenia's independence. In Bosnia–Herzegovina, however, Serb paramilitary units waged a ferocious war of ethnic cleansing against both Muslims and Croats in a war that lasted until 1995. At that time, a complex political arrangement created a Serb republic and a Muslim–Croat federation, both ruled by the same legislature and president. Croatia also fought a devastating but successful war of independence against Serbian nationalists.

Kosovo, in the south of Serbia (as the former Yugoslavia was now called), was another trouble spot, with long-standing tensions between Serbs and Muslims. Although Kosovo had enjoyed differing degrees of autonomy within the former Yugoslavia, this autonomy was withdrawn by Belgrade in 1990 to protect the Serb minority population. Not surprisingly, the Muslim Kosovar rebels responded by proclaiming Kosovo's independence in 1991. This act was resisted vigorously by Serbia, which responded with a violent ethnic-cleansing program designed to oust the Muslims and make Kosovo a pure Serbian province. As warfare escalated in Kosovo, NATO (which included the United States) began bombing Belgrade in 1999 to force Serbia to accept a negotiated settlement. From 1999 until 2008, Kosovo was administered by the UN as a protectorate, an arrangement enforced by some 50,000 peacekeepers from 30 different countries.

Although Serbia remains steadfast about reclaiming Kosovo and only grudgingly recognizes its independence, in most other matters its government is more moderate and less nationalistic. This has led to Serbia's being reinstated in the UN and the Council of Europe and becoming an official EU candidate.

Devolution in Contemporary Europe

As noted in Chapter 1, the concept of geopolitical **devolution** refers to a decentralization of power away from a central authority. This takes many forms in Europe, ranging from a central government sharing power with small states within a larger union to full-on calls for separatism and independence from a larger political entity, commonly by

EXPLORING GLOBAL CONNECTIONS

The New Cold War

Google Earth (MG) Virtual Tour Video
http://goo.gl/GOP0Z7

Connect the dots: Russia annexes Crimea, and U.S. fighter jets appear in Lithuania; Norway sells a former naval base to the highest bidder to be used for arctic exploration, and Russian submarines move in; Lithuania wants to reduce its electricity costs by building an underwater cable to Sweden, and the Russian navy appears in the Baltic; Greece can't pay its fiscal bailout debt to Germany, and Russian banks offer to help; Britain protests the Russian army's meddling in Ukraine, and Russian bombers hold "exercises" off England's coast; fictitious Twitter reports warning Louisiana residents to shelter in place because of a huge chemical spill are tied to Russian hackers in St. Petersburg; Swiss authorities arrest corrupt world soccer executives, and the Kremlin accuses the United States of plotting to discredit Russia's hosting of the 2018 World Cup.

What's going on? It's the new Cold War between Russia and the West, a possible return to the scary days of the 1960s when the world's superpowers pushed each other to the brink of global destruction.

Expansion of NATO After the thawing of the first Cold War that saw several decades of demilitarization in Europe and the former Soviet Union, tensions between Russia and the West increased considerably from 2009 to 2013 as nine former Soviet satellite countries (Poland, the Czech Republic, Slovakia, Hungary, Bulgaria, Romania, Lithuania, Latvia, and Estonia) joined the EU. Even more distressing to Russia was the fact that all nine of these former Warsaw Pact countries joined the North Atlantic Treaty Organization (NATO), the international military organization created in 1949 to contain Soviet ambitions in Europe. Instead of Russia's longed-for buffer zone between the West and its heartland, it now shared a border with three Baltic NATO countries (Estonia, Latvia, and Lithuania) and one formerly trusted but now West-leaning ally, Ukraine. Even though there had been considerable cooperation and even joint military exercises between NATO and Russia in previous decades, this interaction was "suspended" in April 2014 in response to the Russian army's intervention in Ukraine (see Chapter 9).

Figure 8.4.1 Russian Bomber in British Airspace Russia seems increasingly willing to test NATO's European defenses by flying its airplanes into national airspace and sending Russian ships into national coastal waters. Here a British fighter plane escorts a Russian bomber out of British airspace.

Recent Incidents Since then, Cold War 2.0 has escalated, with "dangerous or sensitive" incidents reaching levels not seen since the 1960s; Russian aircraft are testing western Europe's air defenses (Figure 8.4.1), and the Russian navy is asserting its rights in international waters off Norway's coast and in the Baltic Sea. In November 2015 NATO member country Turkey shoots down a Russian bomber for ostensibly violating its national airspace; then, following strong Russia protests, Macedonia and several other Balkan countries are invited to become new NATO members. In the Baltic Sea Russian ships assemble in the construction area of the new underwater cable between Lithuania and Sweden, threatening the cable by "inadvertently" dragging ship anchors across the area. That Lithuania is building this new link to Sweden's hydropower while cutting back on its consumption of Russian natural gas is indicative of the many different aspects of Cold War 2.0.

1. In what European countries or areas would the people in the street be affected by the new Cold War? In what ways?
2. In what other world regions are there expressions of the new Cold War?

minority groups having their own mother tongue or distinctive cultural identity who reside within a defined part of a country. Germany and France, which share power between their national government and regional governments (in France) and *Länder* (the German equivalent to U.S. states, in Germany), illustrate one end of the spectrum. Spain is another example of power sharing; its 1978 constitution recognizes 17 autonomous communities within the nation-state.

At the extreme other end of the devolution spectrum was Scotland's historic 2014 referendum on independence from the United Kingdom. In this election, Scottish voters were asked simply to vote yes or no on the question "Should Scotland be an independent country?" (Figure 8.28). With the highest voter turnout in the UK's long election history, the no vote won by 55 percent. Notable is that the opposition to Scottish independence received a last-minute boost from London when all three major UK parties pledged to devolve "extensive new powers" to the Scottish Parliament by the end of 2015. In Spain, Catalonia—which accounts for 19 percent of the country's economy—is also calling for a referendum on independence, a move opposed by the Spanish government.

Different, but similarly complex is the movement in many European countries to withdraw—or at least regain powers—from the European Union. The current UK prime minister, David Cameron, for example, was reelected on a platform that included a referendum on EU membership by the end of 2017. While EU politicians are understandably against UK withdrawal (as is the Obama administration), just about every EU member country has an anti-EU party of some sort—usually associated with ultraconservative ideologies, such as France's anti-immigrant National Front. The shared complaint is that local and national power has been usurped by the EU's agenda of European unity and standardization.

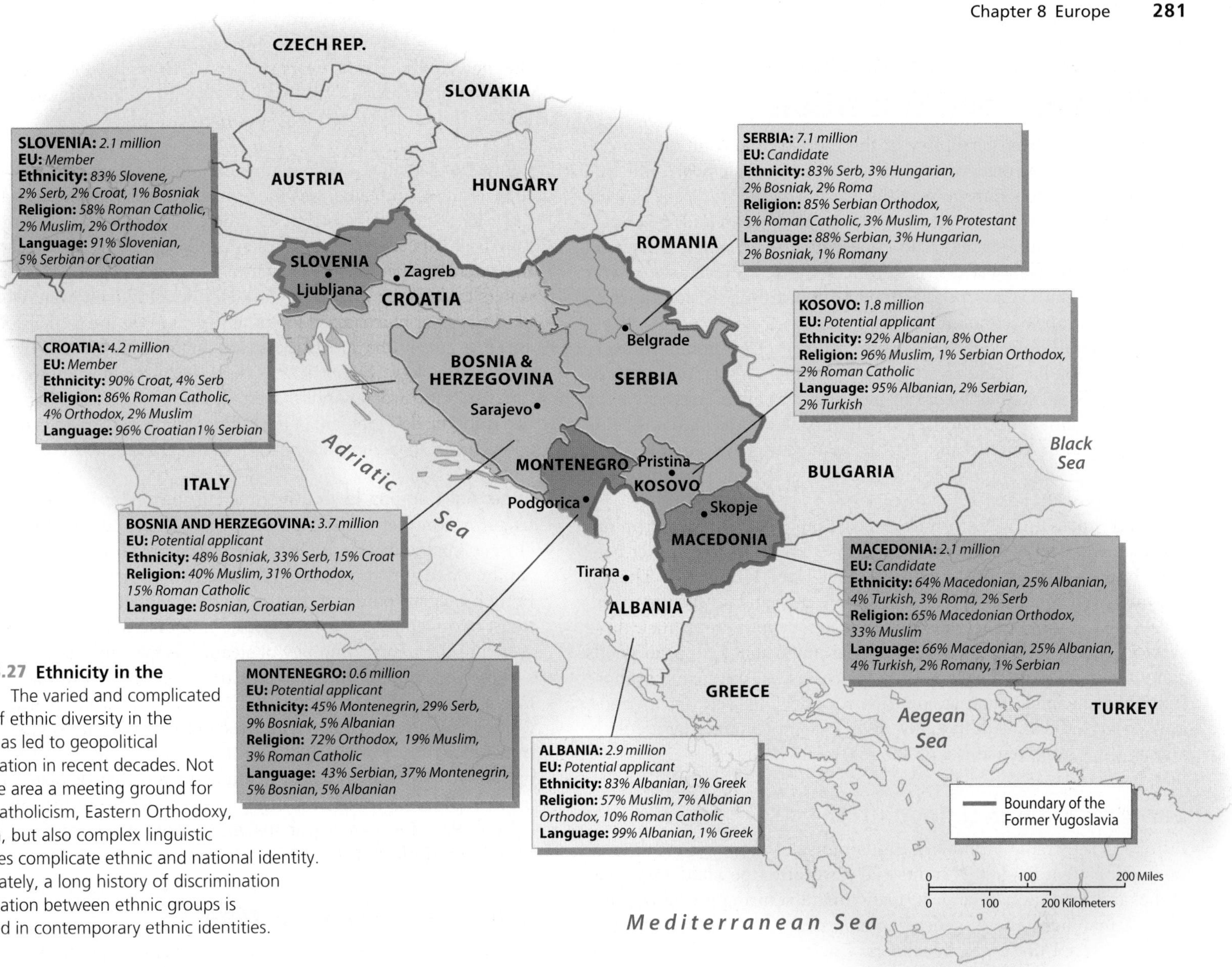

CZECH REP.
SLOVAKIA
AUSTRIA
HUNGARY
ROMANIA
SLOVENIA
Ljubljana
Zagreb
CROATIA
BOSNIA & HERZEGOVINA
Sarajevo
Belgrade
SERBIA
MONTENEGRO
Podgorica
Pristina
KOSOVO
Skopje
MACEDONIA
Tirana
ALBANIA
BULGARIA
ITALY
GREECE
TURKEY
Adriatic Sea
Black Sea
Aegean Sea
Mediterranean Sea
Boundary of the Former Yugoslavia
0 100 200 Miles
0 100 200 Kilometers

SLOVENIA: *2.1 million*
EU: *Member*
Ethnicity: *83% Slovene, 2% Serb, 2% Croat, 1% Bosniak*
Religion: *58% Roman Catholic, 2% Muslim, 2% Orthodox*
Language: *91% Slovenian, 5% Serbian or Croatian*

CROATIA: *4.2 million*
EU: *Member*
Ethnicity: *90% Croat, 4% Serb*
Religion: *86% Roman Catholic, 4% Orthodox, 2% Muslim*
Language: *96% Croatian1% Serbian*

BOSNIA AND HERZEGOVINA: *3.7 million*
EU: *Potential applicant*
Ethnicity: *48% Bosniak, 33% Serb, 15% Croat*
Religion: *40% Muslim, 31% Orthodox, 15% Roman Catholic*
Language: *Bosnian, Croatian, Serbian*

MONTENEGRO: *0.6 million*
EU: *Potential applicant*
Ethnicity: *45% Montenegrin, 29% Serb, 9% Bosniak, 5% Albanian*
Religion: *72% Orthodox, 19% Muslim, 3% Roman Catholic*
Language: *43% Serbian, 37% Montenegrin, 5% Bosnian, 5% Albanian*

SERBIA: *7.1 million*
EU: *Candidate*
Ethnicity: *83% Serb, 3% Hungarian, 2% Bosniak, 2% Roma*
Religion: *85% Serbian Orthodox, 5% Roman Catholic, 3% Muslim, 1% Protestant*
Language: *88% Serbian, 3% Hungarian, 2% Bosniak, 1% Romany*

KOSOVO: *1.8 million*
EU: *Potential applicant*
Ethnicity: *92% Albanian, 8% Other*
Religion: *96% Muslim, 1% Serbian Orthodox, 2% Roman Catholic*
Language: *95% Albanian, 2% Serbian, 2% Turkish*

MACEDONIA: *2.1 million*
EU: *Candidate*
Ethnicity: *64% Macedonian, 25% Albanian, 4% Turkish, 3% Roma, 2% Serb*
Religion: *65% Macedonian Orthodox, 33% Muslim*
Language: *66% Macedonian, 25% Albanian, 4% Turkish, 2% Romany, 1% Serbian*

ALBANIA: *2.9 million*
EU: *Potential applicant*
Ethnicity: *83% Albanian, 1% Greek*
Religion: *57% Muslim, 7% Albanian Orthodox, 10% Roman Catholic*
Language: *99% Albanian, 1% Greek*

Figure 8.27 Ethnicity in the Balkans The varied and complicated pattern of ethnic diversity in the Balkans has led to geopolitical fragmentation in recent decades. Not only is the area a meeting ground for Roman Catholicism, Eastern Orthodoxy, and Islam, but also complex linguistic boundaries complicate ethnic and national identity. Unfortunately, a long history of discrimination and retaliation between ethnic groups is embedded in contemporary ethnic identities.

Figure 8.28 Scotland's Independence Vote. Although there are many different separatist movements throughout Europe, Scotland took the issue much further in September 2014 by having its citizens vote on independence from the United Kingdom. Despite 55 percent of the electorate voting against independence, the movement for complete separation from England remains strong. In this photo a woman waves the Scottish flag, referred to as the saltire, in support of the vote for independence.

Review

8.14 Describe briefly how the map of Europe changed with the Treaty of Versailles in 1918.

8.15 What European countries were considered Soviet satellites during the Cold War?

8.16 What countries made up the former Yugoslavia?

8.17 Discuss, with examples, the different forms of political devolution in contemporary Europe.

KEY TERMS nation-state, irredentism, Cold War, Iron Curtain, buffer zone, North Atlantic Treaty Organization (NATO), Warsaw Pact, balkanization, devolution

Economic and Social Development: Integration and Transition

As the acknowledged birthplace of the Industrial Revolution, Europe in many ways invented the modern economic system of industrial capitalism. Though Europe was the world's industrial leader in the early 20th century, it was later eclipsed by both Japan and the United States as the region struggled to cope with the effects of two world wars, a decade of global depression, and the more recent Cold War and its aftermath. Currently, a drawn-out fiscal crisis continues to challenge Europe's economic structure.

In general, however, the last half-century of economic recovery and integration has been largely successful (Table A8.2). In fact, western Europe's success at blending national economies has given the world a new model for regional cooperation, an approach that is being imitated in Latin America and Asia. Eastern Europe has fared less well because the results of four decades of Soviet economic planning were, at best, mixed.

Accompanying western Europe's economic boom has been an unprecedented level of social development as measured by worker benefits, health services, education, and literacy (see Table A8.2). Though the improved social services set an admirable standard for the world, today cost-cutting politicians and businesspeople argue that these services increase the cost of business so much that European goods cannot compete in the global marketplace. As a result, many of those traditional benefits, such as job security and long vacation periods, have been eroded. Additionally, high unemployment, particularly among the young, remains a problem.

Europe's Industrial Revolution

Europe is the cradle of modern industrialism, with two fundamental innovations that made this **Industrial Revolution** possible: First, machines replaced human labor in many manufacturing processes, and, second, inanimate energy sources (water, steam, electricity, and petroleum) powered these new machines. England was the birthplace of this new system in the years between 1730 and 1850, but by the late 19th century this new industrialism had spread throughout Europe and, within decades, to the rest of the world.

Figure 8.29 Hearth Area of the Industrial Revolution Europe's Industrial Revolution began on the flanks of England's Pennine Mountains, where swift-running streams were used to power mechanized looms to weave cotton and wool. Later, once the railroads were developed, many of these early factories switched to coal power.

Centers of Change England's textile industry, located on the flanks of the Pennine Mountains, was the center of the earliest industrial innovation. The county of Yorkshire, on the eastern side of the Pennines, had long been a hearth area for woolen textiles, drawing raw materials from the extensive sheep herds of that region and using the clean mountain waters to wash the wool before it was spun. Originally, waterwheels were used to power mechanized looms at the rapids and waterfalls of Pennine streams, but by the 1790s the steam engine had become the preferred source of energy (Figure 8.29). These steam engines, however, needed fuel, and local wood supplies were quickly exhausted. Progress was stifled until the development of the railroad in the early 19th century, which allowed coal to be moved long distances at a reasonable cost.

Industrial Regions in Continental Europe The first industrial districts in continental Europe began appearing in the 1820s, located close to coalfields (Figure 8.30). The first area outside Britain was the Sambre-Meuse region, named for the two river valleys straddling the French-Belgian border. Like the English Midlands, it had a long history of cottage-based woolen textile manufacturing that quickly converted to the new technology of steam-powered mechanized looms.

By 1850, the dominant industrial area in all Europe (including England) was the Ruhr district in northwestern Germany, near the Rhine River. Rich coal deposits close to the surface powered the Ruhr's transformation from a small textile region to one of heavy industry, particularly of iron and steel manufacturing. Decades later, the Ruhr industrial region became synonymous with the industrial strength behind Nazi Germany's war machine, resulting in its being bombed heavily in World War II.

Rebuilding Postwar Europe

The leader of the industrial world in 1900, Europe produced 90 percent of the world's manufactured output. However, by 1945, after four decades of war and economic chaos, industrial Europe was in shambles, with many of its cities and industrial areas in ruins and much of the region's population dispirited, homeless, and hungry. Clearly, a new pathway for postwar Europe had to be forged to provide economic, political, and social security.

Evolution of the EU In 1950, the leaders of western Europe began discussing a new form of economic integration that would avoid the historical pattern of nationalistic independence that duplicated industrial effort. Robert Schuman, France's farsighted foreign minister, proposed the radical idea that instead of rebuilding separate iron and steel facilities in each country, Europe should share its natural resources. In 1952, France, Germany, Italy, the Netherlands, Belgium, and Luxembourg ratified a treaty that joined them together in the European Coal and Steel Community (ECSC). Five years later, because of the resounding success of the ECSC, these six states agreed to work toward further integration by creating a larger European common market that would encourage the free movement of goods, labor, and capital. In 1957, the Treaty of Rome was signed, establishing the European Economic Community (EEC).

Figure 8.30 Industrial Regions of Europe From England, the Industrial Revolution spread to continental Europe, starting with the Sambre-Meuse region on the French-Belgian border and then diffusing to the Ruhr area in Germany. Readily accessible surface coal deposits powered these new industrial areas. Early on, iron ore for steel manufacture came from local deposits, but later ore was imported from Sweden and other areas in the shield country of Scandinavia. Today, because of the dense truck, railroad, and barge transportation network, the newer industrial areas are closely linked to urban areas where access to skilled labor, rather than raw materials, is a major factor.

Europe expanded its goals once again and in 1991 became the European Union (EU), at which time the supranational organization moved even further into the region's affairs by, for example, establishing regulations on agricultural products and pricing. In 2004, the EU expanded beyond its core membership in western Europe by adding 10 new states, including a cluster of former Soviet-controlled communist satellites from eastern Europe. These new members—Latvia, Estonia, Lithuania, Poland, Slovakia, the Czech Republic, Hungary, Slovenia, Malta, and Cyprus—brought the total to 25. Bulgaria and Romania were admitted in 2007 and Croatia in 2013, resulting in the current 28 EU countries. Iceland, Serbia, Montenegro, Macedonia, Albania, and Turkey have all been accepted as official applicants to the EU, while Kosovo and Bosnia-Herzegovina are potential candidates as of 2015 (Figure 8.31).

Economic Disintegration and Transition in Eastern Europe

Eastern Europe has historically been less developed economically than its western counterpart. This is partially explained by the fact that this region is simply not as rich in natural resources as western Europe

Figure 8.31 The European Union The driving force behind Europe's economic and political integration has been the EU, which was formed in the 1950s as an organization with six members focused solely on rebuilding the region's coal and steel industries. As of 2015, the EU had 28 members. Besides the official applicants of Turkey, Serbia, Macedonia, and Iceland, several Balkan countries are preparing applications to the EU. Note that Norway is not a member of the EU, primarily because membership would inhibit that country's fishing industry. **Q: Why is Switzerland not a member of the EU?**

and what resources there are have long been exploited by outside interests rather than being developed internally. This pattern began in the 19th century, when the Ottoman and Hapsburg empires dominated eastern Europe and the Balkans. Later, eastern Europe was exploited by Nazi Germany and, more recently, by the Soviet Union's postwar centralized planning.

Even though Soviet economic planning was ostensibly an attempt to develop eastern Europe's economy by coordinating resource usage, these efforts were, in fact, designed to serve Soviet homeland interests. This system worked with mixed results for more than 40 years; however, eastern European countries were plunged into economic and social chaos when the Soviet Union itself collapsed in 1991. Recovery and development since that time have been difficult, with some countries making the transition more rapidly and more fully than others, resulting in a geographic patchwork of wealth and hardship throughout eastern Europe (Figure 8.32).

Change Since 1991 In place of Soviet coordination and subsidies came a painful period of economic transition that was outright chaotic in many eastern European countries. As the Soviet Union turned its attention to its own economic and political turmoil, it stopped exporting cheap natural gas and petroleum to eastern Europe. Instead, Russia sold these fuels on the open global market to gain hard currency. Without cheap energy, many eastern European industries could not operate and were forced to close, idling millions of workers. For example, in the first two years of the transition (1990–1992), industrial production fell 35 percent in Poland and 45 percent in Bulgaria. In addition, the guaranteed Soviet markets for eastern European products simply evaporated, further aggravating the collapse of eastern European economies.

To recover, these countries began redirecting their economies toward western Europe. But doing this meant moving from a socialist-based economy of state ownership and control to a capitalist economy of private ownership and free markets. Without Soviet price supports

(a)

(b)

Figure 8.32 The Old and the New in Poland (a) Warsaw's Old Town, on the banks of the Vistula River, was originally established in the 13th century. Today it's not only a symbol of historical Poland but also testimony to the country's pride and resiliency. In 1938 the Old Town was destroyed by Nazi German bombers during Hitler's invasion of Poland. Polish citizens rebuilt the Old Town as a symbol of resistance, but then it was systematically destroyed once again by the Nazis after the Warsaw Uprising of 1944. Most recently the Old Town was rebuilt during the post-war period and became a World Heritage site in 1994. (b) Ten minutes walking distance from Warsaw's Old Town is the country's new commercial hub in Warsaw's new downtown. On the right is the Warsaw Financial Center, built in 1999; in the center is the 48-floor hotel Warsaw Intercontinental; and on the left is Zlota 44, a 54-floor residential building completed in 2014. Three other major skyscrapers are currently under construction in the downtown area, indicative of economic vitality in this former Soviet satellite country.

and subsidies, the countries of eastern Europe had to construct completely new economic systems, ones that could compete favorably in the new global marketplace. While some countries—Poland, the Czech Republic, Slovenia, and Slovakia—made the transition quickly, others—primarily the Balkan states—are taking longer (Figure 8.33).

Promise and Problems of the Eurozone

A traditional aspect of a country's sovereignty is the ability to control its own monetary system, and indeed this was the case in Europe until two decades ago. During the 20th century, the currency of Germany was deutschmarks, with francs in France, lira in Italy, pesetas in Spain, and so on. Today, however, most European countries have replaced national currencies with a common currency, the *euro*. This shift began in 1999 when 11 of the then 15 EU member states joined together to form the **Economic and Monetary Union (EMU)**. In 2002, new euro coins and bills replaced national currencies in the EMU countries, creating an economic subregion of Europe commonly called the **Eurozone**. Today 19 of the EU's 28 member states use the euro.

By adopting a common currency, Eurozone members sought to increase the efficiency of both domestic and international business by eliminating the costs associated with payments made in different currencies. Although many traditional economists had (and still have) misgivings about this common-currency system, the political goal of enhancing European unity through a common currency won out. In 1999, several EU member countries resisted relinquishing control over their national monetary system and opted out of the Eurozone; consequently, the UK still uses its traditional currency, the pound, as does Denmark with the krone and Sweden with the krona.

The non-euro EU member states of Bulgaria, Croatia, the Czech Republic, Hungary, Poland, and Romania are legally obligated to join the Eurozone at some future date. However, this expansion process has slowed recently because of a lengthy and controversial fiscal crisis that has illuminated EMU shortcomings.

The 2015 Greek Debt Crisis and Its Implications for the Eurozone

Europe has long consisted of a diverse mosaic of richer and poorer countries. Historically, this disparity was between the richer industrial heartland—England, France, Germany, the Netherlands, Belgium, and northern Italy—and the poorer Mediterranean agricultural periphery—Spain, Portugal, southern Italy, the Balkans, and Greece. To lessen this economic disparity, a founding assumption of the EU was that unity would emerge only if the richer countries helped the poorer ones through financial loans and subsidies. The EMU's creation in 1999 furthered this process by making relatively cheap loans available in the new common currency.

Initially, the process seemed to be working as Greece, Portugal, and Spain borrowed money to pay existing debts and start new projects that reduced unemployment and expanded the consumer base. But as shockwaves from the 2008 global financial crisis rippled through the Eurozone, it became painfully apparent that the borrowing countries were hopelessly in debt. At that point, the lending countries—led by Germany and the Netherlands, two rich countries noted for their thrifty cultures—demanded that the borrowing countries implement a program of austerity by reducing government spending and raising revenue through increased taxes. As a result, what had previously been a banker's financial abstraction now became a humanitarian crisis as jobs disappeared, pensions shrank, social services evaporated, taxes increased, and prices for food and other necessary goods soared.

Before the EMU, countries that found themselves in financial trouble would adjust their currency through devaluation, a common pathway out of indebtedness. But this was not allowed under the EMU, since the euro was a common currency controlled by the European Central Bank (ECB). Consequently, the suffering continued. Thousands of Portuguese and Spaniards took advantage of the new mobility under Schengen and left their home countries for jobs in Germany, the Netherlands, England, and Belgium. In Greece, people took to the streets in riots and demonstrations and changed governments frequently (Figure 8.34). By early summer 2015, Greece was on the brink of economic collapse and became the focus of the euro crisis. With 3.5 billion euros in payments due in July and no way for the government to pay, banks closed, ATMs ran dry, crucial imports

Figure 8.33 Development Issues in Europe: Economic Disparity This map shows that economic disparity between rich and poor countries still exists in contemporary Europe, with stark differences between northern Europe and the Mediterranean (compare Norway and Portugal), and traditional western Europe and the former Soviet satellite countries of eastern Europe (compare France and Germany with Romania and Bulgaria). However, also of note is that some former Soviet satellite countries (Poland, Czech Republic, Slovenia) are doing better than others (Latvia, Bulgaria, Romania). **Q. What factors could explain the contemporary economic disparity between the former Soviet satellite countries?**

like food and medicine stalled, and the tourists that were the mainstay of Greece's summer economy stayed away.

Greece's options were limited: The country could either borrow more money from the ECB and hope for the best or, alternatively, declare bankruptcy, leave the Eurozone, reinvent its traditional currency (the drachma), and muddle through somehow. Negotiations to borrow more money from the ECB were ugly: Germany, which had loaned Greece 57 billion euros, was furious and accused the Greeks of profligacy (a polite word for squandering money) and not only refused to lend more money, but also suggested Greece be thrown out of the Eurozone; France, to whom Greece owed 43 billion euros, was slightly more understanding, suggesting things could be worked out; the Netherlands, Finland, and Ireland—countries that had successfully balanced their books through austerity—supported Germany's hard-line stance; Poland, Bulgaria, and Romania—potential euro countries—put their Eurozone membership on hold; and even Russia got in the act by offering to bail out Greece (even though experts said Russia lacked the financial means), an offer that sent chills through NATO headquarters because of the geopolitical implications.

Finally, after months of contentious negotiations, Greece and the Eurozone came to an agreement in August 2015 for a third round of bailout funds that many believe is only a temporary solution.

Figure 8.34 Euro Crisis Protests Greece suffered deeply because of the euro crisis, and its population has protested vehemently against the fiscal policies resulting from its bailout. Here, in Athens, Greece, labor unions protest against those policies. The term *troika* is originally a Russian word used to describe the three-part leadership of the Soviet era and is used here in an unflattering reference to the three organizations administering the bailout funds to Greece: the EU, the European Central Bank, and the International Monetary Fund (IMF).

Regardless, these drawn-out negotiations have changed the conversation about the Eurozone's future in several ways:

- The assumption that Europe's richer countries will subsidize the region's poorer countries must be revised.
- If the Eurozone is to survive, it, too, must be modified to allow individual countries more flexibility in managing their internal finances.
- What was once unthinkable—Greece leaving the Eurozone—now is accepted in the conversation about the Eurozone's future, as the costs of such a move for the EU have been revised downward.
- The International Monetary Fund (IMF), a persistent critic of Eurozone indebtedness in the past, became the strongest advocate for actually reducing Greece's debt because it sees the current solution as unsustainable. This new stance by the major international lender may have a significant influence on future Eurozone actions on indebtedness.

Social Development in Europe: Gender Issues

Despite the visibility of female political leaders and the fact that Europe is considered one of the most developed regions of the world, gender equity issues persist in government, business, and domestic life. For the EU countries as a whole, for example, male employment is 21 percent higher than female employment, and women who work generally make 25 percent less than men.

However, given the complexity of Europe, with its mixture of national, urban, rural, and migrant cultures, the nature and extent of gender issues differ widely among countries and regions. To illustrate, within the 28 EU countries about a quarter of the parliamentary offices are held by women. Sweden has the highest representation, with more than half of its ministers being female, whereas Cyprus has absolutely none. Similarly, in the business world, only 11 of Europe's largest companies have women in top management, yet women make up almost a third of top management in Norway, compared to just 1 percent in Luxembourg (Figure 8.35).

One interesting pattern is that female participation in the workforce is generally higher in the countries of eastern Europe and the Balkans. Two interrelated factors explain this. First, women were expected to work in the communist economies of these countries from 1945 to 1990. Second, families often needed two incomes to survive during the difficult economic transition that followed the collapse of the Soviet Union in 1991. Regardless of cause, the results are startling. Today Bulgaria has the highest percentage of female CEOs (21 percent) of any EU country, and Slovenia, formerly part of socialist Yugoslavia, is the country with the least income disparity between men and women.

Women are well represented in both government and business in the Scandinavian countries, but for very different reasons than in eastern Europe and the Balkans. It is generally agreed that the foundation of Scandinavia's gender equity comes from a combination of comprehensive child care, liberal family benefits that guarantee job security and career advancement after maternity and paternity leaves, and a tax code that does not punish dual-income families. As a result, Norway and Sweden have the highest percentage of females in the workforce. Portugal has the third-highest number at 71 percent; because of its struggling economy, women usually work out of necessity rather than choice, with grandparents and other family members providing child care—unlike the government-sponsored child care common in Scandinavia.

Not to be overlooked are the extraordinarily complex gender issues within Europe's large migrant cultures and their host cultures. We mentioned earlier how France's national policies have become entangled with Muslim gender and cultural preferences. Other examples of these complexities can be found in Germany as the state finds itself embroiled in cultural tensions. These range from increasing women's freedom beyond the family household to prosecuting Turkish honor killings, where young women have paid with their lives for behaviors, such as dating and marrying without parental consent, that are common in German culture but unacceptable in traditional Turkish culture.

Review

8.18 What geographic factors explain the locations of early industry in Europe?

8.19 Describe the origin and evolution of the European Union in terms of its goals.

8.20 Why and how has eastern Europe reinvented its economy in the last two decades?

8.21 Discuss the factors explaining the highly variable geography of women in Europe's workforce.

KEY TERMS Industrial Revolution, Economic and Monetary Union (EMU), Eurozone

Figure 8.35 Women in Europe's Business World The employment of women in Europe's workforce differs significantly among different countries and regions. Scandinavia, for example, has the highest percentage of women in upper management positions; this contrasts with the lowest percentage in the Mediterranean countries. This photo was taken at a business meeting in Berlin, Germany.

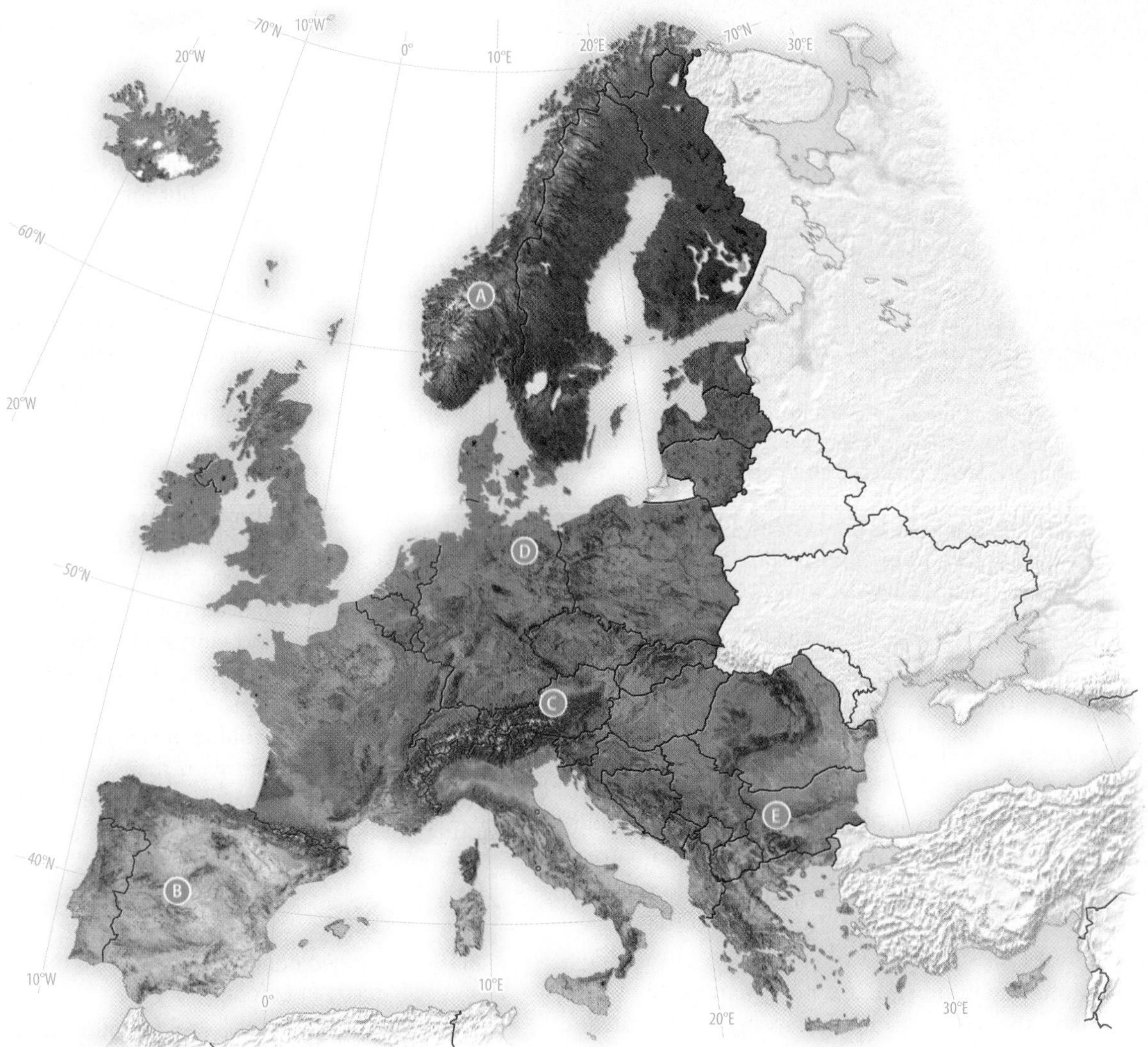

Review

Physical Geography and Environmental Issues

8.1 Describe the topography, climate, and hydrology of Europe.

8.2 Identify the major environmental issues in Europe as well as measures taken to resolve those problems.

Because of its immense latitudinal stretch, Europe includes a diverse array of climates and landscapes, from arctic tundra to dry-summer Mediterranean shrublands. As for environmental issues, the EU's leadership has been important in resolving trans-boundary water, air, and toxic hazard problems in both western and eastern Europe. Europe has also shown leadership in reducing atmospheric emissions through its cap-and-trade program.

1. What are the dominant landforms along the coast of Norway, and what geologic process created them?

2. The EU proposes to reduce its CO_2 emissions to 20 percent below 1990 emission levels by 2020. How will this be done?

Population and Settlement

8.3 Provide examples of countries with different rates of natural growth.

8.4 Describe both the patterns of internal migration within Europe and the geography of foreign migration to the region.

With the notable exception of Ireland, all European countries are below replacement level in terms of natural growth. Unless this changes through family-friendly policies, future population growth (or decline) will be determined solely by in-migration.

3. Explain the reasons behind the differing population densities in this map of Spain.

4. Using data from Table A8.1 (population), construct a simple map showing which countries are gaining and which are losing population from migration.

Cultural Coherence and Diversity

8.5 Describe the major languages and religions of Europe and locate these on a map.

Historically, Europe's diverse cultural geography was primarily a product of language and religion; today, however, it has become more complex because of diffuse global influences interacting with migrant cultures. The result is a complex mixture of traditional, global, and ethnic cultures. An example is how France tries to retain its traditional French language in the face of U.S. English–speaking media and street-level African migrant speech.

5. What does this photo of the cultural landscape of Salzburg, Austria, tell you about the cultural geography of that country?

6. Find information on birth and death rates for the Muslim population of France. How do these compare with the rates for the native population of that country? How might this influence France's cultural geography?

C

Geopolitical Framework

8.6 Summarize how the map of European states has changed in the last 100 years.

8.7 Explain why and how Europe was divided during the Cold War and what the geographic implications are today.

Europe's borders changed often during the 20th century because of two world wars, the Cold War's end in 1990, and the devolution of the former Yugoslavia. Today there are different degrees of separatism, ranging from the autonomy given Basques and Catalonians in Spain to the very real possibility Scotland will separate from the United Kingdom.

7. This sign is in an icon of Berlin. Why?

8. Search the Internet for information on the 2014 Scottish independence vote to determine how an independent Scotland would address matters of finance, trade, and defense.

D

Economic and Social Development

8.8 Describe Europe's economic and political integration as driven by the EU.

8.9 Identify the major characteristics of Europe's current economic and social crisis.

After decades of postwar economic growth, Europe is now mired in a troublesome fiscal crisis that calls into question the EU's common-currency program and, in many ways, aggravates the economic and social disparities between Europe's rich and not-so-rich countries.

9. After reviewing data in Tables A8.1 and A8.2 on Bulgaria's social and economic conditions, write a short description of how the activities along this street in Sofia might have changed since 1990.

10. Discuss and critique the arguments for and against the UK staying in the European Union.

E

KEY TERMS

asylum laws (p. 265)
balkanization (p. 279)
buffer zone (p. 278)
Cold War (p. 278)
continental climate (p. 260)
Cyrillic alphabet (p. 271)
devolution (p. 279)
Economic and Monetary Union (EMU) (p. 285)
European Union (EU) (p. 283)
Eurozone (p. 285)
family-friendly policies (p. 264)
fjord (p. 257)
Industrial Revolution (p. 282)
Iron Curtain (p. 278)
irredentism (p. 275)
marine west coast climate (p. 258)
maritime climate (p. 260)
medieval landscape (p. 267)
Mediterranean climate (p. 258)
nation-state (p. 275)
North Atlantic Treaty Organization (NATO) (p. 278)
Renaissance–Baroque period (p. 267)
Schengen Agreement (p. 264)
Warsaw Pact (p. 278)

DATA ANALYSIS

http://goo.gl/dSEMCq

The flood of extralegal migrants seeking political asylum in Europe is overwhelming the region. This exercise is designed to give you a better understanding of this migration crisis.

1. Go to EU's Eurostat page (http://ec.europa.eu), access the asylum statistics, and acquaint yourself with the data tables in the right-hand column and the text content on the left. Note that asylum seekers come from many different countries and file their asylum applications in all the different EU countries.
2. Now make either a map or a bar chart linking the different migrant source countries to the different European countries. This can be done most simply by mapping one or two leading source countries for each EU country. Summarize your findings in a paragraph or two.
3. Next, go to the data table that shows whether the migrant's asylum application was successful or not—that is, whether the migrant was granted asylum and stayed in Europe or was sent back to his or her country of origin. Once again, you can present your findings in either map or table form.
4. Which countries are most likely and which least likely to approve asylum applications from which source countries? Consider social factors such as the country's wealth (measured as gross domestic product per capita), unemployment rates, natural birth rates, and cultural factors such as each country's religion and language. Write a summary essay explaining your findings and reasoning.

MasteringGeography™

Looking for additional review and test prep materials? Visit the Study Area in MasteringGeography™ to enhance your geographic literacy, spatial reasoning skills, and understanding of this chapter's content by accessing a variety of resources, including MapMaster interactive maps, geoscience animations, videos, *In the News* RSS feeds, flashcards, web links, self-study quizzes, and an eText version of *Globalization and Diversity.*

Author's Blog

Scan to visit the **Author's Blog for field notes, media resources, and chapter updates**

https://gad4blog.wordpress.com/category/europe/

Scan to visit the **GeoCurrents Blog**

http://www.geocurrents.info/category/place/europe

9 The Russian Domain

PHYSICAL GEOGRAPHY AND ENVIRONMENTAL ISSUES

Many areas within the Russian domain suffered severe environmental damage during the Soviet era (1917–1991). Today air, water, toxic chemical, and nuclear pollution plague large portions of the region.

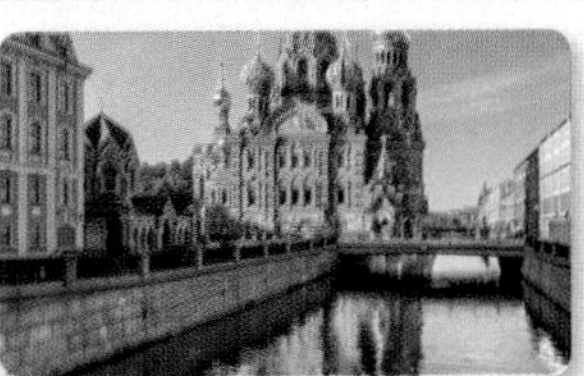

POPULATION AND SETTLEMENT

Urban landscapes within the Russian domain reflect a fascinating mix of imperial, socialist, and post-communist influences. Many larger urban areas within the region are showing trends toward sprawl and decentralization similar to those seen in North America and western Europe.

CULTURAL COHERENCE AND DIVERSITY

Although Slavic cultural influences dominate the region, many non-Slavic minorities, including a variety of indigenous peoples in Siberia and a complex collection of ethnic groups in the Caucasus Mountains, shape the cultural and political geography of the domain.

GEOPOLITICAL FRAMEWORK

Centralization of Russian political power under President Vladimir Putin has had widespread consequences within the region, sparking ongoing tensions with neighboring Ukraine as well as limiting democratic freedoms within Russia itself.

ECONOMIC AND SOCIAL DEVELOPMENT

The region's economy has recently been hard hit by falling energy prices, global economic sanctions against Russia, and war within Ukraine.

◀ The eastern Ukrainian city of Debaltseve has been devastated by the recent civil war in that country. Many residents fled the city as it became a battleground between Russian rebels and Ukrainian military forces.

The devastated streets of Debaltseve in eastern Ukraine suggest the magnitude of Europe's latest political unraveling. Closed factories and schools, abandoned homes, and the threat of violence have haunted many portions of eastern Ukraine since civil war erupted there in 2014. Russian-backed rebels, especially in the Luhansk and Donetsk regions, declared their independence from Kiev, setting up new "People's Republics." Cities such as Luhansk, Donetsk, and Debaltseve have been caught in the crossfire as the rebels—with Russian help—hope to secure their political autonomy and fight off Ukrainian troops bent on reasserting national authority in these breakaway regions. The result is a war-torn landscape more reminiscent of World War II than of 21st-century Europe.

The recent instability is nothing new. In 1991, the Russian domain witnessed the complete collapse of the **Soviet Union** (or the Union of Soviet Socialist Republics [USSR]), a sprawling communist state that had dominated the region since 1917. In its place stood 15 former "republics" once united under the USSR. Now independent, each republic has tried to make its way in a post-Soviet world. The Russian Republic remained dominant in size and political influence and thus came to form the core of a new Russian domain. Today the domain includes not only Russia, but also the nations of Ukraine, Belarus, Moldova, Georgia, and Armenia (Figure 9.1). The term *domain* suggests the persistent Russian influence on the five other nations included in the region. Slavic Russia, Ukraine, and Belarus make up the region's core. Nearby Moldova and Armenia broadly remain within Russia's geopolitical orbit. Relations between Russia and Georgia remain strained, and recent conflicts in Georgia brought Russian troops into that small country.

New patterns of foreign investment and redefined migration flows demonstrate the unpredictable nature of the Russian domain's post-Soviet global connections.

Two significant areas once part of the Soviet Union are excluded from the domain. The mostly Muslim republics of Central Asia and the Caucasus (Kazakhstan, Uzbekistan, Kyrgyzstan, Turkmenistan, Tajikistan, and Azerbaijan), despite continued ties to Russia, have become aligned with the Central Asia world region (see Chapter 10), while the Baltic republics (Estonia, Latvia, and Lithuania) are best grouped with Europe (see Chapter 8).

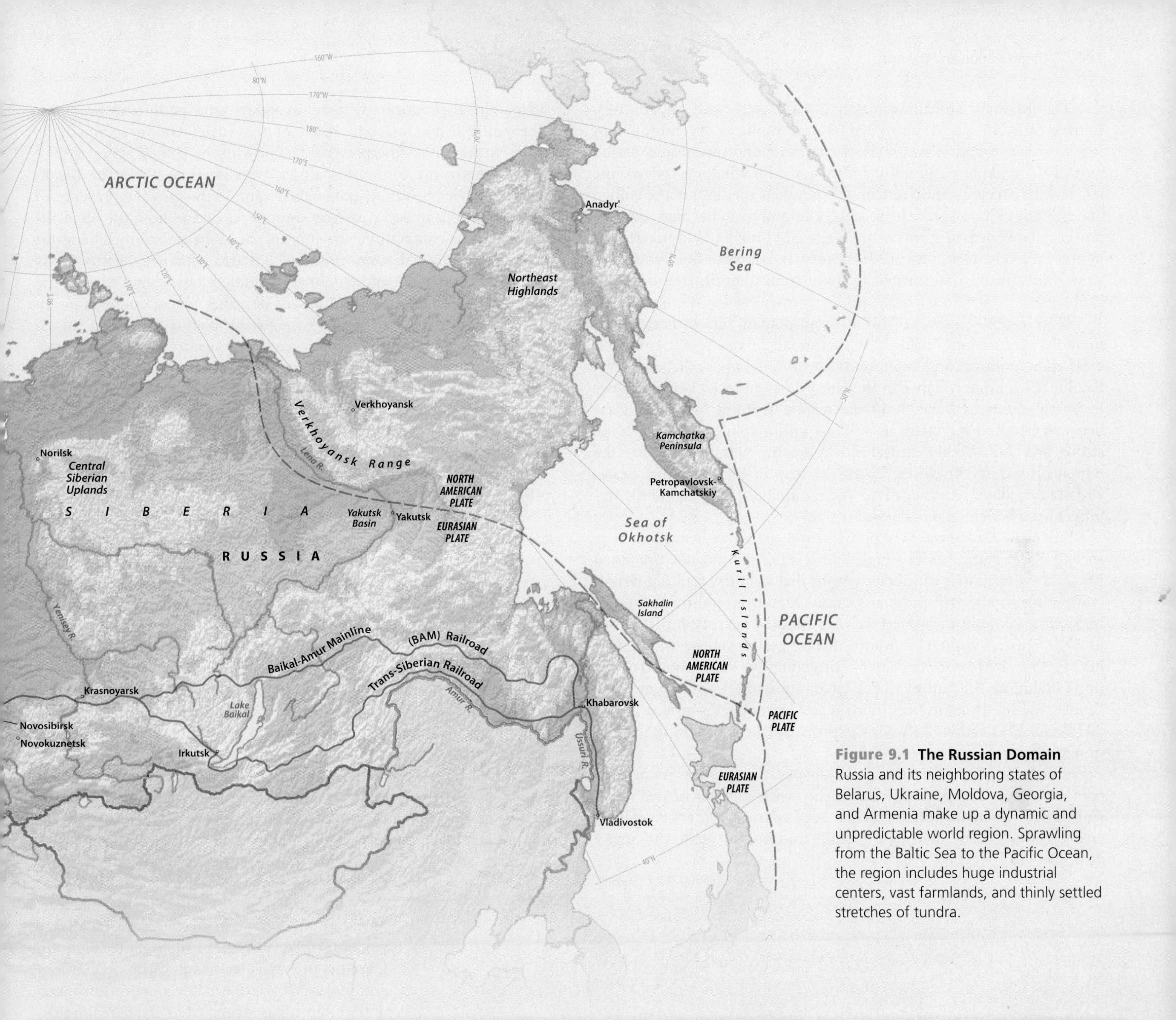

Figure 9.1 The Russian Domain Russia and its neighboring states of Belarus, Ukraine, Moldova, Georgia, and Armenia make up a dynamic and unpredictable world region. Sprawling from the Baltic Sea to the Pacific Ocean, the region includes huge industrial centers, vast farmlands, and thinly settled stretches of tundra.

Learning Objectives

After reading this chapter you should be able to:

9.1 **Explain the close connection among latitude, regional climates, and agricultural production in Russia.**

9.2 **Describe the major environmental issues affecting the region and suggest how climate change might impact high-latitude areas.**

9.3 **Identify the region's major migration patterns, in both the Soviet and the post-Soviet eras.**

9.4 **Explain major urban land-use patterns in a large city such as Moscow.**

9.5 **Describe the major phases of Russian expansion across Eurasia.**

9.6 **Identify the key regional patterns of linguistic and religious diversity.**

9.7 **Summarize the historical roots of the region's modern geopolitical system.**

9.8 **Provide examples of recent geopolitical conflicts in the region and indicate how these reflect persistent cultural differences.**

9.9 **Identify key ways in which natural resources, including energy, have shaped economic development in the region.**

9.10 **Describe key sectors of the region's economy in the Soviet and post–Soviet eras and discuss how recent geopolitical events affect prospects for future economic growth.**

The Russian domain remains a land rich with superlatives: Endless Siberian spaces, unlimited natural resources, legends of ruthless Cossack warriors, and tales of epic wars and revolutions are all part of the region's geographic and historical mythology. Indeed, the rise of Russian civilization remarkably parallels the story of the United States. Both cultures grew from small beginnings to become imperial powers that benefited from the fur trade, gold rushes, and transcontinental railroads during the 19th century. In addition, both countries were dramatically transformed by industrialization during the 20th century.

More recently, however, the Russian domain has witnessed particularly breathtaking change. With the fall of the Soviet Union in 1991, new political and economic institutions have reshaped everyday life. Economic collapse in the late 1990s produced steep declines in living standards throughout the region. Political instability grew between neighboring states as well as within countries. After 2000, strong and increasingly centralized leadership within Russia set the region on a different course. With the help of higher energy prices (Russia is a major exporter of oil and natural gas), real economic improvements benefited most of the region.

Since 2013, new waves of political and economic instability have rocked the region. Russian President Vladimir Putin has tightened his hold on executive authority within that country, limiting democratic reforms and free press autonomy. Moreover, political instability within Ukraine prompted Russia's illegal annexation of Crimea in 2014. Russia's military support of rebel groups in eastern Ukraine has further increased tensions. Subsequently, many western nations including the United States have imposed economic sanctions against Russia. In addition, dramatically lower energy prices in late 2014 and 2015 reduced Russia's exports, further deepening its economic woes.

Processes of globalization have shaped the Russian domain in complex ways. The region's relationship with the rest of the world shifted during the last decade of the 20th century. Under centralized Soviet control, large increases in industrial output made the area a major global producer of steel, weaponry, and petroleum products. The political and military reach of the Soviet Union spanned the globe, making it a superpower on a par with the United States.

Suddenly, the communist order evaporated in 1991. The breakdown of Soviet control exposed the region to the growing presence of western European and American influences and to both the opportunities and the competitive pressures of the global economy. The result was a world region that saw its global linkages redefined. New patterns of foreign investment and redefined migration flows demonstrated the unpredictable nature of the Russian domain's global connections in the post-Soviet period (Figure 9.2). Recent conflicts between Russia and Ukraine, escalating economic sanctions against Russia, and gyrating energy prices suggest more uncertainty ahead.

Slavic Russia (population 142 million) dominates the region. Although only about three-quarters the size of the former Soviet Union, Russia's dimensions still make it the largest state on Earth. Its area of 6.6 million square miles (17 million square kilometers) dwarfs even Canada, and its nine time zones are a reminder that dawn in Vladivostok on the Pacific Ocean is still only evening in Moscow.

The demise of the Soviet Union ended almost 75 years of communist rule. After a decade of political and economic instability (1991–2000), Russia made impressive progress early in the new century. Russian President Putin built a reputation for strong leadership (2000–2008, 2012–), and he has stressed Russia's economic growth. Much of that growth has come to Russian cities, where expanding middle and professional classes are enjoying better living standards. Many rural areas, however, remain deeply mired in poverty. Another concern is that Putin's desire for wealth and power has been matched by his need for more centralized political control inside Russia. His close cooperation with the country's **oligarchs**, a small group of wealthy, very private businessmen who control (along with organized crime) important aspects of the Russian economy, also raises suspicions about Putin's actions. In addition, recent Russian aggression in Ukraine, as well as new agreements to build closer economic ties with neighboring nations (such as Belarus and Kazakhstan), suggests that Russia's desire for more-centralized authority did not end with the demise of the Soviet Union.

The bordering states of Ukraine, Belarus, Moldova, Georgia, and Armenia are inevitably yoked to the evolution of their giant neighbor. Ukraine, in particular, has the size, population, and resource base to become a major European nation, but it has struggled to create real political and economic change since independence. With 43 million people and a rich storehouse of resources, Ukraine's size of 233,000 square miles (604,000 square kilometers) is similar to that of France. Many years of political instability recently escalated in the country, however, prompting Russia's takeover of Crimea and fomenting a rebel-led civil war (with Russia's support) in the eastern portion of the country.

Figure 9.2 Moscow's International Business Center Also known as Moscow City, the International Business Center dominates the capital's western skyline and features some of Europe's tallest buildings.

Figure 9.3 Chisinau, Moldova With a population of almost 1 million, the Moldovan capital is in the most economically affluent part of this small eastern European nation. Its landscape still reflects the influence of Soviet-era planning and urban design.

Nearby Belarus is smaller (80,000 square miles [208,000 square kilometers]), and its population of about 9.5 million remains closely tied economically and politically to Russia. Presently, its strikingly authoritarian and antiforeign leadership reflects many aspects of the old Soviet empire.

Moldova, with 3.5 million people, shares cultural links with Romania, but its economic and political connections have kept it tied to the Russian domain (Figure 9.3). South of Russia and beyond the bordering Caucasus Mountains, the Transcaucasian countries of Armenia and Georgia are similar in size to Moldova. Their populations differ culturally from that of their Slavic neighbor to the north. In addition, these two nations face significant political challenges: Armenia shares a hostile border with Azerbaijan (see Chapter 10), and Georgia's ethnic diversity and contentious relations with neighboring Russia threaten its political stability.

Physical Geography and Environmental Issues: A Vast and Challenging Land

The region's physical geography continues to shape its economic prospects in fundamental ways. For example, Russia's vast size poses special challenges, but its rich store of natural resources has offered unique economic benefits. At the same time, the Soviet period (1917–1991) witnessed unparalleled, unrestrained economic development that damaged the region's environment. The region still bears the scars of that legacy.

A Diverse Physical Setting

The Russian domain's northern latitudinal position shapes its basic geographies of climate, vegetation, and agriculture (see Figure 9.1). Indeed, the Russian domain provides the world's largest example of a high-latitude continental climate, where seasonal temperature extremes and short growing seasons limit human settlement (Figure 9.4). In terms of latitude, Moscow is positioned as far north as Ketchikan, Alaska, and even the Ukrainian capital of Kiev (Kyiv) sits farther north than the Great Lakes in Canada. Thus, apart from a subtropical zone near the Black Sea, the region experiences a classic continental climate with hard, cold winters and marginal agricultural potential.

The European West An airplane flight over the western portions of the Russian domain reveals a vast, barely changing landscape. European Russia, Belarus, and Ukraine cover the eastern portions of the vast European Plain that runs from southwest France to the Ural Mountains. One major geographic advantage of European Russia is that different river systems, all now linked by canals, flow into four separate drainage basins. The result is that trade goods can easily flow in many directions. The Dnieper and Don rivers flow into the Black Sea; the West and North Dvina rivers drain into the Baltic and White seas, respectively; and the Volga, the longest river in Europe, runs to the Caspian Sea.

Most of European Russia experiences cold winters and cool summers by North American standards. Moscow, for example, is about as cold as Minneapolis in January, but it is not nearly as warm in July. In Ukraine, Kiev is milder, however, and Simferopol, near the Black Sea, offers wintertime temperatures that average more than 20°F (11°C) warmer than those of Moscow (see the climographs in Figure 9.4).

Three distinctive environments shape agricultural potential in the European west (Figure 9.5). North of Moscow and St. Petersburg, poor soils and cold temperatures severely limit farming. The region's boreal forests have been extensively logged.

Belarus and central portions of European Russia possess longer growing seasons, but acidic **podzol soils**, typical of northern forest environments, limit this region's ability to support a productive agricultural economy. The diversified agriculture that does occur includes grain (rye, oats, and wheat) and potato cultivation, swine and meat production, and dairying.

South of 50° latitude, agricultural conditions improve across much of southern Russia and Ukraine. Forests gradually give way to steppe environments dominated by grasslands and by fertile "black earth" **chernozem soils**. These have proven valuable for commercial wheat, corn, and sugar beet cultivation and for commercial meat production (Figure 9.6).

The Ural Mountains and Siberia The Ural Mountains (see Figure 9.1) physically separate European Russia from Siberia. Topographically, the Urals are not a particularly impressive range. Still, the ancient rocks of these mountains contain valuable mineral resources, and the mountains themselves traditionally marked European Russia's eastern cultural boundary. In early 2013, the city of Chelyabinsk in the southern Urals achieved momentary global notoriety when a large meteor exploded just above the city, injuring hundreds and damaging thousands of buildings (see *Exploring Global Connections: Russian Meteorite Fragments Go Global*).

East of the Urals, Siberia unfolds across the landscape for thousands of miles. The great Arctic-bound Ob, Yenisey, and Lena rivers (see Figure 9.1) drain millions of square miles of northern country, including the flat West Siberian Plain, the hills and plateaus of the Central Siberian Uplands, and the rugged and isolated Northeast Highlands. Along the Pacific, the Kamchatka Peninsula offers spectacular

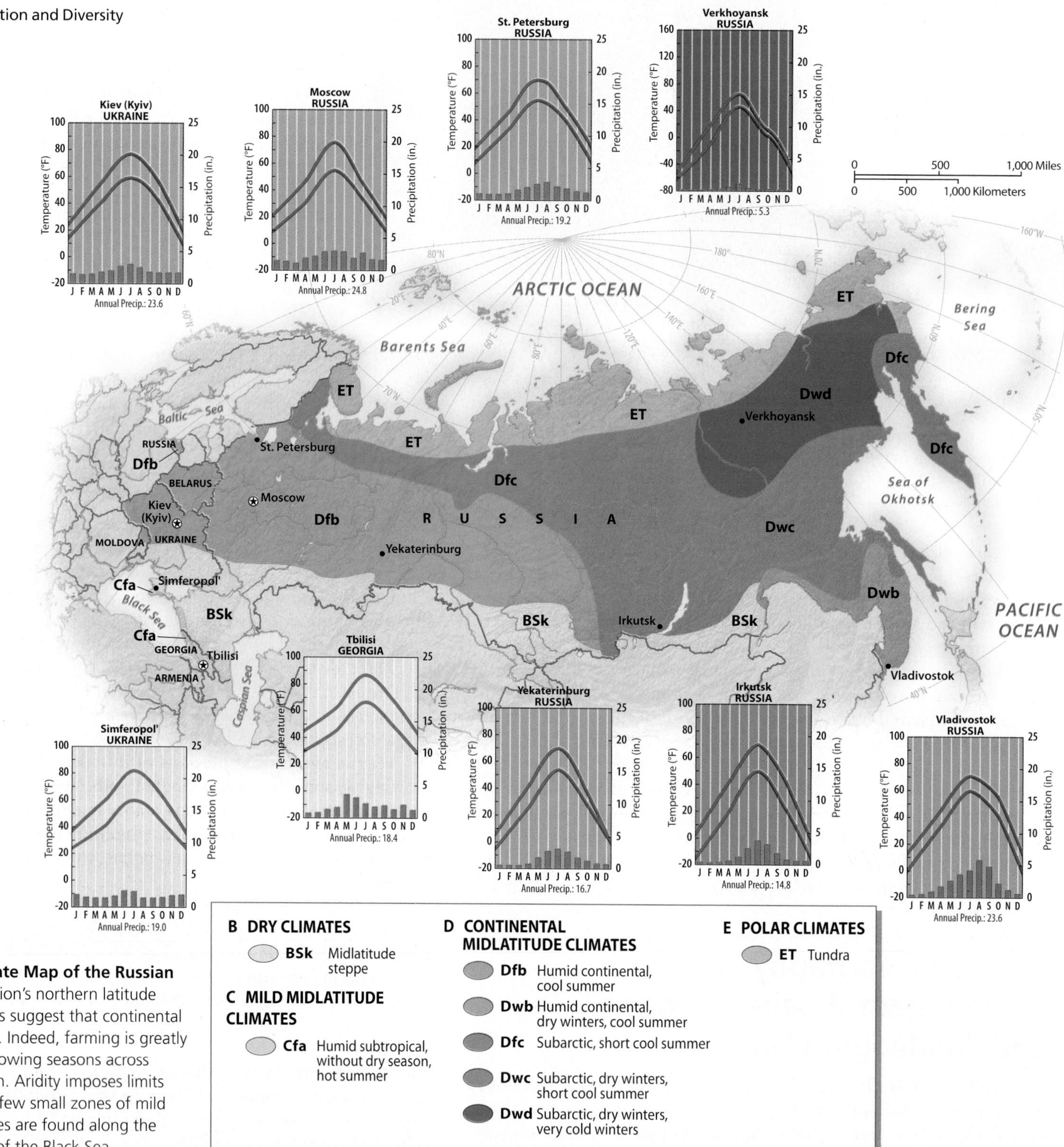

Figure 9.4 Climate Map of the Russian Domain The region's northern latitude and large landmass suggest that continental climates dominate. Indeed, farming is greatly limited by short growing seasons across much of the region. Aridity imposes limits elsewhere. Only a few small zones of mild midlatitude climates are found along the temperate shores of the Black Sea.

volcanic landscapes (Figure 9.7). Wintertime climatic conditions, however, are severe across the entire region.

Siberian vegetation and agriculture reflect the climatic setting. The north is too cold for tree growth and instead supports tundra vegetation, characterized by mosses, lichens, and a few ground-hugging flowering plants. Much of the tundra region is associated with **permafrost**, a cold-climate condition of unstable, seasonally frozen ground underlain by a permanently frozen layer that limits vegetation growth and causes problems for railroad construction. South of the tundra, the Russian **taiga**, or coniferous forest zone, dominates a large portion of the Russian interior. With the huge demand for lumber from nearby Japan and China, the eastern taiga zone is threatened by both authorized logging and illegal timber poaching.

The Russian Far East The Russian Far East is a distinctive subregion characterized by proximity to the Pacific Ocean, more southerly latitude, and fertile river valleys, such as the Amur and the Ussuri. Located at about the same latitude as New England, the region features longer growing seasons and milder climates than those found

Figure 9.5
Agricultural Regions Harsh climate and poor soils combine to limit agriculture across much of the Russian domain. Better farmlands are found in Ukraine and in European Russia south of Moscow. Portions of southern Siberia support wheat production but yield marginal results. In the Russian Far East, warmer climates and better soils make possible higher agricultural productivity.
Q: Describe the relationships between major agricultural zones and patterns on the climate map (Figure 9.4).

AGRICULTURAL REGIONS
Diversified agriculture
Large-scale grain production
Urban truck farming
Humid subtropical specialized agricultural production
Tundra
Taiga
Drylands
Mountains

to the west or north. Here the continental climates of the Siberian interior meet the seasonal monsoon rains of East Asia. It is a fascinating zone of ecological mixing: Conifers of the taiga mingle with Asian hardwoods, and reindeer, Siberian tigers, and leopards also find common ground.

The Caucasus and Transcaucasia In European Russia's extreme south, the Caucasus Mountains stretch between the Black and Caspian seas (Figure 9.8). They mark Russia's southern boundary and are characterized by major earthquakes. Farther south lies Transcaucasia and the distinctive natural settings of Georgia and Armenia. Patterns of both climate and terrain in the Caucasus and Transcaucasia are very complex. Rainfall is higher in the west, while eastern valleys are semi-arid. In areas where rainfall is adequate or where irrigation is possible, agriculture can be quite productive. Georgia in particular produces fruits, vegetables, flowers, and wines (Figure 9.9).

A Devastated Environment

The breakup of the Soviet Union and subsequent opening of the region to international public examination revealed some of the world's most severe environmental degradation (Figure 9.10). Seven decades of intense and rapid Soviet industrialization caused environmental problems that extend across the entire region.

Air and Water Pollution Toxic environments still plague much of the region. Poor air quality affects hundreds of cities and industrial

Figure 9.6 Ukrainian Steppe Akin to North America's Great Plains, much of southern Ukraine is a region of low relief and extensive commercial grain production.

Figure 9.7 Kamchatka Peninsula Russia's Pacific Coast offers many spectacular natural settings, including a variety of volcanic landscape features.

EXPLORING GLOBAL CONNECTIONS

Russian Meteorite Fragments Go Global

Google Earth (MG) Virtual Tour Video http://goo.gl/BDsepG

It was literally the sound heard around the world. At 9:20 in the morning on February 15, 2013, the largest object (between 7000 and 11,000 tons) to enter Earth's atmosphere in the last century (the Tunguska area of Siberia experienced a similar event in 1908) exploded above Chelyabinsk, Russia, in the southern Urals (Figure 9.1.1). Thousands of windows shattered near Chelyabinsk, and more than 1200 people were injured, mostly by flying glass. The fireball stunned local residents, and video footage of the event immediately went viral to every corner of the globe.

Treasures in the Snow In the days following the explosion, fragments of the meteorite, which were eagerly recovered by local residents, began their own worldwide journeys. The word was out: Schoolchildren were digging through the snow, finding small black pebbles everywhere. One woman found a fist-sized stone in her woodshed that had come through the roof. Within a few days, strangers began showing up, their fat wallets filled with rubles and euros. They began quietly buying up the treasures (Russian authorities discourage the black market in meteorites) that had fallen from the sky. Quickly, eBay auctions were offering the fragments, an online meteorite sales site (www.star-bits.com) featured fresh finds from Chelyabinsk, and the International Meteorite Collectors Association (IMCA) website was getting new hits from around the world (Figure 9.1.2).

A Long History of Collecting The Chelyabinsk phenomenon is nothing new. People have collected meteorites for thousands of years. Egyptian hieroglyphs refer to "iron from heaven," and some archaeologists believe ancient Egyptians made artifacts from iron- and nickel-rich meteorites found in the desert. In fact, desert environments make for great meteorite hunting. In 2008, an Italian geologist browsing Google Earth thought he detected an unusual feature in the North African desert. Sure enough, it was a 150-foot-wide meteor crater, and he discovered thousands of meteorite fragments at the isolated site. But when he returned the following year, he found the site had been disturbed. Soon the Egyptian fragments were on sale at a collectors' show in France.

Figure 9.1.1 Meteor in the Sky Above Chelyabinsk, 2013 Stunned residents of Chelyabinsk looked skyward on the morning of February 15, 2013, as a huge meteorite exploded above the city.

Figure 9.1.2 Meteorite Fragments Found near Chelyabinsk Many local residents of this region in the southern Urals cashed in on the lucrative global market for meteorite fragments.

The Chelyabinsk meteorite fragments are now redistributed around the world. The membership list of the IMCA reads like a roll call of the United Nations, with collectors in Germany, New Zealand, Morocco, China, the Philippines, Brazil, Ireland, and, not surprisingly, Russia. What came out of the sky and fell to Earth that February morning quickly acquired human meaning and monetary value and became, in their own small and magical ways, fragments of a much larger global economy.

1. What other examples of collectible artifacts and materials would have their own patterns of global movement and redistribution?
2. Given its relatively small population and isolated character, why is the interior of Russia so well known for its periodic encounters with Earth-bound meteors?

complexes throughout the Russian domain. The Soviet policy of building large clusters of industrial processing and manufacturing plants in concentrated areas, often with minimal environmental controls, produced a collection of polluted cities stretching from Belarus to Russian Siberia (see Figure 9.10). Siberia's northern mining and smelting city of Norilsk is one of Russia's dirtiest urban areas, earning it the dubious distinction of appearing on the Blacksmith Institute's list of "ten most polluted places in the world" (Figure 9.11). In addition, a large swath of larch-dominated forest has died in a huge zone of contamination that extends more than 75 miles (120 km) east of the city. Norilsk Nickel (the area's major industrial polluter) hopes to dramatically cut harmful sulfur dioxide emissions between 2015 and 2020. Elsewhere, growing rates of private car ownership have greatly increased automobile-related pollution. Today 90 percent of Moscow's air pollution is linked to the city's growing automobile traffic.

Degraded water is another hazard. Urban water supplies are vulnerable to industrial pollution, raw sewage, and demands that increasingly exceed capacity. Oil spills have harmed thousands of square miles in the tundra and taiga of the West Siberian Plain and along the Ob River. Water pollution has also affected the Volga River, the Black Sea, portions of the Caspian Sea shoreline, the waters of the Arctic Ocean off Russia's northern coast, and Siberia's Lake Baikal, the world's largest freshwater reserve (Figure 9.12).

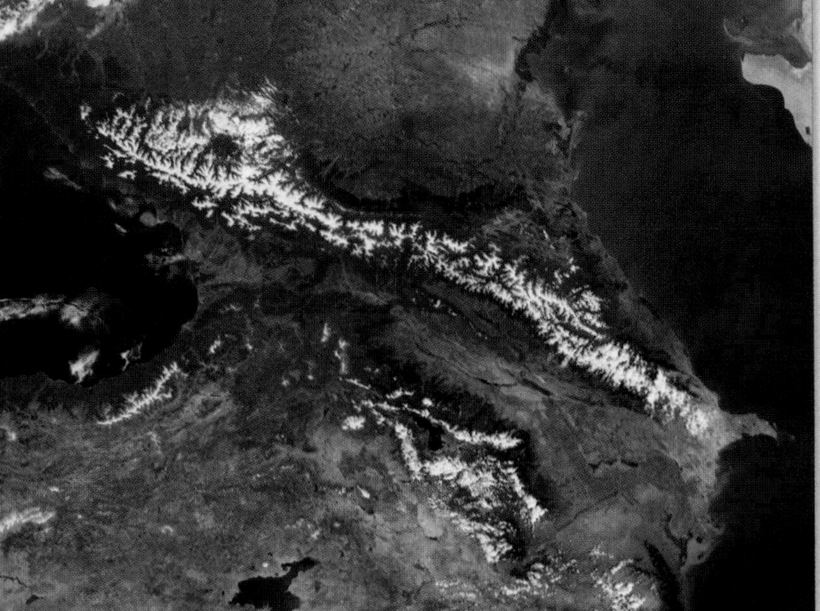

Figure 9.8 Satellite Image of the Caucasus Mountains Aligned between the Black Sea (left) and the Caspian Sea (right), the rugged, snow-capped Caucasus Mountains prevent easy movement across the culturally diverse and politically contested borderlands between southern Russia and Georgia.

The Nuclear Threat The nuclear era brought added dangers to the region. The Soviet Union's nuclear weapons and energy programs often ignored issues of environmental safety. Siberia, for example, suffered regular nuclear fallout when tests were conducted in the atmosphere. Nuclear explosions also were used to move earth in dam-building projects. The once-pristine Russian Arctic has been poisoned: The area around the island of Novaya Zemlya served as an unregulated dumping ground for nuclear wastes in the Soviet era. Aging nuclear reactors dot the region's landscape, often contaminating nearby rivers with plutonium leaks. Nuclear pollution is particularly pronounced in northern Ukraine, where the Chernobyl nuclear power plant suffered a catastrophic meltdown in 1986.

Addressing the Environmental Crisis

Regional leaders are beginning to respond to the environmental crisis. In the case of Chernobyl, plans call for completing a huge protective roof over the entire reactor complex by 2017 or 2018 (see *Working Toward Sustainability: Putting a Lid on Chernobyl*). In Siberia, the successful cleanup of Lake Baikal is a sign of greater environmental awareness in the region (see Figure 9.12). The lake, home to about 20 percent of Earth's unfrozen fresh surface water, suffered during the later Soviet period. Large pulp and paper mills were located along the lakeshore in the 1950s and 1960s. Unfortunately, these industries discharged pollutants into the lake and surrounding atmosphere. Since the early 1990s, stricter regulations have reduced industrial pollution, and the lake's water quality has improved. Indeed, the lake has become *the* national "poster child" of the Russian environmental movement.

Elsewhere, in the Land of the Leopard National Park, along Russia's Amur River, another success story is unfolding. A 2015 census of Amur leopards, the world's rarest big cat, revealed its remarkable comeback after facing extinction less than a decade earlier. Animal counts of the rare cat more than doubled in eight years, suggesting the Russian national park is providing valuable protected habitat for breeding. The next step may be the creation of a jointly managed trans-boundary nature reserve that involves similar habitat in China.

Climate Change in the Russian Domain

Given its latitude and continental climates, the Russian domain is often cited as a world region that would benefit from a warmer global climate. But such an interpretation oversimplifies the complex natural and human responses to global climate change, some of which are already occurring across the region.

Potential Benefits Optimists point to economic benefits that may result from warmer Eurasian climates. Some models predict the northern limit of spring cereal cultivation in northwestern Russia will shift 60–90 miles (100–150 km) poleward for every 1°C (1.8°F) of warming. Less severe winters may make energy and mineral development in arctic settings less costly. About 15 percent of the world's undiscovered oil reserves (and 30 percent of its undiscovered natural gas reserves) are probably located in these settings, and Russia has staked large claims to this area. In the Arctic Ocean and Barents Sea, warmer temperatures and less sea ice are translating into better commercial fishing, easier navigation, more high-latitude commerce, and more ice-free days in northern Russian ports. Since 2010, a growing number of commercial vessels have negotiated the **northern sea route** along Siberia's northern coast. Northern ports such as Murmansk may reap the benefits of this arctic warming (Figure 9.13).

Potential Hazards Even with such rosy scenarios, might long-term regional and global costs outweigh the benefits? First, hotter summers may increase the risk of wildfire. In the summer of 2010, in what could be a sign of things to come, hundreds of blazes broke out in Russia, mostly south and east of Moscow. The fires scorched more than 484,000 acres (196,000 hectares), burning wheat fields and hundreds of structures and filling the skies of Moscow with smoke as visitors and residents experienced record heat.

Second, changes in ecologically sensitive arctic and subarctic ecosystems are already leading to major disruptions in wildlife and indigenous human populations in those settings of northern Russia.

Figure 9.9 Grape Harvest, Subtropical Georgia The moderating influences of the Black Sea and a more southern latitude produce a small zone of humid subtropical agriculture in Georgia.

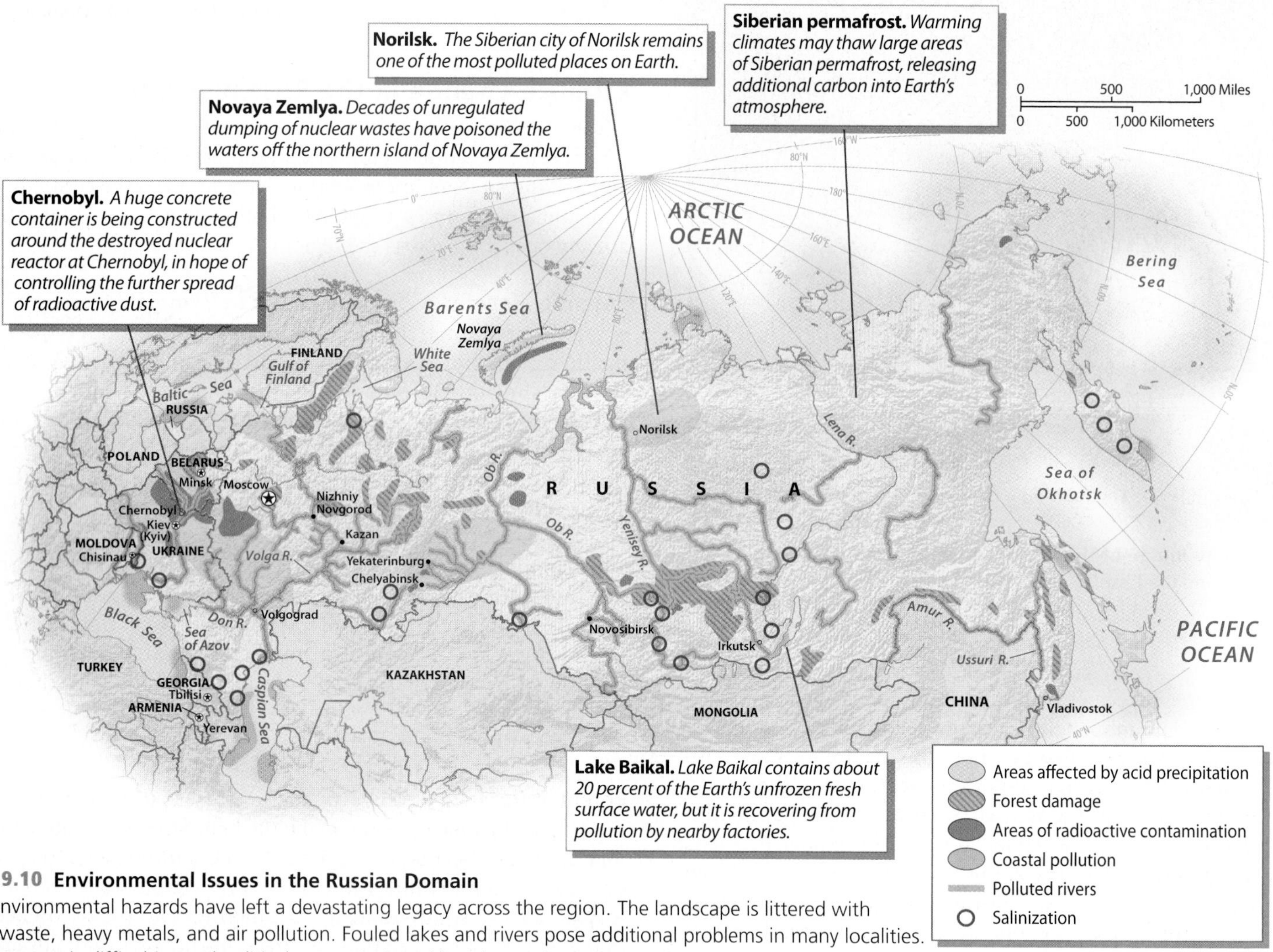

Norilsk. *The Siberian city of Norilsk remains one of the most polluted places on Earth.*

Siberian permafrost. *Warming climates may thaw large areas of Siberian permafrost, releasing additional carbon into Earth's atmosphere.*

Novaya Zemlya. *Decades of unregulated dumping of nuclear wastes have poisoned the waters off the northern island of Novaya Zemlya.*

Chernobyl. *A huge concrete container is being constructed around the destroyed nuclear reactor at Chernobyl, in hope of controlling the further spread of radioactive dust.*

Lake Baikal. *Lake Baikal contains about 20 percent of the Earth's unfrozen fresh surface water, but it is recovering from pollution by nearby factories.*

0 500 1,000 Miles
0 500 1,000 Kilometers

ARCTIC OCEAN
Barents Sea
Novaya Zemlya
White Sea
FINLAND
Gulf of Finland
Baltic Sea
RUSSIA
POLAND
BELARUS
Minsk
Moscow
Chernobyl
Kiev (Kyiv)
MOLDOVA
Chisinau
UKRAINE
Volga R.
Nizhniy Novgorod
Kazan
Yekaterinburg
Chelyabinsk
Volgograd
Don R.
Black Sea
Sea of Azov
TURKEY
GEORGIA
Tbilisi
ARMENIA
Yerevan
Caspian Sea
KAZAKHSTAN
Ob R.
Norilsk
R U S S I A
Yenisey R.
Novosibirsk
Irkutsk
MONGOLIA
Lena R.
Amur R.
Ussuri R.
CHINA
Vladivostok
Sea of Okhotsk
Bering Sea
PACIFIC OCEAN

Areas affected by acid precipitation
Forest damage
Areas of radioactive contamination
Coastal pollution
Polluted rivers
Salinization

Figure 9.10 Environmental Issues in the Russian Domain
Varied environmental hazards have left a devastating legacy across the region. The landscape is littered with nuclear waste, heavy metals, and air pollution. Fouled lakes and rivers pose additional problems in many localities. Present economic difficulties and political uncertainties add to the costly challenge of improving the region's environmental quality in the 21st century.

Figure 9.11 Norilsk The sprawling Norilsk Nickel plant dominates this portion of the city of Norilsk. Extensive air and water pollution is the undesirable consequence of this industrial operation.

Figure 9.12 Lake Baikal Southern Siberia's Baikal is one of the world's largest deep-water lakes. Industrialization devastated water quality after 1950 as pulp and paper factories poured wastes into the lake. Recent cleanup efforts have helped, but environmental threats remain.

Take the example of the polar bear. The shrinking volume of arctic sea-ice habitat for the bears means that they are forced to widen their search for food, bringing them into closer contact with arctic villages and disrupting traditional hunting practices. Poachers have also profited, increasing their illegal harvests.

Third, rising global sea levels will hit low-lying areas of the Black and Baltic seas particularly hard. Officials in St. Petersburg, Russia's second-largest city, are already contemplating significant costs associated with controlling the Baltic's rising waters.

Finally, the largest potential change, with global implications, relates to the thawing of Siberian permafrost. Substantial areas of northern Russia are covered with permafrost that is already close to thawing. This same region has witnessed some of the most persistent large-scale global warming since 1950. Thus, even minor increases in temperature could have significant and irreversible consequences for the region, including major changes in topography (mud flows, slumping, erosion, and craters caused by the release of methane), drainage (lake coverage and rivers), and vegetation. Existing fish and wildlife populations will need to adjust in order to survive. Human infrastructure such as buildings, roads, and pipelines will also require substantial modification.

However, the greatest potential global impact may come with the huge release of carbon that is currently stored in existing permafrost environments. Permafrost soils contain large amounts of organic material that decomposes quickly when thawed. Most of the planet's permafrost could release its carbon reservoir within the next century, the equivalent of 80 years of burning fossil fuels. Such a contribution to the world's carbon budget, which would probably further warm Earth, is only beginning to be incorporated into models of global climate change. Thus, the survival of the Siberian permafrost may be a key to slowing or quickening further global warming.

The region's boreal forests also play a pivotal role in global climate change (Figure 9.14). Russia possesses about 20 percent of world forest reserves, which absorb huge amounts of carbon dioxide. But currently, unsustainable forest management policies and illegal timber poaching are contributing to expanded logging. By 2045, the beneficial impacts of these forests may be negligible, leading to more rapid buildup of carbon dioxide in the atmosphere.

Review

9.1 Compare the climate, vegetation, and agricultural conditions of Russia's European west with those of Siberia and the Russian Far East.

9.2 Describe the high environmental costs of industrialization within the Russian domain, and cite a recent example of efforts to address some of these problems.

KEY TERMS Soviet Union, oligarch, podzol soil, chernozem soil, permafrost, taiga, northern sea route

Figure 9.13 Murmansk The northern Russian port of Murmansk may see healthy growth and expansion of its harbor facilities as global warming brings more ice-free travel to sea lanes in the Arctic Ocean.

Figure 9.14 Boreal Forest Northern Russia's boreal forests are being harvested for the world's commercial lumber markets. Their long-term depletion may lead to further carbon dioxide buildup in Earth's atmosphere.

WORKING TOWARD SUSTAINABILITY

Putting a Lid on Chernobyl

Google Earth (MG) Virtual Tour Video https://goo.gl/C3EK7O

It is a sobering remembrance: April 26, 2016, marks the 30th anniversary of the Chernobyl nuclear plant disaster in northern Ukraine (Figure 9.2.1). Chernobyl was one of the world's worst nuclear nightmares and greatest environmental disasters ever. The explosion at the power plant and the meltdown of the reactor exposed tens of millions of Europeans to elevated doses of radiation, killed thousands of nearby residents, and contaminated a vast landscape surrounding the facility. Long-term effects on soils, livestock, wildlife, and human health are still being tabulated. A study published in 2010 by the New York Academy of Sciences, for example, estimates that the cumulative health effects of Chernobyl may have contributed to more than 985,000 deaths worldwide. An "exclusion zone" of about 1000 square miles (2600 square kilometers) still exists around the plant, where weed-filled city streets and abandoned farmland remain a 21st-century no-man's-land. Right after the disaster, a concrete structure (called the *sarcophagus*) was hastily constructed to contain the radioactive materials on site, but it was never meant to be a long-term solution to the problem.

Figure 9.2.2 Arch Construction near the Reactor Site Chernobyl's new Safe Confinement structure is being assembled in two halves. This view (in 2015) shows the arch being constructed near the reactor site.

Finally, however, a more sustainable approach is being taken to confine the long-term risks at the site (Figure 9.2.2). A remarkable 32,000-ton steel and concrete arch, costing more than $1.5 billion, is being assembled nearby. If construction plans remain on schedule, the otherworldly structure, more than 300 feet (90 meters) high, will be moved over the damaged plant in 2017 or 2018. Once in place, the ends of the arch will be closed off, effectively containing future radioactive dust, especially if the unsafe and aging structure within collapses. The safeguarded plant grounds will also allow for a more comprehensive cleanup of areas surrounding the site. Engineers hope that the arch can stand for at least 100 years.

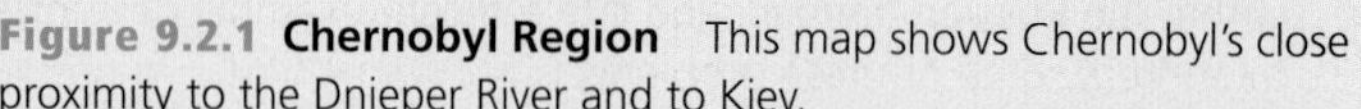

Figure 9.2.1 Chernobyl Region This map shows Chernobyl's close proximity to the Dnieper River and to Kiev.

Many challenges remain. Providing safe conditions for construction workers is a high priority. Maintaining the arch is also problematic. Most steel-supported structures such as this are painted every 15 years to protect them from rusting, but that task would introduce additional health risks. Instead, expensive rustproof stainless steel is being used, and special dehumidifiers will keep bolts and key parts from becoming too moist. Longer term, remaining fuel at the site must be removed to ensure radiation does not leak into groundwater, potentially endangering nearby Kiev. Perhaps most daunting is the country's unstable political environment, making it more difficult and potentially dangerous for global companies to participate in the containment effort. Many ask who will pay for maintaining the arch if Ukraine suffers an economic meltdown, a real possibility given its current challenges. Still, the arch is a testament to human ingenuity and the ability to address a catastrophe that reshaped this corner of the world three decades ago.

1. Is nuclear power a safe energy source today? Defend your answer.
2. Find a map of your local area focused on the college campus, and draw a 1000-square-mile (10-mile x 10-mile) zone around the facility to understand the size of Chernobyl's "exclusion zone." What local areas would be included?

Population and Settlement: An Urban Domain

The Russian domain is home to about 200 million residents (see Table A9.1). Although they are widely dispersed across Eurasia, most live in cities. The region's population geography has been influenced by the distribution of natural resources and by government policies that have encouraged migration out of the traditional centers of population in the western portions of the domain.

Population Distribution

The favorable agricultural setting of the European west offered a home to more people than did the inhospitable conditions found across central and northern Siberia. Although Russian efforts over the past century have encouraged a wider dispersal of the population, it remains heavily concentrated in the west (Figure 9.15). European Russia is home to more than 100 million people, while Siberia, although far larger, holds only about 40 million. If you add the 60 million inhabitants of Belarus, Moldova, and Ukraine, the imbalance between east and west becomes even more striking.

The European Core Sprawling Moscow and its nearby urbanized region dominate the settlement landscape with a metropolitan area containing more than 16 million people, with the majority living outside of the city's inner ring of development (Figure 9.16). Unofficial estimates—including undocumented immigrants—are even higher. Clearly Russia's primate city, Moscow produces about 20 percent of the entire nation's wealth. Moscow is projected to grow at a healthy pace. In 2014, government officials embraced an ambitious urban plan that emphasized continued decentralization and automobile-oriented sprawl, especially to the south (Figure 9.17). Officials hope this takes pressure off the city's crowded core, making it more livable.

On the shores of the Baltic Sea, St. Petersburg (4.9 million people) has traditionally had a great deal of contact with western Europe. Between 1712 and 1917, it served as the capital of the Russian Empire. Its handsome buildings, bridges, and canals give it an urban landscape many have compared to the great cities of western Europe (Figure 9.18).

Other urban clusters are located along the lower and middle stretches of the Volga River, including the cities of Kazan, Samara, and Volgograd. This highly commercialized river corridor, also containing important petroleum reserves, supports a diverse industrial base. Nearby, the resource-rich Ural Mountains include the gritty industrial landscapes of Yekaterinburg (1.4 million) and Chelyabinsk (1.1 million).

Other population clusters are found in Belarus and Ukraine (see Table A9.1). The Belarusian capital of Minsk (1.8 million) dominates that country, and its landscape recalls the drab Soviet-style architecture of an earlier era. Ukraine's capital, Kiev (Kyiv, 2.8 million), straddles the Dnieper River. Kiev's urban landscape offers a mix of traditional architecture and sprawling high-rise apartment buildings (Figure 9.19). Ukraine's major eastern cities have witnessed widespread depopulation and economic chaos because of the civil war that decimated that portion of the country in 2014 and 2015.

Siberian Hinterlands A passenger leaving the southern Urals on a Siberia-bound train becomes aware that the land ahead is ever more sparsely settled (see Figure 9.15). The distance between cities grows, and

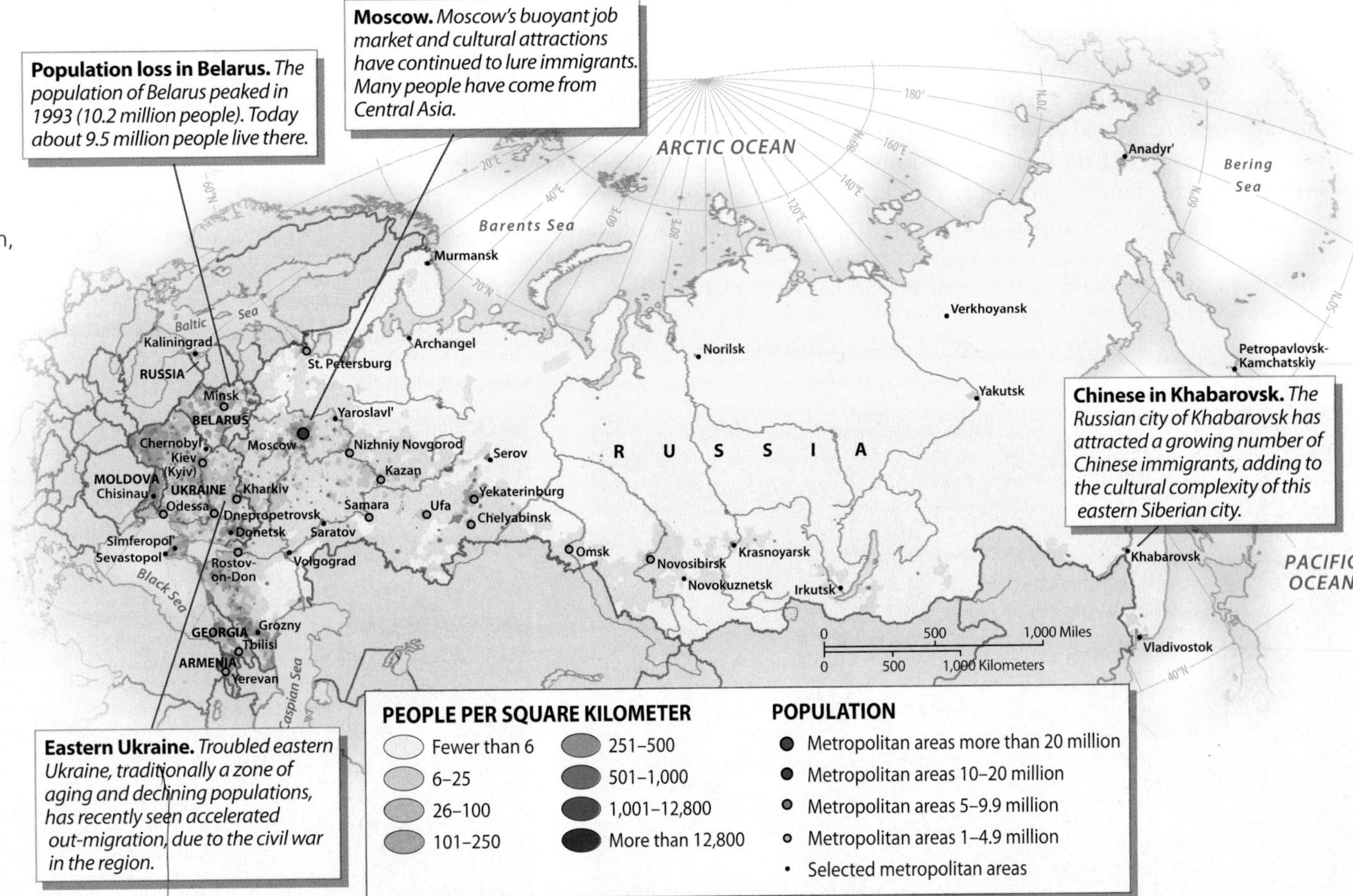

Figure 9.15
Population Map of the Russian Domain The region's population is largely clustered west of the Ural Mountains. Dense agricultural settlements, extensive industrialization, and large urban centers are found of Ukraine, in much of Belarus, and across western Russia south of St. Petersburg and Moscow.

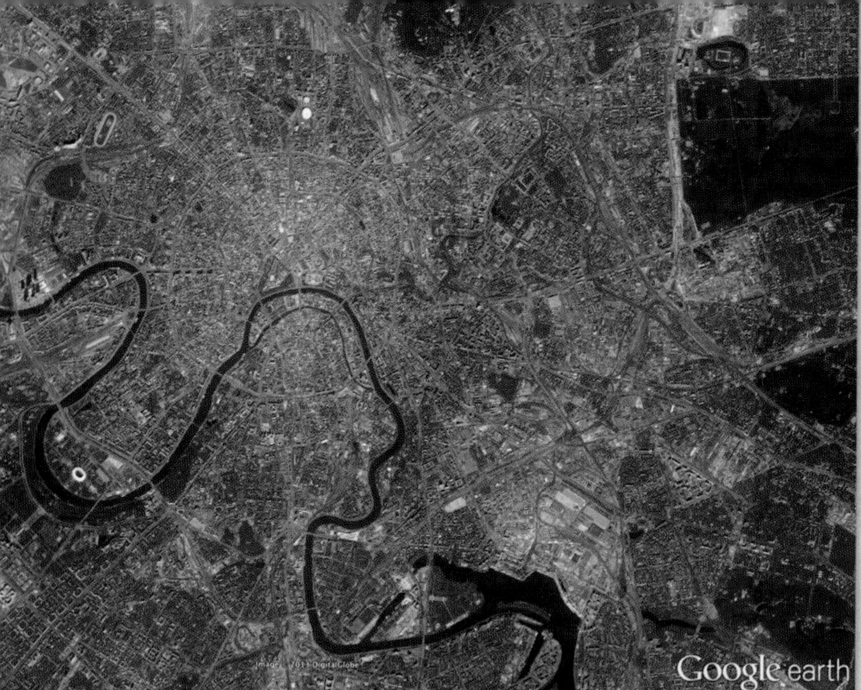

Figure 9.16 Metropolitan Moscow Sprawling Moscow extends more than 50 miles (80 kilometers) beyond the city center, on both sides of the Moscow River. The larger metropolitan area is home to more than 16 million people, and the relative strength of its urban economy continues to attract migrants from elsewhere in the country, thus putting more pressure on its infrastructure.

Figure 9.17 Moscow Traffic The capital's Ring Road (the MKAD) features North American–style interchanges and traffic jams.

the intervening countryside reveals a landscape shifting gradually from farms to forest. To the south, a collection of isolated, but sizable, urban centers follows the **Trans-Siberian Railroad**, a key railroad passage to the Pacific completed in 1904. The eastbound traveler encounters Omsk (1.1 million people) as the rail line crosses the Irtysh River, Novosibirsk (1.5 million) at its junction with the Ob River, and Irkutsk (600,000) near Lake Baikal. The port city of Vladivostok (600,000) marks the end of the Trans-Siberian Railroad and provides access to the Pacific. To the north, a thinner sprinkling of settlements appears along the **Baikal–Amur Mainline (BAM) Railroad** (completed in 1984), which parallels the older line but runs north of Lake Baikal to the Amur River. From the BAM line to the Arctic, the almost empty spaces of central and northern Siberia dominate the scene, interrupted only rarely by small settlements, often oriented around natural resource extraction (see Figure 9.11) (see *Geographers at Work: Exploring Arctic Russia's Changing Urban Landscape*).

Regional Migration Patterns

Over the past 150 years, millions of people within the Russian domain have been on the move. These major migrations, both forced and voluntary, reveal sweeping examples of human mobility that rival the great movements from Europe and Africa or the transcontinental spread of settlement across North America.

Eastward Movement Just as settlers of European descent moved west across North America, exploiting natural resources and displacing native peoples, European Russians moved east across the Siberian frontier. Although these migrations into Siberia began several centuries earlier, the pace accelerated in the late 19th century as the Trans-Siberian Railroad was being constructed. Peasants were attracted to the region by its agricultural

Figure 9.18 St. Petersburg Often called Russia's most beautiful city, St. Petersburg's urban design features a varied mix of gardens, open space, waterways, and bridges. This view shows the ornate architecture (center) of the Church of the Savior of Spilled Blood, completed in 1907.

Figure 9.19 Kiev and the Dnieper River The ornate domes of the Kiev Petchersk Lavra (one of the historical centers of Eastern Orthodox Christianity) frame this view of the city, which includes many new buildings (in the distance) constructed on the east bank of the Dnieper River.

GEOGRAPHERS AT WORK

Exploring Arctic Russia's Changing Urban Landscape

Figure 9.3.1 Geographer Dmitry Streletskiy services a weather station that measures permafrost temperature near Deadhorse, Alaska.

"I was always into the outdoors and hiking," says **Dmitry Streletskiy**, who examines the impacts of climate change on urban settlements in the Russian Arctic. "I got into geography because I thought it was all about the field work, all about being outside." He was also interested in travel: "Tourists pay money to go to places, but geographers are professionals who get paid to go places!" After finishing a master's degree in geography from Moscow University, Streletskiy ventured to the United States to complete his doctoral studies in climatology at the University of Delaware. Now based at George Washington University, Streletskiy studies how warming permafrost affects human activities in Alaska and Russian Arctic.

Effects of Climate Change Streletskiy's work, supported by the National Science Foundation, has taken him to various settlements in Arctic Russia where he examines the impact of climate change in communities such as Igarka and Norilsk (Figure 9.3.2). Most arctic residents actually live in densely settled urban areas, working in energy and natural resource–related industries. Streletskiy found that a majority of building foundations constructed in the 1970s were not designed to account for damage due to climate change impacts, such as thawing permafrost leads to unstable building foundations, and affecting roads and other types of infrastructure. He also found strong geographic variability in permafrost degradation, depending on local site and climatic conditions.

Streletskiy partners with experts in climate change, geographic information systems, urban sustainability, and engineering to assess future impacts given different climate change scenarios and to propose ways to improve urban sustainability in high-latitude environments. Although much of his work keeps him indoors, Streletskiy still enjoys being outdoors and taking in the landscape with a geographer's eye: "I know why the river flows this way, I know why the mountains are there … I like to figure out how things work."

Figure 9.3.2 Degraded Permafrost Reshapes the Siberian Landscape The Siberian town of Igarka has lost much of its population as nearby resource industries have declined. Many of its buildings have also been damaged by processes associated with permafrost degradation.

1. In addition to the effects of human-induced climate change, how else might permafrost in arctic settlements become unstable or damaged?
2. In your local setting, describe a climatic hazard that could have short- or long-term impacts on buildings or infrastructure in your area.

opportunities (in the south) and by greater political freedoms than they traditionally enjoyed under the **tsars** (or czars; Russian for *Caesar*), the authoritarian leaders who ruled the pre-1917 Russian Empire.

Political Motives Political motives shaped migration patterns in the Russian domain. Particularly in the case of Russia, both the imperial and the Soviet-era leaders forcibly relocated people, extending Russian political and economic power into the Eurasian interior. Political dissidents and troublemakers in the Soviet era were exiled to the region's **Gulag Archipelago**, a vast collection of political prisons in which inmates often disappeared or spent years far removed from their families and communities.

Russification, the Soviet policy of resettling Russians into non-Russian portions of the Soviet Union, also changed the region's human geography. Millions of Russians were given economic and political incentives to move elsewhere in the Soviet Union in order to increase Russian dominance in many outlying portions of the country. By the end of the Soviet period, Russians made up significant minorities within former Soviet republics such as Kazakhstan (30 percent Russian), Latvia (30 percent), and Estonia (26 percent).

Since 2014, another wave of politically motivated migration has rocked the region. War-torn areas of eastern Ukraine, caught in the crossfire between Russian rebel and Ukrainian forces, have lost about one-third of their population. About half of these displaced peoples have moved elsewhere within Ukraine, and the other half (largely ethnic Russians) have spilled into nearby Russia, overwhelming border zones with new refugee populations.

New International Movements Other recent migrations cross international borders (Figure 9.20). In the post-Soviet era, Russification has often been reversed. Several of the newly independent non-Russian countries imposed rigid language and citizenship requirements, which encouraged Russian residents to leave. In other settings, ethnic Russians experienced varied forms of discrimination. The Russian government also has promoted a repatriation program for ethnic Russians worldwide, offering incentives for Russian-speaking migrants to return

Figure 9.20 Recent Migration Flows in the Russian Domain Ethnic Russians have returned from former Soviet republics, while other Russians are emigrating from the domain for economic, cultural, and political reasons. Within Russia, both political and economic forces are also at work, encouraging people to be on the move. Note how much of this activity, both inflow and outflow, is centered on Moscow.

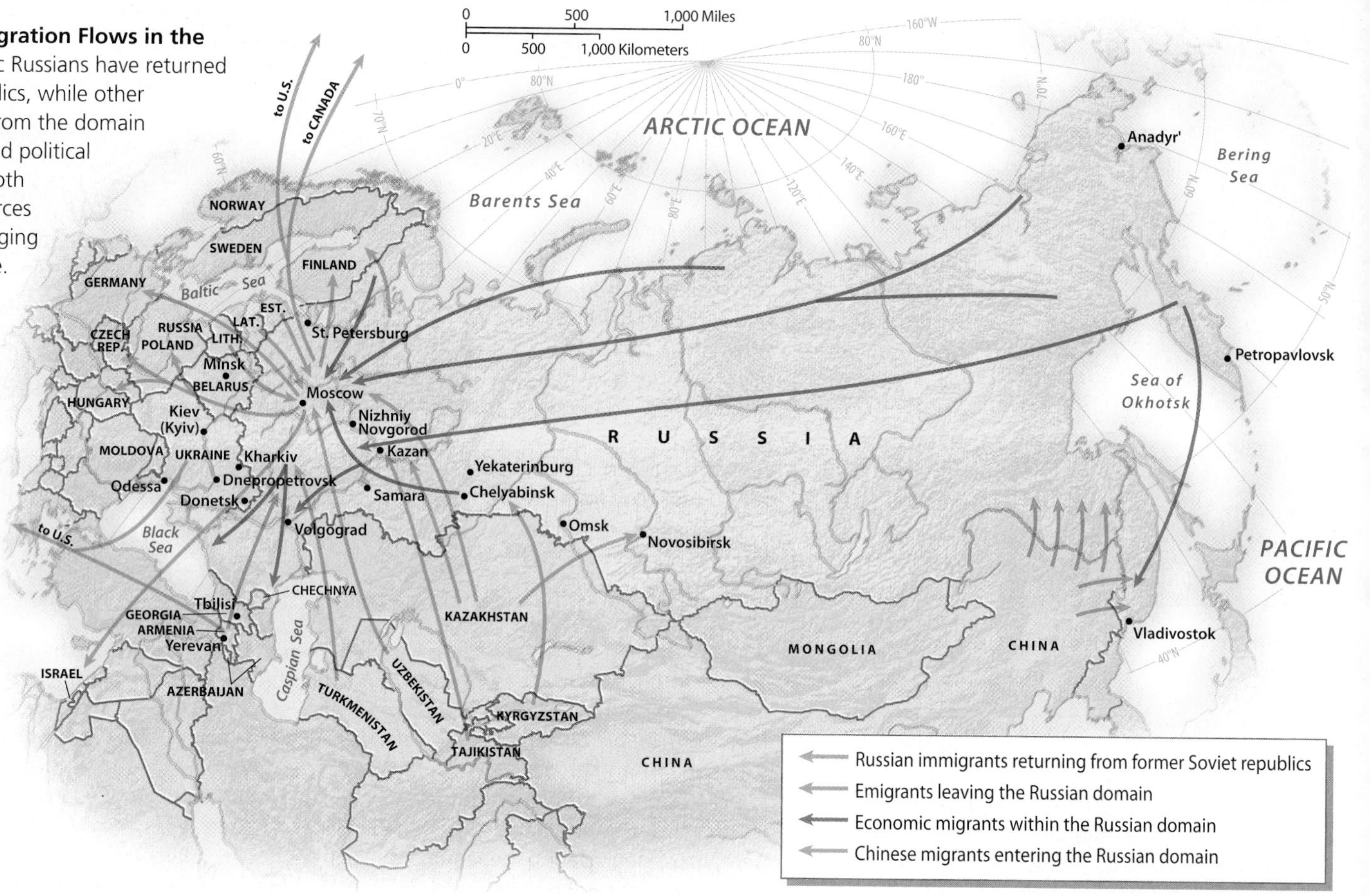

or move to their cultural homeland. As a result, Central Asian and Baltic countries, once part of the Soviet Union, have seen their Russian populations decline significantly since 1991, often by 20 to 35 percent.

Russia has experienced a growing immigrant population, many of them undocumented migrants drawn to the country for work. The story has a familiar ring to it. More than 11 million undocumented immigrants are suspected to be in the country. Most are young males who come for better-paying jobs. Russia's government recently implemented tighter controls to restrict border crossings and enacted tougher penalties against businesses that hire illegal immigrants. There is a growing national debate concerning how many legal foreign workers should be allowed in the country and whether or not undocumented immigrants should be granted amnesty.

The actual flows of undocumented immigrants are complex. Most immigrants—legal and undocumented—come from portions of the former Soviet Union, especially from ethnically non-Slavic regions in Central Asia (Figure 9.21). Almost one-third of the country's undocumented immigrants may live in the job-rich Moscow metropolitan region. Many send money they earn in Russia back home to their native lands. One estimate suggests about 20–30 percent of Tajikistan's economy is based on funds received from émigrés living and working in Russia.

In addition, immigration into Russia's Far East, principally from northern China, is reshaping the economic and cultural geographies of that region. Walk through the Russian cities of Vladivostok and Khabarovsk, and you will see that street signs feature both Russian and Chinese lettering. Entire neighborhoods are dominated by immigrants. Chinese children are learning Russian in school, and Russians find themselves working for Chinese entrepreneurs. A significant nationalist backlash has occurred: Chinese are sometimes attacked, Chinese shopkeepers complain of being rousted by Russian police, and recent legislation has made it harder for Chinese to operate businesses. On the other hand, many younger Russians have welcomed their Chinese counterparts, a growing number of joint Russian–Chinese companies have appeared in the region, and there is more intermarrying between the two groups.

The domain's more open borders make it easier for other residents to leave the region (see Figure 9.20). Deteriorating economic conditions and the region's unpredictable politics have encouraged many to emigrate. The "brain drain" of young, well-educated, upwardly mobile Russians has been considerable. Sometimes, ethnic links play a part. For example, many Russian-born ethnic Finns have moved to nearby Finland, much

Figure 9.21 Central Asian Immigrants in Moscow These migrants from Tajikistan work at an outdoor market near Moscow's Kiev Railway Station.

Figure 9.22 Downtown Moscow Shoppers stroll pleasant Arbat Street in downtown Moscow. Upscale shops, restaurants, and entertainment venues attract both visitors and local residents.

Figure 9.23 New Single-Family Housing, Moscow Suburbs This newer upscale housing development in the Moscow suburbs was built by INKOM-Nedvizhimost. The large, neatly fenced lots and spacious homes build on North American traditions of suburban taste and design.

to the consternation of the Finnish government. Russia's Jewish population also continues to fall, a pattern begun late in the Soviet period. These emigrants have flocked mostly to Israel or the United States.

Inside the Russian City

Today most people in the Russian domain live in cities, the product of a century of urban migration and growth (see Table A9.1). Large Russian cities possess a core area, or center, featuring superior transportation connections; the best-stocked, upscale department stores and shops; desirable housing; and the most important offices (both government and private) (Figure 9.22). The largest urban centers, such as Moscow and St. Petersburg, also feature extensive public spaces and examples of monumental architecture. Within the city, there is usually a distinctive pattern of circular land-use zones, each built at a later date moving outward from the center. Such a ring-like urban morphology is not unique. However, as a result of the extensive power of Soviet-era government planners, this urban form is probably more highly developed here than in most parts of the world.

The cores of many older cities predate the Soviet Union. Pre-1900 stone buildings often dominate older city centers. Some of these are former private mansions that were converted to government offices or subdivided into apartments during the communist period but are now being privatized again. Many of these older buildings, however, are being leveled in rapidly growing districts such as downtown Moscow. Retail malls have replaced many of these older structures. Nearby nightclubs and bars are filled with pleasure seekers as the city's professional elites mingle with foreign visitors and tourists.

Farther out from the city centers are **mikrorayons**: large, Soviet-era housing projects of the 1970s and 1980s. Mikrorayons are typically massed blocks of standardized apartment buildings, ranging from 9 to 24 stories in height. The largest of these supercomplexes contain up to 100,000 residents. Soviet planners hoped that mikrorayons would foster a sense of community, but most now serve as anonymous bedroom communities for larger metropolitan areas.

Some of Russia's most rapid urban growth has occurred on the metropolitan periphery, paralleling the North American experience. Moscow, for example, is increasingly oriented around the automobile, and its urban reach has expanded far beyond the city center. Land prices and tax rates are also lower there than in the central city. New suburban shopping malls and housing districts featuring single-family homes are popping up on the urban fringe, allowing upscale residents to live and shop without having to visit the city center (Figure 9.23).

The Demographic Crisis

Russia has identified population loss as a key issue of national importance. Some government and UN estimates indicate that Russia's population could fall by a startling 45 million by 2100. Similar conditions are affecting the other countries within the region (see Table A9.1). Beginning with World War II, large numbers of deaths combined with low birth rates to produce sizable population losses. Although the population increased in the 1950s, growth slowed by 1970, and death rates began exceeding birth rates in the early 1990s. Two population pyramids tell the troubling tale (Figure 9.24). The first shows that, whereas adult-age populations are prominently represented (with the exception of older males and the birth crash of World War II), relatively few children are being added.

President Putin has declared that demographic decline is Russia's "most acute problem." He has pushed for programs aimed at raising birth rates and challenging the one-child family norm that has become widely accepted. Under the plan, mothers of multiple children receive cash payments, extended maternity leave, and extensive day-care subsidies. Some Russian cities have even sponsored competitions, encouraging couples to have babies in the hope of winning prizes such as automobiles.

Recently, birth rates rose slightly within Russia, perhaps a function of changing government policies. Russian birth rates are now significantly higher than those of Germany, Italy, and Japan. In particular, rates among the country's ethnic non-Russians (for example, in the Caucasus region and portions of Siberia) have been significantly above those of ethnic Russians.

Still, many uncertainties remain concerning future demographic growth. The current economic decline and political instability in the region may depress birth rates. Moreover, residents still face significantly higher death rates than much of the developed world, and

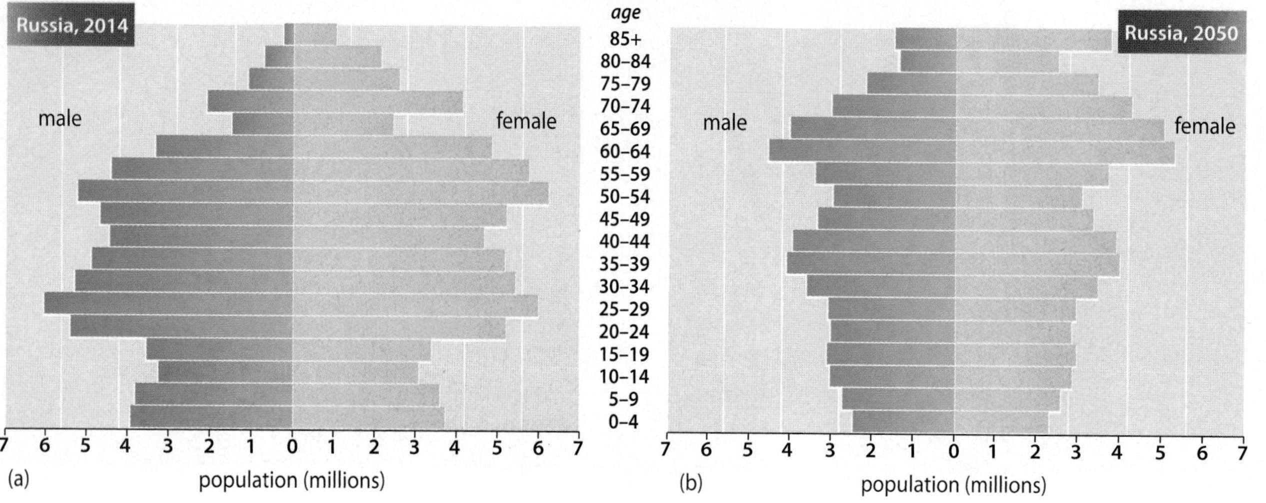

Figure 9.24 Russia's Changing Population Two population pyramids provide (a) a recent glimpse (2014) and (b) a predicted pattern (2050) of Russia's population structure. Present trends suggest that Russia's population will continue to age, with relatively fewer young people supporting a relatively large elderly population. **Q: Among Russian males, what is the evidence of the impact of earlier wars and higher death rates?**

they also have daunting health-care challenges. In addition, the region's demographic structure forecasts fewer women of childbearing age in the near future, making the recent increases difficult to maintain.

Review

9.3 Discuss how major river and rail corridors have shaped the geography of population and economic development in the region. Provide specific examples.

9.4 Contrast Soviet and post-Soviet migration patterns within the Russian domain, and identify the changing forces at work.

9.5 Describe some of the major land-use zones in the modern Russian city, and suggest why it is important to understand the impact of Soviet-era planning within such settings.

KEY TERMS Trans-Siberian Railroad, Baikal–Amur Mainline (BAM) Railroad, tsar, Gulag Archipelago, Russification, mikrorayon

Cultural Coherence and Diversity: The Legacy of Slavic Dominance

For hundreds of years, Slavic peoples speaking the Russian language expanded their influence from an early homeland in central European Russia. Russian cultural patterns and social institutions spread widely during this Slavic expansion, influencing non-Russian ethnic groups that continued to live under the rule of the Russian Empire. The legacy of this diffusion continues today, offering Russians a rich historical identity. It also provides a framework for examining how present-day Russians deal with forces of globalization and how non-Russian cultures have evolved in the region.

The Heritage of the Russian Empire

The Russian Empire's expansion paralleled similar events in western Europe. As Spain, Portugal, France, and Britain carved out overseas empires, Russia expanded eastward and southward across Eurasia. The origin of the Russian Empire lies in the early history of the **Slavic peoples**, a northern branch of the Indo-European ethnolinguistic family. Slavic political power grew by 900 CE as these people intermarried with southward-moving warriors from Sweden known as *Varangians*, or *Rus*. Within a century, the state of Rus extended from Kiev (the capital) to near the Baltic Sea. The new Kiev-Rus state interacted with the Greek Byzantine Empire, and this influence brought Christianity and the Cyrillic alphabet to the region. Even as Russians converted to **Eastern Orthodox Christianity**, a form of Christianity linked to eastern Europe and with church leaders in Constantinople (modern Istanbul), their Slavic neighbors to the west (Poles, Czechs, Slovaks, Slovenians, and Croatians) accepted Catholicism. This early Russian state then faltered, splitting into several principalities later ruled by invading Mongols and Tatars.

In the 14th century, northern Slavic peoples overthrew Tatar rule and established an expanding Slavic state (Figure 9.25). The new Russian Empire's core lay near the eastern fringe of the old state of Rus. Gradually, this area's language diverged from that spoken in the new core, and *Ukrainians* and Russians developed into two separate peoples. A similar development took place among northwestern Russians, who experienced several centuries of Polish rule and were transformed into a distinctive group known as *Belarusians*.

The Russian Empire expanded remarkably in the 16th and 17th centuries. Former Tatar territories in the Volga Valley (near Kazan) were incorporated into the Russian state in the mid-1500s. The Russians also allied with the seminomadic **Cossacks**, Slavic-speaking Christians who had earlier migrated to the region seeking freedom in the ungoverned steppes. This alliance smoothed the way for Russian expansion into Siberia during the 17th century. Furs and precious metals were the region's chief attractions.

Westward expansion was slow and halting. When Tsar Peter the Great (1682–1725) defeated Sweden in the early 1700s, he obtained a foothold on the Baltic Sea. There he built the new capital city of St. Petersburg, designed to give the empire better access to western Europe. Later in the 18th century, Russia defeated both the Poles and the Turks and gained all of modern-day Belarus and Ukraine. Tsarina Catherine the Great (1762–1796) was especially important in colonizing Ukraine and bringing the Russian Empire to the warm-water shores of the Black Sea.

The 19th century witnessed the Russian Empire's final expansion, mostly into Central Asia, where once-powerful Muslim states could no longer resist the Russian army. The mountainous Caucasus region proved a greater challenge, as residents used the rugged terrain in defending their lands. South of the Caucasus, however, Christian Armenians and Georgians accepted Russian power with little struggle because they found it preferable to rule by the Persian or the Ottoman empire.

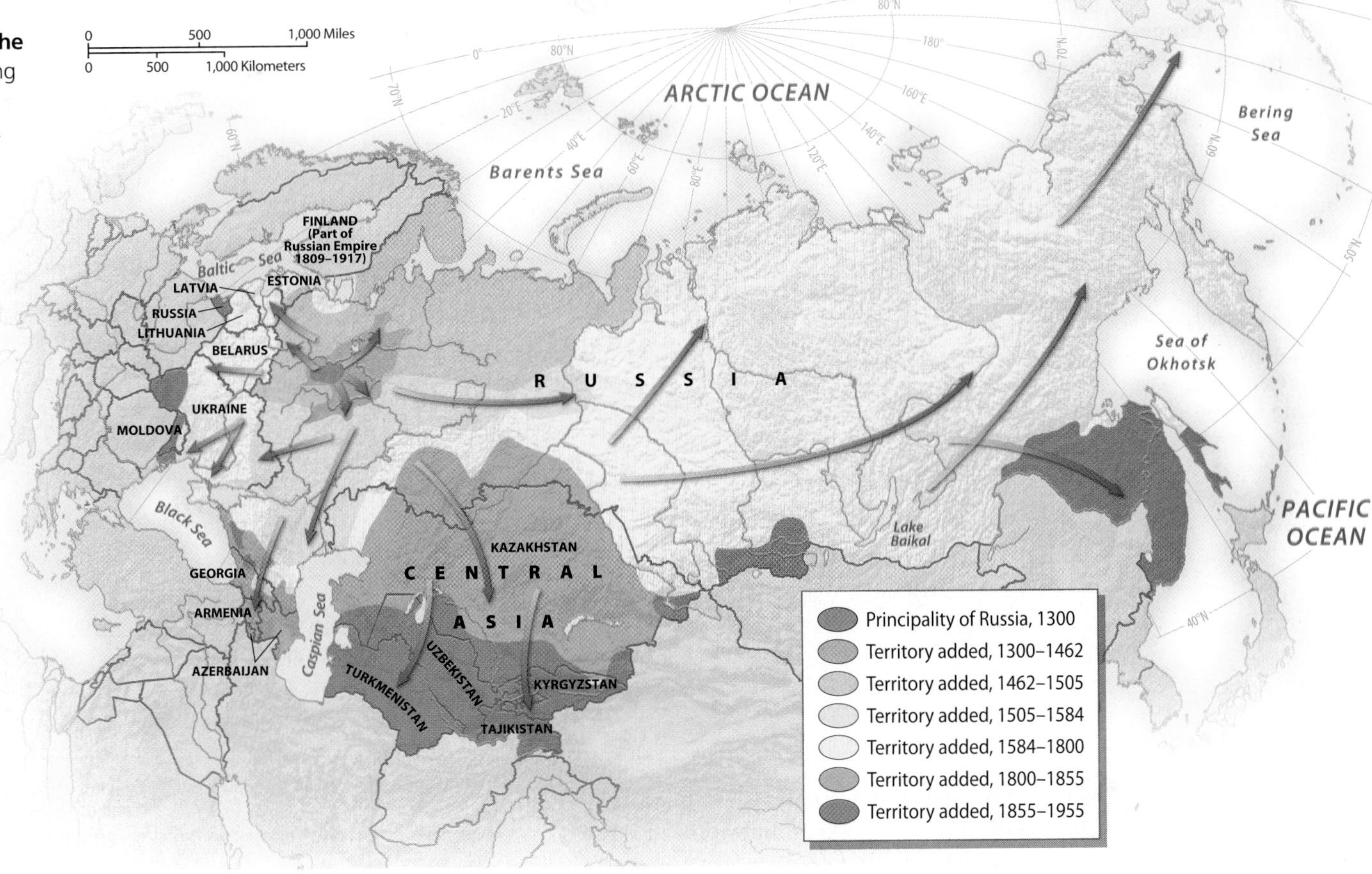

Figure 9.25 Growth of the Russian Empire Beginning as a small principality in the vicinity of modern Moscow, the Russian Empire took shape between the 14th and 16th centuries. After 1600, Russian influence stretched from eastern Europe to the Pacific Ocean. Later, portions of the empire were added in the Far East, in Central Asia, and near the Baltic and Black seas.

Geographies of Language

Slavic languages dominate the region (Figure 9.26). The geographic pattern of the Belarusian people is relatively simple: Most Belarusians reside in Belarus, and most people in Belarus are Belarusians. Ukraine's situation is much more complex (Figure 9.27). Russian speakers historically dominate large parts of eastern Ukraine and form the foundation of rebel resistance forces in that part of the country (along with Russian military support). Western Ukraine, however, contains fewer Russian speakers. Similarly, the Crimean Peninsula, while still considered part of Ukraine, has long ethnic and historical connections to Russia, facilitating that country's occupation of the peninsula in 2014.

Russians inhabit most of European Russia and make up about 80 percent of its population. The Russian zone extends across southern Siberia to the Sea of Japan. In sparsely settled central and northern Siberia, Russians share territory with varied native peoples.

In Moldova, Romanian (a Romance language) speakers dominate, although ethnic Russians and Ukrainians each make up about 13 percent of the country's population. Now that Russia, Ukraine, and Moldova are separate countries with a heightened sense of national distinction, ethnic minorities have become a significant source of tension across the region.

Other non-Slavic peoples also shape the region's cultural geography. The Finno-Ugric peoples include Finnish-speaking settlers who dominate sizable portions of the non-Russian north. Altaic speakers further complicate the country's linguistic geography. They include Volga Tatars, centered on the city of Kazan. While retaining their ethnic identity, the Turkic-speaking Tatars have extensively intermarried with their Russian neighbors. Yakut and Evenki peoples of northeast Siberia also speak Turkic languages (Figure 9.28). In the east, over 400,000 Buryats live near Lake Baikal and are tied to the cultures and history of Central Asia.

The plight of many native peoples in central and northern Siberia parallels the situation in the United States, Canada, and Australia. Poor, rural indigenous peoples remain distinct from dominant European cultures. These peoples are internally diverse, often divided into unrelated linguistic groups. Many Siberian peoples have seen traditional ways challenged by Russification, just as native peoples elsewhere have been subjected to similar pressures of cultural and political assimilation.

Transcaucasia offers a bewildering variety of languages (Figure 9.29). From Russia, along the north slopes of the Caucasus, to Georgia and Armenia east of the Black Sea, a convoluted history and rugged physical setting combine to produce one of the world's most complex linguistic maps. Several language families are found in a region smaller than Ohio, and individual languages are spoken by small, isolated cultural groups.

Geographies of Religion

Most Russians, Belarusians, and Ukrainians share an Eastern Orthodox Christian heritage. For hundreds of years, Eastern Orthodoxy served as a central cultural presence in the Russian Empire. Indeed, church and state were tightly fused until the empire's demise in 1917. Under the Soviet Union, all religion was discouraged and persecuted. However, with the downfall of the Soviet Union, a religious revival swept much of the Russian domain. Now about 75 million Russians are members of the Orthodox Church, including almost 500 monastic orders dispersed across the country.

Other forms of Christianity are present. Western Ukrainians, who experienced several hundred years of Polish rule, eventually joined

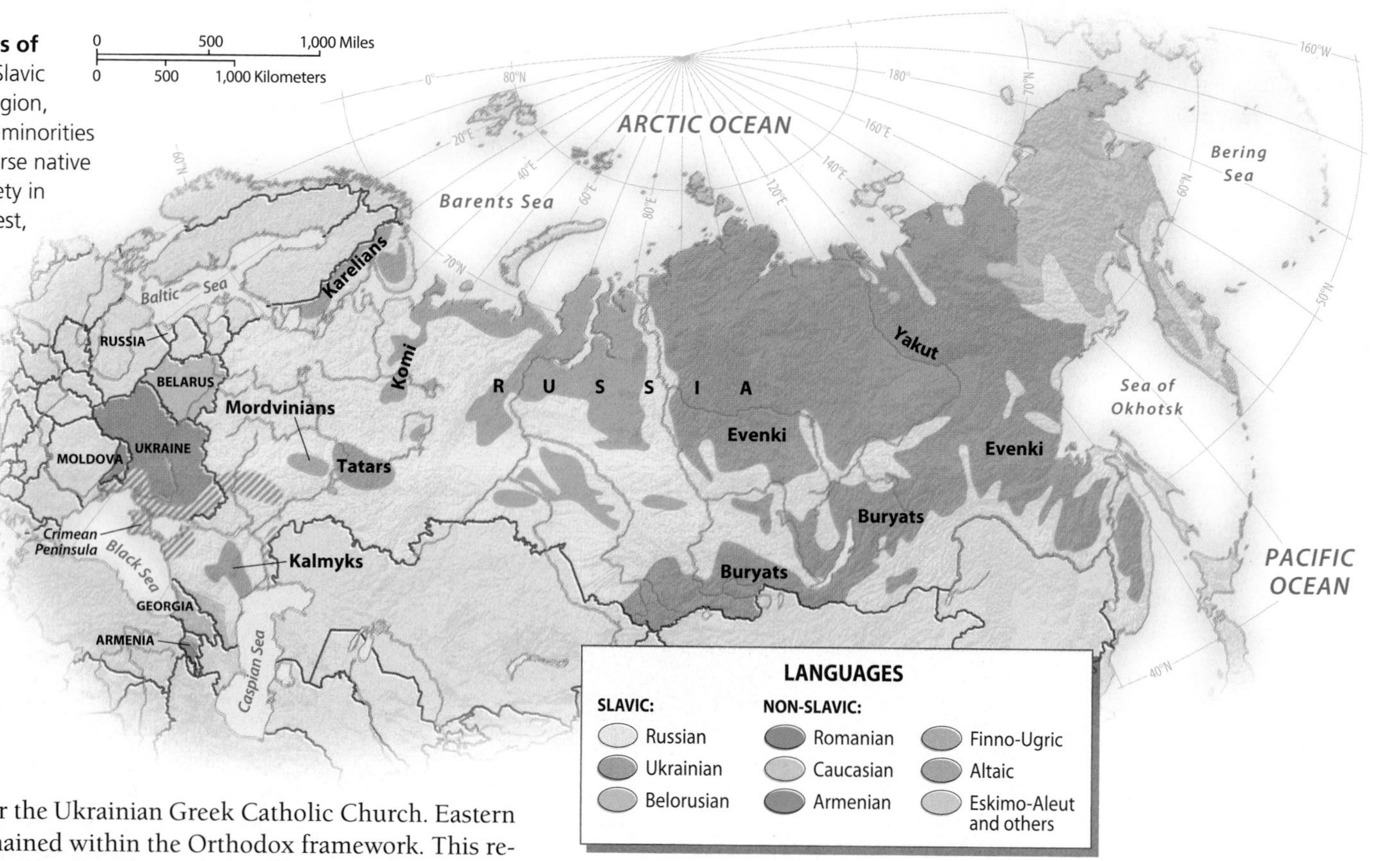

Figure 9.26 Languages of the Russian Domain Slavic Russians dominate the region, although many linguistic minorities are present. Siberia's diverse native peoples add cultural variety in that area. To the southwest, the Caucasus Mountains and the lands beyond contain the region's most complex linguistic geography. Ukrainians and Belarusians, while sharing a Slavic heritage with their Russian neighbors, add further variety in the west.

LANGUAGES
SLAVIC: Russian, Ukrainian, Belorusian
NON-SLAVIC: Romanian, Caucasian, Armenian, Finno-Ugric, Altaic, Eskimo-Aleut and others

the Catholic Church or the Ukrainian Greek Catholic Church. Eastern Ukraine, however, remained within the Orthodox framework. This religious split reinforces cultural differences between eastern and western Ukrainians. Elsewhere, Armenia has a long Christian tradition, but it differs somewhat from both Eastern Orthodox and Catholic practices. Evangelical Protestantism is also on the rise.

In addition, non-Christian religions shape regional cultural geographies. Islam is the largest non-Christian religion. Russia has some 7000 mosques and approximately 20 million adherents (Figure 9.30). Most are Sunni Muslims, and they include peoples in the North Caucasus, the Volga Tatars, and Central Asian peoples near the Kazakhstan border. Growth rates among Russia's Muslim populations are three times those of the non-Muslim populations.

Islamic fundamentalism has grown more popular, particularly among Muslims in the Caucasus region, who increasingly resist what they see as strong-arm tactics and repressive actions on the part of the Russian government. Elsewhere, a growing number of young, conservative Muslim Salafists living in Russian Tatarstan draw sharp distinctions between themselves and traditional Sunnis in the region, raising ethnic and religious tensions. Local authorities, as well as

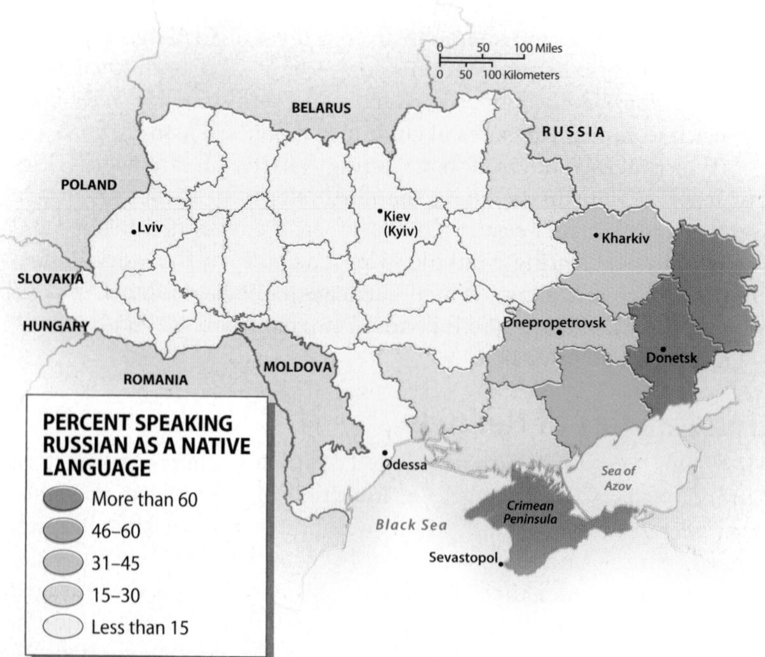

Figure 9.27 The Russian Language in Ukraine Both eastern Ukraine and the Crimean Peninsula retain large numbers of Russian speakers, a function of long cultural and political ties that continue to complicate Ukraine's contemporary human geography.

Figure 9.28 Minority Evenki Russia's indigenous Evenki population speaks an Altaic language and shares many of the challenges faced by native peoples in other world regions. **Q: What social and economic problems might the Evenki share with many indigenous peoples in North America and Australia?**

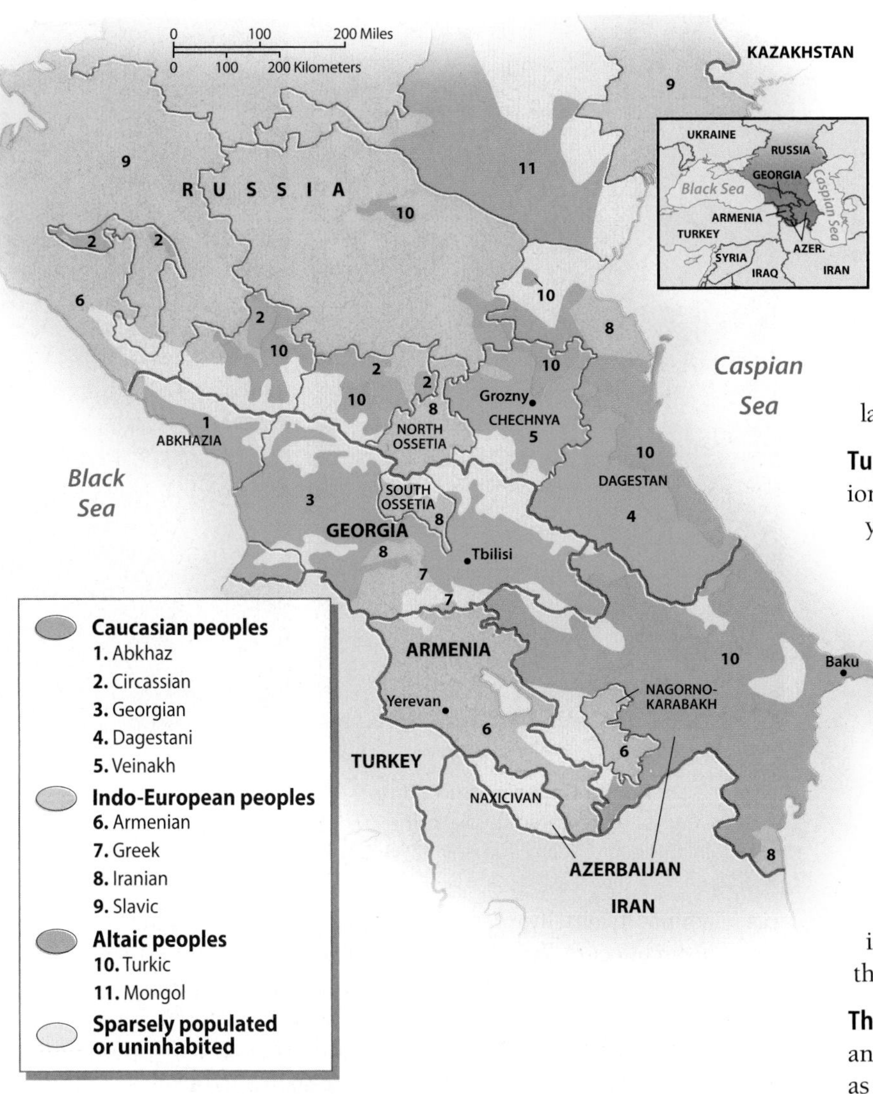

Figure 9.29 Peoples of the Caucasus Region Residents of the Caucasus region speak a complicated mosaic of Caucasian, Indo-European, and Altaic languages. Political problems have periodically erupted in the region as local populations struggle for more autonomy.

Moscow leaders, have drafted laws designed to limit the rising influence of the Salafists, but many observers feel this will only add to their appeal among the region's disaffected youth.

Russia, Belarus, and Ukraine are home to more than 1 million Jews, who are especially numerous in the larger cities of the European domain. Jews suffered severe persecution under both the tsars and the communists. Recent out-migrations, prompted by new political freedoms, have further reduced their numbers in these countries. Buddhists also are represented in the region, associated with the Kalmyk and Buryat peoples of the Russian interior. Indeed, Buddhism has enjoyed a recent renaissance and now claims approximately 1 million practitioners, mostly in Asiatic Russia.

Russian Culture in Global Context

Russian culture has developed its own distinctive traditions and symbols, and it has also been influenced greatly by western Europe. By the 19th century, even as Russian peasants interacted rarely with the outside world, Russian high culture had become thoroughly Westernized, and Russian composers, novelists, and playwrights gained considerable fame in Europe and the United States.

Soviet Days During the Soviet period, new cultural influences shaped the socialist state. Initially, European-style modern art flourished in the Soviet Union, encouraged by Marxist rhetoric. By the late 1920s, however, Soviet leaders turned against modernism, viewing it as the decadent expression of the capitalist world. Instead, state-sponsored Soviet artistic productions centered on **socialist realism**, a style devoted to the realistic depiction of workers heroically challenging nature or struggling against capitalism. Still, traditional high arts, such as classical music and ballet, received generous state subsidies, and to this day Russian artists regularly achieve worldwide fame.

Turn to the West By the 1980s, it was clear that the attempt to fashion a new Soviet culture based on communist ideals had failed. The younger generation was more inspired by fashion and rock music from the West. American mass-consumer culture proved particularly attractive.

After the fall of the Soviet Union, global cultural influences grew, particularly in larger cities such as Moscow. Western books and magazines flooded shops, residents sought advice about home mortgages and condominium purchases, and they enjoyed the newfound pleasures of fake Chanel handbags and McDonald's hamburgers. People embraced the world that their former leaders had warned them about for generations. Not all cultural influences streaming into the country were Western in inspiration. Films from Hong Kong and Mumbai (Bombay), as well as the televised romance novels (*telenovelas*) of Latin America, for example, proved even more popular in the Russian domain than in the United States.

The Music Scene Younger residents have embraced popular music, and their enthusiasm for American and European performers, as well as their support of a budding home-grown music industry, symbolizes the changing values of a post-Soviet generation. Today Russian MTV

Figure 9.30 Moscow Mosque Moscow's growing immigrant population includes many Muslims from portions of the former Soviet Union, particularly from Central Asia. Today their presence is an increasingly visible element in the capital's cultural landscape.

reaches most of the nation's younger viewers. Major global media companies such as Sony Music Entertainment established Russian operations in the post-Soviet era. Universal also opened the way for Russian performers to go global.

Moscow has hosted the Eurovision awards, a regional competition highlighting new musical talent from various European countries. In 2015, the latest generation of regional musicians made major appearances at the 60th annual Eurovision Song Contest in Vienna, Austria (Figure 9.31). Russia's representative was Polina Gagarina, a Moscow native whose "A Million Voices" became a widely popular single and music video (in translation) around the English-speaking world. Another strong contender was Ukrainian heartthrob Eduard Romanyuta. Representing Moldova in the competition (Ukraine withdrew given its political unrest), the Ukrainian singer and actor performed "I Want Your Love," an American-style pop ballad that proved popular with teen girls across the region and beyond.

(a)

(b)

Figure 9.31 Polina Gagarina and Eduard Romanyuta (a) Gagarina, a Moscow native, represented Russia in the 2015 Eurovision Song Contest. (b) Ukrainian singer Eduard Romanyuta sang for Moldova.

 **Review**

9.6 What were the key phases of colonial expansion during the rise of the Russian Empire, and how did each enlarge the reach of the Russian state?

9.7 Identify some of the key ethnic minority groups (as defined by language and religion) within Russia and neighboring states.

KEY TERMS Slavic people, Eastern Orthodox Christianity, Cossack, socialist realism

Geopolitical Framework: Growing Instability Across the Region

The geopolitical legacy of the former Soviet Union still weighs on the Russian domain. After all, the bold lettering of the "Union of Soviet Socialist Republics" dominated the Eurasian map for much of the 20th century, and the country's political influence affected every part of the world. Under President Putin, Russia's post-2000 political resurgence signaled a return to its Soviet-era geopolitical dominance within the region (Figure 9.32). Recent events in Ukraine and in neighboring countries demonstrate Russia's willingness to reassert itself on the regional stage, but sharp global reaction in the West suggests that the country could isolate itself from mainstream global political initiatives and institutions. In response, Russia may pivot to the east, forging closer relations with China.

Geopolitical Structure of the Former Soviet Union

The Soviet Union rose from the ashes of the Russian Empire, which collapsed abruptly in 1917. The Russian tsars did little to modernize the country or improve the lives of the peasant population. After the tsars fell, a government representing several political groups assumed authority. Soon, however, the **Bolsheviks**, a faction of Russian communists representing industrial workers, seized power. They espoused the doctrine of **communism**, a belief based on the writings of Karl Marx that promoted the overthrow of capitalism by the workers, large-scale elimination of private property, state ownership and central planning of major sectors of the economy (both agricultural and industrial), and one-party authoritarian rule. The leader of these Russian communists was Vladimir Ilyich Ulyanov, who adopted the name Lenin. Lenin became the architect of the Soviet Union. The new communist state radically reconfigured Eurasian political and economic geography.

Creating a Political Structure Lenin and other Soviet leaders were aware that they faced major challenges in organizing the new state and designed a geopolitical solution that maintained their country's territorial boundaries and recognized in theory the rights of non-Russian citizens. Each major nationality received its own "union republic,"

Figure 9.32 Russian President Vladimir Putin Since 2000, Putin's influence across the Russian domain has been immense. Within Russia, Putin (as both president and prime minister) has managed an impressive economic recovery, while limiting civil liberties. Beyond Russia, Putin has reestablished the region's geopolitical presence on the global stage.

provided it was situated on one of the nation's external borders (Figure 9.33). Eventually, 15 such republics were established, creating the Soviet Union. So-called **autonomous areas** within these republics gave special recognition to smaller ethnic homelands. Even with these internal republics and autonomous areas, by the late 1920s the Soviet Union was a highly centralized state, with important decisions made in the Russian capital of Moscow.

The chief architect of this political consolidation was Joseph Stalin, who did everything he could to centralize power and assert Russian authority. The Stalin period (1922–1953) also saw the enlargement of the Soviet Union. Victorious in World War II, the country acquired Pacific islands from Japan, the Baltic republics (Estonia, Latvia, and Lithuania), and portions of eastern Europe. One strategic addition on the Baltic Sea was the northern portion of East Prussia (the port of Kaliningrad), previously part of Germany. It still forms a small but strategic Russian **exclave**, defined as a portion of a country's territory that lies outside its contiguous land area.

After World War II, the Soviet Union expanded its influence across eastern Europe. In the words of British leader Winston Churchill, the Soviets extended an **Iron Curtain** between their eastern European allies and the more democratic nations of western Europe. As eastern Europe retreated behind the Iron Curtain, the Soviet Union and the United States became antagonists in a global **Cold War** of military competition that lasted from 1948 to 1991.

End of the Soviet System Ironically, Lenin's system of republics based on cultural differences sowed the seeds of the Soviet Union's demise. Even though the republics were never allowed real freedom, they provided a political framework that encouraged the survival of distinct cultural identities. Contrary to expectations of Soviet leaders, ethnic nationalism intensified in the post–World War II era as the Soviet system grew less repressive. When Soviet President Mikhail Gorbachev initiated his policy of **glasnost**, or greater openness, during the 1980s, several republics—most notably the Baltic states of Lithuania, Latvia, and Estonia—demanded independence. In addition, Gorbachev's policy of **perestroika**, or restructuring of the planned centralized economy, was an admission that the Soviet economy increasingly lagged behind those of western Europe and the United States. A failed war in Afghanistan and growing political protests in eastern Europe added to Gorbachev's problems.

By 1991, Gorbachev saw his authority slip away amid rising pressures for political decentralization and economic reforms. During that summer, Gorbachev's regime was further endangered by the popular election of reform-minded Boris Yeltsin as head of the Russian Republic. By late December, all of the country's 15 constituent republics had become independent states, and the Soviet Union ceased to exist.

Current Geopolitical Setting

The political geography of post-Soviet Russia and the nearby independent republics changed dramatically after the Soviet Union's collapse (Figure 9.34). All of the former republics have struggled to establish stable political relations with their neighbors.

Russia and the Former Soviet Republics For a time, it seemed that a looser political union of most of the former republics, called the *Commonwealth of Independent States (CIS)*, would emerge from the ruins of the Soviet Union. But the CIS faded in importance. More recently, Russia has backed the growth of the **Eurasian Economic Union (EEU)**, a customs union (paralleling the European Union [EU]) designed to encourage trade as well as closer political ties between member states. Formed in 2015, the EEU contains five member states (Russia, Belarus, Kazakhstan, Armenia, and Kyrgyzstan).

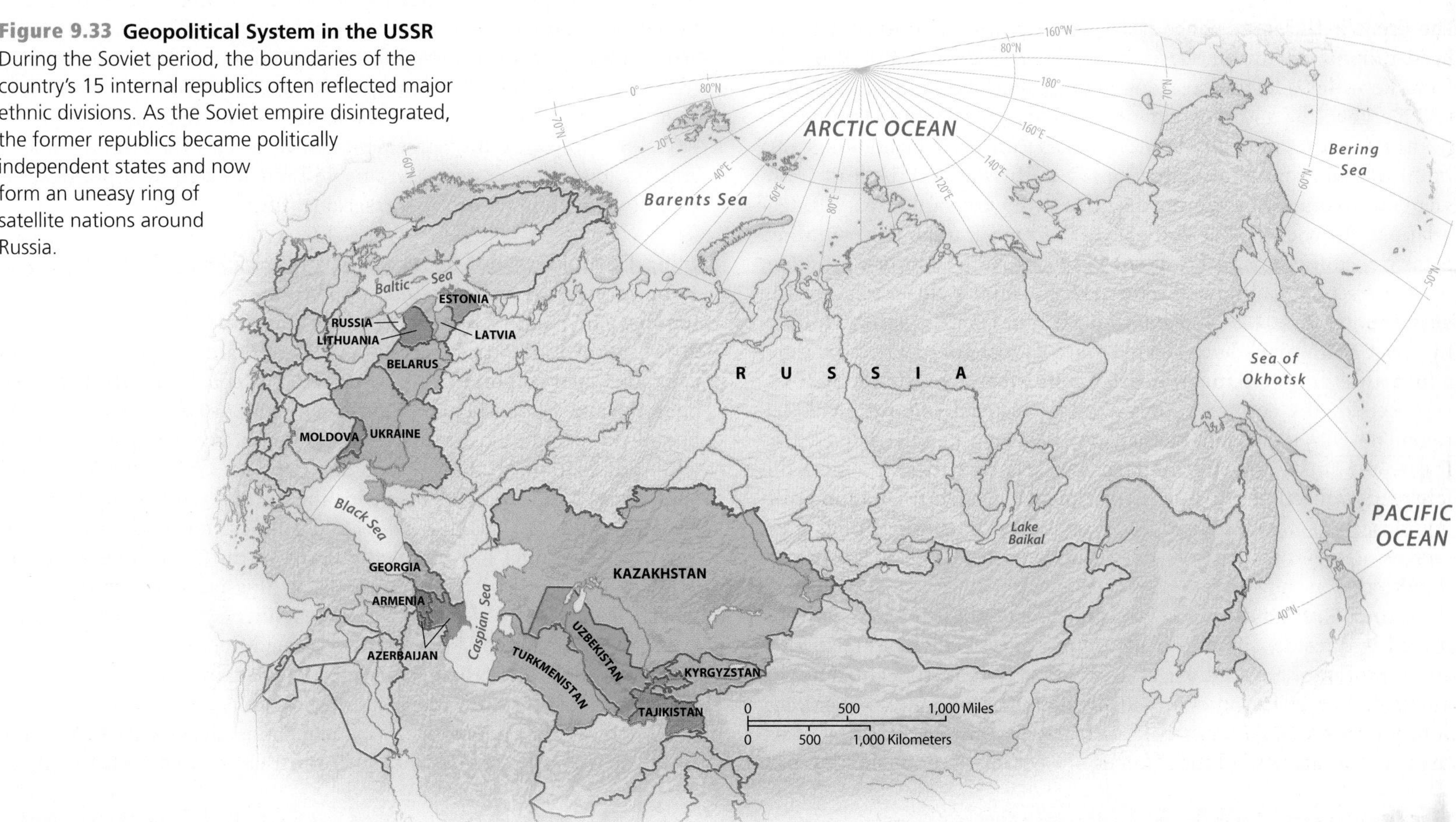

Figure 9.33 Geopolitical System in the USSR During the Soviet period, the boundaries of the country's 15 internal republics often reflected major ethnic divisions. As the Soviet empire disintegrated, the former republics became politically independent states and now form an uneasy ring of satellite nations around Russia.

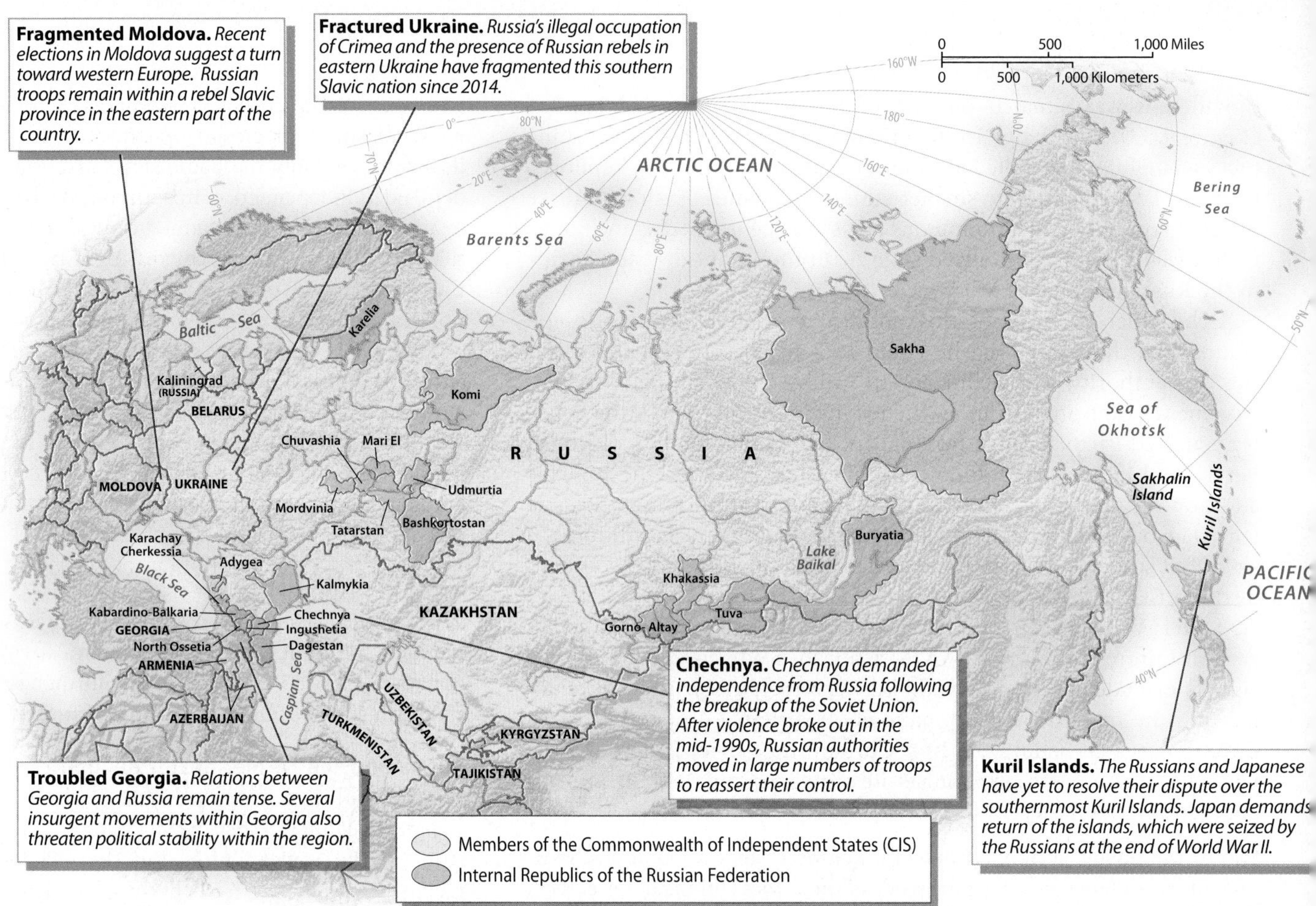

Figure 9.34 Geopolitical Issues in the Russian Domain The Russian Federation Treaty of 1992 created a new internal political framework that acknowledged many of the country's ethnic minorities. Recently, however, Russian authorities have moved to centralize power and limit regional dissent. Russia's relations with several nearby states, especially Ukraine, remain strained. **Q: Cite some similarities between Russia's internal republics shown here and the regional map of languages (Figure 9.26).**

Some observers argue that Russia will use the organization to reconstruct in more formal ways many elements of a larger "Soviet-style" empire that extends from the Baltic to Central Asia.

The Crisis in Ukraine Since 2014, the region's geopolitical map has been dominated by the crisis in Ukraine (Figure 9.35). Why did Ukraine unravel? Three reasons help explain its plight. First, since its emergence as an independent country in the 1990s, Ukraine has been *a divided, politically unstable state*, torn between its long-time political and economic connections with Russia (and the former Soviet Union) and a burgeoning desire to drift westward into a stronger relationship with NATO and the EU. Viktor Yanukovich, who had close ties to Russia and President Putin, became Ukraine's president in 2010. When Yanukovich refused to consider a closer relationship with the EU in 2013, protestors forced him from power the following year. Ukrainians later elected Petro Poroshenko, who advocated a turn away from too much Russian influence, as president in 2014.

Second, the country's long-standing *regional cultural divide* between its Ukrainian and Russian populations, broadly defined (see Figure 9.27), has often reinforced and reflected these political differences, with ethnic Russians in Crimea and eastern Ukraine favoring closer ties with Moscow. Many residents in these settings grew increasingly dissatisfied as events in Kiev seemed to signal a turn to the West.

Third, *Russia's expansive geopolitical ambitions* in the region, reflected by President Putin's bellicose response to the Ukrainian situation, prompted Russia's illegal occupation of the Crimean Peninsula in 2014. Russia has also provided ongoing economic and military support (more than 10,000 Russian troops are estimated to have been in the country) of rebel forces in eastern Ukraine who advocate separation from Kiev, either to create independent states or to be a part of an expanded Russian union (see Figure 9.35).

The result has been a violent, economically catastrophic civil war in eastern Ukraine. By mid-2015, more than 6000 people had died in the conflict, and about 1.5 million residents, mostly in war-torn eastern Ukraine, had been forced from their homes. Rebel-held zones in the east, especially in the disputed Luhansk and Donetsk areas, have been partially depopulated as residents flee the fighting (see Figure 9.35). Several cease-fire agreements failed to end the conflict, and it appears likely that eastern Ukraine will remain a zone of political instability and conflict for the foreseeable future.

Other Geopolitical Hot Spots In Belarus, leaders have been slow to embrace political and economic opportunities in western and central Europe. The country remains firmly within Russia's political orbit. In 2010, the two countries pledged to move toward a "union state" and expanded their common military maneuvers. In 2012, Belarusian President Alexander Lukashenko (often called Europe's last dictator) adopted one of the world's most restrictive policies on free use of the Internet within the nation, strongly discouraging the use of any "foreign" websites. Many political dissidents inside the country remain imprisoned.

Tiny Moldova has also witnessed political tensions in the post-Soviet era. Conflict has repeatedly flared in the Transdniester region in the eastern part of the country, where Russian troops remain and where Slavic separatists push for independence from a central government dominated by Romanian-speaking Moldovans. Complicating matters, Moldovan parliamentary elections late in 2014 signaled a potential political turn toward western Europe.

In addition, Transcaucasia remains unstable. Since 2003, the Georgian government has moved toward closer ties with the West.

Figure 9.35 Ukraine's Geopolitical Hot Spots This map shows where Russian troops occupied the Crimean Peninsula in 2014. Disputed areas in eastern Ukraine, under the influence of Russian-backed rebel forces, are also shown.

In 2008, Russia invaded the country when Georgia attempted to reassert its control over Abkhazia and South Ossetia, two breakaway regions dominated by Russian sympathizers. Almost 1000 people died in the conflict, and more than 30,000 people were displaced from their homes. The situation remains tense today, with Georgia still claiming control over Abkhazia and South Ossetia and with Russia declaring both microstates independent.

In nearby Armenia, territories claimed by Christian Armenians and Muslim Azeris interpenetrate one another in a complex fashion. The far southwestern portion of Azerbaijan (Naxicivan) is actually separated from the rest of the country by Armenia, while the important Armenian-speaking district of Nagorno-Karabakh is officially an autonomous portion of Azerbaijan. After Armenia successfully occupied much of Nagorno-Karabakh in 1994, fighting between the countries diminished. No final peace treaty has been signed, however, and Azerbaijan demands the return of the territory.

Geopolitics Within Russia Within Russia, further pressures for devolution, or more localized political control, produced the March 1992 signing of the Russian Federation Treaty. The treaty granted Russia's internal autonomous republics and its lesser administrative units greater political, economic, and cultural freedoms, including more control of their natural resources and foreign trade (see the map of internal republics, Figure 9.34). Conversely, it weakened Moscow's centralized authority to collect taxes and to shape policies within its varied hinterlands. Defined essentially along ethnic lines, 21 regions possess status as republics within the federation and now have constitutions that often run counter to national mandates.

Since 2000, Russian leaders, especially Putin, have pushed for more centralized control. Putin's prominence has been enduring: He served two terms as president (2000–2008) and one term as prime minister (2008–2012) and then was reelected president in March 2012. A former internal security agent with the KGB during the Soviet period, Putin consolidated power in the country, pushed for strong economic growth oriented around Russia's energy economy, and reasserted Russia's political and military role, both on the world stage and as a dominant regional power. Putin's occupation of Crimea in 2014 proved widely popular in Russia, boosting his domestic standing even as the country slipped into an economic recession.

Russian Challenge to Civil Liberties Still, periodic public protests since 2009 have challenged Putin's authority. After his 2012 election, thousands of protesters marched in the streets of Moscow and St. Petersburg, and hundreds were arrested. Disgruntled members of the urban middle class (Putin received less than half of the vote in Moscow), human rights groups, and opposition political parties resent Putin's grip on power. Protestors demand a freer press, more democracy, more open elections, and a broader commitment to economic growth. They have criticized Putin's strong-arm leadership style and his increasingly close ties to the **siloviki**, members of the nation's military and security forces.

These protests are in part a response to the central government's crackdown on civil liberties. Immediately after the fall of the Soviet Union, Russia enjoyed a genuine flowering of democratic freedoms. A multiparty political system, independent media, and a growing array of locally and regionally elected political officials signaled real change from the authoritarian legacy of the Soviet period.

Since 2002, however, many hard-won civil liberties have slipped away, victims of Putin's campaign to consolidate political power, increase central authority, limit press freedoms, and silence critics. For example, the Russian president now has more direct control of candidate nominations for dozens of Russian governorships and mayoral positions. Many Russian media outlets have lost their autonomy and are now under more direct government ownership and influence. Outspoken journalists critical of the government (such as Anna Politkovskaya) have died under suspicious circumstances or have been murdered. Recently, Russian officials have increased their surveillance and regulation of the Internet, another move to silence opposition.

In 2012, Pussy Riot, a Russian female punk rock band, made global headlines with its performance of a song critical of President Putin in Moscow's Christ Savior Cathedral, one of Russia's largest houses of worship (Figure 9.36). Band members—decked out in

Figure 9.36 Pussy Riot The members of this Russian punk rock band went to jail for hooliganism when they appeared in a prominent Russian Orthodox Church and sang a song protesting President Putin's rule.

fluorescent stockings as they sang "Mother of God, blessed Virgin, drive out Putin!"—were accused of "hooliganism" and jailed. Although the women claimed their actions were a political protest, they were given multiple-year sentences (one was later released), igniting a flurry of public reactions against the government. Even more disturbing, one of Putin's leading critics, Boris Nemtsov, was gunned down in Moscow near Red Square in early 2015, sparking more protests across the country.

The Shifting Global Setting

Since 1991, complicated regional political relationships, both to the east and to the west, have challenged the Russians. In East Asia, Russia is working to build a closer political relationship with China to counter faltering ties to the West. Russia also plays an important role in containing North Korea's nuclear ambitions, pressuring its Far East neighbor to limit uranium enrichment and weapons development projects. Territorial disagreements—specifically, a dispute over the Kuril Islands—continue to affect Russia's relationship with Japan.

To the west, Russia worries about the expansion of NATO. Although most Russian leaders accepted the inevitable inclusion of Poland, Hungary, and the Czech Republic in NATO, they strongly opposed adding the Baltic republics (Estonia, Latvia, and Lithuania) to the increasingly powerful organization. The exclave of Kaliningrad (population 950,000) is a related sore point. Since 2004, the area has been surrounded by NATO and EU nations (Poland and Lithuania), much to Russia's frustration.

Today President Putin is attempting to reassert his nation's global political status. Russia retains a permanent seat on the UN Security Council. The country's nuclear arsenal, while reduced in size, remains a powerful counterpoint to American and western European interests. Russia often acts as a counterweight to the United States in international maneuverings, and Putin has taken an increasingly anti-American slant in his foreign policy, a political strategy that has proven quite popular in Russia. Indeed, the long geopolitical history of the region suggests that Russia's recent reemergence on the global stage as a more powerful, centralized state is a sign of things to come.

 **Review**

9.8 How do current geopolitical conflicts reflect long-standing cultural differences within the region?

9.9 Describe how Vladimir Putin has played a key role in consolidating Russia's power since 2000, both within the country and beyond.

KEY TERMS Bolshevik, communism, autonomous area, exclave, Iron Curtain, Cold War, glasnost, perestroika, Eurasian Economic Union (EEU), siloviki

Economic and Social Development: Coping with Growing Regional Challenges

The economic future of the Russian domain remains difficult to predict (see Table A9.2). In the middle 2010s, much of the region faced enormous economic headwinds and a steep recession. Ukraine was torn by civil war, and Russia's energy-based export economy was adjusting to lower prices, sharply lower revenues, and a weakened currency. In addition, widespread economic sanctions and growing political tensions with the West hurt trade and curtailed foreign investment in the region.

The Legacy of the Soviet Economy

During the communist period, much of the present economic infrastructure was established, including new urban centers, industrial developments, and the modern network of transportation and communication linkages. As communist leaders such as Stalin consolidated power in the 1920s and 1930s, they nationalized Russian industries and agriculture, creating a system of **centralized economic planning**, in which the state controlled production targets and industrial output. The Soviets emphasized heavy basic industries (steel, machinery, chemicals, and electricity generation), rather than consumer goods. By the late 1920s, Stalin shifted agricultural land into large collectives and state-controlled farms.

Much of the Russian domain's basic infrastructure—its roads, rail lines, canals, dams, and communications networks—originated during the Soviet period (Figure 9.37). Dam and canal construction turned many rivers into a virtual network of interconnected reservoirs. The Volga–Don Canal (completed in 1952) connected those two river systems and greatly eased the movement of raw materials and manufactured goods (Figure 9.38). The Trans-Siberian rail line was modernized and complemented by the addition of the BAM link across central Siberia. Farther north, the Siberian Gas Pipeline was built to link the Arctic's energy-rich fields with growing demand in Europe. Overall, the postwar period produced real economic and social improvements for the Soviet people.

Despite the successes, problems increased during the 1970s and 1980s. Soviet agriculture remained inefficient. Manufacturing quality failed to match Western standards. Equally troubling, the Soviet Union failed to participate fully in technological revolutions transforming the United States, Europe, and Japan. Disparities also grew between the Soviet elite and ordinary people who enjoyed few personal freedoms. By the late 1980s, the Soviet Union had reached both an economic and a political impasse.

The Post-Soviet Economy

Fundamental economic changes have shaped the Russian domain since 1991. Particularly within Russia itself, much of the highly centralized state-controlled economy has been replaced by a mixed economy of state-run operations and private enterprise. The collapse of the Soviet Union also meant that economic relationships between the former Soviet republics were no longer controlled by a single, centralized government.

Redefining Regional Economic Ties Since the breakup of the Soviet Union, Russia has worked to maintain many economic ties with other former Soviet republics. The recent expansion of the EEU in 2015 is designed to counterbalance the growth of the EU, reduce trade barriers within the region, and encourage more economic cooperation between member states (Russia, Belarus, Kazakhstan, Armenia, and Kyrgyzstan). Ukraine has followed a different path, turning to the West for more economic aid and trading opportunities. Georgia has followed a similar path, actively exploring membership in the EU.

Privatization and State Control The post-Soviet era has brought a great deal of economic uncertainty. Russia's government initiated a massive program to transform its economy in 1993, opening portions

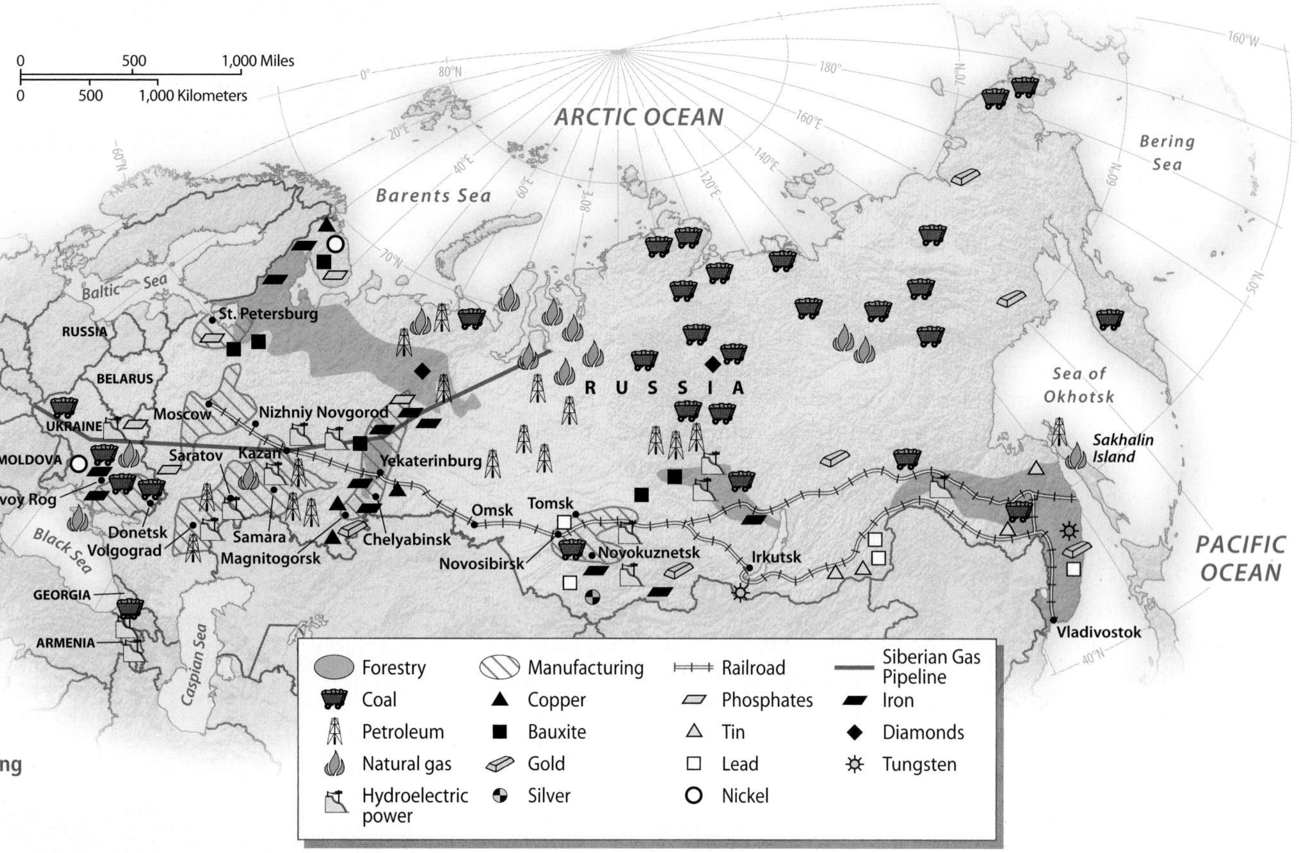

Figure 9.37 Major Natural Resources and Industrial Zones The Russian domain's varied natural resources and chief industrial zones are widely distributed. Fossil fuels are abundant, although their distance from markets often imposes special costs. In southern Siberia, rail corridors offer access to many mineral resources. In the mineral-rich Urals and eastern Ukraine, proximity to natural resources sparked industrial expansion, while Moscow's industrial might is related to its proximity to markets and capital. **Q: Looking at the map, why might it be argued that Russia's size is both a blessing and a curse?**

of the economy to more private initiative and investment. Unfortunately, the lack of legal and financial safeguards invited abuses and often resulted in mismanagement and corruption in the new system.

Almost 90 percent of Russian farmland was privatized by 2003, with many farmers forming voluntary cooperatives to work the same acreage as under the Soviet system. Thousands of private retailing establishments also appeared and now dominate that portion of the economy. In addition, the long-established "informal economy" continues to flourish. Even during the Soviet era, millions of citizens earned extra money by informally selling Western consumer goods, manufacturing food and vodka, and providing skilled services such as computer and automobile repairs. Today these barter transactions and informal cash deals form a huge part of an economy never reported to government authorities.

Figure 9.38 Volga–Don Canal Built during the Soviet era, the Volga-Don canal remains a key commercial link that facilitates the economic integration of southern Russia. This view near Volgograd suggests the canal's enduring economic importance.

The natural resource and heavy industrial sectors of the economy were initially privatized in Russia, but in recent years, under Putin's management, state-run enterprises took back more control of the nation's energy assets and infrastructure. Gazprom, the huge Russian natural gas company, was privatized in 1994, but since 2005 state control of its activities has increased.

Especially in Russia—and particularly in its cities—the successes of the new economy are increasingly visible on the landscape. Luxury malls, office buildings, and more fashionable housing subdivisions are now part of the urban scene as the middle class grows in settings such as Moscow. On the other hand, the gap has grown between increasing urban affluence and grinding rural poverty. In 2015, the *Wall Street Journal* reported that 110 people controlled 35 percent of Russia's wealth, while half the population had a total average household wealth of less than $875.

The Challenge of Corruption Throughout the Russian domain, corruption remains widespread. Doing business often means lining the pockets of government officials, company insiders, or trade union representatives. Organized crime remains pervasive in Russia. Many ties also remain between organized crime and Russian intelligence agencies. Much of the country's real wealth has been exported to foreign bank accounts. Various local and regional crime organizations divide up much of the economy and have links to illegal global business operations. Violence and gangland-style murders still unfold on the streets of Moscow, much to the embarrassment of government officials.

Problems of Health Care and Alcoholism Health care is another major social problem within the Russian domain. Health-care expenditures remain only a fraction of what they were during the Soviet period. To put the problem in global perspective, Russians typically survive on less than 13 percent of what most Americans spend annually on health care

($1043 vs. $7960). Mortality rates for Russian men are especially grim. One in three Russian men dies before retirement (age 60). Cardiovascular disease, often related to high-fat diets and physical inactivity, is a key contributor to these elevated death rates. Smoking remains widely popular (54 percent of physicians smoke). HIV-AIDS is also a major problem. In Russia, more than 700,000 people live with the disease.

Alcohol use in Russia (more than 15 liters [4 gallons] of pure alcohol per person annually) remains far above the global average (6.13 liters [1.6 gallons]). Russian leaders initiated an antidrinking campaign in 2010, calling their country's plight "a national disaster." An ambitious goal was set to reduce alcohol consumption by 50 percent within the next decade. Meanwhile, however, binge drinking and chronic high levels of alcohol consumption continue to threaten the lives of millions of people throughout the region.

Gender, Culture, and Politics

Women still struggle for basic rights within the conservative, patriarchal societies that characterize the Russian domain. While often better educated than men, women earn substantially less money performing the same work. Women are also underrepresented in positions of corporate and political power, often faring worse than their western European and American counterparts. Violence against women has been widely reported in the post-Soviet era. Beatings and rapes are common.

In addition, **human trafficking** (a practice in which women are lured or abducted into prostitution) is a widespread problem. Armenia, Ukraine, Moldova, and rural districts of Russia are major sources of young women who are forced into prostitution in Europe and the Middle East. It is a multibillion-dollar business, involving the large-scale participation of organized crime and hundreds of thousands of young women. Some estimates suggest that in addition to Ukraine's large domestic sex tourism industry, thousands more women have emigrated as sex workers, to both western Europe and the United States. Since 2008 FEMEN, a Ukrainian feminist organization, has made headlines around the world with its members' high-profile topless protests of that country's sex industry (their slogan is "Ukraine is not a brothel") (Figure 9.39). The region is also a major global source for Internet brides and dating, practices that invite additional violence against women.

The Russian Domain in the Global Economy

The relationship between the Russian domain and the world beyond has shifted greatly since the end of communism. During the Soviet era, the region was relatively isolated from the world economic system. But links with the global economy multiplied after the downfall of the Soviet Union. Recent economic sanctions against Russia, however, have thrown the path of its future global economic connections into doubt.

Figure 9.39 FEMEN Protest This group of Ukrainian feminists has vigorously protested the exploitation of women in their own country and around the world.

EVERYDAY GLOBALIZATION

How the Russian Domain Shapes the Virtual World

As every American college student knows, the video- and online-gaming landscape has changed dramatically since Russian Alexie Pajitnov invented *Tetris* at the Soviet Academy of Sciences in 1984. Less apparent, however, is the enduring connection between the Russian domain and the multibillion-dollar video-gaming industry (Figure 9.4.1). Russia dominates the game. Boris Nuraliev, one of the corporate founders of the movement in the early 1990s, created the 1C Company—often called Russia's Microsoft—which moved from the rather ordinary world of business software into the extraordinary world of gaming (*Theater of War, Kings Bounty: The Legend, Pacific Fighters,* etc.). Today 1C, based in Moscow, employs almost 1000 people (including 250 internal game developers) and is the largest game publisher and developer in the region. Dozens of other Russian innovators, including contributors from Belarus and Ukraine, have helped to make our everyday virtual worlds what they are.

1. Find a popular video game that you or your friends play. Who developed the game, and where did it originate?
2. For that video game, describe how its "place" (real or imaginary) is depicted in the game and what role it plays in setting the stage.

Figure 9.4.1 Russia's Blossoming Virtual World These Russian youngsters eagerly explore the gaming cyberworld at a GameWorld interactive entertainment exhibition in Moscow.

More-Globalized Consumers Most visibly, since the fall of the Soviet Union, a barrage of consumer imports has transformed the lives of residents. All of the symbols of global capitalism can be seen in the heart of Moscow and, increasingly, in many other settings throughout the Russian domain. Western luxury goods have found a small but enthusiastic, highly visible market among the Russian elite, a group noted for its devotion to BMW automobiles, Rolex watches, and other status emblems. It is also a two-way street: The region's software engineers and video game developers have had a large impact on these industries around the world (see *Everyday Globalization: How the Russian Domain Shapes the Virtual World*).

Changing Flows of Foreign Investment Post–Soviet era foreign investment has ebbed and flowed with the region's political stability. After President Putin took office in 2000, he successfully encouraged growing foreign investment, especially from the United States, Japan, and

Figure 9.40 Russia's Expanding Pipelines New and planned oil pipelines are designed to expand Russia's presence in the global petroleum economy. (a) Projects near the Caspian Sea take pipelines through politically unstable portions of the region. (b) Projects in the Russian Far East would benefit nearby China and Japan.

western Europe. For many years, the success of Russia's equity markets and the relative stability of its financial sector were encouraging. Since 2014, however, a sliding currency in Russia (the ruble), war in eastern Ukraine, and a growing list of economic sanctions have dramatically slowed foreign investment. Sanctions include freezing overseas Russian bank assets and forbidding the sale of many European and American goods to Russia (especially in energy- and technology-related industries). Capital flight from Russia also dramatically increased in 2014, because many Russian investors feared more instability ahead.

In response to sanctions, Russia banned many imported goods from the United States and Europe, including many food and consumer items. Falling currencies in the region have made the purchase of any dollar- or euro-denominated items more expensive, further dampening consumer demand and stoking inflation.

Globalization and Russia's Petroleum Economy Russia's enormous oil and gas industry has not escaped tougher economic times. For much of the post-Soviet era, the energy economy was a real boon to the region. The statistics are impressive: Russia's energy production makes up more than one-quarter of its economic output and two-thirds of its exports. Russia has 26 percent of the world's natural gas reserves (mostly in Siberia) and is the world's largest gas exporter. As for oil, Russia possesses more than 75 billion barrels of proven reserves and has large producing fields in Siberia, the Volga Valley, the Far East, and the Caspian Sea region. The primary destination for Russian petroleum products has overwhelmingly shifted to western Europe.

Figure 9.41 Sakhalin Island Large-scale investments by both foreign and Russian interests have concentrated on energy-rich Sakhalin Island. The area promises to be a producer of both oil and natural gas in the years to come.

The Siberian Gas Pipeline already connects distant Asian fields with western Europe via Ukraine (see Figure 9.37). Those connections are supplemented by lines through Belarus (the Yamal–Europe Pipeline) and Turkey (the Blue Stream Pipeline). Underwater pipelines beneath the Baltic Sea (Nord Stream—completed in 2011 and 2012) deliver gas to northern Europe. To the south, a large petroleum export terminal opened at Novorossiysk (on the Black Sea) in 2001, delivering Caspian Sea oil supplies to the world market via a pipeline passing through troubled Chechnya (Figure 9.40). Nearby, oil pipelines between Baku (on the Caspian Sea) and the Black and Mediterranean seas cross Azerbaijan and Georgia.

China and Japan are lobbying hard for more pipeline projects. Russians are building a large new Siberian Pacific Pipeline to link the Siberian fields to Asian markets. China wants Russian oil to flow to Daqing, where it can be refined for national and regional markets. The Japanese prefer a large new facility at the Pacific port of Nakhodka, well positioned to supply Japan and offering Russia easy access to global markets via the Pacific Ocean. Other links connect the system with developments on Sakhalin Island, where several major energy projects have been completed (see Figure 9.41).

But Russia's energy sector faces numerous challenges. First, plunging oil and gas prices in 2014 reduced revenues dramatically. Second, economic sanctions made new investments in the sector more difficult as capital and equipment deals dried up. Big Russian energy companies such as Rosneft have been hit hard. Third, President Putin, frustrated by Western sanctions and low energy prices, has trimmed back new energy-related investments, such as the South Stream Gas Pipeline across the Black Sea, which he abruptly canceled late in 2014. Given the new political realities, Putin has aggressively shifted the focus of his energy deals to the east, forging agreements with China both in fossil fuel projects and in renewable energy initiatives.

Review

9.10 Describe how centralized planning created a new economic geography across the former Soviet Union. What is its lasting impact?

9.11 Briefly summarize the key strengths and weaknesses of the post-Soviet Russian economy, and suggest how globalization has shaped its evolution.

KEY TERMS centralized economic planning, human trafficking

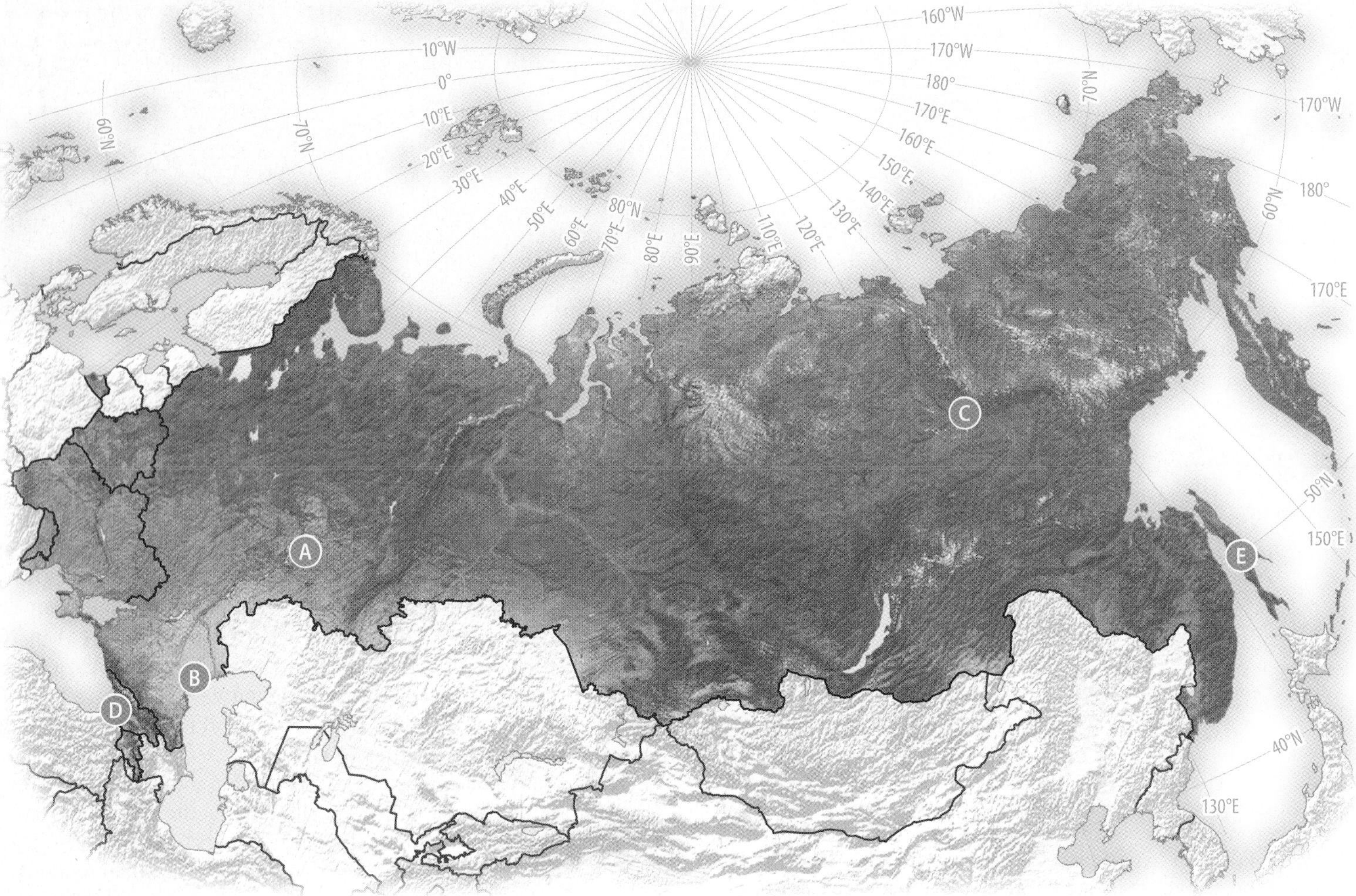

Review

Physical Geography and Environmental Issues

9.1 Explain the close connection among latitude, regional climates, and agricultural production in Russia.

9.2 Describe the major environmental issues affecting the region and suggest how climate change might impact high-latitude areas.

Huge environmental challenges remain for the Russian domain. The legacy of the Soviet era includes polluted rivers and coastlines, poor urban air quality, and a frightening array of toxic wastes and nuclear hazards.

1. Why is the Volga River often referred to as Russia's version of the Mississippi?

2. Join with a group of students to debate another student group on the question of whether Russia's natural environment is one of its greatest assets or one of its greatest liabilities.

Population and Settlement

9.3 Identify the region's major migration patterns, in both the Soviet and the post-Soviet eras.

9.4 Explain major urban land-use patterns in a large city such as Moscow.

Declining and aging populations are part of the sobering reality for much of the region. Although some localities see modest population growth related to in-migration (mostly toward expanding urban areas), many rural areas and less competitive industrial zones are likely to see continued outflows of people and very low birth rates.

3. Traditionally, why is the large area of Russia that is located south of Volgograd so sparsely populated?

4. Given recent economic developments near the Caspian Sea, why might population in this area increase in the future?

Cultural Coherence and Diversity

9.5 Describe the major phases of Russian expansion across Eurasia.

9.6 Identify the key regional patterns of linguistic and religious diversity.

Much of the Russian domain's underlying cultural geography was formed centuries ago from the complex mix of Slavic languages, Orthodox Christianity, and numerous ethnic minorities that continue to complicate the scene today. Further changing the region are new global influences—products, technologies, and attitudes that often clash with traditional cultural values.

5. Where in the region would you be most likely to encounter Yakut-speaking peoples?

6. Cite some key similarities and differences you might observe in comparing the lifestyle of the Yakut with those of native North American populations.

C

Geopolitical Framework

9.7 Summarize the historical roots of the region's modern geopolitical system.

9.8 Provide examples of recent geopolitical conflicts in the region and indicate how these reflect persistent cultural differences.

The region's political legacy is rooted in the Russian Empire, a land-based system of colonial expansion that greatly enlarged Russian influence after 1600 and then reappeared as the Soviet Union expanded its influence. Only remnants of that empire survive, but it still shapes the language of Russian nationalism and has stamped the geopolitical character of the region in lasting ways.

7. Why is this area of South Ossetia troublesome for Georgia's government?

8. Why does Russia recognize South Ossetia as an independent nation?

D

Economic and Social Development

9.9 Identify key ways in which natural resources, including energy, have shaped economic development in the region.

9.10 Describe key sectors of the region's economy in the Soviet and post–Soviet eras and discuss how recent geopolitical events affect prospects for future economic growth.

The region's future economic geography, particularly in Russia, remains tied to the fortunes of the unpredictable global energy economy. Recent economic sanctions imposed on Russia and the ongoing conflict in Ukraine have dimmed near-term economic prospects.

9. What global energy markets are most likely to be served by oil and natural gas produced on Russia's Sakhalin Island?

10. What are some of the key environmental and cultural challenges zones of rapid energy development such as Sakhalin face?

E

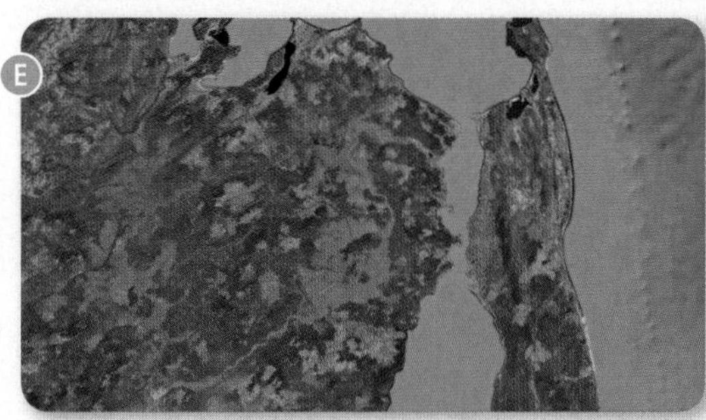

KEY TERMS

autonomous area *(p. 313)*
Baikal–Amur Mainline (BAM) Railroad *(p. 304)*
Bolshevik *(p. 312)*
centralized economic planning *(p. 316)*
chernozem soil *(p. 295)*
Cold War *(p. 313)*
communism *(p. 312)*
Cossack *(p. 308)*
Eastern Orthodox Christianity *(p. 308)*
Eurasian Economic Union (EEU) *(p. 313)*
exclave *(p. 313)*
glasnost *(p. 313)*
Gulag Archipelago *(p. 305)*
human trafficking *(p. 318)*
Iron Curtain *(p. 313)*
mikrorayon *(p. 307)*
northern sea route *(p. 299)*
oligarch *(p. 294)*
perestroika *(p. 313)*
permafrost *(p. 296)*
podzol soil *(p. 295)*
Russification *(p. 305)*
siloviki *(p. 315)*
Slavic people *(p. 308)*
socialist realism *(p. 311)*
Soviet Union *(p. 292)*
taiga *(p. 296)*
Trans-Siberian Railroad *(p. 304)*
tsar *(p. 305)*

DATA ANALYSIS

http://goo.gl/mqXCPu

Foreign direct investment (FDI) can be a valuable gauge of economic activity in a country. Data are often gathered for both incoming FDI (investment capital entering a country) and outgoing FDI (investment capital leaving a country). The Organisation for Economic Co-operation and Development (OECD) keeps annual statistics on FDI. Go to its website (www.oecd-ilibrary.org) and access the statistical profile for the Russian Federation.

1. Make a simple chart showing incoming and outgoing flows of FDI for the years shown in the data table.
2. Briefly summarize the general patterns and trends you observe. How would you explain some of the major changes that occurred during the period?
3. Given Russia's current economic and political situation, what patterns of incoming and outgoing FDI are likely for the current year and the next three years? Explain your answer.
4. How might changing rates of incoming FDI affect patterns of internal movement or international migration? Why?

MasteringGeography™

Looking for additional review and test prep materials? Visit the Study Area in MasteringGeography™ to enhance your geographic literacy, spatial reasoning skills, and understanding of this chapter's content by accessing a variety of resources, including MapMaster interactive maps, geoscience animations, videos, *In the News* RSS feeds, flashcards, web links, self-study quizzes, and an eText version of *Globalization and Diversity*.

Authors' Blogs

Scan to visit the **Author's Blog for field notes, media resources, and chapter updates**

http://gad4blog.wordpress.com/category/the-russian-domain/

Scan to visit the **GeoCurrents Blog**

http://geocurrents.info/category/place/russia-ukraine-and-caucasus

10 Central Asia

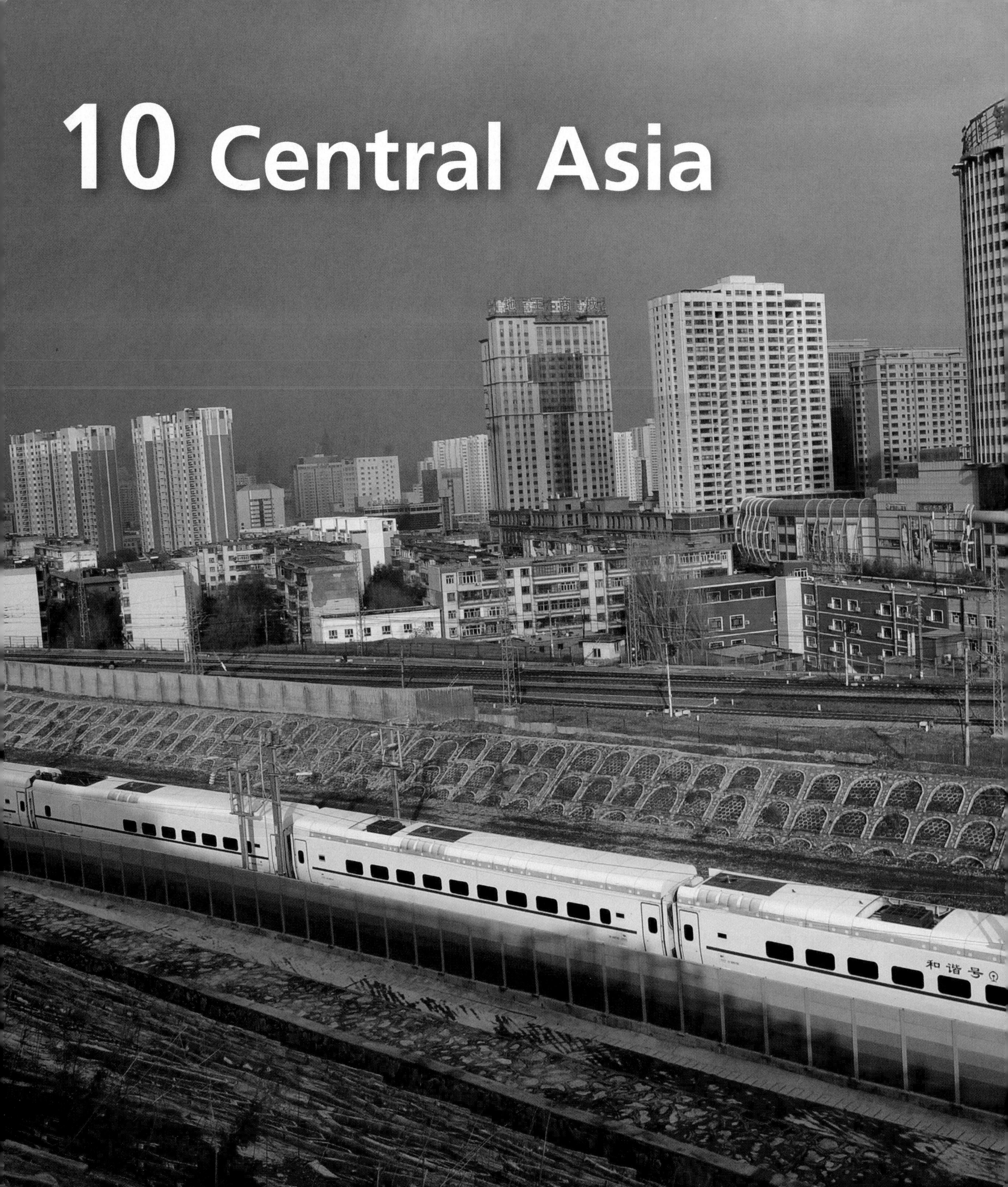

PHYSICAL GEOGRAPHY AND ENVIRONMENTAL ISSUES

Intensive agriculture along the rivers that flow into the deserts of Central Asia has led to serious water shortages and the drying up of many of the region's lakes and wetlands.

POPULATION AND SETTLEMENT

Pastoral nomadism, the traditional way of life across much of Central Asia, is gradually disappearing as people settle in towns and cities.

CULTURAL COHERENCE AND DIVERSITY

In much of eastern Central Asia, the growing Han Chinese population is sometimes seen as a threat to the long-term survival of the indigenous cultures of the Tibetan and Uyghur peoples.

GEOPOLITICAL FRAMEWORK

Afghanistan and its neighbors to the north are frontline states in the struggle between radical Islamic fundamentalism and secular governments.

ECONOMIC AND SOCIAL DEVELOPMENT

Despite its abundant resources, Central Asia remains a poor region, although much of it enjoys relatively high levels of social development.

◀ Chinese investment in infrastructure has benefited Central Asian lands in recent years. This train sped through Urumqi city during its test run on November 11, 2014. This route is a part of the 1,776-kilometer Lanzhou-Xinjiang high-speed railway.

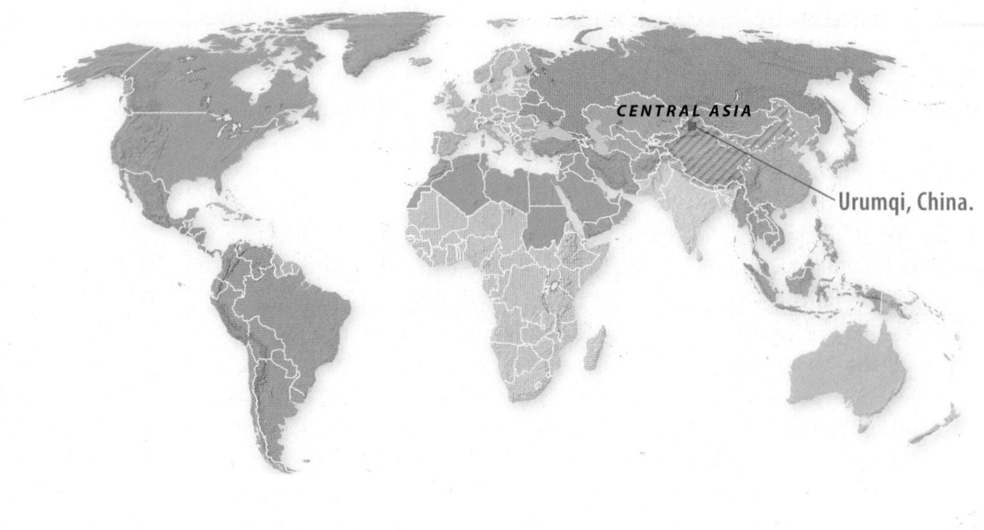

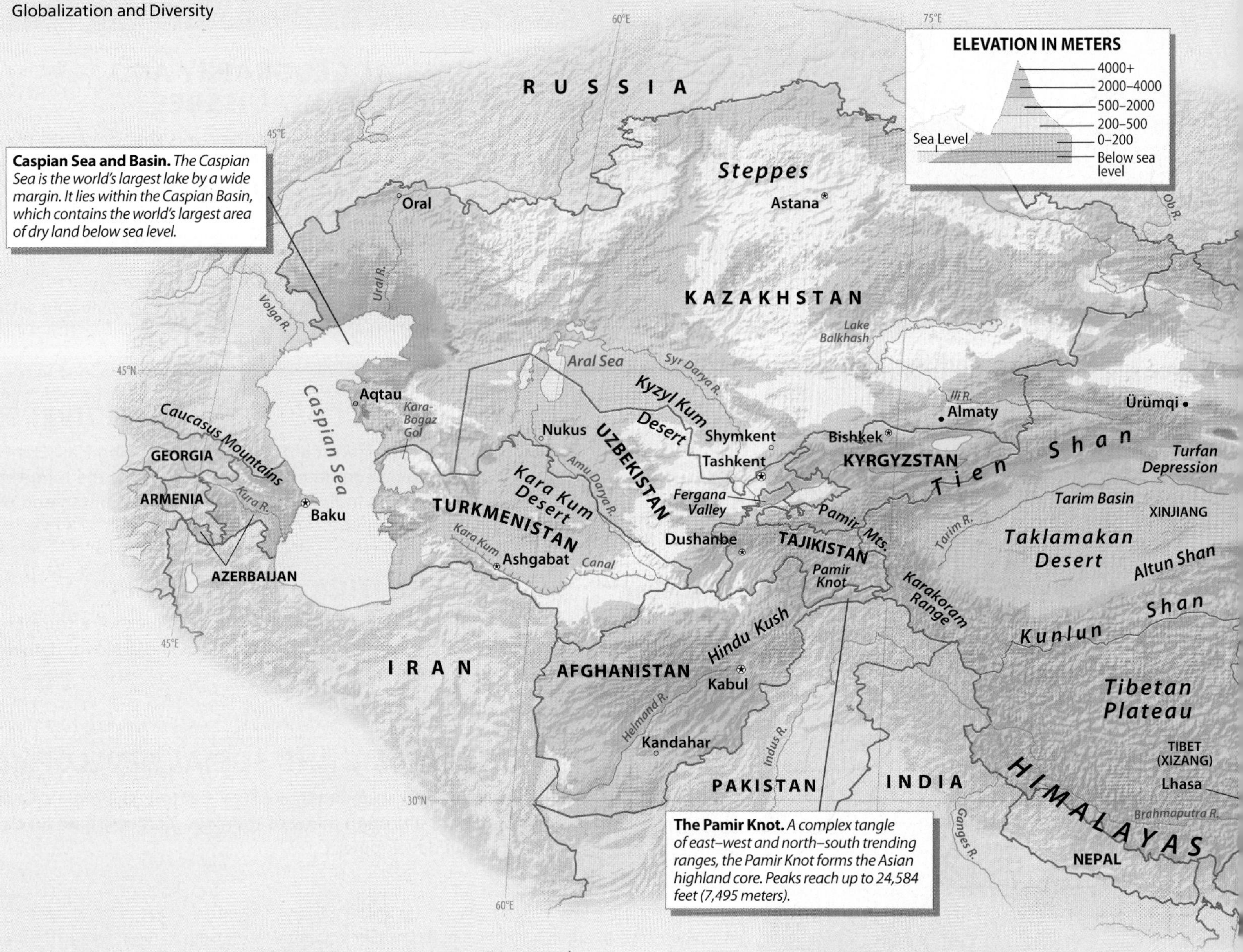

Caspian Sea and Basin. *The Caspian Sea is the world's largest lake by a wide margin. It lies within the Caspian Basin, which contains the world's largest area of dry land below sea level.*

The Pamir Knot. *A complex tangle of east–west and north–south trending ranges, the Pamir Knot forms the Asian highland core. Peaks reach up to 24,584 feet (7,495 meters).*

For hundreds of years, the **Silk Road** was a network of paths and facilities that allowed trade goods to flow from China to Southwest Asia and Europe, crossing Central Asia in the process. In the early modern period (1500–1800), as maritime trade gradually replaced overland transportation, the Silk Road declined and all but disappeared. Currently, however, China is partnering with Russia, Kazakhstan, Kyrgyzstan, and several other Central Asian countries to build a "New Silk Road" based on railways and modern highways.

Most global trade still goes by sea, due largely to cost. Shipping a container of goods from China to Europe costs around $4000, as opposed to $9000 if sent by truck or train. But the cost of overland transport is falling—and will drop further if the projects planned in Central Asia are completed. But price is often less important than timing, as fashions change rapidly and consumers demand new products as quickly as possible. Sending goods from China to Europe by sea currently takes about 60 days, as opposed to 14 days by rail. Chinese officials hope to reduce the latter figure to 10 days within a few years.

But despite both the rapid growth of trans-Eurasian rail lines and the great ambitions of China, Russia, and Kazakhstan, the New Silk Road faces major obstacles. Economic growth in Russia and Kazakhstan, as well as in several other Central Asian countries, has been threatened by low oil prices. Critics also argue that high levels of corruption might undermine the project.

> Although the new Silk Road will link China to Europe through Central Asia, defining "Central Asia" is not an easy matter.

Although the New Silk Road is designed to link China to Europe through Central Asia, defining "Central Asia" is not an easy matter. Most authorities agree that the region includes five former Soviet republics: Kazakhstan, Kyrgyzstan, Uzbekistan, Tajikistan, and Turkmenistan. This chapter adds another former Soviet state, Azerbaijan, as well as Mongolia and Afghanistan. In addition, the autonomous regions of western China (Tibet and Xinjiang) are counted as parts of both Central Asia and East Asia, and several

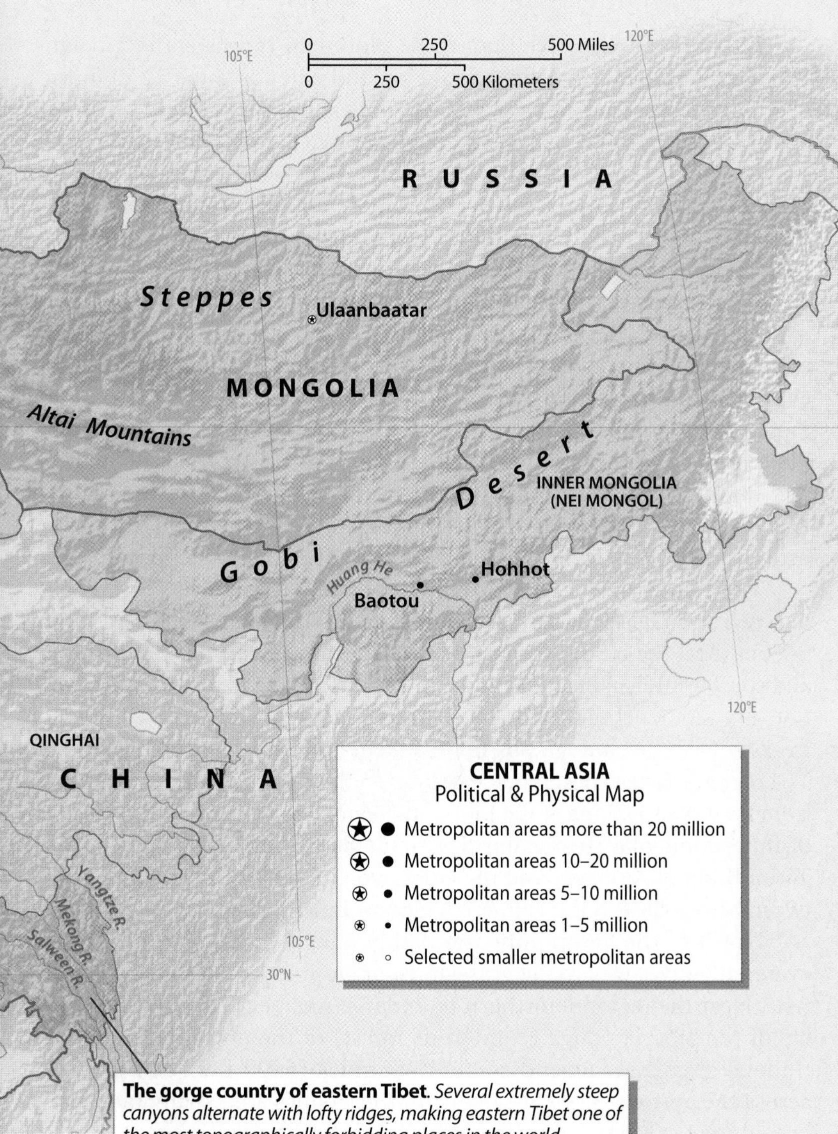

Figure 10.1 Central Asia Extending across the interior of the Eurasian continent, Central Asia is dominated by arid plains and basins, along with high mountain ranges and plateaus. Eight independent countries—Kazakhstan, Turkmenistan, Uzbekistan, Kyrgyzstan, Tajikistan, Azerbaijan, Afghanistan, and Mongolia—form Central Asia's core. The region also includes China's lightly populated far west, which is culturally and environmentally similar to the rest of Central Asia.

other parts of western China, such as Nei Mongol (Inner Mongolia), are occasionally discussed in this chapter as well (Figure 10.1).

Including these additional territories in Central Asia is controversial. Azerbaijan is often classified with its neighbors in the Caucasus (Georgia and Armenia), western China fully belongs to East Asia on political grounds, and Mongolia is also often considered East Asian because of its location and historical connections with China. Afghanistan can also be located within either South Asia or Southwest Asia.

However, there are solid reasons for defining Central Asia as we have. The region has deep historical bonds and similar environmental and economic conditions. Azerbaijan, for example, is more closely linked, both culturally and economically, to Central Asia than to Armenia and Georgia. Central Asia is also increasingly seen as a geopolitical unit, as its various countries face similar political challenges. But it is also not clear whether Central Asia will remain a justifiable world region. Continuing Han Chinese migration into Tibet and especially into Xinjiang, for example, could place these areas in an East Asian cultural framework.

Learning Objectives *After reading this chapter you should be able to:*

10.1 **Identify the key environmental differences among Central Asia's desert areas, its mountain and plateau zone, and its steppe belt, and link these differences to human settlement and economic development.**

10.2 **Summarize the reasons why water resources are of such great importance in Central Asia.**

10.3 **Provide reasons for the Aral Sea's disappearance, and outline the economic and environmental consequences of the loss of this once-massive lake.**

10.4 **Explain why Central Asia's population is so unevenly distributed, with some areas densely settled and others essentially uninhabited.**

10.5 **Describe the differences between Central Asia's historical cities and those that have been established within the past 100 years.**

10.6 **Outline the ways in which religion divides Central Asia and describe how religious diversity has influenced the history of the region.**

10.7 **Identify how cultural globalization affects different parts of Central Asia in distinct ways, and explain why cultural globalization is controversial in much of the region.**

10.8 **Summarize the geopolitical roles played in Central Asia by Russia, China, and the United States, and explain why the region has been the site of pronounced geopolitical tension over the past several decades.**

10.9 **Describe how ethnic conflict has contributed to instability in Afghanistan, and assess the potential of ethnic tension to destabilize the rest of the region.**

10.10 **Explain the role of oil and natural gas production and transport, as well as oil price fluctuations, in generating uneven levels of economic and social development across Central Asia.**

Physical Geography and Environmental Issues: Steppes, Deserts, and Threatened Lakes

Central Asia is unique among world regions in that it is completely landlocked. Located at the core of the world's largest landmass, Central Asia has a distinctly continental climate characterized by extremes in temperature between summer and winter. Cut off from maritime moisture, it is also a relatively dry region, containing several barren deserts. But Central Asia is also environmentally diverse, noted for its lofty, often snow-covered mountains. These highlands are the source areas of the rivers that bring life to the region's arid zones.

Central Asia's Physical Regions

To understand Central Asia's environmental diversity, the region's physical geography must be examined in more detail. In simple terms, Central Asia is characterized by high plateaus and mountains in the south-central and southeastern areas, grassland plains (steppes) in the north, and desert basins in the southwestern and central areas.

Central Asian Highlands The highlands of Central Asia originated in one of the great geological events of Earth's history: the Indian subcontinent's collision with the Asian mainland. This ongoing tectonic impact has created the world's highest mountains, the Himalayas, along the boundary of South Asia and Central Asia. To the northwest, the Himalayas merge with the Karakoram Range and then the Pamir Mountains. From the so-called Pamir Knot—a complex tangle of mountains located where Pakistan, Afghanistan, China, and Tajikistan meet—towering ranges spread outward in several directions. The Hindu Kush curves to the southwest through central Afghanistan, the Kunlun Shan extends to the east, and the Tien Shan swings out to the northeast into China's Xinjiang Autonomous Region. All these ranges have peaks higher than 20,000 feet (6000 meters) in elevation.

Figure 10.2 Tibetan Plateau Alpine grasslands and tundra, interspersed with rugged mountains and saline lakes, dominate the Tibetan Plateau. In summer, the sparse vegetation offers forage for the herds of nomadic Tibetan pastoralists. Much of northern Tibet, however, is too high to support pastoralism and is therefore uninhabited.

Much more extensive than these mountain ranges is the Tibetan Plateau (Figure 10.2). This massive upland extends some 1250 miles (2000 km) from east to west and 750 miles (1200 km) from north to south. Its elevation is as remarkable as its size; almost the entire area is more than 12,000 feet (3700 meters) above sea level. Most of the Tibetan Plateau lies near the maximum elevation at which human life can exist. Rather than a flat surface, the plateau has numerous east–west mountain ranges alternating with basins. Although the southeastern sections of the plateau receive adequate rainfall, most of Tibet is arid (Figure 10.3). Winters on the Tibetan Plateau are cold, and while summer afternoons can be warm, summer nights remain chilly (note the climograph for Lhasa).

Plains and Basins Although the mountains of Central Asia are higher and more extensive than any others in the world, most of the region is characterized by plains and basins. This lower-lying zone features a central belt of deserts and a northern strip of semiarid steppes.

The Tien Shan and Pamir Mountains divide Central Asia's desert belt into two separate segments. To the west lie the arid plains of the Caspian Sea and Aral Sea basins, located primarily in Turkmenistan, Uzbekistan, and southern Kazakhstan (Figure 10.4). The climate here is continental: hot, dry summers, and winters with average temperatures well below freezing (see the climographs for Tashkent and Almaty in Figure 10.3). The region's eastern desert belt extends for almost 2000 miles (3200 km) from far western China at the foot of the Pamirs to the southeastern edge of Inner Mongolia. Two major deserts lie here: the Taklamakan, in the Tarim Basin of Xinjiang, and the Gobi, which runs along the border between Mongolia and the Chinese region of Inner Mongolia.

North of the desert zone, rainfall gradually increases, and desert eventually gives way to the grasslands, or **steppes**, of northern Central Asia. Near the region's northern boundary, trees begin to appear, outliers of the Siberian taiga (coniferous forest) of the north. Nearly continuous grasslands extend some 4000 miles (6400 km) east to west across the entire region. Summers on the steppes are usually pleasant, but winters can be extremely cold.

Major Environmental Issues

Much of Central Asia has a relatively clean environment, largely due to its generally low population density. Industrial pollution, however, is a serious problem in the larger cities, such as Tashkent (in Uzbekistan) and Baku (in Azerbaijan). Elsewhere, the typical problems of arid environments plague the region: **desertification** (the spread of deserts resulting from poor land-use practices), salinization (the accumulation of salt in the soil), and desiccation (the drying up of lakes and wetlands) (Figure 10.5).

Aral Sea Destruction One of the great environmental tragedies of the late 20th century was the destruction of the Aral Sea, a vast, salty lake (until recently, larger than Lake Michigan) located on the boundary of Kazakhstan and Uzbekistan. The Aral's only sources of water are the Amu Darya and Syr Darya rivers flowing out of the distant Pamir Mountains. Both rivers have been intensively used for irrigation for thousands of years, but the scale of water diversion greatly expanded after 1950. The valleys of the two rivers were some of the Soviet Union's only lands for growing such warm-season crops as rice and cotton, and Soviet planners favored huge engineering projects to deliver water to arid lands to "make the deserts bloom."

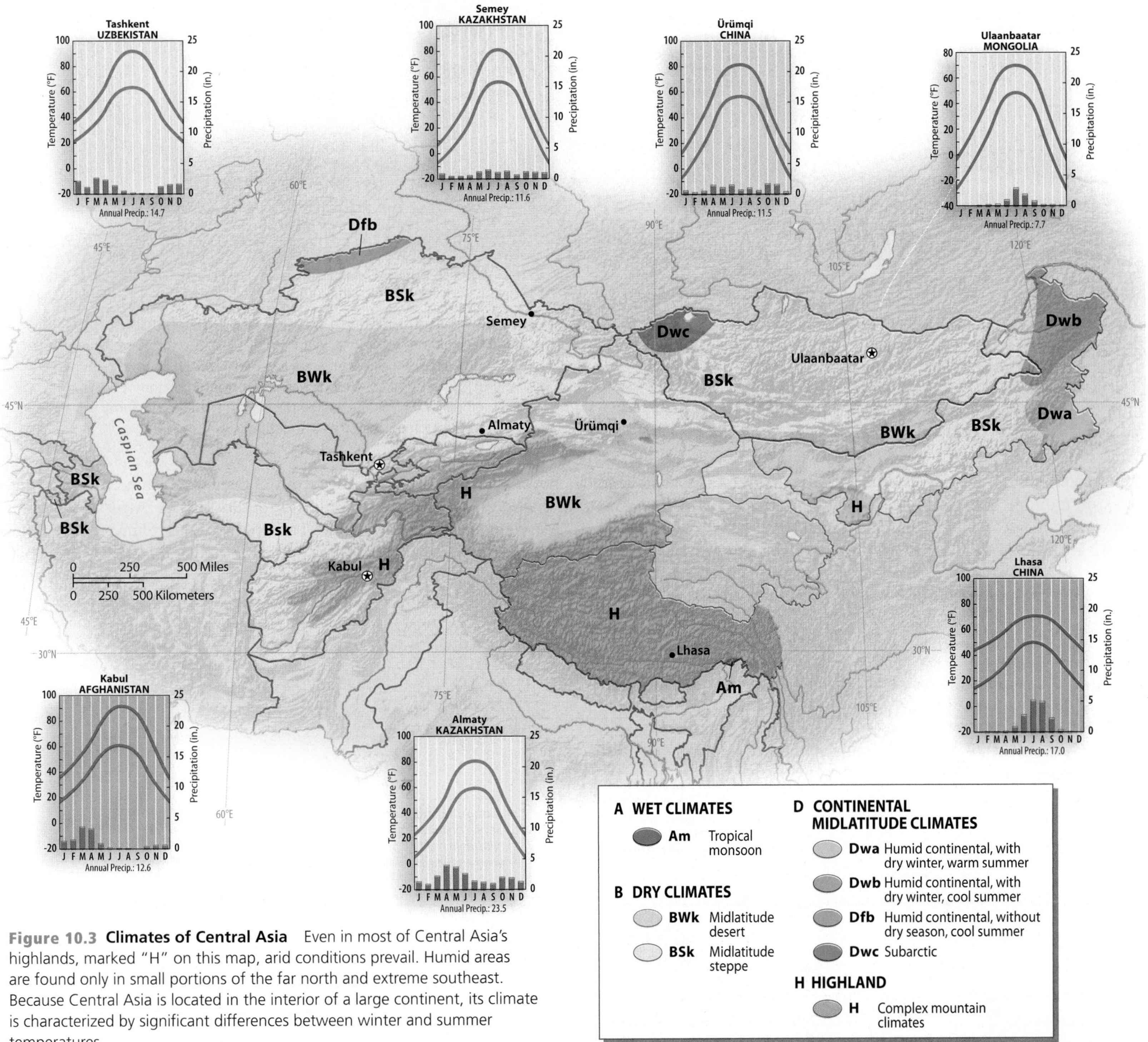

Figure 10.3 Climates of Central Asia Even in most of Central Asia's highlands, marked "H" on this map, arid conditions prevail. Humid areas are found only in small portions of the far north and extreme southeast. Because Central Asia is located in the interior of a large continent, its climate is characterized by significant differences between winter and summer temperatures.

Unfortunately, more water diverted to irrigate crops meant less freshwater for the Aral Sea. As river flow was reduced, the shallow Aral began to recede. With less freshwater flowing into the lake, it also grew increasingly salty, destroying fish stocks. New islands began to emerge, and the Aral Sea split into two separate lakes in 1987 and then into three lakes in the early 2000s (Figure 10.6).

The shrinking Aral Sea has resulted in economic and cultural damage as well as ecological devastation. Fisheries formerly employing 40,000 workers have closed down, and agriculture has suffered. The retreating lake has left large salt flats on its exposed beds. Windstorms pick up the salt, along with agricultural chemicals that had accumulated in the lake's shallows, and deposit these in nearby fields. As a result, farm yields have declined, public health has been threatened, and the local climate has become colder in the winter and hotter in the summer. Equally damaging is desertification of the lake bed and the once-green periphery.

Efforts to save what is left of the Aral Sea are currently focused on the small northern remnant. A series of dikes and dams, financed by the World Bank and the government of Kazakhstan, has managed to raise water levels over 26 feet (8 meters). Improved water quality has also partially revived wildlife and reestablished commercial fishing. In 2014, 5595 metric tons of fish were harvested in the so-called Small Aral Sea of the north. However, the southern Aral Sea, the largest part

Figure 10.4 Central Asian Desert Much of Central Asia is dominated by deserts and other arid lands. The Kara Kum Desert of Turkmenistan is especially dry, supporting little vegetation.

of the basin, has seen no such recovery. In 2014, the Aral Sea's south-eastern basin went completely dry for the first time in 600 years.

Other Fluctuating Lakes Western Central Asia contains several large lakes in addition to the Aral Sea because it forms a low-lying basin, without drainage to the ocean, surrounded by mountains and more humid areas. For example, Lake Balqash in southeastern Kazakhstan, is the world's 15th-largest lake. Like the Aral Sea, Balqash has become smaller and saltier over the past several decades.

The story of the Caspian Sea, located on the region's western boundary, is more complicated. The world's largest lake, the Caspian receives most of its water from the large rivers of the north—namely, the Ural and the Volga, which drain much of European Russia. Extensive irrigation development in the lower Volga Basin reduced the volume of freshwater reaching the Caspian in the second half of the 20th century, eventually exposing large expanses of the former lake bed. Decreased water volume and increased salinity disrupted the ecosystem and devastated fisheries.

The Caspian Sea's water level reached a low point in the late 1970s and then began to rise, probably because of higher-than-normal

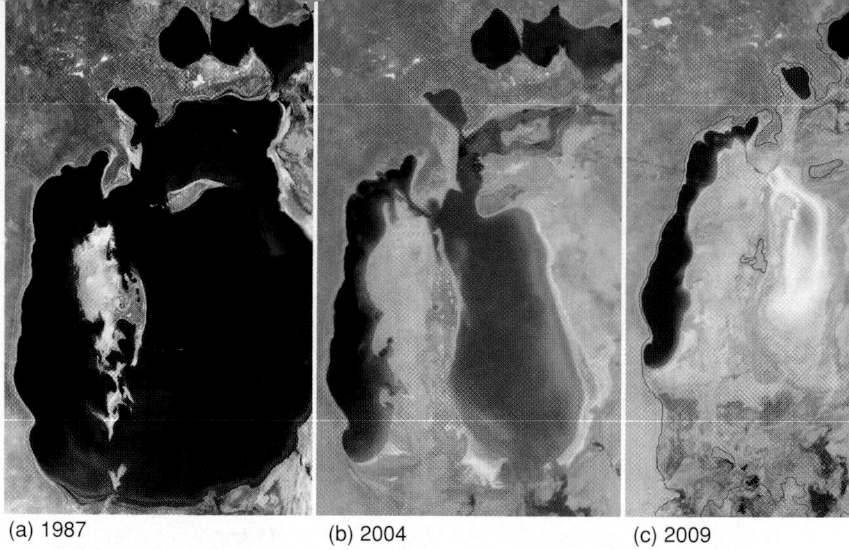

Figure 10.6 The Shrinking of the Aral Sea These satellite images show the steady shrinkage of the Aral Sea from (a) 1987, to (b) 2004, to (c) 2009. The Aral was a massive lake as recently as the 1970s. It is now divided into several much smaller lakes. **Q: Why has the northern portion of the Aral Sea retained its water volume, unlike the rest of the lake, and why has the southeastern portion disappeared altogether?**

Figure 10.5 Environmental Issues in Central Asia Desertification is perhaps more widespread in Central Asia than in any other world region. Soil erosion and overgrazing have led to the advance of desert-like conditions in much of western China and Kazakhstan. In western Central Asia, the most serious environmental problems are associated with the diversion of river water for irrigation and the corresponding desiccation of lakes.

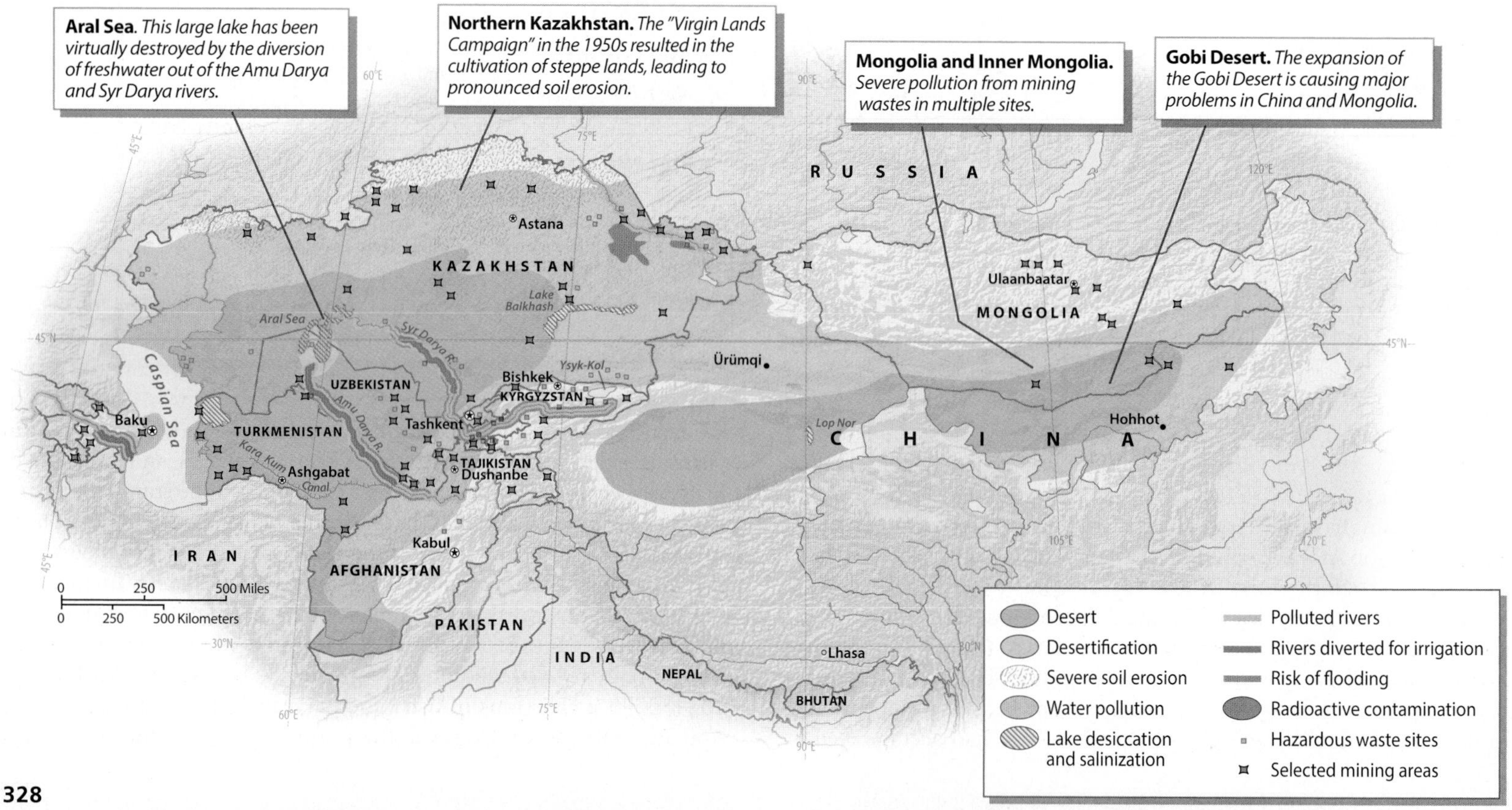

Aral Sea. *This large lake has been virtually destroyed by the diversion of freshwater out of the Amu Darya and Syr Darya rivers.*

Northern Kazakhstan. *The "Virgin Lands Campaign" in the 1950s resulted in the cultivation of steppe lands, leading to pronounced soil erosion.*

Mongolia and Inner Mongolia. *Severe pollution from mining wastes in multiple sites.*

Gobi Desert. *The expansion of the Gobi Desert is causing major problems in China and Mongolia.*

RUSSIA
KAZAKHSTAN
Astana
Lake Balkhash
Aral Sea
Syr Darya R.
Caspian Sea
Baku
UZBEKISTAN
TURKMENISTAN
Amu Darya R.
Kara Kum Canal
Ashgabat
Tashkent
Bishkek
KYRGYZSTAN
Ysyk-Kol
TAJIKISTAN
Dushanbe
Kabul
AFGHANISTAN
IRAN
PAKISTAN
INDIA
NEPAL
BHUTAN
Lhasa
Ürümqi
Lop Nor
CHINA
Hohhot
MONGOLIA
Ulaanbaatar

0 250 500 Miles
0 250 500 Kilometers

Desert
Desertification
Severe soil erosion
Water pollution
Lake desiccation and salinization
Polluted rivers
Rivers diverted for irrigation
Risk of flooding
Radioactive contamination
Hazardous waste sites
Selected mining areas

WORKING TOWARD SUSTAINABILITY

The Greening of the Inner Mongolian Desert

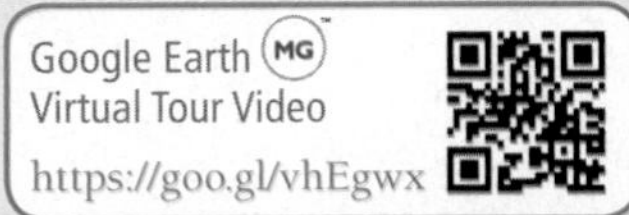

Google Earth (MG) Virtual Tour Video https://goo.gl/vhEgwx

Desertification has long been an extreme problem in China's Nei Mongol Autonomous Region, also known as Inner Mongolia. This vast region has an arid to semiarid climate and was once mostly covered by grasslands. Because of overgrazing and other forms of misuse, however, sand dunes and barren salt flats gradually expanded. Sand dunes have been particularly damaging, as they often spread into and then smother pastures and croplands. The negative effects of desertification in Inner Mongolia have extended well beyond the region itself, as winds whipping across the bare lands generate damaging dust storms as far away as Beijing and beyond.

Halting Desertification in Inner Mongolia In recent years, China's government, along with local authorities and businesspeople, has turned the corner on desertification across much of the region. From 2010 to 2015, China treated 10 million hectares (24.7 million acres) of desertified land. The key process here has been the planting of drought-adapted vegetation on the dunes. If successful, the plants will send out roots to hold the sand in place, preventing dune movement. Such plants, however, usually need some help to become established. The best technique is to cover the sand with a mesh of wheat straw, creating squares roughly the size of a house. The straw temporarily stabilizes the sand until the new plants take root and provides essential plant nutrients as it decays.

Success in the Hobq Desert One of the most successful examples of reversing desertification is found in the Hobq Desert of Inner Mongolia (Figure 10.1.1). A key participant is the private Chinese company Elion Resources Group, which has been working with local and regional governments since the mid-1990s to revegetate vast expanses of barren land. The company originally extracted minerals found in the salt flats, but blowing sand kept covering the roads to these mineral deposits, halting transportation. After much trial and error, Elion researchers learned how to stabilize the sands by planting drought-adapted shrubs.

One plant that was soon thriving on the former dunes was licorice. Before long, Elion Resources learned that it could profit from the sale of licorice root, used in the pharmaceutical industry as well as in candy making. Eventually, the firm began to invest in a wide array of sustainable projects in the area, using solar power extensively and building greenhouses, efficient drip-irrigation systems, and research laboratories. Even a successful tourism industry was established. An estimated 100,000 jobs have emerged, increasing the annual income of local residents 10-fold. In 2012, the company's chair, Wang Wenbiao, was honored with the UN Environment and Development Award.

1. What other parts of the world have experienced severe desertification? Can the techniques pioneered in China be successfully applied in other world regions?
2. List some of the advantages and disadvantages of relying on private firms to combat desertification.

Figure 10.1.1 Fighting Desertification in Inner Mongolia Desertification is a major problem across much of Central Asia. In northern China's Inner Mongolia Autonomous Region, extensive tree-planting campaigns are being conducted in order to halt the spread of sand dunes and thus preserve agricultural fields and grazing grounds. Here villagers are planting trees to stabilize sand dunes in the Hobq Desert.

precipitation in its drainage basin. By the late 1990s, it had risen some 8.2 feet (2.5 meters). This enlargement, too, caused problems, including the flooding of some newly reclaimed farmland in the Volga Delta. But then, from 2007 to 2015, the lake began to shrink again, its level dropping by roughly half a meter. Currently, however, the Caspian's most serious environmental threat is probably pollution from the oil industry, rather than fluctuation in size.

Desertification Shrinking Central Asian lakes create new desert lands in the dried-up lake beds. Desertification also results from overgrazing and poor farming practices. In the eastern part of the region, the Gobi Desert has gradually spread southward, encroaching on densely settled lands in northeastern China proper. The Chinese have tried to prevent the march of desert with massive tree- and grass-planting campaigns, as the roots of such plants stabilize the soil and help to keep sand dunes from moving (see *Working Toward Sustainability: The Greening of the Inner Mongolian Desert*).

Dam Building and Water Conflicts Water resources in Central Asia are very unevenly distributed, and most of the freshwater in western Central Asia is controlled by mountainous Kyrgyzstan and Tajikistan. Both countries are engaged in massive dam-building projects designed to control water resources and generate electricity (Figure 10.7). Tajikistan's Rogun Dam, currently under construction with Russian aid, could be the world's tallest dam when completed. Uzbekistan, Kazakhstan, and Turkmenistan object to the construction of these dams, fearing that Kyrgyzstan and Tajikistan will release too much water during flood periods and withhold too much during dry spells. Leaders from the five countries have met repeatedly to hammer out an agreement on water sharing, but so far have been unable to do so.

Climate Change and Central Asia

Most climate experts expect Central Asia to be hard hit by global warming. The Tibetan Plateau has already seen marked temperature increases, resulting in melting permafrost and the retreat of mountain glaciers.

Figure 10.7 Nurek Dam, Tajikistan The world's second-highest dam, Nurek Dam in Tajikistan stands at 984 feet (300 meters). Completed in 1980 under the Soviet Union, Nurek Dam is designed primarily to supply hydroelectric power, but it is also used for irrigation and for flood control.

The retreat of Central Asia's glaciers is especially worrisome because of their role in providing water for the region's rivers. Deposition of soot from China's coal-burning factories and power plants on Tibetan glaciers has also hastened shrinkage in some areas. Some specialists think that the region's major rivers will decline by roughly 25 percent over the next 20 years.

Climate change is also likely to reduce precipitation in the arid lowlands of western Central Asia. Prolonged and devastating droughts have recently struck Afghanistan, indicating a possible shift to a drier climate. As a result, the UN Intergovernmental Panel on Climate Change has predicted a 30 percent crop decline for Central Asia as a whole by the middle of the 21st century. A 2012 report linked warmer weather to the spread of new strains of wheat rust, already reported to have reduced wheat yields in parts of the region. However, as is true elsewhere in the world, global warming will not affect all parts of Central Asia in the same way. Some areas, including the Gobi Desert and the Tibetan Plateau, could see increased precipitation.

Review

10.1 Why does Central Asia have such large lakes, and why are many of these lakes so severely threatened?

10.2 How does Central Asia's location near the center of the world's largest landmass and at the junction of colliding tectonic plates influence the region's climate and landforms?

KEY TERMS Silk Road, steppe, desertification

Population and Settlement: Densely Settled Oases Amid Vacant Lands

Most of Central Asia is sparsely populated (Figure 10.8). Large areas are either too arid or too high to support much human life. Even many of the most favorable areas are populated only by scattered groups of nomadic **pastoralists** (people who raise livestock for subsistence purposes).

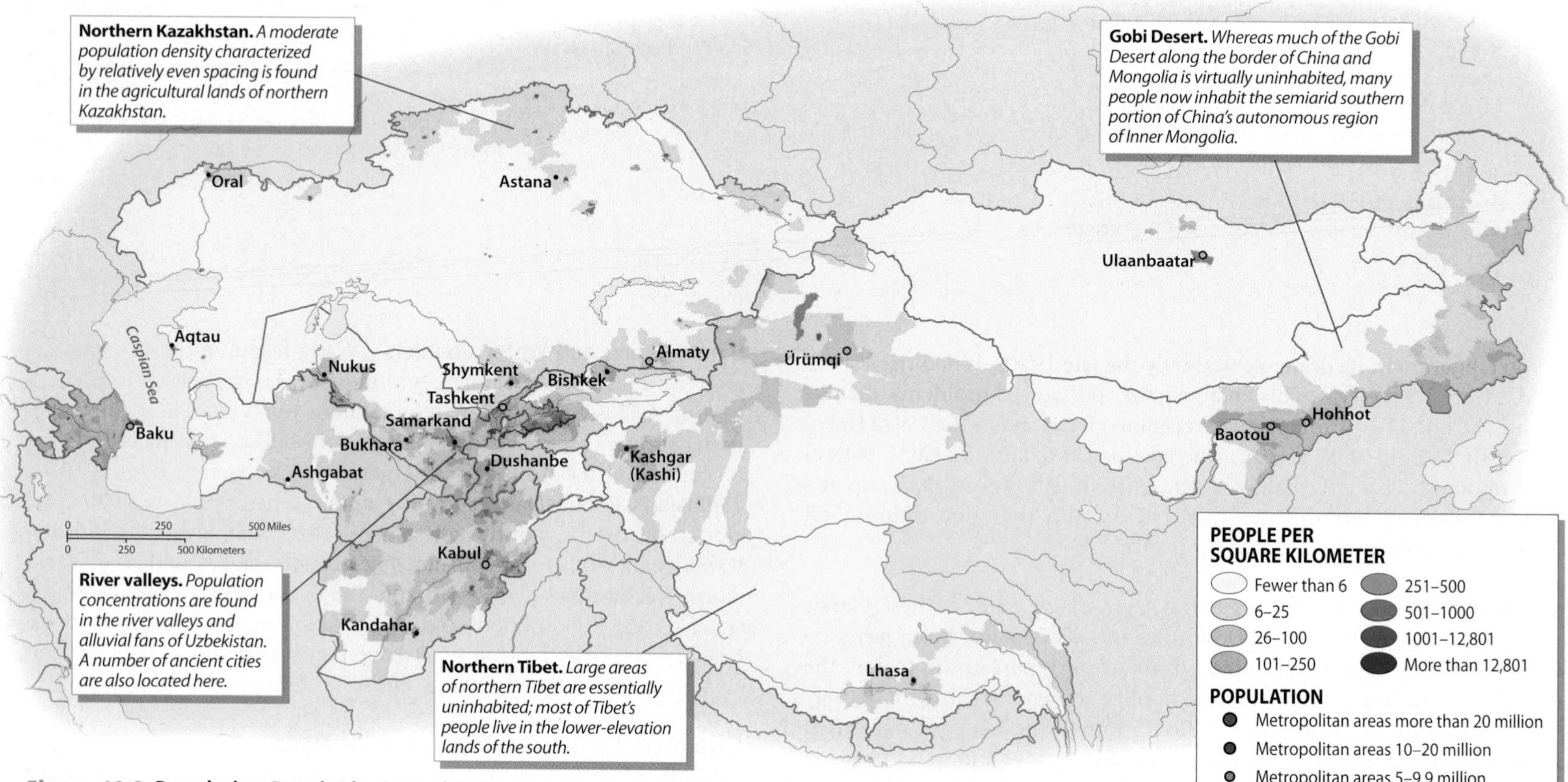

Northern Kazakhstan. *A moderate population density characterized by relatively even spacing is found in the agricultural lands of northern Kazakhstan.*

Gobi Desert. *Whereas much of the Gobi Desert along the border of China and Mongolia is virtually uninhabited, many people now inhabit the semiarid southern portion of China's autonomous region of Inner Mongolia.*

River valleys. *Population concentrations are found in the river valleys and alluvial fans of Uzbekistan. A number of ancient cities are also located here.*

Northern Tibet. *Large areas of northern Tibet are essentially uninhabited; most of Tibet's people live in the lower-elevation lands of the south.*

PEOPLE PER SQUARE KILOMETER: Fewer than 6; 6–25; 26–100; 101–250; 251–500; 501–1000; 1001–12,801; More than 12,801

POPULATION: Metropolitan areas more than 20 million; Metropolitan areas 10–20 million; Metropolitan areas 5–9.9 million; Metropolitan areas 1–4.9 million; Selected smaller metropolitan areas

Figure 10.8 Population Density in Central Asia As a whole, Central Asia remains one of the world's most sparsely populated regions, although it does contain distinct clusters of higher population density. Most of the region's large cities are located near the region's periphery or in its major river valleys.

Mongolia, which is more than twice the size of Texas, has only 3 million inhabitants. But as is common in arid environments, those few lowland settings with good soil and dependable water supplies are thickly settled.

Highland Population and Subsistence Patterns

The environment of the Tibetan Plateau is particularly harsh. Only sparse grasses and herbaceous plants can survive in this high-altitude climate, making human subsistence difficult. The only feasible way of life here is nomadic pastoralism based on the yak, an altitude-adapted relative of the cow. Several hundred thousand people manage to make a living in such a manner, roaming with their herds over vast distances.

Although most of the Tibetan Plateau can support only nomadic pastoralism, the majority of Tibetans are sedentary farmers. Farming in Tibet is possible only in the few locations that are *relatively* low in elevation and that have good soils and either adequate rainfall or dependable irrigation based on local streams. The main zone of sedentary settlement lies in the south and the southeast, where protected valleys offer these favorable conditions.

The mountains of Central Asia are vitally important for people living in the adjacent lowlands, whether they are migratory pastoralists or settled farmers. Many herders use the highlands for summer pasture; when the lowlands are dry and hot, the high meadows provide rich grazing. The Kyrgyz (of Kyrgyzstan) are noted for their traditional economy based on **transhumance**, moving their flocks from lowland pastures in the winter to highland meadows in the summer.

Lowland Population and Subsistence Patterns

Most of the inhabitants of Central Asian deserts live in the narrow belt where the mountains meet the basins and plains. Here water supplies are adequate and soils are neither salty nor alkaline, as is often the case in the basin interiors. For example, the population distribution pattern of China's Tarim Basin forms a ring-like structure (Figure 10.9). Streams flowing out of the mountains are diverted to irrigate fields and orchards in the fertile band along the basin's edge.

The population of former Soviet Central Asia is also concentrated in the transitional zone between the highlands and the plains. A series of **alluvial fans** (fan-shaped deposits of sediments dropped by streams flowing out of the mountains) has long been devoted to intensive cultivation (Figure 10.10). **Loess**, a fertile, silty soil deposited by the wind, is plentiful in this region. Several large valleys in this area also offer fertile, easily irrigated farmland. The large and densely populated

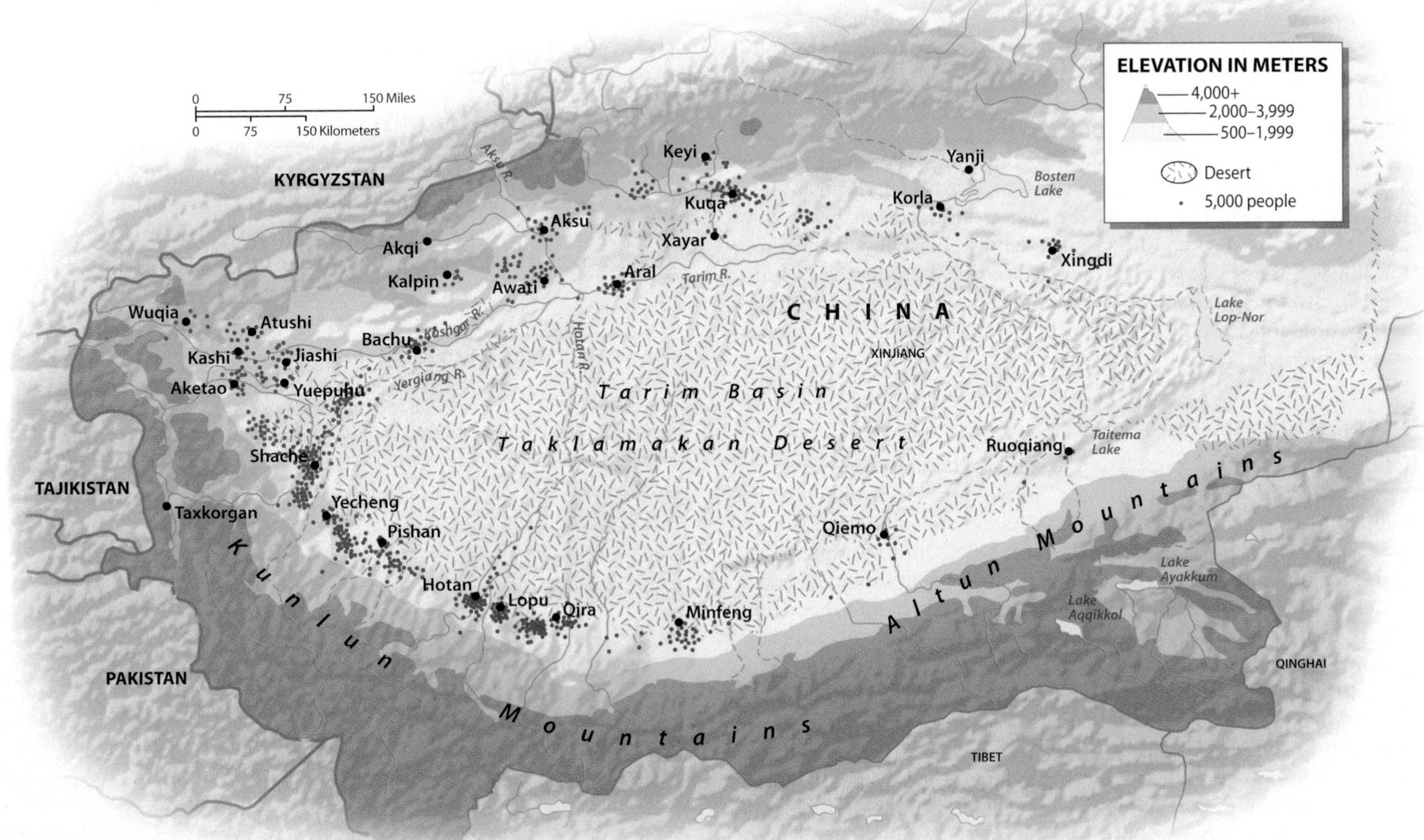

Figure 10.9 Population Patterns in Xinjiang's Tarim Basin The central portion of the Tarim Basin is a nearly uninhabited expanse of sand dunes and salt flats. Along the edge of the basin, however, dense agricultural and urban settlements are located where streams running out of the surrounding mountains allow for intensive irrigation. The largest of these oasis communities are found along the southwestern fringe of the basin.

FIGURE 10.10 Densely Settled Alluvial Fan The Tente River in Kazakhstan forms an alluvial fan where it emerges from the mountains and begins to flow across a relatively flat landscape. The fertile soils of this broad fan, measuring 12 miles (20 kilometers) across at its widest point, are intensively cultivated. Several towns and villages are visible along the fan's outer edge. Railroad tracks form a straight feature that cuts across the northeastern portion of the fan.

Fergana Valley of the upper Syr Darya River is shared by three countries: Uzbekistan, Kyrgyzstan, and Tajikistan.

The steppes of northern Central Asia are the classical land of nomadic pastoralism. Until the 20th century, almost none of this area had ever been plowed and farmed. To this day, pastoralism remains a common way of life across the grasslands, particularly in Mongolia (Figure 10.11). Throughout the region, however, many pastoral peoples have been forced to adopt sedentary lifestyles. In northern Kazakhstan, the Soviet regime converted the most productive pastures into wheat farms in the mid-1900s in order to increase the country's supply of grain. Consequently, northern Kazakhstan has the highest population density in the steppe region.

Population Issues

Central Asia has a low population density overall, but some areas are growing at a moderately rapid pace. In western China, much of the population growth over the past 30 years has stemmed from the migration of Han Chinese into the area. In 2006, China completed the construction of a railway from Beijing to the Tibetan capital of Lhasa, which has increased the flow of migrants and tourists into the region. The trains pass through such high elevations that passengers are supplied with supplemental oxygen.

Most of the former Soviet republics of Central Asia exhibit moderate population expansion due to their own fertility patterns. In contrast to Tibet and Xinjiang, this zone has experienced substantial outmigration since 1991. At first, emigrants were mostly ethnic Russians returning to the Russian homeland. Later, as the Russian economy boomed after 2000, hundreds of thousands of men from Central Asia sought work in Russia (see *Exploring Global Connections: Tajikistan's Remittance-Dependent Economy*).

Human fertility patterns vary substantially from one part of Central Asia to another (Table A10.1). Afghanistan, the least developed, most war-torn, and most male-dominated country of the region, has the highest birth rate by a wide margin. Fertility rates are in the middle range throughout most of the former Soviet area. Kazakhstan's birth rate was well below replacement level as recently as 2000, but it has rebounded and is now over 2.5. Fertility here is much lower for ethnic Russians than it is for the region's indigenous peoples. Azerbaijan's fertility rate, in contrast, is slightly below replacement level; as a result, the population pyramids of Azerbaijan and Afghanistan make an especially striking contrast (Figure 10.12).

Urbanization in Central Asia

Although the steppes of northern Central Asia had no real cities before the modern age, the river valleys have been partially urbanized for thousands of years. Such cities as Samarkand and Bukhara in Uzbekistan, which were key nodes on the old Silk Road, have long been famous for their lavish architecture (Figure 10.13). As a result, both cities have been designated UNESCO World Heritage Sites. These two cities contrast sharply with those built during the period

Figure 10.11 Mongolian Pastoralism Many Mongolians still make their livings as migratory herders, roaming over large areas of the great Mongolian grasslands. Such people typically live in collapsible tents known as gers (or yurts), two of which are visible in this photograph. **Q: Why is nomadic pastoralism so much more widespread in Mongolia than in most other countries?**

EXPLORING GLOBAL CONNECTIONS

Tajikistan's Remittance-Dependent Economy

Google Earth (MG) Virtual Tour Video http://goo.gl/K3j0Zx

Tajikistan, a poor, remote, and mountainous republic, is generally regarded as the world's most remittance-dependent country. In 2013, 42 percent of its earnings came from money sent by migrants working abroad. Kyrgyzstan, in second place, derived 31.5 percent of its economic production from remittances. In both cases, the vast majority of remitted income flowed from workers employed in Russia. Over 1 million of Tajikistan's 8.2 million people are believed to live and work in Russia (Figure 10.2.1).

Tajikistan relies so heavily on remittances in part because the country exports very little to the rest of the world. It does have abundant hydropower, which allows it to refine aluminum for export and to sell some electricity directly to neighboring countries. Overall, however, Tajikistan's import bill is almost three times its export receipts. Only remittances allow the country to import the goods that its citizens and its industries require.

The reliance of Tajikistan on remittances from Russia places the country in an uncomfortable situation. To begin with, it gives Russia great leverage over political decisions. When the Tajik government does something that irritates Russia, Russian leaders not uncommonly threaten to deport Tajik workers. Such threats were leveled in 2013, for example, when Tajikistan delayed renegotiating the lease on a Russian military base located in the country.

Tajikistan's dependence on remittances also means that the country's economy suffers deeply whenever Russia experiences economic difficulties. As falling oil prices, along with Western sanctions, sent Russia into a recession in late 2014, its demand for labor from foreign countries quickly declined. As a result, the World Bank estimated that the flow of remittance money to Tajikistan dropped by 40 percent in 2015. Tajikistan's exports to Russia have also declined, which places further strains on the country.

Figure 10.2.1 Tajik Workers in Russia The economy of Tajikstan is highly dependent on the remittance of wages by Tajik workers living in Russia. Such workers, however, are often treated harshly and during economic downturns many of them are forced to return home. In this photograph, migrant workers from Tajikistan are relaxing on the roof of their shelter after working at a local market near Moscow.

Tajikistan's major response to the decline in remittances from Russia is to seek closer economic ties with China. In late 2014, China announced that it would invest $6 billion in the country over the next three years, which would be more than 10 times Tajikistan's average annual foreign direct investment in a three-year period. Most of this Chinese investment is targeted at gas pipelines, railroads and other transportation projects, and cement manufacturing. It is unlikely, however, that such projects will make up for the decline in remittances. This investment also leads to fears that Tajikistan will come under the economic domination of China.

1. Why do Tajikistan and Kyrgyzstan rely so heavily on remittances? Relate this situation to their geographical location and their historical development.
2. Is such remittance dependency a major problem for these countries? If so, how might they try to reduce such dependency?

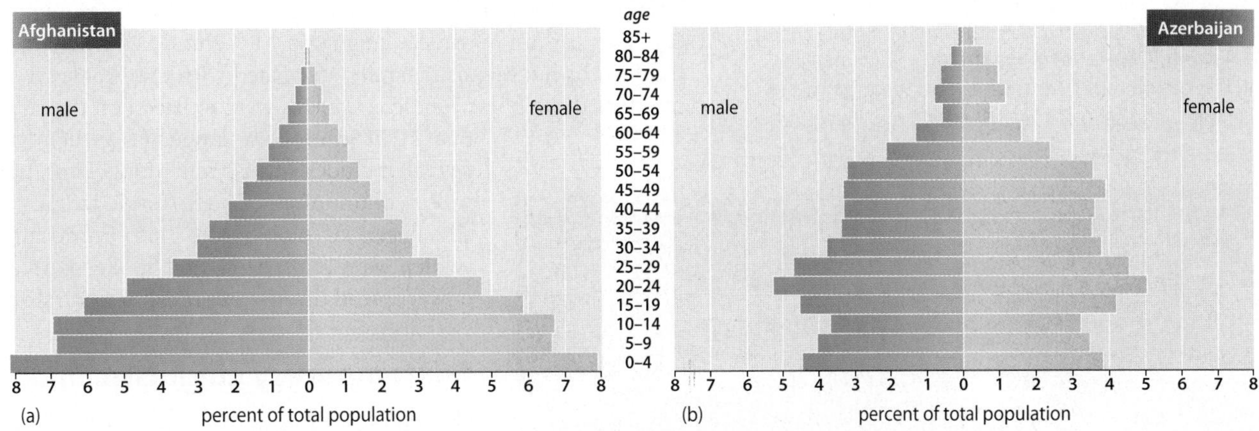

Figure 10.12 Demographic Structure of Afghanistan and Azerbaijan The relatively balanced population pyramid for Azerbaijan reflects the fact that it is close to population stability. Azerbaijan's birth rate has increased in recent years, however, as is indicated by the wider bar for ages "0–4" than for ages "10–14." Note also that women outlive men in (b) Azerbaijan, which is the normal pattern. In contrast, (a) Afghanistan's high birth rates coupled with high death rates have created a more triangular pyramid. In this poor, male-dominated country, elderly men and elderly women are equally few in numbers. **Q: In Afghanistan, why are there so many more people in the "0–4" age bracket than in the "5–9" age bracket?**

Figure 10.13 Traditional Architecture in Samarkand This Silk Road city in Uzbekistan is famous for its lavish Islamic architecture, some of it dating back to the 1400s. Samarkand owes its rich architectural heritage in part to the fact that it was the capital of the empire created by the great medieval conqueror Tamerlane.

of Russian/Soviet rule. Tashkent in Uzbekistan, for example, is largely a Soviet creation and thus looks quite different from the more traditional Bukhara. Several major cities, such as Kazakhstan's former capital of Almaty, were minor settlements before Russian colonization. In cities throughout the former Soviet region, the effects of centralized Soviet urban planning and design are easily seen.

Parts of Central Asia have experienced substantial urbanization in recent decades. A new major city, Astana, was designated Kazakhstan's capital in 1997. Noted for its futuristic architecture, Astana is a good example of a **forward capital**, which is a new capital city established in what had previously been a peripheral area in order to strengthen political and economic power there (Figure 10.14).

Other Central Asian cities are also expanding rapidly. Even Mongolia now has more people living in cities than in the countryside. Its capital, Ulaanbaatar (Ulan Bator), was recently in the midst of a major building boom. In the highlands of Central Asia, however, cities remain relatively few and far between. Only about one-quarter of the people of Tajikistan, for example, are urban residents. Tibet similarly remains predominantly rural, although change is occurring as Han Chinese migrants move into its cities.

New development projects, coupled with Han migration, are transforming cities throughout Chinese Central Asia. Local controversies often follow. Between 2010 and 2014, for example, the Chinese government demolished some 65,000 houses in Kashgar (Kashi) in Xinjiang, once a major station on the Silk Road. The destruction of most of the old city of Kashgar was officially justified on the grounds of earthquake safety. Such actions, however, infuriate most local residents and harm tourism, which China wants to encourage. An international petition drive to have UNESCO declare central Kashgar a World Heritage Site has thus far had little success.

Figure 10.14 Astana, Capital of Kazakhstan Astana, a forward capital, has grown rapidly in recent years, emerging as a new metropolis near the middle of the vast country. Oil wealth has supported Astana's expansion.

Review

10.3 Why are large parts of Central Asia so sparsely settled, while others have dense populations?

10.4 Why is the urban environment of Central Asia changing so rapidly?

KEY TERMS pastoralist, transhumance, alluvial fan, loess, forward capital

Cultural Coherence and Diversity: A Meeting Ground of Different Traditions

Although large areas of Central Asia are environmentally similar, the region's cultural traditions are more diverse. Western Central Asia is largely Muslim and is often classified as part of Southwest Asia, but in Mongolia and Tibet most people traditionally follow Tibetan Buddhism. Tibet is culturally linked to both South and East Asia, and Mongolia is historically associated with China, but neither fits easily within any other world region.

Historical Overview: Changing Languages and Populations

The river valleys and oases of Central Asia were early sites of agricultural communities. Archaeologists discovered abundant evidence of farming villages dating back to Neolithic times (beginning around 8000 BCE) in the Amu Darya and Syr Darya valleys. After the domestication of the horse around 4000 BCE, nomadic pastoralism emerged in the steppes. Eventually, pastoral peoples gained power over the entire region.

The earliest recorded languages of Central Asia belonged to the Indo-European linguistic family, associated with the people who first domesticated the horse. These languages were replaced on the steppe more than 1000 years ago by languages in the Altaic family (which includes Turkish and Mongolian). In the river valley communities as well, Turkic languages gradually began to replace the Indo-European tongues (which were closely related to Persian) as Turkic power spread throughout most of Central Asia.

Contemporary Linguistic and Ethnic Geography

Today people speaking Turkic and Mongolian languages inhabit most of Central Asia (Figure 10.15). A few native Indo-European languages are confined

GEOGRAPHERS AT WORK

Kazakh Migration in Mongolia

"Geography's links with other disciplines comes to life in Mongolia," notes **Holly Barcus**, a geographer at Macalester College in Minnesota. Barcus studies the minority Kazakh population of Mongolia in collaboration with anthropologist Cynthia Werner of Texas A&M University. "Geography allows conversations, not only across the globe, but across disciplines," says Barcus. "Geographers enter these conversations with an understanding of history, politics, the natural environment, drawing on our background in each area to pursue a question." Barcus and Werner want to understand why roughly half of Mongolia's Kazakhs migrated to Kazakhstan after the end of the Cold War, why many later returned to Mongolia, and why still others simply stayed put (Figure 10.3.1).

Studying Migration Barcus's passion for geography was sparked by "the real-world linkages between topics discussed in class and being able to understand the process creating those outcomes—but also my own role in that process. At that moment, what I was studying became active rather than passive learning—something I enjoyed outside the classroom as well as in the classroom."

In 2004 Barcus initiated a long-term project on transnational migration in western Mongolia. She and Werner found that ethnic Kazakhs were not pushed out of Mongolia after 1991, but were instead enticed by the Kazakhstan government, which wanted to increase the proportion of ethnic Kazakhs in its population and had enough mineral and fossil fuel wealth to subsidize migrants. Migration later became more circular, as roughly a third of the 60,000-strong group that had left in the first wave returned to Mongolia. Some returnees complained of discrimination for being too rural and for not speaking Russian, still an important language in Kazakhstan. More recently, younger Kazakhs have left Mongolia for Kazakhstan for better education and career opportunities.

Barcus and Werner discovered that Mongolian Kazakhs generally frame their decisions around geographically informed narratives. The desire to stay in Mongolia is reinforced by ties to one's place of birth, while movement to Kazakhstan stems from the notion of the "original homeland." "The environment is so intrinsically related to our self-identity and how that identity plays into nationalism and migration decisions," says Barcus.

Barcus encourages her own students to travel: "People are moving around, and changes are occurring in the global economy. There are endless possibilities for going out and engaging." She also tells students that geography "is a practical skill that complements a whole range of occupations. At the undergraduate level you can't go wrong with geography. It forms the baseline for anything you want to do."

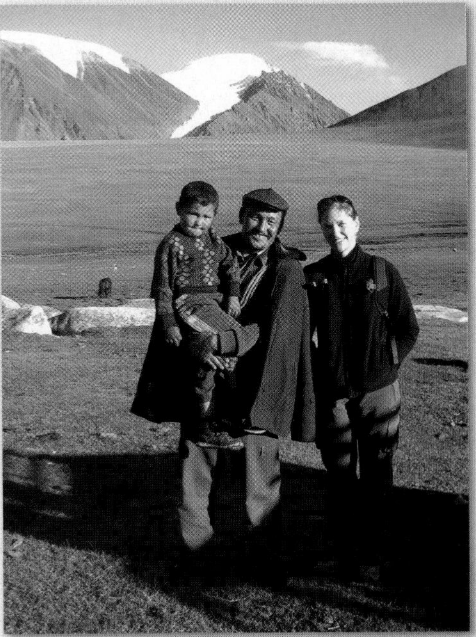

Figure 10.3.1 The Kazakhs of Mongolia Geographer Holly Barcus visits an ethnic Kazakh family in western Mongolia. Dr. Barcus has been studying migration in this region since 2004.

1. What cultural, historical, and environmental factors might be linked to the relative ease of movement of people from Mongolia to Kazakhstan and back again to Mongolia?
2. Do you know immigrants who have returned or want to return to their home country? List push or pull factors that might influence their decision.

to the southwest, while Tibetan remains the main language of the plateau. Russian is also widely spoken in the west, often as a second or third language, while Chinese is increasingly important in the east.

Tibetan Tibetan is usually placed in the Sino-Tibetan language family, suggesting a shared linguistic history between the Chinese and the Tibetan peoples. Some students of Tibetan, however, argue that no definite relationship between the two languages has been established.

The Tibetan language itself is divided into several highly distinctive forms, usually called dialects, but regarded by most linguists as separate languages. Such "Tibetic" languages are spoken over almost the entire Tibetan Plateau. Over 90 percent of the 3 million people who live in the Tibet Autonomous Region speak a Tibetic language, while another 3 million or so Tibetic speakers live in the highland portions of the Chinese provinces of Qinghai and Sichuan. Smaller Tibetic-speaking communities are found elsewhere in China and in adjoining parts of India, Nepal, and Bhutan as well.

Mongolian The Mongolian language includes a cluster of closely related dialects spoken by approximately 5 million people. The standard Mongolian of both the independent country of Mongolia and China's Inner Mongolia is called *Khalkha.* Mongolian has its own distinctive script, but Mongolia itself adopted Russia's Cyrillic alphabet in 1941. Efforts are now being made in Mongolia to revive the old script, which is still used by Mongolian speakers in northern China.

Mongolian speakers form about 90 percent of the population of Mongolia, although Kazakhs, who speak a Turkic language, are the largest minority group (see *Geographers at Work: Kazakh Migration in Mongolia*). In China's Inner Mongolia Autonomous Region, Han Chinese who migrated to the area over the past 100 years now outnumber Mongolian speakers, and only about 17 percent of the 25 million residents are Mongolian. But due to its relatively large population, Inner Mongolia still has a much larger population of Mongolian speakers than does Mongolia itself.

Turkic Languages Far more Central Asians speak Turkic languages than Mongolian and Tibetan combined. The Turkic linguistic sphere extends from Azerbaijan in the west through China's Xinjiang Autonomous Region in the east.

Five of the six countries of the former Soviet Central Asia—Azerbaijan, Uzbekistan, Turkmenistan, Kyrgyzstan, and Kazakhstan—are

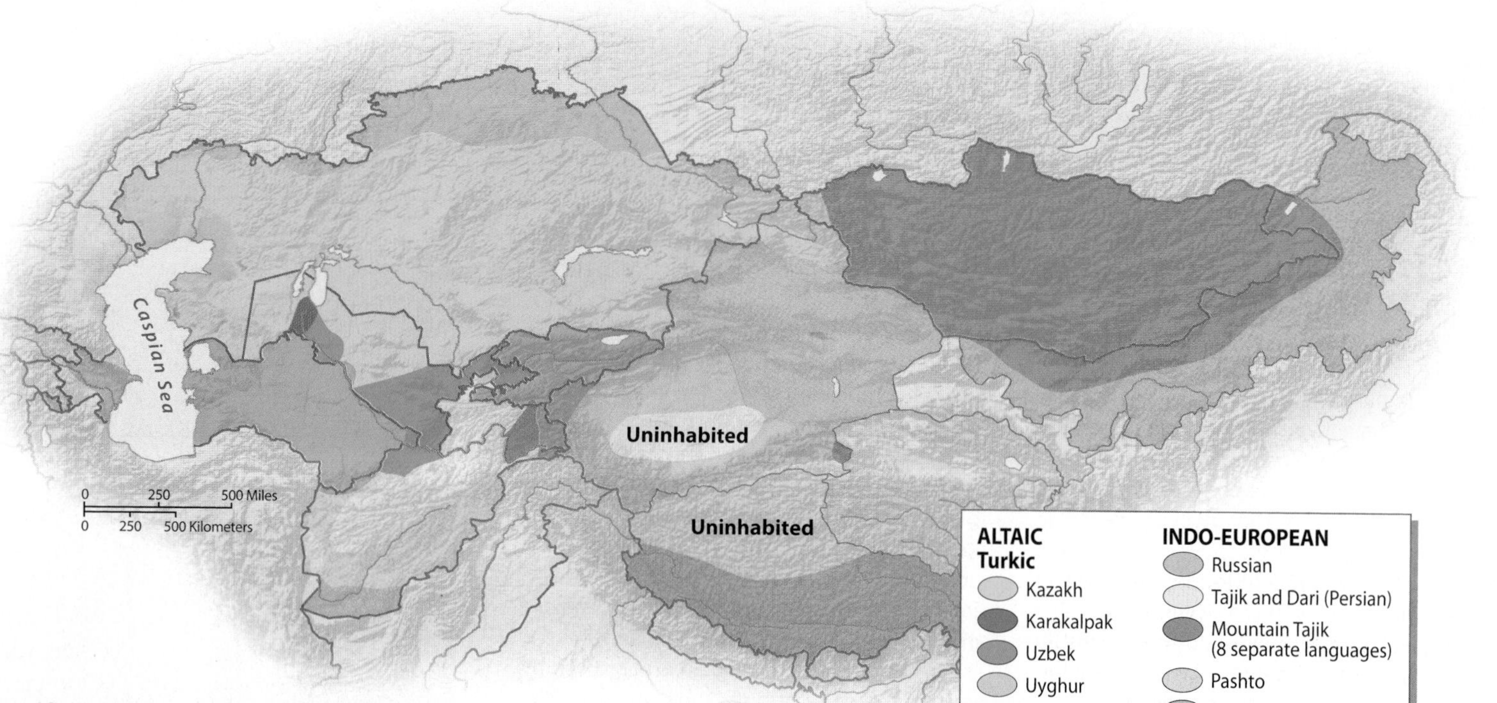

Caspian Sea

Uninhabited

Uninhabited

0 250 500 Miles

0 250 500 Kilometers

ALTAIC

Turkic

- Kazakh
- Karakalpak
- Uzbek
- Uyghur
- Kyrgyz
- Turkmen
- Azeri

Mongolian

- Mongolian

INDO-EUROPEAN

- Russian
- Tajik and Dari (Persian)
- Mountain Tajik (8 separate languages)
- Pashto
- Baluchi

SINO-TIBETAN

- Tibetic languages
- Mandarin Chinese

OTHER

- Other and uninhabited

Figure 10.15 Language Map of Central Asia Most of the region is dominated by languages in the Altaic family, which includes both the Turkic languages (spoken throughout most of the central and western parts of the region) and Mongolian (spoken in the northeast). Several Indo-European languages, however, are spoken in both the far northwestern and the south-central regions, while Tibetic languages cover most of the Tibetan Plateau in the southeast.

named after the Turkic languages of their dominant populations. In Azerbaijan, 90 percent of the people speak Azeri, whereas 74 percent of the people of Uzbekistan speak Uzbek and 72 percent of the people of Turkmenistan speak Turkmen. With more than 24 million speakers, Uzbek is the most widely spoken Central Asian language. In the Amu Darya Delta in the far north of Uzbekistan, however, most people speak a different Turkic language called *Karakalpak*, while Kazakh speakers are found in the sparsely populated Uzbek deserts. Roughly 72 percent of the inhabitants of Kyrgyzstan speak Kyrgyz as their native language, while around 63 percent of the people of Kazakhstan speak Kazakh. More than a quarter of Kazakhstan's residents speak European languages, particularly Russian and Ukrainian. In general, the Kazakhs predominate in the center and south of the country, while Russians live in the agricultural districts of the north and in the cities of the southeast.

Uyghur, spoken in China's Xinjiang Autonomous Region, dates back almost 2000 years. The Uyghur people number about 11 million, most of whom live in Xinjiang (Figure 10.16). As recently as 1953, the Uyghur made up about 80 percent of Xinjiang's population, but that number fell to 46 percent by 2009, mostly due to Han Chinese migration. According to official statistics, the Han Chinese make up around 39 percent of Xinjiang's population, but Uyghur activists think that number is much larger. About 1.5 million Kazakh speakers also live in Xinjiang.

Linguistic Complexity in Tajikistan and Afghanistan The sixth republic of the former Soviet Central Asia, Tajikistan, is dominated by Indo-European language speakers. Tajik is so closely related to Persian that it is often considered to be a Persian (or Farsi) dialect. Roughly 80 percent of the people of Tajikistan speak Tajik as their first language. In the remote mountains of eastern Tajikistan, people speak a variety of distinctive Indo-European languages, sometimes collectively referred to as "Mountain Tajik." The remainder of Tajikistan's population speaks either Uzbek or one of several minor languages.

The linguistic geography of Afghanistan is even more complex than that of Tajikistan (Figure 10.17). Afghanistan became a country in the 1700s under the leadership of the Pashtun people. Afghanistan's rulers, however, never tried to build a nation-state around Pashtun identity, in part because about half of the Pashtun people (who speak the Pashto language) live in what is now Pakistan. In Afghanistan itself, 40–50 percent of the population is estimated to be Pashto speakers, most living to the south of the Hindu Kush.

Figure 10.16 Uyghur Mosque The Uyghur people of Xinjiang in northwestern China speak a Turkic language and follow Sunni Islam. The Id Kah Mosque in the city of Kashgar, pictured here, is China's largest mosque, accommodating up to 10,000 people.

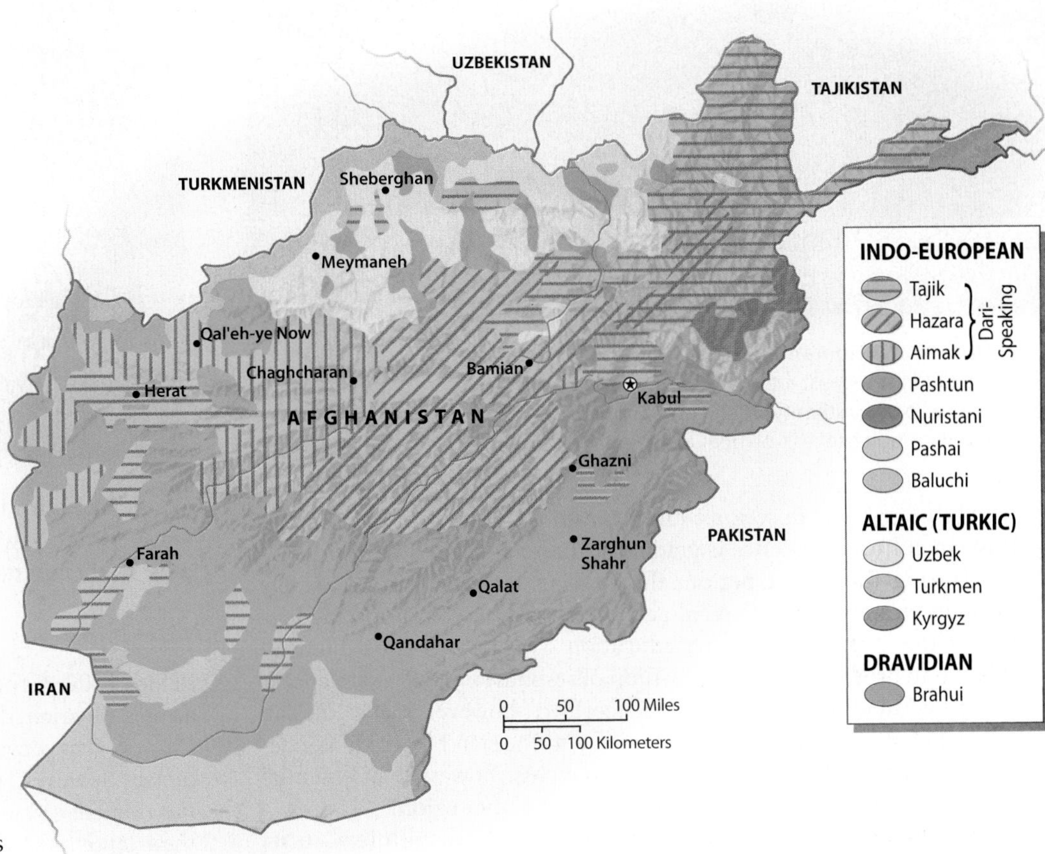

Figure 10.17 Afghanistan's Ethnolinguistic Patchwork Afghanistan is one of the world's most ethnically complex countries. Its largest ethnic group is the Pashtuns, who live in most of the southern portion of the country as well as in the adjoining borderlands of Pakistan. Northern Afghanistan is mostly inhabited by Uzbeks, Tajiks, and Turkmens—whose main population centers are located in Uzbekistan, Tajikistan, and Turkmenistan, respectively. The Hazaras of Afghanistan's central mountains, like the Tajiks, speak a form of Persian. However, the Hazaras are considered a separate ethnic group in part because they, unlike other Afghans, follow Shiite rather than Sunni Islam.

Roughly half of the people of Afghanistan speak Dari, the local form of Persian, primarily in northern Afghanistan. Dari speakers in the west and the far north are considered to be of Tajik ethnicity, whereas those in the central mountains are Hazaras, said to descend from Mongol conquerors who arrived in the 12th century. Finally, another 10 percent or so of the people of Afghanistan speak Turkic languages, mainly Uzbek.

Geography of Religion

At one time, Central Asia was noted for its religious complexity. Major overland trading routes crossed the region, giving easy access to both merchants and missionaries. By 1500, however, the region was divided into two spiritual camps: Islam predominated in the west and center, and Tibetan Buddhism spread in Tibet and Mongolia.

Islam in Central Asia Different Central Asian peoples are known for their different interpretations of Islam. The Pashtuns of Afghanistan are noted for their strict Islamic ideals—although critics contend that many Pashtun rules, such as prohibiting women's faces to be seen in public, are based more on ethnic than religious customs. The traditionally nomadic groups of the northern steppes, such as the Kazakhs and Kyrgyzs, are considered to be more relaxed about religious beliefs and practices. While most of the region's Muslims are Sunnis, Shiism is dominant among the Hazaras of central Afghanistan and the Azeris of Azerbaijan.

Under the communist regimes of China, the Soviet Union, and Mongolia, all forms of religion were discouraged. Chinese authorities and student radicals attempted to suppress Islam in Xinjiang during the Cultural Revolution of the late 1960s and early 1970s. Chinese Muslims now enjoy basic freedom of worship, but the state still closely monitors religious expression out of fear that it will lead to resistance against Chinese rule. In 2014, officials in Urumqi, the capital of Xinjiang, banned the wearing of the Islamic face veil in public.

Periodic persecution of Muslims also occurred in Soviet Central Asia, and until the 1970s many observers thought that Islam would slowly disappear from the region. Religious expression was not, however, so easily repressed. Interest in Islam began to grow in former Soviet Central Asia in the 1970s and 1980s. In the post-Soviet period, Islam continues to revive as people seek their cultural roots. In Xinjiang, Islam has served as a focus of social identity among the Uyghur people. Most Uyghur leaders, however, insist that their religious beliefs are not political. Radical Islamic fundamentalism, however, has emerged as a powerful political movement in Afghanistan, parts of Tajikistan, and the Fergana Valley (mostly in Uzbekistan).

Tibetan Buddhism Mongolia and especially Tibet stand apart from the rest of Central Asia in their peoples' practice of Tibetan Buddhism, sometimes called Lamaism. Buddhism entered Tibet from India many centuries ago and merged with the native religion of the area, called *Bon*. The resulting mix is more oriented toward mysticism than are other forms of Buddhism and is more tightly organized. The head of Tibetan Buddhist society is the Dalai Lama. Until the Chinese conquest, Tibet was essentially a **theocracy** (religious state), with the Dalai Lama holding political as well as religious authority.

Tibetan Buddhists suffered persecution after 1959, when China invaded Tibet, and Tibetan Buddhism inspired many Tibetans to resist Chinese rule. The Dalai Lama, who fled Tibet for India in 1959, has been a powerful advocate for the Tibetan cause in international circles. During the 1960s and 1970s, an estimated 6000 Tibetan Buddhist monasteries were destroyed; the number of active monks today is only about 5 percent of what it was before the Chinese occupation. China has more recently allowed many monasteries to reopen, but their activities remain limited and closely watched (Figure 10.18).

Central Asian Culture in Global Context

In cultural terms, western Central Asia is closely linked to Russia, whereas eastern Central Asia is more closely tied to China. In the east, the main cultural issue is the migration of Han Chinese into the area,

Figure 10.18 Tibetan Buddhist Monastery Tibet is well known for its large Buddhist monasteries, which at one time served as seats of political as well as religious authority. The Ganden Monastery is one of the three great "university monasteries" of Tibet.

Figure 10.19 U.S.-Style Restaurant in Baku A vibrant and increasingly cosmopolitan city, Baku, Azerbaijan, is closely linked to global economic and cultural networks. This coffee shop and wine bar caters to both local and international customers. **Q: Why does the city of Baku support more international-style restaurants than most other Central Asian cities?**

which has resulted in serious ethnic tensions. In the west, in contrast, Russian cultural influence is gradually diminishing.

During the Soviet period, the Russian language spread widely through western Central Asia, serving both as a common language and as the language of higher education. People had to be fluent in Russian in order to reach any position of responsibility. Russian speakers settled in all the major cities. Since 1991, however, many Russian speakers have migrated back to Russia, and Russian has gradually declined in favor of local languages. It remains, however, an essential language of interethnic communication across the region.

Though remote and poorly integrated into global culture, Central Asia is hardly immune to the forces of globalization. The increased usage of English throughout Central Asia shows its close connections with global culture. Such influences have been especially marked in the oil cities of the Caspian Basin, such as Baku, which have seen an influx of foreign oil workers (Figure 10.19). Although their number is small, English speakers in Central Asia, especially those with computer skills, are increasingly valued as the region strives to find a place and a voice in the global community.

Review

10.5 Explain how patterns of religious affiliation divide Central Asia into distinct subregions and why religion is a source of increasing tensions in the region.

10.6 How have the patterns of linguistic geography in Central Asia been transformed over the past two decades?

KEY TERMS theocracy

Geopolitical Framework: Political Reawakening

Central Asia has played a minor role in global political affairs for the past several hundred years. Before 1991, the entire region, except Mongolia and Afghanistan, was under direct Soviet or Chinese control. Mongolia, moreover, had been a close Soviet ally, and even Afghanistan came under Soviet domination in the late 1970s. Today southeastern Central Asia firmly belongs to China. And although the breakup of the Soviet Union saw the emergence of six new Central Asian countries, most of them retain close political ties to Russia (Figure 10.20).

Partitioning of the Steppes

Before 1500, Central Asia was a power center, a region whose mobile armies threatened the far more populous, sedentary states of Asia and Europe. The development of new weapons changed the balance of power, however, allowing the wealthier agricultural states to conquer the nomads. By the 1700s, the nomad armies had been defeated and their lands taken. The winners in this struggle were the two largest states bordering the steppes: Russia and China.

By the mid-1700s, the Chinese empire (under the Manchu, or Qing, dynasty) had reached its greatest territorial extent, including Mongolia, Xinjiang, Tibet, and a slice of modern Kazakhstan. From the height of its power in the late 1700s, China declined rapidly. When the Manchu dynasty fell in 1912, Mongolia became independent, functioning as a **buffer state** between China and Russia, although China still ruled the extensive borderlands of Inner Mongolia (Nei Mongol). Tibet had earlier gained effective independence, although China did not recognize this status.

In the mid-1800s, Russia began to push south of the steppe zone, in part to keep the British out, conquering most of western Central Asia by 1900. Britain attempted to conquer Afghanistan, but failed to do so. Consequently, Afghanistan's position as an independent buffer state between the Russian Empire (later the Soviet Union) and British India remained secure.

Central Asia Under Communist Rule

Western Central Asia came under communist rule not long after the emergence of the Soviet Union in 1917. Mongolia followed in 1924. After the Chinese revolution of 1949, the communist system was also imposed on Xinjiang and Tibet.

Soviet Central Asia The newly established Soviet Union retained all the Central Asian territories of the Russian Empire. The new regime sought to build a Soviet society that would eventually knit together all the massive territories of the Soviet Union. Central Asia's leaders were replaced by Communist Party officials loyal to the new state, Russian immigration was encouraged, and local languages had to be written in Cyrillic (Russian) rather than Arabic script.

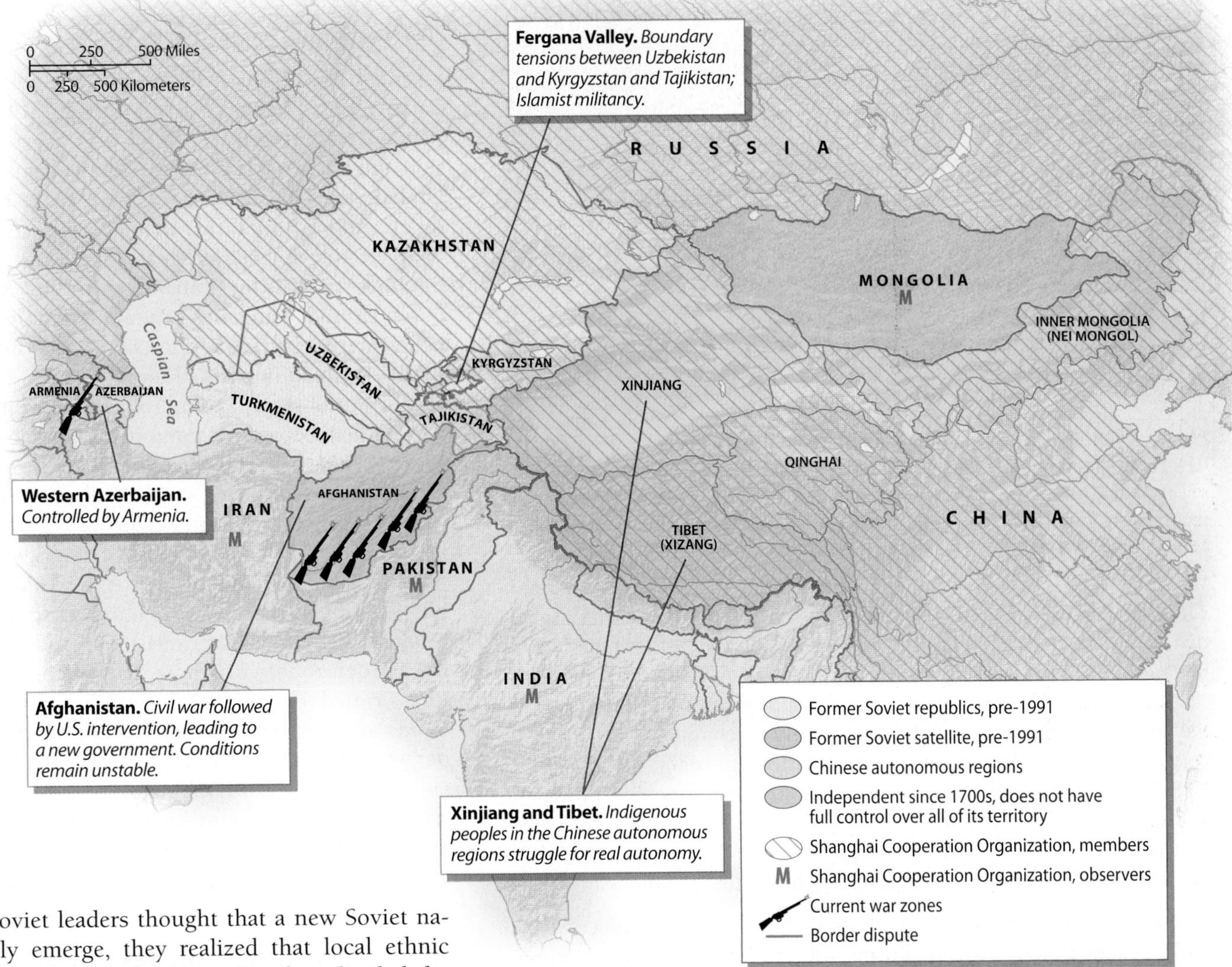

Figure 10.20 Central Asian Geopolitics Six of the eight independent states of Central Asia came into existence in 1991, with the breakup of the Soviet Union. In eastern Central Asia, the most serious geopolitical problems stem from China's maintenance of control over areas in which the native peoples are not Chinese. Afghanistan, the scene of a prolonged and brutal war, has experienced the most extreme forms of geopolitical tension in the region.

Although the early Soviet leaders thought that a new Soviet nationality would eventually emerge, they realized that local ethnic diversity would not disappear overnight. They therefore divided the Soviet Union into a series of nationally defined "union republics," in which a certain degree of cultural autonomy would be allowed. In the 1920s, Kazakhstan, Kyrgyzstan, Tajikistan, Uzbekistan, Turkmenistan, and Azerbaijan assumed their present shapes as Soviet republics.

Contrary to official intentions, the new republics encouraged the development of local national identities more than a broader Soviet identity. Also undercutting Soviet unity was the fact that cultural and economic gaps separating Central Asians from Russians did not decrease as much as planned. In addition, the region's higher birth rates led many Russians to fear that the Soviet Union risked being dominated by Turkic-speaking Muslims, contributing to the breakup of the Soviet Union.

The Chinese Geopolitical Order After decades of political and economic chaos, China reemerged as a united country in 1949. Its new communist government quickly reclaimed most of the Central Asian territories that had slipped out of China's grasp in the early 1900s. China's new leaders promised the non-Chinese peoples a significant degree of political self-determination and cultural autonomy and thus found much local support in Xinjiang. Gaining control over Tibet was more difficult. China occupied Tibet in 1950, but the Tibetans launched a rebellion in 1959. When this was crushed, the Dalai Lama and some 100,000 followers sought refuge in India.

Loosely following the Soviet model, China established autonomous regions in areas occupied primarily by non–Han Chinese peoples, including Xinjiang, Tibet proper (called *Xizang* in Chinese), and Inner Mongolia. Such autonomy, however, often meant little, and it did not prevent Han Chinese immigration to these areas. Nor were all parts of Chinese Central Asia granted autonomous status. The large and historically Tibetan and Mongolian province of Qinghai, for example, remained an ordinary Chinese province.

Current Geopolitical Tensions

Although western Central Asia made the transition to independence rather smoothly, several conflicts still affect the region. Much of China's Central Asian territory is also troubled, but China keeps a strong political hold on the area. Afghanistan, unfortunately, continues to suffer from a particularly brutal war.

Independence in Former Soviet Lands The disintegration of the Soviet Union in 1991 generally proceeded peacefully in Central Asia (Figure 10.21), although a civil war that broke out in Tajikistan lasted until 1997. The six newly independent countries, however, had been dependent on the Soviet system, and it was not easy for them to chart their own courses. In most cases, authoritarian rulers, rooted in the old order, retained power and undermined opposition groups.

The new countries also faced several border conflicts, due in part to the complicated political divisions established by the Soviet Union. Such problems have been particularly severe in the Fergana Valley, a large, fertile, densely populated lowland basin shared by Uzbekistan, Tajikistan, and Kyrgyzstan (Figure 10.22). Boundaries here meander in a complex manner and remain a source of disagreement among the countries in question. Tajikistan and Kyrgyzstan, moreover, have expressed anger at Uzbekistan's uncompromising attitudes about border delineation.

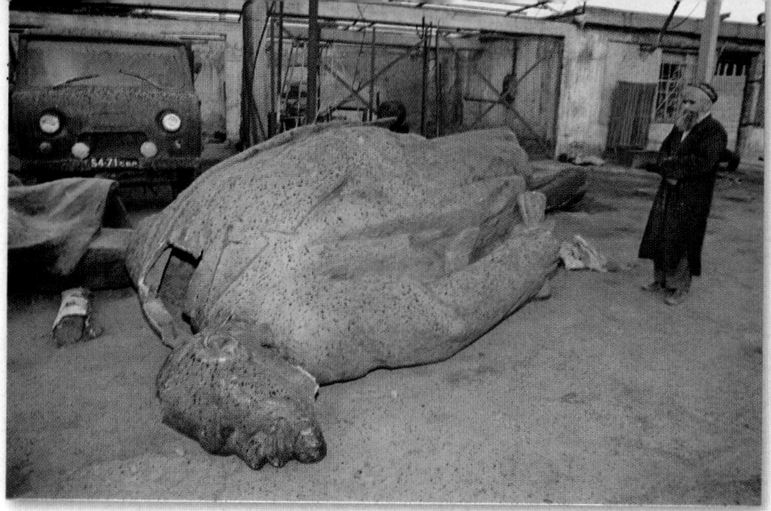

Figure 10.21 The End of Soviet Rule in Central Asia An elderly Tajik man looks over a toppled and decapitated statue of Lenin, founder of the Soviet Union, in Dushanbe. When the Soviet Union collapsed in 1991, emblems of the old regime were immediately destroyed in many areas. **Q: Why were so many statues of Lenin erected in the former Soviet republics of Central Asia in the first place?**

This has resulted in tense relations between Uzbekistan and the other two countries. Tajikistan has further objected to the fact that many ethnic Tajiks in Uzbekistan have been forced to take on an Uzbek identity.

More recently, however, geopolitical tensions in the Fergana Valley area have eased slightly, due mainly to common security concerns associated with extremist interpretations of Islam. In early 2015, Uzbekistan announced as a goodwill gesture the opening of a new commercial airline route to Dushanbe, Tajikistan's capital. Uzbekistan also resumed selling natural gas to energy-poor Kyrgyzstan, which had been suspended due to economic and political disagreements.

Strife in Western China Many of the indigenous inhabitants of Tibet and Xinjiang in western China want independence, or at least real political autonomy. China maintains that all its Central Asian lands are essential parts of its national territory, and it treats separatist groups harshly. In Xinjiang, several groups form the Eastern Turkistan Independence Movement. Some of these have a radical Islamist orientation and are considered terrorist organizations by both China and the United States, but most are more secular, basing their claims on ethnicity and territory rather than religion.

Periodic violence by separatist groups in Xinjiang typically results in quick reprisals by China. In July 2009, severe ethnic rioting broke out in Urumqi, Xinjiang's capital, prompting China to temporarily suspend Internet service and restrict cell-phone coverage (Figure 10.23). In 2014, China intensified its restrictions on Muslim religious observances in the region, going so far as to prohibit fasting during Ramadan. Violent separatists continue to launch deadly attacks against the Chinese state and Han Chinese civilians, provoking harsh retaliation by the Chinese military. In 2014 alone, an estimated 500 people died in the so-called Xinjiang conflict.

Anti-China protests in Tibet have also been severely repressed, but the Tibetans have brought the world's attention to their struggle. China maintains several hundred thousand troops in the region because of both Tibetan resistance and the strategic importance of the border zone with India. In March 2008, Tibetan protestors in Lhasa rioted, burning many Chinese-owned businesses. In response, China arrested hundreds of Tibetan monks and forced others to undergo "patriotic education." The Dalai Lama now agrees that Tibet should remain part of China, but Chinese leaders reject his call for true political autonomy. More recently, a number of Tibetan monks and nuns have burned themselves to death to protest Chinese policies. As of April 2015, about 135 Tibetans had engaged in deadly self-immolation.

War in Afghanistan No other Central Asian conflict compares in intensity to the struggle being waged in Afghanistan. The country's

Figure 10.22 Political Boundaries Around the Fergana Valley Some of the world's most complex political boundaries can be found in the vicinity of the Fergana Valley. The central portion of the valley belongs to Uzbekistan, which is otherwise separated from it by high mountains. The lower valley, on the other hand, is part of Tajikistan, the core area of which is likewise separated from it by highlands. The Fergana's upper periphery belongs to Kyrgyzstan. Note also the small enclaves of Uzbekistan within Kyrgyzstan.

Figure 10.23 Ethnic Tension in Xinjiang Several times over the past few years ethnic rioting pitting the indigenous Uyghurs against Han Chinese immigrants has broken out in the major cities of Xinjiang. Chinese security forces have reacted harshly to Uyghur protestors.

troubles began in 1978 when a Soviet-supported military "revolutionary council" seized power. The new socialist government began to suppress religion, leading to widespread resistance. When the government was about to collapse, the Soviet Union responded with a massive invasion. Despite its power, the Soviet military was never able to control the more rugged parts of the country. Pakistan, Saudi Arabia, and the United States, moreover, ensured that anti-Soviet forces remained well armed. When the exhausted Soviets finally withdrew their troops in 1989, brutal local warlords grabbed power across most of the country.

In 1995, a new movement called the Taliban arose in Afghanistan. Founded by young Muslim religious students, the Taliban believed in the strict enforcement of Islamic law. The Taliban model attracted large numbers of soldiers, and by September 2001 only far northeastern Afghanistan lay outside Taliban power. By 2000, however, most of Afghanistan's people were turning against the group, primarily because of the severe restrictions the Taliban imposed on daily life. These restrictions were more pronounced for women, but even men were compelled to obey the Taliban's numerous decrees. Most forms of recreation were simply outlawed, including television, films, and music—and even kite flying.

The attacks of September 11, 2001, in the United States completely changed the balance of power in Afghanistan. The United States and Britain, working with Afghanistan's anti-Taliban forces, attacked the Taliban government and soon defeated it. Although a democratic Afghan government was established within a few years, peace did not return. The new government proved corrupt and ineffective, failing to establish security in most parts of the country.

By 2004, the Taliban had regrouped, operating from safe havens in Pakistan. Afghanistan's new government has had to rely on the military power of the International Security Assistance Force (ISAF), led by the North Atlantic Treaty Organization (NATO). But despite its tens of thousands of troops, the ISAF was unable to stop the Taliban resurgence (Figure 10.24). In 2009, the United States responded by sending an additional 17,000 troops to Afghanistan and has continued to target high-level Taliban leaders, often by bombing their compounds, many of which are located across the border in Pakistan. This strategy has weakened the Taliban command structure, but has also resulted in many civilian casualties, reducing local support for the war effort. Although many areas of northern Afghanistan are relatively secure, the south remains a battle-scarred war zone.

In 2014, the U.S. government announced that it would gradually reduce its military force in Afghanistan over the next few years, as would its coalition partners. By 2016, only a small body of troops was scheduled to remain to guard the U.S. embassy. Such plans could

Figure 10.24 Taliban Fighters in Afghanistan Taliban militants continue to fight against the government of Afghanistan over much of the country despite having relatively poor equipment. Fighters often move from one part of the country to another on motorcycles, like these militants gathering in an undisclosed city in southern Afghanistan in 2008.

quickly change, however, if the Taliban were to make rapid gains. Although public opinion polls show that few Afghans want the Taliban back in power, the country's weak national government could have difficulty withstanding a Taliban military assault.

International Dimensions of Central Asian Tension

With the collapse of the Soviet Union in 1991, Central Asia emerged as a key arena of geopolitical tension. China, Russia, Pakistan, Iran, India, and the United States have competed for influence in the region. The political revival of Islam has also generated international geopolitical issues.

Russia's economic and military ties to Central Asia did not vanish when the Soviet Union collapsed. Russia maintains military bases in Tajikistan and Kyrgyzstan and continues to be one of the main markets for Central Asian exports. Furthermore, western Central Asia's rail links and gas and oil pipelines are closely tied in with those of Russia.

Russia and China became concerned about the growing influence of the United States in Central Asia after the U.S. military established bases in Uzbekistan, Kyrgyzstan, and Afghanistan following the attacks of September 11, 2001. As a result, the two countries formed the **Shanghai Cooperation Organization (SCO)**, composed of China, Russia, Kazakhstan, Kyrgyzstan, Tajikistan, and Uzbekistan. The SCO seeks cooperation on such security issues as terrorism and separatism and also aims to enhance trade. The SCO works with the **Collective Security Treaty Organization (CSTO)**, a Russian-led military association that includes Belarus, Armenia, Kazakhstan, Kyrgyzstan, Tajikistan, and Uzbekistan. Mounting local opposition forced the United States to shut down its military base in Uzbekistan in 2005, and in 2014 it turned over its base in Kyrgyzstan to the Kyrgyz military (Figure 10.25).

Turkmenistan has not joined the SCO or the CSTO. Turkmenistan's relative isolation stems from the policies of its former president, Saparmurat Niyazov, who forced Turkmens to treat him as a heroic savior-figure. After Niyazov's death in 2006, many observers thought that the country would become more open, but thus far such hopes have been mostly frustrated. Currently, the government of Turkmenistan is spending billions of dollars to turn its capital, Ashgabat, into a "white city," resurfacing most of its public buildings in marble. In early 2015, it went so far as to ban imports of black cars.

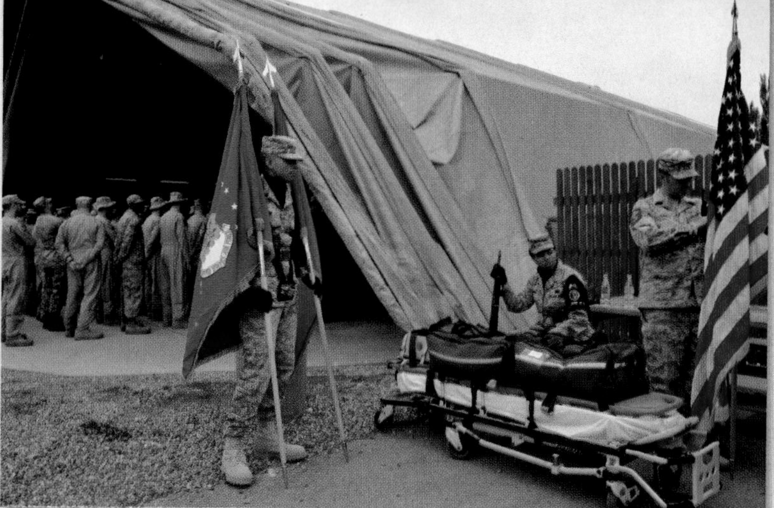

Figure 10.25 Handover of U.S. Military Base in Kyrgyzstan In 2014, a U.S. military base in Kyrgyzstan, officially known as the Transit Center at Manas, was returned to the control of the Kyrgyz government. U.S. servicemen are preparing for the change of command ceremony. From 2001 to 2014, this base provided key services for the war effort in Afghanistan.

The chaotic situation in Afghanistan is generally considered Central Asia's most serious geopolitical issue, as local leaders fear that Islamic radicalism could spread from Afghanistan into their own countries. India is also concerned about Afghanistan and has sought to gain influence in the region. Pakistan, however, views such moves with alarm, fearing that India is trying to encircle its territory.

In early 2015, authorities in Uzbekistan claimed that ISIL (Islamic State of Iraq and the Levant; also known as ISIS) has been plotting attacks on its territory, noting that over 300 Uzbek citizens had joined the extremist group fighting in Syria and Iraq. As a result, Uzbekistan vowed to enhance the security of its borders, especially with Afghanistan. Kyrgyzstan's government made similar charges, claiming that ISIL has allocated US$70 million to finance attacks in the Fergana Valley.

Review

10.7 How did the collapse of the Soviet Union in 1991 change the geopolitical structure of Central Asia, and how has Russia attempted to maintain influence in the region?

10.8 How does Afghanistan's geopolitical situation differ from those of the other countries of the region, and why is its political history so different from those of its neighbors?

KEY TERMS buffer state, Shanghai Cooperation Organization (SCO), Collective Security Treaty Organization (CSTO)

Economic and Social Development: Abundant Resources, Troubled Economies

Central Asia is not a wealthy region. Afghanistan in particular stands near the bottom of almost every list of economic and social indicators, with by far the world's highest infant mortality rate. In the early years of the 21st century, however, several Central Asian countries underwent an economic boom, with annual growth rates exceeding 10 percent. The global economic crisis of 2008–2009 hit the region hard, but most countries bounced back quickly. More recently, the decline in the price of oil and other commodities has slowed growth in most Central Asian countries, hitting Kazakhstan particularly hard.

Post-communist Economies

During the communist period, state planners sought to spread economic development widely across the Soviet Union. This effort required building large factories even in remote areas of Central Asia, regardless of the costs. Such Central Asian industries relied heavily on subsidies from the Soviet government. When those subsidies ended, the region's industrial base collapsed, leading to a huge drop in the standard of living. As is true elsewhere in the former Soviet Union, however, some individuals became very wealthy after the fall of communism.

Since the beginning of the 21st century, Kazakhstan, Azerbaijan, and Turkmenistan have benefited from their energy resources. Today Kazakhstan is Central Asia's most developed country, with a productive agricultural sector, the world's 12th-largest reserves of oil, and its 18th-largest reserves of natural gas. The region's oldest fossil fuel industry is located in Azerbaijan (Figure 10.26), which has attracted much international investment in recent years. Although its economy has grown rapidly, Azerbaijan remains a poor country, burdened by inadequate infrastructure. Turkmenistan, despite having the world's fourth-largest reserves of natural gas, is poorer still, burdened by its heavy-handed government policies.

Uzbekistan's fossil fuel reserves are much smaller than those of Kazakhstan and Turkmenistan, but they are still substantial. However, the country is also much more densely populated, and many people live in overcrowded farmlands that are suffering from environmental stress. An authoritarian country, Uzbekistan retains many aspects of the old Soviet-style command economy (state-run industries, rather than market-oriented private firms). It is, however, the world's fourth-largest cotton exporter, and its large deposits of gold and other mineral resources help keep its economy afloat.

Cotton is also a major crop in Turkmenistan and Tajikistan. As we saw earlier in the chapter, the massive level of irrigation required by cotton cultivation is responsible for many of the region's severe environmental problems, including the destruction of the Aral Sea.

Figure 10.26 Oil Development in Azerbaijan Although oil has brought a certain amount of wealth to Azerbaijan, it has also produced extensive pollution. Because most of the petroleum is located either near or under the Caspian Sea, this sea—actually the world's largest lake—is becoming increasingly polluted, and its fisheries are declining.

Unfortunately, cotton production in Central Asia is also associated with human rights abuses. In Uzbekistan, over 1 million students, teachers, and government employees are forced to pick cotton every harvest season for minimal payment and under appalling conditions. Those who refuse often lose their jobs or are expelled from school (Figure 10.27).

Central Asia's two most mountainous countries, Tajikistan and Kyrgyzstan, hold most of the region's water resources, which they are developing by building massive dams. Tajikistan, however, is burdened by its remote location, rugged topography, and lack of infrastructure. With a per capita gross national income of only around $2000 (Table A10.2), Tajikistan is one of Eurasia's poorest countries, with two-thirds of its people living in poverty.

Kyrgyzstan also remains very poor, relying heavily on agriculture and gold mining. A single facility, the Kumtor Gold Mine, accounts for nearly 12 percent of Kyrgyzstan's gross domestic product and roughly half of its industrial output. Despite several market-oriented reforms in the 1990s, the country has experienced little sustained economic growth. Kyrgyzstan was the only former Soviet Central Asian country belonging to the World Trade Organization until 2013, when Tajikistan joined. In 2015, the Kyrgyz government announced a major new initiative to reduce the size of its underground economy, which accounts for roughly 40 percent of the goods and services produced in the country.

Mongolia, formerly a communist ally of the Soviet Union, also suffered an economic collapse in the 1990s after Soviet subsidies were eliminated. Conditions worsened from 2000 until 2002, when a combination of severe winters and summer droughts devastated livestock, the traditional mainstay of the Mongolian economy. Mongolia does, however, possess vast reserves of copper, gold, and other minerals, and its economy grew by more than 10 percent a year from 2010 to 2013 as the mining sector took off. But in 2014, the mineral boom weakened due to reduced demand from China. By early 2015, the Mongolian government had to request support from the International Monetary Fund to help stabilize its economy.

The Economy of Chinese Central Asia Unlike the rest of the region, the Chinese portions of Central Asia did not experience an economic crash in the 1990s (see *Everyday Globalization: Rare Earths from Inner Mongolia*). China as a whole had one of the world's fastest-growing economies, although its centers of economic dynamism are all located in the eastern zone. However, since China had been much poorer than the Soviet Union before 1991, it is not surprising that poverty in Chinese Central Asia remains widespread.

Tibet in particular remains impoverished, its economy dominated by subsistence agriculture. China has been rapidly building roads and railroads into both Xinjiang and Tibet through its official Western China Development strategy, greatly expanding tourism in the process (Figure 10.28). According to official statistics, Tibet's economy grew by over 12 percent a year from 2010 through 2015, but Tibetan critics

FIGURE 10.27 Child Worker in Uzbek Cotton Field Uzbekistan has been harshly criticized for its reliance on forced labor, particularly that of children, for bringing in its all-important cotton crop. A Uzbek boy is carrying a sack of raw cotton on a field near Yallama, roughly 44 miles (70 kilometers) southwest of Tashkent. **Q: Why is forced labor, especially of children, used so extensively in Uzbekistan's cotton harvest?**

EVERYDAY GLOBALIZATION

Rare Earths from Inner Mongolia

Rare earth elements form crucial components of the high-tech economy, impacting daily life across the world in many ways. Dysprosium, for example, improves high-powered magnets and lasers, yttrium is used in microwave filters for communications equipment and in energy-efficient lightbulbs, and gadolinium has medical and computer-memory applications.

As their name suggests, the 17 rare earth elements are not easy to extract. India, Brazil, South Africa, and the United States were once major producers, but by 2010 China accounted for 95 percent of total world output. Most Chinese rare earths come from Inner Mongolia's Bayan Obo Mining District, near the Mongolian border (Figure 10.4.1). Bayan Obo is regarded as an environmental disaster zone, as huge quantities of dust, toxins, and radioactive waste are produced in the mining process. China's near monopoly on rare earth production has led other countries, including the United States, Canada, Brazil, Tanzania, Australia, Vietnam, and Malaysia, to try to develop their own reserves.

1. Is it important for the United States to mine its own rare earth deposits? If so, how might the U.S. government encourage this?
2. What objects do you own that probably contain rare earth elements?

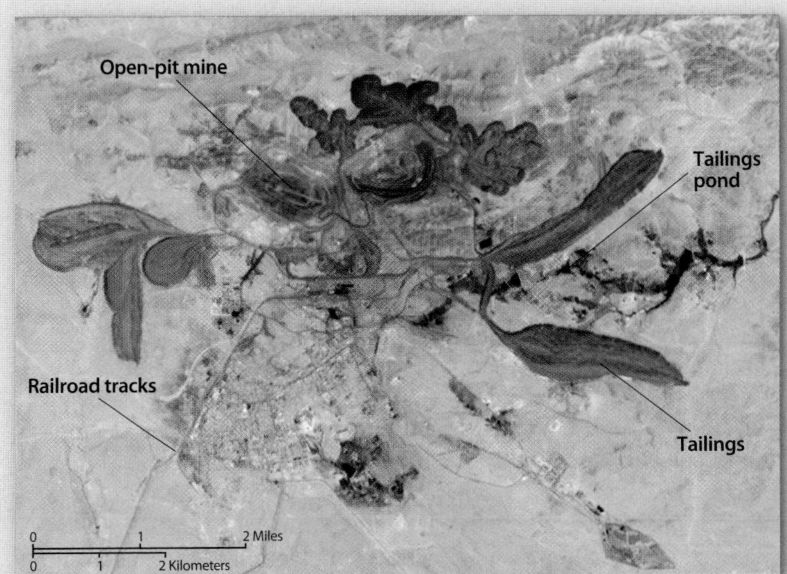

Figure 10.4.1 Rare Earth Mining in Inner Mongolia China dominates the mining of rare earth elements, several of which are extremely important in high-tech manufacturing. Roughly half of China's rare earth production comes from the mines of Bayan Obo, located in Inner Mongolia. In this false-color satellite image of Bayan Obo, vegetation appears red, grassland is light brown, rocks are black, and water surfaces are green. Two circular open-pit mines are visible, as well as a number of tailing ponds and tailings piles.

Figure 10.28 Qinghai–Tibet Railway China is engaged in a massive program to extend roads and railroads into the western half of the country. The main connection with the Tibetan Plateau is the recently completed Qinghai–Tibet Railway, shown here. This railway has greatly increased tourism in Tibet. **Q: Why have the railroad tracks been elevated so high above the valley floor?**

claim that most of these recent economic gains went to Han Chinese immigrants, rather than to indigenous Tibetans.

Xinjiang has tremendous mineral wealth, including China's largest oil reserves. Its agricultural sector, while small, is highly productive. The new highway and rail links bring economic benefits to the region, but also encourage migration from eastern China. In 2014, Xinjiang's economy grew by more than 10 percent, a figure significantly higher than that of China as a whole. Yet many local Muslim peoples believe, like the Tibetans, that the government and Han Chinese migrants take much of the region's wealth.

Afghanistan's Misery Though rich in natural gas and minerals, Afghanistan is a deeply impoverished country. Foreign aid and the export of handwoven rugs constitute its legitimate economic mainstays. Although official statistics indicate that the Afghan economy expanded rapidly from 2011 to 2013, war, corruption, crime, and poor infrastructure have prevented most people from experiencing any real gains. In 2015, moreover, the Afghan economy went into a severe recession, due in part to major cuts in foreign aid and investment.

By the late 1990s, Afghanistan had emerged as the leading producer of one major commodity for the global market: illegal drugs. Most of the world's opium, used to produce heroin, is grown here. In much of southern and western Afghanistan, opium is the main cash crop. Not only is it highly profitable and easy to transport, but also it uses relatively little water. As much as US$100 million a year in narcotics profits have supposedly gone to the Taliban, which operates mainly in the opium-growing areas of the country. As a result, both NATO and the Afghan government have tried to convince villagers to grow alternative crops, and opium production in Afghanistan declined from 2007 to 2010. But by 2012, it was again expanding. The 2014 harvest of 6400 metric tons set a new record.

Afghanistan does have substantial mineral resources. China has already invested heavily in the Ainak Copper Mine, projected to become one of the world's largest mines. When fully operational, the mine could employ an estimated 20,000 Afghans. Taliban insurgents, however, often target such foreign operations, greatly reducing international investment in the country.

Central Asian Economies in Global Context Many foreign countries are drawn to Central Asia's natural resources, but vast pipeline systems must be built to make the region's fossil fuel fields economically viable (Figure 10.29). The inadequate existing pipelines mostly pass through Russia, which charges high transit fees. Seeking an alternative route, an international consortium pushed through the massive Baku–Tbilisi–Ceyhan pipeline, which transports oil from the Caspian Basin to a Turkish port on the Mediterranean Sea. Opened in 2006 at a cost of $3.9 billion, it is the world's second-longest oil pipeline.

China, greatly interested in Central Asia's natural resources, took the lead in commissioning the ambitious Central Asia–China gas pipeline, which links wells in Turkmenistan, Uzbekistan, and Kazakhstan to the China market. The third line of this ambitious project was completed in 2014, and construction of a fourth line is expected to begin within a few years. China has also invested heavily in regional infrastructure, as Kazakhstan in particular hopes to upgrade its decaying transportation network. China is now the top trading partner of Kazakhstan, Kyrgyzstan, Tajikistan, and Turkmenistan.

Russia hopes to keep its former Soviet republics in Central Asia within its sphere of influence. One of its main strategies to do so is the **Eurasian Economic Union**, a tightly connected trading bloc that came into effect in 2015. This union currently includes Russia, Belarus, Armenia, Kazakhstan, and Kyrgyzstan. Tajikistan has also been in negotiations to join the union, but both Uzbekistan and Turkmenistan have rejected the idea.

Social Development in Central Asia

Social conditions in Central Asia vary more than economic conditions. Afghanistan falls at the bottom of the scale on almost every aspect of social development, but in the former Soviet territories, levels of health and education remain relatively high. Social conditions have improved recently in Tibet and Xinjiang, although they are probably not keeping pace with the progress made in China proper. Mongolia is noted for its relatively high levels of social development as well as for its pronounced gender equity. Women have outnumbered men in Mongolian universities since the 1980s, and many have reached high positions in the government. One prominent example is Sanjaasuren Oyun, Mongolia's environmental minister, who was recently named president of the UN Environment Assembly (UNEP).

Social Conditions and the Status of Women in Afghanistan Afghanistan's average life expectancy is about 45 years, one of the lowest figures in the world, and its infant mortality level is the world's highest. Afghanistan endures almost constant warfare, and its rugged topography hinders the provision of basic social and medical services. Illiteracy is commonplace, especially for women. Afghanistan's 12 percent adult female literacy rate is reportedly the lowest in the world (see Table A10.2).

Women in traditional Afghan society—especially Pashtun society—have very little freedom. Restrictions on female activities intensified in the 1990s under Taliban control. Taliban forces prohibited women from working or attending school and often even restricted medical care. After the Taliban lost power, its fighters continued to target female education, destroying schools for girls throughout southern Afghanistan.

Under Afghanistan's new constitution, established in 2004, 25 percent of parliamentary seats are reserved for women (Figure 10.30). Much evidence, however, indicates that the social position of women has improved little outside the capital city of Kabul. Over 80 percent of Afghan women are said to suffer from domestic violence. Throughout Afghanistan and especially in Kabul, many women fear the Taliban will again take control, eliminating the few gains that they have made.

Gender Issues in Mongolia and Former Soviet Central Asia In traditional Central Asia, the social position of women varied considerably.

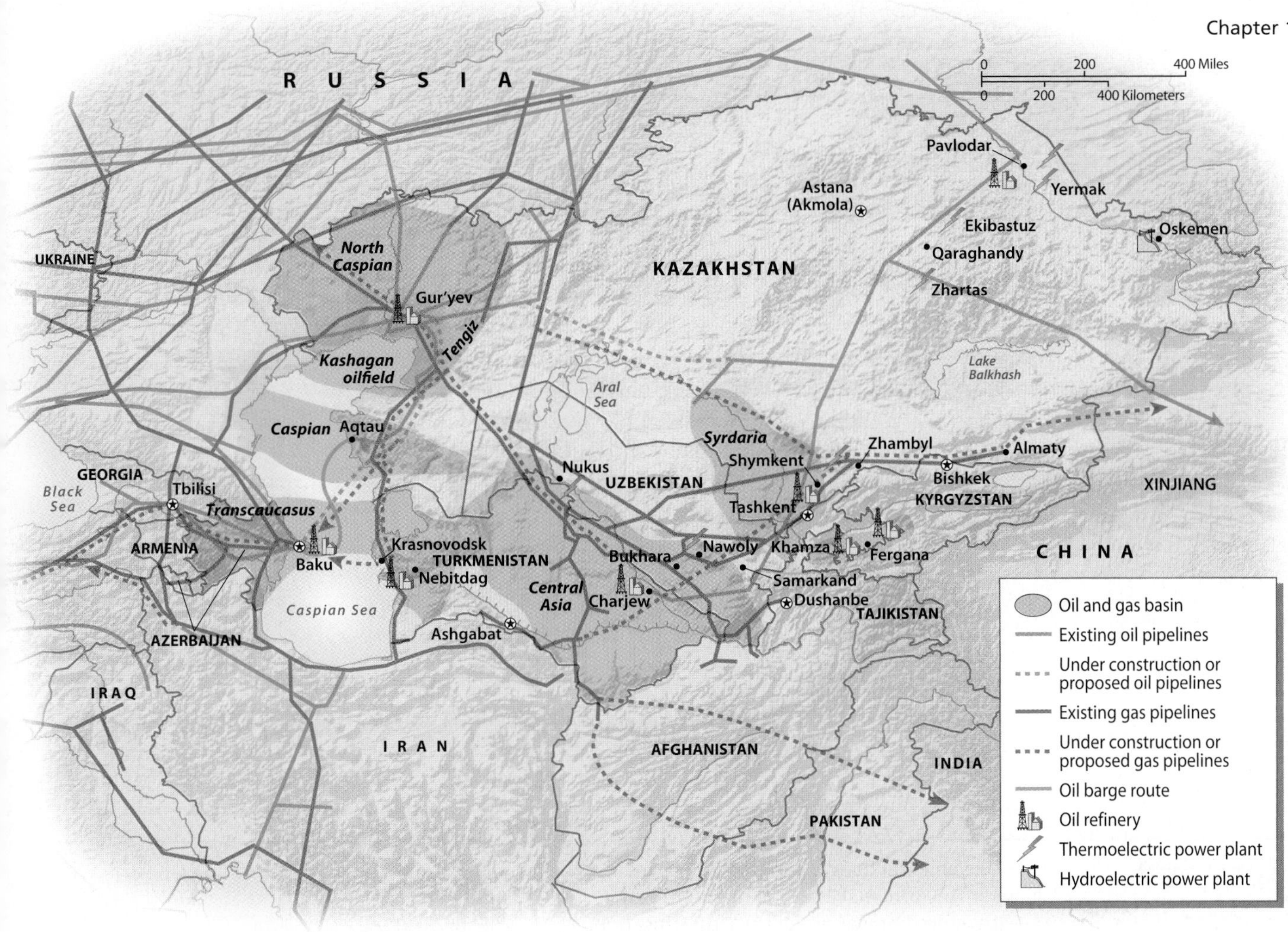

Figure 10.29 Oil and Gas Pipelines Central Asia holds some of the world's largest oil and natural gas deposits and recently has emerged as a major center for drilling and exploration. Because of its landlocked location, Central Asia cannot easily export its petroleum products. Pipelines have been built to solve this problem, and several others are currently being planned. Pipeline construction is a contentious issue, however, as several potential pathways lie across Iran, a country that remains under U.S. sanctions.

In general, women had more autonomy in the pastoral societies of the northern steppes than they did in the agricultural and urban societies located farther to the south.

Traditional Kyrgyz society, however, practiced "bride abduction," in which kidnapped women were forced to marry their abductors. Although many such kidnappings took place with the consent of the bride-to-be, others were true abductions. A recent resurgence in this practice has led to outrage among Kyrgyz women's groups. In early 2013, Kyrgyzstan passed a law to increase the maximum prison term for bride kidnapping to seven years, but according to one report, some 12,000 Kyrgyz young women and girls were kidnapped for this purpose in 2014 alone.

During the Soviet era, efforts were made to educate Central Asian women and place them in the workforce, although women often encountered limits on career advancement. Women in these countries still have relatively high levels of education, and according to the World Economic Forum's *Global Gender Gap Report 2012*, female professional workers and college students outnumber men in Kazakhstan, Kyrgyzstan, and Mongolia.

Much of this achievement can be attributed to Soviet education investment, though doubts have arisen as to whether the region's relative equality can persist into the future. Central Asian countries have attempted to address such concerns by installing quotas for women in the government. By law, 30 percent of Kyrgyz and Uzbek parliament members must be female. Local women's groups complain, however, that their actual power in government is limited.

Demographic issues in Central Asia also influence the position of women. In Tajikistan especially, so many men have relocated to Russia for work that many households are now led by women. At the same time, local women are under pressure to follow more traditional gender roles due to the growth of more fundamentalist forms of Islam. In Kazakhstan, nationalist parties have tried to ban contraception for Kazakh women, as they hope to increase both the population of the country and the proportion of its ethnic Kazakh population.

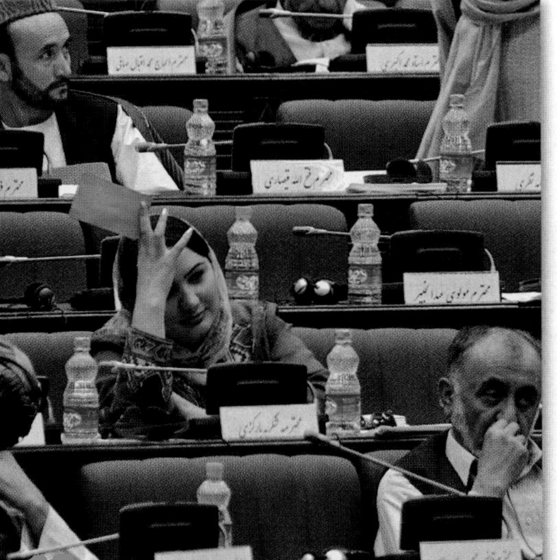

Figure 10.30 Afghan Women in Parliament Women in Afghanistan generally have a low social position and suffer from many restrictions, especially in rural areas. They do have a number of reserved seats in the country's parliament, although critics contend that female legislators have little real power.

Review

10.9 How does the distribution of fossil fuels influence economic and social development in Central Asia?

10.10 Why do the social positions of women vary so much across Central Asia, and why is Afghanistan particularly problematic in this regard?

KEY TERMS Eurasian Economic Union

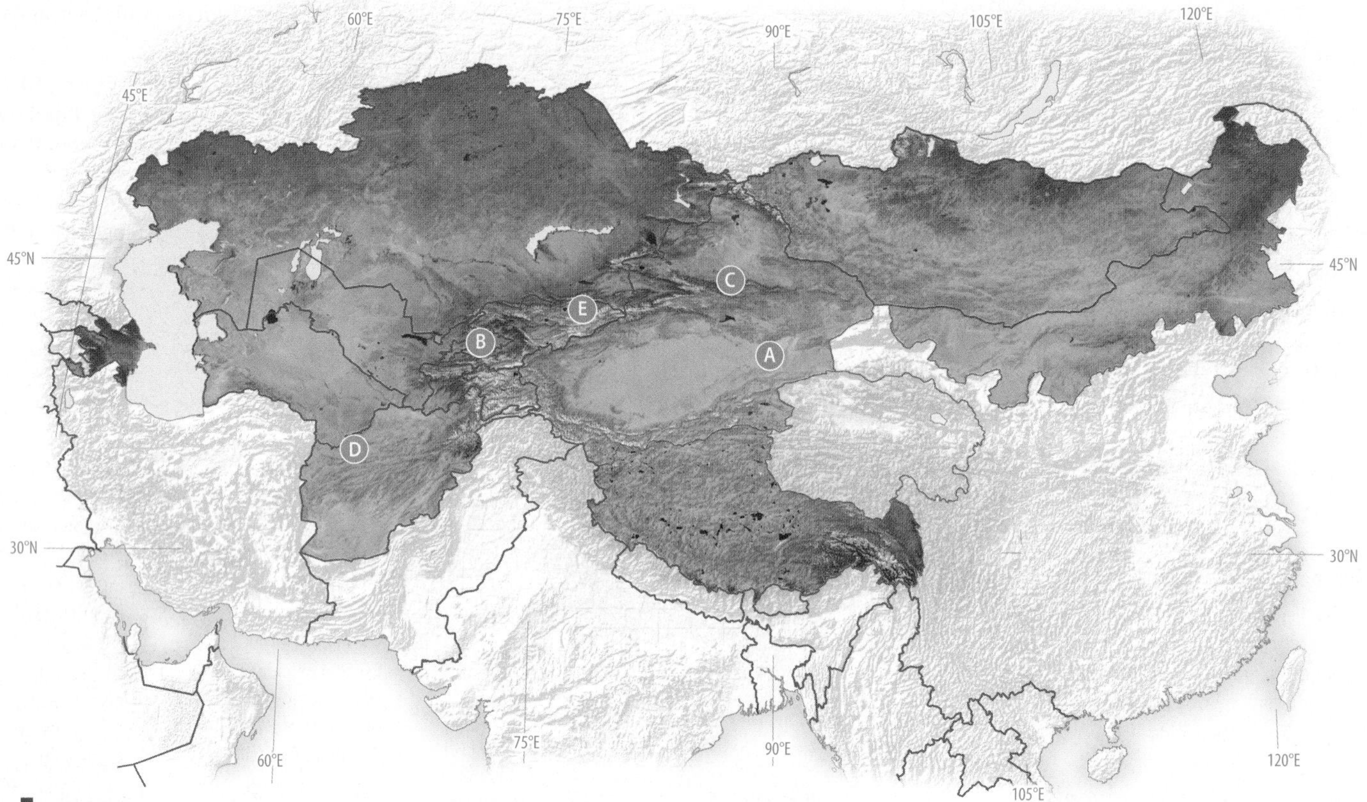

Review

Physical Geography and Environmental Issues

10.1 Identify the key environmental differences among Central Asia's desert areas, its mountain and plateau zone, and its steppe belt, and link these differences to human settlement and economic development.

10.2 Summarize the reasons why water resources are of such great importance in Central Asia.

10.3 Provide reasons for the Aral Sea's disappearance, and outline the economic and environmental consequences of the loss of this once-massive lake.

Environmental problems, among others, have brought the region to global attention. The destruction of the Aral Sea is one of the worst environmental disasters the world has seen, and many other lakes in the region have experienced similar problems. Desertification has devastated many areas, and the booming oil and gas industry has created its own environmental disasters. Successful efforts to stop desertification, however, are under way in some areas.

1. Lop Nur in China's Xinjiang Autonomous Region was once a lake, but is now a dried-out salt flat, used for nuclear tests and containing the world's largest potash fertilizer plant. Why did the lake disappear, and why has China used the dried-up lake bed for nuclear tests and potash processing?

2. What strategies might be used to prevent the desiccation of other lakes in Central Asia?

Population and Settlement

10.4 Explain why Central Asia's population is so unevenly distributed, with some areas densely settled and others essentially uninhabited.

10.5 Describe the differences between Central Asia's historical cities and those that have been established within the past 100 years.

Large movements of people characterize many parts of Central Asia. Han Chinese immigration into Tibet and Xinjiang is steadily turning the indigenous peoples of these areas into minority groups. The movement of Russian speakers out of the former Soviet areas of Central Asia has also resulted in major transformations. Hundreds of thousands of migrant workers from Tajikistan and Kyrgyzstan currently live in Russia. In Afghanistan, war and continuing chaos have generated large refugee populations.

3. An area of very high population density is found near the center of this map. Why is this particular region so crowded, and why is it surrounded by areas of much lower population density?

4. How might such steep population gradients influence political tension and economic development in the region?

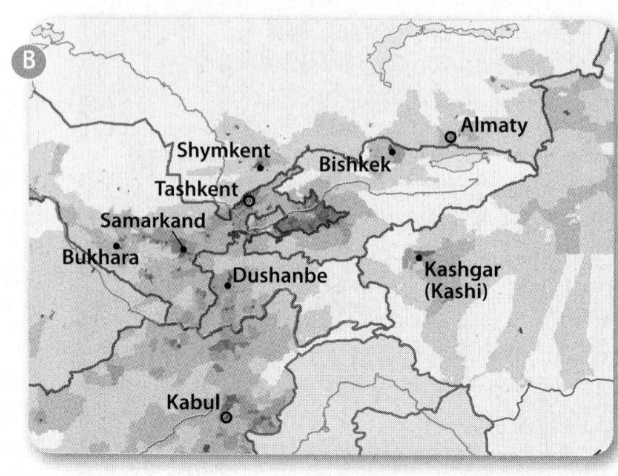

Cultural Coherence and Diversity

10.6 Outline the ways in which religion divides Central Asia, and describe how religious diversity has influenced the history of the region.

10.7 Identify how cultural globalization affects different parts of Central Asia in distinct ways and explain why cultural globalization is controversial in much of the region.

Religious tension has recently emerged as a major cultural issue throughout much of western Central Asia. Radical Islamic fundamentalism remains a potent force in southern and western Afghanistan and, to a somewhat lesser extent, in the Fergana Valley of Uzbekistan, Tajikistan, and Kyrgyzstan. Although moderate forms of Islam prevail in most of the region, Central Asian leaders have used the fear of religiously inspired violence to maintain repressive policies.

5. China recently prohibited women from wearing the Islamic face veil in Urumqi, the capital of Xinjiang. Why did the Chinese government enact this policy, and what might some of its consequences be?

6. Have similar policies been enacted in other parts of the world? If so, how have they been received by local populations?

C

Geopolitical Framework

10.8 Summarize the geopolitical roles played in Central Asia by Russia, China, and the United States, and explain why the region has been the site of pronounced geopolitical tension over the past several decades.

10.9 Describe how ethnic conflict has contributed to instability in Afghanistan, and assess the potential of ethnic tension to destabilize the rest of the region.

China maintains a firm grip on Tibet and Xinjiang, but the rest of Central Asia has become a key area of geopolitical competition. Russia and China in particular are interested in maintaining influences over the region. Afghanistan continues to be a war-torn country, with little chance for an enduring peace settlement. Overall, political regimes throughout most of Central Asia remain largely authoritarian.

7. The Soviet invasion and subsequent occupation of Afghanistan (1979–1989) proved very costly to the Soviet Union. Why did Soviet leaders feel compelled to invade Afghanistan, and why did their occupation of the country ultimately prove unsuccessful?

8. Is it reasonable to draw connections between the Soviet and the U.S. experiences in Afghanistan? In what ways have they been similar, and in what ways have they been different?

D

Economic and Social Development

10.10 Explain the role of oil and natural gas production and transport, as well as oil price fluctuations, in generating uneven levels of economic and social development across Central Asia.

Central Asian economies are gradually opening up to global connections, largely because of their substantial mineral reserves and the construction of the so-called New Silk Road. However, Central Asia is likely to face serious economic difficulties for some time, especially since the region is not a significant participant in global trade and attracts little foreign investment outside the fossil fuel sector. Opium cultivation and heroin manufacture remain a serious problem in Afghanistan.

9. The Canadian-owned Kumtor Gold Mine in Kyrgyzstan is essential to the country's economy, yet it is increasingly controversial. Why do so many people in Kyrgyzstan oppose this particular kind of economic development?

10. What are some of the potential disadvantages faced by a country that focuses much of its economic development planning on exploitation of a particular natural resource such as gold?

E

KEY TERMS

alluvial fan *(p. 331)*
buffer state *(p. 338)*
Collective Security Treaty Organization (CSTO) *(p. 341)*
desertification *(p. 326)*
Eurasian Economic Union *(p. 344)*
forward capital *(p. 334)*
loess *(p. 331)*
pastoralist *(p. 330)*
Shanghai Cooperation Organization (SCO) *(p. 341)*
Silk Road *(p. 324)*
steppe *(p. 326)*
theocracy *(p. 337)*
transhumance *(p. 331)*

DATA ANALYSIS

http://goo.gl/eu7rtk

Spectacular mountain scenery and rich history provide Central Asia with great tourism potential. The well-preserved cities of the Silk Road, such as Uzbekistan's Samarkand, are particularly well suited for tourism. However, international tourism has been slow to take off in Central Asia. Visit the World Bank website (http://data.worldbank.org) to access data on international tourism receipts.

1. Write down the tourism-receipt figure for each Central Asian country for the most recent year available. Note if data are unavailable for any particular country. Use the data to construct a bar graph for the Central Asian countries for the year 2010 and the most recent year. Do the same for Turkey, a country outside the region that shares historical and cultural features with several Central Asian countries.
2. Suggest possible reasons for the differences you see on the graphs, based on environmental, population, cultural, geopolitical, and economic factors.
3. Could Central Asian countries significantly increase their tourism? How could they make this possible?

MasteringGeography™

Looking for additional review and test prep materials? Visit the Study Area in MasteringGeography™ to enhance your geographic literacy, spatial reasoning skills, and understanding of this chapter's content by accessing a variety of resources, including MapMaster interactive maps, geoscience animations, videos, *In the News* RSS feeds, flashcards, web links, self-study quizzes, and an eText version of *Globalization and Diversity*.

Authors' Blogs

Scan to visit the
Author's Blog for field notes, media resources, and chapter updates

http://gad4blog.wordpress.com/category/central-asia/

Scan to visit the
GeoCurrents Blog

http://geocurrents.info/category/place/central-asia-places

11 East Asia

PHYSICAL GEOGRAPHY AND ENVIRONMENTAL ISSUES

China has long experienced severe deforestation and soil erosion, but its current economic boom is generating some of the world's worst pollution problems. Japan, South Korea, and Taiwan, however, have extensive forests and relatively clean environments.

POPULATION AND SETTLEMENT

China is currently undergoing a major transformation as tens of millions of peasants move from impoverished villages in the interior to booming coastal cities. Low birth rates and aging populations are found throughout East Asia.

CULTURAL COHERENCE AND DIVERSITY

Despite several unifying cultural features, East Asia in general and China in particular are divided along striking cultural lines. Historically, however, the entire region was linked by Mahayana Buddhism, Confucianism, and the Chinese writing system.

GEOPOLITICAL FRAMEWORK

China's growing power is generating tension with other East Asian countries, while Korea remains a divided nation, with South Korea set against North Korea. As China's global influence grows, Japan, South Korea, and Taiwan are responding by strengthening ties with the United States.

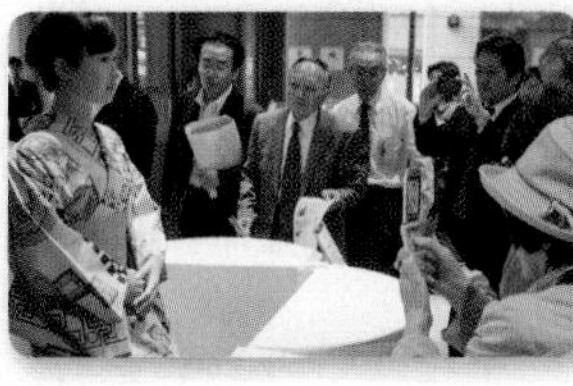

ECONOMIC AND SOCIAL DEVELOPMENT

Over the past several decades, East Asia has emerged as a core area of the world economy, with China undergoing one of the most rapid economic expansions the world has ever seen. North Korea, however, remains desperately poor, plagued by widespread malnutrition.

◀ Located to the south of the Korean Peninsula, Jeju Island is a favored tourist destination, especially for South Koreans. Jeju is noted for its subtropical climate, its beautiful mountains and beaches, and its unique cultural practices. Hikers here are approaching Seongsan Ilchulbong Peak on Jeju Island.

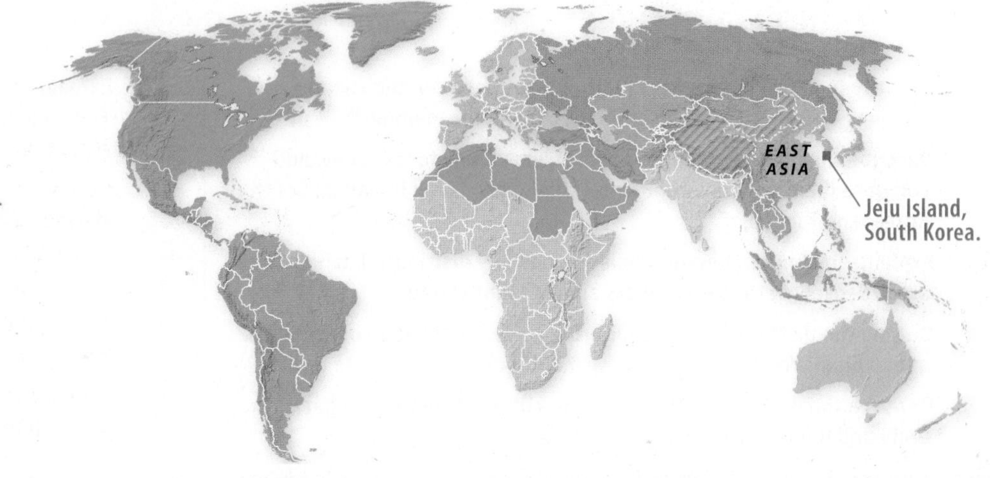

South Korea's Jeju Island has long been noted for its mild climate, its striking volcanic landscapes, and the social prominence of its women. In earlier times, local poverty led many men to seek work on the mainland, resulting in a female-dominated family structure. Jeju was also famous for its *haenyeo*, or "sea women," professional divers who made good livings by gathering abalone and conch deep below the ocean waves. More recently, Jeju's reputation has turned to tourism, with more than 6 million visitors each year. Korean honeymooners flock to the island, many of whom visit Jeju Loveland, its famous outdoor sculpture garden devoted to sex education.

Currently, however, up to half of Jeju's visitors come from China. Many Chinese have purchased property on the island, and Chinese businesspeople are investing heavily in the local economy. Both the local government of Jeju and the national government of South Korea have encouraged such foreign investments, allowing condominium buyers permanent resident status as well as access to South Korean medical and employment benefits. In 2014, the construction of a US$2.4 billion casino-resort project was announced, which will also include a theme park, shopping mall, and three hotels. A joint project of a company from Singapore and another from Hong Kong, the complex is designed to appeal to Chinese tourists.

Although Chinese tourism and investments bring much money to Jeju, they have also prompted local opposition. A 2014 survey indicated that roughly two-thirds of the local residents view the Chinese influx negatively, and people selling land to Chinese buyers have been insulted as "national traitors." But as China's economy grows and its citizens increasingly travel abroad, the flow of both tourists and investments will probably increase, putting more pressure on Jeju. Such issues, however, are by no means unique to the island, as China's rapid rise has transformed international relations across East Asia and, to some extent, throughout the rest of the world.

China's rapid rise has transformed international relations across East Asia and, to some extent, throughout the rest of the world.

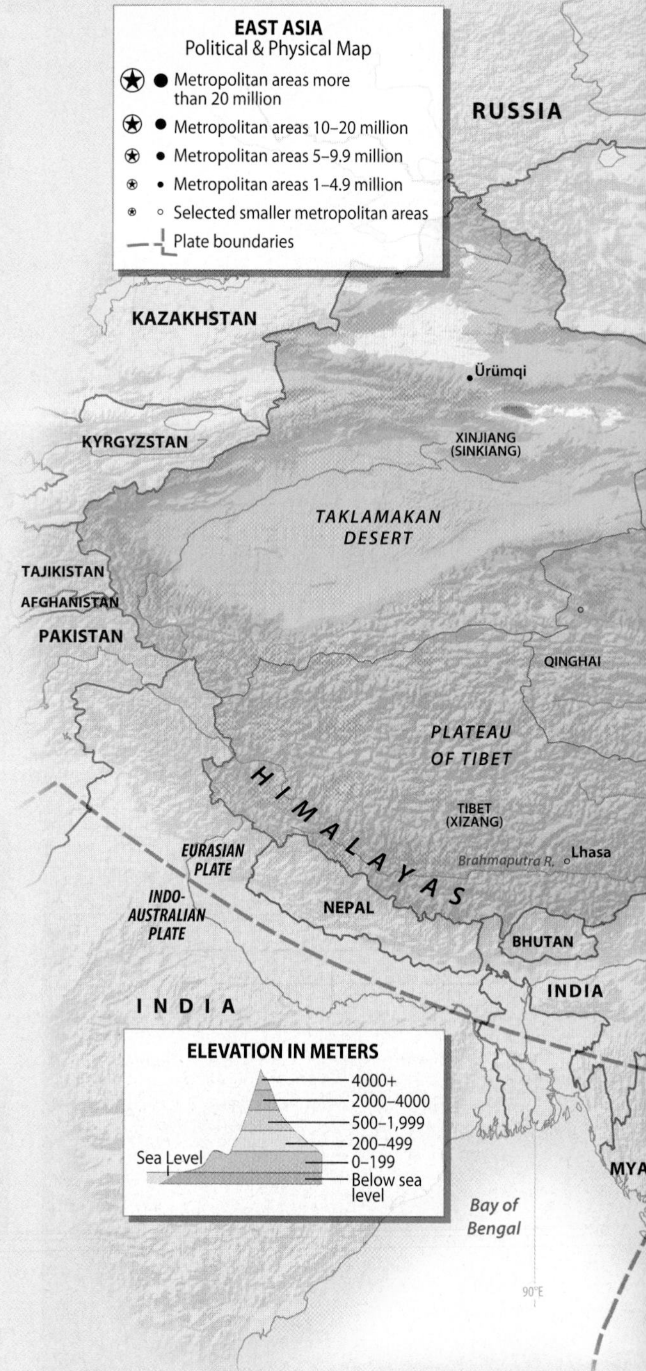

Learning Objectives

After reading this chapter you should be able to:

11.1 Identify the key environmental differences between the island portions of East Asia (Japan and Taiwan) and the mainland.

11.2 Describe the main environmental problems China faces today, and compare them with environmental challenges faced by Japan, South Korea, and Taiwan.

11.3 Explain why China's population is so unevenly distributed, with some areas densely settled and others almost uninhabited.

11.4 Outline the distribution of major urban areas on the map of East Asia and explain why the region's largest cities are continuing to grow.

11.5 Describe the ways in which religion and other systems of belief both unify and divide East Asia.

11.6 Explain the distinction between, and geographical distribution of, the Han Chinese and the other ethnic groups of China, paying particular attention to language.

11.7 Outline the geopolitical division of East Asia during the Cold War period, and explain how the division of that period still influences East Asian geopolitics.

11.8 Identify the main reasons behind East Asia's rapid economic growth in recent decades, and discuss possible limitations to continued expansion at such a rate.

11.9 Summarize the geographical differences in economic and social development found in China and across East Asia as a whole.

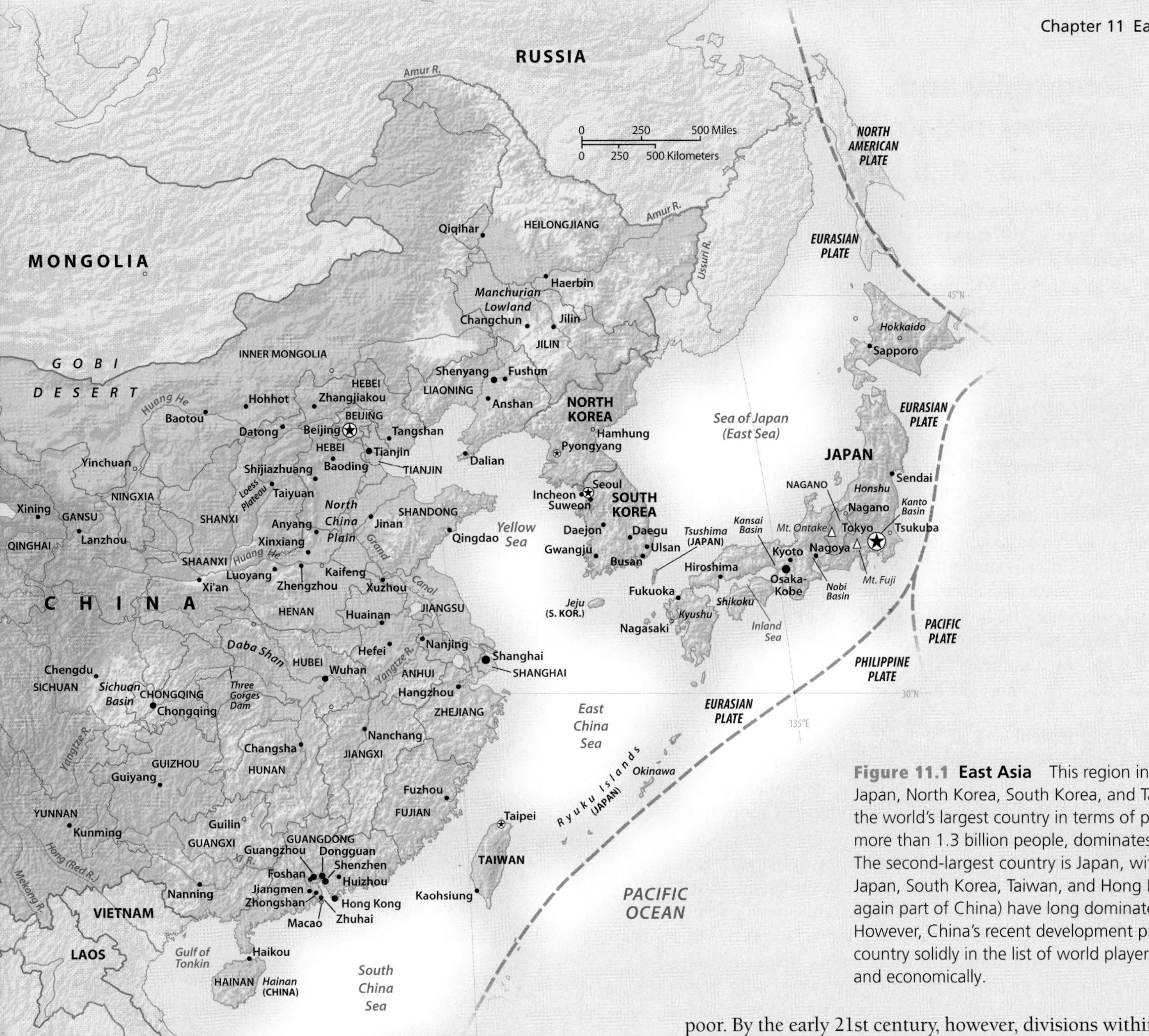

Figure 11.1 East Asia This region includes China, Japan, North Korea, South Korea, and Taiwan. China, the world's largest country in terms of population, with more than 1.3 billion people, dominates East Asia. The second-largest country is Japan, with 127 million. Japan, South Korea, Taiwan, and Hong Kong (now once again part of China) have long dominated economically. However, China's recent development places that country solidly in the list of world players, both politically and economically.

East Asia, which consists of China, Taiwan, Japan, South Korea, and North Korea (Figure 11.1), is a core area of the world economy, due to both its advanced economies and its huge population. China alone, with more than 1.3 billion inhabitants, has more people than any other world region except South Asia. Japan, South Korea, and Taiwan are among the world's key trading states. China has more recently emerged as a key global trading power and has recently emerged as a major military power as well. According to some measurements, China's rapidly growing economy is already almost as large as that of the United States.

Although historically unified by cultural features, in the second half of the 20th century East Asia was politically divided, with the capitalist economies of Japan, South Korea, Taiwan, and Hong Kong separated from the communist bloc of China and North Korea. As Japan became a leader in the global economy, much of China remained poor. By the early 21st century, however, divisions within East Asia had been reduced. China is still governed by the Communist Party but has been on a path of mostly capitalist development since the 1980s. Yet relations between North Korea and South Korea remain hostile, and China's rapid rise is generating concerns across the rest of the region.

In certain respects, East Asia can be easily defined by the territorial extent of its constituent countries. Taiwan's political status, however, is ambiguous. Although Taiwan is, in effect, an independent country, China (officially, the People's Republic of China) claims it as part of its own territory. As a result, Taiwan is recognized as a sovereign state by only a handful of other countries. In cultural terms, the territorial extent of East Asia is a more complicated issue, especially with regard to the western half of China. This is a huge, but lightly populated space; some 95 percent of the residents of China live in the eastern part of the country (sometimes referred to as **China proper**). Culturally and historically, the indigenous peoples of western China are more closely connected to Central Asia than to the rest of East Asia. As a result, western China is covered most extensively in Chapter 10; in this chapter, we will examine this region only to the extent that it is politically part of China.

Physical Geography and Environmental Issues: Resource Pressures in a Crowded Land

Many environmental problems in East Asia stem from a combination of the region's large population, its massive industrial development, and its physical geography. Steep slopes and heavy rainfall make many areas vulnerable to soil erosion and mudslides, and a seismically active environment generates earthquake threats. Careful examination of the region's physical geography thus sheds light on the environmental issues that it faces.

East Asia's Physical Geography

East Asia is situated in the same general latitudinal range as the United States, although it extends considerably farther north and south. China's northernmost tip lies as far north as central Québec, while its southernmost point is at the same latitude as Mexico City. The climate of southern China is thus roughly comparable to that of the Caribbean, while the climate of northern China is similar to that of south-central Canada (Figure 11.2; note particularly the climographs for Hong Kong, Chongqing, Beijing, and Shenyang). Although far smaller than China, Japan also extends over a wide range of latitude. As a result, Japan's extreme south, in southern Kyushu and the Ryukyu Archipelago, is subtropical, while northern Hokkaido is almost subarctic.

Japan's Physical Environment Most of Japan has a temperate climate with pronounced seasonal differences, much like that of the eastern United States. Tokyo's climate, for example, is similar to that of Washington, DC—although Tokyo receives significantly more rain (see the climograph in Figure 11.2).

The Pacific coast of Japan is separated from the Sea of Japan coast by a series of steep mountain ranges (Figure 11.3). Japan is one of the world's most rugged countries, with mountainous terrain covering some 85 percent of its territory. Most of these uplands are heavily forested (Figure 11.4). Japan owes its lush forests to its mild, rainy climate, its long history of conservation, and its import of timber and wood pulp from other parts of the world. For hundreds of years, both the Japanese state and its village communities have enforced strict forest-conservation rules, ensuring that timber and firewood extraction is balanced by tree growth.

Small alluvial plains located along parts of Japan's coastline are interspersed among its mountains. These areas have long been cleared and drained for intensive agriculture. The largest Japanese lowland is the Kanto Plain north of Tokyo, but even it is only some 80 miles wide and 100 miles long (130 km wide and 160 km long). The country's other main lowland basins are the Kansai, located around Osaka, and the Nobi, centered on Nagoya.

Korean Landscapes The well-defined Korean Peninsula is partially cut off from northeast China by rugged mountains and sizable rivers. The far north, which just touches Russia's far east, has a climate similar to that of Maine, whereas the southern tip is more like the Carolinas. Korea, like Japan, is a mountainous land with scattered alluvial basins. The lowlands of the southern portion of the peninsula are more extensive than those of the north, giving South Korea an agricultural advantage over North Korea. The latter, however, has far more abundant natural resources. The uplands of North Korea are heavily deforested (Figure 11.5), whereas those of South Korea have seen extensive reforestation since the end of World War II.

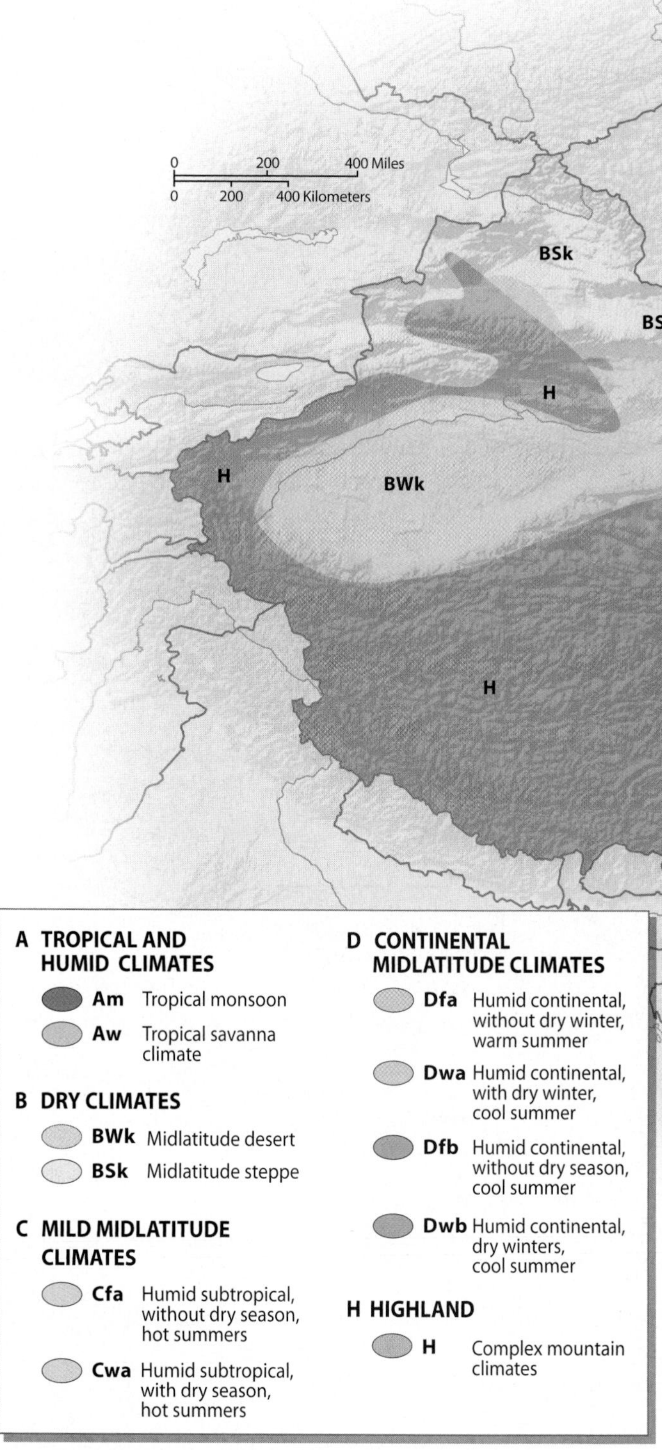

Taiwan's Environment Taiwan, a small, mountainous, and mostly wooded country, is somewhat similar to Japan and Korea in terms of its physical geography. An island about the size of Maryland, Taiwan sits at the edge of the continental landmass. To the west, the Taiwan Strait is only about 200 feet (60 meters) deep; to the east, ocean depths of many thousands of feet lie 10 to 20 miles (15 to 30 km) offshore.

Taiwan's central and eastern regions are rugged and mountainous, while the west is dominated by a rather narrow alluvial plain. Bisected by the Tropic of Cancer, Taiwan has a mild winter climate, but it is often hit by typhoons in the early autumn. Taiwan still has extensive forests, especially in its remote central and eastern uplands.

The Diverse Environments of China Even if its far western provinces are excluded, China is still a vast country with diverse environmental regions.

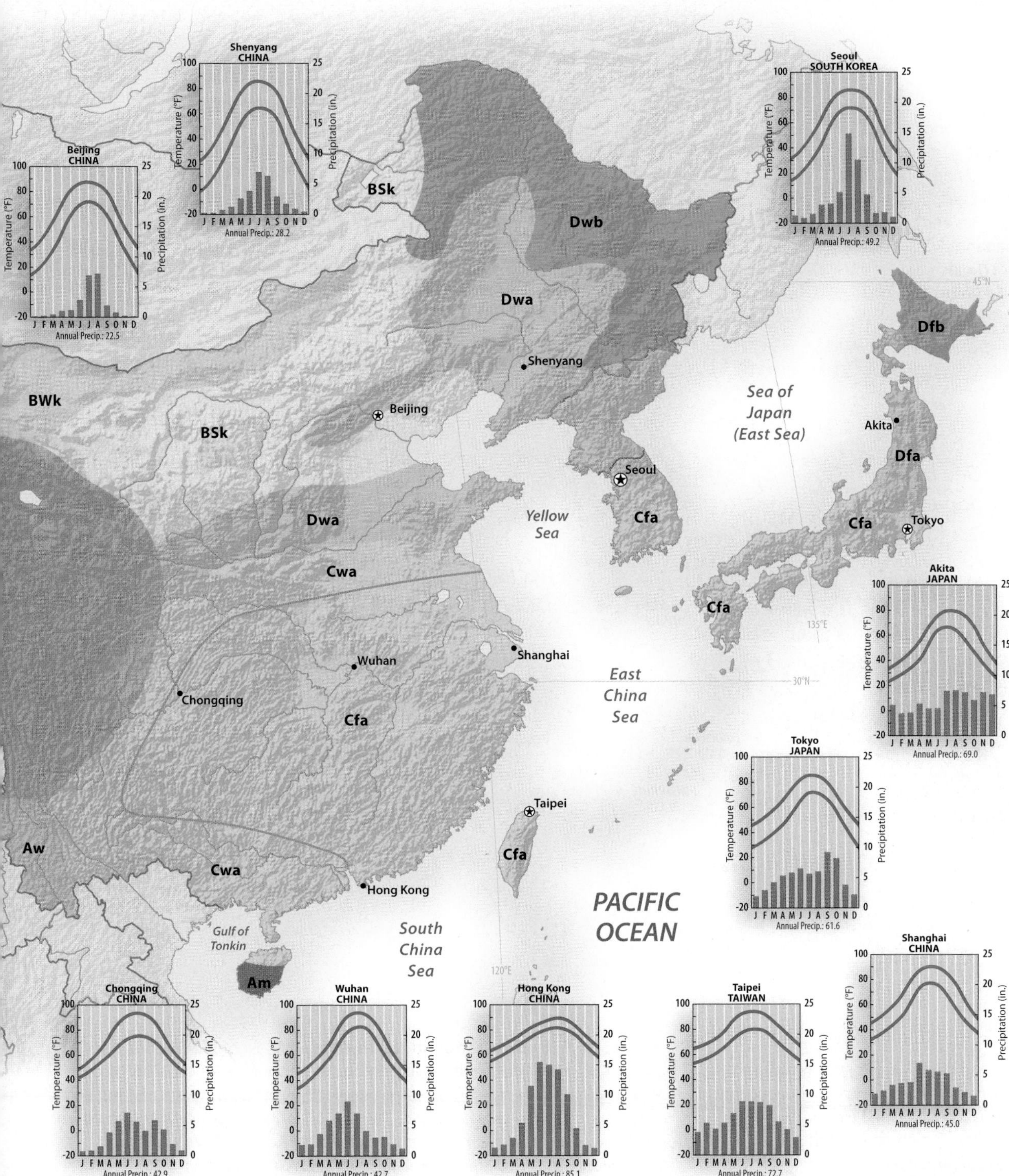

Figure 11.2 Climate Map of East Asia As China, Japan, and Korea are located in roughly the same latitudinal zone as North America, there are climatic parallels between these parts of the world. The northernmost tip of China lies at about the same latitude as Québec and has a similar climate, whereas southern China approximates the climate of Florida. In Japan, maritime influences produce a milder climate.

It can be broadly divided into two main areas: the drier zone lying to the north of the Yangtze River (or Chang Jiang) Valley and the more humid region that includes the Yangtze and all areas to the south.

Large valleys and moderate-elevation plateaus are found in the far south, where the climate is tropical or subtropical. Southeast China's coastal areas are rugged and offer limited agricultural opportunities (Figure 11.6). North of the Yangtze River Valley, the climate is both colder and drier. Summer rainfall is generally abundant, but the other seasons tend to be dry. The North China Plain, a large, flat area of fertile soil crossed by the Huang He (Yellow River), is usually cold and dry in winter and hot and humid in summer. Overall precipitation is somewhat low and unpredictable, and much of the North China Plain is threatened by **desertification** (the degradation of once-arable land in dry areas into desert), which often results in sand- and dust storms. Seasonal water shortages are becoming severe through much of the region as withdrawals for irrigation and industry increase (Figure 11.7). As a result, China is building large diversion projects to move water from the Yangtze River into the North China Plain.

Northeast China, generally known as Manchuria to English speakers, is dominated by a broad, fertile lowland sandwiched between mountains and uplands stretching along China's borders with North Korea, Russia, and Mongolia. Although winters here can be brutally cold, summers are usually warm and moist. Manchuria's upland areas are home to some of China's best-preserved forests and wildlife refuges.

Earthquakes and Tsunamis Much of East Asia is geologically active, with frequent earthquakes. Such hazards are most common in Japan, but they are usually more destructive in China due to lax construction standards. Almost 70,000 people died in the May 2008 Sichuan earthquake in central China, which left over 4 million people homeless. A relatively small Chinese earthquake in 2014 resulted in 367 deaths.

Figure 11.3 Japan's Physical Geography (a) Japan has several lowland plains, primarily along the coastline, which are mixed in with rugged mountains and uplands. (b) Because of its location at the convergence of three major tectonic plates, Japan commonly experiences both earthquakes and volcanic eruptions. In addition, much of Japan's coast is vulnerable to devastating tsunamis (tidal waves) caused by earthquakes in the Pacific Basin.

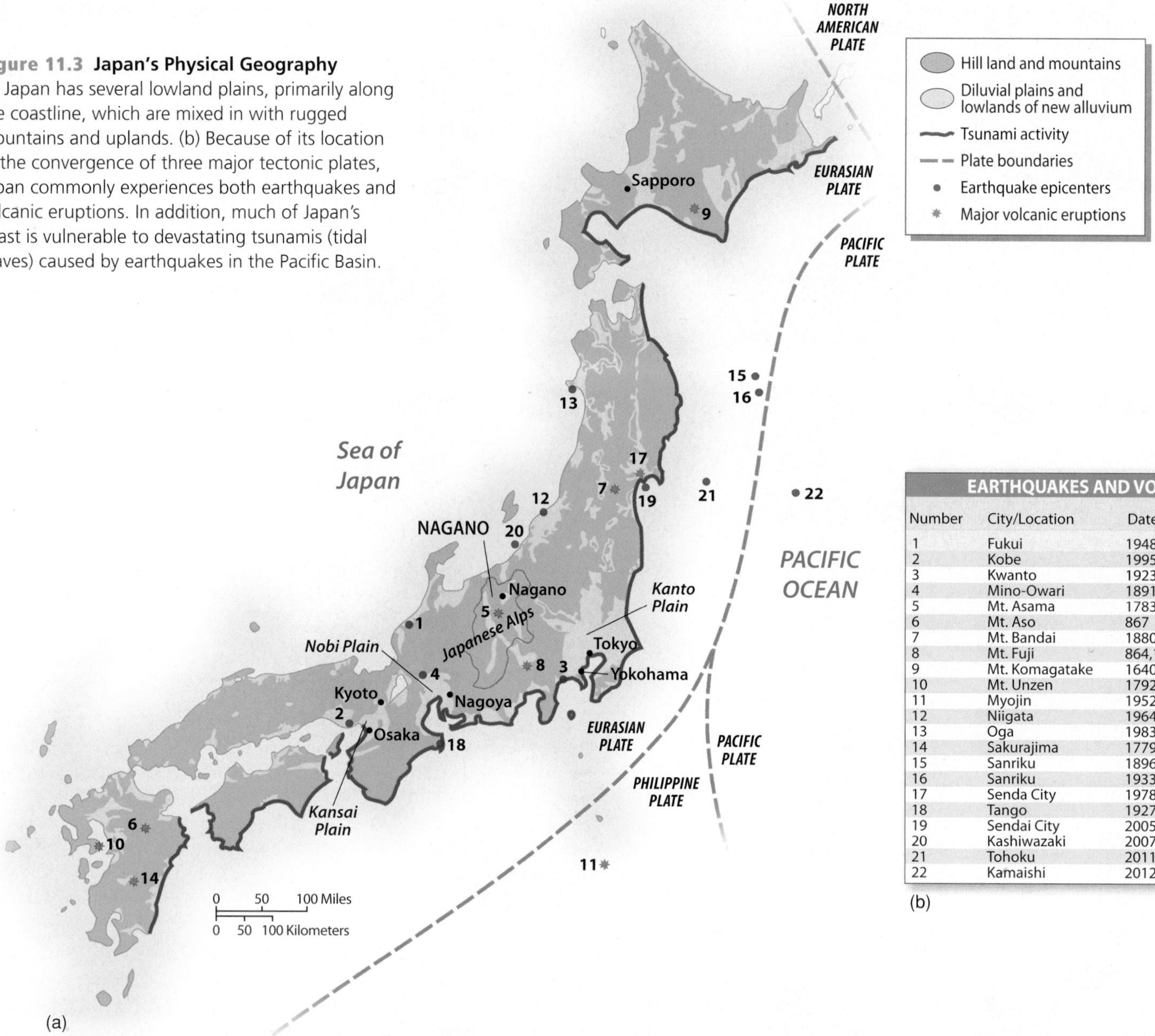

EARTHQUAKES AND VOLCANOES			
Number	City/Location	Date	Richter Magnitude
1	Fukui	1948	7.3
2	Kobe	1995	7.2
3	Kwanto	1923	7.9
4	Mino-Owari	1891	8.4
5	Mt. Asama	1783,1982	
6	Mt. Aso	867	
7	Mt. Bandai	1880	
8	Mt. Fuji	864,1707	
9	Mt. Komagatake	1640	
10	Mt. Unzen	1792,1991	
11	Myojin	1952	
12	Niigata	1964	7.7
13	Oga	1983	7.7
14	Sakurajima	1779,1914	
15	Sanriku	1896	7.6
16	Sanriku	1933	8.5
17	Senda City	1978	
18	Tango	1927	8.0
19	Sendai City	2005	7.2
20	Kashiwazaki	2007	6.8
21	Tohoku	2011	9.0
22	Kamaishi	2012	7.3

(b)

Figure 11.4 Japanese Forests Japan is one of the world's most mountainous countries, and most of its uplands are heavily forested. A combination of abundant rainfall, mild temperatures, and a long history of conservation helps support abundant tree growth across most of the country. This photograph shows picturesque Kegon Falls in Tochigi Prefecture in central Japan.

Figure 11.5 Deforestation in North Korea Most of the uplands of North Korea have been extensively deforested through a long history of overexploitation. Hills and mountains across much of the country are covered with scrub and stunted trees, as can be seen in Hwanghae Province in the southern portion of the country. The contrast with South Korea, where extensive reforestation has occurred, is striking. (AFP/AFP/Getty Images)

Figure 11.6 The Fujian Coast The rugged coastal province of Fujian lies in southeast China. Here the coastal plain is narrow and the shoreline deeply indented, producing a striking landscape. Because of limited agricultural opportunities along this rugged coastline, many residents of Fujian work in maritime activities.

Figure 11.7 Drought in Northern China The Yellow River (or Huang He) is sometimes called the "cradle of Chinese civilization" owing to its historical importance. However, due to the increasing extraction of water for agriculture and industry, the river now often runs dry in its lower reaches. Here children are playing on a boat in the dried-up riverbed in China's Henan Province.

Owing to both its long coastline and its position near the intersection of three tectonic plates, Japan is particularly vulnerable to **tsunamis**, huge sea waves usually produced by underwater earthquakes. The 2011 Tohoku earthquake and tsunami in northeastern Japan destroyed a number of towns and resulted in over 15,000 deaths (Figure 11.8). It also caused severe damage to the Fukushima Daiichi nuclear power plant, releasing significant quantities of radiation and forcing the evacuation of over 200,000 people from the area.

East Asia's Environmental Challenges

The region's most serious environmental issues are found in China, owing to its large population, its rapid industrial development, and the unique features of its physical geography. China suffers from some of the world's most severe air and water pollution. Japan, South Korea, and Taiwan, on the other hand, have invested heavily in environmental protection, resulting in much cleaner natural environments (Figure 11.9).

Urban Pollution As China's industrial base has expanded, urban environmental problems have grown increasingly severe. The burning of high-sulfur coal results in serious air pollution, which is exacerbated by the growing number of automobiles. Most Chinese cities are covered by a heavy shroud of smog, especially during the winter when the air often becomes stagnant. A recent report found that air pollution causes more than 350,000 premature deaths in China every year. Pollutants from China, moreover, regularly reach not only Japan and the Koreas, but also the U.S.

Figure 11.8 Tohoku Tsunami The March 11, 2011, Tohoku earthquake and tsunami were one of the world's most devastating natural disasters of recent times. At its extreme, the tsunami wave was over 40 meters (133 feet) high, giving it enough power to wash away entire villages. This photograph shows the tsunami breeching an embankment and flowing into the city of Miyako in Iwate prefecture. More than 15,000 people died as a result of this disaster. AFP PHOTO / JIJI PRESS **Q: Why did the 2011 Tohoku earthquake and tsunami generate such profound consequences for the Japanese economy and political system?**

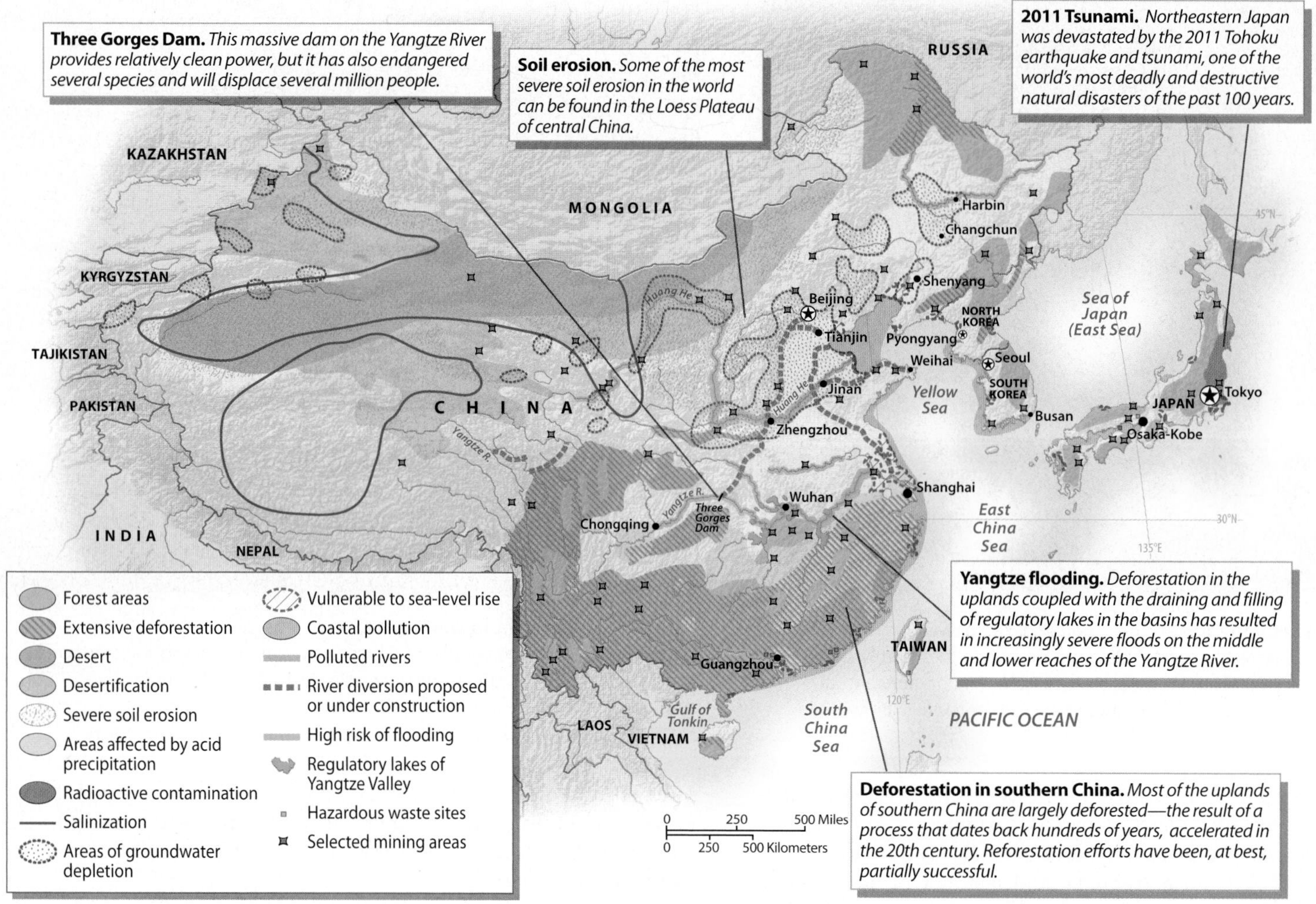

Figure 11.9 Environmental Issues in East Asia This huge world region has been almost completely transformed from its natural state and continues to have serious environmental problems. In China, some of the most pressing environmental issues involve deforestation, flooding, water control, and soil erosion.

West Coast (Figure 11.10). Water pollution, another huge problem, is reportedly responsible for some 60,000 deaths a year.

China is, however, making some progress in addressing pollution. Sulfur dioxide emissions peaked in 2006, and China's major cities are now less polluted than those of India. A hard-hitting video about air pollution in the country, called *Under the Dome*, went viral in both China and abroad in 2015, with several hundred million viewings within a few days. The fact that this highly critical video was not initially blocked by China's censors led many to conclude that the country's leaders were finally regarding pollution as a national emergency. The documentary's popularity, however, generated its own official concerns, and within a week it was removed from all Chinese video sites.

Japan, South Korea, and Taiwan have all made significant progress in controlling pollution, with relatively clean environments considering their high levels of population density and intensive industrialization. In the 1950s and 1960s—Japan's most intensive period of industrial growth—the country suffered from some of the world's worst water and air pollution. Soon afterward, it passed strict environmental laws. Similar developments occurred in Taiwan and South Korea several decades later.

Japan's environmental cleanup was aided by its location, as winds usually carry smog-forming chemicals out to sea. Equally important has been the phenomenon of **pollution exporting**. Japan's high cost of production and strict environmental laws have led to many Japanese companies relocating their dirtiest factories to other areas, especially China and Southeast Asia. This practice, which has also been followed by the United States and western Europe, means that Japan's pollution is partially displaced to poorer countries.

Forests and Deforestation Most of the uplands of China and North Korea support only grass, scrubby vegetation, and stunted trees. China lacks the historical tradition of forest conservation that characterizes Japan. In much of southern China, sweet potatoes, maize, and other crops have been grown on steep and easily eroded hillsides for several hundred years. After centuries of exploitation, many upland areas have lost so much soil that they cannot easily support forests.

Although the Chinese government has started large-scale reforestation programs, only a few have been truly successful. Today substantial forests are found only in China's far north, where a cool climate prevents fast growth, and along the eastern slopes of the Tibetan Plateau, where rugged terrain restricts commercial forestry. As a result, China suffers a severe shortage of forest resources. As its economy continues to expand, China has become a major importer of lumber, pulp, and paper.

Figure 11.10 Chinese Air Pollution China has such severe air pollution that it can easily be seen from space. In this NASA spectroradiometer image from October 28, 2009, thick gray-brown haze covers the western half of the North China Plain. Such pollution also streams across the Yellow Sea into Korea and Japan. Some airborne contaminants from China can even cross the Pacific to reach the western United States.

Flooding, Dams, and Soil Erosion in China

Historically, China's most severe environmental problems are those caused by flooding and soil erosion. Its government is now trying hard to address these problems, with mixed success. One of its most controversial responses has been the construction of the Three Gorges Dam on the Yangtze River.

The Yangtze River and the Three Gorges Controversy The Yangtze River is one of the most important physical features of East Asia. This river, the world's third largest (by volume), emerges from the Tibetan highlands onto the rolling lands of the Sichuan Basin, passes through a magnificent canyon in the Three Gorges area (Figure 11.11), and then meanders across the lowlands of central China before entering the sea in a large delta near the city of Shanghai. The Yangtze has been the main transportation corridor into the interior of China for millennia and has long been famous in Chinese literature for its beauty and power.

China's government is trying to control the Yangtze for two main reasons: to prevent flooding and to generate electricity. To do so, it built a series of large dams, the largest of which is a massive structure in the Three Gorges area, completed in 2006. This $39 billion structure is the largest hydroelectric dam in the world, forming a reservoir 350 miles (563 km) long. It has jeopardized several endangered species (including the Yangtze River dolphin), flooded a major scenic attraction, and displaced more than 1 million people.

The Three Gorges Dam generates large amounts of electricity, supplying roughly 1.7 percent of China's huge demand. As China industrializes, its need for power is increasing rapidly. Most of China's energy supply currently comes from burning coal, resulting in severe air pollution. According to China's government, the Three Gorges Dam prevented the release of 100 million tons of carbon into the atmosphere in 2014 alone. Most environmentalists, however, argue that the costs exceed the benefits. Chinese government officials disagree, claiming the dam has brought additional benefits, such as reducing the threat of flooding in the middle and lower Yangtze River Valley. Dam-building on the Huang He in northern China, however, has not been able to solve that region's water control problems.

Flooding in Northern China The North China Plain has historically suffered from both drought and flooding. This area is dry most of the year, yet it often experiences heavy downpours in summer.

Figure 11.11 The Three Gorges of the Yangtze (a) The spectacular Three Gorges landscape of the Yangtze River is the site of (b) a controversial dam that has displaced over 1 million people. Not only are the human costs high to those displaced, but also there may be significant ecological costs to endangered aquatic species.

(a)

(b)

Figure 11.12 Massive Discharge on China's Yellow River The Yellow River is considered to be the world's muddiest major river. Silt deposits increase the risk of flooding and clog up reservoirs. Visitors look at a massive gush of water released from the Xiaolangdi Reservoir in order to clear out the accumulated silt deposits.

Since ancient times, levees and canals have both controlled floods and allowed irrigation. But no matter how much effort has been put into water control, disastrous flooding has never been completely prevented.

The worst floods in northern China are caused by the Huang He, which cuts across the North China Plain. As a result of upstream erosion, the river carries a huge **sediment load** (the amount of suspended clay, silt, and sand in the water), making it the muddiest major river in the world (Figure 11.12). When the river enters the low-lying plain, its velocity slows, and its sediments begin to settle and accumulate in the riverbed. As a result, the level of the river gradually rises above that of the surrounding lands, and the river must eventually break free of its course to find a new route to the sea over lower-lying ground. In the period of recorded history, the Huang He has changed course 26 times. The resulting floods have been known to kill several million people in a single episode. While the river has not changed its course since the 1930s, most geographers agree that another change is inevitable.

Erosion on the Loess Plateau The Huang He's sediment load comes from the eroding soils of the Loess Plateau, located to the west of the North China Plain. **Loess** is a fine, windblown material that was deposited on this upland area during the last ice age. Here loess deposits accumulated to depths of up to several hundred feet. Loess forms fertile soil but washes away easily when exposed to running water. Cultivation requires plowing, which leads to soil erosion. As the region's population gradually increased, woodland and grassland diminished, leading to ever-greater rates of soil loss. Continued erosion cut great gullies across the plateau, steadily reducing the extent of productive land.

Major efforts are under way to stop the degradation of the Loess Plateau, several of which have achieved marked success. The construction of terraces and the planting of trees have proved crucial in halting soil erosion and preserving the region's farmlands (see *Geographers at Work: China's Agricultural Transformation*).

Climate Change and East Asia

East Asia occupies a central position in climate change debates, largely because of China's extremely rapid increase in carbon emissions. From levels only about half those of the United States in 2000, China's total production of greenhouse gases became the largest in the world in 2007. This staggering rise stems from both China's explosive economic growth and its reliance on coal to generate most of its electricity (Figure 11.13).

The potential effects of climate change in China have serious implications globally. A recent report predicted that the country's wheat, corn, and rice production could fall by as much as 37 percent if average temperatures increase 3.5–5.5°F (2–3°C) over the next 50 to 80 years. Increased evaporation rates could greatly intensify the water shortages that already plague much of northern China. In the wet zones of southern China, climate change concerns center on the probability of more intense storms in this already flood-prone region.

In June 2007, China released its first national plan on climate change, which called for major gains in energy efficiency as well as a partial transition to renewable energy sources. As a result, the Chinese government is heavily subsidizing the manufacture of solar panels, which it sees as a key energy technology of the 21st century. China's climate change strategy also includes a major expansion of both nuclear and wind power, along with ambitious reforestation efforts. China is currently the world's leading producer of renewable energy.

Figure 11.13 Energy Use in China Most of China's surging energy demand is being met by burning coal and oil.

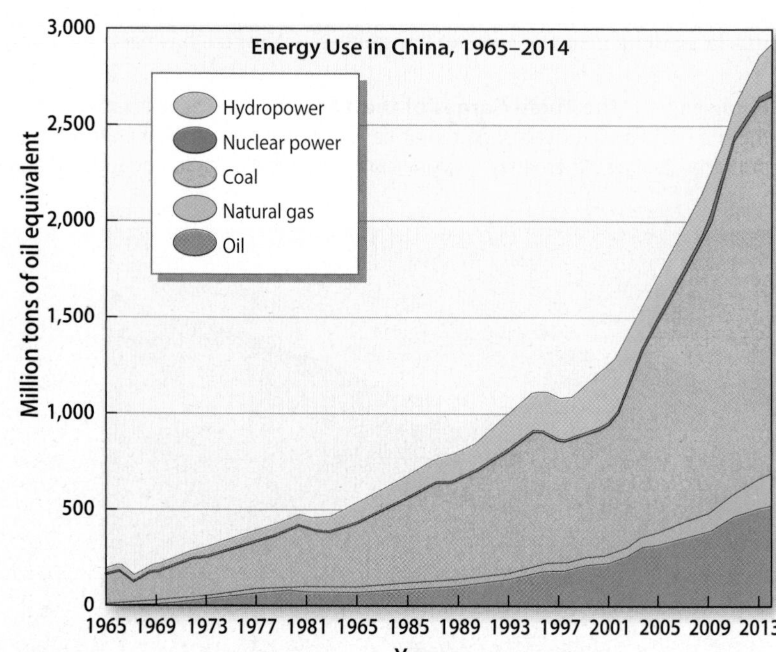

GEOGRAPHERS AT WORK

China's Rural Transformation

A relative latecomer to geography, **Gregory Veeck** of Western Michigan University was working for the U.S. Department of Agriculture when he discovered the geographic perspective, and went on to earn a PhD in the field. "What is great is the synthetic nature of the discipline," he notes. Veeck studies rural China's economy and environment in the wake of changes in that country's agricultural policies, and has conducted in-depth fieldwork in some of the richest and poorest parts of China. When working with Chinese colleagues, he explains, "I'm the glue, in a sense. I know what everyone is doing—the hydrologists, the economic development person—I see the big picture. Then we put the information on a map, which is the language for everyone involved."

Tools of the Geographer Veeck says his interest in geography made sense: "All my life I've liked maps. There's something wonderful about them. And I like talking to people, sitting around in rural China, spending an hour talking about grain prices. Learning Chinese allowed me to do that . . . geographers like languages more than people in other disciplines." He was initially terrified of learning Chinese, but now "I'll roll it out at the drop of a dime. Now is a wonderful time to learn a language."

Veeck also relies on survey data and statistical analysis. His study of off-farm income among farmers in Jiangsu Province, for example, cross-correlated variables related to income, farm size, and family size to demonstrate the importance of location: Farmers living near industrial cities could prosper by augmenting their income through part-time factory work.

More recently, Veeck and colleagues combined remote sensing with in-depth interviews with herders to assess the quality of China's grasslands and pastures (Figure 11.1.1). Poor pasture management threatens both herder incomes and the ecology of these extensive grasslands, and China's government is making efforts to mitigate damages and protect this segment of its economy. Veeck enjoys "old-school fieldwork," interviewing people to make sense of the remote sensing data. "Geography offers all these skills—quantitative and qualitative methods. We have more skills than our sister disciplines. We bring more tools to problem solving."

Figure 11.1.1 Field Work in China's Highlands Gregory Veeck visits Dagu glacier in Heishui County, western Sichuan Province, while studying yak husbandry and pastureland changes in the Sichuan highlands.

1. Why is it important to understand how Chinese farmers gain their income? How could such information be useful for Chinese officials and planners?
2. Think of ways that a geographer could contribute to a research project in your community.

In late 2014, China reached a deal with the United States in which it promised that its carbon emissions would peak around 2030 and then begin to decline. The Chinese government also pledged to fill 20 percent of its energy needs from renewable sources by the same year. Most climate change experts, however, think China should start cutting its greenhouse gas emissions much earlier, and many fear that the country will not actually honor its commitments on this issue.

China's contribution to climate change overshadows those of other East Asian countries. Japan, South Korea, and Taiwan are major greenhouse gas emitters, but they also have energy-efficient economies. Several Japanese and South Korean companies, moreover, are world leaders in energy efficiency and other forms of environmentally responsible technology. Japan's carbon dioxide emissions, however, spiked upward from 2011 to 2014 after it shut down its nuclear reactors due to the Tohoku earthquake and tsunami and the subsequent Fukushima power plant disaster. Although citizen opposition remains pronounced, Japan announced in early 2015 that it would soon restart its nuclear program.

REVIEW

11.1 Why does China suffer much more from soil erosion, floods, desertification, and deforestation than Japan, Taiwan, and South Korea?

11.2 Why has China become the world's largest emitter of greenhouse gases, and what is its government doing about this problem?

KEY TERMS China proper, desertification, tsunami, pollution exporting, sediment load, loess

Population and Settlement: A Realm of Crowded Lowland Basins

East Asia is densely populated (Figure 11.14 and Table A11.1). The lowlands of Japan, the Korean Peninsula, and China are among the most intensely used portions of Earth, containing not only the major cities, but also most of the agricultural lands of these countries. The region's population growth rate, however, has declined dramatically (Figure 11.15). Japan's current concern is population loss along with its aging society. In 2013, Japan's population dropped by almost 250,000, and if current trends continue, the country's population will dip from 127 million to 87 million by 2060. Within a few decades, South Korea and Taiwan will probably face similar issues. Although China's huge population is still expanding, it, too, will begin to decline within a few decades if current trends persist.

Japanese Settlement and Agricultural Patterns

Japan is a highly urbanized country, noted for its large cities. In 2014, the United Nations announced that Tokyo will almost certainly remain the world's largest metropolitan area until at least 2030. But Japan is also one of the world's most mountainous countries, with lightly populated uplands. Agriculture must therefore share the limited lowlands with cities and suburbs, resulting in extremely intensive farming practices and densely packed cities. Such concentrated development is especially pronounced in Japan's core area, extending from Tokyo south and west through Nagoya and Osaka to the northern coast of Kyushu. Most of Japan's more remote areas, in contrast, are currently losing population as younger people move to the urban areas.

Japanese agriculture is largely limited to its coastal plains and interior basins. Rice is Japan's major crop, and irrigated rice demands flat land. Japanese rice farming has long been one of the most productive forms of agriculture in the world, helping to support a large population on a limited amount of rugged land. Although rice is grown in almost all Japanese lowlands, the country's premier rice-growing districts lie along the Sea of Japan's coast of central and northern Honshu. Vegetables are also grown intensively in all the lowland basins, even on tiny patches within urban neighborhoods (Figure 11.16). The valleys of central and northern Honshu are famous for their temperate-climate fruit, while citrus comes from the milder southwestern districts. Crops that thrive in a cooler climate, such as potatoes, are grown mainly in Hokkaido and northern Honshu.

Settlement and Agricultural Patterns in Korea and Taiwan

The Korean Peninsula is also densely populated, containing some 76 million people (25 million in North Korea and 51 million in South Korea) in an area smaller than Minnesota. South Korea's population density is significantly higher than that of Japan. Most Koreans crowd into the alluvial plains and basins of the west and south. The highland spine, extending from the far north to northeastern South Korea, remains relatively sparsely settled. As in Japan, South Korean agriculture is dominated by rice. North Korea, in contrast, relies heavily on corn and other upland crops that do not require irrigation.

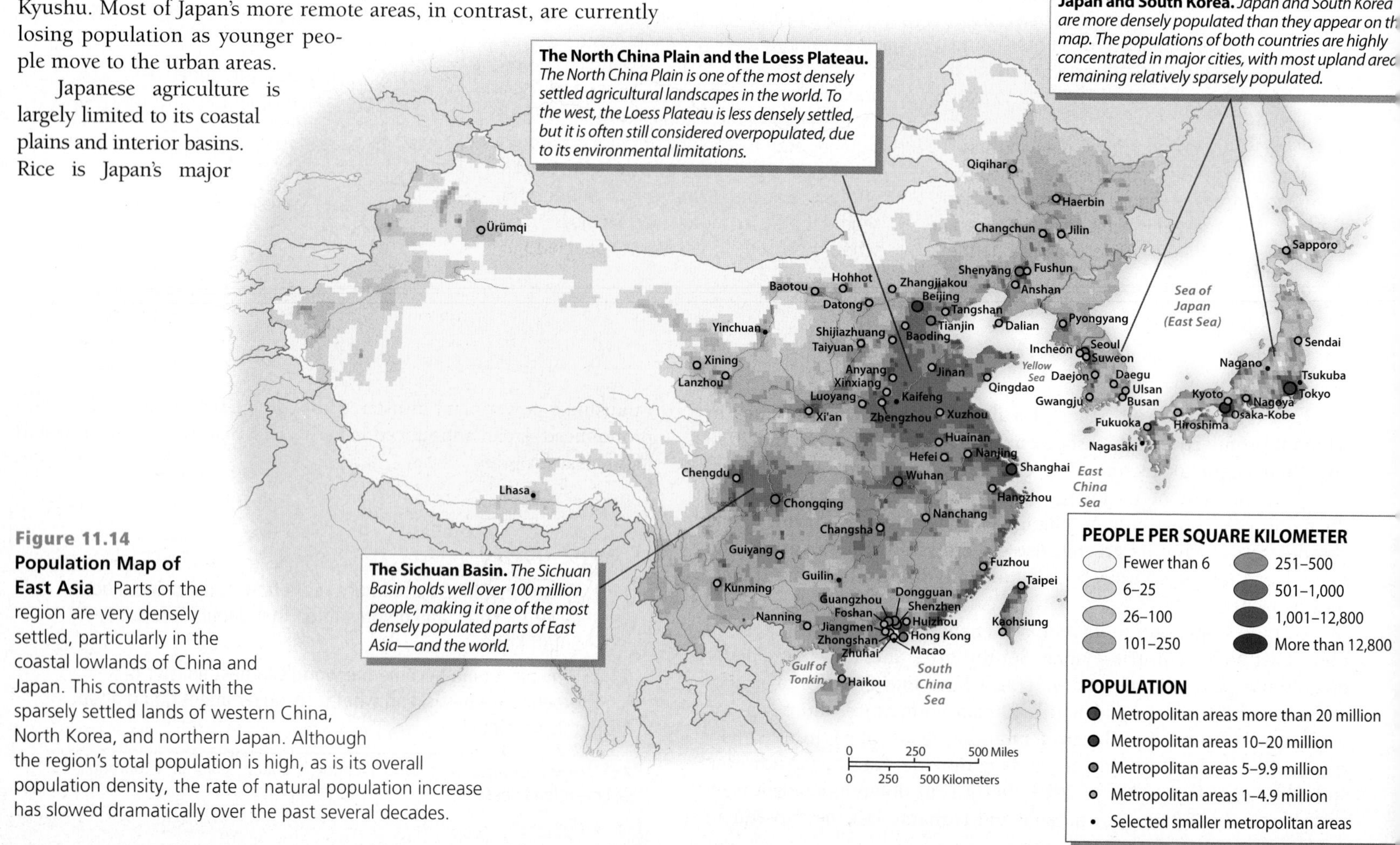

Figure 11.14 Population Map of East Asia Parts of the region are very densely settled, particularly in the coastal lowlands of China and Japan. This contrasts with the sparsely settled lands of western China, North Korea, and northern Japan. Although the region's total population is high, as is its overall population density, the rate of natural population increase has slowed dramatically over the past several decades.

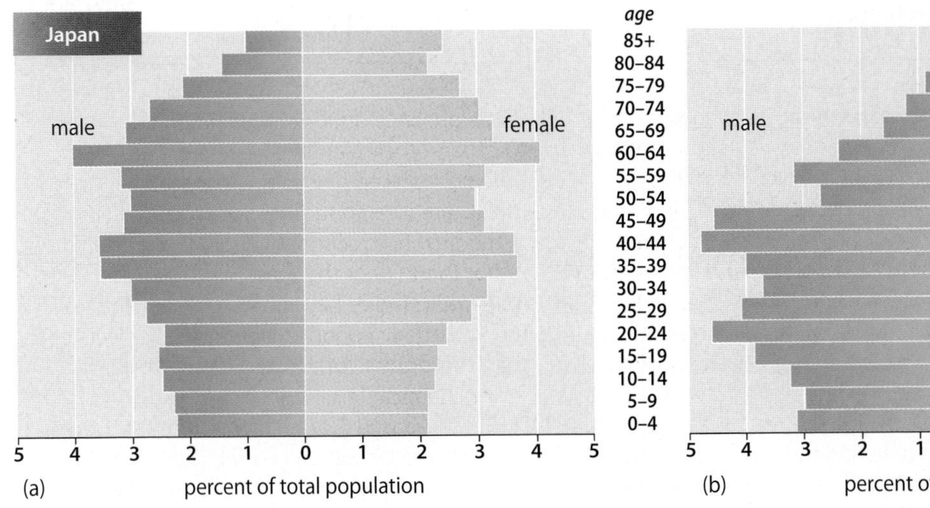

Figure 11.15 Population Pyramids for Japan and China (a) Japan has one of the world's oldest, and most rapidly aging, populations, due to both its low birth rate and its high life expectancy. The large Japanese population in the over-60 category places a great burden on the country's economy. (b) China has a more balanced demographic profile, but it also has a low birth rate, which could lead to similar problems in another 10 to 20 years.

Roughly the size of the Netherlands, Taiwan contains more than 23 million inhabitants. Its overall population density is the highest in East Asia and one of the highest in the world. Because mountains cover most of central and eastern Taiwan, virtually the entire population is concentrated in the narrow lowland belt in the north and west. In this area, large cities and numerous factories are scattered amid lush farmlands.

Settlement and Agricultural Patterns in China

Unlike Japan, Taiwan, and South Korea, China still has a vast rural population, with almost half its population living in the countryside. Although rural China is characterized by significant diversity, a line drawn just to the north of the Yangtze River Valley divides the vast country into two main agricultural regions. To the south, rice is the dominant crop. To the north, wheat, millet, and sorghum are the most common.

Southern and central China's population is highly concentrated in the broad lowlands, which are famous for their fertile soil and intensive agriculture. Planting and harvesting occurs year-round in most of southern and central China; summer rice alternates with winter barley or vegetables, and in many areas two rice crops, as well as one winter crop, are grown. Southern China also produces a wide variety of tropical and subtropical crops, and moderate slopes throughout the area produce sweet potatoes, corn, and other upland crops. In the plateau zone of Yunnan, tea, coffee, rubber, and other tropical plantation crops are widely cultivated (see *Working Toward Sustainability: Tea and Coffee in Yunnan, China*).

Figure 11.16 Japanese Urban Farm Japanese landscapes often combine dense urban settlement with small patches of intensively farmed land. Here a small rice farm is immediately adjacent to an urban neighborhood.

In northern China, population distribution is more variable. The North China Plain has long been one of the world's most thoroughly **anthropogenic landscapes** (that is, it is a landscape heavily transformed by human activities). Virtually its entire extent is either cultivated or occupied by houses, factories, and other structures of human society. Manchuria, on the other hand, was a lightly populated frontier zone as recently as the mid-1800s. Today, with a population of more than 100 million, its central plain is thoroughly settled. The Loess Plateau is also relatively crowded, despite its challenging environment (Figure 11.17).

East Asian Agriculture and Resources in Global Context

Although East Asian agriculture is highly productive, it cannot feed the huge number of people living in the region. Japan, Taiwan, and South Korea are major food importers, and China has recently moved in the same direction. Other resources from all quarters of the world are drawn in by the powerful economies of East Asia.

Japan is mostly self-sufficient in rice, but it is still one of the world's largest food importers. Japan imports much of its meat and the feed

Figure 11.17 Loess Settlement China's Loess Plateau is densely settled, considering its harsh environment. Over much of the plateau, subterranean dwellings are carved out of the soft sediments. This area is prone to major earthquakes that exact a high toll on the local population because of dwelling collapses.

WORKING TOWARD SUSTAINABILITY

Tea and Coffee in Yunnan, China

Google Earth (MG) Virtual Tour Video https://goo.gl/ZcdB99

Yunnan Province in south-central China has much more biological diversity than any other part of the country. It is also a major tourist destination, owing to its mild climate, spectacular scenery, and cultural diversity. But Yunnan is one of China's poorest provinces, beset by soil erosion, deforestation, and other environmental problems. Over the past several decades, the expansion of plantations growing rubber, coffee, tea, and other crops has generated extensive environmental damage (Figure 11.2.1). More recently, however, substantial progress has been made in developing sustainable cropping techniques.

Toward Sustainable Tea Sustainable tea production in Yunnan has been supported by the Rainforest Alliance, a New York–based organization devoted to preserving biodiversity and ensuring local livelihoods by changing land-use patterns and consumer behavior. The Rainforest Alliance recently gave its coveted sustainability certification to a large estate in Yunnan that produces roughly half a million kilograms (1,100,000 pounds) of tea a year.

To gain this status, the farm managers had to prove that they had reduced deforestation, ensured habitat availability for wildlife, and improved both soil and water conservation. They also had to show that they provided a safe working environment and gave fair wages, decent housing, and education and health-care benefits to workers and their families.

Figure 11.2.1 Harvesting Sustainable Coffee in Yunnan, China Several sustainable coffee-growing initiatives have been established in China's ethnically diverse Yunnan Province. Here, farmers of the Miao ethnic group harvest coffee beans at a plantation near Baoshan city in Yunnan in 2010.

Coffee Plantations Although tea has long been cultivated locally, coffee is a relatively new crop for the region. But as China's thirst for coffee has grown, so has production in Yunnan, one of the few parts of the country with a climate suitable for the crop. Between 2008 and 2011, coffee bean prices in the province doubled, encouraging farmers to expand their plantings. In some areas, tea is being replaced by coffee, and elsewhere forests are cleared for the crop. Most coffee in the province is currently grown in the sun, rather than under the shade of large trees. As numerous studies have demonstrated, shade-grown coffee supports much more biodiversity than sun-grown coffee, while also requiring fewer chemical fertilizers.

Seattle-based Starbucks is at the forefront of Yunnan's sustainable coffee production. It has done so through its Coffee and Farm Equity Practices (C.A.F.E.) program, which emphasizes both environmental conservation and social equity. Starbucks has encouraged its Yunnan-based suppliers to follow its C.A.F.E guidelines and in 2011 began to certify selected growers. To gain such certification, growers must show that they are conserving water and otherwise cultivating in a sustainable manner. Wastewater used in processing coffee beans, moreover, must be disposed of in a nonpolluting manner. More shade trees are now appearing in Yunnan's coffee plantations, and they are improving the quality of the soil, reducing fertilization requirements, and helping preserve local wildlife.

1. What environmental features make Yunnan particularly suitable for such plantation crops as tea and coffee?
2. Why are corporations like Starbucks so interested in environmental sustainability in China? How might such projects impact its profits?

used in its domestic livestock industry from the United States, Brazil, Canada, and Australia. These same countries supply soybeans and wheat. Japan has one of the highest rates of fish consumption in the world, and Japanese fishing fleets must scour the world's oceans to meet demand. Japan also depends on imports to supply its demand for forest resources. Although its own forests produce high-quality cedar and cypress logs, Japan buys most of its construction lumber and pulp (for papermaking) from western North America and Southeast Asia.

South Korea has followed Japan in obtaining food and forestry resources from abroad. In 2008, as global grain prices jumped, large South Korean companies began to negotiate long-term leases for vast tracts of farmland in poor tropical countries. South Korean companies also invest heavily in natural resource extraction in other parts of the world, particularly in Central Asia and Africa (see *Exploring Global Connections: South Korean Investments and Aid in Africa*).

Unlike South Korea, North Korea relies on its own farmland for most of its food production. North Korea's inefficient, state-run agricultural system, however, cannot supply adequate foods, resulting in periodically devastating famines. In recent years, however, North Korean leaders have started to allow farmers to cultivate their own small plots and sell surplus goods in small, open markets. These changes have generated a much larger food supply, although it remains uncertain whether they will be allowed to persist.

Through the 1980s, China remained self-sufficient in food, despite its huge population and crowded lands. But growing wealth has brought about a huge increase in the consumption of meat, which requires large amounts of imported feed grain. China now consumes five times more pork per person than it did in the late 1970s, and some experts claim that by the 2020s, as much as half of the world's feed-grain production will be devoted to feeding Chinese hogs. Economic growth has also resulted in a loss of agricultural lands to residential, commercial, and industrial development, requiring more food imports from abroad. But China is also a major exporter of high-value crops, including many fruits, nuts, and vegetables.

China's current demands for agricultural and mineral products are reordering patterns of global trade. Many Latin American, African, and Asian countries now greatly depend on the Chinese market. China continues to invest heavily in infrastructure, farming, and mining projects in Africa, Latin America, and elsewhere.

Figure 11.18 Traditional Chinese Courtyard These women are eating in an interior courtyard. Houses in most parts of China were traditionally built around courtyards, which served as an extension of the home. In China's booming cities, most of these old-style dwellings have been replaced by modern apartment complexes, but many survive in villages and smaller cities.

The Urban Environment of East Asia

China has one of the world's oldest urban foundations. In medieval and early modern times, East Asia as a whole possessed a well-developed system of cities that included some of the largest settlements on the planet. In the early 1700s, Tokyo, then called Edo, probably overshadowed all other cities, with a population of more than 1 million.

Despite this early urbanization, East Asia was overwhelmingly rural at the end of World War II. Some 90 percent of China's people then lived in the countryside, and even Japan was only about 50 percent urbanized. But as the region's economy grew after the war, so did its cities. Japan, Taiwan, and South Korea are now between 78 and 86 percent urban (see Table A11.1), which is typical for advanced industrial countries. A little more than half of the people of China now live in cities, a number that rose from just 26 percent in 1990. The current movement of Chinese people from rural to urban areas is one of the largest population transfers the world has ever seen.

Chinese Cities Traditional Chinese cities were clearly separated from the countryside by defensive walls. Most were planned in accordance with strict geometric principles, with straight streets meeting at right angles. The old-style Chinese city was dominated by low buildings. Houses were typically built around courtyards, and narrow alleyways served both commercial and residential functions (Figure 11.18).

China's cities began to change as Europeans started to gain power in the region in the 1800s. Several port cities were taken over by European interests, which proceeded to build Western-style buildings and modern business districts. By far the most important of these semicolonial cities was Shanghai, built near the mouth of the Yangtze River, the main gateway to interior China. Although Shanghai declined after the Communist Party came to power in 1949, it has experienced a major revival. Migrants are now pouring into Shanghai, and building cranes crowd the skyline (Figure 11.19). Official statistics put the population of the metropolitan area at 24 million.

Beijing was China's capital during the Manchu period (1644–1912), a status it regained in 1949. Under communist rule, Beijing was radically transformed; old buildings were razed, and broad avenues plowed through old neighborhoods. Crowded residential districts gave way to large blocks of apartment buildings and massive government offices. Some historically significant buildings were saved; those of the Forbidden City, for example, where the Manchu rulers once lived, survive as a complex of museums.

In the 1990s, Beijing and Shanghai vied for primary position among Chinese cities, with Tianjin, serving as Beijing's port, coming in a close third. All three cities have historically been removed from the regular provincial structure of the country and granted their own metropolitan governments. In 1997, another major city, Hong Kong, passed from British to Chinese control and was granted a distinctive status as a self-governing "special administrative region." The greater metropolitan area of the Xi Delta—composed of Hong Kong, Shenzhen, and Guangzhou (called Canton in the West)—is now one of China's premier urban areas (see Figure 11.14).

China's Urban Expansion China's government expects to see more than 100 million rural people move to the country's cities by 2020. It is building huge new housing developments and infrastructural projects to accommodate this growth and avoid the emergence of slums. According to official plans, every city with more than 200,000 people will be connected to the rest of the country by railroads and expressways. The central government would like to see such urban expansion proceed in an orderly and regionally balanced manner and has long been frustrated by the fact that the most rapid growth has occurred in the economically booming coastal belt. Its response has been a harsh system of urban residence registration, or ***hukou***, which denies unregistered urban migrants access to education and social services. In 2014, as part of an urban modernization effort, China announced that it would significantly relax such restrictions.

Figure 11.19 Shanghai Skyline Over the past quarter-century, the skyline of Shanghai has been completely transformed. None of the tall buildings visible in this photograph existed before 1990. **Q: Among all Chinese cities, why has Shanghai in particular seen such extraordinary levels of urban development over the past two decades?**

EXPLORING GLOBAL CONNECTIONS

South Korean Investments and Aid in Africa

Google Earth (MG) Virtual Tour Video http://goo.gl/5HbMu7

China's massive presence in Sub-Saharan Africa has gained much attention in recent years. Less often noted is South Korea's growing engagement in the region. South Korean firms not only have increased their exports to the region, but also have been investing, building local factories and distribution centers. The South Korean government is also enhancing its presence in the region. The Korean Initiative for African Development has overseen a huge increase in official developmental assistance, and the official Korea–Africa Industry Cooperation Forum encourages industrial cooperation. Diplomatic ties are also growing; between 2007 and 2014, Kenya, Angola, Senegal, Rwanda, Ethiopia, Sierra Leone, and Zambia all opened embassies in Seoul.

Products for the African Market From 2000 to 2010, South Korea's investments in Sub-Saharan Africa increased roughly 10-fold, and its exports surged ahead as well. More than 75 percent of exports to the region are sophisticated manufactured products, including electronic equipment, phones, vehicles, and construction equipment. Samsung Corporation has been a leading player. From 2010 to 2012, it invested over US$150 million in the region.

Samsung is now pursuing a "build for Africa" strategy based on making products that are well suited to the African market (Figure 11.3.1). Its smartphones are a vital component of this strategy and now command over half of the regional market. Samsung also plans to build local assembly plants for refrigerators, washing machines, and other durable consumer goods. Other South Korean firms involved include LG Electronics, which recently announced that it will invest more than US$2 billion to develop products suited to East and Central Africa.

Other South Korean firms focus instead on construction, building essential infrastructure in Sub-Saharan Africa. Daewoo Corporation, for example, recently announced the construction of a major power plant in Kenya, and another Korean firm is working on an oil refinery in Gabon. Overall, South Korea's construction contracts in the region rose from US$1.5 billion in 2008 to US$2.2 billion in 2011.

Figure 11.3.1 Samsung's Africa Initiative Pedestrians pass a large advertisement for Samsung Electronics Co. mobile phones in Nairobi, Kenya in 2013. Samsung, like several other South Korean companies, is making major efforts to reach the African market.

Local Opposition Not all of South Korea's ventures in Sub-Saharan Africa, however, have been welcomed. Several South Korean firms have leased huge tracts of African farmland in order to enhance Korean food security. Such projects, however, often anger local people, who typically receive few benefits and may even be forced off their lands. In 2009, an agreement allowing Daewoo to lease over a million hectares (2,470,000 acres) of farmland in Madagascar generated anger that led to the downfall of the country's government and the ouster of its president.

Another problem in South Korean–African relations is that of Korean fleets illegally fishing off the shores of West Africa. In 2013, a ship was caught off-loading some 4000 boxes of illicit fish from Sierra Leone in the South Korean port of Busan. Liberia recently detained a number of South Korean ships for illegally fishing in its waters. Unless such problems can be solved, South Korea's engagement with Africa may remain relatively limited.

1. List some of the reasons why South Korean firms are so interested in investing in Sub-Saharan Africa. What features particular to this region might make this investment attractive?
2. Is it a sound policy for South Korea to fish extensively in foreign waters and acquire farmland abroad? What possible advantages and disadvantages could result from such strategies?

Unfortunately, China's urbanization drive has in many areas led to massive overinvestment in apartment blocks and other forms of housing. Local authorities encouraged building in order to generate high levels of economic growth, but the units constructed are often too expensive for most people. As a result, empty new apartment complexes, some large enough to seem like modern ghost towns, are now a common feature of many Chinese cities. A 2014 study by a Chinese university reported that 20 percent of urban homes were vacant.

Although most of China's new city dwellers live in large apartment complexes, a significant amount of urban sprawl is also occurring. Affluent Chinese often prefer to live in single-family houses and pursue an automobile-dependent lifestyle. Critics contend that China is too crowded and polluted to support such an urban pattern. A 2014 report by the World Bank and an official research branch of the Chinese government claimed that the country would save US$1.4 trillion in infrastructure spending if it could build cities as dense as those of South Korea.

City Systems of Japan and South Korea The urban structures of Japan and South Korea are quite different from those of China. South Korea is noted for its pronounced **urban primacy** (the concentration of total urban population in a single city), whereas Japan is the center of a new urban phenomenon—the **superconurbation**, also called a **megalopolis** (a huge zone of coalesced metropolitan areas). (We saw another example of a megalopolis in Chapter 3: the Boston/NewYork/Philadelphia/Baltimore/Washington, DC corridor.)

Seoul, the capital, is by far the largest city in South Korea. The city is home to more than 10 million people, and its greater metropolitan

Figure 11.20 Gangnam District of Seoul The capital of South Korea, Seoul is also the country's economic hub and cultural center. Seoul's Gangnam District is noted for its expensive shops and luxury apartments.

Figure 11.21 Urban Concentration in Japan (a) The inset map shows the rapid expansion of Tokyo in the postwar decades. Today the greater Tokyo metropolitan area is home to almost 30 million people. The map of Japan as a whole shows the cluster of urban settlements along Japan's southeastern coast. The major area of urban concentration is between Tokyo and Osaka, a distance of some 300 miles, known as the *Tokaido corridor.* (b) Roughly 65 percent of Japan's population lives in this area.

TOKYO'S GROWTH
1914
1946
1975
2011

10 km
20 km
30 km
40 km
50 km
Tokyo Bay
(a)
Sea of Japan (East Sea)
PACIFIC OCEAN
Hokkaido
Honshu
Area enlarged
JAPAN
Tokyo
Yokohama
Nagoya
Kobe
Kyoto
Osaka
Hiroshima
Kitakyūshū
Fukuoka
Shikoku
Kyushu
0 50 100 Miles
0 50 100 Kilometers

Primary area of population concentration (Tokkaido corridor)
Secondary area of population concentration
Rail trunk line

POPULATION OF JAPAN'S MAIN URBAN CONCENTRATIONS

Region	Population
Tokyo	36.9 million
Osaka/Kobe/Kyoto	19.3 million
Nagoya	9.1 million

65% of Japan's total population lives in the Tokkaido corridor

(b)

area contains some 40 percent of South Korea's total population (Figure 11.20). All of South Korea's major government, economic, and cultural institutions are concentrated there. However, Seoul's explosive and generally unplanned growth has resulted in serious congestion.

Japan has traditionally been characterized by urban "bipolarity," rather than urban primacy. Until the 1960s, Tokyo, the capital and main business and educational center, together with the neighboring port of Yokohama, was balanced by the trading center of Osaka and its port, Kobe (Figure 11.21). Kyoto, the former imperial capital and the traditional center of elite culture, is also located near Osaka. As Japan's economy boomed in the 1960s through the 1980s, however, so did Tokyo, outpacing all other urban areas in almost every urban function. The greater Tokyo metropolitan area today contains up to 38 million persons, depending on how its boundaries are defined.

Japanese cities sometimes strike foreign visitors as rather gray and dull places, lacking historical interest. Little of the country's premodern architecture remains intact. Traditional Japanese buildings were made of wood, which survives earthquakes much better than stone or brick. However, fires have therefore been a long-standing hazard, and during World War II U.S. forces firebombed most Japanese cities. (Hiroshima and Nagasaki were, of course, completely destroyed by U.S. nuclear bombs.) However, Kyoto, the old imperial capital, was spared devastation. As a result, Kyoto is famous for its beautiful temples, which ring the basin in which central Kyoto lies.

REVIEW

11.3 Why does East Asia get so much of its food and so many of its natural resources from other parts of the world?

11.4 Describe how the urban landscape of China is currently changing.

KEY TERMS anthropogenic landscape, hukou, urban primacy, superconurbation (megalopolis)

Cultural Coherence and Diversity: A Confucian Realm?

East Asia is in some respects one of the world's most culturally unified regions. Although different parts of East Asia have their distinct cultures, the entire region shares certain historically rooted ways of life and systems of ideas that can be traced back to ancient Chinese civilization. Chinese culture emerged roughly 4000 years ago, largely in isolation from the Eastern Hemisphere's other early centers of civilization in the valleys of the Indus, Tigris–Euphrates, and Nile rivers.

Unifying Cultural Characteristics

The most important unifying cultural characteristics of East Asia are related to religious and philosophical beliefs. Throughout the region, Buddhism and especially Confucianism have shaped both individual beliefs and social and political structures. Although the role of traditional belief systems has been seriously challenged, especially in China, traditional cultural patterns remain.

Figure 11.22 Confucian Temple in China Confucianism is usually considered to be a philosophy rather than a religion, but it does have religious aspects. In Confucian temples, which are found across many parts of East Asia, the spirit and philosophy of Confucius are honored and often worshipped. **Q: To what extent can Confucianism be said to unify East Asia, both in the past and in the contemporary period?**

The Chinese Writing System The clearest distinction between East Asia and the world's other cultural regions is found in written language. Writing systems elsewhere in the modern world are based on the alphabetic principle, in which each symbol represents a distinct sound. East Asia evolved an entirely different system of **ideographic writing**, in which each symbol (or ideograph—often called a *character*) usually represents an idea, rather than a sound.

The East Asian writing system can be traced to the dawn of Chinese civilization. As the Chinese Empire expanded, the Chinese writing system spread. Japan, Korea, and Vietnam all came to use the same system, although in Japan it was substantially modified, while in Korea it was later mostly replaced by an alphabetic system. (Modern written Japanese combines Chinese characters, called *kanji*, with two separate systems representing syllables.) A major disadvantage of Chinese writing is its difficulty; to be literate, a person must memorize thousands of characters. Its main benefit is that two literate persons do not have to speak the same language to be able to communicate, because the written symbols they use to express their ideas are the same. Hence, speakers of different Chinese dialects who may not understand one another's speech can still read the same newspapers, books, and websites.

The Confucian Legacy Just as a common writing system helped build cultural linkages throughout East Asia, so, too, the idea system of **Confucianism** (the philosophy based on the teachings of Confucius) came to occupy a significant position in the region. Indeed, so strong is the heritage of Confucius that some writers refer to East Asia as the "Confucian realm" (Figure 11.22).

China's premier philosopher, Confucius (or Kung Fu Zi, in Mandarin Chinese), was born in 551 BCE, a period of political instability. Confucius's goal was to create a philosophy that could generate social stability. While Confucianism is sometimes considered a religion, Confucius himself was far more interested in how to lead a correct life and organize a proper society. He stressed obedience to authority, but he thought that those in power must act in a caring manner. Confucian philosophy also emphasizes education. The most basic level of the traditional Confucian moral order is the family unit, considered the bedrock of society.

The significance of Confucianism in East Asian development has long been debated. In the early 1900s, many observers believed that this conservative philosophy, based on respect for tradition and authority, was responsible for the economically backward position of China and Korea. But because East Asia has more recently enjoyed the world's fastest rates of economic growth, such a position is no longer supportable. New arguments claim that Confucianism's respect for education and the social stability that it generates give East Asia an advantage. It must also be recognized, however, that Confucianism has lost much of the hold it once had on public morality throughout East Asia.

Religious Unity and Diversity in East Asia

Certain religious beliefs have worked alongside Confucianism to unite the region. The most important culturally unifying beliefs are associated with Mahayana Buddhism, though other religious practices challenge this cultural unity.

Mahayana Buddhism Buddhism, a religion that stresses the quest to escape an endless cycle of rebirths and reach union with the cosmos (a state called nirvana), originated in India in the 6th century BCE. It reached China by the 2nd century CE and within a few hundred years had spread throughout East Asia. Today Buddhism remains widespread everywhere in the region, although it is far less significant here than in mainland Southeast Asia, Sri Lanka, and Tibet (Figure 11.23).

The variety of Buddhism practiced in East Asia—Mahayana, or Great Vehicle—is distinct from the Therevada Buddhism of Southeast Asia. Most important, Mahayana Buddhism simplifies the quest for nirvana, in part by putting forward the existence of beings who refuse divine union for themselves in order to help others spiritually. Mahayana Buddhism also permits its followers to practice other religions. Thus, many Japanese are both Buddhists and Shintoists, while many Chinese consider themselves both Buddhists and Taoists (as well as Confucianists).

Shinto The Shinto religion is so closely bound to the idea of Japanese nationality that it is questionable whether a non-Japanese person can follow it. Shinto began as the worship of nature spirits, but it was gradually refined into a subtle set of beliefs about the harmony of nature

Figure 11.23 The Buddhist Landscape Mahayana Buddhism has traditionally been practiced throughout East Asia. This Golden Buddha statue is located in Baomo Park in Chi Lei Village, Guangdong Province, China.

Figure 11.24 South Korean Mega-church South Korea has seen a major expansion of Christianity over the past half-century. Many South Koreans attend large churches that use modern technology to connect ministers to their congregations. This 2007 service at Yoido Soon-Bok-Eum (Full Gospel) church, the largest in South Korea, focused on praying for the safe return of several South Koreans who had been kidnapped in Afghanistan.

and its connections with human existence. Shinto is still a place- and nature-centered religion. Certain mountains, particularly Mount Fuji, are considered sacred. Major Shinto shrines, often located in scenic places, attract numerous religious pilgrims. The most notable of these is the Ise Shrine south of Nagoya, devoted to the emperor of Japan.

Taoism and Other Chinese Belief Systems Similar to Shinto, the Chinese religion Taoism (or Daoism) is rooted in nature worship. Also like Shinto, it stresses spiritual harmony. Taoism is indirectly associated with **feng shui**, the Chinese and Korean practice of designing buildings in accordance with the spiritual powers that supposedly flow through the local topography. Even in hypermodern Hong Kong, skyscrapers worth millions of dollars have occasionally gone unoccupied because their construction failed to follow feng shui principles.

Minority Religions Followers of virtually all world religions can be found in East Asia. More than 1 million Japanese, for example, belong to Christian churches. Christianity has spread much more extensively in South Korea; between 25 and 35 percent of the people are adherents (Figure 11.24). South Korea now sends more Christian missionaries abroad than any other country except the United States. Christianity is also spreading rapidly in China, causing Beijing's communist leadership some concern. A 2012 Chinese survey estimated the Christian population at 30 to 40 million, but some independent sources think that it could be significantly higher.

China's Muslim community is much more deeply rooted than its Christian population. In addition to the Uighurs of Xinjiang (see Chapter 10), roughly 10 million Chinese-speaking Muslims, called Hui, are concentrated in Gansu and Ningxia in the northwest and in Yunnan Province along the south-central border. Smaller clusters of Hui, often separated in their own villages, live in almost every province of China.

Secularism in East Asia Despite the varied forms of religious expression, East Asia is one of the most secular regions of the world. Although a small portion of Japan's population is highly religious, most people only occasionally observe Shinto or Buddhist rituals. Japan also has several "new religions," a few of which are noted for their strong beliefs. But for Japanese society as a whole, religion is not particularly important.

After the communist regime took power in China in 1949, all forms of religion and traditional philosophy—including Confucianism—were severely discouraged. Under the new regime, based on a specific version of **Marxism**, atheism became the official belief system. With the easing of communism during the 1980s and 1990s, however, many forms of religious expression began to return. In North Korea, communist beliefs gradually gave way to a Korean official ideology of *juche*, or "self-reliance." Ironically, *juche* demands absolute loyalty to the country's repressive political leaders.

Linguistic and Ethnic Diversity in East Asia

Written languages may have helped unify East Asia, but the same cannot be said for spoken languages (Figure 11.25). Japanese and Mandarin Chinese partially share a system of writing, but the two languages have no direct relationship. Like Korean, however, Japanese has adopted many words of Chinese origin.

Language and National Identity in Japan Japanese, according to most linguists, is not related to any other language. Korean is also usually classified as the only member of its language family. Some linguists, however, think that Japanese and Korean should be classified together based on shared grammatical features. The Japanese dialect spoken in the southern Ryukyu Islands is so distinct that most linguists considered it a separate language. Many Ryukyu people believe that they are not considered full members of the Japanese nation, and they have historically suffered from some discrimination.

Although the Japanese form one of the world's most homogeneous peoples, in earlier centuries the country was divided between two very different peoples: the Japanese living to the south and the Ainu inhabiting the north. The Ainu are physically distinct from the Japanese and possess their own language (Figure 11.26). For centuries, the two groups competed for land, and by the 10th century CE the Ainu were mostly restricted to the northern island of Hokkaido. Today only about 25,000 Ainu remain, and their language is almost extinct.

Approximately 600,000 people of Korean descent living in Japan today have also felt discrimination. Most were born in Japan and speak Japanese, rather than Korean. Despite their deep bonds to Japan, however, such individuals cannot easily obtain Japanese citizenship. This treatment has led many Japanese Koreans to hold radical political views and support North Korea, angering many other residents of Japan.

Starting in the 1980s, other immigrants began to arrive in Japan, mostly from the poorer countries of Asia. Because Japan severely restricts the flow of immigrants, many do not have legal status. Men from China and southern Asia typically work in construction; women from Thailand and the Philippines often work as entertainers and sometimes as prostitutes. Overall, immigration is less pronounced in Japan than in most other wealthy countries, and relatively few migrants acquire permanent residency, let alone citizenship. In 2014, however, the Japanese government announced that it would seek to increase the flow of immigrants in order to bolster the country's declining population.

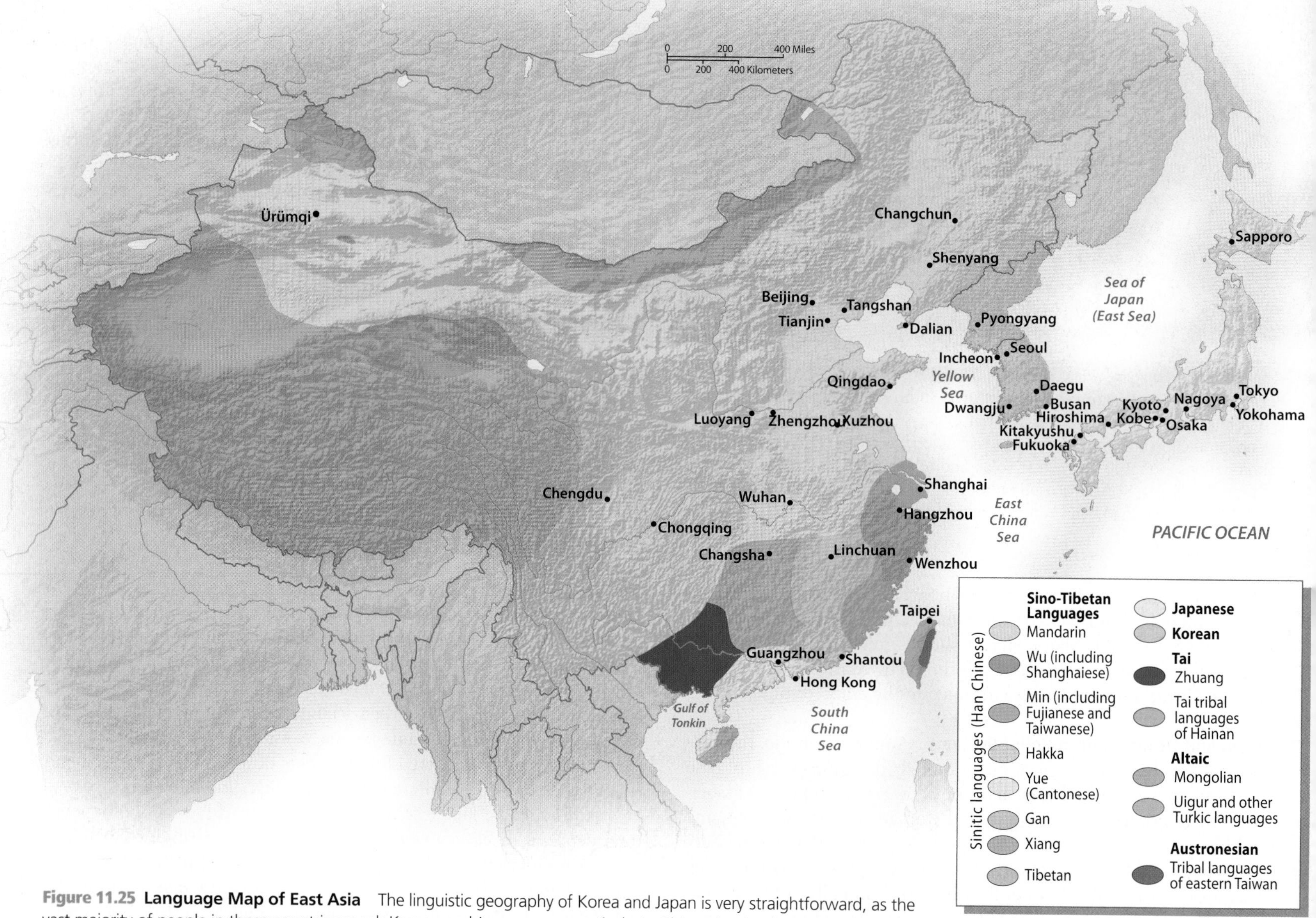

Figure 11.25 Language Map of East Asia The linguistic geography of Korea and Japan is very straightforward, as the vast majority of people in those countries speak Korean and Japanese, respectively. In China, the dominant Han Chinese speak a variety of closely related *Sinitic* languages, the most important of which is Mandarin Chinese. In the peripheral regions of China, a large number of languages—belonging to several different linguistic families—are spoken.

Language and Identity in Korea Koreans are also a relatively homogeneous people. The vast majority of people in both North and South Korea speak Korean and consider themselves to be members of the Korean nation. However, strong regional identities persist, which can be traced back to the medieval period when the peninsula was divided into three separate kingdoms.

Not all Koreans live in Korea. Several million reside directly across the border in northern China. Desperately poor North Koreans often try to sneak across the border to join these Korean-speaking Chinese communities, but China's government regards such migrants as a security threat and returns them to North Korea when it can. A more recent Korean **diaspora** (scattering of a particular group of people over a vast geographical area) has brought hundreds of thousands of people to the United States, Canada, Australia, New Zealand, the Philippines, and other countries.

Figure 11.26 Ainu Men The indigenous Ainu people of northern Japan are much reduced in population, but they still maintain some of their cultural traditions. Here Ainu men participate in the Marimo Festival on the northern Japanese island of Hokkaido.

Figure 11.27 Tribal Villages in South China Non-Han people are usually classified as "tribal" in China, which assumes they have a traditional social order based on autonomous village communities. This is a Miao village in the Xiangxi Tujia and Miao Autonomous Prefecture, Hunan Province.

Language and Ethnicity Among the Han The geography of language and ethnicity in China is more complex than that of Korea or Japan. This is true even if we consider only the densely populated eastern half of the country (see Figure 11.25). The most important distinction is that separating the Han Chinese from the non-Han peoples. The Han, who form the vast majority, are those people who have historically been incorporated into the Chinese cultural and political systems and whose languages are expressed in the Chinese writing system. They do not, however, all speak the same language.

Northern, central, and southwestern China—a vast area extending from Manchuria through the middle and upper Yangtze River Valley to the valleys of Yunnan in the far south—constitute a single linguistic zone. The spoken language here is called Mandarin Chinese in English. Throughout the rest of China, Mandarin (*Putonghua,* or "common language") is the national tongue.

In southeastern China, from the Yangtze Delta to China's border with Vietnam, several separate but related languages are spoken. Traveling from south to north, we hear Cantonese (*Yue*) spoken in Guangdong, Fujianese (alternatively Hokkienese, or *Min*) spoken in Fujian, and Shanghaiese (*Wu*) spoken in and around Shanghai. Linguistically speaking, these are true languages because they are not mutually intelligible. They are usually called dialects, however, because they have no distinctive written form. Despite their many differences, all the languages of the Han Chinese are closely related, belonging to the Sinitic language subfamily.

The Non-Han Peoples Many remote upland districts of China proper are inhabited by various groups of non-Han peoples speaking non-Sinitic languages. Such peoples are often classified as **tribal**, implying that they have a traditional social order based on self-governing village communities. Such a view is not entirely accurate, however, because some of these groups once had their own kingdoms. All are now subject to the Chinese state (Figure 11.27).

As many as 11 million Manchus live in Manchuria. The Manchu language, however, is almost extinct, having been abandoned in favor of Mandarin Chinese. This is ironic because the Manchus ruled the entire Chinese Empire from 1644 to 1912. Until the later part of this period, the Manchus prevented the Han from settling in central and northern Manchuria. Once Han Chinese were allowed to move to Manchuria in the 1800s, the Manchus soon found themselves vastly outnumbered.

Much larger communities of non-Han peoples are found in south-central China, especially in Guangxi, Guizhou, and Yunnan. Although most residents of Yunnan are Han Chinese, the remote areas of the province are inhabited by a wide array of indigenous peoples (Figure 11.28). Because most of its inhabitants in

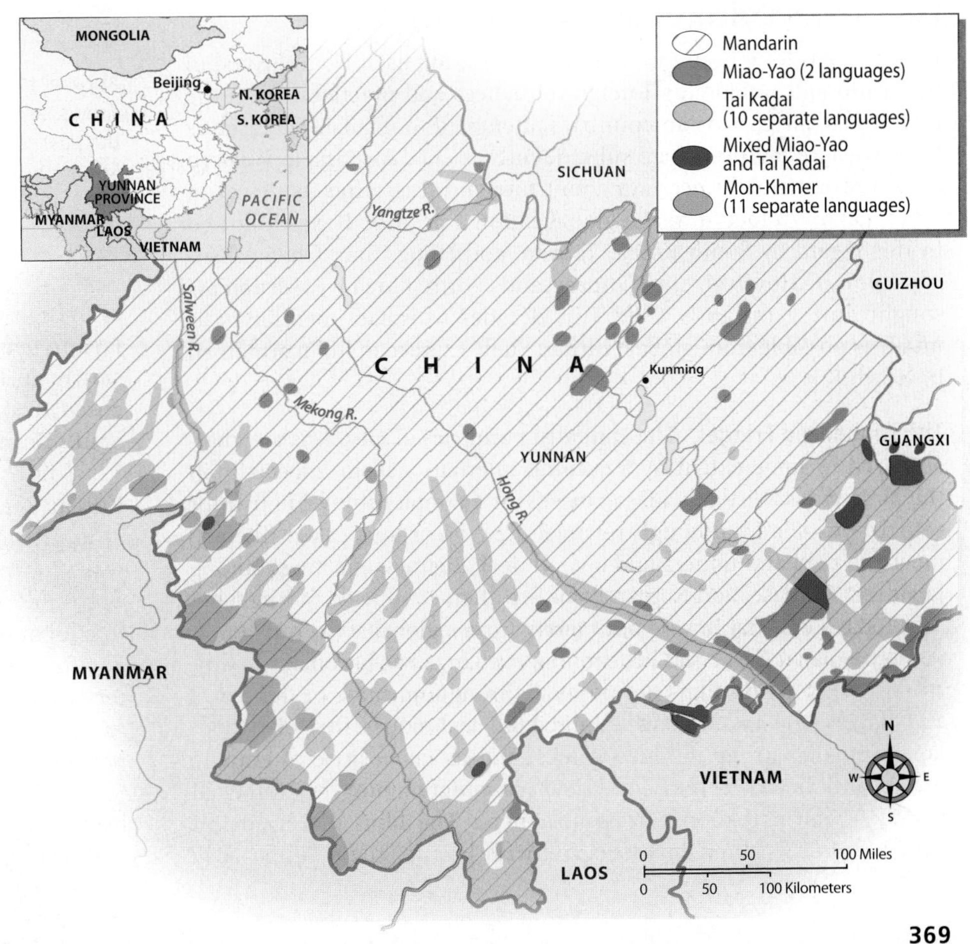

Figure 11.28 Language Groups in Yunnan China's Yunnan Province is the most linguistically complex area in East Asia. In Yunnan's broad valleys and relatively level plateau areas, as well as in its cities, most people speak Mandarin Chinese. In the hills, mountains, and steep-sided valleys, however, a wide variety of tribal languages from several linguistic families are spoken. In certain areas, several different languages can be found in very close proximity.

the uplands and remote valleys speak languages of the Tai family, Guangxi has been designated an **autonomous region**. Critics contend, however, that little real autonomy has ever existed. (In addition to Guangxi, there are four other autonomous regions in China. Three of these—Xizang [Tibet], Nei Mongol [Inner Mongolia], and Xinjiang—are located in Central Asia and are thus discussed at length in Chapter 10. The fourth, Ningxia, located in northwestern China, is distinguished by its large concentration of Hui [Mandarin-speaking Muslims].)

Language and Ethnicity in Taiwan Taiwan is noted for its linguistic and ethnic complexity. In the island's mountainous eastern region, a few small groups of "tribal" peoples speak Austronesian languages related to those of Indonesia. These peoples resided throughout Taiwan before the 16th century. At that time, however, Han migrants began to arrive in large numbers. Most of the newcomers spoke Fujianese dialects, which eventually evolved into the distinctive language called Taiwanese.

Taiwan was transformed almost overnight in 1949, when China's nationalist forces, defeated by the communists, sought refuge on the island. Most of the nationalist leaders spoke Mandarin, which they made the official language. Taiwan's new leadership discouraged Taiwanese, viewing it as a local dialect. As a result, considerable tension developed between the Taiwanese and Mandarin communities. Only in the 1990s did Taiwanese speakers begin to reassert their linguistic identity. At present, proponents of the Taiwanese language tend to advocate formal independence from China, whereas those who favor Mandarin more often hope for eventual reunification.

Figure 11.29 Korean Star Kim Soo-Hyun South Korean films, television shows, and music have become extremely popular across much of Asia over the past several decades. This photograph shows South Korean actor Kim Soo-Hyun attending a press conference on March 21, 2014, in Taipei, Taiwan. Kim Soo-Hyun is most famous for his role in the Korean TV drama "Love from the Star."

East Asian Cultures in Global Context

East Asia has long been torn between separating itself from the rest of the world and welcoming foreign influences and practices. Until the mid-1800s, all East Asian countries attempted to insulate themselves from Western culture. Japan subsequently opened its ports to Western trade but remained uncertain about foreign ideas. After its defeat in 1945, Japan decided to make globalization a priority. It was followed in this regard by South Korea, Taiwan, and Hong Kong (then a British colony). However, the Chinese and North Korean governments sought during the early Cold War decades to isolate themselves as much as possible from global culture. Such a stance is still maintained in North Korea.

The Globalized Fringe The capitalist countries of East Asia are characterized to some extent by a cultural internationalism, especially in the large cities. Virtually all Japanese, for example, study English for 6 to 10 years, and although relatively few learn to speak it fluently, most can read and understand a good deal. Business meetings among Japanese, Chinese, and Korean firms are often conducted in English.

Today's cultural flow is not merely from a globalist West to a previously isolated East Asia. Instead, the exchange has become two-way. Hong Kong's action films are popular throughout most of the world and have come to influence filmmaking techniques in Hollywood. Japan remains strong in video games, and its *anime* style of graphic novels and animated film and television programming has followed karaoke bars in their overseas movement. The films of Hayao Miyazaki and other directors associated with Japan's Studio Ghibli have been highly popular and influential across the world. Employment in Japan's *anime* industry, however, peaked around 2005 and has been slowly declining ever since. Critics claim that the industry is too focused on niche audiences and thus suffers from competition from South Korea and other Asian countries.

South Korea's popular culture industry continues to thrive, as Korean music, movies, and television shows have become popular around the world, a phenomenon known as ***hallyu***, or "Korean-wave." K-pop, a South Korean musical genre based on rock, hip-hop, and electropop styles, has been especially fashionable abroad, its global reach enhanced by its fans' use of Facebook, Twitter, and YouTube to publicize their favorite artists (Figure 11.29). The international appeal of Korean popular culture reached a high point in 2012, when the music video "Gangnam Style" by the artist PSY became the third most often viewed video in YouTube history. South Korean cultural officials now hope to refresh *hallyu* exports by linking entertainment to other Korean products such as food and fashion.

The Chinese Heartland In one sense, Japan is more culturally predisposed to cosmopolitanism than is China. Before the modern era, the Japanese borrowed heavily from other cultures (particularly from China itself), whereas the Chinese have historically been more self-sufficient. The southern coastal Chinese, however, have long had a stronger orientation toward foreign lands.

In most periods of Chinese history, the internal orientation of the core prevailed over the external orientation of the southern coast. After the communist victory of 1949, only the small British enclave of Hong Kong maintained close international cultural connections. In the rest of the country, a grim and puritanical cultural order was rigidly enforced. Once China began to liberalize its economy and open its doors to foreign influences in the late 20th century, however, the southern coastal region suddenly assumed a new prominence. Through its coastal cities, global cultural patterns began to penetrate the rest of the country. The result has been the emergence of a vibrant and somewhat flashy Chinese urban popular culture containing such global features as nightclubs, karaoke bars, fast-food franchises, and theme parks (Figure 11.30).

North Korea's Isolation Unlike the rest of East Asia, North Korea tries to insulate its people as much as possible from global culture. But foreign films and television shows, especially those from South Korea, are increasingly popular. DVDs are smuggled into the country in large numbers, even though smugglers can face public execution if caught. North Korea also uses intimidation to try to control how the county is depicted in the foreign media. In 2014, it threatened "merciless action" against the United States due to the scheduled release of *The Interview*, a Sony Pictures comedy that depicts the fictional assassination of its leader. Sony Pictures was later hacked, and a huge trove of sensitive material was released to the public. Some experts think that North Korea was responsible for the attack, but the evidence remains unclear.

Figure 11.30 Chinese Theme Park As China's economy grows, its people are spending increasing amounts of money on entertainment. Theme parks, which now number over 2000, are particularly popular.

Review

11.5 How has the geography of religion changed in East Asia since the end of World War II?

11.6 What features mark the Han Chinese as an ethnic group distinct from the other peoples of China?

KEY TERMS ideographic writing, Confucianism, feng shui, Marxism, diaspora, tribal, autonomous region, *hallyu*

Geopolitical Framework: The Imperial Legacies of China and Japan

Much of East Asia's political history revolves around the centrality of China and the ability of Japan to remain outside China's grasp. The traditional Chinese conception of geopolitics was based on the idea of a universal empire: All territories were either part of the Chinese Empire, paying tribute to it and acknowledging its supremacy, or outside the system altogether. When China could no longer maintain its power in the face of European aggression, the East Asian political system began to collapse. As European power declined in the 1900s, China and Japan competed for regional leadership. After World War II, East Asia was split by larger Cold War rivalries (Figure 11.31).

The Evolution of China

The original core of Chinese civilization was the North China Plain and the Loess Plateau. For many centuries, periods of unification alternated with times of division into competing states. The most important episode of unification occurred in the 3rd century BCE and lasted until the 3rd century CE. During this period, China expanded vigorously to the south of the Yangtze River Valley. Subsequently, the ideal of a united China triumphed, helping to join Han Chinese into a single people. Although periods of disunity followed the collapse of China's various dynasties, reunification always followed.

Various Chinese dynasties attempted to conquer Korea, but the Koreans resisted. Eventually, China and Korea worked out an arrangement whereby Korea paid token tribute and acknowledged the supremacy of the Chinese Empire. In return, Korea received trading privileges and retained independence. When foreign armies invaded Korea—as did those of Japan in the late 1500s—China sent troops to support its "vassal kingdom."

The Qing Dynasty of the Manchus In 1644, the Manchus toppled the Ming Dynasty and replaced it with the Qing (also spelled Ch'ing) Dynasty. As earlier conquerors did, the Manchus retained the Chinese bureaucracy and made few institutional changes. Their strategy was to adapt themselves to Chinese culture, while at the same time preserving their own identity as an elite military group. This system functioned well until the mid-19th century, when the Chinese Empire began to crumble at the hands of European and, later, Japanese power.

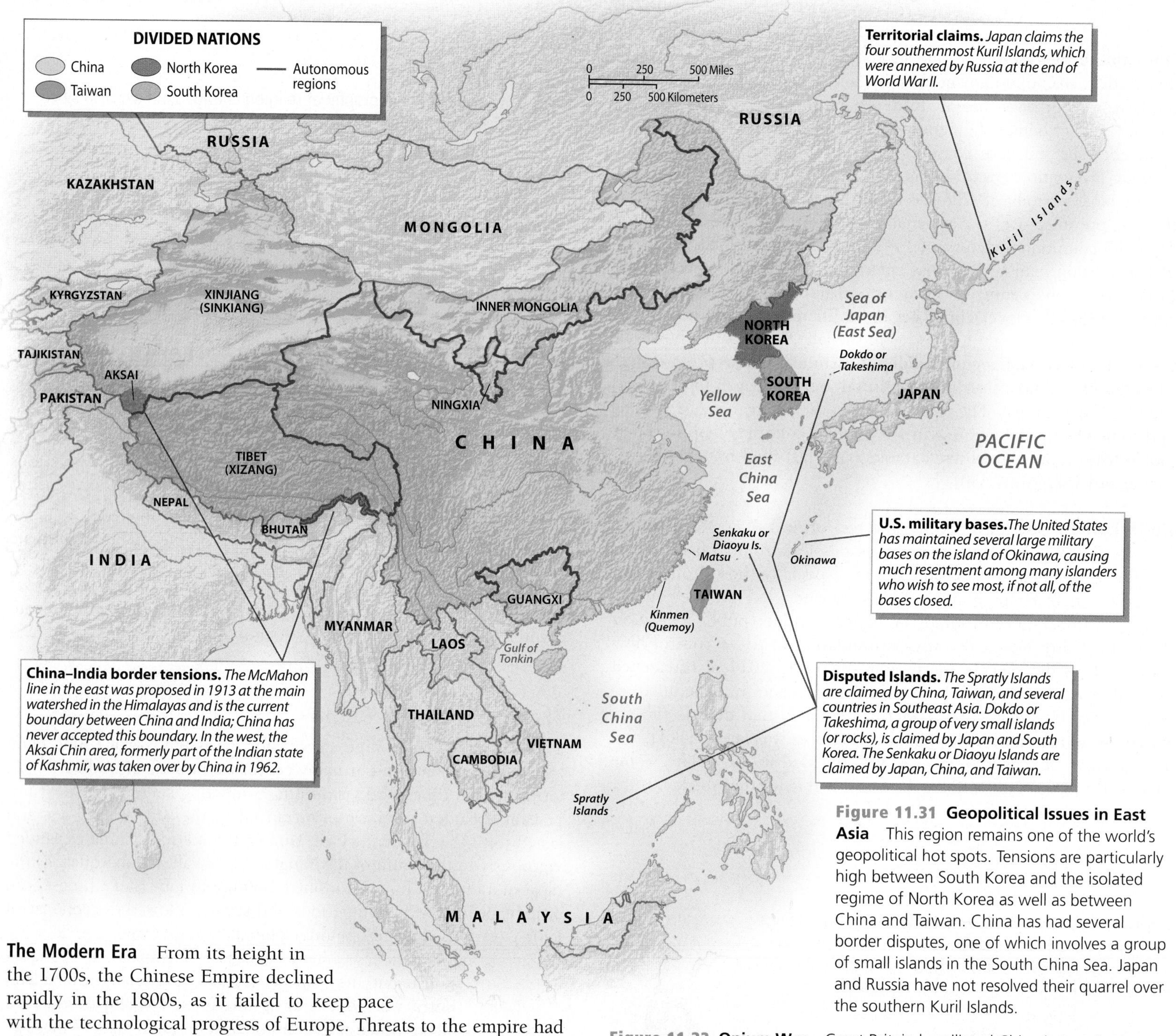

Figure 11.31 Geopolitical Issues in East Asia This region remains one of the world's geopolitical hot spots. Tensions are particularly high between South Korea and the isolated regime of North Korea as well as between China and Taiwan. China has had several border disputes, one of which involves a group of small islands in the South China Sea. Japan and Russia have not resolved their quarrel over the southern Kuril Islands.

The Modern Era From its height in the 1700s, the Chinese Empire declined rapidly in the 1800s, as it failed to keep pace with the technological progress of Europe. Threats to the empire had always come from the north, and imperial officials saw little danger from European merchants operating along their coast. But the Europeans were distressed by the amount of silver needed to obtain Chinese silk, tea, and other products. In response, the British began to sell opium, which Chinese authorities rightfully viewed as a threat. When the imperial government tried to suppress the opium trade in the 1840s, Britain attacked and quickly prevailed (Figure 11.32).

This first "opium war" sparked a century of political and economic chaos in China. The British demanded and received trade privileges in selected Chinese ports. As European businesses penetrated China and weakened local economic interests, anti-Manchu rebellions broke out. At first, all such uprisings were crushed—but not before causing tremendous destruction. Meanwhile, European power continued to advance. In 1858, Russia annexed the northernmost reaches of Manchuria, and by 1900 China had been divided into separate **spheres of influence** (Figure 11.33) in which each colonial power had no formal

Figure 11.32 Opium War Great Britain humiliated China in two "opium wars" in the mid 1800s, forcing the much larger country to open its economy to foreign trade and to grant Europeans extraordinary privileges. This image shows the East India Company steamer *Nemesis* destroying Chinese war junks in January 1841.

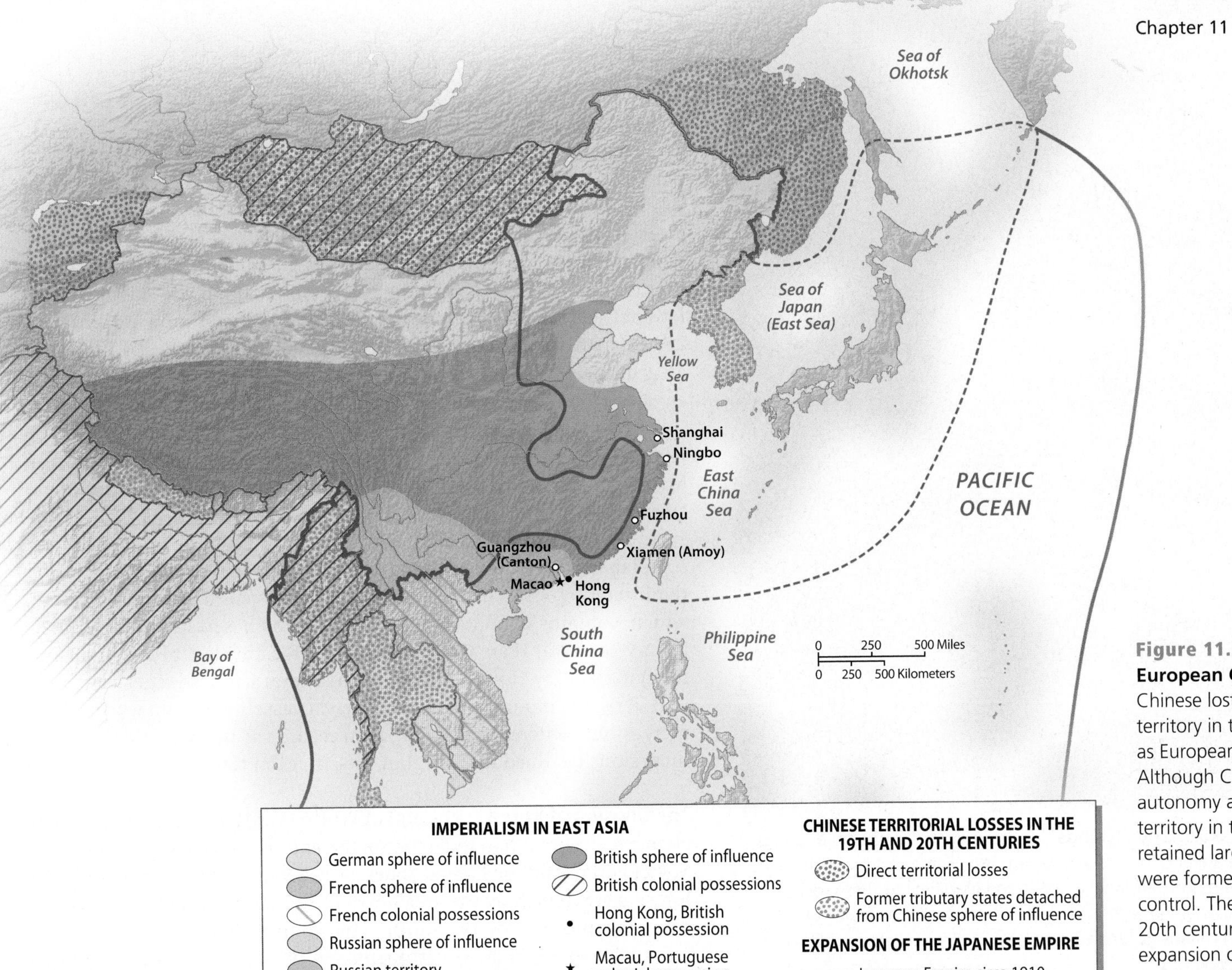

Figure 11.33 19th-Century European Colonialism The Chinese lost influence and territory in the 19th century as European power expanded. Although China regained its autonomy and most of its territory in the 1900s, Russia retained large areas that were formerly under Chinese control. The first half of the 20th century saw the rapid expansion of the Japanese Empire, which ended with the defeat of Japan in World War II.

political authority, but much informal influence and tremendous economic clout.

A successful rebellion in 1911 finally toppled the Manchus and destroyed the empire, but subsequent efforts to establish a unified Chinese Republic were not successful. In many parts of the country, local military leaders ("warlords") grabbed power for themselves. By the 1920s, it appeared that China might be completely torn apart. The Tibetans had gained autonomy; Xinjiang was under Russian influence; and in China proper, Europeans and local warlords vied with the weak Chinese Republic for power. Japan was also increasing its demands and seeking to expand its territory.

The Rise of Japan

Japan did not emerge as a state until the 7th century, some 2000 years later than China. From its earliest days, Japan looked to China for intellectual and political models. Its offshore location, however, insulated Japan from Chinese rule. Between 1000 and 1580, Japan had no real unity, being divided into several small, warring states.

The Closing and Opening of Japan By the early 1600s, Japan had been reunited by the armies of the Tokugawa **shogunate** (a **shogun** was the top military leader who was the true power behind the throne). At this time, Japan isolated itself from the rest of the world. Until the 1850s, Japan traded with China mostly through the Ryukyu Islanders and with Russia through Ainu go-betweens. The only Westerners allowed to trade in Japan were the Dutch, and their activities were strictly limited.

Japan remained largely closed to foreign commerce and influence until U.S. gunboats sailed into Tokyo Bay in 1853 to demand trade access. Aware that China was losing power, Japanese leaders set about modernizing their economic, administrative, and military systems. This effort accelerated when the Tokugawa shogunate was toppled in 1868 by the Meiji Restoration (so called because it restored the emperor to the throne, but did not give the emperor any real power). Unlike China, Japan successfully strengthened its government and economy.

The Japanese Empire Japan's new rulers realized that their country remained threatened by European imperial powers. They decided that the only way to meet the challenge was to expand their own territory. Japan soon took over Hokkaido and began to move farther north. In 1895, the Japanese government tested its newly modernized army against China, winning a quick victory that gave it control of Taiwan. Tensions then mounted with Russia as the two countries competed for power in Manchuria and Korea. The Japanese defeated the Russians in 1905, giving Japan considerable influence in northern China. With no strong rival in the area, Japan annexed Korea in 1910.

Figure 11.34 Senkaku/Diaoyu Islands These rugged, uninhabited islands in the East China Sea are administered by Japan, but claimed by both China and Taiwan. As China has increasingly emphasized its claims to these rocky islands, tensions with Japan have mounted. **Q: Why have these small and relatively insignificant islands generated such potentially dangerous conflict over the past several years?**

The 1930s brought a global depression, greatly reducing world trade and putting a resource-dependent Japan in a difficult situation. The country's leaders sought a military solution, and in 1931 Japan conquered Manchuria. In 1937, Japanese armies occupied the North China Plain and the coastal cities of southern China, angering the United States. When the United States cut off the export of scrap iron, Japan began to experience a resource crunch. In 1941, Japan's leaders decided to destroy the American Pacific fleet in order to clear the way for the conquest of resource-rich Southeast Asia. Their grand strategy was to unite East and Southeast Asia into a "Greater East Asia Co-Prosperity Sphere" ruled by Japan.

Postwar Geopolitics

With Japan's defeat at the end of World War II, East Asia became dominated by the rivalry between the United States and the Soviet Union. American interests initially prevailed in Japan, South Korea, and Taiwan, while Soviet interests advanced on the mainland.

Japan's Revival Japan lost its colonial empire in 1945, its territory reduced to the four main islands plus the Ryukyu Archipelago. In general, the Japanese government agreed to this loss of land, but a remaining territorial conflict concerns the four southernmost islands of the Kuril chain, which were taken by the Soviet Union in 1945. Japan still claims these islands, but Russia maintains that they are part of its territory and refuses to negotiate.

Japan's military power was limited by the constitution imposed on it by the United States, forcing Japan to rely in part on the U.S. military for its defense needs. Many Japanese citizens, however, believe that their country ought to provide its own defense. Slowly but steadily, Japan's military has emerged as a strong regional force, despite the limits imposed on it. North Korean nuclear bomb making and missile testing have raised security concerns in Japan, as has China's growing power. In 2014, Japan accelerated its military buildup, and in early 2015 it signed an arms deal with France in which the two countries agreed to cooperate on drones, underwater surveillance, and other military technologies.

Tensions between China and Japan have grown over the Senkaku Islands to the northeast of Taiwan (called the Diaoyu Islands in Mandarin Chinese) (Figure 11.34). Although Japan controls these small, uninhabited islands, China also claims them, as well as the surrounding waters—which may contain substantial oil resources. Anti-Japanese feelings in China, and in Korea as well, are occasionally reinforced by visits of Japanese prime ministers to Yasukuni Shrine, which contains a military cemetery in which several war criminals from World War II are buried. In 2012, large anti-Japanese demonstrations marked by rioting broke out in several Chinese cities. The next year, tensions mounted as Chinese naval vessels and military airplanes increasingly entered areas that Japan regards as its own territory. Such incursions declined in 2014, but the situation remains tense.

The Division of Korea The end of World War II brought much greater changes to Korea than to Japan. As the end of the war approached, the Soviet Union and the United States agreed to divide the country; Soviet forces were to occupy the area north of the 38th parallel, whereas U.S. troops would occupy the south. This soon resulted in the establishment of two separate governments. In 1950, North Korea invaded South Korea, seeking to reunify the country. The United States, backed by the United Nations, supported the south, while China aided the north. The war ended in a stalemate, and Korea remains divided, its two governments still technically at war.

Large numbers of U.S. troops remained in South Korea after the war, as it was a poor country that could not defend itself. Subsequently, however, the south emerged as a wealthy trading nation, while the fortunes of the north declined. By the late 1990s, South Korea's government came to favor a softer approach to North Korea. As a result, South Korean firms invested substantial funds in joint economic endeavors in North Korea, and South Korean tourists were allowed to cross the border for closely monitored trips to famous locations. Despite these peaceful moves by South Korea, North Korea remained hostile, going so far as to detonate small nuclear bombs in 2006 and 2009.

The international community has made many efforts to persuade North Korea to abandon its quest for nuclear weapons. The 2011 death of North Korea's leader Kim Jong-il, and the succession of his untested 28-year-old son, Kim Jong-un, led to heightened uncertainty (Figure 11.35). International talks on North Korea's nuclear program collapsed in 2012, and tensions escalated further in 2013 when the UN condemned North Korean missile launches. This action prompted North Korea to threaten South Korea, Japan, and the United States with a "nuclear catastrophe." The situation eased slightly later that year after North Korea agreed to reestablish its diplomatic hotline with South Korea that allows the leaders of the two countries to communicate directly.

Figure 11.35 KIM JONG-UN The North Korean political system relies heavily on a "personality cult" constructed around the country's leader. In this propaganda photograph, adoring women are surrounding supreme leader Kim Jong-un as he guides the multiple-rocket launching drill of an all-women military contingent.

The Division of China World War II brought tremendous destruction and loss of life to China. Before the war began, China had already experienced armed conflicts between nationalist and communist forces. After Japan invaded China proper in 1937, the two camps cooperated, but as soon as Japan was defeated, China resumed its civil war. In 1949, the communists proved victorious, forcing the nationalists to retreat to Taiwan. The mainland was then renamed the People's Republic of China, while the nationalist government on Taiwan retained the name the Republic of China.

A dormant state of war between China and Taiwan persisted for decades after 1949. The Beijing government still claims Taiwan as an integral part of China and vows that it will eventually reclaim it. Nationalists in Taiwan long insisted that they represented the true government of China and that Taiwan was merely one province of a temporarily divided country. By the end of the 20th century, however, almost all Taiwanese had given up on the idea of taking over China itself, and many began to press openly for Taiwan's formal independence.

The ideal of Chinese unity continues to be influential both in China and abroad. In the 1950s and 1960s, the United States recognized Taiwan as the only legitimate government of China, but its policy changed after U.S. leaders decided that it would be more useful to recognize mainland China. Soon, China entered the United Nations, and Taiwan found itself diplomatically isolated. Taiwan continues to be recognized, however, as the legitimate government of China by several small countries in Africa, the Americas, and the Pacific, most of which receive Taiwanese economic aid in return.

The geopolitical status of Taiwan continues to be a controversial issue in Taiwan itself. When a supporter of political separation was elected president of Taiwan in 2000, China threatened to invade if the island were to declare formal independence. Military tensions remained high for several years, but during the same period economic connections continued to strengthen. Taiwanese voters grew dissatisfied with an independence movement that had generated geopolitical tension without delivering substantial benefits. In Taiwan's presidential elections of 2008 and 2012, the old nationalist party won clear victories, in part by promising to maintain good relations with the mainland. The growing power of mainland China, however, has more recently helped to revitalize the independence movement.

The Chinese Territorial Domain Despite the fact that it has been unable to regain Taiwan, China has successfully retained most of the territories that the Manchus controlled. In the case of Tibet, this has required considerable force. Resistance by the Tibetans compelled China to launch a full-scale invasion in 1959. The Tibetans, however, continue to struggle for real autonomy, if not actual independence, as they fear that the Han Chinese now moving to Tibet will eventually outnumber them and undermine their culture.

The postwar Chinese government also retained control over Xinjiang in the northwest as well as Inner Mongolia (Nei Mongol), a vast territory stretching along the Mongolian border. Like Tibet, Nei Mongol and Xinjiang are classified as autonomous regions. The native peoples of Xinjiang are asserting their religious and ethnic identities, calling the region Eastern Turkistan to emphasize its Turkish heritage, and separatist attitudes are common. Most Han Chinese, however, see Nei Mongol and Xinjiang as integral parts of their country and regard any talk of independence as treasonous.

One territorial issue was finally resolved in 1997, when China reclaimed Hong Kong from Britain. In the isolationist 1950s, 1960s, and 1970s, Hong Kong acted as China's window on the outside world, and it grew wealthy as a capitalist city. As Chinese relations with the outer world opened in the 1980s, Britain decided to honor its treaty provisions and return Hong Kong to China. China, in turn, promised that Hong Kong would become a **special administrative region**, retaining its fully capitalist economic system for at least 50 years under the "one country, two systems" model. Many civil liberties not enjoyed in China itself remain protected in Hong Kong.

Despite Hong Kong's substantial autonomy, the central government continues to intervene in its affairs, angering many of its people. Interference by Beijing in Hong Kong's local elections in 2014 generated a massive student-led protest movement. At its height, more than 100,000 people occupied the streets of Hong Kong to demand democracy, virtually shutting down the city (Figure 11.36). China's leaders denounced the movement, claiming that it was instigated by Europe and the United States. Eventually, the protests died down without any meaningful concessions by the Chinese government. Critics contend that civil liberties and academic freedom in Hong Kong have suffered as a result of the protest movement.

In 1999, Macau, the last colonial territory in East Asia, was returned to China, becoming the country's second special administrative region. This small former Portuguese enclave, located across the estuary from Hong Kong, has functioned largely as a gambling refuge. In 2008, gambling revenues in Macau surpassed those of Las Vegas, making it the world capital of commercial wagering (Figure 11.37). In 2014, however, Macau's economy experienced a sharp decline, essentially halting the construction of new casinos. Most experts attribute this reversal to China's ongoing anticorruption campaign, which is linked to increasing opposition to gambling among the country's leaders.

The Global Dimension of East Asian Geopolitics

In the early 1950s, East Asia was divided into two hostile Cold War camps: China and North Korea were allied with the Soviet Union, while Japan, Taiwan, and South Korea were linked to the United States.

Figure 11.36 2014 Hong Kong Protests Massive pro-democracy protests led by students broke out in Hong Kong in the autumn of 2014. Protestors block a major road during a rally outside government headquarters on October 10, 2014.

The China–Soviet Union alliance soon deteriorated into mutual hostility, however, and in the 1970s China and the United States found that they could work with each other, sharing as they did a common enemy in the Soviet Union.

The end of the Cold War, coupled with China's rapid economic growth, again altered the balance of power in East Asia. The United States no longer needed China to offset the Soviet Union, and the U.S. military has become increasingly worried about the growing power of the rapidly modernizing Chinese army. From 2000 through 2015, China's military budget grew at an average annual rate of about 10 percent. Although China's military spending is much less than that of the United States, it is much greater than that of any other country. Several of China's neighbors have become concerned about its growing strength. As a result, South Korea and especially Japan have been eager to maintain close military ties with the United States. Currently, some 35,000 U.S. troops are stationed in Japan, with another 28,000 in South Korea.

China is also working to develop its "soft," or nonmilitary, power on the global stage. In 2013, for example, it announced the development of the Asian Infrastructure Investment Bank to serve as a counterweight to the Western-dominated World Bank and International Monetary Fund. The United States opposes the expansion of this Chinese-led bank, and in 2015 it expressed disappointment when the United Kingdom, Italy, Germany, and France announced that they would participate in its activities.

Figure 11.37 Macau Casino A special administrative region of China with its own legal system, Macau is now the world's foremost center of casino gambling. Prominent buildings visible here include the Bank of China, the Grand Lisboa Hotel-Casino, and Wynn Hotel-Casino.

China is thus coming of age as a major force in global politics. Whether it is a force to be feared by other countries is a matter of considerable debate. Chinese leaders insist that they have no intention of interfering in the internal affairs of other countries. They do, however, view concerns expressed by the United States and other countries about their human rights record, as well as about their activities in Tibet, as excessive meddling in their internal affairs. Foreign critics worry that China is growing much more assertive under the leadership of President Xi Jinping, who has been able to acquire much more power than his immediate predecessors. Chinese claims over islands in the South China Sea are a particular international concern (see Chapter 13).

Review

11.7 How did the decline of China during the 1800s affect the geopolitical structure of East Asia?

11.8 How has the geopolitical environment in East Asia changed since the end of the Cold War?

KEY TERMS sphere of influence, shogunate, shogun, special administrative region

Economic and Social Development: A Core Region of the Global Economy

East Asia contains vast disparities in economic and social well-being (Table A11.2). Japan's urban belt has one of the world's greatest concentrations of wealth, whereas many of China's interior districts remain relatively poor. Overall, East Asia has experienced rapid economic growth since the 1970s, but this growth has not been evenly distributed. North Korea, for example, saw its living standards decline sharply between 1990 and 2010.

Japan's Economy and Society

Japan was the pacesetter of the world economy in the 1960s, 1970s, and 1980s. In the early 1990s, however, the Japanese economy experienced a major setback, and growth has remained slow ever since. Despite its recent problems, Japan is still the world's third-largest economic power.

Boom and Bust Although Japan's heavy industrialization began in the late 1800s, most of its people remained poor. The 1950s, however, saw the beginnings of the Japanese "economic miracle." With its empire gone, Japan was forced to export manufactured products. Beginning with inexpensive consumer goods, Japanese industry moved to more sophisticated products, including automobiles, cameras, electronics, machine tools, and computer equipment. By the 1980s, Japan was the leader in many segments of the global high-tech economy.

The early 1990s saw the collapse of Japan's inflated real estate market, leading to a banking crisis. At the same time, many Japanese companies relocated factories to Southeast Asia and China. As a result, Japan's economy stagnated for years. The Japanese government has repeatedly tried to revitalize the economy through massive state spending, leading to huge deficits.

Despite its economic problems, Japan remains a core country of the global economic system. Its economic influence spans the globe, as

Figure 11.38 Japanese Robots Japan is a world leader in robotics. Japanese researchers are now working hard to create useful humanoid robots. Here we see shoppers taking photographs of a humanoid robot named "Aiko Chihira" at the information desk of a Mitsukoshi department store in Tokyo, Japan. This particular robot was installed as a receptionist at the upscale Mitsukoshi department store to give directions and information to shoppers.

Japanese multinational firms invest heavily in production facilities in North America and Europe as well as in developing countries. Japan remains a world leader in a range of high-tech fields, including robotics, optics, and machine tools for the semiconductor industry (Figure 11.38).

The global economic crisis of 2008–2009 caused much damage in Japan, as export markets collapsed. Just as its economy was beginning to recover, Japan was hit by the massive Tohoku earthquake and tsunami of 2011, resulting in an economic decline of roughly 1 percent. Rebuilding in the devastated area of northeastern Japan, moreover, has been slower than anticipated, due in part to bureaucratic obstacles. Several recent government plans did little to boost the economy, but the major drop in oil prices in 2014 and 2015 had more positive effects.

Living Standards and Social Conditions in Japan Despite high levels of economic development, Japanese living standards remain somewhat lower than those of the United States. Housing, food, transportation, and services are particularly expensive in Japan. But the Japanese also enjoy many benefits unknown in the United States. Unemployment remains low, health care is provided by the government, and crime is rare. By such social measures as literacy, infant mortality, and average life expectancy, Japan surpasses the United States by a comfortable margin. Critics contend, however, that Japan suffers from a significant degree of hidden poverty, which largely remains outside of public view.

Women in Japanese Society Critics often point out that Japanese women have not shared the benefits of their country's success. Advanced career opportunities remain limited, especially for women who marry and have children. Mothers are expected to devote themselves to their families and to their children's education. Japanese businessmen often work or socialize with coworkers until late every evening and thus contribute little to child care. An estimated 70 percent of Japanese women drop out of the workforce for at least a decade after having their first child, in contrast to only 30 percent of U.S. women.

One response to the limitations faced by Japanese women has been declining marriage rates. Japan has seen an even more dramatic drop in its fertility rate. Whether this is due to the domestic difficulties faced by Japanese women or merely a result of the pressures of a postindustrial society is an open question. Fertility rates have, after all, dropped even lower in parts of Europe. But regardless of the cause, a shrinking population means an aging population, and increasing numbers of Japanese retirees will have to be supported by smaller numbers of workers.

The Thriving Economies of South Korea, Taiwan, and Hong Kong

The Japanese path to development was successfully followed by its former colonies, South Korea and Taiwan. Hong Kong also emerged in the 1960s and 1970s as a newly industrialized economy, although its economic and political systems are different.

South Korea's Rise The postwar rise of South Korea was even more remarkable than that of Japan (see *Everyday Globalization: East Asia's Domination of Shipbuilding*). During Japanese occupation, Korean industrial development was concentrated in the north, which is rich in natural resources. The south, in contrast, remained a densely populated, poor, agrarian region.

In the 1960s, the South Korean government began a program of export-led economic growth. It guided the economy with a heavy hand and denied basic political freedom to the Korean people. By the 1970s, such policies had proved highly successful in the economic realm. Huge Korean industrial conglomerates, known as ***chaebol***, moved from exporting inexpensive consumer goods to heavy industrial products and then to high-tech equipment. As the South Korean middle class expanded, pressure for political reform increased, and in the 1980s South Korea made the transition from an authoritarian to a democratic country.

Through the 1980s, South Korean firms remained dependent on the United States and Japan for basic technology. By the 1990s, however, that was no longer the case. South Korea invested heavily in education, which has served it well in the global high-tech economy. Increasingly, South Korean companies are themselves becoming multinational, building new factories in low-wage countries elsewhere in Asia and Latin America as well as in the United States and Europe. South Korea has also signed free-trade agreements with a number of countries and seeks to create a major financial center in Seoul. Such ambitions were bolstered in 2015 when Samsung Electronics announced that it would begin to trade Chinese currency directly with South Korean currency, rather than using the U.S. dollar as an intermediary.

South Korea's political and social development has not been as smooth as its economic progress. Issues of economic globalization often provoke serious political tensions. For example, a 2011 free-trade agreement between South Korea and the United States resulted in heated protests in Seoul (Figure 11.39). Another concern is the power of the *chaebols*, which many South Koreans regard as excessive. Anti-*chaebol* sentiment surged in 2014 when the daughter of the chairman of Korean Air caused an international flight delay over the way that a bag of nuts had been served.

Like other trade-dependent countries, South Korea experienced a major recession in 2008–2009. Economic recovery in South Korea, however, occurred more rapidly than in most other industrialized countries. From 2011 through 2014, its economy expanded at a relatively healthy rate of around 3 percent a year. Still, South Korean economists and business leaders regard such a pace of expansion as sluggish, and some worry that their country could follow Japan into a low-growth economic trap. South Korea faces some of the same social challenges that suppress economic

EVERYDAY GLOBALIZATION

East Asia's Domination of Shipbuilding

Economic globalization depends heavily on maritime shipping, as the vast majority of exported and imported goods—whether oil or iPhones—travel by sea. Today almost all large ships, with the exception of those used by the military, are made in East Asia. China has the lead with 45 percent of the global market, while South Korea is second at 29 percent and Japan follows with 18 percent.

Sixty years ago, shipbuilding was concentrated in North America and Europe, with Scotland and Northern Ireland occupying prime positions. Japan then took the lead, outcompeting its Western rivals, but by the 1980s South Korea gained the top position. Although China is now the largest producer, South Korea still makes most of the world's truly massive ships, such as cruise liners, supertankers, and large cargo carriers. The world's biggest shipyard is located in Ulsan, South Korea, which is also the headquarters of the world's largest shipbuilding company, Hyundai Heavy Industries (Figure 11.4.1).

1. What are some of the factors that helped South Korea develop such a large and profitable shipbuilding industry?
2. What percentage of the items you are now wearing or carrying were shipped to the United States from an overseas location?

Figure 11.4.1 Hyundai Heavy Industries Ulsan Shipyard Headquartered in Ulsan, South Korea, Hyundai builds a range of container vessels and tankers. Other South Korean shipbuilders include Samsung Heavy Industries and Daewoo Shipbuilding & Marine Engineering.

growth in Japan, including an aging population and restricted opportunities for professional women.

Taiwan and Hong Kong Taiwan and Hong Kong have also experienced rapid economic growth since the 1960s. Taiwan's government, like those of South Korea and Japan, guided the country's economic development, especially during the early phase of industrial expansion.

Hong Kong, unlike its neighbors, has been characterized as one of the most **laissez-faire** economic systems in the world (laissez-faire refers to market freedom with little governmental control). Government involvement has been minimal, which is one reason why the city's business elite were nervous about the transition to Chinese rule. Hong Kong traditionally functioned as a trading center, but in the 1960s and 1970s it became a major producer of textiles, toys, and other consumer goods. By the 1980s, however, such cheap products could no longer be made in such an expensive city. Hong Kong industrialists subsequently began to move their plants to southern China, while Hong Kong itself increasingly specialized in business services, banking, telecommunications, and entertainment.

Both Taiwan and Hong Kong have close overseas economic connections. Linkages are particularly tight with Chinese-owned firms located in Southeast Asia and North America. Taiwan's high-tech businesses are intertwined with those of Silicon Valley in the United States. Hong Kong's economy is also closely bound with that of the United States (as well as those of Canada and Britain), but its closest connections are with the rest of China.

Such high levels of globalization resulted in major recessions in both economies during the global economic crisis of 2008–2009. However, as was the case in South Korea, the economy of Taiwan bounced back quite rapidly once the crisis passed. Growth was bolstered by a 2010 agreement with China that allowed Taiwanese financial firms more freedom to operate on the mainland, just as it allowed mainland firms to invest more freely in Taiwan. As with Japan and South Korea, low birth rates and aging populations are major concerns in both Hong Kong and Taiwan.

Chinese Development

China dwarfs all of the rest of East Asia in both physical size and population. Its economic takeoff is thus reshaping the economy of the entire world. Despite its recent growth, however, China's economy has several weaknesses. For example, much of the vast interior remains poor, many of its heavy industries are not globally competitive, and a number of massive housing projects are deeply in debt and largely unoccupied. The future of China's economy is one of the biggest uncertainties facing both East Asia and the world as a whole.

China Under Communism More than a century of war, invasion, and near chaos in China ended in 1949, when the communist forces led

Figure 11.39 Protests in South Korea Massive political protests are common in South Korea. In October 2011, heated protests accompanied the ratification of the United States–South Korea Free Trade Agreement.

by Mao Zedong seized power. The new government, inheriting a weak economy, set about nationalizing private firms and building heavy industries. These plans were most successful in Manchuria, where a large amount of heavy industrial equipment had been left by the Japanese.

In the late 1950s and 1960s, however, China experienced two economic disasters. The first, ironically called the "Great Leap Forward," hinged on the idea that small-scale village workshops could produce the large quantities of iron needed for sustained industrial growth. Communist Party officials forced these inefficient workshops to meet unreasonably high production quotas. The result was a horrific famine that may have killed 20 million people. The early 1960s saw a return to more practical policies, but toward the end of that decade a new wave of radicalism swept through China. This "Cultural Revolution" aimed at mobilizing young people to rid the country of "undesirable" traditional social values and replace them with communist ideology. Thousands of experienced industrial managers and college professors were expelled from their positions. Many were sent to villages to be "reeducated" through hard physical labor; others were simply killed. The economic consequences were devastating.

Toward a Postcommunist Economy When Mao Zedong, who had been revered as an almost superhuman being, died in 1976, China faced a crucial turning point. Its economy was nearly stagnant and its people desperately poor. However, the economy of Taiwan, its rival, was booming. This led to a political struggle between pragmatists hoping for change and dedicated communists. The pragmatists emerged victorious, and by the late 1970s it was clear that China would embark on a different economic path. The new China sought closer connections with the world economy and took a modified capitalist road to development.

China did not, however, transform itself into a fully capitalist country. The state continued to run most heavy industries, and the Communist Party held on to political power. Instead of suddenly abandoning the communist model, as the former Soviet Union did, China allowed cracks to appear in which capitalist businesses could take root and thrive.

Industrial Reform An important early industrial reform involved opening **Special Economic Zones (SEZs)**, in which foreign investment was welcome and state interference was minimal. The Shenzhen SEZ, adjacent to Hong Kong, proved particularly successful (Figure 11.40). Additional SEZs were soon opened, mostly in the coastal region. The basic strategy was to attract foreign investment that could generate exports, the income from which could supply China with capital to build its infrastructure (roads, electrical and water systems, telephone exchanges, and the like). China has more recently used the SEZ model to bring economic development to the interior, declaring the entire city of Kashgar in western Xinjiang an SEZ in 2010.

China enacted other capitalistic reforms in the 1980s and 1990s. Former agricultural cooperatives were allowed to produce for the market, and many proved highly successful. From the early 1990s until around 2010, the Chinese economy grew at roughly 10 percent a year, perhaps the fastest rate of expansion the world has ever seen. Seeking to strengthen its connections with the global economic system, in 2001 China joined the World Trade Organization, a body designed to facilitate free trade and provide ground rules for international economic exchange.

Despite China's reliance on trade, it weathered the global economic crisis of 2008–2009 relatively well. But its pace of economic growth eased off considerably after the crisis, averaging only a little more than 7 percent a year from 2011 to 2015. China's leaders have responded by trying to increase domestic consumption, and thus decrease reliance on exports, and by engaging in a massive drive against corruption. Another strategy is the creation of Free Trade Zones, which are similar to the SEZs, but supposedly characterized by more extensive deregulation and global access. The first such Free Trade Zone was created in Shanghai in 2013, and three more followed in 2015 in Tianjin, Guangdong, and Fujian. Some evidence suggests, however, that these Free Trade Zones offer few new advantages, and thus far investments in them have been modest.

Figure 11.40 Shenzhen Adjacent to Hong Kong, Shenzhen was one of China's first Special Economic Zones. It has recently emerged as a major city in its own right. **Q: What political, economic, and spatial factors account for Shenzhen's extraordinarily rapid rise as a major city?**

Criticism of the Chinese Economic Model China's economic expansion has created tensions with several other countries, especially the United States. China exports far more to the United States than it imports, creating economic imbalances. Foreign critics accuse China of unfairly keeping labor costs low in order to enhance exports and of enacting numerous barriers to imports. China's large and growing holdings of U.S. Treasury bonds, however, make it difficult for the United States to exert much pressure on China's economy.

Critics also focus on the mistreatment of labor in factories, especially those producing electronic goods for export. Employees in such plants are usually exposed to hazardous chemicals and are often forced to work overtime under harsh conditions without additional pay. In recent years, the global labor-rights movement led major protests against Foxconn and Catcher Technologies, two Taiwan-based firms that manufacture iPads, iPhones, Kindles, and other U.S.-designed devices in Chinese factories. A 2013 undercover investigation by a group called China Labor Watch found numerous safety and health violations, in addition to harsh working conditions, at a factory that contracts with Apple. In 2014, a Hong Kong student group organized a large protest outside of Foxconn's local headquarters to publicize that fact that 14 of its employees had recently committed suicide, supposedly because of the brutal labor conditions within its factories.

Analysts also point out that political freedom has not accompanied China's more open, market-driven economy. Opponents of the state still face imprisonment, and freedom of the press is highly limited. Reporters Without Borders, a global organization that advocates freedom of the press, rates China as one of the world's 12 "enemies of the Internet." In 2014, the same organization ranked China as 175th out of 180 countries in regard to freedom of the press in general. Intellectual freedom in China's universities, moreover, was curtailed in

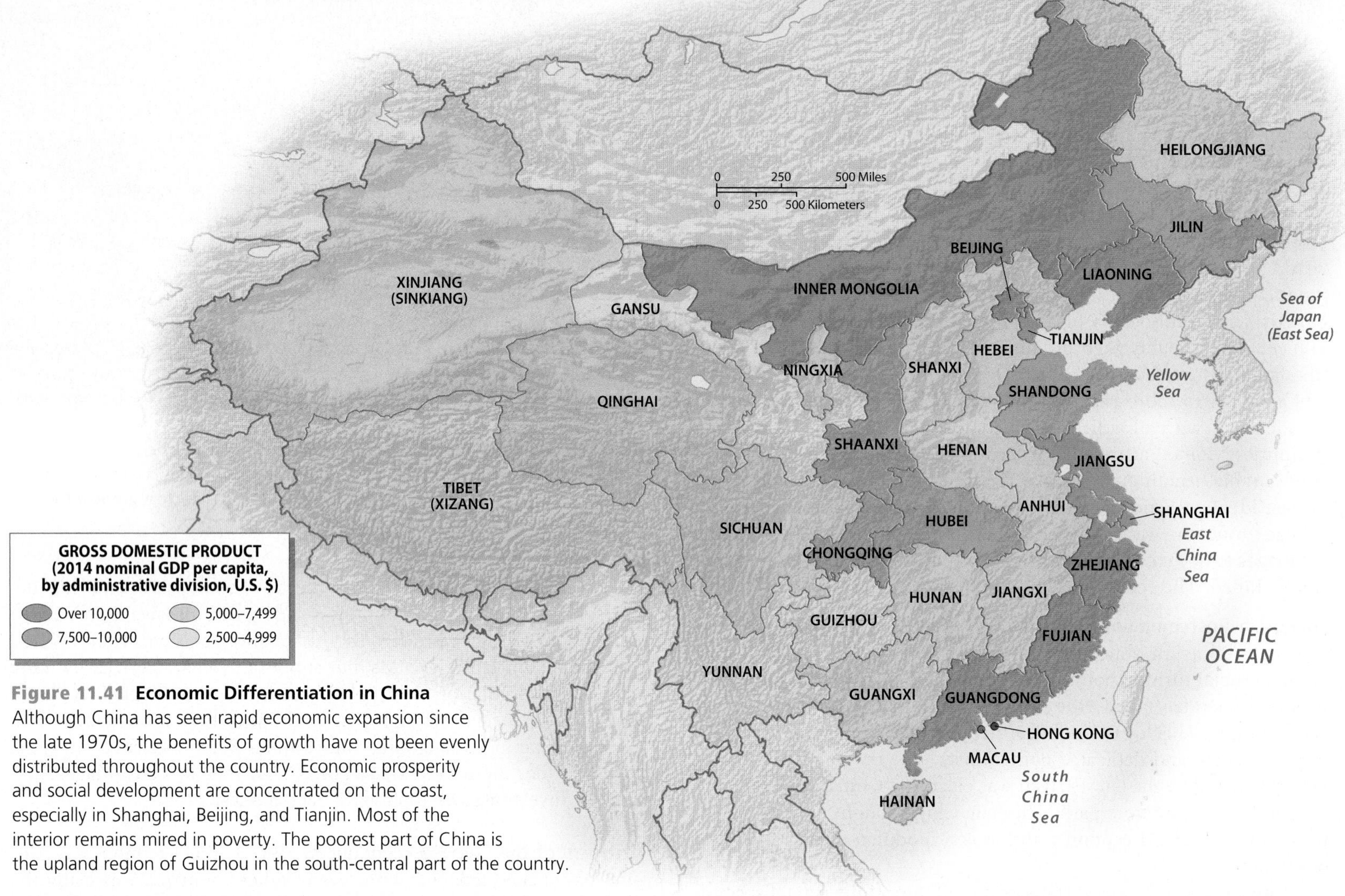

Figure 11.41 Economic Differentiation in China Although China has seen rapid economic expansion since the late 1970s, the benefits of growth have not been evenly distributed throughout the country. Economic prosperity and social development are concentrated on the coast, especially in Shanghai, Beijing, and Tianjin. Most of the interior remains mired in poverty. The poorest part of China is the upland region of Guizhou in the south-central part of the country.

2014 as the government began to reemphasize Marxism, despite the capitalistic nature of the country's economy.

Social and Regional Differentiation The Chinese economic surge brought about by the reforms of the late 1970s and 1980s resulted in growing **social and regional differentiation**. In other words, certain groups of people—and certain portions of the country—prospered, while others experienced much less development (Figure 11.41). Despite its socialist government, the Chinese state encouraged the formation of an economic elite, having concluded that only wealthy individuals can adequately transform the economy. The least fortunate Chinese citizens were sometimes left without work, and many millions migrated from rural villages to seek employment in the booming coastal cities.

Though still officially a communist country, China now has a much more unequal distribution of wealth than its officially capitalist neighbors—Japan, South Korea, and Taiwan. According to some statistical measurements, wealth is now more evenly distributed in the United States than in China.

China's Booming Coastal Region Most of the benefits from China's economic transformation have flowed to the coastal region and to the capital city of Beijing. The southern provinces of Guangdong and Fujian were the first to benefit, profiting from their close connections with the overseas Chinese communities of Southeast Asia and North America. (The vast majority of overseas Chinese emigrants came from Guangdong and Fujian.) Their proximity to Taiwan and especially Hong Kong also proved helpful.

By the 1990s, the Yangtze Delta, centered on Shanghai, reemerged as China's economic core. The Chinese government has encouraged the development of huge industrial, commercial, and residential complexes, hoping to take advantage of the region's vitality. The Suzhou Industrial Park is now a hypermodern city of more than a half million people, thanks largely to a $20 billion investment, most of it from Singapore. Shanghai's Pudong industrial development zone has attracted $10 billion, much of it for construction of a new airport and subway system. Shanghai now has the world's largest container port.

The Beijing–Tianjin region has also played a major role in China's economic boom, based on its proximity to political power and its position as the gateway to northern China. In addition, the other coastal provinces of northern China have done relatively well.

Interior and Northern China The interior and northern parts of China, in contrast, have seen less economic expansion than the rest of the country. Manchuria remains relatively well-off as a result of fertile soil and early industrialization, but it has not participated much in the recent boom. Many of the state-owned heavy industries of the Manchurian **rust belt**, or zone of decaying factories, are relatively inefficient. In recent years, northeastern China has seen much slower economic growth than the rest of the country.

Most of the interior provinces lagged well behind the coast in the boom period of the 1980s and 1990s. To counter such regional differences, China is building roads and railway lines and undertaking other projects in the interior through its so-called Great Western Development Strategy. From 2008 to 2014, China's interior provinces grew more rapidly than its coastal provinces, reducing both economic disparities and the flow of migrants. By 2014, economic growth rates across the country had largely converged at around 7.5 percent.

Figure 11.42 China's Gender Imbalance China has one of the world's most imbalanced gender ratios, with boys greatly outnumbering girls in most parts of the country. This photograph shows a classroom at a primary school in Danzhou City on Hainan Island. Danzhou has the worst gender imbalance in China, with 170 males born for every 100 females, according to figures from the most recent Chinese national census.

Interior provinces are now competitive in the low-value, low-wage manufacturing sector, but the coastal provinces do much better in higher-value manufacturing.

Social Conditions in China

Despite pockets of persistent poverty, China has made significant progress in social development. Since coming to power in 1949, the communist government made large investments in medical care and education, and today China has impressive health and longevity figures. The literacy rate remains lower than those of Japan, South Korea, and Taiwan, but because almost all children attend elementary school, it will rise substantially in the coming years.

China's Population Quandary Population policy remains an unsettling issue for China. With more than 1.3 billion people highly concentrated in less than half of its territory, China is a densely populated country. In 1978, its government had become so concerned that it instituted the infamous "one-child policy." Under this plan, couples in normal circumstances are expected to have only one offspring and can suffer financial and other penalties if they do not comply. Forced abortions and other human rights abuses have occurred as well.

A combination of economic development and the one-child policy successfully reduced China's fertility, which currently stands well below the replacement level at 1.6. Now the government is concerned about a rapidly aging population and possible future labor shortages. Such concerns led China to relax its one-child policy in 2013, allowing couples to have two children if one of the parents is an only child. In 2015, China announced that the one-child would be abandoned in favor of a two-child policy for the entire country. Some experts think that this change will have only a minor effect on China's fertility rate.

Gender Issues in China An equally pressing demographic issue in China is the gender imbalance. There are now 33 million more Chinese men than women, and in 2014, 116 boys were born for every 100 girls (Figure 11.42). This gap reflects the cultural practice of honoring ancestors; because family lines are traced through male offspring, only a male heir can maintain the family's lineage. One result is illegal gender-selective abortion; if ultrasound reveals a female fetus, the pregnancy is sometimes terminated. Poor couples not uncommonly abandon baby girls, and young boys are occasionally kidnapped and sold to wealthy couples without a son. Increasingly, however, the government is taking action, sentencing to death a doctor who sold seven infants to human traffickers in 2014 and vowing to crack down on the practice of pre-birth gender determination in 2015.

Women historically occupied a low position in Chinese society, as is true in most other civilizations. In the 20th century, both the nationalist and the communist governments sought to equalize the relations between the sexes. As a result, women now have a relatively high rate of participation in the Chinese workforce. But as is still true throughout East Asia, few women achieve positions of power in either business or government. According to the World Economic Forum's 2014 Global Gender Gap Report, China ranks 87th out of 142 countries in regard to the social position of women.

The Failure of Development in North Korea

At the end of the Korean War in 1953, North Korea had a higher level of industrial development than South Korea, as it possessed most of the peninsula's mines and factories. But as South Korea experienced export-led industrialization starting in the 1970s, North Korea remained devoted to a state-led economy that rejected globalization and failed to generate development. North Korea did maintain close economic relations with the Soviet Union, however, and when Soviet power collapsed in 1991, it experienced a severe blow. No longer able to afford sufficient fertilizer, its agriculture system sharply declined, leading to severe famines. A 2013 UN report indicated that roughly 25 percent of North Korean children suffer from chronic food insecurity and hunger. Electricity and other basic services, moreover, remain scarce.

North Korea's government has made limited efforts to enhance economic growth. In 2002, it legalized farmers' markets, but several years later it shut most of them down. A more ambitious move toward private farming and food selling began in 2012, with some success. Following China's lead, North Korea also opened a number of Special Economic Zones in which foreign firms, mostly South Korean and Chinese, produce goods for the world market using inexpensive local labor. In 2013 alone, 14 such zones were established. North Korea is also increasingly exporting its own people to work in foreign countries under official government contracts. Partly as a result of such policies, the country's economy finally began expanding at a rapid rate in 2014.

Whether such growth is sustainable is uncertain, owing to North Korea's erratic policies and hostile relations with most of the rest of the world. New human rights abuses have been uncovered as the country attempts to boost its growth. A 2015 UN report claimed that tens of thousands of North Koreans work under slavelike conditions in labor camps in China, Russia, and the Arabian Peninsula.

Review

11.9 How has the process of economic development been similar in Japan, South Korea, Taiwan, and China since the end of World War II, and how has it been different in each country?

11.10 Why do levels of social and economic development vary so extensively from the coastal region of China to the interior portions of the country?

KEY TERMS chaebol, laissez-faire, Special Economic Zone (SEZ), social and regional differentiation, rust belt

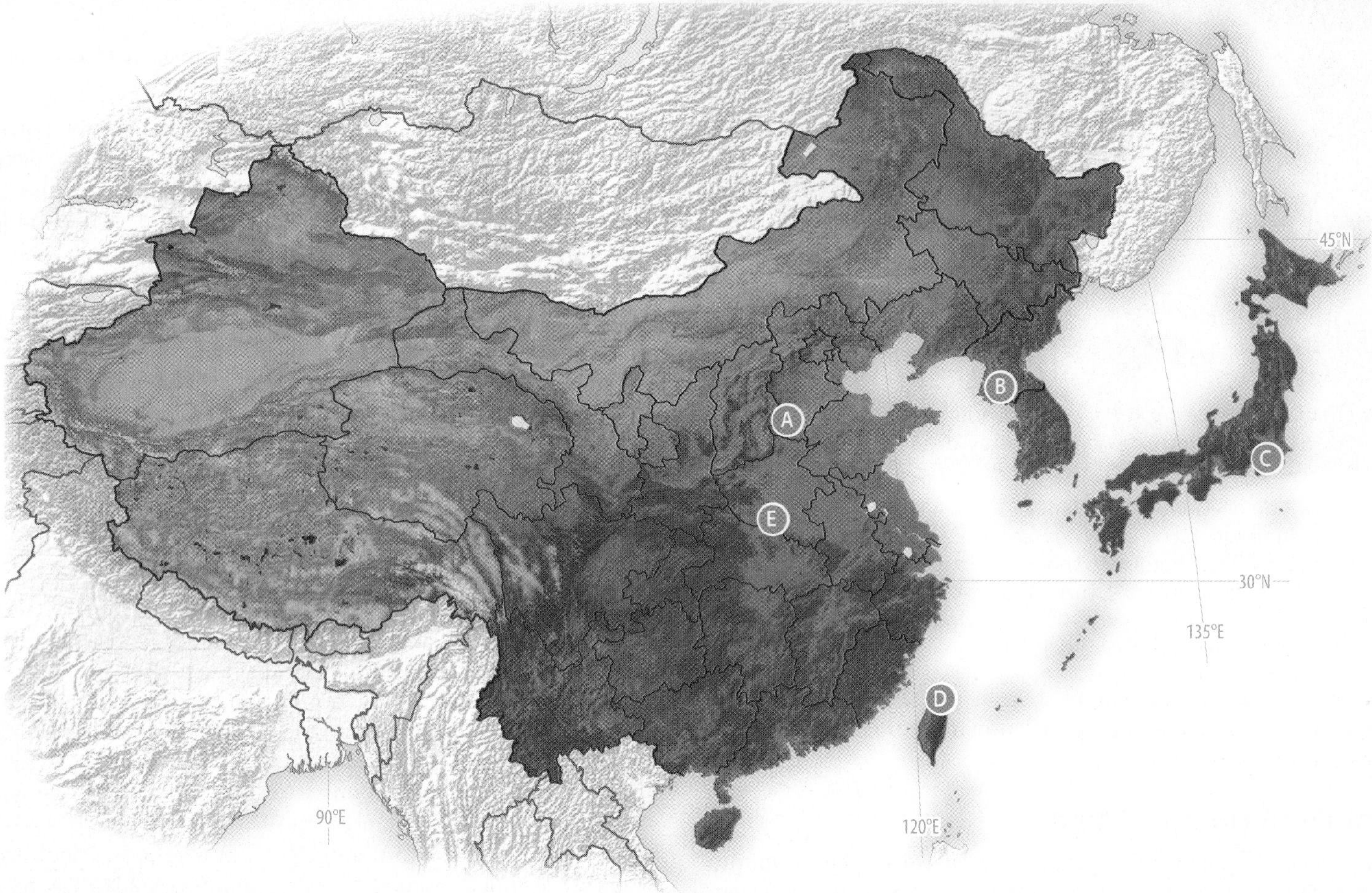

Review

Physical Geography and Environmental Issues

11.1 Identify the key environmental differences between the island portions of East Asia (Japan and Taiwan) and the mainland.

11.2 Describe the main environmental problems China faces today, and compare them with environmental challenges faced by Japan, South Korea, and Taiwan.

The economic success of East Asia has been accompanied by severe environmental degradation. Japan, South Korea, and Taiwan have responded by enacting strict environmental laws and by moving many of their most polluting industries overseas. The major environmental problem in the region today stems from the rapid growth of China's economy. Pollution in Chinese cities is so serious that it has major impacts on human health, and much of the Chinese countryside suffers from environmental problems such as soil erosion and desertification. China is responding, however, with major programs for renewable energy and soil conservation.

1. This map detail shows an area characterized by severe soil erosion, a strong risk of flooding, and a new water transfer project. Are these phenomena related, and if so, how?

2. Describe the potential advantages and disadvantages of diverting water flows from one river basin to another.

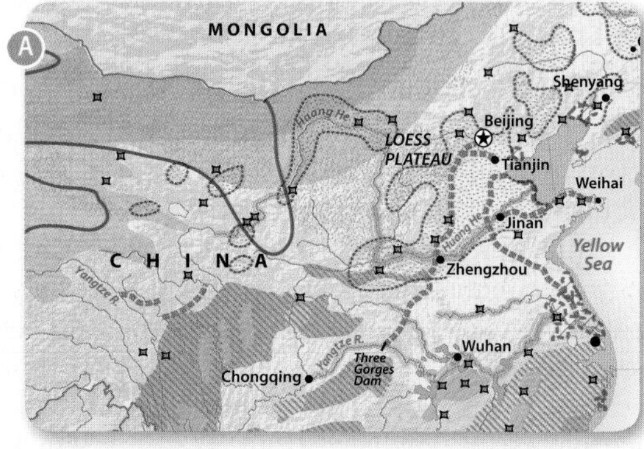

Population and Settlement

11.3 Explain why China's population is so unevenly distributed, with some areas densely settled and others almost uninhabited.

11.4 Outline the distribution of major urban areas on the map of East Asia and explain why the region's largest cities are continuing to grow.

East Asia is a densely populated region, but its birth rates have plummeted in recent decades. Japan is experiencing population decline, which is putting pressure on the Japanese economy. China's biggest demographic challenge results from the massive movement of people from the interior to the coast and from rural villages to rapidly expanding cities. China has been trying to redirect development toward the interior, with mixed results.

3. Why is animal power used so often in North Korean agriculture, and what implications does this have for the North Korean population?

4. What kinds of consequences is a country likely to face in pursuing a policy of food self-sufficiency?

Cultural Coherence and Diversity

11.5 Describe the ways in which religion and other systems of belief both unify and divide East Asia.

11.6 Explain the distinction between, and geographical distribution of, the Han Chinese and the other ethnic groups of China, paying particular attention to language.

East Asia is unified by deep cultural and historical bonds. China is the greatest influence on East Asia because, at one time or another, it ruled most of the region. Although Japan was never under Chinese rule, it has profound historical connections to Chinese civilization. The more prosperous parts of East Asia have welcomed cultural globalization over the past several decades.

5. Mt. Fuji is a national symbol of Japan. How does the Japanese religion of Shinto contribute to the cultural significance of this mountain?

6. How does the natural world figure into beliefs and practices found among other religions?

C

Geopolitical Framework

11.7 Outline the geopolitical division of East Asia during the Cold War period and explain how the division of that period still influences East Asian geopolitics.

East Asia has been characterized by much strife since the end of World War II. China and Korea are still suspicious of Japan, while Japan is concerned about China's growing military power and North Korea's nuclear arms and missiles. Territorial disputes over islands complicate relations between East Asian neighbors, especially China and Japan.

7. If Taiwan is already an independent country, why are the Taiwanese protestors in this photo demanding Taiwanese independence?

8. Should the international community acknowledge an independent Taiwan, or is the "one China policy" necessary for regional and global stability?

D

Economic and Social Development

11.8 Identify the main reasons behind East Asia's rapid economic growth in recent decades, and discuss possible limitations to continued expansion at such a rate.

11.9 Summarize the geographical differences in economic and social development found in China and across East Asia as a whole.

With the notable exception of North Korea, all East Asian countries experienced rapid economic growth after World War II. In the 2000s, the most important story was the rise of China. China's economic expansion has reduced poverty nationwide, but it has also generated serious tensions between the wealthier, more globally oriented coastal regions and the less-prosperous interior provinces. The rise of China also has global implications, as many countries in all regions of the world have profited by exporting the raw materials needed by China's booming industries.

9. China has been rapidly expanding its urban system, yet in some areas huge new apartment blocks are unoccupied. Why were such projects built, and what consequences might they have for China's economy?

10. Describe the advantages and disadvantages of massive, planned residential developments.

E

KEY TERMS

anthropogenic landscape *(p. 361)*
autonomous region *(p. 370)*
chaebol *(p. 377)*
China proper *(p. 351)*
Confucianism *(p. 366)*
desertification *(p. 353)*
diaspora *(p. 368)*
feng shui *(p. 367)*
hallyu ("Korean Wave") *(p. 370)*
hukou *(p. 363)*
ideographic writing *(p. 366)*
laissez-faire *(p. 378)*
loess *(p. 358)*
Marxism *(p. 367)*
pollution exporting *(p. 356)*
rust belt *(p. 380)*
sediment load *(p. 358)*
shogun *(p. 373)*
shogunate *(p. 373)*
social and regional differentiation *(p. 380)*
special administrative region *(p. 375)*
Special Economic Zone (SEZ) *(p. 379)*
sphere of influence *(p. 372)*
superconurbation (megalopolis) *(p. 364)*
tribal *(p. 369)*
tsunami *(p. 355)*
urban primacy *(p. 364)*

DATA ANALYSIS

http://goo.gl/ZwguC7

China is known for its rapid economic growth and its large regional economic disparities. Chinese officials seek to close the development gap. Are policies to balance economic development more evenly across China successful? Visit the National Bureau of Statistics of China website (http://data.stats.gov.cn/) to examine the gross regional product data and compare the relative economic growth of China's provinces and province-level municipalities. A pull-down menu lets you select a province and view statistical measurements. At the bottom of the list, find the data for "Per Capita Gross Regional Product (yuan/person)."

1. Write down these figures for the last 10 years for the following coastal regions: Tianjin, Shanghai, Zhejiang, Fujian, and Guangdong. Graph these five data sets. Then do the same for the interior provinces of Gansu, Guizhou, Sichuan, Yunnan, and Tibet.
2. Compare the two sets of graphs, noting both current levels of economic disparity and those recorded for 2005. Based on your observations, are regional economic differences in China becoming more pronounced, becoming less pronounced, or staying roughly the same?
3. Do your findings correlate with China's desire to create a geographically balanced national economy? Suggest reasons for your results.

MasteringGeography™

Looking for additional review and test prep materials? Visit the Study Area in MasteringGeography™ to enhance your geographic literacy, spatial reasoning skills, and understanding of this chapter's content by accessing a variety of resources, including MapMaster interactive maps, geoscience animations, videos, *In the News* RSS feeds, flashcards, web links, self-study quizzes, and an eText version of *Globalization and Diversity.*

Authors' Blogs

Scan to visit the **Author's Blog for field notes, media resources, and chapter updates**

http://gad4blog.wordpre ss.com/category/east-asia/

Scan to visit the **GeoCurrents Blog**

http://geocurrents.info/category/place/east-asia/

12 South Asia

PHYSICAL GEOGRAPHY AND ENVIRONMENTAL ISSUES

The arid parts of South Asia suffer from water shortages and soil salinization, whereas the humid areas often experience devastating monsoon floods.

POPULATION AND SETTLEMENT

South Asia will soon become the most populous region of the world. Birth rates have, however, decreased substantially in recent years.

CULTURAL COHERENCE AND DIVERSITY

South Asia is one of the most culturally diverse parts of the world, with India alone having more than a dozen official languages as well as numerous adherents of most major religions.

GEOPOLITICAL FRAMEWORK

The South Asian region is burdened not only by several violent secession movements, but also by the ongoing struggle between India and Pakistan, both armed with nuclear weapons.

ECONOMIC AND SOCIAL DEVELOPMENT

Although South Asia is one of the world's poorest regions, certain areas are experiencing rapid economic growth and technological development.

◀ Much of Pakistan experienced an exceptionally severe heat wave in June 2015. Residents of the metropolis of Karachi cool themselves off in the shallow coastal waters of the Arabian Sea.

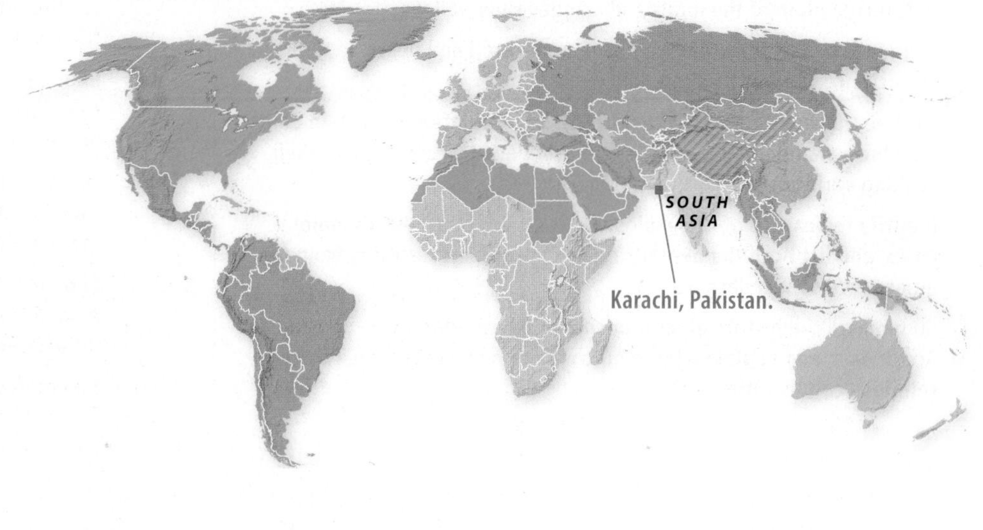

Pakistan has been hard-hit by several extreme weather events in recent years. Monsoon floods in 2010 and 2011 submerged much of the country's most fertile farmland. In June 2015, the opposite problem struck when a lethal heat wave resulted in more than 1400 deaths, mostly from dehydration and heatstroke. Although people perished across southern Pakistan, the crisis was most severe in the great southern metropolis of Karachi, home to more than 20 million people.

Several factors contributed to Karachi's high mortality figure. The city is always hot and humid in June, a condition usually moderated by sea breezes. These winds failed in 2015, allowing temperatures to soar to 113°F (45°C). Equally significant was the recurring failure of the city's electricity supply, which prevented the operation of fans and air conditioners and led to water shortages when pumps stopped working. Cultural factors played a role as well; the 2015 heat wave occurred during the Muslim holy month of Ramadan, when believers are prohibited from eating and drinking between sunrise and sunset. Many poor people who had to work outdoors in the oppressive heat thus experienced dehydration.

Environmentalists also think that human actions have changed the area's climate, making extreme weather more common. Global climate change is linked to both higher temperatures and greater fluctuations. More locally, the destruction of the mangrove forests that once lined the sea near Karachi is also linked to higher summer temperatures.

Pakistan is not the only South Asian country to have experienced extreme weather in recent years. A heat wave one month earlier in India resulted in more than 2500 deaths. Experts predict that such events will become more common, due both to global climate change and to the fact that more people live in huge cities with little shade or greenery. Successful development, however, could partially counteract such effects. Where electricity is reliable, fans and air conditioning can significantly reduce death rates. But generating electricity by burning coal, as is done in much of India, places additional burdens on the global climate. Such are the paradoxes of climate change and development in modern South Asia.

Development itself is a highly uneven process in South Asia. Most of the region has experienced substantial economic and social progress over the past several decades, and some areas have emerged as world-class business centers, often specializing in high-tech fields. But other places have failed to make such gains, and across the region rampant environmental degradation threatens both human communities and natural ecosystems. Population growth adds further pressure; given its current rate of expansion, South Asia will soon surpass East Asia as the world's most populous region.

> Given its current rate of expansion, South Asia will soon surpass East Asia as the world's most populous region.

Geopolitical tensions also persist both within and among the various countries of the region. Since gaining independence in 1947, India and Pakistan have fought several wars and remain locked in conflict. Political tensions have reached such heights that some experts consider South Asia the most likely site for nuclear war. Religious divisions contribute to the geopolitical turmoil, for India is primarily a Hindu country (with a large Muslim minority), while neighboring Pakistan and Bangladesh are both predominantly Muslim.

Learning Objectives

After reading this chapter you should be able to:

- **12.1** **Describe the geological relationship between the Himalayas and the flat, fertile plain of the Indus and Ganges river valleys.**
- **12.2** **Explain how the monsoon winds are generated and describe their importance for South Asia.**
- **12.3** **Outline the ways in which the patterns of population growth in South Asia have changed over the past several decades, and explain their striking variation across the region.**
- **12.4** **Identify the causes of the explosive growth of South Asia's major cities, and list both the benefits and the problems resulting from the emergence of such large cities.**
- **12.5** **Summarize the historical relationship between Hinduism and Islam in South Asia, and explain why so much tension exists between the two religious communities today.**
- **12.6** **Compare and contrast the ways in which India and Pakistan, both containing numerous distinctive language groups, have dealt with the issue of national cohesion.**
- **12.7** **Explain why South Asia was politically partitioned following British rule and how the legacies of partition continue to generate political and economic difficulties in the region.**
- **12.8** **Describe the challenges that India, Pakistan, and Sri Lanka have faced from insurgency movements that seek to carve out new independent states from their territories.**
- **12.9** **Explain why European merchants were eager to trade in South Asia in the 16th, 17th, and 18th centuries and how their activities influenced the region's later economic development.**
- **12.10** **Summarize the variations in economic and social development across the different parts of South Asia, and explain why such variability is so pronounced.**

South Asia as a whole forms a distinct landmass separated from the rest of the Eurasian continent by a series of sweeping mountain ranges, including the Himalayas—the highest mountains in the world. It is often called the **Indian subcontinent**, in reference to its largest country. South Asia also includes the island countries of Sri Lanka and the Maldives as well as the Indian territories of the Lakshadweep, Andaman, and Nicobar islands (Figure 12.1).

India, Pakistan, and Bangladesh dominate the South Asia landmass. India is by far the region's largest country, both in size and in population. Covering more than 1 million square miles

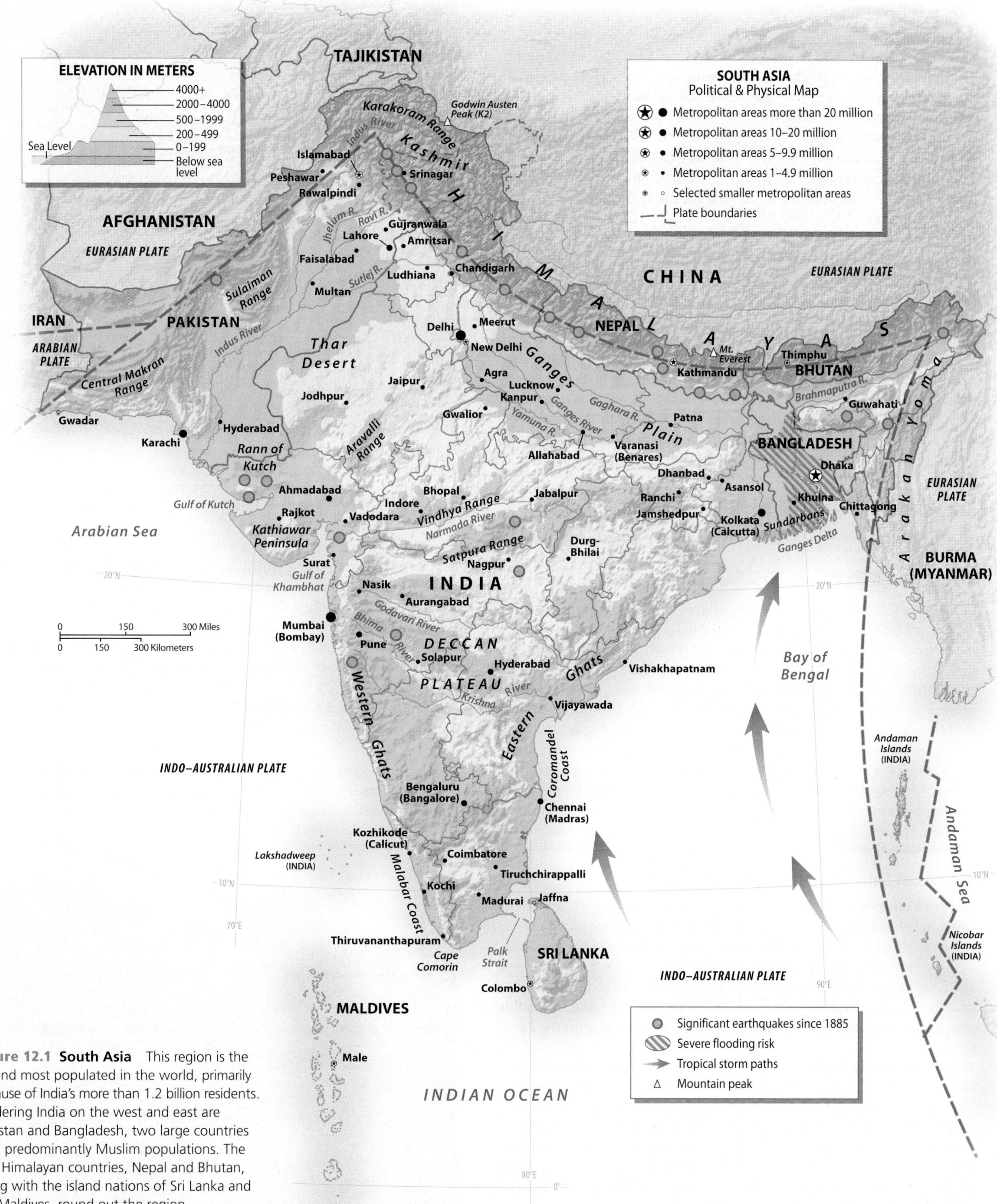

Figure 12.1 South Asia This region is the second most populated in the world, primarily because of India's more than 1.2 billion residents. Bordering India on the west and east are Pakistan and Bangladesh, two large countries with predominantly Muslim populations. The two Himalayan countries, Nepal and Bhutan, along with the island nations of Sri Lanka and the Maldives, round out the region.

(2.6 million square kilometers) from the Himalayan crest to the southern tip of the peninsula, India is the world's seventh-largest country in terms of area and, with more than 1.2 billion inhabitants, second only to China in population. Pakistan, the next largest country, is less than one-third the size of India. Stretching from the high northern mountains to the arid coastline on the Arabian Sea, its population exceeds 182 million. Bangladesh, on India's eastern shoulder, was originally created as East Pakistan in the hurried division of India in 1947 and achieved independence after a brief civil war in 1971. Although small in area (54,000 square miles [140,000 square kilometers]), Bangladesh is one of the world's most densely populated places, with 157 million people living in an area about the size of Wisconsin.

The other countries of South Asia have much smaller populations than the three demographic giants. Nepal and Bhutan are both located in the Himalaya Mountains, sandwiched between India and the Tibetan Plateau of China. Nepal, with some 31 million people, is much larger than Bhutan, which has fewer than 1 million inhabitants. Sri Lanka (formerly Ceylon) and the Maldives round out South Asia. Sri Lanka is a large island with more than 21 million inhabitants, whereas the Maldives is a collection of tiny atoll islands that together support only about 300,000 people.

Figure 12.2 South Asia from Space The four physical subregions of South Asia are clearly seen in this satellite image, from the snow-clad Himalaya Mountains in the north to the islands of the south. The irrigated lands of the Indus River Valley in Pakistan are clearly visible in the upper left.

Physical Geography and Environmental Issues: Diverse and Stressed Landscapes

South Asia's diverse environmental geography ranges from the world's highest mountains to densely populated delta islands barely above sea level; from some of the wettest places on Earth to scorching deserts; and from tropical rainforests to eroded scrublands. Dense human population and rapid industrialization are generating severe environmental problems throughout the region. The rapid increase in automobile use in particular produces serious smog problems in urban areas. Delhi, one of India's largest cities, is now widely considered the world's worst city for air pollution. South Asia has also suffered from some of the world's most severe environmental disasters. Today, across the region, environmental activists, government agencies, and industrial producers are responding to the crisis, often in innovative ways.

Physical Subregions of South Asia

To better understand its diverse environments, we can divide South Asia into four physical subregions, starting with the high mountain ranges of its northern edge and extending to the tropical islands of the far south. Lying south of the mountains are the extensive river lowlands that form the heartland of both India and Pakistan. Between these river lowlands and the island countries is the vast area of peninsular India, extending more than 1000 miles (1600 km) from north to south (Figure 12.2).

Mountains of the North South Asia's northern rim of mountains is dominated by the great Himalayan range, forming the northern borders of India, Nepal, and Bhutan. More than two dozen peaks exceed 25,000 feet (7600 meters), including the world's highest mountain, Everest, on the Nepal–China (Tibet) border (see *Geographers at Work: The Himalayan Environment*). To the east are the lower Arakan Yoma Mountains, forming the border between India and Burma (Myanmar) and separating South Asia from Southeast Asia.

These mountain ranges are a result of the dramatic collision of northward-moving peninsular India with the Asian landmass. The entire region is still geologically active, putting northern South Asia in serious earthquake danger. A 2013 earthquake in Pakistan's Balochistan Province killed at least 825 people. This was a relatively minor event, however, when compared with Nepal's 2015 earthquake, which took more than 9000 lives and resulted in economic losses roughly equal to 25 percent of the country's gross domestic product (GDP; Figure 12.3). It also caused a series of avalanches on Mount Everest that killed at least 19 climbers.

Figure 12.3 2015 Nepal Earthquake Two major earthquakes struck Nepal in April and May of 2015, killing thousands of people and resulting in severe economic losses. Local residents walk past destroyed buildings in the town of Chautara in northeastern Nepal.

GEOGRAPHERS AT WORK

The Himalayan Environment

Over the past 40 years, **P. P. Karan** of the University of Kentucky has proved to be one of the world's most accomplished and versatile geographers (Figure 12.1.1). Karan studies the links between economic development and both the natural and cultural realms, not just in South Asia but also in Tibet, the United States, and especially Japan.

Karan majored in economics as an undergraduate, but later studied geography because it "seemed more 'real world,' whereas economics dealt with theoretical things." He never looked back: "I see geography in everything—what we eat, what we do, what we use. It's even more relevant now with GPS and smartphones. Geography draws from different physical and social sciences and the humanities to explain how distinctive a place is."

Understanding the Himalayas Within South Asia, Karan has worked on India's position in the global community, the impact of the 2004 Indian Ocean tsunami, and environmental movements across the region. He has primarily focused, however, on the Himalayas. For his book, *Himalaya: Life on the Edge of the World*, Karan and coauthor David Zurick conducted fieldwork in the Himalayas, but also employed geological records, scientific reports, and an array of official documents. These multidisciplinary lines of research allowed Karan and Zurick to trace the great mountain range back to its very geological beginnings, and to outline the relations between human societies and local environments over more than a millennium.

Previous studies emphasized deforestation and erosion, often placing blame mainly on overpopulation and the subsistence activities of local villagers, but Karan and Zurick's book showed that this model is simplistic. Conservation efforts such as well-designed ecotourism initiatives and old and new sustainable agriculture techniques have achieved notable successes. Says Karan, "Geographers look at different attributes, and come up with how to improve an area."

Figure 12.1.1 P.P. Karan The geographer is shown here conducting fieldwork in Tibet on the environmental and economic transformation of the region.

1. Why is it so difficult to determine the extent of environmental degradation over a large area like the Himalayas?
2. List the benefits and drawbacks of studying several different parts of the world rather than focusing on a single region.

Although most of South Asia's northern mountains are too rugged and high to support dense human settlement, major population clusters are found in the Kathmandu Valley of Nepal, situated at 4400 feet (1300 meters), and the Valley, or Vale, of Kashmir in northern India, at 5200 feet (1600 meters).

Indus–Ganges–Brahmaputra Lowlands South of the mountain rim lie large lowlands created by three major river systems, which have deposited sediments to build huge alluvial plains of fertile and easily farmed soils. These densely settled lowlands constitute the core population areas of Pakistan, India, and Bangladesh.

The Indus River, which flows 1980 miles (3180 km) from the Himalayas through Pakistan to the Arabian Sea, provides much-needed irrigation for Pakistan's southern deserts. More famous, however, is the Ganges, which flows southeasterly some 1500 miles (2400 km) and empties into the Bay of Bengal. The Ganges provides the fertile alluvial soil that has made northern India one of the world's most densely settled areas. Given the central role of this important river throughout Indian history, it is understandable why Hindus consider the Ganges sacred. Finally, the Brahmaputra River, which rises on the Tibetan Plateau, flows more than 1700 miles (2700 km) before joining the Ganges in central Bangladesh and spreading out over the world's largest river delta.

Peninsular India Extending southward from the river lowlands is peninsular India, made up primarily of the Deccan Plateau, which is bordered on each side by narrow coastal plains backed by north–south mountain ranges. On the west are the higher Western Ghats, generally about 5000 feet (1500 meters) in elevation; to the east, the Eastern Ghats are lower and less continuous. On both coastal plains, fertile soils and an adequate water supply support population densities comparable to those of the Ganges lowland to the north.

Soil quality ranges from fair to poor over much of the Deccan Plateau, but in the state of Maharashtra lava flows have produced particularly fertile black soils. Unfortunately, much of the area does not have a reliable water supply for agriculture. The western portion of the plateau lies in the rain shadow of the Western Ghats, giving it a semiarid climate. For centuries, small reservoirs have collected monsoon rainfall for use during the dry season. Today, deep wells and powerful pumps extract groundwater to support more widespread irrigation.

Partly because of the overuse of groundwater, India's government is building a series of large dams to provide for irrigation. Dam-building, however, is controversial because the resulting reservoirs displace large numbers of rural residents. The Sardar Sarovar Dam project on the Narmada River in the state of Madhya Pradesh has dislodged more than 100,000 people. Local residents and activists throughout India have joined forces in opposition, but farmers in neighboring Gujarat, who are reaping the benefits, strongly support the project. In 2014, the Indian government permitted an increase in dam height that will allow the irrigation of an additional 6900 square miles (18,000 square kilometers) of farmland.

The Southern Islands At the southern tip of peninsular India lies the island of Sri Lanka. Sri Lanka is ringed by extensive coastal plains and

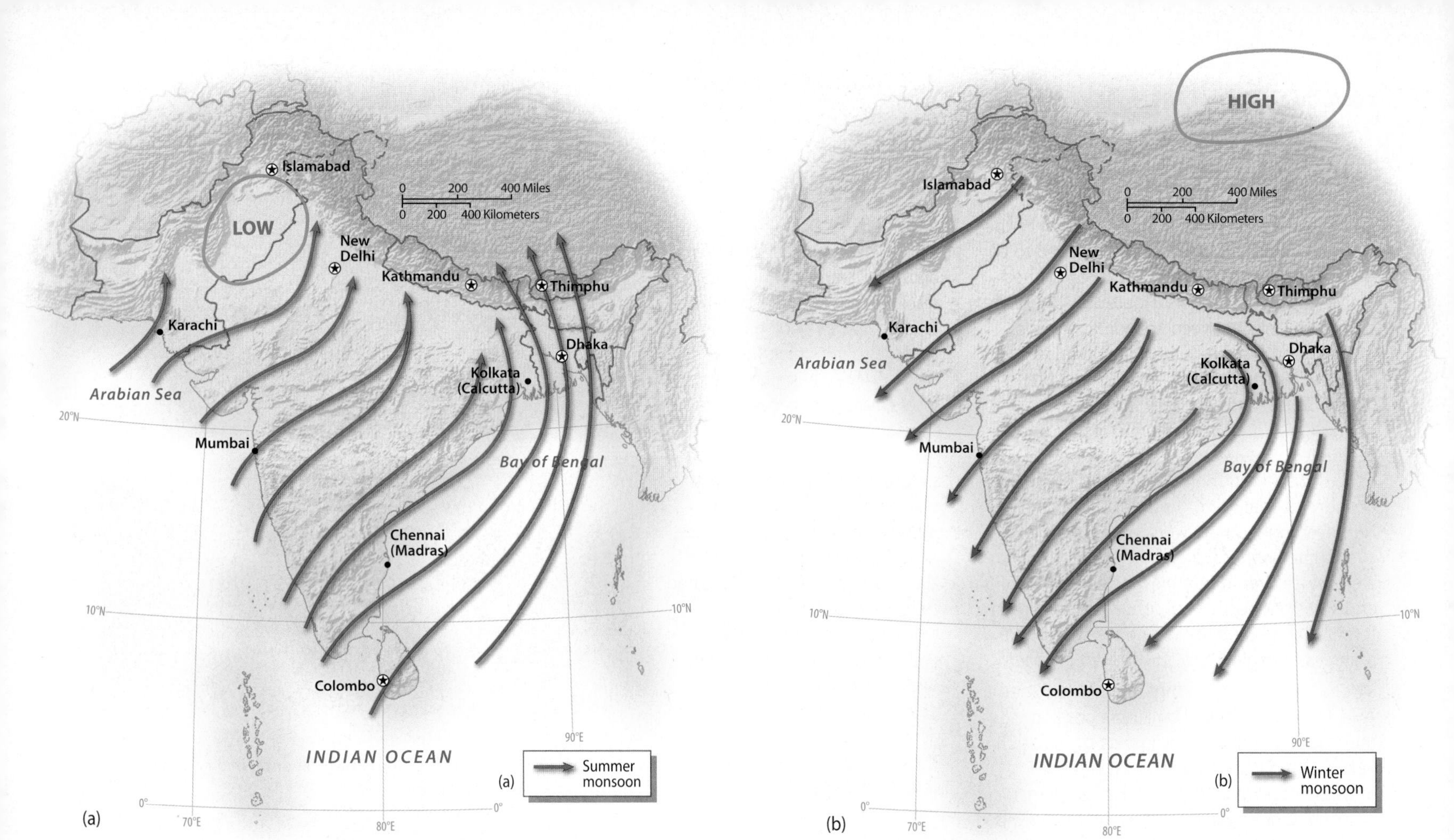

Figure 12.4 The Summer and Winter Monsoons (a) During the summer, low pressure centered over South and Southwest Asia draws in warm, moist air masses that bring heavy monsoon rains to most of the region. Usually, these rains begin in June and last for several months. (b) During the winter, high pressure forms over northern Asia. As a result, winds are reversed from those of the summer. During this season, only a few coastal locations in eastern India and Sri Lanka receive substantial rain.

low hills, but mountains reaching more than 8000 feet (2400 meters) occupy the southern interior, providing a cool, moist climate. Because the main monsoon winds arrive from the southwest, that portion of the island is much wetter than the rain-shadow areas of the north and east.

The Maldives, a chain of more than 1200 islands off the southwestern tip of India, has a combined area of only 116 square miles (290 square kilometers), and only one-quarter of the islands are inhabited. Like many South Pacific islands, the Maldives are low coral atolls, with a maximum elevation just over 6 feet (2 meters) above sea level.

South Asia's Monsoon Climates

The dominant climatic factor for most of South Asia is the **monsoon**, the seasonal change of wind direction that corresponds to wet and dry periods (Figure 12.4). During the winter, a large high-pressure system forms over the cold Asian landmass. Because winds flow from high to low pressure, cool, dry winds flow outward from the continental interior across South Asia. The resulting dry season extends from November until February. As winter turns to spring, these winds diminish, resulting in the hot, dry season of March through May. Eventually, the buildup of heat over South Asia and Southwest Asia produces a large low-pressure cell. By early June, this low-pressure cell is strong enough to cause a shift in wind direction so that warm, moist air from the Indian Ocean moves toward the continental interior. This signals the onset of the warm and rainy season of the southwest monsoon that lasts from June through October (Figure 12.5).

Orographic rainfall is caused by the uplifting and cooling of moist monsoon winds over the Western Ghats and the Himalayan foothills. As a result, some areas receive more than 200 inches (508 cm) of rain during the four-month wet season (Figure 12.6). Cherrapunji, in northeastern India, is one of the world's wettest places, with rainfall averaging 450 inches (1130 cm); see the climograph in Figure 12.6.

Figure 12.5 Monsoon Rain During the summer monsoon, some Indian cities, such as Mumbai, receive more than 70 inches (178 cm) of rain in just three months. These daily torrents cause floods and power outages, but when the monsoon downpours finally arrive, people often react with joy.

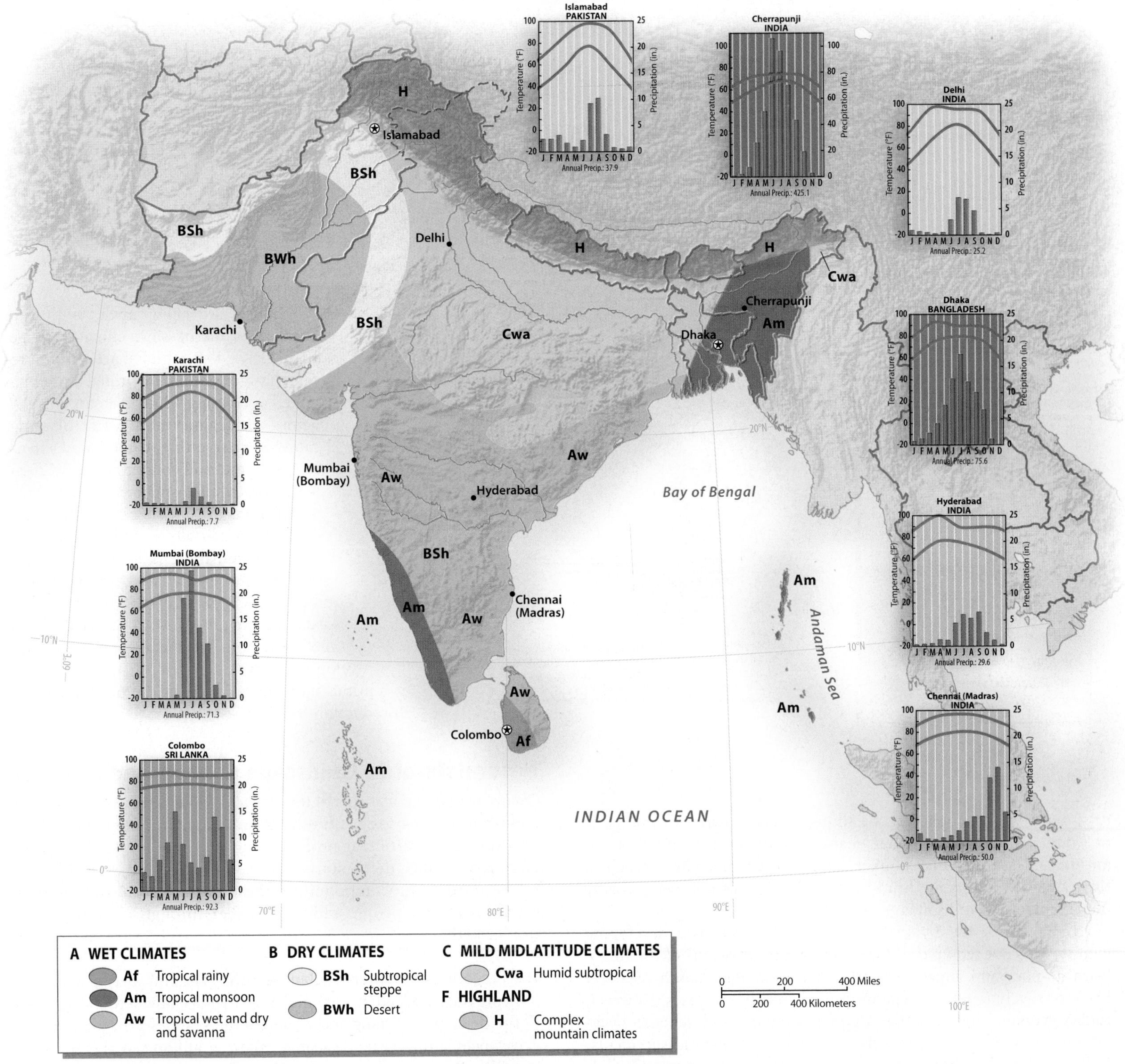

Figure 12.6 Climates of South Asia Except for the extensive Himalayas, South Asia is dominated by tropical and subtropical climates. Many of these climates show a distinct summer rainfall season that is associated with the southwest monsoon. The climographs for Mumbai and Delhi are excellent illustrations.

On the Deccan Plateau, however, rainfall is dramatically reduced by a strong **rain-shadow effect** (see Chapter 2). As winds move downslope, the air becomes warmer, and dry conditions usually prevail; see the climographs for Hyderabad, Delhi, and Karachi in Figure 12.6.

Climate Change and South Asia

According to the UN's Intergovernmental Panel on Climate Change (IPCC) 2014 report, the effects of climate change will be especially severe in South Asia. Even a minor rise in sea level will inundate large areas of the Ganges–Brahmaputra Delta in Bangladesh. Already, more than 18,500 acres (7500 hectares) of swampland in the Sundarbans region are submerged. If the most severe sea-level forecasts are realized, the Maldives will simply vanish beneath the waves. The region's vital rice crops are also threatened by rising sea levels, whereas increased heat could reduce the wheat harvest an estimated 50 percent by 2100.

Climate change could result in major changes to the region's water resources. Many Himalayan glaciers are retreating, threatening the

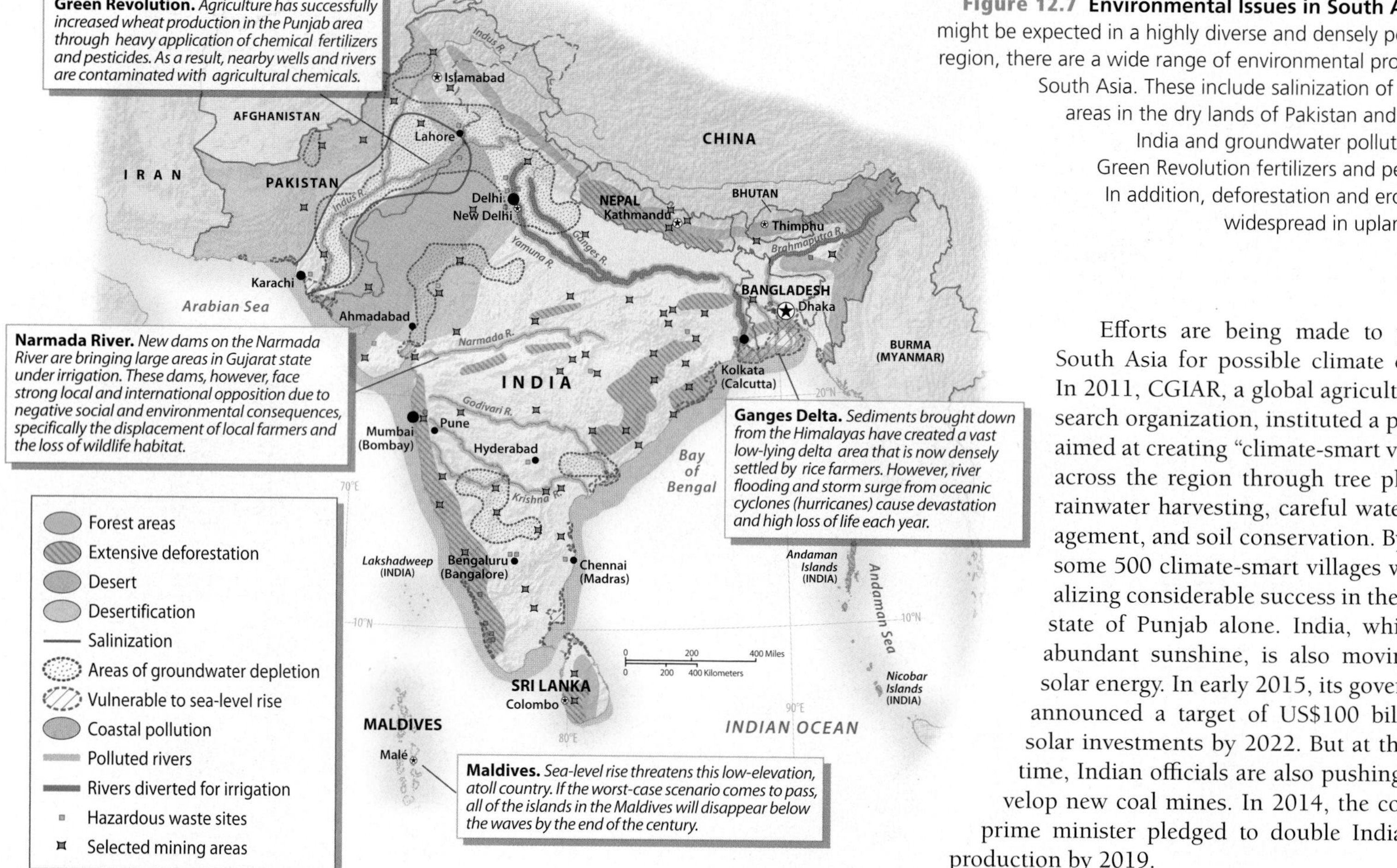

Figure 12.7 Environmental Issues in South Asia As might be expected in a highly diverse and densely populated region, there are a wide range of environmental problems in South Asia. These include salinization of irrigated areas in the dry lands of Pakistan and western India and groundwater pollution from Green Revolution fertilizers and pesticides. In addition, deforestation and erosion are widespread in upland areas.

dry-season water supplies of the Indus–Ganges Plain. In some parts of South Asia, however, climate change could increase rainfall due to an intensification of the summer monsoon. Unfortunately, much of this rainfall would come from intense cloudbursts and would likely increase flooding and soil erosion.

India signed the Kyoto Protocol in 2002, but as a developing country it is not required to follow the main provisions of the treaty. With its poor and largely nonindustrial economies, South Asia still has a relatively low per capita output of greenhouse gases. However, India's economy in particular is growing rapidly and depends heavily on coal to generate electricity. According to some measurements, India is now the world's third-largest emitter of carbon dioxide, following only China and the United States. Official estimates indicate that India must triple or even quadruple its primary energy supply in order to maintain an economic growth rate of 8 percent over the next quarter-century.

Climate change could put Pakistan in a particular bind. As much as 90 percent of the irrigation water for this mostly desert country comes from the mountains of Kashmir, a region divided between Pakistan and India. Because Pakistan already experiences periodic water shortages, many experts believe that it must cooperate with India to develop and conserve the region's water resources. Considering the geopolitical tension between the two countries, which is focused on Kashmir, any such agreement seems unlikely.

Efforts are being made to prepare South Asia for possible climate change. In 2011, CGIAR, a global agricultural research organization, instituted a program aimed at creating "climate-smart villages" across the region through tree planting, rainwater harvesting, careful water management, and soil conservation. By 2015, some 500 climate-smart villages were realizing considerable success in the Indian state of Punjab alone. India, which has abundant sunshine, is also moving into solar energy. In early 2015, its government announced a target of US$100 billion in solar investments by 2022. But at the same time, Indian officials are also pushing to develop new coal mines. In 2014, the country's prime minister pledged to double India's coal production by 2019.

Natural Hazards, Landscape Change, and Pollution

Owing to specific features of its physical geography, South Asia suffers from some severe natural hazards and environmental problems (Figure 12.7). Particular problems include flooding in the region's large river deltas, deforestation in the uplands, and desertification in the northwest. Compounding these problems are the immense numbers of new people added each year through natural population growth.

The Precarious Situation of Bangladesh The link between population pressure and environmental problems is nowhere clearer than in the delta area of Bangladesh, where the search for fertile farmland drives people into hazardous areas, putting millions at risk from seasonal flooding as well as from the powerful cyclones (tropical storms) that form over the Bay of Bengal. For thousands of years, drenching monsoon rains have eroded huge quantities of sediment from the Himalayan slopes, which is then transported to the sea by the Ganges and Brahmaputra rivers, gradually building this low-lying delta environment. Scientists estimate that in an average year approximately 10,000 square miles (26,000 square kilometers) of the country flood, resulting in some 5000 deaths.

Although periodic flooding is a natural, even beneficial, phenomenon that enlarges deltas by depositing fertile river-borne sediment, floods are still a huge problem. In September 1998, more than 22 million Bangladeshis lost their homes when water covered two-thirds of the country. Equally serious flooding in August 2007 caused much

WORKING TOWARD SUSTAINABILITY

Community Development and Mangrove Conservation in Sri Lanka

Mangrove swamp forests, which thrive in shallow coastal waters in many tropical regions, are endangered across southern Asia. In some areas, they are being cleared to make room for export-oriented fish and shrimp ponds, and elsewhere the trees are cut for charcoal or firewood. But mangrove forests are vital components of coastal ecosystems, serving as nurseries for many fish and crustacean species. They also protect the shoreline from storm surges and can trap heavy metals and other pollutants under their roots, keeping them out of the water. As a result, mangrove conservation has become a high priority.

Loss of Mangroves Sri Lanka has already lost many of its mangroves. The most extreme destruction took place in the northwestern corner of the island, where swamp forests were converted into shrimp farms in the 1990s. That development, however, did not prove sustainable, as diseases later wiped out most of the shrimp. In 2015, Sri Lanka's government embarked on a new strategy, becoming the first country in the world to officially protect all of its remaining mangrove forests.

A New Initiative Sri Lanka's new mangrove-protection scheme is attracting considerable global attention. Jointly run by the government, a local coastal-protection foundation, and a U.S.-based conservation group called Seacology, this US$3.4 million initiative combines environmental protection with rural development and women's empowerment (Figure 12.2.1). A central feature is the provision of small low-interest loans of around US$100 to local residents, most of whom are women. Such loans, in turn, are generally used to start small businesses. Funding is also earmarked for planting fast-growing tree species on dry land to provide alternative sources of fuel.

In return for their loans, recipients are organized into 1500 groups of around 10 persons that are tasked with overseeing a specific area of mangroves. They are also responsible for teaching others about the importance of the mangrove ecosystem and for helping replant 9600 acres (3885 hectares) with mangrove seedlings. The groups must work with government rangers who patrol the 21,782 acres (8815 hectares) of mangroves now under official protection. In these areas, mangroves cannot be lawfully cut for commercial purposes.

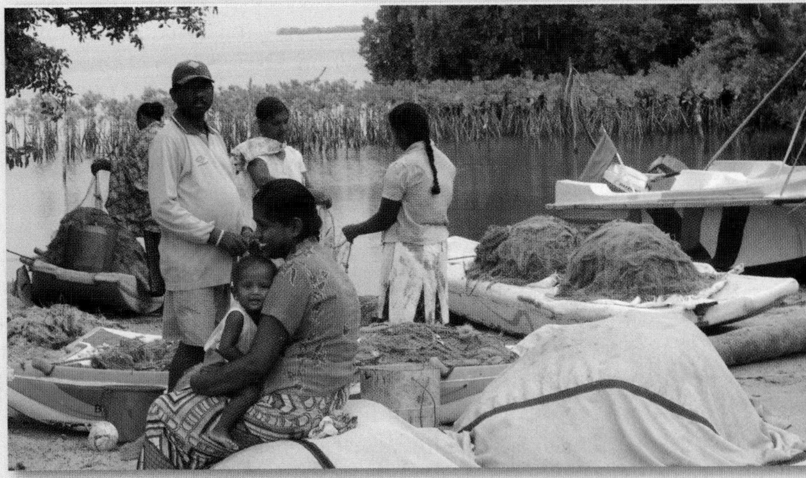

Figure 12.2.1 New Mangrove Trees Sri Lanka has made great efforts to preserve its remaining mangrove forests. Community-oriented mangrove reforestation schemes have proved effective in many areas. Recently replanted mangroves thrive in Puttalam lagoon.

Targeted Poverty Reduction Sri Lanka's new mangrove initiative seeks to reduce rural poverty as much as to protect nature. As a result, it focuses on the most vulnerable members of local communities. Half of all loan recipients must be widows, and the other half must be school dropouts, whether male or female. Thus far, the program appears to be quite successful in both ecological and economic terms. Of the nearly 2000 loans made to local women thus far, the repayment rate has been over 96 percent.

1. Why have mangrove forests in particular been selected for conservation programs in Sri Lanka and elsewhere in the region? How do they differ from other forests in terms of their own vulnerability and their benefits to local communities?
2. Why has Sri Lanka decided to link a social-development program to its mangrove conservation initiative?

Google Earth (MG) Virtual Tour Video
https://goo.gl/OFskkZ

less damage, largely because of an internationally funded development program that elevated hundreds of thousands of houses above the 1998 high-water mark. With Bangladesh's expanding population, however, flooding might take higher tolls in the coming decades as farmers continue to move into the hazardous lower floodplains.

The summers of 2010 and 2011 saw heavy rains across northern South Asia and renewed flooding in Bangladesh. During the 2010 event, however, damage in Pakistan was much more extensive. With losses estimated at US$43 billion, more than 2000 deaths, and more than 20 million people affected, the 2010 Pakistan flood is commonly regarded as that country's worst natural disaster. Less deadly but still destructive floods hit Pakistan in 2012, 2013, and 2014.

Forests and Deforestation Many scholars link the catastrophic floods periodically experienced in much of South Asia with deforestation in the region's uplands. As trees are cleared from the mountains and foothills of the Himalayas, runoff increases in the adjacent lowlands of Bangladesh, Pakistan, and northern India. As a result, efforts to protect remaining forest coverage in the region have intensified. Coastal forest resources have also been targeted for conservation (see *Working Toward Sustainability: Community Development and Mangrove Conservation in Sri Lanka*).

Tropical monsoon forests and savanna woodlands once covered most of South Asia, except for the desert areas in the northwest. In most areas, however, tree cover has vanished as a result of human activities. The Ganges River Valley and coastal plains of India, for example, were largely deforested hundreds of years ago to make room for agriculture. Elsewhere, forests were cleared more gradually for agricultural, urban, and industrial expansion. More recently, hill slopes in the rugged landscapes of eastern and northern South Asia have been logged for commercial purposes. Extensive forests can still be found, however, in some remote areas.

As a result of deforestation, many South Asian villages suffer from a shortage of fuel wood for household cooking, forcing people to burn dung cakes from cattle. This low-grade fuel provides adequate heat, but prevents manure from being used as fertilizer. Where wood is available, collecting it often involves many hours of female labor because the remaining sources of wood are often far from the villages.

Wildlife Although the overall environmental situation in South Asia is somewhat bleak, wildlife protection inspires some optimism. The region has managed to retain a diverse assemblage of wild animals despite heavy population pressure and intense poverty. The only remaining Asiatic lions live in India's state of Gujarat, and even Bangladesh retains a viable population of tigers in the Sundarbans, the mangrove forests of the Ganges Delta. Wild elephants still roam several large reserves in India, Sri Lanka, and Nepal.

Wildlife protection in India far exceeds that in most other parts of Asia. Project Tiger, which operates 42 preserves, is often credited with preserving the species. Surveys in 2006, however, indicated that the tiger population had dropped below 2000, prompting the Indian government to pledge $153 million in further funding. Some success was realized, as a 2015 national tiger census came up with a figure of 2226 (Figure 12.8). Unfortunately, some villagers have been relocated to make more room for tiger populations. Some advocates for the rural poor argue that such initiatives place the needs of wildlife above those of people. Increasingly, however, wildlife advocates are working with local people, who are often surprisingly willing to share their lands with large wild animals, regardless of the dangers that they present.

Pollution Like other developing regions, South Asia suffers from high levels of air and water pollution. According to a 2014 World Health Organization (WHO) study, 13 of the world's 20 dirtiest cities are located in India. An especially severe problem in many South Asian cities is the prevalence of small particulates, a highly unhealthful form of air pollution. The study found that the particulate level in Delhi is six times greater than the WHO's recommended maximum. Partly as a result of these findings, India's government promised to create a comprehensive air-quality index and to enact stricter pollution regulations.

Figure 12.8 Human–Tiger Interactions Tourists photograph a tiger in Ranthambhore National Park in the Indian state of Rajasthan. Ranthambhore is one of the best places in India to see wild tigers. **Q: Why has India, despite its poverty and high population density, been able to maintain healthy populations of wild tigers and other large and potentially dangerous animals?**

Review

12.1 Why is the monsoon so crucial to life in South Asia?

12.2 Why is flooding such an important environmental issue in Bangladesh and adjacent areas of northeastern India?

KEY TERMS Indian subcontinent, monsoon, orographic rainfall, rain-shadow effect

Population and Settlement: The Demographic Dilemma

South Asia will soon surpass East Asia as the world's most populous region (Figure 12.9). India alone is home to over 1.2 billion people, while Pakistan and Bangladesh rank among the world's 10 most populous countries (Table A12.1). Furthermore, much of South Asia is still experiencing rapid population growth. Although South Asia has made remarkable agricultural gains over the past several decades, widespread concern still persists about its ability to feed itself.

India's total fertility rate (TFR) has dropped rapidly, from 6 in the 1950s to the current rate of 2.5. In western and southern India, fertility rates are now generally at or below replacement level. In much of northern India, however, birth rates remain high; the average woman in the poor state of Bihar gives birth to 3.5 children. A distinct cultural preference for male children is found in most of South Asia, a tradition that further complicates family planning.

Pakistan has seen a rapid reduction in its TFR in recent years, but at 3.26 it is still well above replacement level. As a result, Pakistan's population will probably be over 250 million by 2050. This is a worryingly high number, considering the country's arid environment, underdeveloped economy, and political instability (Figure 12.10). Bangladesh has been more successful than Pakistan in reducing its birth rate. As recently as 1975, its TFR was 6.3, but dropped to 2.21 by 2012. The success of family planning can be partly attributed to strong government support, advertised through radio, billboards, and even postage stamps.

Migration and the Settlement Landscape

South Asia is one of the least urbanized world regions, with around one-third of its people living in cities. Most South Asians reside in compact rural villages, but increased mechanization of agriculture, along with the expansion of large farms at the expense of subsistence cultivation, pushes as many people to the region's rapidly growing urban areas as are drawn by employment opportunities in the city.

The region's most densely settled areas are those with fertile soils and dependable water supplies. The highest rural population densities are found in the core area of the Ganges and Indus river valleys and on India's coastal plains. Settlement is less dense on the Deccan Plateau and is relatively sparse in the highlands of the far north and the arid lands of the northwest.

Many South Asians have migrated in recent years from poor and densely populated areas to less densely populated or wealthier areas. Migrants are often attracted to large cities such as Mumbai (Bombay), but those from Bangladesh are settling in large numbers

Figure 12.9 Population Map of South Asia Except for the desert areas of the west and the high mountains of the north, South Asia is a densely populated region. Particularly high densities of people are found on the fertile plains along the Indus and Ganges rivers and in India's coastal lowlands. In rural areas, people typically cluster in villages, often located near water sources, such as streams, wells, canals, or small tanks that store water between monsoon rains.

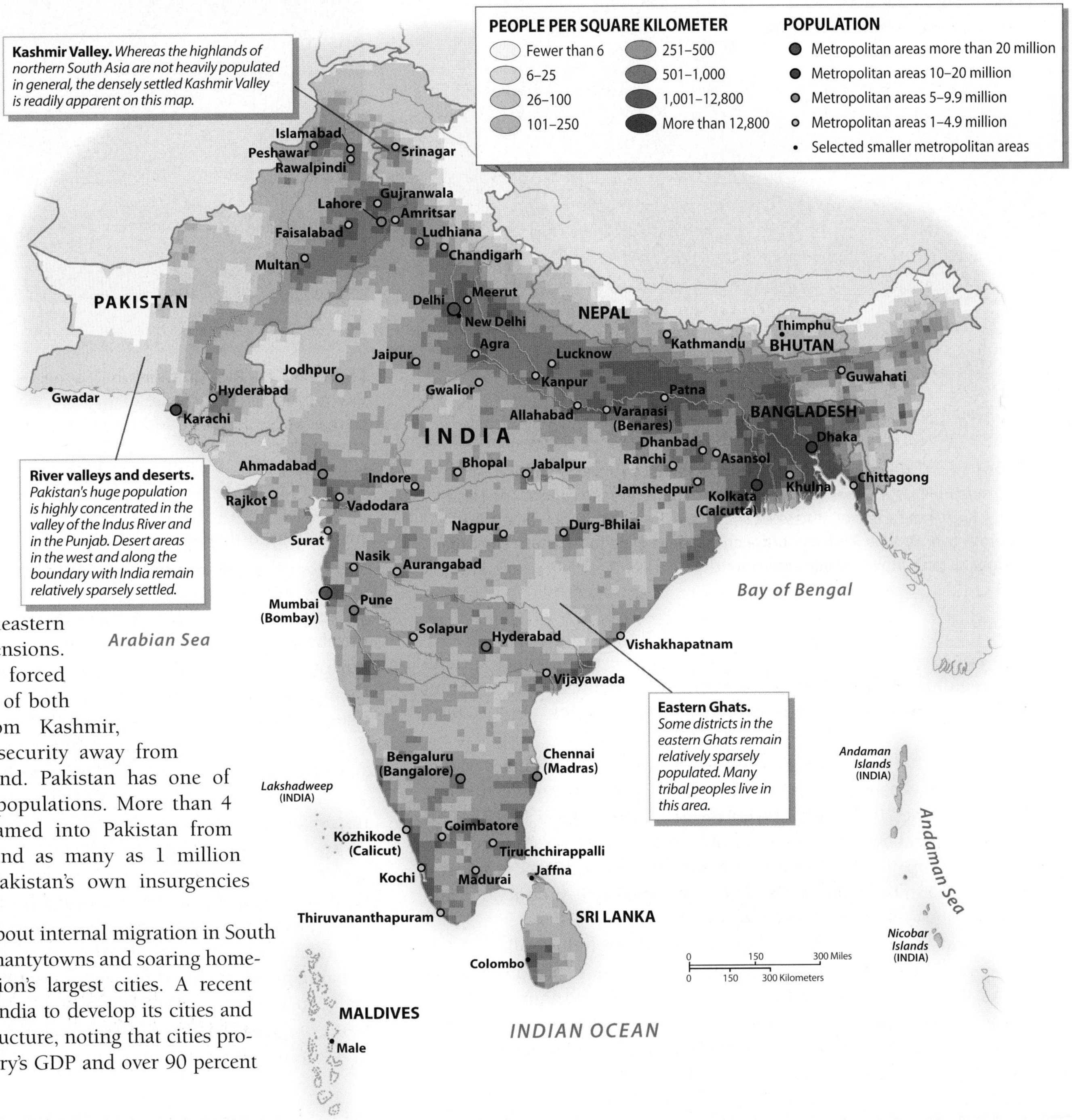

Kashmir Valley. *Whereas the highlands of northern South Asia are not heavily populated in general, the densely settled Kashmir Valley is readily apparent on this map.*

River valleys and deserts. *Pakistan's huge population is highly concentrated in the valley of the Indus River and in the Punjab. Desert areas in the west and along the boundary with India remain relatively sparsely settled.*

Eastern Ghats. *Some districts in the eastern Ghats remain relatively sparsely populated. Many tribal peoples live in this area.*

PEOPLE PER SQUARE KILOMETER
Fewer than 6
6–25
26–100
101–250
251–500
501–1,000
1,001–12,800
More than 12,800

POPULATION
Metropolitan areas more than 20 million
Metropolitan areas 10–20 million
Metropolitan areas 5–9.9 million
Metropolitan areas 1–4.9 million
Selected smaller metropolitan areas

in rural portions of northeastern India, creating ethnic tensions. Sometimes migrants are forced out by war; a large number of both Hindus and Muslims from Kashmir, for example, have sought security away from their battle-scarred homeland. Pakistan has one of the world's largest refugee populations. More than 4 million refugees have streamed into Pakistan from neighboring Afghanistan, and as many as 1 million have been displaced by Pakistan's own insurgencies (Figure 12.11).

Experts are concerned about internal migration in South Asia, which results in huge shantytowns and soaring homeless populations in the region's largest cities. A recent World Bank report advised India to develop its cities and invest more in urban infrastructure, noting that cities produce two-thirds of the country's GDP and over 90 percent of its government revenues.

Agricultural Regions and Activities

South Asian agriculture has historically been relatively unproductive, especially compared with that of East Asia. Since the 1970s, however, agricultural production has grown rapidly. Unfortunately, many South Asian farmers have gone deeply into debt, threatening future agricultural gains.

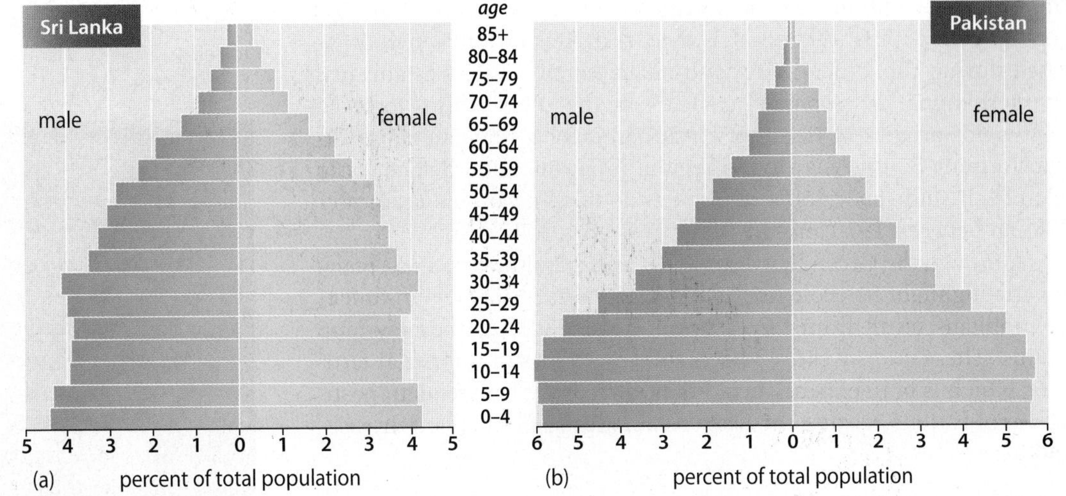

Figure 12.10 Population Pyramids of Sri Lanka and Pakistan (a) Sri Lanka has had a relatively low birth rate for decades, as well as relatively long life expectancies, and thus has a well-balanced population pyramid. (b) Pakistan, in contrast, has a much higher birth rate, as well as a lower average life span, and thus has a bottom-heavy pyramid.

Figure 12.11 Afghan Refugee Camp in Pakistan Several million Afghans still reside in Pakistan, many of them living in grim refugee camps. The Nasir Bagh Camp, near the city of Peshawar, is now closed, but many others like it remain active. **Q: Why have most Afghan refugees gone to Pakistan rather than to Afghanistan's other neighboring countries?**

Figure 12.12 Rice Cultivation Farmers plant rice seedlings in Sri Lanka. Rice requires large quantities of irrigation water. It is the main crop in the lower Ganges Valley and Delta, along the lower Indus River of Pakistan, and in India's coastal plains.

Crop Zones South Asia can be divided into several distinct agricultural regions, all with different problems and potentials. These regions are based on the production of three subsistence crops—rice, wheat, and millet.

Rice is the main crop and foodstuff in the lower Ganges Valley, along India's coastal lowlands, in the delta lands of Bangladesh, along Pakistan's lower Indus Valley, and in Sri Lanka (Figure 12.12). This distribution reflects the large volume of irrigation water needed to grow the crop. The amount of rice grown in South Asia is impressive: India ranks behind only China in world rice production, and Bangladesh is the fourth-largest producer.

Wheat is the principal crop in the northern Indus Valley and in the western half of India's Ganges Valley. South Asia's "breadbasket" includes the northwestern Indian states of Haryana and Punjab and adjacent areas in Pakistan. Here the so-called Green Revolution has been particularly successful in increasing grain yields. In less fertile areas of central India, millet and sorghum are the main crops, along with root crops such as manioc. Wheat and rice are the preferred staples throughout South Asia, but poorer people must often subsist on rougher crops.

The Green Revolution The main reason South Asian agriculture has kept up with population growth is the **Green Revolution**, which originated during the 1960s in agricultural research stations established by international development agencies. By the 1970s, efforts to breed high-yield varieties of rice and wheat had reached their initial goals, transforming South Asia from a region of chronic food deficiency to one of self-sufficiency. India more than doubled its annual grain production between 1970 and the mid-1990s (Figure 12.13).

Although the Green Revolution was an agricultural success, many experts highlight its ecological and social costs. Serious environmental problems result from the chemical dependency of the new crop strains. These crops typically need large quantities of industrial fertilizer, which is both expensive and polluting, as well as frequent pesticide applications because they lack natural resistance to plant diseases and insects.

Social problems have also followed the Green Revolution. In many areas, only the more prosperous farmers can afford the new seed strains, irrigation equipment, farm machinery, fertilizers, and pesticides. Poorer farmers have often been forced to serve as wage laborers for their more successful neighbors or to migrate to crowded cities. To purchase the necessary inputs, moreover, most farmers have had to borrow large sums of money. As fluctuating crop prices often prevent them from repaying their debts, many farmers feel trapped. Approximately 300,000 Indian farmers killed themselves between 1995 and 2015, mostly by hanging or drinking pesticides. This tragedy represents the largest wave of suicides in world history.

The Green Revolution has fed South Asia's expanding population over the past several decades, but whether it can continue to do so remains unclear. One alternative is to expand water delivery systems (through either canals or wells), as many fields are not irrigated. Irrigation, however, brings its own problems. In much of Pakistan and northwestern India, where irrigation has been practiced for generations, soil **salinization**, or the buildup of salt in fields, is a major problem. In addition, groundwater is being depleted, especially in Punjab, India's breadbasket. On the other hand, optimists point out that India's agricultural production continues to grow and that famines, once common, are now a thing of the past.

Figure 12.13 Green Revolution Farming Because of "miracle" wheat strains that have increased yields in the Punjab area, this region has become the breadbasket of South Asia. India has more than doubled its wheat production in the past 25 years and has moved from continual food shortages to self-sufficiency. Here a man is scattering fertilizer in a Green Revolution rice field in India.

Figure 12.14 Mumbai Hutments Hundreds of thousands of people in Mumbai live in crude hutments, with no sanitary facilities, built on formerly busy sidewalks. Hutment construction is forbidden in many areas, but wherever it is allowed, sidewalks quickly disappear.

Urban South Asia

Although South Asia remains a largely rural society, many of its cities are large and growing quickly. India alone has more than 45 metropolitan areas with over 1 million inhabitants. According to a 2014 report, India's urban population will increase from 340 million in 2008 to 600 million in 2031. Because of this rapid growth, South Asian cities have serious problems with homelessness, poverty, congestion, water shortages, air pollution, and sewage disposal. Throughout South Asia, sprawling squatter settlements, or **bustees**, are rapidly expanding, providing meager shelter for many migrants.

Clean water and sanitation are major problems in urban slums throughout South Asia. A 2010 study estimated that only 54 percent of India's city dwellers had access to sewers or other modern sanitation facilities. In Bangladesh, a 2015 report claimed that the country's slum dwellers, those living without piped water or other basic amenities, had increased by 60 percent in the previous 17 years. Yet some progress is being made, as the proportion of urban Indians with access to an improved water source increased from 72 percent in 1990 to over 88 percent in 2008.

Mumbai South Asia's largest city, Mumbai (often called by its former colonial name, Bombay), is India's financial, industrial, and commercial center. Mumbai itself contains roughly 14 million people, while its metropolitan area is home to more than 22 million. Mumbai is responsible for much of India's foreign trade, has long been a manufacturing center, and is the focus of India's film industry. Mumbai's economic vitality draws people from all over India, resulting in simmering ethnic tensions.

Limited space means that most of Mumbai's growth has taken place to the north and east of the historic city. Building restrictions in the downtown area have led to skyrocketing commercial and residential rents, which are some of the highest in the world. Even the city's thriving middle class has difficulty finding adequate housing. Hundreds of thousands of less-fortunate immigrants live in "hutments," crude shelters built on formerly busy sidewalks (Figure 12.14). The least fortunate sleep on the street or in simple plastic tents, often placed along busy roadways.

Mumbai's notorious road congestion eased somewhat in 2009, after the completion of a massive eight-lane, $340 million bridge linking the central city to its northern suburbs. The Mumbai Metro, an ambitious rapid-transit system, is scheduled for completion in 2021, promising further improvements. In 2015, local authorities announced an ambitious new initiative, "Mumbai Next," to build the infrastructure necessary to turn Mumbai into a global financial, commercial, and entertainment hub.

Kolkata To many, Kolkata (more often called by its old name, Calcutta) symbolizes the problems faced by rapidly growing cities in developing countries. About 1 million people here sleep on the streets every night. And with approximately 15 million people in its metropolitan area, Kolkata falls far short of supplying the rest of its residents with water, power, and sewage treatment. Electrical power is woefully inadequate, and during the wet season many streets are routinely flooded.

With rapid growth as migrants pour in from the countryside, a mixed Hindu–Muslim population that generates ethnic tension, a decayed economic base, and an overloaded infrastructure, Kolkata faces a troubled future. Yet it remains a culturally vibrant city, noted for its fine educational institutions, theaters, and publishing firms. Kolkata is currently trying to nurture an information technology industry, but it remains to be seen whether this effort will prove successful.

Karachi Pakistan's largest urban area and commercial core, Karachi is one of the world's fastest-growing cities. Its metropolitan population, already over 20 million, is expanding at about 5 percent per year (Figure 12.15). Karachi served as Pakistan's capital until 1963, when the new city of Islamabad was created in the northeast. Karachi suffered relatively little from the departure of government functions; it is still Pakistan's most cosmopolitan city, its main streets lined with businesses and high-rise buildings.

Karachi does suffer from political and ethnic tensions that have periodically turned parts of the city into armed camps. In the early decades of Pakistan's independence, Karachi's main conflict was between the Sindis, the region's native inhabitants, and the Muhajirs, Muslim refugees from India who settled in the city after Pakistan's separation from India in 1947. More recently, clashes between Sunni and Shiite Muslims have intensified, as have those between Pashtun migrants from northwestern Pakistan and other residents. According to the Human Rights Commission of Pakistan, 2909 people were killed in Karachi in 2014, including 142 law enforcement personnel and 134 political activists.

Figure 12.15 Karachi Street Scene Karachi, Pakistan's largest city and main port, is noted for its economic power, its ethnic violence, and its congested and colorful streets. British influences can be seen at the Karachi Empress Market and bus station.

Review

12.3 Why has the Green Revolution been so controversial, considering the fact that it has greatly increased South Asia's food supply?

12.4 What are the major advantages and disadvantages of the growth of South Asia's megacities?

KEY TERMS Green Revolution, salinization, bustee

Cultural Coherence and Diversity: A Common Heritage Undermined by Religious Rivalries

Historically, South Asia is a well-defined cultural region. A thousand years ago, virtually the entire area was united by the Hindu faith. The subsequent arrival of Islam added a new religious element, but it did not undermine the region's cultural unity. British imperialism later brought other cultural features, from the widespread use of English to a passion for cricket.

India has been a secular state since its creation. Since the 1980s, this political tradition has come under pressure from the growth of **Hindu nationalism**, which promotes Hindu values as the foundation of Indian society. In several high-profile cases, Hindu mobs demolished Muslim mosques that had allegedly been built on the sites of ancient Hindu temples. Since 2000, however, the Hindu nationalist movement has moderated somewhat. Bitter divisions persist, but efforts are being made to promote religious understanding (Figure 12.16).

In Pakistan, Islamic fundamentalism is a divisive issue. Powerful fundamentalist leaders want to make Pakistan a religious state under Islamic law, a plan rejected by the country's secular intellectuals and international businesspeople. The government has attempted to mediate between the two groups, but with little success.

Origins of South Asian Civilizations

Many scholars think that the roots of South Asian culture extend back to the Indus Valley civilization, which flourished 4500 years ago in what is now Pakistan. This remarkable urban-oriented society vanished almost entirely around 1800 BCE, probably due to environmental change. By 800 BCE, however, a new focus of civilization had emerged in the middle Ganges Valley.

Hindu Civilization This early Ganges Valley civilization gave birth to **Hinduism**, a complex faith that lacks a single system of belief. Certain deities are recognized, however, by all believers, as is the notion that these various gods are all expressions of a single divine essence (Figure 12.17). Hindus also share a common set of epic stories, usually written in **Sanskrit**, the sacred language of their religion. Hinduism is noted for its mystical tendencies, which have long inspired many to seek an ascetic lifestyle, renouncing property and sometimes all regular human relations. One of its hallmarks is a belief in the transmigration of souls from being to being through reincarnation. Hinduism is also associated with India's **caste system**, the strict division of society into hereditary groups that are ranked as ritually superior or inferior to one another.

Figure 12.16 Ayodhya Mosque Controversy The dismantling of the Babri Mosque in Ayodhya by Hindu nationalists in 1992 resulted in intense religious conflicts in many parts of India. More recently, Hindu–Muslim brotherhood ("Bhai Bhai") groups have tried to enhance understanding and mutual respect. A rally of one such group is depicted here.

Buddhism Ancient India's caste system was challenged from within by Buddhism. Siddhartha Gautama, the Buddha, was born in 563 BCE in an elite caste. He rejected the life of wealth and power, however, and sought instead to attain enlightenment, or mystical union with the universe. He preached that the path to such enlightenment (or *nirvana*) was open to all, regardless of social position. His followers eventually established Buddhism as a new religion. Buddhism spread throughout South Asia and later expanded into East, Southeast, and Central Asia. But it never fully replaced Hinduism in India and was disappearing from the Indian peninsula by 500 CE.

Arrival of Islam The next major challenge to Hindu society—Islam—came from the outside. Around the year 1000, Turkic-speaking Muslims began to enter the region from Central Asia. By the 1300s, most of South Asia lay under Muslim power, although Hindu kingdoms persisted in southern India. During the 16th and 17th centuries, the **Mughal** (or **Mogul**) **Empire**, the most powerful of the Muslim states, dominated much of the region from its power center in the Indus–Ganges Basin (Figure 12.18).

At first, Muslims formed a small ruling elite, but over time increasing numbers of Hindus converted to the new faith, particularly in the northwest and northeast where the areas now known as Pakistan and Bangladesh became predominantly Muslim.

Figure 12.17 Hindu Temple The Akshardham temple complex in Delhi opened in 2005. Hindu temples are a prominent feature of the Indian landscape, and as India's economy grows, lavish new temples continue to be built.

EVERYDAY GLOBALIZATION

India and the International Day of Yoga

Yoga's global popularity prompted the United Nations to declare June 21 as the International Day of Yoga, an initiative heavily promoted by India's Prime Minister Narendra Modi and supported by 175 countries. To celebrate the first Yoga Day in 2015, Modi joined more than 35,000 fellow practitioners, including dignitaries from 84 countries, to perform yoga poses in New Delhi. In India, yoga instruction is now officially supported in schools across the country (Figure 12.3.1).

Yet the creation of an International Day of Yoga generated controversy, both locally and abroad. Some Indian Muslim clerics and organizations denounced yoga as a Hindu religious practice and therefore an un-Islamic activity. Elsewhere, similar objections were raised by strict Muslim and Christian groups. Proponents argue that, despite its roots in Hindu culture, yoga long ago evolved into a nonreligious set of exercises—an age-old practice with modern appeal to health-conscious practitioners worldwide.

1. Is yoga necessarily associated with Hinduism, or is it merely a set of exercises that just happen to have originated in a mostly Hindu country?
2. What are some other health or exercise practices that you engage in that originated in other cultures?

Figure 12.3.1 International Yoga Day Yoga, which consists of a set of Indian spiritual exercises, is increasingly popular across much of the world. India is now encouraging the practice and spread of yoga. India's Prime Minister Narendra Modi practices yoga along with many others during a celebration marking the International Day of Yoga on June 21, 2015 in New Delhi.

The Caste System Caste is one of the historically unifying features of South Asia, as certain aspects of caste organization are found even among the region's Muslim and Christian populations. *Caste* is actually a rather clumsy term for the complex social order of the Hindu world. It combines two distinct concepts: *varna* and *jati*. *Varna* refers to the ancient fourfold social hierarchy, which distinguishes the Brahmins (priests), Kshatriyas (warriors), Vaishyas (merchants), and Sundras (farmers and craftsmen), in declining order of ritual purity. Standing outside this traditional order are the so-called untouchables, now usually called **Dalits**, whose ancestors held "impure" jobs, such as leather working or trash collection. *Jati*, on the other hand, refers to the hundreds of local endogamous ("marrying within") groups that exist at each *varna* level. Different *jati* groups are often called *subcastes*.

India's caste system is in a state of flux today. Its original occupational structure has long been undermined by the necessities of a modern economy, and various social reforms have chipped away at the discrimination that it embodies. The Dalit community itself has produced several notable national leaders who have waged partially successful political struggles. Owing to such efforts, the very concept of "untouchability" is illegal in India. India's central government also reserves a significant percentage of university seats and government jobs for students from low-caste backgrounds, while a number of Indian states have set higher quotas. Such "reservations," as they are called, are controversial, as many people think that they unfairly penalize people of higher-caste background.

But despite the gains made, caste remains an important feature of Indian social organization, and marriage across caste lines is still relatively rare. Dalits continue to suffer from many forms of oppression, particularly in the poor, rural areas of north-central India. Gang rapes of Dalit women are not uncommon, and Dalit students who break caste barriers are sometimes severely beaten. Radical Hindu activists also pressure Dalit communities by trying to coerce Christians and Muslims of Dalit backgrounds to reconvert to Hinduism.

Figure 12.18 The Red Fort Delhi's Red Fort, completed in 1648, was the power center of the Mughal Empire. Today this massive fortification, one of the largest in the world, is a major tourist destination.

Contemporary Geographies of Religion

In the simplest terms, South Asia has a Hindu heritage overlain by a significant Muslim presence. Such a picture fails, however, to capture the enormous diversity of religion in today's South Asia (Figure 12.19).

Hinduism Less than 1 percent of the people of Pakistan are Hindu, and in Bangladesh and Sri Lanka, Hinduism is a minority religion. However, in India and Nepal, Hinduism is clearly the majority faith. In most of central India, more than 90 percent of the population is Hindu. But Hinduism is itself a geographically complicated religion, with different aspects of faith varying across different parts of India. Some cultural practices originally associated with Hinduism, moreover, have lost their religious significance and have therefore spread widely to other parts of the world (see *Everyday Globalization: India and the International Day of Yoga*).

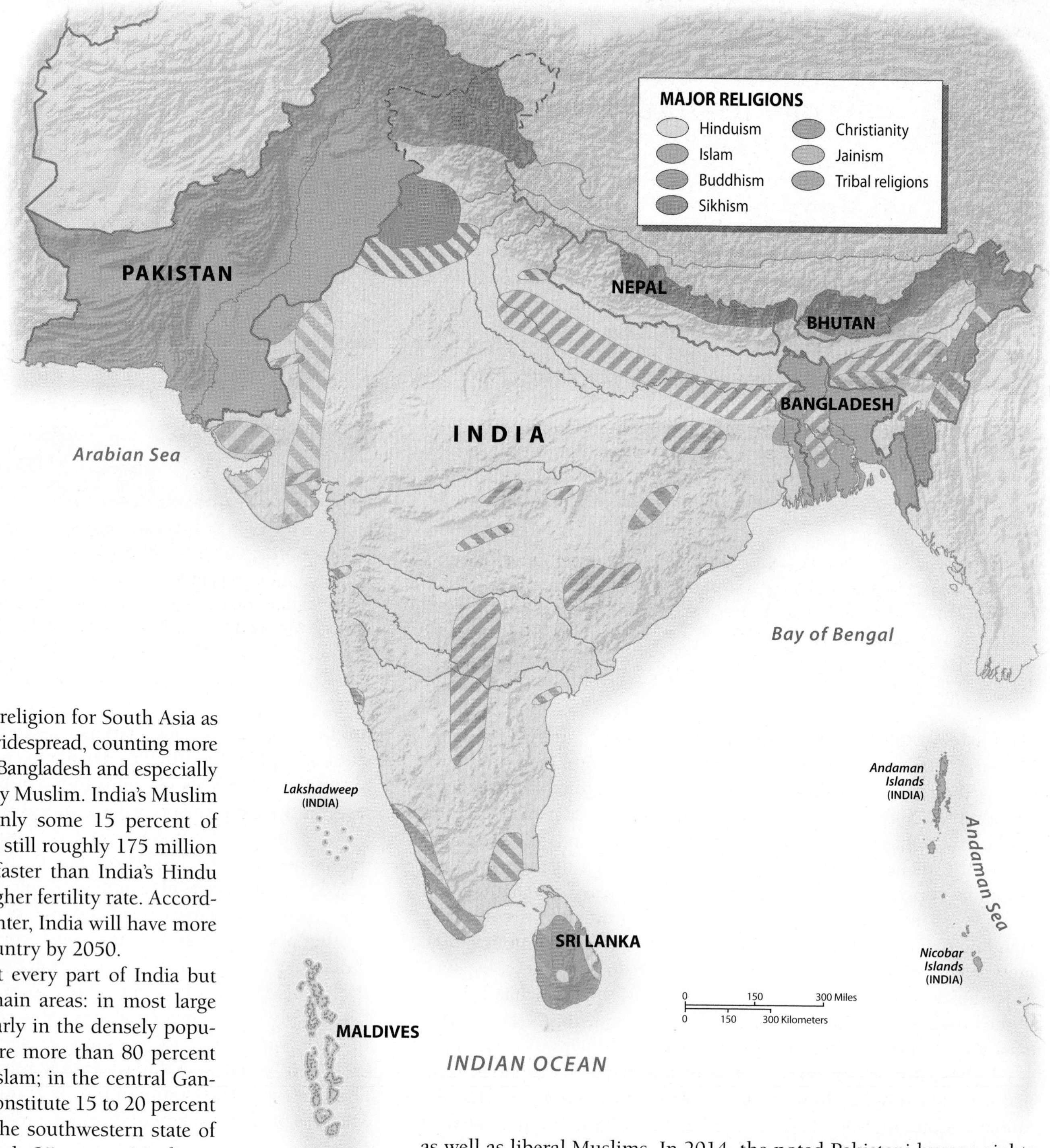

Figure 12.19 Religious Geography of South Asia Hindu-dominated India is bordered by the predominantly Muslim Pakistan and Bangladesh. More than 150 million Muslims, however, live within India, making up roughly 15 percent of the total population, particularly in northwest Kashmir and in the Ganges Valley. Sikhs form the majority population in India's state of Punjab. Also note the Buddhist populations in Sri Lanka, Bhutan, and northern Nepal; the areas of tribal religion in the east; and the centers of Christianity in the southwest.

Islam Though a minority religion for South Asia as a whole, Islam is still very widespread, counting more than 500 million followers. Bangladesh and especially Pakistan are overwhelmingly Muslim. India's Muslim community, constituting only some 15 percent of the country's population, is still roughly 175 million strong. It is also growing faster than India's Hindu population, thanks to its higher fertility rate. According to the Pew Research Center, India will have more Muslims than any other country by 2050.

Muslims live in almost every part of India but are concentrated in four main areas: in most large cities; in Kashmir, particularly in the densely populated Vale of Kashmir, where more than 80 percent of the population follows Islam; in the central Ganges Plain, where Muslims constitute 15 to 20 percent of the population; and in the southwestern state of Kerala, which is approximately 25 percent Muslim.

Interestingly, Kerala was one of the few parts of India that never experienced prolonged Muslim rule. Islam in Kerala was historically connected to trade across the Arabian Sea. Kerala's Malabar Coast supplied spices and other luxury products to Southwest Asia, encouraging many Arab traders to settle there. Gradually, many of Kerala's native residents converted to the new religion. The same trade routes brought Islam to Sri Lanka, which is approximately 9 percent Muslim, and to the Maldives, which is almost entirely Muslim.

In predominantly Muslim Pakistan, rising Islamic fundamentalism has generated severe conflicts. Radical fundamentalist leaders want to make Pakistan a fully religious state, a plan rejected by most of the country's citizens. The government has attempted to intercede between the two groups, but it is often viewed as biased toward the Islamists. Antiblasphemy laws, for example, have been used to persecute members of Pakistan's small Hindu and Christian communities as well as liberal Muslims. In 2014, the noted Pakistani human rights lawyer Rashid Rehman was murdered merely for defending a professor of English who had been charged with blasphemy, which can itself carry the death penalty.

Sikhism The tension between Hinduism and Islam in northern South Asia gave rise to a new religion, **Sikhism**, which originated in the 1400s in the Punjab near the modern boundary between India and Pakistan. The Punjab was the site of intense religious competition at the time; Islam was gaining converts, and Hinduism was on the defensive. The new faith combined elements of both religions. Many orthodox Muslims viewed Sikhism as dangerous because it incorporated elements of their own religion in a manner contrary to accepted beliefs. Periodic persecution led the Sikhs to adopt a militantly defensive stance. Even today many Sikh men work as soldiers and bodyguards, both in India and abroad (Figure 12.20).

At present, the Indian state of Punjab is approximately 60 percent Sikh. Small, but often influential groups of Sikhs are scattered

Figure 12.20 Sikh Soldiers India's Sikh community developed military traditions when their faith was persecuted in earlier centuries. Today many Sikh men serve in the Indian military. Here members of a Sikh regiment march in Delhi to celebrate the 51st anniversary of India's transition to a republic.

across the rest of India. Devout Sikh men are immediately visible because they do not cut their hair or their beards. Instead, they wear their hair wrapped in a turban and sometimes tie their beards close to their faces.

Buddhism and Jainism Although Buddhism virtually disappeared from India in medieval times, it persisted in Sri Lanka. Among the island's dominant Sinhalese people, Theravada Buddhism developed into a national religion. In the high valleys of the Himalayas, the Tibetan form of Buddhism emerged as the majority faith. The town of Dharamsala in the northern Indian state of Himachal Pradesh is the seat of Tibet's government-in-exile and of its spiritual leader, the Dalai Lama, who fled Tibet in 1959 after an unsuccessful revolt.

At roughly the same time as the birth of Buddhism (circa 500 BCE), another religion emerged in northern India: **Jainism**. This religion also stressed nonviolence, taking this creed to its ultimate extreme. Jains are forbidden to kill any living creatures, and as a result the most devoted members of the community wear gauze masks to prevent them from inhaling small insects. Jainism forbids agriculture because plowing can kill small creatures. As a result, most Jains look to trade for their livelihoods and today form a relatively prosperous community concentrated in northwestern India.

Other Religious Groups The Parsis, concentrated in Mumbai, form a tiny but influential religious group. Followers of Zoroastrianism, the ancient faith of Iran, Parsi refugees fled to India in the 7th century (Figure 12.21). The Parsis prospered under British rule, forming some of India's first modern industrial companies, such as the Tata Group of industries. Intermarriage and low fertility, however, now threaten the survival of this small community.

Indian Christians are more numerous than either Parsis or Jains. Their religion arrived some 1700 years ago as missionaries from Southwest Asia brought Christianity to India's southwestern coast. Today roughly 20 percent of the people of Kerala follow Christianity. Several Christian sects are represented, with the largest affiliated with the Syrian Christian Church of Southwest Asia. Another stronghold of Christianity is the small Indian state of Goa, a former Portuguese colony. Here Roman Catholics make up roughly half of the population.

During the colonial period, British missionaries went to great efforts to convert South Asians to Christianity. They had little success, however, in Hindu, Muslim, and Buddhist communities. The remote tribal districts of British India proved to be more receptive to missionary activity, especially those in the northeast. The Indian states of Nagaland, Meghalaya, and Mizoram now have Christian majorities, with more than 75 percent of the people of Nagaland belonging to the Baptist Church.

Geographies of Language

South Asia's linguistic diversity rivals its religious diversity. In northern South Asia, most languages belong to the Indo-European language family, the world's largest. The languages of southern India, on the other hand, belong to the **Dravidian language family**, which is found only in South Asia. Along the region's mountainous northern rim, a third linguistic family, Tibeto-Burman, dominates. Within these broad divisions are many different languages, each associated with a distinct culture. In many parts of South Asia, several languages are spoken within the same area, and the ability to speak several languages is common everywhere (Figure 12.22).

Each of the major languages of India is associated with an Indian state, as the country deliberately structured its political subdivisions along linguistic lines after attaining independence. As a result, the Gujarati language is spoken in Gujarat, Marathi in Maharashtra, Oriya in Odisha, and so on (see Figure 12.37 for a map of India's states). Two of these languages, Punjabi and Bengali, extend into Pakistan and Bangladesh, respectively, as the political borders were established on religious rather than linguistic lines. Dialects closely related to Nepali, the national language of Nepal, are also spoken in many of the mountainous areas of northern India. Minor languages abound in most of the more remote areas.

The Indo-European North The most widely spoken language of South Asia is **Hindi** (not to be confused with the Hindu religion). With more than 500 million native speakers, Hindi is by some measurements the world's second most widely spoken language. It plays a prominent role in present-day India, both because so many people speak it and because it is the main language of the Ganges Valley. Hindi is an official language in 10 Indian states and is widely studied throughout the country.

Bengali, the second most widely spoken language in South Asia, is the official language of Bangladesh and the Indian state of West Bengal, spoken by roughly 200 million people. It also has an extensive literature, as West Bengal (and particularly its capital city, Kolkata [Calcutta])

Figure 12.21 Parsi Temple India's small but influential Parsi community follows the ancient Iranian religion of Zoroastrianism. This Parsi temple in Mumbai is decorated in the style of ancient Iran.

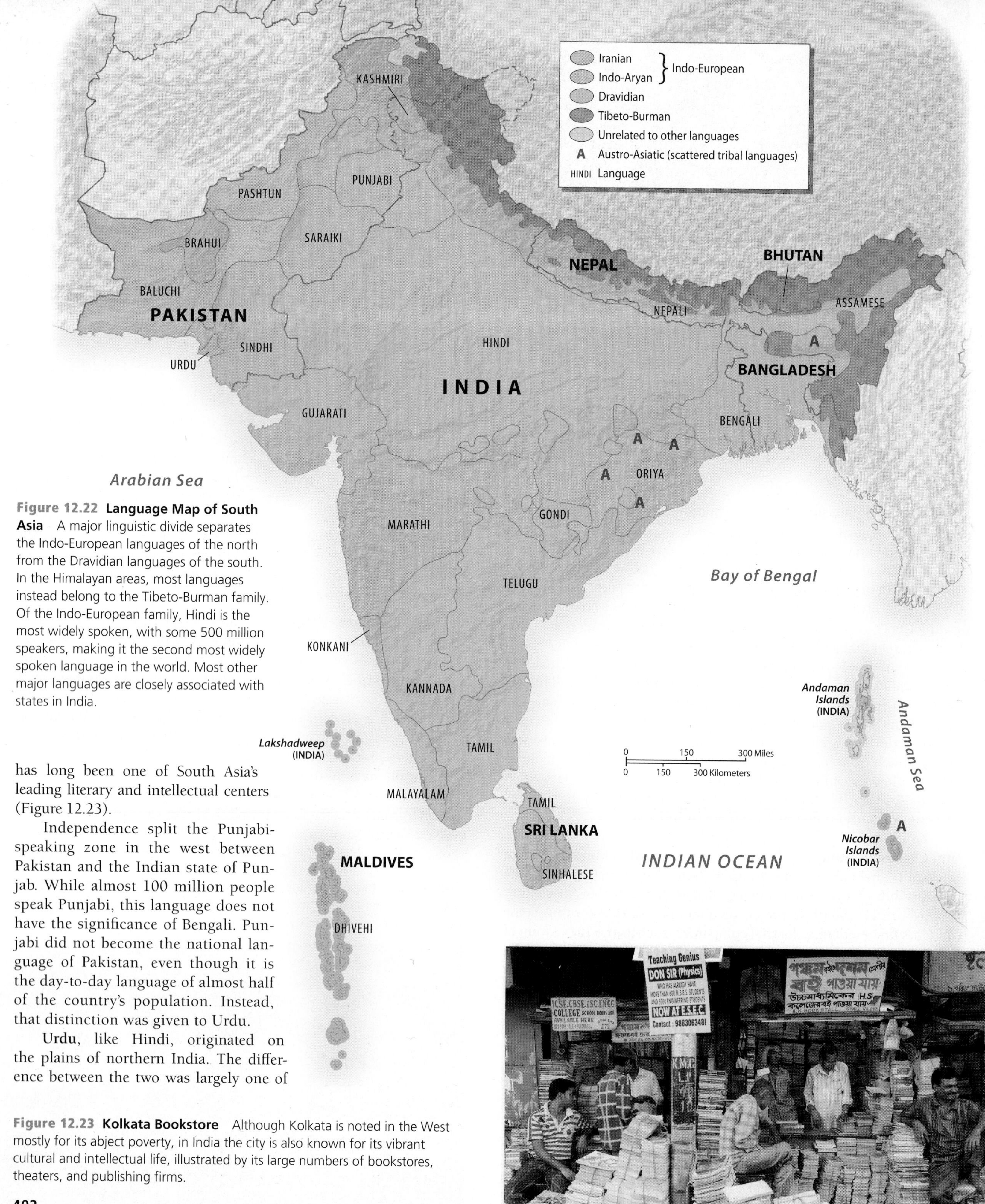

Figure 12.22 Language Map of South Asia A major linguistic divide separates the Indo-European languages of the north from the Dravidian languages of the south. In the Himalayan areas, most languages instead belong to the Tibeto-Burman family. Of the Indo-European family, Hindi is the most widely spoken, with some 500 million speakers, making it the second most widely spoken language in the world. Most other major languages are closely associated with states in India.

has long been one of South Asia's leading literary and intellectual centers (Figure 12.23).

Independence split the Punjabi-speaking zone in the west between Pakistan and the Indian state of Punjab. While almost 100 million people speak Punjabi, this language does not have the significance of Bengali. Punjabi did not become the national language of Pakistan, even though it is the day-to-day language of almost half of the country's population. Instead, that distinction was given to Urdu.

Urdu, like Hindi, originated on the plains of northern India. The difference between the two was largely one of

Figure 12.23 Kolkata Bookstore Although Kolkata is noted in the West mostly for its abject poverty, in India the city is also known for its vibrant cultural and intellectual life, illustrated by its large numbers of bookstores, theaters, and publishing firms.

religion: Hindi was the language of the Hindu majority, Urdu that of the Muslim minority. Because of this distinction, Hindi and Urdu are written differently—Hindi in the Devanagari script (derived from Sanskrit) and Urdu in the Arabic script. Although Urdu borrows many words from Persian, its basic grammar and vocabulary are almost identical to those of Hindi. With independence in 1947, millions of Urdu-speaking Muslims from the Ganges Valley fled to Pakistan. Because Urdu had a higher status than Pakistan's native tongues, it was quickly established as the new country's official language. Although only about 8 percent of the people of Pakistan learn Urdu as their first language, more than 90 percent are able to speak and understand it.

Languages of the South The four main Dravidian languages are confined to southern India and northern Sri Lanka. As in the north, each language is closely associated with one or more Indian states: Kannada in Karnataka, Malayalam in Kerala, Telugu in Andhra Pradesh and Telangana, and Tamil in Tamil Nadu. Tamil is often considered the most important member of the family because it has the longest history and the largest literature. Tamil poetry dates back to the 1st century CE, making it one of the world's oldest written languages.

Although Tamil is spoken in northern Sri Lanka, the country's majority population, the Sinhalese, speak an Indo-European language. Apparently, the Sinhalese migrated from northern South Asia several thousand years ago, settling primarily on the island's fertile southwestern coast and central highlands. These same people also migrated to the Maldives, where the national language, Dhivehi, is essentially a Sinhalese dialect. The drier north and east of Sri Lanka, on the other hand, were settled mainly by Tamils from southern India. Some Tamils later moved to the central highlands, where they were employed as tea-pickers on British-owned estates.

Linguistic Dilemmas Multilingual Sri Lanka, Pakistan, and India are all troubled by linguistic conflicts. Such problems are most complex in India, simply because India is so large and has so many different languages.

Indian nationalists have long dreamed of a national language that could unify their country. But **linguistic nationalism**, or the linking of a specific language with political goals, is often resisted by certain groups. The obvious choice for a national language is Hindi, and Hindi was indeed declared as such in 1947. Raising Hindi to this position, however, angered many non-Hindi speakers, especially in the Dravidian south. It was eventually decided that both Hindi and English would serve as official languages of India as a whole, but that each Indian state could select its own official language. As a result, more than 20 separate Indian languages now have that status.

Despite such opposition, Hindi is expanding, especially in the Indo-European north, where local languages are closely related to Hindi. Hindi is spreading through education as well as television and movies. Films and television programs are made in several northern languages, but Hindi remains primary. In a poor but modernizing country such as India, where many people experience the wider world largely through moving images, the influence of a national film and television culture can be substantial.

Despite its spread, Hindi remains foreign to much of India, and protests occasionally erupt over plans to expand the use of Hindi in non-Hindi-speaking areas (Figure 12.24). National-level communication is thus conducted mainly in English. Although many Indians want to deemphasize English, others advocate it as a neutral national language because all parts of the country have an equal stake in it. Furthermore, English gives substantial international benefits. English-medium schools abound throughout South Asia, and many children of the elite learn this global language well before they begin school.

Figure 12.24 The Use of English in India English serves as a common language across most of India, and its use is increasingly necessary for career advancement. As a result, specialized English-language schools are found throughout the country. An advertisement for such a school in Muzaffarpur, India, located in the impoverished state of Bihar, is partially submerged due to devastating flooding in the region in 2007.

South Asia in Global Cultural Context

The widespread use of English in South Asia has not only helped global culture spread throughout the region, but also helped South Asians' cultural production reach a global audience. The worldwide spread of South Asian literature, however, is nothing new. As early as the turn of the 20th century, Rabindranath Tagore gained international acclaim for his poetry and fiction, earning the Nobel Prize for Literature in 1913. More recently, "Bollywood" films produced in Mumbai gained popularity across much of the world (see *Exploring Global Connections: The Indian Film Industry's International Reach*).

The expansion of South Asian culture abroad has been accompanied by the spread of South Asians themselves. Migration from South Asia during the time of the British Empire led to the establishment of large communities in such distant places as eastern Africa, Fiji, and the southern Caribbean (Figure 12.25). Subsequent migration has been aimed more at the developed world; several million people of South Asian origin now live in Britain, and a similar number are found in North America. Many present-day migrants to the United States are doctors, software engineers, and members of other professions, making Indian Americans the country's wealthiest and best-educated ethnic group.

Within South Asia, cultural globalization has brought severe tensions. Traditional Hindu and Muslim religious norms frown on any overt display of sexuality—a staple feature of global popular culture. Religious leaders often criticize Western films and television shows as being immoral. Although India is a relatively free country, both the national and the state governments periodically ban films and books considered too sexual. In 2012, for example, *The Girl with the Dragon Tattoo* was banned because of its adult scenes. Other films have been barred from India for other reasons.

EXPLORING GLOBAL CONNECTIONS

The Indian Film Industry's International Reach

Google Earth (MG) Virtual Tour Video http://goo.gl/tXcDgE

India's film industry—the world's largest—has long had an impressive global reach. Indian movies are increasingly marketed abroad, and they are starting to influence filmmaking techniques in other countries. Foreign observers usually equate Indian cinema with *Bollywood*, the Hindi-language film business centered in Mumbai (formerly Bombay, hence "Bollywood," a play on "Hollywood"). But movies made in other cities using India's regional languages generate more money overall than those of Bollywood. The Tamil-language films of "Kollywood" and the Telugu-language movies of "Tollywood" have recently become influential both in India and abroad.

Figure 12.4.1 ***Baahubali: The Beginning*** Indian films often rely heavily on posters for publicity, and those for epic historical films tend to be dramatic. The 2015 bilingual (Telugu and Tamil) film *Baahubali: The Beginning* was the first non-Bollywood Indian film to gross over US $90 million worldwide.

International Appeal Bollywood's global reach became apparent to many observers in 2014, when it held its premier awards event in Tampa, Florida. The United States was selected in part because it is the most important foreign market for Hindi-language films. Indian movies are now even being set in the United States, such as the recent blockbuster *Dhoom 3*, set largely in Chicago. Indian filmmakers and distributors are drawn to the United States in part by the 3-million-strong Indian American community. But Indian Americans do not make up the entire audience, as Indian films now routinely open at more than 200 American theaters. Bollywood is the only foreign film industry that distributes its own movies in the United States, rather than relying on Hollywood firms or art-house distributors.

Bollywood films have been widely viewed in other foreign markets for decades. They gained popularity in the former Soviet Union during the Cold War, when Western films were seldom shown. West Africa, Japan, Southeast Asia, and Central Asia have long been important markets. More recently, many Germans have taken to Bollywood; 2006 saw the launch of *Ishq*, a glossy German-language Bollywood magazine. In 2015, a video of German girls doing a dubsmash of Bollywood dialogue became a viral sensation in India.

The highly musical Indian filmmaking tradition has also played a role in the development of other film industries, particularly that of Nigeria, dubbed "Nollywood." Hollywood itself has also been influenced. The 2001 film *Moulin Rouge!*, for example, was directly inspired by Indian musicals, and its success is said to have encouraged a revival of musicals in the United States.

The Rise of Southern Indian Films In India itself, Bollywood films account for only about a third of the total market. Increasingly, southern Indian movies are dubbed into other Indian languages and distributed throughout the country. In 2007, the Tamil film *Sivaji The Boss* surprised many observers by becoming a national hit. In 2015, the southern Indian film *Baahubali: The Beginning* did even better, earning over US$47 million in a mere eight days, an impressive number for India (Figure 12.4.1). Not surprisingly, Tamil and Telugu movies are beginning to have an international reach as well. As a result, in 2011 Walt Disney Pictures coproduced the highly successful Telugu fantasy-adventure *Anaganaga O Dheerudu*.

1. Why does India have so many distinctive regional filmmaking industries?
2. What are some possible reasons for the growing success of Indian films in the United States and other foreign countries?

Indiana Jones and the Temple of Doom, for example, was banned in 1984 for its racist portrayals of Indians.

Still, the pressures of internationalization are hard to resist. In the tourism-oriented Indian state of Goa, such tensions are on full display. There German and British sun worshipers often wear nothing but skimpy swimwear, whereas Indian women tourists go into the ocean fully clothed. Young Indian men, for their part, often simply walk the beach and gawk at the semi-naked foreigners (Figure 12.26).

Review

12.5 Why has religion become such a contentious issue in South Asia over the past several decades?

12.6 How have India and Pakistan tried to foster national unity in the face of ethnic and linguistic fragmentation?

KEY TERMS Hindu nationalism, Hinduism, Sanskrit, caste system, Mughal (Mogul) Empire, Dalit, Sikhism, Jainism, Dravidian language family, Hindi, Urdu, linguistic nationalism

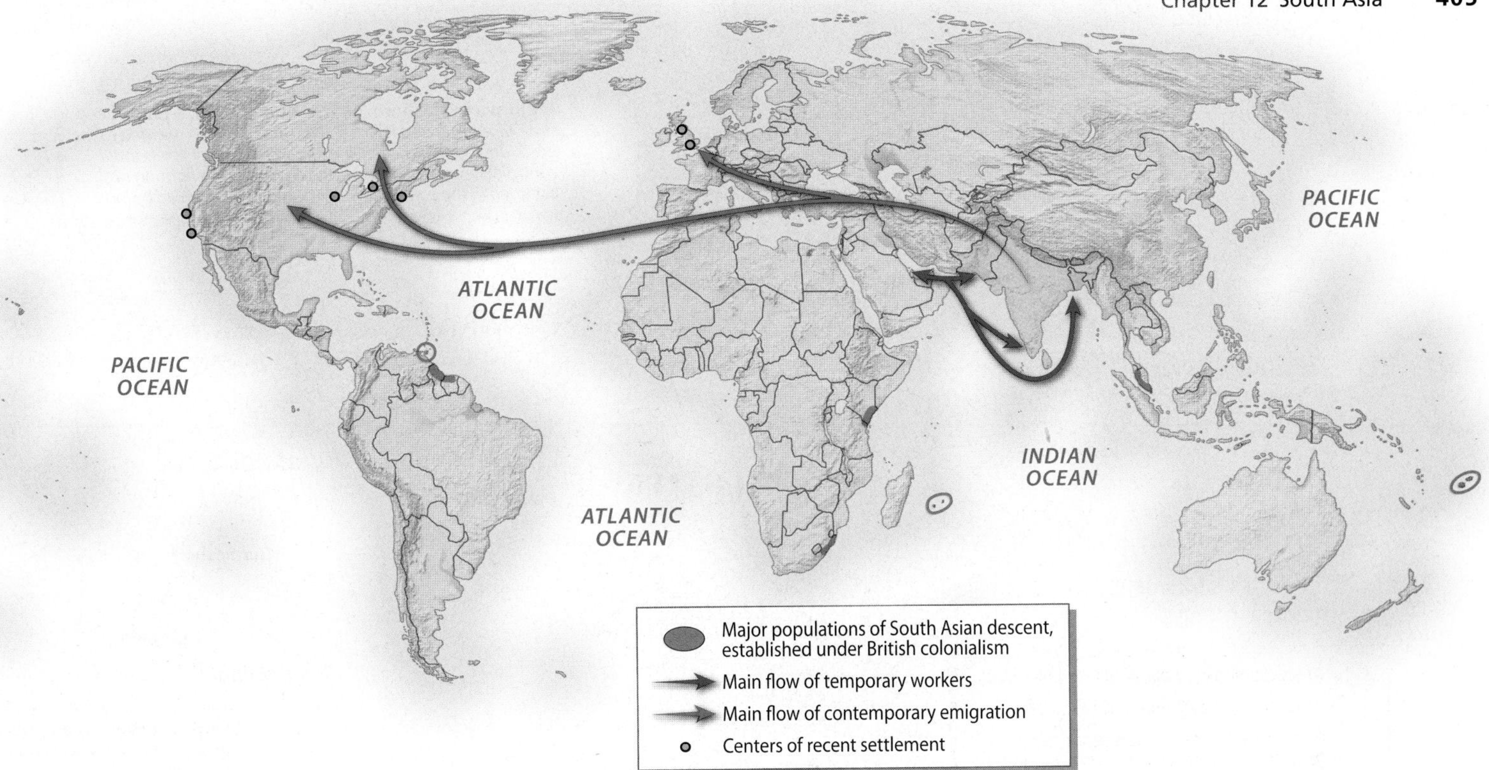

Figure 12.25 The South Asian Global Diaspora During the British imperial period, many South Asian workers migrated to other colonies. Today roughly 50 percent of the population of such places as Fiji and Mauritius is of South Asian descent. More recently, significant populations of South Asians have settled, and are still settling, in Europe (particularly Britain) and North America. Large numbers of temporary workers, both laborers and professionals, are employed in the wealthy oil-producing countries of the Persian Gulf.

Figure 12.26 Goa Beach Scene The liberal Indian state of Goa, formerly a Portuguese colony, is now a major destination for tourists, both from within India and from Europe and Israel. European tourists come in the winter for sunbathing and for "Goan rave parties," where ecstasy and other drugs are widely available. Indian tourists typically find the scantily clad foreigners unusual, if not bizarre. **Q: Why has Goa in particular emerged as a major center of beach-oriented tourism and "party culture"?**

Geopolitical Framework: A Deeply Divided Region

Prior to British imperialism, South Asia had never been politically united. At times, powerful empires ruled most of the subcontinent, but none covered its entire extent. The British, however, brought the entire region under their power by the middle of the 19th century. Independence in 1947 led to separating Pakistan from India; in 1971, Pakistan itself was divided with the independence of Bangladesh, formerly East Pakistan. Today serious geopolitical issues continue to plague the region (Figure 12.27).

South Asia Before and After Independence

During the 1500s, when Europeans first arrived, most of northern South Asia was ruled by the Muslim Mughal Empire (Figure 12.28), while southern India remained under the control of the Hindu kingdom of Vijayanagara. European merchants, eager to obtain spices, textiles, and other Indian products, established coastal trading posts. The Portuguese carved out an enclave in Goa, while the Dutch gained control over much of Sri Lanka, but neither was a significant threat to the Mughals. In the early 1700s, however, the Mughal Empire weakened rapidly, with competing states emerging in its former territories.

The British Conquest The unsettled conditions of the 1700s provided an opening for European imperialism. The British and French, having largely displaced the Dutch and Portuguese, competed for trading posts. Before the Industrial Revolution, Indian cotton textiles were the finest in the world, and European merchants needed large quantities for their global trading networks. After Britain's victory over France in the Seven Years' War (1756–1763), the French retained only a few minor coastal cities. Elsewhere, the **British East India Company**, the private organization that acted as an arm of the British government, was free to carve out its own South Asian empire.

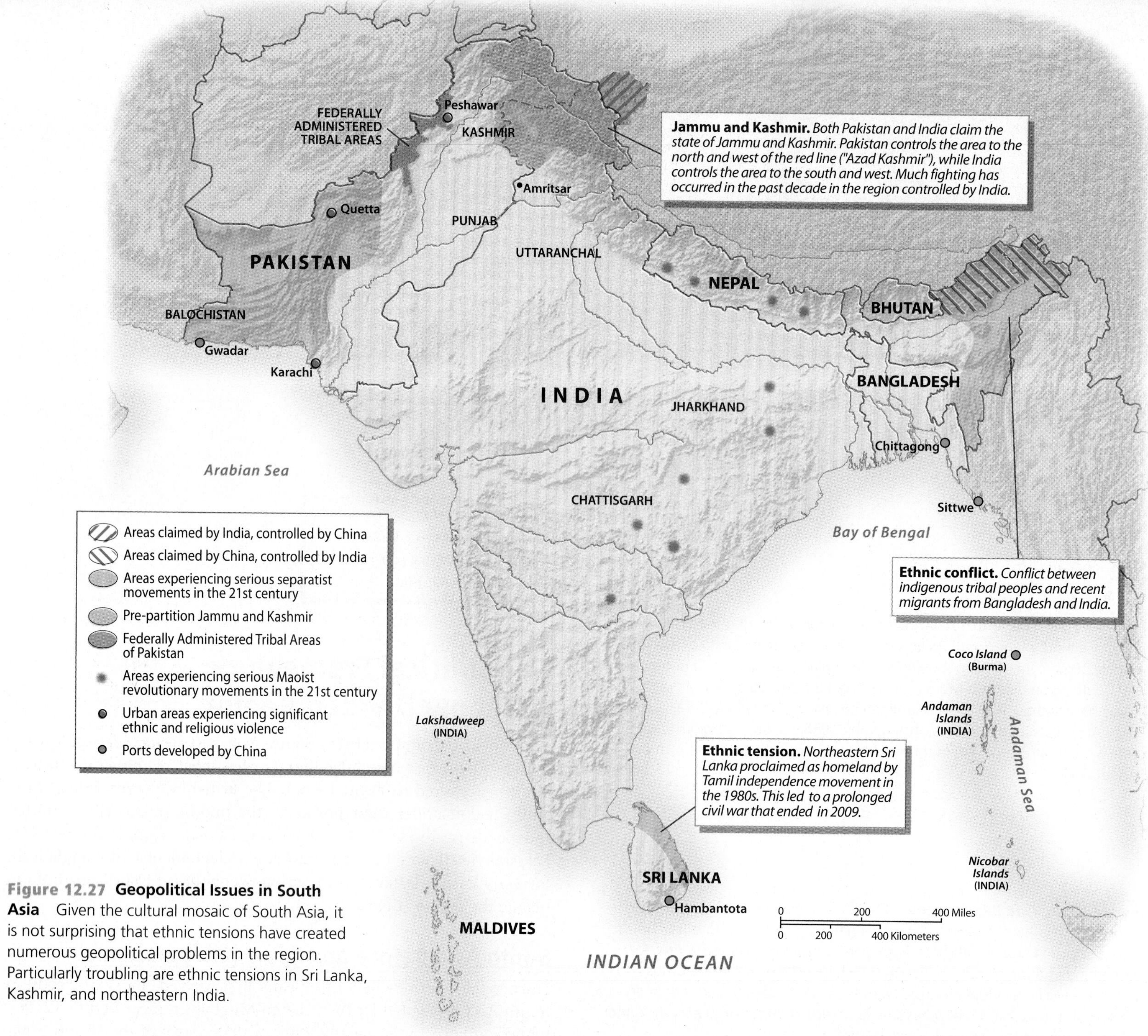

Figure 12.27 Geopolitical Issues in South Asia Given the cultural mosaic of South Asia, it is not surprising that ethnic tensions have created numerous geopolitical problems in the region. Particularly troubling are ethnic tensions in Sri Lanka, Kashmir, and northeastern India.

British control over South Asia was essentially complete by the 1840s, but valuable local allies were allowed to maintain power, provided that they did not threaten British interests. The territories of these indigenous (or "princely") states were gradually reduced, while British advisors increasingly dictated their policies.

The continual expansion of British power led to a rebellion in 1856 across much of South Asia. When this uprising (often called the *Sepoy Mutiny*) was finally crushed, a new political order was implemented. South Asia was now under the authority of the British government, with the Queen of England as its head of state. Britain enjoyed direct control over the region's most productive and densely populated areas. In more remote areas, Britain ruled indirectly, with native rulers retaining their thrones.

British officials, concerned about threats to their immensely profitable Indian colony, particularly from the Russians advancing across Central Asia, attempted to secure their boundaries. In some cases, this merely required making alliances with local rulers. In such a manner, Nepal and Bhutan retained their independence. In the extreme northeast, the British Empire took over some small states and tribal territories that had largely been outside of the South Asian cultural sphere. A similar policy was conducted on the vulnerable northwestern frontier.

Independence and Partition The framework of British India began to unravel in the early 20th century, as South Asians increasingly demanded independence. The British, however, were determined to stay, and by the 1920s South Asia was caught up in massive political protests.

The leaders of the rising nationalist movement faced a dilemma in imagining an independent country. Many leaders, including Mohandas Gandhi—the father figure of Indian independence—favored a unified state that would include all British territories in mainland South Asia. Most Muslim leaders, however, argued for dividing British India into two new countries: a Hindu-majority India and a Muslim-majority Pakistan.

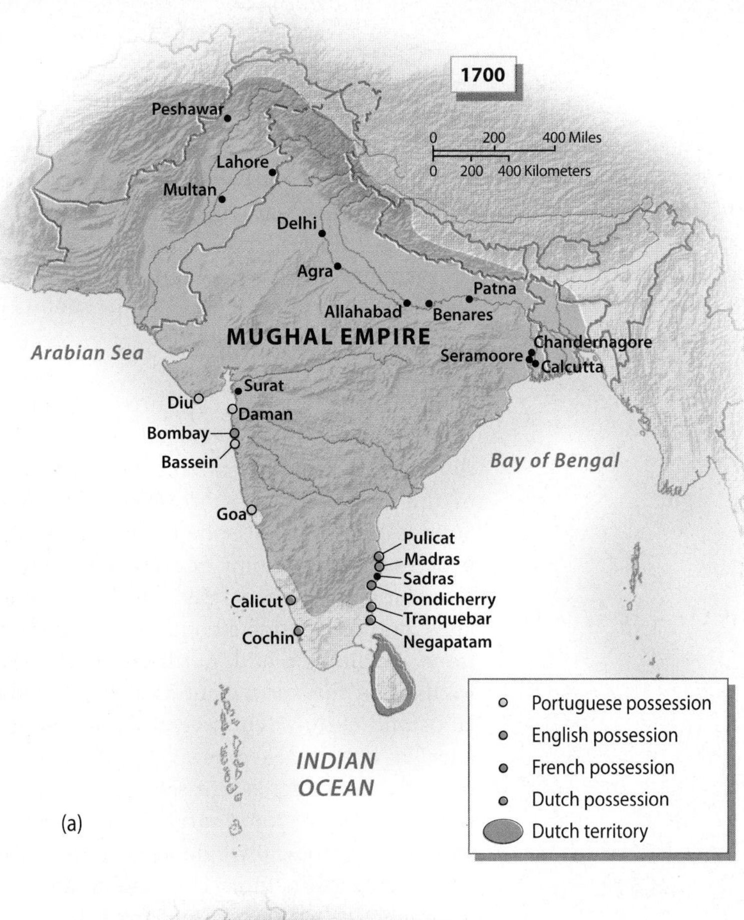

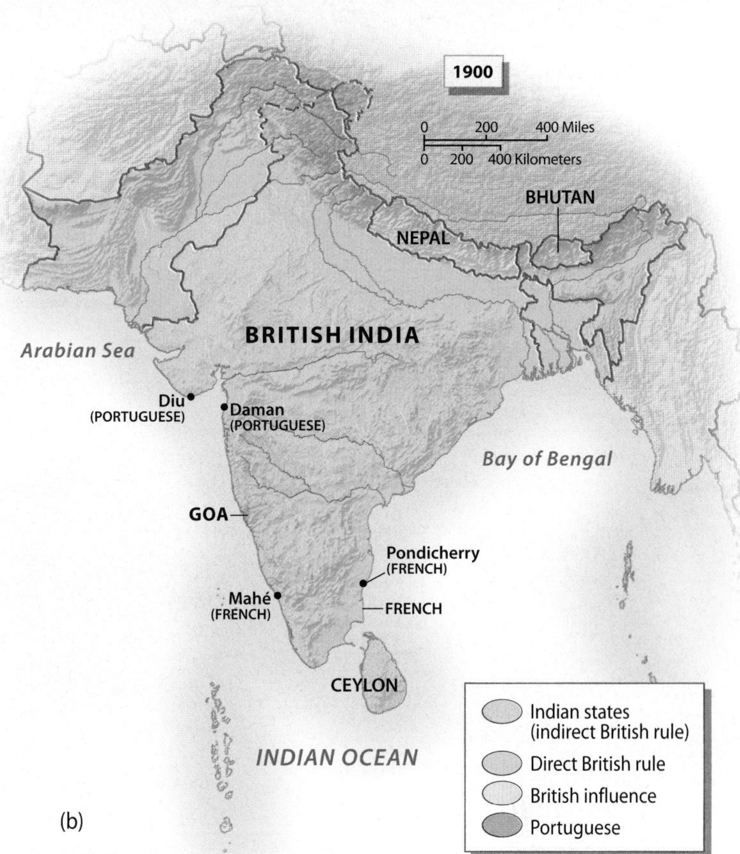

Figure 12.28 Geopolitical Change (a) At the onset of European colonialism before 1700, much of South Asia was dominated by the powerful Mughal Empire. Britain directly ruled the wealthiest parts of the region, but other lands remained under the partial authority of indigenous rulers. (b) Independence for the region came after 1947, when the British abandoned their extensive colonial territory. (c) Bangladesh, which was formerly East Pakistan, gained its independence in 1971 after a short struggle against centralized Pakistani rule from the west.

In several parts of northern South Asia, however, Muslims and Hindus were settled in roughly equal numbers. The fact that areas of clear Muslim majority were on opposite sides of the subcontinent, in present-day Pakistan and Bangladesh, was another problem.

When the British finally withdrew in 1947, South Asia was indeed divided into India and Pakistan. Partition was a horrific event. Not only were some 14 million people displaced, but also roughly 1 million were killed. Hindus and Sikhs fled from Pakistan, to be replaced by Muslims fleeing India.

The Pakistan that emerged from partition was for several decades a clumsy two-part country, with its western section in the Indus Valley and its eastern portion in the Ganges Delta. The Bengalis, occupying the poorer eastern section, complained that they were treated as second-class citizens. In 1971, they launched a rebellion and, with the help of India, quickly prevailed. Bangladesh then emerged as a new country. This second partition did not solve Pakistan's problems, however, as the country remained politically unstable and prone to military rule. Pakistan retained the British policy of allowing almost full autonomy to the Pashtun tribes living along its border with Afghanistan, a relatively lawless area marked by clan fighting. This area would later lend much support to Afghanistan's Taliban regime and to Osama bin Laden's Al Qaeda organization.

Ethnic Conflicts in South Asia

After India and Pakistan gained independence, ethnic and religious tensions continued to plague many parts of South Asia. Some of these conflicts receive little international attention despite their persistent nature, a prime example being the Baloch insurgency in southwestern Pakistan. The region's most complex—and perilous—struggle is that in Kashmir, because it involves both India and Pakistan.

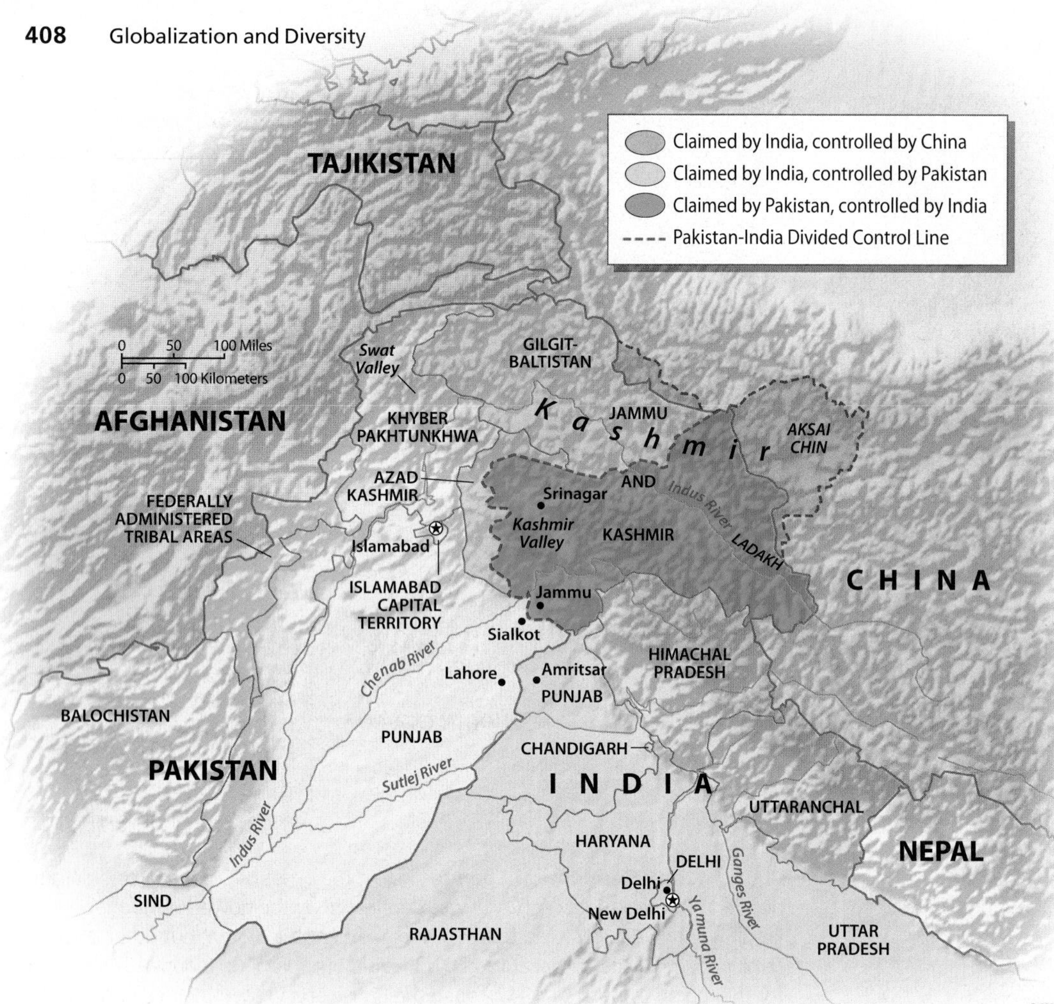

Figure 12.29 Conflict in Kashmir Unrest in Kashmir maintains hostility between the two nuclear powers of India and Pakistan. Under the British, this region of predominantly Muslim population was ruled by a Hindu maharaja, who managed to join the province to India upon partition. Today many Kashmiris wish to join Pakistan, while many others argue for an independent state.

Kashmir Relations between India and Pakistan were hostile from the start, and the situation in the Indian state of Jammu and Kashmir has kept the conflict burning (Figure 12.29). During the British period, Kashmir was a large princely state with a primarily Muslim core joined to a Hindu district in the south (Jammu) and a Tibetan Buddhist district in the northeast (Ladakh). Kashmir was ruled by a Hindu **maharaja**, a local king subject to British advisors. During partition, Kashmir came under severe pressure from both India and Pakistan. After troops linked to Pakistan gained control of western Kashmir, the maharaja decided to join India. But neither Pakistan nor India would accept the other's control over any portion of Kashmir, and they have fought several wars over the issue.

Although the Indo-Pakistani boundary has remained fixed, fighting in Kashmir has continued, reaching a peak in the 1990s. Many Muslim Kashmiris would like to join their homeland to Pakistan, others prefer that it remain a part of India, and a large portion would rather see it become an independent country. Indian nationalists are determined to keep Kashmir, while militants from Pakistan continue to cross the border to fight the Indian army, ensuring that tensions between the two countries remain high.

The Vale of Kashmir, with its lush fields and orchards nestled among some of the world's most spectacular mountains, was once one of South Asia's premier tourist destinations (Figure 12.30). By 2014, conditions seemed to have improved, as a successful election was held and tourists were beginning to reappear. In October of that year, however, Pakistani and Indian soldiers fired on each other across the border, resulting in at least four civilian deaths, again intensifying the conflict.

The Northeast Fringe A relatively obscure series of ethnic conflicts emerged in the 1980s in the uplands of India's extreme northeast. Much of this area was not historically part of the South Asian cultural sphere, and many of its peoples want autonomy or even independence. Northeastern India is relatively lightly populated and has attracted millions of migrants from Bangladesh and northern India. Many locals view this movement as a threat to their lands and culture. On several occasions, local guerillas have attacked newcomer villagers and, in turn, have suffered reprisals from the Indian military. The region's varied ethnic militias, moreover, periodically fight against each other.

Northeastern India is a remote area, and relatively little information from it reaches the outside world. The South Asia Terrorism Portal estimates that fighting here led to about 5000 deaths between 2005 and 2014. India's government is increasingly concerned about the situation, because it would like to build closer connections across the border into Burma. Although fighting has significantly declined since the early 2000s, in early 2015 India was actively conducting military operations in the states of Assam, Manipur, Nagaland, and Tripura.

Figure 12.30 The Vale of Kashmir Surrounded by high mountains, Kashmir Valley is one of the most picturesque locations in South Asia, and it remains an important tourist destination despite its political tensions. Dal Lake is located in the heart of Srinagar, the summer capital of the Indian state of Jammu and Kashmir. The numerous boats on Dal Lake are used for both fishing and sightseeing.

Tensions in the northeast complicate India's relations with Bangladesh. India accuses Bangladesh of giving separatists sanctuary on its side of the border and objects to continuing Bangladeshi emigration. As a result, India is building a 2500-mile (4000-km), $1.2 billion fence along the border (Figure 12.31). One problem with this barrier, however, has been the presence of 111 small Indian **exclaves** located on the Bangladeshi side of the border and 51 Bangladeshi exclaves on the Indian side. In 2015, the two countries finally agreed to swap these little pieces of territory to simplify the division between the two countries.

Sri Lanka Pronounced ethnic violence in Sri Lanka stems from both religious and linguistic differences. Northern Sri Lanka is dominated by Hindu Tamils, whereas the island's majority group is Buddhist in religion and Sinhalese in language. Relations between the two communities have historically been fairly good, but tensions mounted after independence (Figure 12.32). Sinhalese nationalists have favored a centralized government, with some calling for an officially Buddhist state. Most Tamils want political and cultural autonomy and have accused the government of discriminating against them.

In 1983, war erupted when the rebel force informally known as the "Tamil Tigers" attacked the Sri Lankan army. By the 1990s, most of northern Sri Lanka was under the control of the Tamil rebels. In 2007, Sri Lanka's government abandoned negotiations and launched an all-out offensive, crushing the rebel army and killing its leaders in 2009. Two years later, a UN report found evidence that both the Sri Lankan military and the Tamil Tigers had engaged in war crimes.

Figure 12.31 India–Bangladesh Fence India began building a border fence between its territory and that of Bangladesh in 2003 in order to reduce illegal immigration and to stop the influx of militants. These members of the Indian Border Security Force are patrolling a segment of the border fence.

Figure 12.32 Civil War in Sri Lanka The majority of Sri Lankans are Sinhalese Buddhists, many of whom maintain that their country should be a Buddhist state. A Tamil-speaking Hindu minority in the northeast strongly resists this idea. Tamil militants waged war against the Sri Lankan government for several decades, hoping to create an independent country in their northern and eastern homeland. They were decisively defeated by the Sri Lankan military, however, in 2009.

As of late 2014, an estimated 160,000 troops, almost entirely Sinhalese, were stationed in the Tamil areas of northern Sri Lanka. Local residents increasingly complain that the military has been grabbing land for itself, often to develop it for tourism. Several golf courses and resort hotels in the area are now reportedly run by the Sri Lankan army.

The Maoist Challenge

Not all of South Asia's conflicts are rooted in ethnic or religious differences. Poverty, inequality, and environmental degradation in east-central India, for example, have fueled a revolutionary movement inspired by former Chinese leader Mao Zedong. Mao, unlike most communist leaders, thought that peasant farmers and not just industrial workers could constitute a revolutionary force. In the area impacted by this Maoist movement, fighting is sporadic but persistent. More than 13,000 people lost their lives in this struggle between 1996 and early 2015. But after peaking in 2010 at 1177, the annual death toll declined to 372 in 2014.

Maoism has been an even greater challenge to the government of Nepal. Nepalese Maoists, infuriated by the lack of development in rural areas, emerged as a significant force in the 1990s. By 2005, they controlled over 70 percent of the country. At the same time, Nepal's urban population turned against the country's monarchy, launching massive protests. In 2008, the king stepped down and Nepal became a republic, with the leader of the former Maoist rebels serving as prime minister.

The end of the monarchy has not brought stability to Nepal. Several governments have been formed and then disbanded since 2008. Although a stable government finally seemed to emerge in 2014, arguments over proposed revisions to the constitution resulted in violent demonstrations. Much of this tension stems from plans to divide Nepal into ethnically based states. The indigenous people of Nepal's southern lowlands are particularly distressed by the migration of settlers from the more densely populated hill country and are therefore pushing for local autonomy.

International Geopolitics

As Kashmir shows, South Asia's major international geopolitical problem is the continuing cold war between India and Pakistan. The stakes are extremely high; both India and Pakistan have around 100 nuclear weapons and are currently expanding their arsenals. Although low-level fighting persists, the leaders of the two countries have been trying to reduce tensions. In 2012, India agreed to remove restrictions on Pakistani investments in Indian companies. Strains seemed to be easing in early 2015 when India's prime minister accepted an invitation to visit his Pakistani counterpart in Islamabad, but later that year planned peace talks collapsed before they could even begin.

The China Question During the global Cold War, Pakistan allied itself with the United States, while India leaned slightly toward the Soviet Union. Such alliances began to fall apart with the end of the superpower conflict in the early 1990s. Since then, Pakistan has moved closer to China, while India has done the same with the United States.

China's military connection with Pakistan is rooted in its own tensions with India. In 1962, China defeated India in a brief war, gaining control over the virtually uninhabited territory of Aksai Chin in northern Kashmir. Although growing trade has brought the two countries closer together in some respects, the fact that China continues to control Aksai Chin and to claim the entire northeastern Indian state of Arunachal Pradesh ensures that relations between the two massive countries remain frosty. In 2013, India accused China of sending troops over the "line of control" established after the 1962 war, resulting in a three-week standoff. India is also concerned about China's development of ports in India's neighbors, which could potentially be used as naval bases. A particular worry is Pakistan's massive new port of Gwadar, constructed, financed, and managed by Chinese firms (Figure 12.33).

Although India and China continue to discuss a possible border resolution, tensions persist. A 2013 poll found that 83 percent of Indians view China as a security threat. As a result, India is slowly building military relations with other countries. In 2015, the U.S. secretary of defense visited India to forge closer security ties, and India agreed to participate in naval exercises with the United States and Japan in the Indian Ocean.

Pakistan's Complex Geopolitics The conflict between India and Pakistan became more complex after the attacks of September 11, 2001. Although Pakistan had supported Afghanistan's Taliban regime, it soon agreed to help the United States in its struggle against the Taliban in exchange for military and economic aid. This decision came with large risks, as both Al Qaeda and the Taliban are supported by many of the Pashtun people of northwestern Pakistan. Before long, an Islamist-inspired insurgency broke out over much of the region.

Figure 12.33 Port of Gwadar China has invested heavily in expanding the facilities at the excellent deep-water port of Gwadar, located in Pakistan's province of Balochistan. This photo shows the port of Gwadar as it was being developed in 2007. Pakistani and Chinese authorities hope that the port will enhance trade between the two countries. **Q: Why was Gwadar selected for the development of a major port, considering the fact that it is located in a sparsely populated part of Pakistan?**

Following several military reversals, Pakistan sought peace through negotiations and even went so far as to give radical Islamists virtual control over sizable areas. From these bases, militants launched numerous attacks on U.S. forces in Afghanistan and attempted to gain control over broader swaths of Pakistan's territory. The United States responded by using drone aircraft to attack insurgent leaders, resulting in many civilian casualties and generating strong anti-American sentiment throughout the country. Relations with the United States further deteriorated in May 2011 after a U.S. raid killed Osama bin Laden deep in Pakistan's territory; an official Pakistani report later claimed that this action constituted an "American act of war against Pakistan." But despite these tensions, the United States continues to support the Pakistani military with weapons and financial aid.

Most Pakistanis are weary of the fighting, and many still hope for a negotiated settlement. In 2015, as relations between Pakistan and Afghanistan improved, negotiations were initiated in Pakistan between the Afghan government and Taliban forces operating in its territory. Political leaders in Pakistan cited that development in calling for renewed talks with Pakistan's own radical Islamists, but others warned against negotiating with possible terrorists. Some outside observers remain skeptical of all such maneuvers, arguing that Pakistan's military, particularly its extremely powerful Directorate for Inter-Services Intelligence (ISI), has been infiltrated by radical Islamist elements.

Review

12.7 How have relations between India and Pakistan influenced South Asian geopolitical developments over the past several decades?

12.8 Why has South Asia experienced so many insurgencies since the end of British rule in 1947?

KEY TERMS British East India Company, maharaja, exclave, Maoism

Economic and Social Development: Rapid Growth and Rampant Poverty

South Asia is one of the world's poorest regions, yet it is also the site of great wealth. Many of South Asia's scientific and technological accomplishments are world-class, but the area also has some of the world's highest illiteracy and malnutrition rates. And although South Asia's high-tech businesses are closely integrated with the global economy, the South Asian economy as a whole was until recently one of the world's most isolated.

South Asian Poverty

One of the clearest measures of human well-being is nutrition, and by this measure South Asia ranks very low. No other region has so many chronically undernourished people. According to a 2015 report, 39 percent of India's children suffer from stunted growth due mainly to poor diets. Over half of the people of India live on less than $2 a day, and Bangladesh is poorer still (Table A12.2; Figure 12.34). Sanitation is another major problem in the region. An estimated 70 percent of rural Indians have no access to toilets or even latrines and must relieve themselves in open fields, a practice that often spreads disease.

Despite such deep and widespread poverty, South Asia should not be regarded as a zone of misery. More than 300 million Indians are rated by local standards as members of the "middle class" who can buy such modern goods as televisions, motor scooters, and washing machines. By the early years of the 21st century, India's economy was growing at a rapid rate. Growth slumped during the global economic crisis of 2008–2009, but soon ramped back up. In early 2015, India's annual economic expansion rate was 7.3 percent, exceeding that of China. Major campaigns, moreover, are under way across the region to improve nutrition, education, and sanitation.

Figure 12.34 Poverty in India India's rampant poverty results in a significant amount of child labor. Here a 10-year-old boy is moving a large burden of plastic waste by bicycle.

Geographies of Economic Development

Following independence, the governments of South Asia attempted to build new economic systems to benefit their own people rather than foreign countries or corporations. Planners initially stressed heavy industry and economic self-sufficiency. Some gains were realized, but the overall pace of development remained slow. Since the 1990s, however, governments in the region have gradually opened their economies to the global economic system. In the process, core areas of development and social progress have emerged, surrounded by large peripheral zones that have lagged behind.

The Himalayan Countries Both Nepal and Bhutan are disadvantaged by their rugged terrain and remote locations and by their relative isolation from modern technology and infrastructure. Until recently, Bhutan remained purposely disconnected from the modern world economy, allowing its small population to live in a relatively pristine natural environment. Although it now allows direct international flights, cable television, and the Internet, Bhutan still charts its own course, emphasizing "gross national happiness" over economic growth. Bhutan has also invested heavily in hydropower dams and exports large amounts of electricity to India, boosting its economic growth. Nepal, on the other hand, is more heavily populated and suffers much more severe environmental degradation. Nepal has long relied heavily on international tourism, but its tourist industry has contracted due to political instability (Figure 12.35).

Bangladesh By several measures, Bangladesh is the region's poorest country. Environmental degradation and colonialism have contributed to Bangladesh's poverty, as did the partition of 1947. Most of prepartition Bengal's businesses were located in the west, which went to India. Slow economic growth and a rising population in the first several decades after independence meant that Bangladesh remained poor. Roughly three-quarters of its people live on less than $2 a day.

Economic conditions in Bangladesh have, however, improved in recent years. Bangladesh is internationally competitive in clothing manufacture, in part because its wage rate is so low. Low-interest credit provided by the internationally acclaimed Grameen Bank has given hope to many poor women in the country, allowing the emergence of small-scale enterprises (Figure 12.36). Information technology is now being used effectively to improve micro-loan programs and other antipoverty schemes. Bangladesh has also discovered substantial reserves of natural gas, although it has been slow to develop them. Political instability and environmental degradation, however, cloud the country's economic future.

Pakistan Like Bangladesh, Pakistan suffered deeply from partition in 1947. Even so, for several decades after independence, Pakistan maintained a more productive economy than did India. The country has a strong agricultural sector, as it shares the fertile Punjab with India.

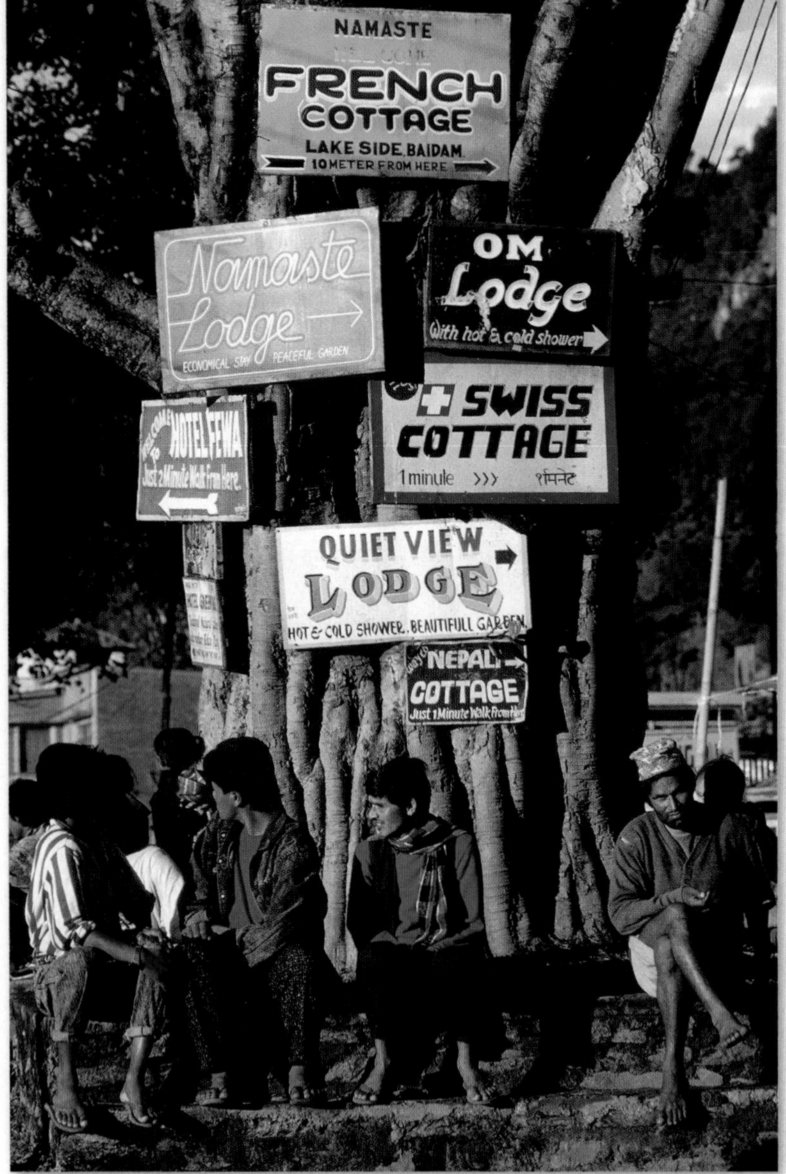

Figure 12.35 **Tourism in Nepal** Although Nepal has long been one of the world's prime destinations for adventure tourism, business has suffered greatly in recent years due to the country's Maoist insurgency. Many tourists in Nepal stay in inexpensive lodges, seen advertised in this photo.

Pakistan also boasts a large textile industry, based on its huge cotton crop. Pakistan's economy, however, is less dynamic than India's and is burdened by high levels of defense spending. In addition, a small but powerful landlord class controls much of its best agricultural lands, yet pays virtually no taxes.

Pakistan's economic growth accelerated in the early 2000s after the country was granted concessions by the international community for its role in combating global terrorism. In 2008, however, its high inflation rate and huge budget deficits required a bailout from the International Monetary Fund. Electricity shortages, a lack of foreign investment, insurgencies, and political instability contribute to Pakistan's unsettled economic future. The drop in the price of oil in late 2014, however, gave Pakistan a boost, and by early 2015 its economy was growing at an annual rate of 4.7 percent, its best figure in eight years.

Sri Lanka and the Maldives Sri Lanka's economy is by several measures the most highly developed in South Asia. The country's exports are concentrated in textiles and agricultural products such as rubber and tea. By global standards, however, Sri Lanka is still a poor country, its progress hampered by ethnic strife and political struggles. But Sri Lanka does enjoy high levels of education, abundant natural resources, and tremendous tourist potential. The end of its civil war in 2009 has boosted growth in recent years, as have investments from China.

Figure 12.36 **Grameen Bank** The internationally famous Grameen Bank supplies low-interest micro-loans to the women of Bangladesh. Here funds are being dispensed at a Grameen branch meeting. **Q: Why do micro-finance organizations such as Grameen Bank focus on women for most of their loans?**

The Maldives is the most prosperous South Asian country based on per capita economic output, but its total economy, like its population, is very small. Most of the country's revenues come from fishing and international tourism. Critics claim that most of the money from the tourist economy goes to a very small segment of the population and that tourism is vulnerable to sea-level rise and political tensions. In 2014, the Maldivian government created several special economic zones to help diversify its economic base.

India's Less Developed Areas India's per capita gross national income is in the same league as that of Pakistan, but its total economy is much larger. As the region's biggest country, India has far more internal variation in economic development (Figure 12.37). The most basic economic division is that between the more prosperous south and west and the poorer districts in the north and east. In some respects, however, India's biggest divide is between its rural and urban areas, as the countryside lags behind the cities. A recent survey found that roughly half of rural Indian households own little or no land, relying on poorly paid casual labor to survive.

India's least developed area has long been Bihar, a state of 104 million people located in the lower Ganges Valley. Bihar's per capita level of economic production is less than one-third that of India as a whole. Neighboring Uttar Pradesh, India's most populous state, is also extremely poor. Both states are also noted for their socially conservative outlooks and caste tensions. Other north-central states, including Madhya Pradesh, Jharkhand, Chhattisgarh, and Odisha, also lag behind India as a whole. In recent years, however, economic growth has been picking up in most of these states, leading to renewed hope for genuine development.

High levels of corruption have hindered development across most of South Asia, but the problem is especially severe in the poorer states of north-central India. A massive grassroots anticorruption campaign, however, has become increasingly influential and now is having a major impact on Indian politics. India is also battling corruption by using its technical expertise to create a massive database of all of its residents, using biometric information and assigning a unique identification number to each person. Critics, however, fear that this system will undermine privacy and give the government too much power.

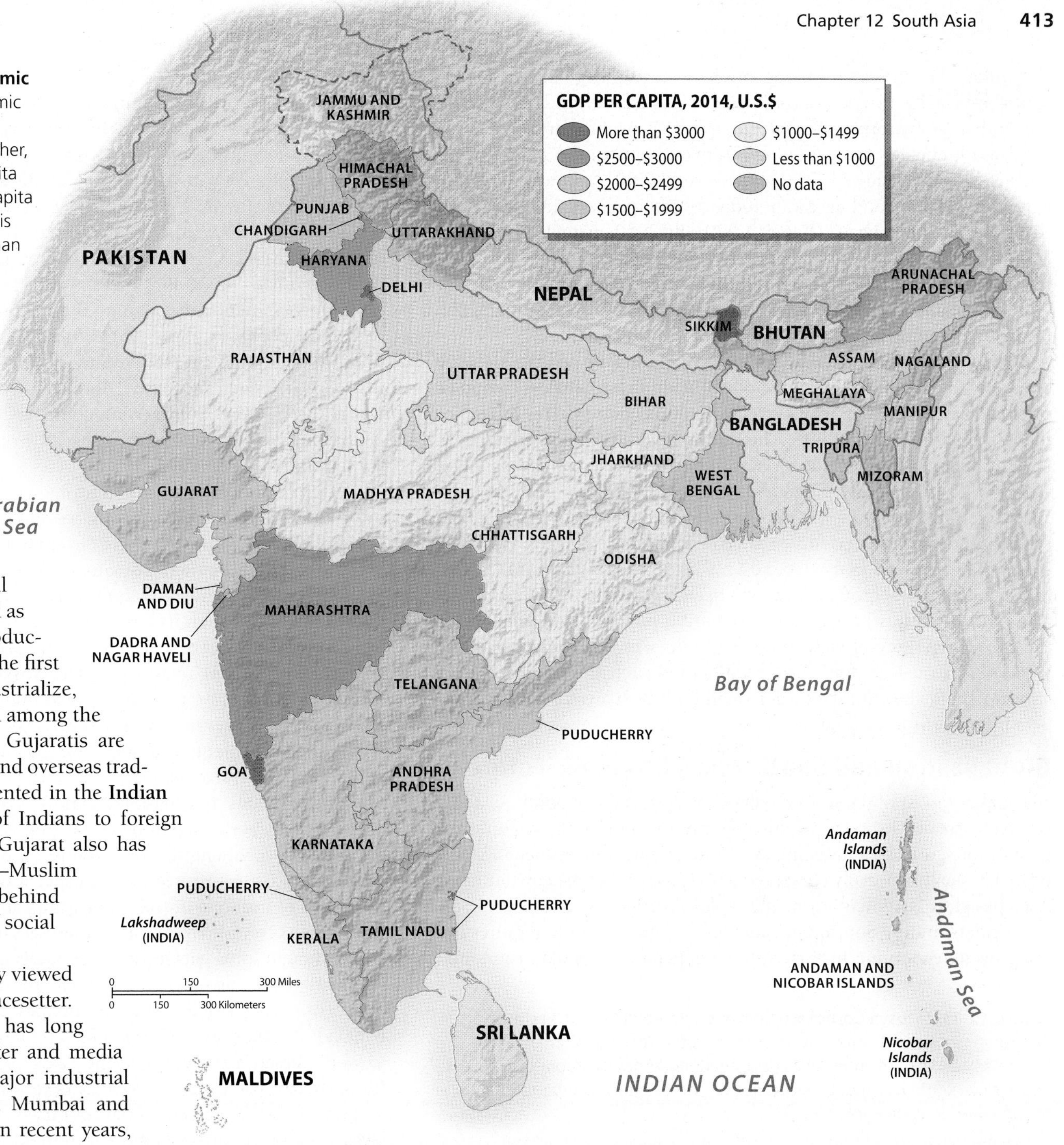

Figure 12.37 India's Economic Disparities Levels of economic development vary profoundly from one part of India to another, as seen in this map of per capita gross domestic product. Per capita economic production in Delhi is roughly seven times greater than that in Bihar.

GDP PER CAPITA, 2014, U.S.$
More than $3000
$2500–$3000
$2000–$2499
$1500–$1999
$1000–$1499
Less than $1000
No data

India's Growth Centers The west-central states of Gujarat and Maharashtra are noted for their industrial and financial power as well as for their agricultural productivity. Gujarat was one of the first parts of South Asia to industrialize, and its textile mills are still among the region's most productive. Gujaratis are well known as merchants and overseas traders and are heavily represented in the **Indian diaspora**, the migration of Indians to foreign countries. Unfortunately, Gujarat also has some of the worst Hindu–Muslim relations in India and lags behind southern India in terms of social development.

Maharashtra is usually viewed as India's economic pacesetter. Its huge city of Mumbai has long been India's financial center and media capital (Figure 12.38). Major industrial zones are located around Mumbai and in cities in Maharashtra. In recent years, Maharashtra's economy has grown more quickly than those of most other Indian states; production per capita is roughly 50 percent greater than that of India as a whole.

In the northwestern states of Punjab and Haryana, showcases of the Green Revolution, economic output per capita is also relatively high. Their economies depend largely on agriculture, but recent investments have gone to food processing and other industries. On Haryana's eastern border lies the National Capital Territory of Delhi, where much of India's political power and wealth is concentrated. This territory, in turn, is split into nine districts, one of which—New Delhi—is India's official capital.

Figure 12.38 Central Business District of Mumbai Formerly called Bombay, Mumbai is the financial and business center of India, famous for its modern and rapidly expanding skyline. But despite its wealth, Mumbai has extensive slums and millions of impoverished residents in close proximity to lavish skyscrapers.

India's fast-growing high-technology sector lies farther to the south, especially in Bengaluru (formerly Bangalore) and Hyderabad. The Indian government selected the upland Bengaluru area, noted for its pleasant climate, for technological investments in the 1950s. Other businesses soon followed. In the 1980s and 1990s, a quickly growing computer software and hardware industry emerged, earning Bengaluru the label "Silicon Plateau" (Figure 12.39). By 2000, many U.S. software, accounting, and data-processing jobs were being transferred, or "outsourced," to Bengaluru and other Indian cities. The southern states of Tamil Nadu and Kerala have also seen rapid growth in recent years, due in part to booming information technology industries.

India has proved especially competitive in software because software development does not require a sophisticated infrastructure—computer code can be exported by wireless telecommunication systems instead of modern roads or port facilities. The Indian government now stresses Internet connectivity to provide services and enhance electronics manufacturing, with several new programs announced in a 2015 "Digital India Week" celebration. India has also invested heavily in mobile telephone services and currently has more than 875 million wireless connections.

India's new economy depends on its abundant technical talent. Many Indian social groups have long been highly committed to education, and India has been a major scientific power for decades. With the growth of the software industry, India's brainpower has finally begun to create economic gains. Whether such developments can spread benefits beyond the rather small high-tech areas they presently occupy remains to be seen.

Globalization and South Asia's Economic Future

Throughout most of the second half of the 20th century, South Asia was relatively isolated from the world economy. Even today the region's volume of foreign trade and its influx of foreign direct investment are still relatively small, especially compared with those of East Asia or Southeast Asia. But globalization is advancing rapidly, particularly in India.

Understanding South Asia's low level of globalization requires examining its economic history. After independence, India's economic policy was based on widespread private ownership combined with government control of planning, resource allocation, and heavy industries. India also established high trade barriers to protect its economy from global competition. This mixed socialist-capitalist system encouraged industrial development and allowed India to become nearly self-sufficient. By the 1980s, however, problems with this model were becoming apparent. Slow economic growth meant that the percentage of Indians living in poverty remained almost constant. At the same time, countries such as Malaysia and Thailand were experiencing rapid development after opening their economies to globalization.

In response to these difficulties, India's government began to open its economy in 1991. Many regulations were eliminated, tariffs were reduced, and partial foreign ownership of local businesses was allowed. Other South Asian countries followed a somewhat similar path. Pakistan, for example, began to privatize many of its state-owned industries in 1994.

Overall, India's economic reforms have proved successful. Indian information technology firms are world-class, and many are expanding globally. Social media apps developed in India, for example, are gaining a global market. Zomato, an Indian food and restaurant locator app, now operates in more than 150 countries, including the United States, through its Urbanspoon subsidiary.

Growth has been so rapid in this sector that some companies struggle to find and retain qualified workers; as a result, wages are rapidly rising. Recent growth has also highlighted India's need to improve its infrastructure, but it is not clear how India will pay for the necessary investments in roads, railroads, and facilities for electricity generation and transmission.

The gradual internationalization and deregulation of India's economy has generated substantial opposition. Foreign competitors are seriously challenging some domestic firms. Cheap manufactured goods from China are seen as an especially serious threat. Moreover, hundreds of millions of Indian peasants and slum dwellers have seen few benefits from their country's rapid economic growth.

Although India gets most of the media attention, other South Asian countries have also experienced significant economic globalization in recent years. Besides exporting textiles and other consumer goods, Bangladesh, Pakistan, and Sri Lanka send large numbers of their citizens to work abroad, particularly in the Persian Gulf countries. Remittances from foreign workers are Bangladesh's second-largest source of income, accounting for 8.2 percent of the country's GDP in 2014. On a per capita basis, remittances are even more important for Sri Lanka. Out of a total population of 21 million, roughly 1.5 million Sri Lankans work abroad.

Figure 12.39 Infosys Corporate Campus Based in the Indian city of Bangalore (Bengaluru), Infosys is one of the world's leading information technology companies. Infosys has developed modernistic corporate campuses in several major India cities. Employees stroll and bike through the company's headquarters in Bangalore, India.

Social Development

South Asia has relatively low levels of health and education, which is not surprising considering its poverty. People in the more developed areas of western and southern India are healthier, live longer, and are better educated, on average, than people in poorer areas, such as the lower Ganges Valley. Bihar, with a female literacy rate of only 53 percent, is at the bottom of most social-development rankings, while Kerala, Punjab, and Maharashtra are near the top. Several key measurements of social welfare are higher in India than in Pakistan (see Table A12.2).

Several oddities stand out when we compare South Asia's map of economic development with its map of social well-being. Portions of India's extreme northeast, for example, show relatively high literacy rates despite their poverty, largely because of the educational efforts of Christian missionaries. For example, Mizoram's literacy rate is over

90 percent. Overall, however, southern South Asia outpaces other parts of the region in terms of social development.

The Educated South Southern South Asia's relatively high levels of social welfare are clearly visible in Sri Lanka. Given its widespread poverty and prolonged civil war, Sri Lanka is a social-development success, with an average life expectancy of 75 years and 98 percent literacy. The Sri Lankan government has achieved these results through universal primary education and inexpensive medical clinics.

On the mainland, Kerala in southwestern India has realized even more impressive results. Kerala is extremely crowded and has long had difficulty feeding its population; despite strong economic growth in recent years, its per capita economic output is only a little above average for India. Kerala's level of social development, however, is India's highest (Figure 12.40): 94 percent of Kerala's residents are literate, the average life expectancy is 75 years, and several diseases, such as malaria, have been eliminated.

Some observers attribute Kerala's social successes to its state policies. Kerala has often been led by a socialist party that stresses education and community health care. While this is undoubtedly important, some researchers point to the relatively high social position of women in Kerala as another factor at work.

Gender Relations and the Status of Women It is often argued that South Asian women have a very low social position in both the Hindu and the Muslim traditions. Throughout most of India, women traditionally leave their own families shortly after puberty to join those of their husbands. As outsiders, often in distant villages, young brides have little freedom and few opportunities. In Pakistan, Bangladesh, and such Indian states as Rajasthan, Bihar, and Uttar Pradesh, female literacy lags far behind male literacy.

An even more disturbing statistic is that of gender ratios, the relative proportion of males and females in the population. A 2011 study found that for India as a whole, only 914 girls are born for every 1000 boys; in parts of northern India, the ratio is as low as 824 to 1000. An imbalance of males over females often results from differences in care. In poor families, boys typically receive better nutrition and medical care than girls. An estimated 10 million girls, moreover, have supposedly been lost in northern India due to sex-selective abortion over the past 20 years.

Economics plays a major role in this situation. In rural households, boys are usually viewed as a blessing because they typically remain with and work for the well-being of their families. Among the poor, elderly people (especially widows) subsist largely on what their sons provide. Girls, on the other hand, are seen as an economic liability, marrying out of their families at an early age and requiring a dowry.

Figure 12.40 Education in Kerala India's southwestern state of Kerala, which has virtually eliminated illiteracy, is South Asia's most highly educated region. It also has the lowest fertility rate in South Asia. Because of this, many argue that women's education and empowerment are the most effective form of contraception.

Figure 12.41 The Gulabi Gang A grass-roots feminist organization active across northern India, the Gulabi Gang is dedicated to women's empowerment and the struggle against oppression. Its members are noted for their pink saris and bamboo sticks that can be used for self-defense.

Efforts to improve India's sex ratio include sting operations in some areas to catch doctors who perform illegal sex-selective abortions. In 2015, Punjab launched a program in which officials greet the parents of newborn girls in the most male-dominated districts and provide them with valuable gifts. Some of these initiatives have achieved significant results. A few villages in the highly traditional northern state of Rajasthan actually saw more female than male births in 2014.

Sexual violence is another huge problem across in the region, particularly in the Ganges Valley. In 2012, almost 25,000 incidents of rape were reported in India, yet activists estimate that 90 percent of rapes go unreported. Women from low-caste backgrounds are the most common victims, but all social groups are vulnerable. Social protests over sexual violence, however, have been gaining strength, as have several women's empowerment movements. One such organization is the Gulabi Gang, whose members wear pink saris and carry bamboo sticks for self-protection (Figure 12.41). The Gulabi Gang was recently the subject of a major Indian motion picture and has gained the corporate sponsorship of several information technology firms.

The social position of women is improving throughout South Asia, especially in the more prosperous areas where employment opportunities outside the family are emerging. But even in many middle-class households, women still experience deep discrimination. Indeed, dowry demands have increased in some areas, and several well-publicized murders of young brides have occurred after their families failed to deliver an adequate supply of goods. Although the social bias against women across the north is striking, it is much less evident in southern South Asia, especially Kerala and Sri Lanka.

Review

12.9 How has the economy of India been transformed since the reforms of 1991?

12.10 Why do levels of social and economic development vary so much across South Asia?

KEY TERMS Indian diaspora

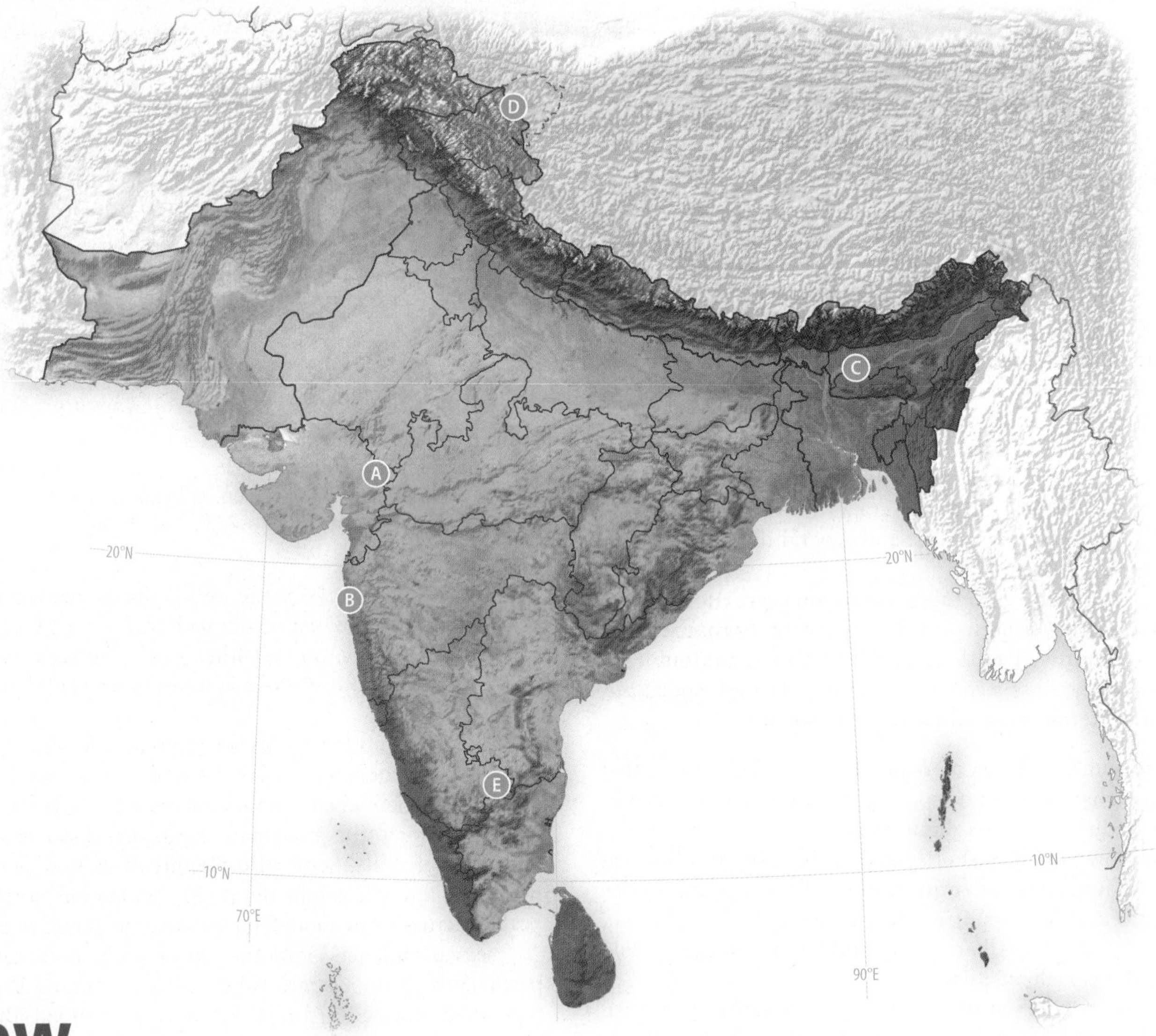

Review

Physical Geography and Environmental Issues

12.1 Describe the geological relationship between the Himalayas and the flat, fertile plain of the Indus and Ganges river valleys.

12.2 Explain how the monsoon winds are generated and describe their importance for South Asia.

Environmental degradation and instability pose particular problems for South Asia. The monsoon climate causes both floods and droughts, which tend to be more severe here than in most other world regions. Rising sea level associated with climate change directly threatens the low-lying Maldives and may seriously affect the monsoon-dependent agricultural systems of India, Pakistan, and Bangladesh.

1. Why is India so eager to build massive canals, such as the one visible here?

2. Why do some Indians deeply oppose such dam- and canal-building projects, and what potential problems are encountered when irrigation water is brought into desert areas?

A

Population and Settlement

12.3 Outline the ways in which the patterns of population growth in South Asia have changed over the past several decades, and explain their striking variation across the region.

12.4 Identify the causes of the explosive growth of South Asia's major cities, and list both the benefits and the problems resulting from the emergence of such large cities.

Continuing population growth in this densely populated region demands attention. Fertility rates have declined in recent years, but Pakistan, northern India, and Bangladesh cannot easily meet the demands imposed by their expanding populations. Increasing social and political instability may result as cities mushroom in size and rural areas grow more crowded.

3. Why have shantytowns like the one in this photograph grown so rapidly in and around the large cities of India in recent years?

4. How might India's government reduce the problem of shantytown growth and create better living conditions for people residing in such slums?

B

Cultural Coherence and Diversity

12.5 Summarize the historical relationship between Hinduism and Islam in South Asia, and explain why so much tension exists between the two religious communities today.

12.6 Compare and contrast the ways in which India and Pakistan, both containing numerous distinctive language groups, have dealt with the issue of national cohesion.

South Asia's diverse cultural heritage, shaped by peoples speaking several dozen languages and following several major religions, makes for a particularly rich social environment. Unfortunately, cultural differences sometimes generate tensions. In India, religious strife between Hindus and Muslims persists, and in Pakistan and Bangladesh, people with radical interpretations of Islam often clash with others. Debates over official languages and the role of English occur over much of the region.

5. What historical and geographical features account for the fact that the far northeastern part of India, visible in this map detail, has such linguistic and cultural diversity?

6. What kinds of problems are associated with the cultural diversity found in this part of India?

C

Geopolitical Framework

12.7 Explain why South Asia was politically partitioned following British rule and how the legacies of partition continue to generate political and economic difficulties in the region.

12.8 Describe the challenges that India, Pakistan, and Sri Lanka have faced from insurgency movements that seek to carve out new independent states from their territories.

Geopolitical tensions within South Asia are particularly severe. The long-standing feud between Pakistan and India escalated dangerously in the late 1990s, leading some observers to fear the onset of a nuclear war. Although tensions between the two countries have been reduced, the underlying sources of conflict—particularly the struggle in Kashmir—remain unresolved. India, Pakistan, and Sri Lanka, moreover, have all faced major internal insurgencies over the past several decades. India's territorial conflict with China also remains unresolved, leading India to seek closer relations with the United States and Japan and leading Pakistan to develop closer ties with China.

7. Why is India concerned about the presence of Chinese troops, like those visible in this photograph, in disputed territories?

8. How might India and China resolve their long-standing territorial dispute? Do any potential compromises seem feasible?

D

Economic and Social Development

12.9 Explain why European merchants were eager to trade in South Asia in the 16th, 17th, and 18th centuries and how their activities influenced the region's later economic development.

12.10 Summarize the variations in economic and social development across the different parts of South Asia and explain why such variability is so pronounced.

Although it is one of the poorest world regions, much of South Asia has seen rapid economic growth in recent years. India in particular seems well positioned to take advantage of economic globalization. Large segments of its huge labor force are well educated and speak excellent English, the major language of global commerce. But will these global connections help the vast numbers of India's poor? Advocates of free markets and globalization tend to see a bright future, whereas skeptics more often see growing problems.

9. Why is India investing so much money in providing a unique identification number to each of its citizens, based on such biological information as photographs, fingerprints, and iris scans?

10. Why is this program considered controversial, both in India and abroad?

E

KEY TERMS

British East India Company (*p. 405*)
bustee (*p. 397*)
caste system (*p. 398*)
Dalit (*p. 399*)
Dravidian language family (*p. 401*)
exclave (*p. 409*)
Green Revolution (*p. 396*)
Hindi (*p. 401*)
Hinduism (*p. 398*)
Hindu nationalism (*p. 398*)
Indian diaspora (*p. 413*)
Indian subcontinent (*p. 386*)
Jainism (*p. 401*)
linguistic nationalism (*p. 403*)
maharaja (*p. 408*)
Maoism (*p. 409*)
monsoon (*p. 390*)
Mughal (or Mogul) Empire (*p. 398*)
orographic rainfall (*p. 390*)
rain-shadow effect (*p. 391*)
salinization (*p. 396*)
Sanskrit (*p. 398*)
Sikhism (*p. 400*)
Urdu (*p. 402*)

DATA ANALYSIS

http://goo.gl/HzkEZO

India, particularly northwestern India, is noted for its male-biased population, which generally indicates a low social standing for women. This bias is partly due to sex-selective abortion (illegal in India, but still widely practiced), but other factors are at work, such as different levels of nutrition and medical care provided for boys and girls. Go to the website for India's Planning Commission (http://planningcommission.nic.in/) and access the sex ratio page.

1. Use these data to construct several graphs showing changes in sex ratios from 1901 to 2011 for India as a whole and for the following Indian states: Haryana, Kerala, Himachal Pradesh, and Bihar.
2. Considering the fact that sex-selective abortion only became widespread in the 1970s, what do the data suggest about the role that this procedure plays in maintaining unbalanced sex ratios in India? Can you see any effect of the campaigns to reduce sex-selective abortion and improve care for baby girls over the past several decades?
3. Compare the trend lines for the four states you graphed. Write a paragraph explaining the differences found across these four states, based on what you know about India's cultural geography and its economic and social development.

MasteringGeography™

Looking for additional review and test prep materials? Visit the Study Area in MasteringGeography™ to enhance your geographic literacy, spatial reasoning skills, and understanding of this chapter's content by accessing a variety of resources, including MapMaster interactive maps, geoscience animations, videos, *In the News* RSS feeds, flashcards, web links, self-study quizzes, and an eText version of *Globalization and Diversity*.

Authors' Blogs

Scan to visit the **Author's Blog for field notes, media resources, and chapter updates**

http://gad4blog.wordpress.com/category/south-asia/

Scan to visit the **GeoCurrents Blog**

http://geocurrents.info/category/place/south-asia

13 Southeast Asia

PHYSICAL GEOGRAPHY AND ENVIRONMENTAL ISSUES

The rainforests of Southeast Asia are vital centers of biological diversity, but they are much diminished due to commercial logging and agricultural expansion.

POPULATION AND SETTLEMENT

As is true for many other regions, Southeast Asia's river valleys, deltas, and areas of fertile volcanic soil tend to be densely populated, whereas most of its upland areas are lightly settled.

CULTURAL COHERENCE AND DIVERSITY

This region is noted for both its linguistic and its religious diversity. Many areas of Southeast Asia, however, are plagued by ethnic conflicts and religious tensions.

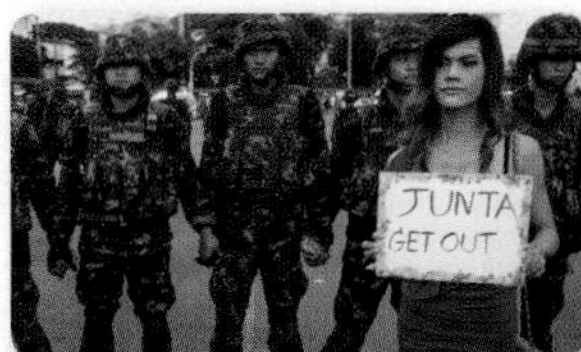

GEOPOLITICAL FRAMEWORK

Southeast Asia is one of the most geopolitically united regions of the world, with all but one of its countries belonging to the Association of Southeast Asian Nations (ASEAN).

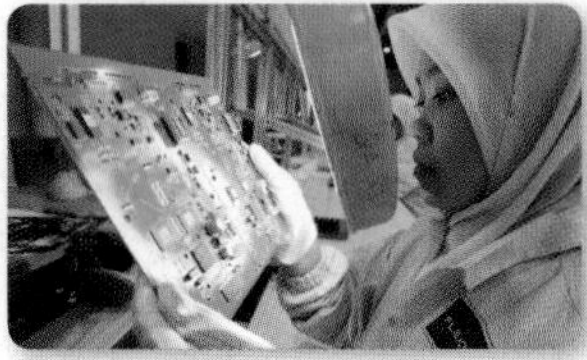

ECONOMIC AND SOCIAL DEVELOPMENT

A region of startling economic contrasts, Southeast Asia contains some of the world's most globalized and dynamic economies, as well as some of its most isolated and impoverished.

◀ The wealthy city-state of Singapore is noted for its modern architecture and lavish shopping malls. Tourists enjoy boating on a canal inside an enclosed mall called The Shoppes at Marina Bay Sands.

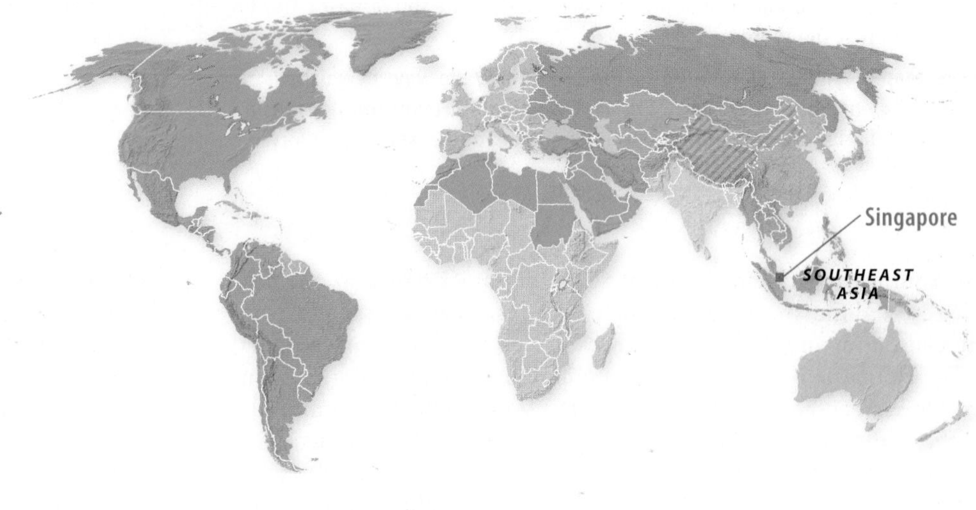

When some people think of Southeast Asian cities, they picture grim squatter settlements and sprawling slums. But others imagine something altogether different: luxury shopping. The malls of Singapore are world famous, catering to a global clientele and focusing intensively on wealthy Chinese tourists. The four-story Mandarin Gallery, located in a five-star hotel, is a case in point. Situated on Orchard Road, Singapore's retail and entertainment hub, the Mandarin Gallery's website describes it as "a retail haven for the discerning fashionista offering an amazing spread of quintessential boutiques." And Singapore is not the only major Southeast Asian city to support high-end shops. Across Southeast Asia as a whole, the luxury retail market expanded by 11 percent from 2012 to 2013.

Of course, upscale retailing is merely one aspect of the Southeast Asian urban environment. With the exception of Singapore, every major city of the region does contain grim slums. Southeast Asia, in short, is a land of extreme developmental differences. This is true, of course, in all world regions, but in Southeast Asia the disparities of wealth are particularly pronounced. Singapore is one of the world's richest and most technologically sophisticated countries, whereas Laos remains one of Asia's poorest countries.

Yet even the region's poorest countries—including Burma (Myanmar), Cambodia, and East Timor as well as Laos—have experienced rapid economic development in recent years. Much of their growth stems from increasing international trade and foreign investments, especially with and from China. Over the past 30 years, Southeast Asia has been one of the most globally connected parts of the developing world, demonstrating both the promise and the perils of globalization.

Southeast Asia, one of the most globally connected parts of the developing world, demonstrates both the promise and the perils of globalization.

Learning Objectives

After reading this chapter you should be able to:

13.1 **Describe how the region's tectonic activity and dominant climate types have influenced Southeast Asian landscapes, human settlement, and development.**

13.2 **Explain the driving forces behind deforestation and habitat loss in the different subregions of Southeast Asia, and list other environmental problems affecting the region.**

13.3 **Show how the differences among plantation agriculture, rice growing, and swidden cultivation in Southeast Asia have molded settlement patterns.**

13.4 **Describe the role of primate cities and other massive urban centers in the development of Southeast Asia, and locate the largest cities on a map of the region.**

13.5 **Map the major religions of the region to show the ways in which religions from other parts of the world spread through Southeast Asia and discuss the role of religious diversity in Southeast Asian history.**

13.6 **Identify the controversies surrounding cultural globalization in Southeast Asia, explaining why some people in the region welcome the process, whereas others resist it.**

13.7 **List the ASEAN countries and explain how this organization has influenced geopolitical relations in the region.**

13.8 **Identify the major ethnic conflict zones on the map of Southeast Asia, and explain why certain countries in the region have such deep problems in this regard.**

13.9 **Explain why levels of economic and social development vary so widely across the Southeast Asian region, giving an example from a mainland country and from an insular country.**

Southeast Asia's involvement with the larger world is not new. Chinese and Indian connections date back many centuries. Later, commercial ties across the Indian Ocean opened the doors to Islam, and today Indonesia is the world's most populous Muslim-majority country. More recently came the impact of the West, as Britain, France, the Netherlands, and the United States gained control of Southeast Asian colonies. Southeast Asia's resources and its strategic location made it a major battlefield during World War II. Yet long after peace was restored in 1945, warfare of a different sort continued in this region. As colonial powers were replaced by newly independent countries, Southeast Asia became a Cold War battleground for world powers and their competing economic systems.

Although communism eventually prevailed in Vietnam, Laos, and Cambodia, all of these countries later opened their economies to the global market. Today the struggle between capitalism and communism has given way to other issues. Relations among the various countries of the region are now generally good. The **Association of Southeast Asian Nations (ASEAN)**, which includes every country in the region except East Timor (Timor-Leste), has created a generally effective system of regional cooperation.

Southeast Asia is a clearly defined region consisting of 11 countries that vary widely in spatial extent, population, and cultural traits (Figure 13.1). Geographers generally divide these countries into those on the Asian continent—mainland Southeast Asia—and those on islands—insular Southeast Asia. The mainland includes Burma,

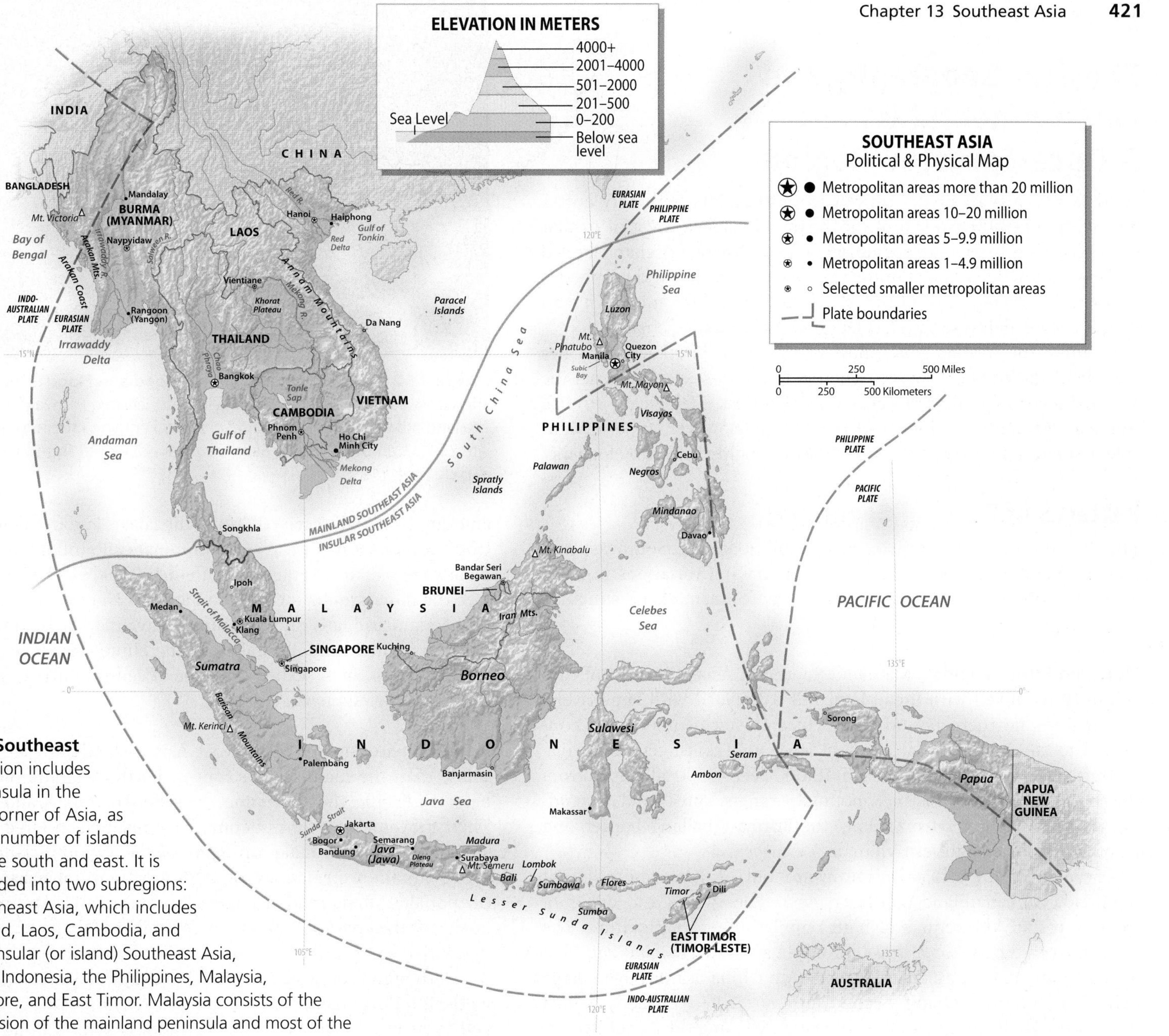

Figure 13.1 Southeast Asia This region includes the large peninsula in the southeastern corner of Asia, as well as a large number of islands scattered to the south and east. It is commonly divided into two subregions: mainland Southeast Asia, which includes Burma, Thailand, Laos, Cambodia, and Vietnam; and insular (or island) Southeast Asia, which includes Indonesia, the Philippines, Malaysia, Brunei, Singapore, and East Timor. Malaysia consists of the southern extension of the mainland peninsula and most of the northern part of the island of Borneo.

Thailand, Cambodia, Laos, and Vietnam. Although Burma has the largest mainland territory, Vietnam has the largest population of the mainland states, with 91 million people.

Insular Southeast Asia includes the large countries of Indonesia, the Philippines, and Malaysia and the small countries of Singapore, Brunei, and East Timor. Although classified as part of the insular realm because of its cultural and historical background, Malaysia actually splits the difference between the continent and the islands. Part of its national territory is on the mainland's Malay Peninsula and part is on the huge island of Borneo, some 300 miles (480 km) distant. Borneo also includes Brunei, a small, oil-rich country of roughly 400,000 people. Singapore is essentially a city-state, occupying an island just to the south of the Malay Peninsula.

Indonesia is an island nation, stretching 3000 miles (4800 km), or about the same distance as from New York to San Francisco. From Sumatra in the west to New Guinea in the east, Indonesia contains more than 13,000 separate islands. Not only does it dwarf all other Southeast Asian states in size, but also it is by far the largest in population. With roughly 252 million people, Indonesia is the world's fourth most populated country. Lying north of the equator is the Philippines, a country of about 100 million people spread over some 7000 islands, both large and small.

Controversies surround the names of two Southeast Asian countries. Since 1989, the government of Burma has insisted that the country's English name is *Myanmar*. Both terms are used in the country itself, although in Burmese they are rendered as "Bama" and "Myanma." Burma's democratic opposition movement, as well as the governments of the United States, the United Kingdom, and Canada, have continued to use the term *Burma*, as does this book. If the country's government continues on its current path of reform, we may switch to *Myanmar* in future editions. Less controversial is the official name of East Timor: Timor-Leste. As *Leste* derives from the Portuguese word for "east," most English-language sources, including this book, continue to use the more familiar term *East Timor*.

Physical Geography and Environmental Issues: A Once-Forested Region

Southeast Asia is an almost entirely tropical region, with only northern Burma extending north of the Tropic of Cancer. Most of insular Southeast Asia lies in the equatorial zone and hence is characterized by heavy rainfall over most of the year. As a result, tropical rainforests historically covered most of the island zone. Since the mid-20th century, however, deforestation has led to rapid landscape change.

Unlike the insular zone, most of mainland Southeast Asia lies outside of the equatorial zone, where a long dry season prevents the establishment of rainforest vegetation. The less ecologically diverse tropical "wet and dry" forests of this region, however, are also valuable, containing teak and other tree species highly valued by loggers.

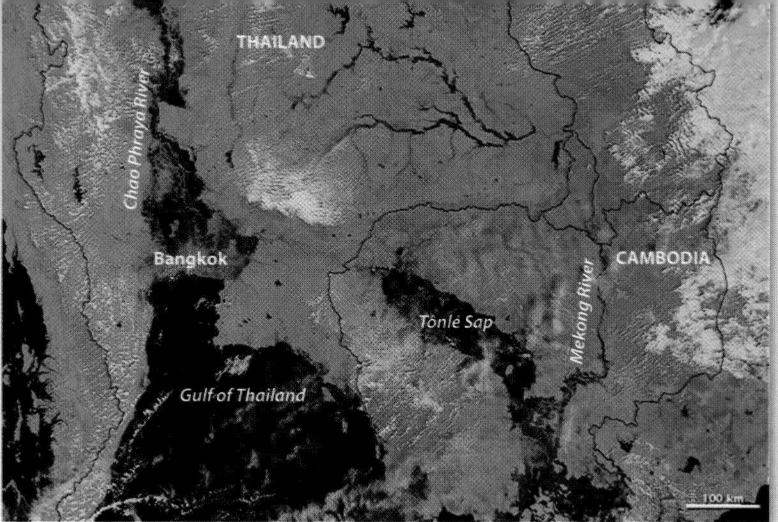

Figure 13.2 Central Thailand and Cambodia in Flood The flat, fertile, and well-watered delta and lower valley of the Chao Phraya River in central Thailand are visible in this satellite image, taken when the river was flooding in late 2011. Cambodia's Tonlé Sap (Great Lake), swollen from the same floods, is also clearly evident.

Patterns of Physical Geography

The different kinds of forests of insular and mainland Southeast Asia mostly reflect differences in climate, but striking variations in landforms and other aspects of the physical environment also distinguish the subregions.

Mainland Environments Mainland Southeast Asia is an area of rugged uplands mixed with broad lowlands associated with large rivers. The region's northern boundary lies in a cluster of mountains connected to the highlands of western Tibet and south-central China. In Burma's far north, peaks reach 18,000 feet (5500 meters). From this point, a series of distinct mountain ranges spread out, extending through western Burma, along the Burma–Thailand border, and through Laos into southern Vietnam.

Several large rivers flow southward out of Tibet and adjacent highlands into mainland Southeast Asia. The valleys and deltas of these rivers are the centers of both population and agriculture. The longest river, the Mekong, flows through Laos and Thailand and then across Cambodia before entering the South China Sea through a large delta in southern Vietnam. Second longest is the Irrawaddy, which flows through Burma's central plain before reaching the Bay of Bengal. Two smaller rivers are equally significant: the Red River, which forms a heavily settled delta in northern Vietnam, and the Chao Phraya, which has created the fertile alluvial plain of central Thailand. The lower Chao Phraya Valley is easily seen in satellite images, especially when the river is flooding (Figure 13.2).

The centermost area of mainland Southeast Asia is Thailand's Khorat Plateau. Neither a rugged upland nor a fertile river valley, this low sandstone plateau averages about 500 feet (175 meters) in height and is noted for its thin, poor soils. Water shortages and periodic droughts pose challenges throughout this extensive area.

The Influence of the Monsoon Almost all of mainland Southeast Asia is affected by the seasonally shifting winds known as the *monsoon*. The climate of this area is characterized by a distinct warm and rainy season from May to October, followed by dry, but still generally hot conditions from November to April (Figure 13.3). Only the central highlands of Vietnam and a few coastal areas receive significant rainfall during this period.

Two types of tropical climates dominate mainland Southeast Asia. Although both are affected by the monsoon, they differ in the total amount of rainfall received. Along the coasts and in the highlands, the tropical monsoon climate (Am) dominates. Rainfall totals for this climate zone usually average more than 100 inches (250 ccm) each year (see the climographs for Rangoon and Da Nang). The greater portion of the mainland falls into the tropical savanna (Aw) climate type, with half the annual rainfall totals (see the climographs for Bangkok and Vientiane). In the so-called dry zone of central Burma, rainfall is even lower, and droughts are common.

Insular Environments Southeast Asia's islands are less geologically stable than the mainland, as four of Earth's tectonic plates converge here: the Pacific, the Philippine, the Indo-Australian, and the Eurasian. Their movements over millennia have resulted in large, often explosive volcanoes that make up many of the region's islands. This still-active situation creates several additional natural hazards, including earthquakes, toxic mud volcanoes, and **tsunamis**. An undersea earthquake off the coast of Sumatra in December 2004 generated a tsunami that killed some 230,000 people in Southeast and South Asia.

Although Indonesia contains thousands of islands, it is dominated by the four large landmasses of Sumatra, Kalimantan (or Borneo), Java, and Sulawesi. A string of active volcanoes extends along the length of eastern Sumatra, across Java, and into the Lesser Sunda Islands. From late 2013 through February 2015, formerly inactive Mount Sinabung in northern Sumatra erupted numerous times, forcing the government to relocate thousands of villagers, many permanently (Figure 13.4). In the Philippines, the two largest and most important islands are Luzon (about the size of Ohio) in the north and Mindanao (the size of South Carolina) in the south. Sandwiched between them are the Visayan Islands, which number roughly a dozen. The topography of the Philippines includes rugged upland landscapes as well as numerous volcanoes.

The climates of insular Southeast Asia are more complex than those of the mainland. Most of Indonesia lies in the equatorial zone, resulting in high levels of precipitation evenly distributed throughout the year (see the climograph for Padang). Southeastern Indonesia and East Timor, however, experience a prolonged dry season from June to October. Most of the Philippines experiences dry conditions from November to April (see the climograph for Manila).

The Typhoon Threat The coastal areas of mainland Southeast Asia and the Philippines are highly vulnerable to tropical cyclones, or **typhoons**

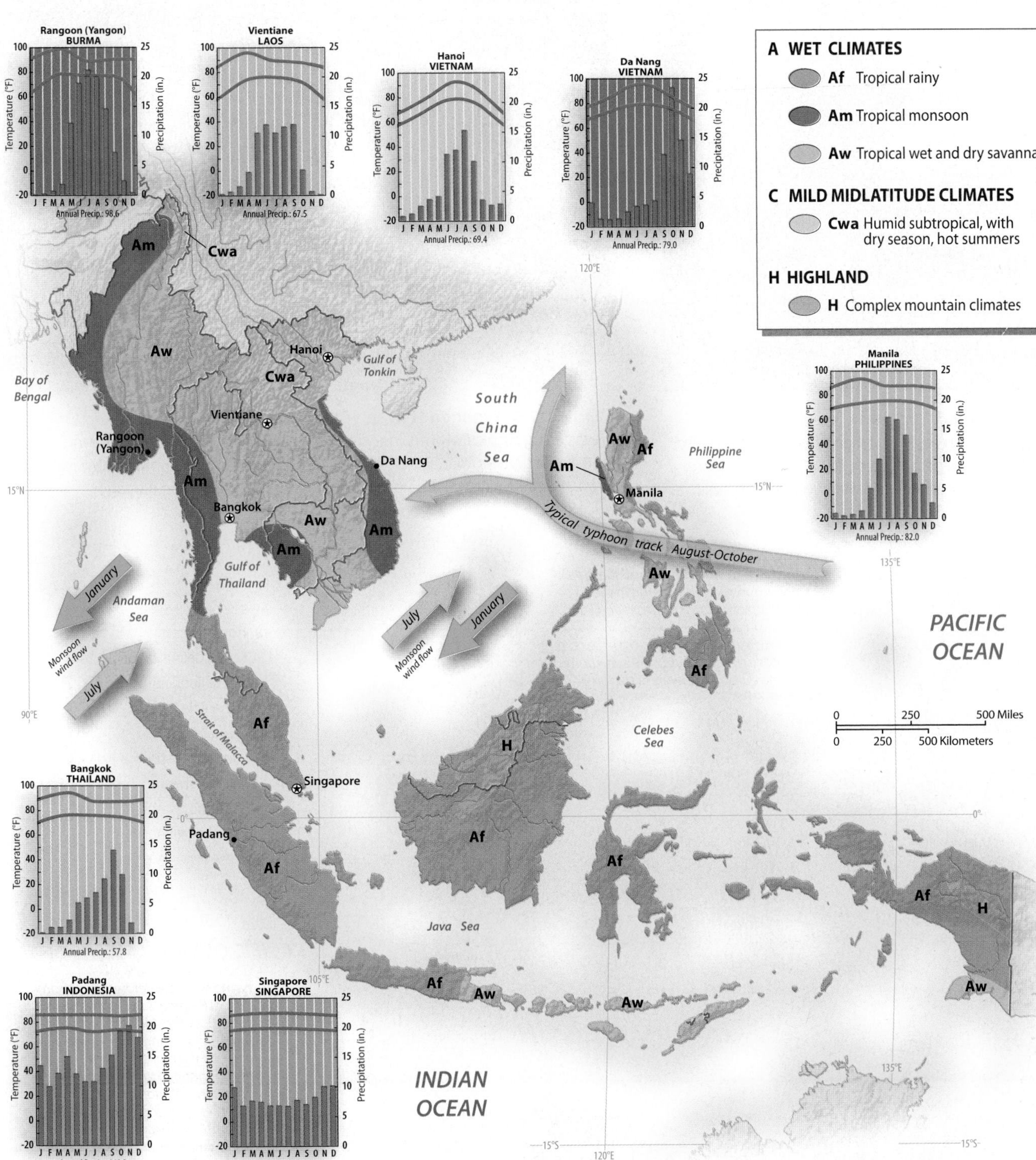

Figure 13.3 Climate Map of Southeast Asia Most of insular Southeast Asia is characterized by the constantly hot and humid climates of the equatorial zone. Mainland Southeast Asia, on the other hand, has the seasonally wet and dry climates of the tropical monsoon and tropical savanna types. Only the far north features subtropical climates, with relatively cool winters. The northern half of the region is strongly influenced by the seasonally shifting monsoon winds. Northeastern Southeast Asia—and especially the Philippines—often experiences typhoons from August to October

as they are called locally. These strong storms bring devastating winds and torrential rain. Each year several typhoons hit Southeast Asia, causing heavy damage through flooding and landslides. Deforestation and farming on steep hillsides intensify the problem.

The Philippines is particularly vulnerable to tropical cyclones. Typhoon Haiyan, which hit the central Philippines in November 2013, was the strongest storm ever to make landfall (Figure 13.5). It killed more than 6300 people, harmed more than 14 million, and caused damage in excess of US$12 billion. After the storm passed, the Philippine government adopted a US$3.7 billion reconstruction plan aimed at allowing a sustainable, long-term recovery. According to current plans, as many as 1 million people will be moved away from the most vulnerable coastal areas.

Environmental Problems: Deforestation, Pollution, and Dams

Deforestation is usually considered the main environmental problem in Southeast Asia, but it is not the only one. As is true over most of the developing world, air and water pollution impacts the lives of hundreds of millions of people. A more recent environmental controversy in Southeast Asia concerns dam-building.

Patterns of Deforestation

Deforestation and related environmental problems are major issues throughout most of Southeast Asia (Figure 13.6). Although countries such as Indonesia have transformed large areas of forests into croplands to

Figure 13.4 Mount Sinabung Erupting Indonesia is famous for its numerous volcanic eruptions, which often result in extensive damage but also generate highly fertile soils. In 2014 and again in early 2015, Mount Sinabung volcano in North Sumatra violently exploded, forcing many villagers to seek safety elsewhere.

Figure 13.5 Typhoon Haiyan The Philippines has suffered several devastating tropical storms in recent years, including Typhoon Bobha in 2012 and Typhoon Haiyan in 2013. **Q: What features of the Philippines' physical and human geography make it particularly vulnerable to cyclone damage?**

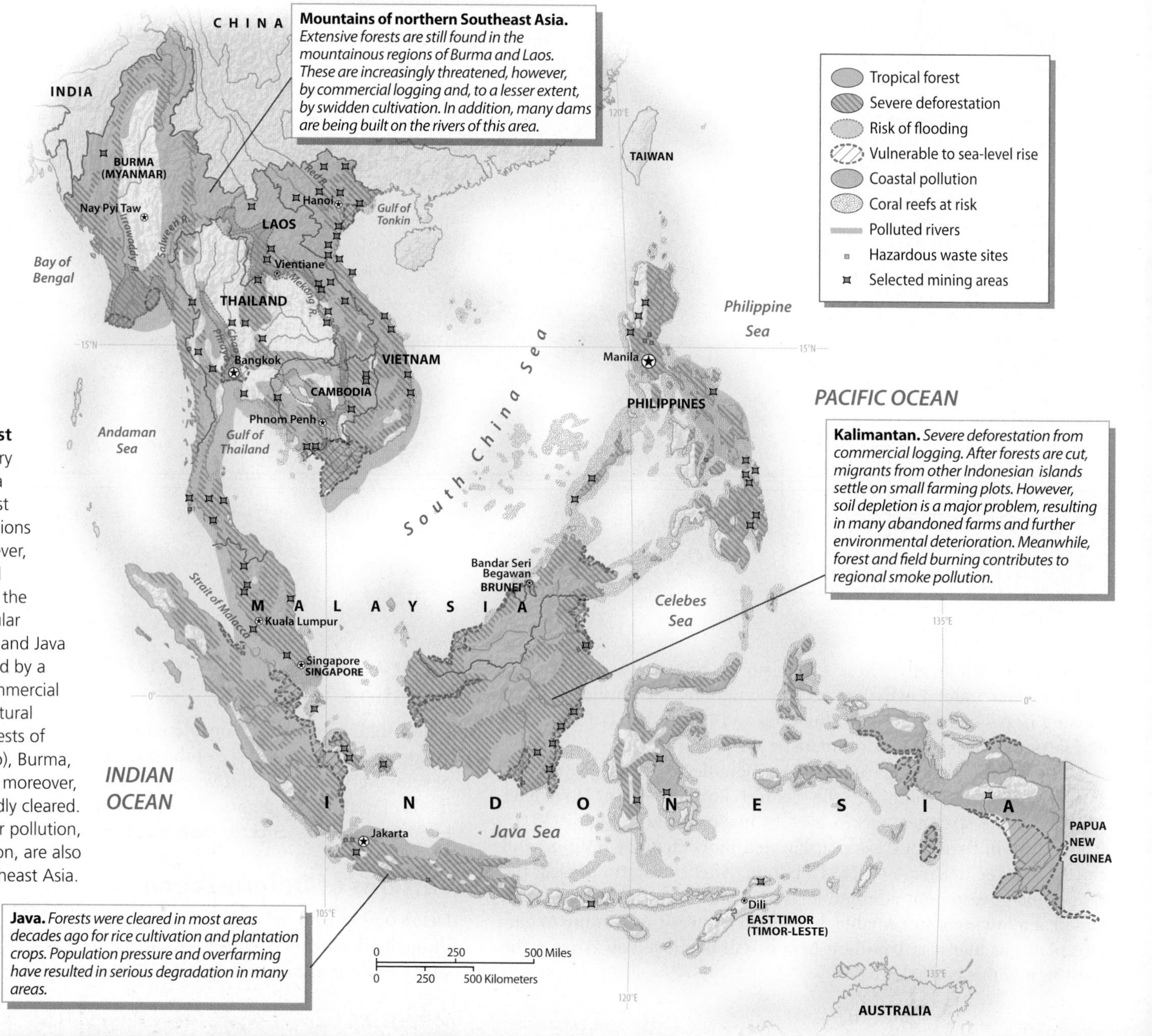

Mountains of northern Southeast Asia. *Extensive forests are still found in the mountainous regions of Burma and Laos. These are increasingly threatened, however, by commercial logging and, to a lesser extent, by swidden cultivation. In addition, many dams are being built on the rivers of this area.*

Tropical forest
Severe deforestation
Risk of flooding
Vulnerable to sea-level rise
Coastal pollution
Coral reefs at risk
Polluted rivers
Hazardous waste sites
Selected mining areas

Kalimantan. *Severe deforestation from commercial logging. After forests are cut, migrants from other Indonesian islands settle on small farming plots. However, soil depletion is a major problem, resulting in many abandoned farms and further environmental deterioration. Meanwhile, forest and field burning contributes to regional smoke pollution.*

Java. *Forests were cleared in most areas decades ago for rice cultivation and plantation crops. Population pressure and overfarming have resulted in serious degradation in many areas.*

Figure 13.6 Environmental Issues in Southeast Asia A half century ago, Southeast Asia was one of the most heavily forested regions of the world. However, most of the tropical forests of Thailand, the Philippines, peninsular Malaysia, Sumatra, and Java have been destroyed by a combination of commercial logging and agricultural settlement. The forests of Kalimantan (Borneo), Burma, Laos, and Vietnam, moreover, are now being rapidly cleared. Water and urban air pollution, as well as soil erosion, are also widespread in Southeast Asia.

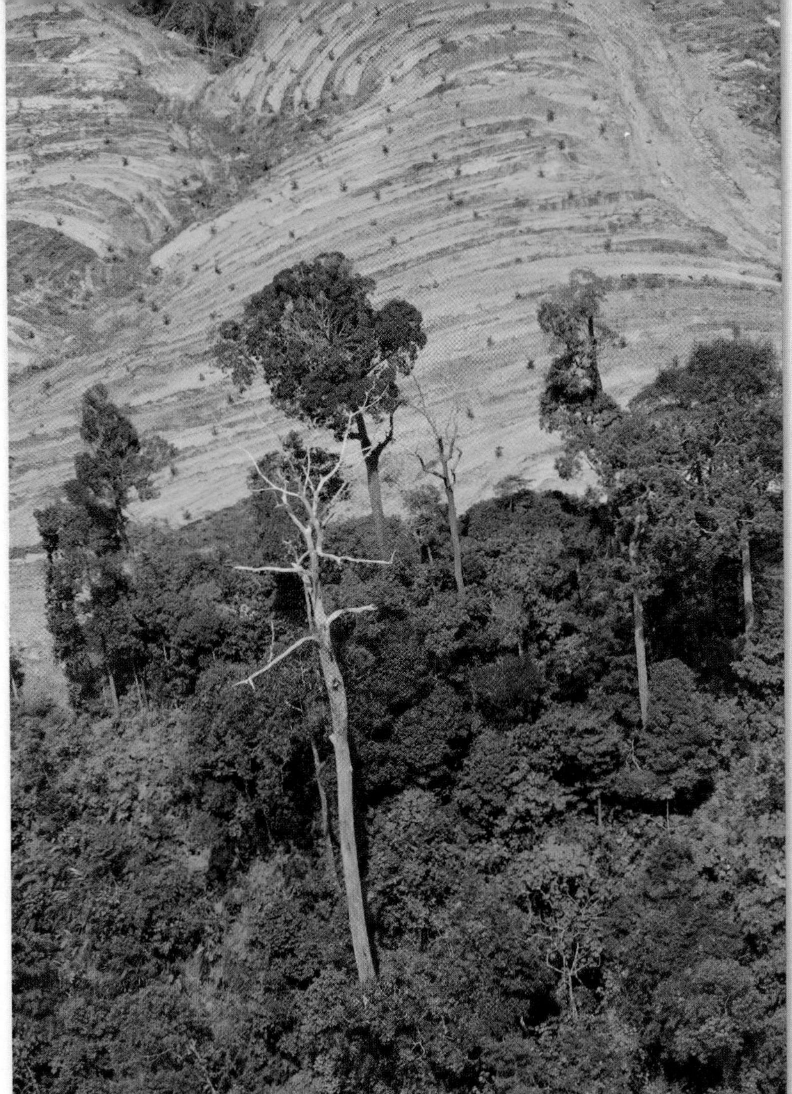

Figure 13.7 Forest Clearance Insular Southeast Asia supports some of the world's most ecologically diverse tropical rainforests. Unfortunately, many of the region's forests have been cleared. After logging, forests are often replaced with oil palm plantations, as seen here in Indonesia's Papua province.

help feed their expanding populations, population growth is not the main cause of deforestation. Most forests are cleared so that the wood products can be exported to other parts of the world. Initially, Japan, Europe, and the United States were the main importers, but as China industrializes, its demand has grown. After loggers move through, cleared lands are often planted with oil palms and other export-oriented crops (Figure 13.7).

Malaysia has long been a leading exporter of tropical hardwoods from Southeast Asia. Mainland Malaysia was mostly deforested by 1985, when a cutting ban was imposed. Since then, logging has been concentrated in the states of Sarawak and Sabah on the island of Borneo. Granting logging concessions to Malaysian and foreign firms has caused considerable problems with local tribal people by disrupting their traditional resource base.

After Thailand cleared more than half of its forests between 1960 and 1980, a series of logging bans virtually eliminated commercial forestry by 1995. Damage to the landscape, however, was severe; flooding increased in lowland areas, and hillslope erosion led to silt accumulation in irrigation works and hydroelectric facilities. Many of these cutover lands are being reforested with fast-growing Australian eucalyptus trees, which do not support local wildlife. The Thai forestry ban, moreover, has resulted in increased logging—much of it illegal—in remote areas of Laos, Cambodia, and Burma.

Indonesia, the largest country in Southeast Asia, has fully two-thirds of the region's forest area, including about 10 percent of the world's true tropical rainforests. Most of Sumatra's forests have been removed, however, and those on Borneo are receding. Indonesia's last major forestry frontier is on the island of New Guinea, where forests are still extensive. Indonesia is trying, however, to preserve the remaining areas of habitat elsewhere in the country. Kutai National Park on Borneo, for example, covers more than 741,000 acres (300,000 hectares). Conservation officials hope that this and other protected areas will allow animals a chance to survive in the wild, including the orangutan, which now lives only in restricted areas of Borneo and northern Sumatra.

Smoke and Air Pollution

At one time, most of Southeast Asia's residents seemed unconcerned about the widespread air pollution created by a combination of urban smog and smoke from forest clearing. Then, late in the 1990s, the region suffered from two consecutive years of disastrous air pollution that served as a wake-up call (Figure 13.8). Although the situation has improved, damaging blankets of haze continue to descend on the region. In 2013, fires in northern Sumatra generated smoke plumes that extended to Borneo, causing record high levels of air pollution in Singapore and across much of Malaysia.

Several factors combined to produce the region's air pollution disaster of the late 1990s and early 2000s. First, large portions of Southeast Asia suffer from periodic extreme drought, often caused by El Niño (see Chapter 4), which can turn the normally wet tropical forests into tinderboxes. Drought, along with draining for agriculture, also dries out the peat bogs of coastal wetlands, which can continue to burn for months (see *Working Toward Sustainability: New Efforts to Create an Environmentally Responsible Palm Oil Industry* on pages 428–429). Second, commercial forest cutting has been responsible for many fires, as the leftover slash (branches, small trees, and so forth) is often burned to clear out the land. A third factor is Southeast Asia's rapidly growing cities, where cars, trucks, and factories emit huge quantities of air pollutants.

Figure 13.8 Urban Air Pollution The rapidly industrializing mega-cities of Southeast Asia experience high levels of air pollution, particularly in Bangkok, Manila, and Jakarta. People sometimes resort to using face masks to filter out soot and other forms of particulate matter. Forest fires, which often follow logging, add to the problem.

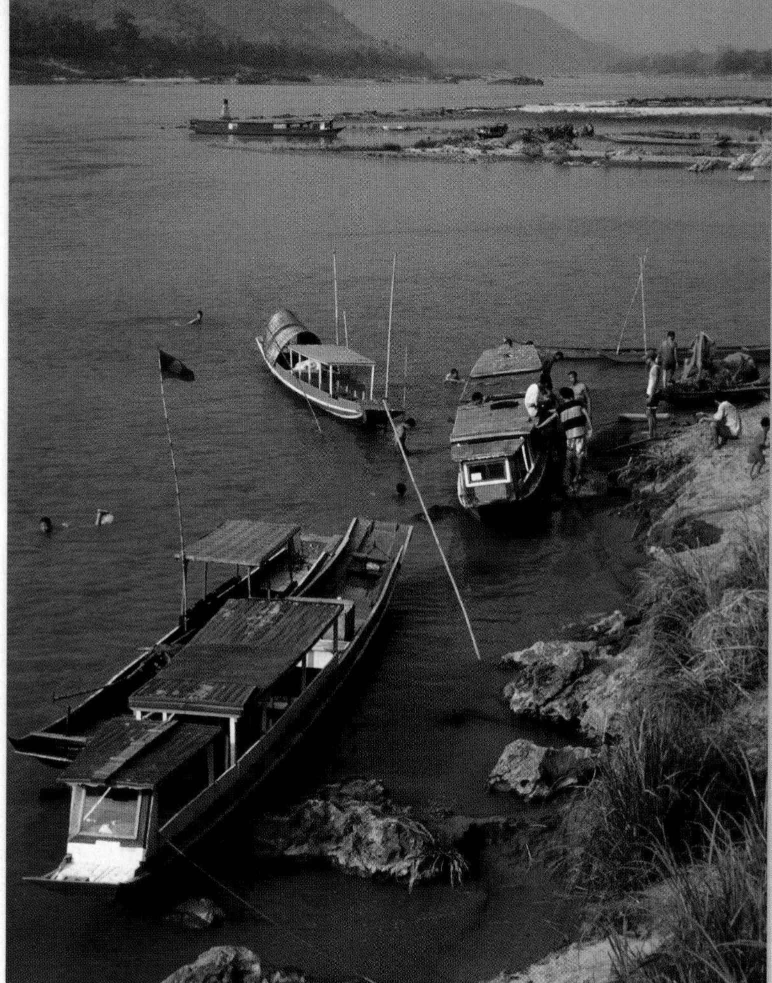

Figure 13.9 Mekong River Flowing through Mainland Southeast Asia, the Mekong is one of the world's largest and most ecologically productive rivers. It is increasingly threatened by dam construction.

Over the past decades, several Southeast Asian cities have built rail-based public transportation systems to reduce traffic and vehicular emissions. Bangkok in particular has made great strides, substantially reducing its levels of ozone and sulfur dioxide. In the Philippines, Manila still suffers from deteriorating air quality, but new transportation initiatives promise to improve the situation. Efforts have also been made to reduce forest fires and the resulting smoke crises. Particularly important is the 2002 ASEAN Agreement on Transboundary Haze Pollution, which was finally ratified by Indonesia in 2014.

Dam-Building in Southeast Asia

Dam-building has been an especially controversial matter in Laos, Burma, and Cambodia, where China has been financing and building large dams, in part to meet its own energy needs. Laos currently plans to expand its hydroelectric output from 3200 megawatts in 2014 to 12,000 megawatts in 2030, which will require 30 new dams, 12 of which will be on the Mekong River. Environmental groups fear that such dams will reduce biological diversity and threaten the livelihoods of local people. Freshwater fisheries are particularly threatened by damming the massive Mekong River, the delta of which is already receding (Figure 13.9). Scientists have warned that as many as 20 fish species could be wiped out by the project, including the Mekong giant catfish, which can reach 660 pounds (300 kilograms).

The Laotian government has responded to such criticism by redesigning dams to allow some fish migration and to reduce sedimentation, although not all plans have been released to the public. But Laos is determined to proceed with its ambitious dam-building plans, claiming that this is the best way to lift the country out of poverty and to generate electricity without emitting massive quantities of carbon dioxide. Most Southeast Asian economies show strong growth, increasing the region's energy demands and carbon emissions at a rapid pace.

Climate Change and Southeast Asia

According to a 2014 UN report, climate change will hit Southeast Asian cities and agricultural areas particularly hard. The region is highly vulnerable to the rise in sea level associated with global warming. Most of Southeast Asia's people live along the coast, and periodic flooding is already a major problem in many of the region's low-lying cities, particularly Bangkok and Manila. Southeast Asian farmland is also concentrated in delta environments and thus could suffer from saltwater intrusion and higher storm surges. Environmentalists also fear that higher temperatures themselves could reduce rice yields throughout the region. Other possible problems include reduced fish stocks from ocean acidification and a decline in tourism in coastal resorts.

Changes in precipitation across Southeast Asia due to climate change remain uncertain. Some experts foresee an intensification of the monsoon pattern, which could bring increased rainfall to much of the mainland, but reduce it in the equatorial zone. While enhanced precipitation would likely result in more destructive floods, it could bring some agricultural benefits to dry areas such as Burma's central Irrawaddy Valley. Others think that both dry spells and flooding could be intensified. In 2010, northern mainland Southeast Asia saw one of the worst droughts in recorded history. Then, in the summer of 2011, Thailand experienced massive flooding that inundated 14.8 million acres (6 million hectares) of land.

Southeast Asia's overall carbon emissions from conventional sources remain low by global standards. When greenhouse gas output associated with deforestation is factored in, however, Southeast Asia's role in global climate change becomes much larger. By some estimates, Indonesia is the world's third-largest contributor to the problem, after China and the United States. Much of Indonesia's carbon emissions stem from the burning and oxidation of peat in deforested swamps (Figure 13.10).

Figure 13.10 Burning Peatlands The soil of most wetland areas in Southeast Asia is composed largely of peat, an organic substance that can burn when dry. Draining for agricultural expansion, as well as drought, often results in extensive peat fires. Here a C130 airplane is dropping water on burning peat in Lampung, Sumatra.

Review

13.1 Why do the mainland and insular regions of Southeast Asia have such distinctive climates and landforms, and how have these differences affected the human communities of these two regions?

13.2 How have people changed the physical landscapes of Southeast Asia over the past 50 years? How is global climate change impacting these landscapes now, and how might it further impact them in the future?

KEY TERMS Association of Southeast Asian Nations (ASEAN), tsunami, typhoon

Population and Settlement: Subsistence, Migration, and Cities

Southeast Asia's population issue is quite different from those of East Asia and South Asia. With over 600 million people, Southeast Asia is still *relatively* sparsely settled. Throughout the region, extensive tracts of land with infertile soil and rugged topography remain thinly inhabited. In contrast, dense populations live in the region's deltas, coastal areas, and zones of fertile volcanic soil (Figure 13.11). Southeast Asia experienced rapid population growth in the second half of the 20th century. More recently, birthrates have dropped quickly, especially in the wealthier countries of the region.

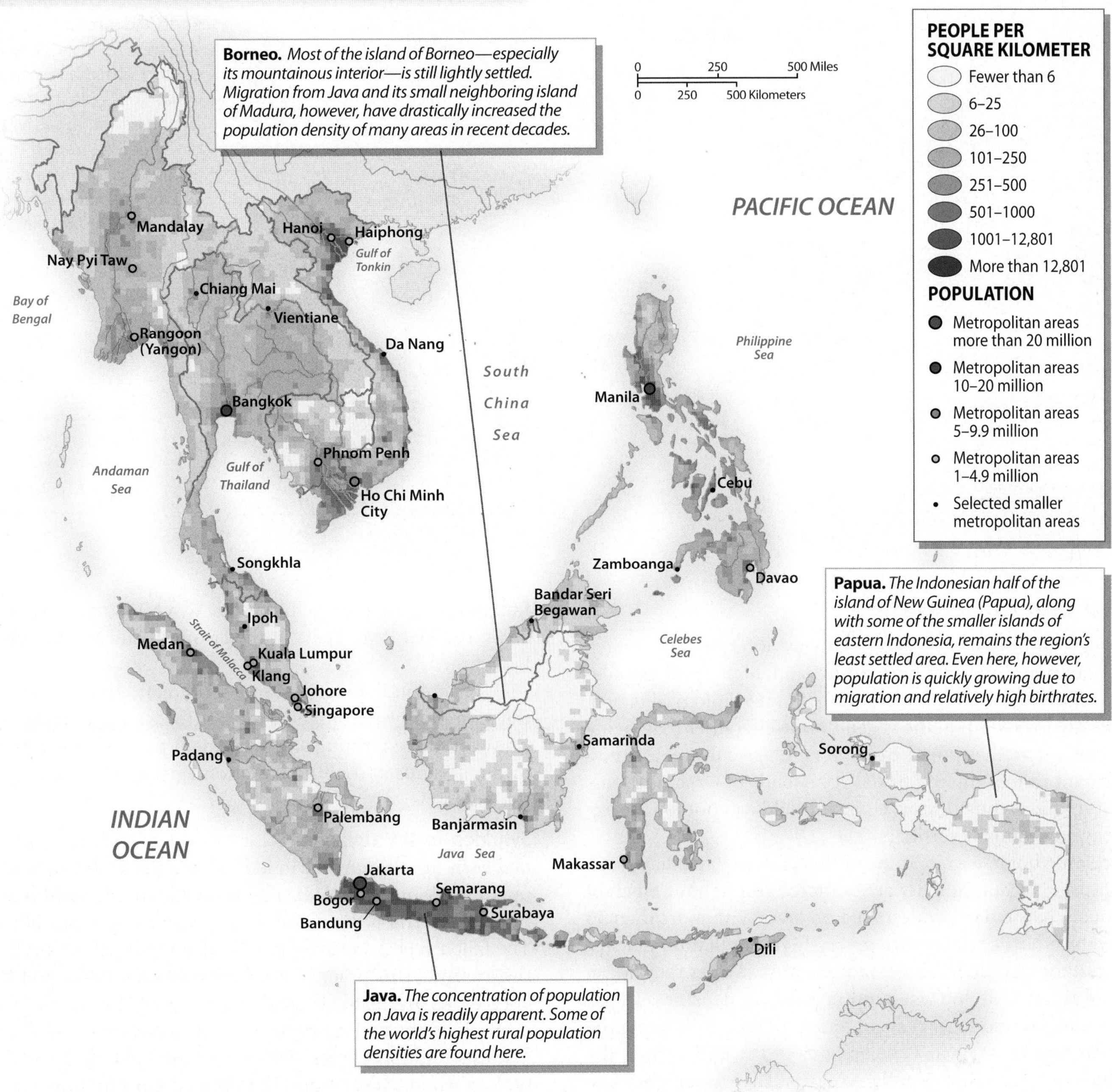

Figure 13.11 Population Map of Southeast Asia In mainland Southeast Asia, population is concentrated in the valleys and deltas of the region's large rivers. In the uplands, population density remains relatively low. In Indonesia, density is extremely high on Java, an island noted for its fertile soil and large cities. Some of Indonesia's outer islands, especially those of the east, remain lightly settled. Overall, population density is high in the Philippines, especially in central Luzon. **Q: Why is the population of Vietnam so unevenly distributed? What are some of the political and economic consequences of such uneven distribution?**

WORKING TOWARD SUSTAINABILITY

New Efforts to Create an Environmentally Responsible Palm Oil Industry

Palm oil is an inexpensive vegetable oil, useful for both baking and frying, that is extracted from the fruit of the African oil palm tree. The global demand for palm oil is expected to double by 2030. Most of the crop is produced on large plantations in Indonesia and Malaysia (Figure 13.1.1).

Problems with Palm Oil Palm oil is also controversial. Some critics claim that it is unhealthful, leading to high cholesterol levels and heart disease. Palm oil's negative effects on the environment have been more clearly confirmed. Plantations are often established where tropical rainforests have been removed, resulting in habitat loss that threatens endangered species, including orangutans, Sumatran tigers, and Sumatran elephants. Even more destructive is the practice of draining and clearing wetland forests to plant oil palms. This process leads to oxidation of the organic-based peat soils of the former wetlands, releasing huge quantities of carbon dioxide. Dried-out peat also burns easily, and the resulting fires periodically spread devastating air pollution over much of insular Southeast Asia.

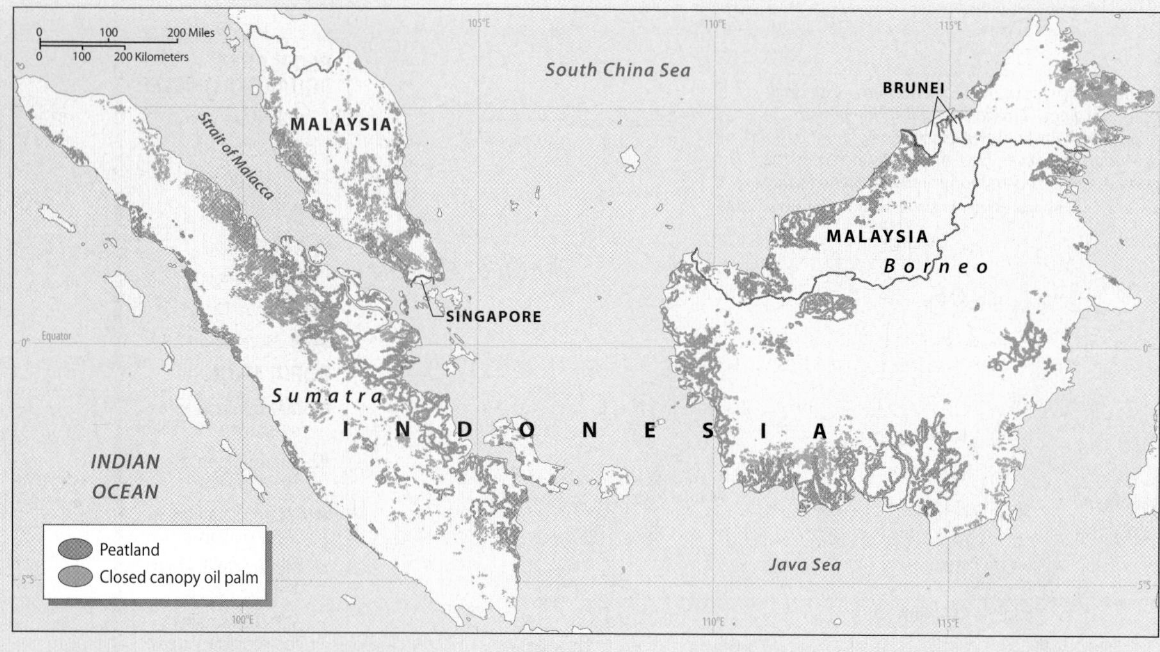

Figure 13.1.1 Palm Oil and Peatland Areas Swamplands with organic peat soils are widespread in eastern Sumatra, Borneo, and the Malay Peninsula. Many of these areas, along with other local forestlands, have been converted to oil-palm plantations in recent decades.)
Q: Why are vulnerable peatlands so often used for oil-palm plantations?

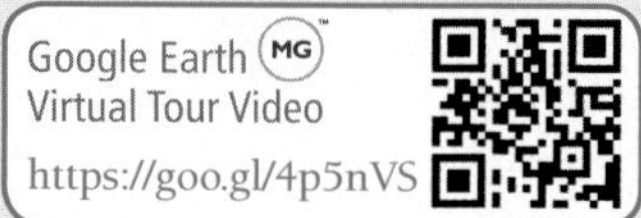

Settlement and Agriculture

Much of insular Southeast Asia has relatively infertile soil, which cannot easily support intensive agriculture and high rural population densities. Although island forests are lush and biologically rich, plant nutrients are locked up in the vegetation itself, rather than being stored in the soil where they could easily benefit agriculture. Furthermore, the constant rain of the equatorial zone tends to wash nutrients away. Agriculture must thus rely on constant field rotation or the application of huge amounts of fertilizer.

There are some notable exceptions to this pattern. Unusually rich soils deriving from volcanic activity are scattered throughout much of the region, but are particularly widespread on Java. With more than 50 volcanoes, Java is a fertile island with a very high population density. Roughly 143 million people live on Java in an area smaller than Iowa, giving it one of the world's highest rural population densities. Dense populations are also found in pockets of fertile alluvial soils along the coasts of insular Southeast Asia. The core area of the Philippines, the central lowlands of Luzon near Manila, is particularly densely settled.

Mainland Southeast Asia's population is concentrated in the agriculturally intensive valleys and deltas of the large rivers, whereas most upland areas remain relatively lightly settled. The population core of Thailand is formed by the valley and delta of the Chao Phraya River, just as Burma's is focused on the Irrawaddy River. Vietnam has two distinct core areas: the Red River Delta in the far north and the Mekong Delta in the far south.

Agricultural practices and settlement forms vary widely across the complex environments of Southeast Asia. Generally speaking, however, three farming and settlement patterns are apparent: swidden in the uplands and both plantation agriculture and rice cultivation in the lowlands.

Swidden in the Uplands Also known as *shifting cultivation* or *slash-and-burn agriculture*, swidden is practiced throughout the rugged uplands of Southeast Asia (Figure 13.12). In the **swidden** system, small plots of several acres of forest or brush are periodically cut by hand. The fallen vegetation is then burned, and the ash naturally spreads over the ground. This practice transfers nutrients to the soil before subsistence crops are planted. Yields remain high for several years and then drop off as the soil nutrients are exhausted and insect pests and plant diseases multiply. These plots are abandoned after a few years and return to woody vegetation. The cycle of cutting, burning, and planting then moves to another small plot not far away.

Swidden is a sustainable form of agriculture when population densities remain relatively low. Today, however, the swidden system is increasingly threatened. With higher population densities, the rotation

Figure 13.1.2 Palm Oil Harvest The oil palm is an extremely valuable crop that produces inexpensive cooking oil. Grown across much of Insular Southeast Asia, it is associated with extensive environmental damage. This worker is lifting recently harvested palm fruit onto a wheelbarrow at a plantation in North Sumatra, Indonesia.

Toward a Solution? These environmental problems have led to efforts to establish sustainable palm oil production. In 2004, the Roundtable on Sustainable Palm Oil was established by international environmental groups and local producers to encourage and certify environmentally responsible palm oil. Critics argued, however, that as of 2011 only about 12 percent of the crop was actually sustainably produced.

More recently, large international businesses that trade palm oil have intensified their actions to create a more environmentally responsible industry. In 2014, Minnesota-based Cargill, the largest privately held corporation in the United States, agreed it would no longer trade palm oil that came from deforested lands or peat-rich wetlands or that involved the exploitation of indigenous peoples. At the 2014 UN Climate Summit, three major palm oil producers joined Cargill in its pledge. Other major food companies, including Unilever and Nestlé, have also agreed to strengthen their environmental safeguards in regard to palm oil. These companies are now working with Southeast Asian farmers in sustainable agricultural production.

Although a number of environmental groups have helped companies to establish guidelines, many environmentalists remain skeptical. A significant issue is accreditation, as even producers who have signed sustainability agreements often purchase oil from illegal farms that continue to clear forests and drain wetlands. Cargill and Unilever are thus working on how to trace out their supply chains to ensure that all producers follow the agreed-upon practices, but the process will not be easy (Figure 13.1.2).

1. What other widely consumed foodstuffs generate similar environmental controversies? How do these controversies differ from those associated with palm oil?
2. What responsibility do multinational corporations like Cargill bear for the environmental damage caused by local farmers who supply their raw materials?

period must be shortened, which damages the soil. Swidden farming is also harmed by commercial logging, which both displaces farmers and removes soil nutrients from the ecosystem as logs are exported.

When swidden can no longer support the population, upland people sometimes adapt by switching to cash crops that will allow them to participate in the commercial economy.

Figure 13.12 Swidden Agriculture In the uplands of Southeast Asia, swidden (or slash-and-burn) agriculture is widely practiced. When done by tribal peoples with low population densities, swidden is not environmentally harmful. When practiced by large numbers of immigrants from the lowlands, however, swidden can cause deforestation and extensive soil erosion.

EXPLORING GLOBAL CONNECTIONS

The Opium Resurgence in Northern Southeast Asia

Google Earth (MG) Virtual Tour Video https://goo.gl/QESZOh

Until late in the 20th century, the Golden Triangle of northern Southeast Asia, focused on Burma's Shan State, was the global center of opium growing and heroin production. In 1990, however, the Burmese army defeated several ethnic rebellions that were financed by narcotics production. These victories, along with antidrug campaigns by local governments and the United Nations, resulted in a massive opium decline (Figure 13.2.1). At the same time, the crop surged ahead in war-torn Afghanistan. By 2007, over 85 percent of the world's illicit opium originated in Afghanistan.

The Rise of Methamphetamine As opium growing declined in northern Southeast Asia, other drugs took its place, since both drug lords and ethnically based militias were eager to retain their profits. Of particular importance was methamphetamine, a synthetic drug that is easy to make with the right chemicals. A lively trade thus developed, with "meth" flowing north into China and south into Thailand and with chemicals moving in the opposite directions. In 2014 alone, 36 tons of methamphetamine were seized by Southeast Asian officials. Even the United States found itself involved. In 2005, a New York grand jury indicted eight leaders of Burma's United Wa State Army on charges of trafficking amphetamines and other drugs.

Opium's Comeback Moreover, 2008 saw a resurgence in the growing of opium (Figure 13.2.2). Production in Afghanistan dropped, reducing competition, but more important was the rise of China as a new major market. According to the UN Office on Drugs and Crime (UNDOC), 55 percent of Asia's estimated 3.3 million heroin users currently reside in China. The opium fields of the Golden Triangle not only are much closer to this market than those of Afghanistan, but also produce a higher-quality product. A recent survey found that the area devoted to opium growing in Burma and Laos increased by 6000 acres (2600 hectares) in 2014 alone. In that year, the two countries produced an estimated 762 metric tons of opium, generating

Figure 13.2.1 The Fight Against Opium The opium crop in northern mainland Southeast Asia results in widespread economic corruption and social damage. As a result, both governmental forces and ethnic militias have been involved in eradication programs. These soldiers from the Ta'ang (or Palaung) National Liberation Army in Burma are destroying an opium field in the ethnic Palaung area.

In the mountains of northern Southeast Asia, also called the **Golden Triangle**, one of the main cash crops historically has been opium, grown by local farmers for the global drug trade (see *Exploring Global Connections: The Opium Resurgence in Northern Southeast Asia*).

Plantation Agriculture With European colonization, Southeast Asia became a focus for plantation agriculture, growing high-value specialty crops ranging from coconuts to rubber. Even in the 19th century, Southeast Asia was linked to world trade flow through the plantation system. Forests were cleared and swamps drained to make room for commercial farms; labor was supplied by local people or by workers brought in from India or China.

Plantations are still an important part of Southeast Asia's geography and economy. Most of the world's natural rubber is produced in Malaysia, Indonesia, and Thailand. Sugarcane has long been a major crop of the Philippines and Thailand, although it is no longer very profitable; as a result, growing it in the Philippines is associated with intense rural poverty. Indonesia is the region's leading producer of tea, and Vietnam dominates coffee production (Figure 13.13). In recent years, oil palm plantations have been spreading through much of the region, often at the expense of tropical rainforests. Coconuts are widely grown in the Philippines, Indonesia, and elsewhere.

Rice in the Lowlands The lowland basins of mainland Southeast Asia are largely devoted to intensive rice cultivation. Through most of the region, rice is the preferred staple food. Rice harvests are increasingly traded to meet the needs of expanding markets throughout the world. Three delta areas have been the focus for commercial rice cultivation: the Irrawaddy in Burma, the Chao Phraya in Thailand, and the Mekong in Vietnam. The use of agricultural chemicals and high-yield crop varieties has allowed production to keep pace with population growth, although at the cost of significant environmental damage.

As of 2014, the world's top rice exporter was Thailand, with Vietnam in the third position. In 2008, these two countries, along with Burma, Cambodia, and Laos, joined together to form the Organization of Rice Exporting Countries (OREC), which seeks to maintain high and even prices. The Philippines has strongly criticized the organization, as it is the world's top rice-importing country. Although agriculture in much of the Philippines is dominated by rice, Filipino farmers cannot keep pace with the country's rapidly growing population.

Recent Demographic Changes

Both population density and fertility patterns vary widely across Southeast Asia. Although birth rates have dropped sharply in recent years, in several countries they remain well above the replacement level.

76 metric tons of heroin. Nearly 90 percent of this production comes from Burma's Shan State.

With the resurgence of opium growing in the Golden Triangle, new eradication programs are currently under way. Officials realize, however, that law enforcement efforts alone will not be adequate, as the value of opium is too high. A small field that would produce a rice crop worth US$30 could yield an opium crop worth US$585. Therefore, UNDOC officials argue that broad-based economic and social development in the region's rugged uplands is needed. In particular, they emphasize the building of an effective transportation system to allow lower-value crops to be brought to market.

1. What are some of the political, cultural, and environmental factors that make the Golden Triangle a major source of illicit drugs?
2. Compare the possible advantages and disadvantages of an approach to drug production in the region based on economic and social development rather than law enforcement.

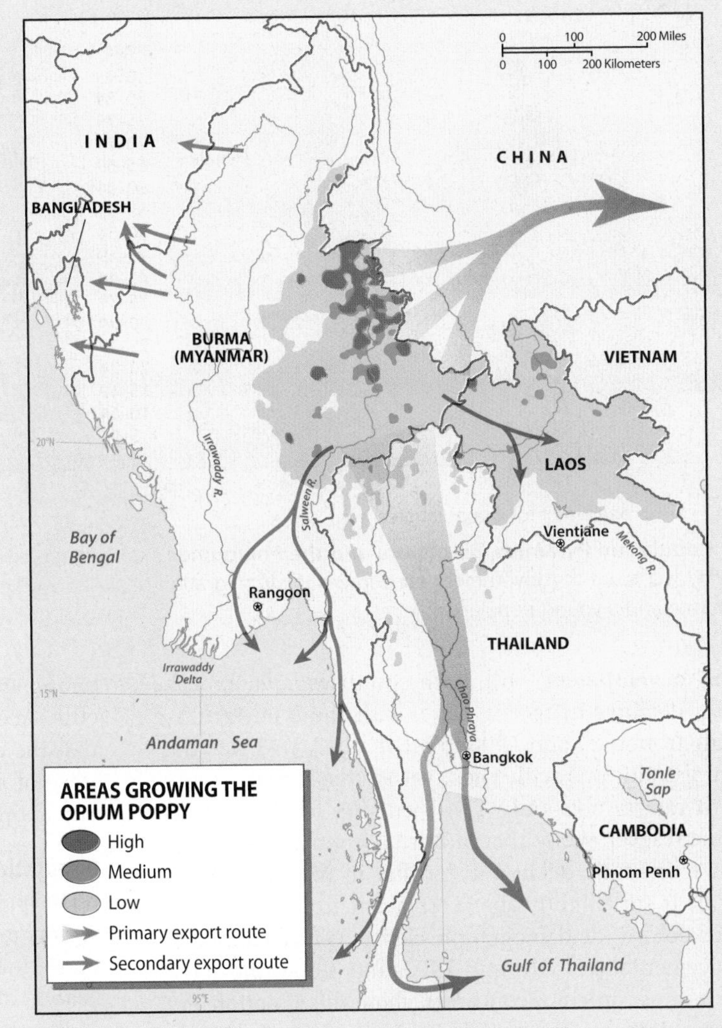

Figure 13.2.2 Opium Growing Areas The opium poppy has long been commercially cultivated in the northern region of Mainland Southeast Asia. Although production declined significantly in the 1990s and early 2000s, it has recently made a dramatic comeback.

East Timor in particular has had an elevated birth rate—over five children per woman—that did not begin to decline until 2000. Other Southeast Asian states are moving toward population stabilization, and in a few, low fertility levels could lead to demographic decline (Figure 13.14).

Population Contrasts The Philippines, the second most populous country in Southeast Asia, has a relatively high total fertility rate of 3.1 (Table A13.1). Effective family planning here has been difficult to establish. When a popular democratic government replaced a dictatorship in the 1980s, the Philippine Roman Catholic Church pressured the new government to cut funding for family-planning programs. A combination of rapid growth and economic stagnation forced many Filipinos to migrate (Figure 13.15). Up to 13 million Filipinos now live and work abroad, with over 1 million in Saudi Arabia alone.

Elevated fertility rates are also found in Laos and Cambodia, two countries of Buddhist religious tradition marked by low levels of

Figure 13.13 Tea Harvesting in Indonesia Plantation crops, such as tea, are major sources of exports for several Southeast Asian countries. Coconut, rubber, palm oil, and coffee are other major cash crops. Many of these crops require large amounts of labor, particularly at harvest time.

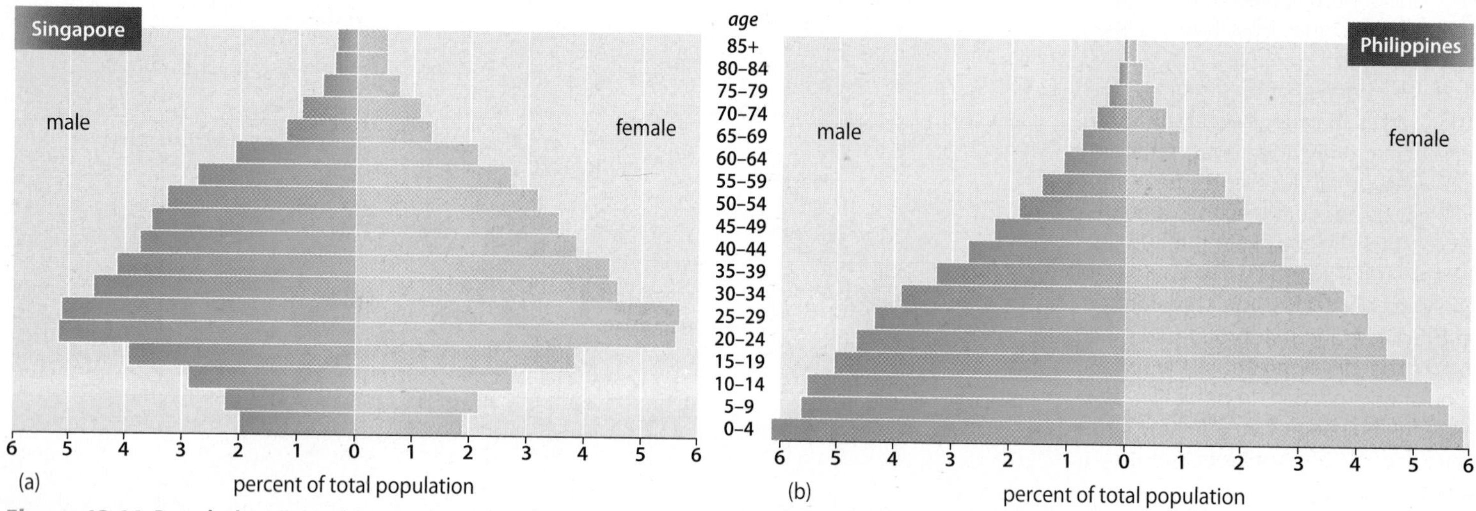

Figure 13.14 Population Pyramids, Singapore and the Philippines (a) Singapore has a very low birth rate and an aging population; as a result, its government encourages the immigration of skilled workers. (b) The Philippines, by contrast, has a high birth rate and a young population.

economic and social development. Still, both countries have experienced major fertility declines in recent years, with the total fertility rate of Laos dropping from over 6 in 1990 to 3.1 in 2012. In Thailand, which shares cultural traditions with Laos, yet is considerably more developed, the birth rate is now only 1.4. Vietnam's fertility rate is also below replacement level, while that of Burma is close to it. Indonesia has also seen a dramatic decline in fertility in recent decades, although its birth rate is still slightly above replacement level.

Singapore stands out on the demographic charts, as its fertility rate dropped below replacement level in the mid-1970s and is now one of the lowest in the world. Its government is concerned about this situation and is actively promoting childbearing, particularly among the most highly educated segment of its population. Under Singapore's official programs, middle-income Singaporean couples can receive various incentives worth up to US$120,000 if they have two children. The government has also sponsored pro-fertility advertisement and video campaigns; one 2012 YouTube hit included these lyrics: "I'm a patriotic husband, you're my patriotic wife, let's do our civic duty and manufacture life!"

Yet because of migration, Singapore's population has been steadily increasing. The government especially encourages highly skilled people to relocate to the city-state, offering high wages and other benefits. According to an official 2013 report, Singaporean officials hope that the country's population will exceed 6.5 million by 2030. The release of this report generated a public protest, a rare event in orderly Singapore, as many citizens think that the island is too crowded as it is.

Figure 13.15 Overseas Employment Fair in the Philippines Due to a combination of poor economic conditions at home, rapid population growth, and good English-language skills, many Filipinos seek employment abroad. Here job seekers crowd an overseas employment fair.

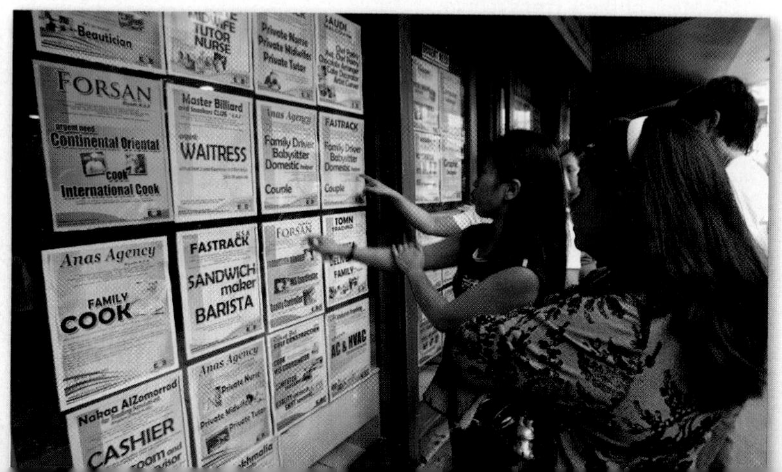

Population Density and Migration Indonesia has long had an explicit policy of **transmigration**, or relocation of people from one region to another within its national territory. Primarily because of migration from densely populated Java, the population of the outer islands of Indonesia has grown rapidly since the 1970s. The province of East Kalimantan, for example, experienced an annual growth rate of 30 percent from 1980 to 2000.

High social and environmental costs have accompanied these relocation programs. Javanese peasants, accustomed to the fertile soils of their home island, often fail when farming the former rainforest of Borneo. In some areas, farmers have little choice but to adopt a semi-swidden form of cultivation, a process associated with deforestation and conflicts with indigenous peoples. Partly because of these problems, the Indonesian government significantly reduced its official transmigration program in 2000. It still helps around 60,000 people migrate to the outer islands every year, however, and many others borrow money or use their personal savings to do the same (Figure 13.16). Indonesia is also creating a major new agricultural project in the Merauke region of south-central New Guinea, which could entail the transfer of over half a million people.

Urban Settlement

Despite the relatively high level of economic development found in parts of Southeast Asia, the region is not heavily urbanized. Even Thailand's population is roughly two-thirds rural, which is unusual for a country that has experienced so much industrialization. But cities are growing rapidly throughout the region, increasing the rate of urbanization.

Many Southeast Asian countries have **primate cities**—single, large urban settlements that overshadow all others. Thailand, for example, is dominated by Bangkok, just as Manila far surpasses all other cities in the Philippines (Figure 13.17). Both have grown recently into

Figure 13.16 Transmigration in Indonesia Over the past several decades, people have been moving in large numbers from the crowded island of Java to Indonesia's outer islands. The migrant settlement visible in this image is on the Indonesian portion of the island of New Guinea.

megacities with more than 11 million residents in their respective metropolitan areas.

In Manila, Bangkok, Jakarta, and other massive cities, explosive growth has led to housing shortages, congestion, and pollution. Bangkok suffers from some of the worst traffic in the world, although the Thai government has responded with large-scale highway and mass-transit construction programs. It is estimated that more than half of Manila's people live in squatter settlements, often without basic water and electricity service. Most large Southeast Asian cities lack parks and other public spaces, which is one reason why massive shopping malls have become so popular. Bangkok's Paragon Mall has recently emerged as a major urban focus, complete with a conference center and a concert hall.

Urban primacy is not encountered in all Southeast Asian countries. Vietnam, for example, has two main cities: Ho Chi Minh City (formerly Saigon) in the south and Hanoi, the capital, in the north. The metropolitan area of Jakarta, Indonesia's capital, has more than 20 million residents, but the country has several other large and growing cities, including Bandung and Surabaya. Yangon (formerly Rangoon) remains the primate city of Burma, with more than 5 million people, but it is no longer the country's capital; the government was moved to inaccessible Naypyidaw in 2006 because of security concerns.

Kuala Lumpur, the largest city in Malaysia, has received heavy investments from both the national government and the global business community. This has produced a modern, forward-looking city that is largely free of most of the traffic, water, and slum problems that plague most other Southeast Asian cities. Although Kuala Lumpur has only 1.6 million residents, its greater metropolitan area tops 7 million, making it the primate city of Malaysia.

The independent republic of Singapore is essentially a city-state of 5.4 million people on an island of 274 square miles (710 square kilometers), about three times the size of Washington, DC (Figure 13.18). While space is at a premium, Singapore has been very successful in building high-density housing. Unlike most other Southeast Asian cities, Singapore has no squatter settlements or slums.

Figure 13.17 Bangkok Thailand's capital saw the development of an impressive skyline during its boom years from the late 1970s through the late 1990s. Unfortunately, transportation development did not keep pace with population and commercial growth, resulting in one of the most congested and polluted urban landscapes in the world.

Review

13.3 Why does Southeast Asia have such distinctive forms of agriculture?

13.4 Explain why different parts of Southeast Asia vary so much in regard to both population density and population growth.

KEY TERMS swidden, Golden Triangle, transmigration, primate city

Cultural Coherence and Diversity: A Meeting Ground of World Cultures

Unlike many other world regions, Southeast Asia lacks the historical dominance of a single civilization. Instead, the region has been a meeting ground for cultural influences from South Asia, China, the Middle East, Europe, and North America. Abundant natural resources, along with the region's strategic location on oceanic trading routes connecting major landmasses, have long made Southeast Asia attractive to outsiders.

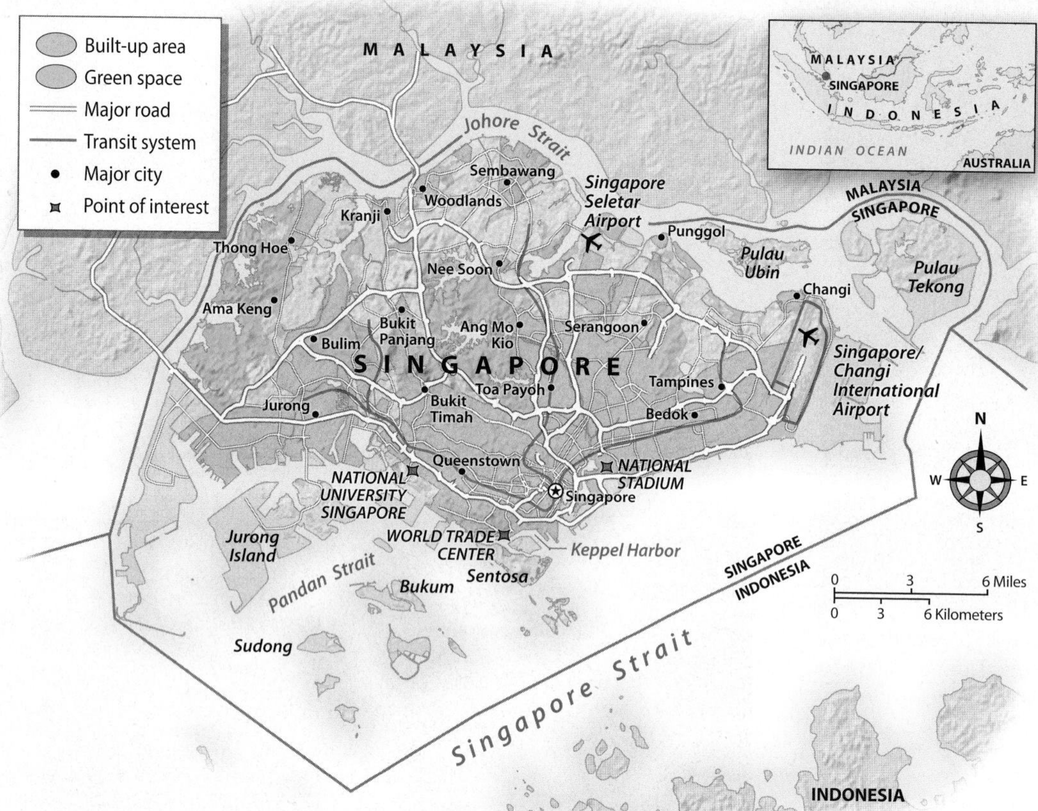

Figure 13.18 Singapore The economic and technological hub of Southeast Asia, Singapore is famous for its clean, efficiently run, and very modern urban environment. Some residents complain, however, that Singapore has lost much of its charm as it has developed.

The Introduction and Spread of Major Cultural Traditions

In Southeast Asia, contemporary cultural diversity is related to the historical influence of the major religions of the region: Hinduism, Buddhism, Islam, and Christianity (Figure 13.19).

South Asian Influences The first major external religious influence arrived from South Asia some 2000 years ago when small numbers of educated migrants from what is now India helped local leaders establish Hindu kingdoms, with some Buddhist influences, in the lowlands of Burma, Thailand, Cambodia, Malaysia, and western Indonesia. Although Hinduism later faded away in most locations, it is still the dominant religion on the Indonesian island of Bali.

A second wave of South Asian religious influence reached mainland Southeast Asia in the 13th century in the form of Theravada Buddhism, which spread from Sri Lanka. Almost all of the people in lowland Burma, Thailand, Laos, and Cambodia converted to Buddhism at that time, and today this religion forms the foundation for their social institutions. Saffron-robed monks, for example, are a common sight, and Buddhist temples abound.

Chinese Influences Unlike most other mainland peoples, the Vietnamese were not heavily influenced by South Asian civilization. Instead, their early connections were to East Asia. Vietnam was a province of China until about 1000 CE, when the Vietnamese rejected Chinese political rule and established their own kingdom. But the Vietnamese retained many features of Chinese culture, and their traditional religious and philosophical beliefs are centered on Mahayana Buddhism and Confucianism.

East Asian cultural influences in many other parts of Southeast Asia are directly linked to the more recent immigration of southern Chinese. This migration reached a peak in the 19th and early 20th centuries (Figure 13.20). China was then a poor and crowded country, which made sparsely populated Southeast Asia appear to be a place of opportunity. Eventually, distinct Chinese settlements were established in every Southeast Asian country, especially in urban areas. In Malaysia, the Chinese minority now constitutes roughly one-third of the population, whereas in Singapore some three-quarters of the people are of Chinese ancestry.

In many places in Southeast Asia, relationships between the Chinese minority and the native majority are strained. Even though their ancestors arrived generations ago, many Chinese are still considered resident aliens because they maintain their Chinese identities. A more significant source of tension is the fact that most Chinese communities in Southeast Asia are relatively wealthy. Many Chinese immigrants prospered as merchants, an occupation avoided by most local people. As a result, they have acquired substantial economic influence, which is often resented by others.

The Arrival of Islam Muslim merchants from South and Southwest Asia arrived in Southeast Asia hundreds of years ago and soon began converting many of their local trading partners. From an initial core established around 1200 CE in northern Sumatra, Islam spread through much of insular Southeast Asia. By 1650, it had largely replaced Hinduism and Buddhism throughout Malaysia and Indonesia, with the exception of the small but fertile island of Bali.

Some 88 percent of Indonesia's inhabitants follow Islam, making it the world's most populous Muslim country (Figure 13.21). This figure, however, hides significant internal religious diversity. In some parts of Indonesia, such as northern Sumatra (Aceh), highly orthodox forms of Islam took root. In others, such as central and eastern Java, a more relaxed form of worship emerged that included certain Hindu and even animistic beliefs. Islamic reformers, however, have long tried to instill more mainstream forms of the faith among the Javanese, with increasing success in recent years.

Christianity Islam was still spreading eastward through insular Southeast Asia when the Europeans arrived in the 16th century. When the Spanish claimed the Philippine Islands in the 1570s, they found the southwestern portion of the archipelago to be thoroughly Islamic. To this day, the southwest Philippines is still largely Muslim, but the rest of the country is mostly Roman Catholic. East Timor, long a Portuguese colony, is also a predominantly Roman Catholic country.

Christian missions spread through other parts of Southeast Asia in the late 19th and early 20th centuries, when European colonial powers controlled most of the region. Although French priests converted many Vietnamese to Catholicism, missionaries had little influence in other lowland areas. They were more successful in highlands inhabited by tribal peoples worshiping nature spirits and their ancestors. The general name for such religions is **animism**. While many modern hill tribes

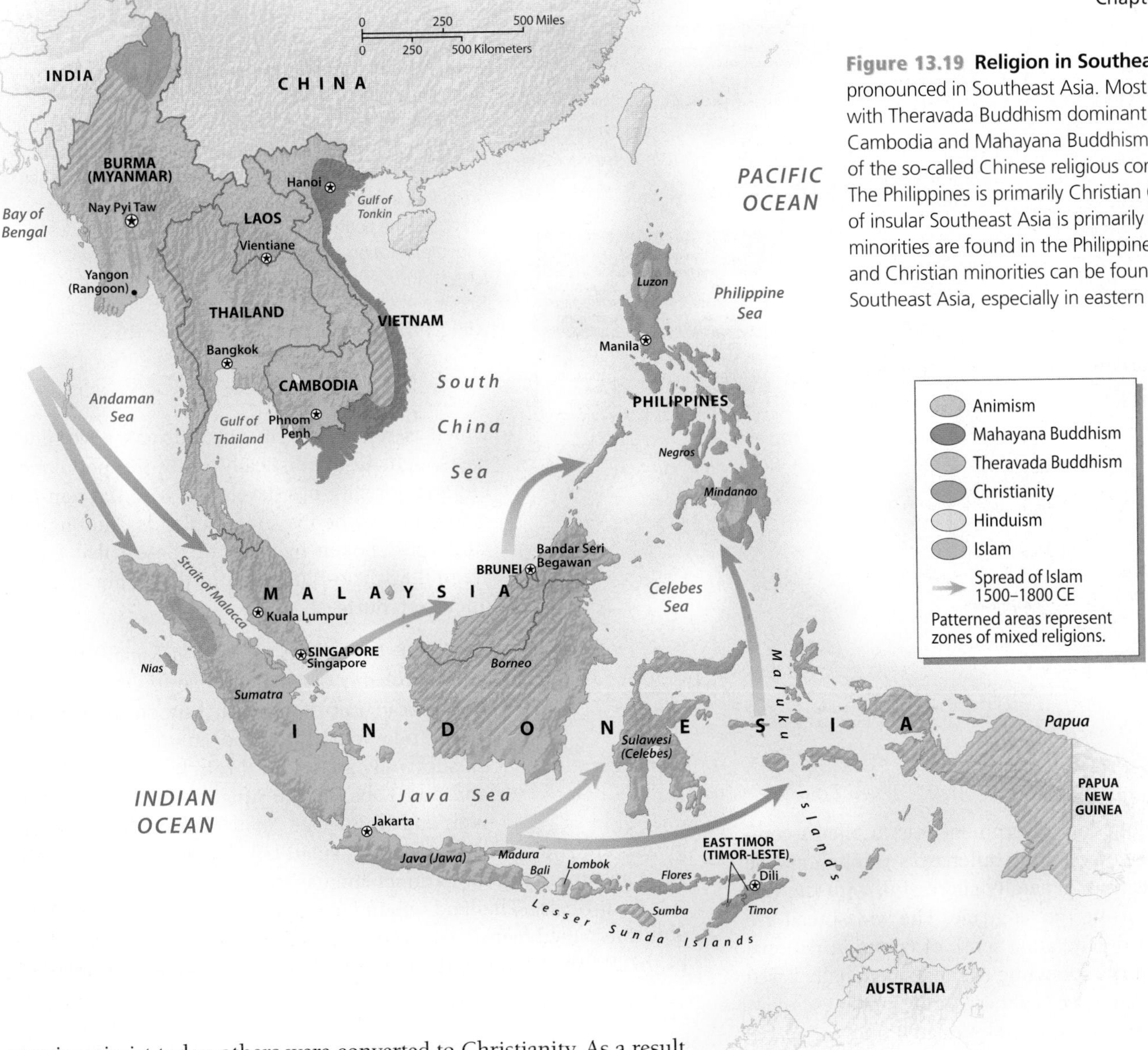

Figure 13.19 Religion in Southeast Asia Religious diversity is pronounced in Southeast Asia. Most of the mainland is Buddhist, with Theravada Buddhism dominant in Burma, Thailand, Laos, and Cambodia and Mahayana Buddhism (combined with other elements of the so-called Chinese religious complex) prevailing in Vietnam. The Philippines is primarily Christian (Roman Catholic), but the rest of insular Southeast Asia is primarily Muslim. Substantial Muslim minorities are found in the Philippines, Thailand, and Burma. Animist and Christian minorities can be found in remote areas throughout Southeast Asia, especially in eastern Indonesia.

remain animist today, others were converted to Christianity. As a result, significant Christian concentrations are found in several upland areas of Indonesia, Vietnam, and Burma.

Religious Persecution Religious persecution has recently become a serious issue in parts of Southeast Asia. Vietnam's communist government is struggling against a revival of faith among the country's Buddhist majority and its 8 million Christians, many of whom live in the central highlands. In 2014, a UN special report condemned the country for failing to respect the freedom of belief. Burma's government has generally supported Buddhism, but when Buddhist monks led massive demonstrations against the government in late 2007, it cracked down, killing 30 to 40 monks. Burma has also severely repressed its Rohingya Muslim minority, driving hundreds of thousands out of the country (Figure 13.22). In 2014, the Burmese government told the country's estimated 1 million Rohingyas that they would be placed in detention camps and face deportation unless they could prove that their family had resided in the country for more than 60 years. Indonesia has sought to ban the Ahmaddiya Muslim sect, a group regarded as heretical by many mainstream Muslims.

Geography of Language and Ethnicity

The linguistic geography of Southeast Asia is complicated (Figure 13.23). The several hundred distinct languages of the region are all placed into five major linguistic families. Four of these families are discussed below, whereas the fifth, Papuan, is discussed in Chapter 14.

The Austronesian Languages One of the world's most widespread language families is Austronesian, which extends from Madagascar to Easter Island in the eastern Pacific. Today almost all insular Southeast Asian languages belong to the Austronesian family. But despite this common linguistic grouping, more than 50 distinct languages are spoken in Indonesia alone. In far eastern Indonesia, various languages fall into the completely separate family of Papuan, closely associated with New Guinea.

The Malay language overshadows all others in insular Southeast Asia. Malay is native to the Malay Peninsula, eastern Sumatra, and coastal Borneo and was spread historically by merchants and seafarers. As a result, it became a common trade language, or **lingua franca**, throughout much of the insular realm. When Indonesia gained independence in 1949, its leaders decided to use the lingua franca version of Malay as the basis for a new national language called *Bahasa Indonesia* (or simply *Indonesian*). Although Indonesian is slightly different from the Malaysian spoken in Malaysia, they form a single, mutually understandable language. Both are written in the Roman script.

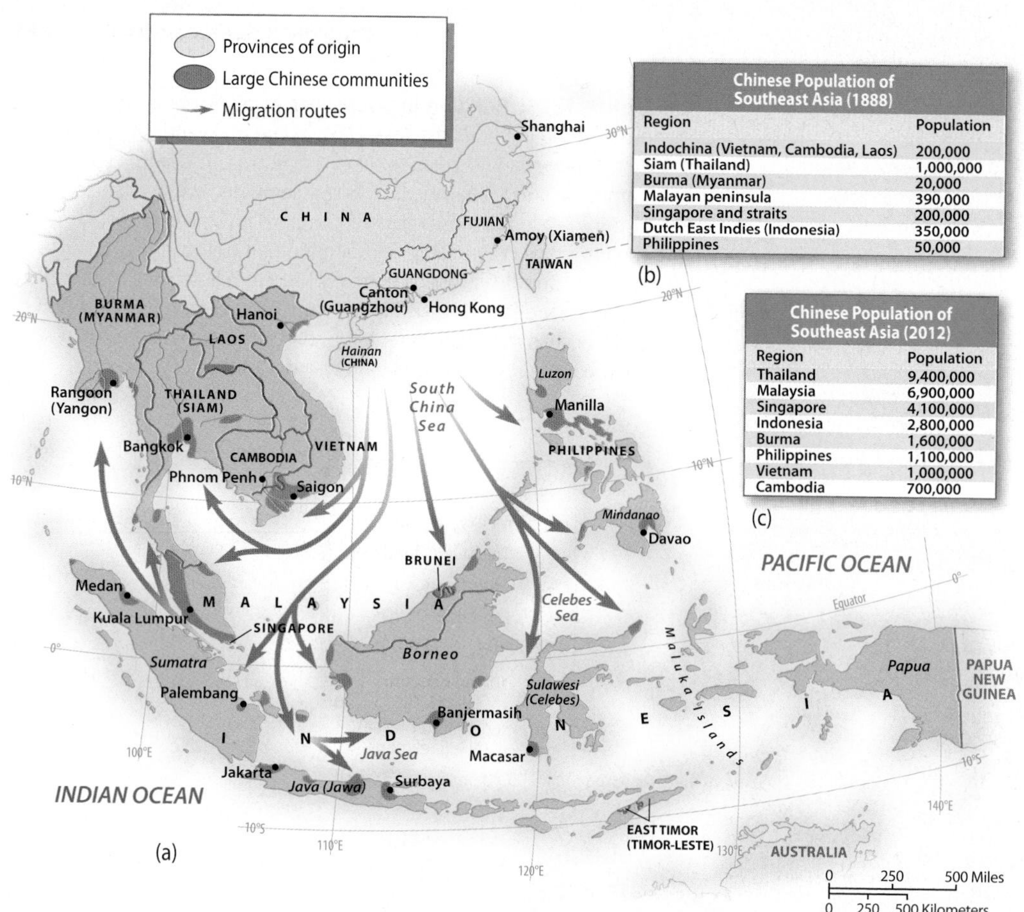

Chinese Population of Southeast Asia (1888)

Region	Population
Indochina (Vietnam, Cambodia, Laos)	200,000
Siam (Thailand)	1,000,000
Burma (Myanmar)	20,000
Malayan peninsula	390,000
Singapore and straits	200,000
Dutch East Indies (Indonesia)	350,000
Philippines	50,000

Chinese Population of Southeast Asia (2012)

Region	Population
Thailand	9,400,000
Malaysia	6,900,000
Singapore	4,100,000
Indonesia	2,800,000
Burma	1,600,000
Philippines	1,100,000
Vietnam	1,000,000
Cambodia	700,000

Figure 13.20 Chinese in Southeast Asia (a) People from the southern coastal region of China have been migrating to Southeast Asia for hundreds of years, a process that reached a peak in (b) the late 1800s and early 1900s. Most Chinese migrants settled in the major urban areas, but in peninsular Malaysia large numbers were drawn to the countryside to work in the mining industry and in plantation agriculture. (c) Today Malaysia has the largest number of people of Chinese ancestry in the region. Singapore, however, is the only Southeast Asian country with a Chinese majority.

The goal of the new Indonesian government was to offer a common language that could overcome ethnic differences throughout the huge country. This policy has been generally successful, with the vast majority of Indonesians now using the language. The widespread use of Bahasa Indonesia has helped generate a sense of common national identity over most of the country's sprawling territory. Regionally based languages, however, such as Javanese, Balinese, and Sundanese, continue to be the primary languages of most Indonesian homes.

The Philippines contains eight major languages and several dozen minor languages, all of which are closely related. Despite more than 300 years of colonialism by Spain, Spanish never became a unifying force for the islands. During the American period (1898–1946), English became the language of government and education. After independence, Philippine nationalists selected Tagalog, the language of the Manila area, to replace English and help unify the new country. After Tagalog was standardized and modernized, it was renamed Pilipino (or Filipino) and has gradually become the unifying national language through its use in education, television, and films.

Tibeto-Burman Languages Each country of mainland Southeast Asia is closely identified with the national language spoken in its core territory. This does not mean, however, that all residents of these countries speak these official languages on a daily basis. In the mountains and other remote districts, local languages are commonly used. This linguistic diversity reinforces ethnic differences, often presenting challenges for programs designed to build national unity.

A good example of such linguistic challenges is Burma. Its national language, spoken by some 32 million people, is Burmese, which is related to Tibetan. Burma's government has sought to linguistically unify the population, but a major split has developed with the non-Burman peoples who live in the uplands. Although the languages spoken by most of these tribal groups are in the Tibeto-Burman family, they are quite distinct from Burmese.

Tai-Kadai Languages The Tai-Kadai linguistic family probably originated in southern China and then spread into Southeast Asia starting around 1100. Today closely related languages within the Tai subfamily are found throughout most of Thailand and Laos, in the uplands of northern Vietnam, and in Burma's Shan Plateau. Most Tai languages are spoken by members of small tribal groups, but two, Thai and Lao, are important national languages.

Historically, Thailand's main language, Siamese (just as the kingdom was called Siam), was restricted to the lower Chao Phraya Valley. In the 1930s, however, the country changed its name to Thailand to

Figure 13.21 Indonesian Mosque Indonesia is often said to be the world's largest Muslim country, as more Muslims reside here than in any other country in the world. Islam was first established in northern Sumatra, which is still the most devoutly Islamic part of Indonesia. Acehnese people are shown praying in front of the Baiturrahman Grand Mosque in the city of Banda Aceh. **Q: What are some of the political consequences of the high level of religious devotion found among most of the people of Aceh in northern Sumatra?**

Figure 13.22 Rohingya Refugee Camp Tens of thousands of Muslim Rohingyas have been driven out of Burma. Most of these refugees live in grim camps in neighboring Bangladesh.

emphasize the unity of all the peoples speaking the closely related Tai languages within its territory. Siamese was similarly renamed Thai and gradually became the country's unifying language. There is still much variation in dialect, however, with northern Thai sometimes considered a separate language. Somewhat more distinctive is Lao, the Tai language that became the national tongue of Laos. In Thailand's Khorat Plateau, most people speak Isan, a dialect closer to Lao than to standard Thai. Even Isan food is distinct from standard Thai food, as it is spicier and is based on glutinous ("sticky") rice.

Mon-Khmer Languages The Mon-Khmer language family probably once covered virtually all of mainland Southeast Asia. It contains two major languages—Vietnamese and Khmer (Cambodia's national language)—as well as a large group of minor languages spoken by hill peoples and a few lowland groups. Because of the historical Chinese influence in Vietnam, the Vietnamese language was written with Chinese characters until the French colonial government imposed the Roman alphabet, which remains in use today. Khmer, on the other hand, is—like Lao, Thai, and Burmese—written in its own Indian-derived script.

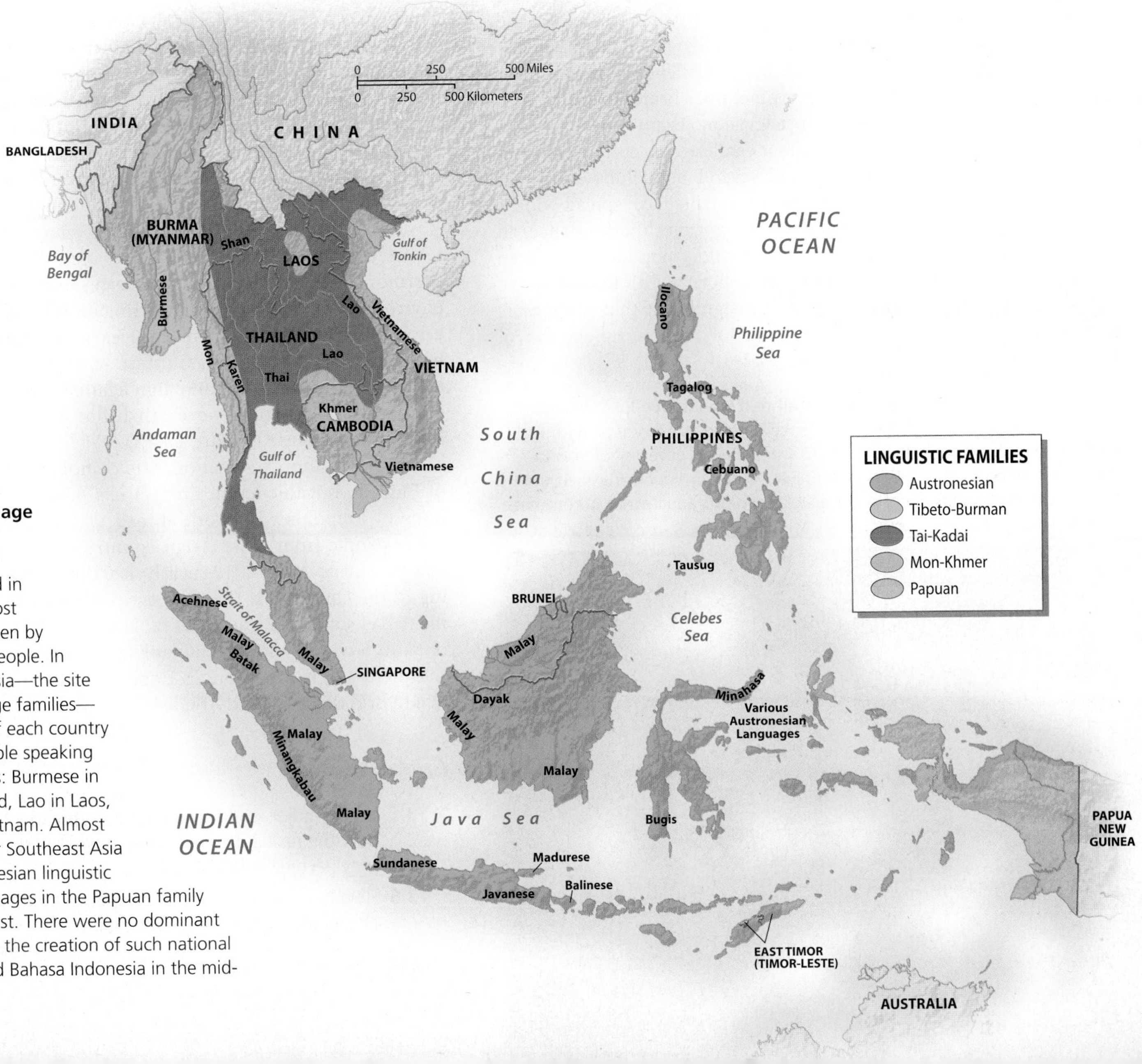

FIGURE 13.23 Language Map of Southeast Asia A huge number of languages are found in Southeast Asia, but most are tribal tongues spoken by only a few thousand people. In mainland Southeast Asia—the site of three major language families—the central lowlands of each country are dominated by people speaking the national languages: Burmese in Burma, Thai in Thailand, Lao in Laos, and Vietnamese in Vietnam. Almost all languages in insular Southeast Asia belong to the Austronesian linguistic family, although languages in the Papuan family are found in the far east. There were no dominant languages here before the creation of such national tongues as Filipino and Bahasa Indonesia in the mid-20th century.

The most important aspect of mainland Southeast Asia's linguistic geography is the fact that each country's national language is spoken mainly in the core lowlands, whereas the uplands are populated by tribal peoples speaking separate languages. In Vietnam, for example, Vietnamese speakers occupy less than half of the national territory, even though they constitute a sizable majority of the country's population. Ethnic tensions here have recently mounted, as Vietnamese speakers, aided by the country's major road-building program, have begun moving into the sparsely populated highlands.

Southeast Asian Culture in Global Context

Several Southeast Asian countries have been quite receptive to global cultural influences. This is particularly true of the Philippines, where U.S. colonialism encouraged the country to embrace many forms of popular Western culture. As a result, Filipino musicians and other entertainers are in demand elsewhere in Asia. Several aspects of indigenous Southeast Asian culture have also successfully spread to other parts of the world. Thai, Vietnamese, and Indonesian styles of cooking, for example, are popular in many urban areas of North America and Europe. Sports are also undergoing similar forms of globalization. Thailand's *muay thai*, or Thai kickboxing—noted for its wide variety of punches, kicks, elbow jabs, and knee thrusts—is currently gaining popularity across much of the world.

Cultural globalization has also been challenged in Southeast Asian countries. The Malaysian government has been especially critical of American films and satellite television. Islamic revivalism in Indonesia and Malaysia also presents a challenge to cultural globalization. Islamic radicals have attacked several nightclubs and other tourist destinations, and anti-Western sentiments have spread widely. Risqué musicians like Lady Gaga have had to cancel performances in Indonesia. But it is also true that several wildly popular Indonesian performers, most notably Jupe (Julia Perez, born as Yuli Rachmawati), are equally sexually suggestive. It is also significant that Indonesia's popular president, Joko Widodo (called Jokowi), is a huge fan of hard-core heavy-metal music (Figure 13.24).

Figure 13.24 Joko Widodo: First Metalhead President Joko Widodo ("Jokowi"), the president of Indonesia, is well known as a fan of heavy-metal music. In this photograph, taken in 2013 when he was governor of Jakarta, Jokowi is holding a maroon bass guitar given to him by Robert Trujillo of the U.S. band Metallica. Jokowi often wears black T-shirts and leather jackets when attending heavy-metal concerts.

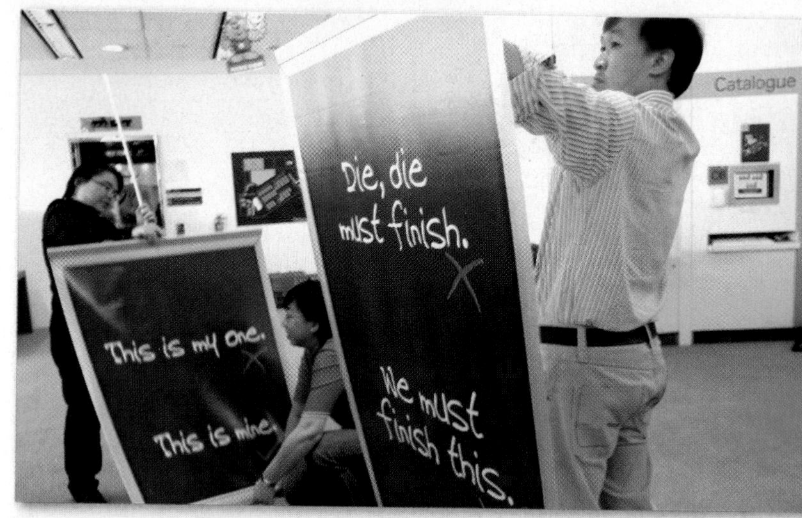

Figure 13.25 Anti-Singlish Campaigns in Singapore English is one of Singapore's four official languages, but most Singaporeans speak an informal version of English called "Singlish" that is highly influenced by Malay and the languages of southern China. As a result, the government of Singapore has been trying to encourage proper English and discourage the use of Singlish. Here library personnel are setting up posters to promote the use of correct spoken English.

The use of English as the global language also causes controversy in Southeast Asia. As the language of popular global culture, it is opposed by some conservatives, yet it must be mastered if citizens are to participate in global business and politics. In Malaysia, the widespread use of English became controversial in the 1980s as nationalists stressed the importance of the native tongue. This worried both the business establishment and the influential Chinese community.

In Singapore, the situation is more complex. Mandarin Chinese, English, Malay, and Tamil are all official languages. Furthermore, the languages of southern China are common in home environments, as 75 percent of Singapore's population is of southern Chinese ancestry. In recent years, the Singapore government has encouraged Mandarin Chinese over the southern Chinese dialects. It also launched a campaign against "Singlish," a popular form of speech that is based on English but employs many words from Malay and Chinese (Figure 13.25).

In the Philippines, nationalists complain about the common use of English, but fluency in English has proved beneficial for the millions of Filipinos who work abroad or for international businesses. Many people from other Asian countries come to the Philippines to study English. As of 2012, nearly 100,000 South Koreans were residing in the country for this purpose. Although the Philippine government has sought to gradually replace English with Filipino, English remains a widespread official language. At the same time, Filipino itself increasingly incorporates English words and phrases, giving rise to a hybrid dialect called "Taglish."

Review

13.5 Which major world religions have spread into Southeast Asia over the past 2000 years, and in what parts of the region did they become established?

13.6 How have different Southeast Asian countries reacted to the challenges of cultural globalization?

KEY TERMS animism, lingua franca

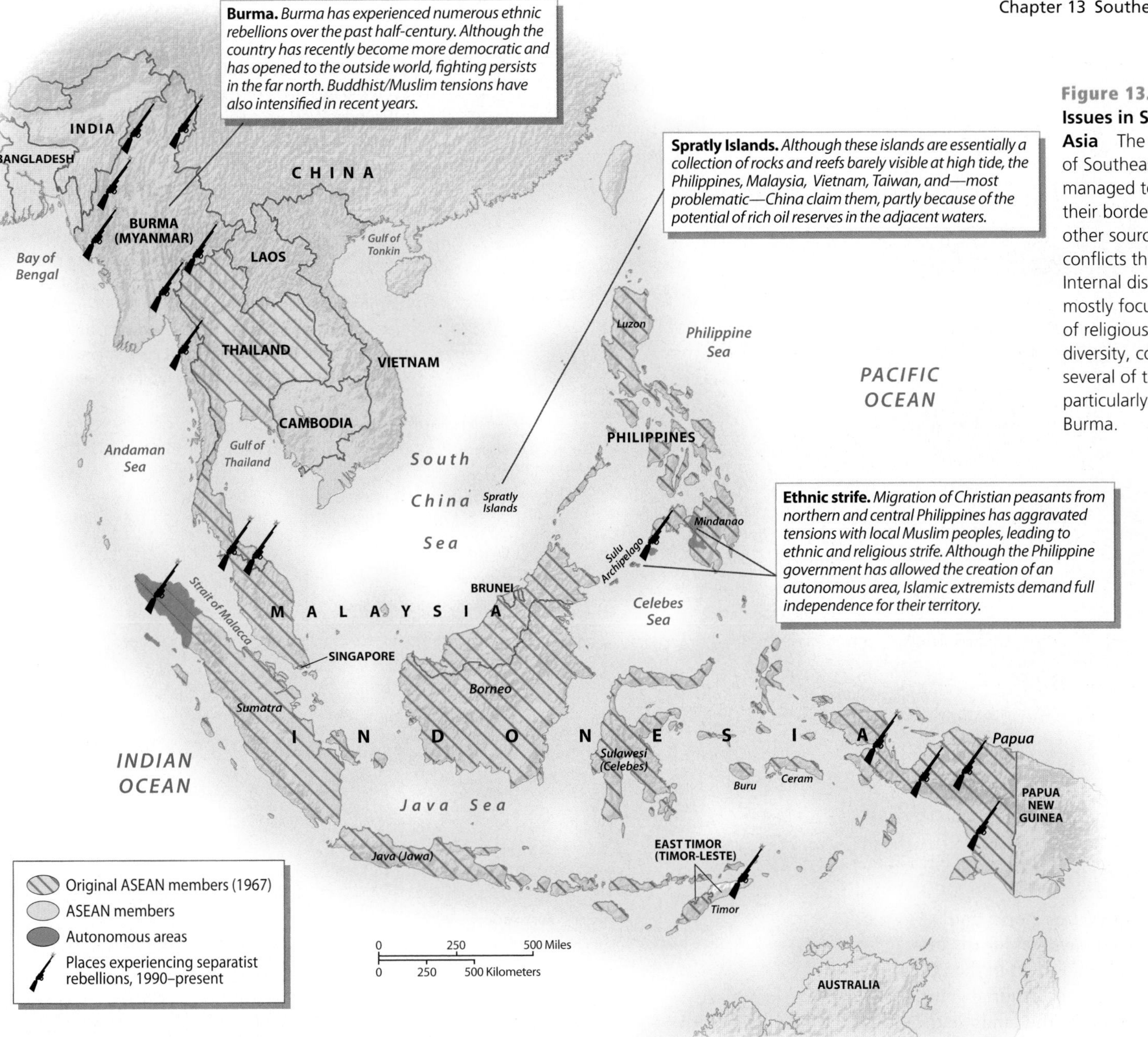

Figure 13.26 Geopolitical Issues in Southeast Asia The countries of Southeast Asia have managed to solve most of their border disputes and other sources of potential conflicts through ASEAN. Internal disputes, however, mostly focused on issues of religious and ethnic diversity, continue to trouble several of the region's states, particularly Indonesia and Burma.

Geopolitical Framework: War, Ethnic Strife, and Regional Cooperation

Southeast Asia is sometimes defined as the geopolitical grouping of 10 countries that have joined together as ASEAN (Figure 13.26). Although East Timor is not a member, its government has applied to join and in April 2015 claimed that it had fulfilled all obligations for membership; other countries in the organization, however, insist that East Timor first show that it has the financial capability to participate in ASEAN programs. ASEAN has significantly reduced geopolitical problems among its member states, while giving Southeast Asia as a whole a greater degree of regional coherence. But despite ASEAN's successes, many Southeast Asian countries still experience internal ethnic conflicts as well as tensions with their neighbors.

Before European Colonialism

The modern countries of mainland Southeast Asia all existed in one form or another as kingdoms before European colonialism. Cambodia emerged first, over 1000 years ago, and by the 1300s, independent states had been established by the Burmese, Siamese, Lao, and Vietnamese people.

Insular Southeast Asia's premodern map, on the other hand, was completely different from that of the modern nation-states. Many kingdoms existed on the Malay Peninsula and on the islands of Sumatra, Java, and Sulawesi, but few were territorially stable. Indonesia, the Philippines, and Malaysia thus largely owe their territorial shapes and later state boundaries to European colonialism (Figure 13.27).

The Colonial Era

The Portuguese were the first Europeans to arrive (around 1500), lured mainly by the spices of the Maluku Islands in eastern Indonesia. In the late 1500s, the Spanish conquered most of the lowland Philippines, which they used as a base for their silver trade between China and the Americas. By the 1600s, the Dutch began to establish trading bases, followed by the British. With superior naval weapons, the Europeans

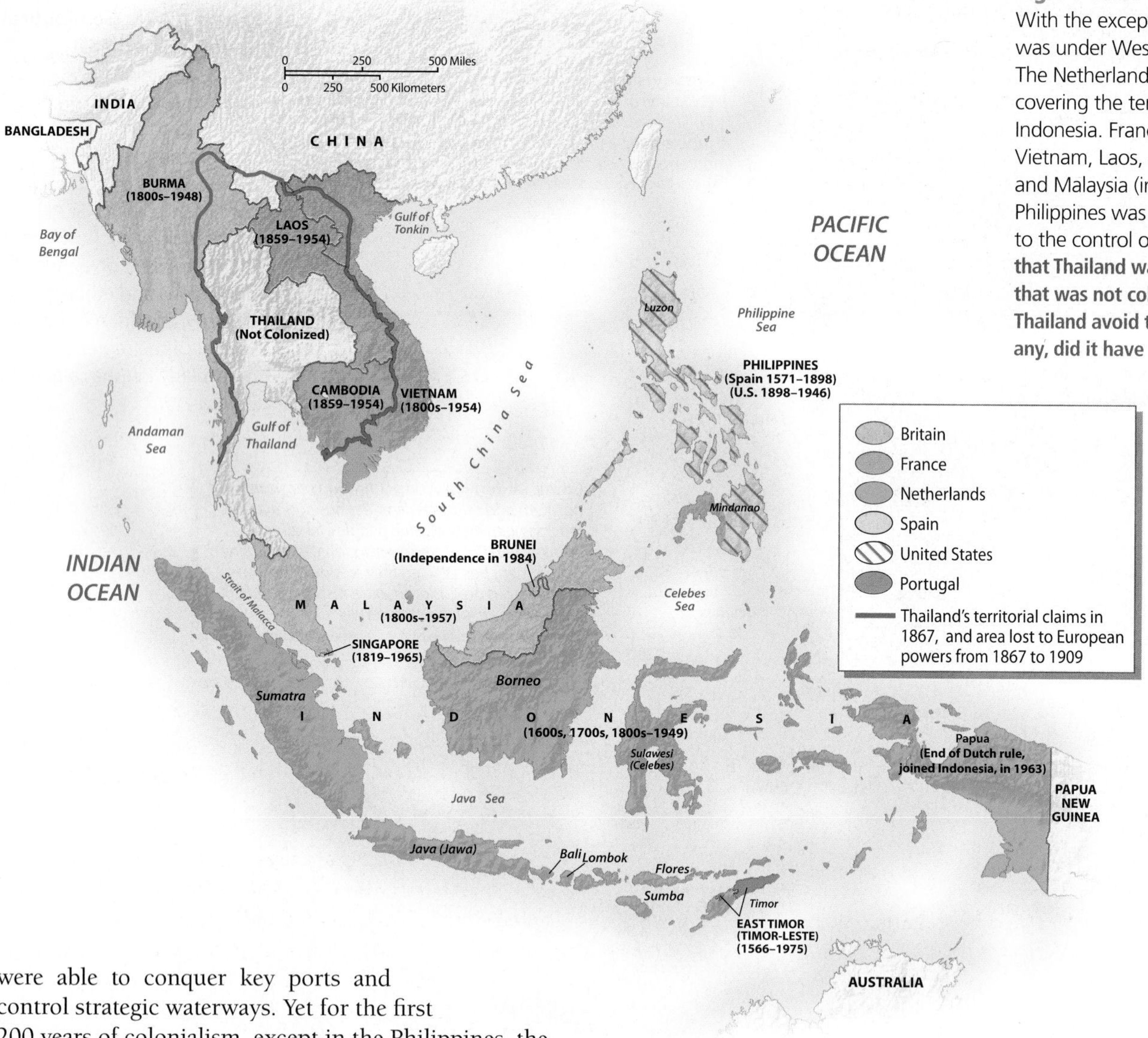

Figure 13.27 Colonial Southeast Asia With the exception of Thailand, all of Southeast Asia was under Western colonial rule by the early 1900s. The Netherlands had the largest empire in the region, covering the territory that was later to become Indonesia. France maintained a large colonial realm in Vietnam, Laos, and Cambodia, as did Britain in Burma and Malaysia (including Singapore and Brunei). The Philippines was initially colonized by Spain, but passed to the control of the United States in 1898. **Q: Note that Thailand was the only country in Southeast Asia that was not colonized by Western powers. How did Thailand avoid this fate, and what consequences, if any, did it have for the country's future development?**

were able to conquer key ports and control strategic waterways. Yet for the first 200 years of colonialism, except in the Philippines, the Europeans made no major territorial gains.

By the 1700s, the Netherlands had become the most powerful naval force in the region. As a result, a Dutch Empire in the "East Indies" began appearing on world maps. This empire continued to grow into the early 20th century, when it defeated its last major enemy, the Islamic state of Aceh in northern Sumatra. Later the Netherlands divided the island of New Guinea with Germany and Britain, rounding out its Indonesian colony.

The British, preoccupied with India, concentrated their attention on the sea lanes linking South Asia to China. As a result, they established several fortified outposts along the Strait of Malacca, the most important of which was Singapore, founded in 1819. To avoid conflict, the British and Dutch agreed that the British would limit their attention to the Malay Peninsula and the northern portion of Borneo. The British allowed local rulers to retain limited powers, much as they had done in India.

In the 1800s, European colonial power spread through most of mainland Southeast Asia. The British conquered the kingdom of Burma and extended their power into the nearby uplands. During the same period, the French moved into Vietnam's Mekong Delta, gradually expanding their territorial control into Cambodia and northward to China's border. Thailand was the only country to avoid colonial rule, although it did lose territories to the British and French. The final colonial power to enter the region was the United States, which took the Philippines first from Spain and then from Filipino nationalists between 1898 and 1902.

Organized resistance to European rule began in the 1920s, but it took the Japanese occupation of World War II to show that colonial power was vulnerable. Between 1942 and 1945, Japan occupied virtually the entire region. After Japan's surrender in 1945, pressure for independence intensified throughout Southeast Asia. Britain withdrew from Burma in 1948 and began to pull out of its colonies in insular Southeast Asia in the late 1950s, although Brunei did not gain independence until 1984. Singapore briefly joined Malaysia, but it was soon expelled and then gained independence in 1965. In the Philippines, the United States granted long-promised independence on July 4, 1946, although it retained military bases for several decades. The Dutch attempted to reestablish colonial rule after World War II, but they were forced to acknowledge Indonesia's independence in 1949.

Figure 13.28 Ho Chi Minh Trail North Vietnam supplied communist insurgents in South Vietnam with weapons and other goods that they moved over the ill-marked Ho Chi Minh Trail, running through Laos and Cambodia. The United States extensively bombed the trail and sprayed chemical defoliants on the surrounding forests, but it was never able to shut down this supply route.

The Vietnam War and Its Aftermath

After World War II, France was determined to regain control of its Southeast Asian colonies. Resistance to French rule was organized primarily by communist groups based mainly in northern Vietnam. Open warfare between French soldiers and the communist forces continued until 1954, when France withdrew after a major military defeat. An international peace council then divided Vietnam into a communist North Vietnam, allied with the Soviet Union, and a capitalist-oriented South Vietnam, with close ties to the United States.

The peace accord did not, however, end the fighting. Communist guerrillas in South Vietnam fought to overthrow the new government and unite it with the north. North Vietnam sent troops and war materials across the border to aid the rebels. Most of these supplies reached the south over the Ho Chi Minh Trail, a confusing network of forest passages through Laos and Cambodia, thus steadily drawing these two countries into the conflict (Figure 13.28). In Laos, the communist Pathet Lao forces challenged the government, while in Cambodia the **Khmer Rouge** guerrillas gained considerable power.

The **domino theory** guided U.S. foreign policy. According to this notion, if Vietnam fell to the communists, then so would Laos, Cambodia, and the other countries of the region. Fearing such an outcome, the United States was drawn ever deeper into the war. By 1965, thousands of U.S. troops were fighting to support the government of South Vietnam. But despite superiority in arms and troops, U.S. forces gradually lost control over much of the countryside. As casualties mounted and the antiwar movement back home strengthened, the United States began secret talks toward a negotiated settlement. Subsequent U.S. troop withdrawals began in earnest in the early 1970s.

With the withdrawal of U.S. forces, the noncommunist government began to collapse. Saigon fell in 1975, and in the following year Vietnam was officially reunited under the government of the north. Reunification was a traumatic event in southern Vietnam. Hundreds of thousands of people fled from the new regime, with many settling in the United States.

Vietnam proved fortunate compared to Cambodia. There the Khmer Rouge installed one of the most brutal regimes the world has ever seen. Cities were largely evacuated as urban residents were forced into the countryside to become peasants and most wealthy and educated people were executed. The Khmer Rouge's goal was to create an agriculturally self-sufficient society that would eventually provide the foundation for industrialization. After several years of horrific bloodshed, neighboring Vietnam invaded and installed a far less brutal, but still repressive, regime. Fighting between different groups continued until 1991, when a comprehensive peace settlement was finally reached.

Geopolitical Tensions in Contemporary Southeast Asia

In several parts of Southeast Asia, local ethnic groups have been struggling against national governments that inherited their territory from former colonial powers. Tensions have also emerged where tribal groups attempt to preserve their homelands from logging, mining, or migrant settlers (see Figure 13.26).

Conflicts in Indonesia When Indonesia gained independence in 1949, it included all of the former Dutch possessions in the region except western New Guinea (Papua). In 1962, the Netherlands organized an election to see whether the people of this area wished to join Indonesia or form an independent country. The vote went for union, but many observers believe that Indonesia's government rigged the election. As a result, many of the local people began to rebel. Indonesia is determined to maintain control of western New Guinea, in part because it is the site of the country's largest source of tax revenue—the highly polluting Grasberg mine, run by Phoenix-based Freeport-McMoRan Corporation (Figure 13.29).

In 2000, the Indonesian government granted a slight degree of autonomy to Papua, which reduced support for the rebellion. But a decade later, tensions again mounted, and violence continues to flare. In December 2014, five Papuan high school students were killed when Indonesian security forces fired into a crowd of demonstrators, prompting *Time* magazine to call Papua a "killing field." One underlying source of conflict is the continuing migration of people to Papua from Java and neighboring islands.

The island of Timor has also experienced political bloodshed in recent years. The eastern half of this poor and rather dry island had been a Portuguese colony and had evolved into a largely Christian society. The East Timorese expected independence when Portugal finally withdrew in 1975. Indonesia, however, viewed the area as its own and immediately invaded. A brutal war followed, won in part when the Indonesian army blocked food from reaching the province.

After a severe economic crisis in 1997, Indonesia's power in the region slipped. A new Indonesian government promised an election in 1999 to determine whether the East Timorese still wanted independence. At the same time, however, the Indonesian army began

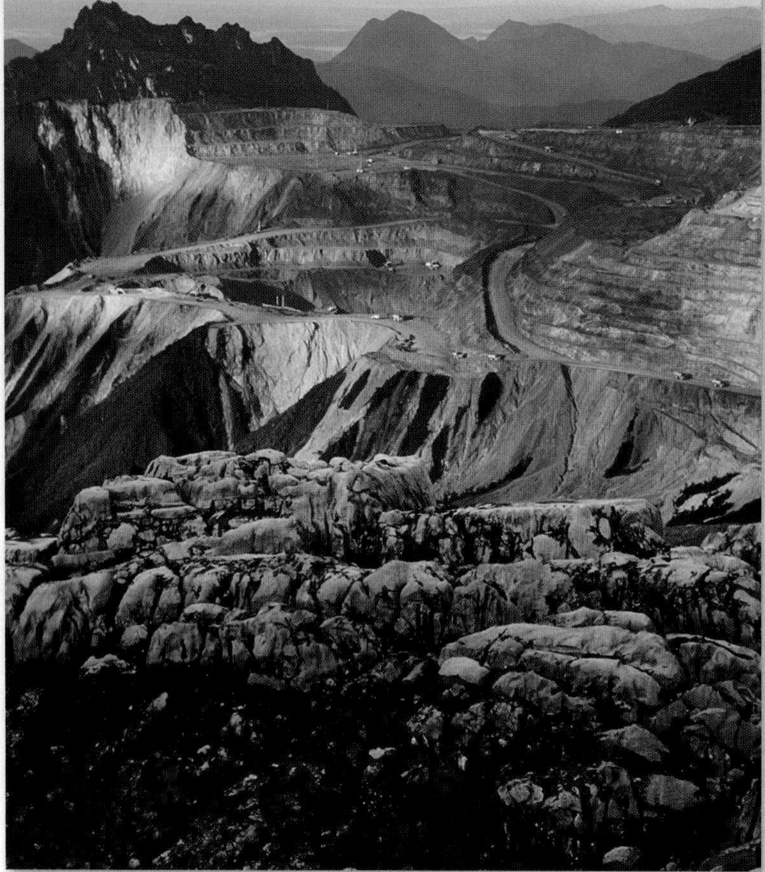

Figure 13.29 Grasberg Mine Located in the Indonesian province of Papua, Grasberg is the world's largest gold mine and third-largest copper mine. Employing more than 19,000 people, Grasberg is of great economic importance to Indonesia. It is also highly polluting, generating 253,000 tons (230,000 metric tons) of tailings per day, and is thus opposed by many local inhabitants.

to organize militias to intimidate the East Timorese into voting to remain with Indonesia. When it was clear that the vote would be for independence, the militias began rioting, looting, and killing civilians. Under international pressure, Indonesia finally withdrew, and the East Timorese began to build a new country, with independence formalized in 2002.

The Aceh region of northern Sumatra has also experienced prolonged political violence, as local rebels have sought to create an independent Islamic state. The Indonesian government has given Aceh "special autonomy," but did so to prevent independence. Ironically, the devastation of the December 2004 tsunami seems to have generated a solid peace, as local needs were so great that the separatist fighters agreed to lay down their weapons. Tensions persist in the area, however, as Aceh's autonomous government seeks to enforce strict Islamic law. In 2014, these laws were extended to cover the region's 90,000 non-Muslims. Adultery, sex before marriage, and homosexuality are now severely punished, and even the possession of alcohol can result in cane-strokes, imprisonment, and large fines.

Regional Tensions in the Philippines The Philippines has also suffered from regional political violence. Its most persistent problem is the Islamic southwest, where rebels have long demanded independence. After successful negotiations with the main separatist group in 1989, the government created the Autonomous Region in Muslim Mindanao (ARMM). The more radical Islamist groups, however, rejected the settlement and have continued to fight, setting off bombs, attacking military patrols, and kidnapping civilians. In 2014, governmental officials and Islamist rebels signed a new peace treaty based on creation of an expanded autonomous region to be called Bangsamoro. The agreement has not yet been ratified, however, and fighting continues.

The Muslim southwest does not present the Philippines' only political problem. A revolutionary communist group called the New People's Army operates in most parts of the country and still controls many rural districts. Furthermore, the country's national government, although democratic, is far from stable, suffering from periodic coup threats, corruption scandals, mass protests, and impeachment efforts.

Burma's Many Problems Of all the countries of Southeast Asia, Burma has experienced the most extreme ethnic conflicts. Burma's simultaneous wars have pitted the central government, dominated by the Burmese-speaking ethnic group (Burmans), against the country's varied non-Burman societies. Fighting intensified gradually after independence in 1948, and by the 1980s almost half of the country's territory had become a combat zone.

In the 1990s, however, successful Burmese army offensives, combined with cease-fire agreements, resulted in a marked decrease in warfare. In early 2012, the Burmese government finally signed a cease-fire with the Karen ethnic rebels, though at the same time it intensified a military drive against the Kachin insurgents of the far north (Figure 13.30). From 2011 to 2014, an estimated 100,000 civilians were displaced in the resulting Kachin War. Burma is also plagued by religious tensions, as discussed earlier in this chapter in connection with the Rohingyas. In June 2014, the country's second-largest city, Mandalay, was shaken by severe ethnic riots after a Muslim tea-shop owner was accused of raping a Buddhist girl.

Due to its repressive government, Burma has long suffered from trade restrictions imposed by the United States and the European Union. In 2010, however, Burma held elections and began to create a more open society, and two years later it allowed noted opposition

Figure 13.30 Women Soldiers of the Kachin Liberation Army Although Burma has recently made peace with many insurgent groups, the Kachin Liberation Army has been fighting the Burmese government for decades. Like several other ethnic insurgent groups in Burma, it employs many female fighters.

Figure 13.31 2014 Thai Military Coup The government of Thailand has moved between military rule and democracy on several occasions in recent decades. Many civilians protested the most recent military take-over of the country, which occurred in May 2014. Demonstrators hold signs as soldiers stand guard during a protest against Victory Monument in central Bangkok on May 26, 2014.

leader and 2001 Nobel Peace Prize winner Aung San Suu Kyi to run for parliament. The United States has reestablished diplomatic relations with Burma, and in 2013 the European Union dropped its trade sanctions. In November 2014, however, U.S. President Barack Obama accused the country's leaders of backsliding on their reforms and hinted that sanctions could be reimposed.

Thailand's Troubles Unlike Burma, Thailand has enjoyed basic human freedoms for many years. Most observers thought that it had become a stable democracy, but in 2006, Thai Prime Minister Thaksin Shinawatra was overthrown by a military coup that apparently had the blessing of Thailand's revered king. Thaksin, a wealthy businessman, had gained the support of Thailand's poor in part by setting up a national health-care system, but he infuriated the middle and upper classes by taking too much power into his own hands. Huge protests and counterprotests followed for several years. In 2011, Thaksin's sister, Yingluck Shinawatra, became prime minister after winning a landslide victory. In May 2014, however, Thailand's military overthrew the elected government, dissolved the senate, and repealed the constitution (Figure 13.31). A military-dominated national legislature was then established. Many observers believe Thailand will eventually return to democratic rule, but the process will probably not be easy.

Thailand's recent political chaos makes it difficult for the country to deal with its main political threat, the murky insurgency that plagues its southernmost provinces. Far southern Thailand is relatively poor, and its people are mostly Malay in language and Muslim in religion. They have long resented being ruled by Thailand, and periodic rebellions have flared up for decades. After 2004, violence sharply increased, resulting in more than 5000 deaths by early 2015. One odd feature of the insurgency in southern Thailand is the fact that no group has emerged to claim responsibility for the violence, and hence no political demands have been made.

International Dimensions of Southeast Asian Geopolitics

Despite ASEAN's general successes in encouraging peaceful relations among its member countries, border disputes continue to flare up. In late 2008 and early 2009, a conflict between Thailand and Cambodia over the area around the 11th-century Preah Vihear Temple resulted in 40 deaths. In January 2015, Thailand sent 200 troops to the border to protest Cambodian construction activities in the disputed area.

A more complicated territorial dispute centers on the Spratly and Paracel islands in the South China Sea, two groups of tiny islands and reefs that may sit over substantial oil reserves (Figure 13.32). The Philippines, Malaysia, Brunei, and Vietnam have all claimed territory there, as have China and Taiwan. International tensions in the South China Sea began to heat up in 2010, as China increased its naval presence in the area. In May 2014, China set up an oil-drilling rig in waters claimed by Vietnam, resulting in several collisions involving Chinese and Vietnamese boats. As a result of such tensions, both Vietnam and the Philippines began to seek closer military cooperation with the United States.

One of the biggest problems faced by ASEAN leaders has been the establishment of radical Islamist networks in the region. The largest of these is Jemaah Islamiya (JI), a militant group dedicated to establishing an Islamic state that would contain all Muslim areas within Southeast Asia. JI agents are believed to have detonated several deadly bombs in Java and Bali in the early 2000s. Indonesia responded by creating an elite counterterrorism squad (Detachment 88) and by establishing a "deradicalization program." By 2008, most observers had concluded that these programs were so successful that JI was no longer a threat. In September 2014, however, the group claimed responsibility for bombing a public monument in front of the city hall of General Santos in the southern Philippines.

Piracy, especially in the vital shipping lanes in the Strait of Malacca, is another international security concern in Southeast Asia. Between April and December 2014, six oil tankers were completely emptied of their cargos while their crews were held captive. These attacks have led Malaysia, Singapore, Indonesia, and Thailand to form a joint antipiracy naval force called the Malacca Straits Patrol.

Figure 13.32 The Spratly Islands The struggle over the Spratly Islands, some or all of which are claimed by China, Taiwan, Vietnam, Malaysia, Brunei, and the Philippines, has been heating up in recent years. In order to bolster its claims, China has been building facilities, as well as actual islands, in the area. This March 2015 photograph shows Chinese construction and dredging under way at Mischief Reef.

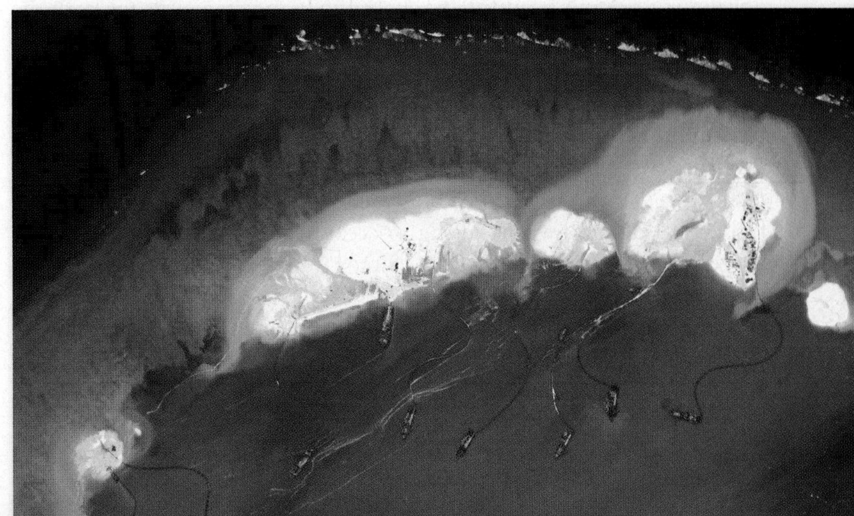

Review

13.7 How did European colonization influence the development of the modern countries of Southeast Asia, and how did that process differ in the insular and the mainland regions?

13.8 How did the emergence and spread of ASEAN reduce geopolitical tensions in Southeast Asia, and why could certain conflicts in the region not be solved by the ASEAN process?

KEY TERMS Khmer Rouge, domino theory

Figure 13.33 Philippine Call Center The Philippines competes with India for the position as the world's leading international call center. Almost a million Filipinos now work at these centers, providing information and advice to English-speaking consumers in North America and Europe.

Economic and Social Development: The Roller-Coaster Ride of Developing Economies

Over the past few decades, Southeast Asia has experienced major economic fluctuations (Table A13.2). An economic boom between 1980 and 1997 was followed by a major recession. Southeast Asian economies began to grow quickly after 2000, but the global economic crisis of 2008–2009 resulted in another blow. Within a few years, however, the region had fully recovered. Overall, Southeast Asia accounted for some of the world's fastest-growing economies during the first 15 years of the new millennium. Some observers fear that a slowdown in China, along with the recent decline in global commodity prices, could hurt the region, but others think that the reduced price of oil will give it an economic boost.

Uneven Economic Development

While the region as a whole has experienced pronounced ups and downs, parts of Southeast Asia have done much better in the global economy than have others. Oil-rich Brunei and technologically sophisticated Singapore rank among the world's more prosperous countries, whereas Cambodia, Laos, Burma, and East Timor are among the poorer countries of Asia.

The Philippine Decline and Recovery In the 1950s, the Philippines was the most highly developed Southeast Asian country. By the late 1960s, however, Philippine development had been derailed. Through the 1980s and early 1990s, the country's economy failed to outpace its population growth, resulting in declining living standards for both the poor and the middle class. Filipinos are still well educated and reasonably healthy by world standards, but even the country's educational and health systems declined during this period.

Why did the Philippines fail despite its earlier promise? While there are no simple answers, it is clear that dictator Ferdinand Marcos (who ruled from 1968 to 1986) wasted—and perhaps even stole—billions of dollars, while failing to create conditions that would lead to genuine development. The Marcos regime instituted a kind of **crony capitalism**, in which the president's friends were given huge economic favors, while those believed to be enemies had their properties taken.

By 2010, however, the economy of the Philippines finally began to recover. From 2012 to 2014, it expanded by almost 7 percent a year. A particularly bright spot in the Philippine economy has been the growth of business outsourcing operations, attracted by the country's educated, English-speaking population. The recent development of modern communications systems in the region has been particularly beneficial. It is estimated that by 2017, 1.3 million Filipinos will work in international call centers, handling telephone inquiries from customers in the United States and other wealthy countries (Figure 13.33). The fact that most Filipinos are familiar with American culture gives their country an advantage in this rapidly growing industry.

The Regional Hub: Singapore Singapore and Malaysia have been Southeast Asia's major developmental successes. Singapore has transformed itself from an **entrepôt**—a place where goods are imported, stored, and then transshipped—to one of the world's wealthiest and most modern states. Singapore is the communications and financial hub of Southeast Asia as well as a thriving high-tech manufacturing center. Its government has played an active role in the development process, encouraging investment by multinational companies and investing itself in housing, education, and some social services (Figure 13.34). The country is currently encouraging growth in the pharmaceutical and medical technology industries. Singapore's government, however, remains only partly democratic, as the ruling party maintains a firm grip on government and restricts free speech.

The Malaysian Boom Although not nearly as well-off as Singapore, Malaysia has also experienced rapid economic growth. Development was initially concentrated in agriculture and natural resources, focusing on tropical hardwoods, plantation products, and tin mining. More recently, manufacturing, especially in labor-intensive high-tech sectors, has become the main engine of growth (Figure 13.35). Malaysia, however, is also an oil exporter, and lower oil prices may take a toll on its economy.

The modern economy of Malaysia is not uniformly distributed across the country. One difference is geographical: Most industrial development has occurred on the west side of peninsular Malaysia. More important, however, are differences based on ethnicity: Malaysia's industrial wealth has been concentrated in the Chinese community. Ethnic Malays remain less prosperous than Chinese-Malaysians, and those of South Asian descent are poorer still.

The income gap between majority populations and local Chinese communities is a feature of most Southeast Asian countries. The problem

Figure 13.34 Public Housing in Singapore Despite its free-market approach to economics, the government of Singapore has invested heavily in public housing. Most Singaporeans live in buildings similar to these apartments. **Q: Singapore is in general a conservative country devoted to free-market economics. Considering this fact, why has its government invested so heavily in public housing?**

is particularly acute in Malaysia, however, because its Chinese minority is so large. The government's response has been one of aggressive "affirmative action," by which economic power is transferred to the dominant Malay, or **Bumiputra** ("sons of the soil"), community. This policy has been reasonably successful. Since the economy as a whole has expanded significantly, the Chinese community can thrive even as its relative share of wealth has declined. But opposition to such policies is growing, as even many ethnic Malay professionals want a more open and competitive society. According to a 2008 survey, 71 percent of Malaysians claim that the country's affirmative action programs are obsolete.

Thailand's Ups and Downs Thailand, like Malaysia, climbed rapidly during the 1980s and 1990s into the ranks of the world's newly industrialized countries. Japanese companies were leading players in the Thai boom, attracted by Thailand's low-wage, yet reasonably well-educated, workforce. Thailand experienced major downturns, however, in the late 1990s and again in 2008–2009. The Thai economy has seen some renewed vitality in recent years, but continued political instability has reduced its rate of growth.

Thailand's economic growth over the past several decades has by no means benefited the entire country to an equal extent. Most industrial development has occurred in the historical core, especially in Bangkok and surrounding areas. The vulnerability of such industrial concentration was demonstrated in the summer of 2011, when extensive flooding in central Thailand disrupted not only the Thai national economy, but also global supply chains in the automobile and consumer electronics sectors.

Thailand's Lao-speaking northeast (the Khorat Plateau) and Malay-speaking far south remain the country's poorest regions. Because of the poverty of their homeland, northeasterners are often forced to seek employment in Bangkok. Men typically find work in the construction industry, while women not uncommonly make their living as sex workers. In the far south, poverty and unemployment have contributed to the region's brutal insurgency.

Unstable Economic Expansion in Indonesia At the time of independence (1949), Indonesia was one of the world's poorest countries. The Indonesian economy finally began to expand in the 1970s. Oil exports fueled the early growth, as did the logging of tropical forests. But Indonesia continued to grow as its reserves were depleted and it became a net oil importer. Like Thailand and Malaysia, it has attracted multinational companies looking for a low-wage and relatively well-educated workforce. Large Indonesian firms, many of them owned by local Chinese families, have also capitalized on the country's human and natural resources.

Despite its recent economic growth, Indonesia remains a poor country. Its pace of economic expansion has seldom matched those of Singapore and Malaysia, and it still depends more on the unsustainable exploitation of natural resources. But the global economic crisis of 2008–2009 was not as severe in Indonesia as it was in the wealthier countries of Southeast Asia, and from 2010 to 2015 its economy recorded solid growth. In early 2015, the global fall in the price of oil allowed the Indonesian government to reduce its expensive fuel subsidies, a move that many economists think could both boost its economic performance and allow more spending on social programs.

The Recent Rise of Vietnam and Cambodia The former French colonial zone was long noted as one of the poorest and least globalized parts of Southeast Asia. From the time of the Vietnam War through the early 1990s, Vietnam experienced relatively little economic development. Frustrated with their country's economic performance, Vietnam's leaders began to follow China by embracing market economics while retaining the political forms of a communist state. The Vietnamese economy subsequently began to expand quickly, experiencing annual rates of growth of up to 8 percent in the early 2000s. But Vietnam is still a poor country, and more recently its annual rate of economic growth has declined to around 5.5 percent. Critics complain that the country's banking sector is poorly developed and is burdened by too many nonperforming loans.

Figure 13.35 High-Tech Manufacturing in Malaysia Many foreign companies have established high-tech manufacturing facilities in Malaysia. Most of their factories are located in the western part of peninsular Malaysia.

Figure 13.36 Cambodian Casino The Cambodian border city of Poipet has recently emerged as a major gambling center. Most of the investments, and most of the tourists, come from Thailand, angering many Cambodians. Organized crime is also a major problem in the city.

Vietnam now welcomes multinational corporations, drawn by its low wages and relatively well-educated workforce. Japanese and South Korean companies now often favor Vietnam over other Southeast Asian countries. Local businesses, however, complain of harassment by state officials, and development remains geographically uneven. Southern Vietnam is still more entrepreneurial and capitalistic than the north, while deep and persistent poverty remains entrenched in many rural areas, particularly in the tribal highlands.

Cambodia's recent economic history is somewhat similar to that of Vietnam, only more extreme. Long burdened by war and corruption, Cambodia was one of Asia's poorest countries, its economy focused on subsistence agriculture. The discovery of oil and other mineral resources in 2005, however, combined with a thriving tourist economy and large-scale international investments to generate an economic boom. The garment industry did especially well and now accounts for more than half of Cambodia's total exports.

Cambodia's recent economic expansion has resulted in problems as well as opportunities. Workers in garment factories, mainly young women, are very poorly paid and must work long hours under harsh conditions. A property boom in Phnom Penh, moreover, saw thousands of poor people being forced out of their homes to make way for new projects. Tourism has also proved to be a mixed blessing. Several Cambodian border towns, most notably Poipet, have set themselves up as gambling centers, attracting investment—and criminal activities—from Thai underworld figures (Figure 13.36). Many Cambodians fear that their country is coming under the economic domination of Thailand and Vietnam.

Turnarounds in Laos and East Timor? Like Cambodia, Laos has long been dominated by subsistence agriculture, which employs roughly three-quarters of its workforce. Laos has particular economic difficulties owing to its rough terrain and relative isolation; outside of its few cities, paved roads and reliable electricity are rare. As a result, it remains heavily dependent on foreign aid. More recently, however, the Laotian economy has begun to grow rapidly.

The Laotian government is pinning its economic hopes on hydropower development, mining, tourism, and investment from Thailand, Vietnam, and China. Hydropower is particularly important, as the country is mountainous and has many rivers and could therefore generate and export large quantities of electricity. Laos also benefits from the increasing volume of barge traffic going up the Mekong River to China and by Chinese-financed railroad building. In 2014, a minister in the Laotian government claimed that his country would secure its economic place by serving as a bridge between China and the ASEAN bloc.

East Timor is also notable for its economic problems. This small country took many years to recover from the devastation that accompanied its independence, and it has been further weakened by the gradual withdrawal of international aid agencies. But discoveries of massive offshore oil and natural gas deposits have significantly boosted East Timor's prospects, and from 2011 to 2014 it posted one of the world's fastest rates of economic expansion. It remains to be seen whether such growth is sustainable and whether it will significantly increase the meager living standards of the East Timorese people.

Burma's Troubled but Emerging Economy Burma is at the bottom of the scale of Southeast Asian economic development. For all of its many problems, however, Burma remains a land of great potential. It has abundant natural resources (including oil and other minerals, water, and timber) as well as a large expanse of fertile farmland. Its population density is moderate, and its people are reasonably well educated. But despite these advantages, Burma's economy remained relatively stagnant for the first 60 years after independence in 1948.

Since 2011, however, Burma has opened its economy while simultaneously reforming its political system. Trade with the rest of the world has boomed, and foreign investments have begun to pour into the country. Chinese firms have been building major oil and gas pipelines across the country to bring fuel supplies from the Indian Ocean into southwestern China. Anticorruption measures have also been passed, and the country has at long last created a stable currency. In 2014, Burma achieved an economic growth rate of over 7 percent.

Globalization and the Southeast Asian Economy

Southeast Asia as a whole has rapidly integrated into the global economy (Figure 13.37). Singapore has staked its future on the success of multinational capitalism, as have several other countries. Even communist Laos and once-isolationist Burma have opened their doors to the global system. The region is also one of the world's premier tourist destinations, noted for its beautiful beaches, tropical climate, rich culture, and moderate prices. Tourist arrivals have been growing by more than 10 percent a year since 1995. Although tourism has been

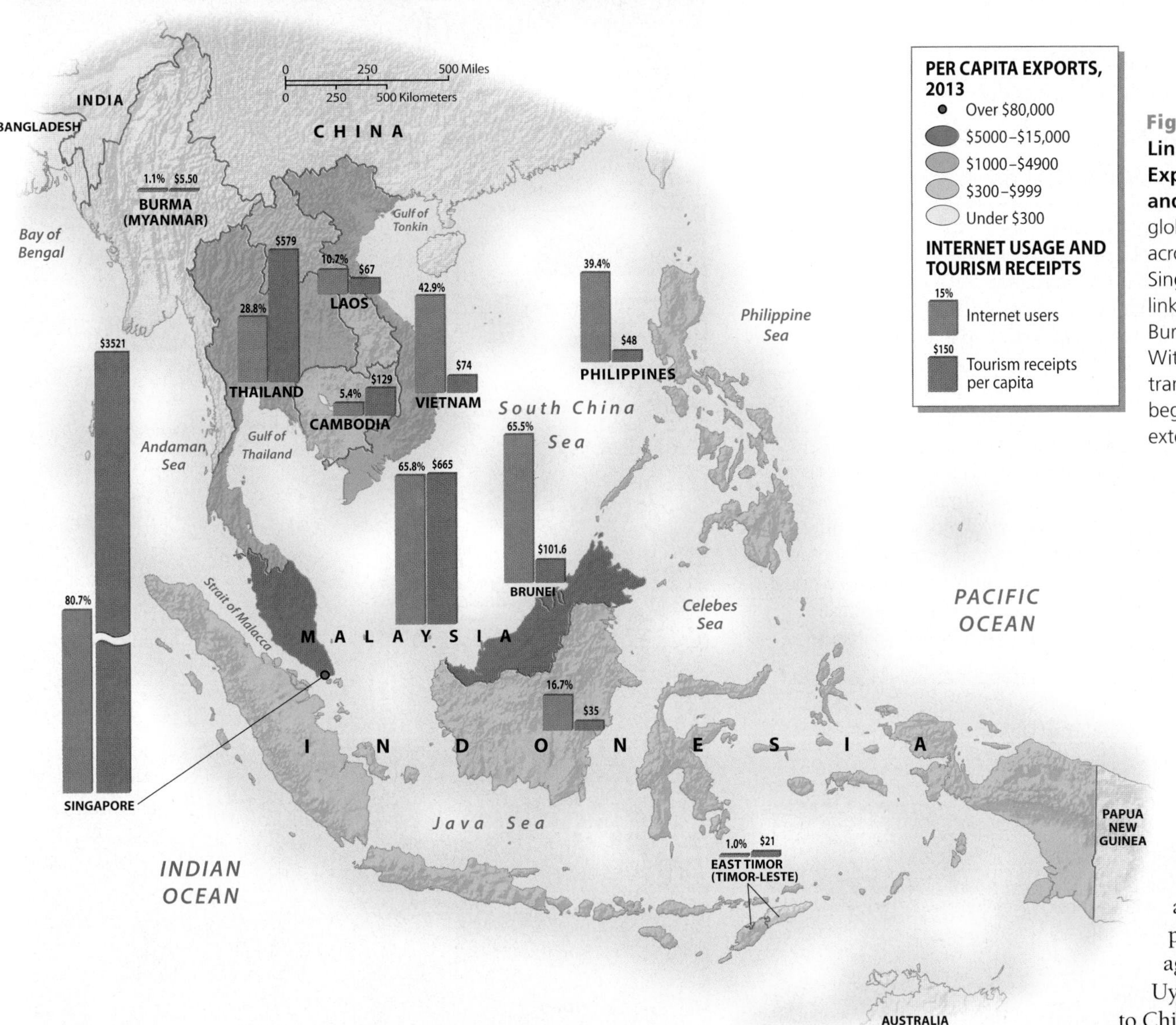

Figure 13.37 Global Linkages in Southeast Asia: Exports, Internet Usage, and Tourism Levels of globalization vary tremendously across Southeast Asia. Whereas Singapore and Malaysia are closely linked to the global economy, Burma is still relatively isolated. With Burma's recent political transformation, however, it is beginning to experience more extensive globalization.

associated with a number of environmental and social problems, sustainable tourism now forms the fastest-growing segment of the market.

Globalization Controversies Global economic integration has undoubtedly brought significant development to Singapore, Malaysia, Thailand, and even Indonesia. Outside of Singapore and Malaysia, however, economic expansion has generally been based on labor-intensive manufacturing, in which workers are paid low wages and subjected to harsh discipline. Movements have thus begun in Europe, the United States, and elsewhere to pressure both multinational corporations and Southeast Asian governments to improve working conditions for laborers in the export-oriented industries (see *Everyday Globalization: Thailand's Troublesome Seafood Exports*).

Much of the labor in Southeast Asia's exporting companies comes from women, who are often paid much less than men doing the same work (see *Geographer at Work: Female Migrant Workers in Southeast Asia*). One study of factories in central Java found that most workers were young, single women from poor, landless families. A 2013 report claimed that even in relatively prosperous Malaysia, female factory workers earn 22 percent less than male factory workers. In Thailand, impoverished migrant women from Burma form a large, but mostly hidden labor force.

China figures prominently in the recent globalization of Southeast Asian economies. It has invested heavily in infrastructural projects, most notably in Laos, Burma, and Cambodia. Dams, highways, railroads, and ports in the region have all received large infusions of money from China. Southeast Asian leaders are generally pleased with Chinese investments, but local residents are often angered over environmental degradation and the loss of land. Political strings are also sometimes tied to these projects. In 2010, Cambodia agreed to deport 20 ethnic Uyghur asylum-seekers back to China; two days later the Beijing government agreed to release $1.2 billion for Cambodian infrastructure improvements.

Issues of Social Development

As might be expected, most key indicators of social development in Southeast Asia are closely linked to levels of economic development. Singapore thus ranks among the world leaders in regard to health and education, while Laos, Cambodia, Burma, and East Timor lag well behind. The people of Vietnam, however, are somewhat healthier and better educated than might be expected based on their country's overall economic performance.

Life Expectancy and Education Overall, Southeast Asia has achieved relatively high levels of social welfare. In the poorer countries of the region, however, life expectancy is still below 70 years, and the female literacy rates are below 70 percent in Laos and Cambodia. But even the least developed countries of the region have made steady improvements on these issues over the past several decades, in part due to international assistance. In 2014, for example, the World Bank announced a US$2 billion program to help Burma provide health care to all of its citizens and deliver electricity throughout the country.

Most Southeast Asian governments place a high priority on basic education. Literacy rates are relatively high in most countries, even Burma. Gains in university and technical education have been slower however, primarily because of a lack of investment by the countries

EVERYDAY GLOBALIZATION

Thailand's Troublesome Seafood Exports

Southeast Asia's economic globalization links it to the United States in various ways. For example, up to 90 percent of the seafood consumed in the United States is imported, and three of the top six source countries are Thailand, Indonesia, and Vietnam.

Thailand's seafood export industry is ranked third in the world (after China and Norway). But it is also controversial. In 2014, Thailand's shrimp exports plunged by 32 percent as a virulent disease spread through artificial shrimp ponds. Environmentalists argue that much of the country's seafood industry depends on illegal fishing and the unsustainable conversion of coastal wetlands into fishponds (Figure 13.3.1). Human rights issues also loom large. Many people employed in fishing and fish processing are undocumented foreign migrants who are often brutally treated. The U.S. State Department reports that the Thai shrimp industry is a major violator, forcing tens of thousands of people to work as virtual slaves. Such allegations have led U.S. supermarket groups to create a task force to address forced labor in shrimp-exporting countries.

1. List some of the reasons why the Thai seafood industry is associated with environmental and labor controversies.
2. Go to a local market that sells seafood. Determine how much of the fish being sold originated in another country.

Figure 13.3.1 Thai Shrimp Pond Shrimp are cultivated in this pond in Chumporn Province in southern Thailand. Shrimp farming has emerged as a lucrative industry in coastal Southeast Asia, but it is associated with environmental damage and the exploitation of labor.

of the region. As Southeast Asian economies continue to grow, this educational gap is beginning to have negative consequences, forcing many students to study abroad for higher degrees.

Gender Equity and the Sex Trade In many ways, Southeast Asian societies exhibit contradictory views on the social position of women. Historically speaking, Southeast Asia has long been noted for its gender equity. Women in the region have played important economic roles as household managers and market vendors. Early European visitors to the region were often shocked at how freely men and woman mixed and at how much authority women exercised. Such patterns have not completely disappeared. Southeast Asia has had a significant number of female leaders, including two recent presidents of the Philippines and

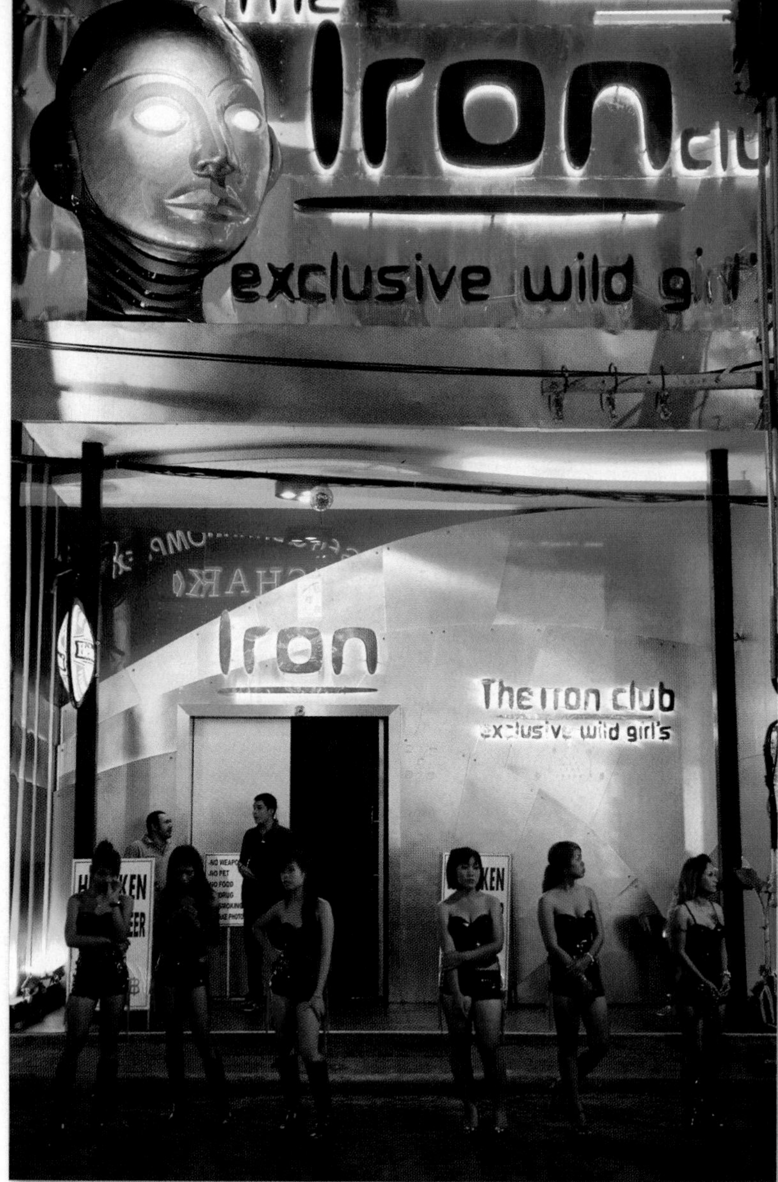

Figure 13.38 Commercial Sex in Pattaya The beach city of Pattaya was once a sleepy fishing village near an American military base. It is now a major center of the global sex trade, which employs an estimated 200,000 people. Russian interests figure prominently in the city's business community.

Burma's world-famous opposition leader and Nobel Peace Prize laureate, Aung San Suu Kyi. Some anthropologists have gone so far as to describe the Minangkabau society of western Sumatra as a "modern matriarchy," as Minangkabau women have traditionally controlled their large households, which are based on descent from female ancestors.

Yet Southeast Asia is also the site of some of the world's most extensive sexual exploitation. Commercial sex is a huge business in Thailand. Despite its massive scope, prostitution is technically illegal in Thailand, which means that it is a major source of corruption. Other Southeast Asian countries—particularly the Philippines, Vietnam, and Cambodia—are also centers of a globally oriented commercial sex trade. Many workers in Southeast Asian brothels are underage, and many have been coerced into the activity. Young women, girls, and boys are frequently trafficked from the poorer parts of the region, often in connection with the drug trade.

Two of the main centers of commercial sex in Southeast Asia developed around U.S. military bases during the Cold War: Angeles City in the Philippines and Pattaya in Thailand (Figure 13.38).

GEOGRAPHERS AT WORK

Female Migrant Workers in Southeast Asia

"Geography provides a really unique lens on contemporary world issues," notes **Rachel Silvey** of the University of Toronto. Silvey has studied the intersection of labor migration, gender roles, and economic development in Indonesia. Her recent research examines the movement of Indonesian workers to Persian Gulf countries, so in addition to Indonesia, Silvey has conducted fieldwork in Dubai and plans to travel to Saudi Arabia, allowing her to look at migration from different vantage points.

Connecting to Place and People

Silvey "fell in love with Indonesia" well before she discovered geography, spending her junior year in college there as part of a Volunteers in Asia program. She learned the language and lived in a rural community, studying gender and agriculture, and "felt like I had something to add to academic conversations when I came back." Her undergraduate thesis advisor asked her if she would consider geography for graduate studies. "I hadn't even heard of geography! But I found that geographers get to learn about the world firsthand through travel, and that it was really the disciplinary home for my diverse interests."

Silvey's research relies heavily on both household surveys and in-depth interviews (Figure 13.4.1). As Silvey relates, her graduate advisor, geographer Victoria Lawson, "encouraged me to make the most of my fieldwork and the connection to place and the people that I'd gotten to know so deeply from my year abroad." Her interviews extend well beyond fact-finding: "To do a good interview is to have a good conversation with someone . . . when it works, you've not only collected data, you've developed a relationship with the person that then continues over the course of your life and theirs."

Figure 13.4.1 **Indonesian Workers Share Their Experiences** Geographer Rachel Silvey gets to know her subjects well through in-depth conversations about their work and their communities.

International Labor Migration Silvey's research on labor migration and gender looks beyond Southeast Asia to consider the position of Indonesian women working in the Gulf states, many of whom provide "care labor" for the elderly as well as children. She is currently part of a research team under the Centre for Global Social Policy (http://www.cgsp.ca/people), studying care pathways. The multidisciplinary team integrates social work, gender and aging issues, migration paths, and public policy—an effort for which Silvey's geographic perspective and interview skills are invaluable. Says Silvey about her interviews: "There are histories . . . that need to be explored, along with the changing landscapes and politics of migration."

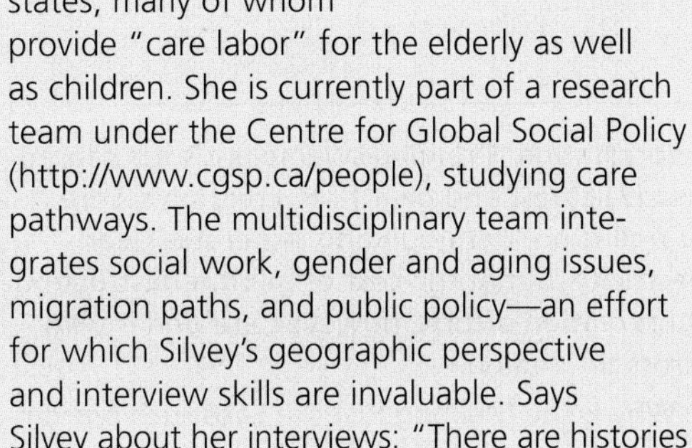

1. How might patterns of female labor migration in Indonesia differ from those of mainland Southeast Asian countries, such as Burma or Thailand?
2. Explain how interviewing immigrants could benefit a government or nongovernmental organization that tracks migration data.

The military bases are gone, but both cities have expanded their economies by focusing on tourism, much of it sex-related. Pattaya now supports an estimated 20,000 sex workers. Here the high end of the business includes thousands of Russian and Ukrainian women. As a result, Pattaya now has a major Russian presence, attracting hundreds of thousands of Russian tourists every year in addition to wealthy Russian investors. Russian organized crime now plays a major role in the city, illustrating one of the seamier aspects of globalization in modern Southeast Asia.

A large Southeast Asian business connected with the sex trade is surrogacy services, in which infertile couples desiring a child pay to have embryos implanted in local women to carry to term. In Thailand, the cross-border surrogacy market is now worth hundreds of millions of dollars a year. In 2014, however, the Thai government declared the practice exploitive and vowed to restrict and perhaps prohibit it.

Review

13.9 Explain why some Southeast Asian countries have experienced sustained economic growth and social development, whereas others have experienced stagnation in the same period.

13.10 Why has Southeast Asia emerged as a major center of the global sex trade, and what problems does this cause for the region?

KEY TERMS crony capitalism, entrepôt, Bumiputra

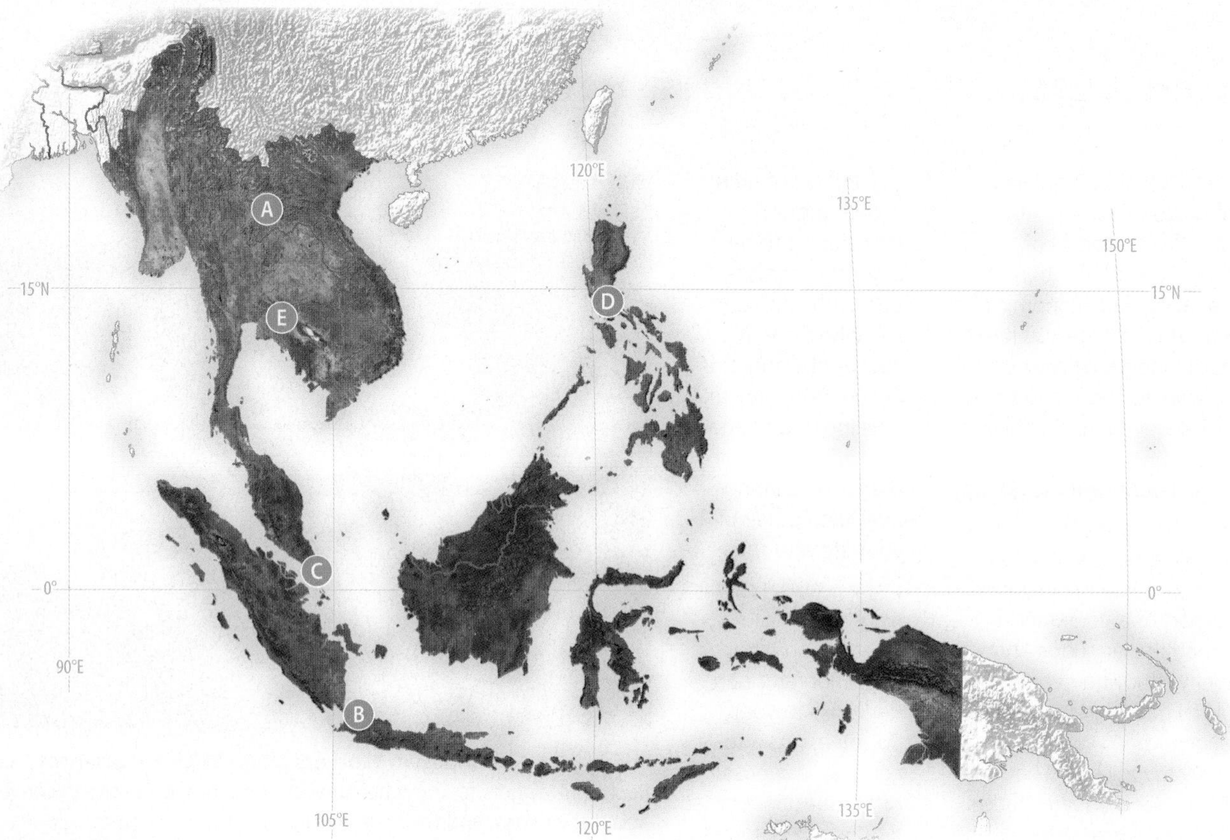

Review

Physical Geography and Environmental Issues

13.1 Describe how the region's tectonic activity and dominant climate types have influenced Southeast Asian landscapes, human settlement, and development.

13.2 Explain the driving forces behind deforestation and habitat loss in the different subregions of Southeast Asia, and list other environmental problems affecting the region.

Some of the most serious problems created by globalization in Southeast Asia are environmental. Commercial logging and plantation agriculture have led to extensive deforestation. Draining swamplands has caused massive forest and peat fires, creating severe air pollution. Dam-building generates clean electricity, but at the cost of habitat destruction. Conservation efforts, however, are under way across the region.

1. Why is the Mekong River in Laos such a promising site for generating hydroelectricity?

2. List the major advantages and disadvantages of dam-building and hydroelectricity development in Southeast Asia. Organize a class debate on this topic, taking into account concerns about global climate change.

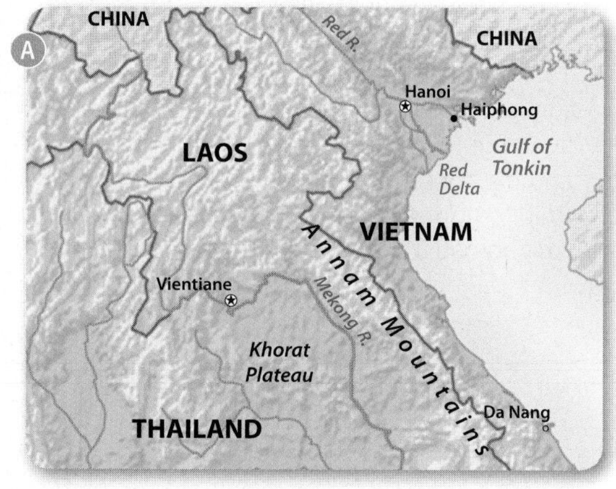

Population and Settlement

13.3 Show how the differences among plantation agriculture, rice growing, and swidden cultivation in Southeast Asia have molded settlement patterns.

13.4 Describe the role of primate cities and other massive urban centers in the development of Southeast Asia, and locate the largest cities on a map of the region.

As people move from densely populated, fertile lowland areas into remote uplands, both environmental damage and cultural conflicts often follow. Population movements in Southeast Asia also have a global dimension. Most countries of the region have seen major reductions in their birth rates, but the Philippines and East Timor continue to experience rapid population gains. Cities are growing rapidly throughout the region, and the Jakarta, Bangkok, and Manila metropolitan areas are now some of the world's largest urban aggregations.

3. What do the patterns in this image tell us about the development of the Jakarta metropolitan area?

4. Why do you think Jakarta has grown much more rapidly over the past several decades than other Indonesian cities? List some of the advantages and disadvantages of urban concentration in Southeast Asia.

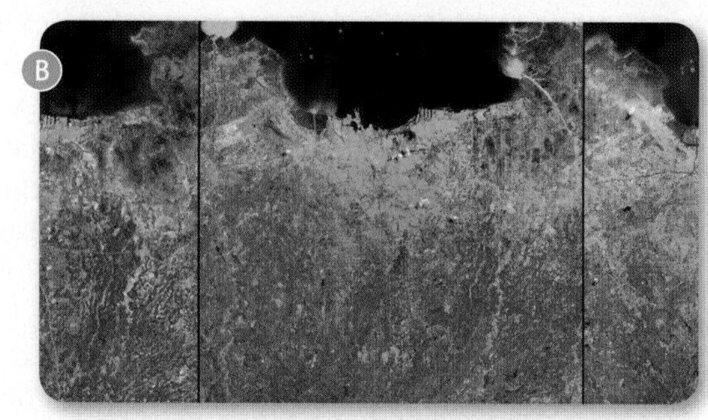

Cultural Coherence and Diversity

13.5 Map the major religions of the region to show the ways in which religions from other parts of the world spread through Southeast Asia and discuss the role of religious diversity in Southeast Asian history.

13.6 Identify the controversies surrounding cultural globalization in Southeast Asia, explaining why some people in the region welcome the process, whereas others resist it.

Southeast Asia is characterized by tremendous cultural diversity. Most of the world's major religions, for example, are represented in the region. In recent decades, conflicts over language and religion have caused serious problems in several Southeast Asian countries. However, the region has found a sense of regional identity as expressed through the Association of Southeast Asian Nations (ASEAN). Cultural globalization has also long been pronounced in many parts of Southeast Asia—particularly Singapore, Thailand, and the Philippines.

5. Why does the government of Singapore work hard to teach its citizens proper spoken English when it is already considered to be an English-speaking country?

6. Should English serve as a global language? Should Southeast Asian countries try to teach everyone to speak English? Organize a class debate on this topic.

C

Geopolitical Framework

13.7 List the ASEAN countries and explain how this organization has influenced geopolitical relations in the region.

13.8 Identify the major ethnic conflict zones on the map of Southeast Asia and explain why certain countries in the region have such deep problems in this regard.

The relative success of ASEAN has not resolved all of Southeast Asia's political tensions. Many of its countries still argue about geographical, political, and economic issues, while insurgencies remain active in the Philippines and Thailand. Both the Philippines and Indonesia have established autonomous areas in order to reduce the desire for secession. Cambodia, Laos, and Burma have been held back by repressive governments, although reforms have recently been enacted, especially in Burma.

7. Some of the signs carried by protestors in this rally demand that China stop "poaching" in Philippine waters. Why is concern about illegal fishing by other countries so pronounced in the Philippines?

8. Why are the tiny Spratly Islands such a controversial issue in Southeast Asia? What does this say about relations between China and Southeast Asia?

D

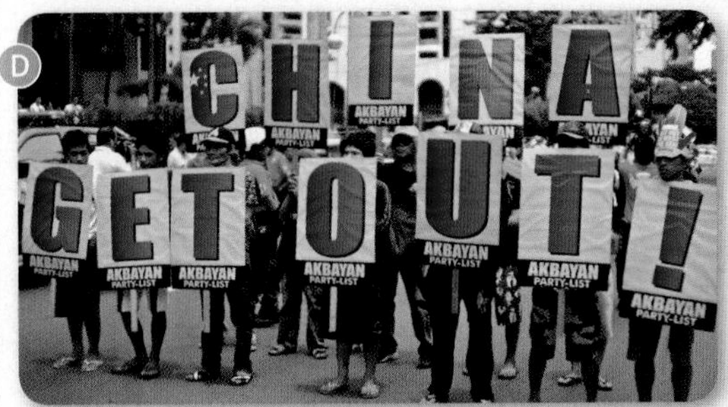

Economic and Social Development

13.9 Explain why levels of economic and social development vary so widely across the Southeast Asian region, giving an example from a mainland country and from an insular country.

Although ASEAN plays an economic as well as a political role, its economic successes have been more limited. Most of the region's trade is still directed outward toward the traditional centers of the global economy: North America, Europe, and East Asia. A significant question for Southeast Asia's future is whether the region will develop an integrated regional economy. A more important issue is whether social and economic development can lift the entire region out of poverty instead of benefiting just the more fortunate areas.

9. How has tourism affected the economy and social relations of different parts of Southeast Asia?

10. What are some of the advantages and disadvantages of tourism-related economic development?

E

KEY TERMS

animism *(p. 434)*
Association of Southeast Asian Nations (ASEAN) *(p. 420)*
Bumiputra *(p. 445)*
crony capitalism *(p. 444)*
domino theory *(p. 441)*
entrepôt *(p. 444)*
Golden Triangle *(p. 430)*
Khmer Rouge *(p. 441)*
lingua franca *(p. 435)*
primate city *(p. 432)*
swidden *(p. 428)*
transmigration *(p. 432)*
tsunami *(p. 422)*
typhoon *(p. 422)*

Data Analysis

http://goo.gl/rRLxiY

Southeast Asia is noted for its numerous indigenous languages. Many languages, however, are declining or endangered, and a number have died out in recent years, threatening cultural diversity. The *Ethnologue: Languages of the World* website (https://www.ethnologue.com) maintains a living language database. Access the site's "Browse Country" page and click on each Southeast Asian country. You will find a summary page listing the number of languages as well as the number of "extinct," "in trouble," and "dying" languages in that country.

1. Construct two graphs, one showing the total number of languages in each country and the other showing the number of "in trouble" and "dying" languages for each country.
2. Write a paragraph describing the patterns you see in the graphs. Does the total number of languages spoken in each country correlate with the number of endangered languages? Suggest reasons why the number of languages, and the number of threatened languages, vary so much from country to country.
3. Based on the graphs and what you know about the forces of globalization working in the region, how might Southeast Asia's linguistic geography change over the next 50 years?

MasteringGeography™

Looking for additional review and test prep materials? Visit the Study Area in MasteringGeography™ to enhance your geographic literacy, spatial reasoning skills, and understanding of this chapter's content by accessing a variety of resources, including MapMaster interactive maps, geoscience animations, videos, *In the News* RSS feeds, flashcards, web links, self-study quizzes, and an eText version of *Globalization and Diversity.*

Authors' Blogs

Scan to visit the **Author's Blog for field notes, media resources, and chapter updates**

http://gad4blog.wordpress.com/category/southeast-asia/

Scan to visit the **GeoCurrents Blog**

http://geocurrents.info/category/place/southeast-asia

14 Australia and Oceania

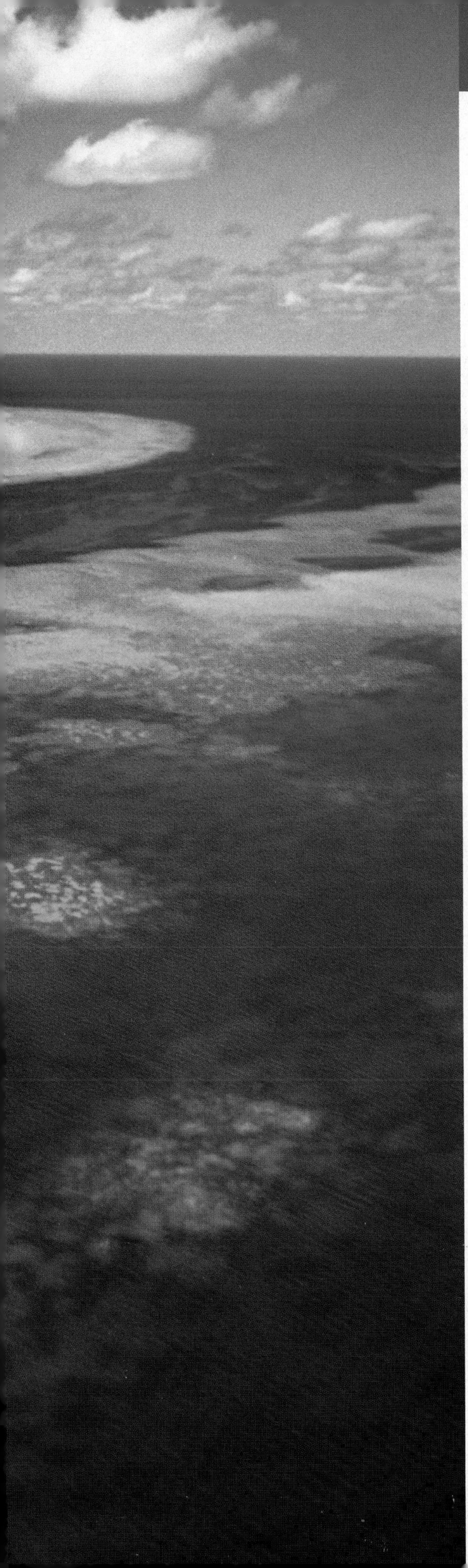

PHYSICAL GEOGRAPHY AND ENVIRONMENTAL ISSUES

Diverse environments characterize this huge region, which includes a continent-sized landmass as well as thousands of small oceanic islands. Global climate change and rising sea levels threaten the very survival of some low-lying countries within the region.

POPULATION AND SETTLEMENT

Growing, dense cities punctuate the sparse rural settlement pattern of Oceania, with urban places as the magnets attracting migrants from both within and outside of the region.

CULTURAL COHERENCE AND DIVERSITY

Both Australia and New Zealand, originally products of European culture, are seeing new cultural geographies take shape because of immigrants from other parts of the world as well as their own native peoples, the Aborigines and Maori.

GEOPOLITICAL FRAMEWORK

A heritage of colonial geographies overlaying native cultures is being replaced by contemporary power struggles among global powers, dominated by the tensions between China and the United States.

ECONOMIC AND SOCIAL DEVELOPMENT

While Australia and New Zealand are relatively wealthy because of world trade, most of island Oceania struggles economically. Even Hawaii has troubles during global downturns with its high cost of living and its boom-or-bust tourist economy.

◀ Off the eastern coast of Australia, the Great Barrier Reef stretches for more than 1400 miles (2300 km) through the Coral Sea.

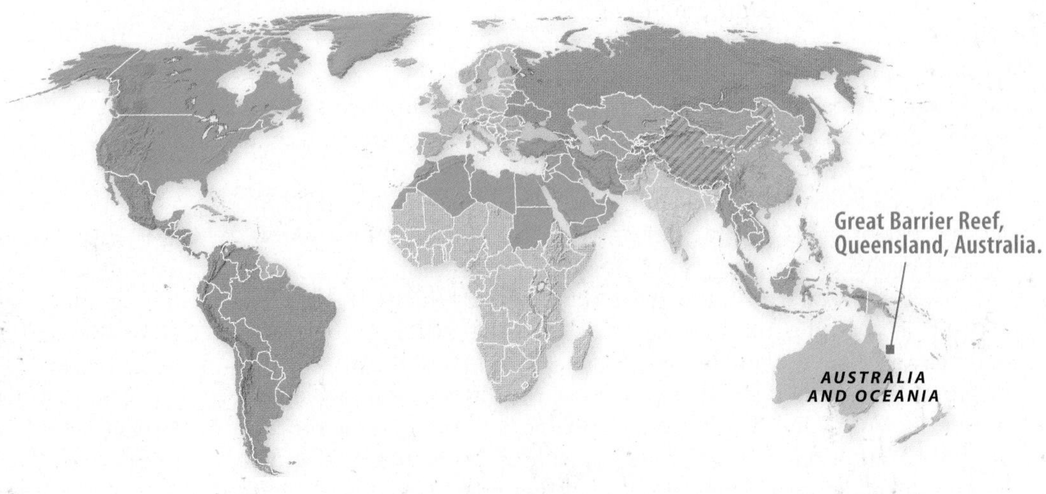

Scientists call it the world's single largest expression of a living organism. Stretching through the azure- and turquoise-tinted waters of the Coral Sea for more than 1400 miles (2300 km) off the coast of Queensland, the Great Barrier Reef (GBR) includes more than 900 small islands and a myriad of underwater coral reefs. This remarkable ecosystem is home to 1500 species of fish, 400 species of coral, whales, dolphins, sea turtles, sea eagles, terns, and plant species found nowhere else on Earth. Taking more than 10,000 years to form, the reef has been a UN World Heritage Site since 1981, and much of the area is protected by Australia's Great Barrier Reef Marine Park.

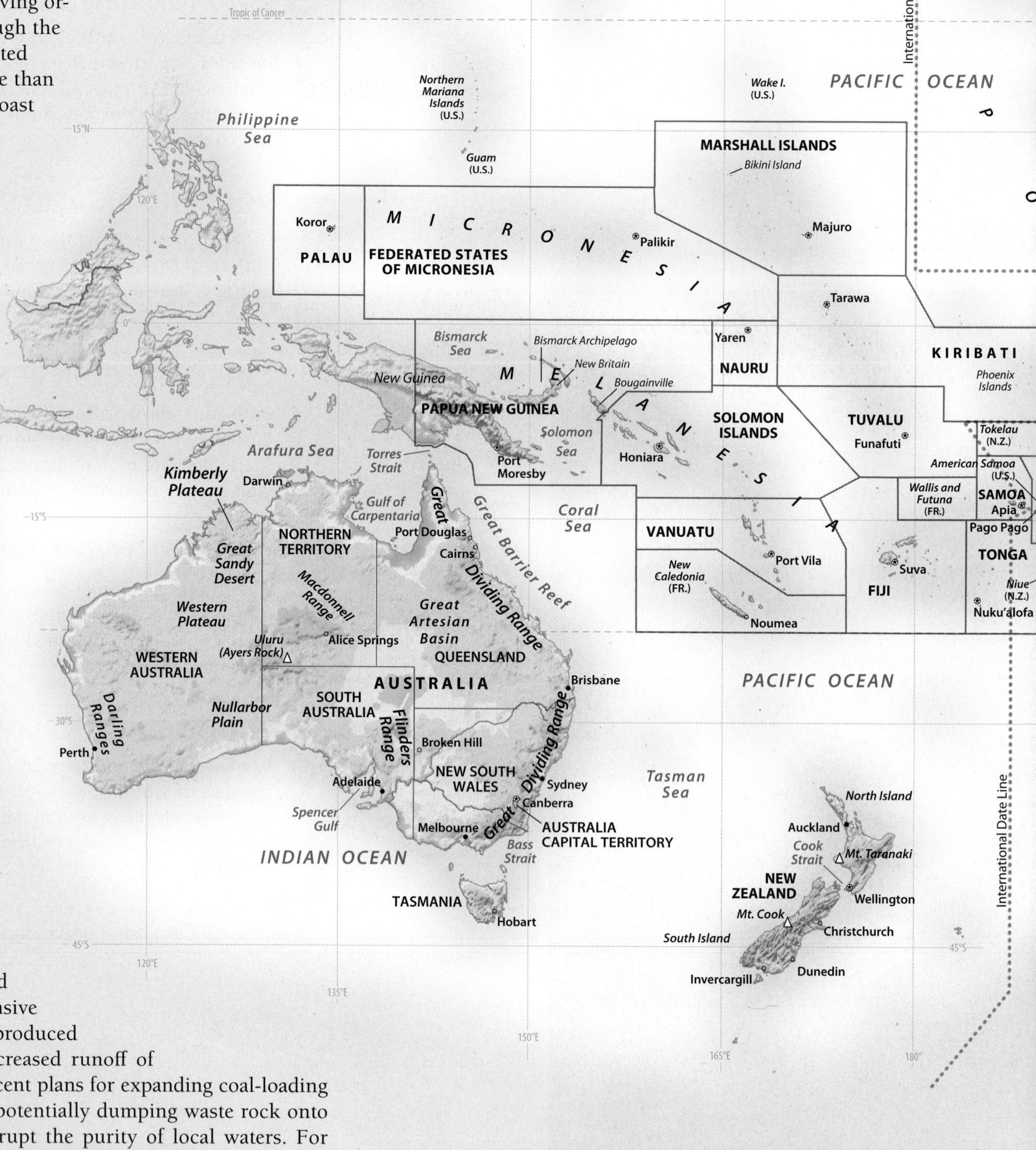

Today, however, the GBR is in the fight of its life. Thanks to global climate change, the reef has lost more than half its coral cover since 1985, much of it damaged by warmer ocean temperatures that have accelerated rates of seawater acidification and coral bleaching. Coastal development has added to its watery woes: More intensive agriculture in Queensland has produced ocean-bound sediment and increased runoff of toxic agricultural chemicals. Recent plans for expanding coal-loading depots at Abbot Point include potentially dumping waste rock onto the reef, a practice sure to disrupt the purity of local waters. For now, the reef's survival hangs in the balance, a giant poster child for a long list of damaging human impacts that threaten the environmental health of the entire South Pacific.

This vast world region includes the island continent of Australia as well as **Oceania**, a collection of islands that reaches from New Guinea and New Zealand to the U.S. state of Hawaii in the mid-Pacific (Figure 14.1). Although native peoples settled the area long ago, more recent European and North American colonization began the process of globalization that is now producing new and sometimes unsettled environmental, cultural, and political geographies. The region's colonial legacy still dominates many agricultural and urban landscapes, and a number of political entities remain territories, still closely tied to distant colonial rule. In the past 30 years, however, growing linkages with Asia have produced major shifts in migration to the region as well as new economic ties. Today Chinese tourists, South Asian immigrants, and Southeast Asian workers are all part of increasingly close connections between this region and its giant neighbor to the north and west.

The vast distances of the Pacific stretching from New Guinea to Hawaii help define the boundaries of this world region, but many of the national boundaries were born from political convenience during an earlier period of colonial globalization. Australia (or "southern land"), often thought of as a continent with its population of more than 23 million people, forms a coherent political unit and subregion.

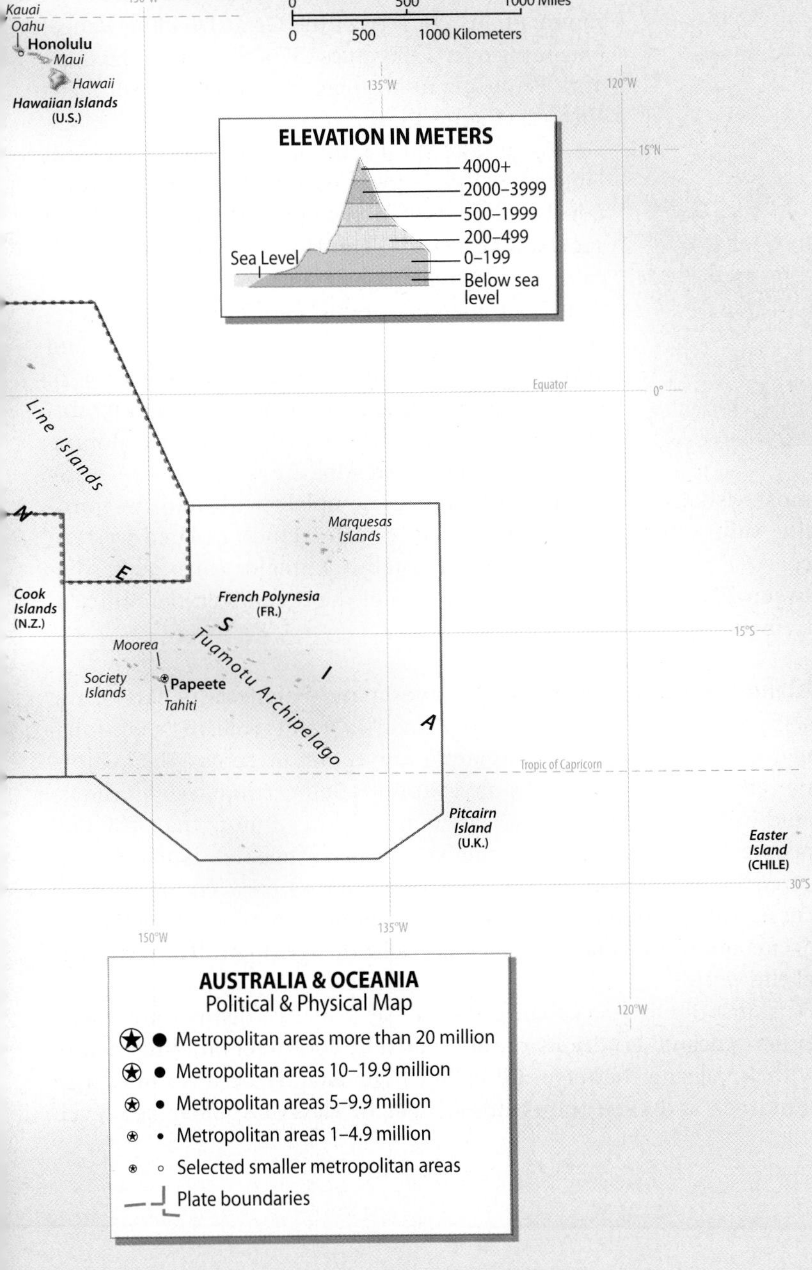

Figure 14.1 Australia and Oceania More water than land, the Australia and Oceania region sprawls across the vast reaches of the western Pacific Ocean. Australia dominates the region, both in its physical size and in its economic and political clout. Along with New Zealand, Australia represents largely Europeanized settlement in the South Pacific. Elsewhere, however, the island subregions of Melanesia, Micronesia, and Polynesia contain large native populations that have mixed in varied ways with later European, Asian, and North American arrivals.

Growing linkages with Asia have produced major shifts in migration to Australia and Oceania, as well as new economic ties.

To the east, New Zealand is a three-hour flight from Australia, has a much smaller population of just over 4.5 million, and is linked to Australia by shared historical ties to Britain. However, New Zealand is considered part of **Polynesia** ("many islands") because of its native Maori people.

Hawaii, 4400 miles (7100 km) northeast of New Zealand, shares the same Polynesian heritage as New Zealand. Hawaii is thought of as the northeastern boundary of Oceania, while the region's southeastern boundary is usually delimited by the Polynesian islands of Tahiti, 3000 miles (4400 km) to the southeast (Figure 14.2).

Four thousand miles (6400 km) west of French Polynesia, well across the International Date Line, lies the island of New Guinea, the accepted, yet sometimes confusing boundary between Oceania and Asia. Today an arbitrary boundary line bisects the island, dividing Papua New Guinea (the eastern half, which is usually considered part of Oceania) from neighboring Papua and West Papua (the western half, which, as part of Indonesia, is usually considered part of Southeast Asia). This western part of Oceania is sometimes called **Melanesia** (meaning "dark islands") because early explorers considered local peoples to be darker-skinned than those in Polynesia.

Finally, the more culturally diverse region of **Micronesia** (meaning "small islands") lies north of Melanesia and west of Polynesia. It includes microstates such as Nauru and the Marshall Islands as well as the U.S. territory of Guam.

Learning Objectives *After reading this chapter you should be able to:*

14.1 Describe the physical geographic characteristics of the region known as Oceania.

14.2 Identify major environmental issues in Australia and Oceania as well as pathways toward solving those problems.

14.3 Use a map to identify and describe major migration flows to (and within) the region.

14.4 Describe historical and modern interactions between native peoples and Anglo-European migrants in Australia and Oceania and their impacts on the region's cultures.

14.5 Identify and describe different pathways to independence taken by countries in Oceania.

14.6 List several geopolitical tensions that persist in Australia and Oceania.

14.7 Describe the diverse economic geographies of Oceania.

14.8 Explain the positive and negative interactions of Australia and Oceania with the global economy.

Figure 14.2 Tahiti, Outlier of Polynesia The islands of Tahiti were first settled between 300 and 800 CE and form the southeastern corner of Polynesia some 3000 miles (4800 kilometers) from New Zealand to the west and roughly the same distance from Hawaii to the north. This is a view of Cook's Bay on the island of Moorea.

Physical Geography and Environmental Issues: Varied Landscapes and Habitats

Australia is made up of a vast semiarid interior—the **Outback**, a dry, sparsely settled land of scrubby vegetation—fringed by tropical environments in its far north and by hilly topography with summer-dry Mediterranean climates to the east, west, and south (Figure 14.3). In contrast, New Zealand is known for its green rolling foothills and rugged snow-capped mountains, landscapes that result from tectonic activity and more humid, cooler climates. Surrounding Australia and New Zealand is the true island realm of Oceania, consisting of a varied array of both high, volcano-created islands and low-lying coral atolls.

Regional Landforms and Topography

Three major topographic regions dominate Australia's physical geography. The vast, irregular Western Plateau, averaging only 1000 to 1800 feet (300 to 550 meters) in elevation, occupies more than half of the continent. To the plateau's east, the basins of the Interior Lowlands stretch for more than 1000 miles (1600 km), from the swampy coastlands of the Gulf of Carpentaria in the north to the valleys of the Murray and Darling rivers in the south, Australia's largest river system. Farthest east is the forested and mountainous country of the Great Dividing Range, extending over 2300 miles (3700 km) from the Cape York Peninsula in northern Queensland to southern Victoria (Figure 14.4).

As part of the Pacific Ring of Fire, New Zealand owes its geologic origins to volcanic mountain-building, which produced its two rugged and spectacular islands. The North Island's active volcanic peaks, reaching heights of more than 9100 feet (2800 meters), and its geothermal features reveal the country's fiery origins. Even higher and more rugged mountains comprise the western spine of the South Island. Mantled by high mountain glaciers and surrounded by steeply sloping valleys, the Southern Alps are some of the world's most visually spectacular mountains, complete with narrow, fjord-like valleys that indent much of the South Island's isolated western coast (Figure 14.5). In nearby Christchurch, major earthquakes between 2010 and 2012 were reminders of the region's vulnerability to tremors.

Island Landforms Much of Melanesia and Polynesia is part of the seismically active Pacific Ring of Fire. As a result, volcanic eruptions, major earthquakes, and tsunamis are common across the region. For example, in 1994, volcanic eruptions and earthquakes on the island of New Britain (Papua New Guinea) forced more than 100,000 people from their homes. Four years later a massive tsunami triggered by an offshore earthquake swept across New Guinea's north coast, killing 3000 residents and destroying numerous villages. Such events are unfortunately a part of life in this geologically active part of the world.

Most of Oceania's islands were created by two distinct processes: either volcanic eruptions or, alternatively, coral reef-building. Those with a volcanic heritage are called **high islands** because most rise hundreds and even thousands of feet in elevation above sea level.

Figure 14.3 The Australian Outback Arid and generally treeless, the vast lands of the Australian Outback resemble some of the dry landscapes of the American West. This view is near Pilbara, Western Australia.

Figure 14.4 Grampian National Park This park, located in Australia's Great Dividing Range west of Melbourne, was added to the national heritage list in 2006 because of its natural beauty and rich indigenous rock art sites.

Figure 14.6 Bora Bora The jewel of French Polynesia, Bora Bora displays many of the classic features of Pacific high islands. As the island's central volcanic core erodes, surrounding coral reefs produce a mix of wave-washed sandy shores and shallow lagoons.

The Hawaiian Islands are good examples, with volcanic peaks of more than 13,000 feet (4000 meters) on the Big Island of Hawaii. Tonga, Samoa, Bora Bora, and Vanuatu provide other examples of high islands (Figure 14.6). Even larger and more geographically complex are the *continental high islands* of New Guinea, New Zealand, and the Solomon Islands.

In contrast, **low islands**, as the name suggests, are formed from coral reefs, making the islands not just lower, but also flatter and usually smaller, than high islands. Further, because the soil on these islands originated as coral, it is generally less fertile than the volcanic soil of high islands and supports less varied plant life. Low islands often begin as barrier reefs around or over sunken volcanic high islands, resulting in an **atoll** (Figure 14.7). The world's largest atoll, Kwajalein in Micronesia's Marshall Islands, is 75 miles (120 km) long and 15 miles (25 km) wide. Low islands dominate the countries of Tuvalu, Kiribati, and the Marshall Islands. Clearly, these low islands are the most vulnerable to rising sea levels associated with climate change.

Figure 14.5 The New Zealand Alps The dominant topographic feature of the South Island, these picturesque mountains, known locally as the Southern Alps, rise to heights over 12,000 feet (3600 meters).

Regional Climate Patterns

Zones of higher precipitation encircle Australia's arid heartland (Figure 14.8). In the tropical low-latitude north, seasonal changes are dramatic and unpredictable. For example, Darwin can experience drenching monsoonal rains in the Southern Hemisphere summer, December to March, followed by bone-dry winters from June to September (see the climograph for Darwin in Figure 14.8).

Along the coast of Queensland, precipitation is high (60–100 in., or 150–250 cm), but rainfall diminishes rapidly in the state's western interior. Rainfall in central Australia, such as the Northern Territory's Alice Springs, averages less than 10 inches (25 cm) annually. South of Brisbane, more midlatitude influences dominate eastern Australia's climate. Coastal New South Wales, southeastern Victoria, and Tasmania experience the country's most dependable year-round precipitation, averaging 40–60 inches (100–150 cm) per year; winter snow frequently covers the nearby mountains. Even here, however, extreme summer heat has added to recent wildfires during the dry season, threatening suburban settings in both Sydney and Melbourne (Figure 14.9). Farther west, summers are hot and dry in much of South Australia and in the southwest corner of Western Australia. These zones of Mediterranean climate produce scrubby eucalyptus woodlands known as **mallees**.

(a)

(b)

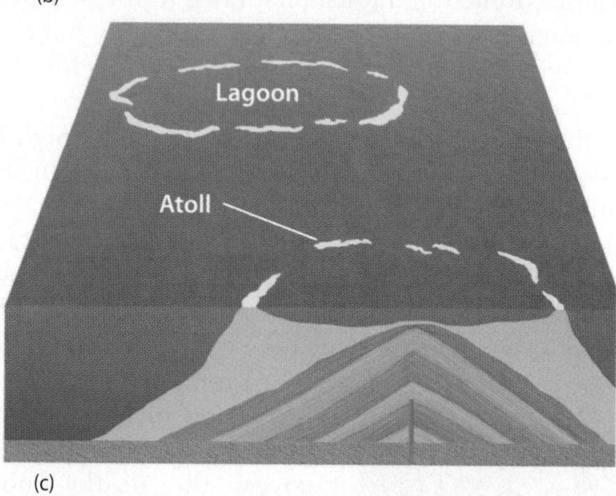

(c)

Figure 14.7 Evolution of an Atoll (a) Many Pacific low islands begin as rugged volcanoes with fringing coral reefs. (b) However, as the extinct volcano erodes and subsides, the coral reef expands, becoming a larger barrier reef. (The term *barrier reef* comes from the hazards that these features pose to navigators approaching the island from the sea.) (c) Finally, all that remains is a coral atoll surrounding a shallow lagoon.

Climates in New Zealand are influenced by three factors: latitude, the moderating effects of the Pacific Ocean, and proximity to local mountain ranges. Most of the North Island is distinctly subtropical (see Figure 14.8); the coastal lowlands near Auckland, for example, are mild and wet year-round. On the South Island, conditions become distinctly cooler as you move closer to the South Pole. Indeed, the island's southern edge feels the seasonal breath of Antarctic chill, as it lies more than 46° south of the equator, at latitudes similar to Portland, Oregon. Mountain ranges on New Zealand's South Island also display incredible local variations in precipitation: West-facing slopes are drenched with more than 100 inches (250 cm) of precipitation annually, whereas lowlands to the east average only 25 inches (65 cm) per year. The Otago region, inland from Dunedin, sits partially in the rain shadow of the Southern Alps, and its rolling, open landscapes resemble the semiarid expanses of North America's West (Figure 14.10).

Island Climates The Pacific high islands usually receive abundant precipitation because of the orographic effect, resulting in dense tropical forests and vegetation. On the Hawaiian island of Kauai, Mt. Walaleale may be one of the wettest spots on Earth, receiving an average annual rainfall of 470 inches (1200 cm). In contrast to the high islands, low islands receive less precipitation, typically less than 100 inches (250 cm) annually. As a result, water shortages are common.

Unique Plants and Animals

Because of the Australian continent's long geologic history of separation and isolation from other landmasses, its bioregions contain an array of plants and animals found nowhere else in the world. More specifically, 83 percent of its mammals and 89 percent of its reptiles are unique to that country. Best known are the country's **marsupials** (mammals that raise their young in a pouch)—the kangaroo, koala, possum, wombat, and Tasmanian devil. Fully 70 percent of the world's known marsupials are found in Australia. Also unique is the platypus, an egg-laying mammal.

Exotic Species The introduction of **exotic species**—nonnative plants and animals—has caused problems for endemic (native) species throughout the Pacific region. In Australia, rabbits brought from Europe successfully multiplied in an environment that lacked the diseases and predators that elsewhere checked their numbers. Before long, rabbit populations had reached plague-like proportions, stripping large sections of land of its vegetation. The animals were brought under control only through the purposeful introduction of the rabbit disease myxomatosis.

Exotic plants and animals in Oceania's island environments have had similar effects. For example, many small islands possessed no native land mammals, and their native bird and plant species proved vulnerable to the ravages of introduced rats, pigs, and other animals. The larger islands of the region, such as those of New Zealand, originally supported several species of large, flightless birds; the largest, called moas, were substantially larger than ostriches. During the first wave of human settlement in New Zealand some 1500 years ago, moa numbers fell rapidly, as they were hunted by humans and their eggs were consumed by invading rats. By 1800, moas had become extinct.

The spread of nonnative species continues today. In Guam, the brown tree snake, which arrived accidentally by cargo ship from the Solomon Islands in the 1950s, has taken over the landscape (Figure 14.11). The island's forest areas now contain more than 10,000 snakes per square mile. The snakes have wiped out nearly all the native bird species and also cause frequent power outages as they crawl along electrical wires. The brown tree snake has already done its damage to Guam, but it threatens other islands as well because it readily hides in cargo containers shipped elsewhere.

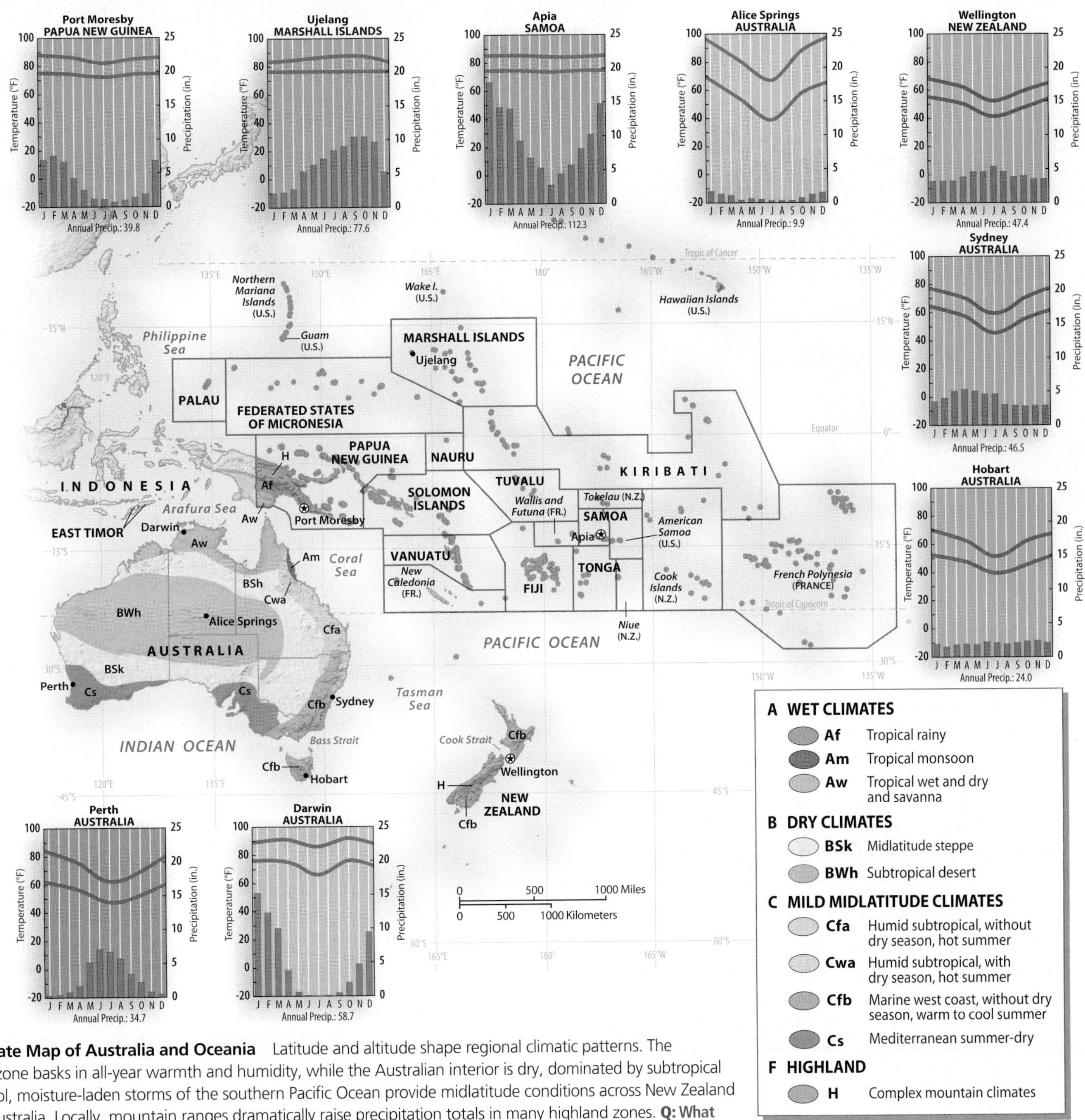

Figure 14.8 Climate Map of Australia and Oceania Latitude and altitude shape regional climatic patterns. The equatorial Pacific zone basks in all-year warmth and humidity, while the Australian interior is dry, dominated by subtropical high pressure. Cool, moisture-laden storms of the southern Pacific Ocean provide midlatitude conditions across New Zealand and portions of Australia. Locally, mountain ranges dramatically raise precipitation totals in many highland zones. **Q: What part of the United States might have a climate that most resembles the climate of Perth? Alice Springs?**

Complex Environmental Issues

Globalization has exacted an environmental toll on Australia and Oceania (Figure 14.12). Specifically, the region's considerable natural resources base has been opened to development, much of it by outside interests. Although benefiting from global investment, the region has also paid a considerable price for encouraging development, and the result is an increasingly threatened environment.

Historically, the region's peripheral economic and political status has often been environmentally costly. When the United States and France required atomic testing grounds for their nuclear weapons programs, the South Pacific was chosen as an ideal location (Figure 14.13). The environmental consequences have been long-lasting. Residents of Bikini Atoll in the Marshall Islands and across various parts of French Polynesia have been forced to evacuate their islands for decades. Elevated levels of toxic radioactive substances remain concentrated in soils, forever disrupting these island settings. While the U.S.-backed Marshall Islands Nuclear Claims Tribunal has awarded more than $2 billion in health- and land-related claims to residents, many argue additional payments are necessary. In French Polynesia, more than 200 nuclear tests were conducted between 1966 and 1996. French government documents declassified in 2013 reported that islands such as Tahiti were exposed to

Figure 14.9 Australian Wildfires Huge and savage dry-season wildfires (known as bushfires in Australia) threaten both rural settlements and sprawling city suburbs in the southeast. This 2009 blaze, just 70 miles from the heart of Melbourne, was the worst fire disaster in 25 years and may be a harbinger of even more damaging fires accompanying climate change.

500 times more radiation than recommended, and increased cancers in the region have been traced to the testing. Now regional authorities are asking for more payments from the French government to address these persisting issues.

Major global mining operations also greatly affect Australia, Papua New Guinea, New Caledonia, and Nauru. Some of Australia's largest gold, silver, copper, and lead mines are located in sparsely settled portions of Queensland and New South Wales, polluting watersheds in these semiarid regions. In Western Australia, huge open-pit iron mines dot the landscape, unearthing ore that is bound for global markets, particularly China and Japan. To the northeast, gold mining is transforming the Solomon Islands, while an even larger gold-mining venture has raised environmental concerns on the island of New Guinea. Elsewhere, Micronesia's tiny Nauru has been virtually turned inside out as much of the island's jungle cover has been removed to get at some of the world's richest phosphate deposits.

Figure 14.10 Otago Valley on New Zealand's South Island Many travelers have compared the scenery and open landscapes of New Zealand's Otago region to those of the American West.

Figure 14.11 Island Pest The brown tree snake, brought into Guam accidentally in the 1950s, has now taken over large parts of the island's forestlands and killed off most native bird species. Because these snakes, which reach 10 feet in length, climb along electrical wires, they frequently cause power outages throughout Guam.

Other environmental threats are found across the region. Vast stretches of Australia's eucalyptus woodlands have been destroyed to create pastures; in the Outback, overgrazing has increased desertification and salinization, resulting in saltier water tables and less productive soils. In addition, coastal rainforests in Queensland cover only a fraction of their original area, although a growing environmental movement in the region is fighting to save the remaining forest tracts. Tasmania has also been an environmental battleground, particularly given the biodiversity of its midlatitude forest landscapes. The island's earlier European and Australian development featured many logging and pulp mill operations, but more than 20 percent of the land is now protected by national parks.

Many high islands in Oceania are also threatened by deforestation. With limited land areas, islands are subject to rapid tree loss, which, in turn, often leads to soil erosion. Although rainforests still cover 70 percent of Papua New Guinea, more than 37 million acres (15 million hectares) have been identified as suitable for logging (Figure 14.14). Some of the world's most biologically diverse environments are being threatened in these operations, but landowners see the quick cash sales to loggers as attractive, even though this nonsustainable practice is contrary to their traditional lifestyles.

Climate Change in Oceania

Even though Oceania contributes relatively little atmospheric pollution to the global atmosphere, the harbingers of climate change are widespread and problematic in this region. New Zealand mountain glaciers are melting. Recently, Australia has suffered from frequent droughts, exceptionally destructive heat waves (which set records in both 2013 and 2014), and devastating wildfires. Warmer ocean waters have caused widespread bleaching of Australia's Great Barrier Reef as microorganisms die, and rising sea levels threaten low-lying island nations. UN projections for the future are also highly disturbing: Sea levels may rise 4 feet (1.4 meters) by century's end, and island inhabitants will suffer from changed coastal resources as the ocean continues

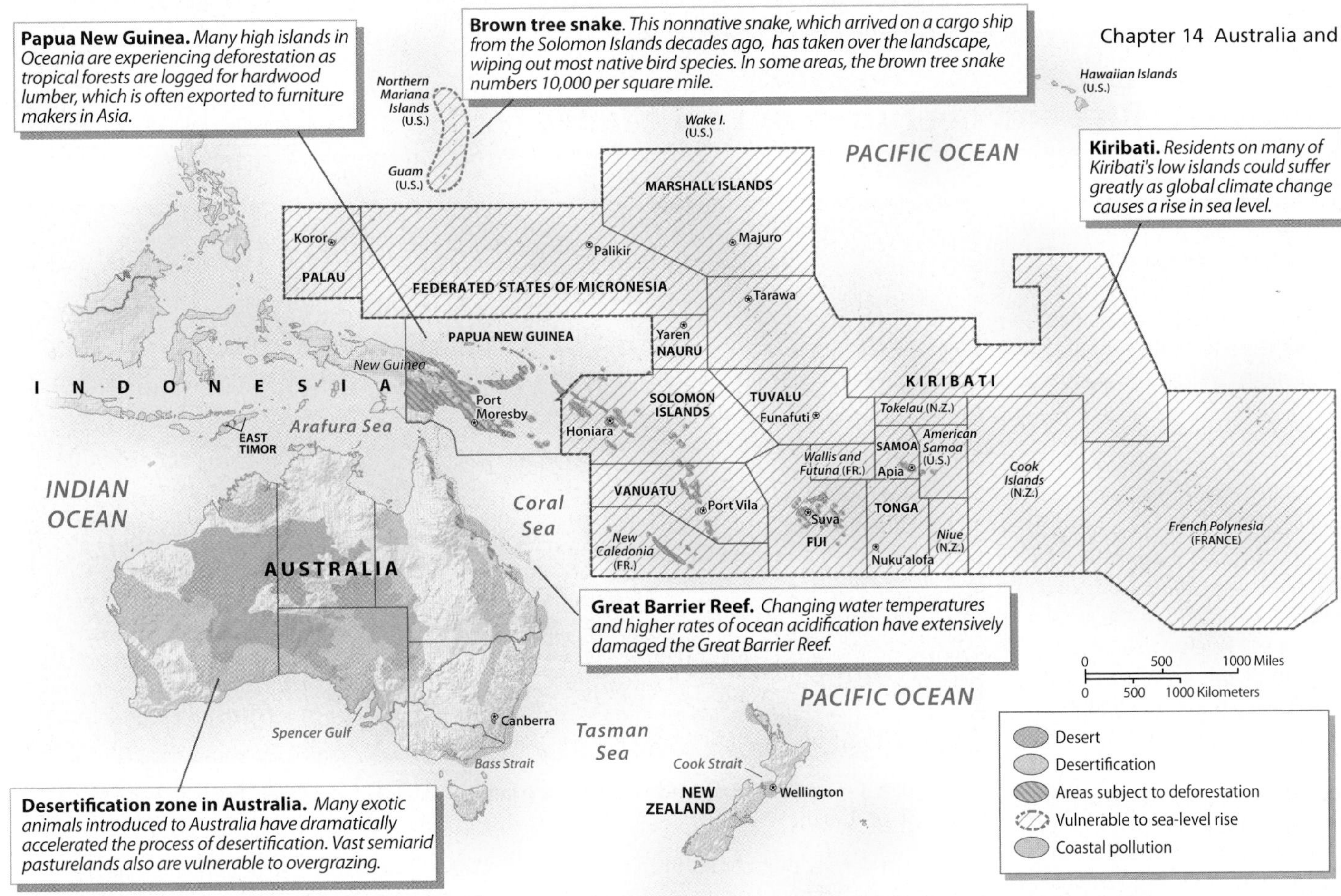

Figure 14.12 Environmental Issues in Australia and Oceania Modern environmental problems belie the myth that the region is an earthly paradise. Tropical deforestation, extensive mining, and a long record of nuclear testing by colonial powers have brought severe challenges to the region. Human settlements have also extensively modified the pattern of natural vegetation. Future environmental threats loom for low-lying Pacific islands as sea levels rise due to global climate change.

to warm (see *Working Toward Sustainability: Sea-Level Rise and the Future of Low Islands*). Furthermore, stronger tropical cyclones could devastate Pacific islands, with widespread damage to land and life. Tropical cyclone Pam, a category 5 storm, struck Vanuatu in 2015, wiping out crops and leaving more than 100,000 islanders homeless.

In response to these threats, the actions taken and policies implemented by Oceania's countries vary, depending on their susceptibility to climate change, the source and magnitude of their atmospheric emissions, and the state of their local economies. Australia, with its 23 million people, generates the most carbon emissions in this large region, primarily because almost 80 percent of its electricity is generated from fossil fuels such as coal. Recent government proposals moved toward a carbon tax program (that would have accelerated emission reductions),

Figure 14.13 Bikini Atoll Atomic Testing The "Baker" atomic explosion, detonated by the United States in 1946, contributed to the long-term radioactive contamination of Bikini Atoll in the South Pacific.

Figure 14.14 Logging in Oceania Much of Oceania's tropical forest is being destroyed by logging. Here, hardwood tree trunks await loading for transfer to Asian mills where they will be made into furniture for China's growing markets.

WORKING TOWARD SUSTAINABILITY

Sea-Level Rise and the Future of Low Islands

Google Earth (MG) Virtual Tour Video https://goo.gl/E48xOc

Low islands have very little topographic relief, since they are basically coral atolls that have grown from the sea as barrier reefs. For example, the highest point in the Marshall Islands, home to nearly 71,000 people, is about 30 feet (9 meters). Because the sustainability of Oceania's low islands appears questionable, given the array of challenging environmental problems in their future, Pacific islanders are working on what many consider inevitable—abandoning their homeland.

Rising Seas Sea levels, according to many estimates, may rise 4 feet (1.4 meters) by 2100 (Figure 14.1.1). These estimates, however, are at the conservative end of the scale. Some scientists think sea levels could be considerably higher by then, perhaps closer to 10 feet (3.3 meters). These are average figures; global sea-level rise could be higher or lower depending on local factors such as ocean basin geology and ocean temperatures (recall that warm water expands). Additionally, this average sea-level rise does not take into account the tidal range for specific locations. In the South Pacific, seasonal extreme high tides can add another 10 feet (3.3 meters) to the average sea level.

Pacific islanders say they are already experiencing the effects of climate change, and they are causing numerous problems. Flooding is more common during high tides, and flooding brings related problems such as saltwater contamination of precious freshwater supplies and damage to low-lying farm fields. Also, warmer ocean temperatures are causing detrimental changes to local reef ecologies, impacting marine life that is crucial for islanders' sustenance. Even the deep-water tuna are leaving as they seek cooler waters elsewhere.

Aggravating the problems of sea-level rise due to climate change is the existing high population density that comes from decades of relatively rapid population growth and little vacant land. With growing island populations, there is no option but to leave. Tragically, that's what many people are preparing for.

Figure 14.1.1 Sea-Level Rise and Pacific Low Islands This is Funafuti Atoll, home to half the population of Tuvalu.

Migration with Dignity Kiribati, for example, has adopted a "migration with dignity" program that emphasizes education and vocational training so its people can find jobs elsewhere. A new maritime training college has been created to help locals gain jobs with global shipping firms. Australian aid has created a nurse training program for islanders seeking jobs off-island in that understaffed field. Regionally, Kiribati is also one of five South Pacific nations (including Samoa, the Solomon Islands, Tuvalu, and Vanuatu) that qualify for UN assistance in developing National Adaptation Programs of Action (NAPAs). These adaptation schemes are designed to develop long-term options for island residents as sea levels rise.

These programs are mainly designed for younger people to give them the training and skills needed to migrate and to compete in other countries. But what about the older people, too old for retraining and less able to adapt to life in a foreign land? How do they maintain their dignity under those conditions? No one seems to have the answer to that question.

1. Which island (or parts of an island) do you think is most vulnerable to flooding from rising sea levels? Why?
2. What areas of the United States face problems from sea-level rise similar to those of Oceania?

but strong opposition to the plan, especially from mining and industrial interests, led the government to scrap it in 2014.

New Zealand has been more politically committed to addressing the global climate change issue. More than half of the country's own energy comes from hydroelectric power, primarily generated in the high, wet mountains of the South Island. Wind and solar—particularly wind—supply fully 13 percent of the country's power (Figure 14.15). The government is now proposing a 10–20 percent reduction in

Figure 14.15 New Zealand Wind Power Given New Zealand's location in the breezy middle latitudes, many areas are excellent settings for the commercial development of wind power.

greenhouse gas (GHG) emissions by 2020. To address New Zealand's livestock emissions of methane (a potent GHG), livestock specialists are experimenting with grass and grain fodder mixtures that could reduce these levels in the future.

Oceania's low islands are already facing the most dramatic consequences of global climate change, especially as measured by rising sea levels. Many Pacific nations—most notably Tuvalu, Kiribati, and the Marshall Islands—maintain they are already experiencing problems from sea-level rise, coral bleaching, and degraded fishery resources because of warming ocean waters related to climate change. As a result, these nations have banded together into a strident political union lobbying for a global solution to climate change. In the climate change debate, these small island nations have been outspoken in their demands that developed nations such as the United States, Japan, and the countries of western Europe provide the island nations with financial aid to mitigate damage from climate change.

 **Review**

14.1 How are high islands, coral atolls, and barrier reefs formed?

14.2 Describe the different climate regions found in Australia and New Zealand. What climate controls produce those regions?

14.3 List three key environmental issues for this region, and suggest why these have global importance.

KEY TERMS Oceania, Polynesia, Melanesia, Micronesia, Outback, high island, low island, atoll, mallee, marsupial, exotic species

Population and Settlement: Booming Cities and Empty Spaces

Modern population patterns reflect both indigenous and European settlement. In New Zealand, Australia, and the Hawaiian Islands, Anglo-European migration has structured the distribution and concentration of contemporary populations. Elsewhere in Oceania, population geographies are determined by the needs of native island peoples (Figure 14.16). Migration patterns to both Australia and New Zealand have also brought more residents from Asia. This pattern is coupled with an increase of intraregional migration as people move about for complex reasons, including the push forces of unemployment, resource depletion, and even the threat of flooding associated with climate change.

Contemporary Population Patterns

Despite the stereotypes of life in the Outback, modern Australia has one of the most urbanized populations in the world (Table A14.1). About 90 percent of the country's residents live within either the Sydney or the Melbourne metropolitan area. Indeed, Australia's eastern and southern coasts are home to the majority of its 23 million people.

Inland, population densities decline as rapidly as the rainfall: Semiarid hills west of the Great Dividing Range still contain significant rural settlement, but Queensland's southwestern periphery remains sparsely populated. New South Wales is the country's most heavily populated state; its sprawling capital city of Sydney (4.4 million people), focused around one of the world's most magnificent natural harbors, is the largest metropolitan area in the entire South Pacific (Figure 14.17). In the nearby state of Victoria, Melbourne (with 4.3 million residents) has long competed with Sydney for status as Australia's premiere city, claiming cultural and architectural supremacy over its slightly larger neighbor. Since 2010, employment growth in Melbourne has made it Australia's fastest-growing metropolitan area, and it may soon surpass Sydney in population. Located between these metropolitan giants, the much smaller federal capital of Canberra (population 380,000) represents a classic geopolitical compromise, in the same spirit that created Washington, DC, midway between the populous southern and northern portions of the eastern United States.

Outside of the Australian core, Aboriginal populations are widely but thinly scattered inland across Western Australia and South Australia as well as in the Northern Territory, creating smaller but regionally important centers of settlement. In the far southwest, sprawling Perth dominates the urban scene with its 1.9 million residents. Some planners fear the rapidly growing metropolitan area (projected 2050 population: 4.6 million) may become the "biggest low-density city on earth" because of its suburban fringe spreading into surrounding rural lands (Figure 14.18).

More than 70 percent of New Zealand's 4.5 million residents live on the North Island, with the Auckland metropolitan area (over 1.4 million) dominating the scene in the north and the capital city, Wellington (400,000), anchoring settlement along the Cook Strait in the south. Settlement on the South Island is mostly located in the somewhat drier lowlands and coastal districts east of the mountains, with Christchurch (375,000) serving as the largest urban center (Figure 14.19). Elsewhere, rugged and mountainous terrain on both the North and the South islands produces much lower densities.

In Papua New Guinea, less than 15 percent of the country's population is urban, with most people living in the isolated interior highlands. The nation's largest city is the capital, Port Moresby (400,000), located along the narrow coastal lowland in the far southeastern corner of the country. In stark contrast to Papua New Guinea, the largest urban area on the northern margin of Oceania is Honolulu (1 million), on the island of Oahu. Here rapid metropolitan growth since World War II is due to U.S. statehood and Hawaii's scenic attractions.

Historical Geography

The region's remoteness from the world's early population centers meant that it lay beyond the dominant migratory paths of earlier peoples. Even so, prehistoric settlers eventually found their way to the isolated Australian interior and even the far reaches of the Pacific. Later, once Europeans had explored the region and identified its resource potential, the pace of new in-migrations increased.

Peopling the Pacific The large islands of New Guinea and Australia, relatively near to the Asian landmass, were settled much earlier than the more distant islands of the Pacific. By 60,000 years ago, the ancestors of today's native Australians, or **Aborigines**, were making their way from Southeast Asia into Australia (Figure 14.20). The first Australians most likely arrived on some kind of watercraft, but because such boats were probably not capable of more lengthy voyages, the more distant islands remained inaccessible to settlement for tens of thousands of years. During the last glacial period, however, sea levels were much lower than they are now, allowing easier movement to Australia across relatively narrow spans of water from what is now called Southeast Asia. It is not known whether the original

Figure 14.16 Population of Australia and Oceania Less than 40 million people occupy this world region. Although Papua New Guinea and many Pacific islands feature mainly rural settlements, most residents of the region live in the large urban areas of Australia and New Zealand. Sydney and Melbourne account for almost half of Australia's population, and most New Zealand residents live on the North Island, home to the cities of Auckland and Wellington.

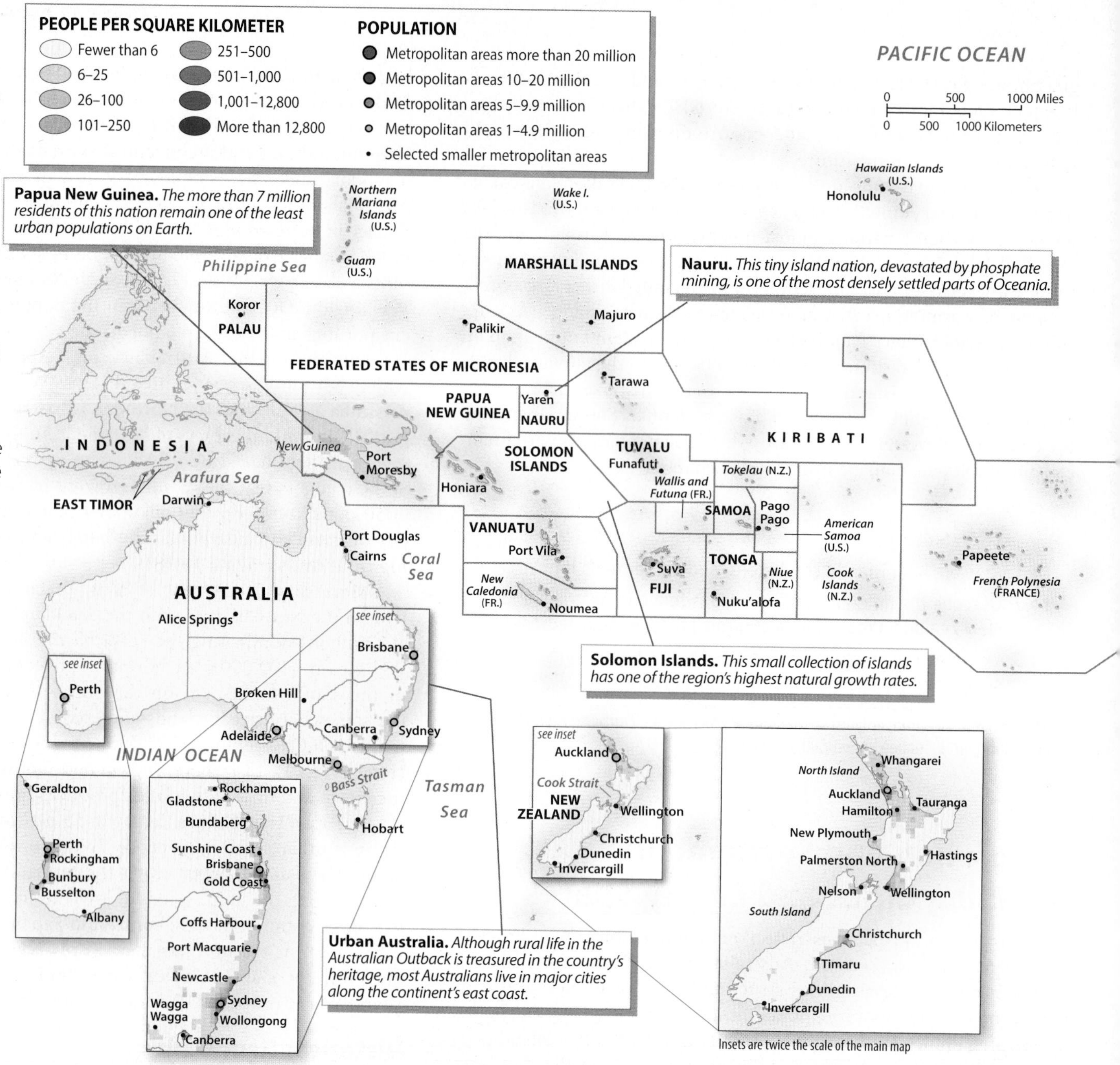

PEOPLE PER SQUARE KILOMETER
Fewer than 6
6–25
26–100
101–250
251–500
501–1,000
1,001–12,800
More than 12,800
POPULATION
Metropolitan areas more than 20 million
Metropolitan areas 10–20 million
Metropolitan areas 5–9.9 million
Metropolitan areas 1–4.9 million
Selected smaller metropolitan areas
PACIFIC OCEAN
0 500 1000 Miles
0 500 1000 Kilometers
Hawaiian Islands (U.S.)
Honolulu
Papua New Guinea. The more than 7 million residents of this nation remain one of the least urban populations on Earth.
Northern Mariana Islands (U.S.)
Wake I. (U.S.)
Guam (U.S.)
Philippine Sea
MARSHALL ISLANDS
Nauru. This tiny island nation, devastated by phosphate mining, is one of the most densely settled parts of Oceania.
Koror
PALAU
Palikir
Majuro
FEDERATED STATES OF MICRONESIA
Tarawa
PAPUA NEW GUINEA
Yaren
NAURU
KIRIBATI
INDONESIA
New Guinea
Port Moresby
SOLOMON ISLANDS
TUVALU
Funafuti
Tokelau (N.Z.)
Arafura Sea
Honiara
Wallis and Futuna (FR.)
SAMOA
Pago Pago
American Samoa (U.S.)
EAST TIMOR
Darwin
VANUATU
Port Vila
Port Douglas
Cairns
Coral Sea
Suva
FIJI
TONGA
Niue (N.Z.)
Cook Islands (N.Z.)
Papeete
French Polynesia (FRANCE)
AUSTRALIA
New Caledonia (FR.)
Noumea
Nuku'alofa
Alice Springs
see inset
Brisbane
Solomon Islands. This small collection of islands has one of the region's highest natural growth rates.
see inset
Perth
Broken Hill
Adelaide
Canberra
Sydney
INDIAN OCEAN
Melbourne
Bass Strait
Tasman Sea
Hobart
see inset
Auckland
Cook Strait
NEW ZEALAND
Wellington
Christchurch
Dunedin
Invercargill
Geraldton
Perth
Rockingham
Bunbury
Busselton
Albany
Rockhampton
Gladstone
Bundaberg
Sunshine Coast
Brisbane
Gold Coast
Coffs Harbour
Port Macquarie
Newcastle
Sydney
Wagga Wagga
Wollongong
Canberra
Urban Australia. Although rural life in the Australian Outback is treasured in the country's heritage, most Australians live in major cities along the continent's east coast.
North Island
Whangarei
Auckland
Hamilton
Tauranga
New Plymouth
Palmerston North
Hastings
Nelson
Wellington
South Island
Christchurch
Timaru
Dunedin
Invercargill
Insets are twice the scale of the main map

Figure 14.17 Sydney Harbor, Australia This aerial view, taken above North Sydney, shows Australia's most spectacular and famous harbor. Note the Opera House near the water on the opposite shore.

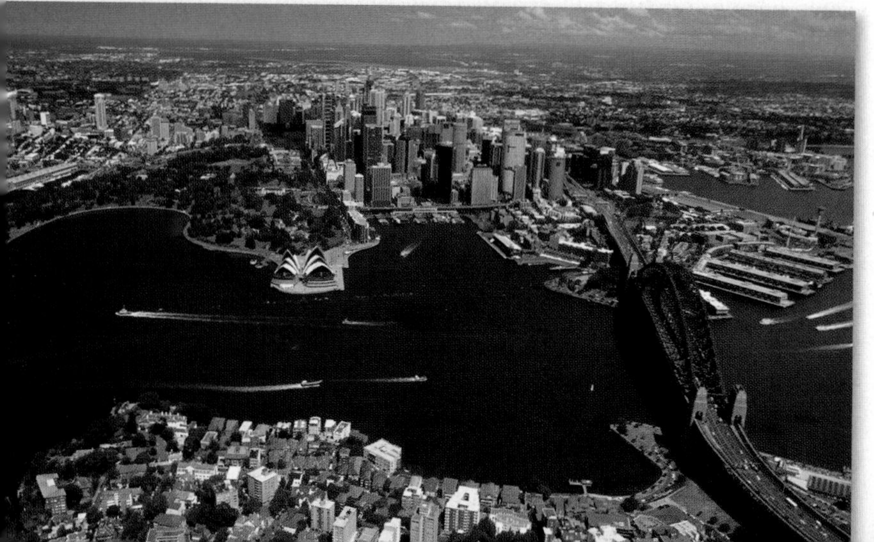

Figure 14.18 Perth Suburbs To accommodate rapid suburban growth, the southwestern Australian city of Perth has witnessed a boom in the construction of new homes and apartment houses.

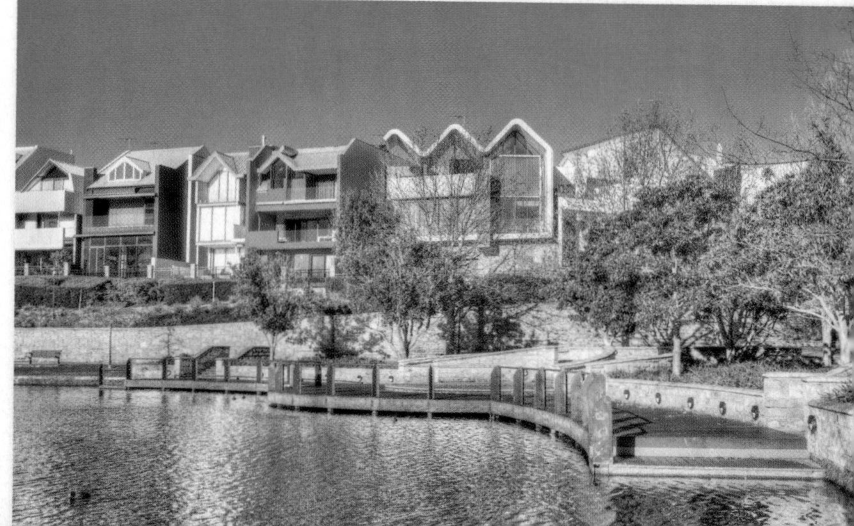

Figure 14.19 Christchurch, New Zealand This image of Christchurch shows the downtown area and nearby Hagley Park. Note the sprawling Canterbury Plain bordered by the snow-clad Southern Alps in the distance.

Australians arrived in one wave of people or in many, but the available evidence suggests that they soon occupied large portions of the continent, including Tasmania, which was once connected to the mainland by a land bridge because of the lower sea level.

Eastern Melanesia was settled much later than Australia and New Guinea. By 3500 years ago, some Pacific peoples had mastered long-distance sailing and navigation, which eventually opened the entire oceanic realm to human habitation. In that era, people gradually moved east to occupy New Caledonia, the Fiji Islands, and Samoa. From there, later movements took seafaring folk north into Micronesia, with the Marshall Islands occupied around 2000 years ago.

Continuing movements from Asia further complicated the story of these migrating Melanesians. Some migrants mixed culturally and eventually reached western Polynesia, where they formed the core population of the Polynesian people. By 800 CE, they had reached such distant places as Tahiti, Hawaii, and Easter Island. Prehistorians hypothesize that population pressures may have quickly reached crisis stage on the relatively small islands, leading people to attempt dangerous voyages to colonize other Pacific islands. Equipped with sturdy outrigger sailing vessels and ample supplies of food, the Polynesians were quickly able to colonize most of the islands they discovered.

European Colonization About six centuries after the Maori people brought New Zealand into the Polynesian realm, Dutch navigator

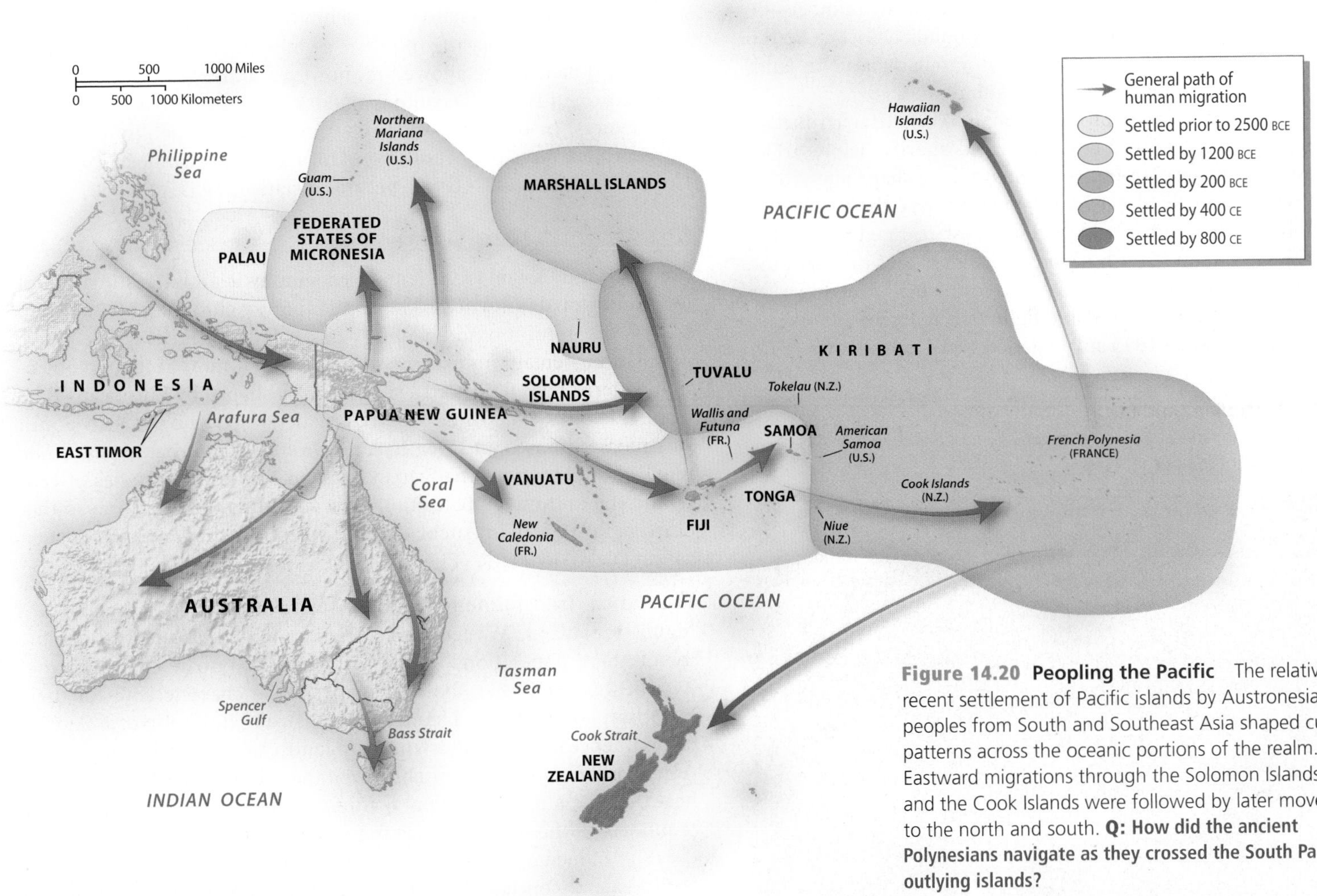

Figure 14.20 Peopling the Pacific The relatively recent settlement of Pacific islands by Austronesian peoples from South and Southeast Asia shaped cultural patterns across the oceanic portions of the realm. Eastward migrations through the Solomon Islands, Fiji, and the Cook Islands were followed by later movements to the north and south. **Q: How did the ancient Polynesians navigate as they crossed the South Pacific to outlying islands?**

Abel Tasman spotted the islands on his global exploration of 1642, marking the beginning of a new chapter in the human occupation of the South Pacific. British sea captain James Cook surveyed the shorelines of both New Zealand and Australia between 1768 and 1780, with the belief that these distant lands might be worthy of European development. By 1800, other European expeditions were exploring the Pacific, placing most of Oceania's island groups on colonial maps.

European colonization of the region began in Australia when the British needed a remote penal colony to which it could exile convicts. Australia's southeastern coast was selected as an appropriate site, and in 1788 the First Fleet arrived with 750 prisoners in Botany Bay, near what is now Sydney. Other fleets and more convicts soon followed, as did boatloads of free settlers. Before long, free settlers outnumbered the convicts, who were gradually being released after serving their sentences. The growing population of English-speaking people soon moved inland and also settled other favorable coastal areas. British and Irish settlers were attracted by the agricultural and stock-raising potential of the distant colony as well as by the lure of gold and other minerals. A major gold rush occurred in Australia during the 1850s, paralleling historical developments in western North America (Figure 14.21).

These new settlers clashed with the Aborigines almost immediately after arriving. No treaties were signed, however, and in most cases Aborigines were simply displaced from their lands. In some places—most notably, Tasmania—they were hunted down and killed. In mainland Australia, the Aborigines were greatly reduced in numbers by disease, removal from their lands, and pure economic hardship. By the mid-19th century, Australia was primarily an English-speaking land, as the native peoples had been driven into submission.

The lush and fertile lands of New Zealand also attracted British settlers. European whalers and sealers arrived shortly before 1800, with more permanent agricultural settlement taking shape after 1840, when the British formally declared sovereignty over the region. As new arrivals grew in number and the scope of planned settlement colonies on the North and South islands expanded, tensions with the native Maori population increased. Organized in small kingdoms, or chiefdoms, the Maori were formidable fighters (Figure 14.22). Widespread Maori wars began in 1845 and engulfed New Zealand until 1870.

Figure 14.21 Australian Gold Rush This late-19th-century sketch of Victoria shows how the landscape was dramatically modified by miners as they searched for gold near Melbourne.

Figure 14.22 Maori Warrior Body ornamentation, including tattoos, was common in traditional Maori culture, particularly among the high-status warrior class. Before battle, Maori warriors would perform a ceremonial dance called the *haka*, which included fierce facial contortions with the tongue and eyes, as shown by this Maori dancer. *Haka* dances (along with the facial contortions) are now a common part of New Zealand culture, particularly among sports teams.

The British eventually prevailed, however, and as in Australia, the native Maori lost most of their land.

Native Hawaiians also lost control of their lands to immigrants. Hawaii emerged as a united and powerful kingdom in the early 1800s, and for many years its native rulers limited U.S. and European claims to their islands. Increasing numbers of missionaries and settlers from the United States were allowed in, however, and by the late 19th century control of Hawaii's economy had largely passed to foreign plantation owners. In 1893, U.S. interests were strong enough to overthrow the Hawaiian monarchy, resulting in formal political annexation to the United States in 1898. Hawaii became a state in 1959.

Settlement Landscapes

Australia and Oceania's settlement geography is an interesting mix of local and global influences. The contemporary cultural landscape still reflects the imprint of indigenous peoples in those areas where native populations remain numerically dominant, but more recent colonization patterns have produced a scene mainly shaped by Europeans. The result includes everything from German-owned vineyards in South Australia to houses on New Zealand's South Island that appear to be plucked directly from the British Isles. In addition, processes of economic and cultural globalization have structured urbanization so that cities such as Perth or Auckland look strikingly similar to places such as San Diego or Seattle (Figure 14.23).

The Urban Transformation Both Australia and New Zealand are highly urbanized, Westernized societies, and thus the vast majority of their populations live in city and suburban environments. As cities evolved, they took on many of the characteristics of their European counterparts, blended with a strong dose of North American influences. The result is an urban landscape in which many North Americans are quite comfortable, even though the varied local accents heard on the street and many features of the metropolitan scene are reminders of the strong and lasting attachments to British traditions.

The affluent Western-style urban environments of Australia and New Zealand contrast dramatically with the urban landscapes found in less developed countries in the region. The streets of Port Moresby,

Papua New Guinea's capital and largest commercial center, reveal evidence of the large gap between rich and poor within Oceania (Figure 14.24). Rapid growth here has produced many of the classic problems of urban underdevelopment: There is a shortage of adequate housing, the building of roads and schools lags far behind the need, and street crime and alcoholism are rising. Elsewhere, urban centers such as Suva (Fiji), Noumea (New Caledonia), and Apia (Samoa) also reflect the economic and cultural tensions generated as local populations are exposed to Western influences. Rapid urban growth is a common problem in the smaller cities of Oceania because native people from rural areas and nearby islands gravitate toward the available job opportunities. In the past 50 years, the huge global growth of tourism in places such as Fiji and Samoa has also transformed the urban scene, replacing traditional village life with a landscape of souvenir shops, honking taxicabs, and crowded seaside resorts.

Figure 14.24 Port Moresby, Papua New Guinea Urban poverty and high crime afflict the city of Port Moresby, capital of Papua New Guinea. The city's slums, many built out on the water, reflect the stresses of recent urban growth as rural residents emigrate from even poorer nearby highlands.

Rural Australia and New Zealand Rural landscapes across Australia and the Pacific region express a complex mosaic of cultural and economic influences. In some settings, Australian Aborigines can still be found in their familiar homelands, their traditional way of life and settlements barely changed from pre-European times. Yet such settlement landscapes are becoming rare. Global influences penetrate the scene as the cash economy, foreign tourism and investment, and the currents of popular culture flow from city to countryside.

Most of the rural Australian interior is too dry for farming and features range-fed livestock. Sheep and cattle dominate Australia's livestock economy, and rural landscapes in the interior of New South Wales, Western Australia, and Victoria are oriented around isolated sheep stations—ranch operations that move the flocks from one large pasture to the next. Cattle can sometimes be found in these same areas, although many of the more extensive range-fed cattle operations are concentrated farther north, in Queensland (Figure 14.25). Other isolated interior settings remain home to Aboriginal peoples who still pursue their traditional forms of hunting and gathering.

Croplands also vary across the region. A band of commercial wheat farming sometimes mingles with the sheep country across southern Queensland; the moister interiors of New South Wales, Victoria, and South Australia; and a swath of favorable land east and north of Perth. Specialized sugarcane operations thrive along the narrow, warm, and humid coastal strip of Queensland. To the south and west, productive irrigated agriculture has developed in places such as the Murray River Basin, where orchard crops and vegetables are grown. **Viticulture**, or grape cultivation, dominates South Australia's Barossa Valley, the Riverina district in New South Wales, and Western Australia's Swan Valley.

New Zealand's rural settlement landscape includes a variety of agricultural activities. Ranching clearly dominates, with the vast majority of agricultural land devoted to livestock production, particularly sheep grazing and dairying. Commercial livestock outnumber people in New Zealand by a ratio of more than 20 to 1, and this is apparent throughout the countryside. Dairy operations are present mostly in the lowlands of the north, where they sometimes mingle with suburban landscapes in the vicinity of Auckland.

Rural Oceania The rural landscape varies considerably elsewhere in Oceania. On high islands with more water, denser populations take advantage of diverse agricultural opportunities; on the more barren low islands, fishing is often more important. Several types of rural

Figure 14.23 Downtown Auckland Apart from motorist New Zealand's cities, like this view of busy Queen Street, resemble many North American cities.

Figure 14.25 Cattle Ranch, Queensland Interior This motorized cowboy is herding cattle at Longreach, deep in the interior Queensland Outback.

settlement can be identified across the island region. In rural New Guinea, village-centered shifting cultivation dominates: Farmers clear a patch of forest and then, after a few years, shift to another patch, thus practicing the swidden agriculture common in Southeast Asia. Subsistence foods such as sweet potatoes, taro (another starchy root crop), coconut palms, bananas, and other garden crops are often found in the same field, an agricultural practice known as **intercropping**.

Commercial plantation agriculture has also made its mark in many of the more accessible rural settings. Unlike subsistence agriculture, these commercial operations generally feature a form of monoculture, where only one crop is grown in a field. In these places, settlements consist of worker housing near crops that is typically controlled by absentee landowners. For example, copra (coconut), cocoa, and coffee operations have transformed many agricultural settings in places such as the Solomon Islands and Vanuatu. Sugarcane and taro plantations have reshaped other island settings, particularly in Fiji and Hawaii (Figure 14.26).

Figure 14.26 Commercial Taro Cultivation While taro is a traditional subsistence crop throughout Oceania, it is also grown on plantations in Hawaii for commercial sales throughout the region. This taro plantation is on the island of Kauai.

Diverse Demographic Paths

Various population-related issues face residents of the region today. Australian and New Zealand populations grew rapidly (mostly from natural increases) in the 20th century, but today's low birth rates parallel the pattern in North America, where population growth stems from immigration (Figure 14.27).

Different demographic challenges grip many less-developed island nations of Oceania. High population growth rates of over 2 percent per year are not uncommon, as in Vanuatu and the Marshall Islands (see Table A14.1). The larger islands of Melanesia contain some room for settlement expansion, but competitive pressures from commercial mining and logging operations limit the availability of new agricultural land. On some of the smaller island groups in Micronesia and Polynesia, population growth is a more pressing problem. Tuvalu (north of Fiji), for example, has just over 11,000 inhabitants crowded onto a land area of about 10 square miles (26 square kilometers), making it one of the world's most densely populated countries.

Out-migration from several island nations is very high. For example, in Tonga and Samoa the lack of employment is a considerable push force. In contrast, Australia and New Zealand remain attractive to migrants, as does New Caledonia because of its recent mining boom.

Review

14.4 Compare and contrast the populations of Australia and New Zealand in terms of size, density, and level of urbanization.

14.5 Trace on a map the prehistoric peopling of the Pacific.

14.6 Describe the rural settlement patterns for Australia and the island countries, and explain why these differ.

KEY TERMS Aborigine, viticulture, intercropping

Cultural Coherence and Diversity: A Global Crossroads

The Pacific world offers excellent examples of how culture is transformed as different groups migrate to a region, interact with one another, and evolve over time. As Europeans and other outsiders arrived in Oceania, colonization forced native peoples to resist or adjust. More recently, worldwide processes of globalization have redefined the region's cultural geography.

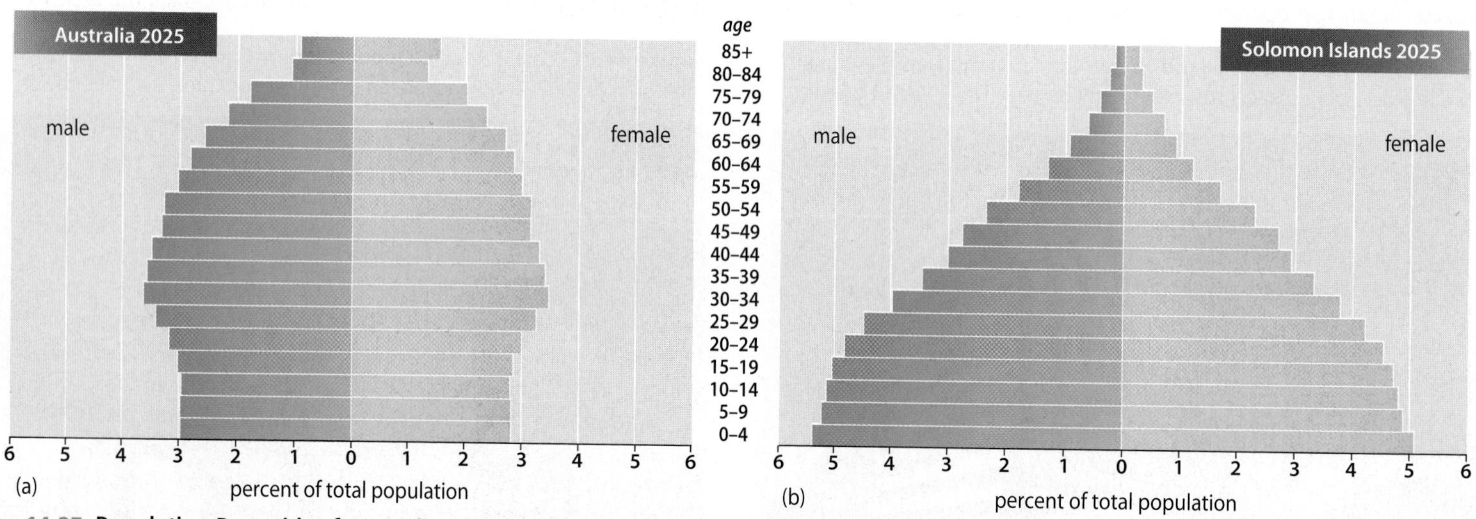

Figure 14.27 Population Pyramids of Australia and the Solomon Islands (2025) (a) Like many developed countries, Australia has very low natural growth, as shown by the forecast for 2025. (b) In contrast, like that of many developing countries, the Solomon Islands forecast shows the classic pyramidal shape of a young and growing population.

Multicultural Australia

Australia's cultural patterns illustrate globalization processes at work. Today, while the country is still dominated by European roots, its multicultural character is increasingly visible as native peoples assert their cultural identities and as immigrant populations play larger roles in society, particularly in metropolitan areas.

Aboriginal Imprints For thousands of years, Australia's indigenous Aborigines dominated the continent. They practiced hunting and gathering, a way of life that persisted up to the European conquest. Settlement densities remained low, tribal groups were often isolated from one another, and the overall population probably never exceeded 300,000 inhabitants. Cultural geographies were diverse and fragmented, producing many local languages. Even today about 50 indigenous languages can still be found.

Europeans brought radical demographic and cultural changes. Aboriginal populations were decimated in the process. The geographic results of colonization were striking: Indigenous residents were relocated to the sparsely settled interior, particularly in northern and central Australia, where fewer Europeans competed for land. Historically, the European attitude toward the Aboriginal population was often more discriminatory than it was toward the native peoples of North America.

Today Aboriginal culture perseveres in Australia, and a native people's movement is growing, similar to what is happening in the Americas (Figure 14.28). Indigenous people account for approximately 2 percent (or 430,000) of Australia's population, but their geographic distribution has changed dramatically over the past century. Aborigines account for almost 30 percent of the Northern Territory's population (many located near Darwin), and other large native reserves are located in northern Queensland and Western Australia. Most native people, however, live in the same urban areas that dominate the country's overall population geography. Indeed, more than 70 percent of Aborigines live in cities, and very few of them still practice traditional hunting-and-gathering lifestyles. Processes of cultural assimilation are clearly at work: Urban Aborigines are frequently employed in service occupations, Christianity has often replaced traditional animist religions, and only 13 percent of Aborigines still speak a native language.

Still, there is a growing Aboriginal interest in preserving traditional cultural values, particularly in the Outback, where indigenous languages remain strong. Cultural leaders also are preserving Aboriginal spiritualism, and these religious practices often link local populations to surrounding places and natural features that are considered sacred. In fact, a growing number of these sacred locations are at the center of land-use controversies (such as mining on sacred lands) between Aboriginal populations and Australia's European majority.

Figure 14.28 Australian Aborigines The native people of Australia inhabited the continent before European colonization. Relocated to less desirable land by the European immigrants, they continue to struggle for equal rights and decent living conditions. This family lives in a traditional community near Alice Springs in the Northern Territory, where Aborigines make up 20 percent of the population.

A Land of Immigrants Most Australians reflect the continent's more recent European-dominated migration history, but even these patterns have become more complex as more Asian immigrants arrive. Overall, more than 70 percent of Australians claim a British or Irish cultural heritage. These groups dominated 19th- and early-20th-century migrations into the country, and their close cultural ties to the British Isles remain strong.

European plantation owners along the Queensland coast imported inexpensive farm workers from the Solomons and New Hebrides (now Vanuatu) in the late 19th century. These Pacific island laborers, known as **kanakas**, were spatially and socially segregated from their Anglo employers, but they further diversified the cultural mix of Queensland's "sugar coast." Historically, however, nonwhite immigration was strictly limited by what is termed the **White Australia Policy**, in which government guidelines after 1901 promoted European and North American immigration at the expense of other groups. The policy was not dismantled until 1973.

Recent migration trends feature more diverse inflows of new workers, adding to the country's multicultural character. Since the 1970s, the government's migration program has selected people on the basis of their educational background and potential for succeeding economically in Australian society. Many families have arrived from such places as China, India, Malaysia, and the Philippines. Refugees from troubled parts of the world, such as Southeast Asia, the Middle East, and the Balkans, are also coming to Australia, creating a real asylum problem for the country (see *Exploring Global Connections: Asylum Seekers Arrive in Australia*).

The result is a diverse foreign-born population (Figure 14.29). Indeed, today about 25 percent of Australia's people are foreign-born, reflecting the country's global popularity as a migration destination. In the period 2000–2010, almost 40 percent of arriving settlers were from Asia. Major cities offer particularly attractive possibilities: Both Melbourne and Sydney are about 20 percent Asian and feature increasingly heterogeneous neighborhoods, businesses, and cultural institutions that cater to these new ethnic groups (Figure 14.30).

Cultural Patterns in New Zealand

New Zealand's cultural geography broadly reflects the patterns seen in Australia, although the precise cultural mix differs slightly. Native **Maori** people are more numerically important and culturally visible in New Zealand than their Aboriginal counterparts are in Australia. While British colonization clearly mandated the dominance of Anglo cultural traditions by the late 19th century, the Maori survived, even though they lost most of their land in the process. After an initial decline, the native population rebounded in the 20th century, and today self-identified Maori account for more than 15 percent of the country's

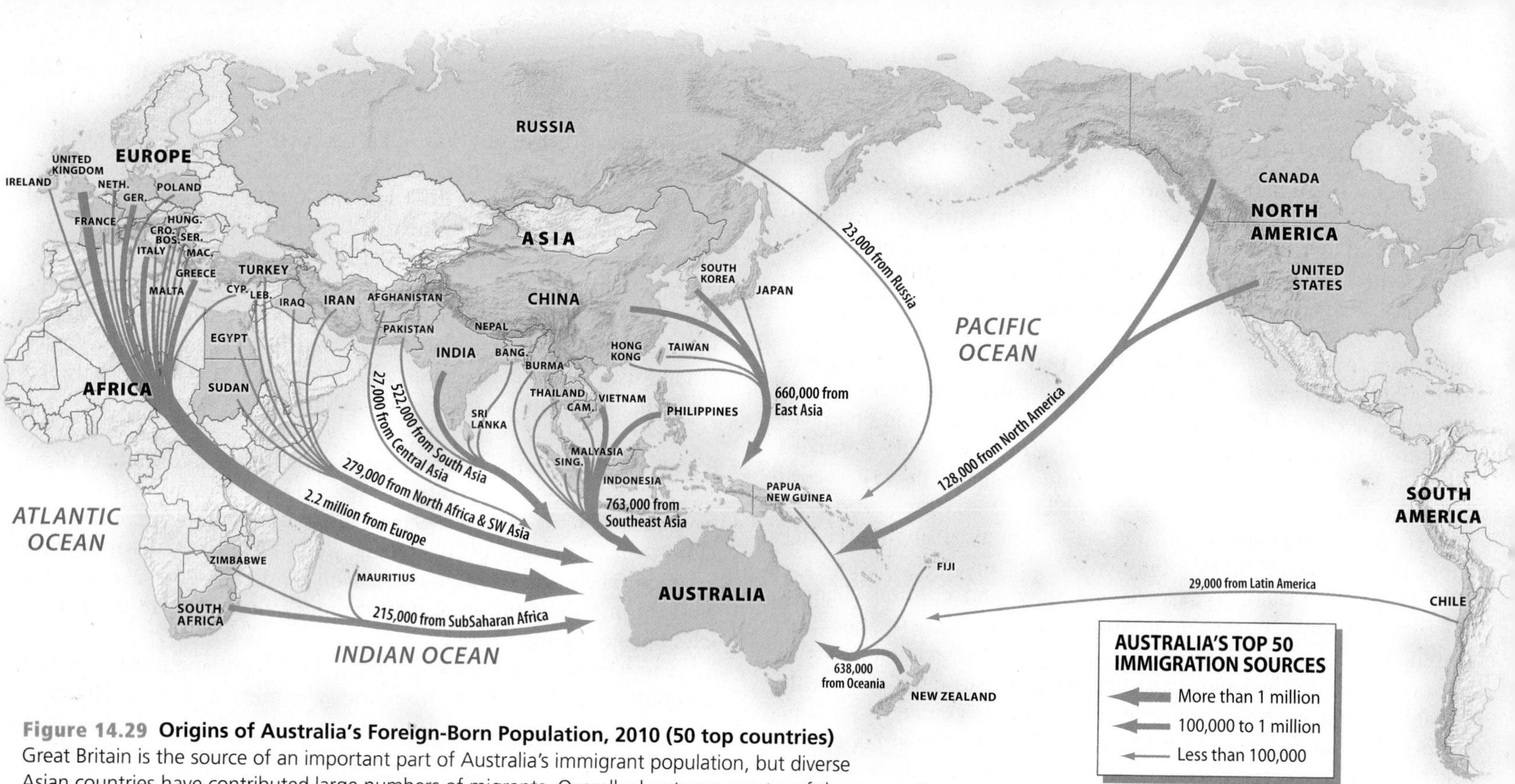

Figure 14.29 Origins of Australia's Foreign-Born Population, 2010 (50 top countries) Great Britain is the source of an important part of Australia's immigrant population, but diverse Asian countries have contributed large numbers of migrants. Overall, about one-quarter of the country's population is foreign-born. **Q: What different factors explain the changing patterns of immigration to Australia?**

4.5 million residents. Geographically, the Maori remain most numerous on the North Island, including a sizable concentration in metropolitan Auckland. While urban living is on the rise, many Maori are committed to preserving their religion, traditional arts, and Polynesian culture. In addition, Maori joins English as an official language of New Zealand.

Although many New Zealanders still identify with their largely British heritage, the country's cultural identity has increasingly separated from its British roots. In many ways, popular culture ties the country ever more closely to Australia, the United States, and continental Europe, a function of increasingly global mass media. Several major movies, for example, have been filmed in New Zealand, including *The Adventures of Tintin, The Hobbit, Avatar,* the *Lord of the Rings* trilogy, and *Whale Rider*.

Paralleling the Australian pattern, many immigrants have also moved to New Zealand, especially to larger cities such as Auckland. Solid economic growth and construction-related jobs have contributed to growing employment opportunities for immigrants. In 2013, the New Zealand census reported that about one in four residents of New Zealand was foreign-born. Since 2010, the largest source nations for these migrants have been the United Kingdom, China, India, and the Philippines.

Figure 14.30 Asian-Australian Neighborhood, Melbourne Little Bourke Street in Melbourne has long been home to a thriving community of Chinese-Australian residents.

The Mosaic of Pacific Cultures

Native and colonial influences produce a range of cultures across the islands of the South Pacific. In more isolated places, traditional cultures are largely insulated from outside influences. In most cases, however, modern life in the islands revolves around an intricate cultural and economic interplay of local and Western influences.

Language Geography A modern language map reveals some significant cultural patterns that both unite and divide the region (Figure 14.31). Most native languages of Oceania belong to the **Austronesian** language family, which is found across wide expanses of the Pacific, much of insular Southeast Asia, and, somewhat surprisingly, Madagascar. Linguists hypothesize that the first prehistoric wave of oceanic mariners spoke Austronesian languages and thus spread them throughout this vast realm of islands and oceans. Within this broad language family, the Malayo-Polynesian subfamily includes most of the related languages of Micronesia and Polynesia, suggesting a common cultural and migratory history for these widespread peoples.

Melanesia's language geography is more complex. Although coastal peoples often speak languages brought to the region by the

EXPLORING GLOBAL CONNECTIONS

Asylum Seekers Arrive in Australia

Google Earth (MG) Virtual Tour Video
http://goo.gl/1kl02h

In 2014, almost 17 million people sought asylum from political and social persecution, according to the UN. Many of them left their homelands and sought refuge in a neighboring country. Others traveled longer distances, seeking out a specific country for their refuge. Of these homeless masses, some have sought out Australia's shores, seeing the land Down Under as a potential path to a new life. In 2013, for example, more than 20,000 asylum seekers applied for admission to the country. While many Australians are proud of permitting refugees to relocate there, critics note that in 2014 more than $1.2 billion was spent to operate detention centers where refugees are processed and housed while awaiting decisions regarding asylum.

The Pacific Solution One of the most controversial aspects of the refugee-asylum topic is Australia's so-called Pacific Solution, a government policy designed to dissuade refugees from targeting Australia as their goal. Originally instigated in 2001, the Pacific Solution had several components. The Australian navy is charged with intercepting refugee boats before they can make landfall on Australian territory (Figure 14.2.1). Australia also relinquished its ownership of small offshore islands where refugees often landed and where they could plead amnesty under Australian law.

In addition, detention centers created on Nauru, Christmas Island, and Manus (an island in Papua New Guinea) are designed to keep refugees off Australian territory and deprive them of Australian law and civil rights while their claims are processed. These centers have proven expensive and problematic: Through an agreement with Papua New Guinea, for example, the Manus Island facility has grown to more than 1000 detainees (Figure 14.2.2). While some Manus residents applaud the influx of people and cash, others point to potential environmental and social issues as asylum seekers try to resettle in nearby areas.

Figure 14.2.2 Manus Island Detention Center, Papua New Guinea Many people who seek asylum in Australia end up being processed and resettled in Papua New Guinea. This view shows some of the initial accommodations at the Manus Island Detention Center.

Refugee Source Areas How do these asylum seekers get to Australia, and where do they come from? Most arrive in overloaded boats of questionable seaworthiness that probably departed most recently from some part of Indonesia, Australia's closest neighbor. These craft, however, may have started their voyages much farther away, in Sri Lanka, Bangladesh, Pakistan, or even China. Fiji and Malaysia have also contributed growing numbers of refugees. In some cases, the actual nationality of refugees is unclear. For example, many people reaching the shores of Australia are originally from Afghanistan but may have spent time in refugee camps in Pakistan, Iran, or Iraq. Because of their time in these other countries, the question of whether or not they have been persecuted becomes complex. Another complication is that, if the decision is made to return these refugees to their country of origin, where do they go? For now, Australia is struggling with its role as a home for refugee populations, and given current global political instability, no relief appears to be on the horizon.

1. A key element in Australia's amnesty policy is whether or not refugees were "persecuted" in their homeland. List the reasons why Australian refugees might have been persecuted in these different countries—Afghanistan, Sri Lanka, Pakistan, and Iran.
2. What other countries and regions of the world that you've read about in this book are magnets for refugees, and what sort of amnesty policies do they have?

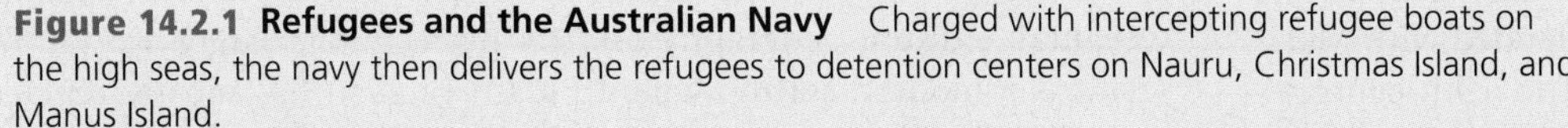

Figure 14.2.1 Refugees and the Australian Navy Charged with intercepting refugee boats on the high seas, the navy then delivers the refugees to detention centers on Nauru, Christmas Island, and Manus Island.

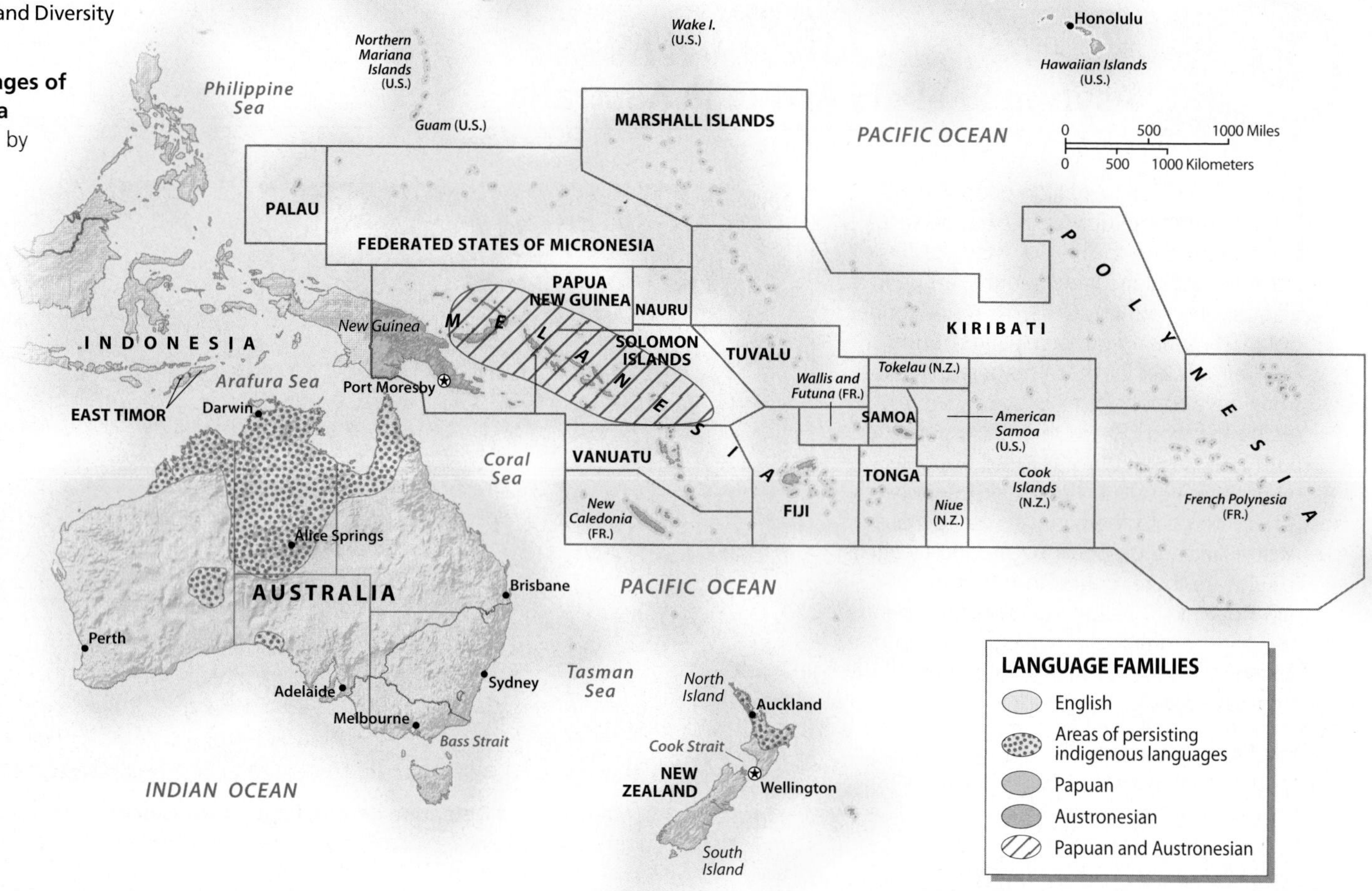

Figure 14.31 Languages of Australia and Oceania While English is spoken by most residents, native peoples and their linguistic traditions remain an important cultural and political force in both Australia and New Zealand. Elsewhere, traditional Papuan and Austronesian languages dominate Oceania. The French colonial legacy also persists in some Pacific locations. Tremendous linguistic diversity has shaped the cultural geography of Melanesia, and more than 1000 languages have been identified in Papua New Guinea alone.

seafaring Austronesians, more isolated highland cultures, particularly on the island of New Guinea, speak varied Papuan languages. In fact, more than 1000 languages have been identified across the mountainous interior, creating such linguistic complexity that many experts question whether they even constitute a unified "Papuan family" of related languages. Some scholars estimate that half of New Guinea's languages are spoken by fewer than 500 people.

Given the frequency of contact between different island cultures, it is no surprise that people have generated new forms of intercultural communication. For example, several forms of **Pidgin English** (also known simply as *Pijin*) are found in the Solomons, Vanuatu, and New Guinea, where it is the major language used between ethnic groups. In Pijin, a largely English vocabulary is reworked and blended with Melanesian grammar. Pijin's origin is commonly traced to 19th-century Chinese sandalwood traders ("pijin" is the Chinese pronunciation of the word for "business"). While of historical origin, Pijin has become a globalized language of sorts in Oceania as trade and political ties have developed between different native island groups. About 300,000 people in Oceania speak Pidgin on a regular basis, and it forms an important element of cultural identity among native Hawaiians, who use a version of Pidgin in their everyday vernacular language.

Village Life Traditional patterns of social life are as complex and varied as the language map. Across much of Melanesia, including Papua New Guinea, most people live in small villages often occupied by a single clan or family group. Many of these traditional villages contain fewer than 500 residents, although some larger communities may house more than 1000 people. Life often revolves around the gathering and growing of food, annual rituals and festivals, and complex networks of kin-based social interactions (Figure 14.32).

Traditional Polynesian culture also focuses on village life, although strong class-based relationships often exist between local elites (who are often religious leaders) and ordinary residents. Polynesian villages are also more likely to be linked to other islands by wider cultural and political ties. Despite the Western stereotype of Polynesian communities as idyllic and peaceful, violent warfare was actually quite common across much of the region prior to European contact.

Interactions with the Larger World

While traditional cultures persist in some areas, most Pacific islands have witnessed tremendous cultural transformations in the past 150 years. Settlers from Europe, the United States, and Asia brought new values and technological innovations that have forever changed Oceania's cultural geography and its place in the larger world. The result is a modern setting where Pidgin English has mostly replaced native languages, Hinduism is practiced on remote Pacific islands, and people from traditional fishing communities now work at resort hotels and golf course complexes.

Colonial Connections Anglo-European colonialism transformed the cultural geography of the Pacific world. The region's cultural makeup also was changed by new people migrating into the Pacific islands. Hawaii illustrates the pattern. By the mid-19th century, Hawaii's King Kamehameha was already entertaining assorted whalers, Christian missionaries, traders, and navy officers from Europe and the United States. A small, elite group of **haoles**, light-skinned Europeans and Americans, successfully profited from sugarcane plantations and Pacific shipping contracts. Labor shortages on the islands led to the importing of Chinese, Portuguese, and Japanese workers, who further complicated the region's cultural geography. By 1900, Japanese immigrants dominated the island workforce.

The United States formally annexed the islands in 1898, and the cultural mix revealed in the Hawaiian census of 1910 suggests the

Figure 14.32 Highlands Village, Papua New Guinea Residents of this village in the Papua New Guinea Highlands live in a loosely clustered array of homes created from local building materials.

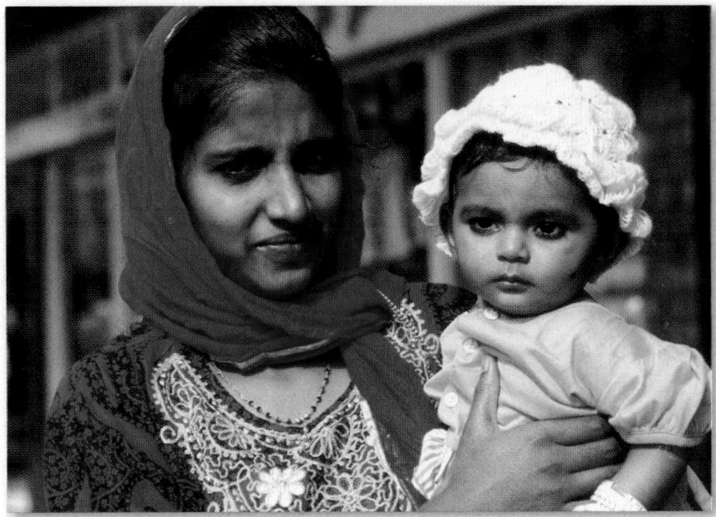

Figure 14.33 South Asians in Fiji This young Indian woman and her daughter live on Vanua Levu, one of Fiji's largest islands.

magnitude of culture change: More than 55 percent of the population was Asian (mostly Japanese and Chinese), native Hawaiians made up another 20 percent, and about 15 percent (mostly imported European workers) were white. By the end of the 20th century, however, the Asian population was less dominant, as about 40 percent of Hawaii's residents were white. In addition, native Hawaiians were joined by diverse migrants from other Pacific islands. Ethnic mixing among these groups has produced a rich mosaic of Hawaiian cultures, showing a unique blend of North American, Asian, Pacific, and European influences.

Hawaii's story has also played out on many other Pacific islands. In the Mariana Islands, Guam was absorbed into the United States' Pacific empire at the conclusion of the Spanish-American War in 1898. Thereafter, the native people were influenced not only by Americanization (the island remains a self-governing U.S. territory today), but also by the thousands of Filipinos who were moved in to supplement Guam's modest labor force. To the southeast, the British-controlled Fiji Islands offered similar opportunities for redefining Oceania's cultural mix. The same sugar-plantation economy that spurred changes in Hawaii prompted the British to import thousands of South Asian laborers to Fiji. The descendants of these Indians (most practicing Hinduism) now constitute almost half the island country's population and sometimes come into sharp conflict with native Fijians (Figure 14.33). In French-controlled portions of the region, small groups of traders and plantation owners filtered into the Society Islands (Tahiti), but a larger group of French colonial settlers (many originally part of a penal colony) had a major impact on the cultural makeup of New Caledonia. Still a French colony, New Caledonia is more than one-third French, and its capital city of Noumea reveals a cultural setting forged from French and Melanesian traditions.

Sports and Globalization Like all aspects of culture, colonial influences have left their mark on Oceania's playing fields, particularly in those former English colonies where cricket, soccer, and rugby dominate the sporting scene. Netball, an English version of basketball played with seven players, is one of the most popular women's sports in both Australia and New Zealand, followed by field hockey. In New Zealand, where rugby is arguably the nation's favorite sport, the national team, the All Blacks (named for their uniform color, not their ethnicity), has integrated Polynesian warrior rituals and dances into its pre- and postgame activities, to international acclaim (Figure 14.34).

In recent years, many young men from the Pacific islands have looked to American football as their ticket to a better future, one that includes a U.S. college scholarship and perhaps even a chance to play in the National Football League (NFL). Many players come from American Samoa because they are American citizens and need no visas, although football players from Tonga and Fiji are also recruited. Coaches note that in addition to physical size and agility, these players often possess a strong work ethic. For many Polynesian athletes, however, success at the professional level remains elusive. But given the lack of home-based job opportunities, young Pacific island men will likely remain on the rosters of many college athletic teams.

Review

14.7 Compare the Australian Aborigines and the New Zealand Maori in terms of their initial encounters with Europeans and the challenges they face today.

14.8 How has Australia's immigrant flow changed in the last 50 years?

14.9 Describe Hawaiian cultural changes over the last century.

KEY TERMS kanaka, White Australia policy, Maori, Austronesian, Pidgin English, haole

Figure 14.34 New Zealand All Blacks Rugby Team Before and after each rugby game, the All Blacks perform the Maori *haka*, a traditional ritual dance associated with war, contests, or tribal challenges.

Geopolitical Framework: Diverse Paths to Independence

Pacific geopolitics reflects a complex interplay of local, colonial-era, and current global-scale forces (Figure 14.35). These complexities become apparent in the story of Micronesia's Marshall Islands. This sprinkling of islands and atolls (covering 70 square miles, or 180 square kilometers, of land) historically consisted of many ethnic groups that made up small political units. In 1914, the Japanese moved into the islands, and the area remained under their control until 1944, when U.S. troops occupied the region. Following World War II, a UN trust territory (administered by the United States) was created across a wide swath of Micronesia, including the Marshall group. Demands for local self-government grew during the 1960s and 1970s, resulting in a new constitution and independence for the Marshall Islands by the early 1990s. Today, still benefiting from U.S. aid, government officials in the modest capital city on Majuro Atoll struggle to unite island populations, protect large maritime sea claims, and resolve a generation of legal and medical problems that grew from U.S. nuclear bomb testing in the region. Similar stories are typical across the Pacific region, suggesting a 21st-century political geography that is still very much in the making.

Roads to Independence

The newness and fluidity of the region's political boundaries are remarkable. The region's oldest independent states are Australia and New Zealand, both 20th-century creations still considering whether to complete their formal political separation from the British Crown. Elsewhere, political ties between colony and mother country are closer and perhaps more enduring. Even many of the newly independent Pacific **microstates**, with their tiny overall land areas, keep special political and economic ties to countries such as the United States.

Independent Australia (1901) and New Zealand (1907) gradually created their own political identities, yet both still struggle with the final shape of these identities. Although Australia became a commonwealth in 1901 and New Zealand finally broke formal legislative links with the mother country in 1947, both nations still acknowledge the British Crown as the symbolic head of government (Figure 14.36).

Colonial ties were cut even more slowly in other parts of the region, and the process has not yet been completed. New Caledonia remains affiliated (as a "special collectivity") with France after the island nation's citizens voted down a referendum for independence in 2014. In the 1970s, Britain and Australia began giving up their own colonial empires in the Pacific. Fiji (Great Britain) gained independence in 1970,

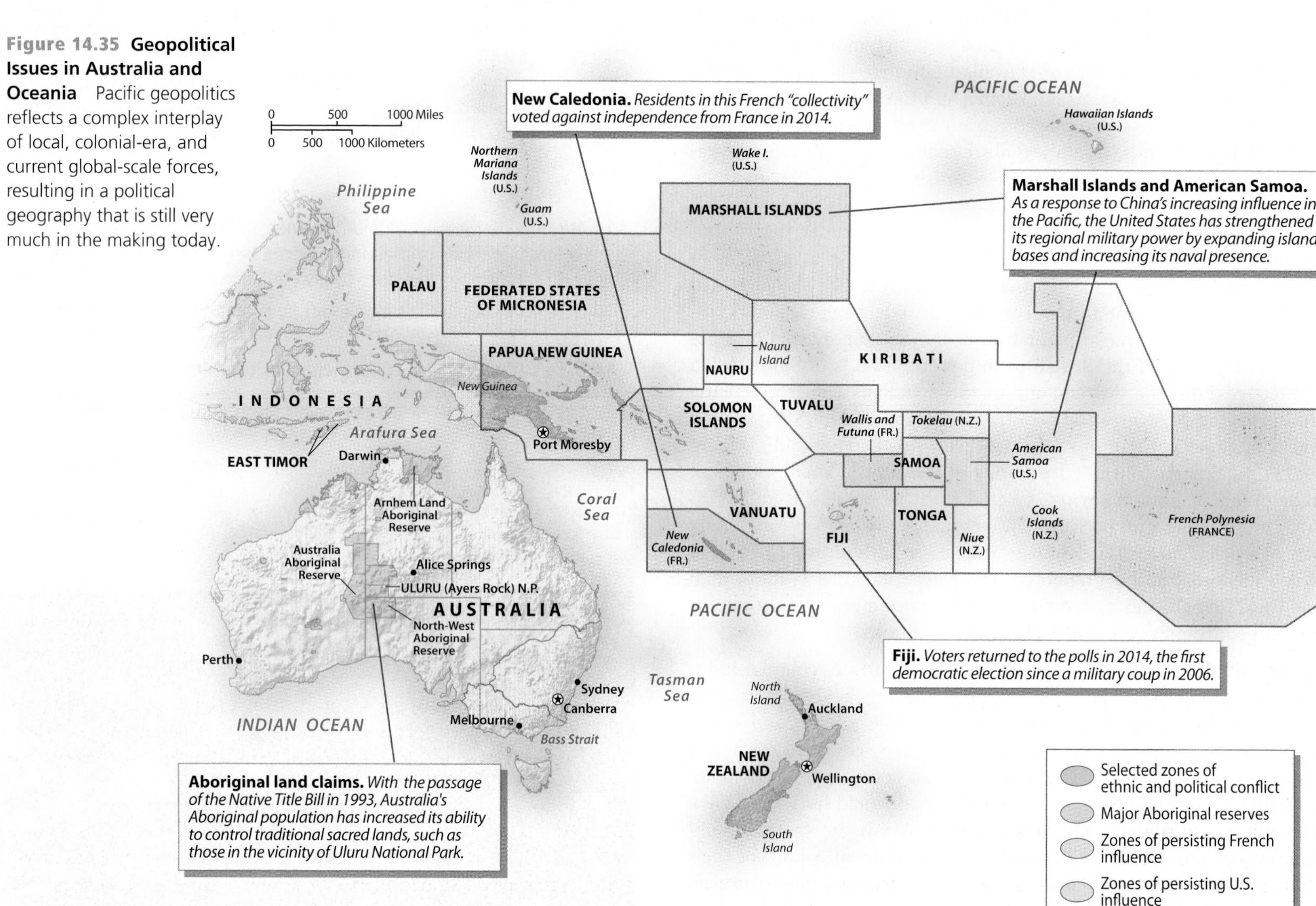

Figure 14.35 Geopolitical Issues in Australia and Oceania Pacific geopolitics reflects a complex interplay of local, colonial-era, and current global-scale forces, resulting in a political geography that is still very much in the making today.

Figure 14.36 Prince William and the Duchess of Cambridge Visit Australia in 2014 Close ties between Australia and the British Crown are reflected on this visit to Sydney's Taronga Zoo by Prince William and the Duchess of Cambridge.

Figure 14.37 The Streets of Pago Pago, American Samoa This view of downtown Pago Pago shows a small shopping area and a nearby church.

followed by Papua New Guinea (Australia) in 1975 and the Solomon Islands (Great Britain) in 1978. The small island nations of Kiribati and Tuvalu (Great Britain) also became independent in the late 1970s. Independence, however, has not necessarily guaranteed political stability: Fiji experienced a military coup in 2006 and only recently returned to a representative and elected government, while Papua New Guinea suffered a constitutional crisis in 2012 as its judicial and legislative branches battled for power following a disputed election.

The United States has recently turned over most of its Micronesian territories to local governance, while still maintaining a large influence in the area. After gaining these islands from Japan in the 1940s, the U.S. government provided aid to the islanders and utilized a number of islands for military purposes. Bikini Atoll was destroyed by nuclear tests, and the large lagoon of Kwajalein Atoll was used as a giant missile target. A major naval base was established in Palau, the westernmost archipelago of Oceania.

By the early 1990s, the Marshall Islands and the Federated States of Micronesia (including the Caroline Islands) gained independence. Their ties to the United States remain close. Several other Pacific islands remain under U.S. administration. Palau is a U.S. "trust territory," which gives Palauans some local autonomy. The people of the Northern Marianas chose to become a "self-governing commonwealth in association with the United States," a rather vague political position that allows them to become U.S. citizens. Residents of self-governing Guam and American Samoa are also U.S. citizens (Figure 14.37). Hawaii became a full-fledged U.S. state in 1959, yet debate continues regarding native land claims and sovereignty.

Other colonial powers appear less inclined to give up their oceanic possessions. New Zealand still controls substantial territories in Polynesia, and France has even more extensive holdings in the region. In addition to New Caledonia in Melanesia, France's territories include French Polynesia, a large expanse of mid-Pacific territory, as well as the much smaller territory of Wallis and Futuna to the west.

Persistent Geopolitical Tensions

Cultural diversity, colonial legacy, youthful states, and a rapidly changing political map contribute to ongoing geopolitical tensions in the Pacific world. Indeed, some of these conflicts have consequences that extend far beyond the boundaries of the region. Others are more locally based, but they are still reminders of the difficulties that occur as political space is redefined in varied natural and cultural settings.

Native Rights in Australia and New Zealand Indigenous peoples in Australia and New Zealand have used the political process to gain more control over land and resources in their two countries, paralleling similar efforts in North America and elsewhere. Australia's Aboriginal groups discovered newfound political power from effective lobbying during a time when a more sympathetic federal government was interested in rectifying historical discrimination that left native peoples with no legal land rights. In recent years, the Australian government established several Aboriginal reserves, particularly in the Northern Territory, and expanded Aboriginal control over sacred national parklands such as Uluru (called Ayers Rock by the British settlers) (Figure 14.38). Further concessions to indigenous groups were made in 1993 when the government passed the **Native Title Bill**, which compensated Aborigines for lands already given up and gave them the right to gain title to unclaimed lands they still occupied. The bill also provided Aborigines with legal standing to deal with mining companies in native-settled areas.

Efforts to expand Aboriginal land rights, however, have met strong opposition. In 1996, an Australian court ruled that pastoral leases (the form of land interest held by ranchers who control most of the Outback) do not necessarily negate or replace Aboriginal land rights. Grazing interests were infuriated, which led the government to respond that Aboriginal claims allow them to visit sacred sites and do some hunting and gathering, but do not give them complete economic control over the land (Figure 14.39).

In New Zealand, native land claims are even more contentious because the Maori constitute a larger proportion of the overall population and the lands they claim are more valuable than rural Aboriginal lands in Australia. Recent protests have included civil disobedience, ever-increasing Maori land claims over much of the North and South islands, and a call to return the country's name to the indigenous *Aotearoa*, "Land of the Long White Cloud." Overall, the government response has been to increasingly recognize Maori land and fishing rights.

Native Rights in Hawaii Native Hawaiians, who call themselves *Kanaka maoli*, also have issues with the U.S. government concerning human rights, access to ancestral land, and the political standing of native people. Native Hawaiians are descendants of Polynesian people who arrived in

Figure 14.38 Aborigines near Uluru Rock An indigenous ranger patrols Uluru-Kata Tjuta National Park in Australia's Northern Territory. Uluru Rock is in the background.

the Hawaiian Islands about 1000 years ago. Until 1898, when the islands were annexed by the United States, Hawaiians lived in an independent and sovereign state that was recognized by major foreign powers. The legality of U.S. annexation is still contested today and underlies native Hawaiian demands for a return of their historical sovereignty.

Today many Hawaiian nationalists advocate a form of Polynesian sovereignty similar to that of the Navajo Nation in Arizona, Utah, and New Mexico. Extremists, however, reject the idea of native Hawaiians having the same standing as U.S. Indian tribes. Instead, they demand that the UN invalidate the 1898 annexation and grant native Hawaiians complete sovereignty, creating a separate country within the U.S. state of Hawaii (Figure 14.40). This extreme solution is resisted by private and corporate landowners in Hawaii, since it calls into question the legality of their ownership of everything from residential properties in the Honolulu suburbs to hotel parcels along Waikiki Beach.

Figure 14.39 Applications for Native Land Claims in Australia This map shows the applications for native land claims in Australia filed by different Aboriginal groups as of 2013. Note that these are applications only, not government-approved claims. Nevertheless, the widespread extent of the claims shows why the topic is so contentious and controversial. **Q: What might account for the relative lack of native land claims in the center of Australia?**

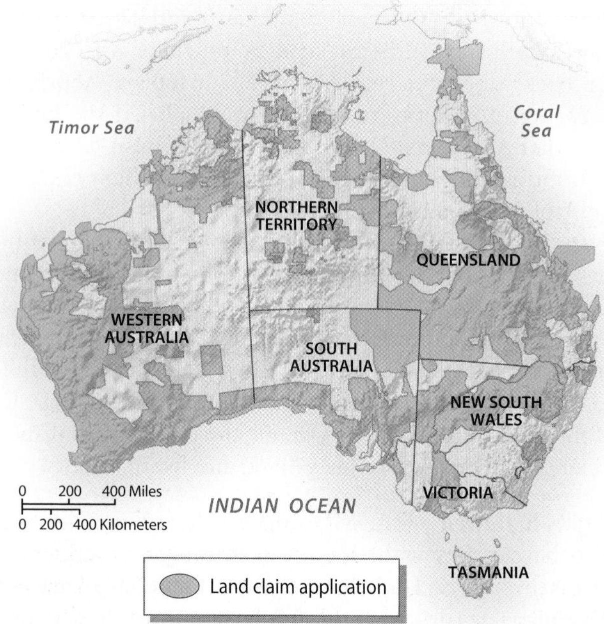

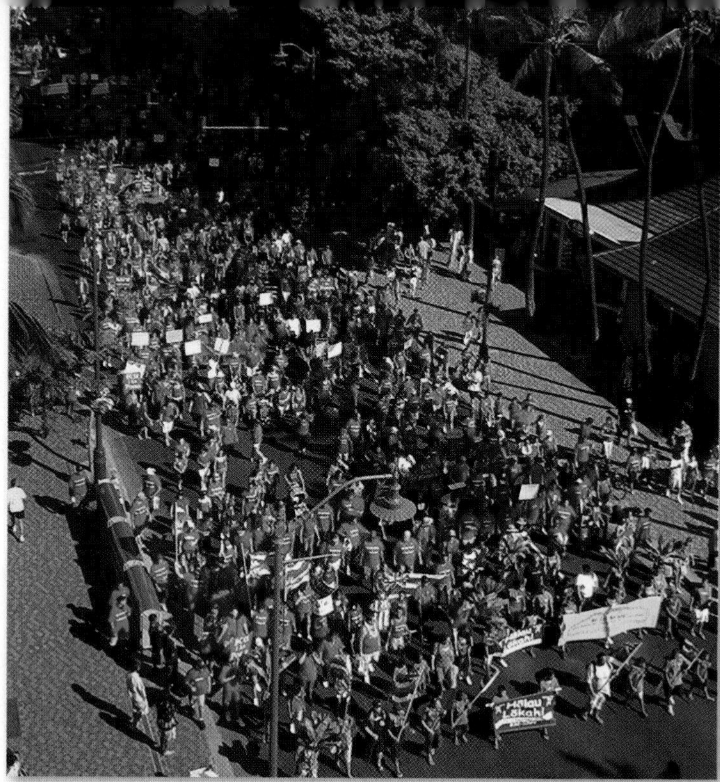

Figure 14.40 Native Hawaiian Nationalism Thousands of native Hawaiians march in protest down Honolulu's Kalakaua Avenue on Waikiki Beach. They are showing their support for the greater legal recognition of Native Hawaiian rights in Hawaii.

The Strategic Pacific As was the case during World War II, Oceania is once again a strategic global region where numerous countries seek to expand their influence. The major players include the reigning superpowers, the United States and China; the Pacific Rim countries of Japan, South Korea, Russia, Taiwan, and Indonesia; France, as a former colonial master; and the two most powerful locals, Australia and New Zealand.

For several decades now, disputes between Taiwan and the giant People's Republic of China (PRC) have inflamed tensions in Oceania. Each country has vied for influence over island communities, particularly those countries with votes in the UN. Samoa, Tonga, and Fiji, for example, have received vast amounts of aid from the PRC and have often voted in the UN to support China's controversial policy toward Tibet. In late 2014, Chinese President Xi Jinping reinforced these connections by holding a summit of allied island nations in Fiji and by promising more economic aid to friendly Pacific countries supportive of Chinese policies in the region. In contrast, Taiwan relies on the UN votes of Palau, Kiribati, and the Marshall Islands (all of them recent recipients of Taiwanese aid) to neutralize China's ambitions to reclaim Taiwan.

Because of the PRC's growing influence in Oceania, coupled with North Korea's continued hostility toward the West, the United States is increasing its diplomatic and military presence in the South Pacific. The term *Asia Pivot* is commonly used to describe the Obama administration's proposed shift in America's foreign and military policy away from Iraq and Afghanistan and toward the Asia–Pacific world. When China protested that this U.S. shift was a thinly veiled policy of containment, Washington's reply was that the U.S. policy was simply one of "Asia management."

Semantics aside, the Asia Pivot includes the establishment of numerous small military bases scattered around the Pacific (and, for that matter, around the world), commonly called "lily-pad" outposts. The recent—and controversial—agreement with Australia to post 2500 U.S. Marines near Darwin in northern Australia by 2016 is an example

Figure 14.41 U.S. Marines in Australia As part of the Asia Pivot, the United States reached an agreement with Australia to station up to 2500 Marines on the country's north coast near Darwin. Although the United States claims that northern Australia is simply a good place to train troops, China and other observers believe that the Marines are part of a strategy to contain Chinese influence in the region.

(Figure 14.41). Although the U.S. military understandably does not publicize its lily-pad bases, reportedly there are new or recently revitalized military bases in Australia's Cocos Islands, American Samoa, Tinian Island, the Marshall Islands, and the Marianas. Andersen Air Force Base in Guam has expanded its facilities, and similar expansion is taking place on existing U.S. military bases in Japan, the Philippines, and South Korea. The geopolitical goal behind this Asia Pivot seems clear enough: to reestablish the United States as a major political and military player in Oceania.

Since the policy emerged around 2011, the United States has successfully joined the East Asia Summit, an important regional discussion forum, but geopolitical distractions elsewhere in the world (the Middle East, Ukraine, etc.) have continued to dominate its daily geopolitical agenda. In addition, confronting China—politically and economically—in its own backyard in the western Pacific is proving to be daunting and expensive. Indeed, China's recent commitment to increased aid to sympathetic Pacific nations underscores the potentially high price tag of the Asia Pivot policy for the United States.

Not to be overlooked are Australia and New Zealand, which still play key political roles in the South Pacific. Although these two countries sometimes disagree on strategic and military matters, their size, wealth, and collective political influence in the region make them important forces for political stability. Also, special colonial relationships still connect these two nations with many Pacific islands. Australia maintains close political ties to its former colony of Papua New Guinea, and New Zealand's continuing control over Niue, Tokelau, and the Cook Islands in Polynesia confirms that its political influence extends well beyond its borders. How these two countries will respond to China's growing influence, however, is unclear.

Australia's number one trade partner is China, yet it maintains unquestioned political ties to North America. Australia could be showing Oceania that a middle pathway is possible, that Pacific island countries do not necessarily have to choose between the United States and China (or between China and Taiwan). Small island states, however, may find such a conciliatory stance more problematic and less likely to produce supportive flows of economic aid and low-interest loans from their chosen allies.

14.10 Describe the colonial history of the Pacific islands, and contrast it to that of Australia and New Zealand.

14.11 What are the arguments for and against sovereignty for native Hawaiians?

14.12 Explain the characteristics of and the motivations behind the United States' Asia Pivot policy.

KEY TERMS microstate, Native Title Bill

Economic and Social Development: Growing Asian Connections

As with all other world regions, the Pacific region contains a diversity of economic situations creating both wealth and poverty. Even within affluent Australia and New Zealand, pockets of pronounced poverty occur, and large economic disparities exist as well between those Pacific countries with global trade ties and small island nations lacking resources and external trade. While tourism offers some relief from abject poverty, the tourist economy can be fickle, prone to unpredictable booms and busts. Overall, the economic future of the Pacific region remains uncertain and highly variable because of its small domestic markets, peripheral position in the global economy, and diminishing resource base (Table A14.2).

Australian and New Zealand Economies

Much of Australia's economic wealth has been built historically on the cheap extraction and export of abundant raw materials. Export-oriented agriculture, for example, has long been one of the key supports of Australia's economy. Australian agriculture is highly productive in terms of labor input and produces varied temperate and tropical crops as well as huge quantities of beef and wool for world markets. Although farm exports are still important to the economy, the mining sector has grown more rapidly since 1970, making Australia one of the world's mining superpowers.

Mining's recent growth is primarily due to increased trade with China, which has made Australia the world's largest exporter of iron and coal. Besides these two resources, Australia produces an assortment of other materials—namely, bauxite (aluminum ore), copper, gold, nickel, lead, and zinc. As a result, the New South Wales–based Broken Hill Proprietary Company (BHP) is one of the world's largest mining corporations. China's appetite for natural resources is huge—equaled, apparently, only by Australia's willingness to sell its resources. In return, Chinese consumer goods are eagerly purchased by Australian shoppers (Figure 14.42). Half a million Chinese tourists also visit Australia annually and add to the local economy by keeping busy Australian lifeguards, blackjack dealers, and real estate brokers. In addition, 88,000 Chinese students currently attend Australian universities.

With growing numbers of Asian immigrants and expanding economic links with Asian markets, Australia's economic future is promising. In addition, an expanding tourism industry is helping to diversify the economy. More than 7 percent of the nation's workforce is now devoted to serving the needs of more than 6.6 million tourists each year. Popular destinations include Melbourne and Sydney as well as Queensland's resort-filled Gold Coast, the Great Barrier Reef, and

Figure 14.42 Australian Trade with China Containers from Asia testify to the recent explosion of two-way trade between Australia and China. Raw materials, mainly iron ore, are exported to China, and consumer goods flow from China into Australia, making that country a key beneficiary of China's recent economic growth.

the vast, arid Outback. Along the Gold Coast, many luxury hotels are owned by Japanese firms and provide a bilingual resort experience for their Asian clientele (Figure 14.43).

New Zealand is also a wealthy country, although somewhat less well-off than Australia. Before the 1970s, New Zealand relied heavily on exports to Great Britain—primarily agricultural products such as wool and butter. These colonial trade linkages became problematic in 1973 when Britain joined the European Union, with its strict agricultural protection policies. Unlike Australia, New Zealand lacked a rich base of mineral resources to export to global markets. As a result, the country saw its economy stagnate in the 1980s. Eventually, the New Zealand government enacted drastic neoliberal reforms. Tax rates fell, and many state-run industries were privatized. As a result, New Zealand has been transformed into one of the most market-oriented countries in the world.

Oceania's Divergent Development Paths

Varied economic activities characterize the Pacific island nations. One way of life is oriented around subsistence-based economies, such as shifting cultivation or fishing. In other places, a commercial extractive economy dominates, and large-scale plantations, mines, and timber activities compete with the traditional subsistence sector for both land and labor (Figure 14.44). Elsewhere, the huge growth in global tourism has transformed the economic geographies of many island settings, forever changing how people make a living. In addition, many island nations benefit from direct subsidies and economic aid from present and former colonial powers, assistance that is designed to promote development and stimulate employment.

Melanesia is the least developed and poorest part of Oceania because these countries have benefited the least from tourism and from subsidies from wealthy colonial and ex-colonial powers. Today many Melanesians still live in remote villages isolated from the modern economy. The Solomon Islands, for example, with few industries other than fish canning and coconut processing, has a per capita gross national income (GNI) below $3000 per year. In contrast, Fiji is a more affluent Melanesian country, largely because of its tourist economy, reflecting its popularity with Chinese and Japanese visitors. Recent investments in Papua New Guinea's energy sector may speed economic growth in that country. Beginning in 2014, a huge multibillion-dollar project to produce liquefied natural gas, mostly for consumption in East Asia, is raising hopes for more economic development there. One key challenge for the government will be translating that increased energy-related wealth into an improved standard of living for the nation's 7.5 million residents.

Throughout Micronesia and Polynesia, economic conditions depend on both local subsistence economies and economic linkages to the wider world. Many islands export food products, but native populations survive mainly on fish, coconuts, bananas, and yams. Some island groups enjoy large subsidies from either France or the United States, even though such support often comes with a political price. China is also increasingly involved with economic development plans that often imply political support.

Other Polynesian island groups have been completely transformed by tourism. In Hawaii, more than one-third of the state's income flows directly from tourist dollars. With more than 8 million visitors annually (including more than 1.2 million from Japan), Hawaii represents all the classic benefits and risks of the tourist economy. Job creation and economic growth have reshaped the islands, but congested highways, high prices, and the unpredictable spending habits of tourists have put the region at risk for future problems (see *Geographers at Work: Planning for the Future Across the Pacific Basin*). Elsewhere, French Polynesia has long been a favored destination of the international jet set. More than 20 percent of French Polynesia's GNI is derived from tourism, making it one of the wealthiest areas of the Pacific. More recently, Guam has seen a resurgence of Japanese and Korean tourists, especially those on honeymoons.

The South Pacific Tuna Fishery The South Pacific is home to the world's largest tuna fishery, and this resource contributes significantly to local island economies, both directly and indirectly. Tuna fishing and processing account for about 10 percent of all wage employment in southern Oceania, although 90 percent of the working tuna boats come from far-flung Pacific nations—namely China, Japan, South Korea, and the United States. These foreign boats are charged access fees for fishing in each island's offshore territory (Figure 14.45).

International law allows the extension of sovereign rights 200 miles (320 km) beyond a country's coastline in an **Exclusive Economic Zone (EEZ)**. Within the EEZ, that country has economic

Figure 14.43 Queensland's Gold Coast Many of these luxury hotels in the Surfer's Paradise section of the Gold Coast are owned by Japanese firms specializing in accommodations for Asian tourists. **Q: What are the environmental implications of this Gold Coast development?**

GEOGRAPHERS AT WORK

Planning for the Future Across the Pacific Basin

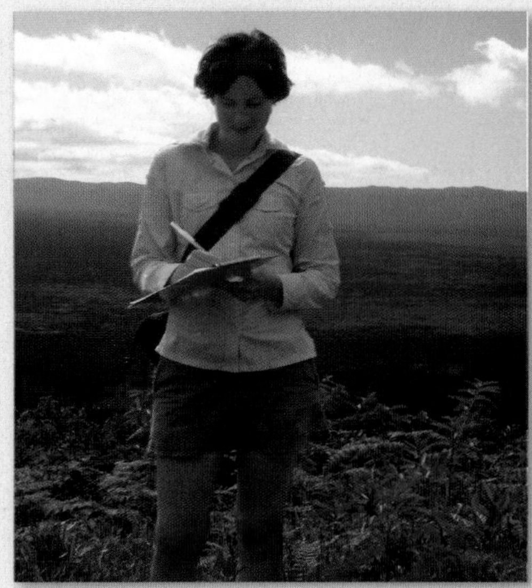

Figure 14.3.1 **Laura Brewington, East-West Center, Hawaii** Trained in conservation work, Brewington has turned her fieldwork skills and geographic tools toward meeting the challenges of climate change in island environments.

"I don't know what it is about islands, but my whole life path has taken me from one to another," says **Laura Brewington**, a Research Fellow at the University of Hawaii's East–West Center (Figure 14.3.1). Brewington has worked in New Zealand, Thailand, Iceland, and the Galapagos Islands, and currently focuses on developing environmental policy and land-use guidelines that integrate traditional lifeways, economic development, and unfolding climate change scenarios across fragile Pacific islands.

GIS and Climate Change Brewington's earlier work was set in the eco-sensitive Galapagos Islands and included mapping environmental vulnerabilities and developing ways to contain invasive species across the region as it becomes increasingly exposed to the effects of globalization. More recently, Brewington's attention shifted to climate change adaptations in the state of Hawaii and U.S.-affiliated Pacific Islands. She partners with climate change scientists, the U.S. Geological Survey, water conservation groups, farmers, developers, and local residents to develop a range of land- and water-use possibilities (all produced in a series of GIS layers) to inform policy decisions. Brewington says, "The Pacific region is huge and people very disparate, so our goal is to get those voices heard, but also to get the right climate science information into the right decision-makers' hands."

Brewington, a biostatistician in the public health sector, discovered geography when her work required her to learn spatial statistics and GIS. She is passionate spokesperson for the discipline: "I love getting students engaged in hands-on experiences. I tell them to choose a career path in line with what they find exciting. I didn't do that until much later, and I wish I'd had that piece of advice."

1. Explain how land use and resource mapping can help a government or nongovernment agencies make policy decisions.
2. How might potential climate change scenarios shape local planning initiatives in your own community?

control over resources such as fisheries as well as retaining ocean-floor mineral rights. In Oceania, because island countries are often composed of a series of islands and because each country's EEZ is delimited from its outermost points, the South Pacific tuna fishery is composed of a patchwork of intersecting EEZs, each with different fishing regulations and fees. Complicating matters, as tuna become more scarce (and more valuable as a resource), many island nations have closed their EEZs to outsiders, reserving the tuna catch for local boats and processing operations. Whether this fragmentation will aid or injure the sustainability of the tuna fishery is not clear.

Mining in the South Pacific Mining dominates the economy of New Caledonia. Its nickel reserves, the world's second largest, are both a blessing and a curse. Although they currently sustain much of the island's export economy, income from nickel mining will decrease in the near future as reserves dwindle. Dramatic price fluctuations for this commodity also hamper economic planning for the French colony.

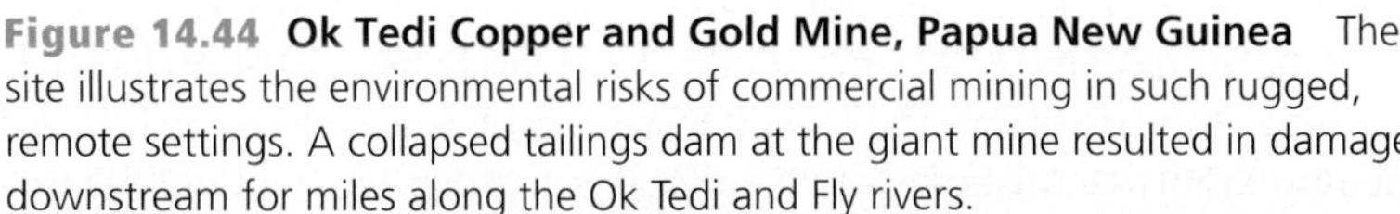

Figure 14.44 **Ok Tedi Copper and Gold Mine, Papua New Guinea** The site illustrates the environmental risks of commercial mining in such rugged, remote settings. A collapsed tailings dam at the giant mine resulted in damage downstream for miles along the Ok Tedi and Fly rivers.

Figure 14.45 **Trouble in the Tuna Industry** These Samoan fishermen returned to traditional subsistence fishing in their outriggers after the economic collapse of the Samoan tuna fishing and canning industry. The collapse was due to competition from countries in East Asia that pay lower wages.

EVERYDAY GLOBALIZATION

Wine from Down Under Gains Global Appeal

Americans have enjoyed Australian wines since the 1990s, but the region's appeal is broadening: Australian wine bottles are found on dinner tables from China to India, and Americans are embracing additional offerings from New Zealand.

Thirsty Chinese consumers with a taste for European-style wines recently spurred an 8 percent increase in annual wine imports from Australia, now the world's fourth-largest wine exporter (Figure 14.4.1). A new bilateral trade agreement will end Chinese import tariffs on Australian wine by 2018, likely prompting more consumption. Australia is also the second-largest wine supplier to nearby India.

At the same time, U.S. wine imports from New Zealand have leaped since 2008, growing 13 percent in 2014. More than 80 percent of New Zealand's wine exports are its prize-winning sauvignon blanc varieties, but its pinot noir also appeals to Americans. The shift toward Kiwi wine is enhanced by the fact that savvy American growers, familiar with the challenges of importing alcoholic beverages into the United States (rules vary from state to state) and with marketing to North Americans, have bought up several major New Zealand vintners.

1. In addition to wine, what food products or other consumer goods from Australia or New Zealand do you most commonly see for sale in your local community? Explain why.
2. Where do the beverages you consume come from? Keep a simple beverage journal for five days and summarize the patterns you find.

Figure 14.4.1 Barossa Valley Wine Country, South Australia These vineyards in South Australia's beautiful Barossa Valley now produce wine for global export.

To the north, the tiny island of Nauru has already seen its mining economy wither as phosphate deposits have dwindled, leaving its people with almost no job prospects. In Papua New Guinea, however, gold and copper mining has dramatically transformed the landscape and remains an important part of the nation's economy. Large open-pit mine operations have largely been financed by foreign corporations based in Australia, the United States, and elsewhere. Major projects are active in many districts on the main island as well as on the nearby island of Bougainville.

Deep-sea mining may be the region's next economic frontier, although environmentalists question its long-term effects. Many resources are a geological product of the Pacific Ring of Fire: Underwater volcanism and mineralization have produced vast deposits of commercially valuable mining resources. Since 2014, for example, several New Zealand mining interests have applied for permission to dredge deep-sea minerals such as iron ore and phosphate nodules from the sea floor within that country's EEZ. Groups such as Kiwis Against Seabed Mining strongly oppose the applications, which still await approval. Off the Cook Islands, cobalt and manganese nodules the size of lettuce heads have also whetted the appetites of mining companies and government officials, although some in the area worry that mining could damage the region's pristine waters and large local tourist economy. However, Fiji, the Solomon Islands, Tonga, and Vanuatu have already moved forward with issuing exploration licenses. Similarly, by 2016, a Canadian company hopes to initiate the world's first operational deep-sea mining project in the Bismarck Sea, off the coast of New Guinea, at the site of volcanic hot springs rich in gold and copper deposits. Time will tell whether there is a steep environmental price to be paid for these submarine riches.

Oceania in Global Context

Many international trade flows link the region to the far reaches of the Pacific and beyond. Australia and New Zealand dominate global trade patterns in the region (see *Everyday Globalization: Wine from Down Under Gains Global Appeal*). In the past 30 years, ties to the United Kingdom, the British Commonwealth, and Europe have weakened in comparison with growing trade links to Japan, East Asia, the Middle East, and the United States. Australia, for example, now imports more manufactured goods from China, Japan, and the United States than it does from Britain and Europe. Both Australia and New Zealand also participate in the **Asia–Pacific Economic Cooperation Group (APEC)**, an organization designed to encourage economic development in Southeast Asia and the Pacific Basin.

To promote more regional economic integration, Australia and New Zealand signed the **Closer Economic Relations (CER) Agreement** in 1982, slashing trade barriers between the two countries. New Zealand now benefits from larger Australian markets for its exports, and Australian corporate and financial interests have gained access to New Zealand business opportunities. Since the CER Agreement's signing, trade between the two countries has steadily grown. More than 20 percent of New Zealand's imports and exports come from Australia, and this pattern of regional free trade is likely to strengthen in the future.

Smaller nations of Oceania, while often closely tied to countries such as China, Taiwan, Japan, the United States, and France, also benefit from their proximity to Australia and New Zealand. More than half of Fiji's imports come from those two nearby nations, and countries such as Papua New Guinea, Vanuatu, and the Solomon Islands enjoy a similarly close trading relationship with their more developed Pacific neighbors.

Continuing Social Challenges

Australians and New Zealanders enjoy high levels of social welfare but face some of the same challenges evident elsewhere in the developed world (see Table A14.2). Life spans average about 80 years in both countries, and rates of child mortality have fallen greatly since 1960. Like North America and Europe, however, the two countries count cancer and heart disease as the leading causes of death, and alcoholism is a continuing social problem, particularly in Australia. Furthermore, Australia's rate of skin cancer is among the world's highest, the consequence of a largely fair-skinned, outdoors-oriented population from

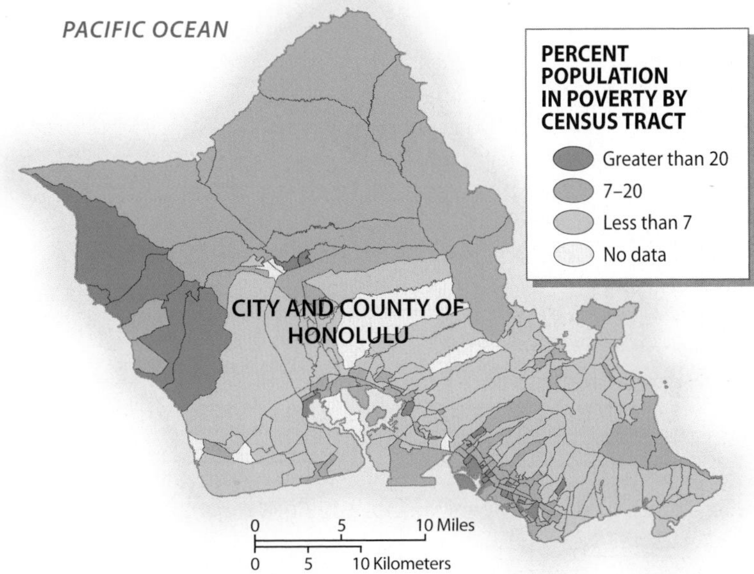

Figure 14.46 Poverty in Hawaii The distribution of poverty on the Hawaiian island of Oahu shows that the highest levels of poverty are in the urban pockets around Honolulu in the southeast as well as in the western census tracts, which have the highest numbers of native Hawaiians.

northwest Europe living in a sunny, low-latitude setting. Overall, high-quality medical care is offered by Australia's Medicare program (initiated in 1984) and by New Zealand's system of social services.

Not surprisingly, the social conditions of the Aborigines and Maori are much less favorable than those of the population overall. Schooling is irregular for many native peoples, and levels of postsecondary education for Aborigines (12 percent) and Maori (14 percent) remain far below national averages (32–34 percent). Other social measures reflect the pattern. Less than one-third of Aboriginal households own their own homes, while more than 70 percent of white Australians are homeowners. Discrimination against native peoples continues in both countries, a situation aggravated and publicized with the recent assertion of indigenous political rights and land claims. As with African-American, Hispanic, and Native American populations in North America, simple social policies have not yet solved these lasting problems.

Even in Hawaii, the proportion of native Hawaiians living below the poverty level is much higher than those of other ethnic groups (Figure 14.46). This group also has the shortest life expectancy and the highest infant mortality rate; rates of death from cancer and heart disease are almost 50 percent higher than for other groups in the United States; and native Hawaiian women have the highest rate of breast cancer in the world. Further, 55 percent of native Hawaiians do not complete high school, and only 7 percent have college degrees.

In other parts of Oceania, levels of social welfare are higher than might be expected based on the islands' economic situations. Sizable investments in health and education services have achieved considerable success. For example, the average life expectancy in the Solomon Islands, one of the world's poorer countries as measured by per capita GNI, is a respectable 67 years. By other social measures as well, the Solomon Islands and several other Oceania states have reached higher levels of human well-being than exist in most Asian and African countries with similar levels of economic output. This is partly a result of successful policies, but it also reflects the relatively healthy natural environment of Oceania; many tropical diseases so troublesome in Africa simply do not exist in the Pacific region.

Gender, Culture, and Politics

Both Australia and New Zealand were early supporters of female suffrage, empowering women to vote in national elections in 1893 (New Zealand) and 1902 (Australia). In some cases, women were allowed to vote even earlier in local and state elections, although not until decades later could females actually hold office. Since that time, both countries have elected women as national leaders, with New Zealand's Helen Clark serving as prime minister between 1999 and 2008 and Australia's Julia Gillard serving in a similar office between 2010 and 2013. Yet both countries still have considerable gender gaps in terms of female representation in government, employment salaries, and social support services. As a result, in the global gender index New Zealand ranks 7th and Australia is 24th, both countries scoring lower than selected western European countries.

Indigenous gender roles were transformed by colonialism in New Zealand. Before European colonization, Maori women and men were equal in social status and power. This was because of the overarching Maori principle of equity and balance in their nonhierarchical society. For example, the Maori language had no gender distinctions in personal pronouns (no "his" or "hers"). Further evidence for this gender equity comes from the prominent role women played in Maori proverbs and legends.

However, English colonial society was troubled by this gender-neutral native culture and sought to deprive Maori women of their status. In 1909, New Zealand law required Maori women to undergo legal marriage ceremonies that emphasized male ownership of all property—not just of land and livestock, but also of the wife. Missionary schools for Maori women had also long reinforced English notions of female domesticity where women were subject to male authority. As a result, gender roles in contemporary Maori society now reflect these colonial notions of male dominance. To further their European values, missionaries rewrote Maori proverbs and legends to emphasize male characters with heroic warrior attributes. Today Maori women never participate in the haka war dances; instead, female participation is limited to subservient songs and dances.

Across Australia, traditional Aboriginal society has distinct gender roles, with clear-cut distinctions between "women's business" and "men's business." These distinctions are played out not just in daily affairs, but also in the Dreamtime, an abstract parallel universe of central importance for Aboriginal people. In physical life as well as in Dreamtime, women are responsible for the vitality and resilience of family lives, while men's business centers on the larger group or tribe. These distinct gender roles also involve the landscape, with certain areas and locales linked closely to either men or women, but rarely to both. As a result, the Aboriginal territory is also highly gendered.

Review

14.13 Explain what an Exclusive Economic Zone (EEZ) is and why it is important to Pacific island economies.

14.14 What is the Closer Economic Relations (CER) Agreement, and why is it important in understanding the economics of Oceania?

14.15 Describe the major challenges to social development in Oceania.

KEY TERMS Exclusive Economic Zone (EEZ), Asia–Pacific Economic Cooperation Group (APEC), Closer Economic Relations (CER) Agreement

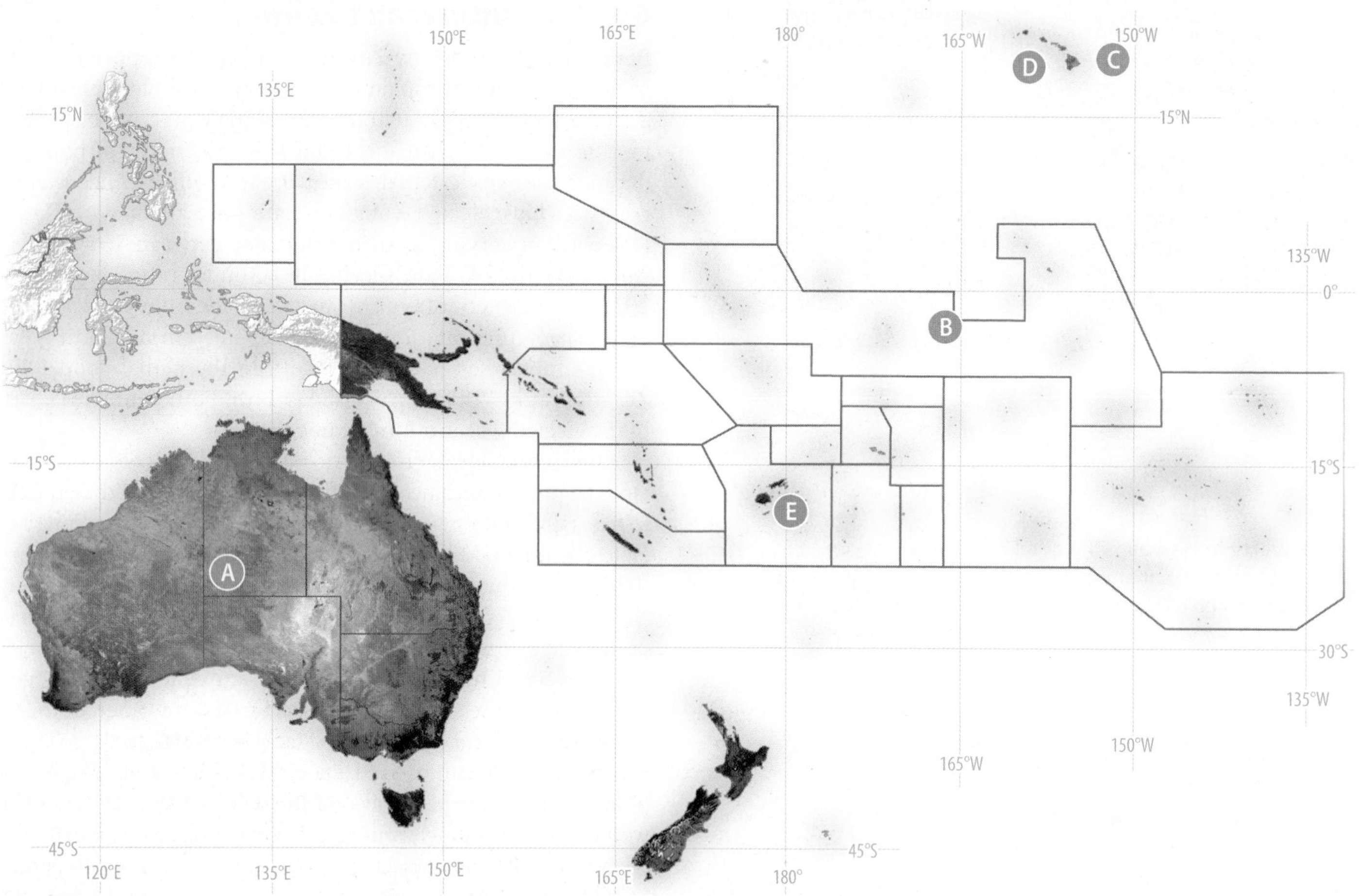

Review

Physical Geography and Environmental Issues

14.1 Describe the physical geographic characteristics of the region known as Oceania.

14.2 Identify major environmental issues in Australia and Oceania as well as pathways toward solving those problems.

The natural environment has undergone accelerated change in the past 50 years. Urbanization, tourism, extractive economic activity, exotic species, and, most recently, global climate change have altered the landscapes of Oceania.

1. Imagine a line from Darwin in the north through Alice Springs to Sydney in the southeast; then, using the climographs for these three places, describe in a narrative essay the different climates found along this transect in terms of rainfall, temperatures, and seasonality.

2. Given the different climates you've discussed, describe the different agricultural activities you might expect to find along this north–south transect.

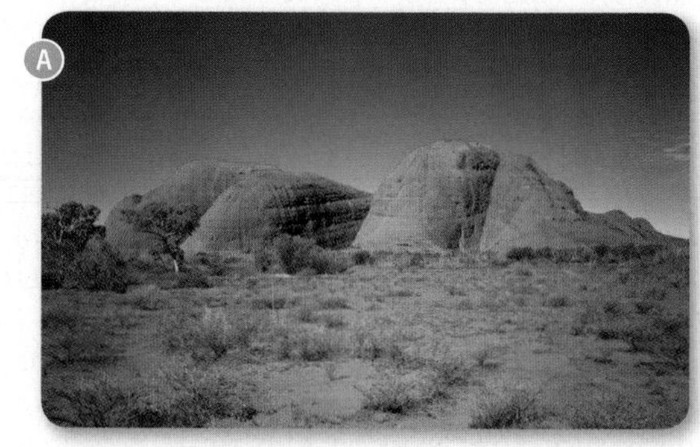

Population and Settlement

14.3 Use a map to identify and describe major migration flows to (and within) the region.

Migration is a major theme in Oceania, beginning with the earliest Aboriginal peoples who came from the Asian mainland, to prehistoric island-hopping Polynesian people, to more recent European emigrants who populated Australia and New Zealand.

3. Make a list of what Polynesian sailors would have to take along on their voyages to set up new settlements on distant islands.

4. Today, because of sea-level rise due to climate change, many low-island people see an unsustainable future and are discussing migrating elsewhere. What can you identify as some of the key challenges these migrants might face as they relocate to an unfamiliar setting elsewhere in Oceania, in Australia, or beyond?

Cultural Coherence and Diversity

14.4 Describe historical and modern interactions between native peoples and Anglo-European migrants in Australia and Oceania and their impacts on the region's cultures.

In Australia and New Zealand, the Aborigines and Maori work to preserve their minority cultures in the face of dominating Anglo influences. In different ways, the same is true of Oceania's peoples as tourists and foreign workers become increasingly common on once-remote islands.

5. Native Hawaiians are roasting a pig in the traditional manner. What other aspects of Polynesian culture have become tourist attractions in Hawaii? What about in New Zealand?

6. In a group of three, have one person each represent the issues facing Australian Aborigines, New Zealand Maori, and native Hawaiians in their different countries. Then, as a group, compare the similarities and differences. Predict where each native group might be in 10 years.

C

Geopolitical Framework

14.5 Identify and describe different pathways to independence taken by countries in Oceania.

14.6 List several geopolitical tensions that persist in Australia and Oceania.

Oceania has become entangled in the world's larger geopolitical tensions as China and the United States jockey for economic and political influence in the region. Some small island nations feel pressure to align themselves with one superpower or the other, but economically powerful Australia has forged a different solution, maintaining continued political allegiance to the United States while continuing to develop economic ties to its major trading partner, China.

7. This Google Earth image of Pearl Harbor, Hawaii, includes the huge U.S. naval base that is the headquarters for the Pacific fleet. Drawing on what you've learned about tropical island environments in this chapter, make a list of the aspects of this coastal environment that have been changed since 1900.

8. Working with several other students, each of whom will represent a specific country in Oceania (including Australia), discuss the geopolitical and economic ties and tensions between that country and China. How might these change in the next five years?

D

Economic and Social Development

14.7 Describe the diverse economic geographies of Oceania.

14.8 Explain the positive and negative interactions of Australia and Oceania with the global economy.

Australia is the economic powerhouse of the region because of its rich mineral resources, with coal and iron dominating the list of trade commodities. In contrast, island nations, from Fiji to Hawaii, are largely at the mercy of global tourism, with boom-and-bust cycles linked to the vitality of the global economy.

9. Make a list of the economic and social advantages and liabilities for a traditional island community or a large tourist resort, such as the one pictured, which is in Fiji.

10. You are part of a team responsible for creating a comprehensive economic and social development plan for a small South Pacific island. Choose a specific island, learn its needs and resources, define your goals, and describe a plan for achieving them within five years.

E

KEY TERMS

Aborigine *(p. 463)*
Asia–Pacific Economic Cooperation Group (APEC) *(p. 480)*
atoll *(p. 457)*
Austronesian *(p. 470)*
Closer Economic Relations (CER) Agreement *(p. 480)*
Exclusive Economic Zone (EEZ) *(p. 478)*
exotic species *(p. 458)*
haole *(p. 472)*
high island *(p. 456)*
intercropping *(p. 468)*
kanaka *(p. 469)*
low island *(p. 457)*
mallee *(p. 457)*
Maori *(p. 469)*
marsupial *(p. 458)*
Melanesia *(p. 455)*
Micronesia *(p. 455)*
microstate *(p. 474)*
Native Title Bill *(p. 475)*
Oceania *(p. 454)*
Outback *(p. 456)*
Pidgin English *(p. 472)*
Polynesia *(p. 455)*
viticulture *(p. 467)*
White Australia Policy *(p. 469)*

DATA ANALYSIS

http://goo.gl/9ZC29D

At the broad scale of our regional climate maps, local detail for New Zealand's South Island (see Figure 14.8) is difficult to see. Yet a more detailed look at the island's diverse microclimates reveals an amazing and complex picture. Visit the website for Te Ara: The Encyclopedia of New Zealand (http://www.teara.govt.nz) and examine the map of annual precipitation for the South Island.

1. Briefly describe and explain today's pattern of precipitation across the South Island.
2. Given your knowledge of New Zealand geography, write a paragraph about how this precipitation pattern may have shaped historical and contemporary geographies of settlement, agriculture, and economic development across the island.
3. Visit NIWA's website (https://www.niwa.co.nz) and access a map of projected changes for 2080–2099, based on global climate change models and predictions. See Figure 5 in the document. Briefly summarize what you see, and suggest possible social and economic implications of these changes for future residents and government planners.

MasteringGeography™

Looking for additional review and test prep materials? Visit the Study Area in MasteringGeography™ to enhance your geographic literacy, spatial reasoning skills, and understanding of this chapter's content by accessing a variety of resources, including MapMaster interactive maps, geoscience animations, videos, *In the News* RSS feeds, flashcards, web links, self-study quizzes, and an eText version of *Globalization and Diversity.*

Authors' Blogs

Scan to visit the **Author's Blog for field notes, media resources, and chapter updates**

www.gad4blog.wordpress.com/category/australia-and-oceania/

Scan to visit the **GeoCurrents Blog**

www.geocurrents.info/category/place/australia-and-pacific

Appendix: Population & Development Indicator Tables

Table A3.1: NORTH AMERICA POPULATION INDICATORS

Country	Population (millions) 2013	Population Density (per square kilometer)[1]	Rate of Natural Increase (RNI)	Total Fertility Rate	Percent Urban	Percent < 15	Percent > 65	Net Migration (Rate per 1000)
Canada	35.8	4	0.4	1.6	80	16	16	6
United States	321.2	35	0.5	1.9	81	19	15	3

Source: Population Reference Bureau, *World Population Data Sheet, 2015.*

[1]World Bank, *World Development Indicators,* 2015.

Table A3.2: NORTH AMERICA DEVELOPMENT INDICATORS

Country	GNI per capita, PPP 2013	GDP Average Annual %Growth 2009–13	Human Development Index (2013)[1]	Percent Population Living Below $2 a Day	Life Expectancy (2015)[2]	Under Age 5 Mortality Rate (1990)	Under Age 5 Mortality Rate (2013)	Youth Literacy (% pop ages 15–24)	Gender Inequality Index (2013)[3,1]
Canada	42,120	2.3	.902	–	81	8	5	–	0.136
United States	53,750	2.1	.914	–	79	11	7	–	0.262

Source: World Bank, *World Development Indicators*, 2015.

[1]United Nations, *Human Development Report, 2014.*

[2]Population Reference Bureau, *World Population Data Sheet, 2015.*

[3]Gender Inequality Index—A composite measure reflecting inequality in achievements between women and men in three dimensions: reproductive health, empowerment and the labor market that ranges between 0 and 1. The higher the number, the greater the inequality.

Table A4.1: LATIN AMERICA POPULATION INDICATORS

Country	Population (millions) 2015	Population Density (per square kilometer)[1]	Rate of Natural Increase (RNI)	Total Fertility Rate	Percent Urban	Percent < 15	Percent > 65	Net Migration (Rate per 1000)
Argentina	42.4	15	1.0	2.2	93	24	11	0
Bolivia	10.5	10	1.9	3.2	69	31	6	–1
Brazil	204.5	24	0.9	1.8	86	24	7	0
Chile	18.0	24	0.8	1.8	90	21	10	2
Colombia	48.2	44	1.3	1.9	76	27	7	–1
Costa Rica	4.8	95	1.1	1.9	73	23	7	2
Ecuador	16.3	63	1.6	2.6	70	31	7	0
El Salvador	6.4	306	1.3	2.0	67	31	7	–8
Guatemala	16.2	144	2.0	3.1	52	40	5	–1
Honduras	8.3	72	1.9	2.7	54	34	5	–2
Mexico	127.0	63	1.4	2.3	79	28	7	–2
Nicaragua	6.3	51	1.8	2.4	59	32	5	–4
Panama	4.0	52	1.4	2.7	78	28	8	2
Paraguay	7.0	17	1.7	2.8	64	33	5	–1
Peru	31.2	24	1.5	2.5	79	29	6	–1
Uruguay	3.6	19	0.4	1.9	93	21	14	–1
Venezuela	30.6	34	1.5	2.5	94	28	6	0

Source: Population Reference Bureau, *World Population Data Sheet, 2015.*

[1]World Bank, *World Development Indicators,* 2015.

Table A4.2: LATIN AMERICA DEVELOPMENT INDICATORS

Country	GNI per capita, PPP 2013	GDP Average Annual %Growth 2009–13	Human Development Index (2013)[1]	Percent Population Living Below $2 a Day	Life Expectancy (2015)[2]	Under Age 5 Mortality Rate (1990)	Under Age 5 Mortality Rate (2013)	Youth Literacy (% pop ages 15–24)	Gender Inequality Index (2013)[3,1]
Argentina	–	5.2	.808	2.9	77	28	13	99	0.381
Bolivia	5,750	5.3	.667	12.7	67	120	39	99	0.472
Brazil	14,750	3.1	.744	6.8	75	11	14	99	0.441
Chile	21,060	5.3	.822	< 2	79	19	8	99	0.355
Colombia	11,960	4.9	.711	12.0	75	34	17	98	0.460
Costa Rica	13,570	4.6	.763	3.1	79	17	10	99	0.344
Ecuador	10,720	5.5	.711	8.4	75	52	23	99	0.429
El Salvador	7,490	1.8	.662	8.8	73	60	16	97	0.441
Guatemala	7,130	3.5	.628	29.8	73	78	31	94	0.523
Honduras	4,270	3.6	.617	29.2	74	55	22	95	0.482
Mexico	16,020	3.6	.756	7.5	75	49	15	99	0.376
Nicaragua	4,510	4.8	.614	20.8	75	66	24	87	0.458
Panama	19,300	9.1	.765	8.9	78	33	18	98	0.506
Paraguay	7,670	6.2	.676	7.7	72	53	22	99	0.457
Peru	11,160	6.6	.737	8.0	75	75	17	99	0.387
Uruguay	18,940	5.8	.790	< 2	77	23	11	99	0.364
Venezuela	17,900	2.9	.764	12.9	75	31	15	99	0.464

Source: World Bank, *World Development Indicators, 2015.*

[1]United Nations, *Human Development Report, 2014.*

[2]Population Reference Bureau, *World Population Data Sheet, 2015.*

[3]Gender Inequality Index—A composite measure reflecting inequality in achievements between women and men in three dimensions: reproductive health, empowerment and the labor market that ranges between 0 and 1. The higher the number, the greater the inequality.

Table A5.1: THE CARIBBEAN POPULATION INDICATORS

Country	Population (millions) 2015	Population Density (per square kilometer)[1]	Rate of Natural Increase (RNI)	Total Fertility Rate	Percent Urban	Percent < 15	Percent > 65	Net Migration (Rate per 1000)
Anguilla*	0.02	173	–	1.8	100	24	8	13
Antigua and Barbuda	0.09	205	0.8	1.5	30	24	8	0
Bahamas	0.4	38	0.9	1.9	85	26	7	1
Barbados	0.3	662	0.3	1.7	46	20	13	2
Belize	0.4	15	1.7	2.4	44	36	4	4
Bermuda*	0.07	1,301	–	2.0	100	18	16	2
Cayman*	0.05	244	–	1.9	100	19	11	15
Cuba	11.1	106	0.3	1.7	75	17	13	–2
Dominica	0.07	96	0.5	2.1	68	22	10	–5
Dominican Republic	10.5	215	1.5	2.5	72	31	6	–3
French Guiana	0.3	–	2.3	3.5	77	34	5	5
Grenada	0.1	311	0.9	2.1	41	26	7	–2
Guadeloupe	0.4	–	0.6	2.2	98	21	14	–2
Guyana	0.7	4	1.4	2.6	29	27	6	–7
Haiti	10.9	374	1.9	3.2	59	35	4	–3
Jamaica	2.7	251	1.1	2.3	52	24	9	–5
Martinique	0.4	–	0.3	1.9	89	19	17	–10
Montserrat*	0.005	51	–	1.3	14	26	6	0
Puerto Rico	3.5	408	0.2	1.5	99	18	17	–15
St. Kitts and Nevis	0.05	208	0.6	1.8	32	21	8	1
St. Lucia	0.2	299	0.6	1.5	15	22	9	0
St Vincent and the Grenadines	0.1	280	0.9	2.0	49	26	7	–9
Suriname	0.6	3	1.1	2.3	71	28	6	–2
Trinidad and Tobago	1.4	261	0.6	1.7	15	21	9	–1
Turks and Caicos*	0.05	50	–	1.7	93	22	4	15

Source: Population Reference Bureau, *World Population Data Sheet, 2015*

**Additional data from the CIA Factbook*

[1]World Bank, *World Development Indicators,* 2015.

Table A5.2: THE CARIBBEAN DEVELOPMENT INDICATORS

Country	GNI per capita, PPP 2013	GDP Average Annual %Growth 2009–13	Human Development Index (2013)[1]	Percent Population Living Below $2 a Day	Life Expectancy (2015)[2]	Under Age 5 Mortality Rate (1990)	Under Age 5 Mortality Rate (2013)	Youth Literacy (% pop ages 15–24)	Gender Inequality Index (2013)[3,1]
Anguilla	12,200*	–		–	81*	–	–	–	–
Antigua and Barbuda	20,490	–0.9	.774	–	77	27	9	–	–
Bahamas	22,700	1.1	.789	–	74	22	13	–	0.316
Barbados	15,090	0.4	.776	–	75	18	14	–	0.350
Belize	7,870	2.7	.732	22.0	74	44	17	–	0.435
Bermuda	66,430	–3.4	–	–	81*	–	–	–	–
Cayman	43,800*	–	–	–	81*	–	–	99	–
Cuba	18,520	2.5	.815	–	78	13	6	100	0.350
Dominica	10,060	–0.4	.717	–	75	17	11	–	–
Dominican Republic	11,630	4.2	.700	8.8	73	58	28	97	0.505
French Guiana	–	–	–	–	80	–	–	–	–
Grenada	11,230	0.3	.744	–	76	21	12	–	–
Guadeloupe	–	–	–	–	81	–	–	–	–
Guyana	6,610	5.0	.638	18.0	66	63	37	93	0.524
Haiti	1,720	2.2	.471	77.5	64	143	73	72	0.599
Jamaica	8,490	–	.715	5.9	74	35	17	96	0.457
Martinique	–	–	–	–	82	–	–	–	–
Montserrat	8,500*	–	–	–	74*	–	–	–	–
Puerto Rico	23,840	–2.0	–	–	79	–	–	99	–
St. Kitts and Nevis	20,990	0.3	.750	–	75	28	10	–	–
St. Lucia	10,290	–0.4	.714	40.6	79	23	15	–	–
St Vincent and the Grenadines	10,440	–0.2	.719	–	71	27	19	–	–
Suriname	15,960	4.1	.705	27.2	71	52	23	98	0.463
Trinidad and Tobago	26,220	0.3	.766	13.5	75	37	21	100	0.321
Turks and Caicos	29,100*	–	–	–	80*	–	–	–	–

Source: World Bank, *World Development Indicators, 2015.*

[1]United Nations, *Human Development Report, 2014.*

[2]Population Reference Bureau, *World Population Data Sheet, 2015.*

[3]Gender Inequality Index—A composite measure reflecting inequality in achievements between women and men in three dimensions: reproductive health, empowerment and the labor market that ranges between 0 and 1. The higher the number, the greater the inequality.

*Additional data from the *CIA World Fackbook*, 2015

Table A6.1: SUB-SAHARAN AFRICA POPULATION INDICATORS

Country	Population (millions) 2015	Population Density (per square kilometer)[1]	Rate of Natural Increase (RNI)	Total Fertility Rate	Percent Urban	Percent < 15	Percent > 65	Net Migration (Rate per 1000)
WESTERN AFRICA								
Benin	10.6	92	2.7	4.9	45	45	3	0
Burkina Faso	18.5	62	3.3	6.0	27	45	2	–1
Cape Verde	0.5	124	1.5	2.4	62	31	6	–2
Gambia	2.0	183	3.2	5.6	57	46	2	–1
Ghana	27.7	114	2.5	4.2	51	39	5	–2
Guinea	11.0	48	2.6	5.1	36	42	3	0
Guinea–Bissau	1.8	61	2.4	4.9	49	43	3	–1
Ivory Coast	23.3	64	2.3	4.9	50	41	3	0
Liberia	4.5	45	2.7	4.7	47	42	3	–1
Mali	16.7	13	2.9	5.9	39	47	3	–4
Mauritania	3.6	4	2.5	4.2	59	40	3	–1
Niger	18.9	14	3.9	7.6	22	52	4	0
Nigeria	181.8	191	2.5	5.5	50	43	3	0
Senegal	14.7	73	2.9	5.0	45	42	4	–1
Sierra Leone	6.5	84	2.3	4.9	41	41	3	–1
Togo	7.2	125	2.7	4.8	38	42	3	0
EASTERN AFRICA								
Burundi	10.7	396	3.3	6.2	10	46	3	0
Comoros	0.8	395	2.4	4.3	28	41	3	–3
Djibouti	0.9	38	1.8	3.4	77	34	4	–3
Eritrea	5.2	63	3.0	4.4	21	43	2	–5
Ethiopia	98.1	94	2.3	4.1	17	41	4	0
Kenya	44.3	78	2.3	3.9	24	41	3	0
Madagascar	23.0	39	2.7	4.4	33	41	3	0
Mauritius	1.3	620	0.3	1.4	41	20	9	–1
Reunion	0.9	–	1.2	2.4	94	24	10	–3
Rwanda	11.3	477	2.3	4.2	28	41	3	–1
Seychelles	0.09	194	0.9	2.4	54	22	8	6
Somalia	11.1	17	3.2	6.6	38	47	3	–7
South Sudan	12.2	152*	2.4	6.9	17	42	3	11
Tanzania	52.3	56	3.0	5.2	30	45	3	–1
Uganda	40.1	188	3.1	5.9	18	48	2	–1
CENTRAL AFRICA								
Cameroon	23.7	47	2.6	4.9	52	43	3	0
Central African Republic	5.6	7	2.9	6.2	39	45	3	0
Chad	13.7	10	3.4	6.5	22	48	2	1
Congo	4.8	13	2.7	4.8	64	41	3	–8
Dem. Rep. of Congo	73.3	30	3.0	6.6	42	46	3	0
Equatorial Guinea	0.8	27	2.4	5.1	39	39	3	5
Gabon	1.8	7	2.3	4.1	86	38	5	1
São Tomé and Principe	0.2	201	2.9	4.3	67	42	4	–6
SOUTHERN AFRICA								
Angola	25.0	17	3.2	6.1	62	47	2	1
Botswana	2.1	4	1.8	2.9	57	33	5	2
Lesotho	1.9	68	1.1	3.3	27	36	5	–5
Malawi	17.2	174	2.6	5.0	16	44	3	0
Mozambique	25.7	33	3.2	5.9	31	45	3	0
Namibia	2.5	3	2.2	3.6	46	35	4	0
South Africa	55.0	44	1.2	2.6	62	30	6	3
Swaziland	1.3	73	1.6	3.3	21	37	4	–1
Zambia	15.5	20	3.0	5.6	40	46	3	0
Zimbabwe	17.4	37	2.4	4.3	33	43	3	–3

Source: Population Reference Bureau, *World Population Data Sheet, 2015.*

[1]World Bank, *World Development Indicators,* 2015.

*Combined data from the *World Population Data Sheet,* 2013 and the *World Development Indicators,* 2015.

Table A6.2: SUB-SAHARAN AFRICA DEVELOPMENT INDICATORS

Country	GNI per capita, PPP 2013	GDP Average Annual %Growth 2009–13	Human Development Index (2013)[1]	Percent Population Living Below $2 a Day	Life Expectancy (2015)[2]	Under Age 5 Mortality Rate (1990)	Under Age 5 Mortality Rate (2013)	Youth Literacy (% ages 15–24)	Gender Inequality Index (2013)[3,1]
WESTERN AFRICA									
Benin	1,780	4.2	.476	74.3	59	177	85	42	0.614
Burkina Faso	1,680	7.7	.388	72.4	56	208	98	39	0.607
Cape Verde	6,210	2.0	.636	34.7	75	58	26	98	–
Gambia	1,610	2.6	.441	55.9	59	165	74	69	0.624
Ghana	3,900	10.2	.573	51.8	61	121	78	86	0.549
Guinea	1,160	3.2	.392	72.7	60	228	101	31	–
Guinea-Bissau	1,410	2.9	.396	78.0	54	210	124	74	–
Ivory Coast	3,090	3.8	.452	< 2	51	151	100	48	0.645
Liberia	790	10.3	.412	94.9	60	241	71	49	0.655
Mali	1,540	2.3	.407	78.8	53	257	123	47	0.673
Mauritania	2,850	5.5	.487	47.7	63	125	90	56	0.644
Niger	890	6.4	.337	76.1	60	314	104	24	0.674
Nigeria	5,360	5.4	.504	82.2	52	214	117	66	–
Senegal	2,210	3.1	.485	60.3	65	136	55	66	0.537
Sierra Leone	1,690	5.5	.374	82.5	50	267	161	63	0.643
Togo	1,180	5.1	.473	72.8	57	147	85	80	0.579
EASTERN AFRICA									
Burundi	770	4.1	.389	93.5	59	183	83	89	0.501
Comoros	1,490	2.8	.488	65.0	61	122	78	86	–
Djibouti	–	4.4	.467	41.2	62	122	70	–	–
Eritrea	1,180	5.4	.381	–	63	138	50	91	–
Ethiopia	1,380	10.5	.435	72.2	64	198	64	55	0.547
Kenya	2,780	6.0	.535	67.2	62	98	71	82	0.548
Madagascar	1,370	1.9	.498	95.1	65	161	56	65	–
Mauritius	17,730	3.6	.771	< 2	74	24	14	98	0.375
Reunion	–	–	–	–	80	–	–	–	–
Rwanda	1,450	7.4	.506	82.3	65	156	52	77	0.410
Seychelles	23,730	5.4	.756	< 2	73	17	14	99	–
Somalia	–	–	–	–	55	180	146	–	–
South Sudan	1,860	–	–	–	55	217	99	–	–
Tanzania	2,430	6.6	.488	73.0	62	158	52	75	0.553
Uganda	1,630	5.9	.484	62.9	59	178	66	87	0.529
CENTRAL AFRICA									
Cameroon	2,770	4.4	.504	53.2	57	145	95	81	0.622
Central African Republic	600	–5.3	.341	80.1	50	169	139	36	0.654
Chad	2,010	6.1	.372	60.5	51	208	148	49	0.707
Congo	4,600	4.6	.564	57.3	58	119	49	81	0.617
Dem. Rep. of Congo	740	7.3	.338	95.2	50	181	119	66	0.669
Equatorial Guinea	23,270	1.2	.556	*	57	190	96	98	–
Gabon	17,230	6.3	.674	20.9	63	94	–	–	0.508
São Tomé and Principe	2,950	4.4	.558	73.1	66	96	51	80	–
SOUTHERN AFRICA									
Angola	7,000	4.8	.526	67.4	52	243	167	73	–
Botswana	15,640	6.0	.683	27.8	64	53	47	96	0.486
Lesotho	3,160	5.3	.486	73.4	44	88	98	83	0.557
Malawi	750	4.2	.414	88.1	61	227	68	72	0.591
Mozambique	1,100	7.3	.393	82.5	54	226	87	67	0.657
Namibia	9,490	5.3	.624	43.2	64	73	50	87	0.450
South Africa	12,530	2.7	.658	26.2	61	62	44	99	0.461
Swaziland	6,060	1.3	.530	59.1	49	83	80	94	0.529
Zambia	3,810	7.3	.561	86.6	53	193	87	64	0.617
Zimbabwe	1,690	9.9	.492	–	61	79	89	91	0.516

Source: World Bank, *World Development Indicators, 2015.*

[1]United Nations, *Human Development Report, 2014.*

[2]Population Reference Bureau, *World Population Data Sheet, 2015.*

[3]Gender Inequality Index—A composite measure reflecting inequality in achievements between women and men in three dimensions: reproductive health, empowerment and the labor market that ranges between 0 and 1. The higher the number, the greater the inequality.

Table A7.1: SOUTHWEST ASIA AND NORTH AFRICA POPULATION INDICATORS

Country	Population (millions) 2015	Population Density (per square kilometer)[1]	Rate of Natural Increase (RNI)	Total Fertility Rate	Percent Urban	Percent < 15	Percent > 65	Net Migration (Rate per 1000)
Algeria	39.3	17	2.0	3.0	73	28	6	–1
Bahrain	1.4	1,753	1.3	2.1	100	21	2	5
Egypt	89.1	82	2.5	3.5	43	31	4	0
Gaza and West Bank	4.5	693	2.8	4.1	83	40	3	–2
Iran	78.5	48	1.4	1.8	71	24	5	–1
Iraq	37.1	77	2.7	4.2	71	41	3	2
Israel	8.4	372	1.6	3.3	91	28	11	1
Jordan	8.1	73	2.2	3.5	83	37	3	3
Kuwait	3.8	189	1.5	2.3	98	23	2	22
Lebanon	6.2	437	1.0	1.7	87	26	6	31
Libya	6.3	4	1.7	2.4	78	29	5	–11
Morocco	34.1	74	1.6	2.5	60	25	6	–2
Oman	4.2	12	1.8	2.9	75	22	3	45
Qatar	2.4	187	1.1	2.0	100	15	1	28
Saudi Arabia	31.6	13	1.6	2.9	81	30	3	5
Sudan	40.9	21	2.9	5.2	33	43	3	–2
Syria	17.1	124	1.6	2.8	54	33	4	–26
Tunisia	11.0	70	1.3	2.1	68	23	8	–1
Turkey	78.2	97	1.2	2.2	77	24	8	3
United Arab Emirates	9.6	112	1.3	1.8	83	16	1	8
Western Sahara	0.6	–	1.4	2.4	82	26	3	9
Yemen	26.7	46	2.6	4.4	34	41	3	1

Source: Population Reference Bureau, *World Population Data Sheet, 2015.*

[1]World Bank, *World Development Indicators,* 2015.

Table A7.2: SOUTHWEST ASIA AND NORTH AFRICA DEVELOPMENT INDICATORS

Country	GNI per capita, PPP 2013	GDP Average Annual %Growth 2009–13	Human Development Index (2013)[1]	Percent Population Living Below $2 a Day	Life Expectancy (2015)[2]	Under Age 5 Mortality Rate (1990)	Under Age 5 Mortality Rate (2013)	Youth Literacy (% pop ages 15–24)	Gender Inequality Index (2013)[3,1]
Algeria	13,070	3.1	.717	22.8	74	66	25	92	0.425
Bahrain	36,290	3.6	.815	–	76	21	6	98	0.253
Egypt	10,790	2.6	.662	15.4	71	86	22	89	0.580
Gaza and West Bank	5,300	6.0	.686	< 2	73	43	22	99	–
Iran	15,610	1.7	.749	8.0	74	61	17	98	0.510
Iraq	14,930	8.1	.642	21.2	69	46	34	82	0.542
Israel	31,780	4.0	.888	–	82	12	4	100	0.101
Jordan	11,660	2.6	.745	< 2	74	37	19	99	0.488
Kuwait	84,800	5.7	.814	–	74	17	10	99	0.288
Lebanon	17,400	3.0	.765	–	77	33	9	99	0.413
Libya	–	–8.6	.784	–	71	44	15	100	0.215
Morocco	7,000	3.9	.617	14.2	74	81	30	82	0.460
Oman	52,780	3.5	.783	–	77	48	11	98	0.348
Qatar	128,530	10.2	.851	–	78	20	8	99	0.524
Saudi Arabia	53,640	6.6	.836	–	74	43	16	99	0.321
Sudan	3,230	–4.6	.473	44.1	62	123	77	88	0.628
Syria	–	–	.658	16.9	70	36	15	96	0.556
Tunisia	10,610	2.4	.721	4.5	76	51	15	97	0.265
Turkey	18,570	5.9	.759	2.6	77	72	19	99	0.360
United Arab Emirates	59,890	4.2	.827	–	77	22	8	95	0.244
Western Sahara	–	–	–	–	68	–	–	–	–
Yemen	3,820	–2.7	.500	37.3	65	126	51	87	0.733

Source: World Bank, *World Development Indicators, 2015.*

[1]United Nations, *Human Development Report, 2014.*

[2]Population Reference Bureau, *World Population Data Sheet, 2015.*

[3]Gender Inequality Index—A composite measure reflecting inequality in achievements between women and men in three dimensions: reproductive health, empowerment and the labor market that ranges between 0 and 1. The higher the number, the greater the inequality.

Table A8.1: EUROPE POPULATION INDICATORS

Country	Population (millions) 2015	Population Density (per square kilometer)[1]	Rate of Natural Increase (RNI)	Total Fertility Rate	Percent Urban	Percent < 15	Percent > 65	Net Migration (Rate per 1000)
WESTERN EUROPE								
Austria	8.6	103	0.1	1.5	67	14	18	6
Belgium	11.2	369	0.1	1.8	99	17	18	5
France	64.3	120	0.4	2.0	78	19	18	0
Germany	81.1	231	–0.3	1.5	73	13	21	5
Ireland	4.6	67	0.9	2.0	60	22	13	–5
Luxembourg	0.6	210	0.4	1.5	90	17	14	19
Netherlands	16.9	498	0.1	1.7	90	17	17	2
Switzerland	8.3	205	0.2	1.5	74	15	18	11
United Kingdom	65.1	265	0.3	1.9	80	18	17	4
SOUTHERN EUROPE								
Albania	2.9	106	0.5	1.8	56	19	12	–6
Bosnia & Herzegovina	3.7	75	–0.2	1.2	40	15	16	0
Croatia	4.2	76	–0.3	1.5	56	15	18	–2
Cyprus	1.2	124	0.6	1.4	67	17	12	–12
Greece	11.5	86	–0.1	1.3	78	15	21	–1
Italy	62.5	205	–0.2	1.4	68	14	22	2
Kosovo	1.8	168	0.9	2.3	38	28	7	–12
Macedonia	2.1	84	0.1	1.5	57	17	13	0
Montenegro	0.6	46	0.2	1.6	64	18	14	–1
Portugal	10.3	114	–0.2	1.2	61	14	19	–3
Serbia	7.1	82	–0.5	1.6	60	14	18	–2
Slovenia	2.1	102	0.1	1.6	50	15	18	0
Spain	46.4	93	0.0	1.3	77	15	18	–2
NORTHERN EUROPE								
Denmark	5.7	132	0.1	1.7	87	17	19	7
Estonia	1.3	31	–0.2	1.5	68	16	19	–1
Finland	5.5	18	0.0	1.7	85	16	20	3
Iceland	0.3	3	0.7	1.9	95	20	14	3
Latvia	2.0	32	–0.3	1.6	68	15	19	–4
Lithuania	2.9	47	–0.3	1.7	67	15	18	–4
Norway	5.2	14	0.4	1.8	80	18	16	7
Sweden	9.8	24	0.3	1.9	84	17	20	8
EASTERN EUROPE								
Bulgaria	7.2	67	–0.6	1.5	73	14	20	0
Czech Republic	10.6	136	0.0	1.5	74	15	17	2
Hungary	9.8	109	–0.4	1.4	69	15	18	–3
Poland	38.5	126	0.0	1.3	60	15	15	0
Romania	19.8	87	–0.4	1.3	54	16	17	–4
Slovakia	5.4	113	0.1	1.4	54	15	14	0
MICRO STATES								
Andorra	0.08	169	0.5	1.3	86	15	18	–7
Liechtenstein	0.04	231	0.2	1.5	15	15	16	4
Malta	0.4	1,323	0.2	1.4	95	15	16	3
Monaco	0.04	18,916	–0.1	1.4	100	13	24	13
San Marino	0.03	524	0.1	1.5	94	15	18	5
Vatican City	–	–	–	–	–	–	–	–

Source: Population Reference Bureau, *World Population Data Sheet, 2015*.

[1]World Bank, *World Development Indicators*, 2015.

Table A8.2: EUROPE DEVELOPMENT INDICATORS

Country	GNI per capita, PPP 2013	GDP Average Annual %Growth 2009–13	Human Development Index (2013)[1]	Percent Population Living Below $2 a Day	Life Expectancy (2015)[2]	Under Age 5 Mortality Rate (1990)	Under Age 5 Mortality Rate (2013)	Youth Literacy (% pop ages 15–24)	Gender Inequality Index (2013)[3,1]
WESTERN EUROPE									
Austria	45,040	1.6	.881	–	81	9	4	–	0.056
Belgium	41,160	1.1	.881	–	80	10	4	–	0.068
France	38,180	1.2	.884	–	82	9	4	–	0.080
Germany	45,010	2.0	.911	–	80	9	4	–	0.046
Ireland	38,870	0.7	.899	–	81	9	4	–	0.115
Luxembourg	57,830	2.1	.881	–	82	8	2	–	0.154
Netherlands	46,260	0.1	.915	–	81	8	4	–	0.057
Switzerland	59,610	1.8	.917	–	83	8	4	–	0.030
United Kingdom	37,970	1.4	.892	–	81	9	5	–	0.193
SOUTHERN EUROPE									
Albania	9,950	2.3	.716	3.0	78	41	15	99	0.245
Bosnia & Herzegovina	9,660	0.6	.731	< 2	75	19	7	100	0.201
Croatia	20,810	–1.3	.812	< 2	77	13	5	100	0.172
Cyprus	27,630	–1.5	.845	–	80	11	4	100	0.136
Greece	25,660	–6.4	.853	–	81	13	4	99	0.146
Italy	35,220	–0.6	.872	–	83	10	4	100	0.067
Kosovo	9,090	3.3	–	–	77	–	–	–	–
Macedonia	11,520	1.9	.732	4.2	75	38	7	99	0.162
Montenegro	14,410	1.3	.789	< 2	77	18	5	99	–
Portugal	27,190	–1.5	.822	–	80	15	4	99	0.116
Serbia	12,480	0.7	.745	< 2	75	29	7	99	–
Slovenia	28,650	–0.6	.874	< 2	81	10	3	100	0.021
Spain	32,870	–1.1	.869	–	83	11	4	100	0.100
NORTHERN EUROPE									
Denmark	45,300	0.4	.900	–	81	9	4	–	0.056
Estonia	24,920	4.7	.840	< 2	77	20	3	100	0.154
Finland	39,860	0.7	.879	–	81	7	3	–	0.075
Iceland	41,090	1.1	.895	–	82	6	2	–	0.088
Latvia	22,510	3.8	.810	2.0	74	21	8	100	0.222
Lithuania	24,530	3.8	.834	< 2	74	17	5	100	0.116
Norway	65,450	1.5	.944	–	82	8	3	–	0.068
Sweden	46,170	2.2	.898	–	82	7	3	–	0.054
EASTERN EUROPE									
Bulgaria	15,210	1.1	.777	3.9	75	22	12	98	0.207
Czech Republic	26,970	0.7	.861	59.1	79	14	4	–	0.087
Hungary	22,660	0.6	.818	< 2	76	19	6	99	0.247
Poland	22,830	3.0	.834	< 2	78	17	5	100	0.139
Romania	18,390	1.3	.785	8.8	75	37	12	99	0.320
Slovakia	25,970	2.5	.830	< 2	76	18	7	–	0.164
MICRO STATES									
Andorra	–	–	.830	–	83*	8	3	–	–
Liechtenstein	–	–	.889	–	82	10	–	–	–
Malta	27,020	2.2	.829	–	82	11	6	98	0.220
Monaco	–	–	–	–	90	8	4	–	–
San Marino	–	–	–	–	83	12	3	–	–
Vatican City	–	–	–	–	–	–	–	–	–

Source: World Bank, *World Development Indicators, 2015*.

[1]United Nations, *Human Development Report, 2014*.

[2]Population Reference Bureau, *World Population Data Sheet, 2015*.

[3]Gender Inequality Index—A composite measure reflecting inequality in achievements between women and men in three dimensions: reproductive health, empowerment and the labor market that ranges between 0 and 1. The higher the number, the greater the inequality.

*Additional data from the *CIA World Fackbook*, 2015

Table A9.1: THE RUSSIAN DOMAIN POPULATION INDICATORS

Country	Population (millions) 2015	Population Density (per square kilometer)[1]	Rate of Natural Increase (RNI)	Total Fertility Rate	Percent Urban	Percent < 15	Percent > 65	Net Migration (Rate per 1000)
Armenia	3.0	105	0.5	1.5	63	19	11	–6
Belarus	9.5	47	0.0	1.7	76	16	14	2
Georgia	3.8	78	0.2	1.7	54	17	14	–2
Moldova	4.1	124	0.0	1.3	42	16	10	–1
Russia	144.3	9	0.0	1.8	74	16	13	2
Ukraine	42.8	79	–0.4	1.5	69	15	15	1

Source: Population Reference Bureau, *World Population Data Sheet, 2015.*

[1]World Bank, *World Development Indicators,* 2015.

Table A9.2: THE RUSSIAN DOMAIN DEVELOPMENT INDICATORS

Country	GNI per capita, PPP 2013	GDP Average Annual %Growth 2009–13	Human Development Index (2013)[1]	Percent Population Living Below $2 a Day	Life Expectancy (2015)[2]	Under Age 5 Mortality Rate (1990)	Under Age 5 Mortality Rate (2013)	Youth Literacy (% pop ages 15–24)	Gender Inequality Index (2013)[3,1]
Armenia	8,180	4.7	.730	15.5	75	47	16	100	0.325
Belarus	16,950	3.9	.786	< 2	73	17	5	100	0.152
Georgia	7,020	5.9	.744	31.3	75	47	13	100	–
Moldova	5,180	5.0	.663	2.8	72	35	15	100	0.302
Russia	24,280	3.5	.778	< 2	71	27	10	100	0.314
Ukraine	8,970	2.8	.734	< 2	71	19	10	100	0.326

Source: World Bank, *World Development Indicators, 2015.*

[1]United Nations, *Human Development Report, 2014.*

[2]Population Reference Bureau, *World Population Data Sheet, 2015.*

[3]Gender Inequality Index—A composite measure reflecting inequality in achievements between women and men in three dimensions: reproductive health, empowerment and the labor market that ranges between 0 and 1. The higher the number, the greater the inequality.

Table A10.1: CENTRAL ASIA POPULATION INDICATORS

Country	Population (millions) 2015	Population Density (per square kilometer)[1]	Rate of Natural Increase (RNI)	Total Fertility Rate	Percent Urban	Percent < 15	Percent >65	Net Migration (Rate per 1000)
Afghanistan	32.2	47	2.6	4.9	25	45	2	2
Azerbaijan	9.7	114	1.2	2.2	53	22	6	0
Kazakhstan	17.5	6	1.7	3.0	53	25	7	0
Kyrgyzstan	6.0	30	2.1	4.0	36	32	4	–1
Mongolia	3.0	2	2.2	3.1	68	27	4	–1
Tajikistan	8.5	59	2.6	3.8	27	36	3	–3
Turkmenistan	5.4	11	1.3	2.3	50	28	4	–1
Uzbekistan	31.3	71	1.8	2.4	51	28	4	–1

Source: Population Reference Bureau, *World Population Data Sheet, 2015.*

[1]World Bank, *World Development Indicators,* 2015.

Table A10.2: CENTRAL ASIA DEVELOPMENT INDICATORS

Country	GNI per capita, PPP 2013	GDP Average Annual %Growth 2009–13	Human Development Index (2013)[1]	Percent Population Living Below $2 a Day	Life Expectancy (2015)[2]	Under Age 5 Mortality Rate (1990)	Under Age 5 Mortality Rate (2013)	Youth Literacy (% pop ages 15–24)	Gender Inequality Index (2013)[3,1]
Afghanistan	1,960	8.1	.468	–	61	192	97	47	0.705
Azerbaijan	16,180	2.8	.747	2.4	74	95	34	100	0.340
Kazakhstan	20,680	6.4	.757	< 2	70	57	16	100	0.323
Kyrgyzstan	3,080	3.7	.628	21.1	70	70	24	100	0.348
Mongolia	8,810	12.5	.698	–	69	107	32	98	0.320
Tajikistan	2,500	7.2	.607	27.4	67	114	48	100	0.383
Turkmenistan	12,920	11.6	.698	49.7	65	94	55	100	–
Uzbekistan	5,290	8.2	.661	–	68	75	43	100	–

Source: World Bank, *World Development Indicators, 2015.*

[1]United Nations, *Human Development Report, 2014.*

[2]Population Reference Bureau, *World Population Data Sheet, 2015.*

[3]Gender Inequality Index—A composite measure reflecting inequality in achievements between women and men in three dimensions: reproductive health, empowerment and the labor market that ranges between 0 and 1. The higher the number, the greater the inequality.

Table A11.1: EAST ASIA POPULATION INDICATORS

Country	Population (millions) 2015	Population Density (per square kilometer)[1]	Rate of Natural Increase (RNI)	Total Fertility Rate	Percent Urban	Percent < 15	Percent > 65	Net Migration (Rate per 1000)
China	1,371.9	145	0.5	1.7	55	17	10	0
Hong Kong	7.3	6,845	0.3	1.2	100	11	15	3
Japan	126.9	349	−0.2	1.4	93	13	26	1
North Korea	25.0	207	0.5	2.0	61	22	10	0
South Korea	50.7	516	0.4	1.2	82	14	13	3
Taiwan	23.5	652	0.2	1.2	73	14	12	1

Source: Population Reference Bureau, *World Population Data Sheet, 2015.*

[1]World Bank, *World Development Indicators,* 2015.

Table A11.2: EAST ASIA DEVELOPMENT INDICATORS

Country	GNI per capita, PPP 2013	GDP Average Annual %Growth 2009–13	Human Development Index (2013)[1]	Percent Population Living Below $2 a Day	Life Expectancy (2015)[2]	Under Age 5 Mortality Rate (1990)	Under Age 5 Mortality Rate (2013)	Youth Literacy (% pop ages 15–24)	Gender Inequality Index (2013)[3,1]
China	11,850	8.7	.719	18.6	75	49	13	100	0.202
Hong Kong	54,270	3.8	.891	–	84	–	–	–	–
Japan	37,550	1.6	.890	–	83	6	3	–	0.138
North Korea	–	–	–	–	70	45	27	100	–
South Korea	33,360	3.7	.891	–	82	8	4	–	0.101
Taiwan	–	–	–	–	80	–	–	–	–

Source: World Bank, *World Development Indicators, 2015.*

[1]United Nations, *Human Development Report, 2014.*
[2]Population Reference Bureau, *World Population Data Sheet, 2015.*
[3]Gender Inequality Index—A composite measure reflecting inequality in achievements between women and men in three dimensions: reproductive health, empowerment and the labor market that ranges between 0 and 1. The higher the number, the greater the inequality.

Table A12.1: SOUTH ASIA POPULATION INDICATORS

Country	Population (millions) 2015	Population Density (per square kilometer)[1]	Rate of Natural Increase (RNI)	Total Fertility Rate	Percent Urban	Percent < 15	Percent > 65	Net Migration (Rate per 1000)
Bangladesh	160.4	1,203	1.4	2.3	23	33	5	−3
Bhutan	0.8	20	1.1	2.2	38	31	5	2
India	1314.1	421	1.4	2.3	32	29	5	−1
Maldives	0.3	1,150	1.9	2.2	45	26	5	0
Nepal	28.0	194	1.5	2.4	18	33	6	−1
Pakistan	199.0	236	2.3	3.8	38	36	4	−2
Sri Lanka	20.9	327	1.2	2.3	18	25	8	−4

Source: Population Reference Bureau, *World Population Data Sheet, 2015.*

[1]World Bank, *World Development Indicators,* 2015.

Table A12.2: SOUTH ASIA DEVELOPMENT INDICATORS

Country	GNI per capita, PPP 2013	GDP Average Annual %Growth 2009–13	Human Development Index (2013)[1]	Percent Population Living Below $2 a Day	Life Expectancy (2015)[2]	Under Age 5 Mortality Rate (1990)	Under Age 5 Mortality Rate (2013)	Youth Literacy (% pop ages 15–24)	Gender Inequality Index (2013)[3,1]
Bangladesh	3,190	6.2	.558	76.5	71	139	41	80	0.529
Bhutan	6,920	6.6	.584	15.2	68	138	36	74	0.495
India	5,350	6.9	.586	59.2	68	114	53	81	0.563
Maldives	9,900	4.5	.698	12.2	74	105	10	99	0.283
Nepal	2,260	4.2	.540	56.0	67	135	40	82	0.479
Pakistan	4,840	3.1	.537	50.7	66	122	86	71	0.563
Sri Lanka	9,470	7.4	.750	23.9	74	29	10	98	0.383

Source: World Bank, *World Development Indicators, 2015.*

[1]United Nations, *Human Development Report, 2014.*
[2]Population Reference Bureau, *World Population Data Sheet, 2015.*
[3]Gender Inequality Index—A composite measure reflecting inequality in achievements between women and men in three dimensions: reproductive health, empowerment and the labor market that ranges between 0 and 1. The higher the number, the greater the inequality.

Table A13.1: SOUTHEAST ASIA POPULATION INDICATORS

Country	Population (millions) 2015	Population Density (per square kilometer)[1]	Rate of Natural Increase (RNI)	Total Fertility Rate	Percent Urban	Percent < 15	Percent > 65	Net Migration (Rate per 1000)
Burma (Myanmar)	52.1	82	1.0	2.3	34	24	5	–1
Brunei	0.4	79	1.4	1.6	77	25	5	1
Cambodia	15.4	86	1.8	2.7	21	31	6	–2
East Timor	1.2	79	2.8	5.7	32	42	5	–9
Indonesia	255.7	138	1.5	2.6	54	29	5	–1
Laos	6.9	29	2.1	3.1	38	37	4	–3
Malaysia	30.8	90	1.2	2.0	74	26	6	3
Philippines	103.0	330	1.7	2.9	44	34	4	–1
Singapore	5.5	7,713	0.5	1.3	100	16	11	14
Thailand	65.1	131	0.4	1.6	49	18	11	0
Vietnam	91.7	289	1.0	2.4	33	24	7	0

Source: Population Reference Bureau, *World Population Data Sheet, 2015.*

[1]World Bank, *World Development Indicators,* 2015.

Table A13.2: SOUTHEAST ASIA DEVELOPMENT INDICATORS

Country	GNI per capita, PPP 2013	GDP Average Annual %Growth 2009–13	Human Development Index (2013)[1]	Percent Population Living Below $2 a Day	Life Expectancy (2015)[2]	Under Age 5 Mortality Rate (1990)	Under Age 5 Mortality Rate (2013)	Youth Literacy (% pop ages 15–24)	Gender Inequality Index (2013)[3]
Burma (Myanmar)	–	–	.524	–	65	107	51	96	0.430
Brunei	–	1.5	.852	–	79	12	10	100	–
Cambodia	2,890	7.0	.584	41.3	64	117	38	87	0.505
East Timor	7,670	11.0	.620	71.1	68	180	55	80	–
Indonesia	9,270	6.2	.684	43.3	71	82	29	99	0.500
Laos	4,550	8.2	.569	62.0	68	148	71	84	0.534
Malaysia	22,530	5.7	.773	2.3	75	17	9	98	0.210
Philippines	7,840	6.1	.660	41.7	69	57	30	98	0.406
Singapore	76,860	6.3	.901	–	83	8	3	100	0.090
Thailand	13,430	4.2	.722	3.5	75	35	13	97	0.364
Vietnam	5,070	5.8	.638	12.5	73	50	24	97	0.322

Source: World Bank, *World Development Indicators, 2015.*

[1]United Nations, *Human Development Report, 2014.*

[2]Population Reference Bureau, *World Population Data Sheet, 2015.*

[3]Gender Inequality Index—A composite measure reflecting inequality in achievements between women and men in three dimensions: reproductive health, empowerment and the labor market that ranges between 0 and 1. The higher the number, the greater the inequality.

Table A14.1: AUSTRALIA AND OCEANIA POPULATION INDICATORS

Country	Population (millions) 2015	Population Density (per square kilometer)[1]	Rate of Natural Increase (RNI)	Total Fertility Rate	Percent Urban	Percent < 15	Percent > 65	Net Migration (Rate per 1000)
Australia	23.9	3	0.6	1.9	89	19	15	8
Fed. States of Micronesia	0.1	152	1.9	3.5	22	34	4	–14
Fiji	0.9	48	1.3	3.1	51	29	5	–6
French Polynesia	0.3	76	1.1	2.0	56	24	7	0
Guam	0.2	306	1.5	2.9	93	26	8	–6
Kiribati	0.1	126	2.1	3.8	54	36	4	–1
Marshall Islands	0.06	292	2.6	4.1	74	41	3	–17
Nauru	0.01	503	2.7	3.9	100	37	1	–9
New Caledonia	0.3	14	0.9	2.3	70	24	9	4
New Zealand	4.6	17	0.6	1.9	86	20	15	11
Palau	0.02	45	0.2	1.7	84	20	6	0
Papua New Guinea	7.7	16	2.3	4.3	13	39	3	0
Samoa	0.2	67	2.4	4.7	19	39	5	–28
Solomon Islands	0.6	20	2.5	4.1	20	39	3	0
Tonga	0.1	125	2.0	3.9	23	37	6	–19
Tuvalu	0.01	329	1.6	3.2	59	33	5	0
Vanuatu	0.3	21	2.8	4.2	24	39	4	0

Source: Population Reference Bureau, *World Population Data Sheet, 2015.*

[1]World Bank, *World Development Indicators,* 2015.

Table A14.2: AUSTRALIA AND OCEANIA DEVELOPMENT INDICATORS

Country	GNI per capita, PPP 2013	GDP Average Annual %Growth 2009–13	Human Development Index (2013)[1]	Percent Population Living Below $2 a Day	Life Expectancy (2015)[2]	Under Age 5 Mortality Rate (1990)	Under Age 5 Mortality Rate (2013)	Youth Literacy (% pop ages 15–24)	Gender Inequality Index (2013)[3]
Australia	42,110	2.7	.933	–	82	9	4	–	0.113
Fed. States of Micronesia	3,680	0.4	.630	44.7	70	56	36	–	–
Fiji	7,590	2.6	.724	22.9	70	30	24	–	–
French Polynesia	–	–	–	–	77	–	–	–	–
Guam	–	–	–	–	79	–	–	–	–
Kiribati	2,780	2.2	.607	–	65	88	58	–	–
Marshall Islands	4,630	3.2	–	–	72	52	38	–	–
Nauru	–	–	–	–	66	–	–	–	–
New Caledonia	–	–	–	–	77	–	–	100	–
New Zealand	30,970	2.1	.910	–	81	11	6	–	0.185
Palau	14,540	3.9	.775	–	72	32	18	100	–
Papua New Guinea	2,510	8.3	.491	57.4	62	88	61	71	0.617
Samoa	5,560	1.8	.694	–	74	30	18	100	0.517
Solomon Islands	1,810	6.8	.491	–	70	42	30	–	–
Tonga	5,450	1.9	.705	–	76	25	12	99	0.458
Tuvalu	5,260	2.2	–	–	70	58	29	–	–
Vanuatu	2,870	1.6	.616	–	71	39	17	95	–

Source: World Bank, *World Development Indicators, 2015.*

[1]United Nations, *Human Development Report, 2014.*

[2]Population Reference Bureau, *World Population Data Sheet, 2015.*

[3]Gender Inequality Index—A composite measure reflecting inequality in achievements between women and men in three dimensions: reproductive health, empowerment and the labor market that ranges between 0 and 1. The higher the number, the greater the inequality.

Glossary

Aborigine An indigenous inhabitant of Australia.

acid rain A harmful form of precipitation high in sulfur and nitrogen oxides. Caused by industrial and auto emissions, acid rain damages aquatic and forest ecosystems in regions such as eastern North America and Europe.

adiabatic lapse rate The rate a moving air mass cools or warms with changes in elevation, which is usually around 5.5 degrees F per 1000 feet or 1 degree C per 100 meters. Contrast with environmental lapse rate.

African diaspora The forced removal of Africans from their native area and their resettlement throughout the world, especially in the Americas.

African Union (AU) A mostly political body that has tried to resolve regional conflicts. Founded in 1963, the organization grew to include all the states of the continent except South Africa, which finally was asked to join in 1994. In 2004, the body changed its name from the Organization of African Unity to the African Union.

agrarian reform A popular but controversial strategy to redistribute land to peasant farmers. Throughout the 20th century, various states redistributed land from large estates or granted title from vast public lands in order to reallocate resources to the poor and stimulate development. Agrarian reform occurred in various forms, from awarding individual plots or communally held land to creating state-run collective farms.

agribusiness The practice of large-scale, often corporate farming in which business enterprises control closely integrated segments of food production, from farm to grocery store.

alluvial fan A fan-shaped deposit of sediments dropped by a river or stream flowing out of a mountain range.

Altiplano The largest intermontane plateau in the Andes, which straddles Peru and Bolivia and ranges in elevation from 10,000 to 13,000 feet (3,000 to 4,000 meters).

altitudinal zonation The relationship between elevation, temperature, and changes in vegetation that result from the environmental lapse rate (average 3.5°F for every 1,000 feet [6.5°C for every 1,000 meters]). In Latin America, four general altitudinal zones exist: tierra caliente, tierra templada, tierra fria, and tierra helada.

animism A wide variety of tribal religions based on the worship of nature's spirits and human ancestors.

anthropogenic An adjective for human-caused change to a natural system, such as the atmospheric emissions from cars, industry, and agriculture that are causing global warming.

anthropogenic landscape A landscape heavily transformed by human agency.

apartheid The policy of racial separateness that directed the separate residential and work spaces for white, blacks, coloureds, and Indians in South Africa for nearly 50 years. It was abolished when the African National Congress came to power in 1994.

Arab League A regional political and economic organization focused on Arab unity and development.

Arab Spring A series of public protests, strikes, and rebellions in the Arab countries, often facilitated by social media, that have called for fundamental government and economic reforms.

areal differentiation The geographic term for description and analysis of how physical or human traits differ within a spatial unit or area on the surface of Earth.

areal integration The geographic term for description and analysis of how different places or points on Earth interact with each other.

Asia–Pacific Economic Cooperation Group (APEC) A regional organization designed to encourage economic development in Southeast Asia and the Pacific Basin.

Association of Southeast Asian Nations (ASEAN) A supranational geopolitical group linking together the 10 different states of Southeast Asia.

asylum laws Protection for refugees who are victims of ethnic, religious, or political persecution in other parts of the world.

atoll A low, sandy island made from coral. Atolls are often oriented around a central lagoon.

Austronesian A language family that encompasses wide expanses of the Pacific, insular Southeast Asia, and Madagascar.

autonomous areas Minor political subunits created in the former Soviet Union and designed to recognize the special status of minority groups within existing republics.

autonomous region In the context of China, provinces that have been granted a certain degree of political and cultural autonomy, or freedom from centralized authority, due to the fact that they contain large numbers of non-Han Chinese people. Critics contend that they have little true autonomy.

Baikal–Amur Mainline (BAM) Railroad Key central Siberian railroad connection completed in the Soviet era (1984), which links the Yenisey and Amur rivers and parallels the Trans-Siberian Railroad.

balkanization A geopolitical term and concept to describe the breaking up of large political units into smaller ones, the type example being the replacement of the former Yugoslavia with smaller independent states such as Bosnia, Macedonia, Kosovo, and so on.

Berlin Conference A 1884 conference that divided Africa into -European colonial territories. The boundaries created in Berlin satisfied European ambition but ignored indigenous cultural affiliations. Many of Africa's civil conflicts can be traced to ill-conceived territorial divisions crafted in 1884.

biodiversity The array of species, both flora and fauna, found in an ecosystem or bioregion.

biofuels Energy sources derived from plants or animals. Throughout the developing world, wood, charcoal, and dung are primary energy sources for cooking and heating.

bioregion A spatial unit or region of local plants and animals adapted to a specific environment, such as a tropical savanna.

Bolsa Familia This is a Brazilian conditional cash transfer program created to reduce extreme poverty. Families who qualify receive a monthly check from the government as long as they keep their children in school and take them for regular health checkups.

Bolshevik A member of the Russian Communist movement led by Lenin that successfully took control of the country in 1917.

boreal forest A coniferous forest found in a high-latitude or mountainous environment in the Northern Hemisphere.

brain drain Migration of the best-educated people from developing countries to developed nations where economic opportunities are greater.

brain gain The potential of return migrants to contribute to the social and economic development of a home country with the experiences they have gained abroad.

British East India Company A private trade organization that acted as an arm of colonial Britain in ruling most of South Asia until 1857, when it was abolished and replaced by full governmental -control.

buffer state A country that is situated between much stronger countries, and which is intended to reduce conflicts between those more powerful countries by preventing them from sharing a common border.

buffer zone An array of nonaligned or friendly states that "buffer" a larger country from invasion. In Europe, keeping a buffer zone has been a long-term policy of Russia (and also of the former Soviet Union) to protect its western borders from European invasion.

Bumiputra The name given to native Malay (literally, "sons of the soil"), who are given preference for jobs and schooling by the Malaysian government.

bustees Settlements of temporary and often illegal housing in Indian cities, caused by rapid urban migration of poorer rural people and the inability of the cities to provide housing for this rapidly expanding population.

capital leakage The gap between the gross receipts an industry (such as tourism) brings into a developing area and the amount of capital retained.

Caribbean Community and Common Market (CARICOM) A regional trade organization established in 1972 that includes former English colonies in the Caribbean Basin as its members.

Caribbean diaspora The economic flight of Caribbean peoples across the globe.

caste system The complex division of South Asian society into different hierarchically ranked hereditary groups. The caste system is most explicit in Hindu society but is also found in other cultures to a lesser degree.

Central American Free Trade Association (CAFTA) A trade agreement between the United States and Guatemala, El Salvador, Nicaragua, Honduras, Costa Rica, and the Dominican Republic to reduce tariffs and increase trade between member countries.

centralized economic planning An economic system in which the state sets production targets and controls the means of production.

chaebol A very large South Korean business conglomerate that is composed of numerous smaller companies.

chernozem soils A Russian term for dark, fertile soil, often associated with grassland settings in southern Russia and Ukraine.

China proper The eastern half of the country of China, where the Han Chinese form the dominant ethnic group. The vast majority of China's population is located in China proper.

choke point Strategic setting where narrow waterways or other narrow passages are vulnerable to military blockade disruption.

choropleth map A thematic map in which areas are colored or shaded to depict differences in whatever is being mapped.

clan A social unit that is typically smaller than a tribe or an ethnic group but larger than a family, based on supposed descent from a common ancestor.

climate The average weather conditions for a place, usually based upon 30 years of weather measuresments.

climate change The measured change in climate from a previous state, contrasted with normal variability.

climograph A graph of average annual temperature and precipitation data by month and season.

Closer Economic Relationship (CER) Agreement An agreement signed in 1982 between Australia and New Zealand, designed to eliminate all economic and trade barriers between the two countries.

Cold War An ideological struggle between the United States and the Soviet Union that was conducted between 1946 and 1991.

Collective Security Treaty Organization (CSTO) A Russian-led military association that includes Belarus, Armenia, Kazakhstan, Kyrgyzstan, Tajikistan, and Uzbekistan. The CSTO and SCO work together to address military threats, crime, and drug smuggling.

colonialism Formal, established (mainly historical) rule over local peoples by a larger imperialist government for the expansion of political and economic empire.

coloured A racial category used throughout South Africa to define people of mixed European and African ancestry.

communism A belief based on the writings of Karl Marx

conflict diamonds that promoted the overthrow of capitalism by the workers, the large-scale elimination of private property, state ownership and central planning of major sectors of the economy (both agricultural and industrial), and one-party authoritarian rule.

Confucianism A philosophical system based on the ideas of Confucius, a Chinese philosopher who lived in the 6th century bce. Confucianism stresses education and the importance of respecting authority figures, as well as the importance of authority figures acting in a responsible manner. Confucianism is historically significant throughout East Asia.

connectivity The degree to which different locations are linked with one another through transportation and communication infrastructure.

continental climate A climate region in a continental interior, removed from moderating oceanic influences, characterized by hot summers and cold winters. In such a climate, at least one month must average below freezing.

convergent plate boundary Areas where two tectonic plates moving in opposing directions meet and converge.

core–periphery model A conceptualization of the world into two economic spheres. The developed countries of western Europe, North America, and Japan form the dominant core, with less-developed countries making up the periphery. Implicit in this model is that the core gained its wealth at the expense of peripheral countries.

Cossacks Highly mobile Slavic-speaking Christians of the southern Russian steppe who were pivotal in expanding Russian influence in 16th- and 17th-century Siberia.

counterinsurgency The suppression of a rebellion or insurgency by both military and political means, which includes not just armed warfare but also winning the support of local peoples by improving local infrastructure (schools, roads, etc.).

creolization The blending of African, European, and some Amerindian cultural elements into the unique sociocultural systems found in the Caribbean.

crony capitalism A system in which close friends of a political leader are either legally or illegally given business advantages in return for their political support.

cultural assimilation A process in which immigrants are culturally absorbed into the larger host society.

cultural homeland A culturally distinctive settlement in a well-defined geographic area, whose ethnicity has survived over time, stamping the landscape with an enduring personality.

cultural imperialism The active promotion of one cultural system over another, such as the implantation of a new language, school system, or bureaucracy. Historically, cultural imperialism has been primarily associated with European colonialism.

cultural landscape A physical or natural landscape that has been changed considerably by the influences of human settlement.

cultural nationalism A process of protecting, either formally (with laws) or informally (with social values), the primacy of a certain cultural system against influences (real or imagined) from another culture.

cultural syncretism or hybridization The blending of two or more cultures, which produces a synergistic third culture that exhibits traits from all cultural parents. Also called *cultural hybridization*.

culture hearth An area of historical cultural innovation.

culture Learned and shared behavior by a group of people that gives them a distinct "way of life." Culture is made up of both material (technology, tools, etc.) and abstract (speech, religion, values, etc.) components.

Cyrillic alphabet An alphabet based on the Greek alphabet and used by Slavic languages heavily influenced by the Eastern Orthodox Church. It is attributed to the missionary work of St. Cyril in the 9th century.

Dalit The currently preferred term used to denote the members of India's most discriminated against ("lowest") caste groups, those people previously referred to as "untouchables."

decolonialization The process of a former colony's gaining (or regaining) independence over its territory and establishing (or reestablishing) an independent government.

demographic transition model A five-stage model of population change derived from the historical decline of the natural rate of -increase as a population becomes increasingly urbanized through industrialization and economic development.

desertification The spread of desert conditions into semiarid areas due to improper management of the land.

devolution The breakng apart or separation within a political unit such as a nation-state.

diaspora The scattering of a particular group of people over a vast geographic area. Originally, the term referred to the migration of Jews out of their homeland, but now it has been generalized to refer to any ethnic dispersion.

divergent plate boundary A geologic boundary where tectonic plates move away from each other, in opposite diections, thereby creating either a rift zone, which is a depression, or, in other places, a ridge built of volcanic material.

diversity Refers to the state of having different landscapes, cultures, or ideas, as well as the inclusion of distinct peoples in a particular society.

dollarization An economic strategy in which a country adopts the U.S. dollar as its official currency. A country can be partially dollarized, using U.S. dollars alongside its national currency, or fully dollarized, in which case the U.S. dollar becomes the only medium of exchange and the country gives up its own national currency. Panama fully dollarized in 1904; more recently, Ecuador fully dollarized in 2000.

domestication The purposeful selection and breeding of wild plants and animals for cultural purposes.

domino theory A U.S. geopolitical policy of the 1970s that stemmed from the assumption that if Vietnam fell to the Communists, the rest of Southeast Asia would soon follow.

Dravidian language family A strictly South Asian language family that includes such important languages as Tamil and Telugu. Once spoken through most of the region, Dravidian languages are now largely limited to southern South Asia.

Eastern Orthodox Christianity A loose confederation of self-governing churches in eastern Europe and Russia that are historically linked to Byzantine traditions and to the primacy of the patriarch of Constantinople (Istanbul).

Economic and Monetary Union (EMU) Created by the European Union (EU) in 1999 to facilitate economic matters amongst member states, including usage of a common currency.

Economic Community of West African States (ECOWAS) ECOWAS is an intergovernmental organization that promotes economic integration and security among its 15 member states in West Africa. It was founded in 1975.

economic convergence The notion that globalization will result in the world's poorer countries gradually catching up with more advanced economies.

edge city Suburban node of activity that features a mix of peripheral retailing, industrial parks, office complexes, and entertainment facilities.

El Niño An abnormally large warm current that appears off the coast of Ecuador and Peru in December. During an El Niño year, torrential rains can bring devastating floods along the Pacific coast and drought conditions in the interior continents of the Americas.

entrepôt A city and port that specializes in transshipment of goods.

environmental lapse rate The decline in temperature as one ascends higher in the atmosphere. On average, the temperature declines 3.5°F for every 1,000 feet of elevation, or 6.5°C for every 1,000 meters. Not to be confused with the adiabatic lapse rate.

ethnic religion A religion closely identified with a specific ethnic or tribal group, often to the point of assuming the role of the major defining characteristic of that group. Normally, ethnic religions do not actively seek new converts.

ethnicity A shared cultural identity held by a group of people with a common background or history, often as a minority group within a larger society.

Eurasian Economic Union (EEU) A customs union (paralleling the European Union [EU]) designed to encourage trade as well as closer political ties between member states. Formed in 2015, the EEU contains five member states (Russia, Belarus, Kazakhstan, Armenia, and Kyrgyzstan).

European Union (EU) The current association of 28 European countries that are joined together in an agenda of economic, political, and cultural integration.

Eurozone The common monetary policy and currency of the European Union; those countries of Europe using the euro as its currency and who are members of the EU's common monetary system, contrasted to those countries having a national currency and a monetary system. France is an example of the former, and the United Kingdom of the latter.

exclave A portion of a country's territory that lies outside its contiguous land area.

Exclusive Economic Zone (EEZ) An area of the ocean decreed by international law where one local country has more rights to fishing and mineral rights than do other countries.

exotic river A river that issues from a humid area and flows into a dry area otherwise lacking streams.

exotic species Nonnative plants and animals.

fair trade An international certification movement to identify primary commodities exported from the developing world in which farmers earn a better price for their product. Commodities such as coffee, tea, and forest products are certified "fair trade" when small-scale producers earn more for their product and production methods are viewed as environmentally and socially sustainable.

family-friendly policies Public policies that encourage higher birth rates. An example would be extended maternity and paternity leaves for parents of a newborn.

federal state Nations that allocate considerable political power to units of government beneath the national level.

feng shui Literally translated into English as "wind-water," feng shui refers to a set of Chinese beliefs based on harmonizing human activities and buildings with the spiritual forces found in the natural environment. It is closely associated with the religion of Daoism.

Fertile Crescent An ecologically diverse zone of lands in Southwest Asia that extends from Lebanon eastward to Iraq and that is often associated with early forms of agricultural domestication.

fjord Flooded, glacially carved valley. In Europe, fjords are found primarily along Norway's western coast.

formal region A geographic concept used to describe an area where a static and specific trait (such as a language or a climate) has been mapped and described. A formal region contrasts with a functional region.

forward capital A capital city deliberately positioned near a contested territory, signifying the state's interest and presence in this zone of conflict.

fossil water Water supplies that were stored underground during wetter climatic periods.

fracking A set of drilling technologies that injects a mix of water, sand, and chemicals underground in order to release and enhance the removal of natural gas and oil.

free trade zone (FTZ) A duty-free and tax-exempt industrial park created to attract foreign corporations and create industrial jobs.

functional region A geographic concept used to describe the spatial extent dominated by a specific activity. The circulation area of a newspaper is an example, as is the trade area of a large city.

gender The social and cultural expressions of male- and femaleness, which contrasts with sex, which is the biological distinction -between male and female.

gender gap The difference in parity or equity between males and females in a specific social or cultural context. A term often used to describe gender differences in salary, working conditions, or political power.

gender inequality Refers to unequal treatment or perceptions of people based on their gender. Typically these are socially constructed gender differencs such as access to education, differences in pay, or political participation.

gender roles How female and male behavior differs in a specific cultural context.

gentrification A process of urban revitalization in which higher-income residents displace lower-income residents in central city neighborhoods.

geographic information system (GIS) A computerized mapping and information system that analyzes vast amounts of data that may include many layers of specific kinds of information, such as microclimates, hydrology, vegetation, or land-use zoning regulations.

geography The spatial science that describes and explains physical and cultural phenomena on Earth's surface.

geopolitics The relationship between politics and space and territory.

glasnost A policy of greater political openness initiated during the 1980s by then Soviet President Mikhail Gorbachev.

global positioning system (GPS) Originally used to describe a very accurate satellite-based location system, but now also used in a general sense to describe smartphone location systems that may use cell phone towers as a subsitute for satellites.

globalization The increasing interconnectedness of people and places throughout the world through converging processes of economic, political, and cultural change.

glocalization Combines the idea of globalization with that of local considerations or practices. For example, a global corporation might market its goods differently depending upon local culture.

Golden Triangle The world's second largest opium and heroin producing area, located in northern Laos, Thailand, and Burma.

graphic or linear scale A ruler-like symbol on a map that translates the map's cartographic scale into visual terms.

grassification The conversion of tropical forest into pasture for cattle ranching. Typically, this process involves introducing species of grasses and cattle, mostly from Africa.

Great Escarpment A landform that rims southern Africa from Angola to South Africa. It forms where the narrow coastal plains meet the elevated plateaus in an abrupt break in elevation.

Great Rift Valley The valleys and lakes of East Africa that form on the divergent plate boundary that extends north to south across East Africa. In this area African Plate is in the slow process of splitting in two.

Greater Antilles The four large Caribbean islands of Cuba, Jamaica, Hispaniola, and Puerto Rico.

Greater Arab Free Trade Area (GAFTA) An organization created in 2005 by 17 members of the Arab League that is designed to eliminate all intraregional trade barriers and spur economic cooperation.

Green Revolution Highly productive agricultural techniques developed since the 1960s that entail the use of new hybrid plant varieties combined with large applications of chemical fertilizers and pesticides. The term is generally applied to agricultural changes in developing countries, particularly India.

greenhouse effect The natural process of lower atmospheric heating that results from the trapping of incoming and reradiated solar energy by water moisture, clouds, and other atmospheric gases.

Greenhouse gases (GHGs) Those atmospheric gases, both natural and human-caused, that trap reradiated solar energy, warming the lower layers of the atmosphere. Human-generated GHGs such as carbon dioxide and methane are causing Earth to warm, resulting in changes to the planet's climates.

gross domestic product (GDP) The total value of goods and services produced within a given country (or other geographical unit) in a single year.

gross national income (GNI) The value of all final goods and services produced within a country's borders (gross domestic product) plus the net income from abroad (formerly referred to as gross national product).

gross national income (GNI) per capita The figure that results from dividing a country's GNI by the total population.

Group of Eight (G8) A collection of powerful countries—United States, Canada, Japan, Great Britain, Germany, France, Italy, and -Russia—that confers regularly on key global economic and political issues.

Gulag Archipelago A collection of Soviet-era labor camps for political prisoners, made famous by writer Aleksandr Solzhenitsyn.

Hajj An Islamic religious pilgrimage to Makkah. One of the five essential pillars of the Muslim creed to be undertaken once in life, if an individual is physically and financially able to do it.

***hallyu* ("Korean Wave")** Literally meaning the "flow of Korea," hallyu refers to the popularity of South Korean music, films, and television shows in other Asian countries and increasingly across the rest of the world as well.

haoles Light-skinned Europeans or U.S. citizens in the Hawaiian Islands.

high islands Large, elevated islands, often focused around recent volcanic activity.

Hindi An Indo-European language with more than 480 million speakers, making it the second-largest language group in the world. In India, it is the dominant language of the heavily populated north, specifically the core area of the Ganges Plain.

Hindu nationalism A contemporary "fundamental" religious and political movement that promotes Hindu values as the essential—and exclusive—fabric of Indian society. As a political movement, Hindu nationalism appears to be less tolerant of India's large Muslim minority than do other political movements.

Hinduism Refers to the main religion of India and Nepal, which developed in the South Asian subcontinent over the past several thousand years. As many beliefs and religious practices very significantly from one Hindu community to another, some scholars think of Hinduism more as a family of closely related religions than as a single faith.

Horn of Africa The northeastern corner of Sub-Saharan Africa that includes the states of Somalia, Ethiopia, Eritrea, and Djibouti. Drought, famine, and ethnic warfare in the 1980s and 1990s resulted in political turmoil in this area.

hukou An official record used in China to identify a specific person as a resident of a particular place. The hukou system is used in China to control the movement of people, particularly from rural areas to cities.

human Development Index (HDI) For the past three decades, the United Nations has tracked social development in the world's countries through the Human Development Index (HDI), which combines data on life expectancy, literacy, educational attainment, gender equity, and income.

human geography The branch of geography aligned with the social sciences. It deals with the human settlement of the earth, its peoples, settlement patterns, cultures, economies, social systems and inteactions with the environment across space and at different scales.

human trafficking A practice in which women are lured or abducted into prostitution.

hurricane A storm system with an abnormally low-pressure center, sustaining winds of 75 miles per hour (121 km/hour) or higher. Each year during hurricane season (July–October), a half dozen to a dozen hurricanes form in the warm waters of the Atlantic and Caribbean, bringing destructive winds and heavy rain.

hydraulic fracturing (fracking) A set of drilling technologies that injects a mix of water, sand, and chemicals underground in order to release and enhance the removal of natural gas and oil.

hydropolitics The interplay of water resource issues and politics.

ideographic writing A writing system in which each symbol represents not a sound but a concept.

indentured labor Foreign workers (generally South Asians) contracted to labor on Caribbean agricultural estates for a set period of time, often several years. Usually the contract stipulated paying off the travel debt incurred by the laborers. Similar indentured labor arrangements have existed in most world regions.

Indian diaspora The historical and contemporary propensity of Indians to migrate to other countries in search of better opportunities. This has led to large Indian populations in South Africa, the Caribbean, and the Pacific islands, along with western Europe and North America.

Indian subcontinent The name frequently given to South Asia in reference to its largest country. It forms a distinct landmass separated from the rest of the Eurasian continent by a series of sweeping mountain ranges, including the Himalayas—the highest mountains in the world.

Industrial Revolution That period of time in the 18th century when European factories first changed from using animate power (human and animals) to inanimate power (water and coal) to power machines.

informal sector A much-debated concept that presupposes a dual economic system consisting of formal and informal sectors. The informal sector includes self-employed, low-wage jobs that are usually unregulated and untaxed. Street vending, shoe shining, artisan manufacturing, and self-built housing are considered part of the informal sector. Some scholars include illegal activities such as drug smuggling and prostitution in the informal economy.

insolation A measure of solar radiation often expressed in units of solar energy received over a specific area (square foot or square meter) over a specific period of time.

insurgency A political rebellion or uprising.

Intercropping The practice of planting multiple crops in the same field.

internally displaced persons (IDPs) Groups and individuals who flee an area due to conflict or famine but still remain in their country of origin. These populations often live in refugee-like conditions but are difficult to assist because they technically do not qualify as -refugees.

Iron Curtain A term coined by British leader Winston Churchill during the Cold War to define the western border of Soviet power in Europe. The notorious Berlin Wall was a concrete manifestation of the Iron Curtain.

irredentism A state or national policy of reclaiming lost lands or those inhabited by people of the same ethnicity in another nation-state.

ISIL (Islamic State of Iraq and the Levant; also ISIS or Islamic State) A violent Sunni extremist organization that has expanded its influence in Iraq, Syria, and elsewhere as it attempts to create a new religious state (a caliphate) in the region.

Islamic fundamentalism A movement within both the Shiite and Sunni Muslim traditions to return to a more conservative, -religious-based society and state. Often associated with a rejection of Western culture and with a political aim to merge civic and religious authority.

Islamism A political movement within the religion of Islam that challenges the encroachment of global popular culture and blames colonial, imperial, and Western elements for many of the region's problems. Adherents of Islamism advocate merging civil and religious authority.

isolated proximity A concept that explores the contradictory position of the Caribbean states, which are physically close to North America and economically dependent upon that region but also have strong loyalties to locality and limited economic opportunity.

Jainism A religious group in South Asia that emerged as a protest against orthodox Hinduism around the 6th century bce. Its ethical core is the doctrine of noninjury to all living creatures. Today, Jains are noted for their nonviolence, which prohibits them from taking the life of any animal.

kanakas Melanesian agricultural workers imported to Australia, historically concentrated along Queensland's "Sugar Coast."

Khmer Rouge Literally, "Red (or Communist) Cambodians," the left-wing insurgent group that overthrew the royal Cambodian government in 1975 and subsequently created one of the most brutal political systems the world has ever seen.

kleptocracy A state where corruption is so institutionalized that politicians and bureaucrats siphon off a huge percentage of a country's wealth.

Kyoto Protocol An international treaty to limit greenhouse gas emissions. It was enacted in 1997 and expired in 2015 when it was replaced by the Paris Treaty of 2015.

laissez-faire An economic system in which the state has minimal involvement and in which market forces largely guide economic activity.

latifundia A large estate or landholding in Latin America.

latitude (parallels) The angular distance north or south from the equator of a point on the earth's surface. It is measured from 0 at the equator to 90 at the poles.

legend Most maps have legends that define the meaning of the symbols and colors used on a map to represent some aspect of the real world.

Lesser Antilles The arc of small Caribbean islands from St. Maarten to Trinidad.

Levant The eastern Mediterranean region.

lingua franca An agreed-upon common language to facilitate communication on specific topics such as international business, politics, sports, or entertainment.

linguistic nationalism The promotion of one language over others that is, in turn, linked to shared notions of nationalism. In India, some Hindu nationalists promote Hindi as the national language, yet this is resisted by many other groups in which that language is either not spoken or does not have the same central cultural role as in the Ganges Valley. The lack of a national language in India remains problematic.

lithosphere The outer layer of Earth's interior where the convection cells that drive plate tectonics are located.

location factor The various influences that explain why an economic activity takes place where it does.

loess A fine, wind-deposited sediment that makes fertile soil but is very vulnerable to water erosion.

longitude (meridians) Longitude lines or meridians run nothe-south, from pole to pole, and measure distance east or west from the Prime Meridian (0 degrees) located in Greenwich, England near London.

low islands Flat, low-lying islands formed by coral reefs, and contrasting with high islands that were formed from volcanic eruptions.

Maghreb A region in northwestern Africa that includes portions of Morocco, Algeria, and Tunisia.

maharaja Regional Hindu royalty, usually a king or prince, who ruled specific areas of South Asia before independence but who was usually subject to overrule by British colonial advisers.

mallee A tough and scrubby eucalyptus woodland of limited economic value that is common across portions of interior Australia.

Maoism Refers to the specific variety of Marxism that was developed by the Chinese leader Mao Zedong in the mid twentieth century. Unlike mainstream Marxism, Maoism regards peasants rather than industrial workers as forming the main potentially revolutionary class that can help create in a Communist society.

Maori Indigenous Polynesian people of New Zealand.

map projections The cartographic and mathematical solution to translating the surface of a rounded globe (usually Earth) to a flat surface (usually a piece of paper) with a minimum of distortion.

map scale The relationship between distances on a mapped object such as Earth and depection of that space on a map. Large scale maps cover small areas in great detail, whereas small scale maps depict less detail but over large areas.

maquiladora Assembly plants on the Mexican border built by foreign capital. Most of their products are exported to the United States.

marine west coast climate A moderate climate with cool summers and mild winters that is heavily influenced by maritime conditions. Such climates are usually found on the west coasts of continents between the latitudes from 45 to 50 degrees.

maritime climate A climate moderated by proximity to oceans or large seas. It is usually cool, cloudy, and wet and lacks the temperature extremes of continental climates.

maroons Runaway slaves who established communities rich in African traditions throughout the Caribbean and Brazil.

marsupial A class of mammals found primarily in the Southern Hemisphere with the distinctive characteristic of carrying their young in a pouch. Kangaroos are perhaps the best-known marsupial, with wallabies, koalas, wombats, and the Tasmanian Devil also found in Oceania.

Marxism A philosophy developed by Karl Marx, the most important historical proponent of communism. Marxism, which has many variants, presumes the desirability and, indeed, the necessity of a socialist economic system run through a central planning agency.

medieval landscape An urban landscape from 900 to 1500 CE, characterized by narrow, winding streets, and three- or four-story structures (usually in stone, but sometimes wooden), with little open space except for the market square. These landscapes are still found in the centers of many European cities.

medina The original urban core of a traditional Islamic city.

Mediterranean climate A unique climate, found in only five locations in the world, that is characterized by hot, dry summers with very little rainfall. These climates are located on the west side of continents, between 30 and 40 degrees latitude.

Megalopolis A large urban region formed as multiple cities grow and merge with one another. The term is often applied to the string of cities in eastern North America that includes Washington, DC; Baltimore; Philadelphia; New York City; and Boston.

Melanesia A Pacific Ocean region that includes the culturally complex, generally darker-skinned peoples of New Guinea, the Solomon Islands, Vanuatu, New Caledonia, and Fiji.

Mercosur The Southern Common Market, established in 1991, which calls for free trade among member states and common external tariffs for nonmember states. Argentina, Paraguay, Brazil, -Uruguay, and Venezuela are members; Chile, Peru, Bolivia, Ecuador and -Colombia are associate members.

mestizo A person of mixed European and Indian ancestry.

Micronesia A Pacific Ocean region that includes the culturally diverse, generally small islands north of Melanesia. Micronesia includes the Mariana Islands, Marshall Islands, and Federated States of Micronesia.

microstates Usually independent states that are small in both area and population.

mikrorayon A large, state-constructed urban housing project built during the Soviet period in the 1970s and 1980s.

Millennium Development Goals A program of the United Nations, in collaboration with the World Bank, that aims to reduce extreme poverty by focusing resources on improving basic education, health care, and access to clean water in developing countries. The targeted goals are based on 1990 baselines and are supposed to be reached by 2015. Many countries in the developing world will reach their targets; it appears that many Sub-Saharan African countries will not.

minifundia A small landholding farmed by peasants or tenants who produce food for subsistence and the market.

mono crop production Agriculture based on a single crop.

monotheism A religious belief in a single God.

Monroe Doctrine A proclamation issued by U.S. President James Monroe in 1823 that the United States would not tolerate European military action in the Western Hemisphere. Focused on the Caribbean as a strategic area, the doctrine was repeatedly invoked to justify U.S. political and military intervention in the region.

monsoon The seasonal pattern of changes in winds, heat, and moisture in South Asia and other regions of the world that is a product of larger meteorological forces of land and water heating, the resultant pressure gradients, and jet-stream dynamics. The monsoon produces distinct wet and dry seasons.

monsoon wind Continental-scale winds that flow from high to low pressure. In South, Southeast Asia and in North America's Southwest monsoon winds are associated with rainy weather.

Mughal (or Mogul) Empire The powerful Muslim state that ruled most of northern South Asia in the 1500s and 1600s. The last vestiges of the Mughal dynasty were dissolved by the British following the rebellion of 1857.

nation-state A relatively homogeneous cultural group (a nation) with its own political territory (the state).

Native Title Bill A bill by the Australian legislation signed in 1993 that provides Aborigines with enhanced legal rights regarding land and resources within the country.

neocolonialism Economic and political strategies by which powerful states indirectly (and sometimes directly) extend their influence over other, weaker states.

neoliberalism Economic policies widely adopted in the 1990s that stress privatization, export production, and few restrictions on imports.

neotropics Tropical ecosystems of the Americas that evolved in relative isolation and support diverse and unique flora and fauna.

net migration rate A statistic that depicts whether more people are entering or leaving a country.

new urbanism An urban design movement stressing higher density, mixed-use, pedestrian-scaled neighborhoods where residents might be able to walk to work, school, and local entertainment.

nonmetropolitan growth A pattern of migration in which people leave large cities and suburbs and move to smaller towns and rural areas.

nonrenewable energy Those energy sources such as oil and coal with finite reserves.

North American Free Trade Agreement (NAFTA) An agreement made in 1994 between Canada, the United States, and Mexico that established a 15-year plan for reducing all barriers to trade among the three countries.

North Atlantic Treaty Organization (NATO) Initially NATO was a group of North Atlantic and European allies who came together in 1949 to counter the Soviet threat to western Europe.

northern sea route An ice-free channel along Siberia's northern coast that will grow in importance given sustained global warming.

novel ecosystems New and previously unseen assemblages of plants and animals resulting from human modification of the environment, including climate change.

Oceania A major world subregion that is usually considered to include New Zealand and the major island regions of Melanesia, Micronesia, and Polynesia.

offshore banking Financial services offered by islands or microstates that are typically confidential and tax exempt. As part of a global financial system, offshore banks have developed a unique niche, offering their services to individual and corporate clients for set fees. The Bahamas and the Cayman Islands are leaders in this sector.

oligarchs A small group of wealthy, very private businessmen who control (along with organized crime) important aspects of the Russian economy.

Organization of American States (OAS) Founded in 1948 and headquartered in Washington, DC, an organization that advocates hemispheric cooperation and dialogue. Most states in the Americas, except Cuba, belong to the OAS.

Organization of the Petroleum Exporting Countries (OPEC) An international organization (formed in 1960) of 12 oil-producing nations that attempts to influence global prices and supplies of oil. Algeria, Gabon, Indonesia, Iran, Iraq, Kuwait, Libya, Nigeria, Qatar, Saudi Arabia, the United Arab Emirates, and Venezuela are members.

orographic effect The influence of mountains on weather and climate, usually referring to the increase of precipitation on the windward side of mountains, and a drier zone (or rain shadow) on the leeward or downwind side of the mountain.

orographic rainfall Enhanced precipitation over uplands that results from lifting and cooling of air masses as they are forced over mountains.

Ottoman Empire A large, Turkish-based empire (named for Osman, one of its founders) that dominated large portions of southeastern -Europe, North Africa, and Southwest Asia between the 16th and 19th centuries.

Outback Australia's large, generally dry, and thinly settled interior.

outsourcing A business practice that transfers portions of a company's production and service activities to lower-cost settings, often located overseas.

Palestinian Authority (PA) A quasi-governmental body that represents Palestinian interests in the West Bank and Gaza.

Pangaea The supercontinent of tightly clustered tectonic plates that existed 250 million years ago.

pastoral nomadism A traditional subsistence agricultural system in which practitioners depend on the seasonal movements of livestock within marginal natural environments.

pastoralists Nomadic and sedentary peoples who rely on livestock (especially cattle, camels, sheep, and goats) for sustenance and livelihood.

perestroika A program of partially implemented, planned economic reforms (or restructuring) undertaken during the Gorbachev years in the Soviet Union and designed to make the Soviet economy more efficient and responsive to consumer needs.

permafrost A cold-climate condition in which the ground remains permanently frozen.

physical geography The other major subfield of geography more closely aligned with the natural sciences. It studies the processes and patterns in the natural world in terms of climate, vegetation, landforms, and hydrosphere, and how humans modify these systems.

physiological density A population statistic that relates the number of people in a country to the amount of arable land.

Pidgin English A version of English that also incorporates elements of other local languages, often utilized to foster trade and basic communication between different culture groups.

plantation America A cultural region that extends from midway up the coast of Brazil, through the Guianas and the Caribbean, and into the southeastern United States. In this coastal zone, European-owned plantations, worked by African laborers, produced agricultural products for export.

plate tectonics A geophysical theory that postulates Earth's surface is made up of numerous large segments (tectonic plates) that move very slowly about. The tensions between these tectonic plates, through long periods of time, have shaped Earth's surface and topography. In the shorter term plate pressures and movement cause some areas of Earth to be more prone to earthquakes and volcanoes.

podzol soil A Russian term for an acidic soil of limited fertility, typically found in northern forest environments.

polar jet streams A powerful atmospheric wind located in the higher latitudes of both hemispheres, north and south. These winds are caused by Earth's rotation.

pollution exporting The process of exporting industrial pollution and other waste material to other countries. Pollution exporting can be direct, as when waste is simply shipped abroad for disposal, or indirect, as when highly polluting factories are constructed abroad.

Polynesia A Pacific Ocean region, broadly unified by language and cultural traditions, that includes the Hawaiian Islands, Marquesas Islands, Society Islands, Tuamotu Archipelago, Cook Islands, American Samoa, Samoa, Tonga, and Kiribati.

population density The population of an area as measured by people per spatial unit, usually people per sqare mile or square kilometer.

population pyramid The structure of a population measuring the percentage of young and old, presented graphically as a pyramid-shaped graph. This graph plots the percentage of all different age groups along a vertical axis that divides the population into male and female.

postindustrial economy An economy in which the tertiary and quaternary sectors dominate employment and expansion.

prairie An extensive area of grassland in North America. In the more humid eastern portions, grasses are usually longer than in the drier western areas, which are in the rain shadow of the Rocky Mountain range.

primate city The largest urban settlement in a country that dominates all other urban places, economically and politically. Often, but not always, the primate city is also the country's capital. Primate cities are usually three to four times larger than the next largest city in a country.

prime meridian Zero degrees longitude, from which locations east and west are measured in a system of latitude and longitude. Currently, the most used Prime Merdian is that established in 1851 at the Naval Observatory in Greenwich, England (in southeastern London). Before that other countries and cultures established their own prime meridians upon which to base their maps and navigation systems.

proven reserves The amount of a non-renewable energy source (oil, coal, and gas) still in the ground that is feasible to exploit under current market conditions.

purchasing power parity (PPP) An important qualification to these GNI per capita data is the concept of adjustment through PPP, an adjustment that takes into account the strength or weakness of local currencies.

Quran (or Koran) A book of divine revelations received by the prophet Muhammad that serves as a holy text in the religion of Islam.

rain shadow A drier area of precipitation, usually on the leeward or downwind side of a mountain range, that receives less rain and snowfall than the windward or upwind side.

rain-shadow effect Caused by the warming of air as it descends down a mountain range; this warming increases the ability of an air mass to hold moisture.

rate of natural increase (RNI) The standard statistic used to express natural population growth per year for a country, a region, or the world, based on the difference between birthrates and death rates. RNI does not consider population change from migration. Though most often a positive figure (such as 1.7 percent), RNI can also be expressed as a negative (such as –.08 percent figure) for no-growth countries.

refugee A person who flees his or her country because of a well-founded fear of persecution based on race, ethnicity, religion, ideology, or political affiliation.

region A geographic concept of areal or spatial similarity, large or small.

regional geography Regional geography focuses on areal units that share particular characteristics. This book examines 12 world regions and outlines processes that unite these areas as well as divides them.

remittances Monies sent by immigrants working abroad to family members and communities in their countries of origin. For many countries in the developing world, remittances often amount to billions of dollars each year. For small countries, remittances can equal 5 to 10 percent of a country's gross domestic product.

remote sensing A method of digitally photographing Earth's surface from satellites or high altitude aircraft so that the information captured can be manipulated by computers to translate information into certain electromagnetic bandwidths, which, in turn, emphasizes certain features and patterns on Earth's surface.

Renaissance–Baroque period A historical period, dated roughly from the 16th to the 19th century, characterized by certain urban planning designs and architectural styles that are still found today in many European cities: wide, ceremonial boulevards; large monumental structures (palaces, public squares, churches, and so on); and ostentatious housing for the urban elite.

renewable energy Energy sources, such as solar, wind, and hydro, that are replenished by nature at a faster rate than they are used or consumed.

replacement rate The average number of children born to women in order to have a stable population. The global replacement rate is 2.1.

representative fraction The cartographic and mathematical expression of the relationship between distance on a map and that on Earth's surface which is expressed as a fraction depicting the scale of the map. For example, a common scale is that of one inch on a map depicting one mile on the surface, which is then rounded off to the representative fraction of 1/62,500.

reradiate The process of heat exchange in the lower atmosphere from Earth surfaces originally warmed by direct solar radiation.

rift valley Deep trenches or valleys on Earth's surface created where tectonic plates diverge from each other.

rimland The mainland coastal zone of the Caribbean, beginning with Belize and extending along the coast of Central America to northern South America.

Russification A policy of the Soviet Union designed to spread Russian settlers and influences to non-Russian areas of the country.

rust belt Regions of heavy industry that experience marked economic decline after their factories cease to be competitive.

Sahel The semidesert region at the southern fringe of the Sahara, and the countries that fall within this region, which extends from Senegal to Sudan. Droughts in the 1970s and early 1980s caused widespread famine and dislocation of population.

salinization The accumulation of salts in the upper layers of soil, often causing a reduction in crop yields, resulting from irrigation using water with high natural salt content and/or irrigation of soils that contain a high level of mineral salts.

Sanskrit The original Indo-European language of South Asia, introduced into northwestern India perhaps 4,000 years ago, from which modern Indo-Aryan languages evolved. Over the centuries, Sanskrit has become the classical literary language of the Hindus and is widely used as a scholarly second language, much like Latin in medieval Europe.

Schengen Agreement The 1985 agreement between some—but not all—European Union member countries to reduce border formalities in order to facilitate freer movements of Europeans between countries for work, study, or tourism. Initially this agreement created a new "Schengenland" of a borderless Europe where one could move freely, for example, between France, Germany, and Italy. By 2015, however, many countries had reestablished border controls in response to the large number of extralegal migrants entering Europe.

sectarian violence Conflicts that divide people along ethnic, religious, and sectarian lines.

sectoral transformation The evolution of a labor force from being highly dependent on the primary sector to being oriented around more employment in the secondary, tertiary, and quaternary sectors.

secularism The term describes both the separation of politics and religion, as well as the non-religious segment of a population. An example of the first usage is the secularism of the United States constitution, which clearly separates State from Church; whereas in the second usage it is common to refer to the growing secularism of Europe, referring to the disinterest in religion of a large part of the population.

sediment load The amount of sand, silt, and clay carried by a river.

Shanghai Cooperation Organization (SCO) Formed in 2001, a geopolitical group composed of China, Russia, Kazakhstan, Kyrgyzstan, Uzbekistan, and Tajikistan that focuses on common security threats and works to enhance economic cooperation and cultural exchange in Central Asia.

shield A large upland area of very old exposed rocks. Shields range in elevation from 600 to 5,000 feet (200 to 1,500 meters). The three major shields in South America are the Guiana, Brazilian, and Patagonian.

Shiite A Muslim who practices one of the two main branches of Islam. Shiites are especially dominant in Iran and nearby southern Iraq.

shogun, shogunate The true ruler of Japan before 1868. In contrast, the emperor's power was merely symbolic.

Sikhism An Indian religion combining Islamic and Hindu elements, founded in the Punjab region in the late 15th century. Most of the people of the Indian state of Punjab currently follow this religion.

Silk Road An historical trade route that extended across Central Asia, linking China with Europe and Southwest Asia.

siloviki Members of military and security forces in the Russian domain.

Slavic peoples A group of peoples in eastern Europe and Russia who speak Slavic languages, a distinctive branch of the Indo-European language family.

social and regional differentiation A process by which certain classes of people, or regions of a country, grow richer when others grow poorer.

socialist realism An artistic style once popular in the Soviet Union that was associated with realistic depictions of workers in their patriotic struggles against capitalism.

Southern African Development Community (SADC) SADC is an intergovernmental organization concerned with the socio-economic cooperation, integration, and security of its 15 member states in southern Africa. It is headquarted in Gaborone, Botswana.

sovereignty The ability of a government to control its territorial state without any interference from outside sources or bodies.

Soviet Union Created in 1917, a sprawling communist state that dominated the region until 1991. Also known as the Union of Soviet Socialist Republics (or USSR).

Spanglish A hybrid combination of English and Spanish spoken by Hispanic Americans.

special administrative region In China, a region of the country that temporarily maintains its own laws and own system of government. When Hong Kong was rejoined with China in 1997, it became a special administrative region, a position that it is scheduled to keep until 2047. In 1999 Macao passed from Portuguese rule to become China's second special administrative region.

Special Economic Zones (SEZs) Relatively small districts in China that have been fully opened to global capitalism.

spheres of influence In countries not formally colonized in the 19th and early 20th centuries (particularly China and Iran), limited areas gained by particular European countries for trade purposes and more generally for economic exploitation and political manipulation.

squatter settlement Makeshift housing on land not legally owned or rented by urban migrants, usually in unoccupied open spaces within or on the outskirts of a rapidly growing city.

steppe Semiarid grasslands found in many parts of the world. Grasses are usually shorter and less dense in steppes than in prairies.

structural adjustment programs Controversial yet widely implemented programs used to reduce government spending, encourage the private sector, and refinance foreign debt. Typically, these International Monetary Fund and World Bank policies trigger drastic cutbacks in government-supported services and food subsidies, which disproportionately affect the poor.

subduction zone A tectonic boundary where one colliding plate slips under another.

subnational organizations Groups that form along ethnic, ideological, or territorial lines that can induce serious internal divisions within a state.

subtropical jet stream A powerful atmospheric wind found in the lower latitudes of both the northern and southern hemispheres. Like the polar jet stream, the subtropical jet is caused by Earth's rotation.

Suez Canal A pivotal waterway connecting the Red Sea and the Mediterranean opened by the French in 1869.

Sunni A Muslim who practices the dominant branch of Islam.

superconurbation (megalopolis) A massive urban agglomeration that results from the coalescing of two or more formerly separate metropolitan areas.

supranational organizations Governing bodies that include several states, such as trade organizations, and often involve a loss of some state powers to achieve organizational goals.

sustainability To manage or to maintain something at a certain level so that it lasts, especially particular resources such as forests or soils.

sustainable agriculture A system of agriculture where organic farming principles, a limited use of chemicals, and an integrated plan of crop and livestock management combine to offer both producers and consumers environmentally friendly alternatives.

swidden Also called slash-and-burn agriculture, a form of cultivation in which forested or brushy plots are cleared of vegetation, burned, and then planted to crops, only to be abandoned a few years later as soil fertility declines. Also often called *shifting cultivation*.

syncretic religions Religions that feature a blending of different belief systems. In Latin America, for example, many animist practices were folded into Christian worship.

taiga The vast coniferous forest of Russia that stretches from the Urals to the Pacific Ocean. The main forest species are fir, spruce, and larch.

terrorism The systematic use of terror to achieve political or cultural goals.

thematic geography (systematic geography) These are the subject areas that geographers organize their research efforts such as environmental, population, cultural, political or economic geography, to name a few.

thematic map A thematic map focuses on a particular topic or subject area such as political boundary changes, refugee flows, or vegetation patterns. Most geographic research lends itself to thematic mapping at different scales of analysis.

theocracy A political state led by religious authorities. Also called a theocratic state.

theocratic state A political state led by religious authorities. Also called a *theocracy*.

total fertility rate (TFR) The average number of children who will be borne by women of a hypothetical, yet statistically valid, population, such as that of a specific cultural group or within a particular country. Demographers consider TFR a more reliable indicator of population change than the crude birthrate.

township Racially segregated neighborhoods created for nonwhite groups under apartheid in South Africa. They are usually found on the outskirts of cities and classified as black, coloured, or South Asian.

transform fault An earthquake fault where the ground on each side of the fault moves in opposite directions because of tectonic forces.

transform plate boundary Areas where two tectonic plates meet, with one plate sliding past the other in a horizontal direction.

transhumance A form of pastoralism in which animals are taken to high-altitude pastures during the summer months and returned to low-altitude pastures during the winter.

transmigration The planned, government-sponsored relocation of people from one area to another within a state territory.

Trans-Siberian Railroad A key southern Siberian railroad connection completed during the Russian empire (1904) that links European Russia with the Russian Far East terminus of Vladivostok.

Treaty of Tordesillas A treaty signed in 1494 between Spain and Portugal that drew a north–south line some 300 leagues west of the Azores and Cape Verde islands. Spain received the land to the west of the line and Portugal the land to the east.

tribal Refers to ethnic groups that historically did not have states or cities of their own and were politically organized only at the local level. In most tribal societies, relations of kinship are particularly important.

tribalism Allegiance to a particular tribe or ethnic group rather than to the nation-state. Tribalism is often blamed for internal conflict in Sub-Saharan states.

tsar A Russian term (also spelled *czar*) for "Caesar," or ruler. Tsars were the authoritarian rulers of the Russian empire before its collapse in the 1917 revolution.

tsetse fly A fly that is a vector for a parasite that causes sleeping sickness (typanosomiasis), a disease that especially affects humans and livestock. Livestock is rarely found in areas of Sub-Saharan Africa where the tsetse fly is common.

tsunami A very large sea wave induced by earthquakes.

tundra Arctic region with a short growing season in which vegetation is limited to low shrubs, grasses, and flowering herbs.

typhoon A large tropical storm, similar to a hurricane, that forms in the western Pacific Ocean in tropical latitudes and can cause widespread damage to the Philippines and coastal Southeast and East Asia.

UNASUR (Union of South American Nations) A supranational organization that seeks to integrate trade and population movements within South America. Created in 2008, it is modeled after the European Union.

unitary state A political system in which power is centralized at the national level.

universalizing religion A religion, usually with an active missionary program, that appeals to a large group of people, regardless of local culture and conditions. This contrasts with ethnic religions. Christianity and Islam both have strong universalizing components.

urban decentralization A process in which cities spread out over a larger geographic area.

urban heat island An effect in built-up areas in which development associated with cities often produces nighttime temperatures some 9 to 14°F (5 to 8°C) warmer than nearby rural areas.

urban primacy A state in which a disproportionately large city (for example, London, New York or Bangkok) dominates the urban system and is the center of economic, political, and cultural life.

urbanized population The percentage of a country's population living in settlements characterized as cities. Usually, high rates of urbanization are associated with higher levels of industrialization and economic development, since these activities are usually found in and around cities. Conversely, lower urbanized populations (less than 50 percent) are characteristic of developing countries.

Urdu One of Pakistan's official languages (along with English), Urdu is very similar to the Indian language of Hindi, although it includes more words derived from Persian and Arabic and is written in a modified form of the Persian Arabic alphabet. Although most Pakistanis do not speak Urdu at home, it is widely used as a second language and is extensively employed in education, the media, and government, thus giving Pakistan a kind of cultural unity.

viticulture Grape cultivation.

Warsaw Pact The Cold War military alliance of eight Soviet-controlled eastern European states created to counter the west's NATO Pact. The Warsaw Pact was formed in 1954 and disbanded in 1991.

water stress A condition where water availability is less than water demand, either currently or projected for the future.

White Australia Policy Before 1973, a set of stringent Australian limitations on nonwhite immigration to the country. It has been largely replaced by a more flexible policy today.

World Trade Organization (WTO) Formed as an outgrowth of the General Agreement on Tariffs and Trade (GATT) in 1995, a large collection of member states dedicated to reducing global barriers to trade.

Credits

Chapter 1 opening photo Rob Crandall **O1.1** Image © 2013 DigitalGlobe **O1.2** Edgar Su/Reuters **O1.4** John Zada/Alamy **O1.5** Marie Price **O1.6** Handout/Alamy **O1.7** Terry Whittaker/Alamy **Figure 1.1** Genevieve Vallee/Alamy **Figure 1.2** Image © 2013 DigitalGlobe **Figure 1.3** Robert Harding World Imagery **Figure 1.6** Rob Crandall **Figure 1.7** Rob Crandall **Figure 1.8** Frans Lemmens/Getty Images **Figure 1.9** Interfoto/Travel/Alamy **Figure 1.11** Noor Khamis/Reuters **Figure 1.12** Jon Arnold/Alamy **Figure 1.13** Edgar Su/Reuters **Figure 1.14** ASK Images/Alamy **Figure 1.15** Jim West/Alamy **Figure 1.16** ZUMA Press, Inc./Alamy **Figure 1.17** Jay Directo/AFP/Getty Images **Figure 1.22** NASA **Figure 1.2.1** AGE Fotostock **Quote, page 19** UN World Commission on Environment and Development, 1987 **Figure 1.25** Tim Graham/Alamy **Table 1.1** Data from: Population Reference Bureau, World Population Data Sheet, 2012 **Figure 1.28** John Zada/Alamy **Figure 1.29** Umit Bektas/Reuters **Figure 1.30** Creatista/YAY Media AS/Alamy **Figure 1.3.1** Marie Price **Figure 1.31** Joao Padua/AFP/Getty Images **Figure 1.33** Rosalrene Betancourt/Alamy **Figure 1.35** ThavornC/Shutterstock **Figure 1.36** Data from: http://www.bbc.com/news/world-?-25927595 and/or Washington Post, 2013 **Figure 1.37** Alexander Ryumin/ZUMA Press/Newscom **Figure 1.38** Gaizka Iroz/AFP/Getty Images **Figure 1.41** Handout/Alamy **Figure 1.4.1** Courtesy of Susan Wolfinbarger **Figure 1.4.2** Photo DigitalGlobe/AAAS/Getty Images **Quote, page 34** Susan Wolfinbarger **Figure 1.42** Rolex Dela Pena/EPA/Newscom **Table 1.2** Data from: World Bank, World Development Indicators, 2012. **Figure 1.46** Jon Arnold Images, Ltd./Alamy **Figure 1.47** Terry Whittaker/Alamy **Figure 1.48** Images&Stories/Alamy **R1.A** Joerg Boethling/Alamy **R1.B** Michele Falzone/JAI/Alloy/Corbis **R1.D** Les Rowntree **R1.E** Philip Ojisua/Afp/Getty Images **R1.F** Liba Taylor/Robert Harding Picture Library Ltd/Alamy **QR code** World Bank Development Indicators for 2015, Table 1.1, http://wdi.worldbank.org/table/1.1?

Chapter 2 opening photo Christian Beier/AGE Fotostock **O2.1** Ivan Alvarado/Reuters **O2.2** Andrew Biraj/Reuters **O2.3** Chris Harris/Glow Images **O2.4** Paul Strawson/Alamy **O2.5** Otmar Smit/Shutterstock **Figure 2.1** Wicaksono Saputra/Alamy **Figure 2.4** Ivan Alvarado/Reuters **Figure 2.5** Ragnar Th. Sigurdsson/Arctic Images/Alamy **Figure 2.6** US Geological Survey, United States Department of the Interior. **Figure 2.8** Greg Vaughn/VW Pics/AGE Fotostock **Figure 2.1.1** Courtesy of M. Jackson **Figure 2.11** NASA **Figure 2.14** National Weather Service, National Oceanic and Atmospheric Administration. http://www.weather.gov/satellite?image=ir#vis **Figure 2.17b** Robbie Shone/Science Source **Figure 2.19** Andrew Biraj/Reuters **Figure 2.2.2** Blickwinkel/Hummel/Alamy **Figure 2.20** Pearson Education, Inc. **Figure 2.21** Duncan McKenzie/Getty Images **Figure 2.23a** Berndt Fischer/AGE Fotostock **Figure 2.23b** Karen Desjardin/Getty Images **Figure 2.24** Kevin Foy/Alamy **Figure 2.25a** 06photo/Fotolia **Figure 2.25b** Tuul and Bruno Morandi/Getty Images **Figure 2.26a** Mshch/Fotolia **Figure 2.26b** SHSPhotography/Fotolia **Figure 2.27** Chris Harris/Glow Images **Figure 2.3.1** Universal Images Group/Getty Images **Figure 2.4.1** Paul Strawson/Alamy **Figure 2.4.2** Koenig/Wello/Splash/Newscom **Figure 2.29** McClatchy-Tribune Content Agency, LLC/Alamy **Figure 2.31** Otmar Smit/Shutterstock **Figure 2.32** Jake Lyell/Alamy **R2.A** Kyodo/Reuters **R2.B** Todd Shoemake/Shutterstock **R2.C** USGS **R2.D** 68/Dinodia Photos/Ocean/Corbis **QR code** World Bank data on CO2 emissions, http://data.worldbank.org/indicator/EN.ATM.CO2E.PC?

Chapter 3 opening photo Rudy Sulgan/Corbis **O3.1** NASA **O3.2** Rob Crandall/The Image Works **O3.3** J. Emilio Flores/La Opinion/Newscom **O3.4** Ian Shive/Aurora Photos/Alamy **O3.5** Angela Peterson/MCT/Landov **Figure 3.2** Don Mason/Corbis/Glow Images **Figure 3.3** Michael Reynolds/EPA/Newscom **Figure 3.4** NASA **Figure 3.5** Caleb Foster/Fotolia **Figure 3.1.2** Pete Mcbride/National Geographic Creative/Corbis **Figure 3.8** Bruce Coleman, Inc./Alamy **Figure 3.10** Stephen Shames/Polaris/Newscom **Figure 3.2.2** 2d Alan King/Alamy **Figure 3.12** Luiz Felipe Castro/Getty Images **Figure 3.15** Mastering Microstock/Shutterstock **Figure 3.17** Rob Crandall/The Image Works **Figure 3.18** Ian Dagnall/Alamy **Figure 3.19** NASA **Figure 3.21** William Wyckoff **Figure 3.25** William Wyckoff **Figure 3.26a** National Geographic Image Collection/Alamy **Figure 3.26b** William Wyckoff **Figure 3.27b** J. Emilio Flores/La Opinion/Newscom **Figure 3.29** Matthieu Paley/Corbis **Figure 3.30** NASA **Figure 3.32** Pat and Rosemarie Keough/Corbis **Figure 3.33** Ian Shive/Aurora Photos/Alamy **Figure 3.35** Jim Noelker/AGE Fotostock **Figure 3.36** Katja Kreder/Image Broker/Alamy **Figure 3.37** Angela Peterson/MCT/Landov **Figure 3.4.1** Courtesy of Lucia Lo **Figure 3.39** David South/Alamy **R3.B** Radius/Corbis **R3.C** Phil Augustavo/Getty Images **R3.D** Geogphotos/Alamy **R3.E** Frontpage/Shutterstock **QR code** Census Bureau data on state populations, http://www.census.gov/population/projections/data/statepyramid.html?

Chapter 4 opening photo Rieger Bertrand/Hemis/Corbis **O4.1** Rob Crandall **O4.2** Rob Crandall **O4.3** Rob Crandall **O4.4** Guillermo Granja/Reuters **O4.5** Keith Dannemiller/Alamy **Figure 4.2** Rob Crandall **Figure 4.3** Ian Trower/Robert Harding/Newscom **Figure 4.4** Danny Lehman/Documentary Value/Corbis **Figure 4.5** Rob Crandall **Figure 4.6** Newscom **Figure 4.8** Rob Crandall **Figure 4.10a** Bernard Francou-IRD **Figure 4.10b** Bernard Francou-IRD **Figure 4.12a** NASA **Figure 4.12b** NASA **Figure 4.13** Rob Crandall **Figure 4.14** Rob Crandall **Figure 4.15** Newscom **Figure 4.1.1** Fernando Vergara/AP Images **Figure 4.1.2** Javier Galeano/AP Images **Figure 4.17** Enrique Castro-Mendivil/Landov **Figure 4.19** David Santiago Garcia/Aurora Photos/Alamy **Figure 4.20a** Rob Crandall **Figure 4.20b** Rob Crandall **Figure 4.20c** Rob Crandall **Figure 4.20d** Rob Crandall **Figure 4.22** Orlando Kissner/AFP/Getty Images **Figure 4.23** Rob Crandall **Figure 4.24** Rob Crandall **Figure 4.26** Alexandro Auler/LatinContent/Getty Images **Figure 4.2.2** Guillermo Granja/Reuters **Figure 4.29** Felipe Trueba/EPA/Newscom **Figure 4.30** Adapted from: The Economist, Nov. 22, 2012 **Figure 4.3.1** Courtesy of Corrie Drummond Garcia **Quote, page 131** Corrie Drummond **Figure 4.31** Paulo Fridman/Corbis **Figure 4.4.1** Kaveh Kazemi/Getty Images **Figure 4.33** Keith Dannemiller/Alamy **Figure 4.35** Rob Crandall **Figure 4.36** Rob Crandall **Figure 4.37** Adapted from: World Bank Development Indicators, 2013 **R4.A** Fernando Vergara/AP Images **R4.B** Luoman/Getty Images **R4.D** Google Earth **R4.E** Frontpage/Shutterstock **QR code** ICO coffee production Figures, 2011-2014, http://www.ico.org/prices/po-production.pdf?

Chapter 5 opening photo Roberto Fumagalli/Alamy **O5.1** Image © 2013 DigitalGlobe. US Dept of State Geographer. © 2013 Google. **O5.2** Margaret S/Alamy **O5.3** Rob Crandall **O5.4** Ana Martinez/Reuters **O5.5** Rob Crandall **Figure 5.2** Rob Crandall **Figure 5.3** Grand Tour/Corbis **Figure 5.4** Hufton Crow/AGE Fotostock **Figure 5.5** Reuters **Crisis Mapping, page 146** Adapted from www.newswatch.nationalgeographic.com, How Crisis Mapping Saved Lives in Haiti, July 2, 2012. **Figure 5.1.1** Ushahidi Haiti Project (UHP) **Figure 5.6** Adapted from: Temperature and precipitation data from E. A. Pearce and C. G. Smith, The World Weather Guide, London: Hutchinson, 1984 **Figure 5.7** Desmond Boylan/Reuters **Figure 5.8** Based on DK World Atlas, London: DK Publishing, 1997, pp. 7, 55 **Figure 5.9** Image © 2013 DigitalGlobe. US Dept of State Geographer. © 2013 Google. **Figure 5.10** Rob Crandall **Figure 5.11** Rob Crandall **Figure 05.14** Data from Barry Levin, Caribbean Exodus, Westport, CT: Praeger Publishers, 1987 **Figure 5.15** Corbis **Figure 5.16** Orlando Barria/Newscom **Figure 5.17** Rob Crandall **Figure 5.18** Margaret S/Alamy **Figure 5.2.1** Google Earth **Figure 5.19** Data based on Philip Curtin, The Atlantic Slave Trade, A Census, Madison: University of Wisconsin Press, 1969, p. 268 **Figure 5.20** Ranu Abhelakh/Reuters/Corbis **Figure 5.21** Rob

Crandall **Figure 5.3.1** Shi Rong Xinhua News Agency/Newscom **Figure 5.23** Rob Crandall **Figure 5.24** Rob Crandall **Figure 5.26** Ana Martinez/Reuters **Figure 5.27** Rob Crandall **Figure 5.29** Frank Heuer/laif/Redux **05.31** Frank Fell/Robert Harding World Imagery **Figure 5.4.1** Courtesy of Sarah Blue **Quote, page 167** Sarah Blue **Figure 5.33** Disability Images/Alamy **R5.A** Google Earth **R5.D** Google Earth **R5.E** Grand Tour/Corbis **QR code** World Bank data on migration and remittances, http://econ.worldbank.org/WBSITE/EXTERNAL/EXTDEC/EXTDECPROSPECTS/0,,contentMDK:22759429~pagePK:64165401~piPK:64165026~theSitePK:476883,00.html?

Chapter 6 opening photo Rob Crandall **O6.1** Robert Caputo/Getty Images **O6.2** Ton Koene/Horizons WWP/Alamy **O6.3** Amar Grover/Getty Images **O6.4** Alexander Joe/Afp/Gettyimages **O6.5** Heiner Heine/Image Broker/Alamy **Figure 6.2** Jake Lyell/Alamy **Figure 6.3** Rob Crandall **Figure 6.4** Afripics. Com/Alamy **Figure 6.5** Robert Caputo/Getty Images **Figure 6.7** Rob Crandall **Figure 6.8** Heeb Christian/Robert Harding World Imagery **Figure 6.10** Daniel Berehulak/Getty Images **Figure 6.1.1** Mike Goldwater/Alamy **Figure 6.12** Akintunde Akinleye/Reuters **Figure 6.13** Rob Crandall **Figure 6.16** Jake Lyell/Alamy **Figure 6.17** Newscom **Figure 6.18** Data from http://www.cdc.gov/vhf/ebola/outbreaks/2014-west-africa/distribution-map.html. **Figure 6.20** Thierry Gouegnon/Reuters **Figure 6.21** Ton Koene/Horizons WWP/Alamy **Figure 6.22** Jeremy Graham/DB Images/Alamy **Figure 6.23** Max Milligan/Getty Images **Figure 6.24a** Google Earth **Figure 6.24b** Google Earth **Figure 6.26** Neil Cooper/Alamy **Figure 6.28** Gavin Hellier/Alamy **Figure 6.30** Amar Grover/Getty Images **Figure 6.2.1** Shashank Bengali/Mct/Newscom **Figure 6.2.2** Face To Face/ZUMA Press/Newscom **Figure 6.31** Streeter Lecka/Getty Images **Figure 6.34** Ulrich Doering/Alamy **Figure 6.35** Data from UNHCR Global Trends, 2013 **Figure 6.3.1** Alexander Joe/Afp/Gettyimages **Figure 6.36** Newscom **Figure 6.37** Jake Lyell/Alamy **Figure 6.38** Pan Siwei Xinhua News Agency/Newscom **Figure 6.39** Solar Reserve **Figure 6.4.1** Courtesy of Fenda Akiwumi **Quote, page 208** Fenda Akiwumi **Figure 6.40** Data from World Development Indicators, 2015, Table 6 **Figure 6.41** Siphiwe Sibeko/Reuters **Figure 6.43** Heiner Heine/Image Broker/Alamy **R6.A** Rob Crandall **R6.B** Google Earth **R6.E** Abenaa/Getty Images **QR code** United Nations AIDS information website, http://aidsinfo.unaids.org/#?

Chapter 7 opening photo Gavin Hellier/Robert Harding World Imagery **O7.1** Alan Carey/Spirit/Corbis **O7.2** DarkGrey/Getty Images **O7.3** Duby Tal/Albatross/Superstock **O7.4** Yin Bogu/Xinhua Press/Corbis **O7.5** Arterra Picture Library/Alamy **Figure 7.2** PixelPro/Alamy **Figure 7.3** Walter Bibikow/Jon Arnold Images Ltd/Alamy **Figure 7.4** NASA **Figure 7.5** Independent Picture Service/Alamy **Figure 7.6** Alan Carey/Spirit/Corbis **Figure 7.8** DeAgostini/Getty Images **Figure 7.9** Galyna Andrushko/Shutterstock **Figure 7.11** Egmont Strigl/Image Broker/AGE Fotostock **Figure 7.12** Muratart/Shutterstock **Figure 7.13** Worldspec/NASA/Alamy **Figure 7.1.1** Jochen Tack/arabianEye/Corbis **Figure 7.14** Patrick Syder/Lonely Planet Images/Getty Images **Figure 7.16** Izzet Keribar/Getty Images **Figure 7.17** Modified from Clawson and Fisher, 2004, World Regional Geography,8th ed. **Figure 7.18** Wigbert Röth/Image Broker/Newscom **Figure 7.19** Peter Horree/Alamy **Figure 7.20** Jalil Bounhar/AP Images **Figure 7.21** Marcia Chambers/dbimages/Alamy **Figure 7.22** Wael Hamdan/Alamy **Figure 7.23** Megapress/Alamy **Figure 7.24** DarkGrey/Getty Images **Figure 7.27** US Census Bureau, http://www.census.gov/population/international/data/idb/region.php?N=%20Results%20&T=12&A=separate&RT=0&Y=2015&R=-1&C=EG, http://www.census.gov/population/international/data/idb/region.php?N=%20Results%20&T=12&A=separate&RT=0&Y=2015&R=-1&C=IR, http://www.census.gov/population/international/data/idb/region.php?N=%20Results%20&T=12&A=separate&RT=0&Y=2015&R=-1&C=AE **Figure 7.28** Source: Modified from Rubenstein, 2005, An Introduction to Human Geography,8th ed., Upper Saddle River, NJ: Prentice Hall **Figure 7.2.1** Data from: Economist, 25 April 2015. **Figure 7.2.2** Fabrizio Villa/Polaris/Newscom **Figure 7.29** Newscom **Figure 7.30** Modified from Rubenstein, 2011, An Introduction to Human Geography,10th ed., Upper Saddle River, NJ: Prentice Hall, and National Geographic Society, 2003,Atlas of the Middle East, Washington, DC **Figure 7.31** Duby Tal/Albatross/Superstock **Figure 7.32** Modified from Rubenstein, 2011, An Introduction to Human Geography, 10th ed., Upper Saddle River, NJ: Prentice Hall and National Geographic Society, 2003, Atlas of the Middle East, Washington, DC **Figure 7.33** Santiago Urquijo/Getty Images **Figure 7.34** Chris Hondros/Staff/Getty Images **Figure 7.35** Fadi Al-Assaad/Reuters **Figure 7.36** Data from: Economist, May 2015 **Figure 7.37** Rex Features/AP Images **Figure 7.38** Modified from Rubenstein, 2011, An Introduction to Human Geography, 10th ed., Upper Saddle River, NJ: Prentice Hall **Figure 7.39** Modified from Rubenstein, 2011, An Introduction to Human Geography, 10th ed., Upper Saddle River, NJ: Prentice Hall **Figure 7.39b** Yin Bogu/Xinhua Press/Corbis **Figure 7.3.1** Courtesy of Karen Culcasi **Figure 7.3.2** Courtesy of Karen Culcasi **Figure 7.41** Modified from Rubenstein, 2011, An Introduction to Human Geography, 10th ed., Upper Saddle River, NJ: Prentice Hall **Figure 7.42** Image Landsat. Image ? 2013 TerraMetrics. Image ? DigitalGlobe. Data SIO, NOAA, U.S. Navy, NGA, GEBCO. **Figure 7.4.1** Adam Reynolds/Bloomberg/Contributor/Getty Images **Figure 7.43a** Xi Li, Wuhan University/University of Maryland/Courtesy of #withSyria **Figure 7.43b** Xi Li, Wuhan University/University of Maryland/Courtesy of #withSyria **Figure 7.45** Arterra Picture Library/Alamy **R7.A** Lukasz Janyst/Shutterstock **R7.C** Johnny Dao/Shutterstock **R7.D** Barry Gregg/Getty Images **R7.E** Google Earth **QR code** WHO data/interactive atlas page on physicians, http://www.who.int/en/?

Chapter 8 opening photo Shahid Khan/Shutterstock **O8.1** Philippe Body/AGE Fotostock **O8.2** Stefania Mizara/Corbis **O8.3** Clynt Garnham Education/Alamy **O8.4** Dylan Martinez/Reuters **O8.5** Photocreo Bednarek/Fotolia **Figure 8.1** Roz Gaizkaroz/AFP/Getty Images **Figure 8.4** Hemis/Alamy **Figure 8.5** Philippe Body/AGE Fotostock **Figure 8.6** Smart.art/Shutterstock **Figure 8.9** Magnus Qodarion/Alamy **Figure 8.10** Sarah Leen/National Geographic Creative **Figure 8.13** Thomson Reuters **Figure 8.1.1** Courtesy of Weronika Kusek **Figure 8.1.2** Courtesy of Weronika Kusek **Figure 8.2.1** Eurasia/Robert Harding World Imagery **Figure 8.2.2** Gabriele Hanke/Image Broker/Alamy **Figure 8.15** Stefania Mizara/Corbis **Figure 8.16** Les Rowntree **Figure 8.17** Clynt Garnham Education/Alamy **Figure 8.3.1** Alex Segre/Alamy **Figure 8.19** Peter Forsberg/EU/Alamy **Figure 8.21** Koray Ersin/Fotolia **Figure 8.22** Jonathan Larsen/Diadem Images/Alamy **Figure 8.25** Peter Jordan/Alamy **Figure 8.26a** Bettmann/Corbis **Figure 8.26b** Picture-Alliance/DPA/Newscom **Figure 8.4.1** Raf/ZUMA Press/Newscom **Figure 8.28** Dylan Martinez/Reuters **Figure 8.29** Ann Pickford/Alamy **Figure 8.32a** Skorpionik00/Shutterstock **Figure 8.32b** Photocreo Bednarek/Fotolia **Figure 8.33** Data from: Sources: http://ec.europa.eu/eurostat/statisticsexplained/index.php/Unemployment_statistics, http://ec.europa.eu/eurosta/statisticsexplained/index.php/Unemployment_statistics_at_regional_level **Figure 8.34** Nikolas Georgiou/Alamy **Figure 8.35** Kerkla/Getty Images **R8.A** CSP Remik44992/Fotosearch LBRF/AGE Fotostock **R8.C** Leonid Serebrennikov/AGE Fotostock **R8.D** Bernd.neeser/Shutterstock **R8.E** Peter Erik Forsberg/AGE Fotostock **QR code** EU's Eurostat page on asylum statistics, http://ec.europa.eu/eurostat/statisticsexplained/index.php/Asylum_statistic?

Chapter 9 opening photo Gleb Garanich/Reuters **O9.1** Yul/Fotolia **O9.2** Yulenochekk/Fotolia **O9.3** Gerner Thomsen/Alamy **O9.4** Sergie Chirkov/EPA/Corbis **O9.5** Ivan Sekretarev/AP Images **Figure 9.2** Andrey Rudakov/Bloomberg/Getty Images **Figure 9.3** Serghei Starus/Alamy **Figure 9.6** Mykola Mazuryk/Shutterstock **Figure 9.7** Masami Goto/Glow Images **Figure 9.1.1** AP Images **Figure 9.1.2** Krasilnikov Stanislav/ZUMA Press/Newscom **Figure 9.8** NASA **Figure 9.9** Irakli Gedenidze/AP Images **Figure 9.11** PhotoXpress/ZumaPress/Newscom **Figure 9.12** Yul/Fotolia **Figure 9.13** Lev Fedoseyev/ITAR-TASS Photo Agency/Newscom **Figure 9.14** Environment Images/UIG/Getty Images **Figure 9.2.2** Volodymyr Shuvayev/AFP/Getty Images **Figure 9.16** Google Earth **Figure 9.17** ITAR-TASS Photo Agency/Alamy **Figure 9.18** Yulenochekk/Fotolia **Figure 9.19** Hemis/Alamy **Figure 9.3.1** Courtesy of Dmitry Streletskiy **Figure 9.3.2** Ilya Naymushin/Reuters **Figure 9.21** Misha Japaridze/AP Images **Figure 9.22** Cindy Miller Hopkins/

Alamy **Figure 9.23** ITAR-TASS Photo Agency/Alamy **Figure 9.28** Gerner Thomsen/Alamy **Figure 9.30** Mikhail Japaridze/AP Images **Figure 9.31a** Schoendorfer/REX Shutterstock/Newscom **Figure 9.31b** Dieter Nagl/AFP/ Getty Images **Figure 9.32** Sergie Chirkov/EPA/Corbis **Figure 9.36** Sergey Ponomarev/AP Images **Figure 9.38** ITAR-TASS Photo Agency/Alamy **Figure 9.39** Sergey Dolzhenko/EPA/Newscom **Figure 9.4.1** Astapkovich Vladimir/ITAR-TASS Photo AgencyPhotos/Newscom **Figure 9.41** Ivan Sekretarev/AP Images **R9.A** Svetlana Bobrova/Shutterstock **R9.C** ITAR-TASS Photo Agency/Alamy **R9.D** ITAR-TASS Photo Agency/Alamy **R9.E** Google Earth **QR code** OECD's annual FDI statistics on Russian Federation, http://www.oecd-ilibrary.org/economics/ country-statistical-profile-russian-federation_20752288-table-rus?

Chapter 10 opening photo ChinaFotoPress/Getty Images **O10.1** Theodore Kaye/Alamy **O10.2** Ilya Postnikov/Fotolia **O10.3** J. Marshall/Tribaleye Images/ Alamy **O10.4** Stringer/Reuters **O10.5** View Stock/Alamy **Figure 10.2** Ru Baile/Fotolia **Figure 10.4** Deidre Sorensen/Photoshot/Newscom **Figure 10.6a** NASA **Figure 10.6b** NASA **Figure 10.6c** NASA **Figure 10.1.1** Phil Micklin **Figure 10.7** Theodore Kaye/Alamy **Figure 10.10** Jesse Allen/Robert Simmon/NASA Earth Observatory **Figure 10.11** Ted Wood/Aurora Photos/ Alamy **Figure 10.2.1** Denis Sinyakov/Reuters **Figure 10.13** GM Photo Images/Alamy **Figure 10.14** Ilya Postnikov/Fotolia **Figure 10.3.1** Courtesy of Holly Barcus **Figure 10.16** Cyrille Gibot/Alamy **Figure 10.18** J. Marshall/ Tribaleye Images/Alamy **Figure 10.19** Matthew Ashton/Alamy **Figure 10.21** Dimitri Borko/AFP/Getty Images **Figure 10.23** Peter Parks/AFP/Getty Images **Figure 10.24** Stringer/Reuters **Figure 10.25** AFP/Getty Images **Figure 10.26** Oliviero Olivieri/Robert Harding/Newscom **Figure 10.27** Shamil Zhumatov/ Reuters **Figure 10.4.1** Jesse Allen/Robert Simmon/NASA Earth Observatory **Figure 10.28** View Stock/Alamy **Figure 10.30** John Costello/MCT/Newscom **R10.A** Cai Chuqing/Shutterstock **R10.C** Mark Ralston/AFP/Getty Images **R10.D** Maximilian Clarke/Getty Images **R10.E** Michal Cerny/Alamy **QR code** World Bank data on international tourism receipts, http://data.worldbank.org/ indicator/ST.INT.RCPT.CD?

Chapter 11 opening photo Topic Photo Agency/AGE Fotostock **O11.1** JIJI Press/AFP/Getty Images **O11.2** B. Lawrence/BL Images/Alamy **O11.3** Iain Masterton/Alamy **O11.4** KCNA/Reuters **O11.5** Chris McGrath/Getty Images **Figure 11.4** Masao Takahashi/AFLO/Glow Images **Figure 11.5** AFP/Getty Images **Figure 11.6** Documentary Value/Corbis **Figure 11.7** Color China Photo/AP Images **Figure 11.8** JIJI Press/AFP/Getty Images **Figure 11.10** NASA **Figure 11.11a** Panorama Images/The Image Works **Figure 11.11b** Meng Liang/ChinaFotoPress/ZUMApress/Newscom **Figure 11.12** China Daily/Reuters **Figure 11.13** Source: http://www.bp.com/en/global/corporate/ about-bp/energy-economics/statistical-review-of-world-energy/statistical-review-downloads. **Figure 11.1.1** Courtesy of Gregory Veeck **Figure 11.16** Skye Hohmann Japan Images/Alamy **Figure 11.17** Liu Xiaoyang/China Images/Alamy **Figure 11.2.1** Yang Zheng/Imaginechina/AP Images **Figure 11.18** Aldo Pavan/Getty Images **Figure 11.19** B. Lawrence/BL Images/Alamy **Figure 11.3.1** Trevor Snapp/Bloomberg/Getty Images **Figure 11.20** Pearson Education, Inc. **Figure 11.22** Iain Masterton/Alamy **Figure 11.23** Rob Crandall **Figure 11.24** Jo Yong-Hak/Reuters **Figure 11.26** Masa Uemura/ Alamy **Figure 11.27** Olaf Schubert/Image Broker/Alamy **Figure 11.29** Ashley Pon/Getty Images **Figure 11.30** China Photos/Stringer/Getty Images **Figure 11.32** Universal Images Group/SuperStock **Figure 11.34** Kyodo/Newscom **Figure 11.35** KCNA/Reuters **Figure 11.36** Bobby Yip/Reuters **Figure 11.37** Lucas Vallecillos/AGE Fotostock **Figure 11.38** Chris McGrath/Getty Images **Figure 11.39** Ahn Young-Joon/AP Images **Figure 11.4.1** Chung Sung-Jun/ Getty Images **Figure 11.40** Yuan Shuiling/Imaginechina/AP Images **Figure 11.41** Source: World Bank https://www.imf.org/external/pubs/ft/weo/2015/01/ weodata/index.aspx **Figure 11.42** Rex/Newscom **R11.B** Adrian Bradshaw/ EPA/Newscom **R11.C** SeanPavonePhoto/Fotolia **R11.D** Patrick Lin/AFP/ Getty Images **R11.E** Ma Jian/Chinafotopress/ZUMA Press/Newscom **QR code** National Bureau of Statistics of China gross regional product data, http://data.stats.gov.cn/english/easyquery.htm?cn=E0103?

Chapter 12 opening photo Abbas Ali/Anadolu Agency/Getty Images **O12.1** Kamal Sachar/AFP/Getty Images **O12.2** Ajit Solanki/AP Images **O12.3** Hemant Chawla/The India Today Group/Getty Images **O12.4** Anupam Nath/AP Images **O12.5** Newscom **Figure 12.2** Science Source **Figure 12.3** Roberto Schmidt/ AFP/Getty Images **Figure 12.1.1** Courtesy of P.P. Karan **Figure 12.5** Kamal Sachar/AFP/Getty Images **Figure 12.7** Data from: http://statisticstimes.com/ economy/gdp-capita-ofindian-states.php **Figure 12.2.1** Mohammed Abidally/ Alamy **Figure 12.8** Aditya "Dicky" Singh/Alamy **Figure 12.11** Education Images/Getty Images **Figure 12.12** Glow Images **Figure 12.13** Ajit Solanki/ AP Images **Figure 12.14** Rob Crandall **Figure 12.15** Alamy **Figure 12.16** Newscom **Figure 12.17** Hemant Chawla/The India Today Group/Getty Images **Figure 12.18** Money Sharma/EPA/Newscom **Figure 12.3.1** Vinod Singh/Anadolu Agency/Getty Images **Figure 12.20** Newscom **Figure 12.21** Newscom **Figure 12.23** Olaf Krüger/Image Broker/Newscom **Figure 12.24** Gideon Mendel/Corbis **Figure 12.4.1** Arka Mediaworks **Figure 12.26** Rob Crandall **Figure 12.30** R. Kiedrowski/Arco Images/Alamy **Figure 12.31** Anupam Nath/AP Images **Figure 12.33** Qadir Baloch/Reuters **Figure 12.34** Newscom **Figure 12.35** Hemis/Alamy **Figure 12.36** Philippe Lissac/Picture-Alliance/Godong/Newscom **Figure 12.38** Chirag Wakaskar/Alamy **Figure 12.39** Namas Bhojani/Bloomberg/Getty Images **Figure 12.40** Rob Crandall **Figure 12.41** Joerg Boethling/Alamy **R12.A** Travel India/Alamy **R12.B** Paul Biris/Getty Images **R12.D** Yawar Nazir/Getty Images **R12.E** Harish Tyagi/Epa/ Newscom **QR code** India's Planning Commission web page on sex ratio, http:// planningcommission.nic.in/data/datatable/data_2312/DatabookDec2014%20 215.pdf?

Chapter 13 opening photo Massimo Borchi/Atlantide Phototravel/Corbis **O13.1** Y. T. Haryono/Anadolu Agency/Getty Images **O13.2** Romeo Gacad/ Newscom **O13.3** Thierry Falise/LightRocket/Getty Images **O13.4** Athit Perawongmetha/Reuters **O13.5** Jonathan Drake/Bloomberg/Getty Imagaes **Quote, page 420** Mandarin Gallery's website, Starwood Hotels and Resorts Worldwide, Inc. http://www.thewestinsingapore.com/en/localareashopping. **Figure 13.2** Google Earth **Figure 13.4** Y. T. Haryono/Anadolu Agency/Getty Images **Figure 13.5** Kevin Frayer/Getty Images **Figure 13.7** Romeo Gacad/ Newscom **Figure 13.8** V. Miladinovic/India Government Tourist Office **Figure 13.9** Philippe Body/Hemis/Alamy **Figure 13.10** Michael S. Yamashita/ Terra/ Corbis **Figure 13.1.2** Dimas Ardian/Bloomberg/Getty Images **Figure 13.12** The Asahi Shimbun Premium/Getty Images **Figure 3.2.1** Thierry Falise/ LightRocket/Getty Images **Figure 13.13** Peter Bowater/Alamy **Figure 13.15** Cheryl Ravelo/Reuters **Figure 13.16** Google Earth **Figure 13.17** Image Source/Superstock **Figure 13.21** EPA/Corbis **Figure 13.22** Thierry Falise/ LightRocket/Getty Images **Figure 13.24** AFP/Getty Images **Figure 13.25** Wong Maya-E/AP Images **Figure 13.28** AP Images **Figure 13.29** Stewart Cohen/ Tetra Images/Alamy **Figure 13.30** Vincent Yu/AP Images **Figure 13.31** Athit Perawongmetha/Reuters **Figure 13.32** DigitalGlobe/ScapeWare3d/Getty Images **Figure 13.33** Erik de Castro/Reuters **Figure 13.34** Roslan Rahman/ Newscom **Figure 13.35** Jonathan Drake/Bloomberg/Getty Imagaes **Figure 13.36** Sukree Sukplang/Reuters **Figure 3.3.1** Think4photop/Shutterstock **Figure 13.38** Heeb Christian/Prisma Bildagentur AG/Alamy **Figure 3.4.1** Courtesy of Rachel Silvey **R13.B** NASA **R13.C** Wong Maya-E/AP Images **R13.D** Ezra Acayan/NurPhoto/Corbis **R13.E** Sukree Sukplang/Reuters **QR code** Ethnologue: Languages of the World, https://www.ethnologue.com/browse/ countries?

Chapter 14 opening photo William Chopart/Mond Image/AGE Fotostock **O14.1** Raga Jose Fuste/Prisma Bildagentur AG/Alamy **O14.2** Don Fuchs/ AGE Fotostock **O14.3** Newscom **O14.4** Lucy Pemoni/Reuters **O14.5** Wayne G. Lawler/Science Source **Figure 14.2** Raga Jose Fuste/Prisma Bildagentur AG/Alamy **Figure 14.3** Paul Mayall/Alamy **Figure 14.4** Jochen Schlenker/ Robert Harding World Imagery/Getty Images **Figure 14.5** Maxine House/ Alamy **Figure 14.6** Chad Ehlers/Getty Images **Figure 14.9** Glenn Nicholls/ AP Images **Figure 14.10** 42pix/Alamy **Figure 14.11** Arco Images/Glow Images **Figure 14.13** US Department of Defense **Figure 14.14** Friedrich

Stark/Alamy **Figure 14.1.1** Torsten Blackwood/Newscom **Figure 14.15** David Wall/Danita Delimont Photography/Newscom **Figure 14.17** Don Fuchs/AGE Fotostock **Figure 14.18** David Steele/Shutterstock **Figure 14.19** Colin Monteath/AGE Fotostock **Figure 14.21** Historical Image Collection by Bildagentur-Online/Alamy **Figure 14.22** Bill Bachmann/Science Source **Figure 14.23** Mago World Image/Pixtal/AGE Fotostock **Figure 14.24** Jackie Ellis/Alamy **Figure 14.25** Egmont Strigl/Image Broker/Alamy **Figure 14.26** Mark A. Johnson/Alamy **Figure 14.28** Newscom **Figure 14.30** Van der Meer Marica/ ArTerra Picture Library/AGE Fotostock **Figure 14.2.1** Rick Rycroft/AP Images **Figure 14.2.2** Australian Department of Immigration and Citizenship/Getty Images **Figure 14.32** Blickwinkel/Alamy **Figure 14.33** Thomas Cockrem/Alamy **Figure 14.34** Phil Walter/Getty Images **Figure 14.36** Danny Martindale/WireImage/Getty Images **Figure 14.37** Education Images/UIG/Getty Images **Figure 14.38** Angela Prati/AGE Fotostock **Figure 14.40** Lucy Pemoni/Reuters **Figure 14.41** Glenn Campbell/The Sydney Morning Herald/Fairfax Media via Getty Images **Figure 14.42** Torsten Blackwood/Getty Images **Figure 14.43** Brisbane Architectual and Landscape Photographer/Getty Images **Figure 14.3.1** Courtesy of Laura Brewington **Figure 14.44** Wayne G. Lawler/Science Source **Figure 14.45** Fred Kruger/ Picture-Alliance/Dumont Bildar/Newscom **Figure 14.4.1** Stephen Andrews/ AGE Fotostock **R14.A** Takashi Hagihara/Corbis **R14.B** Les Rowntree **R14.C** Chris Cheadle/Alamy **R14.D** Stocktrek Images/Getty Images **R14.E** Tatiana Belova/Fotolia **QR code** Map of average annual precipitation for New Zealand's South Island, http://www.teara.govt.nz/en?

CVR WIN-Initiative/Getty Images **Icon.1 Globe icon** Iconspro/Shutterstock **Icon.2 Leaf icon** Helga Pataki/Shutterstock **Icon.3 Computer icon** Cobalt88/Shutterstock **Icon.4 Recycle icon** Zebra-Finch/Shutterstock **Icon.5 Pointer icon** Doungtawan/Shutterstock **Icon.6 Business icon** Route55/Shutterstock?

Index

F

J

K

L

O

P

Q

R

World – Physical

Great Basin	Land features
Caribbean Sea	Water bodies
Aleutian Trench	Underwater features

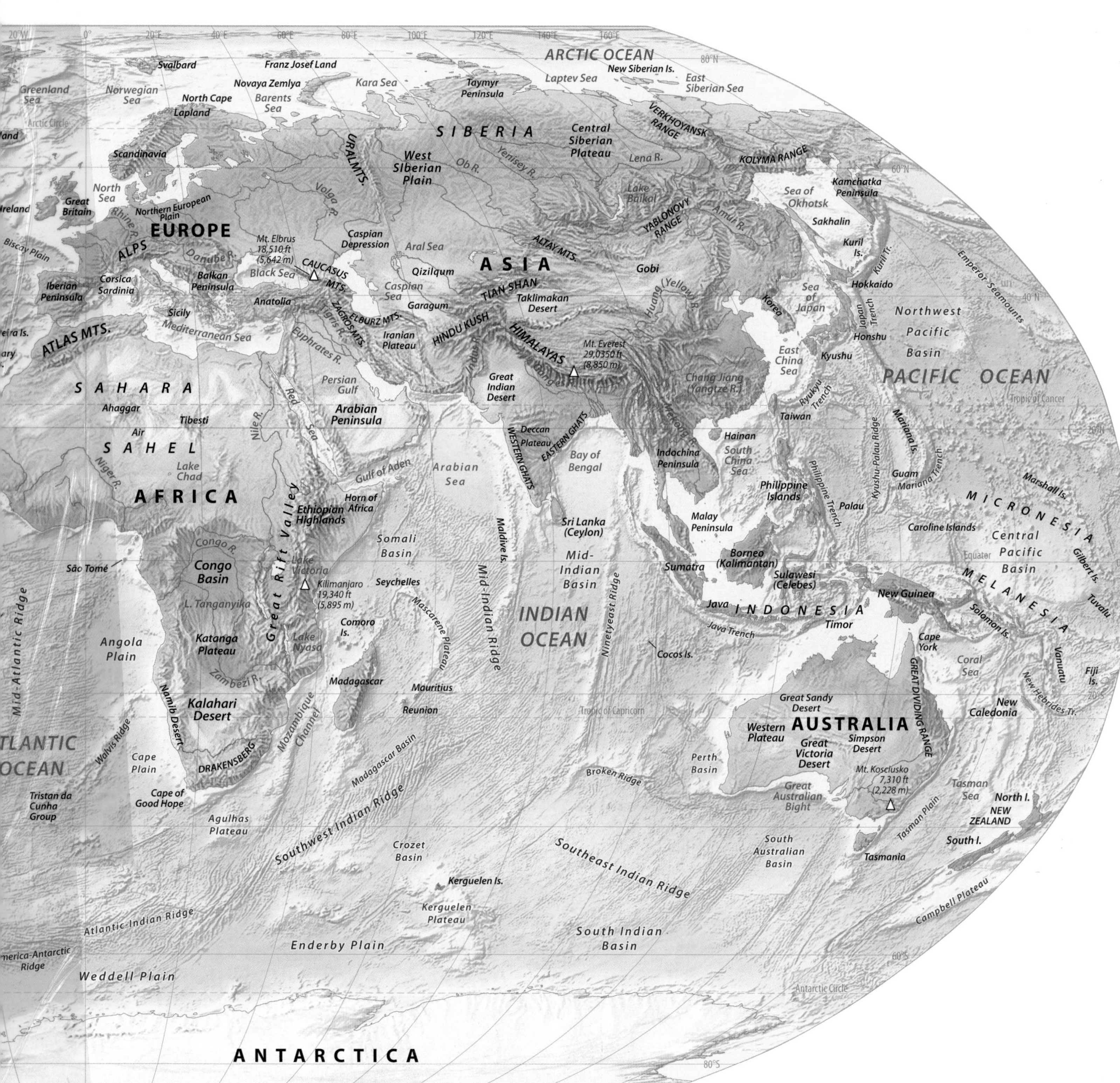

ARCTIC OCEAN
Svalbard
Franz Josef Land
Novaya Zemlya
Kara Sea
Taymyr Peninsula
Laptev Sea
New Siberian Is.
East Siberian Sea
Greenland Sea
Norwegian Sea
North Cape
Barents Sea
Lapland
Arctic Circle
SIBERIA
Central Siberian Plateau
VERKHOYANSK RANGE
Scandinavia
URAL MTS.
West Siberian Plain
Ob R.
Yenisey R.
Lena R.
KOLYMA RANGE
Kamchatka Peninsula
North Sea
Great Britain
Ireland
Rhine R.
Northern European Plain
Volga R.
Lake Baikal
YABLONOVY RANGE
Amur R.
Sea of Okhotsk
Sakhalin
EUROPE
Mt. Elbrus 18,510 ft (5,642 m)
Caspian Depression
Aral Sea
ALTAY MTS.
Kuril Is.
Kuril Tr.
Emperor Seamounts
Biscay Plain
ALPS
Danube R.
CAUCASUS MTS.
Black Sea
Qizilqum
ASIA
Gobi
Hokkaido
Iberian Peninsula
Corsica
Sardinia
Balkan Peninsula
Caspian Sea
TIAN SHAN
Huang (Yellow) R.
Sea of Japan
Korea
Japan Trench
Northwest Pacific Basin
Anatolia
Garagum
Taklimakan Desert
Sicily
ELBURZ MTS.
ZAGROS MTS.
Tigris R.
Iranian Plateau
HINDU KUSH
HIMALAYAS
Honshu
ATLAS MTS.
Mediterranean Sea
Euphrates R.
Mt. Everest 29,035 ft (8,850 m)
East China Sea
Kyushu
Indus R.
Ganges R.
Chang Jiang (Yangtze R.)
PACIFIC OCEAN
SAHARA
Persian Gulf
Great Indian Desert
Ryukyu Trench
Tropic of Cancer
Ahaggar
Red Sea
Arabian Peninsula
Tibesti
Nile R.
Taiwan
Air
Deccan Plateau
EASTERN GHATS
WESTERN GHATS
Mekong R.
Hainan
SAHEL
Bay of Bengal
Indochina Peninsula
South China Sea
Philippine Trench
Kyushu-Palau Ridge
Mariana Is.
Mariana Trench
Lake Chad
Niger R.
Gulf of Aden
Arabian Sea
Philippine Islands
Guam
Marshall Is.
AFRICA
Ethiopian Highlands
Horn of Africa
Palau
MICRONESIA
Great Rift Valley
Maldive Is.
Sri Lanka (Ceylon)
Malay Peninsula
Caroline Islands
Central Pacific Basin
Congo R.
Somali Basin
Borneo (Kalimantan)
Equator
São Tomé
Congo Basin
Lake Victoria
Mid-Indian Basin
Sumatra
Gilbert Is.
Kilimanjaro 19,340 ft (5,895 m)
Seychelles
Sulawesi (Celebes)
MELANESIA
L. Tanganyika
Mid-Indian Ridge
New Guinea
Mascarene Plateau
INDIAN OCEAN
Ninetyeast Ridge
Java
INDONESIA
Tuvalu
Mid-Atlantic Ridge
Comoro Is.
Java Trench
Timor
Solomon Is.
Angola Plain
Katanga Plateau
Lake Nyasa
Cape York
Cocos Is.
Zambezi R.
Coral Sea
Vanuatu
Madagascar
GREAT DIVIDING RANGE
Fiji Is.
Namib Desert
Mauritius
Kalahari Desert
Mozambique Channel
Reunion
Great Sandy Desert
New Caledonia
Tropic of Capricorn
New Hebrides Tr.
Walvis Ridge
Western Plateau
AUSTRALIA
ATLANTIC OCEAN
Cape Plain
DRAKENSBERG
Madagascar Basin
Perth Basin
Great Victoria Desert
Simpson Desert
Broken Ridge
Mt. Kosciusko 7,310 ft (2,228 m)
Tristan da Cunha Group
Cape of Good Hope
Southwest Indian Ridge
Great Australian Bight
Tasman Sea
North I.
NEW ZEALAND
Agulhas Plateau
Tasman Plain
Crozet Basin
Southeast Indian Ridge
South Australian Basin
South I.
Tasmania
Kerguelen Is.
Kerguelen Plateau
Campbell Plateau
Atlantic-Indian Ridge
South Indian Basin
Enderby Plain
America-Antarctic Ridge
Weddell Plain
Antarctic Circle
ANTARCTICA